Plant Invasion

Plant Invasion

Topic Editors

Bruce Osborne
Panayiotis G. Dimitrakopoulos

MDPI
Basel • Beijing • Wuhan • Barcelona • Belgrade • Novi Sad • Cluj • Manchester

Topic Editors
Bruce Osborne
UCD School of Biology and Environmental Science and UCD Earth Institute
University College Dublin
Dublin
Ireland

Panayiotis G. Dimitrakopoulos
Department of Environment
University of the Aegean
Mytilene
Greece

Editorial Office
MDPI AG
Grosspeteranlage 5
4052 Basel, Switzerland

This is a reprint of the Topic, published open access by the journals *Agronomy* (ISSN 2073-4395), *Diversity* (ISSN 1424-2818), *Forests* (ISSN 1999-4907), *International Journal of Plant Biology* (ISSN 2037-0164) and *Plants* (ISSN 2223-7747) , freely accessible at: https://www.mdpi.com/topics/Plant_Invasion.

For citation purposes, cite each article independently as indicated on the article page online and as indicated below:

Lastname, A.A.; Lastname, B.B. Article Title. *Journal Name* **Year**, *Volume Number*, Page Range.

ISBN 978-3-7258-6177-4 (Hbk)
ISBN 978-3-7258-6178-1 (PDF)
https://doi.org/10.3390/books978-3-7258-6178-1

Contents

About the Editors

Bruce Osborne

Bruce Osborne is a Professor of Plant Ecophysiology at the School of Agriculture and Food Science, University College Dublin. He carried out his PhD at the University of Nottingham, UK, after which he held postdoctoral positions at the University of Dundee and Queens University of Belfast. He has wide-ranging research interests spanning whole-plant to ecosystem levels of organization, encompassing the effects of climate/environmental change on greenhouse gas emissions and their mitigation. He has a special interest in the Gunnera–Nostoc symbiosis, including Gunnera invasions, for which he is internationally recognized. He was formerly President of the European Society of Plant Biology and a member of the Board of the European Plant Science Organization. He is the editor of the local journal *Biology and Environment* and serves as an associate editor of five other journals. He has published over 200 peer-reviewed articles.

Panayiotis G. Dimitrakopoulos

Panayiotis G. Dimitrakopoulos is Professor of Functional Ecology in the Department of Environment of the University of the Aegean, Greece. He obtained his Ph.D. from the University of the Aegean in 2001. He carried out his postdoctoral research at the University of Zurich, Switzerland, funded by the European Science Foundation (LINKECOL program). His research focuses on functional plant ecology, community ecology, biodiversity conservation, and conservation policy. He has served as President of the Hellenic Ecological Society (2012–2014), a member of the "Natura 2000" National Committee (2010–2020), and Honorary Visiting Senior Fellow at Anglia Ruskin University, Cambridge (2019–2022). He serves as the Section Editor-in-Chief of Ecology in *Biology* (Basel) and is a member of the editorial board of eight journals and a referee for more than 70 journals. He has published more than 100 papers in peer-reviewed journals.

Preface

The topic *Plant Invasion* collects research exploring the causes, consequences, and management of invasive species across diverse ecosystems and geographical regions. It synthesizes studies on the mechanisms underlying invasion success and the functional traits of invasive plants, alongside applied research on distribution modeling, ecosystem impacts, and mitigation strategies. The motivation for creating this topic is the growing recognition that biological invasions, as a key component of global environmental change, exert profound effects on biodiversity and ecosystem resilience worldwide. Through an interdisciplinary perspective, we seek to promote collaboration and knowledge exchange among researchers, practitioners, and policymakers. We believe that this will help readers understand the processes of plant invasion and the strategies proposed to mitigate their impact on the natural and man-made environment.

Bruce Osborne and Panayiotis G. Dimitrakopoulos
Topic Editors

International Journal of
Plant Biology

MDPI

Review

Review of Invasive Plant Functional Traits and Management Using Remote Sensing in Sub-Saharan Africa

Fredrick Ojija [1,2,*], Francesco Petruzzellis [2] and Giovanni Bacaro [2]

[1] Department of Earth Sciences, College of Science and Technical Education, Mbeya University of Science and Technology, Mbeya P.O. Box 131, Tanzania

[2] Department of Life Sciences, University of Trieste, Via L. Giorgieri 10, 34127 Trieste, Italy; fpetruzzellis@units.it (F.P.); gbacaro@units.it (G.B.)

* Correspondence: fredrick.ojija@yahoo.com

Abstract: Biodiversity and sustainable development in Sub-Saharan Africa (SSA) are considerably impacted by invasive alien plants (IAPs). Increasing plant invasions in SSA threaten agricultural productivity, biodiversity conservation, and other socioeconomic activities, which in turn put the United Nations Sustainable Development Goals (SDGs) in peril. In order to effectively combat IAPs, understanding their functional traits (morphological, physiological, and phenological traits) and integrating them into remote sensing (RS) is vital. While functional traits influence IAPs' fitness to invade and establish in a new geographical range, RS aids in studying them remotely, delineating and mapping them, and predicting their potential invasions. The information on this study topic was gathered by reviewing various existing studies published between 2000 and 2024. Based on this review, it was deduced that the majority of IAPs are fast-growing (or acquisitive), with a shorter leaf lifespan, bigger leaves, and higher plant height, ultimately resulting in a higher resource acquisition ability. We established further that in SSA, there are limited studies on IAP functional traits and their integration in RS. Many studies conducted in the region focus mostly on IAP distribution. Evidence from prior studies revealed that functional trait remote sensing (FTRS)-based research not only improves detection and mapping but also predicts whether a certain alien plant can become invasive or expand its distribution range. Thus, using the FTRS approach could help IAP management in SSA, ultimately achieving the SDGs. Our review discusses IAP implications in SSA (e.g., Angola, Tanzania, Benin, Kenya, Uganda, Rwanda, Zambia, Burundi, Zimbabwe, Botswana, Malawi, etc.) and for the achievement of SDGs; functional traits and their impact on alien invasions; and the importance of incorporating functional traits into RS.

Keywords: aliens; biodiversity; conservation; ecosystem; functional diversity; invasion; plant traits; SDGs

1. Introduction

Invasive alien plants (hereafter IAPs) are considered environmental pollutants because, like other pollutants, they are threatening human well-being and livelihoods [1,2], biodiversity, and natural ecosystems [3–5]. They cause biodiversity loss and functional changes [6,7], endangering and/or suppressing native (local or indigenous) species [3,8], and forming novel plant communities [1,9]. They are one of the major factors inducing biotic homogenization [9], which is defined by an increase in genetic, taxonomic, or functional similarities across different sites over a predetermined period of time [10,11]. They alter the community structure and composition of recipient environments by displacing or suppressing the abundance and/or species richness of native species [12,13]. A study conducted by Forey et al. [9] in a Fijian rainforest provides evidence in favor of this, showing that IAP *Pinanga coronata* (Blume) Blume decreased native species richness and Shannon diversity by 50% and 33%, respectively. Another study in southern Poland by Stefanowicz et al. [14] showed that IAP (*Solidago gigantean* Aiton) decreased soil microbe biomass and microbial activity. Also, it was reported that IAPs can change the biocenotic conditions and life cycles of animals [12,15]. For instance, invasive moss (*Campylopus introflexus* (Hedw.) Brid) changed beetle feeding preference, spider hunting mode, and body size [12]. They added that in areas with moss invasion, the proportions of web-building spiders and phytophagous beetles were lower.

In addition to the loss of native flora and fauna species [16–18], the IAPs can also decrease the functional diversity of native plant [19,20] and fauna communities [12] leading to functional homogenization. Functional diversity can be defined as a measure of an organism's functional traits that affect one or more aspects of an ecosystem's functioning. It usually quantifies the abundance and dispersion of living things occupying a specific niche. Examples of functional diversity are functional divergence, evenness, and richness [19–21]. Considering plants, Tordoni et al. [21] found that the presence of alien plants decreased the functional diversity of the co-occurring native plant communities in coastal dunes, and similar results were found in other ecosystems [22–24], as well as in urban environments [25]. Another study revealed that moss invasion reduced the functional diversity of carabid beetles while increasing spiders' functional similarity [12]. This occurred along with the dramatic change in species composition, i.e., abundance, richness, and diversity [12], ecosystem structure [13], and the functioning of an invaded ecosystem [5,16,20]. Because of this, it is claimed that biological invasion is one of the major causes of biodiversity loss and changes in ecosystem services provision globally [4,18,20,26]. Nevertheless, the impacts of IAPs and biodiversity loss are deemed significant based on how species traits interact with the recipient environment [5,9].

In the last decades, many studies addressed plant invasion under a functional point of view, i.e., by measuring plant functional traits, which are defined as any morpho-physio-phenological characteristics that impact plant fitness indirectly via their effects on growth, reproduction, and survival [27,28]. This approach allowed us to answer several questions related to plant invasion, from assessing the effects of IAP invasion on ecosystems structure and functionality [21,29] to highlighting the determinants of IAPs' invasiveness [30–32]. In particular, several studies performed multispecies comparison between native and/or non-invasive alien species vs. IAPs [30–32], revealing consistent patterns in traits differences between IAPs and native/non-invasive alien species, with some exceptions. Generally, the IAPs' ability to adapt to a variety of environmental conditions and disturbances, as well as their competitive capacity, enhance their invasiveness [8]. Indeed, in many cases, IAPs outperformed native plants in terms of traits related to growth, e.g., height, size, relative growth rate, leaf area, photosynthetic rates, etc., and reproduction, e.g., fecundity, seed number, germination rate [16,32–37]. However, despite the great number of studies and substantial meta-analytic efforts, a unique set of traits responsible for IAP invasion has not been identified to date. For instance, Daehler [30] found that growth rate, competitive ability, or fecundity did not differ between IAPs and native species, concluding that differences between the two groups of species often depend on growing conditions. Similar results were found in Mediterranean ecosystems [38–40], suggesting that invasion is a context-dependent process and that multiple suites of traits could promote invasiveness in different environments [32,40]. In this light, different hypotheses have been proposed to explain the success of IAPs. The "try harder" hypothesis [41] suggests that successful aliens deal better with the local conditions than resident species, expressing a set of functional traits different from those of native species. On the other hand, the "join the locals" hypothesis [41] predicts similarities among alien invasive and native species.

Another feature that might be associated with invasiveness in alien species is phenotypic plasticity [42,43], defined as the ability of an organism to develop different phenotypes under different environmental conditions [44]. A higher phenotypic plasticity could help IAPs to adjust and cope faster with new environments and eventually expand the distribution range and ecological niche [34,45]. However, only a few studies reported higher phenotypic plasticity in IAPs, while in others no differences between IAPs and native species were detected [43,46–48]. Besides functional traits, IAP invasion generally occurs when niches become unoccupied due to disturbances that increase resource availability or when competitors and natural enemies reduce the performance of native species [4,49]. Anthropogenic pressure and elevation [8,18], and the absence of natural antagonists (e.g., natural enemies or predators) to suppress them, also foster their invasion [5,7,50].

Considering the deleterious impacts of IAPs on the environment, biodiversity, and global economy [3,17], there is a need for effective countermeasures for their invasions [51,52]. In Article 8(h), the Convention on Biodiversity Diversity (CBD) underlines preventing the introduction, management, or eradication of alien species that threaten habitats, ecosystems, or species [16]. Additionally, the UN Decade on Ecosystem Restoration, led by Resolution 73/284, seeks to stop, prevent, and reverse ecosystems' degradation on our planet caused by alien species (https://www.decadeonrestoration.org/, accessed on 27 November 2023). This can be effectively achieved through the integration of functional traits into remote sensing (RS). For instance, by detecting and utilizing high-resolution imagery and hyperspectral imaging, scientists can identify, monitor, and manage IAPs with unprecedented accuracy and efficiency. Furthermore, the use of a functional trait remote sensing (FTRS) research approach can further improve our knowledge of the dynamics, control, mechanisms, and impacts of alien invasions [53–55]. For studying alien invasions, RS is the most thorough and practical method available as it demonstrates great potential to detect functional traits in IAPs [56–61]. However, Sub-Saharan Africa (SSA) still has significant gaps in its application when it comes to studying and managing alien plant invasions.

Despite the increasing research on IAPs, most studies in SSA focus on the distribution of IAPs and a few functional traits, e.g., allelochemicals and chemical defenses that facilitate invasion [1,50,62–64]. In general, functional trait-based analyses on SSA IAPs are very scarce and have seldom been integrated into RS. As such, much less is known in SSA about how IAP functional traits play a role in plant invasions and the application of RS in invasion control. Since functional traits and RS are useful in mapping, predicting, and managing plant invasions [65–68], integrating them into research in SSA is important. It is noteworthy, however, that a combination of multiple traits—rather than a single trait—is crucial for a more accurate prediction of invasion of alien plants [17,69]. Because of the inadequate knowledge of IAPs' impact in developing countries, particularly in SSA, and limited FTRS-research in the region, this review paper discusses the (i) implications of IAPs in SSA and for the achievement of the United Nations Sustainable Development Goals (UN SDGs), (ii) functional traits and their impact on invasions, and (iii) importance of incorporating functional traits into RS; and lastly, it presents (iv) the lessons learned and ways forward.

2. Literature Review Method

The published peer-reviewed original and review research articles on biological invasions, specifically on IAPs, were retrieved from international scientific databases and publishers. These include Springer, Elsevier, Taylor & Francis, SAGE, Scopus, Wiley-Blackwell, PLOS ONE, Hindawi, MDPI, the Directorate of Open Access Journals, African Journal Online, and the Web of Science (Table 1). We reviewed one hundred and ten (110) articles focusing on biological invasions, alien species, plant functional traits, remote sensing, and the impacts of IAPs on biodiversity and the environment. We specifically looked for the words "invasive plants", "alien species", "plant invasion", "biological invasion", "plant traits", "plant functional traits", "allelopathy", "allelochemicals", "impact of invasive", "biodiversity loss", "mycorrhizal plants", "UN SDGs", "biological invasion in Sub-Sahara", "functional diversity", "remote sensing", and "biological invasion and UN SDG" in the keywords, titles, and abstracts of the articles. We did not use pre-print articles and unpublished research materials.

Table 1. The number of literature sources that were processed.

Publisher	Number of Reviewed Articles	Indexing Database	Number of Reviewed Articles
Springer	20	Scopus	33
Elsevier	32	Directorate of Open Access Journals	17
Taylor & Francis	10	African Journal Online	11
SAGE	8	Web of Science	49
Wiley-Blackwell	6		
PLOS ONE	9		
MDPI	21		
Hindawi	4		
Total reviewed articles	110		110

3. Invasive Plants of Sub-Saharan Africa and Their Deleterious Impacts

There are many IAPs that have been reported in SSA countries [6,70–72]. Table 2 provides examples of detrimental IAPs found in SSA. They are widely distributed throughout the region and have a major impact on the natural environment, biodiversity, and livelihoods [70,72,73]. Numerous IAPs emit secondary compounds, or allelopathic chemicals, which can change the soil chemistry or physicochemical characteristics, including microbial populations, soil organic matter content, and the availability of nutrients [74–77]. For instance, *Acacia mearnsii* and *Acacia melanoxylon* have been reported to increase soil nitrogen levels and alter soil nutrient cycling [74,75]. Increased commodity traffic, the movement of people and tourists from different parts of the world, anthropogenic activities, and climate change led by globalization are responsible for the influx of IAPs and their spread [16,71,74]. SSA has the highest IAP vulnerability among all the countries due to increasing anthropogenic activities and climate change [71].

The widespread alien plant invasion of natural systems in SSA has led to a decline in the biomass of some native species, the disruption of ecosystem services and functions, and the loss of the aesthetic and economic values of ecosystems [70,77]. Also, the spread of IAPs is increasing disease transmission in SSA countries, as some species have been reported to be host disease vectors or pathogens [70,78]. Agha et al. [70] state in their review that certain IAPs can interact directly with the vectors, thereby increasing the risk of certain arbovirus transmission. And Nyasembe et al. [78] found that female malaria vector *Anopheles gambiae* increased fitness and energy reserves when fed on *Parthenium hysterophorus* L., *Ricinus communis* L., and *Bidens pilosa*.

Table 2. IAPs in Sub-Saharan Africa, their native range, and their negative impacts.

Scientific Name	Common Name(s)	Native Range	Impacts	References
Acacia mearnsii De Wild (Family: Fabaceae)	Green wattle, Black wattle, or Late black wattle	South-eastern Australia and Tasmania	Averts native species growth, decreases rangeland productivity and surface water, raises the amount of soil nitrogen, and alters its physicochemical characteristics.	[16,74]
Acacia melanoxylon R. Br (Family: Fabaceae)	Blackwood acacia, or Blackwood	South eastern Australia	Drives away native plant species, and modifies soil nutrients by adding nitrogen.	[16,70,74]
Argemone mexicana L. (Family: Papaveraceae)	Mexican poppy or Mexican prickly poppy	Central America and the Caribbean	Poisonous to livestock; it is rarely eaten, causes health disorders, and exerts allelopathic effects on native plants.	[16,64,70,74]
Bidens pilosa L. (Family: Asteraceae)	Blackjack	Tropical America	Hinders the growth and establishment of native plant species. Also, it competes with crops for resources (water, nutrients, light, and spaces)	[16,74]
Caesalpinia decapetala (Roth) Alston (Family: Fabaceae)	Cat's claw, Mauritius, or Mysore thorn	Asia mainly India	Its impenetrable thickets prevent peoples' and animals' free movement; its massive spines on the stems hinder the management of forests, also harm wildlife, livestock, and people.	[16,74]

Table 2. *Cont.*

Scientific Name	Common Name(s)	Native Range	Impacts	References
Calotropis procera (Aiton) W.T.Aiton (Family: Apocynaceae)	Sodom apple, king's crown, rubber bush, or rubber tree	South and Western Asia, North Africa, and Tropical Africa	Displaces native plants, grows into dense thickets, and its sap irritates the eyes severely. When consumed, it makes people sick.	[16,74]
Chromolaena odorata (L.) R. M. King & H. Rob. (Family: Asteraceae)	Siam weed, Rouge plant, Christmas bush, or Devil weed	South and North America	Lowers rangelands productivity, suppresses native plants leading homogenization, toxic to animals including human, and intensifies fires.	[16,74]
Clidemia hirta (L.) D. Don (Family: Melastomataceae)	Clidemia, Soapbush, or Koster's curse	Tropical America	Harmful to livestock, suppresses native vegetation, and forms thick or dense stands.	[16,74]
Datura stramonium L. (Family: Solanaceae)	Thorn apple, Jimson weed, devil's trumpet, or devil's weed	Central, South and North America	Forms monospecific dense stands that replace native species, and it is harmful to animals and plants.	[16,70,74]
Lantana camara L. (Family: Verbenaceae)	Lantana	South America and Central America	Lowers the production of fodder, inhibits the growth of vegetation, and destroys or leads to biodiversity loss.	[16,74,76]
Leucaena leucocephala (Lam.) de Wit (Family: Fabaceae)	White leadtree, River tamarind, Pearl wattle, or Jumbay	Southern Mexico and Northern Central America	Displaces native flora and fauna species, changes the ecosystems structure, disrupts primary succession processes, and decreases environmental quality.	[16,74]
Mimosa diplotricha C. Wright ex Sauvalle (Family: Fabaceae)	Giant false sensitive plant or Giant sensitive plant	Tropical America	Produces shadows that stop light-demanding plant species from regenerating; thick stands make it difficult for animals and wildlife to roam freely. Additionally, it poisons animals.	[16,70,74]
Mimosa pigra L. (Family Fabaceae)	Giant sensitive tree	South America	Decreases native biodiversity, blocks the open habitats used by wildlife, modifies the ecosystem, and reduces native resources and grazing areas for livestock and wildlife.	[16,72,74]
Parthenium hysterophorus L. (Family: Asteraceae)	Carrot weed, or Whitetop weed	North and South America	Toxic invasive plant; it rapidly suppresses native vegetation through allelopathy and resource competition. Alters native plant community structure to monospecific stands, reduces rangeland productivity and crop yields, and causes health problems to people and animals.	[33,50,72,78,79]
Opuntia ficus-indica (L.) Mill. (Family: Cactaceae)	Sweet prickly pear, India fig opuntia, Barbary fig, or Cactus pear	North America	Its spines hinder access to pasture and harm humans, animals, and wildlife, and drives out native species.	[16,74]
Opuntia stricta (Haw.) Haw. (Family: Cactaceae)	Erect prickly pear	Tropical America	Its spines hinder access to pasture and injure people, animals, and wildlife, and displaces native species.	[16,74]
Pinus patula Schiede ex Schltdl. & Cham. (Family: Pinaceae)	Patula pine or Spreading-leaved pine	Central America, e.g., Mexico	Its dense stands displace and/or inhibit the growth and germination of native plants and reduce drainage or water run-off.	[16,74]
Pistia stratiotes L. (Family: Araceae)	Water cabbage, Nile cabbage, Water lettuce, or Shellflower	Probably Tropical America or Africa	Impedes fishing, obstructs waterways, slows down water flow, destroys fish rookeries (breeding colonies), increases nutrient loading and siltation rates (thus, depresses water quality), and threatens fish and other species survival.	[16,74]
Prosopis juliflora (Sw.) DC. (Family: Fabaceae)	Mesquite	Tropical America	Depletes groundwater, lowers ecosystems and rangelands ability to support wildlife, and eradicates native species from invaded areas.	[16,72,74]
Psidium guajava L. (Family: Myrtaceae)	Common guava	Central America and the Caribbean	Makes thick stands that hinder or displaces native species; uses allelopathy to negatively impact plants and crops.	[16,58,74]
Ricinus communis L. (Family: Euphorbiaceae)	Castor bean	East Africa	Forms thick stands that, especially in riparian areas, supplant native plants.	[16,74]
Rubus niveus Thunb. (Family: Rosaceae)	Ceylon raspberry, Mysore raspberry, or hill raspberry	East and South Asia, Australia, or the Himalayas	Transforms the plant community and forms dense thickets that interfere with or impede the growth and rejuvenation of native plants.	[16,72,74]
Senna spectabilis (DC.) H.S. Irwin & Barneby (Family: Fabaceae)	Golden wonder tree	Tropical America	Inhibits the growth and rejuvenation of native plants using allelopathy.	[16,74]

Table 2. *Cont.*

Scientific Name	Common Name(s)	Native Range	Impacts	References
Tagetes minuta L. (Family: Asteraceae)	Wild marigold	South America	Forms monospecific stands by suppressing the growth and germination of native plant species.	[16,72,74]
Tephrosia vogelli Hook. f. (Family: Fabaceae)	Fish bean or Fish-poison bean	Tropical Africa	Its leaves are toxic to animals including fishes, worms, insects, molluscs, toads, and frogs.	[16,70,74]
Tithonia diversifolia (Hemsl.) A. Gray (Family: Asteraceae)	Mexican sunflower, Tree marigold, or Japanese sunflower	Mexico and Central America	Decreases the productivity of rangelands, modifies the organization of plant communities, and leads to the loss of some native species.	[16,74]

Furthermore, as IAPs' invasions continue to expand in the region, so do the costs of eliminating the IAPs [17]. According to the CBD, IAPs and other invasive species ruin the global economy by billions of dollars every year [16,55]. Thus, because of the IAPs' negative effects on agricultural production [71], biodiversity conservation, human health [70,78], and other socioeconomic development activities in SSA [16,36,77,80], the UN SDGs are in peril. Subject to socioeconomic factors, anthropogenic activities, and climate change [81], IAP invasions are likely to increase further in the future, mainly in arable and urban environments in SSA [72,76]. Therefore, in order to effectively manage IAPs, strategic management plans combining their functional traits and RS should be taken into account. This is because RS can help to detect and utilize high-resolution imagery and functional traits to identify, monitor, and manage IAPs before they spread and cause further negative impact.

4. Implications of IAPs for Sustainable Development Goals in Sub-Saharan Africa

The management and prevention of biological invasions are being widely discussed in a number of international conventions (e.g., CBD, Ramsar Convention, Bern Convention), organizations (e.g., International Union for Conservation of Nature), policies, and initiatives intended to promote sustainable development, biodiversity conservation, human health, and other socioeconomic development challenges [6,16,36]. Therefore, it becomes essential to avert IAPs to ensure a healthy environment that supports sustainable development through biodiversity resources and human health in order to achieve the UN SDGs [36,82]. The UN SDGs and human livelihoods—such as farming and livestock rearing—are supported by healthy, uninvaded ecosystems because they offer ecosystem services such as controlling water flow, foraging, recycling nutrients, producing food—such as seeds, fruits, and vegetables—controlling climate change, and biodiversity conservation [82,83]. However, the invasions of IAPs on pristine ecosystems in SSA could result in the loss of biodiversity and valuable ecosystem services [2,80] with a deleterious impact on human livelihoods (decreases in or losses of crop yields or livestock productivity) and eventually failure to meet the SDGs [72]. In SSA, biological invasions caused by IAPs have proven to be a significant problem toward achieving the UN SDGs [16,36,83]. Some of these SDGs that are directly connected to global biological invasion management efforts to ensure sustainable livelihoods, biodiversity conservation, human health, sanitation, and zero carbon emissions include SDG 1 (ending or eliminating poverty), SDG 2 (zero hunger), SDG 3 (good health and well-being), SDG 6 (clean water and sanitation), SDG 13 (combat climate change), and SDG 15 (life on land).

As IAPs (e.g., *B. pilosa*, *P. hysterophorus*, and *P. stratiotes*) reduce crop yields, it could be difficult for people to achieve food security [78,83]. In developed countries, AIPs are responsible for reducing crop yields by 10% [16]. In Africa, especially in SSA, alien invasive plants alone cause about a 25% decline in yield [16]. In Ethiopia, it was reported that AIP, i.e., *P. hysterophorus*, reduced the yield of finger millet, sorghum, and other vital subsistence crops by 75% [16]. Limited food (and nutrients) could affect other essential livelihoods needed by people, consequently increasing the poverty level and diseases and failing to achieve SDGs 1 and 2 [36]. IAPs can further affect the achievement of SDG 3 because they pose potential harm to human health because some are toxic to humans and animals e.g., *P. hysterophorus*, *A. Mexicana*, *C. hirta*, *T. vogelli*, and *O. stricta* [74,79]. People's health risks might increase if they are exposed to areas invaded with toxic IAPs.

IAPs (e.g., *P. stratiotes*) can cause pollution in aquatic ecosystems, causing a loss of clean water and aquatic species [70,74,78], which affects the achievement of SDG 6. *Pistia stratiotes*, *Salvinia molesta*, and *Pontederia crassipes* (formerly *Eichhornia crassipes*), for instance, have been reported to increase nutrient loading and siltation rates in lakes and rivers, thereby reducing water quality and subsequently reducing fish survival [16,74]. This could lead to diseases and health problems for people who use such water. Moreover, some IAPs (e.g., *P. juliflora*) tend to deplete groundwater, making water unavailable to humans, plants, and animals. This could make it difficult for humans and animals in SSA countries to have good health and well-being (SDG 6).

The spread and effects of IAPs are not only exacerbated by climate change [81] but they also actively contribute to it. Certain native plants that are effective carbon sinks could be suppressed or undergo major changes in their structure, function, and composition [16]. Additionally, IAPs can alter temperature and precipitation patterns and seasonal distributions in ways that benefit IAPs [16]. Thus, invasive plants impede efforts to mitigate climate change and accomplish UN SDG 13 [36,82]. Furthermore, the invasions of IAPs on the environment can negatively impact land-based life [82]. In SSA, for instance, by damaging the habitats (i.e., changing species composition and structure) that provide vital ecosystem services (food, forages, medicine, water, and nutrient cycling) to sustain humans and nonhuman biota, IAPs threaten the lives of many terrestrial native species with extinction [80]. They may impair humans' and other animals' health because some IAPs influence disease transmission by hosting disease vectors or pathogens, for instance the invasive *Parthenium hysterophorus* [70,78]. The disease might cause the loss of keystones and other vulnerable species, which could lead to an ecosystem imbalance or collapse. Because of this, IAPs impede the effort to protect the lives of species on land (UN SDG 15). In SSA countries, the loss of terrestrial biodiversity due to IAPs could further affect the livelihood and economy of millions of people, impeding the achievement of the UN SDGs [76,77,82]. To achieve UN SDGs in SSA, the planning and management of IAPs are important and could be facilitated through integrating invasive functional traits and RS.

5. IAP Functional Traits and Their Ecological Importance

Despite the fact that an IAP's invasion depends on the recipient ecosystem's resilience, it mostly rests on various functional traits [4,5,17,21], which influence the success or failure of an alien plant in the invaded habitat [18,69]. Functional traits (Table 3) define the ecological strategies that shape plants' responses to different environmental conditions [17,84] and are vital in studying biological invasions as they provide insights into the mechanisms that underlie the plant's biological invasion [4,84,85]. They help to understand why specific plant species may become invasive in their new habitats [69,85,86], determining how IAPs respond to environmental filters, changes, and biotic interactions and how they impact ecological processes [27,87]. For researchers in SSA, this information is crucial since it may aid in planning and developing strategies for managing and controlling IAPs. They might also utilize this information to stop new invasions and restore affected ecosystems. However, as already mentioned, multiple suite of traits are thought to make a species invasive and successful [69,87].

Table 3. Functional trait categories associated with IAPs in Sub-Saharan Africa.

IAP Functional Trait Categories	Ecological Importance	References
Reproduction, spread and growth rate (seed mass, seed number, dispersal method, relative growth rate)	IAPs have a higher fecundity (or reproduction rate). They make significant reproductive investments, resulting in a large number of seeds that can swiftly invade new areas. Their high fecundity makes them compete more successfully than native species.	[4,17–21,34,37,84,87,88]

Table 3. *Cont.*

IAP Functional Trait Categories	Ecological Importance	References
Reproduction, spread and growth rate (seed mass, seed number, dispersal method, relative growth rate)	IAPs have efficient and different seed dispersal mechanisms. Propagule size, weight, shape, water dispersal (*hydrochory*), wind dispersal (*anemochory*), dispersal, animal dispersal (*epizoochory*) are examples of seed dispersal traits.	[84,87–89]
	As IAPs devote a greater number of resources to growth and reproduction, they grow faster than native species.	[18,37,84,87,90]
Phenology	In contrast to native species, IAPs can alter their phenological patterns. This makes it possible for them to take advantage of resources when there is little competition. IAPs also begin flowering earlier and continue flowering for longer than native plants.	[19,21,66,91,92]
Resource acquisition and utilization (leaf economic spectrum traits)	IAPs' functional traits related to resource (water, nutrients, and light) acquisition include leaf area (LA), specific leaf area (SLA), vein length per unit area (VLA), leaf mass per area (LMA), leaf area, rapid growth rate, height, and extensive and deep root system. These traits indicate IAP's ability to capture and use resources efficiently, which enables them to surpass native species.	[4,17–21,37]
Hydraulic traits (vein traits and drought tolerance traits) and water use	Most IAPs are able to grow in a variety of environmental conditions because they possess traits associated with severe drought and/or temperature resilience or tolerance. For instance, most IPAs show high venation, leaf water potential at turgor loss point, and leaf osmotic potential at full turgor. But some IAPs may have lower drought resistance than natives.	[4,7,19,21,48,93]
Mycorrhizal associations	IAPs' access to nutrients may be influenced by their mycorrhizal associations. Some IAPs may exhibit flexibility in forging associations or have a variety of mycorrhizal partners.	[7,18,93]
Allelopathy and chemical defense	Some IAPs produce secondary metabolites or allelochemicals that inhibit the growth of native plants. They use these allelochemicals to influence their invasiveness and reduce competition with their nearby native species.	[64,72,89]

Other functional traits that have been investigated in various studies include leaf dry matter content (LDMC), leaf nitrogen (LN), potassium (LP), and phosphorus (LPh) content, carbon-to-nitrogen ratio (C:N) [4,9,21,27], leaf area ratio (LAR), leaf area–sapwood area ratio (LA:SA), saturated water content of the wood (SWC), wood density (WD), leaf lifespans (LL), midday leaf water potential (Ψ_{leaf}) [4,37,94], chlorophyll, anthocyanin, leaf orientation, wood specific gravity (WSG), and carotene concentrations [53,60,95]. Previous comparative studies and meta-analyses between IAPs and native species reported that IAPs tend to share functional traits that favor fast growth, giving IAPs an advantage over native species in resource acquisition and utilization [21,90]. Indeed, IAPs typically have higher values of traits related to resource (water, nutrients, space, and light) capture and use compared with native plant species [4,91]. Most of the acquired resources are relocated for reproduction and growth [90,94], which makes the IAPs have a higher fecundity and rapid growth rate than natives [37,84]. Investing in vegetative growth, for instance, promotes population persistence following establishment [92], while allocating enough resources for reproduction encourages invasions of new environments [84]. Furthermore, IAPs tend to have higher plant height, SLA, seed number, and leaf area than native species (Table 2) [19,34]. Indeed, the majority of IAPs are fast-growing (or possess an acquisitive strategy), with higher LL, photosynthetic rates, and nitrogen content [58,92,95,96], while native plants grow slower, since they typically have a more conservative strategy.

It was reported that alien species display a stronger capacity for resource acquisition than native species, which exhibited a resource conservative strategy [4]. They further established evidence that IAPs exhibit higher values of SLA and VLA, which demonstrate lower carbon investment in the construction of a leaf and higher water transport efficiency, respectively. High photosynthetic rates and water transport efficiency are generally due to a higher venation network in IAP leaves [21]. And higher SLA in IAPs implies that they reduce the expense of leaf construction and maintenance [21]. However, native plant species may behave in the exact opposite way; evidently, they invest greater resources into constructing leaves, thus lowering SLA [4,21]. A study carried out on a mountain in central Argentina revealed that native species *Polylepis australis* switched to more conservative resource-use strategies (lower LA, LAR, SLA, and higher LDMC and WD), while an invasive *Cotoneaster franchetii* maintained acquisitive resource-use strategies (i.e., high SLA, LAR, LA, and low LDMC) along an elevation gradient [37]. A successful water-transport strategy was also demonstrated by *C. franchetii*, which showed low WD and high Ψ_{leaf} and LA:SA [37].

A study conducted by Tordoni et al. [21] using $\delta^{13}C$ as a functional trait found considerably lower values of $\delta^{13}C$ in IAPs, suggesting higher gas exchange rates than natives. This result supported the hypothesis that IAPs have a higher rate of photosynthesis [4,21]. On the other hand, IAPs showed a capacity to resist and survive environmental stressors, i.e., drought, extreme temperatures, and/or salinity, because they possess drought and salinity tolerance and environmental plasticity traits [4,69]. Moreover, the phenological traits of some IAPs provide them with the ability to exploit resources differently from native plant species [4]. These patterns enable them to take advantage of resources when there is low competition. The flexibility and ability of IAPs to exploit patchy resources are also indicated by high SLA [34,61,92], which denotes higher photosynthetic capability, higher leaf nutrient levels, and greater turnover of leaves i.e., shorter leaf lifespan [20,66]. In terms of their ability to adapt to different light levels, IAPs exhibit more plasticity than native plant species, which is important for plant growth, development, and survival [15,26,48]. Given the previous results, trait divergence has been suggested to enhance the invasion success of IAPs [21,97,98], but the direction of trait differences is not always consistent. In fact, the invasion process is considered strongly context-dependent, as in some environments, IAPs were found to exhibit similar traits to native species [38,39].

Furthermore, there is a strong correlation between biological invasion and reproductive dispersal traits such as seed production, seed dry mass, and seed dispersal mechanisms [15,26,84,89]. Alien species with long-range dispersal are more successful in invading larger areas [84,89]. This is due to their increased likelihood of arriving in previously uninvaded areas [84]. High-fecundity (i.e., large number of seeds) and easy seed dispersal mechanisms (wind-dispersed and animal-dispersed) are advantageous for invasive species because they allow the seeds to spread quickly and invade vast areas in a brief period of time [54,84]. Many native plant species do not possess these traits and, thus, have poor dispersal abilities and are unlikely to colonize habitats [53,89]. In addition to investing heavily in a large number of seeds to increase the likelihood of invasions, IAPs are effective seed dispersants [26,89]. Nevertheless, in contrast to native plant species, they achieve this by reducing seed mass and size [84]. The growth form and architecture (woody, forbs, shrub, herb, or height) affect how much light they receive and how IAPs compete with natives for resources [15]. In view of this, Mologni et al. [84] observed that growth forms (woody, forbs, and graminoids) and dispersal traits effectively predicted the effective alien invaders in the northern New Zealand islands.

Moreover, some alien plant species rely on mycorrhiza symbioses (or plant–fungal associations) to establish and thrive in the environment [18,93]. In the case of IAPs, they establish strong mutualistic relationships with mycorrhizal fungi in their novel range [7]. The association helps them to establish and enhance their invasiveness in their newly introduced area [7,93]. With this association, IAPs capture essential nutrients more effectively than native nearby plants [7,93]. It also enhances carbohydrate flow and tolerance to harmful heavy metals, salinity, root pathogens, and herbivory for IAPs [7,18,93]. Further, to ensure that they continue their invasiveness, they produce allelochemicals that inhibit native plant germination and growth [19,64]. Many IAPs employ these strategies to suppress and outcompete native plant species, eventually changing the structure of the vegetative community to monospecific stands [88]. Additionally, IAPs display some trade-offs between several functional traits to increase their invasiveness, for instance, by allocating enough resources to growth [4,88]. This means that such IAPs prioritize growth over defense mechanisms, which would enable them to spread invasively [4,21]. Based on the roles and significance of functional traits in plant invasions, this review recommends employing them in research aiming at the management and prediction of new invasions in SSA. Incorporating functional in RS would help to predict which plant species are more likely to become invasive and to evaluate the possible ecological and management challenges associated with the invasion [85].

6. Importance of Functional Trait Remote Sensing-Based Research in Plant Invasions

Studies show that RS is an efficient tool for mapping IAPs in diverse environments [92,99,100]. There are several RS platforms that provide spectral data with higher spatial and spectral resolutions and have the potential to produce estimations of functional traits with higher accuracy [55,92,99]. These include, for instance, unmanned aerial vehicles (UAVs: e.g., aircraft and drones), field spectrometers, and satellites [56,92,99,100]. The use of RS in the study of plant invasions extends beyond invasion science to biological invasion management [92,99,100]. It is employed in plant invasion mapping, prediction, and monitoring, as well as in the supply of ecological information for studies on impacts, mechanisms, dynamics, and drivers (Figure 1). Phenological, morphological, and physiological plant functional traits have proven useful in mapping IAPs [92].

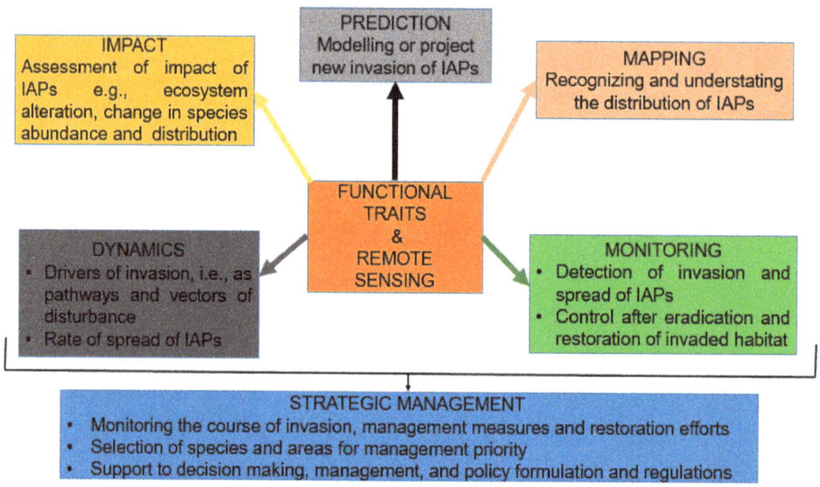

Figure 1. Functional traits and remote sensing's contribution to the knowledge of alien plant invasions and management. Modified from Müllerová et al. [55].

However, its applications for mapping IAPs have often relied on spatial distribution and rarely on functional traits, despite the strong link between IAP invasion and functional traits [4,26,86,101]. Remotely sensed images can be used to detect the physiological, morphological, and chemical qualities of IAPs that correlate to their functional traits [101]. This is due to the fact that variations in spectral band values are also linked to plant biophysic and phenotypic traits [101,102]. Functional traits provide vital information regarding IAPs' success over native species [90,94,103]. Thus, using IAP functional traits in RS (Figure 1) would improve mapping and help predict whether a particular alien plant can become invasive or expand its distribution range [26,92,99]. Given its potential for the mapping and prediction of IAP invasions using function traits, this might be important in SSA, where plant invasions are threatening the UN SDGs and biodiversity conservation [16,36,76,80,83].

Further, the use of an FTRS database can make it possible to continuously and affordably estimate ongoing biological invasions over wide areas in SSA. Such predictions could be useful for extension officers, conservation managers, and farmers in SSA in pre-planning countermeasures for the predicted invasion. Functional traits have been used in RS and various models (i.e., regression and classification trees) to predict the outcomes of alien plant species invasions [56,89,92]. Some studies—the majority of which were conducted in developed countries—have shown how useful it is to study plant invasions using functional traits in RS [61,66–68]. Some of the functional traits of IAPs have been demonstrated, either to aid in mapping and predicting IAPs or to delineate IAPs from noninvasive plant species using RS (Table 4).

Table 4. Functional traits that have been used in RS for mapping and predicting IAPs or to delineate IAPs from noninvasive plants.

Functional Traits	Uses in Remote Sensing	RS Technique	Country	References
Leaves phenology (green or dry leaves)	Monitoring the invasion of *Ligustrum lucidum* W.T. Aiton.	Satellite	Argentina	[66]
SLA, LDMC, and LNC	Assessing impacts of plant invasions (*Impatiens glandulifera* Royle and *S. gigantea*) on ecosystem.	Field spectrometer	Belgium	[61]
Sum of leaf magnesium and calcium contents (leaf Ca + Mg)	Assessing impacts of plant invasions (*I. glandulifera* and *S. gigantea*) on ecosystem.	Field spectrometer	Belgium	[61]
LDMC	Verifying statistical models' predictive power using IAPs *Festuca rubra* L. *Elytriagia atherica* (Link) Kerguélen, and *Puccinellia maritima* (Jacq.) Parl.	Satellite	Netherlands	[65]
Flowering periods (flower phenology)	*Tamarix* spp. invasion detection and mapping using spectral signatures acquired during flowering periods	Satellite	US	[56]
Leaf C:N	Comparing carbon-to-nitrogen ratios of senescent leaf in plants.	Satellite	Bangladesh	[104]
Plant leaf colour (leaf phenology)	Mapping occurrences of IAP *Sapium sebiferum* (L.) Roxb.	UAV (aircraft)	US	[105]
Patches or clumps (Structural traits)	Mapping of IAP *Pennisetum ciliare* (L.) Link	Satellite	Mexico	[106]
Total nitrogen, magnesium, canopy height, potassium, and total chlorophyll (Chla + Chlb)	Mapping IAP *Lespedeza cuneata* (Dum. Cours.) G.Don in grassland ecosystems.	Field spectrometer + UAV (aircraft)	US	[107]
Plant height, inflorescence, flowering, germination, and vegetative growth	Mapping of IAP *Spartina alterniflora* (Loisel.) P.M.Peterson & Saarela	Field spectrometer	China	[96]
Chlorophyll, anthocyanin, and carotene concentrations	Elucidate functional dissimilarity between IAPs (e.g., *Egeria densa Planch.*, *Myriophyllum spicatum* L, etc.) and native (e.g., *Elodea Canadensis Michx.*, *Stuckenia pectinate (L.) Böerner*, etc.).	Field spectrometer + UAV (aircraft)	US	[95]
Canopy structure, senesced leaves, eight, biochemical and biophysical features, and inflorescences	Detecting three different IAP species (*Carpobrotus edulis* (L.) N.E. Br, *Eucalyptus globulus* Labill., and *Cortaderia jubata* (Lem.) Stapf).	UAV (aircraft)	US	[108]
Leaf orientation	Delineating IAP (*Taeniatherum caput-medusae L*) and native plants.	Field spectrometer	US	[60]
Canopy leaf nitrogen content	Assessing impact of IAP (*Morella faya* Ait.) invasion on nitrogen-oxide emissions.	UAV (aircraft)	US	[57]
SLA, LMA, water content, carotenoid (Car) content, Chla, Chlb, total chlorophyll (Chla + Chlb), chlorophyll a:b ratio (Chla/Chlb), carotenoid:total chlorophyll ratio (Car/Chl)	Distinguishing IAPs (*Hovenia dulcis* and *P. guajava*), from native plant species (*Luehea divaricata* Mart and *Psidium cattleianum* Sabine).	Field spectrometer	Brazil	[58]
Canopy, tree diameter, and height	Identification of IAP *Acer negundo* L in forests.	UAV	Poland	[59]

Previous studies affirm that using optically measured functional traits for remote sensing can help ecologists to better understand the mechanisms and effects of plant invasion (Table 3). Functional traits can be utilized in RS not only to map the invasion of IAPs but also to delineate or group IAPs based on how they share effects on ecosystem processes or how they respond to the environment [54,59,60]. Also, Chacón-Madrigal et al. [53] contends that functional traits can be used in RS to determine how closely related IAP species differ in their geographical range sizes. In their study, they used LDMC, SLA, plant height, leaf thickness, WSG, LN, LP, and leaf N:P ratio to determine the geographical range size difference among related plant species. In addition, current studies reveal that plant functional diversity patterns can be detected from the spectral diversity of high-resolution multispectral imagery without the need for satellite data [109,110]. This is because recent developments in algorithms have shown the potential of calculating spectral diversity as a proxy for functional diversity compared with FTRS, which still needs the testing of empirical or theoretical models using data from satellites [109,110]. This could help in highlighting the effects of IAP on an ecosystem scale. Overall, these studies clearly show the relevance of the FTRS research approach to understanding invasive plant behavior and management. Despite the fact that studies that demonstrate the use of FTRS to monitor, predict, and control IAPs are missing or limited in SSA, the functional traits and RS techniques in Table 3 can also be used to study biological invasions of various IAPs in SSA.

Despite the fact that FTRS-based research can be used to detect and predict IAPs and their distribution, it should be understood, however, that without an initial field recognition of the range of a given IAP, remote technology cannot help in monitoring that species. The recognition of species' detailed characteristics (mainly colour and growth forms) in the field can be the key to species determination.

7. Lessons Learned and Way Forward

The alternative way of delineating IAPs from native plants is by using functional traits, as explained by Mallmann et al. [58] and Mielczarek et al. [59]. For instance, in contrast to native plant species, the IAPs tend to have bigger leaves, higher specific leaf areas, minimal defense investments, higher plant heights, and high resource acquisition. Nevertheless, it is important to note that the success of alien species could depend either on a more acquisitive or a more conservative strategy with respect to native species, according to the invasion context. This review underlines further that the utilization of FTRS-based data would advance IAP knowledge and aid in planning and preventing plant invasions in SSA. Even with RS's increased capacity for the assessment of IAPs' functional traits, ecologists in the SSA have failed to keep up with the recent growth in remote-sensing technologies, which has hindered the region from fully utilizing these tools for invasive species mapping, mitigation, and prediction. Given the complexity of mapping invasive species in SSA due to its high diversity of species and structural heterogeneity, the integration of functional traits into RS mapping might be particularly helpful, where IAP species pose a far more serious problem [16,36,58,73,74,76,80]. Hence, the efficiency of biological invasion control may be achieved with better knowledge of invasion dynamics and the identification of their drivers through a FTRS-research approach. In this light, we propose the following objectives to promote the use of functional traits and remote sensing techniques to monitor and contrast IAPs also spread in the SSA region. The first is to promote the use of satellite-derived RS data to study IAP invasion in the SSA region, since a great number of satellite missions (such as LANDSAT and COPERNICUS missions) offer free data that can be easily used by researchers in SSA. Indeed, UAVs and field spectrometer instruments are not cheap, and it might not be easy to use them in countries in SSA. Satellite data (such as LANDSAT and SENTINEL), however, could be analyzed both with medium–high level workstations or under cloud computing frameworks, such as the Google Earth Engine, the usage of which only requires an internet connection and an entry-level laptop. Secondly, there is a need for collaborations between invasion ecologists from developed countries and those from SSA in research-related functional traits and RS to impart FTRS knowledge to researchers in SSA. Furthermore, we recommend that ecologists in the SSA start employing a process-based mapping and biological invasion prediction tool. In doing so, it is expected that the FTRS approach will effectively prevent plant invasions in SSA, improve biodiversity and conservation, and consequently achieve the UN SDGs. However, even though IAPs may share specific functional traits that can be leveraged to improve their detection remotely, it should be noted that the success of detection may vary depending on the species.

Author Contributions: Conceptualization, F.O.; methodology, F.O.; resources, F.O., F.P. and G.B.; writing—original draft preparation, F.O.; writing—review and editing, F.O., F.P. and G.B.; supervision, F.P. and G.B.; project administration, F.O., F.P. and G.B.; funding acquisition, F.O., F.P. and G.B. All authors have read and agreed to the published version of the manuscript.

Funding: This review article is a part of the research visit at Trieste University, in the Department of Science, funded by the Italian Ministry of Foreign Affairs and International Cooperation, and the Italian Agency for Development Cooperation for the TWAS-SISSA-Lincei Research Cooperation Research Visit Programme through UNESCO-TWAS that was funded to F.O. F.P. was supported by the funding PON Ricerca e Innovazione D.M. 1062/21—Contratti di ricerca, from the Italian Ministry of University (MUR).

Institutional Review Board Statement: Not applicable.

Informed Consent Statement: Not applicable.

Data Availability Statement: Data are unavailable due to privacy restrictions.

Acknowledgments: We thank all of our colleagues who provided insightful and practical suggestions. We also thank them for taking the time to polish the manuscript and provide an incisive critique that greatly improved the work.

Conflicts of Interest: The authors declare no conflicts of interest. The funders had no role in the design of the study; in the collection, analyses, or interpretation of data; in the writing of the manuscript; or in the decision to publish the results.

References

1. Borokini, I.T.; Kortz, A.; Anibaba, Q.A.; Witt, A.; Aigbokhan, W.I.; Hejda, M.; Pyšek, P. Alien flora of Nigeria: Taxonomy, biogeography, habitats, and ecological impacts. *Biol. Invasions* **2023**, *25*, 3677–3696. [CrossRef]
2. Sheergojri, I.A.; Rashid, I.; Rehman, I.U.; Rashid, I. Invasive species services-disservices conundrum: A case study from Kashmir Himalaya. *J. Environ. Manag.* **2022**, *309*, 114674. [CrossRef] [PubMed]
3. Jocienė, L.; Krokaitė, E.; Rekašius, T.; Juškaitytė, E.; Ielciu, I.; Galanina, O.; Kupčinskienė, E. The Molecular Evidence for Invasive Climber *Echinocystis lobata* (Michx.) Torr. & A. Gray in Eastern and Central Europe. *Diversity* **2023**, *15*, 1084. [CrossRef]
4. Petruzzellis, F.; Tordoni, E.; Tomasella, M.; Savi, T.; Tonet, V.; Palandrani, C.; Castello, M.; Nardini, A.; Bacaro, G. Functional differentiation of invasive and native plants along a leaf efficiency/safety trade-off. *Environ. Exp. Bot.* **2021**, *188*, 104518. [CrossRef]
5. Pyšek, P.; Jarošík, V.; Hulme, P.E.; Pergl, J.; Hejda, M.; Schaffner, U.; Vilà, M. A global assessment of invasive plant impacts on resident species, communities and ecosystems: The interaction of impact measures, invading species' traits and environment. *Glob. Chang. Biol.* **2012**, *18*, 1725–1737. [CrossRef]
6. Munishi, L.K.; Ngondya, I.B. Realizing UN decade on ecosystem restoration through a nature-based approach: A case review of management of biological invasions in protected areas. *PLoS Sustain Transform* **2022**, *1*, e0000027. [CrossRef]
7. Santolamazza-Carbone, S.; Durán-Otero, M.; Calviño-Cancela, M. Context dependency, co-introductions, novel mutualisms, and host shifts shaped the ectomycorrhizal fungal communities of the alien tree *Eucalyptus globulus*. *Sci. Rep.* **2019**, *9*, 7121. [CrossRef] [PubMed]
8. Bacaro, G.; Maccherini, S.; Chiarucci, A.; Jentsch, A.; Rocchini, D.; Torri, D.; Gioria, M.; Tordoni, E.; Martellos, S.; Altobelli, A.; et al. Distributional patterns of endemic, native and alien species along a roadside elevation gradient in Tenerife, Canary Islands. *Community Ecol.* **2015**, *16*, 223–234. [CrossRef]
9. Forey, E.; Lodhar, S.Y.F.; Galvin, S.D.; Lowry, J.H.; Gopaul, S.; Hanson, G.; Carboni, M.; Chauvat, M.; Boehmer, H.J. Alien palm invasion leads to selective biotic filtering of resident plant communities towards competitive functional traits. *Biol. Invasions* **2023**, *25*, 1489–1508. [CrossRef]
10. Olden, J.D.; LeRoy Poff, N.; Douglas, M.R.; Douglas, M.E.; Fausch, K.D. Ecological and evolutionary consequences of biotic homogenization. *Trends Ecol. Evol.* **2004**, *19*, 18–24. [CrossRef]
11. Qian, H.; Guo, Q. Linking biotic homogenization to habitat type, invasiveness and growth form of naturalized alien plants in North America. *Divers. Distrib.* **2010**, *16*, 119–125. [CrossRef]
12. Schirmel, J.; Buchholz, S. Invasive moss alters patterns in life-history traits and functional diversity of spiders and carabids. *Biol. Invasions* **2013**, *15*, 1089–1100. [CrossRef]
13. Smith, T.C.; Bishop, T.B.B.; Duniway, M.C.; Villarreal, M.L.; Knight, A.C.; Munson, S.M.; Waller, E.K.; Jensen, R.; Gill, R.A. Biophysical factors control invasive annual grass hot spots in the Mojave Desert. *Biol. Invasions* **2023**, *25*, 3839–3858. [CrossRef]
14. Stefanowicz, A.M.; Stanek, M.; Majewska, M.L.; Nobisb, M.; Zubek, S. Invasive plant species identity affects soil microbial communities in a mesocosm experiment. *Appl. Soil Ecol.* **2019**, *136*, 168–177. [CrossRef]
15. Garcia, R.A.; Clusella-Trullas, S. Impacts of invasive plants on animal diversity in South Africa: A synthesis. *Bothalia-Afr. Biodivers. Conserv.* **2017**, *47*, 1–12.
16. Boy, G.; Witt, A. *Invasive Alien Plants and Their Management in Africa*, 1st ed.; CABI Africa: Nairobi, Kenya, 2013.
17. Oh, M.; Heo, Y.; Lee, E.J.; Lee, H. Major environmental factors and traits of invasive alien plants determine their spatial distribution: A case study in Korea. *J. Ecol. Environ.* **2021**, *45*, 29. [CrossRef]
18. Ribotta, S.; Liccari, F.; Muggia, L.; Pallavicini, A.; Bagnolini, F.; Tordoni, E.; Bacaro, G. Invasion at the Edge: The Case of *Rosa rugosa* (Rosaceae) in Italy. *Diversity* **2021**, *13*, 645. [CrossRef]
19. Milanović, M.; Knapp, S.; Pyšek, P.; Kühn, I. Linking traits of invasive plants with ecosystem services and disservices. *Ecosyst. Serv.* **2020**, *42*, 101072. [CrossRef]
20. Sheppard, C.S. Relative performance of co-occurring alien plant invaders depends on traits related to competitive ability more than niche differences. *Biol. Invasions* **2019**, *21*, 1101–1114. [CrossRef]
21. Tordoni, E.; Petruzzellis, F.; Nardini, A.; Savi, T.; Bacaro, G. Make it simpler: Alien species decrease functional diversity of coastal plant communities. *J. Veg. Sci.* **2019**, *30*, 498–509. [CrossRef]
22. Abadie, J.C.; Machon, N.; Muratet, A.; Porcher, E. Landscape disturbance causes small-scale functional homogenization, but limited taxonomic homogenization, in plant communities. *J. Ecol.* **2011**, *99*, 1134–1142. [CrossRef]
23. Brice, M.; Pellerin, S.; Poulin, M. Does urbanization lead to taxonomic and functional homogenization in riparian forests? *Divers. Distrib.* **2017**, *23*, 828–840. [CrossRef]

24. Lambdon, P.W.; Lloret, F.; Hulme, P.E. Do non-native species invasions lead to biotic homogenization at small scales? The similarity and functional diversity of habitats compared for alien and native components of Mediterranean floras. *Divers. Distrib.* **2008**, *14*, 774–785. [CrossRef]
25. Milanović, M.; Kühn, I.; Pyšek, P.; Knapp, S. Functional diversity changes in native and alien urban flora over three centuries. *Biol. Invasions* **2021**, *23*, 2337–2353. [CrossRef]
26. Kaushik, P.; Pati, P.K.; Khan, M.L.; Khare, P.K. Plant functional traits best explain invasive species' performance within a dynamic ecosystem—A review. *Trees For. People* **2022**, *8*, 100260. [CrossRef]
27. Carmona, C.P.; De Bello, F.; Mason, N.W.H.; Leps, J. Traits without borders: Integrating functional diversity across scales. *Trends Ecol. Evol.* **2016**, *31*, 382–394. [CrossRef] [PubMed]
28. Violle, C.; Navas, M.; Vile, D.; Kazakou, E.; Fortunel, C.; Hummel, I.; Garnier, E. Let the concept of trait be functional! *Oikos* **2007**, *116*, 882–892. [CrossRef]
29. Hejda, M.; Pyšek, P.; Jarošík, V. Impact of invasive plants on the species richness, diversity and composition of invaded communities. *J. Ecol.* **2009**, *97*, 393–403. [CrossRef]
30. Daehler, C.C. Performance comparisons of co-occurring native and alien invasive plants: Implications for conservation and restoration. *Annu. Rev. Ecol. Evol. Syst.* **2003**, *34*, 183–211. [CrossRef]
31. Funk, J.L.; Zachary, V.A. Physiological responses to short-term water and light stress in native and invasive plant species in southern California. *Biol. Invasions* **2010**, *12*, 1685–1694. [CrossRef]
32. van Kleunen, M.; Weber, E.; Fischer, M. A meta-analysis of trait differences between invasive and non-invasive plant species. *Ecol. Lett.* **2010**, *13*, 235–245. [CrossRef] [PubMed]
33. Ojija, F.; Arnold, S.E.J.; Treydte, A.C. Impacts of alien invasive *Parthenium hysterophorus* on flower visitation by insects to co-flowering plants. *Arthropod-Plant Interact.* **2019**, *13*, 719–734. [CrossRef]
34. Ruprecht, E.; Fenesi, A.; Nijs, I. Are plasticity in functional traits and constancy in performance traits linked with invasiveness? An experimental test comparing invasive and naturalized plant species. *Biol. Invasions* **2014**, *16*, 1359–1372. [CrossRef]
35. Pyšek, P.; Richardson, D.M. Traits Associated with Invasiveness in Alien Plants: Where Do we Stand? In *Biological Invasions*; Nentwig, W., Ed.; Springer: Berlin/Heidelberg, Germany, 2008; pp. 97–125.
36. Rangi, D. *Invasive Species: The Hidden Threat to Sustainable Development*, 1st ed.; CABI: Nairobi, Kenya, 2018.
37. Ferrero, M.C.; Tecco, P.A.; Gurvich, D.E. Is intraspecific variability an advantage in mountain invasions? Comparing functional trait variation in an invasive and a native woody species along multiple environmental gradients. *Biol. Invasions* **2022**, *24*, 1393–1412. [CrossRef]
38. Funk, J.L. The physiology of invasive plants in low-resource environments. *Conserv. Physiol.* **2013**, *1*, cot026. [CrossRef] [PubMed]
39. Funk, J.L.; Vitousek, P.M. Resource-use efficiency and plant invasion in low-resource systems. *Nature* **2007**, *446*, 1079–1081. [CrossRef] [PubMed]
40. Funk, J.L.; Standis, R.J.; Stock, W.D.; Valladares, F. Plant functional traits of dominant native and invasive species in mediterranean-climate ecosystems. *Ecology* **2016**, *97*, 75–83. [CrossRef] [PubMed]
41. Tecco, P.A.; Diaz, S.; Cabido, M.; Urcelay, C. Functional traits of alien plants across contrasting climatic and land-use regimes: Do aliens join the locals or try harder than them? *J. Ecol.* **2010**, *98*, 17–27. [CrossRef]
42. Heberling, J.M.; Kichey, T.; Decocq, G.; Fridley, J.D. Plant functional shifts in the invaded range: A test with reciprocal forest invaders of Europe and North America. *Funct. Ecol.* **2016**, *30*, 875–884. [CrossRef]
43. Davidson, A.M.; Jenions, M.; Nicotra, A.B. Do invasive species show higher phenotypic plasticity than native species and, if so, is it adaptive? A meta-analysis. *Ecol. Lett.* **2011**, *14*, 419–431. [CrossRef]
44. Nicotra, A.B.; Atkin, O.K.; Bonser, S.P.; Davidson, A.M.; Finnegan, E.J.; Mathesius, U.; Poot, P.; Purugganan, M.D.; Richards, C.L.; Valladares, F.; et al. Plant phenotypic plasticity in a changing climate. *Trends Plant Sci.* **2010**, *15*, 684–692. [CrossRef] [PubMed]
45. Godoy, O.; Vlladares, F.; Castro-Díez, P. Multispecies comparison reveals that invasive and native plants differ in their traits but not in their plasticity. *Funct. Ecol.* **2011**, *25*, 1248–1259. [CrossRef]
46. Leishman, M.R.; Cooke, J.; Richardson, D.M.; Newman, J. Evidence for shifts to faster growth strategies in the new ranges of invasive alien plants. *J. Ecol.* **2014**, *102*, 1451–1461. [CrossRef] [PubMed]
47. Taylor, K.T.; Maxwell, B.D.; Pauchard, A.; Nunez, M.A.; Rew, L.J. Native versus non-native invasions: Similarities and differences in the biodiversity impacts of *Pinus contorta* in introduced and native ranges. *Divers. Distrib.* **2016**, *22*, 578–588. [CrossRef]
48. Petruzzellis, F.; Peng, G.; Tyree, M.T.; Tonet, V.; Savi, T.; Torboli, V.; Pallavicini, V.; Bacaro, G.; Nardini, A. Plasticity of functional traits of tree of heaven is higher in exotic than in native habitats. *Trees* **2019**, *33*, 411–420. [CrossRef]
49. Thompson, K.; Davis, M.A. Why research on traits of invasive plants tells us very little. *Trends Ecol. Evol.* **2011**, *26*, 155–156. [CrossRef] [PubMed]
50. Ojija, F. Eco-friendly management of *Parthenium hysterophorus*. *Sci. Prog.* **2022**, *105*, 1–15. [CrossRef] [PubMed]
51. Minuto, L. Management of an invasive plant in a Mediterranean protected area: The experience of *Senecio deltoideus* in Italy. *Ann. Bot.* **2021**, *11*, 1–12.
52. Tassin, J.; Kull, C.A. Facing the broader dimensions of biological invasions. *Land Use Policy* **2015**, *42*, 165–169. [CrossRef]
53. Chacón-Madrigal, E.; Wanek, W.; Hietz, P.; Dullinger, S. Traits indicating a conservative resource strategy are weakly related to narrow range size in a group of neotropical trees. *Perspect. Plant Ecol. Evol. Syst.* **2018**, *32*, 30–37. [CrossRef]

54. Lavorel, S.; Garnier, E. Predicting changes in community composition and ecosystem functioning from plant traits: Revisiting the Holy Grail. *Funct. Ecol.* **2002**, *16*, 545–556. [CrossRef]
55. Müllerová, J.; Brundu, G.; Große-Stoltenberg, A.; Kattenborn, T.; Richardson, D.M. Pattern to process, research to practice remote sensing of plant invasions. *Biol. Invasions* **2023**, *25*, 3651–3676. [CrossRef]
56. Evangelista, P.; Stohlgren, T.; Morisette, J.; Kumar, S. Mapping invasive Tamarisk (Tamarix): A comparison of single-scene and time-series analyses of remotely sensed data. *Remote Sens.* **2009**, *1*, 519–533. [CrossRef]
57. Hall, S.J.; Asner, G.P. Biological invasion alters regional nitrogen-oxide emissions from tropical rainforests. *Glob. Chang. Biol.* **2007**, *13*, 2143–2160. [CrossRef]
58. Mallmann, C.L.; Filho, W.P.; Dreyer, J.B.B.; Tabaldi, L.A.; Durgante, F.M. Leaf-level field spectroscopy to discriminate invasive species (*Psidium guajava* L. and *Hovenia dulcis* Thunb.) from native tree species in the Southern Brazilian Atlantic Forest. *Remote Sens.* **2023**, *15*, 791. [CrossRef]
59. Mielczarek, D.; Sikorski, P.; Archiciński, P.; Ciezkowski, W.; Zaniewska, E.; Chormanski, J. The use of an airborne laser scanner for rapid identification of invasive tree species *Acer negundo* in riparian forests. *Remote Sens.* **2022**, *15*, 212. [CrossRef]
60. Naupari, J.A.; Vierling, L.A.; Eitel, J.U.H. Delineating native and invasive plant functional groups in shrub-steppe vegetation using bidirectional reflectance. *J. Appl. Remote Sens.* **2013**, *7*, 073563. [CrossRef]
61. Van Cleemput, E.; Van Meerbeek, K.; Helsen, K.; Honnay, O.; Somers, B. Remotely sensed plant traits can provide insights into ecosystem impacts of plant invasions: A case study covering two functionally different invaders. *Biol. Invasions* **2020**, *22*, 3533–3550. [CrossRef]
62. Bordbar, F.; Meerts, P. Patterns in the alien flora of the Democratic Republic of the Congo: A comparison of Asteraceae and Fabaceae. *Plant Ecol. Evol.* **2020**, *153*, 373–389. [CrossRef]
63. Foxcroft, L.C.; Richardson, D.M.; Rejmánek, M.; Pysek, P. Alien plant invasions in tropical and sub-tropical savannas: Patterns, processes and prospects. *Biol. Invasions* **2010**, *12*, 3913–3933. [CrossRef]
64. Ojija, F. Allelopathic effects of *Sphaeranthus suaveolens* (Forssk.) DC. and *Argemone mexicana* L leaf crude extract on *Zea mays* L germination and growth. *AGBIR* **2023**, *39*, 651–656.
65. Ali, A.M.; Darvishzadeh, R.; Shahi, K.R.; Skidmore, A. Validating the predictive power of statistical models in retrieving leaf dry matter content of a coastal wetland from a Sentinel-2 Image. *Remote Sens.* **2019**, *11*, 1936. [CrossRef]
66. Gavier-Pizarro, G.I.; Kuemmerle, T.; Hoyos, L.E.; Stewart, S.I.; Huebner, C.D.; Keuler, N.S.; Radeloff, V.C. Monitoring the invasion of an exotic tree (*Ligustrum lucidum*) from 1983 to 2006 with Landsat TM/ETM+ satellite data and Support Vector Machines in Córdoba, Argentina. *Remote Sens. Environ.* **2012**, *122*, 134–145. [CrossRef]
67. Hantson, W.; Kooistra, L.; Slim, P.A. Mapping invasive woody species in coastal dunes in the Netherlands: A remote sensing approach using LIDAR and high-resolution aerial photographs. *Appl. Veg. Sci.* **2012**, *15*, 536–547. [CrossRef]
68. Maruthi Sridhar, B.B.; Vincent, R.K.; Clapham, W.B.; Sritharan, S.I.; Osterberg, J.; Neale, C.M.U.; Watts, D.R. Mapping saltcedar *Tamarix ramosissima* and other riparian and agricultural vegetation in the Lower Colorado River region using multi-spectral Landsat TM imagery. *Geocarto Int.* **2010**, *25*, 649–662. [CrossRef]
69. Devin, S.; Beisel, J.-N. Biological and ecological characteristics of invasive species: A gammarid study. *Biol. Invasions* **2006**, *9*, 13–24. [CrossRef]
70. Agha, S.B.; Alvarez, M.; Becker, M.; Fèvre, E.M.; Junglen, S.; Borgemeister, C. Invasive alien plants in Africa and the potential emergence of mosquito-borne arboviral diseases—A review and research outlook. *Viruses* **2020**, *13*, 32. [CrossRef] [PubMed]
71. Paini, D.R.; Sheppard, A.W.; Cook, D.C.; De Barro, P.J.; Worner, S.P.; Thomas, M.B. Global threat to agriculture from invasive species. *Proc. Natl. Acad. Sci. USA* **2016**, *113*, 7575–7579. [CrossRef] [PubMed]
72. Uyi, O.; Mukwevho, L.; Ejomah, A.J.; Toews, M. Invasive alien plants in Sub-Saharan Africa: A review and synthesis of their insecticidal activities. *Front. Agron.* **2021**, *3*, 725895. [CrossRef]
73. Weidenhamer, J.D.; Callaway, R.M. Direct and indirect effects of invasive plants on soil chemistry and ecosystem function. *J. Chem. Ecol.* **2010**, *36*, 59–69. [CrossRef]
74. Witt, A.; Beale, T.; Van Wilgen, B.W. An assessment of the distribution and potential ecological impacts of invasive alien plant species in eastern Africa. *Trans. R. Soc. S. Afr.* **2018**, *73*, 217–236. [CrossRef]
75. Afzal, M.R.; Naz, M.; Ashraf, W.; Du, D. The Legacy of Plant Invasion: Impacts on Soil Nitrification and Management Implications. *Plants* **2023**, *12*, 2980. [CrossRef]
76. Goncalves, E.; Herrera, I.; Duarte, M.; Bustamante, R.O.; Lampo, M.; Velasquez, G.; Sharma, G.P.; Garcıa-Rangel, S. Global invasion of *Lantana camara*: Has the climatic niche been conserved across continents? *PLoS ONE* **2014**, *9*, e111468. [CrossRef]
77. Kilawe, C.J.; Baltazary, I.S.; Malila, B.P.; Lyimo, P.J.; Mwakalukwa, E.E. Replacement of native trees by the neotropical invasive tree *Cedrela odorata* L. in the Kimboza Forest Reserve, Tanzania. *Biol. Invasions* **2023**, *25*, 3697–3710. [CrossRef]
78. Nyasembe, V.O.; Cheseto, X.; Kaplan, F.; Foster, W.A.; Teal, P.E.A.; Tumlinson, J.H.; Borgemeister, C.; Torto, B. The invasive american weed *Parthenium hysterophorus* can negatively impact malaria control in Africa. *PLoS ONE* **2015**, *10*, e0137836. [CrossRef] [PubMed]
79. Ojija, F.; Ngimba, C. Suppressive abilities of legume fodder plants against the invasive weed *Parthenium hysterophorus* (Asteraceae). *Environ. Sustain. Indic.* **2021**, *10*, 100111. [CrossRef]
80. Landmann, T. Invasive plants and food security in Africa: The potential of earth observation data. *S. Afr. Inst. Int. Aff.* **2017**, *1*, 1–4.

81. Pellegrini, E.; Buccheri, M.; Martini, F.; Boscutti, F. Agricultural land use curbs exotic invasion but sustains native plant diversity at intermediate levels. *Sci. Rep.* **2021**, *11*, 8385. [CrossRef]
82. Egoh, B.N.; Ntshotsho, P.; Maoela, M.A.; Blanchard, R.; Ayompe, L.M.; Rahlao, S. Setting the scene for achievable post-2020 convention on biological diversity targets: A review of the impacts of invasive alien species on ecosystem services in Africa. *J. Environ. Manag.* **2020**, *261*, 110171. [CrossRef]
83. Diagne, C.; Turbelin, A.J.; Moodley, D.; Novoa, A.; Leroy, B.; Angulo, E.; Adamjy, T.; Dia, C.A.K.M.; Taheri, A.; Tambo, J.; et al. The economic costs of biological invasions in Africa: A growing but neglected threat? *NeoBiota* **2021**, *67*, 11–51. [CrossRef]
84. Mologni, F.; Bellingham, P.J.; Cameron, E.K.; Dinh, K.; Wright, A.E.; Burns, K.C. Functional traits explain non-native plant species richness and occupancy on northern New Zealand islands. *Biol. Invasions* **2022**, *24*, 2135–2154. [CrossRef]
85. Van Kleunen, M.; Dawson, W.; Dostal, P. Research on invasive-plant traits tells us a lot. *Trends Ecol. Evol.* **2011**, *26*, 317. [CrossRef] [PubMed]
86. Korpelainen, H.; Pietiläinen, M. What makes a good plant invader? *Life* **2023**, *13*, 1596. [CrossRef] [PubMed]
87. Mathakutha, R.; Steyn, C.; Le Roux, P.C.; Blom, I.J.; Chown, S.L.; Daru, B.H.; Ripley, B.S.; Louw, A.; Greve, M. Invasive species differ in key functional traits from native and non-invasive alien plant species. *J. Veg. Sci.* **2019**, *30*, 994–1006. [CrossRef]
88. Lamarque, L.J.; Delzon, S.; Lortie, C.J. Tree invasions: A comparative test of the dominant hypotheses and functional traits. *Biol. Invasions* **2011**, *13*, 1969–1989. [CrossRef]
89. Moravcová, L.; Pyšek, P.; Jarošík, V.; Pergl, J. Getting the right traits: Reproductive and dispersal characteristics predict the invasiveness of herbaceous plant species. *PLoS ONE* **2015**, *10*, e0123634. [CrossRef]
90. Smith, M.D.; Knapp, A.K. Physiological and morphological traits of exotic, invasive exotic, and native plant species in tallgrass prairie. *Int. J. Plant Sci.* **2001**, *162*, 785–792. [CrossRef]
91. Milanović, M.; Knapp, S.; Pyšek, P.; Kühn, I. Trait–environment relationships of plant species at different stages of the introduction process. *NeoBiota* **2020**, *58*, 55–74. [CrossRef]
92. Niphadkar, M.; Nagendra, H. Remote sensing of invasive plants: Incorporating functional traits into the picture. *Int. J. Remote Sens.* **2016**, *37*, 3074–3085. [CrossRef]
93. Moyano, J.; Dickie, I.A.; Rodriguez-Cabal, M.A.; Nuñez, M.A. Patterns of plant naturalization show that facultative mycorrhizal plants are more likely to succeed outside their native Eurasian ranges. *Ecography* **2020**, *43*, 648–659. [CrossRef]
94. Díaz De León Guerrero, S.D.; González-Rebeles Guerrero, G.; Ibarra-Montes, T.M.; Bastarrachea, A.R.; Cobos, R.S.; Bullock, S.H.; Sack, L.; Mendez-Alonzo, R. Functional traits indicate faster resource acquisition for alien herbs than native shrubs in an urban Mediterranean shrubland. *Biol. Invasions* **2020**, *22*, 2699–2712. [CrossRef]
95. Santos, M.J.; Hestir, E.L.; Khanna, S.; Ustin, S.L. Image spectroscopy and stable isotopes elucidate functional dissimilarity between native and nonnative plant species in the aquatic environment. *New Phytol.* **2012**, *193*, 683–695. [CrossRef] [PubMed]
96. Ouyang, Z.-T.; Gao, Y.; Xie, X.; Guo, H.; Zhang, T.; Zhao, B. Spectral Discrimination of the Invasive Plant *Spartina alterniflora* at multiple phenological stages in a Saltmarsh Wetland. *PLoS ONE* **2013**, *8*, e67315. [CrossRef] [PubMed]
97. Divíšek, J.; Chytrý, M.; Beckage, B.; Gotelli, N.J.; Lososová, Z.; Pyšek, P.; Richardson, D.M.; Molofsky, J. Similarity of introduced plant species to native ones facilitates naturalization, but differences enhance invasion success. *Nat. Commun.* **2018**, *9*, 4631. [CrossRef] [PubMed]
98. Marchini, G.L.; Maraist, C.A.; Cruzan, M.B. Trait divergence, not plasticity, determines the success of a newly invasive plant. *Ann. Bot.* **2019**, *123*, 667–679. [CrossRef] [PubMed]
99. Gränzig, T.; Clasen, A.; Fassnacht, F.E.; Cord, A.; Förster, M. Combining remote sensing, habitat suitability models and cellular automata to model the spread of the invasive shrub Ulex europaeus. *Biol. Invasions* **2023**, *25*, 3711–3736. [CrossRef]
100. Vaz, A.S.; Alcaraz-Segura, D.; Campos, J.C.; Vicente, J.R.; Honrado, J.P. Managing plant invasions through the lens of remote sensing: A review of progress and the way forward. *Sci. Total Environ.* **2018**, *642*, 1328–1339. [CrossRef] [PubMed]
101. Beccari, E.; Carmona, C.P.; Tordoni, E.; Petruzzellis, F.; Martinucci, D.; Casagrande, G.; Pavanetto, N.; Rocchini, D.; D'Antraccoli, M.; Ciccarelli, D.; et al. Plant spectral diversity from high-resolution multispectral imagery detects functional diversity patterns in coastal dune communities. *J. Veg. Sci.* **2024**, *35*, e13239. [CrossRef]
102. Hacker, P.W.; Coops, N.C.; Laliberté, E.; Michaletz, S.T. Variations in accuracy of leaf functional trait prediction due to spectral mixing. *Ecol. Indic.* **2022**, *136*, 108687. [CrossRef]
103. Tordoni, E.; Petruzzellis, F.; Nardini, A.; Bacaro, G. Functional divergence drives invasibility of plant communities at the edges of a resource availability gradient. *Diversity* **2020**, *12*, 148. [CrossRef]
104. Rahman, M.M.; Zhang, X.; Ahmed, I.; Iqbal, Z.; Zeraatpisheh, M.; Kanzaki, M.; Xu, M. Remote sensing-based mapping of senescent leaf C:N ratio in the sundarbans reserved forest using machine learning tech-niques. *Remote Sens.* **2020**, *12*, 1375. [CrossRef]
105. Ramsey, E.W.; Nelson, G.A.; Sapkota, S.K.; Seeger, E.B.; Martella, K.D. Mapping Chinese Tallow with Color-Infrared Photography. *Photogramm. Eng.* **2002**, *68*, 251–255.
106. Brenner, J.C.; Christman, Z.; Rogan, J. Segmentation of landsat thematic mapper imagery improves buffelgrass (*Pennisetum ciliare*) pasture mapping in the Sonoran Desert of Mexico. *Appl. Geogr.* **2012**, *34*, 569–575. [CrossRef]
107. Gholizadeh, H.; Friedman, M.S.; McMillan, N.A.; Hammond, W.M.; Hassani, K.; Sams, A.V.; Charles, M.D.; Garrett, D.R.; Joshi, O.; Hamilton, R.G.; et al. Mapping invasive alien species in grassland ecosystems using airborne imaging spectroscopy and remotely observable vegetation functional traits. *Remote Sens. Environ.* **2022**, *271*, 112887. [CrossRef]

108. Underwood, E.C.; Ustin, S.L.; Ramirez, C.M. A Comparison of Spatial and Spectral Image Resolution for Mapping Invasive Plants in Coastal California. *Environ. Manag.* **2007**, *39*, 63–83. [CrossRef] [PubMed]
109. Van Cleemput, E.; Adler, P.B.; Suding, K.N. Making remote sense of biodiversity: What grassland characteristics make spectral diversity a good proxy for taxonomic diversity? *Glob. Ecol. Biogeogr.* **2023**, *32*, 2177–2188. [CrossRef]
110. Schweiger, A.K.; Cavender-Bares, J.; Townsend, P.A.; Hobbie, S.E.; Madritch, M.D.; Wang, R.; Tilman, D.; Gamon, J.A. Plant spectral diversity integrates functional and phylogenetic components of biodiversity and predicts ecosystem function. *Nat. Ecol. Evol.* **2018**, *2*, 976–982. [CrossRef]

agronomy

MDPI

Article

A Method of Invasive Alien Plant Identification Based on Hyperspectral Images

Xi Qiao [1,2,3,†], Xianghuan Liu [1,2,†], Fukuan Wang [1,2,†], Zhongyu Sun [4], Long Yang [4], Xuejiao Pu [1], Yiqi Huang [2], Shuangyin Liu [3] and Wanqiang Qian [1,*]

1 Shenzhen Branch, Guangdong Laboratory of Lingnan Modern Agriculture, Genome Analysis Laboratory of the Ministry of Agriculture and Rural Affairs, Agricultural Genomics Institute at Shenzhen, Chinese Academy of Agricultural Sciences, Shenzhen 518120, China
2 College of Mechanical Engineering, Guangxi University, Nanning 530004, China
3 Guangzhou Key Laboratory of Agricultural Products Quality & Safety Traceability Information Technology, Zhongkai University of Agriculture and Engineering, Guangzhou 510225, China
4 Guangzhou Institute of Geography, Guangdong Academy of Sciences, Guangzhou 510070, China
* Correspondence: qianwanqiang@caas.cn
† These authors contributed equally to this work.

Abstract: Invasive alien plants (IAPs) are considered to be one of the greatest threats to global biodiversity and ecosystems. Timely and accurate detection technology is needed to identify these invasive plants, helping to mitigate the damage to farmland, fruit trees and woodland. Hyperspectral technology has the potential to identify similar species. However, the challenge remains to simultaneously identify multiple invasive alien plants with similar colors based on image data. The spectral images were collected by a hyperspectral camera with a spectral range of 450–998 nm, and the raw spectra were extracted by Cubert software. First derivative (FD), Savitzky-Golay (SG) smoothing and standard normal variate (SNV) were used to preprocess the raw spectral data, respectively. Then, on the basis of preprocessing, principal component analysis (PCA) and ant colony optimization (ACO) were used for feature dimensionality reduction, and the reduced features were used as input variables for later modeling. Finally, a combination of both dimensionality reduction and non-dimensionality reduction is used for identification using support vector machines (SVM) and random forests (RF). In order to determine the optimal recognition model, a total of 18 combinations of different preprocessing methods, dimensionality reduction methods and classifiers were tested. The results showed that a combination of SG smoothing and SVM achieved a total accuracy (A) of 89.36%, an average accuracy (AA) of 89.39% and an average precision (AP) of 89.54% with a test time of 0.2639 s. In contrast, the combination of SG smoothing, the ACO, and SVM resulted in weaker performance in terms of A (86.76%), AA (86.99%) and AP (87.22%), but with less test time (0.0567 s). The SG-SVM and SG-ACO-SVM models should be selected considering accuracy and time cost, respectively, for recognition of the seven IAPs and background in the wild.

Keywords: hyperspectral analysis; invasive plants; data preprocessing; dimension reduction; classification

1. Introduction

Invasive alien plants (IAPs) have become one of the greatest threats to biodiversity, ecosystems and agriculture around the world, causing very large ecological and economic losses [1]. Furthermore, the threat caused by IAPs is growing due to the globalization of human activities and economics [2]. Various projects have been adopted to restrain and eliminate IAPs in recent years. During this time management has played a limited role and the threat is expanding [3,4]. More recently, accurate identification and monitoring, which can provide the detailed distribution of IAPs, has been considered an efficient and low-cost approach in IAP control [5].

Traditionally, manual inspection has been the main method of monitoring for IAPs, which is considered an accurate but inefficient measure [6]. Moreover, various factors, such as the growth of plants, the inability to assess a large spatial coverage, the subjective judgment of scientists, and the inaccessible terrain have restricted monitoring [7]. Hyperspectral remote sensing is regarded as an effective monitoring method and has been successfully used in some IAP monitoring, such as for *Mikania micrantha* Kunth [8], *Spartina alterniflora* Loisel [9–11], *Psidium cattleianum* [12], *Lantana camara* L. [13], *Common milkweed* [14] and *Ailanthus altissima* [15]. These studies only identify single invasive plants and did not attempt to classify multiple species. However, the simultaneous identification of multiple IAPs based on imaging data remains challenging due to similar phenotypic colors. Hyperspectral images can provide rich spectral information and have the potential to simultaneously identify the colors of similar species [16,17]. This paper aims to identify multiple IAPs based on hyperspectral images.

Generally, the full-spectrum curve information of objects is extracted from hyperspectral images and used for identification through machine learning methods, such as support vector machine (SVM) and random forest (RF) [18]. There is no denying that capturing certain bands from hyperspectral images may result in the loss of important spectral information. However, seeking key information from the full spectral curve is indispensable since the hyperspectral image contains numerous invalid data, which will interfere with the identification and decrease the time-validity [19]. Furthermore, it is certain that some noise will occur in the wild and interfere with the hyperspectral information. Therefore, image preprocessing and the selection of critical spectral bands are essential [20].

Researchers have proposed a number of spectral preprocessing methods and have applied them in many studies. Sun et al. applied five preprocessing methods, including standard normal variate (SNV), multiple scattering correction, Savitzky–Golay (SG) smoothing, normalization and first derivative (FD), to pretreat the spectral data, and then a SVM model was used for the determination of moisture content in barley seeds [21]. Pang et al. proposed a method of detecting the vigor of variabilis seeds. In this research, using SG smoothing and second-order derivative (SD) preprocessing combined with least squares support vector machines (LS-SVM) achieved 98.81% accuracy in the test set [22]. Shen et al. explored an application of hyperspectral imaging technology to estimate the soil organic matter, no denoising, SG smoothing denoising and wavelet domain denoising were used to eliminate the noise in hyperspectral data [23].

Meanwhile, it is necessary to select the information that is valuable for the experiment, as hyperspectral data are always redundant and much information is useless [24]. The methods for selecting the information, which also means reducing the dimensionality of hyperspectral data, have been widely applied in many studies [25,26]. Principal component analysis (PCA) and ant colony optimization (ACO) have shown potential in their application for reducing dimensionality.

This research aims to explore the feasibility of classifying Mikania micrantha Kunth, *Sphagneticola calendulacea* (L.) Pruski, *Ageratum conyzoides* L., *Mimosa pudica* Linn., *Lantana camara* L., *Lpomoea cairica* (L.) Sweet, *Bidens pilosa* L. and background in the wild by hyperspectral data. Selecting a proper method for processing hyperspectral data and distinguishing the seven IAPs is a great challenge. Moreover, challenges are manifested in the variability of the hyperspectral data of IAPs in a complex field environment, the lack of prior knowledge and background interference.

To meet these challenges, hyperspectral preprocessing methods FD, SG smoothing, SNV and the dimensionality reduction algorithms PCA and ACO were applied for processing the hyperspectral data of IAPs and background. In the final classification, SVM and RF were used. Finally, preprocessing algorithms, dimensionality reduction algorithms and classifiers were randomly combined to study and explore an optimal identification method for IAPs in the field.

2. Materials and Methods

2.1. Sample Preparation

In this paper, we manually applied the high-speed imaging spectrograph S185 manufactured Cubert (Germany) to collect hyperspectral images of seven species of IAP in the wild. The spectrometer simultaneously recorded 138 bands in the spectral range of 450–950 nm, with a sampling interval of 4 nm. The collection sites were Jiulong town (23°22′29.5″ N, 113°29′52.9″ E; 23°22′13.4″ N, 113°28′46.9″ E) (collected *M. micrantha*, *S. calendulacea*, *A. conyzoides*, *M. pudica*, *L. camara* and *B. Pilosa*), Conghua district (23°28′25.8″ N, 113°29′09.9″ E) (collected *S. calendulacea*, *A. conyzoides*, *M. pudica*, *L. camara* and *L. cairica*), Guangzhou, China, and Sanjiao town (22°42′24.6″ N, 113°25′41.7″ E) (collected *M. micrantha*, *A. conyzoides*, *M. pudica*, *L. cairica* and *B. pilosa*), Zhongshan, China.

These seven IAPs pose a huge threat to native species and biodiversity. They are similar in color, phenotype, and appear together in small plots, which is why hyperspectral techniques were used to identify seven species simultaneously. The images were captured between 10 am and 3 pm around solar noon on between 21 and 25 November 2020, when the weather was cloudy. This heightened the chances of suitable weather conditions for the measurements [27,28]. Before image acquisition, dark reference (by closing the camera lens) and white reference (using a white plate) images were collected to calibrate the spectrometer according to the following equation [29]:

$$I_C = (I_R - I_D)/(I_W - I_D) \tag{1}$$

where I_C is the calibrated image, I_R is the raw image, I_W is the white reference, and I_D is the dark reference.

Then, the spectrometer's lens was aimed directly at the surface of the invasive plant and manually focused to the middle of the leaves. Eighty images of vegetation were collected using the hyperspectral camera and were used in this study. Partial samples of seven invasive plants are shown in Figure 1. The images contain leaves, flowers, stalks, other plants and the no-plant cover area. Among them, flowers, stalks and the no-plant cover area belong to background. The leaves and background feature points were extracted manually from the original images using Cubeware software (approximately 43 per image, 38 of which are leaves feature points), as shown in Figure 2A; the spectral curves are shown in Figure 2B and saved in ASCII format. In this work, 3359 raw spectral data samples (*M. micrantha*: 428, background: 409, *S. calendulacea*: 428, *A. conyzoides*: 425, *M. pudica*: 419, *L. camara*: 408, *L. cairica*: 418, *B. pilosa*: 424) were collected, including a random partition a training set (2400 samples, each category has 300 samples) and testing set (959 samples).

Figure 1. Partial samples of seven invasive plants and the calibration plate. (**A**) *Mikania micrantha* Kunth, (**B**) *Sphagneticola calendulacea* (L.) Pruski, (**C**) *Ageratum conyzoides* L., (**D**) *Mimosa pudica* Linn., (**E**) *Lantana camara* L., (**F**) *Ipomoea cairica* (L.) Sweet, (**G**) *Bidens pilosa* L.

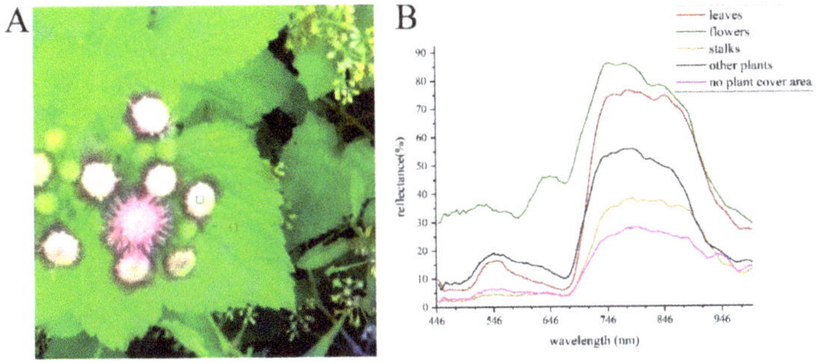

Figure 2. Extraction of raw spectral data. (**A**) Pixels of hyperspectral images used for raw spectra extraction, (**B**) raw spectral data.

2.2. Preprocessing Methods

Hyperspectral data have a large amount of interference redundancy information, which seriously affects the accuracy of the identification. Studies have shown that the interference of noise, astigmatism and baseline drift caused by human factors and background can be reduced by preprocessing [30]. At present, preprocessing methods are relatively mature and can be divided into derivative algorithms, smoothing algorithms, and light scattering correction algorithms. FD, SG smoothing and SNV were used in this paper.

2.2.1. First Derivative

First derivatives (FD) can reduce hyperspectral mutual interference and reduce noise in the analysis of hyperspectral data [31], and they are often used for baseline correction and background elimination in hyperspectral analysis [32]. The formula for spectral reflectance for the FD at wavelength point k can be solved as follows:

$$x_{i,FD} = \frac{x_i - x_{i+k}}{k} \tag{2}$$

In the above formula, k is the window width, $k = 9$.

2.2.2. Savitzky–Golay Smoothing

The smoothing methods include moving window smoothing and Savitzky–Golay (SG) convolution smoothing [33]. Compared with the moving window smoothing method, SG convolution smoothing can achieve a better denoising effect because moving window smoothing is limited by the number of windows. If the number of windows is too large or too small, the denoising effect will be affected. The two smoothing methods are similar, but the difference is that SG convolution smoothing focuses on the calculation of the central value, while moving window smoothing focuses on the number of windows. That is, SG smoothing is an algorithm based on polynomial smoothing, which smooths the spectral curve through moving windows combined with least squares fitting. Its calculation formula is as follows:

$$Y_{k,SG} = \frac{1}{N} \sum_{i=-m}^{m} (Y_{k+i} A_i) \tag{3}$$

where A_i is the smoothing coefficient, $Y_{k,SG}$ is the smoothed value of k points, Y_{k+i} is the unsmoothed value of k, i is the number of smoothed data points from this point to the left, m refers to the number of smoothed data points from this point to the right, and N is the normalization factor.

2.2.3. Standard Normal Variable Transformation

The standard normal variate (SNV) transformation is mainly used to eliminate the influence of solid particle size, surface scattering and optical path changes on the reflected spectrum [34]. SNV is the same as the standardized algorithm, except that the standardized algorithm addresses a set of spectra (based on the column of the spectral array), while SNV addresses a spectrum (based on the row of the spectrum) [35].

$$x_{snv} = (x - \omega) / \sqrt{\frac{\sum_{k=1}^{m} (x_k - \omega)^2}{(m-1)}} \tag{4}$$

on the type of $\omega = \frac{\sum_{k=1}^{m} x_k}{m}$, m is the number of wavelength points.

2.3. Dimension Reduction

The raw and preprocessing spectral data are 138-dimensional, and a high number of dimensions may lead to low efficiency of hyperspectral modeling and poor model performance. Therefore, it is necessary to reduce the dimensionality of hyperspectral data and retain most of the dataset information. Many studies have shown that the dimensionality reduction processing of hyperspectral data can not only shorten the time but can also improve the recognition efficiency to some extent [36]. In this paper, PCA and ACO were used to reduce the dimensionality of hyperspectral data of invasive plants.

2.3.1. Principal Component Analysis

PCA is a dimensionality reduction method that transforms multidimensional data features into a few comprehensive features and uses variable information represented by a few variable features [37,38]. The PCA dimension reduction transformation process is as follows: suppose that the n-dimensional preprocessing result variable $(x_1, x_2, x_3, \cdots, x_n)$ to perform the linear transformation of Formula (5) on variable x_i $(i = 1, 2, 3, \cdots, n)$ to obtain the comprehensive variable $(y_1, y_2, y_3, \cdots, y_n)$ is,

$$\begin{cases} y_1 = a_{11}x_1 + a_{12}x_2 + a_{13}x_3 + \cdots + a_{1n}x_n \\ y_2 = a_{21}x_1 + a_{22}x_2 + a_{23}x_3 + \cdots + a_{2n}x_n \\ \vdots \\ y_n = a_{n1}x_1 + a_{n2}x_2 + a_{n3}x_3 + \cdots + a_{nn}x_n \end{cases} \tag{5}$$

Equation (5) linear transformation satisfies the following conditions:
(1) $y_1, y_2, y_3, \cdots, y_n$ are linearly independent;
(2) $a_{i1}^2 + a_{i2}^2 + a_{i3}^2 \cdots + a_{in}^2 = 1, (i = 1, 2, 3, \cdots, n)$;
(3) the composite variables are arranged according to their variances.

The process of PCA is based on the contribution rate of a single principal component and the cumulative contribution rate. When the cumulative contribution rate of the first m principal components reaches 85%, the first m principal components can represent most of the information of the n-dimensional data to achieve dimensionality reduction. The calculation formulas of the principal component contribution rate and cumulative contribution rate are as follows:

$$\eta = \frac{\lambda_i}{\sum_{i=1}^{n} \lambda_i} \tag{6}$$

$$\rho = \frac{\sum_{k=1}^{m} \lambda_k}{\sum_{i=1}^{n} \lambda_i} \ (m = 1, 2, 3, \cdots, n) \tag{7}$$

In Formulas (6) and (7), λ_i is the principal component contrast.

2.3.2. Ant Colony Optimization

ACO is a probabilistic dimension reduction method for spectral feature selection of hyperspectral data [38]. Based on ACO, the main idea is to feature the selection problem of modeling for solving the minimum cost path search graph, thus creating a node of the search space and design procedures to find the manners of solution. ACO states the path that the ant chooses of the choice probability of the feature subset, based on the pheromone and heuristic information to compute the ant's choice. In this paper, the Fisher criterion is used to calculate heuristic information, and the ant's transition from one node to another node is determined by probability. Assume that the transfer probability P_{ij}^{k} of ant k from node i to node j at time t is calculated by the following formula:

$$P_{ij}^{k}(t) = \begin{cases} \frac{\tau_{ij}^{\alpha}(t)\gamma_{ij}^{\beta}(t)}{\sum_{s \in allowed_k} \tau_{is}^{\alpha}(t)\gamma_{is}^{\beta}(t)}, j \in allowed_k \\ 0, else \end{cases} \tag{8}$$

When the ant completes a cycle, the pheromone on each path is calculated as follows:

$$\tau_{ij}(t+n) = (1+\rho)\tau_{ij} + \Delta\tau_{ij} \tag{9}$$

τ_{ij} refers to the pheromone concentration on the path from node i to node j at time t, and γ_{ij} is the heuristic function, which refers to the expectation of ant k choosing to remove node j from node i. The calculation formula is as follows:

$$\gamma_{ij} = {}^{1}/_{d_{ij}} \tag{10}$$

where d_{ij} is the distance from node i to node j. The value of $allowed_k$ = 1, 2, ..., n − 1, n is the total number of nodes, $allowed_k$ represents the nodes that ant k can select next, α, β is the weight parameter, which plays a role in the importance of pheromones and heuristics, ρ is the pheromone evaporation coefficient, and $\Delta\tau_{ij}$ is the pheromone concentration increment on the path from node i to node j.

2.4. Identification Models

The typical algorithms based on hyperspectral image data classification include the neighborhood algorithm, SVM, RF, and neural network. Domestic and foreign researchers have accomplished many achievements on invasive plants based on hyperspectral data, most of which are based on the spatial distribution [39], area extraction [40], and species identification of invasive plants [41]. According to the characteristics of the samples, SVM and RF were selected to identify seven invasive plants.

SVM is a classical machine learning algorithm based on traditional theory; the core content of this is to transform the object of our study into a high-dimensional feature space by nonlinear transformation and to construct a linear decision function in the high-dimensional space to realize the nonlinear decision function of the original space [42]. In this paper, the kernel function selected by SVM classifier is the radical basis function (RBF). For each SVM model, a grid search combined with a 5-fold cross validation method was used to optimize parameters. The search interval of the penalty coefficient and kernel parameter was set as 2^{-5}~2^{5}, the step size was 0.1.

RF is a combination of classification trees that implements randomization [43]. In the 1980s, Breiman et al. invented the classification tree algorithm [44]. It is calculated by repeating dichotomous data. In the same year, Breiman merged classification trees into random forests. RF generate multiple classification trees by randomizing the data and variables, and then obtaining statistical results. The main parameters of RF are the number of trees and the number of nodes. In this paper, the number of trees was set to 1000, and the number of nodes was set to 10 for all of the RF models.

2.5. Performance Evaluation

The performance of all of the combined models was evaluated based on four statistical parameters, namely, accuracy (A), average accuracy (AA), average precision (AP) and test time, which were often used to evaluate the performance of the classification models and calculated from Equations (11) and (12) [45].

$$Accuracy = \sum_{i=1}^{8} TP_i/N \times 100\% \tag{11}$$

$$Average\ Precision = \sum_{i=1}^{8} \left(\frac{TP_i}{TP_i + FP_i}\right)/8 \tag{12}$$

where *TP* is the true positive sample, *TN* is the true negative sample, *FN* is the false negative sample, *FP* is the false positive sample, *N* is the total sample number, *i* is the number of categories.

All of the aforementioned steps were coded and developed in the environment of MATLAB R2020a (The Math Works Inc., Natick, MA, USA). The GPU of the PC is a NVIDIA GTX 3060.

3. Results

3.1. Preprocessing

Raw and preprocessing spectral data of seven invasive plants and background are shown in Figure 3A–D. In Figure 3A, *M. micrantha* has the highest reflectivity in the wavelength range of approximately 700 to 900 nm, followed by *B. pilosa*, *M. pudica*, *L. cairica*, *A. conyzoides*, *S. calendulacea* and *L. camara*. In the range of approximately 450 to 680 nm, the background reflectance is relatively scattered, while in the range of approximately 700 nm to 998 nm, the background overlaps with the spectral reflectance of seven invasive species, and the overlap region between species was large. Therefore, it is challenging to classify seven invasive species simultaneously. To facilitate the identification of raw spectral data, three processing methods were used to reduce data noise and highlight the distribution patterns of reflectivity with wavelength. The original spectral data processed by SG smoothing are shown in Figure 3C. Compared with Figure 3A, the curve of Figure 3C is relatively smooth, and small fluctuations in reflectance over the whole wavelength range are eliminated. The other two processing methods, FD and SNV, remove other noise from the original spectral data. Figure 3B presents the FD processing of direct difference analysis of the raw spectral data, that is, noise reduction is performed at both ends of the spectral band. Compared with Figure 3B, the raw spectral data are obviously interfered with by artificial and systematic noises, and the intraclass difference of spectral data after FD processing is less than that of the original spectral data. Figure 3D displays the preprocessing data of SNV, and the intraclass difference in spectral data is reduced.

In summary, all three preprocessing methods can eliminate part of the spectral noise. SG smoothing reduced the small range of spectral fluctuations, FD and SNV reduced the intra-class differences in species, but none of the processing methods significantly reduced the inter-class differences in species and background. Therefore, it is necessary to determine the inter-class differences of the seven species through subsequent treatments. To determine a more appropriate pretreatment method, the next step is to analyze the impact of each processing method combined with dimension reduction.

3.2. Dimension Reduction by PCA

Based on the preprocessing results, dimension reduction analysis was conducted by PCA, as shown in Figure 4A,C,E, which are the point states drawn in horizontal (first principal component scores) and vertical (second principal component scores) coordinates. In this way, we can preliminarily judge the convergence degree of FD, SG smoothing and SNV. According to Figure 4A,C,E, it is obvious that FD-PCA has a better sample clustering effect, but *S. calendulacea*, *L. camara* and *B. pilosa* overlap clearly. In Figure 4C, the seven invasive plants are relatively scattered; that is, after SG-PCA treatment, the sample points of invasive plants have almost no similar convergence. In terms of clustering performance, FD ranked first, SNV ranked second, and SG ranked third. The worst clustering performance of SG smoothing preprocessing is predictable because SG smoothing reduces noise but does not change details and overall trends. Therefore, more components need to be considered later.

The cumulative contribution rate of the first 138 principal components were calculated as shown in Figure 4B,D,F. It is clear that at around the first 22 principal components, all curves tended to be smooth and the cumulative principal component contribution rate approaches 100%. Therefore, the principal component characteristics of the first 22 dimensions can fully represent the spectral information of 138 dimensions. In other words, after PCA, the 138-dimensional spectrum is reduced to 22 dimensions.

3.3. Dimension Reduction by ACO

Based on the preprocessing results, feature selection was carried out by ACO, and the feature bands are shown in Table 1.

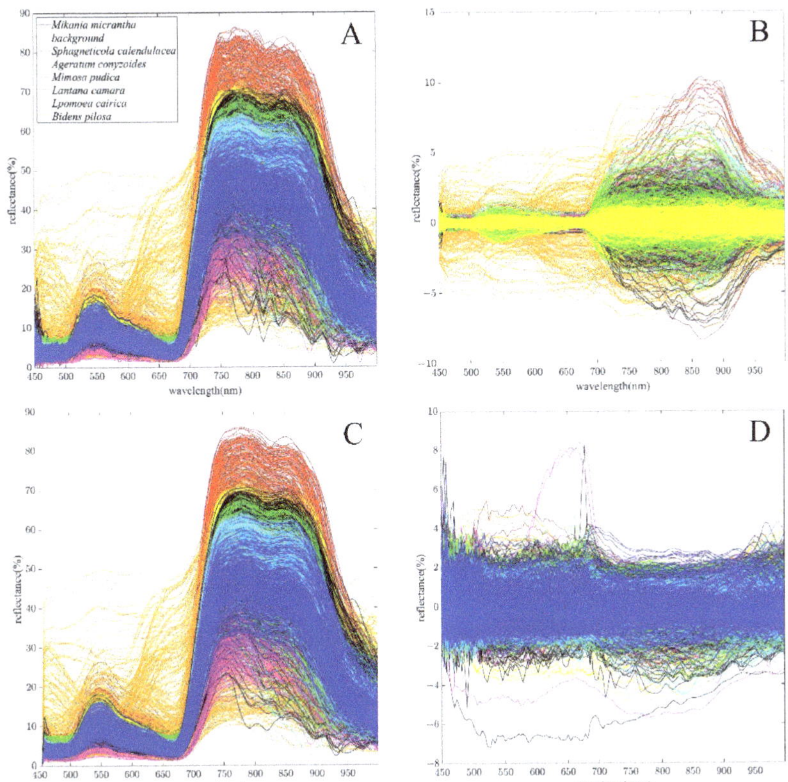

Figure 3. Raw spectral curve of 3359 samples (**A**), raw spectral curve, preprocessed by (**B**) first derivative, (**C**) SG smoothing, (**D**) standard normal variate.

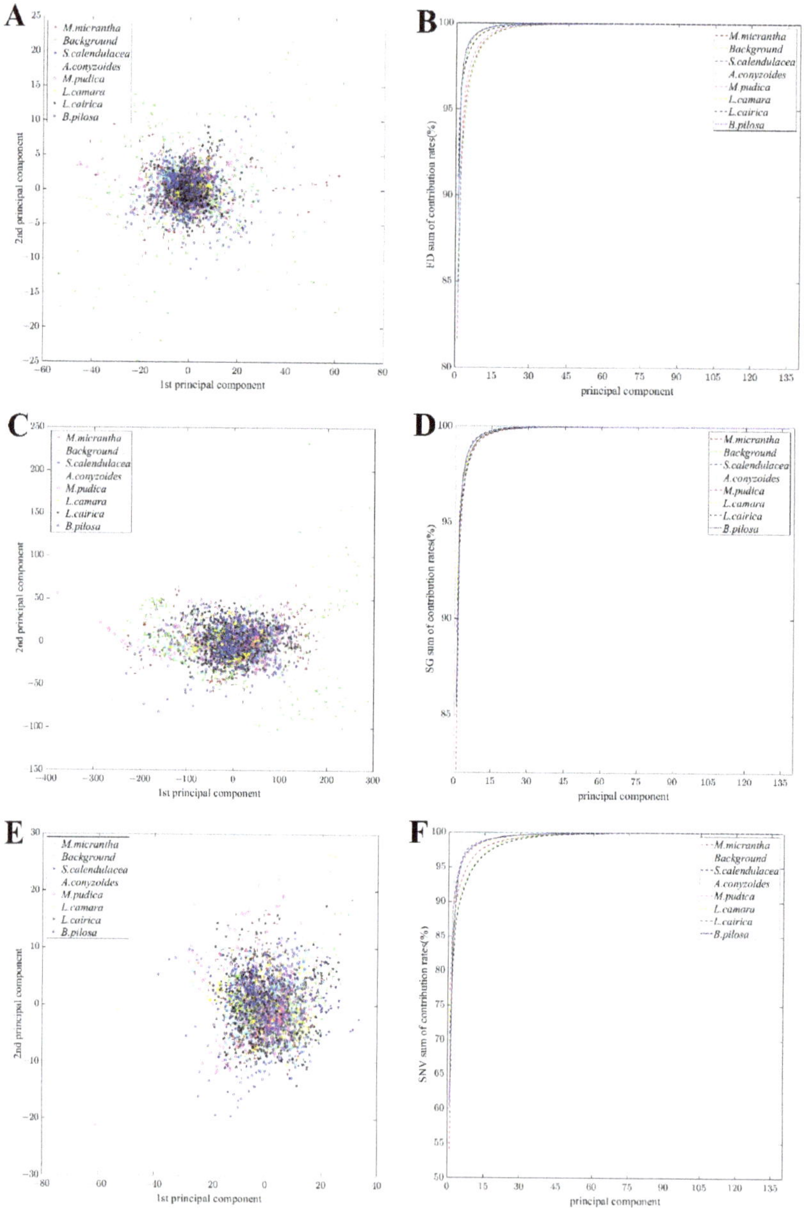

Figure 4. First and second principal component scores and cumulative contribution rate curve of the principal components of 3359 samples based on (**A**,**B**) first derivative, (**C**,**D**) SG smoothing, (**E**,**F**) standard normal variable transformation.

Table 1. Feature bands.

Preprocessing Methods	Number of Feature Bands	Feature Bands (nm)
FD	28	450, 454, 458, 462, 466, 470, 474, 478, 482, 486, 494, 498, 502, 506, 510, 514, 518, 526, 530, 542, 598, 614, 738, 770, 834, 858, 934, 942,
SG smoothing	20	450, 454, 458, 462, 466, 470, 474, 478, 482, 486, 490, 494, 510, 518, 526, 638, 658, 702, 798, 914
SNV	25	450, 454, 458, 462, 466, 470, 474, 478, 482, 486, 494, 498, 502, 506, 510, 514, 518, 526, 530, 542, 578, 662, 774, 822, 950

The number of feature bands selected based on FD was 28, SNV was 25 and SG was 20. Table 1 shows that the feature bands selected based on the three preprocessing methods are concentrated in the range of 450–700 nm. That is, the feature wavelengths are concentrated in the visible range.

Wavelengths of 450, 454, 458, 462, 466, 470, 474, 478, 482, 486, 494, 510, 518 and 526 nm were common to all three preprocessing methods, indicating that these wavelengths were important features in distinguishing the seven species. Based on ACO, the extracted feature wavelengths were relatively continuous between 450–700 nm and relatively scattered between 700–998 nm. Furthermore, the selection of feature wavelengths is related to the quality of the preprocessing. After all, the number of feature wavelengths extracted based on SG is the lowest and the number based on FD is the highest.

3.4. Identification Performance Evaluation of Different Combination Models

Among the 3359 samples, 2400 samples were used for model training and 959 samples were used for model testing. To explore the single species recognition rates and total recognition rates for each combined model, confusion matrices, single species recognition accuracy tables and total accuracy tables are presented, as shown in Figure 5, Tables 2 and 3 (Figure 5 and Table 2 only present the models with the highest accuracy under the same conditions).

Table 2. Recognition accuracy of combined model for single species.

Species \ Models	SNV-PCA-SVM	SG-ACO-SVM	SG-SVM	SNV-PCA-RF	SG-ACO-RF	SG-RF
M. micrantha	86.72%	92.19%	92.97%	80.00%	87.40%	86.44%
Background	91.74%	95.41%	94.50%	95.24%	90.23%	89.08%
S. calendulacea	70.31%	82.81%	89.06%	51.51%	67.18%	81.40%
A. conyzoides	83.20%	79.20%	90.40%	65.25%	81.99%	81.25%
M. pudica	69.75%	84.03%	93.28%	60.66%	82.46%	81.75%
L. camara	91.67%	96.30%	88.89%	63.25%	92.31%	95.12%
L. cairica	67.80%	80.51%	80.51%	56.41%	71.19%	76.27%
B. pilosa	79.03%	85.49%	85.49%	61.72%	80.17%	77.19%

SG-SVM showed outstanding advantages in the identification process among nearly all species. The SG-SVM identified 92.97%, 89.06%, 90.40%, 93.28%, 80.51% and 85.49% accuracy for *M. micrantha*, *S. calendulacea*, *A. conyzoides*, *M. pudica*, *L. cairica* and *B. Pilosa*, respectively, which are far better than those of SNV-PCA-SVM (86.72%, 70.31%, 83.20%, 69.75%, 67.80%, 79.03%), SNV-PCA-RF (80.00%, 51.51%, 65.25%, 60.66%, 56.41%, 61.72%). Meanwhile, the total accuracy (A), the average accuracy (AA), the average precision (AP) of SG-SVM are also excellent, reaching 89.36%, 89.39% and 89.54%, respectively. SG-ACO-SVM (86.76%, 86.99%, 87.22%) cannot be ignored. Compared to SG-SVM, the test time is also shorter, although A, AA and AP are all reduced. When the A of the model was reduced by about 3%, the cost time was nearly 5 times shorter. Moreover, for *L. camara* recognition,

SG-ACO-SVM (96.30%) is superior to SG-SVM (88.89%). SG-SVM and SG-ACO-SVM show advantages in recognition accuracy and time cost, respectively. In practice, when the total accuracy is relatively important, the SG-SVM model is selected, and when the test time cost is relatively important, the SG-ACO-SVM model is used.

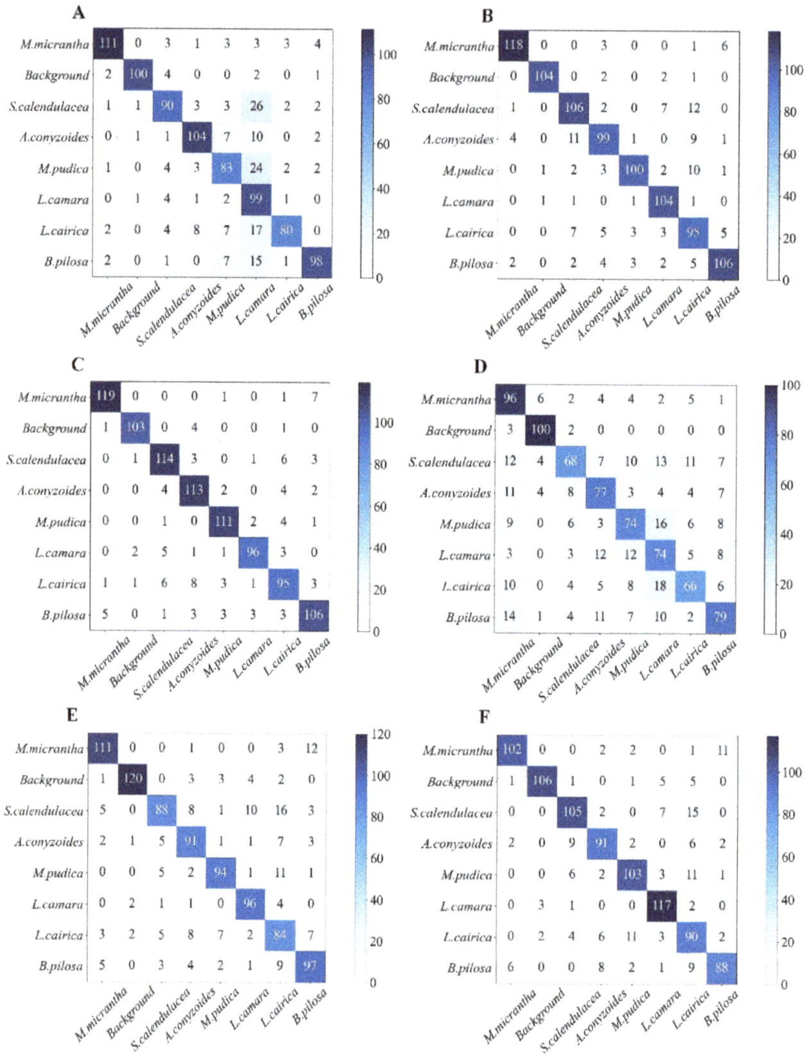

Figure 5. Confusion matrices of the combined models, (**A**) SNV-PCA-SVM, (**B**) SG-ACO-SVM, (**C**) SG-SVM, (**D**) SNV-PCA-RF, (**E**) SG-ACO-RF, (**F**) SG-RF.

Table 3. The total accuracy, average accuracy and average precision of the 18 combined models.

Models	A (%)	AA (%)	AP (%)	Test Time (s)
FD-PCA-SVM	77.89	78.13	79.82	0.09493
SG-PCA-SVM	72.99	73.37	75.63	0.09382
SNV-PCA-SVM	79.77	80.03	82.82	0.08942
FD-ACO-SVM	48.28	48.95	54.09	0.0766
SG-ACO-SVM	86.76	86.99	87.22	0.0567
SNV-ACO-SVM	67.67	68.16	67.76	0.0746
FD-SVM	67.26	67.66	67.70	0.3524
SG-SVM	89.36	89.39	89.54	0.2639
SNV-SVM	84.15	84.32	84.21	0.329
FD-PCA-RF	63.30	63.77	64.70	0.1031
SG-PCA-RF	64.44	64.42	64.36	0.0978
SNV-PCA-RF	66.11	66.76	66.75	0.1035
FD-ACO-RF	61.00	61.34	63.41	0.0841
SG-ACO-RF	81.44	81.61	81.74	0.0788
SNV-ACO-RF	67.36	68.08	68.01	0.0821
FD-RF	62.15	62.40	62.71	0.1407
SG-RF	83.63	83.56	84.15	0.0862
SNV-RF	71.01	71.23	71.11	0.1015

4. Discussion

Currently, manual monitoring is still the mainstay of invasive plant monitoring. However, this makes the monitoring of invasive species both laborious and time-consuming. Given these facts, unmanned aerial vehicle (UAV) with hyperspectral cameras can offer significant opportunities for invasive plant monitoring. Hyperspectral remote sensing, which combines the advantages of imaging and spectroscopy, has been successfully applied to invasive plant monitoring. Previous studies using hyperspectral imaging to monitor the spectral information of invasive plants at different time periods have been very effective. The results obtained in this study show that the spectral distribution of different invasive plants during the same period is different, but the differences are not significant. In general, the lower reflectance of invasive plants occurs in the visible part of the spectrum (400–700 nm), and the absorption of photosynthetic pigments in the region associated with it is largely concentrated in the blue (450-chlorophyll b) and red (650-chlorophyll a) regions.

The highest total accuracy of the combined model in this paper was only 89.39%. On the one hand, only traditional machine learning models are used, on the other hand, SVM and RF is mostly suitable for binary classification [46,47]. After all, SVM and RF can often achieve high accuracy in the application of binary classification [48]. Although the LIBSVM toolbox is capable of solving multiclassification problems [49], the total classification accuracy of samples is not high because the difference of spectral features between various samples is not big, or the difference of features between samples is further reduced after normalization. Dimensionality reduction reduces the training and testing time of the model, but inevitably leads to the loss of some important information, which is the main reason why the accuracy of dimensionality reduction models is lower than that of non-dimensionality reduction models. Of course, the small amount of data is also an important reason. Therefore, it is necessary to expand the datasets or try to use a new type of model to improve the accuracy and efficiency.

Recent studies have used hyperspectral imaging combined with deep learning methods to monitor invasive plants [50,51]. Meanwhile, deep learning is very popular in species classification [52,53], but it does not mean that the recognition effect of deep learning is superior to that of other algorithms. For example, in the recognition of grapevine species, the recognition effect of a SVM is not inferior to that of a convolutional neural network [54]. However, it cannot be denied that deep learning is a current trend in image recognition. To further improve the accuracy of recognition, it is necessary to adopt a deep learning algorithm.

By studying the leaf spectra of the seven invasive plants, 18 combination models were able to distinguish seven invasive plants within a short time interval. By analyzing different combination models, it is found that the total accuracy of SG-SVM is the highest, followed by SG-ACO-SVM. It is worth noting that the spectrum of SG-ACO-SVM was only 20-dimensional, while the spectrum of SG-SVM was 138-dimensional. Although the total accuracy of the SG-ACO-SVM is slightly less, the testing time has been significantly shortened. Of course, the reduction in dimensionality does not necessarily indicate that the models will take less time to test. After all, the SG-PCA-RF and SNV-PCA-RF models are more time consuming than their non-dimensional reduction counterparts. However, the training time for the reduced dimension models is clearly lower than that for the non-dimensional reduction models, and this is true for all of the combined models. While the total accuracy is reduced, a satisfactory recognition accuracy is obtained in a shorter period of time. This is the purpose of using dimensionality reduction and the reason for comparing the 18 combined models.

In this paper, seven invasive plants were identified using 18 models, though promising, in-depth research is necessary before the technique can be applied to the field for accurate monitoring. One aspect that should be considered in future research is the spectral monitoring of invasive plants over their entire period. On the other hand is the consideration of the average spectra of more organs of the invasive plant, such as the flowers, stems, leaves, etc. of the plant. To achieve this, operational challenges have to be overcome. Such as the selection and collection of optimal conditions for hyperspectral images of the plant under test, the determination of the spectral distributions and correlated color temperatures of solar radiations, the calibration of black and white under different atmospheric conditions, the possibility of mounting a hyperspectral sensor on an unmanned aircraft, the calibration of the sensor, the collection, storage and analysis.

The main work of this research was to complete the identification of seven invasive species in four small fields. The image samples used were hyperspectral images taken with a handheld spectrometer. In practical applications, it is often necessary to identify a large range of invasive plants, which requires the installation of a hyperspectral imager on a UAV for image acquisition. Our study provides a reliable reference for hyperspectral image data processing and the classification of a variety of invasive plants.

5. Conclusions

Hyperspectral data of invasive plants in the wild were used as sample data in this work, and three groups of different results were obtained by different preprocessing methods (FD, SG, and SNV). Based on the three preprocessing results, nine groups of data were obtained by combining dimensionality reduction (PCA, ACO) and no dimensionality. The nine groups of data were combined with classification methods (SVM and RF) to establish different classification models to obtain the best recognition method for invasive plants.

The results show that the SG preprocessing method has a better preprocessing effect on hyperspectral data, and most values of the models obtained under this condition are optimal. Based on the preprocessing results, by comparing the results of dimensionality reduction and no dimensionality reduction, the method of dimension reduction may not improve the accuracy of recognition, but it can improve time efficiency. Compared with the RF classifier, the SVM classifier has a better recognition effect on invasive plants. In the dimension reduction method based on PCA and ACO, SVM model recognition based on the ACO dimension reduction method is better. Among the 18 models, the 2 best models are SG-ACO-SVM (AA, 86.99%, AP, 87.22%, TT, 0.0567) and SG-SVM (AA, 89.39%, AP, 89.54%, test time, 0.2639).

Author Contributions: X.Q. and Y.H.: methodology, software, validation, writing. X.L.: methodology, software, validation, writing-original draft, writing-review. F.W.: software. Z.S. and L.Y.: project administration. X.P. and S.L.: project administration. W.Q.: project administration, Writing-original draft, Writing-review and editing. All authors contributed to the article and approved the submitted version. All authors have read and agreed to the published version of the manuscript.

Funding: The work in this paper was supported by the National Natural Science Foundation of China (32272633, 61871475), the Key Research and Development Program of Nanning (20192065), and the Guangdong Science and Technology Planning Project (2017A020216022).

Data Availability Statement: Not applicable.

Acknowledgments: The authors thank the native English-speaking experts for polishing our paper.

Conflicts of Interest: The authors declare no conflict of interest.

References

1. Pan, X.B.; Zhang, J.Q.; Xu, H.; Zhang, X.L.; Zhang, W.; Song, H.H.; Zhu, S.F. Spatial similarity in the distribution of invasive alien plants and animals in China. *Nat. Hazards* **2015**, *77*, 1751–1764. [CrossRef]
2. Liu, Y.J.; Oduor, A.M.O.; Zhang, Z.; Manea, A.; Tooth, I.M.; Leishman, M.R.; Xu, X.L.; Van Kleunen, M. Do invasive alien plants benefit more from global environmental change than native plants? *Glob. Chang. Biol.* **2017**, *23*, 3363–3370. [CrossRef] [PubMed]
3. Miller, K.M.; McGill, B.J.; Weed, A.S.; Seirup, C.E.; Comiskey, J.A.; Matthews, E.R.; Perles, S.; Schmit, J.P. Long-term trends indicate that invasive plants are pervasive and increasing in eastern national parks. *Ecol. Appl.* **2021**, *31*, e02239. [CrossRef] [PubMed]
4. Nguyen, N.A.; Eskelson, B.N.I.; Meitner, M.J.; Murray, T. People's Knowledge and Risk Perceptions of Invasive Plants in Metro Vancouver, British Columbia, Canada. *Environ. Manag.* **2020**, *66*, 985–996. [CrossRef] [PubMed]
5. Qian, W.Q.; Huang, Y.Q.; Liu, Q.; Fan, W.; Sun, Z.Y.; Dong, H.; Wan, F.H.; Qiao, X. UAV and a deep convolutional neural network for monitoring invasive alien plants in the wild. *Comput. Electron. Agric.* **2020**, *174*, 105519. [CrossRef]
6. Carlier, J.; Davis, E.; Ruas, S.; Byrne, D.; Caffrey, J.M.; Coughlan, N.E.; Dick, J.T.A.; Lucy, F.E. Using open-source software and digital imagery to efficiently and objectively quantify cover density of an invasive alien plant species. *J. Environ. Manag.* **2020**, *266*, 110519. [CrossRef]
7. Cohen, J.G.; Lewis, M.J. Development of an Automated Monitoring Platform for Invasive Plants in a Rare Great Lakes Ecosystem Using Uncrewed Aerial Systems and Convolutional Neural Networks. In Proceedings of the International Conference on Unmanned Aircraft Systems (ICUAS), Athens, Greece, 1–4 September 2020; pp. 1553–1564.
8. Ma, Y.; Liu, X.; Liu, M.; Shi, L.; Zhang, Z.; Zhao, N. Feature analysis and model monitoring of different florescences of Mikania micrantha based on hyper-spectrum imaging. *J. Yunnan Univ.-Nat. Sci. Ed.* **2021**, *43*, 290–298. [CrossRef]
9. Qiu, Y.; Lu, J. Dynamic simulation of Spartina alterniflora based on CA-Markov model-a case study of Xiangshan bay of Ningbo City, China. *Aquat. Invasions* **2018**, *13*, 299–309. [CrossRef]
10. Liu, M.; Mao, D.; Wang, Z.; Li, L.; Man, W.; Jia, M.; Ren, C.; Zhang, Y. Rapid Invasion of Spartina alterniflora in the Coastal Zone of Mainland China: New Observations from Landsat OLI Images. *Remote Sens.* **2018**, *10*, 1933. [CrossRef]
11. Ren, G.; Zhou, L.; Liang, J.; Lu, F.; Wang, A.; Wang, J.; Li, X.; Ma, Y. Monitoring the Invasion of Spartina Alterniflora Using Hyperspectral Remote Sensing Image of GF-5. *Adv. Mar. Sci.* **2021**, *39*, 312–326.
12. Barbosa, J.M.; Asner, G.P.; Martin, R.E.; Baldeck, C.A.; Hughes, F.; Johnson, T. Determining Subcanopy Psidium cattleianum Invasion in Hawaiian Forests Using Imaging Spectroscopy. *Remote Sens.* **2016**, *8*, 33. [CrossRef]
13. Khare, S.; Latifi, H.; Ghosh, S.K. Multi-scale assessment of invasive plant species diversity using Pleiades 1A, RapidEye and Landsat-8 data. *Geocarto Int.* **2018**, *33*, 681–698. [CrossRef]
14. Gosselin, H.; Sagan, V.; Maimaitiyiming, M.; Fishman, J.; Belina, K.; Podleski, A.; Maimaitijiang, M.; Bashir, A.; Balakrishna, J.; Dixon, A. Using Visual Ozone Damage Scores and Spectroscopy to Quantify Soybean Responses to Background Ozone. *Remote Sens.* **2020**, *12*, 93. [CrossRef]
15. Tarantino, C.; Casella, F.; Adamo, M.; Lucas, R.; Beierkuhnlein, C.; Blonda, P. Ailanthus altissima mapping from multi-temporal very high resolution satellite images. *Isprs J. Photogramm. Remote Sens.* **2019**, *147*, 90–103. [CrossRef]
16. Xu, Y.; Zhang, C.; Jiang, R.; Wang, Z.; Zhu, M.; Shen, G. UAV-based hyperspectral images and monitoring of canopy tree diversity. *Biodivers. Sci.* **2021**, *29*, 647–660. [CrossRef]
17. Zheng, D.; Shen, G.; Wang, B.; Dai, G.; Lin, F.; Hu, J.; Ye, J.; Fang, S.; Hao, Z.; Wang, X.; et al. Classification of dominant species in coniferous and broad-leaved mixed forest on Changbai Mountain based on UAV-based hyperspectral image and deep learning algorithm. *Chin. J. Ecol.* **2022**, *41*, 1024–1032.
18. Sabat-Tomala, A.; Raczko, E.; Zagajewski, B. Comparison of Support Vector Machine and Random Forest Algorithms for Invasive and Expansive Species Classification Using Airborne Hyperspectral Data. *Remote Sens.* **2020**, *12*, 516. [CrossRef]
19. Liu, Y.; Chang, M.; Xu, J. High-Resolution Remote Sensing Image Information Extraction and Target Recognition Based on Multiple Information Fusion. *IEEE Access* **2020**, *8*, 121486–121500. [CrossRef]

20. Yan, R.; Peng, J.; Ma, D. Dimensionality reduction based on parallel factor analysis model and independent component analysis method. *J. Appl. Remote Sens.* **2019**, *13*, 014532. [CrossRef]
21. Sun, H.; Zhang, L.; Rao, Z.H.; Ji, H.Y. Determination of moisture content in barley seeds based on hyperspectral imaging technology. *Spectrosc. Lett.* **2020**, *53*, 751–762. [CrossRef]
22. Pang, L.; Xiao, J.; Ma, J.J.; Yan, L. Hyperspectral imaging technology to detect the vigor of thermal-damaged Quercus variabilis seeds. *J. For. Res.* **2021**, *32*, 461–469. [CrossRef]
23. Shen, L.; Gao, M.; Yan, J.; Li, Z.L.; Duan, S.B.J.R.S. Hyperspectral Estimation of Soil Organic Matter Content using Different Spectral Preprocessing Techniques and PLSR Method. *Remote Sens.* **2020**, *12*, 1206. [CrossRef]
24. Xu, H.; Zhang, H.; He, W.; Zhang, L. Superpixel-based spatial-spectral dimension reduction for hyperspectral imagery classification. *Neurocomputing* **2019**, *360*, 138–150. [CrossRef]
25. Garcia-Salgado, B.P.; Ponomaryov, V.I.; Sadovnychiy, S.; Reyes-Reyes, R. Efficient dimension reduction of hyperspectral images for big data remote sensing applications. *J. Appl. Remote Sens.* **2020**, *14*, 032611. [CrossRef]
26. Yang, Q.L.; Wan, X.X.; Xiao, G.S. Multispectral Dimension Reduction Algorithm Based on Partial Least Squares. *Laser Optoelectron. Prog.* **2020**, *57*, 013003. [CrossRef]
27. Papp, L.; van Leeuwen, B.; Szilassi, P.; Tobak, Z.; Szatmari, J.; Arvai, M.; Meszaros, J.; Pasztor, L. Monitoring Invasive Plant Species Using Hyperspectral Remote Sensing Data. *Land* **2021**, *10*, 29. [CrossRef]
28. Bellia, L.; Blaszczak, U.; Fragliasso, F.; Gryko, L. Matching CIE illuminants to measured spectral power distributions: A method to evaluate non-visual potential of daylight in two European cities. *Sol. Energy* **2020**, *208*, 830–858. [CrossRef]
29. Huang, Y.; Li, J.; Yang, R.; Wang, F.; Li, Y.; Zhang, S.; Wan, F.; Qiao, X.; Qian, W. Hyperspectral Imaging for Identification of an Invasive Plant Mikania micrantha Kunth. *Front. Plant Sci.* **2021**, *12*, 626516. [CrossRef]
30. Ge, X.-Y.; Ding, J.-L.; Wang, J.-Z.; Sun, H.-L.; Zhu, Z.-Q. A New Method for Predicting Soil Moisture Based on UAV Hyperspectral Image. *Spectrosc. Spectr. Anal.* **2020**, *40*, 602–609. [CrossRef]
31. Saberioon, M.; Cisar, P.; Labbe, L.; Soucek, P.; Pelissier, P. Spectral imaging application to discriminate different diets of live rainbow trout (*Oncorhynchus mykiss*). *Comput. Electron. Agric.* **2019**, *165*, 104949. [CrossRef]
32. Golhani, K.; Balasundram, S.K.; Vadamalai, G.; Pradhan, B. Estimating chlorophyll content at leaf scale in viroid-inoculated oil palm seedlings (*Elaeis guineensis* Jacq.) using reflectance spectra (400 nm–1050 nm). *Int. J. Remote Sens.* **2019**, *40*, 7647–7662. [CrossRef]
33. Feng, J.; Liu, Y.H.; Shi, X.W.; Wang, Q.Q. Potential of hyperspectral imaging for rapid identification of true and false honeysuckle tea leaves. *J. Food Meas. Charact.* **2018**, *12*, 2184–2192. [CrossRef]
34. Asaari, M.S.M.; Mishra, P.; Mertens, S.; Dhondt, S.; Inze, D.; Wuyts, N.; Scheunders, P. Close-range hyperspectral image analysis for the early detection of stress responses in individual plants in a high-throughput phenotyping platform. *Isprs J. Photogramm. Remote Sens.* **2018**, *138*, 121–138. [CrossRef]
35. Liu, J.B.; Han, J.C.; Chen, X.; Shi, L.; Zhang, L. Nondestructive detection of rape leaf chlorophyll level based on Vis-NIR spectroscopy. *Spectrochim. Acta Part A-Mol. Biomol. Spectrosc.* **2019**, *222*, 117202. [CrossRef]
36. Hamim, M.; El Moudden, I.; Pant, M.D.; Moutachaouik, H.; Hain, M. A Hybrid Gene Selection Strategy Based on Fisher and Ant Colony Optimization Algorithm for Breast Cancer Classification. *Int. J. Online Biomed. Eng.* **2021**, *17*, 148–163. [CrossRef]
37. Weng, S.; Tang, P.; Yuan, H.; Guo, B.; Yu, S.; Huang, L.; Xu, C. Hyperspectral imaging for accurate determination of rice variety using a deep learning network with multi-feature fusion. *Spectrochim. Acta Part A-Mol. Biomol. Spectrosc.* **2020**, *234*, 118237. [CrossRef]
38. Bonah, E.; Huang, X.; Yi, R.; Aheto, J.H.; Yu, S. Vis-NIR hyperspectral imaging for the classification of bacterial foodborne pathogens based on pixel-wise analysis and a novel CARS-PSO-SVM model. *Infrared Phys. Technol.* **2020**, *105*, 103220. [CrossRef]
39. Masemola, C.; Cho, M.A.; Ramoelo, A. Assessing the Effect of Seasonality on Leaf and Canopy Spectra for the Discrimination of an Alien Tree Species, Acacia Mearnsii, From Co-Occurring Native Species Using Parametric and Nonparametric Classifiers. *IEEE Trans. Geosci. Remote Sens.* **2019**, *57*, 5853–5867. [CrossRef]
40. Kopec, D.; Zakrzewska, A.; Halladin-Dabrowska, A.; Wylazlowska, J.; Kania, A.; Niedzielko, J. Using Airborne Hyperspectral Imaging Spectroscopy to Accurately Monitor Invasive and Expansive Herb Plants: Limitations and Requirements of the Method. *Sensors* **2019**, *19*, 2871. [CrossRef]
41. Kganyago, M.; Odindi, J.; Adjorlolo, C.; Mhangara, P. Evaluating the capability of Landsat 8 OLI and SPOT 6 for discriminating invasive alien species in the African Savanna landscape. *Int. J. Appl. Earth Obs. Geoinf.* **2018**, *67*, 10–19. [CrossRef]
42. Wang, C.X.; Wang, S.L.; He, X.G.; Dong, H. The Identification of Beef Varieties by Fusing Image Information Based on Hypersepctral Image Technology. *Spectrosc. Spectr. Anal.* **2020**, *40*, 911–916. [CrossRef]
43. Huang, H.; Liu, J.; Liu, S.; Wu, T.; Jin, P. A method for classifying tube structures based on shape descriptors and a random forest classifier. *Measurement* **2020**, *158*, 107705. [CrossRef]
44. Breiman, L. Random forests. *Mach. Learn.* **2001**, *45*, 5–32. [CrossRef]
45. Dash, J.P.; Watt, M.S.; Paul, T.S.H.; Morgenroth, J.; Pearse, G.D. Early Detection of Invasive Exotic Trees Using UAV and Manned Aircraft Multispectral and LiDAR Data. *Remote Sens.* **2019**, *11*, 1812. [CrossRef]
46. Hsu, C.W.; Lin, C.J. A comparison of methods for multiclass support vector machines. *IEEE Trans. Neural Netw.* **2002**, *13*, 415–425. [CrossRef]

47. Xia, S.; Chen, B.; Wang, G.; Zheng, Y.; Gao, X.; Giem, E.; Chen, Z. mCRF and mRD: Two Classification Methods Based on a Novel Multiclass Label Noise Filtering Learning Framework. *IEEE Trans. Neural Netw. Learn. Syst.* **2022**, *33*, 2916–2930. [CrossRef]
48. Li, L.; Li, L.; Wu, Y.; Ye, M. Experimental Comparisons of Multi-class Classifiers. *Inform.-J. Comput. Inform.* **2015**, *39*, 71–85.
49. Liu, M. English speech emotion recognition method based on speech recognition. *Int. J. Speech Technol.* **2022**, *25*, 391–398. [CrossRef]
50. Theron, J.; Pryke, J.S.; Latte, N.; Samways, M.J. Mapping an alien invasive shrub within conservation corridors using super-resolution satellite imagery. *J. Environ. Manag.* **2022**, *321*, 116023. [CrossRef]
51. Zhao, H.; Zhong, Y.; Wang, X.; Hu, X.; Luo, C.; Boitt, M.; Piiroinen, R.; Zhang, L.; Heiskanen, J.; Pellikka, P. Mapping the distribution of invasive tree species using deep one-class classification in the tropical montane landscape of Kenya. *Isprs J. Photogramm. Remote Sens.* **2022**, *187*, 328–344. [CrossRef]
52. Gonzalez-Perez, A.; Abd-Elrahman, A.; Wilkinson, B.; Johnson, D.J.; Carthy, R.R. Deep and Machine Learning Image Classification of Coastal Wetlands Using Unpiloted Aircraft System Multispectral Images and Lidar Datasets. *Remote Sens.* **2022**, *14*, 3937. [CrossRef]
53. Lake, T.A.; Runquist, R.D.B.; Moeller, D.A. Deep learning detects invasive plant species across complex landscapes using Worldview-2 and Planetscope satellite imagery. *Remote Sens. Ecol. Conserv.* **2022**. [CrossRef]
54. Fernandes, A.M.; Utkin, A.B.; Eiras-Dias, J.; Cunha, J.; Silvestre, J.; Melo-Pinto, P. Grapevine variety identification using "Big Data" collected with miniaturized spectrometer combined with support vector machines and convolutional neural networks. *Comput. Electron. Agric.* **2019**, *163*, 104855. [CrossRef]

plants

MDPI

Article

Identifying and Managing Areas under Threat in the Iberian Peninsula: An Invasion Risk Atlas for Non-Native Aquatic Plant Species as a Potential Tool

Argantonio Rodríguez-Merino

Independent Researcher, 28019 Madrid, Spain; argantonio.rodriguez@gmail.com

Abstract: Predicting the likelihood that non-native species will be introduced into new areas remains one of conservation's greatest challenges and, consequently, it is necessary to adopt adequate management measures to mitigate the effects of future biological invasions. At present, not much information is available on the areas in which non-native aquatic plant species could establish themselves in the Iberian Peninsula. Species distribution models were used to predict the potential invasion risk of (1) non-native aquatic plant species already established in the peninsula (32 species) and (2) those with the potential to invade the peninsula (40 species). The results revealed that the Iberian Peninsula contains a number of areas capable of hosting non-native aquatic plant species. Areas under anthropogenic pressure are at the greatest risk of invasion, and the variable most related to invasion risk is temperature. The results of this work were used to create the Invasion Risk Atlas for Alien Aquatic Plants in the Iberian Peninsula, a novel online resource that provides information about the potential distribution of non-native aquatic plant species. The atlas and this article are intended to serve as reference tools for the development of public policies, management regimes, and control strategies aimed at the prevention, mitigation, and eradication of non-native aquatic plant species.

Keywords: alien species; biological invasions; MaxEnt; non-native species; species distribution models; virtual atlas

1. Introduction

When non-native species arrive and become established in new areas, there are often environmental and/or economic costs associated with the impacts of the species and the resulting management needs [1–3]. In addition, biological invasions are considered a major threat to natural systems and a key cause of biodiversity loss worldwide [4,5]. They are a problem that shows no signs of abating, given that the number of non-native species has not stopped growing in recent decades and is expected to continue along this global trajectory for the foreseeable future [6–8].

Consequently, predicting the risk of biological invasions remains one of the greatest challenges in conservation [9] and, therefore, it is essential to develop mitigation strategies that target future invasions [10]. In addition, gathering information on invasion risk is a fundamental part of horizon scanning, a technique that can guide management strategies for eradicating or controlling populations of established non-native species and preventing the arrival and establishment of non-native species with invasive potential [11]. Early warning systems can play an extremely important role, given that control and eradication are generally most effective when efforts are deployed during the introduction phase of the invasion before species become established [9].

Species distribution models (SDMs) are currently the most common approach used to explain and predict where species occur in space and time [12]. Using species occurrence data and a set of predictor variables, SDMs can infer species distributions in areas for which no observations are available [13]. They are thus helpful tools for identifying the areas most likely to be colonized by and host a given non-native species; in this way, it is possible to determine which areas are at the greatest risk of invasion [14].

Inland aquatic ecosystems, both freshwater and estuarine, provide innumerable goods and services and are, at the same time, especially vulnerable to biological invasions compared to terrestrial ecosystems for three key reasons. First, they experience a high rate of species introductions because of human activities. Second, they tend to be quite isolated. Third, they host a large number of endemic species [11,15–17]. Some of the most harmful invaders in inland aquatic ecosystems are aquatic plant species, given their ability to alter habitat structure and composition as well as ecosystem biogeochemistry and water quality [18]. Research suggests that the number of non-native aquatic plants that have been introduced to the Iberian Peninsula has been increasing over recent years [19], a pattern also observed in Europe [20]. Such is clearly concerning for inland aquatic habitats and their native species, especially considering that the Iberian Peninsula is particularly species-rich and harbors a large number of endemic taxa [21]. Yet, few resources are currently available for monitoring non-native species, especially inconspicuous groups such as aquatic plant species, which limits knowledge of the potential distribution of invasive species and the ability to develop appropriate preventive measures. In this context, the SDMs could be used to help governmental authorities direct time, resources, and energy to the locations at greatest risk of invasion. Indeed, for the Iberian Peninsula, SDMs have been successfully used with both established non-native aquatic plants [19,20] and with non-native aquatic plant species with invasive potential [22].

Although information on the occurrence and global distribution of non-native species is becoming increasingly available, thanks to databases and platforms such as GBIF (https://www.gbif.org/, accessed on July 2023), it remains difficult to obtain information on the potential distribution and hence invasion risk for certain species. This challenge means that environmental managers and planners lack the full breadth of data they need to develop and implement effective management and conservation strategies.

This study thus aims to fill some of the information gaps that exist regarding the potential distribution of non-native aquatic plant species in the Iberian Peninsula and to create practical resources from this information. The main objective is to produce a map of suitable areas for assessing the invasion risk of invasive and potentially invasive species in the Iberian Peninsula. To achieve this, there are three specific objectives: (1) to establish the methodological basis for mapping potential species distributions and, consequently, determining the invasive potential of non-native aquatic plants; (2) to identify the areas at greatest risk of invasion by non-native aquatic plants; and (3) to characterize these areas based on climatic, topographical, and anthropogenic variables.

2. Results

2.1. Occurrence Data

This study looked at 72 target species: 32 non-native species that have already become established in the Iberian Peninsula and 40 non-native species with invasive potential. After filtering the original data used to create the models, the record number ranged between 68 and 15,072 for the established species (Table S1) and between 22 and 6212 for the species with invasive potential (Table S2). The decision was made to perform the models without *Lemna aequinoctialis* Welw., a species that had been on the list created by the LIFE INVASAQUA Project (LIFE17 GIE/ES/000515) (hereafter Life Invasaqua project) [23]. Although good results were obtained with few data, *L. aequinoctialis* had too few records of occurrence ($n = 3$) once the protocol for cleaning up the distribution data had been applied.

2.2. Distribution of Currently Established Non-Native Aquatic Plant Species

The current known distribution of non-native aquatic plant species in the Iberian Peninsula are represented in Figure 1A. Of the 9039 cells in the regional map, 759 cells contained instances of occurrence for non-native aquatic plants. A given cell was occupied by one to six species (1 species: 555 cells; 2 species: 126 cells; 3 species: 40 cells; 4 species:

17 cells; 5 species: 19 cells; and 6 species: 2 cells). In general, most occupied cells were found in the western half of the Iberian Peninsula, specifically in coastal Portugal, around the mouth of the Tagus River, across a portion of the Guadiana Basin, and the coast of the southwestern Iberian Peninsula. Additionally, cells along the coast of the eastern Peninsula were occupied by a large number of non-native aquatic plant species. Indeed, only two cells containing six different species of non-native aquatic plants were found in this area.

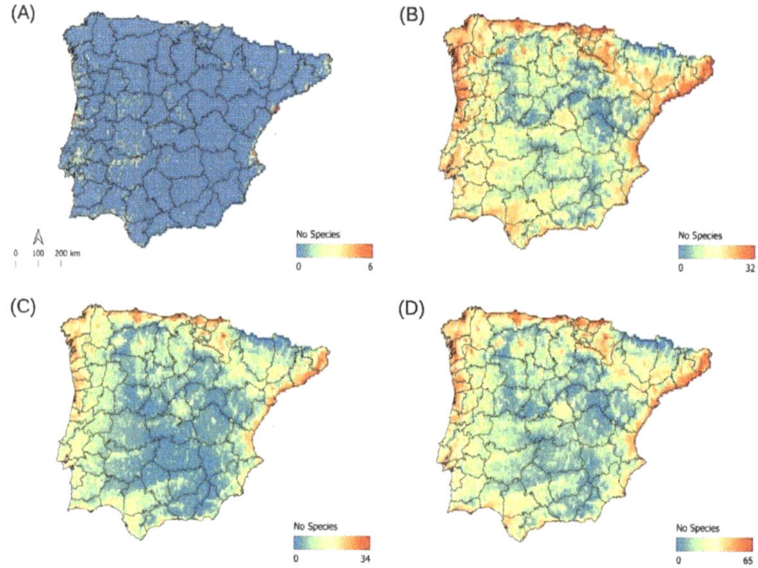

Figure 1. (**A**) Map of observed species richness for non-native aquatic plant species; (**B**) Map of potential species richness for established non-native aquatic plant species; (**C**) Map of potential species richness for non-native aquatic plant species with invasive potential; (**D**) Map of potential species richness for both established species and species with invasive potential.

2.3. Species Distribution Models

Overall, the models yielded highly accurate predictions. For the established species, AUC values ranged from 0.831 to 0.993 (mean AUC ± SD: 0.918 ± 0.058) (Table S1). For the species with invasive potential, the range was 0.810–0.993 (mean AUC ± SD: 0.940 ± 0.043) (Table S2). The lowest AUC value of 0.702 was associated with an established species, *Elodea canadensis* Michx.

For the established species, the most important predictor variables were temperature (combination of Bio5 and Bio6) as well as HFP and precipitation (combination of Bio15 and Bio17) (Figure 2A). For the species with invasive potential, these variables were temperature, precipitation, and HFP, the same variables as for the two groups of species combined (Figure 2A). Altitude was the variable of least importance for the established species, the species with invasive potential, and both groups combined (Figure 2A). For the established species, variable importance independently was ranked as follows: Bio6 > HFP > Bio17 > Bio5 > Altitude > Bio15 (Figures 2B and 3). For the species with invasive potential, the order of ranking was: Bio6 > Bio17 > Bio5 > HFP > Altitude > Bio15 (Figures 2B and 4). For both groups combined, it was: Bio6 > Bio17 > HFP > Bio5 > Altitude > Bio15 (Figure 2B).

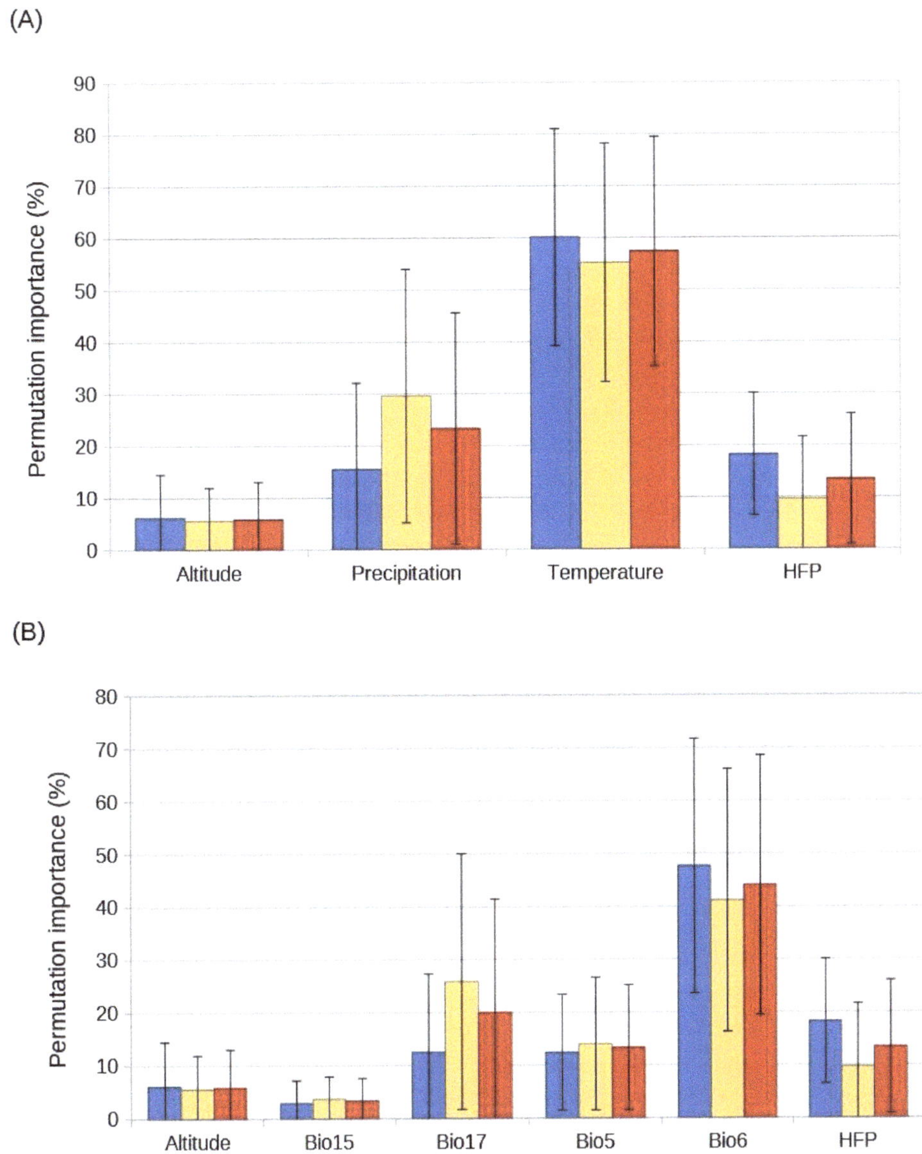

Figure 2. Relative importance (%) of the predictor variables used during SDM development and the standard error for each grouping of the target species: (**A**) Main variable types and (**B**) Independent variables. Bio5 = maximum temperature of the warmest month. Bio6 = minimum temperature of the coldest month. Bio15 = seasonality of precipitation. Bio17 = precipitation of the driest quarter. HFP = human footprint index. Altitude = digital elevation model. Blue bars: established species, yellow bars: potential species, and red bars: all established and potential species.

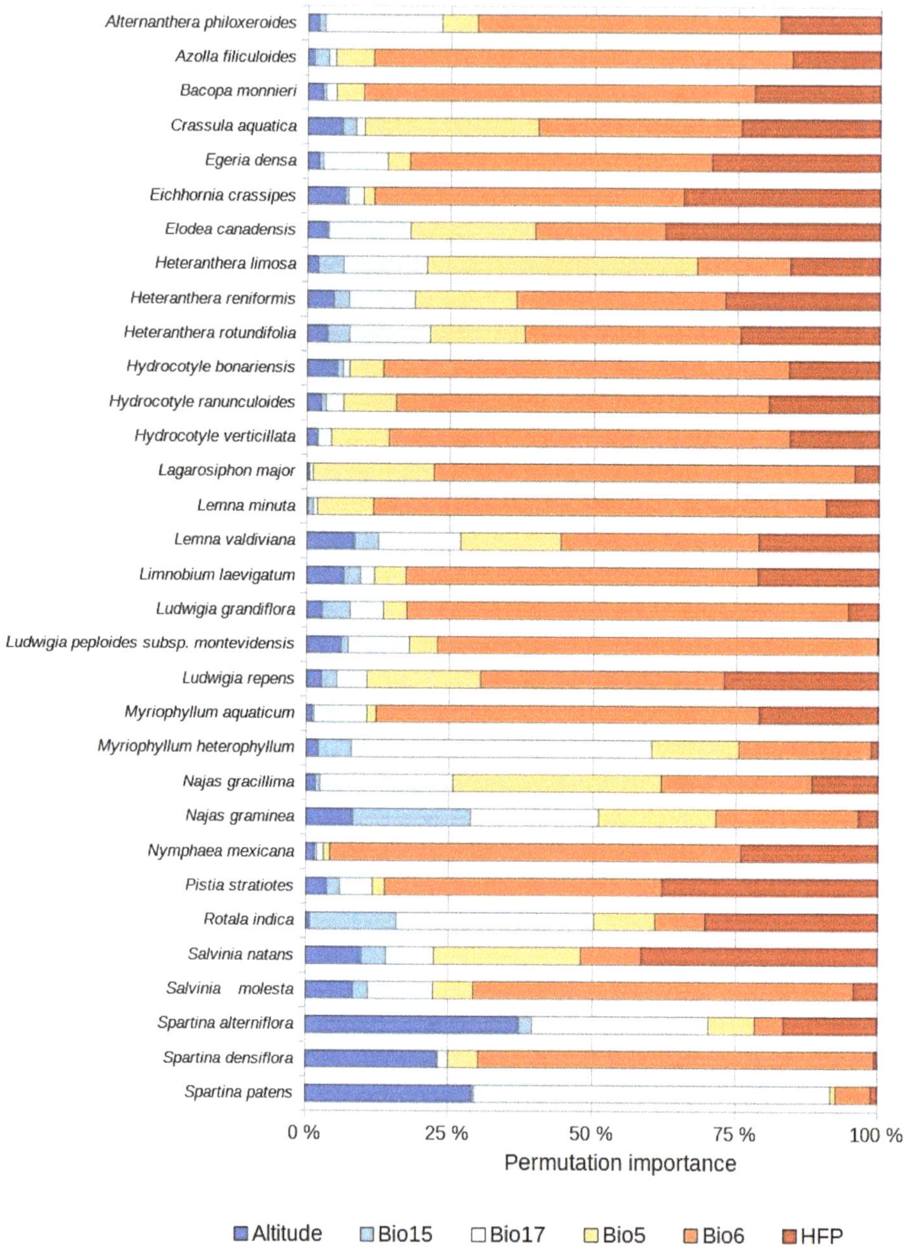

Figure 3. Relative importance (%) of the variables used to model the potential distributions of each established non-native aquatic plant species. Bio5 = maximum temperature of the warmest month. Bio6 = minimum temperature of the coldest month. Bio15 = seasonality of precipitation. Bio17 = precipitation of the driest quarter. HFP = human footprint index. Altitude = digital elevation model.

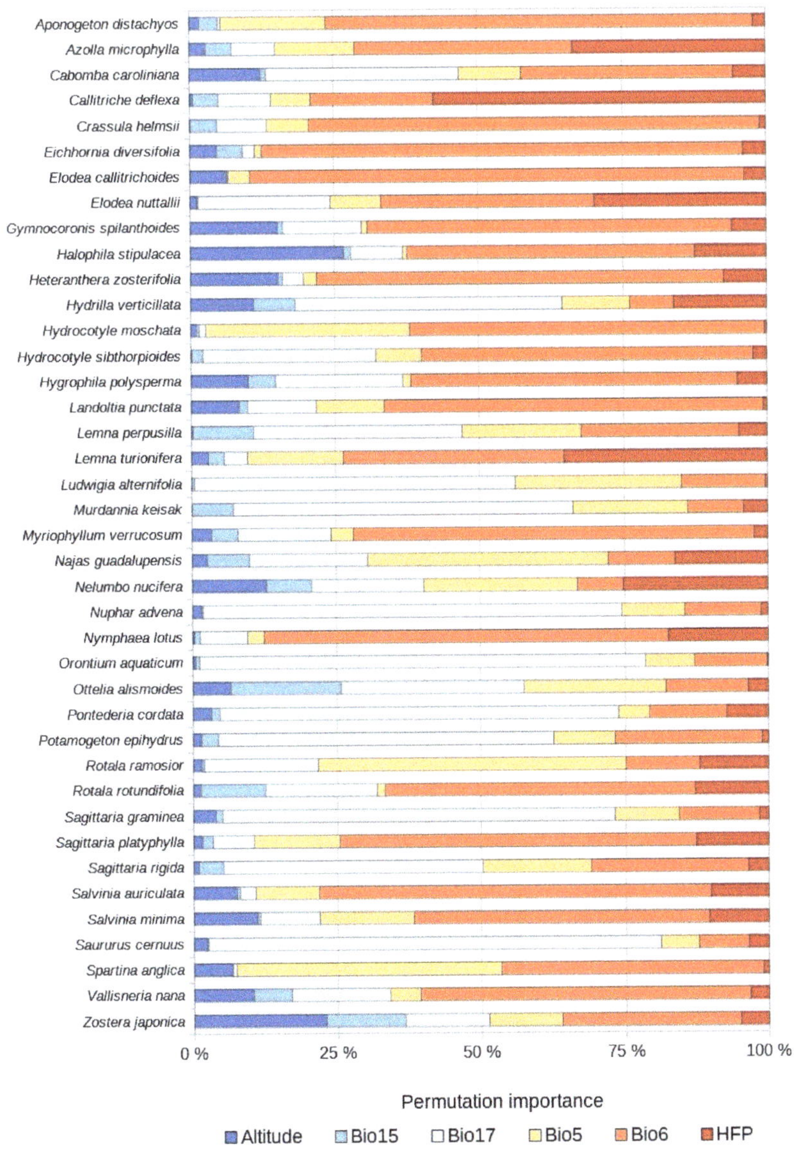

Figure 4. Relative importance (%) of the variables used to model the potential distributions of each non-native aquatic plant species with invasive potential. Bio5 = maximum temperature of the warmest month. Bio6 = minimum temperature of the coldest month. Bio15 = seasonality of precipitation. Bio17 = precipitation of the driest quarter. HFP = human footprint index. Altitude = digital elevation model.

2.4. Potential Species Distributions (Potential Invasion Risk)

The potential distribution of each target species is included in the online Invasion Risk Atlas ([24], Figure 5). Here zones and macroecological patterns associated with the greatest risk of invasion by each target species are indicated, whose distributions vary across the Iberian Peninsula (Figure 6).

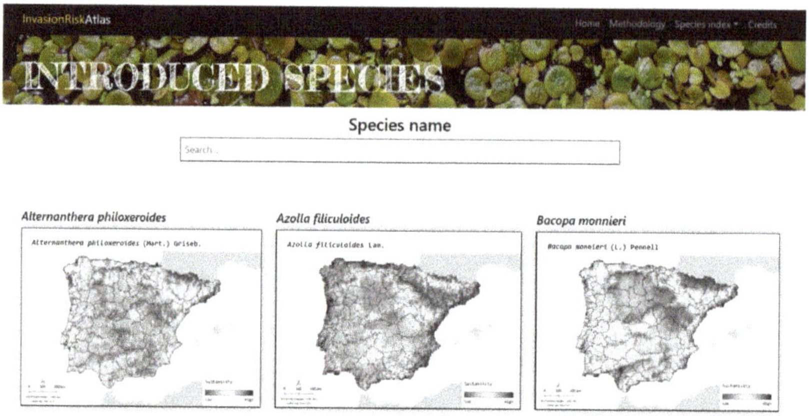

Figure 5. Illustration of the interface for the online Invasion Risk Atlas (https://InvasionRiskAtlas.github.io/, accessed on July 2023).

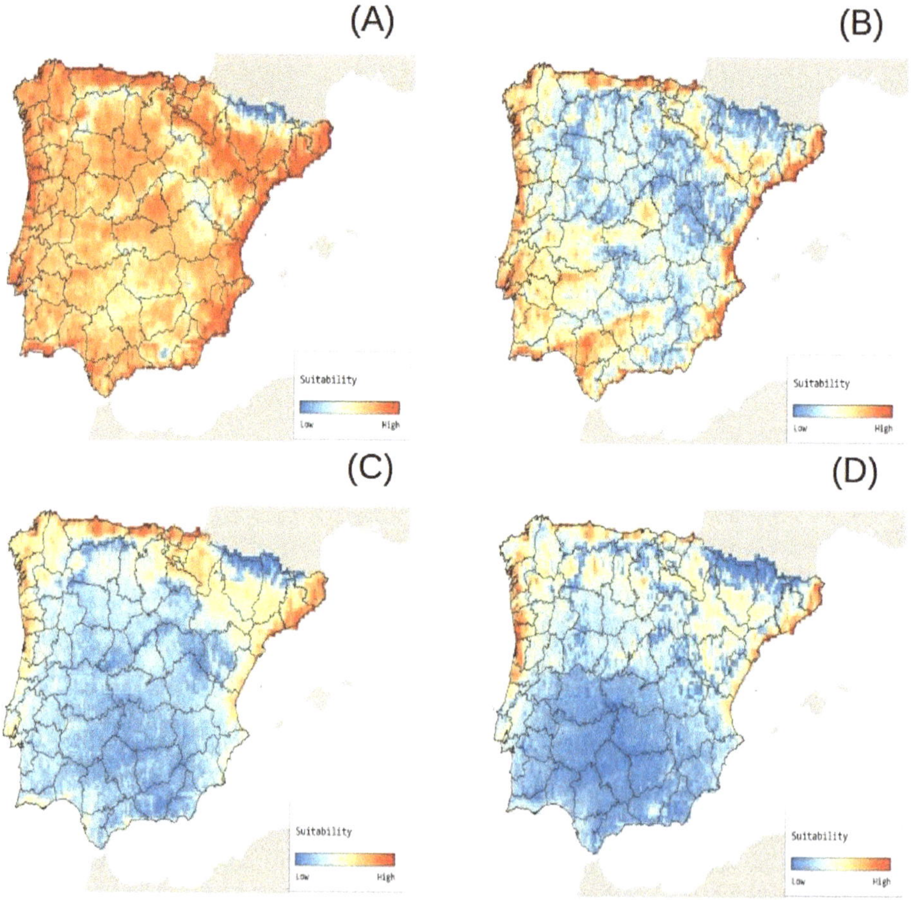

Figure 6. *Cont.*

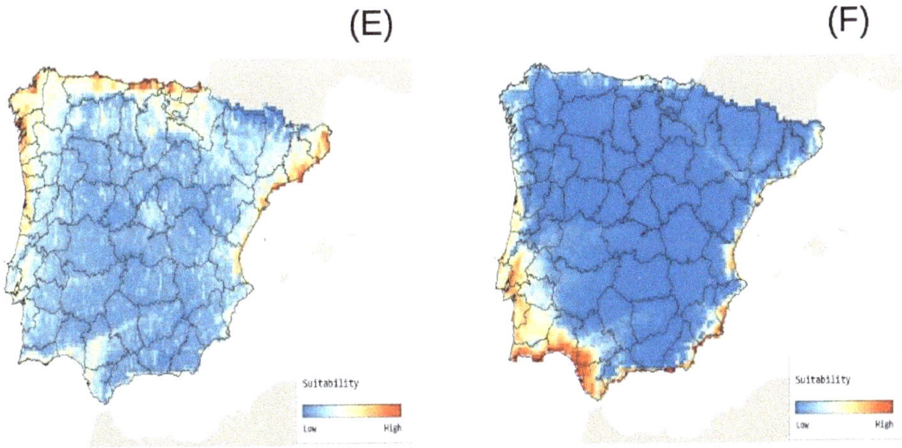

Figure 6. Potential distributions (invasion risk maps) of six non-native aquatic plant species that have been identified by experts as very high risk for Iberian inland waters (Life Invasaqua project; [23]): (**A**) *Azolla filiculoides* Lam. [24]; (**B**) *Eichhornia crassipes* (Mart.) Solms [25]; (**C**) *Ludwigia grandiflora* (Michx.) Greuter & Burdet [26]; (**D**) *Salvinia natans* (L.) All. [27]; (**E**) *Salvinia molesta* D.S. Mitch. [28]; (**F**) *Spartina densiflora* Brongn. [29].

2.5. Potential Richness of Non-Native Aquatic Plant Species

For the established species, the species with invasive potential, and both groups combined, potential species richness displayed similar associations with macroecological patterns (Figure 1B–D). Although the risk of invasion was heightened for the established species versus the species with invasive potential, the areas at greatest risk coincided: with coastal areas, large river basins, and densely populated urban centers. At the lowest risk were the most isolated zones of the Iberian Peninsula, namely high-elevation areas and areas far from large river basins and urban centers (Figure 1B–D).

The established species, the species with invasive potential, and both groups together showed similar responses to the variables used in the SDMs in terms of areas with the highest potential species richness (Figure 7A–F). HFP was positively correlated with potential species richness: the areas experiencing more anthropogenic pressure seem to be at greater risk of invasion by the target species (Figure 7A). Lower elevation areas appear to be more likely to host a larger number of non-native aquatic plant species (Figure 7B). With regards to temperature, species richness increased with Bio6 (Figure 7D) and was higher for Bio5 at values between 20 °C and 25 °C (Figure 7C). There was also a relationship between precipitation and potential species richness. First, areas with greater seasonality in precipitation seem more likely to be invaded by a greater number of species (Figure 7E). Precipitation levels displayed a bimodal association: potential species richness was high in areas with low precipitation (<20 mm) and areas with higher precipitation (100–220 mm); between and above these precipitation levels, potential species richness decreased drastically (Figure 7F).

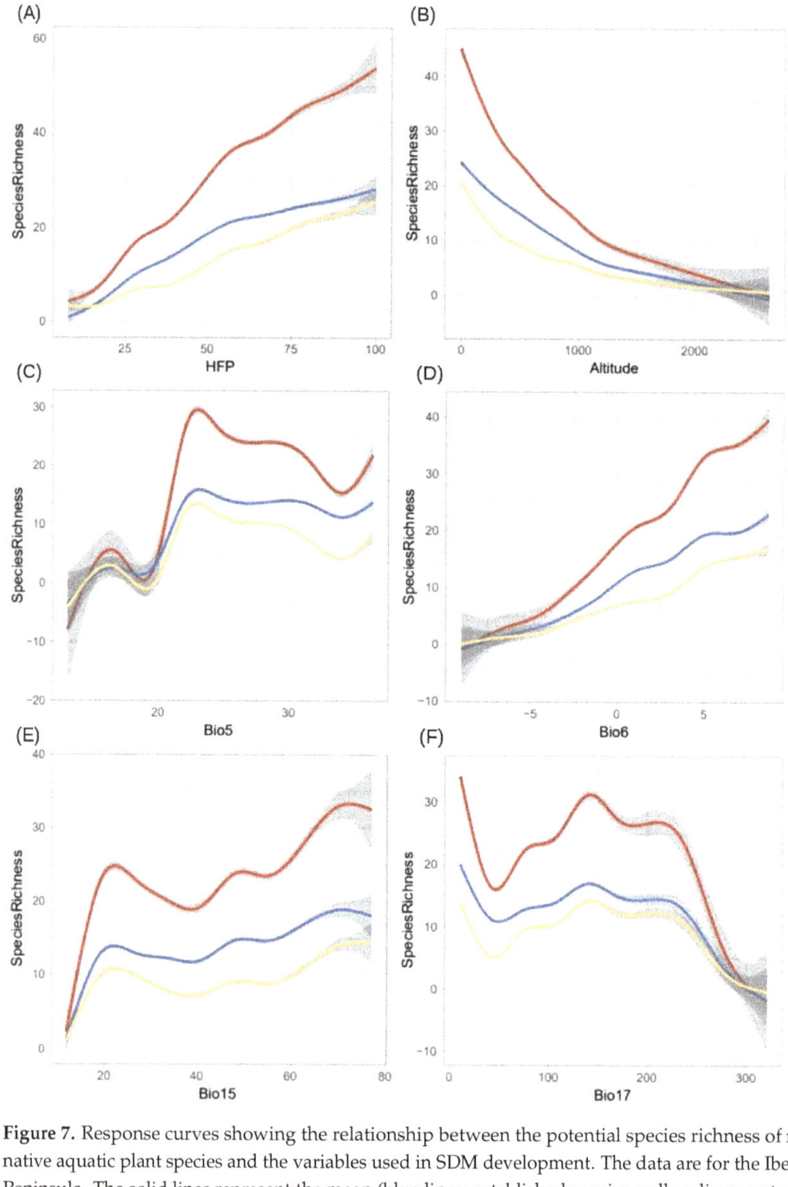

Figure 7. Response curves showing the relationship between the potential species richness of non-native aquatic plant species and the variables used in SDM development. The data are for the Iberian Peninsula. The solid lines represent the mean (blue lines: established species, yellow lines: potential species, and red lines: all established and potential species), and the shaded areas represent the 95% confidence intervals. (**A**) HFP index; (**B**) Digital elevation model (Altitude); (**C**) Maximum temperature of the warmest month (Bio5); (**D**) Minimum temperature of the coldest month (Bio6); (**E**) Seasonality of precipitation (Bio15); (**F**) Precipitation of the driest quarter (Bio17).

3. Discussion

Identifying and understanding the distribution of non-native species is one of the keys to establishing appropriate management strategies aimed at preventing, mitigating, and eradicating invasions. The Iberian Peninsula already hosts a large number of non-native aquatic species and is at substantial risk of future invasions [23]. This study underscores that lack of knowledge about the potential distribution of non-native aquatic plant species is one of the greatest limitations in biological invasion management and, consequently, about the possible threat they pose to the Iberian Peninsula, an important hotspot for aquatic plant species in Europe and around the Mediterranean [30,31].

Although information exists on the distribution of non-native plant species in the Iberian Peninsula, most are scattered across different sources, such as digital platforms and atlases. It also tends to be very general in nature and focused on known species distributions. A few references include Anthos (http://www.anthos.es/, accessed on January 2023), the Flora-On project (https://flora-on.pt/, accessed on January 2023), *Atlas de las plantas alóctonas de España* [32]. However, little information is available about the potential distribution and invasion risk of non-native plant species [33], specifically of non-native aquatic plant species. Such knowledge is available for certain established species [19,34] and for species with invasive potential [22]. When available, this information tends to be very specific and is not always easy to find. For instance, sometimes only a map of overall invasion risks is presented, which conveys information about all species without showing the potential distribution of target species independently. This study serves as an up-to-date and comprehensive source of information on the potential invasion risks of non-native aquatic plant species. It used the results of the Life Invasaqua project [23] to create a list of target species: non-native aquatic plant species currently established in the Iberian Peninsula as well as non-native aquatic plant species with invasive potential. In addition, the study's findings were used to create the online Invasion Risk Atlas [35]. Easy to use, the atlas shows the potential distribution and, consequently, the areas at greatest risk of invasion by each of the target species. Both this study and the atlas represent a major advance with regards to existing resources for understanding the risk of invasion by non-native aquatic plant species in the Iberian Peninsula.

In the Invasion Risk Atlas [35], the maps for the different target species show variable macroecological patterns, which makes sense given the diversity in species types and ecologies (Figure 6). That said, many of the target species display similar patterns with regards to the nature of the areas at greatest risk of invasion. These are the areas that have the capacity to host a large number of non-native aquatic plant species, and they tend to occur near densely populated zones, coastal zones, and large river basins. These results concur with those obtained for other non-native aquatic species in the Iberian Peninsula, e.g., [36,37] and, specifically, for non-native aquatic plant species [19,20]. It is also important to note that zones, where non-native aquatic plant species richness is higher, tend to overlap with zones where native aquatic plant species richness is greatest, a key finding that should inform the conservation of aquatic biodiversity in the Iberian Peninsula [31]. This overlap likely stems from the fact that these areas are characterized by pronounced environmental heterogeneity. The factors that favor biological diversity occur in tandem with anthropogenic forces that facilitate the arrival and establishment of non-native species in communities of native species [38].

As highlighted in previous studies, climatic variables strongly influence potential species distributions [39,40], especially at the scales examined in this study and of ecological importance for aquatic plant species [19,20,31,36,41,42]. In the models, temperature—specifically the minimum temperature of the coldest month—was the variable of greatest importance. Its positive relationship with potential species richness and, consequently, invasion risk may be a consequence of how much temperature can limit species growth and development. In addition, the results showed that potential species richness was higher when the maximum temperature fell between 20 °C and 25 °C. Similar results have been obtained in previous research on non-native aquatic species [36,43] and on aquatic plant species in particular [20,42]. Additionally, precipitation, or the lack thereof, was also an important variable in the SDMs. First, it is an indicator of water availability and levels of soil moisture, both important abiotic variables in the life cycles of the target species. Notably, precipitation is key to species dispersion because it can create connections among water bodies. Second, precipitation levels serve as an indirect indicator of other conditions that could affect species distributions, such as modifications in the water regime or the concentration of nutrients [44–46]. In the case of the Iberian Peninsula, the areas at higher risk of invasion are those where rainfall is lower and displays greater seasonality, a pattern mentioned in past studies [19,20]. These finding underscores that climatically moderate zones may play an important role in facilitating invasions.

Consequently, the shifts in temperature and precipitation brought about by climate change could alter distribution patterns and the risk of invasion by non-native aquatic plant species. They could promote plant development and proliferation by creating new habitats with favorable conditions. Indeed, some currently non-problematic species could start to cause problems, move northwards, and even disappear entirely from their present ranges [42,43,47,48].

The use of non-climatic variables in SDMs can help improve predictions. Here, the inclusion of altitude yielded decent results in SDMs for certain species, such as *Spartina* spp. Schreb., *Halophila stipulacea* (Forssk.) Asch., and *Zostera japonica* Asch. & Graebn. (Figures 3 and 4), which all occur in estuarine environments and marshes. This result is noteworthy because some have suggested that it is questionable to include this variable in SDMs for plant species [49]. Thus, it would appear to be important to account for altitude when modeling the distributions of aquatic plant species, especially those found in habitats where this variable matters, such as marshes and estuaries.

Although it is increasingly common for variables related to water chemistry to be included in SDMs, e.g., [41,50], their use is not always appropriate, as it will depend on the modeling scale. For example, water chemistry can help explain local species distribution patterns [40] but may not make the same contribution to higher-scale models, but see [41,51]. In contrast, variables that provide information about anthropogenic pressures, habitat accessibility, and propagule pressure have proven to be useful in SDMs for non-native species in general [36,39,52] and for non-native aquatic plant species in particular [19,20]. As a result, this study included the HFP index in its SDMs. Indeed, anthropogenic pressures can help explain the presence of certain non-native taxa in climatically unfavorable areas or their absence from climatically favorable areas: successful colonization relies at least in part on factors such as propagule pressure [52].

The results show that the HFP index and altitude present opposite trends in relation to potential species richness (Figure 7A,B). This pattern can be explained by the fact that, in the Iberian Peninsula, high-elevation areas are less accessible and, consequently, more pristine than lower-elevation areas, which experience greater anthropogenic pressure, propagule pressure, and nutrient availability. Other research has found similar results: non-native species are rare in or entirely absent from isolated areas but abundant in accessible areas [43]. In this context, it is worth noting that the richness of non-native species is directly correlated with the degree of anthropogenic activity within regions [36].

At the same time, limited and irregular levels of precipitation in low-lying areas may interact with high levels of anthropogenic activity to increase nutrient concentrations in watersheds [53], which can promote the arrival and establishment of certain non-native aquatic plant species. Such a relationship would support a hypothesis proposed elsewhere [19], which was based on the positive association observed between irrigated farmland and invasion risk in the Iberian Peninsula.

Based on the above, it seems likely that using variables that quantify the degree of anthropogenic pressure could have two important functions: to improve SDM quality, as also highlighted elsewhere [52], and to indirectly incorporate information normally contained in variables absent from the models, such as the level of nutrients in aquatic environments.

Both the established non-native species and the species with invasive potential overlapped in their areas at greatest risk of invasion. That said, there were subtle differences between the two groups (Figure 1B,C). The main reason may be that the models for established species contain data on actual species presence and environmental conditions in both the Iberian Peninsula and other invaded regions. Such is not the case for the species with invasive potential. Consequently, it seems important to include as much occurrence data as possible from a species' native and invasive ranges when modeling potential distributions. Taking this approach will result in more robust models that account for the entire range of environmental conditions in which the species is found [47].

This study is not without its limitations. Sometimes the use of alternative settings is more appropriate for model development than the use of MaxEnt default options [54]. However, this would mean applying individual models for each species, which would go against standardizing and automating the process [55]. This is outside the scope of this work, as the original idea of this project is to implement a common protocol for all species studied, in order to save time and costs in all phases of control and management of non-native aquatic plant species. On the other hand, the lack of specific predictors based on the ecology of the species studied may be a limiting factor in this work [51]. However, as reflected in the manuscript and the references cited therein, the use of climatic variables and the impact of human activities on the area have worked well in the development of SDMs for non-native species [36,41]. It is possible that the use of general predictors may overestimate the potential distribution of the species studied, but it is preferable to overestimate rather than underestimate the potential distribution of species from the perspective of non-native species management [56]. It is important to remember that the resulting models are simplifications of reality, based on a small number of predictors that reflect habitat suitability rather than the absolute limits of species survival [56]. Despite the uncertainty of such spatial representations, SDMs are currently one of the most useful tools for detecting trends in the spread of non-native species.

This work is dynamic in nature and will evolve as new records on species presence are collected. In this way, the models will continue to improve in the future, as new non-native species are uncovered and studied by the scientific community. This characteristic is a strength, not a limitation. Indeed, the function of this article and the Invasion Risk Atlas [35] is to lay a foundation for developing targeted strategies and identifying new areas where non-native aquatic plant species occur. This information will help update the models and improve the resulting predictions.

4. Materials and Methods

4.1. Study Area

The focus of this study was the Iberian Peninsula, which comprises continental parts of Spain and Portugal. The Balearic Islands and the Macaronesian archipelagos were excluded, in accordance with recommendations made by the Life Invasaqua project [23]. Located in the Mediterranean Basin, the Iberian Peninsula is a hotspot of plant biodiversity [57,58]. It harbors a range of highly diverse native aquatic plant species [30,31] that are currently under threat given the arrival and presence of non-native aquatic plant species [19,20].

4.2. Species Occurrence Records

This study's list of target non-native aquatic plant species was taken from the Life Invasaqua project [23] and was expanded using the project's technical reports [59–61]. The list comprises 32 non-native aquatic plant species that have become established in the Iberian Peninsula (Table S1) as well as 41 non-native aquatic plant species with invasive potential (Table S2). The reference nomenclature was adopted from Tropicos (https://tropicos.org, accessed on August 2023).

Occurrence data for all 73 species were downloaded from GBIF (https://www.gbif.org/, accessed on January 2023; Table S3) and EASIN (https://easin.jrc.ec.europa.eu/easin, accessed on January 2023). A database was compiled after incorporating the global occurrence data (i.e., a species' native and non-native distributions). Such an approach makes it possible to include the environmental breadth acquired by species in newly colonized areas, which reflects the species' ability to acclimate [62]. It is therefore possible to produce more reliable maps than when using a more limited set of records [47].

Because data quality determines the model quality [63], a data cleaning protocol was applied. Erroneous data (geographical and taxonomic) were removed, as were data with a low degree of spatial resolution, i.e., zero coordinates and equal latitude and longitude data [64]. In addition, to reduce the influence of sampling bias in the models, duplicates were eliminated, and, to reduce spatial autocorrelation, the degree of data aggregation was reduced by retaining only one occurrence point per pixel with respect to the predictor variables grid [65,66]. Despite these efforts, it is possible that the data still displayed a degree of spatial autocorrelation. It may actually be a natural result of scattering processes and should not be artificially removed [67]. After the data cleaning process, the species records were ready for use in the SDMs and to establish macroecological patterns of species [55]. Species with less than ten records were excluded from the analyses [68].

All the analysis and management of the geographical data was performed using R v. 4.1.3 (https://www.r-project.org/ accessed on April 2023) and QGIS v. 3.28 (https://qgis.org/, accessed on May 2023). Furthermore, the QGIS PyQGIS library and Python (https://www.python.org/, accessed on May 2023) were used to automate repetitive processes.

4.3. Selection of Predictor Variables

The set of 19 bioclimatic predictor variables, which are described in Fick and Hijmans [69], and the digital elevation model (altitude) from the Shuttle Radar Topography Mission were downloaded from WorldClim (https://worldclim.org/, accessed on January 2023).

In addition, to include anthropogenic impacts, accessibility, and propagule pressure, values for the human footprint (HFP) index were downloaded from the Socioeconomic Data and Applications Center (https://sedac.ciesin.columbia.edu/, accessed on January 2023). This index includes eight variables that reflect regional anthropogenic pressures: human constructions, population density, electricity infrastructure, croplands, pastures, roads, railways, and waterways [70].

All the variable values were compiled and associated with the EPSG:3857 reference coordinates system (spatial resolution: 5 arc min). Given the large number of species being modeled, this level of resolution best balances computational space and time with model accuracy [36]. Variables were selected by determining their ecological significance for the target species, drawing on previous studies [41,42,47]. Next, to minimize model overfitting and optimize model parsimony, the number of variables was refined using the variance inflation factor (VIF). This factor expresses the degree of collinearity among variables; here, variables with a VIF value of greater than 5 were eliminated [11] using the R package HH [71]. The variables retained were the digital elevation model (altitude), the maximum temperature of the warmest month (Bio5), the minimum temperature of the coldest month (Bio6), the seasonality of precipitation (Bio15), the precipitation of the driest quarter (Bio17), and the human footprint (HFP) index.

4.4. Species Distribution Models

The SDMs were calibrated using MaxEnt v. 3.4.4 (maximum entropy modeling; https://biodiversityinformatics.amnh.org/open_source/maxent/, accessed on April 2023), an algorithm that combines machine learning with the maximum entropy principle. MaxEnt estimates the probability that a species will be present by determining the maximum entropy distribution based on known species occurrence [72]. MaxEnt was chosen because it can be employed with presence-only data, even if relatively few records are available, and because it is not strongly influenced by spatial sampling errors [54,63,73,74]. Furthermore, MaxEnt performs and predicts potential areas comparatively well against ensemble models [75], making it one of the most reliable techniques for modeling distributions of incomplete datasets [76].

The SDMs were carried out with MaxEnt using the default parameters ('Auto features', convergence = 10–5, the maximum number of iterations = 500, prevalence = 0.5, regularization value $\beta = 1$) [72,77,78], an approach previously used to model the potential distributions of aquatic plant species in general [31] and non-native aquatic plant species in particular [19,20]. As only presence data were used, 10,000 random background points were generated and employed as "pseudoabsences" when running the models. The predictive performance of each model was evaluated using a 10-fold cross-validation procedure, where 80% of the data were employed to train the model and the remaining 20% were utilized to validate the model. The predictive accuracy of the models was determined by analyzing the area under the ROC curve (AUC). Modeling results were considered to express plausible habitat occupation if the AUC was greater than 0.7 [79]. To obtain the most robust estimate of model predictive performance, the average of 10 replicates of each model was used to represent potential species distributions [80].

4.5. Online Invasion Risk Atlas

The resulting maps of invasion risk for each target species have been made available online, resulting in the first invasion risk atlas for the Iberian Peninsula ([35], https://InvasionRiskAtlas.github.io/, accessed on July 2023). The atlas was developed using HyperText Markup Language (HTML), Cascading Style Sheets (CSS), and JavaScript scripting language. In addition, the Bootstrap v. 5 framework (https://getbootstrap.com/, accessed on July 2023) was used to speed up the web development process, which was accomplished using the Visual Studio Code (https://code.visualstudio.com/, accessed on July 2023) editor. The atlas was designed in accordance with responsive web criteria, which means it can be optimally visualized regardless of the viewing device (i.e., cell phone, tablet, computer). The atlas' source code has been deposited in a GitHub repository (https://github.com/InvasionRiskAtlas?tab=repositories, accessed on July 2023), which has been made public via GitHub Pages.

4.6. Potential Richness of Non-Native Aquatic Plant Species

The modeling results, which expressed habitat suitability on a continuous scale, were transformed into maps expressing habitat suitability on a binary scale (presence/absence). This process used the maximum training sensitivity and specificity threshold, which produced the most accurate predictions [81]. The zones of binary suitability were delineated for the Iberian Peninsula and then summed to produce a map of the potential richness of non-native aquatic plant species, which showed the number of species per pixel that were likely to encounter conditions allowing colonization. Consequently, these are the locations where the risk of invasion by non-native aquatic plant species is expected to be highest [20].

In addition, using the ggplot2 package in R [82], response curves were generated for the variables included in the SDMs, with the goal of analyzing the relationships between these variables and the potential richness of non-native species in the Iberian Peninsula. The curves were fitted using generalized additive models (GAMs) because GAMs can handle nonlinear relationships between response and predictor variables [36,83].

5. Conclusions

The SDM results have emphasized the importance of considering climatic and anthropogenic variables when exploring the potential distributions of non-native aquatic plant species, which aligns with the findings of previous research on non-native aquatic plant species [19,20]. From the modeling based on these variables, it is evident that the Iberian Peninsula displays ideal environmental conditions for maintaining established populations of non-native aquatic plant species and for hosting new species with invasive potential. It is hoped that the results here as well as the taxon-specific information in the online Invasion Risk Atlas [35] will become important resources that can guide decision-making by environmental managers. This work should help build biogeographical and macroecological knowledge of aquatic plant species, a taxonomic group whose distribution patterns have been poorly characterized. Furthermore, this study presents an objective methodology that can be used with other taxonomic groups, which contrasts with previous work in this domain. Another advantage is that disseminating information about the potential distributions of non-native species can benefit the general public because knowledge and understanding are the first steps along the path to effective management. In addition, the information provided could help spur citizen science initiatives, for example, and thus contribute to data-gathering efforts and further model improvement. Finally, it is important to develop monitoring programs so that the arrival of non-native aquatic plant species in the Iberian Peninsula can be detected as early as possible. It is also essential to implement or support eradication, control, or mitigation programs in areas where the presence of non-native species can present a risk to the diversity of native species.

Supplementary Materials: The following supporting information can be downloaded at: https://www.mdpi.com/article/10.3390/plants12173069/s1, Table S1: Results from SDMs performed for non-native aquatic plant species that are already established in the Iberian Peninsula. Occurrences = sample size used to run the model at the global scale after cleaning up the original occurrence data. AUC = area under the ROC curve. SD = standard error. MaxTSS = maximum training sensitivity and specificity threshold; Table S2: Results from SDMs performed for non-native aquatic plant species that have the potential to invade the Iberian Peninsula. Occurrences = sample size used to run the model at the global scale after cleaning up the original occurrence data. AUC = area under the ROC curve. SD = standard error. MaxTSS = maximum training sensitivity and specificity threshold. Table S3: List of links to download species occurrence records from the GBIF platform.

Funding: This research received no external funding. This work and the Invasion Risk Atlas began as a personal project while the author was studying a Master's degree in Open Source Geospatial Technologies.

Data Availability Statement: The data are available in the author's personal GitHub repository (https://github.com/InvasionRiskAtlas?tab=repositories, accessed on July 2023).

Acknowledgments: I would like to thank the editors Dimitrakopoulos and Yi for inviting me to submit this manuscript for inclusion in the special issue of MDPI Plant Invasion. I am also grateful to Patricio Soriano for his feedback on the Invasion Risk Atlas, which has helped me to improve its design and web development. Thanks to two anonymous reviewers whose comments and suggestions improved the manuscript considerably.

Conflicts of Interest: The author declare no conflict of interest.

References

1. Cuthbert, R.N.; Pattison, Z.; Taylor, N.G.; Verbrugge, L.; Diagne, C.; Ahmed, D.A.; Leroy, B.; Angulo, E.; Briski, E.; Capinha, C.; et al. Global Economic Costs of Aquatic Invasive Alien Species. *Sci. Total Environ.* **2021**, *775*, 145238. [CrossRef]
2. Cuthbert, R.N.; Diagne, C.; Hudgins, E.J.; Turbelin, A.; Ahmed, D.A.; Albert, C.; Bodey, T.W.; Briski, E.; Essl, F.; Haubrock, P.J.; et al. Biological Invasion Costs Reveal Insufficient Proactive Management Worldwide. *Sci. Total Environ.* **2022**, *819*, 153404. [CrossRef] [PubMed]
3. Diagne, C.; Leroy, B.; Vaissière, A.-C.; Gozlan, R.E.; Roiz, D.; Jarić, I.; Salles, J.-M.; Bradshaw, C.J.A.; Courchamp, F. High and Rising Economic Costs of Biological Invasions Worldwide. *Nature* **2021**, *592*, 571–576. [CrossRef]
4. Bellard, C.; Cassey, P.; Blackburn, T.M. Alien Species as a Driver of Recent Extinctions. *Biol. Lett.* **2016**, *12*, 20150623. [CrossRef]
5. Pyšek, P.; Hulme, P.E.; Simberloff, D.; Bacher, S.; Blackburn, T.M.; Carlton, J.T.; Dawson, W.; Essl, F.; Foxcroft, L.C.; Genovesi, P.; et al. Scientists' Warning on Invasive Alien Species. *Biol. Rev.* **2020**, *95*, 1511–1534. [CrossRef]
6. Seebens, H.; Essl, F.; Dawson, W.; Fuentes, N.; Moser, D.; Pergl, J.; Pyšek, P.; van Kleunen, M.; Weber, E.; Winter, M.; et al. Global Trade Will Accelerate Plant Invasions in Emerging Economies under Climate Change. *Glob. Change Biol.* **2015**, *21*, 4128–4140. [CrossRef]
7. Seebens, H.; Blackburn, T.M.; Dyer, E.E.; Genovesi, P.; Hulme, P.E.; Jeschke, J.M.; Pagad, S.; Pyšek, P.; Winter, M.; Arianoutsou, M.; et al. No Saturation in the Accumulation of Alien Species Worldwide. *Nat. Commun.* **2017**, *8*, 14435. [CrossRef] [PubMed]
8. Seebens, H.; Bacher, S.; Blackburn, T.M.; Capinha, C.; Dawson, W.; Dullinger, S.; Genovesi, P.; Hulme, P.E.; van Kleunen, M.; Kühn, I.; et al. Projecting the Continental Accumulation of Alien Species through to 2050. *Glob. Change Biol.* **2021**, *27*, 970–982. [CrossRef]
9. Crafton, R.E. Modeling Invasion Risk for Coastal Marine Species Utilizing Environmental and Transport Vector Data. *Hydrobiologia* **2015**, *746*, 349–362. [CrossRef]
10. Roy, H.E.; Peyton, J.; Aldridge, D.C.; Bantock, T.; Blackburn, T.M.; Britton, R.; Clark, P.; Cook, E.; Dehnen-Schmutz, K.; Dines, T.; et al. Horizon Scanning for Invasive Alien Species with the Potential to Threaten Biodiversity in Great Britain. *Glob. Change Biol.* **2014**, *20*, 3859–3871. [CrossRef]
11. Fletcher, D.H.; Gillingham, P.K.; Britton, J.R.; Blanchet, S.; Gozlan, R.E. Predicting Global Invasion Risks: A Management Tool to Prevent Future Introductions. *Sci. Rep.* **2016**, *6*, 26316. [CrossRef] [PubMed]

12. Ryo, M.; Angelov, B.; Mammola, S.; Kass, J.M.; Benito, B.M.; Hartig, F. Explainable Artificial Intelligence Enhances the Ecological Interpretability of Black-Box Species Distribution Models. *Ecography* **2021**, *44*, 199–205. [CrossRef]
13. Guisan, A.; Thuiller, W. Predicting Species Distribution: Offering More than Simple Habitat Models. *Ecol. Lett.* **2005**, *8*, 993–1009. [CrossRef] [PubMed]
14. Thuiller, W.; Richardson, D.M.; Pyšek, P.; Midgley, G.F.; Hughes, G.O.; Rouget, M. Niche-Based Modelling as a Tool for Predicting the Risk of Alien Plant Invasions at a Global Scale. *Glob. Change Biol.* **2005**, *11*, 2234–2250. [CrossRef]
15. Gozlan, R.E.; Britton, J.R.; Cowx, I.; Copp, G.H. Current Knowledge on Non-Native Freshwater Fish Introductions. *J. Fish Biol.* **2010**, *76*, 751–786. [CrossRef]
16. Strayer, D.L. Alien Species in Fresh Waters: Ecological Effects, Interactions with Other Stressors, and Prospects for the Future. *Freshw. Biol.* **2010**, *55*, 152–174.
17. Moorhouse, T.P.; Macdonald, D.W. Are Invasives Worse in Freshwater than Terrestrial Ecosystems? *WIREs Water* **2015**, *2*, 1–8. [CrossRef]
18. Gallardo, B.; Clavero, M.; Sánchez, M.I.; Vilà, M. Global Ecological Impacts of Invasive Species in Aquatic Ecosystems. *Glob. Change Biol.* **2016**, *22*, 151–163. [CrossRef]
19. Rodríguez Merino, A.; Fernández Zamudio MD, R.; García Murillo, P. An Invasion Risk Map for Non-Native Aquatic Macrophytes of the Iberian Peninsula. *An. Jard. Bot. Madr.* **2017**, *74*, e055. [CrossRef]
20. Rodríguez-Merino, A.; García-Murillo, P.; Cirujano, S.; Fernández-Zamudio, R. Predicting the Risk of Aquatic Plant Invasions in Europe: How Climatic Factors and Anthropogenic Activity Influence Potential Species Distributions. *J. Nat. Conserv.* **2018**, *45*, 58–71. [CrossRef]
21. Araújo, M.B.; Lobo, J.M.; Moreno, J.C. The Effectiveness of Iberian Protected Areas in Conserving Terrestrial Biodiversity. *Conserv. Biol. J. Soc. Conserv. Biol.* **2007**, *21*, 1423–1432. [CrossRef]
22. Rodríguez-Merino, A. Detección de la distribución potencial de plantas acuáticas emergentes invasoras en la península Ibérica, un reto para la conservación. In *Bosque Mediterráneo y Humedales: Paisaje, Evolución y Conservación: Aportaciones Desde la Biogeografía*; Gosálvez Rey, R.U., Díaz Sanz, M.C., García Rayego, J.L., Serrano de la Cruz Santos-Olmo, M.A., Jerez García, O., Eds.; Almud, Ediciones de Castilla-La Mancha: Ciudad Real, Spain, 2018; Volume 1, pp. 512–522, ISBN 978-84-948075-6-5.
23. Oficialdegui, F.J.; Zamora-Marín, J.M.; Guareschi, S.; Anastácio, P.M.; García-Murillo, P.; Ribeiro, F.; Miranda, R.; Cobo, F.; Gallardo, B.; García-Berthou, E.; et al. A Horizon Scan Exercise for Aquatic Invasive Alien Species in Iberian Inland Waters. *Sci. Total Environ.* **2023**, *869*, 161798. [CrossRef]
24. Rodríguez-Merino, A. *Azolla filiculoides* Lam. Invasion Risk Atlas for Alien Aquatic Plants in the Iberian Peninsula. 2023. Available online: https://InvasionRiskAtlas.github.io/ (accessed on 11 July 2023).
25. Rodríguez-Merino, A. *Eichhornia crassipes* (Mart.) Solms. Invasion Risk Atlas for Alien Aquatic Plants in the Iberian Peninsula. 2023. Available online: https://InvasionRiskAtlas.github.io/ (accessed on 11 July 2023).
26. Rodríguez-Merino, A. *Ludwigia grandiflora* (Michx.) Greuter & Burdet. Invasion Risk Atlas for Alien Aquatic Plants in the Iberian Peninsula. 2023. Available online: https://InvasionRiskAtlas.github.io/ (accessed on 11 July 2023).
27. Rodríguez-Merino, A. *Salvinia natans* (L.) All. Risk Atlas for Alien Aquatic Plants in the Iberian Peninsula. 2023. Available online: https://InvasionRiskAtlas.github.io/ (accessed on 11 July 2023).
28. Rodríguez-Merino, A. *Salvinia molesta* D.S. Mitch. Invasion Risk Atlas for Alien Aquatic Plants in the Iberian Peninsula. 2023. Available online: https://InvasionRiskAtlas.github.io/ (accessed on 11 July 2023).
29. Rodríguez-Merino, A. *Spartina densiflora* Brongn. Invasion Risk Atlas for Alien Aquatic Plants in the Iberian Peninsula. 2023. Available online: https://InvasionRiskAtlas.github.io/ (accessed on 11 July 2023).
30. Chappuis, E.; Ballesteros, E.; Gacia, E. Distribution and Richness of Aquatic Plants across Europe and Mediterranean Countries: Patterns, Environmental Driving Factors and Comparison with Total Plant Richness. *J. Veg. Sci.* **2012**, *23*, 985–997. [CrossRef]
31. Rodríguez-Merino, A.; Fernández-Zamudio, R.; García-Murillo, P. Identifying Areas of Aquatic Plant Richness in a Mediterranean Hotspot to Improve the Conservation of Freshwater Ecosystems. *Aquat. Conserv. Mar. Freshw. Ecosyst.* **2019**, *29*, 589–602. [CrossRef]
32. Sanz Elorza, M.; Sobrino Vesperinas, E.; Dana Sánchez, E.D. *Atlas de las Plantas Alóctonas Invasoras en España*; Ministerio para la Transición Ecológica y el Reto Demográfico: Madrid, Spain, 2004; ISBN 978-84-8014-575-6.
33. Gassó, N.; Sol, D.; Pino, J.; Dana, E.D.; Lloret, F.; Sanz-Elorza, M.; Sobrino, E.; Vilà, M. Exploring Species Attributes and Site Characteristics to Assess Plant Invasions in Spain. *Divers. Distrib.* **2009**, *15*, 50–58. [CrossRef]
34. Rodríguez-Merino, A.; Fernández Zamudio, R.; García Murillo, P. Determinación de las zonas con mayor riesgo de invasión por macrófitos acuáticos exóticos en la Península Ibérica. In *Teledetección: Humedales y Espacios Protegidos*; Bustamante, J., Díaz-Delgado, R., Aragonés, D., Afán, I., García, D., Eds.; XVI Congreso de la Asociación Española de Teledetección; Asociación Española de Teledetección: Sevilla, Spain, 2015; pp. 338–341.
35. Rodríguez-Merino, A. Invasion Risk Atlas for Alien Aquatic Plants in the Iberian Peninsula. 2023. Available online: https://InvasionRiskAtlas.github.io/ (accessed on 11 July 2023).
36. Gallardo, B.; Zieritz, A.; Aldridge, D.C. The Importance of the Human Footprint in Shaping the Global Distribution of Terrestrial, Freshwater and Marine Invaders. *PLoS ONE* **2015**, *10*, e0125801. [CrossRef] [PubMed]
37. Cano-Barbacil, C.; Radinger, J.; García-Berthou, E. The Importance of Seawater Tolerance and Native Status in Mediating the Distribution of Inland Fishes. *J. Biogeogr.* **2022**, *49*, 2037–2049. [CrossRef]

38. Bartomeus, I.; Sol, D.; Pino, J.; Vicente, P.; Font, X. Deconstructing the Native–Exotic Richness Relationship in Plants. *Glob. Ecol. Biogeogr.* **2012**, *21*, 524–533. [CrossRef]
39. Bellard, C.; Leroy, B.; Thuiller, W.; Rysman, J.-F.; Courchamp, F. Major Drivers of Invasion Risks throughout the World. *Ecosphere* **2016**, *7*, e01241. [CrossRef]
40. Pearson, R.G.; Dawson, T.P. Predicting the Impacts of Climate Change on the Distribution of Species: Are Bioclimate Envelope Models Useful? *Glob. Ecol. Biogeogr.* **2003**, *12*, 361–371. [CrossRef]
41. Kelly, R.; Leach, K.; Cameron, A.; Maggs, C.A.; Reid, N. Combining Global Climate and Regional Landscape Models to Improve Prediction of Invasion Risk. *Divers. Distrib.* **2014**, *20*, 884–894. [CrossRef]
42. Gillard, M.; Thiébaut, G.; Deleu, C.; Leroy, B. Present and Future Distribution of Three Aquatic Plants Taxa across the World: Decrease in Native and Increase in Invasive Ranges. *Biol. Invasions* **2017**, *19*, 2159–2170. [CrossRef]
43. Gallardo, B.; Aldridge, D.C.; González-Moreno, P.; Pergl, J.; Pizarro, M.; Pyšek, P.; Thuiller, W.; Yesson, C.; Vilà, M. Protected Areas Offer Refuge from Invasive Species Spreading under Climate Change. *Glob. Change Biol.* **2017**, *23*, 5331–5343. [CrossRef] [PubMed]
44. Evans, K.L.; Warren, P.H.; Gaston, K.J. Species-Energy Relationships at the Macroecological Scale: A Review of the Mechanisms. *Biol. Rev.* **2005**, *80*, 1–25. [CrossRef] [PubMed]
45. Reshetnikov, A.N.; Ficetola, G.F. Potential Range of the Invasive Fish Rotan (*Perccottus glenii*) in the Holarctic. *Biol. Invasions* **2011**, *13*, 2967–2980. [CrossRef]
46. Soomers, H.; Karssenberg, D.; Soons, M.B.; Verweij, P.A.; Verhoeven, J.T.A.; Wassen, M.J. Wind and Water Dispersal of Wetland Plants Across Fragmented Landscapes. *Ecosystems* **2013**, *16*, 434–451. [CrossRef]
47. Rodríguez-Merino, A.; Fernández-Zamudio, R.; García-Murillo, P.; Muñoz, J. Climatic Niche Shift during *Azolla filiculoides* Invasion and Its Potential Distribution under Future Scenarios. *Plants* **2019**, *8*, 424. [CrossRef]
48. Rodríguez-Merino, A. Efecto del clima en la distribución futura de plantas acuáticas exóticas en Europa. In *Bosque Mediterráneo y Humedales: PAISAJE, Evolución y Conservación: Aportaciones Desde la Biogeografía*; Gosálvez Rey, R.U., Díaz Sanz, M.C., García Rayego, J.L., Serrano de la Cruz Santos-Olmo, M.A., Jerez García, O., Eds.; Almud, Ediciones de Castilla-La Mancha: Ciudad Real, Spain, 2018; Volume 1, pp. 523–531, ISBN 978-84-948075-6-5.
49. Hof, A.R.; Jansson, R.; Nilsson, C. The Usefulness of Elevation as a Predictor Variable in Species Distribution Modelling. *Ecol. Model.* **2012**, *246*, 86–90. [CrossRef]
50. Gallardo, B.; Castro-Díez, P.; Saldaña-López, A.; Alonso, Á. Integrating Climate, Water Chemistry and Propagule Pressure Indicators into Aquatic Species Distribution Models. *Ecol. Indic.* **2020**, *112*, 106060. [CrossRef]
51. Gillard, M.B.; Aroviita, J.; Alahuhta, J. Same Species, Same Habitat Preferences? The Distribution of Aquatic Plants Is Not Explained by the Same Predictors in Lakes and Streams. *Freshw. Biol.* **2020**, *65*, 878–892. [CrossRef]
52. Gallardo, B.; Vila, L. La influencia humana, clave para entender la biogeografía de especies invasoras en el Antropoceno. *Cuad. Investig. Geogr.* **2019**, *45*, 61–86. [CrossRef]
53. Kalkhoff, S.J.; Hubbard, L.E.; Tomer, M.D.; James, D.E. Effect of Variable Annual Precipitation and Nutrient Input on Nitrogen and Phosphorus Transport from Two Midwestern Agricultural Watersheds. *Sci. Total Environ.* **2016**, *559*, 53–62. [CrossRef]
54. Merow, C.; Smith, M.J.; Silander Jr, J.A. A Practical Guide to MaxEnt for Modeling Species' Distributions: What It Does, and Why Inputs and Settings Matter. *Ecography* **2013**, *36*, 1058–1069. [CrossRef]
55. García-Roselló, E.; Guisande, C.; Manjarrés-Hernández, A.; González-Dacosta, J.; Heine, J.; Pelayo-Villamil, P.; González-Vilas, L.; Vari, R.P.; Vaamonde, A.; Granado-Lorencio, C.; et al. Can We Derive Macroecological Patterns from Primary Global Biodiversity Information Facility Data? *Glob. Ecol. Biogeogr.* **2015**, *24*, 335–347. [CrossRef]
56. Pérez, G.; Vilà, M.; Gallardo, B. Potential Impact of Four Invasive Alien Plants on the Provision of Ecosystem Services in Europe under Present and Future Climatic Scenarios. *Ecosyst. Serv.* **2022**, *56*, 101459. [CrossRef]
57. Myers, N.; Mittermeier, R.A.; Mittermeier, C.G.; da Fonseca, G.A.B.; Kent, J. Biodiversity Hotspots for Conservation Priorities. *Nature* **2000**, *403*, 853–858. [CrossRef]
58. Fady, B.; Conord, C. Macroecological Patterns of Species and Genetic Diversity in Vascular Plants of the Mediterranean Basin. *Divers. Distrib.* **2010**, *16*, 53–64. [CrossRef]
59. Oliva-Paterna, F.J.; Ribeiro, F.; Miranda, R.; Anastácio, P.; García-Murillo, P.; Cobo, F.; Gallardo, B.; García-Berthou, E.; Boix, D.; Medina, L.; et al. *List of Aquatic Alien Species of the Iberian peninsula (2020). Updated List of Aquatic Alien Species Introduced and Established in Iberian Inland Waters*; Technical Report prepared by LIFE INVASAQUA (LIFE17 GIE/ES/000515); 2021; ISBN 978-84-12-35001-2. Available online: http://dspace.uevora.pt/rdpc/handle/10174/34424 (accessed on 11 April 2023).
60. Oliva-Paterna, F.J.; Ribeiro, F.; Miranda, R.; Anastácio, P.; García-Murillo, P.; Cobo, F.; Gallardo, B.; García-Berthou, E.; Boix, D.; Medina, L.; et al. *List of Potential Aquatic Alien Species of the Iberian Peninsula (2020). Updated List of the Potential Aquatic Alien Species with High Risk of Invasion in Iberian Inland Waters*; Technical Report prepared by LIFE INVASAQUA (LIFE17 GIE/ES/000515); 2021; ISBN 978-84-12-35004-3. Available online: https://lifeinvasaqua.com/wp-content/uploads/2021/05/Lista-especies-acu%C3%A1ticas-potencialmente-invasoras-libro-2-ingles-PDF-INTERACTIVO.pdf (accessed on 11 April 2023).

61. Oliva-Paterna, F.J.; Oficialdegui, F.J.; Anastácio, P.; García-Murillo, P.; Zamora-Marín, J.M.; Ribeiro, F.; Miranda, R.; Cobo, F.; Gallardo, B.; García-Berthou, E.; et al. *Black List and Alert List of The Aquatic Invasive Alien Species of the Iberian Peninsula—Horizon Scanning Exercise Focused on the High-Risk Aquatic Invasive Alien Species for the Iberian Inland Waters;* Technical Report prepared by LIFE INVASAQUA (LIFE17 GIE/ES/000515); 2022; ISBN 978-84-12-35008-1. Available online: https://lifeinvasaqua.com/wp-content/uploads/2023/01/TechRepp_3_INVASAQUA_Complet_ING.pdf (accessed on 11 April 2023).
62. Broennimann, O.; Guisan, A. Predicting Current and Future Biological Invasions: Both Native and Invaded Ranges Matter. *Biol. Lett.* **2008**, *4*, 585–589. [CrossRef]
63. Wisz, M.S.; Hijmans, R.J.; Li, J.; Peterson, A.T.; Graham, C.H.; Guisan, A.; Group, N.P.S.D.W. Effects of Sample Size on the Performance of Species Distribution Models. *Divers. Distrib.* **2008**, *14*, 763–773. [CrossRef]
64. Zizka, A.; Silvestro, D.; Andermann, T.; Azevedo, J.; Duarte Ritter, C.; Edler, D.; Farooq, H.; Herdean, A.; Ariza, M.; Scharn, R.; et al. CoordinateCleaner: Standardized Cleaning of Occurrence Records from Biological Collection Databases. *Methods Ecol. Evol.* **2019**, *10*, 744–751. [CrossRef]
65. Boria, R.A.; Olson, L.E.; Goodman, S.M.; Anderson, R.P. Spatial Filtering to Reduce Sampling Bias Can Improve the Performance of Ecological Niche Models. *Ecol. Model.* **2014**, *275*, 73–77. [CrossRef]
66. Pereira, P.F.; Barbosa, A.M.; Godinho, C.; Salgueiro, P.A.; Silva, R.R.; Lourenço, R. The Spread of the Red-Billed Leiothrix (*Leiothrix lutea*) in Europe: The Conquest by an Overlooked Invader? *Biol. Invasions* **2020**, *22*, 709–722. [CrossRef]
67. Legendre, P.; Legendre, L. *Numerical Ecology*, 3rd ed.; Elsevier: Amsterdam, The Netherlands, 2012.
68. Papeş, M.; Gaubert, P. Modelling Ecological Niches from Low Numbers of Occurrences: Assessment of the Conservation Status of Poorly Known Viverrids (Mammalia, Carnivora) across Two Continents. *Divers. Distrib.* **2007**, *13*, 890–902. [CrossRef]
69. Fick, S.E.; Hijmans, R.J. WorldClim 2: New 1-Km Spatial Resolution Climate Surfaces for Global Land Areas. *Int. J. Climatol.* **2017**, *37*, 4302–4315. [CrossRef]
70. Venter, O.; Sanderson, E.W.; Magrach, A.; Allan, J.R.; Beher, J.; Jones, K.R.; Possingham, H.P.; Laurance, W.F.; Wood, P.; Fekete, B.M.; et al. *Last of the Wild Project, Version 3 (LWP-3): 2009 Human Footprint, 2018 Release;* NASA Socioeconomic Data and Applications Center (SEDAC): Palisades, NY, USA, 2018.
71. Heiberger, R.M. HH: Statistical Analysis and Data Display: Heiberger and Holland 2022. R Package Version 3.1-49. Available online: https://CRAN.R-project.org/package=HH (accessed on 11 April 2023).
72. Phillips, S.J.; Anderson, R.P.; Schapire, R.E. Maximum Entropy Modeling of Species Geographic Distributions. *Ecol. Model.* **2006**, *190*, 231–259. [CrossRef]
73. Elith, J.; Graham, C.H.; Anderson, R.P.; Dudík, M.; Ferrier, S.; Guisan, A.; Hijmans, R.J.; Huettmann, F.; Leathwick, J.R.; Lehmann, A.; et al. Novel Methods Improve Prediction of Species' Distributions from Occurrence Data. *Ecography* **2006**, *29*, 129–151. [CrossRef]
74. Graham, C.H.; Elith, J.; Hijmans, R.J.; Guisan, A.; Townsend Peterson, A.; Loiselle, B.A.; Group, T.N.P.S.D.W. The Influence of Spatial Errors in Species Occurrence Data Used in Distribution Models. *J. Appl. Ecol.* **2008**, *45*, 239–247. [CrossRef]
75. Kaky, E.; Nolan, V.; Alatawi, A.; Gilbert, F. A Comparison between Ensemble and MaxEnt Species Distribution Modelling Approaches for Conservation: A Case Study with Egyptian Medicinal Plants. *Ecol. Inform.* **2020**, *60*, 101150. [CrossRef]
76. Fois, M.; Cuena-Lombraña, A.; Fenu, G.; Bacchetta, G. Using Species Distribution Models at Local Scale to Guide the Search of Poorly Known Species: Review, Methodological Issues and Future Directions. *Ecol. Model.* **2018**, *385*, 124–132. [CrossRef]
77. Phillips, S.J.; Dudík, M. Modeling of Species Distributions with Maxent: New Extensions and a Comprehensive Evaluation. *Ecography* **2008**, *31*, 161–175. [CrossRef]
78. Elith, J.; Phillips, S.J.; Hastie, T.; Dudík, M.; Chee, Y.E.; Yates, C.J. A Statistical Explanation of MaxEnt for Ecologists. *Divers. Distrib.* **2011**, *17*, 43–57. [CrossRef]
79. Pearce, J.; Ferrier, S. Evaluating the Predictive Performance of Habitat Models Developed Using Logistic Regression. *Ecol. Model.* **2000**, *133*, 225–245. [CrossRef]
80. Barnes, M.A.; Jerde, C.L.; Wittmann, M.E.; Chadderton, W.L.; Ding, J.; Zhang, J.; Purcell, M.; Budhathoki, M.; Lodge, D.M. Geographic Selection Bias of Occurrence Data Influences Transferability of Invasive *Hydrilla verticillata* Distribution Models. *Ecol. Evol.* **2014**, *4*, 2584–2593. [CrossRef]
81. Barbet-Massin, M.; Jiguet, F.; Albert, C.H.; Thuiller, W. Selecting Pseudo-Absences for Species Distribution Models: How, Where and How Many? *Methods Ecol. Evol.* **2012**, *3*, 327–338. [CrossRef]
82. Wickham, H.; Chang, W.; Henry, L.; Pedersen, T.L.; Takahashi, K.; Wilke, C.; Woo, K.; Yutani, H.; Dunnington, D.; Posit, P.B.C. Ggplot2: Create Elegant Data Visualisations Using the Grammar of Graphics 2023. R Package Version 3.4.3. Available online: https://CRAN.R-project.org/web/packages/ggplot2 (accessed on 11 April 2023).
83. Wood, S.N. Fast Stable Direct Fitting and Smoothness Selection for Generalized Additive Models. *J. R. Stat. Soc. Ser. B Stat. Methodol.* **2008**, *70*, 495–518. [CrossRef]

diversity

MDPI

Article

Invasive Plant Species Demonstrate Enhanced Resource Acquisition Traits Relative to Native Non-Dominant Species but not Compared with Native Dominant Species

Yingcan Chen [1], Yijie Xie [1], Caihong Wei [1], Si Liu [1], Xiaoyue Liang [1], Jiaen Zhang [1,2] and Ronghua Li [1,2,*]

[1] College of Natural Resources and the Environment, South China Agricultural University, Guangzhou 510642, China; chenyingcan@stu.scau.edu.cn (Y.C.); xyj_student2020@163.com (Y.X.); weicaihong@stu.scau.edu.cn (C.W.); liusi.scau.edu.cn@stu.scau.edu.cn (S.L.); liangxy56@stu.scau.edu.cn (X.L.); jeanzh@scau.edu.cn (J.Z.)

[2] Key Laboratory of Agro-Environment in the Tropics, Ministry of Agriculture and Rural Affairs, South China Agricultural University, Guangzhou 510642, China

* Correspondence: lironghua@scau.edu.cn; Tel.: +86-18664519534

Abstract: Invasive plant species are often characterized by superior resource acquisition capabilities compared with native species, contributing to their success in new environments. However, the dominance of these species varies, and not all invasive species become dominant, nor are all native species uniformly vulnerable to competitive exclusion. In this study, we analyzed 19 functional traits across 144 herbaceous plant species in Guangzhou, China. The studied species included 31 invasive dominant species (IDS), 19 invasive non-dominant species (INS), 63 native dominant species (NDS), and 31 native non-dominant species (NNS). Our findings reveal no significant differences in functional traits between IDS and INS, indicating a broad trait similarity within invasive categories. Pronounced similarities between invasive species and NDS suggest an ecological equivalency that facilitates successful integration and competition in new habitats. Notable differences in several key traits—height, leaf thickness, leaf water content, stoichiometry, photosynthetic rate, water use efficiency, and nitrogen use efficiency—indicate a competitive superiority in resource acquisition and utilization for invasive species over NNS. These distinctions are vital for understanding the mechanisms driving the success of invasive species and are crucial for developing strategies to manage their impact on native ecosystems.

Keywords: invasive species; native species; dominant; leaf functional traits; nutrient content; photosynthesis; seed; height

1. Introduction

Recent studies reveal that China hosts 933 naturalized plant species [1]. Among these, 403 have been classified as invasive alien plants [2]. The annual direct economic losses attributed to 283 invasive species in China are estimated to surpass USD 2397.39 million, with invasive alien plants responsible for 66.4% of these costs. Previous studies utilizing the Global Naturalized Alien Flora Database have revealed that the highest numbers of alien species are found in Australian states and several regions in North America, England, Japan, and New Zealand, and that 3.9% of the existing global vascular flora have become naturalized worldwide due to human activities [3,4]. Furthermore, studies based on the year of alien species records suggest an anticipated increase of approximately 1200 alien vascular plant species in Asia from 2005 to 2050 [5], indicating an ongoing trend without signs of saturation.

Concerns regarding the impact of alien species have significantly heightened interest in identifying the determinants of invasion success. Consequently, the mechanisms driving alien plant species invasions have emerged as a principal focus within the field of biology. A leading hypothesis posits that alien plant species exhibit greater competitiveness compared with native species [6]. The comparative analysis of trait differences between native and alien plant species has gained prominence in the field of plant invasion ecology [7–10]. The variance in traits between native and alien species is often considered a key mechanism explaining the successful invasion of plant communities [11]. To elucidate the traits characteristic of invasive plants, numerous case studies have compared these traits of both native and non-invasive exotic species. Despite these efforts, the findings have been inconclusive or contradictory at times, failing to reveal a consistent pattern [8,9,11–15].

Not every alien species becomes dominant, and not all native species are susceptible to being competitively excluded [9]. In the same region, there are many alien invasive species, but their abundance varies significantly, and the same is true for native species. Therefore, if all alien species are grouped together and all native species are also grouped together for trait comparison, the impact of the abundance differences among the species might be overlooked [16–18]. For example, a study conducted in the United States highlights that invasive grasses intensify the adverse effects of drought on native annual forbs, particularly impacting resource-acquisitive species more severely than resource-conservative ones by altering demographic rates, such as mortality and seed production [19]. Thus, the results of previous studies on competition between alien and native species may be biased by focusing on a single common alien species and a rare native species. Rigorous tests of competition between alien and native species that include both rare and common species are, thus, crucial to understanding the determinants of invasion success and rarity [20].

When comparing trait differences among species groups, the shared evolutionary history, which leads to non-independence of species data, can affect the outcomes. Recognizing the significant influence of phylogenetic relationships on species traits, many researchers have focused on the comparison of functional traits between specific alien invasive species and their native counterparts within the same genus. Nonetheless, the competition involving alien species typically includes multiple coexisting native species in a particular region. Therefore, it becomes important to evaluate trait differences across a broader spectrum of coexisting alien and native species. Yet, such assessments often overlook the phylogenetic context, omitting a critical factor in understanding trait evolution and species interactions. Phylogenetic ANOVA, which considers phylogenetic relationships among species, is crucial for accurately analyzing trait differences among groups [21]. However, integrating functional trait analyses with phylogenetic distance measures, as recommended by Hulme and Bernard-Verdier [11], is rarely done. This oversight indicates a significant gap in research methodologies, emphasizing the need for more nuanced approaches that include phylogenetic and functional traits for a better understanding of species invasions.

Furthermore, while most studies on functional traits have concentrated on a core set of traits that are straightforward to measure, such as plant height, seed weight, and specific leaf area, these traits do not necessarily predict the success or failure of plant species. A review has shown that the influence of traits on species demographic performance depends on both the abiotic and biotic environment [22]. Therefore, our research was dedicated to quantifying the precise characteristics of plant species by incorporating both easily measurable and physiological traits. Specifically, we focused on leaf stoichiometry, carbon assimilation capacity, and nutrient use efficiency, as these traits directly influence the performance and competitive ability of species [23,24].

Located at the junction of tropical and subtropical zones, Guangzhou boasts a rich diversity of plant species and complex ecosystem types. As the political, economic, and transportation hub of southern China, the city engages extensively in international communication and trade. These factors contribute to the notable richness of invasive plant species in Guangzhou. Field surveys combined with a review of relevant literature indicate that there are 131 invasive plant species in the city, of which 114 are terrestrial herbaceous plants [25]. In this study, we analyzed a dataset comprising 144 herbaceous plant species, including 50 alien and 94 native species, identified as either common or rare in Guangzhou.

We constructed a phylogenetic tree to contextualize the evolutionary relationships among the species under investigation. Our analysis focused on 19 functional traits, encompassing both easily measurable and physiological characteristics, to assess potential differences between alien and native species within a phylogenetic framework. We conducted two primary tests: (1) We examined if the functional traits' differences were associated with commonness and rarity among alien and native species. (2) We assessed how these differences varied when considered within a phylogenetic context.

2. Materials and Methods

2.1. Study Area

Guangzhou is located at the southern tip of mainland China (112°57′ to 114°03′ E, 22°26′ to 23°56′ N), covering an area of 7434.4 square kilometers (Figure 1). The region experiences an average annual temperature ranging between 21.5 °C and 22.2 °C. It boasts abundant water resources, with an average annual precipitation of over 1800 mm and approximately 150 rainy days per year. The area's complex topography harbors rich plant resources, creating favorable ecological conditions for plant growth. The study area is situated in a large urban setting, where the predominant land uses are divided between construction land and various forms of vegetation. The potential vegetation of the area primarily consists of woodland, farmland, and grassland. The woodland coverage rate in the region is 42.14%, reflecting the richness of plant species and diversity of natural vegetation.

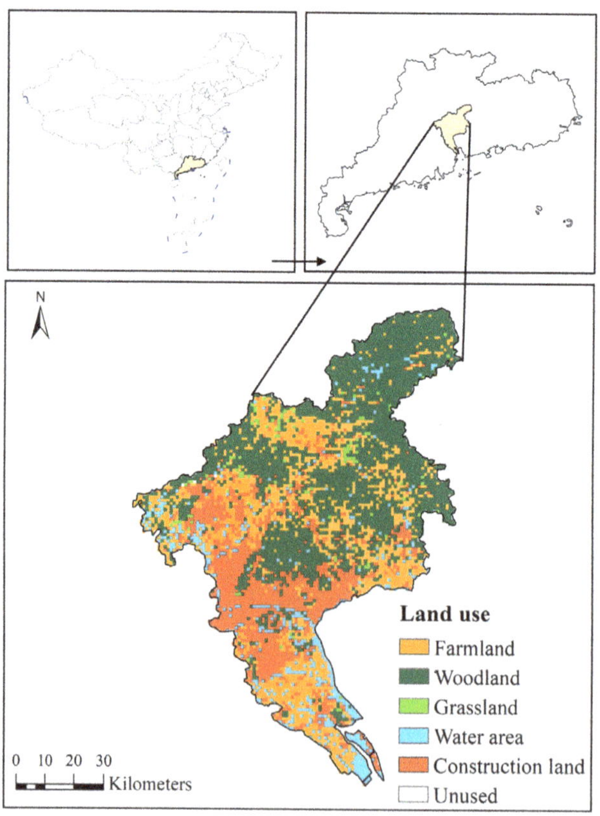

Figure 1. Geographic map and land use pattern of the studied area. Note: the land use pattern of Guangzhou is based on data sourced from the website https://www.resdc.cn/.

In this study, we analyzed a dataset comprising 144 herbaceous plant species in Guangzhou, which includes 50 alien and 94 native species (Table S1). These species primarily inhabit wastelands, ecosystems characterized by herbaceous vegetation interspersed with occasional shrubs or small trees. Their overlapping distributions suggest potential interactions within their natural habitats. The species were categorized as common or rare according to their frequency of occupancy and levels of local abundance. The categories were determined mainly according to the Guangzhou Invasive Plants [25] and Global Biodiversity Information Facility databases (https://www.gbif.org). Specifically, a species was deemed common if it exhibited widespread presence and local abundance across Guangzhou, and rare if it was neither widespread nor locally abundant. Species categorized under invasion levels 1 and 2 are designated as invasive dominant species, whereas those classified under levels 3, 4, and 5 are termed invasive non-dominant species [25]. In total, the 144 species were categorized into 31 invasive dominant species (IDS), 19 invasive non-dominant species (INS), 63 native dominant species (NDS), and 31 native non-dominant species (NNS).

2.2. Functional Traits Measurements

For each of the 144 species, we measured 3–5 individuals of the invasive and native species. The thousand-seed weight (TSW) data were mainly extracted from the seed information database of Kew (https://ser-sid.org/). Seed mass data were also collected from papers published up to December 2023. For data from these sources, when the authors provided a single-seed mass value for a species, that value was adopted. When multiple values were reported for a species from different sources, we used the mean value. Plant height (H) was measured for five mature plants using a ruler. Three to five leaves per plant were used to determine leaf thickness (LT) using a microcalliper. For leaf area (LA) measurements, 20 fully expanded, sun-exposed leaves from the canopy of three to five individuals per species were measured using a Li-3000A leaf area meter (Li-Cor, Lincoln, NE, USA), after removing petioles and/or rachises. Leaves were weighed to obtain their fresh weight and then oven-dried at 70 °C for 48 h to determine their dry mass. Leaf water content (LWC) was calculated as the ratio of the difference between fresh weight and dry weight to fresh weight. Specific leaf area (SLA) for each specimen was calculated as the ratio of total leaf area to leaf dry mass. Subsequently, the oven-dried leaves were ground into a fine powder for chemical analysis. Nitrogen content (N_{mass}, mg g^{-1}) was determined by Kjeldahl analysis, while phosphorus content (P_{mass}, mg g^{-1}) was measured using atomic absorption spectrophotometry. Leaf carbon content (LC) was determined using the external heating method with potassium dichromate. The C:N and N:P ratios were determined by calculating the leaf carbon to nitrogen content ratio and the leaf nitrogen to phosphorus content ratio, respectively. Leaf chlorophyll content was assessed using a SPAD meter (Konica Minolta, Tokyo, Japan) on selected leaves.

To assess light capture strategies, we measured the maximum CO_2 assimilation rate per unit area (A_{area}, µmol m^{-2} s^{-1}) and stomatal conductance per unit area (g_{sa}, mol m^{-2} s^{-1}) using a Li-6400 portable photosynthesis system (Li-Cor, Lincoln, NE, United States) between 9:00 and 11:00 am on sunny days. For each species, 5–10 fully expanded, sun-exposed leaves were measured. The mass-specific CO_2 assimilation rate (A_{mass}, µmol g^{-1} s^{-1}) and stomatal conductance (g_s) were calculated using the formulas A_{mass} = SLA × A_{area}/10,000 and g_s = SLA × g_{sa}/10, respectively.

Photosynthetic nitrogen use efficiency (PNUE, µmol g^{-1} s^{-1}), photosynthetic phosphorus use efficiency (PPUE, µmol g^{-1} s^{-1}), and intrinsic water use efficiency (WUE_i) were calculated to assess resource utilization efficiency. PNUE was determined as the ratio of A_{mass} to leaf nitrogen content per mass (N_m), PPUE as the ratio of A_{mass} to leaf phosphorus content per mass (P_m), and WUE_i as the ratio of A_{mass} to g_s. These calculations allow for a comparative analysis of how plants efficiently utilize nitrogen, phosphorus, and water under similar photosynthetic conditions.

2.3. Data Analyses

All data were analyzed in R 4.3.2 [26]. To account for the effects of shared evolutionary history on species traits, we conducted phylogenetically informed statistical analyses. Firstly, we standardized species names via the R package "Taxonstand", using the Plant List (https://wfoplantlist.org/) as the backbone, to have a taxonomically consistent dataset. Then, the construction of the phylogenetic tree was accomplished using the phylo.maker function within the R package "V.PhyloMaker", as developed by Jin and Qian [27] (Figure S1). We investigated differences in individual trait values among IDS, INS, NDS, and NNS with a phylogenetic one-way analysis of variance (phylogenetic ANOVA) via the R package "phytools".

To examine multivariate associations among the traits, we used principal component analysis (PCA). To better adhere to the assumptions of normality and homogeneous variances, we scaled the trait values. Mean factor loading values of the four species groups on the first two axes were also tested by phylogenetic ANOVA, to examine whether the species groups differed significantly along the PCA axes.

3. Results

3.1. Phenotypic and Reproductive Traits

Figure 2 shows phenotypic and reproductive traits among invasive dominant species (IDS), invasive non-dominant species (INS), native dominant species (NDS), and native non-dominant species (NNS). The analysis reveals statistical equivalence in leaf area (LA) across all groups, indicating that neither invasiveness nor dominance status significantly affects LA (Figure 2a). In terms of leaf thickness, IDS did not significantly differ from INS but varied compared with native species; both NDS and NNS exhibited thicker leaves than IDS, with no significant difference from INS (Figure 2b). Furthermore, there was no significant difference in specific leaf area (SLA) among the groups, suggesting that SLA is unaffected by the species' ecological status or dominance (Figure 2c). Regarding plant height, no differences were observed between IDS and INS, suggesting that invasive species may generally share similar growth heights. However, NNS were significantly shorter than both IDS and INS, and NDS were shorter than INS, suggesting a competitive advantage in height for invasive species (Figure 2d). Finally, no significant difference was noted in thousand-seed weight (TSW) among the groups, indicating that seed weight is not influenced by a species' invasive or native status, nor by its dominance (Figure 2e).

3.2. Leaf Stoichiometry and Leaf Water Content

Figure 3 provides a comparative analysis of leaf stoichiometry and water content among the four groups. For leaf carbon content (LC), NNS exhibited significantly higher values than the other groups, while IDS, INS, and NDS showed no significant differences among themselves (Figure 3a). In terms of leaf nitrogen content (N), INS had significantly higher levels compared with NDS. However, there were no significant differences observed between the invasive species and NNS (Figure 3b). As for leaf phosphorus content (P), NNS were significantly lower compared with the invasive species, but not in comparison to NDS. Among IDS, INS, and NDS, no significant differences were noted for leaf phosphorus content (Figure 3c). Regarding the C:N ratio, the NNS group exhibited a markedly higher C:N ratio when compared with the other groups. Conversely, the IDS, INS, and NDS did not show significant disparities in their C:N ratios (Figure 3d). When examining the N:P, NNS showed significantly higher ratios compared with IDS and NDS but not with INS. No significant differences were detected among IDS, INS, and NDS (Figure 3e). For leaf water content (LWC), NNS were found to have a significantly lower LWC than the other groups. Conversely, IDS, INS, and NDS did not show any significant differences in LWC among themselves (Figure 3f).

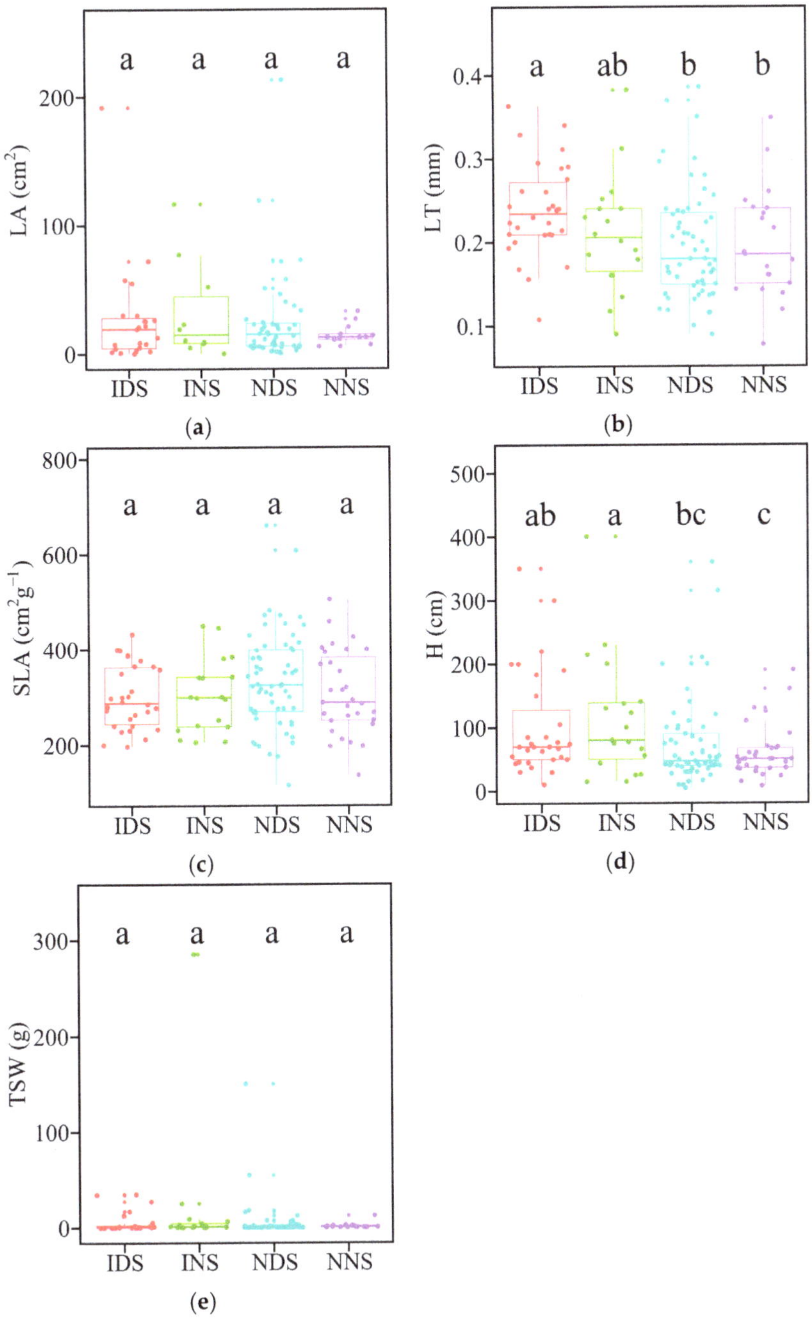

Figure 2. Differences in plant phenotypic and reproductive traits using phylogenetic ANOVA. (**a**) Leaf area (LA, cm^2); (**b**) leaf thickness (LT, mm); (**c**) specific leaf area (SLA, $cm^2\ g^{-1}$); (**d**) height (H, cm); (**e**) thousand-seed weight (TSW, g).

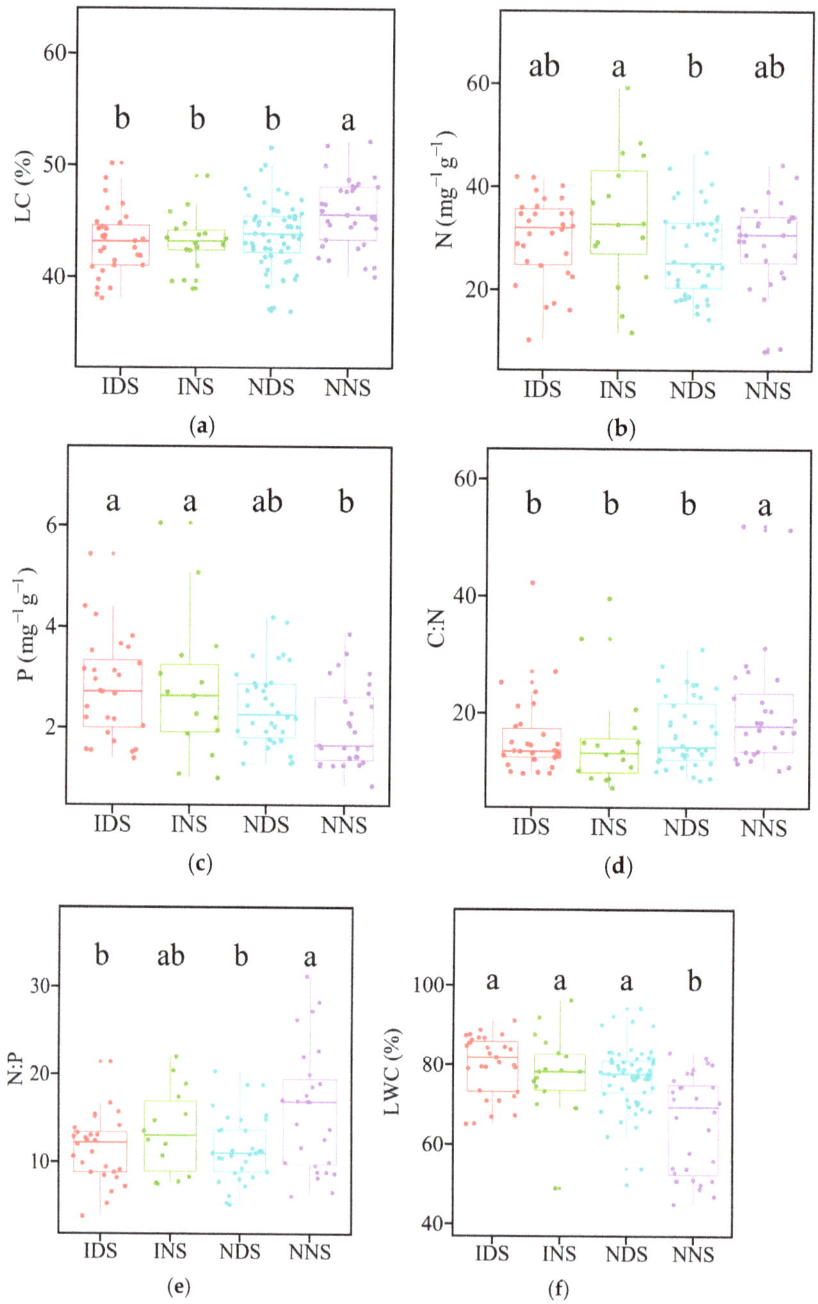

Figure 3. Differences in plant leaf stoichiometry and leaf water content using phylogenetic ANOVA. (**a**) Leaf carbon content (LC, %); (**b**) leaf nitrogen content (N, mg^{-1} g^{-1}); (**c**) leaf phosphorus content (P, mg^{-1} g^{-1}); (**d**) leaf carbon/nitrogen ratio (C:N); (**e**) leaf nitrogen/phosphorus ratio (N:P); (**f**) leaf water content (LWC, %).

3.3. Leaf Chlorophyll Content and Photosynthetic Rate

Figure 4 demonstrates the differences in plant leaf chlorophyll content and photosynthetic rate across the four groups. Leaf chlorophyll content (SPAD) was statistically indistinguishable among IDS, INS, and NDS (Figure 4a). Conversely, the NNS exhibited a significantly lower chlorophyll content. No significant differences were found in photosynthetic rate and stomatal conductance per area among IDS, INS, and NDS, whereas NNS display reduced values in both parameters (Figure 4b,c). Similarly, the photosynthetic rate per mass (A_{mass}) followed the same pattern, with NNS showing significantly lower rates than other groups (Figure 4d). For g_s, NDS had significantly higher stomatal conductance per mass than IDS and NNS (Figure 4e). Thus, the observed trends suggest that native non-dominant species (NNS) tend to exhibit lower values in the assessed physiological parameters, which may imply a reduced capacity for photosynthesis and water transport compared with other groups.

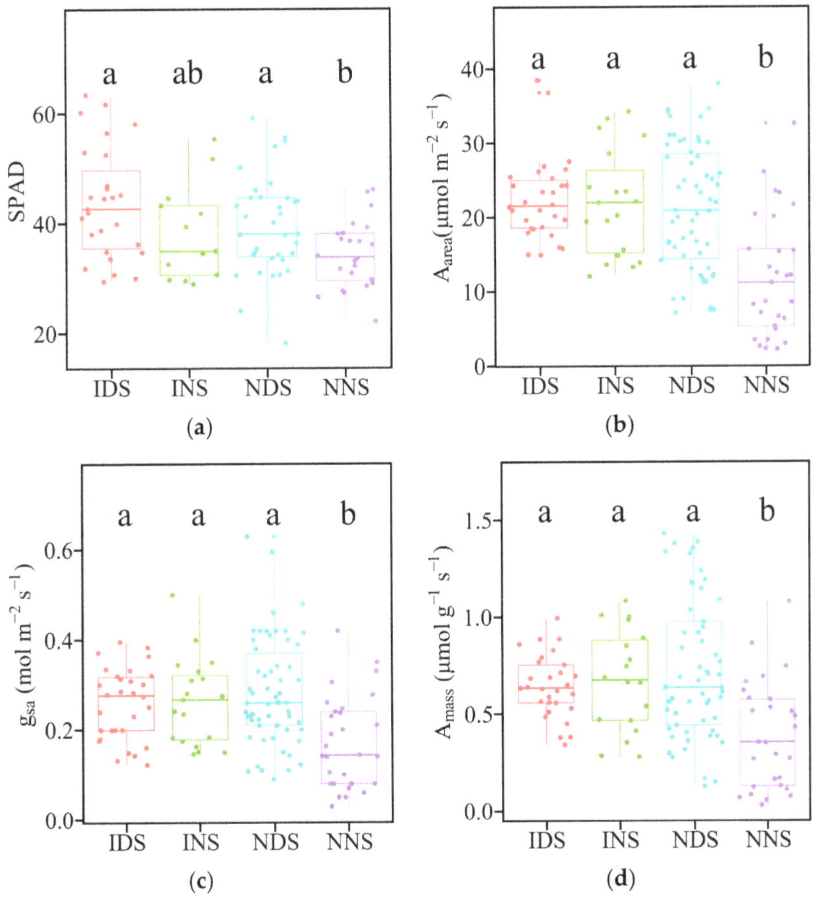

Figure 4. *Cont.*

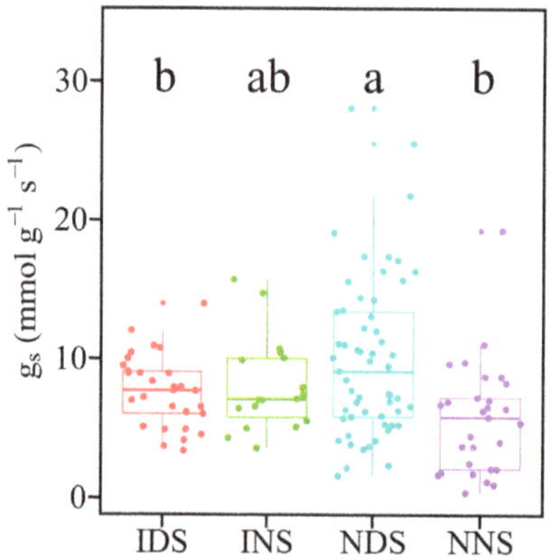

(**e**)

Figure 4. Comparative analysis of leaf chlorophyll content and photosynthetic rates across four groups using phylogenetic ANOVA. (**a**) Leaf chlorophyll content (SPAD); (**b**) maximum photosynthetic rate per area (A_{area}, μmol m^{-2} s^{-1}); (**c**) maximum stomatal conductance per area (g_{sa}, mol m^{-2} s^{-1}); (**d**) maximum photosynthetic rate per mass (A_{mass}, μmol g^{-1} s^{-1}); (**e**) maximum stomatal conductance per mass (g_s, mmol g^{-1} s^{-1}).

3.4. Leaf Water Use Efficiency and Nutrient Use Efficiency

Figure 5 presents an analysis of variance in plant water and nutrient use efficiencies among the four groups. The data on leaf water use efficiency (WUE_i) showed that there was no significant distinction between the IDS and INS (Figure 5a). However, IDS demonstrated higher WUE_i when compared with NDS. In contrast, the comparison between INS and NDS did not yield any significant differences, indicating similar WUE_i between these two groups. The invasive groups, both IDS and INS, exhibited significantly higher WUE_i than the NNS. In terms of leaf nitrogen use efficiency, IDS stood out, with significantly higher efficiency compared with NNS, while the differences between the other groups were not statistically significant (Figure 5b). When analyzing leaf phosphorus use efficiency, the study observed a homogeneous efficiency among all groups, indicating that this trait index did not differ significantly between invasive and native species, nor between dominant and non-dominant species (Figure 5c).

3.5. Impact of Invasiveness and Dominance on Trait Coordination: PCA

The first two principal components (PCs) extracted from the analysis accounted for a considerable proportion of the variability within the dataset, 28.87% for PC1 and 21.84% for PC2 (Figure 6a). The loadings on PCA axis 1 were positive for traits associated with photosynthetic-related traits (A_{area}, g_{sa}, A_{mass}, g_s, PNUE, and PPUE) and leaf water content (LWC), while LT, N:P were loaded at the negative end. The second PCA axis was primarily structured by leaf stoichiometry traits (N, P, LC, and C:N) and leaf area. PC1 significantly separated NNS from IDS, INS, and NDS (Figure 6b,c), while PC2 significantly separated invasive species from NDS and NNS (Figures 6b and 5d). The results indicate that the divergence of multiple functional traits along the first two axes was closely associated with both invasiveness and dominance.

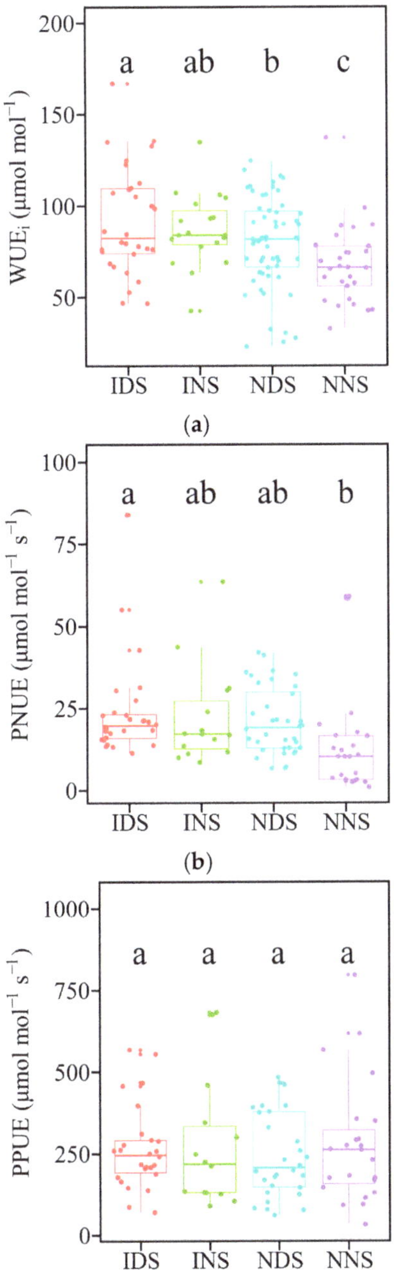

Figure 5. Differences in plant water and nutrient use efficiency traits using phylogenetic ANOVA. (**a**) Intrinsic water use efficiency (WUE_i, μmol mol^{-1}); (**b**) photosynthetic nitrogen use efficiency (PNUE, μmol mol^{-1} s^{-1}); (**c**) photosynthetic phosphorus use efficiency (PPUE, μmol mol^{-1} s^{-1}).

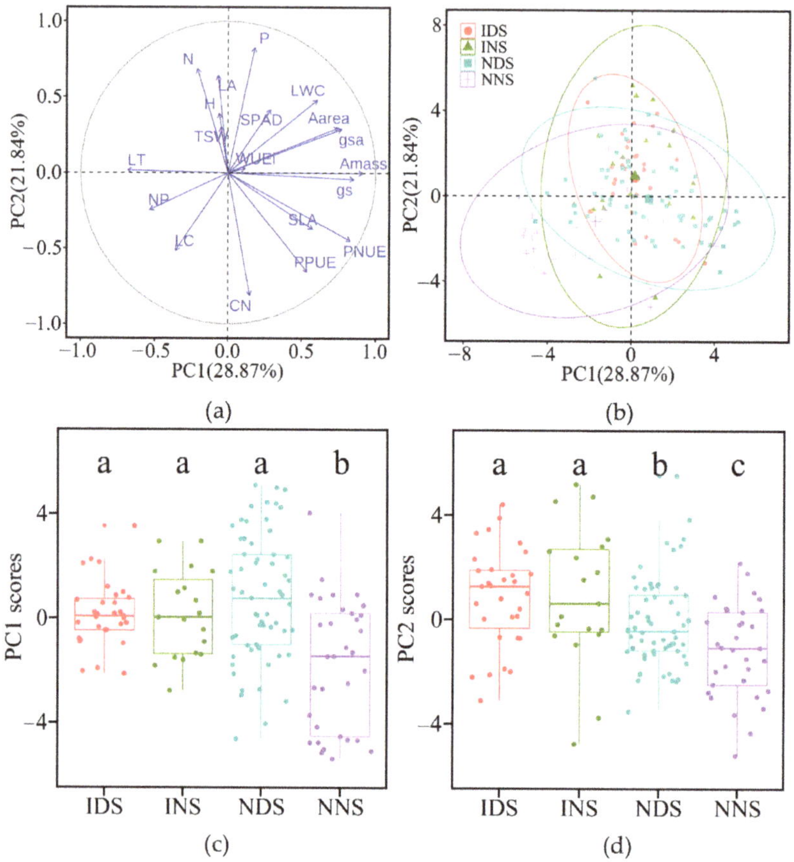

Figure 6. A principal component analysis of 19 functional traits of the 144 study species. (**a**) Loading of the 19 traits on the first two axes; (**b**) species loadings on the first and second axes; (**c**) box-plots of species scores on PC1; (**d**) box-plots of species scores on PC2. The red circles, blue triangles, purple squares, and pink crosses in (**b**) represent invasive dominant species (IDS), invasive non-dominant species (INS), native dominant species (NDS), and native non-dominant species (NNS), respectively. The abbreviations used for the traits are consistent with those mentioned in the text.

4. Discussion

This study reveals significant trait similarities between both dominant and non-dominant invasive species and native dominant species, as well as between dominant and non-dominant invasive species. However, invasive species display unique adaptations compared with native non-dominant species. These include variations in height, leaf thickness, leaf water content, leaf stoichiometry, photosynthetic rate, water use efficiency, and nitrogen use efficiency, suggesting a competitive superiority in resource acquisition and utilization. However, no significant differences in leaf area, specific leaf area, seed weight, and phosphorus use efficiency were found among all groups. Principal component analysis (PCA) further illustrates a systematic differentiation between native non-dominant species and both groups of invasive species. The application of a phylogenetic ANOVA to our dataset reveals that the differences observed among the four groups for most of the traits remained statistically significant, even when accounting for their evolutionary relationships. This analysis underscores the robustness of the trait variations, indicating that they are not merely results of phylogenetic relatedness but likely reflect genuine adaptive differences [21,28]. Such findings are crucial as they suggest that despite sharing a common evolutionary background, the groups have diverged significantly in terms of certain key traits [29].

In this study, we did not find differences in seed weight among all groups. Seed weight is a pivotal trait influencing dispersal capability, germination success, and early-life-stage survival, which in turn affects the establishment and spread of species in new environments. Some studies report that invasive species often have smaller seeds, facilitating wider dispersal over large areas, a trait beneficial for species establishing themselves in new regions [30,31]. Conversely, other research finds invasive species with larger seeds, which might contribute to higher survival rates and competitive advantages in certain environments [9,14,32]. Additionally, many studies have documented similar seed weights between invasive and native species, indicating that seed size alone does not predict invasiveness [14,33]. Our results regarding leaf area diverge from studies conducted in Mediterranean ecosystems, where invasive species exhibit larger leaves than native species, and contrast with findings from the sub-Antarctic region, where invasive species have smaller leaves than their native counterparts [8,34]. For specific leaf area, our study is similar to a study carried out in northeast China [9,14,35] and different from studies indicating that invasive species have a higher SLA [8,36,37]. For leaf phosphorus use efficiency, our results differ from a study reporting that invasive species have higher PPUE [7].

Our results showed no significant differences in any of the 19 functional traits between invasive dominant species (IDS) and invasive non-dominant species (INS). IDS and INS may both employ generalist strategies that enable them to thrive in diverse environmental conditions. The similarity in traits suggests a concept of ecological equivalence among invasive species, where both IDS and INS possess an overlapping set of adaptive traits conducive to invasion success. This suggests that invasiveness and dominance may not always be readily apparent through traditional trait comparisons. Our findings are consistent with other studies reporting that common invasive species exhibit similar functional traits or intrinsic growth rates with rare invasive species [9,12,38]. In summary, the observed lack of significant trait differences between IDS and INS adds an important dimension to our understanding of biological invasions, highlighting the need for comprehensive approaches to studying, managing, and mitigating the impacts of invasive species on native ecosystems. Our results also indicate that, based solely on the functional traits studied, we cannot accurately predict which alien species will become more invasive and cause significant damage in invaded areas. Therefore, it is essential to carefully monitor and manage all alien species during the early stages of invasion.

The research provides compelling evidence for the significant trait similarities between IDS and native dominant species (NDS), as well as between INS and NDS, spanning a wide range of ecological and physiological characteristics. However, certain specific traits—such as leaf thickness, maximum stomatal conductance per mass, intrinsic water use efficiency for IDS compared with NDS, and plant height and leaf nitrogen content for INS versus NDS—were exceptions. This trait convergence suggests a fundamental ecological equivalency that may underlie the ability of invasive species to successfully integrate into and compete within new environments [9,12]. The lack of significant differences among IDS, INS, and NDS could be attributed to their adaptation to disturbance-prone environments. However, these traits do not uniformly confer a competitive advantage, as demonstrated by a study in France, where most invasive dominant species exhibited larger biomass production compared with native species [39]. A study conducted in California, United States, found that dominant invasive species negatively impacted all target species, particularly those that are resource-acquisitive, suggesting that IDS and NDS may share similar traits [19]. Our observation supports the Pre-Adaptation Hypothesis in invasion ecology, which posits that invasive species possess pre-existing traits that are advantageous in new environments, even before exposure to these environments.

IDS manifest an array of superior traits relative to native non-dominant species (NNS), characterized by increased leaf thickness, enhanced plant height, augmented leaf phosphorus and water content, elevated chlorophyll concentrations, and markedly higher rates of photosynthesis and stomatal conductance. Additionally, these species exhibit diminished leaf carbon content and reduced carbon-to-nitrogen (C:N) and nitrogen-to-phosphorus (N:P) ratios. Such phenotypic characteristics underscore a heightened efficiency in water and nitrogen utilization. A congruent pattern was observed between INS and NNS. Our results are consistent with recent studies which found that non-native plant species exhibit significantly enhanced leaf nutrient content and leaf photosynthetic-related traits compared with native species, as well as greater photosynthetic water use efficiency and nitrogen use efficiency [14,40–42]. Invasive species across various communities exhibit leaf traits indicative of an acquisitive resource use strategy, in contrast to the more conservative traits observed in native non-dominant species. This is contrary to another study, which found that both native and invasive species within annual and perennial groups exhibited similar patterns of carbon assimilation and resource use [43]. And it is contrary to a study in the sub-Antarctic area, which found that invasive species had lower plant height, smaller leaf area, and higher specific leaf area than non-invasive aliens, suggesting that these traits are associated with invasiveness [8].

5. Conclusions

Our investigation into the phenotypic, physiological, and reproductive characteristics of both invasive and native plant species, categorized by their dominance status, illuminates the complex dynamics underlying plant competition and survival. While our findings indicate that certain traits such as leaf area, specific leaf area, and thousand-seed weight do not significantly vary across the four studied groups, this suggests a fundamental homogeneity in these traits, which transcends ecological and biological boundaries. Invasive species generally exhibit traits conducive to a competitive advantage over native non-dominant species. These traits underscore that dominance within a habitat is closely linked to superior resource acquisition and utilization capabilities, a trait especially pronounced in invasive dominant species. Such traits not only facilitate their survival and proliferation but also enable them to outcompete other species. This pattern suggests that the capacity for effective resource use is a critical determinant of dominance and success in various ecological settings. Further reflecting on the broader impacts, anthropization and environmental degradation can significantly alter ecological balances, potentially enhancing the invasiveness of certain species by creating niches that exploit their competitive traits. The scientific implications of this work extend to enhance our understanding of invasion biology and the factors that confer competitive advantages to invasive species in various ecological contexts. Practically, this knowledge assists in the development of more effective management strategies that consider both the phenotypic traits and environmental factors driving the success of invasive species, thereby informing conservation efforts and ecological restoration practices for maintaining biodiversity and ecological health.

Supplementary Materials: The following supporting information can be downloaded at https://www.mdpi.com/article/10.3390/d16060317/s1, Table S1: Taxonomic classification and dominance status of the 144 plant species analyzed in the study. Figure S1: Phylogenetic tree of the 144 species studied.

Author Contributions: Conceptualization, R.L.; methodology, R.L., Y.X.; software, R.L and Y.C; formal analysis, R.L.; Y.C and Y.X.; investigation, Y.X., Y.C., C.W. and S.L.; data curation, Y.C., R.L. and Y.C.; writing—original draft preparation, Y.C., X.L. and R.L.; visualization, Y.C. and R.L.; supervision, writing—review and editing, J.Z.; funding acquisition, R.L. All authors have read and agreed to the published version of the manuscript.

Funding: This research was supported by grants from the Guangzhou Basic and Applied Basic Research Foundation (202201010506) and the Guangdong Forestry Science and Technology Innovation Project (2022KJCX018).

Institutional Review Board Statement: Not applicable.

Data Availability Statement: The data presented in this study are available on request from the corresponding author.

Conflicts of Interest: The authors declare no conflicts of interest. The funders had no role in the design of the study; in the collection, analyses, or interpretation of data; in the writing of the manuscript; or in the decision to publish the results.

References

1. Yan, X.; Wang, Z.; Ma, J.; Li, H.; Yan, J. *The Checklist of the Naturalized Plants in China*; Shanghai Scientific and Technical Publishers: Shanghai, China, 2019.
2. Hao, Q.; Ma, J.S. Invasive alien plants in China: An update. *Plant Divers.* **2023**, *45*, 117–121. [CrossRef]
3. Pyšek, P.; Pergl, J.; Essl, F.; Lenzner, B.; Dawson, W.; Kreft, H.; Weigelt, P.; Winter, M.; Kartesz, J.; Nishino, M.; et al. Naturalized alien flora of the world. *Preslia* **2017**, *89*, 203–274. [CrossRef]
4. van Kleunen, M.; Dawson, W.; Essl, F.; Pergl, J.; Winter, M.; Weber, E.; Kreft, H.; Weigelt, P.; Kartesz, J.; Nishino, M.; et al. Global exchange and accumulation of non-native plants. *Nature* **2015**, *525*, 100–103. [CrossRef]
5. Seebens, H.; Bacher, S.; Blackburn, T.M.; Capinha, C.; Dawson, W.; Dullinger, S.; Genovesi, P.; Hulme, P.E.; van Kleunen, M.; Kuhn, I.; et al. Projecting the continental accumulation of alien species through to 2050. *Glob. Chang. Biol.* **2020**, *27*, 970–982. [CrossRef]
6. Huang, K.; Kong, D.; Lu, X.; Feng, W.; Liu, M.; Feng, Y. Lesser leaf herbivore damage and structural defense and greater nutrient concentrations for invasive alien plants: Evidence from 47 pairs of invasive and non-invasive plants. *Sci. Total Environ.* **2020**, *723*, 137829. [CrossRef]
7. Liu, Y.; van Kleunen, M. Responses of common and rare aliens and natives to nutrient availability and fluctuations. *J. Ecol.* **2017**, *105*, 1111–1122. [CrossRef]
8. Mathakutha, R.; Steyn, C.; le Roux, P.C.; Blom, I.J.; Chown, S.L.; Daru, B.H.; Ripley, B.S.; Louw, A.; Greve, M.; Price, J. Invasive species differ in key functional traits from native and non-invasive alien plant species. *J. Veg. Sci.* **2019**, *30*, 994–1006. [CrossRef]
9. Liu, M.; Dong, T.; Feng, W.; Qu, B.; Kong, D.; van Kleunen, M.; Feng, Y. Leaf trait differences between 97 pairs of invasive and native plants across China: Effects of identities of both the invasive and native species. *Neobiota* **2022**, *71*, 1–22. [CrossRef]
10. Zhao, Y.; Liu, M.; Feng, Y.; Wang, D.; Feng, W.; Clay, K.; Durden, L.A.; Lu, X.; Wang, S.; Wei, X.; et al. Release from below- and aboveground natural enemies contributes to invasion success of a temperate invader. *Plant Soil* **2020**, *452*, 19–28. [CrossRef]
11. Hulme, P.E.; Bernard-Verdier, M. Comparing traits of native and alien plants: Can we do better? *Funct. Ecol.* **2018**, *32*, 117–125. [CrossRef]
12. Zhang, Z.; van Kleunen, M. Common alien plants are more competitive than rare natives but not than common natives. *Ecol. Lett.* **2019**, *22*, 1378–1386. [CrossRef]
13. Canessa, R.; Saldana, A.; Rios, R.S.; Gianoli, E. Functional trait variation predicts distribution of alien plant species across the light gradient in a temperate rainforest. *Perspect. Plant Ecol.* **2018**, *32*, 49–55. [CrossRef]
14. Divisek, J.; Chytry, M.; Beckage, B.; Gotelli, N.J.; Lososova, Z.; Pysek, P.; Richardson, D.M.; Molofsky, J. Similarity of introduced plant species to native ones facilitates naturalization, but differences enhance invasion success. *Nat. Commun.* **2018**, *9*, 4631. [CrossRef]
15. Gallagher, R.V.; Randall, R.P.; Leishman, M.R. Trait differences between naturalized and invasive plant species independent of residence time and phylogeny. *Conserv. Biol.* **2015**, *29*, 360–369. [CrossRef]
16. Jeschke, J.M.; Strayer, D.L. Are threat status and invasion success two sides of the same coin? *Ecography* **2008**, *31*, 124–130. [CrossRef]
17. Milanovic, M.; Knapp, S.; Pysek, P.; Kuehn, I. Trait-environment relationships of plant species at different stages of the introduction process. *Neobiota* **2020**, *58*, 55–74. [CrossRef]
18. Iqbal, M.F.; Liu, M.; Iram, A.; Feng, Y. Effects of the invasive plant Xanthium strumarium on diversity of native plant species: A competitive analysis approach in North and Northeast China. *PLoS ONE* **2020**, *15*, e228476. [CrossRef]
19. LaForgia, M.L.; Harrison, S.P.; Latimer, A.M. Invasive species interact with climatic variability to reduce success of natives. *Ecology* **2020**, *101*, e03022. [CrossRef]
20. van Kleunen, M.; Schlaepfer, D.R.; Glaettli, M.; Fischer, M. Preadapted for invasiveness: Do species traits or their plastic response to shading differ between invasive and non-invasive plant species in their native range? *J. Biogeogr.* **2011**, *38*, 1294–1304. [CrossRef]
21. Rohlfs, R.V.; Nielsen, R. Phylogenetic ANOVA: The Expression Variance and Evolution Model for Quantitative Trait Evolution. *Syst. Biol.* **2015**, *64*, 695–708. [CrossRef]
22. Yang, J.; Cao, M.; Swenson, N.G. Why Functional Traits Do Not Predict Tree Demographic Rates. *Trends Ecol. Evol.* **2018**, *33*, 326–336. [CrossRef]
23. Li, R.H.; Zhu, S.D.; Chen, H.; John, R.; Zhou, G.Y.; Zhang, D.Q.; Zhang, Q.M.; Ye, Q. Are functional traits a good predictor of global change impacts on tree species abundance dynamics in a subtropical forest? *Ecol. Lett.* **2015**, *18*, 1181–1189. [CrossRef]
24. Qiu, S.; Liu, S.; Wei, S.; Cui, X.; Nie, M.; Huang, J.; He, Q.; Ju, R.; Li, B. Changes in multiple environmental factors additively enhance the dominance of an exotic plant with a novel trade-off pattern. *J. Ecol.* **2020**, *108*, 1989–1999. [CrossRef]
25. Wang, R. *Guangzhou Invasive Plants*; Guangdong Science & Technology Press: Guangzhou, China, 2019.

26. R Core Team. *R: A Language and Environment for Statistical Computing*; R Foundation for Statistical Computing: Vienna, Austria, 2017.
27. Jin, Y.; Qian, H.V. PhyloMaker: An R package that can generate very large phylogenies for vascular plants. *Ecography* **2019**, *42*, 1353–1359. [CrossRef]
28. Adams, D.C.; Collyer, M.L. Phylogenetic ANOVA: Group-clade aggregation, biological challenges, and a refined permutation procedure. *Evolution* **2018**, *72*, 1204–1215. [CrossRef]
29. Adams, D.C.; Collyer, M.L. Consilience of methods for phylogenetic analysis of variance. *Evolution* **2022**, *76*, 1406–1419. [CrossRef]
30. Rindyastuti, R.; Hapsari, L.; Byun, C. Comparison of ecophysiological and leaf anatomical traits of native and invasive plant species. *J. Ecol. Environ.* **2021**, *45*, 4. [CrossRef]
31. Ni, M.; Deane, D.C.; Li, S.; Wu, Y.; Sui, X.; Xu, H.; Chu, C.; He, F.; Fang, S. Invasion success and impacts depend on different characteristics in non-native plants. *Divers. Distrib.* **2021**, *27*, 1194–1207. [CrossRef]
32. Pik, D.; Lucero, J.E.; Lortie, C.J.; Braun, J. Light intensity and seed density differentially affect the establishment, survival, and biomass of an exotic invader and three species of native competitors. *Community Ecol.* **2020**, *21*, 259–272. [CrossRef]
33. Kunstler, G.; Falster, D.; Coomes, D.A.; Hui, F.; Kooyman, R.M.; Laughlin, D.C.; Poorter, L.; Vanderwel, M.; Vieilledent, G.; Wright, S.J.; et al. Plant functional traits have globally consistent effects on competition. *Nature* **2016**, *529*, 174–204. [CrossRef]
34. Díaz De León Guerrero, S.D.; González-Rebeles Guerrero, G.; Ibarra-Montes, T.M.; Rodríguez Bastarrachea, A.; Santos Cobos, R.; Bullock, S.H.; Sack, L.; Méndez-Alonzo, R. Functional traits indicate faster resource acquisition for alien herbs than native shrubs in an urban Mediterranean shrubland. *Biol. Invasions* **2020**, *22*, 2699–2712. [CrossRef]
35. Luo, J.; Gao, Y.; Feng, W.; Liu, M.; Qu, B.; Zhang, C.; Feng, Y. Stronger ability to absorb nitrate and associated transporters in the invasive plant Xanthium strumarium compared with its native congener. *Environ. Exp. Bot.* **2022**, *198*, 104851. [CrossRef]
36. Castro-Díez, P.; Valle, G.; González-Muñoz, N.; Alonso, Á. Can the Life-History Strategy Explain the Success of the Exotic Trees Ailanthus altissima and Robinia pseudoacacia in Iberian Floodplain Forests? *PLoS ONE* **2014**, *9*, e100254. [CrossRef] [PubMed]
37. Liu, W.W.; Wang, W.W.; Zhang, Y.H. Differences in leaf traits of *Spartina alterniflora* between native and invaded habitats: Implication for evolution of alien species competitive ability increase. *Ecol. Indic.* **2022**, *138*, 108799. [CrossRef]
38. Fenesi, A.; Botta-Dukat, Z.; Miholcsa, Z.; Szigeti, V.; Molnar, C.; Sandor, D.; Szabo, A.; Kuhn, T.; Kovacs-Hostyanszki, A. No consistencies in abundance-impact relationships across herbaceous invasive species and ecological impact metrics. *J. Ecol.* **2023**, *111*, 1120–1138. [CrossRef]
39. Bottollier Curtet, M.; Planty Tabacchi, A.M.; Tabacchi, E.; Güsewell, S. Competition between young exotic invasive and native dominant plant species: Implications for invasions within riparian areas. *J. Veg. Sci.* **2013**, *24*, 1033–1042. [CrossRef]
40. Liu, M.; Kong, D.; Lu, X.; Huang, K.; Wang, S.; Wang, W.; Qu, B.; Feng, Y. Higher photosynthesis, nutrient- and energy-use efficiencies contribute to invasiveness of exotic plants in a nutrient poor habitat in northeast China. *Physiol. Plant.* **2017**, *160*, 373–382. [CrossRef] [PubMed]
41. Abdallah, M.; Douthe, C.; Flexas, J. Leaf morpho-physiological comparison between native and non-native plant species in a Mediterranean island. *Biol. Invasions* **2022**, *24*, 2597–2612. [CrossRef]
42. Barros, V.; Oliveira, M.T.; Santos, M.G. Low foliar construction cost and strong investment in root biomass in *Calotropis procera*, an invasive species under drought and recovery. *Flora* **2021**, *280*, 151848. [CrossRef]
43. Funk, J.L.; Standish, R.J.; Stock, W.D.; Valladares, F. Plant functional traits of dominant native and invasive species in mediterranean-climate ecosystems. *Ecology* **2016**, *97*, 75–83. [CrossRef]

forests

MDPI

Article

Functional Segregation of Resource Utilization Strategies between Invasive and Native Plants and Invasion Mechanisms in the Water Level Fluctuation Zone: A Case Study of Pengxi River in Three Gorges Reservoir, China

Lideng Cheng [1,2], Xingzhong Yuan [1,2,*], Kuo Sun [1,2] and Peiwu Li [1,2]

1 Faculty of Architecture and Urban Planning, Chongqing University, Chongqing 400044, China; 20192001742@cqu.edu.cn (L.C.); peazwoo@gmail.com (P.L.)

2 Research Center for Ecological Restoration and Control of Water Level Fluctuating Zone in the Three Gorges Reservoir, Chongqing University, Chongqing 400044, China

* Correspondence: xzyuan63@cqu.edu.cn

Abstract: The ecosystem of the water level fluctuation (WLF) zone of the Three Gorges Reservoir (TGR) is highly vulnerable and sensitive due to its unique cyclical flooding and drought conditions. The ecological impact of biological invasion in this area is particularly severe, making it crucial to study the differences in resource utilization strategies between invasive plants (IPs) and native plants (NPs) using functional traits to explore the mechanisms of invasion. We selected the WLF zone of Pengxi River in the TGR area and conducted a multi-scale study along the elevation gradient. The results reveal that at the regional scale, IPs have a larger height and specific leaf area, smaller leaf tissue density, and specific root length compared to NPs, showing a preference for enhancing aboveground resource acquisition over leaf defense capabilities. They allocate more tissue construction resources to their roots to withstand environmental pressures, which may be the key to their successful intrusion, highlighting the role of niche differentiation. On the community scale, the H and SLA of IPs and NPs are positively correlated with elevation, while the LTD of IPs shows a negative correlation. At elevations of 175 m and below, IP and NP exhibit functional convergence, while above 175 m, functional divergence was observed. This indicates that although the different resource utilization strategies are crucial for successful IP invasion, the environmental filtering from periodic floods and drought pressures play a significant role in community assembly in the WLF zone, allowing IP to integrate into habitats with similar functional characteristics already inhabited by NP and establish their own communities.

Keywords: invasive plants; functional traits; invasion mechanism; the water level fluctuation zone; Three Gorges Reservoir

1. Introduction

The global concern surrounding invasive species has been growing due to its profound [1–3]. Not only do invasive species threaten global biodiversity, but they also lead to substantial economic losses [4]. Unfortunately, there is no indication that the number of invasive species will decline in the near future [5]. As a result, both scholars and society have come to recognize the importance of addressing the issue of biological invasions. Plant functional traits are observable characteristics that have evolved in individual plants over time to adapt to their specific habitats.

Plant functional traits refer to a set of observable characteristics that have developed within individual plants over time to suit their specific habitats [6]. These traits are crucial for enhancing plants' survival and functioning in a dynamic environment. Furthermore, they serve as indicators for studying how plants respond and adapt to changes in their surroundings [7], which are usually closely related to resource utilization strategies and functional niches [8]. Comparing and analyzing functional trait variations between native and invasive species has become crucial in the field of plant invasion ecology [9]. It is crucial to identify specific traits linked to invasiveness to predict potential invasive species and effectively manage established invaders [10].

In the field of invasion ecology, a central focus is understanding the mechanisms through which invasive plants (IPs) spread. This comprehension aids in predicting their potential to establish populations in specific ecosystems [3,11,12]. Various perspectives have been proposed to elucidate species coexistence mechanisms. Niche differentiation emphasizes interspecies competition, suggesting that functional similarity heightens competition and exclusion, hindering species coexistence [13]. Conversely, environmental filtering underscores the significance of abiotic factors, indicating that similar habitat conditions promote functional similarity among coexisting species in the same habitat [14]. Moreover, the Darwin naturalization hypothesis (DNH) and Darwin pre-adaptation hypothesis (DPH) highlight the role of phylogenetic relationships in the successful invasion of plants [15–17]. The DNH suggests that establishing sustainable populations of alien plants in ecosystems with a high number of native species is challenging [18]. The DPH argues that close phylogenetic relationships imply similar ecological requirements, enhancing the adaptability of alien plants to the local environment. These two hypotheses that seemed contradictory at the time were collectively referred to as the "Darwin's naturalization conundrum" [19,20]. Recent studies have shown that these seemingly contradictory hypotheses may not be entirely incompatible [19,21]. Depending on the specific environment or scale of the study, both hypotheses can effectively predict the outcomes of biological invasions. Research conducted on plants in the United States revealed that at larger spatial scales, environmental filtering traits tend to have a greater impact than competitive traits. This finding has provided a plausible explanation for reconciling Darwin's conflicting hypotheses across different spatial scales [19], highlighting the significance of the scale of research. Differences in traits between invasive and native species are often seen as a key mechanism for successful invasions [22,23]. Understanding these trait-based approaches to invasion success involves comparing the traits of invasive species with those of native species, either within a community or between invasive and native species [24]. Therefore, conducting a multi-scale study based on functional traits is essential for comprehending the mechanism of invasive plants in the WLF zone. Focusing research on understudied and threatened ecosystems also allows us to develop a more holistic understanding of the ecological dynamics of IP [25].

The Three Gorges Reservoir (TGR) in China is widely recognized as the largest hydraulic and hydroelectric project in the world. Its operational strategy, known as "storing clear water and discharging turbid water", leads to fluctuations in the water level within the reservoir area. Specifically, the water level rises to 175 m during winter and decreases to 145 m during summer, creating a significant water level fluctuation (WLF) zone with a vertical difference of 30 m [26]. The TGR underwent experimental impounding from 2006 to 2007, reaching 175 m in 2009. The WLF area of the TGR was officially established in 2010. These annual flooding and drought events have a profound impact on plant communities, requiring the development of unique ecological strategies to thrive in this specific habitat [27]. Consequently, these strategies influence the morphological and physiological adaptive characteristics of these plants [28]. The cyclic changes between land and water levels within the zone have destabilized the plant community structure [29]. Human activities and alterations in the natural water level fluctuations of the Yangtze River's ecosystem have created favorable conditions for the spread of non-native plant species, leading to challenges from invasive species in the TGR area [30,31]. Studies showed a significant increase in alien species in the TGR area between 2003 and 2015 [32]. Failure to address this trend could result in severe damage to local plant resources, reduced genetic diversity, habitat disruption, decreased plant community diversity [33], and pose a substantial threat to the fragile ecosystem. After 14 years of the WLF zone's establishment, the presence of widespread IPs in various parts of the area indicates a challenging invasion scenario. To protect the ecological integrity of this zone and enhance biodiversity, a comprehensive investigation and research on IPs within the WLF zone, along with the implementation of appropriate control measures, are crucial [34]. Recent studies have focused on the impact of habitat characteristics on IPs within the WLF area, such as their proximity to the dam [35], land use composition, surrounding landscape [36], and the distribution patterns of invasive plant species [31]. However, there is a lack of systematic exploration concerning the resource utilization patterns of IPs in this zone and their invasion mechanisms from a functional trait perspective. There is evidence that freshwater wetlands are particularly vulnerable to invasive impacts [37]. It is crucial to analyze the resource acquisition and allocation patterns of IPs in the WLF zone based on functional traits and compare them with native plants (NPs) to enhance our understanding of the adaptation and invasion mechanisms of wetland plant communities in this zone. Current research on the characteristics of invasive plants in China through multi-species comparison is still in its preliminary stage [38], so the results can offer a theoretical foundation for future scientific predictions and the prevention of plant invasions in the zone.

We conducted a multi-scale study on the functional traits of IP and NP in the Pengxi River of the TGR area to investigate the following: (1) the differences in functional traits between IP and NP at regional and community scales under various inundation gradients; (2) the mechanisms of IP invasion in the WLF zone. Our results highlight distinct resource utilization strategies between IPs and NPs, offering new insights into the mechanisms of plant invasion in the WLF zone.

2. Materials and Methods

2.1. Study Area

The study area is located in Pengxi River, Kaizhou District, Chongqing, which lies in the hinterland of TGR (Figure 1). This area is a major tributary on the left bank of the Yangtze River, significantly impacted by the impoundment of the TGR. The terrain is predominantly flat, with extensive alluvial plains on both sides. Consequently, the WLF zone in the Kaizhou District is the largest among all the districts and counties in the TGR area, covering 42.78 km^2, which represents 12.3% of the total area of the WLF zone in the TGR. The study area experiences a subtropical humid monsoon climate, with an average annual temperature of approximately 18.5 °C and annual precipitation of around 1385 mm. To mitigate the effects of the TGR impoundment, the Kaizhou district government built a water level regulation dam downstream of the urban area, reducing the water level drop from 30 m to 4.72 m. This led to the creation of Hanfeng Lake, an urban lake, causing fluctuations in the water levels of Hanfeng Lake and Pengxi River in the research area in line with the TGR water levels. The Three Gorges project and subsequent resettlement efforts have also triggered significant changes in land use around the WLF zone, heightening the risk of invasion by non-native species. The study area is mainly composed of herbaceous plants and sparse shrubs and trees (natural growth and artificial planting).

2.2. Community Surveys and Identification of IP

The survey was conducted in July and August when the water level fluctuation (WLF) zone was fully exposed, and plant growth was at its peak. Eight sampling sites were selected in the Pengxi River Basin, including Wuyangba, Furongba, Shilongchuan Bridge, Toudao River Estuary, Dishuiba, Qukouba, Dalangba, and Baijia Stream (Table S3). These sampling points were evenly distributed in the Pengxi River basin, with similar substrate characteristics and minimal human interference. The elevation of these sites covers most of the elevation range of the WLF zone of the TGR, and water level fluctuations align with this pattern. Parallel sample lines were established at each sample point, ranging from the lowest to the highest elevation. These lines were strategically located at different sites: 6 lines at Wuyangba, 3 lines at Furongba, the Shilongchuan Bridge, Toudao River estuary, and Dishuiba, Dalangba, 4 lines at Qukouba, and 7 lines at the Baijia Stream. Along each sample line, sample plots were placed at different elevations. At each elevation, a single sample plot measuring 10 m × 10 m was established. Within each sample plot, at least five herbaceous sample squares measuring 1 m × 1 m were randomly positioned using the random sampling method. Plant data, including species identification, count, height, coverage, and relevant environmental factors, were meticulously recorded. The identification of IP was based on the "Invasive Alien Species of China" (IASC) database (https://www.iplant.cn/frps, accessed on 9 April 2024), while NP identification followed the website of Flora Reipublicae Populis Sinicae (FRPS) (https://www.iplant.cn/frps; accessed on 21 May 2024) according to the description of each plant's distribution area. In this study, a total of 21 species of IPs belonging to 11 families, 19 genera (Table S1), and 55 species of NPs belonging to 25 families, 43 genera (Table S2), were recorded. The intensity of flooding was inversely related to elevation. Hydrological data obtained from the Yangtze River Hydrological Network (http://www.cjh.com.cn; accessed on 21 May 2024) revealed that the lower section of the Pengxi River WLF zone (elevation 160 m–168 m) experienced flooding depths exceeding 10 m for about 160 days annually, whereas the upper WLF zone (elevation 170 m–175 m) experienced 5 m flood depths for around 70 days per year.

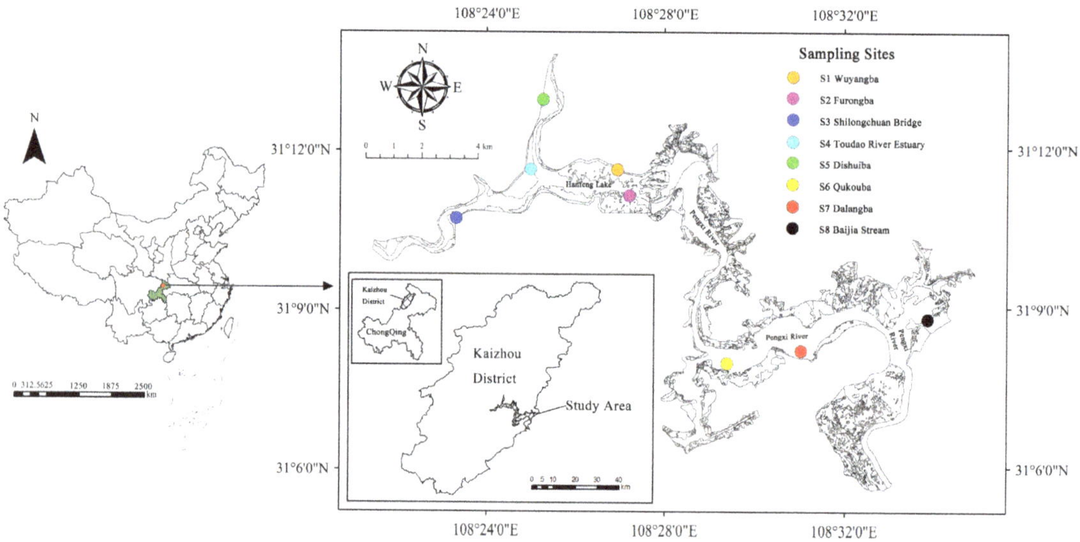

Figure 1. Location of the study area in Pengxi River, Kaizhou District, Chongqing, China.

2.3. Sampling and Measurement of Plant Functional Traits

In each sample area, plants with coverage exceeding 40% were classified as dominant species. In each sample area where dominant IP species were present, 5 intact target individuals that were well-exposed to light, fully developed, and of similar size were selected. Furthermore, 5 companion NPs with a coverage rate exceeding 40% were also included in the selection process. On-site measurements were conducted to assess the characteristics of the sampled plants, including their roots, stems, leaves, and other relevant indicators. Each plant yielded 3–5 undamaged leaves, free from shade and pests. Moreover, primary stems and roots were collected and transported to the laboratory for the determination and calculation of functional traits (Table 1). These selected traits represent plant strategies for acquiring, utilizing, and conserving resources, as well as for exploiting different temporal niches [39,40]. The determination of plant functional trait indicators was referred to as the standard protocol [41]. Leaf area measurements were conducted using ImageJ 1.53c.

Table 1. Method table for calculating functional characteristics.

Functional Traits	Abbreviation	Unit	Plant Characteristics Represented	Definitions and Calculation Methods
Height	H	cm	Ability to acquire sunlight resources	Vertical height of the highest point of the plant from the ground in its natural state
Specific leaf area	SLA	$cm^2 \cdot g^{-1}$	Ability of plants to absorb and store nutrients	Leaf area/leaf dry mass
Leaf dry matter content	LDMC	$g \cdot g^{-1}$	The extent to which dry matter synthesized by plants is put into leaf construction	Leaf dry mass/saturated fresh mass
Stem dry matter content	SDMC	$g \cdot g^{-1}$	The extent to which dry matter synthesized by plants is put into stem construction	Stem dry mass/saturated fresh mass
Specific root length	SRL	$cm \cdot g^{-1}$	The potential absorption rate of water and nutrients by plant root represents morphological indicators of underground competitiveness.	Root length/root dry mass
Leaf tissue density	LTD	$g \cdot cm^{-3}$	The accumulation status of leaf biomass, related to the stretching and defense forces of tissues.	Leaf dry mass/(leaf area × leaf thickness)

2.4. Statistical Analysis

At the regional scale, principal component analysis (PCA) was employed to examine the mean functional trait values of native and invasive plants within the study area. A one-way ANOVA was conducted to determine if significant differences existed between these groups along various axes. The two traits exhibiting the strongest correlation with each primary axis were chosen for further investigation. On the community scale, a comparison was conducted between the character distribution observed in the survey and a null model to detect differences in character composition between actual and random communities. This comparison aided in evaluating the dominant role of ecological processes in community assembly. According to the null hypothesis, the local community reflects the random distribution of individuals and traits derived from the regional species pool across the entire landscape, with weighting based on species abundance at the regional level. This model maintains constant species abundance at the regional scale while allowing for fluctuations at the quadrat level without imposing a fixed number of species in each quadrat. The analysis was carried out using the Picante toolkit in the R software 4.3.2 (999 iterations). The Community Weighted Mean (CWT) was used to quantify the average character value at the community level, weighted by the relative abundance of each species as follows:

$$\mathrm{CWT}_j = \sum_{i}^{n} p_i T_i$$

where p_i is the abundance of species i at point j and T_i is the average eigenvalue of species i at point j.

Our study employed the functional differences index (FD_j) to assess the functional similarity between IPs and NPs. This index quantifies the functional difference between IPs and co-existing NPs by comparing the absolute disparity in weighted trait values within a given plot to the average disparity observed across all plots at the same elevation. The computation of weighted trait values for plant species is derived from the mean trait values of the plant community, which are subsequently weighted based on the relative abundance of each species [6,42]. The FD_j was calculated using the following equation:

$$FD_j = \frac{|\text{CWTalien}_j - \text{CWTnative}_j|}{\text{CWTalien} - \text{CWTnative}}$$

The FD_j represents the disparity between the weighted invasive plant traits (CWTalien_j) and the weighted native plant traits (CWTnative_j) in sample j. CWTalien_j refers to the weighted invasive plant traits observed in sample j, while CWTnative_j represents the weighted native plant traits observed in sample j. CWTalien and CWTnative denote the weighted invasive plant traits and weighted native plant traits, respectively, across all samples. The weighted character value of the community is calculated based on the average character value of the community and weighted by the relative abundance of each species [6,42]. Based on 999 randomizations, we calculated the 95% confidence interval and compared the zero prediction with the observed data by calculating the deviation from the zero prediction. When the observed FD_j value exceeded the 95% confidence interval of the null model, it suggested functional divergence between invasive plants and native plants in terms of physiological and ecological characteristics, indicating niche differentiation (IPs and NPs were more different from each other than expected by chance). When FD_j was observed to occur below the 95% confidence interval, it indicated functional convergence between the IPs and NPs in terms of physiological and ecological characteristics, suggesting the presence of environmental filtering (IPs and NPs were more similar to each other than expected by chance).

The data were tested for normality using the Shapiro–Wilk test and all plant functional traits were log-transformed with a base of 10 to ensure normal distribution before analysis. One-way analysis of variance (ANOVA) was used to analyze the significant differences in plant functional traits at different altitudes at the regional scale. Tukey's method was used for multiple comparisons. All data processing and mapping in this study were performed using Origin 2021 and R 4.3.2.

3. Results

3.1. Differences in Plant Functional Strategies at the Regional Scale

Principal component analysis (PCA) was performed to compare the functional traits of plants between IPs and NPs at various altitudes within the WLF region at the regional scale (Figure 2). The first two principal components out of six accounted for 36.2% and 26.7% of the explained variance, respectively. Axis 1, moving from left to right, showed a gradual increase in LDMC and LTD, along with a gradual decrease in SLA, predominantly representing leaf functional traits. Axis 2, moving from top to bottom, displayed a gradual decrease in SRL and a gradual increase in H and SDMC, which mainly reflected the stem and root traits of the plants. The results highlighted differences between IPs and NPs in terms of their combinations of functional traits. Specifically, IPs showed higher SLA, H, SDMC, and lower LDMC, LTD, and SRL compared to NPs. The notable disparities observed between IPs and NPs along the PCA axis 1 and axis 2 in the WLF area indicate a functional differentiation in their strategies for resource utilization (Figure 3).

3.2. Differences in Plant Functional Traits at the Regional Scale

Based on the results of PCA, SLA and LTD were selected from axis 1, while H and SRL were selected from axis 2 for analysis. The results indicate that IPs exhibit a larger H and SLA compared to NPs, whereas SRL and LTD were lower. Specifically, IPs at elevations ranging from 165 m to 172 m significantly exceeded NP, with SLA at altitudes of 160 m, 168 m, and 175 m also surpassing NP. In contrast, SRL was observed to be lower at 170 m, 172 m, and 178 m, while LTD exhibited lower values at 160 m, 168 m, 170 m, and 175 m ($p < 0.05$; Figure 4).

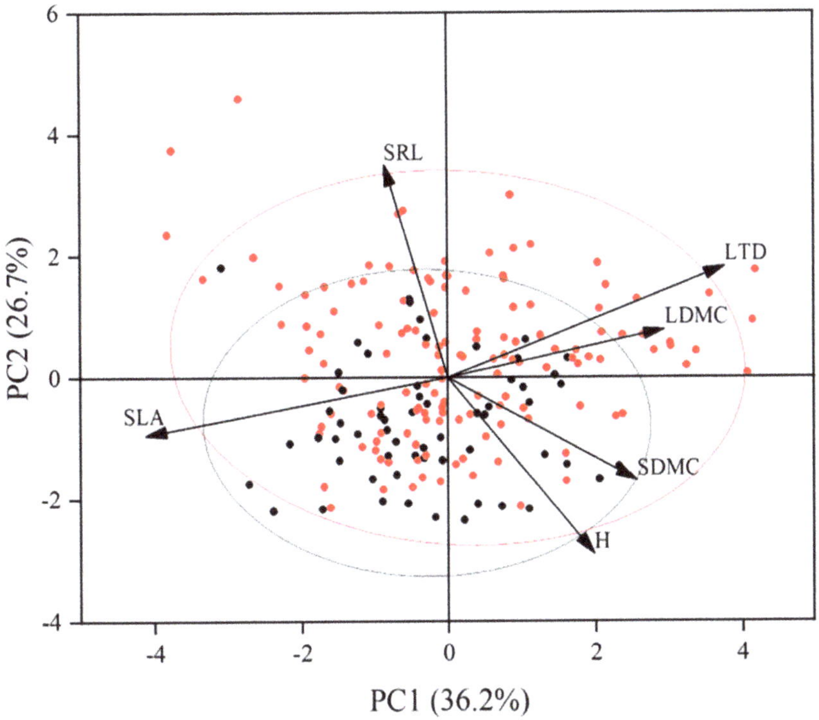

Figure 2. Results of PCA analysis for invasive and native plants at different elevations in the water level fluctuation zone. Black color represents invasive plants and red color represents native plants.

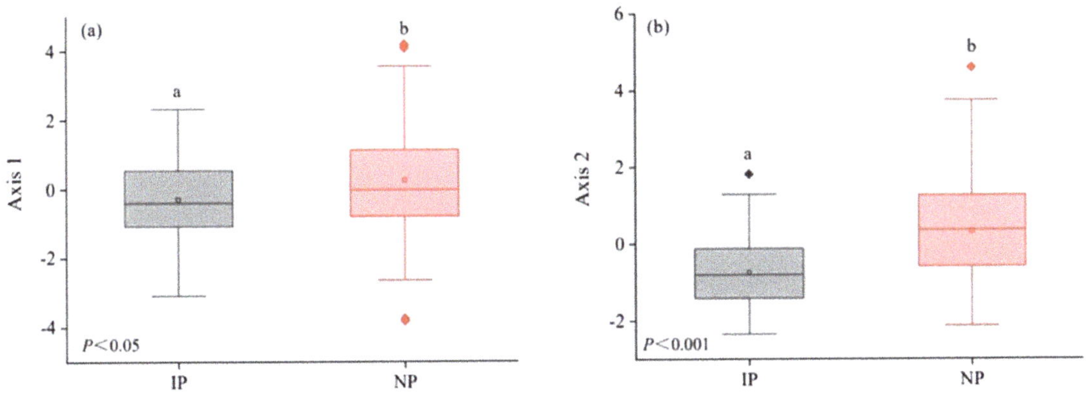

Figure 3. Results of two-tailed *t*-tests describing overall trait differences between invasive plants and native plants across the water level fluctuation zone on PCA axes 1 and axes 2. (**a**) Axis 1; (**b**) axis 2. Different lowercase letters on the boxplot indicate a significant difference.

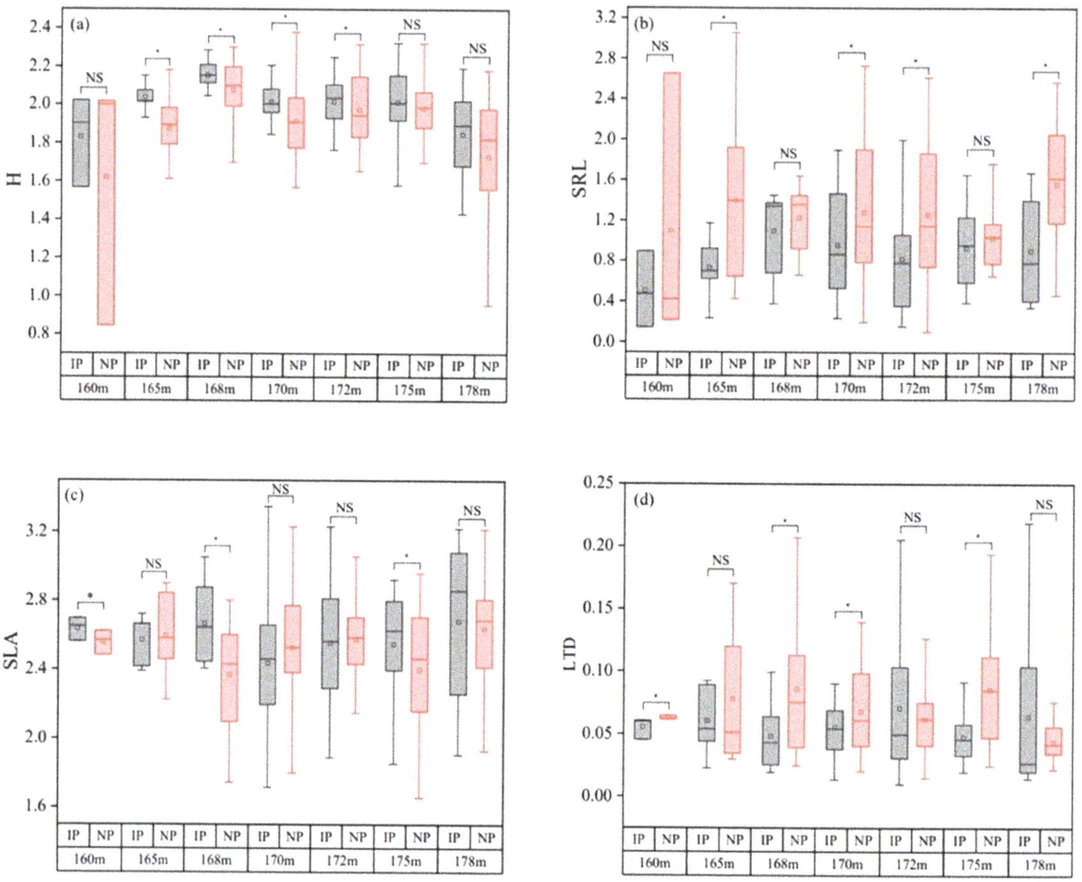

Figure 4. Differences in functional traits of invasive plants and native plants at different elevations of the water level fluctuation zone. Trait values of native plants and invasive plants calculated as the mean value of measured functional traits. (**a**) H, (**b**) SRL, (**c**) SLA, and (**d**) LTD. * indicates significant differences between invasive plants and native plants at different elevations ($p < 0.05$). NS indicates no significant difference between invasive plants and native plants.

3.3. *Response of Functional Traits at the Community Scale to Elevation*

Based on the PCA analysis outcomes, four traits were chosen for examination at the community scale as follows: SLA and LTD of axis 1; H and SRL of axis 2. The results indicate that the IP communities exhibited higher H and SLA and lower SRL and LTD. Both the H and SLA of IPs and NPs have a significant positive correlation with elevation at the community scale. The LTD of IPs was significantly negatively correlated with elevation. The slope absolute values of the H, SLA, and LTD of IPs were higher than those of NPs, while the slope absolute value of the SRL of NPs was higher than that of IPs (Figure 5, Table 2).

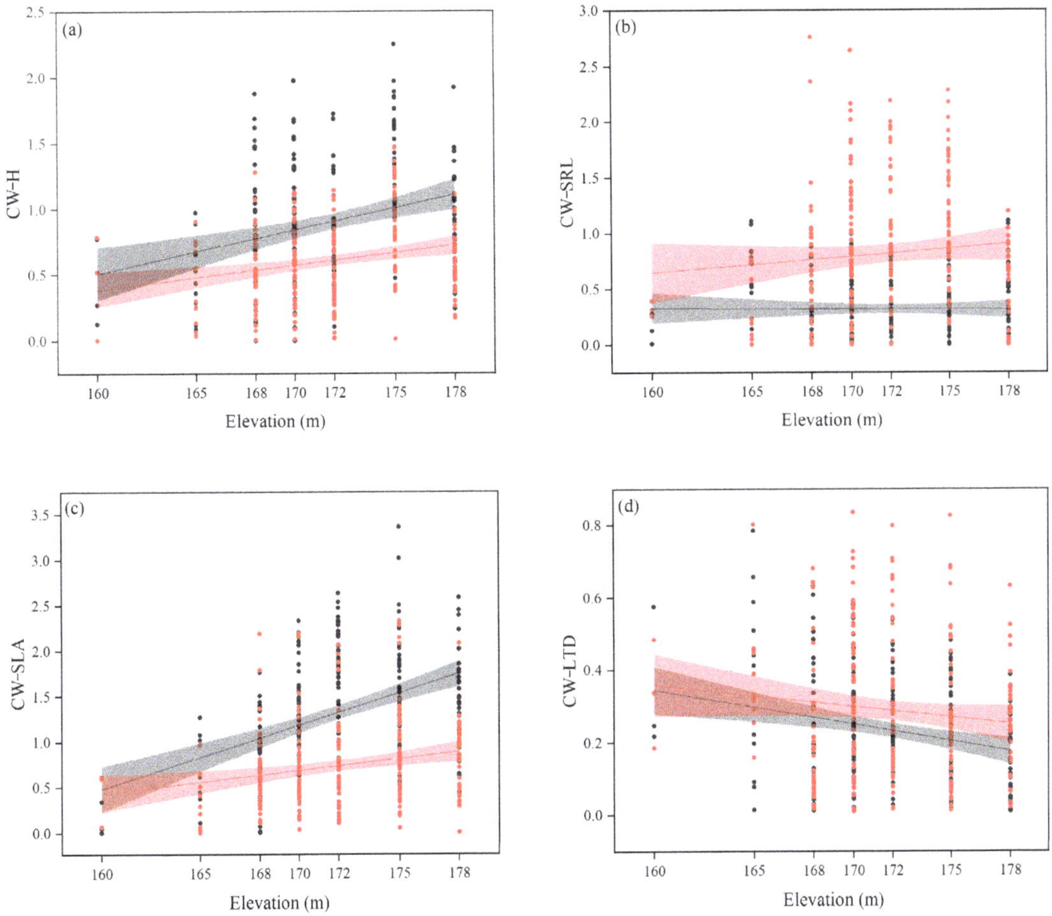

Figure 5. The response of weighted functional traits of invasive communities and native communities in the water level fluctuation to elevation. (**a**) CW−H, (**b**) CW−SRL, (**c**) CW−SLA, and (**d**) CW−LTD. Black color represents invasive plants, and red color represents native plants. The line represents the linear fit curve and the red/black area represents the 95% confidence band.

Table 2. Differences in the weighted functional properties of invasive and native communities in the water level fluctuation zone in relation to elevation.

Functional Traits (at the Community Scale)	Community Type	Slope	R^2	p
CW−H	IP	0.034	0.066	<0.001 ***
	NP	0.019	0.048	<0.001 ***
CW−SRL	IP	<0.001	0.004	0.901
	NP	0.014	0.003	0.190
CW−SLA	IP	0.070	0.170	<0.001 ***
	NP	0.026	0.039	<0.01 **
CW−LTD	IP	−0.009	0.048	<0.05 *
	NP	−0.006	0.008	0.093

***: $p < 0.001$; **: $p < 0.01$; *: $p < 0.05$.

3.4. Functional Differences at the Community Scale

A comparison was made between the functional differences of CW values in the IP and NP communities at various elevations within the WLF area in relation to a null model. The results showed a significant deviation from the null model, indicating a non-random community structure. The functional differences were generally below zero expectations. The functional differences of H, SRL, and SLA on the community scale were significantly positively correlated with elevation. The mean value of FD_j was lower than the 95% confidence interval of the null prediction in the elevation range of 160–175 m and higher than the 95% confidence interval of the null prediction at 178 m elevation. The functional differences of LTD were lower than the 95% confidence interval of the null prediction. The SRL was lower than the 95% confidence interval of spatial prediction in the low-elevation area, except for the lack of a significant difference between the SRL and spatial prediction at 178 m (Figure 6).

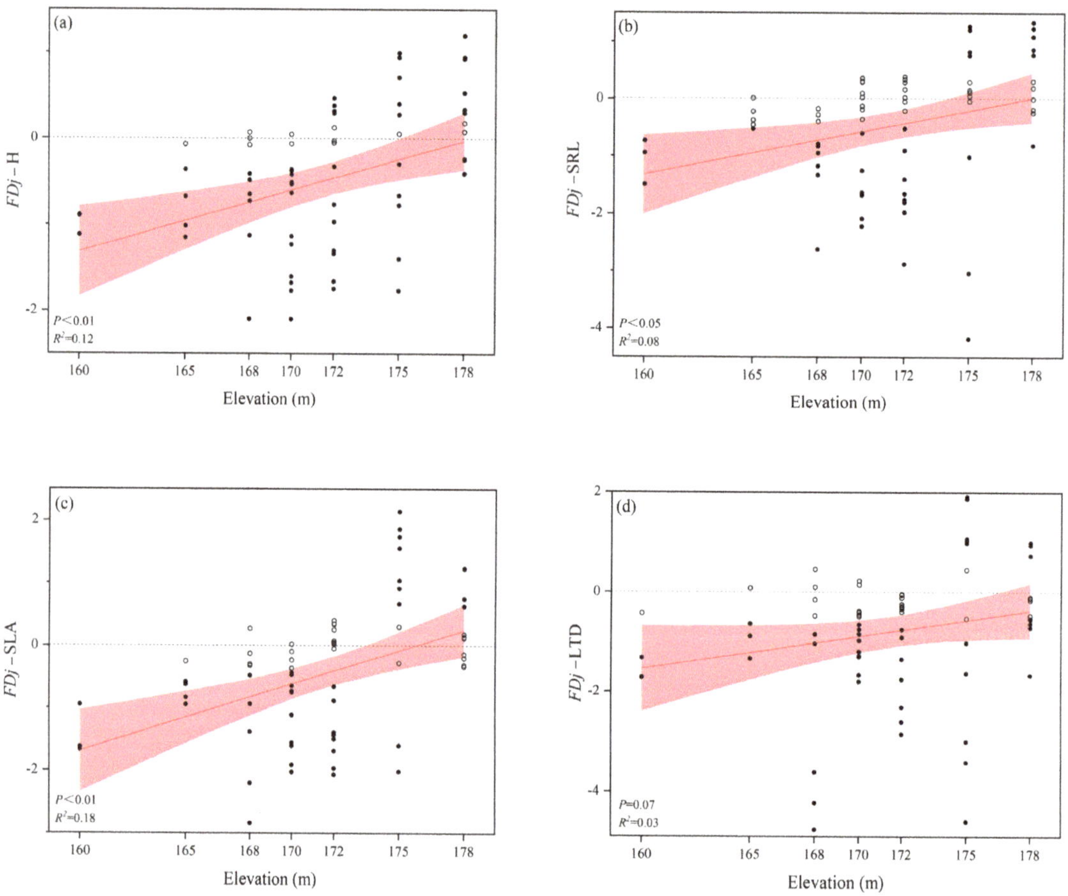

Figure 6. The functional differences index (FD_j) and elevation responses between the invasive community and the native community in the water level fluctuation zone. (**a**) FD_j−H, (**b**) FD_j−SRL, (**c**) FD_j−SLA, and (**d**) FD_j−LTD. When the point is greater than zero, alien species and native species tend to diverge in function. When the point is below zero, the function tends to converge. The solid point is beyond the 95% confidence interval, indicating a significant difference from the null prediction, while the hollow point does not differ from the null prediction. The red line represents the linear fit curve and the red area represents the 95% confidence band.

4. Discussion

4.1. Spatial Differentiation of IP Species in the WLF Zone

Plants have the ability to respond to changes in their habitat and adjust their behavior accordingly. Through evolution, organisms have developed a self-regulating mechanism to adapt to dynamic environments [43]. Our research findings reveal significant differences in the key traits between invasive and native species in the WLF zone of the Pengxi River, providing strong evidence for phenotypic differentiation [10]. These trait variations observed in specific ecosystems between native and invasive species align with previous studies [24,44], suggesting that these distinct attributes may contribute to the spread of invasive species within the WLF zone.

The PCA results indicated that the trait space in the WLF zone is primarily influenced by the following three main trait axes: the leaf economics spectrum, root economics spectrum, and plant main stem (including height). These findings suggest a correlation among different plant characteristics in the WLF zone [45]. Moreover, it has been noted that plant functions are constrained within specific trait combinations [46,47]. Our study identified significant functional differentiation in resource utilization strategies between IPs and NPs across various elevations in the WLF zone, which is consistent with prior research [48,49]. These results suggest that the successful invasion of non-native plants is closely linked to the adoption of distinct resource utilization strategies [50,51]. Specifically, our study highlights the crucial role of the leaf economics spectrum in distinguishing between them. IPs in the WLF zone also exhibit trait values aligned with a 'fast return on investment' strategy, similar to those in other regions [52], which remains consistent across different elevations [53]. These findings imply that IPs may have a competitive edge in resource extraction, especially in the WLF zone, with significant abiotic constraints, while NPs tend to display a more conservative approach to resource absorption [45,54]. In response to varying environmental stresses, IPs allocate resources differently by prioritizing above-ground resources and expanding their ecological niche through significant investments in H and SLA. This shift in resource allocation often leads to a reduction in leaf defense capabilities [10,44]. In the face of periodic flooding and drought-induced challenges in the WLF zone, IPs tend to invest more resources in fortifying their root systems. This resource allocation pattern could be considered a distinguishing feature of their resource utilization strategy in such habitats.

The height of IPs is significantly higher at most elevations. Plant height is indicative of a plant's capacity to obtain light resources and is closely linked to its competitive advantage [55–57]. There is a high level of competition among plants for light and space in the WLF zone. Taller plants have an advantage in accessing light and shading out shorter plants, limiting their growth [58]. This competitive advantage enables them to achieve a higher level of success over invasive species by outcompeting other species by overshadowing them and securing a greater share of light resources [59,60]. The result is consistent with other conclusions in ecosystems experiencing cycles of drought and rewetting [61]. SLA plays a key role in regulating plant functions and resource acquisition [36]. SLA has been found to have a positive impact on plant growth rates, biomass production, and nutrient accumulation efficiency [62]. In the WLF zone, IPs generally exhibit a higher SLA, allowing them to efficiently acquire resources, leading to enhanced growth and faster leaf turnover rates [63–65]. Therefore, traits associated with rapid growth and efficient resource acquisition contribute to the invasiveness of plant species [66].

The growth and acquisition of resources in plants are often balanced by their ability to tolerate abiotic stress [67]. This phenomenon has been observed in the WLF zone, where IPs have been found to have a smaller LDMC and LTD compared to NPs. LDMC and LTD are indicators of how plants allocate dry matter for leaf construction. An increase in LTD, affecting water diffusion from the leaf interior to the surface [68], can reduce water loss and enhance drought stress resistance. IPs invest less in structural compounds and resource allocation for leaf defense, resulting in lower leaf toughness and drought tolerance compared to NPs [44]. This trade-off allows invasive plants to prioritize resource access over tolerance to abiotic stress conditions. Interestingly, this study also found that invasive plants in the floodplain, especially in the upper section, tend to have a smaller SRL. The SRL is a trait sensitive to environmental changes. A larger SRL indicates reduced carbohydrate requirements for root system development and maintenance per unit length, leading to increased energy expenditure for nutrient and water uptake, as well as faster root turnover rates [69,70]. IPs in the WLF zone exhibit an opposite SRL compared to those in other areas [10]. IPs opt to allocate more defensive resources toward the development of root systems, resulting in increased thickness. This investment comes at the cost of higher construction and maintenance expenses in exchange for enhanced soil penetration capabilities and extended longevity [71]. Consequently, IPs in the WLF zone prioritize acquiring more aboveground resources to expand their presence and occupy larger habitat areas. They allocate more resources to their root systems rather than leaves to better withstand harsh environmental conditions (Figure 7).

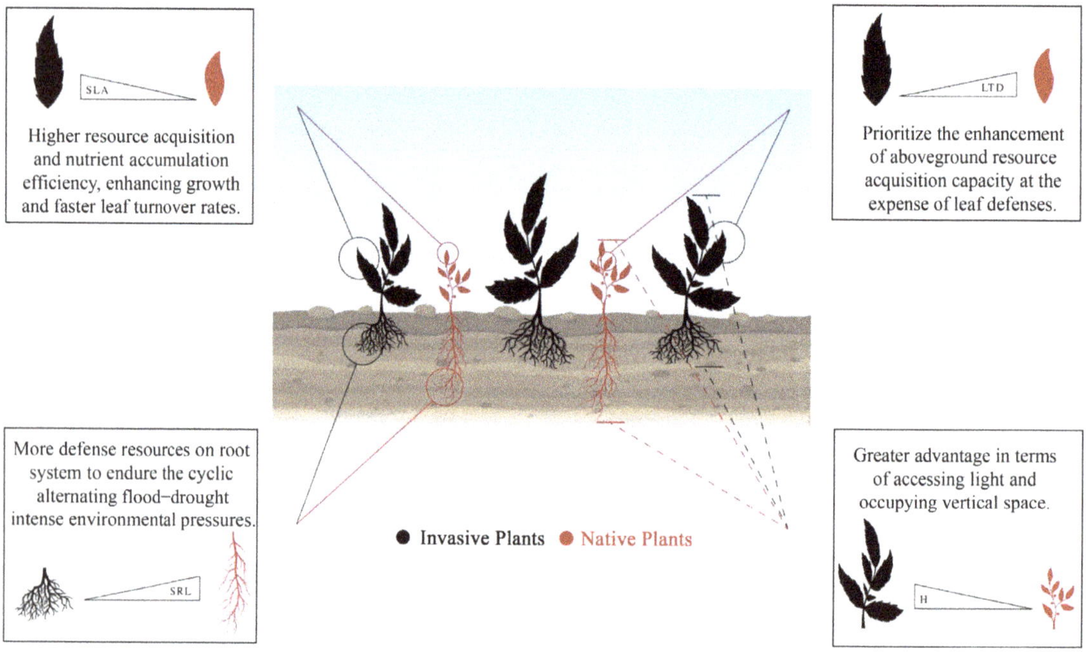

Figure 7. Diagram of functional segregation between invasive plants and native plants in the water level fluctuation zone. The black color represents invasive plants, and the red color represents native plants.

4.2. The Invasion Mechanism of IP in the WLF Zone

Biological invasion provides a valuable opportunity to study the colonization of new species populations [11,72]. Given the unique ecological conditions of intermittent floods and droughts in the WLF region, it is imperative to assess the invasion mechanism of the IP community in this area. This mechanism plays a key role in determining the success or failure of the invasion.

Our findings suggest that at a regional scale, significant differences in features indicate that niche differentiation has a substantial impact. For an alien species to become invasive, it must possess adequate functional similarity to be embraced by the local community while also having enough differences to occupy the edge of the plant functional feature space (unused niche) within the local ecosystem [59]. In most altitude areas of WLF, the unique growth strategy of IP allows it to occupy a vacant niche in the ecosystem of WLF, which may explain its invasive potential. Hence, on a regional scale, the likelihood of invasive success can be predicted by pre-existing characteristics and habitat preferences. Alien species often show significant distinctions from native species at the species level, which may indicate the operation of neutral dynamic and/or density-dependent mechanisms related to limiting similarity, such as competition and evasion/resistance to predators [73,74]. Plants like *Bidens frondosa* and *Bidens pilosa* exhibit flood resistance, possess achene fruits, and bear barb-shaped prickles, making them easily dispersed by animals, including humans, as well as through water and wind due to their lightweight seeds. Their long dispersal distance is also widely used to indicate high invasiveness [75]. Research indicates that IPs of the Asteraceae family can further enhance their community stability by altering the local pollination network [76,77]. This adaptive trait to various habitats, along with advantageous propagules and stress tolerance [78], creates favorable conditions for these plants to invade and inhabit, leading to their widespread presence throughout the WLF zone. Another invasive species, *Alternanthera philoxeroides*, originally from South America and introduced in China nearly a century ago, demonstrates strong habitat adaptability and phenotypic plasticity [79], expanding its ecological niche [80,81] and enabling it to thrive in this zone. Furthermore, the formation of patch-like mosaics post-invasion in specific areas is often attributed to continuous clonal growth by a genealogical strain [82], contributing to the extensive distribution of this species across the entire zone. Our investigation also revealed that native species within the same genus as invasive plants, such as *Bidens tripartita* related to *Bidens frondosa* and *Bidens pilosa* and *Alternanthera sessilis* related to *Alternanthera philoxeroides*, are typically absent in this research area. Consequently, from a phylogenetic standpoint, there exists a considerable genetic distance, which may support the DNH. To gain a deeper understanding of the ecological processes that drive species invasion in the WLF zone, additional comprehensive research is required. This research should encompass intra-species variation, inter-species interactions, and phylogenetic distance within the context of global change in the future.

Significant differences in functional traits were observed between IPs and NPs at the community scale. As elevation increased, these differences gradually increased. Moreover, certain traits showed a notable positive correlation with elevation. In particular, the FD_j between IP and NP communities, which was below 175 m in the WLF area, fell below the 95% confidence interval of the null model, indicating a minimal divergence in physiological and ecological attributes, suggestive of functional convergence. This study highlights the significant role of environmental filtering in community assembly, indicating that foreign species in degraded areas must endure harsh abiotic conditions to establish and sustain the community, resulting in functional convergence with local plant communities [83,84]. As elevation rises, the effects of environmental filtering are mitigated by niche differentiation. The functional differences between communities in the 178 m zone, largely unaffected by flooding, surpass the 95% confidence interval of the null model. These findings highlight the significant impact of environmental filtering, particularly through regular flooding and drought stress, to explain the variations in IP and NP functions within the WLF zone. Although there are challenges between the theory of reconciled coexistence and the concept of environmental filtering because the strength and direction of biotic interactions can be strongly influenced by the abiotic environment, both factors may dynamically interact to drive community patterns [85]. Dominant species in a community typically do not coexist with companion plants of the same genus. This absence of potential competitors or natural enemies for invasive species in the WLF zone may accentuate the importance of habitat filtering. As a result, IPs may successfully integrate into habitats with comparable functional characteristics already occupied by NPs and establish their own communities.

The importance of the research scale when elucidating intrusion mechanisms is highlighted in our study, as research findings can differ depending on the scale considered [86]. Specifically, at the regional scale of the WLF zone, there are notable differences in the functional characteristics of IPs and NPs in response to changes in altitude. IPs typically display more invasive traits due to the homogeneity of the habitat environment at the regional scale, leading to direct interactions and competition among individuals for limited resources. Competitive exclusion among coexisting species can lead to significant variations in functional traits, favoring certain branches and species with competitive advantages [87,88]. On the community scale, in resource-limited environments, environmental factors constrain the variation in physiological and morphological characteristics within the plant community. Environmental filtering emerges as a key factor influencing community assembly, potentially outweighing interspecific interactions at the species level [20,86]. Evidence suggests that strong local competitors may share similarities with strong foreign competitors [89], resulting in convergent character values at the community scale after adjusting for relative abundance. In harsh environmental conditions, the effects of intraspecific competition may be less apparent [19]. Additionally, species loss due to abiotic tolerance or differences in competitive fitness with other residents [88] often leads to the functional traits of coexisting species being more similar than expected [90]. The variations in the results at the two scales may emphasize the distinct roles of abiotic and biotic effects on community assembly and species distribution at different scales.

5. Conclusions

Our multi-scale research method is valuable for comprehending the functional differentiation and invasion mechanisms of local plants and invasive plants in the water level fluctuation zone. At the regional scale, IPs exhibit larger H and SLA and smaller LTD and SRL compared to NP. IPs tend to prioritize enhancing their ability to acquire above-ground resources at the expense of leaf defense capabilities, allocating more tissue construction resources to their roots to withstand the strong environmental pressure in the WLF zone. This resource allocation strategy may be the key to their successful intrusion, highlighting the significant role of niche differentiation. On the community scale, the H and SLA of IP and NP communities show a significant positive correlation with elevation, while the LTD of IP communities shows a significant negative correlation with elevation. The functional differences in H, SRL, and SLA in the IP and NP communities are significantly positively correlated with elevation. Functional convergence is observed at elevations of 175 m and below, with functional divergence at elevations of 175 m and above. This suggests that although different resource utilization strategies drive successful IP invasion, the environmental filtering caused by periodic floods and drought stress leads to similar ecological needs for both IP and NP communities, playing a major role in the community assembly in the WLF zone. IPs may successfully integrate into habitats with similar functional characteristics already occupied by NPs and establish their own communities. This environmental filtering effect decreases notably with increasing elevation. Our results can reflect differences in resource utilization strategies between IPs and NPs, offering new insights into the mechanism of plant invasion in the WLF zone.

Supplementary Materials: The following supporting information can be downloaded at https://www.mdpi.com/article/10.3390/f15060959/s1, Table S1: List of invasive plants in the water level fluctuation zone; Table S2: List of native plants in the water level fluctuation zone; Table S3: Summary table of sampling sites.

Author Contributions: L.C.: Investigation, Software, Visualization, Data curation, Writing—Original draft. X.Y.: Supervision, Writing—Review and Editing. K.S.: Investigation, Data curation. P.L.: Visualization. All authors have read and agreed to the published version of the manuscript.

Funding: This research was funded by the National Natural Science Fund of China (No. 52178031) and the Fundamental Research Funds for the Central Universities (No. 2021CDJQYJC005).

Data Availability Statement: Data will be available upon request to the corresponding author.

Acknowledgments: We thank the Nature Reserve Administration Center of Kaizhou District, Chongqing for supporting our work. We also thank Jiaqi Lin, Wei Cheng, and Chunli Hou for assisting with the investigation.

Conflicts of Interest: The authors declare no conflicts of interest.

References

1. Korpelainen, H.; Pietilainen, M. What Makes a Good Plant Invader? *Life* **2023**, *13*, 1596. [CrossRef] [PubMed]
2. Munro, D.; Steer, J.; Linklater, W. On allegations of invasive species denialism. *Conserv. Biol.* **2019**, *33*, 797–802. [CrossRef]
3. Simberloff, D.; Martin, J.-L.; Genovesi, P.; Maris, V.; Wardle, D.A.; Aronson, J.; Courchamp, F.; Galil, B.; García-Berthou, E.; Pascal, M.; et al. Impacts of biological invasions: What's what and the way forward. *Trends Ecol. Evol.* **2013**, *28*, 58–66. [CrossRef]
4. Lin, W.; Zhou, G.; Cheng, X.; Xu, R. Fast economic development accelerates biological invasions in China. *PLoS ONE* **2007**, *2*, e1208. [CrossRef]
5. Seebens, H.; Blackburn, T.M.; Dyer, E.E.; Genovesi, P.; Hulme, P.E.; Jeschke, J.M.; Pagad, S.; Pysek, P.; Winter, M.; Arianoutsou, M.; et al. No saturation in the accumulation of alien species worldwide. *Nat. Commun.* **2017**, *8*, 14435. [CrossRef] [PubMed]
6. Violle, C.; Navas, M.-L.; Vile, D.; Kazakou, E.; Fortunel, C.; Hummel, I.; Garnier, E. Let the concept of trait be functional! *Oikos* **2007**, *116*, 882–892. [CrossRef]
7. De Bello, F.; Lavorel, S.; Díaz, S.; Harrington, R.; Cornelissen, J.H.C.; Bardgett, R.D.; Berg, M.P.; Cipriotti, P.; Feld, C.K.; Hering, D.; et al. Towards an assessment of multiple ecosystem processes and services via functional traits. *Biodivers. Conserv.* **2010**, *19*, 2873–2893. [CrossRef]
8. Sharma, P.; Rathee, S.; Ahmad, M.; Siddiqui, M.H.; Alamri, S.; Kaur, S.; Kohli, R.K.; Singh, H.P.; Batish, D.R. Leaf functional traits and resource use strategies facilitate the spread of invasive plant Parthenium hysterophorus across an elevational gradient in western Himalayas. *BMC Plant Biol.* **2024**, *24*, 234. [CrossRef]
9. Hulme, P.E.; Bernard-Verdier, M.; Baltzer, J. Comparing traits of native and alien plants: Can we do better? *Funct. Ecol.* **2017**, *32*, 117–125. [CrossRef]
10. Mathakutha, R.; Steyn, C.; le Roux, P.C.; Blom, I.J.; Chown, S.L.; Daru, B.H.; Ripley, B.S.; Louw, A.; Greve, M.; Price, J. Invasive species differ in key functional traits from native and non-invasive alien plant species. *J. Veg. Sci.* **2019**, *30*, 994–1006. [CrossRef]
11. Guisan, A.; Petitpierre, B.; Broennimann, O.; Daehler, C.; Kueffer, C. Unifying niche shift studies: Insights from biological invasions. *Trends Ecol. Evol.* **2014**, *29*, 260–269. [CrossRef]
12. Strayer, D.L. Eight questions about invasions and ecosystem functioning. *Ecol. Lett.* **2012**, *15*, 1199–1210. [CrossRef] [PubMed]
13. Bezeng, B.S.; Savolainen, V.; Yessoufou, K.; Papadopulos, A.S.T.; Maurin, O.; Van der Bank, M. A phylogenetic approach towards understanding the drivers of plant invasiveness on Robben Island, South Africa. *Bot. J. Linn. Soc.* **2013**, *172*, 142–152. [CrossRef]
14. Lososová, Z.; de Bello, F.; Chytrý, M.; Kühn, I.; Pyšek, P.; Sádlo, J.; Winter, M.; Zelený, D. Alien plants invade more phylogenetically clustered community types and cause even stronger clustering. *Glob. Ecol. Biogeogr.* **2015**, *24*, 786–794. [CrossRef]
15. Castro, S.A.; Escobedo, V.M.; Aranda, J.; Carvallo, G.O. Evaluating Darwin's naturalization hypothesis in experimental plant assemblages: Phylogenetic relationships do not determine colonization success. *PLoS ONE* **2014**, *9*, e105535. [CrossRef]
16. Darwin, C. *The Origin of Species by Means of Natural Selection*, 6th ed.; J.M. Dent & Sons Ltd.: London, UK, 1872.
17. Strauss, S.Y.; Webb, C.O.; Salamin, N. Exotic taxa less related to native species are more invasive. *Proc. Natl. Acad. Sci. USA* **2006**, *103*, 5841–5845. [CrossRef] [PubMed]
18. Daehler, C.C. Darwin's naturalization hypothesis revisited. *Am. Nat.* **2001**, *158*, 324–330. [CrossRef] [PubMed]
19. Park, D.S.; Feng, X.; Maitner, B.S.; Ernst, K.C.; Enquist, B.J. Darwin's naturalization conundrum can be explained by spatial scale. *Proc. Natl. Acad. Sci. USA* **2020**, *117*, 10904–10910. [CrossRef] [PubMed]
20. Thuiller, W.; Gallien, L.; Boulangeat, I.; De Bello, F.; Münkemüller, T.; Roquet, C.; Lavergne, S. Resolving Darwin's naturalization conundrum: A quest for evidence. *Divers. Distrib.* **2010**, *16*, 461–475. [CrossRef]
21. Carboni, M.; Munkemuller, T.; Gallien, L.; Lavergne, S.; Acosta, A.; Thuiller, W. Darwin's naturalization hypothesis: Scale matters in coastal plant communities. *Ecography* **2013**, *36*, 560–568. [CrossRef]
22. Leffler, A.J.; James, J.J.; Monaco, T.A.; Sheley, R.L. A new perspective on trait differences between native and invasive exotic plants: Reply. *Ecology* **2015**, *96*, 1152–1153. [CrossRef] [PubMed]
23. Wang, C.; Yu, Y.; Cheng, H.; Du, D. Which factor contributes most to the invasion resistance of native plant communities under the co-invasion of two invasive plant species? *Sci. Total Environ.* **2022**, *813*, 152628. [CrossRef]
24. van Kleunen, M.; Weber, E.; Fischer, M. A meta-analysis of trait differences between invasive and non-invasive plant species. *Ecol. Lett.* **2010**, *13*, 235–245. [CrossRef] [PubMed]
25. Chiu, J.H.; Chong, K.Y.; Lum, S.K.Y.; Wardle, D.A. Trends in the direction of global plant invasion biology research over the past two decades. *Ecol. Evol.* **2023**, *13*, e9690. [CrossRef]
26. Yuan, X.Z.; Zhang, Y.W.; Liu, H.; Xiong, S.; Li, B.; Deng, W. The littoral zone in the Three Gorges Reservoir, China: Challenges and opportunities. *Environ. Sci. Pollut. Res. Int.* **2013**, *20*, 7092–7102. [CrossRef]
27. Sun, R.; Deng, W.q.; Yuan, X.z.; Liu, H.; Zhang, Y.w. Riparian vegetation after dam construction on mountain rivers in China. *Ecohydrology* **2013**, *7*, 1187–1195. [CrossRef]

28. Li, H.; Yu, K.; Ratajczak, Z.; Nippert, J.B.; Tondrob, D.; Xu, D.; Li, W.; Du, G. When variability outperforms the mean: Trait plasticity predicts plant cover and biomass in an alpine wetland. *Plant Soil* **2016**, *407*, 401–415. [CrossRef]
29. Zhang, L.; Yang, L.; Shen, W. Dramatic altitudinal variations in leaf mass per area of two plant growth forms at extreme heights. *Ecol. Indic.* **2020**, *110*, 105890. [CrossRef]
30. Kepa Brian Morgan, T.K.; Sardelic, D.N.; Waretini, A.F. The Three Gorges Project: How sustainable? *J. Hydrol.* **2012**, *460–461*, 1–12. [CrossRef]
31. Wang, Y.; Liu, Y.; Ma, M.; Ding, Z.; Wu, S.; Jia, W.; Chen, Q.; Yi, X.; Zhang, J.; Li, X.; et al. Dam-induced difference of invasive plant species distribution along the riparian habitats. *Sci. Total Environ.* **2022**, *808*, 152103. [CrossRef]
32. Xiong, W.; Wang, H.; Wang, Q.; Tang, J.F.; Bowler, P.A.; Xie, D.; Pan, L.; Wang, Z.X. Non-native species in the Three Gorges Dam Reservoir: Status and risks. *Bioinvasions Rec.* **2018**, *7*, 153–158. [CrossRef]
33. Wen, Y.; Su, X.; Cai, F.; Qian, R.; Bejarano, M.D.; Wu, S.; Yang, Q.; Liu, X.; Zeng, B. The differences in plant invasion in two types of shorelines under flow regulation of the Three Gorges Reservoir. *Sci. Total Environ.* **2024**, *912*, 168892. [CrossRef] [PubMed]
34. Bao, Y.; Gao, P.; He, X. The water-level fluctuation zone of Three Gorges Reservoir—A unique geomorphological unit. *Earth-Sci. Rev.* **2015**, *150*, 14–24. [CrossRef]
35. Wang, C.; Wei, M.; Wang, S.; Wu, B.; Cheng, H. *Erigeron annuus* (L.) Pers. and *Solidago canadensis* L. antagonistically affect community stability and community invasibility under the co-invasion condition. *Sci. Total Environ.* **2020**, *716*, 137128. [CrossRef] [PubMed]
36. Cheng, J.; Chu, P.; Chen, D.; Bai, Y.; Niu, S. Functional correlations between specific leaf area and specific root length along a regional environmental gradient in Inner Mongolia grasslands. *Funct. Ecol.* **2016**, *30*, 985–997. [CrossRef]
37. Mazurczyk, T.; Brooks, R.P. Native biodiversity increases with rising plant invasions in temperate, freshwater wetlands. *Wetl. Ecol. Manag.* **2021**, *30*, 139–160. [CrossRef]
38. Wang, F.; Huang, J.; Zhang, N.; Li, Y.; He, S.; Wen, J.; Yin, L.; Liang, Y. Exploring plant characteristics for constructing a pre-border weed risk assessment for China. *Biol. Invasions* **2024**, *26*, 909–933. [CrossRef]
39. Gross, N.; Suding, K.N.; Lavorel, S. Leaf dry matter content and lateral spread predict response to land use change for six subalpine grassland species. *J. Veg. Sci.* **2007**, *18*, 289–300. [CrossRef]
40. Maire, V.; Gross, N.; Borger, L.; Proulx, R.; Wirth, C.; Pontes, L.D.S.; Soussana, J.F.; Louault, F. Habitat filtering and niche differentiation jointly explain species relative abundance within grassland communities along fertility and disturbance gradients. *New Phytol.* **2012**, *196*, 497–509. [CrossRef]
41. Cornelissen, J.H.C.; Lavorel, S.; Garnier, E.; Diaz, S.; Buchmann, N.; Gurvich, D.E.; Reich, P.B.; ter Steege, H.; Morgan, H.D.; van der Heijden, M.G.A.; et al. A handbook of protocols for standardised and easy measurement of plant functional traits worldwide. *Aust. J. Bot.* **2003**, *51*, 335–380. [CrossRef]
42. Gross, N.; Börger, L.; Duncan, R.P.; Hulme, P.E.; Wilson, S. Functional differences between alien and native species: Do biotic interactions determine the functional structure of highly invaded grasslands? *Funct. Ecol.* **2013**, *27*, 1262–1272. [CrossRef]
43. de Oliveira, A.C.P.; Nunes, A.; Rodrigues, R.G.; Branquinho, C. The response of plant functional traits to aridity in a tropical dry forest. *Sci. Total Environ.* **2020**, *747*, 141177. [CrossRef] [PubMed]
44. Daehler, C.C. Performance comparisons of co-occurring native and alien invasive plants: Implications for conservation and restoration. *Annu. Rev. Ecol. Evol. Syst.* **2003**, *34*, 183–211. [CrossRef]
45. Laughlin, D.C.; Leppert, J.J.; Moore, M.M.; Sieg, C.H. A multi-trait test of the leaf-height-seed plant strategy scheme with 133 species from a pine forest flora. *Funct. Ecol.* **2010**, *24*, 493–501. [CrossRef]
46. Diaz, S.; Kattge, J.; Cornelissen, J.H.; Wright, I.J.; Lavorel, S.; Dray, S.; Reu, B.; Kleyer, M.; Wirth, C.; Prentice, I.C.; et al. The global spectrum of plant form and function. *Nature* **2016**, *529*, 167–171. [CrossRef] [PubMed]
47. Lloret, F.; de la Riva, E.G.; Perez-Ramos, I.M.; Maranon, T.; Saura-Mas, S.; Diaz-Delgado, R.; Villar, R. Climatic events inducing die-off in Mediterranean shrublands: Are species' responses related to their functional traits? *Oecologia* **2016**, *180*, 961–973. [CrossRef] [PubMed]
48. Ordonez, A.; Wright, I.J.; Olff, H. Functional differences between native and alien species: A global-scale comparison. *Funct. Ecol.* **2010**, *24*, 1353–1361. [CrossRef]
49. Tecco, P.A.; DÃ az, S.; Cabido, M.; Urcelay, C. Functional traits of alien plants across contrasting climatic and land-use regimes: Do aliens join the locals or try harder than them? *J. Ecol.* **2010**, *98*, 17–27. [CrossRef]
50. Funk, J.L.; Standish, R.J.; Stock, W.D.; Valladares, F. Plant functional traits of dominant native and invasive species in mediterranean-climate ecosystems. *Ecology* **2016**, *97*, 75–83. [CrossRef]
51. Helsen, K.; Van Cleemput, E.; Bassi, L.; Graae, B.J.; Somers, B.; Blonder, B.; Honnay, O. Inter- and intraspecific trait variation shape multidimensional trait overlap between two plant invaders and the invaded communities. *Oikos* **2020**, *129*, 677–688. [CrossRef]
52. Montesinos, D. Fast invasives fastly become faster: Invasive plants align largely with the fast side of the plant economics spectrum. *J. Ecol.* **2021**, *110*, 1010–1014. [CrossRef]
53. Tordoni, E.; Petruzzellis, F.; Nardini, A.; Savi, T.; Bacaro, G.; Botta-Dukát, Z. Make it simpler: Alien species decrease functional diversity of coastal plant communities. *J. Veg. Sci.* **2019**, *30*, 498–509. [CrossRef]
54. de la Riva, E.G.; Maranon, T.; Violle, C.; Villar, R.; Perez-Ramos, I.M. Biogeochemical and Ecomorphological Niche Segregation of Mediterranean Woody Species along a Local Gradient. *Front. Plant Sci.* **2017**, *8*, 1242. [CrossRef] [PubMed]

55. Pérez-Harguindeguy, N.; Díaz, S.; Garnier, E.; Lavorel, S.; Poorter, H.; Jaureguiberry, P.; Bret-Harte, M.S.; Cornwell, W.K.; Craine, J.M.; Gurvich, D.E.; et al. Corrigendum to: New handbook for standardised measurement of plant functional traits worldwide. *Aust. J. Bot.* **2016**, *64*, 715–716. [CrossRef]
56. Wahl, S.; Ryser, P. Root tissue structure is linked to ecological strategies of grasses. *New Phytol.* **2000**, *148*, 459–471. [CrossRef] [PubMed]
57. Westoby, M.; Falster, D.S.; Moles, A.T.; Vesk, P.A.; Wright, I.J. Plant ecological strategies: Some leading dimensions of variation between species. *Annu. Rev. Ecol. Syst.* **2002**, *33*, 125–159. [CrossRef]
58. Mao, L.; Chen, S.; Zhang, J.; Zhou, G. Altitudinal patterns of maximum plant height on the Tibetan Plateau. *J. Plant Ecol.* **2018**, *11*, 85–91. [CrossRef]
59. Divisek, J.; Chytry, M.; Beckage, B.; Gotelli, N.J.; Lososova, Z.; Pysek, P.; Richardson, D.M.; Molofsky, J. Similarity of introduced plant species to native ones facilitates naturalization, but differences enhance invasion success. *Nat. Commun.* **2018**, *9*, 4631. [CrossRef] [PubMed]
60. Hejda, M.; Pyšek, P.; Jarošík, V. Impact of invasive plants on the species richness, diversity and composition of invaded communities. *J. Ecol.* **2009**, *97*, 393–403. [CrossRef]
61. Zhang, X.; Oduor, A.M.O.; Liu, Y. Invasive plants have greater growth than co-occurring natives in live soil subjected to a drought-rewetting treatment. *Funct. Ecol.* **2022**, *37*, 513–522. [CrossRef]
62. te Beest, M.; Esler, K.J.; Richardson, D.M. Linking functional traits to impacts of invasive plant species: A case study. *Plant Ecol.* **2014**, *216*, 293–305. [CrossRef]
63. Eissenstat, D.M.; Wells, C.E.; Yanai, R.D.; Whitbeck, J.L. Building roots in a changing environment: Implications for root longevity. *New Phytol.* **2000**, *147*, 33–42. [CrossRef]
64. Grotkopp, E.; Rejmanek, M. High seedling relative growth rate and specific leaf area are traits of invasive species: Phylogenetically independent contrasts of woody angiosperms. *Am. J. Bot.* **2007**, *94*, 526–532. [CrossRef] [PubMed]
65. Quan, G.M.; Mao, D.J.; Zhang, J.E.; Xie, J.F.; Xu, H.Q.; An, M. Response of invasive Chromolaena odorata and two coexisting weeds to contrasting irradiance and nitrogen. *Photosynthetica* **2015**, *53*, 419–429. [CrossRef]
66. Gioria, M.; Osborne, B.A. Resource competition in plant invasions: Emerging patterns and research needs. *Front. Plant Sci.* **2014**, *5*, 501. [CrossRef]
67. Herms, D.A.; Mattson, W.J. The Dilemma of Plants—To Grow or Defend. *Q. Rev. Biol.* **1992**, *67*, 478. [CrossRef]
68. Henn, J.J.; Yelenik, S.; Damschen, E.I. Environmental gradients influence differences in leaf functional traits between native and non-native plants. *Oecologia* **2019**, *191*, 397–409. [CrossRef]
69. Geng, Y.; Wang, L.; Jin, D.; Liu, H.; He, J.S. Alpine climate alters the relationships between leaf and root morphological traits but not chemical traits. *Oecologia* **2014**, *175*, 445–455. [CrossRef] [PubMed]
70. Guo, D.; Xia, M.; Wei, X.; Chang, W.; Liu, Y.; Wang, Z. Anatomical traits associated with absorption and mycorrhizal colonization are linked to root branch order in twenty-three Chinese temperate tree species. *New Phytol.* **2008**, *180*, 673–683. [CrossRef]
71. Li, W.; Wang, L.; Qian, S.; He, M.; Cai, X.; Ding, J. Root characteristics explain greater water use efficiency and drought tolerance in invasive Compositae plants. *Plant Soil* **2022**, *483*, 209–223. [CrossRef]
72. Simberloff, D.; Vitule, J.R.S. A call for an end to calls for the end of invasion biology. *Oikos* **2014**, *123*, 408–413. [CrossRef]
73. Cavender-Bares, J.; Keen, A.; Miles, B. Phylogenetic Structure of Floridian Plant Communities Depends on Taxonomic and Spatial Scale. *Ecology* **2006**, *87*, S109–S122. [CrossRef] [PubMed]
74. Loiola, P.P.; de Bello, F.; Chytrý, M.; Götzenberger, L.; Carmona, C.P.; Pyšek, P.; Lososová, Z.; Swenson, N. Invaders among locals: Alien species decrease phylogenetic and functional diversity while increasing dissimilarity among native community members. *J. Ecol.* **2018**, *106*, 2230–2241. [CrossRef]
75. Zhang, B.; Hastings, A.; Grosholz, E.D.; Zhai, L. The comparison of dispersal rate between invasive and native species varied by plant life form and functional traits. *Mov. Ecol.* **2023**, *11*, 73. [CrossRef] [PubMed]
76. Kovács-Hostyánszki, A.; Szigeti, V.; Miholcsa, Z.; Sándor, D.; Soltész, Z.; Török, E.; Fenesi, A. Threats and benefits of invasive alien plant species on pollinators. *Basic Appl. Ecol.* **2022**, *64*, 89–102. [CrossRef]
77. Wang, X.-P.; Fu, X.; Shi, M.-M.; Zhao, Z.-T.; Li, S.-J.; Tu, T.-Y. Invasive Asteraceae plants can enhance community stability by changing pollination network structure, yet cause intense pollen disturbance to native plants in an oceanic island community. *Biol. Invasions* **2023**, *25*, 3603–3618. [CrossRef]
78. Sun, J.; Yuan, X.; Liu, H.; Liu, G. Emergy and eco-exergy evaluation of wetland reconstruction based on ecological engineering approaches in the three Gorges Reservoir, China. *Ecol. Indic.* **2020**, *122*, 107278. [CrossRef]
79. Huang, Y.; Ge, Y.; Wang, Q.; Zhou, H.; Liu, W.; Christie, P. Allelopathic Effects of Aqueous Extracts of *Alternanthera philoxeroides* on the Growth of *Zoysia matrella*. *Pol. J. Environ. Stud.* **2017**, *26*, 97–105. [CrossRef]
80. Brock, M.T.; Weinig, C.; Galen, C. A comparison of phenotypic plasticity in the native dandelion Taraxacum ceratophorum and its invasive congener T. officinale. *New Phytol.* **2005**, *166*, 173–183. [CrossRef]
81. Richards, C.L.; Pennings, S.C.; Donovan, L.A. Habitat range and phenotypic variation in salt marsh plants. *Plant Ecol.* **2005**, *176*, 263–273. [CrossRef]
82. Masoodi, A.; Khan, F. Invasion of alligator weed (*Alternanthera philoxeroides*) in Wular Lake, Kashmir, India. *Aquat. Invasions* **2012**, *7*, 143–146. [CrossRef]
83. Keddy, P.A. Assembly and response rules: Two goals for predictive community ecology. *J. Veg. Sci.* **1992**, *3*, 157–164. [CrossRef]

84. Kraft, N.J.B.; Adler, P.B.; Godoy, O.; James, E.C.; Fuller, S.; Levine, J.M.; Fox, J. Community assembly, coexistence and the environmental filtering metaphor. *Funct. Ecol.* **2014**, *29*, 592–599. [CrossRef]
85. Callaway, R.M.; Brooker, R.W.; Choler, P.; Kikvidze, Z.; Lortie, C.J.; Michalet, R.; Paolini, L.; Pugnaire, F.I.; Newingham, B.; Aschehoug, E.T.; et al. Positive interactions among alpine plants increase with stress. *Nature* **2002**, *417*, 844–848. [CrossRef] [PubMed]
86. Li, S.P.; Cadotte, M.W.; Meiners, S.J.; Hua, Z.S.; Shu, H.Y.; Li, J.T.; Shu, W.S. The effects of phylogenetic relatedness on invasion success and impact: Deconstructing Darwin's naturalisation conundrum. *Ecol. Lett.* **2015**, *18*, 1285–1292. [CrossRef]
87. de Bello, F.; Price, J.N.; Munkemuller, T.; Liira, J.; Zobel, M.; Thuiller, W.; Gerhold, P.; Gotzenberger, L.; Lavergne, S.; Leps, J.; et al. Functional species pool framework to test for biotic effects on community assembly. *Ecology* **2012**, *93*, 2263–2273. [CrossRef]
88. Mayfield, M.M.; Levine, J.M. Opposing effects of competitive exclusion on the phylogenetic structure of communities. *Ecol. Lett.* **2010**, *13*, 1085–1093. [CrossRef] [PubMed]
89. Leishman, M.R.; Thomson, V.P.; Cooke, J. Native and exotic invasive plants have fundamentally similar carbon capture strategies. *J. Ecol.* **2009**, *98*, 28–42. [CrossRef]
90. Park, D.S.; Potter, D. A test of Darwin's naturalization hypothesis in the thistle tribe shows that close relatives make bad neighbors. *Proc. Natl. Acad. Sci. USA* **2013**, *110*, 17915–17920. [CrossRef]

International Journal of *Plant Biology*

MDPI

Article

The Influence of Nitrogen and Phosphorus Addition on Growth of the Invasive C_4 Grass *Saccharum spontaneum*

Justin A. Cummings [1], Ingrid M. Parker [1] and Gregory S. Gilbert [2,*]

[1] Department of Ecology and Evolutionary Biology, University of California, Santa Cruz, CA 95060, USA; jacummin@ucsc.edu (J.A.C.); imparker@ucsc.edu (I.M.P.)
[2] Department Environmental Studies, University of California, Santa Cruz, CA 95064, USA
* Correspondence: ggilbert@ucsc.edu; Tel.: +1-831-459-5002

Abstract: Invasive C_4 grasses can inhibit the natural regeneration of secondary forest in tropical landscapes after the cessation of intensive use for grazing and agriculture. In Panama, invasive *Saccharum spontaneum* forms dense stands that require active management to re-establish forest successional processes. In this region, restoration strategies typically involve clearing grass cover manually and applying fertilizer prior to planting tree seedlings. However, if fertilizers alleviate nutrient limitation and enhance grass competition with tree seedlings, these practices may exacerbate the costs of *Saccharum* control and hamper restoration goals. Here, we evaluated how *S. spontaneum* responds to nitrogen and phosphorus addition in the field to determine whether *S. spontaneum* is nutrient limited in this system. *S. spontaneum* was limited by both nitrogen and phosphorus, as revealed through increased foliar nutrient concentrations. *S. spontaneum* biomass was significantly greater in both nitrogen and phosphorus addition plots after both the first growth period (early rainy season) and second growth period (late rainy season), with stronger effects of nutrient limitation during the second growth period for both N limitation and N and P co-limitation. Nutrient limitation in *S. spontaneum* highlights a potential risk of fertilizer applications during restoration, agriculture, and agroforestry activities in which invasion of this aggressive weed is a challenge to land management.

Keywords: biological invasions; restoration ecology; nutrient limitation; *Saccharum spontaneum*; Panama Canal; invasive grasses; tropical ecology; nitrogen limitation; phosphorus limitation; nutrient co-limitation

1. Introduction

Intensive agriculture and grazing in tropical systems can lead to soil degradation and reduced productivity. When such lands are abandoned, succession is expected to lead to habitat recovery. However, low propagule availability, seed and seedling predation, seasonal drought, fire, competition with non-native weeds, and reduced nutrient availability are barriers that can slow or deter the rate of succession [1–5]. Although tropical soils are diverse, they are generally less fertile than temperate soils and more likely to be limited by phosphorus than nitrogen [6–8]. Tropical nutrient availability is strongly regulated by nutrient cycling through decomposition [9]. Forest clearing and removal of above-ground biomass in agriculture reduce the amount of nutrients that cycle back into the system through decomposition [10], leading to a rapid loss of nutrients after as few as three cycles of shifting agriculture [11]. As productivity decreases, increasing amounts of exposed soils can cause further losses of nitrogen (N), phosphorus (P), potassium (K), and other nutrients through erosion and/or run-off, leading to nutrient limitation and delayed succession [6,9,12,13]. Reduced nutrient availability can impede the establishment of native species [9,14].

Numerous species of non-native C_4 grasses were introduced to the American tropics as pasture grasses and many became weeds [1,3,5,15,16]. Invasive C_4 grasses can inhibit forest succession through accelerated fire regimes [3–5,17–19]. Additionally, in comparison to most native C_3 woody plants, C_4 grasses have high photosynthetic nitrogen use efficiency, allowing them to maintain high photosynthetic rates under the hot, sunny, low-fertility conditions of abandoned agricultural lands. High nitrogen use efficiency gives these grasses an advantage in competition for nutrients and other resources, and it complicates the maintenance or establishment of preferred vegetation [20,21]. Soil degradation via intensive agriculture and grazing has the potential to favor colonization by weedy non-native C_4 grasses [2,3,15,22].

Understanding how soil fertility influences the growth of invasive plants is critical to develop management strategies to restore invaded areas to agricultural use or forest cover. Because fertilizers that are commonly applied as part of reforestation activities to help reestablish trees could preferentially benefit undesired plants, it is important to understand how invasive species will respond to nutrient addition. The same is true for other tropical land use practices that involve grass control, including agriculture.

Saccharum spontaneum L. subsp. *spontaneum* is a large, exotic C_4 grass species, native to Asia and Northern Africa, that invades abandoned agricultural fields in many tropical countries. It is adapted to drought, burns frequently, and is difficult to eradicate, which is why it is considered one of the most serious weeds where it occurs [18,23]. *S. spontaneum* was most likely introduced to the Republic of Panamá in 1939 with a United States Department of Agriculture (USDA) germplasm collection for sugarcane [24] and by the 1960s, had spread widely through disturbed areas [25]. Control of *S. spontaneum* is actively pursued throughout the country and is especially important in the watershed of the Panama Canal, where *S. spontaneum* is widespread and persistent [24,26]. Fires promoted by abundant *S. spontaneum* reduce the germination and species richness of native trees and encourage additional growth and reproduction of *S. spontaneum* [17,18,27]. In Panama, *S. spontaneum* inhibits agriculture, forest restoration, and succession, generating huge costs associated with weed management [23]. The use of fertilizers to support agricultural production or forest restoration may facilitate *S. spontaneum* growth and persistence, thereby reducing seedling establishment and increasing management costs.

The objective of this experiment was to test whether *S. spontaneum* in this system is nutrient-limited, and specifically evaluate the relative importance of nitrogen and phosphorus limitation. We expected that, as for many other tropical systems, *S. spontaneum* growth would be more P-limited than N-limited.

2. Materials and Methods

2.1. Site Description

This project was established in Parque Nacional Soberanía in the Panama Canal Watershed, Republic of Panamá (N 9°6′5.5434″; W 79°36′42.8934″), in abandoned agricultural lands now dominated by *S. spontaneum*. This site experiences a mean rainfall of 2226 mm and about 4 dry months annually from December to early April (defined as months with <100 mm rainfall), with the wettest part of the year occurring in October and November [28]. Soils at this site are characterized as ultisols, predominantly clay and silty clay loams [29]. Moist tropical forest was cleared from this site during the 1960s, and the area was then used for grazing and small-scale agriculture, until it was abandoned around 1993.

2.2. Experimental Design

A full factorial block design was used to test nutrient limitation of growth of *S. spontaneum*. In June of 2011, a homogeneous area of *S. spontaneum* was cleared of aboveground biomass by machete and gas-powered brush cutter from all blocks before applying fertilizers. This experiment consisted of twelve blocks, each with four nutrient treatments: no nutrient addition (control; C), nitrogen addition (N), phosphorus addition (P), and nitrogen + phosphorus combination (N + P). Plots were 5×5 m^2, with a 2 m buffer between plots within a block and a 3 m buffer between blocks. Buffer zones were maintained free of

S. *spontaneum* to inhibit rhizomatous spread between the plots by *S. spontaneum* and reduce the probability of horizontal movement of nutrients between plots.

Dry fertilizer was added by hand. Nitrogen was added as urea ($(NH_2)_2CO_2$), and phosphorus was added as triple super phosphate ($Ca(H_2PO_4)_2 \bullet H_2O$). Nutrient application corresponded to 125 kg ha^{-1} N and 50 kg ha^{-1} P, as recommended for this region [30]. Based on these values, we calculated a nutrient application of 135.63 g triple super phosphate and 339.06 g urea per plot for each application. Nutrients were applied in July after clearing the site and again in October after the first biomass harvest. Soil cores were taken from each plot at the end of the study (December) to compare final soil nutrient availability across treatments.

2.3. Nutrient Concentration Analysis

We compared soil and leaf nutrient concentration data across treatments. In December, we randomly collected and homogenized 10 soil cores at 10 cm depth for each plot; a 20 g subsample was used to extract soil nitrogen and another for phosphorus. We followed the potassium chloride (KCl) [31] and Mehlich [32] standardized protocols to extract available nitrate (NO_3^-), ammonium (NH_4^+), and phosphate (PO_4^{-2}) [33]. Soil samples were placed directly into solution in the field and processed in the lab within 24 h of being collected. For plant tissue nutrient analysis, we collected the third mature leaf from the base of 15 randomly selected individuals in each plot. Leaves were dried for 3 days at 60 °C, and samples were processed at the University of California Santa Cruz. Five leaves were selected from each plot, and leaf N and P were extracted following the Kjeldahl acid digestion protocol [34] using a Lachat BD 46 block digester (Lachat Instruments, Milwaukee, WI, USA).

We assessed *S. spontaneum* performance, in terms of density and above-ground biomass, in September and December 2011. Measurements of both biomass and density are useful because they each affect the establishment and persistence of native species in this system; biomass is an indicator of the overall dominance of *S. spontaneum* in the field; beyond biomass, the density of individual stems influences physical availability of space for other plants to grow. The ability of *S. spontaneum* to spread through underground rhizomes makes it very difficult to differentiate between genetic individuals of *S. spontaneum* in the field. For this reason, we estimated density as the number of stalks; a tiller visibly emerging from a stalk was considered a part of that stalk. In September, we measured density by counting the number of stalks of *S. spontaneum* in 3 randomly placed 1 m^2 quadrats in each plot. *S. spontaneum* height varied across treatments but reached as high as 2 m (personal observation). We randomly collected 10 stalks of *S. spontaneum* from half of the 1 m^2 area used for quantifying *S. spontaneum* density to calculate above-ground biomass. *S. spontaneum* biomass was dried at 60 °C for 3 days and then weighed. We estimated *S. spontaneum* biomass as the product of the number of stalks and mean biomass per stalk. At the end of September, we cleared the plots of *S. spontaneum* to simulate common management practices in areas under restoration. In October, we re-applied nutrients and allowed *S. spontaneum* to grow back until the onset of the dry season in December. Individuals of *S. spontaneum* appeared much smaller during the December harvest, with few individuals reaching heights greater than 2 m and many individuals less than 1 m, suggesting a reduction in above-ground growth after clearing. Data collection in December followed the methods used in September.

2.4. Data Analysis

We used two-way factorial ANOVA to compare the response of soil nutrients, leaf nutrients, and *S. spontaneum* growth to application of nitrogen, phosphorus, or their combination. Blocks were included in the model as a random effect. Soil nutrient response variables included nitrate, ammonium, total N (nitrate + ammonium), phosphate, and N:P ratio (total inorganic nitrogen/phosphate). Plant nutrient variables were leaf N and P concentrations (mg/g), as well as leaf N:P ratios. Plant growth response variables included

S. spontaneum density and biomass measured in both September (July–September growth) and December (October–December growth).

3. Results

Soil nutrient concentrations responded significantly to nutrient additions. Although there was no difference in soil nitrate across treatments, soil ammonium and total N were significantly higher in plots where N was added (Table 1). Nutrient concentrations varied across blocks for all nitrogen measures (Table 1). Soil phosphate was greatest in plots where P was added (Table 1; Figure 1). There was no difference in soil N:P ratios across nutrient treatments (Table 1; Figure 1).

Table 1. Two-way factorial analysis of final soil nutrient availability across treatments, showing the degrees of freedom (df), F ratios (*F*), and probability (p). Nutrients were added at the beginning of the study in July and again after the first harvest in October, and nutrient availability was measured in December. Nutrient additions are treated as fixed effects with blocks as random effects. Bold values represent significant treatment effects.

		Ammonium (mg/kg)		Nitrate (mg/kg)		Total Inorganic N (mg/kg)		Phosphate (mg/kg)		N:P	
Treatment	**df**	*F*	*p*	*F*	*p*	*F*	*p*	*F*	*p*	*F*	*p*
N addition	1.33	9.20	**0.005**	0.10	0.10	9.73	**0.004**	0.11	0.75	0.001	0.98
P addition	1.33	1.77	0.19	0.59	0.59	1.78	0.19	4.14	**0.05**	0.08	0.78
N × P interaction	1.33	0.16	0.70	1.49	0.23	0.03	0.87	0.32	0.35	0.57	0.46
Block	11.33	5.91	**<0.001**	2.51	**0.02**	5.95	**<0.001**	1.17	0.58	1.12	0.38

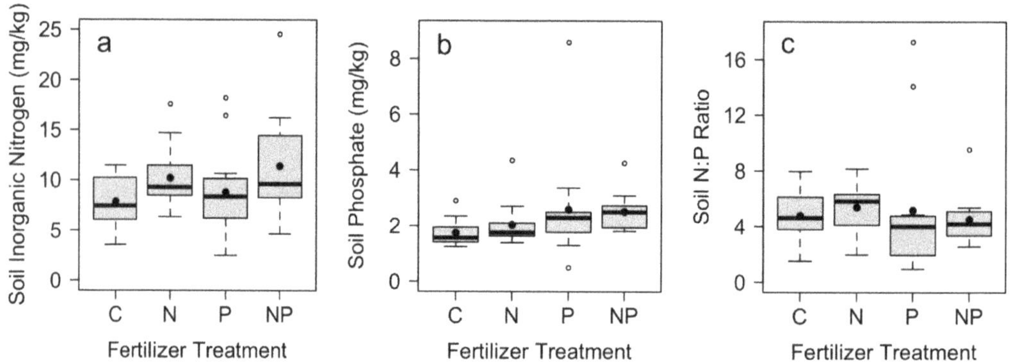

Figure 1. Mean total soil (**a**) inorganic nitrogen and (**b**) phosphate nutrient concentrations, and (**c**) soil N:P across control (C), nitrogen (N), phosphorus (P), and nitrogen and phosphorus (NP) fertilizer treatments (n = 12). Mean soil nutrient concentrations were measured in December at the end of the study, with fertilizer applications in July and October. Box plots indicate median (thick line), 1st and 3rd quartiles (grey box), and range whiskers, with extreme values as open circles. Filled circles indicate the mean. See Table 1 for associated statistics.

The above-ground biomass of *Saccharum spontaneum* showed rapid growth from July to September (Figure 2), averaging approximately 2–3 kg dry biomass per m^2 across the four treatments. The growth from October through to December was nearly an order of magnitude slower than the growth from July to September. However, stem density was higher in the later period (Figure 2).

We found a significant positive effect of nitrogen and phosphorus addition on mean *S. spontaneum* biomass from July to September (Table 2; Figure 2a). From October to December, we found a significant positive effect of nitrogen addition and a significant positive interaction between nitrogen and phosphorus on *S. spontaneum* above-ground biomass (Table 2; Figure 2b). Stem density in September increased by an average of 14% with the addition of N ($p = 0.10$), whereas in December, it increased by 25% ($p = 0.02$) (Table 2; Figure 2c,d). We found no significant effect of P on density in September or December (Table 2; Figure 2c,d).

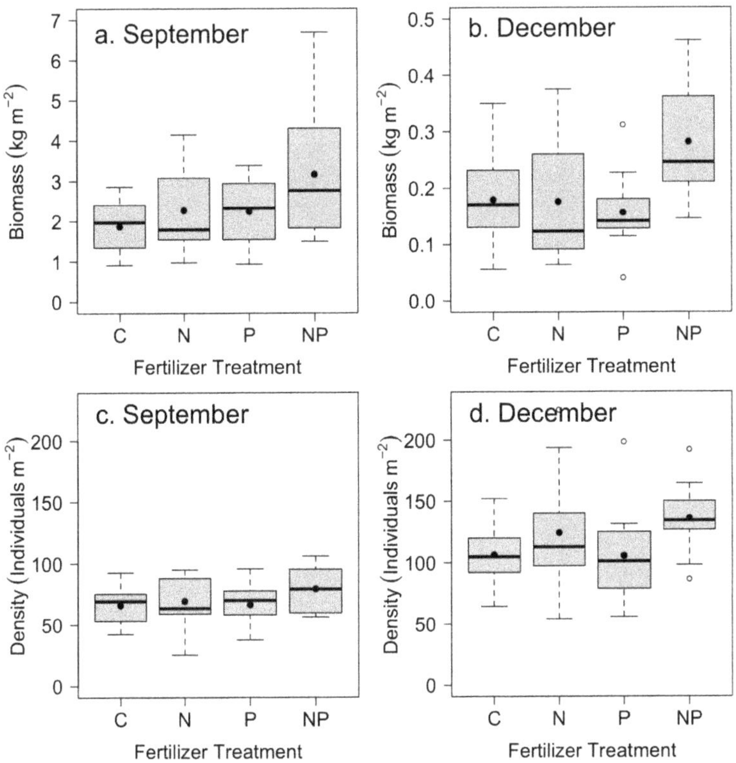

Figure 2. Mean *Saccharum spontaneum* above-ground biomass in (**a**) September and (**b**) December (note different scales) across control (C), nitrogen (N), phosphorus (P), and nitrogen and phosphorus (NP) nutrient treatments ($n = 12$). Mean stem density in (**c**) September and (**d**) December. Box plots indicate median (thick line), 1st and 3rd quartiles (grey box), and range whiskers, with extreme values as open circles. Filled circles indicate the mean. See Table 2 for associated statistics.

Table 2. Two-way factorial analysis of *S. spontaneum* density and biomass in response to nutrient additions, after growth from July to September ("September") and from October to December ("December"). Nutrient additions are treated as fixed effects with blocks as random effects. Columns show the degrees of freedom (df), F ratios (*F*), and probability (*p*) for each treatment. Bold values represent significant treatment effects.

		Density				**Biomass**			
		September		**December**		**September**		**December**	
Treatment	**df**	*F*	*p*	*F*	*p*	*F*	*p*	*F*	*p*
N addition	1.33	2.87	0.10	5.84	**0.02**	6.78	**0.01**	6.88	**0.01**
P addition	1.33	1.17	0.29	0.29	0.59	6.23	**0.02**	3.20	0.08
N × P interaction	1.33	1.02	0.17	0.39	0.31	0.81	0.31	7.74	**0.008**
Block	11.33	1.53	0.32	1.22	0.53	3.36	**0.003**	2.20	**0.04**

S. spontaneum responded to nutrient additions by increasing internal nutrient concentrations. In plots where N was added, we found significantly higher leaf N and no significant effect on leaf P (Table 3; Figure 3). In plots where P was added, we found significantly higher leaf P and no significant increase in leaf N (Table 3; Figure 3). Both leaf N and leaf P also varied by block (Table 3). In addition, there was a significant effect of both N and P fertilization on the leaf N:P ratio, where the leaf N:P was significantly lower in plots where P was added (Table 3; Figure 3). There was no significant interaction between N and P treatments for leaf nutrient concentrations ($p > 0.15$, Table 3).

Table 3. Two-way factorial analysis of *S. spontaneum* leaf nutrient concentrations across treatments, showing the degrees of freedom (df), F ratios (*F*), and probability (*p*). Leaves were collected during the December *S. spontaneum* harvest, following July and October fertilization treatments. Nutrient additions are treated as fixed effects with blocks as random effects. Bold values represent significant treatment effects.

		Leaf N (mg/g)		Leaf P (mg/g)		Leaf N:P	
Treatment	df	*F*	*p*	*F*	*p*	*F*	*p*
N addition	1.33	10.82	**0.002**	0.22	0.64	6.71	**0.01**
P addition	1.33	0.40	0.53	65.20	**<0.001**	126.63	**<0.001**
N × P interaction	1.33	2.10	0.16	0.08	0.77	0.51	0.48
Block	11.33	2.19	**0.04**	3.94	**0.001**	5.14	**<0.001**

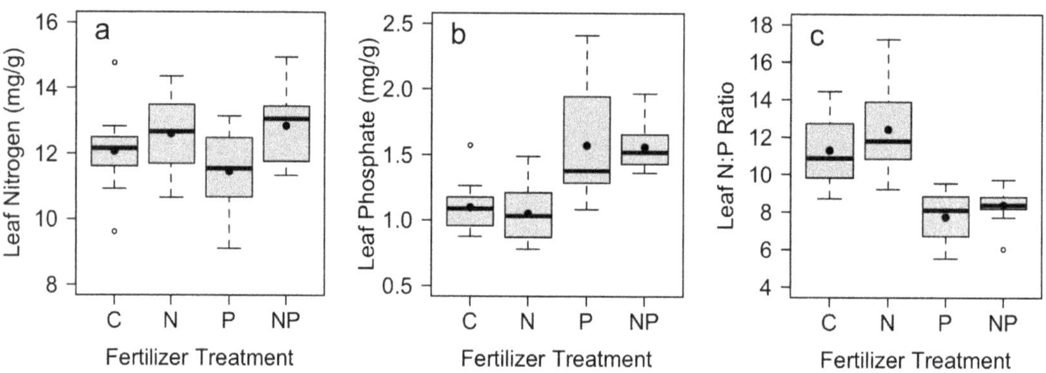

Figure 3. Mean leaf concentrations of (**a**) nitrogen and (**b**) phosphorus, together with (**c**) leaf N:P ratio across control (C), nitrogen (N), phosphorus (P), and nitrogen and phosphorus (NP) fertilizer treatments ($n = 12$). Mean leaf nutrient concentrations were measured in December at the end of the study, with fertilizer applications in July and October. Box plots indicate median (thick line), 1st and 3rd quartiles (gray box) and range whiskers, with extreme values as open circles. Filled circles indicate the mean. See Table 2 for associated statistics.

4. Discussion

Biomass growth of *S. spontaneum* showed a positive response to both N and P addition, suggesting that *S. spontaneum* is both nitrogen and phosphorus was limited in this system. This positive effect is likely linked to increased photosynthetic capacity under increased nutrient availability. The net CO_2 assimilation rate increases nonlinearly with increasing leaf N for eight species of *Saccharum* [35], likely reflecting an increase in the concentration of Rubisco. Similarly, leaf P and growth in C_4 grasses increased with increasing soil P supply in other studies [36,37]. We also found evidence for N and P co-limitation of biomass in *S. spontaneum*, in particular in the later growing period.

Final soil N and P concentrations were significantly higher in plots where N and P were added. This indicates that treatments were effective in increasing nutrient availability throughout the study. Leaf N and P were greater in plots where we added N and P, respectively, indicating that *S. spontaneum* had increased nutrient uptake in fertilized plots.

In our system, nitrogen addition produced a slight increase in stem density. Tillering of both C_3 and C_4 grasses in temperate systems increased in response to increased nitrogen [38–40]. In addition, frequent cutting can increase tillering in *Hyparrhenia rufa*, another tropical invasive C_4 grass [41]. Our results suggest that *S. spontaneum* responds similarly to other grasses by increasing tillering in response to both nitrogen addition and cutting.

Growth in above-ground biomass was an order of magnitude lower in December than September. After the first growing period, the energy reserves in the rhizomes may have been severely depleted, meaning there was little stored carbohydrate left to sustain growth during the second growth period. Alternatively, this growth pattern may suggest that resource allocation varies temporally in *S. spontaneum*. In our study, the period from July to September represents the middle of the wet season, with peak flowering in August–September. The October–December periods include the wettest and cloudiest parts of the rainy season, along with the transition into the dry season. Investment in above-ground growth in the early rainy season may allow the plant to shade competitors and attain heights that optimize pollination and seed dispersal, with a shift in allocation to below-ground growth before the onset of the dry season. Other studies of tropical C_4 grasses have also found a high growth rate at the beginning of the rainy season (e.g., [42]).

In conjunction with resource reallocation, multiple rounds of removal of above-ground biomass may have reduced the growth ability of *S. spontaneum*. Repeated clearing can reduce below-ground carbon stores [43]. For a number of temperate grass species, increased annual above-ground biomass clearing reduced carbohydrate reserves [43]. Our results are consistent with studies that show that repeated clearing of *S. spontaneum* can reduce *S. spontaneum* growth [44]. Future studies should evaluate whether growth after clearing varies seasonally.

5. Conclusions

In summary, we found that nutrient addition increased the growth of *Saccharum spontaneum*, but these effects varied seasonally and differed between biomass and density responses. Given the substantial impacts of *S. spontaneum* within the Panama Canal Watershed [24] and in many other parts of the world where it is a weedy invader [45], there is a great need to understand the management activities that help suppress, or unintentionally promote, its growth. This study highlights a potential risk of fertilizer applications during restoration, agriculture, agroforestry, or other activities in which *S. spontaneum* invasion is a challenge to land management.

Author Contributions: Conceptualization, J.A.C., I.M.P. and G.S.G.; methodology, J.A.C.; formal analysis, J.A.C.; investigation, J.A.C.; data curation, J.A.C.; writing—original draft preparation, J.A.C.; writing—review and editing, J.A.C., I.M.P. and G.S.G.; visualization, J.A.C. All authors have read and agreed to the published version of the manuscript.

Funding: This research was funded by fellowships to J.C. from the NSF Pre-Doctoral Fellowship program, the STRI Short Term Fellowship program, the UCSC Center for Tropical Ecology, Agriculture, and Development, the Graduate Assistance in Areas of National Need (GAANN) program, and the Jean Langenheim Fellowship.

Institutional Review Board Statement: Not applicable.

Informed Consent Statement: Not applicable.

Data Availability Statement: Data are available in a publicly accessible repository at Gilbert, Gregory; Cummings, Justin; Parker, Ingrid (2023), *Saccharum spontaneum* biomass and associated soil and foliar nutrient data, Dryad, Dataset, https://doi.org/10.7291/D1KD69.

Acknowledgments: The authors thank the Smithsonian Tropical Research Institute (STRI) for logistical support and ANAM/MiAmbiente for granting permission to conduct research in the Republic of Panamá. We also thank J.S. Hall (PRORENA) and S.J. Wright (STRI) for serving as mentors to J.C. while conducting work in Panamá, and to Haley Stott and Emelie L. M. McKain for their field assistance.

Conflicts of Interest: The authors declare no conflict of interest.

References

1. Moran, E.F.; Brondizio, E.S.; Tucker, J.M.; da Silva-Forsberg, M.C.; McCracken, S.; Falesi, I. Effects of soil fertility and land-use on forest succession in Amazonia. *For. Ecol. Manag.* **2000**, *139*, 93–108. [CrossRef]
2. Nepstad, D.C.; Uhl, C.; Serrao, E.A. Recuperation of a degraded Amazonian landscape: Forest recovery and agricultural restoration. *Ambio* **1991**, *20*, 248–255.
3. D'Antonio, C.M.; Vitousek, P.M. Biological invasions by exotic grasses, the grass/fire cycle, and global change. *Annu. Rev. Ecol. Syst.* **1992**, *23*, 63–87. [CrossRef]
4. Holl, K.D. Factors limiting tropical rain forest regeneration in abandoned pasture: Seed rain, seed germination, microclimate, and soil. *Biotropica* **1999**, *31*, 229–242. [CrossRef]
5. Aide, T.M.; Cavelier, J. Barriers to lowland tropical forest restoration in the Sierra Nevada de Santa Marta, Colombia. *Restor. Ecol.* **1994**, *2*, 219–229. [CrossRef]
6. Vitousek, P.M. Litterfall, nutrient cycling, and nutrient limitation in tropical forests. *Ecology* **1984**, *65*, 285–298. [CrossRef]
7. Sollins, P. Factors influencing species composition in tropical lowland rain forest: Does soil matter? *Ecology* **1998**, *79*, 23–30. [CrossRef]
8. Townsend, A.R.; Cleveland, C.C.; Asner, G.P.; Bustamante, M.M. Controls over foliar N: P ratios in tropical rain forests. *Ecology* **2007**, *88*, 107–118. [CrossRef] [PubMed]
9. Lu, D.; Moran, E.; Mausel, P. Linking Amazonian secondary succession forest growth to soil properties. *Land Degrad. Devel.* **2002**, *13*, 331–343. [CrossRef]
10. Newman, E. Phosphorus balance of contrasting farming systems, past and present. Can food production be sustainable? *J. Appl. Ecol.* **1997**, *34*, 1334–1347. [CrossRef]
11. Lawrence, D.; D'Odorico, P.; Diekmann, L.; DeLonge, M.; Das, R.; Eaton, J. Ecological feedbacks following deforestation create the potential for a catastrophic ecosystem shift in tropical dry forest. *Proc. Natl. Acad. Sci. USA* **2007**, *104*, 20696–20701. [CrossRef] [PubMed]
12. Baribault, T.W.; Kobe, R.K.; Finley, A.O. Tropical tree growth is correlated with soil phosphorus, potassium, and calcium, though not for legumes. *Ecol. Monogr.* **2012**, *82*, 189–203. [CrossRef]
13. Mercado, L.M.; Patino, S.; Domingues, T.F.; Fyllas, N.M.; Weedon, G.P.; Sitch, S.; Quesada, C.A.; Phillips, O.L.; Aragão, L.E.; Malhi, Y. Variations in Amazon forest productivity correlated with foliar nutrients and modelled rates of photosynthetic carbon supply. *Philos. Trans. R. Soc. Lond. Ser. B* **2011**, *366*, 3316–3329. [CrossRef] [PubMed]
14. Chaer, G.M.; Resende, A.S.; Campello, E.F.C.; de Faria, S.M.; Boddey, R.M. Nitrogen-fixing legume tree species for the reclamation of severely degraded lands in Brazil. *Tree Physiol.* **2011**, *31*, 139–149. [CrossRef] [PubMed]
15. Williams, D.G.; Baruch, Z. African grass invasion in the Americas: Ecosystem consequences and the role of ecophysiology. *Biol. Invasions* **2000**, *2*, 123–140. [CrossRef]
16. Parsons, J.J. Spread of African pasture grasses to the American tropics. *Rangel. Ecol. Manag.* **1972**, *25*, 12–17. [CrossRef]
17. Hooper, E.; Legendre, P.; Condit, R. Barriers to forest regeneration of deforested and abandoned land in Panama. *J. Appl. Ecol.* **2005**, *42*, 1165–1174. [CrossRef]
18. Hooper, E.R.; Legendre, P.; Condit, R. Factors affecting community composition of forest regeneration in deforested, abandoned land in Panama. *Ecology* **2004**, *85*, 3313–3326. [CrossRef]
19. D'Antonio, C.M.; Hughes, R.; Tunison, J. Long-term impacts of invasive grasses and subsequent fire in seasonally dry Hawaiian woodlands. *Ecol. Appl.* **2011**, *21*, 1617–1628. [CrossRef]
20. Edwards, E.J.; Smith, S.A. Phylogenetic analyses reveal the shady history of C_4 grasses. *Proc. Natl. Acad. Sci. USA* **2010**, *107*, 2532–2537. [CrossRef]
21. Lambers, H.; Chapin, F.S.; Pons, T.L. *Plant Physiological Ecology*; Springer: Berlin, Germany, 2008; Volume 2.
22. D'Antonio, C.M.; Hughes, R.F.; Vitousek, P.M. Factors influencing dynamics of two invasive C_4 grasses in seasonally dry Hawaiian woodlands. *Ecology* **2001**, *82*, 89–104.
23. Hammond, B.W. *Saccharum spontaneum* (Gramineae) in Panama: The physiology and ecology of invasion. *J. Sustain. Forest.* **1998**, *8*, 23–38. [CrossRef]
24. Saltonstall, K.; Bonnett, G.D.; Aitken, K.S. A perfect storm: Ploidy and preadaptation facilitate *Saccharum spontaneum* escape and invasion in the Republic of Panama. *Biol. Invasions* **2021**, *23*, 1101–1115. [CrossRef]
25. Craven, D.; Hall, J.; Verjans, J.M. Impacts of herbicide application and mechanical cleanings on growth and mortality of two timber species in *Saccharum spontaneum* grasslands of the Panama Canal Watershed. *Restor. Ecol.* **2009**, *17*, 751–761. [CrossRef]

26. Boeschoten, L.E.; van Breugel, M.; Bailon, M.; Balbuena, J.; Nuñez, M.; Cerezo, A.; Hall, J.S. Framework species approach proves robust in restoring forest on fire prone invasive grass: A case study from Panama. *J. Sustain. For.* **2021**, *40*, 197–215. [CrossRef]
27. Saltonstall, K.; Bonnett, G.D. Fire promotes growth and reproduction of *Saccharum spontaneum* (L.) in Panama. *Biol. Invasions* **2012**, *14*, 2479–2488. [CrossRef]
28. Wishnie, M.; Dent, D.; Mariscal, E.; Deago, J.; Cedeno, N.; Ibarra, D.; Condit, R.; Ashton, P. Initial performance and reforestation potential of 24 tropical tree species planted across a precipitation gradient in the Republic of Panama. *For. Ecol. Manag.* **2007**, *243*, 39–49. [CrossRef]
29. Park, A.; van Breugel, M.; Ashton, M.S.; Wishnie, M.; Mariscal, E.; Deago, J.; Ibarra, D.; Cedeño, N.; Hall, J.S. Local and regional environmental variation influences the growth of tropical trees in selection trials in the Republic of Panama. *For. Ecol. Manag.* **2010**, *260*, 12–21. [CrossRef]
30. Wright, S.J.; Yavitt, J.B.; Wurzburger, N.; Turner, B.L.; Tanner, E.V.; Sayer, E.J.; Santiago, L.S.; Kaspari, M.; Hedin, L.O.; Harms, K.E. Potassium, phosphorus, or nitrogen limit root allocation, tree growth, or litter production in a lowland tropical forest. *Ecology* **2011**, *92*, 1616–1625. [CrossRef] [PubMed]
31. Maynard, D.; Kalra, Y.; Crumbaugh, J. Nitrate and exchangeable ammonium nitrogen. *Soil Sampl. Methods Anal.* **1993**, *1*, 25–38.
32. Mehlich, A. Mehlich 3 soil test extractant: A modification of Mehlich 2 extractant. *Commun. Soil Sci. Plant Anal.* **1984**, *15*, 1409–1416. [CrossRef]
33. Baldeck, C.A.; Harms, K.E.; Yavitt, J.B.; John, R.; Turner, B.L.; Valencia, R.; Navarrete, H.; Davies, S.J.; Chuyong, G.B.; Kenfack, D. Soil resources and topography shape local tree community structure in tropical forests. *Proc. R. Soc. B* **2013**, *280*, 20122532. [CrossRef] [PubMed]
34. Novozamsky, I.; Houba, V.; Van Eck, R.; Van Vark, W. A novel digestion technique for multi-element plant analysis. *Commun. Soil Sci. Plant Anal.* **1983**, *14*, 239–248. [CrossRef]
35. Meinzer, F.C.; Zhu, J. Nitrogen stress reduces the efficiency of the C_4CO_2 concentrating system, and therefore quantum yield, in *Saccharum* (sugarcane) species. *J. Exp. Bot.* **1998**, *49*, 1227–1234. [CrossRef]
36. Rubio, G.; Gutiérrez Boem, F.H.; Lavado, R.S. Responses of C_3 and C_4 grasses to application of nitrogen and phosphorus fertilizer at two dates in the spring. *Grass Forage Sci.* **2010**, *65*, 102–109. [CrossRef]
37. Barger, N.N.; D'Antonio, C.M.; Ghneim, T.; Brink, K.; Cuevas, E. Nutrient limitation to productivity in a secondary savanna in Venezuela. *Biotropica* **2002**, *34*, 493–501. [CrossRef]
38. Lafarge, M. Reproductive tillers in cut tall fescue swards: Differences according to sward age and fertilizer nitrogen application, and relationships with the local dynamics of the sward. *Grass Forage Sci.* **2006**, *61*, 182–191. [CrossRef]
39. Jaramillo, V.J.; Detling, J.K. Small-scale heterogeneity in a semi-arid North American grassland. I. Tillering, N uptake and retranslocation in simulated urine patches. *J. Appl. Ecol.* **1992**, *29*, 1–8. [CrossRef]
40. Moyer, J.; Sweeney, D. Long-term responses in the yield of Eastern Gamagrass [*Tripsacum dactyloides* (L.) L.] to nitrogen fertilizer under two harvest regimes in the United States. *Grass Forage Sci.* **2008**, *63*, 390–397. [CrossRef]
41. Simoes, M.; Baruch, Z. Responses to simulated herbivory and water stress in two tropical C_4 grasses. *Oecologia* **1991**, *88*, 173–180. [CrossRef]
42. Baruch, Z.; Gómez, J.A. Dynamics of energy and nutrient concentration and construction cost in a native and two alien C_4 grasses from two neotropical savannas. *Plant Soil* **1996**, *181*, 175–184. [CrossRef]
43. Klimeš, L.; Klimešová, J. The effects of mowing and fertilization on carbohydrate reserves and regrowth of grasses: Do they promote plant coexistence in species-rich meadows? In Proceedings of the Ecology and Evolutionary Biology of Clonal Plants: Proceedings of Clone-2000, Obergurgl, Austria, 20–25 August 2000; pp. 141–160.
44. Hooper, E.; Condit, R.; Legendre, P. Responses of 20 native tree species to reforestation strategies for abandoned farmland in Panama. *Ecol. Appl.* **2002**, *12*, 1626–1641. [CrossRef]
45. Holm, L.G.; Plucknett, D.L.; Pancho, J.V.; Herberger, J.P. *Saccharum spontaneum* L. In *World Weeds: Natural Histories and Distribution*; University of Hawaii Press: Honolulu, HI, USA, 1977; pp. 693–698.

diversity

MDPI

Article

Mechanisms of Seed-To-Seed Interactions between Dominant Species in the Yangtze River Estuary under Saline Condition

Cheng-Huan Wang [1], Zhen-Lin Yu [1], Yuerenisha Yasenjiang [1], Long Tang [2], Yang Gao [3] and Chun-Jing Zou [1,*]

[1] Shanghai Key Lab for Urban Ecological Processes and Eco-Restoration, School of Life Sciences, East China Normal University, 500 Dongchuan Road, Shanghai 200241, China
[2] School of Human Settlements and Civil Engineering, Xi'an Jiaotong University, 28 Xianning West Road, Xi'an 710049, China
[3] State Key Laboratory of Eco-Hydraulics in Northwest Arid Region of China, Institute of Water Resources and Hydro-Electric Engineering, Xi'an University of Technology, 5 South Jinhua Road, Xi'an 710048, China
* Correspondence: cjzou@bio.ecnu.edu.cn

Abstract: Plant community assembly is the central issue in community ecology. As plant traits differ in different life history stages, the form, intensity and mechanism of interspecific interactions may change with the ontogenetic process of plants. However, our understanding of interspecific interaction mechanisms during germination is still limited. Here, we conducted a laboratory germination experiment using five dominant species in Chongming Dongtan (*Spartina alterniflora*, *Scirpus mariqueter*, *Phragmites australis*, *Suaeda glauca* and *Tripolium vulgare*) to assess their germination performance in control (monoculture), allelopathy and mixture treatments. The results indicated that seeds could affect germination performance of neighbors through both allelopathy and salinity modification. Salinity of the solution in Petri dishes after seed germination decreased significantly in most species combinations in competition treatment, and was negatively correlated with the number of total germinated seeds. Seed leachate of invasive *Spartina alterniflora* significantly accelerated the germination of two native halophytes *Suaeda glauca* and *Tripolium vulgare*, but not *Scirpus mariqueter* and *Phragmites australis*. The salt absorption by *Spartina alterniflora* seeds had inconsistent effects compared with that of its seed leachate. On the other hand, seed leachate of native *Scirpus mariqueter* and *Phragmites australis* significantly slowed down the germination of invasive *Spartina alterniflora*. The effect of salinity modification of *Scirpus mariqueter* on *Spartina alterniflora* was positive, whereas that of other species was neutral. Considering seed-to-seed interactions is an important perspective to understand the underlying mechanisms of community dynamics, species diversity maintenance and invasion of alien species, and can improve the effectiveness in the management of invaded coastal wetlands.

Keywords: allelopathy; interspecific interactions; salinity modification; seed germination

1. Introduction

The spatial distribution pattern of different plants and the underlying mechanism of species coexistence (i.e., plant community assembly mechanisms) have been central issues in community ecology for decades [1,2]. Interspecific interactions are thought of as the driving force of plant community assembly, especially in the later successional stages [3,4]. Positive interactions often occur in stressful environment, and plants can facilitate the survival, growth and reproduction of their neighbors via habitat modification [5]. In relatively benign environment, competition is more common, which can negatively affect the performance of adjacent plant species through resource contention, allelopathy and other mechanisms [6].

Previous studies on interspecific interactions mainly focused on established seedlings and ramets [7–9]. However, as plant traits differ in different life history stages, the form, intensity and mechanism of interspecific interactions may change with the ontogenetic process of plants [10–12]. Researchers have recognized the importance of regeneration on community assembly and investigated some adult-to-seed and seed-to-seed interactions in recent years [13–15], but our understanding of interspecific interaction mechanisms during early life history stage of plants (i.e., germination) is still limited.

Seed germination is vital for plant life cycle. Difference in germination percentage and germination timing under various conditions has a significant impact on recruitment and potential competitive ability of seedlings [16,17], which then determines population distribution and community structure [18]. A great number of studies have demonstrated the effects of environmental factors (e.g., temperature, light, soil moisture and salinity) on seed germination [19–22], but few considered the biotic interactions among seeds. In field sites, seeds of different species often appear in the same place due to the dispersal by wind, tide or animals [23,24], which makes the germination process affected by neighboring conspecific or heterospecific seeds.

Resource competition and direct interference such as allelopathy are well-known underlying mechanisms of interspecific interactions [25,26]. Due to the small size and relatively low resource requirement of seeds, allelopathy has been considered the driving force mediating seed-to-seed interactions in grasslands and croplands [27–30], and the interaction intensity usually change with the concentration of allelochemicals [31]. On the other hand, many ecosystem engineers have indirect effects on adjacent species through habitat modification [32], which may also work during germination. To our knowledge, no study has directly examined the potential effects of microenvironment modification (e.g., salinity absorption) by seeds on germination process.

Chongming Dongtan wetland of international importance is located in the Yangtze River estuary, and plays a crucial part in water purification, biodiversity maintenance and flood control [33]. *Spartina alterniflora* originated in the Atlantic coast of North America, and was introduced to China in 1979 for siltation promotion, beach and levee protection. This invasive species has caused great negative ecosystem impacts through competition with native species and habitat modification these years [34]. *Spartina alterniflora* produces a large number of seeds [35], which disperse in a wide area in coastal salt marshes and often mix with other native species [36]. Although a lot of studies have investigated the competition between invasive *Spartina alterniflora* and dominant native species in Chongming Dongtan, such as *Scirpus mariqueter* and *Phragmites australis* [37–39], few took into account their interspecific interactions during germination (however, see [13]). As a result, we lack enough understanding of the mechanisms of seed-to-seed interactions.

This study tried to assess the interspecific interactions during germination and the underlying mechanisms using five dominant species in Chongming Dongtan (*Spartina alterniflora, Scirpus mariqueter, Phragmites australis, Suaeda glauca* and *Tripolium vulgare*). Specifically, the following two questions are addressed: (1) Do these plants germinate differently in monoculture, in seed leachate of competitors and in mixture with competitor seeds? (2) Does salinity modification by germinating seeds have any impact on the germination process? We predicted that *Spartina alterniflora* seeds may significantly influence the germination of native species through allelopathy and microenvironment modification (i.e., regulation of salinity) as well. The results of this study contribute to our understanding of the regeneration and community assembly mechanisms in salt marshes and provide a basis for wetland ecosystem management and restoration.

2. Materials and Methods

2.1. Study Site

Chongming Dongtan is located at the east end of Chongming Island (121°50′~122°05′ E, 31°25′~31°38′ N). It covers about 320 km^2 and has a subtropical monsoon climate. Dongtan has an annual average rainfall of 1145 mm and annual average temperature of 15.7 °C. The highest temperature is in July with a long-term average temperature of 27.3 °C, and the lowest temperature is in January with a long-term average temperature of 4.2 °C [40]. The salinity of surface soil in Chongming Dongtan usually varies from 2‰ to 15‰, and can exceed 20‰ during salt water intrusion [41].

Major native plant communities in Chongming Dongtan are sedge meadows dominated by *Scirpus mariqueter* (Sm) in middle to low marshes and grass marshes dominated by *Phragmites australis* (Pa) at higher elevations. *Scirpus mariqueter* is a perennial rhizomatous species mainly distributed in the Yangtze River estuary and the Hangzhou Bay. It can reproduce both via seeds and via corms and rhizomes. As a pioneer species, it can colonize stressful habitats where other plants cannot survive [36]. *Phragmites australis* (Pa) is a perennial species that commonly distributes in inland or coastal wetlands [42]. It mainly relies on vegetative reproduction through rhizomes in mature populations, whereas seedlings play important roles in colonization of bare patches. In high marsh, *Suaeda glauca* (Sg) and *Tripolium vulgare* (Tv) often occurs in hypersaline bare patches. *Suaeda glauca* is an annual halophyte which often forms large scale communities around saline-alkaline lakes. It has high salt tolerance and can also grow in intertidal zones frequently affected by tides [43]. *Tripolium vulgare* is also an annual forb which often grows in alkaline lake wetlands with soil pH ranges from 8.10 to 9.15. It can invade *Phragmites australis* community under saline and dry conditions [44]. After the introduction of *Spartina alterniflora* (Sa), this invasive species thrived in Chongming Dongtan from middle to high marsh and formed *Scirpus mariqueter-Spartina alterniflora* and *Spartina alterniflora-Phragmites australis* mixtures [13].

2.2. Experimental Design

The seeds of *Spartina alterniflora* (Sa), *Scirpus mariqueter* (Sm), *Phragmites australis* (Pa), *Suaeda glauca* (Sg) and *Tripolium vulgare* (Tv) were collected in the pure stands of each species in Chongming Dongtan in early October, 2017. The collected seeds were stored in a refrigerator at 4 °C. *Spartina alterniflora* and *Scirpus mariqueter* seeds were immersed in fresh water and the others were kept in dry condition to maintain seed vigor [13]. The germination experiment began on 13 March 2018. All the seeds were washed with distilled water for 3~5 min three times and then placed in a Petri dish (9 cm in diameter) with two pieces of filter paper on the bottom. The Petri dishes were placed in a light incubator (PGX-600B) for germination, with the setting of 30 °C and 100% light intensity (12,000 lux) in the daytime (from 08:00 to 20:00) and 18 °C and 0% light intensity in the night (from 20:00 to 08:00).

For preparation of seed leachate, 50 g of *Spartina alterniflora, Scirpus mariqueter, Phragmites australis, Suaeda glauca* and *Tripolium vulgare* seeds were placed in a 500 mL conical flask, respectively. A total of 400 mL of distilled water was added into each flask. The flasks were sealed and vibrated in a shaker at room temperature for 48 h to extract the maximum concentration of allelochemicals from seeds [45]. After that, NaCl was added into the flasks to reach the salinity of 1%, which is close to the condition of their natural habitats in Chongming Dongtan.

To highlight the effects of invasive plant on native species during germination, this study focused on the interactions between the seeds of invasive species and native species rather than those between different native species. Hence, there were four species combinations: Sa + Sm, Sa + Pa, Sa + Sg and Sa + Tv. We set three different treatments to all species combinations: control (target species seeds in monoculture), allelopathy (target species seeds + competitor seeds leachate) and competition (target species seeds + competitor seeds in mixture). In the control treatment, 20 seeds of a single species were evenly placed in a Petri dish and 6 mL of 1% NaCl solution was added. In the allelopathy treatment, 20 seeds of each species were evenly placed in a Petri dish with 6 mL seed leachate of its competitor seeds. *Spartina alterniflora* seeds were treated by leachate of *Scirpus mariqueter, Phragmites australis, Suaeda glauca* and *Tripolium vulgare*, respectively. Similarly, four native species seeds were treated by *Spartina alterniflora* leachate, respectively. In the competition treatment (Sa + Sm, Sa + Pa, Sa + Sg and Sa + Tv), 20 seeds of each competing species were placed in mixture in each Petri dish and 6 mL of 1% NaCl solution was added. Our experiment followed an additive design to focus on interspecific interactions rather than intraspecific interactions among these species. The two density levels (20 seeds per dish or 3144 seeds m^{-2}, 40 seeds per dish or 6288 seeds m^{-2}) were set according to the soil seed bank density in Chongming Dongtan (250~8500 seeds m^{-2} for *Scirpus mariqueter* and 500~6900 seeds m^{-2} for *Spartina alterniflora*) [36]. There were 17 treatments in all (5 control + 8 allelopathy + 4 competition) and each treatment was replicated four times, making a total of 68 dishes.

Distilled water was added into each Petri dish every day to maintain a shallow layer of water. The number of germinated seeds (where root tips had protruded from the seed coat) of each species was counted and recorded every day, and then, all the Petri dishes were randomly rotated to change its position in the incubator until no seeds germinated within a week. The germination experiment lasted for about a month (13 March to 16 April). After germination, the solution in each Petri dish was diluted to 30 mL, and the salinity was determined by a portable salinometer (SANXIN 5052).

2.3. Data Analysis

Time lag was calculated as the time between the beginning of the experiment and the first germination of each species. Germination period was the time between the first germination and the last germination of each species. Germination percentage was calculated as the proportion of total germinated seeds of each species. Germination index (GI) was calculated as follows [46]:

$$GI = \sum Gi/Di$$

Mean time to germination (MTG) was calculated as follows [47]:

$$MTG = \sum(Gi \times Di)/\sum Gi,$$

where Gi is the number of germinating seeds on the ith day and Di is the number of days after the beginning of the germination experiment.

One-way ANOVA (post hoc Tukey's HSD test) was used to examine the differences of germination traits (time lag, germination period and germination index) of each species in the control treatment. For each competing species in each species combination, one-way ANOVA (post hoc Tukey's HSD test) was used to examine the effects of different treatments on germination percentage, MTG and the final salinity of solution in Petri dishes after germination. The relationships between final salinity and the number of total germinated seeds were analyzed using linear regression for each species combination. The original germination data were log transformed if necessary to meet the assumption of homogeneity of variance. The significance level was set to 0.05 ($p < 0.05$). All the statistical analyses were carried out by STATISTICA 13.5 (TIBCO software).

3. Results

3.1. Germination Traits of Dominant Species

Tripolium vulgare had the highest time lag among the species studied, followed by *Scirpus mariqueter*. *Spartina alterniflora* and *Suaeda glauca* had a significantly lower time lag than *Tripolium vulgare* and *Scirpus mariqueter*, while *Phragmites australis* fell in between (Figure 1a). *Tripolium vulgare* had a significantly higher germination period than *Phragmites australis*, while *Spartina alterniflora*, *Suaeda glauca* and *Scirpus mariqueter* had an intermediate germination period (Figure 1b). The germination index of *Suaeda glauca* and *Spartina alterniflora* was significantly higher than that of *Phragmites australis*, *Scirpus mariqueter* and *Tripolium vulgare* (Figure 1c).

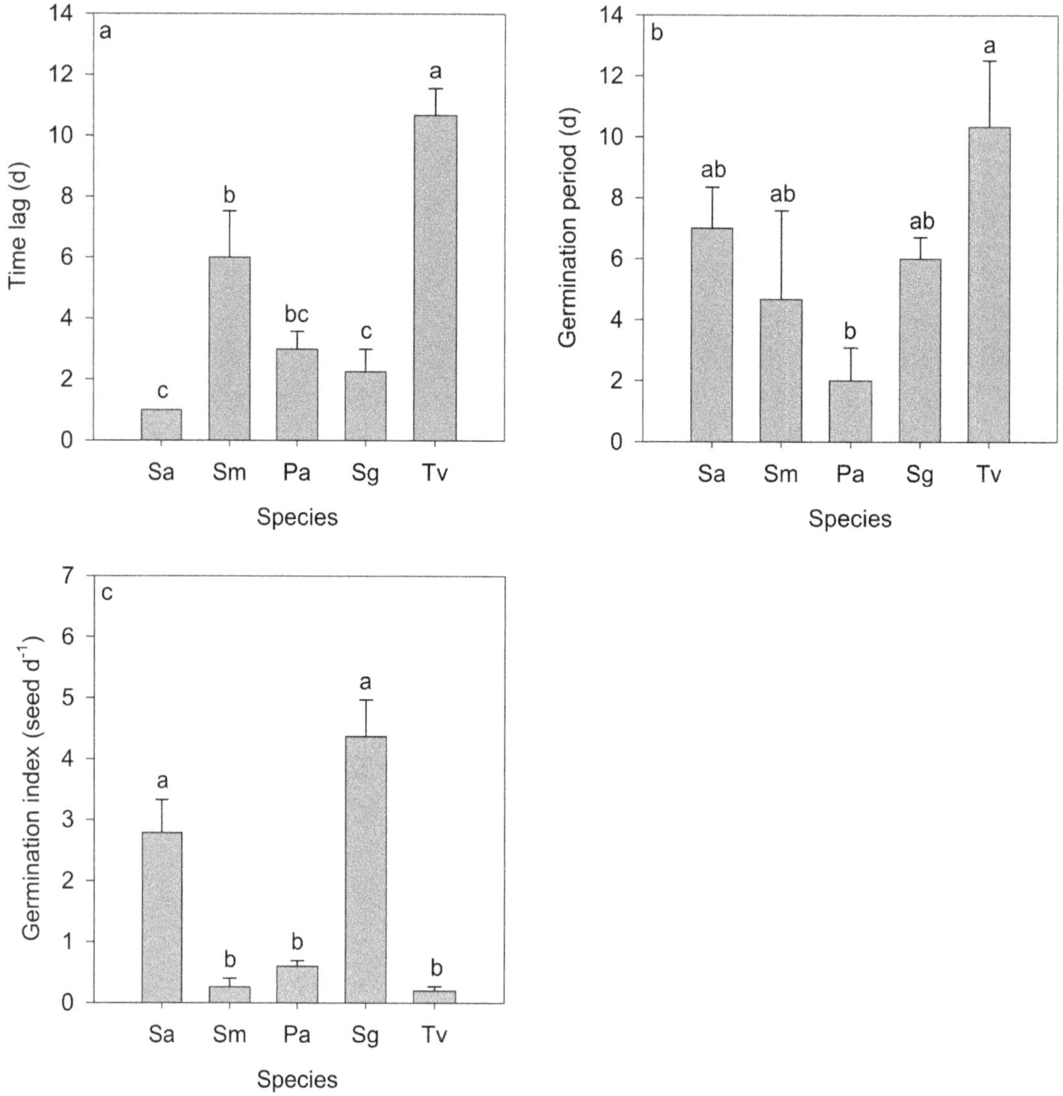

Figure 1. Germination traits of invasive *Spartina alterniflora* (Sa) and four native species (*Scirpus mariqueter*, Sm, *Phragmites australis*, Pa, *Suaeda glauca*, Sg, and *Tripolium vulgare*, Tv) in control treatment (target species in monoculture). (**a**) Time lag before first germination, (**b**) Germination period from the first germination to the last germination, (**c**) Germination index. Different lower cases denote significant differences among different species. Error bars are standard errors.

3.2. Effects of Interspecific Interaction on Germination

In species combination of Sa + Sm, the germination percentage of both species was significantly promoted in competition treatment than in control and allelopathy treatments (Table 1, Figure 2a). The germination percentage of *Phragmites australis* seeds was also significantly affected by interspecific interaction (Table 1). The allelopathy treatment inhibited the germination of *Phragmites australis* seeds, resulting in a significantly lower germination percentage than in the competition treatment, which had a promotion effect

(Figure 2b). Interspecific interactions between *Spartina alterniflora* and native halophytes *Suaeda glauca* and *Tripolium vulgare* did not significantly affect their germination percentage (Table 1, Figure 2c,d).

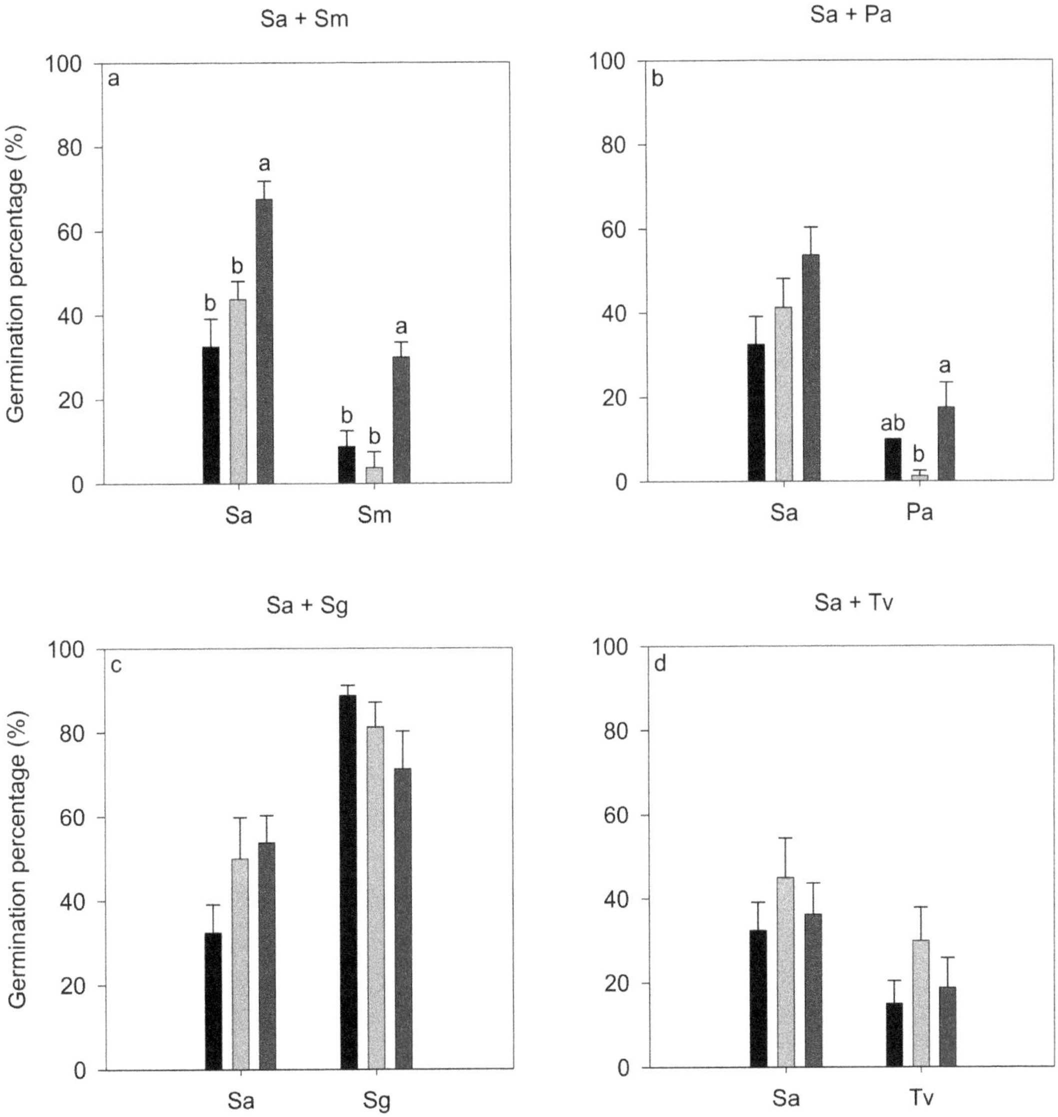

Figure 2. Seed germination percentage of invasive *Spartina alterniflora* (Sa) and four native species (*Scirpus mariqueter*, Sm, *Phragmites australis*, Pa, *Suaeda glauca*, Sg, and *Tripolium vulgare*, Tv) in different interspecific interaction treatments (control: target species in monoculture, allelopathy: target species seeds + competitor seeds leachate, competition: target species seeds + competitor seeds in mixture). (**a**) Species combination of Sa + Sm, (**b**) Species combination of Sa + Pa, (**c**) Species combination of Sa + Sg, (**d**) Species combination of Sa + Tv. Different lower cases denote significant differences among different interspecific interaction treatments of the same species. Error bars are standard errors.

The germination speed of most species (indicated by mean time to germination) was significantly affected by interspecific interaction treatment except for *Phragmites australis* seeds (Table 2). The germination of *Spartina alterniflora* seeds was the fastest (with the lowest mean time to germination) in control treatment and was significantly slowed down in allelopathy treatment with *Scirpus mariqueter* and *Phragmites australis* leachate (Figure 3a,b) and in competition treatment with *Suaeda glauca* and *Tripolium vulgare* seeds (Figure 3c,d). Competition treatment with *Spartina alterniflora* seeds significantly slowed down the germination of *Scirpus mariqueter* (Figure 3a), whereas allelopathy treatment with *Spartina alterniflora* leachate significantly promoted the germination speed of *Suaeda glauca* and *Tripolium vulgare* (Figure 3c,d).

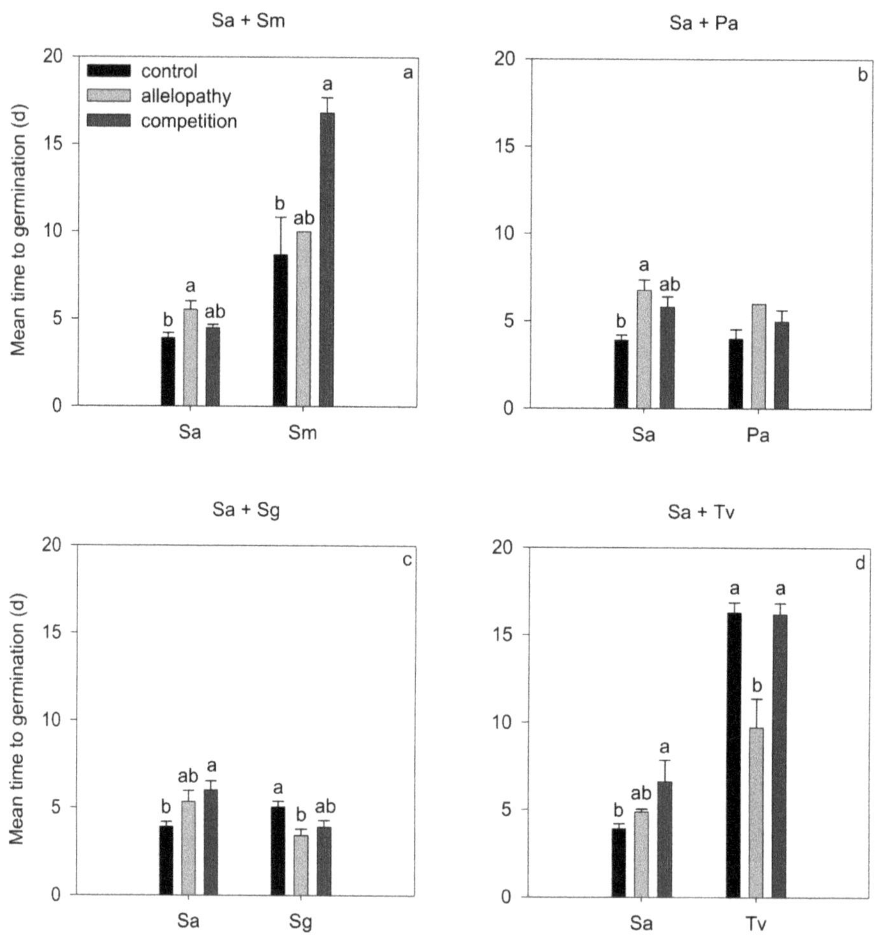

Figure 3. Mean time to seed germination of invasive *Spartina alterniflora* (Sa) and four native species (*Scirpus mariqueter*, Sm, *Phragmites australis*, Pa, *Suaeda glauca*, Sg, and *Tripolium vulgare*, Tv) in different interspecific interaction treatments (control: target species in monoculture, allelopathy: target species seeds + competitor seeds leachate, competition: target species seeds + competitor seeds in mixture). (**a**) Species combination of Sa + Sm, (**b**) Species combination of Sa + Pa, (**c**) Species combination of Sa + Sg, (**d**) Species combination of Sa + Tv. Different lower cases denote significant differences among different interspecific interaction treatments of the same species. Error bars are standard errors.

Table 1. Results of one-way ANOVA testing the effects of interspecific interaction treatment on seed germination percentage of invasive *Spartina alterniflora* (Sa) and four native species (*Scirpus mariqueter*, Sm, *Phragmites australis*, Pa, *Suaeda glauca*, Sg, and *Tripolium vulgare*, Tv) seeds in different species combinations. The abbreviations in brackets after species combinations denote the target species considered. Significant *p* values are in bold.

Source of Variation	Sa + Sm (Sa)			Sa + Pa (Sa)			Sa + Sg (Sa)			Sa + Tv (Sa)		
	Df	F	*p*	Df	F	*p*	Df	F	*p*	Df	F	*p*
treatment	2	11.86	**0.003**	2	2.55	0.133	2	2.11	0.177	2	0.30	0.746
Source of variation	**Sa + Sm (Sm)**			**Sa + Pa (Pa)**			**Sa + Sg (Sg)**			**Sa + Tv (Tv)**		
	Df	F	*p*	Df	F	*p*	Df	F	*p*	Df	F	*p*
treatment	2	14.35	**0.002**	2	4.52	**0.044**	2	1.91	0.204	2	1.28	0.325

Table 2. Results of one-way ANOVA testing the effects of interspecific interaction treatment on the mean time to seed germination of invasive *Spartina alterniflora* (Sa) and four native species (*Scirpus mariqueter*, Sm, *Phragmites australis*, Pa, *Suaeda glauca*, Sg, and *Tripolium vulgare*, Tv) in different species combinations. The abbreviations in brackets after species combinations denote the target species considered. Significant *p* values are in bold.

Source of Variation	Sa + Sm (Sa)			Sa + Pa (Sa)			Sa + Sg (Sa)			Sa + Tv (Sa)		
	Df	F	*p*	Df	F	*p*	Df	F	*p*	Df	F	*p*
treatment	2	5.40	**0.029**	2	8.03	**0.010**	2	4.51	**0.044**	2	5.27	**0.030**
Source of variation	**Sa + Sm (Sm)**			**Sa + Pa (Pa)**			**Sa + Sg (Sg)**			**Sa + Tv (Tv)**		
	Df	F	*p*	Df	F	*p*	Df	F	*p*	Df	F	*p*
treatment	2	8.31	**0.026**	2	1.60	0.291	2	5.32	**0.030**	2	11.02	**0.005**

3.3. Salinity Modification during Germination

The final salinity of the solution in Petri dishes after germination differed significantly among different interspecific interaction treatments for *Spartina alterniflora* in species combinations of Sa + Sm and Sa + Sg, and for *Scirpus mariqueter*, *Phragmites australis* and *Tripolium vulgare* (Table 3). In most species combinations, final salinity was significantly lower in competition treatment than in control and allelopathy treatments, except for *Spartina alterniflora* in species combinations of Sa + Pa and Sa + Tv, and for *Suaeda glauca* in species combination of Sa + Sg (Figure 4).

Table 3. Results of one-way ANOVA, testing the effects of interspecific interaction treatment on the final salinity of the solution in Petri dishes after seed germination in different species combinations. The abbreviations in brackets after species combinations denote the target species considered. Significant *p* values are in bold.

Source of Variation	Sa + Sm (Sa)			Sa + Pa (Sa)			Sa + Sg (Sa)			Sa + Tv (Sa)		
	Df	F	*p*	Df	F	*p*	Df	F	*p*	Df	F	*p*
treatment	2	16.49	**0.001**	2	3.82	0.063	2	7.09	**0.014**	2	1.85	0.212
Source of variation	**Sa + Sm (Sm)**			**Sa + Pa (Pa)**			**Sa + Sg (Sg)**			**Sa + Tv (Tv)**		
	Df	F	*p*	Df	F	*p*	Df	F	*p*	Df	F	*p*
treatment	2	78.77	**<0.001**	2	30.89	**<0.001**	2	2.00	0.191	2	13.33	**0.002**

Sa: *Spartina alterniflora*, Sm: *Scirpus mariqueter*, Pa: *Phragmites australis*, Sg: *Suaeda glauca*, Tv: *Tripolium vulgare*.

In all species combinations, the final salinity of the solution in Petri dishes after germination significantly decreased with an increasing total germinated seed number (Figure 5).

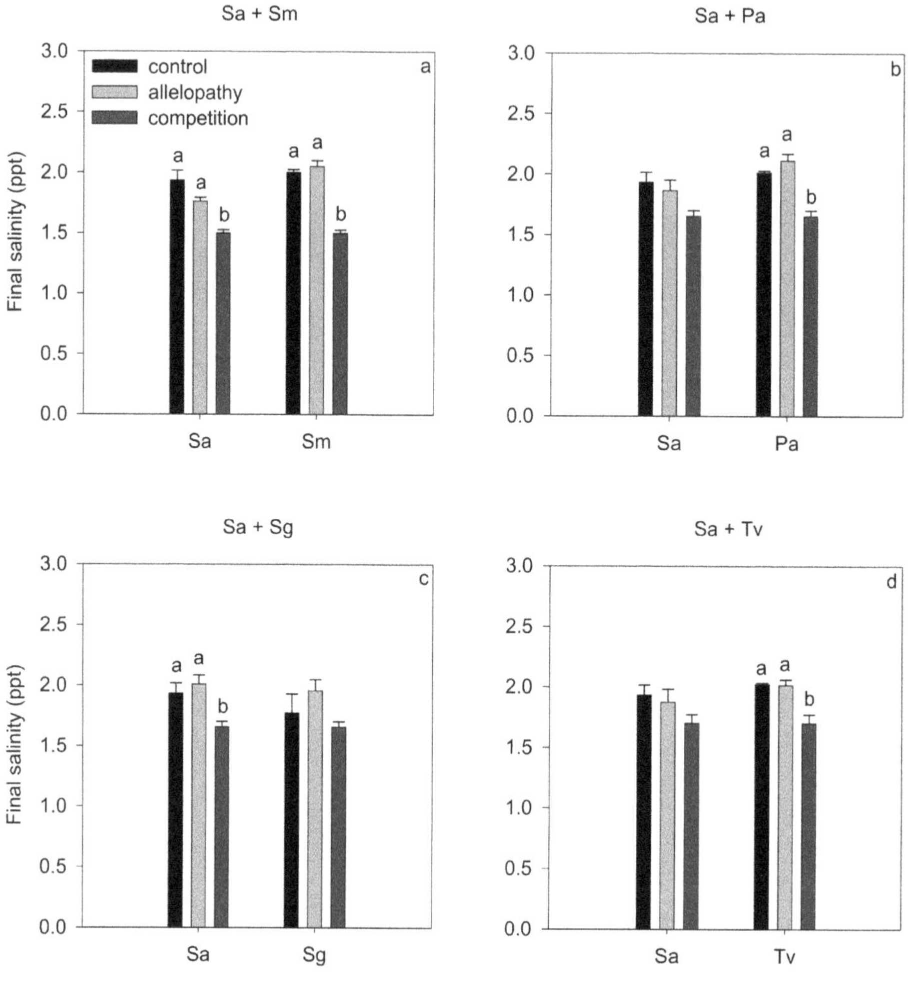

Figure 4. Final salinity of the solution in Petri dishes after germination in different interspecific interaction treatments (control: target species in monoculture, allelopathy: target species seeds + competitor seeds leachate, competition: target species seeds + competitor seeds in mixture). (**a**) Species combination of Sa + Sm, (**b**) Species combination of Sa + Pa, (**c**) Species combination of Sa + Sg, (**d**) Species combination of Sa + Tv. Sa: *Spartina alterniflora*, Sm: *Scirpus mariqueter*, Pa: *Phragmites australis*, Sg: *Suaeda glauca*, Tv: *Tripolium vulgare*. Different lower cases denote significant differences among different interspecific interaction treatments of the same species. Error bars are standard errors.

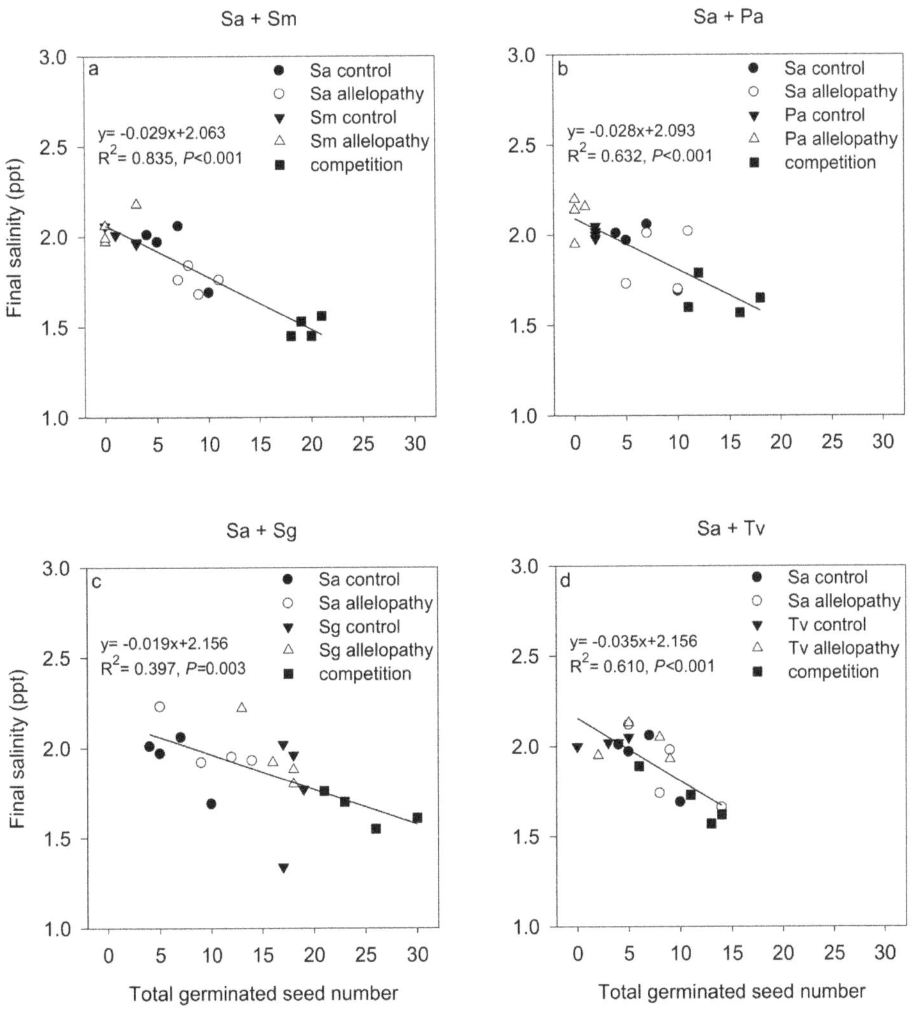

Figure 5. Relationship between final salinity of the solution in Petri dishes after germination and total germinated seed number in different interspecific interaction treatments (control: target species in monoculture, allelopathy: target species seeds + competitor seeds leachate, competition: target species seeds + competitor seeds in mixture). (**a**) Species combination of Sa + Sm, (**b**) Species combination of Sa + Pa, (**c**) Species combination of Sa + Sg, (**d**) Species combination of Sa + Tv. Sa: *Spartina alterniflora*, Sm: *Scirpus mariqueter*, Pa: *Phragmites australis*, Sg: *Suaeda glauca*, Tv: *Tripolium vulgare*.

4. Discussion

Seed germination is an irreversible process and the weakest stage in the life cycle of plants [48], which plays important roles in maintaining plant populations and communities. The findings in this and previous studies [41] suggested that the interaction between seeds is more complicated than we realized before. Our study indicated that the leachate of competitor seeds and the entire competitor seeds had inconsistent effects on germination of target species. This, together with the reduction of solution salinity in Petri dishes, implied that seeds could affect germination performance of neighbors through both allelochemical release and microenvironment modification.

4.1. Effects of Allelopathy on Seed Germination

In a natural environment, soil seed banks usually consist of seeds of different species, which makes seed-to-seed interactions ubiquitous. Adult plants can affect neighbors in a number of ways (e.g., resource competition and direct interference), whereas seeds interact with each other primarily through allelopathy [49]. In addition, the nature and magnitude of interspecific interactions among seeds usually vary with different types and concentration of allelochemicals [50].

In our experiment, the leachate of invasive *Spartina alterniflora* seeds slightly reduced (but not significantly) both the germination percentage and germination speed of native *Scirpus mariqueter* and *Phragmites australis* (Figures 2a,b and 3a,b). That is to say, only a small part of seeds of these species germinated slowly, while the others failed. This effect was probably caused by allelochemicals released from *Spartina alterniflora* seeds. As the shape and anatomical structure of seeds (e.g., seed coat thickness, wax content and embryo position) play important roles in seed response to external conditions [48,51], seed leachate will have different effects on different species. That might be the reason why the germination process of *Suaeda glauca* and *Tripolium vulgare* was significantly accelerated (Figure 3c,d).

The leachate of native *Scirpus mariqueter* and *Phragmites australis* seeds altered the microenvironment and significantly reduced the germination speed of invasive *Spartina alterniflora* seeds (Figure 3a,b). That means that more *Spartina alterniflora* seeds (especially those with poor germination ability) were activated to germination. These asymmetric effects between invasive and native seeds may be due to their substantial mass difference [35,52–54]. During germination, all the seeds may absorb allelochemicals from surrounding leachate, but accumulate them in different concentration, and the allelopathic effects of these chemicals usually depend on the concentration [31,50]. We are not clear what kind of allelochemical was released in the seed leachate of these species and how the interaction effects vary with their concentration, which needs further validation.

4.2. Effects of Salinity Modification on Seed Germination

During germination, seeds can absorb ions from the surrounding solution, which allow them to regulate osmotic potential [55,56]. We found a decrease in salinity after germination in most species combinations in competition treatment (Figure 4), which indicated the importance of microenvironment modification in mediating seed-to-seed interactions. Under the additive design, seed density in competition treatment was twice that of other treatments, which might be the main cause of decreased salinity in mixture. However, the change of total density is also a part of the influence of competing species. Future research following a response surface design can help us understand the effects of both intraspecific and interspecific interactions in different species proportion and total density levels. Notably, seeds in the mixture treatment are affected by allelopathy and microenvironment modification simultaneously, and we are not able to separate them from each other.

In our experiment, invasive *Spartina alterniflora* seeds had inconsistent effects compared with seed leachate and promoted the germination percentage of *Scirpus mariqueter* seeds (Figure 2a,b). Meanwhile, the germination speed of *Scirpus mariqueter* was significantly slowed down (Figure 3a), whereas that of *Phragmites australis* was not affected (Figure 3b). This can be explained by the positive effects of decrease in solution salinity in mixture. *Scirpus mariqueter* seeds are salt sensitive [13]. When the seeds of this native species germinate alone in 1% NaCl treatment, they will suffer from high osmotic and ion-toxicity stress, and only a small proportion of seeds successfully germinate. As *Spartina alterniflora* seeds are more salt tolerant, a great deal of salt ions will be absorbed during germination when this invasive plant is present, which led to a significant decrease in solution salinity. The positive effect of salinity modification by *Spartina alterniflora* on native species seemed to be stronger than the negative effect of its seed leachate, as the germination percentage of *Scirpus mariqueter* seeds was the highest in the mixture (Figure 2a). Conversely, halophyte

Suaeda glauca and *Tripolium vulgare* seeds are much more salt tolerant, and require a certain concentration of salt during germination. Hence, alleviation of salt stress by competing *Spartina alterniflora* seeds in mixture had minor effects, which did not significantly affect the germination percentage (Figure 2c,d) and germination speed of these two species (Figure 3c,d).

For *Spartina alterniflora*, salt-sensitive *Scirpus mariqueter* seeds significantly promoted its germination percentage (Figure 2a), whereas the seeds of halophyte *Suaeda glauca* and *Tripolium vulgare* significantly delayed its germination process (Figure 3c,d). The net effects of microenvironment modification (i.e., regulation of salinity) on this invasive plant differed in different combinations with native species, which was also related to their salt preference. Despite of different species combinations and treatments, the extent of solution salinity decrease in Petri dishes after seed germination is all significantly correlated with the number of total germinated seeds (Figure 5), indicating that the salinity modification effect by competing seeds was mainly due to the change of total density level rather than the biological attributes of competitors. Surprisingly, the Sa + Sg combination had the lowest efficiency of salt absorption per single seed (Figure 5c), even though these two species were the most salt tolerant ones among the five species used in our study. Whether there is another interaction mechanism during seed germination needs further investigation.

4.3. Implications for Community Assembly

Interspecific interactions play important roles in species coexistence and biodiversity. As seeds of different species coexist universally in the field, they may interact with each other through leachate or the modification of microenvironment in the soil. Seeds that germinate earlier usually have an advantage over those germinate later in competition [57]. For invasive plants, high seed yield, high germination percentage and short germination time are important traits which facilitate their invasion into native plant communities [58]. However, sometimes, rapid germination causes seeds to be exposed to a stressful environment, and is bad for subsequent plant survival [59]. Previous studies found that there exists a special mechanism for seeds to "sense" the surrounding environment via allelochemicals, which allow them to assess the optimal germination timing for more competitive advantages [57,60,61]. Among the five species used in our study, *Phragmites australis* had the shortest germination period (Figure 1b), and hence, the lowest potential of germination timing regulation. On the contrary, *Tripolium vulgare* had the longest germination period (Figure 1b), which may improve its adaptability to changing environments. It should be noted that due to the difference in germination timing, seeds that germinate later will not only affected by neighboring seeds, but also by seedlings from early germinated seeds [62]. Future research on the interaction between seeds and seedlings will contribute to our understanding of the role of the germination process in community assembly.

Our study found inconsistent effects of seed leachate and salinity modification on germination of both invasive and native species in Chongming Dongtan. On the whole, *Spartina alterniflora* had more positive effects on *Scirpus mariqueter* and *Phragmites australis* seeds during germination than on halophyte *Suaeda glauca* and *Tripolium vulgare*, and vice versa. Although invasive *Spartina alterniflora* promoted the germination percentage of *Scirpus mariqueter* seeds, it may still gain competitive advantage over native *Scirpus mariqueter*, *Tripolium vulgare* and *Phragmites australis* due to its short time lag (Figure 1a) and high germination index (Figure 1c). Besides, the strong competitive ability of *Spartina alterniflora* in subsequent life stages would promote its invasion into these native plant communities and had caused a series of negative ecosystem impacts [37,63,64]. The competitive mechanisms between invasive *Spartina alterniflora* and native species during germination stage were quite different from those between adult individuals, which highlighted the importance of including regeneration into the framework of community assembly.

In Chongming Dongtan, perennial clonal plants *Spartina alterniflora*, *Phragmites australis* and *Scirpus mariqueter* are heavily dependent on vegetative reproduction in mature patches, whereas seed dispersal and germination play important roles in new habitat colonization [13]. Therefore, eradication of both the aboveground and belowground part of *Spartina alterniflora* before its florescence is critical for control of this invasive species. After the elimination of *Spartina alterniflora*, transplantation of *Phragmites australis* should be implemented due to its relatively low seed vigor in saline water, whereas *Scirpus mariqueter* has the potential to recover from its persistent seed bank [65]. Artificial reduction of sediment salinity will be an effective measure to facilitate the recovery of these native species, but the cost is too high under current conditions. On the other hand, seed dispersal and germination are crucial for the recruitment of annual *Suaeda glauca* and *Tripolium vulgare*, but they are less affected by invasive *Spartina alterniflora* in their hypersaline habitats. Seed addition of these halophytes may also be beneficial to community biodiversity, as there is intransitive competition among the dominant species in Chongming Dongtan during germination stage [66].

Seed-to-seed interactions are an important perspective to understand the underlying mechanisms of community dynamics, species diversity maintenance and invasion of alien species and can improve the effectiveness in the management of invaded coastal wetlands. However, it should be noted that due to the frequent scouring of tide in the field, the concentration of allelochemicals in the soil will be much lower than in laboratory treatment, and seeds will have minor effects of salinity modification. We need to be very careful to extrapolate our laboratory results to natural ecosystems.

Author Contributions: Conceptualization, C.-H.W. and C.-J.Z.; Data curation, Y.Y.; Formal analysis, C.-H.W., L.T. and C.-J.Z.; Investigation, Z.-L.Y.; Methodology, Z.-L.Y. and Y.Y.; Project administration, Y.Y.; Software, C.-H.W. and Y.G.; Supervision, C.-J.Z.; Validation, Y.Y. and Y.G.; Writing—original draft, C.-H.W. and Z.-L.Y.; Writing—review and editing, C.-H.W. and L.T. All authors have read and agreed to the published version of the manuscript.

Funding: This research was funded by the National Natural Science Foundation of China (grant number 31870405, 31872032) and by the Shanghai Municipal Natural Science Foundation (grant number 22ZR1421500).

Institutional Review Board Statement: Not applicable.

Data Availability Statement: Data are contained within the article.

Acknowledgments: We thank the Shanghai Chongming Dongtan National Nature Reserve for providing convenience in seeds collection.

Conflicts of Interest: The authors declare no conflict of interest.

References

1. Rosindell, J.; Hubbell, S.P.; Etienne, R.S. The Unified Neutral Theory of Biodiversity and Biogeography at Age Ten. *Trends Ecol. Evol.* **2011**, *26*, 340–348. [CrossRef] [PubMed]
2. Rosenzweig, M.L. *Species Diversity in Space and Time*; Cambridge University Press: Cambridge, UK, 1995.
3. Purschke, O.; Schmid, B.C.; Sykes, M.T.; Poschlod, P.; Michalski, S.G.; Durka, W.; Kuehn, I.; Winter, M.; Prentice, H.C. Contrasting changes in taxonomic, phylogenetic and functional diversity during a long-term succession: Insights into assembly processes. *J. Ecol.* **2013**, *101*, 857–866. [CrossRef]
4. Xu, Y.; Dong, S.; Li, S.; Shen, H. Research progress on ecological filtering mechanisms for plant community assembly. *Acta Ecol. Sin.* **2019**, *39*, 2267–2281.
5. Wang, X.; Liang, C.; Wang, W. Balance between facilitation and competition determines spatial patterns in a plant population. *Chin. Sci. Bull.* **2014**, *59*, 1405–1415. [CrossRef]
6. Smith, A.L.; Atwater, D.Z.; Kim, W.; Haak, D.C.; Barney, J.N. Invasive plant rhizome production and competitiveness vary based on neighbor identity. *J. Plant Ecol.* **2021**, *14*, 638–647. [CrossRef]
7. Wang, C.-h.; Li, B. Salinity and disturbance mediate direct and indirect plant-plant interactions in an assembled marsh community. *Oecologia* **2016**, *182*, 139–152. [CrossRef]
8. Wehnert, A.; Wagner, S.; Huth, F. Spatio-Temporal Distribution of Carabids Influenced by Small-Scale Admixture of Oak Trees in Pine Stands. *Diversity* **2020**, *12*, 398. [CrossRef]
9. Mikami, O.K.; Kawata, M. Does interspecific territoriality reflect the intensity of ecological interactions—A theoretical model for interspecific territoriality. *Evol. Ecol. Res.* **2004**, *6*, 765–775.
10. Le Roux, P.C.; Shaw, J.D.; Chown, S.L. Ontogenetic shifts in plant interactions vary with environmental severity and affect population structure. *New Phytol.* **2013**, *200*, 241–250. [CrossRef]

11. Schiffers, K.; Tielborger, K. Ontogenetic shifts in interactions among annual plants. *J. Ecol.* **2006**, *94*, 336–341. [CrossRef]
12. Brooker, R.W.; Callaghan, T.V. The balance between positive and negative plant interactions and its relationship to environmental gradients: A model. *Oikos* **1998**, *81*, 196–207. [CrossRef]
13. Dai, W.-l.; Wang, T.-h.; Wang, C.-h. Effects of interspecific interactions on seed germination between dominant species in the Yangtze River Estuary. *Estuar. Coast. Shelf Sci.* **2020**, *232*, 106483. [CrossRef]
14. Margreiter, V.; Walde, J.; Erschbamer, B. Competition-free gaps are essential for the germination and recruitment of alpine species along an elevation gradient in the European Alps. *Alp. Bot.* **2021**, *131*, 135–150. [CrossRef]
15. Elsalahy, H.; Bellingrath-Kimura, S.; Kautz, T.; Doering, T. Effects of mixing two legume species at seedling stage under different environmental conditions. *Peerj* **2021**, *9*, e10615. [CrossRef]
16. Afonso, A.; Castro, S.; Loureiro, J.; Mota, L.; de Oliveira, J.C.; Torices, R. The Effects of Achene Type and Germination Time on Plant Performance in the Heterocarpic *Anacyclus clavatus* (Asteraceae). *Am. J. Bot.* **2014**, *101*, 892–898. [CrossRef]
17. Zacchello, G.; Vinyeta, M.; Agren, J. Strong stabilizing selection on timing of germination in a Mediterranean population of *Arabidopsis thaliana*. *Am. J. Bot.* **2020**, *107*, 1518–1526. [CrossRef]
18. Cheng, H.; Wu, B.; Yu, Y.; Wang, S.; Wei, M.; Wang, C.; Du, D. The allelopathy of horseweed with different invasion degrees in three provinces along the Yangtze River in China. *Physiol. Mol. Biol. Plants* **2021**, *27*, 483–495. [CrossRef]
19. Roman, A.M.; Truta, A.M.; Viman, O.; Morar, I.M.; Spalevic, V.; Dan, C.; Sestras, R.E.; Holonec, L.; Sestras, A.F. Seed Germination and Seedling Growth of *Robinia pseudoacacia* Depending on the Origin of Different Geographic Provenances. *Diversity* **2022**, *14*, 34. [CrossRef]
20. Fernandez-Torquemada, Y.; Luis Sanchez-Lizaso, J. Effects of salinity on seed germination and early seedling growth of the Mediterranean seagrass *Posidonia oceanica* (L.) Delile. *Estuar. Coast. Shelf Sci.* **2013**, *119*, 64–70. [CrossRef]
21. Hayasaka, D.; Nakagawa, M.; Maebara, Y.; Kurazono, T.; Hashimoto, K. Seed germination characteristics of invasive *Spartina alterniflora* Loisel in Japan: Implications for its effective management. *Sci. Rep.* **2020**, *10*, 2116. [CrossRef]
22. Xiaojun, Y.; Shangli, S.; Ruijun, L.; Fang, W.; Benjian, C. Research progress on effects of ecological factors on seed germination. *Pratacultural Sci.* **2006**, *23*, 44–49.
23. Reynolds, C.; Cumming, G.S. Seed traits and bird species influence the dispersal parameters of wetland plants. *Freshw. Biol.* **2016**, *61*, 1157–1170. [CrossRef]
24. Soons, M.B. Wind dispersal in freshwater wetlands: Knowledge for conservation and restoration. *Appl. Veg. Sci.* **2006**, *9*, 271–278. [CrossRef]
25. Partzsch, M. Warming differently affects the inter- and intraspecific interactions among semi-dry grassland species. *Perspect. Plant Ecol. Evol. Syst.* **2019**, *40*, 125481. [CrossRef]
26. Thorpe, A.S.; Aschehoug, E.T.; Atwater, D.Z.; Callaway, R.M. Interactions among plants and evolution. *J. Ecol.* **2011**, *99*, 729–740. [CrossRef]
27. Rashidi, S.; Reza Yousefi, A.; Goicoechea, N.; Pouryousef, M.; Moradi, P.; Vitalini, S.; Iriti, M. Allelopathic Interactions between Seeds of *Portulaca oleracea* L. and Crop Species. *Appl. Sci.* **2021**, *11*, 3539. [CrossRef]
28. Laterra, P.; Bazzalo, M.E. Seed-to-seed allelopathic effects between two invaders of burned Pampa grasslands. *Weed Res.* **1999**, *39*, 297–308. [CrossRef]
29. Suman, A.; Shahi, H.N.; Singh, P.; Gaur, A. Allelopathic influence of *Vigna mungo* (black gram) seeds on germination and radical growth of some crop plants. *Plant Growth Regul.* **2002**, *38*, 69–74. [CrossRef]
30. Zhang, S.; Liu, J.; Bao, X.; Niu, K. Seed-to-seed potential allelopathic effects between *Ligularia virgaurea* and native grass species of Tibetan alpine grasslands. *Ecol. Res.* **2011**, *26*, 47–52. [CrossRef]
31. Mercedes Aguilar-Franco, Z.; Flores-Palacios, A.; Flores-Morales, A.; Perea-Arango, I.; de Jesus Arellano-Garcia, J.; Valencia-Diaz, S. Density-dependent effect of allelopathy on germination and seedling emergence in two Ipomoea species. *Rev. Chil. De Hist. Nat.* **2019**, *92*, 7. [CrossRef]
32. Badano, E.I.; Cavieres, L.A. Ecosystem engineering across ecosystems: Do engineer species sharing common features have generalized or idiosyncratic effects on species diversity? *J. Biogeogr.* **2006**, *33*, 304–313. [CrossRef]
33. Ma, Y.A.; Ma, Z.J. *Chongming Dongtan Wetland of International Importance*; China Forestry Publishing House: Beijing, China, 2006.
34. Xie, B.; Han, G.; Qiao, P.; Mei, B.; Wang, Q.; Zhou, Y.; Zhang, A.; Song, W.; Guan, B. Effects of mechanical and chemical control on invasive *Spartina alterniflora* in the Yellow River Delta, China. *Peerj* **2019**, *7*, e7655. [CrossRef]
35. Xiao, D.; Zhang, L.; Zhu, Z. A study on seed characteristics and seed bank of *Spartina alterniflora* at saltmarshes in the Yangtze Estuary, China. *Estuar. Coast. Shelf Sci.* **2009**, *83*, 105–110. [CrossRef]
36. Wang, C.-H.; Tang, L.; Fei, S.-F.; Wang, J.-Q.; Gao, Y.; Wang, Q.; Chen, J.-K.; Li, B. Determinants of seed bank dynamics of two dominant helophytes in a tidal salt marsh. *Ecol. Eng.* **2009**, *35*, 800–809. [CrossRef]
37. Chen, Z.Y.; Li, B.; Zhong, Y.; Chen, J.K. Local competitive effects of introduced *Spartina alterniflora* on *Scirpus mariqueter* at Dongtan of Chongming Island, the Yangtze River estuary and their potential ecological consequences. *Hydrobiologia* **2004**, *528*, 99–106. [CrossRef]
38. Wang, Q.; Wang, C.H.; Zhao, B.; Ma, Z.J.; Luo, Y.Q.; Chen, J.K.; Li, B. Effects of growing conditions on the growth of and interactions between salt marsh plants: Implications for invasibility of habitats. *Biol. Invasions* **2006**, *8*, 1547–1560. [CrossRef]
39. Yuan, Y.; Wang, K.; Li, D.; Pan, Y.; Lv, Y.; Zhao, M.; Gao, J. Interspecific Interactions between *Phragmites australis* and *Spartina alterniflora* along a Tidal Gradient in the Dongtan Wetland, Eastern China. *PLoS ONE* **2013**, *8*, e53843. [CrossRef]

40. Zhu, Z.; Zhang, L.; Xiao, D. Seed production of *Spartina alterniflora* and its response of germination to temperature at Chongming Dongtan, Shanghai. *Acta Ecol. Sin.* **2011**, *31*, 1574–1581.
41. Wang, Q.; Shi, H.H.; Yu, Z.L.; Wang, T.H.; Wang, C.H. Effects of salinity and interspecific interaction on germination and growth of *Scirpus mariqueter* and *Spartina alterniflora*. *Acta Ecol. Sin.* **2022**, *42*, 8300–8310. [CrossRef]
42. Lissner, J.; Schierup, H.H. Effects of salinity on the growth of *Phragmites australis*. *Aquat. Bot.* **1997**, *55*, 247–260. [CrossRef]
43. Peng, Y.; Xie, T.; Zhou, F.; Wan, H.; Zhang, C.; Zhai, R.; Zheng, Q.; Zheng, C.; Liu, Z. Response of plant growth and photosynthetic characteristics in *Suaeda glauca* and *Atriplex triangularis* seedlings to different concentrations of salt treatments. *Acta Prataculturae Sin.* **2012**, *21*, 64–74.
44. Gong, J.-n.; Wang, K.-y.; Zhang, C.; Ma, Y.-l. Invasion and its effects of xerarch halophytes in reclaimed tidal wetlands. *J. Appl. Ecol.* **2009**, *20*, 33–39.
45. De Souza Coelho, L.C.; Broch Mignoni, D.S.; Barbedo, C.J.; Braga, M.R. Seed leachates of the tropical legume *Sesbania virgata*: Their effects on germination and seedling growth of tomato and rice. *Acta Physiol. Plant.* **2022**, *44*, 96. [CrossRef]
46. Schmer, M.R.; Xue, Q.; Hendrickson, J.R. Salinity effects on perennial, warm-season (C-4) grass germination adapted to the northern Great Plains. *Can. J. Plant Sci.* **2012**, *92*, 873–881. [CrossRef]
47. Brenchley, J.L.; Probert, R.J. Seed germination responses to some environmental factors in the seagrass *Zostera capricorni* from eastern Australia. *Aquat. Bot.* **1998**, *62*, 177–188. [CrossRef]
48. Pellissier, F. Improved germination bioassays for allelopathy research. *Acta Physiol. Plant.* **2013**, *35*, 23–30. [CrossRef]
49. Guido, A.; Hoss, D.; Pillar, V.D. Exploring seed to seed effects for understanding invasive species success. *Perspect. Ecol. Conserv.* **2017**, *15*, 234–238. [CrossRef]
50. Baskin, C.C.; Baskin, J.M.; Chester, D.W.; Smith, M. Ethylene as a possible cue for seed germination of *Schoenoplectus hallii* (Cyperaceae), a rare summer annual of occasionally flooded sites. *Am. J. Bot.* **2003**, *90*, 620–627. [CrossRef]
51. Song, J.; Shi, W.; Liu, R.; Xu, Y.; Sui, N.; Zhou, J.; Feng, G. The role of the seed coat in adaptation of dimorphic seeds of the euhalophyte *Suaeda salsa* to salinity. *Plant Species Biol.* **2017**, *32*, 107–114. [CrossRef]
52. Sun, S.C.; Gao, X.M.; Cai, Y.L. Variations in sexual and asexual reproduction of *Scirpus mariqueter* along an elevational gradient. *Ecol. Res.* **2001**, *16*, 263–274. [CrossRef]
53. Wu, G.-L.; Du, G.-Z.; Shi, Z.-H. Germination strategies of 20 alpine species with varying seed mass and light availability. *Aust. J. Bot.* **2013**, *61*, 404–411. [CrossRef]
54. McKee, J.; Richards, A.J. Variation in seed production and germinability in common reed (*Phragmites australis*) in Britain and France with respect to climate. *New Phytol.* **1996**, *133*, 233–243. [CrossRef]
55. Zhang, H.; Irving, L.J.; McGill, C.; Matthew, C.; Zhou, D.; Kemp, P. The effects of salinity and osmotic stress on barley germination rate: Sodium as an osmotic regulator. *Ann. Bot.* **2010**, *106*, 1027–1035. [CrossRef]
56. Rehman, S.; Harris, P.J.C.; Bourne, W.F.; Wilkin, J. The effect of sodium chloride on germination and the potassium and calcium contents of Acacia seeds. *Seed Sci. Technol.* **1997**, *25*, 45–57.
57. Dyer, A.R.; Fenech, A.; Rice, K.J. Accelerated seedling emergence in interspecific competitive neighbourhoods. *Ecol. Lett.* **2000**, *3*, 523–529. [CrossRef]
58. Williams, L.K.; Kristiansen, P.; Sindel, B.M.; Wilson, S.C.; Shaw, J.D. Quantifying the seed bank of an invasive grass in the sub-Antarctic: Seed density, depth, persistence and viability. *Biol. Invasions* **2016**, *18*, 2093–2106. [CrossRef]
59. Orrock, J.L.; Christopher, C.C. Density of Intraspecific Competitors Determines the Occurrence and Benefits of Accelerated Germination. *Am. J. Bot.* **2010**, *97*, 694–699. [CrossRef]
60. Fenesi, A.; Kelemen, K.; Sandor, D.; Ruprecht, E. Influential neighbours: Seeds of dominant species affect the germination of common grassland species. *J. Veg. Sci.* **2020**, *31*, 1028–1038. [CrossRef]
61. Houseman, G.R.; Mahoney, A.K. Intraspecific seed interactions alter seedling emergence of *Lespedeza cuneata* under field conditions. *Popul. Ecol.* **2015**, *57*, 539–544. [CrossRef]
62. Tielboerger, K.; Prasse, R. Do seeds sense each other? Testing for density-dependent germination in desert perennial plants. *Oikos* **2009**, *118*, 792–800. [CrossRef]
63. Tang, L.; Wolf, A.A.; Gao, Y.; Wang, C.H. Photosynthetic tolerance to non-resource stress influences competition importance and intensity in an invaded estuary. *Ecology* **2018**, *99*, 1327–1337. [CrossRef]
64. Li, B.; Liao, C.-H.; Zhang, X.-D.; Chen, H.-L.; Wang, Q.; Chen, Z.-Y.; Gan, X.-J.; Wu, J.-H.; Zhao, B.; Ma, Z.-J.; et al. *Spartina alterniflora* invasions in the Yangtze River estuary, China: An overview of current status and ecosystem effects. *Ecol. Eng.* **2009**, *35*, 511–520. [CrossRef]
65. Hu, Z.; Ma, Q.; Cao, H.; Zhang, Z.; Tang, C.; Zhang, L.; Ge, Z. A trial study on revegetation of the native *Scirpus mariqueter* population in the coastal wetland of the Yangtze Estuary. *Ecol. Sci.* **2016**, *35*, 1–7.
66. Shi, H.H.; Xue, Q.; Yu, Z.L.; Wang, C.H. Effects of density and species proportion on intraspecific and interspecific interactions between salt marsh plants during germination. *Chin. J. Plant Ecol.* **2022**, *in press*.

forests

MDPI

Article

Ipomoea cairica (L.) from Mangrove Wetlands Acquired Salt Tolerance through Phenotypic Plasticity

Jiatong Zou [1,2,†], Benqi Yuan [1,2,†], Weihua Li [1,2], Xiaoting Xie [1,2], Minghao Chen [1,3,*] and Tiantian Xiong [1,2,*]

1 Guangdong Provincial Key Laboratory of Biotechnology for Plant Development, School of Life Science, South China Normal University, Guangzhou 510631, China; zoujiatong@m.scnu.edu.cn (J.Z.); yuanbenqi@m.scnu.edu.cn (B.Y.); whli@scnu.edu.cn (W.L.); xiexiaoting@m.scnu.edu.cn (X.X.)

2 Guangzhou Key Laboratory of Subtropical Biodiversity and Biomonitoring, School of Life Science, South China Normal University, Guangzhou 510631, China

3 Key Laboratory of Vegetation Restoration and Management of Degraded Ecosystems, South China Botanical Garden, Chinese Academy of Sciences, Guangzhou 510650, China

* Correspondence: xiongtiantian@m.scnu.edu.cn (T.X.); minghaochen@scbg.ac.cn (M.C.); Tel.: +86-20-85211555 or +86-18026310570 (T.X.)

† These authors contributed equally to this work.

Abstract: Palmate-leaved morning glory (*Ipomoea cairica* (L.) Sweet) is a fast-growing perennial herbaceous twining vine that was recently discovered to invade mangrove wetlands in China. To understand the mechanism of its successful invasion, the salt tolerance of a coastal ecotype from Zhuhai and a terrestrial ecotype from Guangzhou were compared under salt stress. The morphological, physiological, and biochemical parameters related to growth, ion homeostasis, photosynthetic pigments, chlorophyll fluorescence parameters, oxidative stress, and apoptosis were measured in both ecotypes. Monitoring apoptosis showed that the protoplasts of the coastal ecotype underwent apoptosis and were later compared with those of the terrestrial ecotype. The coastal ecotype was also found to have higher regenerated stems; less water loss, sodium (Na^+) uptake, and membrane damage; higher salt gland density and area; and better photosynthetic performance than the terrestrial ecotype. The coastal ecotype probably prevented salt-related damage by reducing its water loss and secreting excess Na^+ through its lower stomatal density and higher density and area of salt glands. The coastal ecotype also maintained a better balance of Na^+, potassium ions, nitrogen, and phosphorus under salt stress. Moreover, the coastal ecotype had higher activities of antioxidant enzymes, including superoxide dismutase, peroxidase, and catalase, and a higher content of non-enzymatic antioxidants, including proline and anthocyanins, which indicate a stronger antioxidant ability. Our results suggest that the coastal ecotype adapts to a higher salt tolerance than the terrestrial ecotype by enhancing its exclusion of salt, adjusting its osmolytes, and through photosynthetic efficiency, which could explain its successful invasion in the mangrove wetland ecosystem.

Keywords: antioxidant activity; ecological adaptation; invasive plant; ion homeostasis; morning glory *Ipomoea cairica* (L.) sweet; salt stress

1. Introduction

A rapidly growing number of alien invasive species is causing increasing damage to Chinese agriculture, forestry, and livestock industries [1], with annual economic losses that exceed USD 100 billion [2]. In the hotspots of the introduced species, such as coastal zones, biological invasions are becoming a major driver of the decline in biodiversity and loss of ecological services in communities, such as coastal forests [3,4]. Mangroves are unique higher plant communities that are widely distributed in the intertidal zones of tropical and subtropical regions. They are unique wetland forest ecosystems in the coastal zone. Owing to their high productivity and biodiversity, the abundance of biological resources and species in the mangrove wetland ecosystem is among the highest in the natural environment [5,6]. However, the mangrove wetland ecosystem has been severely degraded, and one of the reasons for the degradation is the invasion of alien species, including smooth cordgrass (*Spartina alterniflora*) and bitter vine (*Mikania micrantha*) [7,8]. Palmate-leaved morning glory (*Ipomoea cairica*) is a typical invasive plant in south China, which has caused severe damage to ecosystems and local economies [9]. This plant has been listed as one of the worst invasive alien species in China [10]. In some previous studies, *I. cairica* was reported to invade mangroves along the southeastern coast of China [9,11]. The vine that invades mangrove wetlands in Zhuhai is a potential salt-tolerant plant. Compared with other habitats, the soil of mangroves is more saline [12]. Therefore, salt tolerance is a limiting factor for invasive plants to invade mangroves.

The most extensive and profound effect of an external salt environment on plants is its inhibition of plant growth and development [13]. For example, salt alters many functional and biochemical processes in plants, which can lead to reduced growth, productivity, and even death. Tissue dehydration, ion-specific toxicity, nutrient shortages, and hormonal/enzymatic imbalances are the primary mechanisms that are affected by salt stress [13,14]. As a result, salt stress leads to a variety of physiological and molecular changes and impedes the growth of plants by impairing photosynthesis, which reduces the availability of resources and suppresses cell division and expansion [15]. Photosynthesis is the primary driver for plant growth and development and is one of the physiological processes of plants that are the most sensitive to stress [16]. Changes in photosynthesis are directly related to the chlorophyll fluorescence parameters of Photosystem II (PSII) [17]. Analyses of the chlorophyll fluorescence parameters can rapidly determine the photochemical efficiency of the plant PSII without damaging the plant, and it is an important method to diagnose whether PSII is functional and analyze the mechanism of the plant's response to stress [18,19]. Therefore, monitoring different chlorophyll fluorescence parameters can provide important information on the changes in plant photosynthetic dynamics under a wide range of environmental stresses [19]. The osmotic potential is another important indicator that is related to salt stress [20]. When the osmotic potential in a plant exceeds that of the external environment, it is difficult for the plant to absorb enough water, which causes a physiological water shortage in the plant; this is also known as osmotic stress [21]. Salt tolerance in plants is implemented at both the organ and cellular levels. The stomata play key roles in the survival of plants in adverse environments. A reduction in the stomatal density may delay the accumulation of toxic ions and signaling molecules that inhibit growth in the leaves by reducing the transpiration flow under salt stress, thus facilitating the adaptation to salinity stress [22,23]. Salt glands actively secrete excessive amounts of ions from the plant cells to avoid toxicity [24].

Previous research on *I. cairica* has primarily focused on its invasion in terrestrial ecosystems. *I. cairica* has high rates of stem growth and photosynthesis, and it can reproduce asexually [9,25,26]. Compared with native companion species, *I. cairica* is more resistant to abiotic stressors, such as heat and drought, which enables it to rapidly invade tropical and subtropical terrestrial ecosystems [27,28]. The resistance of *I. cairica* to stress primarily originates from its high antioxidant capacity and the stability of its photosynthetic system. The levels of expression of *IcSRO1* (SIMILAR TO RCD-One) and *IcOr* (ORANGE) in *I. cairica* under unfavorable conditions were higher than those in the non-invaded plants [11,28]. However, the mechanism of its invasion in coastal wetland ecosystems remains unknown. Therefore, understanding the responses of *I. cairica* to salt stress and its associated mechanisms of tolerance may help to explain its adaptation to mangrove wetlands.

This study investigated coastal ecotypes in the mangrove wetland ecosystem and terrestrial ecotypes in the inland ecosystem and compared their responses to salt stress. The results provide evidence for how *I. cairica* maintains its ability to invade during the process of diffusing from non-salt to salty habitats.

2. Materials and Methods

2.1. Plant Material and Growth Conditions

The terrestrial ecotype of *I. cairica* was collected from the Biological Park of South China Normal University, Tianhe District, Guangzhou City, China (23.13° N, 113.35° E). The coastal ecotype of *I. cairica* was collected from the Qi'ao Mangrove Wetland Park in the Gaoxin District, Zhuhai City, China (22.43° N, 113.62° E). Stems of *I. cairica* that were 10 cm long and had two nodes were selected for cultivation in one-half Hoagland nutrient solution (pH = 5.7) that contained 0.48 mmol/L, 50 mmol/L, and 100 mmol/L of Na^+ [29]. NaCl was used to provide sodium (Na^+) for both control and treatment groups. The solution was changed every 4 days(d). The root lengths were measured after 2 weeks of cultivation. Seedlings that had been cultivated in 0.48 mmol/L Na+ were transplanted into flowerpots that contained clean sand that had been filtered through a 10-mesh sieve. Solely during the root growth experiment, a medium that contained 0.48 mmol/L Na^+ was used for the control, while media that contained 50 mmol/L and 100 mmol/L Na^+ were used for the treatment. For the other experiments, one-half Hoagland nutrient solution (150 mmol/L Na^+) was used for the salt treatments, while 0.48 mmol/L Na^+ was used for the control.

2.2. Plant Harvest and Growth Parameters

Nondestructive growth parameters, including the root length, tendency of the main stem to grow, leaf number during the salt treatment, branch number, and number of flowers, were measured every 10 d for a total of 60 d. In addition, the growth of the main stem was measured with a tape measure (cm), and the root length was measured with a straight edge ruler (cm). The plant harvest and growth parameters were measured as described by Luo et al. [30] and Kumar et al. [31]. After the harvest, the roots were carefully rinsed with tap water; the stems and roots were separated with scissors; the surface water was wiped off with absorbent paper; and the fresh weights of the aboveground parts and roots were measured using an electronic balance. Moreover, two leaves in each replicate were selected and immediately weighed as the fresh weight (FW). The leaves were then immersed in distilled water for 24 h in the dark and weighed to obtain the turgid weight (TW) after the superficial moisture had been dried. The leaves were subsequently placed in a kraft bag, oven-dried at 70 °C for 48 h, and measured as the dry weight (DW). The leaf relative water content (%) was measured as described by Kumar et al. [31] using the following formula:

$$\text{LRWC} = [(\text{FW} - \text{DW})/(\text{TW} - \text{DW})] \times 100$$

Furthermore, the root-to-shoot ratio was calculated using the root dry weight/shoot dry weight as previously described [30]. Each treatment had 12 replicates.

2.3. Salt Rejection and Ionic Balance

2.3.1. Determination of the Contents of Nitrogen, Phosphorus, Potassium, and Sodium in the Roots, Stems, and Leaves

After 60 d of cultivation, the plants were harvested and divided into three parts, including the roots, stems, and leaves. All of the materials were heated at 110 °C for 1.5 h and at 75 °C for 2 d until stable. The shoots and roots were crushed and passed through a 20-mesh sieve. After dehydration, carbonization, and oxidation by digestion with sulfuric acid·hydrogen peroxide ($H_2SO_4 \cdot H_2O_2$) as previously described [32], the organic matter in the sample was destroyed and transformed to inorganic salts. After dilution with acid, an atomic absorption spectrophotometer (AA-7000; Shimadzu, Kyoto, Japan) was used to measure the contents of nitrogen (N), phosphorus (P), potassium (K), and Na^+ in each sample. Each treatment had 12 replicates.

2.3.2. Measurement of the Density and Area of the Salt Glands

In combination with a slight modification of a previous method [33], the back of the leaf blade was coated with clear nail polish. The dried nail polish layer was then peeled off with tweezers and placed on a glass slide. These blotches were then observed under a 5× objective light microscope, and the field of view was photographed numerous times. The number of all the salt glands in the field of view and the area of the salt glands were calculated using ImageJ (NIH, Bethesda, MD, USA), which were used to calculate the density of the salt glands (the total number of salt glands per unit leaf area in the field of view) and their area as previously described [34,35]. The density of salt glands was calculated as the total number of salt glands divided by the per unit leaf area of 10 replicate observations as described by Ding et al. [36] and Leng et al. [37]. There were 150 samples per treatment and line (three epidermal samples per leaf × five fields of view per sample ×10 biological replicates). The total area of at least 10 salt glands at different locations in the field of view was measured using ImageJ. Each treatment had 10 replicates, including three epidermal samples per replicate, five fields of view per sample, and at least 10 salt glands per field of view.

2.4. *Determination of the Relative Electrical Conductivity (REC) and Content of Malondialdehyde (MDA)*

2.4.1. Determination of the Relative Electrical Conductivity

The REC was measured using a DJS-1D conductivity meter (Leica, Shanghai, China). A total of 0.1 g of fresh leaf samples were cut into approximate squares, immersed in 10 mL of deionized water in a test tube, and soaked for 2 h at 25 °C to measure the electrical conductivity, R1. After this, the sample was boiled for 30 min, and once the temperature dropped to room temperature, the conductivity of R2 was measured. The REC was used to represent the cell membrane permeability. The content of REC was calculated using the following formula: R1/R2 × 100% [38]. Each treatment had 12 replicates.

2.4.2. Determination of the Relative Content of Malondialdehyde (MDA)

The content of MDA was measured as described by Hodges et al. [39] with slight modifications. Briefly, 0.1 g of fresh leaves was ground in 2 mL of 5% (*w/v*) trichloroacetic acid (TCA) on ice. The extract was transferred to an Eppendorf centrifuge tube and then centrifuged at 8000 rpm for 15 min at 4 °C. A volume of 1 mL of supernatant was mixed with 1 mL of 0.6% (*w/v*) thiobarbituric acid (TBA) and heated in boiling water for 30 min. After cooling, the mixture was centrifuged at 6000 rpm for 10 min. The supernatant was extracted, and the values of absorbance were measured at three different wavelengths, 600, 532, and 450 nm, using a DU730 UV/Vis Spectrophotometer (Beckman Coulter, Inc., Brea, CA, USA). The concentration of MDA in the mixture (μmol/L) = 6.45 × (A532 − A600) − (0.56 × A450). Each treatment had 12 replicates.

2.5. *Detection of Apoptosis and the Integrity of Root Cell Membranes*

Four-week-old leaves of *I. cairica* were obtained and cut into strips of 0.74–1.25 mm, immersed in an enzymolysis solution, and digested for 5–6 h. The digestive solution that was obtained was passed through a 50 μm sieve, transferred to a 10 mL centrifuge tube, and centrifuged at 100 rpm for 5 min at 4 °C. After the supernatant was removed, the precipitate was washed with 0.5 M mannitol, purified by resuspension in 15% (*w*/(*v*) sucrose, and centrifuged at 178 rpm for 3 min at 4 °C to obtain the protoplasts. The enzyme solution contained the following compounds and enzymes: 2.0% (*w*/(*v*) cellulase R-10 (Yakult Honsha Co., Ltd., Tokyo, Japan), 0.1% (*w*/(*v*) macerozyme R-10 (Yakult Honsha Co., Ltd.), 0.5 M mannitol, 20 mM KCl, and 5 mM $CaCl_2$. The pH was adjusted to 5.5 with KOH. The protoplasts were then resuspended in modified KM8P liquid media [40]. The apoptosis of cells was detected using Annexin V-FITC/PI (Sangon Biotech [Shanghai] Co., Ltd., Shanghai, China) double-staining flow cytometry [41]. Each treatment had 12 replicates.

The membrane integrity in the *I. cairica* root cells was detected using a slight modification of a method described by Lequeux et al. [42]. The roots of both *I. cairica* lines in Guangzhou and Zhuhai were treated with 100 mmol/L NaCl for 12 h, stained with 0.25% Evans blue for 15 min, and then washed three times with ultrapure water before they were photographed. The surfaces of these roots were observed under a microscope (DM3000; Leica, Wetzlar, Germany) after they had been stained with Evans blue. The roots were immersed in 3 mL of pure N,N-dimethylformamide (N,N-dimethylformamide, DMF, analytical grade) (Sigma-Aldrich, St. Louis, MO, USA) for 30 min, and the extract was obtained by filtering the roots. The absorbance of the extract at 600 nm was measured by spectroscopy on a DU730 UV/Vis spectrophotometer (Beckman Coulter, Inc.). Each treatment had 12 replicates.

2.6. *Determination of the Anthocyanin Content and Antioxidant Enzyme Activities*

2.6.1. Assays of the Antioxidant Enzymes

The antioxidant enzymes were extracted as described by Khalid et al. [43], with brief modifications. The main veins of the fresh leaves were removed, and 0.1 g of the sample was weighed and placed in a cold mortar. A volume of 1 mL of grinding buffer (50 mM PBS [pH = 7.8], 0.1 M EDTA-2Na, 0.1% [*v*/*v*] Triton X-100, and 2% [*w/v*] polyvinylpyrrolidone [PVP]) and a small amount of quartz sand were added to the mortar, and the samples were ground to a homogenous paste. The homogenate was then transferred to a centrifuge tube, and the mortar was washed with 1 mL of grinding buffer. The homogenate was centrifuged for 15 min (12,000× *g*, 4 °C), and the supernatant was used for the subsequent enzyme assays. Each treatment had 12 replicates.

The activity of superoxide dismutase (SOD) was assayed as described by Liu et al. [44] and Khalid et al. [43], with minor modifications. A volume of 0.1 mL of the extract was pipetted into a 4 mL centrifuge tube that contained 2.6 mL of the mixture (0.3 mL of 130 mM Met, 1.7 mL of 50 mM PBS [pH = 7.8], 0.3 mL of 750 μmol L^{-1} nitroblue tetrazolium [NBT], and 0.3 mL of 100 μmol L^{-1} EDTA-2Na). Finally, 0.3 mL of 20 μM riboflavin was added. Two negative controls and one positive control were prepared. In the negative control, the enzyme extract was replaced with 50 mm of PBS (pH = 7.8), and the tube was placed in the dark. The positive control entailed the exposure of 50 mm of PBS (pH = 7.8), and the sample tubes were exposed to a 4000 lux fluorescent lamp for 20 min, followed by incubation in the dark. The absorbance of the reaction mixture at 560 nm was measured by spectrophotometry and set to zero by the negative control. A unit of SOD was defined as 50% inhibition of the blue-light chemical reduction in tetrazole min^{-1} g^{-1} of the leaf. Each treatment had 12 replicates.

The activity of peroxidase (POD) was assayed using the guaiacol method [45]. Briefly, 0.1 mL of enzyme and 2.9 mL of reaction mixture (1.875 mL of 50 mM PBS [pH = 7.0], 1 mL of 30 mM H_2O_2, and 0.025 mL of guaiacol) were pipetted into a cuvette, and 50 mM PBS (pH = 7.0) was used as the blank control. Measurements were taken 15 s after the reaction and then every 20 s for a total of nine measurements. Each treatment had 12 replicates.

The catalase (CAT) was assayed as described by Ghalati et al. [46]. The reaction mixture (3 mL) contained 2.9 mL of 30 mM H_2O_2 (Guangzhou Chemical Reagent Factory, Guangzhou, China) and 0.1 mL of enzyme extract. The decrease in absorbance was recorded at 240 nm. One unit of enzyme activity was defined as 1 nmol H_2O_2 dissociated·min^{-1}. Each treatment had 12 replicates.

2.6.2. Determination of the Anthocyanin Content

The anthocyanin content was determined by washing fresh leaves of *I. cairica* with ultrapure water and blotting them with filter paper. A total of 0.1 g of leaves were extracted in 10 mL tubes that contained 4 mL of 1% HCl-methanol at 4 °C for 24 h. The supernatant was then centrifuged at 4500 rpm for 5 min and collected for testing. The absorbance of the supernatant at 530 nm was measured using a DU730 UV/Vis Spectrophotometer

(Beckman Coulter, Inc.). The content of anthocyanin was calculated using a standard curve of cornflower-3-*O*-glucoside [47]. Each treatment had 12 replicates.

2.7. Determination of the Osmotic Adjustment Indicators

2.7.1. Determination of the Content of Proline

The content of Pro was measured as described by Shi et al. [48]. A total of 0.1 g of sample was added to 3 mL of 3% sulfosalicylic acid, incubated for 10 min in a boiling water bath, and then cooled and centrifuged at 6000× *g* for 10 min. A volume of 1 mL of the supernatant was then collected and mixed with 4 mL of extraction buffer (distilled water/glacial acetic acid/acidic ninhydrin = 1:1:2). The mixture was boiled for 1 h and cooled at room temperature. After that, 2 mL of toluene was added, and the phases were left to separate for 2 h. After full extraction, the red toluene phase was subjected to a colorimetric analysis at 520 nm. The content of Pro was calculated from a standard curve. Each treatment had 12 replicates.

2.7.2. Measurement of Stomatal Density

The stomatal density was determined as described by Tao et al. [33]. Clear nail varnish was applied to the abaxial surface of the leaves. The dried layer of the nail varnish was then peeled off using tweezers and placed on a glass slide. The imprints were subsequently observed under an optical microscope with a 10× objective. A certain number of pictures of the field of view were taken, and the number of all the stomata in the field of view was counted using ImageJ. This program was used to calculate the stomatal density (the number of stomata per unit of leaf area in the field of view) with the number of stomata/the area of the field of view. The sample size for each treatment and line was 150 (three epidermal samples per leaf × five fields of view per sample × ten biological replicates). Each treatment had 10 replicates.

2.8. Measurement of the Content of Chlorophyll and Parameters of Chlorophyll Fluorescence

The contents of chlorophyll a (Chl a), chlorophyll b (Chl b), and total chlorophyll were determined as described by Wang and Huang [49]. Approximately 0.1 g of fresh leaves were placed in a 2 mL centrifuge tube and crushed into powder with a small amount of liquid nitrogen. A volume of 2 mL of 80% acetone was then added to the centrifuge tube and extracted for 24 h at 4 °C in the dark. The absorbance of the extracts was measured at 663 nm, 645 nm, and 470 nm using a DU730 UV/Vis spectrophotometer, Beckman Coulter, Inc.), and the concentration and content of each pigment were calculated as described by Wang and Huang [49]. Each treatment had 12 replicates.

After 30 d of treatment, the chlorophyll fluorescence parameters were measured using a portable modulating chlorophyll fluorescence instrument, a PAM-2500 fluorometer (Heinz Walz GmbH, Effeltrich, Germany), as described by Chen et al. [28]. The third and fourth pairs of leaves in the upper part of the plant were acclimated in the dark for at least 20 min before the chlorophyll fluorescence was assessed. The initial fluorescence (F0) was measured in the dark (light intensity < 0.1 μmol m^{-2} s^{-1}). Under saturating light intensity (200 μmol m^{-2} s^{-1}), the stable fluorescence (Fs) and maximum fluorescence (Fm) were measured. The parameters included the maximum efficiency of PSII photochemistry (Fv/Fm), effective quantum yield of PSII (Y[II]), photochemical quenching coefficient (qL), electron transport rate (ETR), and non-photochemical chlorophyll fluorescence quenching (NPQ), which were calculated and processed as described by Baker and Rosenqvis [19]. Each treatment had 12 replicates.

2.9. Statistical Analysis

The data were analyzed using a one-way analysis of variance (ANOVA) or two-way ANOVA. The subsequent multiple comparisons were examined based on the least significant difference (LSD) test, $p < 0.05$ was considered the level of significance. The data were analyzed using SPSS 21.0 (IBM, Inc., Armonk, NY, USA), and all the values were expressed as the mean ± standard error (SE). The figures were prepared using Origin 2019 (OriginLab, Northampton, MA, USA).

3. Results

3.1. Growth of Both Lines under Salt Stress

With the increase in the concentration of salt, the lengths of the roots of the terrestrial ecotype decreased significantly (Figure 1A,B). To track the growth of both ecotypes under salt stress, the seedlings were treated with salt after transplantation, and the stem length, number of leaves, number of branches, and number of flowers of each plant were measured (Figure 1C–F). After 10 d, the stems of the coastal ecotype under the salt treatment were significantly shorter than those in the control. In the control, the stems of the terrestrial ecotype gradually grew longer. At 50 d, the main stem length from longest to shortest was the terrestrial ecotype of the control, the coastal ecotype of the control, the coastal ecotype under the NaCl treatment, and the terrestrial ecotype under the NaCl treatment. Among them, there was a significant difference between the terrestrial ecotype and the coastal ecotype in the control (Figure 1C). After 20 d, there were significantly more leaves in the control than in the salt treatment, but there was no significant difference between the terrestrial ecotype and the coastal ecotype in both the control and salt treatments (Figure 1D). At 20 d, the coastal ecotype in the control flowered first. At the end of the experiment, there were significantly more flowers in the control of the coastal ecotypes than in the salt treatment. In contrast, there was no difference in the number of flowers between the control and salt treatment in the terrestrial ecotype (Figure 1F). There were significantly more branches in the terrestrial ecotype under the salt treatment than in the coastal ecotype. However, there was no significant difference between these two ecotypes in the control (Figure 1E). The results of this analysis indicated that the coastal ecotype and terrestrial ecotype may have adapted to their respective habitats. The terrestrial ecotype is more adapted to the non-salt environment, while the coastal ecotype is more adapted to the saline environment.

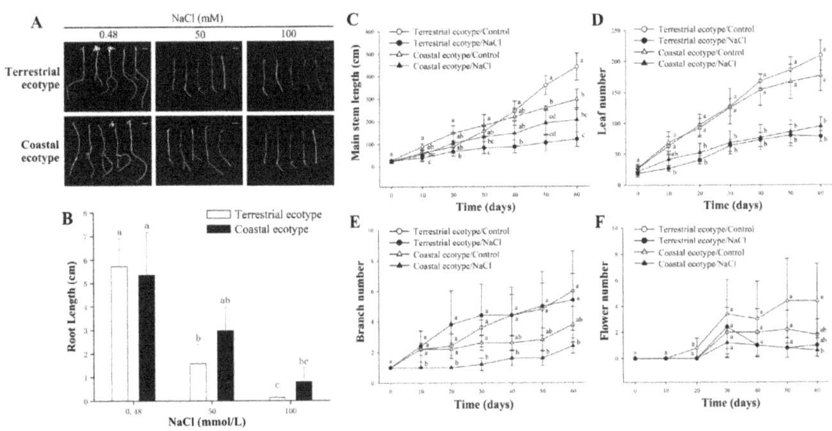

Figure 1. The growth of different ecotypes of *Ipomoea cairica* under salt stress that shows the phenotypic adaptability of the coastal ecotype to salt stress. (**A**,**B**) Root growth after 2 weeks of hydroponics, including (**A**) the morphology of the root (Bar = 1 cm) and (**B**) root length. (**C**–**F**) The growth trend of the two ecotypes within 60 d after transplantation, including (**C**) the main stem length. (**D**) Leaf number during the salt treatment. (**E**) Branch number. (**F**) Flower number. All the data represent the mean ± SE. The different letters in the bar graphs indicate significant differences at $p < 0.05$. SE, standard error.

The elevated concentration of NaCl significantly inhibited the growth of the terrestrial ecotype and coastal ecotype (Figure 2A–D). The coastal ecotype grew better than the terrestrial ecotype in both the control and treatment groups. Compared with the control, the root FW was significantly reduced in both ecotypes under the salt treatment (Figure 2E). In addition, the elevated concentration of NaCl had a more substantial impact on the aboveground biomass and total biomass of the terrestrial ecotype, with rates of inhibition of 43% and 45%, respectively. In contrast, the rates of inhibition of the aboveground biomass and total biomass of the coastal ecotype due to the elevated concentration of NaCl were 31% and 37%, respectively (Figure 2F,G). The coastal ecotype had a significantly higher root/shoot ratio than the terrestrial ecotype after exposure to salt stress. In contrast, there were no detectable differences in the control groups (Figure 2H).

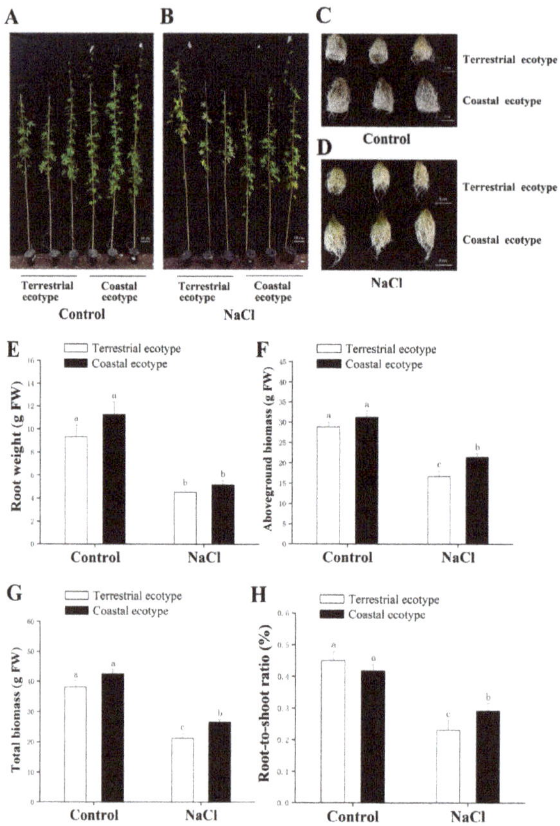

Figure 2. Effect of 150 mmol/L NaCl on the growth of terrestrial ecotype and coastal ecotype of *Ipomoea cairica* (**A–D**) after 60 d. (**E**) Fresh weight of the roots. (**F**) Fresh weight of the aboveground biomass. (**G**) Fresh weight of the total biomass. (**H**) Root/shoot ratio. All the data represent the mean ± SE. Different letters indicate a significant difference at $p < 0.05$. SE, standard error.

3.2. Cell Damage of Both Ecotypes under Salt Stress

Apoptosis of the leaf protoplasts was induced under salt stress in both ecotypes. The terrestrial ecotype entered the apoptosis stage earlier than the coastal ecotype (Figure 3A,B). Evans blue staining was used to explore the effect of salt stress on the membrane integrity of the root cells. After 12 h of treatment, the terrestrial ecotype displayed a darker blue stain on the root tip than the coastal ecotype, which suggested that there was more severe cell damage. The values of absorbance of the decolorizing solution at 600 nm indicated that the terrestrial ecotype suffered greater damage to the cell membrane than the coastal ecotype under salt stress (Figure 3C,D). The contents of MDA and REC of the terrestrial ecotype were higher than those of the coastal ecotype under both the control and salt treatments. Furthermore, the contents of MDA and REC of the terrestrial ecotype were significantly higher than those of the coastal ecotype in the salt treatment (Figure 3E,F). The leaf relative water content was significantly lower in both ecotypes after the salt treatment compared with that of the control (Figure 3G).

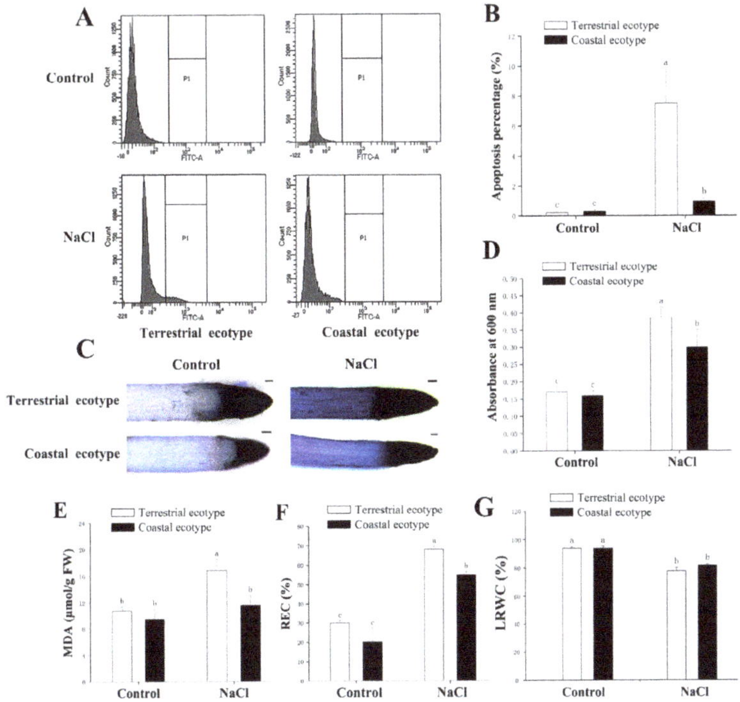

Figure 3. The terrestrial ecotype suffered more severe cellular damage under salt stress than the coastal ecotype. (**A**,**B**) Apoptosis of the protoplasts under the salt treatment detected by Annexin V-FITC/PI double staining flow cytometry. (**A**) Flow cytometric estimation of apoptosis. (**B**) Representative flow cytometry histograms of the apoptosis. (**C**,**D**) Root cell membrane integrity detected by Evans blue staining. A darker stain, lower integrity. (**C**) Dead cells that were stained blue. (**D**) The degree of damage to the cell membrane. (**E**) Malondialdehyde (MDA). (**F**) Relative electrical conductivity (REC). (**G**) Leaf relative water content. Bar = 1 μm. All the data represent the mean ± SE. The different letters in the bar graphs indicate significant differences at $p < 0.05$. SE, standard error.

3.3. Antioxidant Capacity of *I. cairica* in Response to Salt Stress

Under the control treatment, the coastal ecotype appeared to accumulate significantly more anthocyanins than the terrestrial ecotype. The coastal ecotype had a concentration of anthocyanins that was approximately twice as high. However, there was no significant difference between the two ecotypes under the salt treatment since the anthocyanin content of the terrestrial ecotype increased significantly compared with its control (Figure 4A). The SOD activity of the coastal ecotype under both the control and salt treatment was significantly higher than that of the terrestrial ecotype. However, the activity of both ecotypes increased under the salt treatment (Figure 4B). The POD activity of the coastal ecotype in the control group was significantly higher than that of the terrestrial ecotype, and the activity in the salt treatment group increased in both ecotypes (Figure 4C). The CAT activity increased in the salt treatment. The activity of the coastal ecotype was higher in the control and salt treatments than in the terrestrial ecotype, but the difference was not significant (Figure 4D).

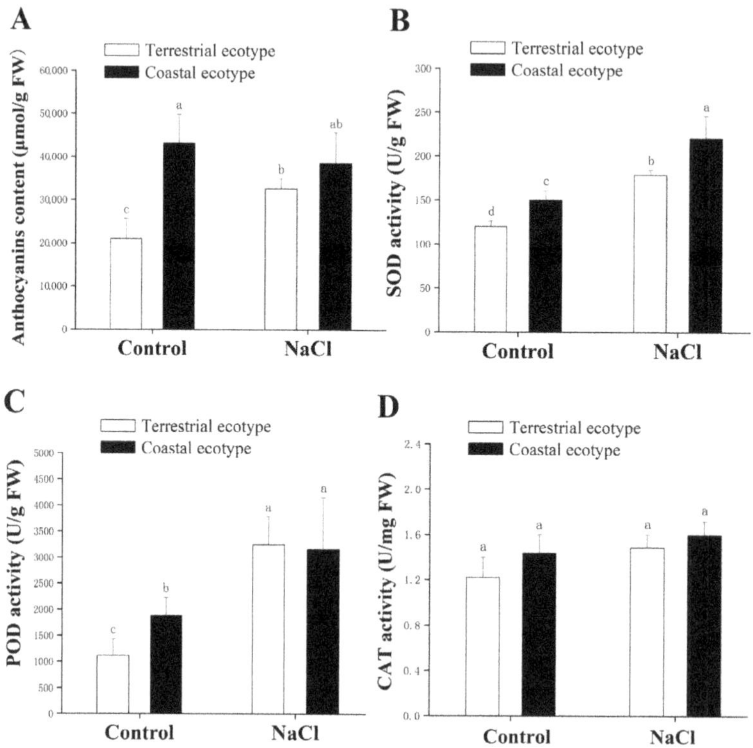

Figure 4. Physiological changes induced by salt in *Ipomoea cairica* leaves. (**A**) Anthocyanin content. (**B**) Superoxide dismutase (SOD). (**C**) Peroxidase (POD). (**D**) Catalase (CAT). All the data represent the mean ± SE. The different letters in the bar graphs indicate significant differences at $p < 0.05$. SE, standard error.

3.4. Salt Rejection and Transport

The appearance of the salt glands of the two ecotypes was examined in detail, and their associated metrics were determined (Figure 5A–D). There was no significant difference in the area of the salt glands between the two ecotypes in either the control or treatments. However, the areas of the salt glands in both ecotypes increased under the salt treatments and were approximately 1.13- and 1.14-fold that of the control, respectively (Figure 5E). The number and density of the salt glands increased in both ecotypes compared with the control. There was a significant difference in these values between the two ecotypes under the salt treatment, with the coastal ecotype being approximately 1.36- and 1.4-fold higher than that of the terrestrial ecotype, respectively. Furthermore, in the control, there was no significant difference in the number and density of salt glands between the two ecotypes, but the values of the coastal ecotype were higher than those of the terrestrial ecotype (Figure 5F,G).

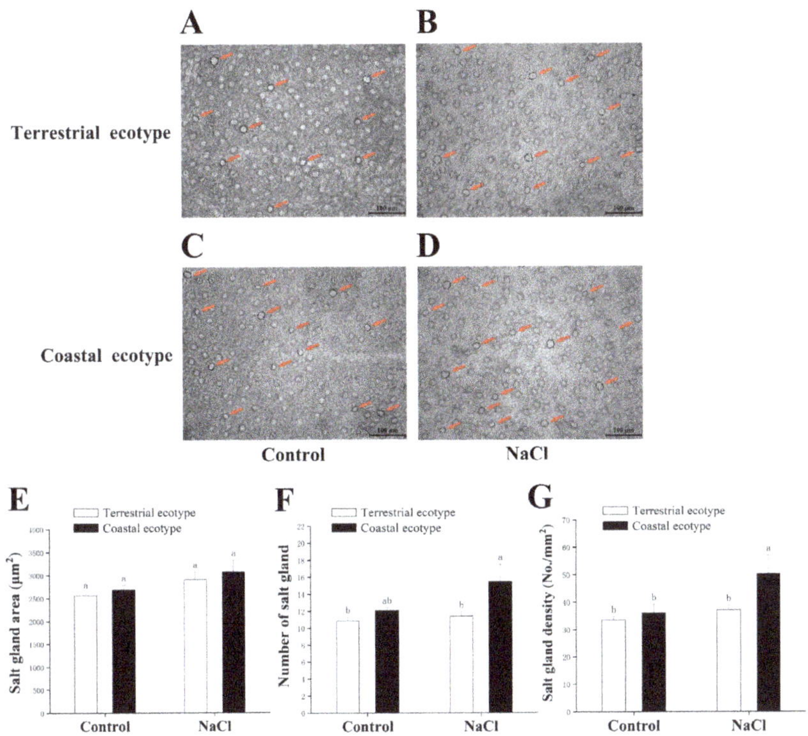

Figure 5. Effect of NaCl concentrations on the salt glands of the two ecotypes after 60 d. (**A–D**) Salt glands, red arrows. (**E**) Salt gland area. (**F**) Number of salt glands. (**G**) Salt gland density. All the data represent the mean ± SE. The different letters in the bar graphs indicate significant differences at $p < 0.05$. SE, standard error.

Under non-saline conditions, significantly less Na^+ accumulated in the roots, stems, and leaves of the coastal ecotype compared with the terrestrial ecotype. When plants were exposed to salinity, excess Na^+ was taken up by the roots and transported upwards to the stems and leaves, which significantly increased the concentration of Na^+ in the roots, stems, and leaves in both ecotypes. Compared to the control, the K^+ contents in the roots, stems, and leaves were significantly reduced in both ecotypes, and the trend was the opposite for Na^+. In the control, the K^+ contents in the leaves and the K^+: Na^+ ratio in the roots, stems, and leaves of the coastal ecotype were significantly higher than those of the terrestrial ecotype. The K^+: Na^+ ratios of the roots, stems, and leaves of the two ecotypes decreased significantly after the salt treatment, but those of the stems and leaves of the coastal ecotype were marginally higher than those of the terrestrial ecotype. Under salt stress, there was a slightly higher content of K^+ in the leaves of the coastal ecotype, while there were lower contents of Na^+ in the stems and leaves (Figure 6A–C).

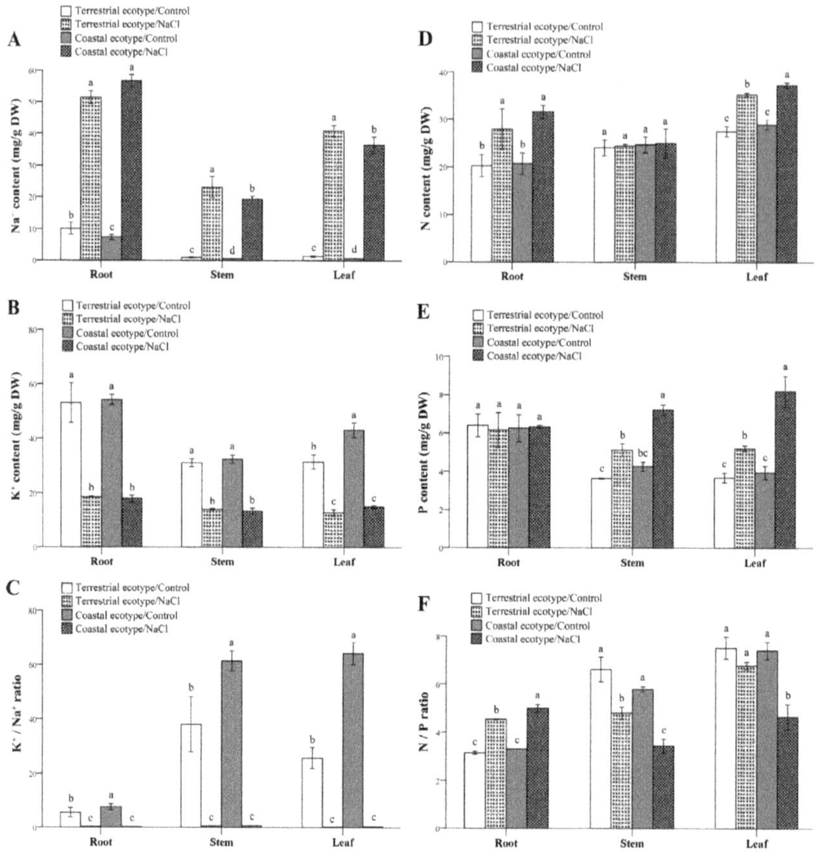

Figure 6. The contents of elements and their allocations in the roots, stems, and leaves of the coastal ecotype were more balanced and conducive to growth than those of the terrestrial ecotype under salt stress. (**A**) Content of sodium ion (Na^+). (**B**) Content of potassium ion (K^+). (**C**) K^+: Na^+ ratio. (**D**) Content of nitrogen (N). (**E**) Content of phosphorus (P). (**F**) N: P ratio. All the data represent the mean ± SE. The different letters in the bar graphs indicate significant differences at $p < 0.05$. SE, standard error.

Salt stress also enhanced the accumulation of N and P in *I. cairica*. Compared with the control, the concentrations of N in the roots and leaves were significantly increased in both ecotypes under the salt treatment. There were significantly higher concentrations of N in the leaves of the coastal ecotype than in the terrestrial ecotype when exposed to salt. The concentrations of P in the stems and leaves were significantly increased in both ecotypes under the salt treatment when compared with the control. And the concentrations of P in the stems and leaves of the coastal ecotype were significantly higher than those of the terrestrial ecotype, respectively. After the salt treatment, the N: P ratio in the stems and leaves of both ecotypes decreased slightly, while that in the roots increased significantly (Figure 6D–F).

3.5. Osmotic Regulation

The appearance of the stomata of the two ecotypes was studied in detail, and their relevant indices were determined (Figure 7A–D). In both the control and salt treatments, the number of stomata and stomatal density were higher in the terrestrial ecotype than in the coastal ecotype, and a significant difference was observed in the salt treatment group

(Figure 7E,F). The content of Pro in the coastal ecotype was higher than that of the terrestrial ecotype under both the control and salt treatments, while the content in the coastal ecotype was significantly higher than that of the terrestrial ecotype in the salt treatment (Figure 7G).

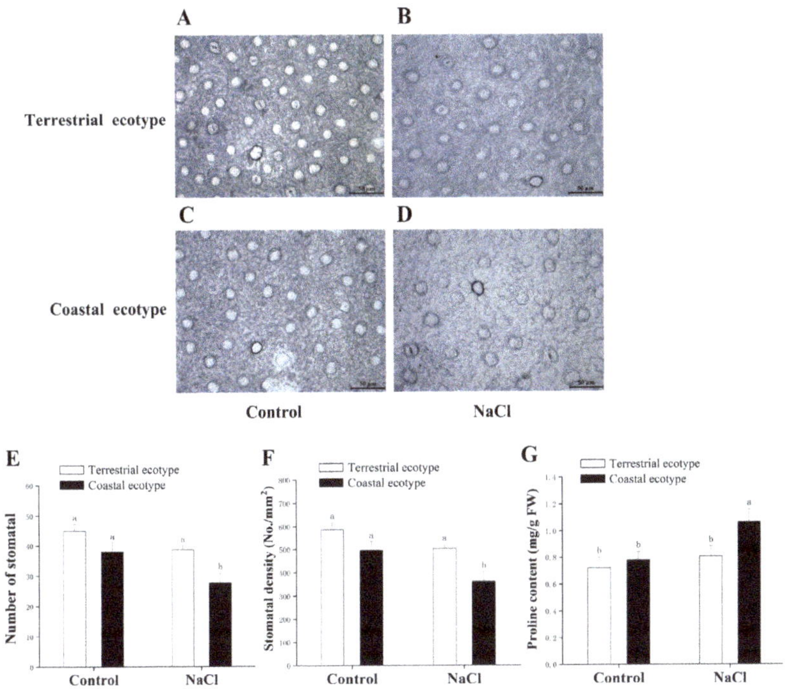

Figure 7. Effect of NaCl concentrations on the stomata and proline content of the two ecotypes after 60 d. (**A–D**) Stomata. (**E**) Number of stomata. (**F**) Stomatal density. (**G**) Proline. All the data represent the mean ± SE. The different letters in the bar graphs indicate significant differences at $p < 0.05$. SE, standard error.

3.6. Photosynthetic Performance

Compared with the control, the content of total Chl decreased by 31% in the terrestrial ecotype and by 27% in the coastal ecotype; the content of Chl a decreased by 33% in the terrestrial ecotype and by 26% in the coastal ecotype, and that of Chl b decreased by 27% in the terrestrial ecotype and by 29% in the coastal ecotype under the salt treatment, respectively (Figure 8A–C). The Chl a/Chl b ratios of both ecotypes were not significantly different under either the control treatment or salt treatment (Figure 8D). As shown in Figure 8E,G, the Fv/Fm and qL of both ecotypes remained stable under the two treatments. Compared with the control, the Y[II] of the terrestrial ecotype and the coastal ecotype decreased by 35.54% and 29.03%, respectively, and the ETR decreased by 35.31% and 28.97%, respectively, under the salt treatment (Figure 8F,I). Compared to the control, the NPQ of both ecotypes increased under the salt treatment (Figure 8H).

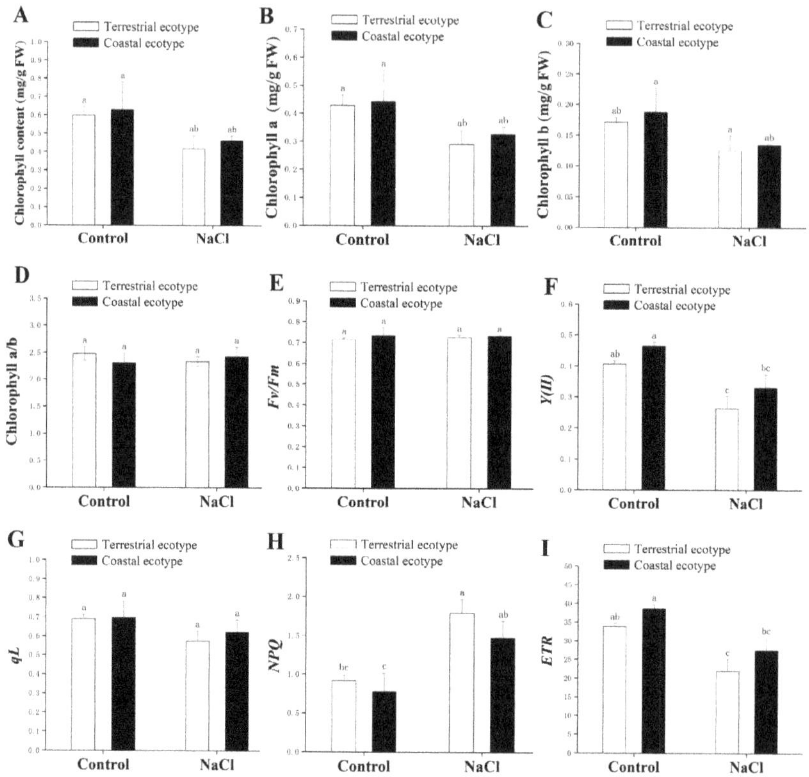

Figure 8. The chlorophyll content and chlorophyll fluorescence parameters of the two ecotypes under salt stress. (**A**) Total chlorophyll (chl a + b). (**B**) Chlorophyll a. (**C**) Chlorophyll b. (**D**) Chlorophyll a/b. (**E**) Maximal efficiency of PSII photochemistry. (**F**) Actual quantum yield of PSII. (**G**) Photochemical quenching coefficient. (**H**) Non-photochemical quenching coefficient. (**I**) Electron transport rate. PSII, photosystem II. All the data represent the mean ± SE. The different letters in the bar graphs indicate significant differences at $p < 0.05$. SE, standard error.

4. Discussion

In most cases, alien species tend to invade and colonize habitats that resemble their native environments [50]. Interestingly, as a species that is generally considered to be a terrestrial ecotype, *I. cairica* was found to invade the saline environment. In this study, by comparing the responses of the coastal ecotype and the terrestrial ecotype of *I. cairica* to salt stress, the coastal ecotype of *I. cairica* was shown to have obtained salt tolerance. To our knowledge, this is the first observation of salt tolerance in this species.

4.1. Morphological Changes in the Two Lines under Salt Stress

Salt tolerance in plants is related to various aspects of plant physiology, biochemistry, morphology, and molecular mechanisms, which are all related to osmosis, photosynthesis, and nutrient imbalances that limit plant growth through salt stress and reduce the biomass of all parts of the plant [51,52]. These results indicate that the coastal ecotype is more effective at growing in saline habitats compared with the terrestrial ecotype. The stems of the coastal ecotype appeared to regenerate more strongly, which suggested that they were more effective at reproducing asexually; this may play a key role in its invasion in saline habitats [53]. The decrease in the number of branches and the advance of the flowering period indicate that the coastal ecotype allocates more energy to stem growth and sexual reproduction for both escape and adaptation [54,55]. Under the salt treatment, the biomass of the terrestrial ecotype was reduced more than that of the coastal ecotype, which indicated that the coastal ecotype was better adapted to salt stress. This result was supported by previous similar studies on different plants [56,57].

4.2. Osmotic Adjustment Responses to Salt Stress

The accumulation of proline is an important adaptive strategy under salt stress that contributes to the increase in salt tolerance [58]. Sweet potato (*Ipomoea batatas* (L.) Lam) plants that accumulated more Pro significantly improved their degree of salt tolerance [59]. An increase in the accumulation of Pro has been associated with the enhancement of salt tolerance in Arabidopsis [60]. In this study, the coastal ecotype had significantly higher contents of Pro under salt stress. The accumulation of more Pro in the coastal ecotype treated with salt may have maintained the osmotic balance between the intracellular and extracellular environments and protected the membrane integrity, which enhanced the degree of salt tolerance. There have also been reports of similar results in several other studies [59,61]. Both the coastal ecotype and terrestrial ecotype improved their water use efficiency by reducing the number and density of stomata under salt stress [33]. The coastal ecotype reduced the density of stomata, decreased their number, reduced water loss, and maintained a higher leaf water content, thereby increasing their tolerance to salt and preserving the water potential and water status in the cells under osmotic stress conditions [62,63]. This is also supported by findings in wheat (*Triticum aestivum* L.) [33] and cogon grass (*Imperata cylindrica* (L.) Raeuschel) [64]. These observations emphasize the importance of reducing the stomatal density as a strategy to minimize the stress of saline conditions. Therefore, it is hypothesized that the coastal ecotype, through the reduction in the stomatal density and the accumulation of more proline, may maintain the osmotic balance between the intracellular and extracellular environments, thereby increasing its tolerance to salt.

4.3. Salt Exclusion and Nutrient Management

Salt glands play a pivotal role in the secretion of salt in recretohalophytes that secrete excess Na^+ to prevent damage from salt [65]. The coastal ecotype appeared to have more salt glands, a higher density, and a larger area of these glands, which enabled the coastal ecotype to secrete excess salt through its salt glands to adapt to saline environments [34,35,66].

Salt stress disturbs the ionic balance of cells and activates the ion channels in the cell membranes, which causes an efflux of K^+ [67]. Therefore, maintaining high levels of K+ and ensuring that the Na^+ in the cytoplasm remains low is a common strategy to facilitate the proper function of enzymes and metabolic activities [68]. The coastal ecotype line exhibits its advantages in terms of a balance of the elements by maintaining a high K^+: Na^+ ratio, which is necessary to maintain enzymatic activity, and is even more important than maintaining low levels of Na^+ [69,70]. With a higher K^+: Na^+ ratio than the terrestrial ecotype, the coastal ecotype obtained a more balanced content of ions and more stable growth [71,72]. Moreover, the coastal ecotype transports large amounts of salt ions into its stems and leaves, thereby reducing the negative impacts of excessive ions on the root system [73]. These findings have led to suggestions to improve salt tolerance by removing Na^+ from the stems and leaves, providing K^+ to the stems, and retaining K^+ in the leaf tissues to maintain an optimal K^+: Na^+ ratio in the stems and leaves.

The accumulation of N, P, and K^+ in the leaves promotes the growth and development of plants [74], while the maintenance of the K^+: Na^+ ratio and N: P ratio in the leaves, the selective transport of K^+, N, and P from the roots to the leaves, and the accumulation of Pro in the leaves helps to reduce the toxic effects of Na^+ and the damage caused by osmotic stress [75,76]. After acclimating to the salt environment, the coastal ecotype allocated more nutrient resources to the leaves for morphogenesis and, ultimately, the enhancement of photosynthesis [77]. P is also related to the structural composition and protein activity of plants [78]. The accumulation of P may improve the integrity of the leaf cells in the coastal ecotype, thereby accelerating the biosynthesis of photosynthetic products.

4.4. Salt Stress Response to the Cell Damage and Antioxidant Capacity

The antioxidant system, which has developed during the evolution of plants, contributes to the homeostasis of reactive oxygen species (ROS) metabolism, which directly reflects the salt tolerance of plants [79]. Under the salt treatment, the lower degree of cell damage of the coastal ecotype was associated with a higher antioxidant capacity. Anthocyanins are natural pigments that protect antioxidant enzymes, scavenge free radicals, regulate cellular osmotic pressure, and stabilize the photosynthetic structure, thus limiting the inhibition of plant growth [80,81]. Antioxidant enzymes are other tools that plants can use to scavenge reactive oxygen species and resist oxidative damage, and they include SOD, POD, and CAT [82]. Increased antioxidant defenses have been positively correlated with a reduction in the oxidative damage in plants under abiotic stress [83]. In this study, both the coastal and terrestrial ecotypes enhanced their antioxidant capacities by increasing their contents of anthocyanin and activities of POD, SOD, and CAT under salt stress. In addition, salt-tolerant plants showed a significant increase in the molecular regulation of the POD and SOD activities under the salt treatment to maintain the balance of reactive oxygen metabolism, which may be related to the high level of expression of *IcSRO1* [11]. In addition, the coastal ecotype had a higher content of anthocyanin and activities of POD, SOD, and CAT under the salt treatment, enhanced ROS scavenging and osmoregulation [84], and a reduction in the levels of REC and MDA, which thereby reduced the contents of MDA and oxidative stress; this led to lipid peroxidation and cellular damage [80]. Consequently, cells of the coastal ecotype entered the apoptotic stage later than those of the terrestrial ecotype, which indicates that the coastal ecotype is better at maintaining photosynthesis and cell homeostasis in a saline environment [85,86].

4.5. Effects of Salt Stress on the Chlorophyll Content and Fluorescence Properties

Chlorophyll fluorescence parameters are often used to examine the relationship between the photosynthesis of plants and the environment in which they live and are important indicators that reflect the physiology of photosynthesis [87]. The Fv/Fm indicates the level of photoinhibition [88]. In this study, the Fv/Fm remained stable across both ecotypes under the two treatments, which indicated that there was little photoinhibition; this suggests that both ecotypes can adapt photosynthetically to salt stress [89,90]. A more substantial increase in NPQ in the terrestrial ecotype was observed under the salt treatment, which protected the photosynthetic apparatus from damage by dissipating excess light energy in the form of heat through non-photochemical processes [90]. Salt stress may disrupt the biochemistry of photosynthesis, which decreases the efficiency of Photosystem I (PSI) and PSII owing to disturbances in the integrity of chloroplasts [91]. Therefore, it can be suggested that the coastal ecotype is more tolerant to salinity stress owing to its ability to maintain the contents of chlorophyll and its fluorescence parameters.

5. Conclusions

This study investigated the growth and physiological and biochemical responses of two lines against salt stress. The results indicated that salt stress is reflected in the obvious impairment of photosynthetic performance and the inhibition of Y(II), NPQ, ETR, and chlorophyll biosynthesis. An increase in the concentration of MDA and REC highlighted the lipid peroxidation and oxidative damage caused to the cell membranes, respectively. However, the adverse effects of salt stress could be mitigated by various internal mechanisms in the cell, such as the accumulation of antioxidant enzymes (SOD, POD, and CAT) and osmolytes (Pro and anthocyanins) (Figure 9). The coastal ecotype of *I. cairica* in Zhuhai and the terrestrial ecotype in Guangzhou appeared to have different physiological responses to salt stress. The coastal ecotype was considerably less inhibited, which demonstrates that it rapidly adjusted to increase its adaptability to salt and enhance its invasiveness in mangrove wetland ecosystems. In addition, this adaptability might be heritable, and it will be investigated further in our subsequent research study.

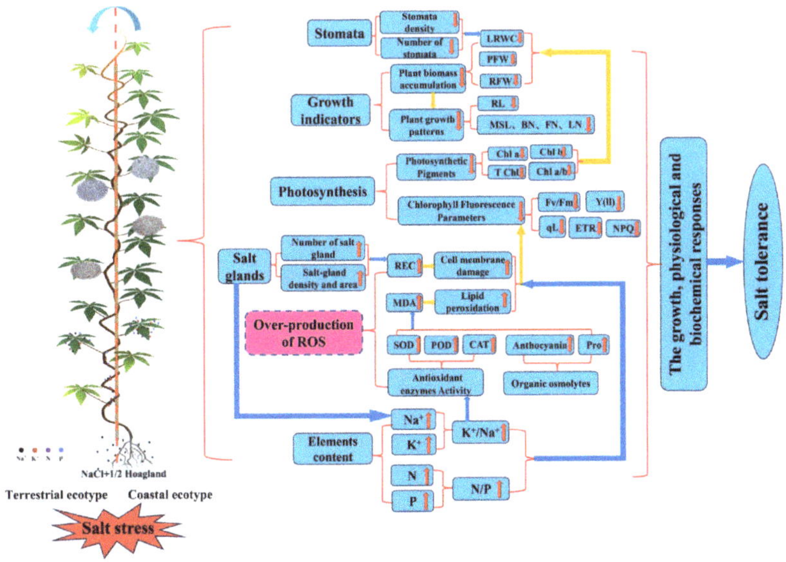

Figure 9. Schematic representation of the inhibition of growth induced by salinity. LRWC, leaf relative water content; PFW, plant fresh weight; RFW, fresh weight of roots; RL, root length; MSL, main stem length; BN, branch number; FN, flower number; LN, leaf number; Chl a, chlorophyll a content; Chl b, chlorophyll b content; TChl, total chlorophyll content; Chl a/b, chlorophyll a/b; Fv/Fm, maximal efficiency of PSII photochemistry; qL, photochemical quenching coefficient; Y(ll), actual quantum yield of PSII; NPQ, non-photochemical quenching coefficient; ETR, electron transport rate; MDA, malondialdehyde; REC, relative electrical conductivity; Pro, proline; SOD, superoxide dismutase; POD, peroxidase. PSII, Photosystem II; Na^+, sodium ion; K^+, potassium ion; K^+/Na^+ sodium and potassium ion ratio; N, nitrogen; P, phosphorus; N/P, nitrogen and phosphorus ratio.

Author Contributions: T.X. and M.C. conceived and designed the research; J.Z. and B.Y. conducted the experiments; T.X., W.L. and X.X. supervised the project administration.; J.Z. and M.C. wrote the manuscript; T.X. and J.Z. revised the manuscript. All authors have read and agreed to the published version of the manuscript.

Funding: This research was funded by the National Natural Science Foundation of China (grant no. 32071618) to Tiantian Xiong, and the Natural Science Foundation of Guangdong Province (grant no. 2023A1515012608) to Tiantian Xiong.

Data Availability Statement: The data that support the findings of this study are available from the corresponding author upon reasonable request.

Acknowledgments: We are grateful for every reviewer's helpful comments. In addition, we thank Paul Giller from the School of Biological, Earth and Environmental Sciences, University College of Cork, Ireland for carrying out English language editing.

Conflicts of Interest: The authors declare that they have no known competing financial interests or personal relationships that could have appeared to influence the work reported in this paper.

References

1. Ding, H.; Xu, H.G.; Wu, J. Status and trends of invasive alien species in China. *World Environ.* **2009**, *3*, 39–40.
2. Gu, Z.Y.; Wu, X.H.; Yang, G.; Chen, J.D. Status and Countermeasures against Invasion of Exotic Organisms in China. *Jiangsu Agric. Sci.* **2006**, *6*, 418–421.

3. Bellard, C.; Cassey, P.; Blackburn, T.M. Alien species as a driver of recent extinctions. *Biol. Lett.* **2016**, *12*, 20150623. [CrossRef] [PubMed]
4. Slingsby, J.A.; Merow, C.; Aiello-Lammens, M.; Allsopp, N.; Hall, S.; Kilroy Mollmann, H.; Turner, R.; Wilson, A.M.; Silander, J.A. Intensifying postfire weather and biological invasion drive species loss in a Mediterranean-type biodiversity hotspot. *Proc. Natl. Acad. Sci. USA* **2017**, *114*, 4697–4702. [CrossRef] [PubMed]
5. Saenger, P.; Ragavan, P.; Sheue, C.R.; López-Portillo, J.; Yong, J.W.H.; Mageswaran, T. Mangrove biogeography of the Indo-Pacific. *Sabkha Ecosyst.* **2019**, *43*, 379–400.
6. Wang, C.W.; Wong, S.L.; Liao, T.S.; Weng, J.H.; Chen, M.N.; Huang, M.Y.; Chen, C.I. Photosynthesis in response to salinity and submergence in two Rhizophoraceae mangroves adapted to different tidal elevations. *Tree Physiol.* **2022**, *42*, 1016–1028. [CrossRef]
7. Ren, H.; Guo, Q.; Liu, H.; Li, J.; Zhang, Q.; Xu, H.; Xu, F. Patterns of alien plant invasion across coastal bay areas in southern China. *J. Coast. Res.* **2014**, *295*, 448–455. [CrossRef]
8. Zhang, Z.; Li, J.; Li, Y.; Liu, W.W.; Chen, Y.X.; Zhang, Y.H.; Li, Y.F. Spatially discontinuous relationships between salt marsh invasion and mangrove forest fragmentation. *For. Ecol. Manag.* **2021**, *499*, 0378–1127. [CrossRef]
9. Liu, G.; Huang, Q.Q.; Lin, Z.G.; Huang, F.F.; Liao, H.X.; Peng, S.L. High tolerance to salinity and herbivory stresses may explain the expansion of *Ipomoea cairica* to salt marshes. *PLoS ONE* **2012**, *7*, e48829. [CrossRef] [PubMed]
10. Bai, F.; Chisholm, R.; Sang, W.; Dong, M. Spatial risk assessment of alien invasive plants in China. *Environ. Sci. Technol.* **2013**, *47*, 7624–7632. [CrossRef] [PubMed]
11. Yuan, B.; Chen, M.; Li, S. Isolation and Identification of *Ipomoea cairica* (L.) Sweet Gene IcSRO1 Encoding a SIMILAR TO RCD-ONE Protein, Which Improves Salt and Drought Tolerance in Transgenic Arabidopsis. *Int. J. Mol. Sci.* **2020**, *21*, 1017. [CrossRef] [PubMed]
12. Morimaru, K.; Mitsutoshi, T.; Yasuo, I.; Kazutoshi, K.; Toshiyuki, O.; Nobuhide, F. High salinity leads to accumulation of soil organic carbon in mangrove soil. *Chemosphere* **2017**, *177*, 51–55. [CrossRef]
13. Polash, M.A.S.; Sakil, M.A.; Hossain, M.A. Plants responses and their physiological and biochemical defense mechanisms against salinity: A review. *Trop. Plant Res.* **2019**, *6*, 250–274. [CrossRef]
14. Wei, L.; Zhang, J.; Wei, S.; Hu, D.; Liu, Y.; Feng, L.; Li, C.; Qi, N.; Wang, C.; Liao, W. Nitric Oxide Enhanced Salt Stress Tolerance in Tomato Seedlings, Involving Phytohormone Equilibrium and Photosynthesis. *Int. J. Mol. Sci.* **2022**, *23*, 4539. [CrossRef]
15. Hao, S.; Wang, Y.; Yan, Y.; Liu, Y.; Wang, J.; Chen, S. A Review on Plant Responses to Salt Stress and Their Mechanisms of Salt Resistance. *Horticulturae* **2021**, *7*, 132. [CrossRef]
16. Allakhverdiev, S.I. Optimising photosynthesis for environmental fitness. *Funct Plant Biol.* **2020**, *47*, iii–vii. [CrossRef]
17. Guo, S.; Zhao, K. Possible Mechanism of NaCl Stress Inhibiting Photosynthesis of Maize Seedlings. *Acta Plant Physiol.* **2001**, *27*, 461–466.
18. Baker, N.R. Chlorophyll fluorescence: A probe of photosynthesis in vivo. *Annu. Rev. Plant Biol.* **2008**, *59*, 89–113. [CrossRef]
19. Baker, N.R.; Rosenqvist, E. Applications of chlorophyll fluorescence can improve crop production strategies: An examination of future possibilities. *J. Exp. Bot.* **2004**, *55*, 1607–1621. [CrossRef] [PubMed]
20. Hessini, K.; Issaoui, K.; Ferchichi, S.; Saif, T.; Abdelly, C.; Siddique, K.H.M.; Cruz, C. Interactive effects of salinity and nitrogen forms on plant growth, photosynthesis and osmotic adjustment in maize. *Plant Physiol. Biochem.* **2019**, *139*, 171–179. [CrossRef] [PubMed]
21. Zhu, J. Abiotic stress signaling and responses in plants. *Cell* **2016**, *167*, 313–324. [CrossRef]
22. Barbieri, G.; Vallone, S.; Orsini, F.; Paradiso, R.; De Pascale, S.; Negre-Zakharov, F.; Maggio, A. Stomatal density and metabolic determinants mediate salt stress adaptation and water use efficiency in basil (*Ocimum basilicum* L.). *J. Plant Physiol.* **2012**, *169*, 1737–1746. [CrossRef] [PubMed]
23. Hasanuzzaman, M.; Zhou, M.; Shabala, S. How Does Stomatal Density and Residual Transpiration Contribute to Osmotic Stress Tolerance? *Plants* **2023**, *12*, 494. [CrossRef] [PubMed]
24. Yuan, F.; Leng, B.; Wang, B. Progress in Studying Salt Secretion from the Salt Glands in Recretohalophytes: How Do Plants Secrete Salt? *Front. Plant Sci.* **2016**, *7*, 977. [CrossRef] [PubMed]
25. Shen, X.Y.; Peng, S.L.; Chen, B.M.; Pang, J.X.; Chen, L.Y.; Xu, H.M.; Hou, Y.P. Do higher resource capture ability and utilization efficiency facilitate the successful invasion of native plants. *Biol. Invasions* **2011**, *13*, 869–881. [CrossRef]
26. Hou, Y.P.; Peng, S.L.; Lin, Z.G.; Huang, Q.Q.; Ni, G.Y.; Zhao, N. Fast-growing and poorly shade-tolerant invasive species may exhibit higher physiological but not morphological plasticity compared with non-invasive species. *Biol. Invasions* **2014**, *17*, 1555–1567. [CrossRef]
27. Tognon, G.B.; Petry, C.; Cuquel, F.L. Response to water deficit of *Ipomoea cairica* (L.) Sweet. *Cienc. E Agrotecnolgia* **2012**, *36*, 318–324. [CrossRef]
28. Chen, M.H.; Mingling Cai, M.L.; Xiang, P.; Qin, Z.F.; Peng, C.L.; Li, S.S. Thermal adaptation of photosynthetic physiology of the invasive vine *Ipomoea cairica* (L.) enhances its advantage over native *Paederia scandens* (Lour.) Merr. in South China. *Tree Physiol.* **2023**, *43*, 575–586. [CrossRef] [PubMed]
29. Hoagland, D.; Arnon, D. *The Water-Culture Method for Growing Plants without Soil*; University of California: Berkeley, CA, USA, 1950; pp. 13–57.
30. Luo, B.; Wang, C.; Wang, X.; Zhang, H.; Zhou, Y.; Wang, W.; Song, P. Changes in photosynthesis and chlorophyll fluorescence in two soybean (*Glycine max*) varieties under NaCl stress. *Int. J. Agric. Biol. Eng.* **2021**, *14*, 76–82. [CrossRef]
31. Marriboina, S.; Sengupta, D.; Kumar, S.; Reddy, A.R. Physiological and molecular insights into the high salinity tolerance of *Pongamia pinnata* (L.) Pierre, a potential biofuel tree species. *Plant Sci.* **2017**, *258*, 102–111. [CrossRef]

32. Smith, V.R. A comparison between the H_2SO_4-H_2O_2-Li_2SO_4-Se oxidation method and alternative digestion procedures for plant nutrient analysis. *Commun. Soil Sci. Plan* **1979**, *10*, 1067–1077. [CrossRef]
33. Tao, R.; Ding, J.; Li, C.; Zhu, X.; Guo, W.; Zhu, M. Evaluating and Screening of Agro-Physiological Indices for Salinity Stress Tolerance in *Wheat* at the Seedling Stage. *Front. Plant Sci.* **2021**, *12*, 646175. [CrossRef]
34. Yuan, F.; Wang, X.; Zhao, B.; Xu, X.; Shi, M.; Leng, B.; Dong, X.; Lu, C.; Feng, Z.; Guo, J.; et al. The genome of the recretohalophyte *Limonium bicolor* provides insights into salt gland development and salinity adaptation during terrestrial evolution. *Mol. Plant* **2022**, *15*, 1024–1044. [CrossRef]
35. Zhao, B.; Zhou, Y.; Jiao, X.; Wang, X.; Wang, B.; Yuan, F. Bracelet salt glands of the recretohalophyte *Limonium bicolor*: Distribution, morphology, and induction. *J. Integr. Plant Biol.* **2023**, *65*, 950–966. [CrossRef]
36. Ding, F.; Chen, M.; Sui, N.; Wang, B.S. Ca^{2+} significantly enhanced development and salt-secretion rate of salt glands of *Limonium bicolor* under NaCl treatment. *S. Afr. J. Bot.* **2010**, *76*, 95–101. [CrossRef]
37. Leng, B.Y.; Yuan, F.; Dong, X.X.; Wang, J.; Wang, B.S. Distribution pattern and salt excretion rate of salt glands in two recretohalophyte species of *Limonium* (Plumbaginaceae). *S. Afr. J. Bot.* **2018**, *115*, 74–80. [CrossRef]
38. Fan, S.; Wu, H.; Gong, H.; Guo, J. The salicylic acid mediates selenium-induced tolerance to drought stress in tomato plants. *Sci. Hortic.* **2022**, *300*, 111092. [CrossRef]
39. Hodges, D.M.; DeLong, J.M.; Forney, C.F.; Prange, R.K. Improving the thiobarbituric acid-reactive substances assay for estimating lipid peroxidation in plant tissues containing anthocyanin and other interfering compounds. *Planta* **1999**, *207*, 604–611. [CrossRef]
40. Kao, K.; Michayluk, M. A method for high-frequency intergeneric fusion of plant protoplasts. *Planta* **1974**, *115*, 355–367. [CrossRef] [PubMed]
41. Dong, X.Y.; Fu, J.; Yin, X.B.; Qu, C.H.; Yang, C.J.; Li, H.Y.; He, G.Q.; Ni, J. Induction of Apoptosis in HepaRG Cell Line by Aloe-Emodin through Generation of Reactive Oxygen Species and the Mitochondrial Pathway. *Cell. Physiol. Biochem.* **2017**, *42*, 685–696. [CrossRef]
42. Lequeux, H.; Hermans, C.; Lutts, S.; Verbruggen, N. Response to copper excess in *Arabidopsis thaliana*: Impact on the root system architecture, hormone distribution, lignin accumulation and mineral profile. *Plant Physiol. Bioch.* **2010**, *48*, 673–682. [CrossRef] [PubMed]
43. Khalid, M.F.; Hussain, S.; Anjum, M.A.; Ahmad, S.; Ali, M.A.; Ejaz, S.; Morillon, R. Better salinity tolerance in tetraploid vs. diploid volkamer lemon seedlings is associated with robust antioxidant and osmotic adjustment mechanisms. *J. Plant Physiol.* **2020**, *244*, 153071. [CrossRef]
44. Liu, C.; Yan, M.; Huang, X.; Yuan, Z. Effects of salt stress on growth and physiological characteristics of pomegranate (*Punica granatum* L.) cuttings. *Pak. J. Bot.* **2018**, *50*, 457–464.
45. Shahid, M.A.; Balal, R.M.; Khan, N.; Simón-Grao, S.; Alfosea-Simón, M.; Cámara-Zapata, J.M.; Mattson, N.S.; Garcia-Sanchez, F. Rootstocks influence the salt tolerance of Kinnow mandarin trees by altering the antioxidant defense system, osmolyte concentration, and toxic ion accumulation. *Sci. Hortic.* **2019**, *250*, 1–11. [CrossRef]
46. Ghalati, R.E.; Shamili, M.; Homaei, A. Effect of putrescine on biochemical and physiological characteristics of guava (*Psidium guajava* L.) seedlings under salt stress. *Sci. Hortic.* **2020**, *261*, 108961–108967. [CrossRef]
47. Hoch, W.; Singsaas, E.; Mccown, B. Anthocyanins facilitate nutrient recovery in autumn by shielding leaves from potentially damaging light levels. *Plant Physiol.* **2003**, *133*, 12961305. [CrossRef] [PubMed]
48. Shi, Y.; Zhang, Y.; Han, W.; Feng, R.; Hu, Y.; Guo, J.; Gong, H. Silicon enhances water stress tolerance by improving root hydraulic conductance in *Solanum lycopersicum* L. *Front. Plant Sci.* **2016**, *7*, 196. [CrossRef] [PubMed]
49. Wang, X.K.; Huang, J.L. *Principles and Techniques of Plant Physiological Biochemical Experiment*, 3rd ed.; Higher Education Press: Beijing, China, 2015.
50. Van Kleunen, M.; Schlaepfer, D.R.; Glaettli, M.; Fischer, M. Preadapted for invasiveness: Do species traits or their plastic response to shading differ between invasive and non-invasive plant species in their native range? *J. Biogeogr.* **2011**, *38*, 1294–1304. [CrossRef]
51. Flowers, T.J.; Colmer, T.D. Salinity tolerance in halophytes. *New Phytol.* **2008**, *179*, 945–963. [CrossRef]
52. Chourasia, K.N.; More, S.J.; Kumar, A.; Kumar, D.; Singh, B.; Bhardwaj, V.; Kumar, A.; Das, S.K.; Singh, R.K.; Zinta, G.; et al. Salinity responses and tolerance mechanisms in underground vegetable crops: An integrative review. *Planta* **2022**, *255*, 68. [CrossRef]
53. Wang, Y.J.; Müller-Schärer, H.; van Kleunen, M.; Cai, A.M.; Zhang, P.; Yan, R.; Dong, B.C.; Yu, F.H. Invasive alien plants benefit more from clonal integration in heterogeneous environments than natives. *New Phytol.* **2017**, *216*, 1072–1078. [CrossRef] [PubMed]
54. Amasino, R. Seasonal and developmental timing of flowering. *Plant J.* **2010**, *61*, 10011013. [CrossRef] [PubMed]
55. Liu, G.; Guan, R.; Chang, R.; Qiu, L. Correlation between Na^+ Contents in Different Organs of Soybean and Salt Tolerance at the Seedling Stage. *Acta Agron. Sin.* **2011**, *37*, 1266–1273. [CrossRef]
56. Menezes, R.V.; Azevedo Neto, A.D.D.; Ribeiro, M.D.O.; Cova, A.M.W. Growth and contents of organic and inorganic solutes in amaranth under salt stress. *Pesqui. Agropecuária Trop.* **2017**, *47*, 22–30. [CrossRef]
57. Sahin, U.; Ekinci, M.; Ors, S.; Turan, M.; Yildiz, S.; Yildirim, E. Effects of individual and combined effects of salinity and drought on physiological, nutritional and biochemical properties of cabbage (*Brassica oleracea* var. capitata). *Sci. Hortic.* **2018**, *240*, 196–204. [CrossRef]
58. Shirazi, M.; Khan, M.; Mahboob, W.; Khan, M.A.; Shereen, A.; Mujtaba, S.; Asad, A. Inconsistency in salt tolerance of some wheat (*Triticum aestivium* L.) genotypes evaluated under various growing environments. *Pak. J. Bot.* **2018**, *50*, 471–479.

59. Zhang, H.; Gao, X.R.; Zhi, Y.H.; Li, X.; Zhang, Q.; Niu, J.B.; Wang, J.; Zhai, H.; Zhao, N.; Li, J.G.; et al. A non-tandem CCCH-type zinc-finger protein, IbC3H18, functions as a nuclear transcriptional activator and enhances abiotic stress tolerance in sweet potato. *New Phytol.* **2019**, *223*, 1918–1936. [CrossRef]
60. Li, G.; Ye, Y.X.; Ren, X.Q.; Qi, M.Y.; Zhao, H.Y.; Zhou, Q.; Chen, X.H.; Wang, J.; Yuan, C.Y.; Wang, F.B. The rice Aux/IAA transcription factor gene OsIAA18 enhances salt and osmotic tolerance in Arabidopsis. *Biol. Plant* **2020**, *64*, 454–464. [CrossRef]
61. Kang, C.; He, S.Z.; Zhai, H.; Li, R.J.; Zhao, N.; Liu, Q.C. A sweetpotato auxin response factor gene (IbARF5) is involved in carotenoid biosynthesis and salt and drought tolerance in transgenic Arabidopsis. *Front. Plant Sci.* **2018**, *9*, 1307. [CrossRef]
62. Munns, R.; Passioura, J.B.; Colmer, T.D.; Byrt, C.S. Osmotic adjustment and energy limitations to plant growth in saline soil. *New Phytol.* **2020**, *225*, 1091–1096. [CrossRef]
63. Zhao, J.; Zhang, W.; da Silva, J.A.T.; Liu, X.; Duan, J. Rice histone deacetylase HDA704 positively regulates drought and salt tolerance by controlling stomatal aperture and density. *Planta* **2021**, *254*, 79. [CrossRef]
64. Hameed, M.; Ashraf, M.; Naz, N. Anatomical adaptations to salinity in cogon grass [*Imperata cylindrica* (L.) Raeuschel] from the Salt Range, Pakistan. *Plant Soil* **2009**, *322*, 229–238. [CrossRef]
65. Yuan, F.; Wang, B. Adaptation of Recretohalophytes to Salinity. In *Handbook of Halophytes*; Springer: Cham, Switzerland, 2020; pp. 1–21.
66. Wei, X.; Yan, X.; Yang, Z. Salt glands of recretohalophyte *Tamarix* under salinity: Their evolution and adaptation. *Ecol. Evol.* **2020**, *10*, 9384–9395. [CrossRef]
67. Shabala, S.; Demidchik, V.; Shabala, L.; Cuin, T.; Smith, S.; Miller, A.; Davies, J.; Newman, I. Extracellular Ca^{2+} ameliorates NaCl-induced K^+ loss from Arabidopsis root and leaf cells by controlling plasma membrane K^+-permeable channels. *Plant Physiol.* **2006**, *141*, 1653–1665. [CrossRef]
68. Zhang, Y.; Fang, J.; Wu, X.; Dong, L. Na^+/K^+ balance and transport regulatory mechanisms in weedy and cultivated rice (*Oryza sativa* L.) under salt stress. *BMC Plant Biol.* **2018**, *18*, 375. [CrossRef]
69. Fang, S.; Hou, X.; Liang, X. Response mechanisms of plants under saline-alkali stress. *Front. Plant Sci.* **2021**, *12*, 667458. [CrossRef]
70. Chen, H.N.; Tao, L.Y.; Shi, J.M.; Han, X.R.; Cheng, X.G. Exogenous salicylic acid signal reveals an osmotic regulatory role in priming the seed germination of *Leymus* chinensis under salt-alkali stress. *Environ. Exp. Bot.* **2021**, *188*, 104498.
71. Rahman, A.; Nahar, K.; Hasanuzzaman, M.; Fujita, M. Calcium Supplementation Improves Na^+/K^+ Ratio, Antioxidant Defense and Glyoxalase Systems in Salt-Stressed Rice Seedlings. *Front. Plant Sci.* **2016**, *7*, 609. [CrossRef] [PubMed]
72. Wu, G.Q.; Li, H.; Zhu, Y.H.; Li, S.J. Comparative physiological response of sainfoin (*Onobrychis viciaefolia*) seedlings to alkaline and saline-alkaline stress. *J. Anim. Plant Sci.* **2021**, *31*, 1028–1035.
73. Qin, Y.; Bai, J.; Wang, Y.; Liu, J.; Hu, Y.; Dong, Z.; Ji, L. Comparative effects of salt and alkali stress on photosynthesis and root physiology of oat at anthesis. *Arch. Biol. Sci.* **2018**, *70*, 329–338. [CrossRef]
74. Wang, R.; Wang, C.; Feng, Q.; Liou, R.M.; Lin, Y.F. Biological inoculant of salt-tolerant bacteria for plant growth stimulation under different saline soil conditions. *J. Microbiol. Biotechnol.* **2021**, *31*, 398–407. [CrossRef]
75. Zelm, E.V.; Zhang, Y.; Testerink, C. Salt Tolerance Mechanisms of Plants. *Annu. Rev. Plant Biol.* **2020**, *71*, 403–433. [CrossRef] [PubMed]
76. Pan, J.; Xue, X.; Huan, C.; Peng, F.; Liao, J.; Ma, S.; You, Q.; Wang, T. Salt Tolerance Strategies of *Nitraria tangutorum* Bobr. and *Elaeagnus angustifolia* Linn. Determine the Inoculation Effects of Microorganisms in Saline Soil Conditions. *Agronomy* **2022**, *12*, 913. [CrossRef]
77. Umar, S.; Diva, I.; Anjum, N.A.; Iqbal, M.; Ahmad, I.; Pereira, E. Potassium-induced alleviation of salinity stress in *Brassica campestris* L. *Cent. Eur. J. Biol.* **2011**, *6*, 1054–1063. [CrossRef]
78. Ratnayake, M.; Leonard, R.; Menge, J. Root exudation in relation to supply of phosphorus and its possible relevance to mycorrhizal formation. *New Phytol.* **1978**, *81*, 543–552. [CrossRef]
79. Yang, Z.; Li, J.L.; Liu, L.N. Photosynthetic regulation under salt stress and salt-tolerance mechanism of sweet sorghum. *Front. Plant Sci.* **2020**, *10*, 1722. [CrossRef] [PubMed]
80. Sun, M.; Feng, X.X.; Gao, J.J.; Peng, R.H.; Yao, Q.H.; Wang, L.J. VvMYBA6 in the promotion of anthocyanin biosynthesis and salt tolerance in transgenic Arabidopsis. *Plant Biotechnol. Rep.* **2017**, *11*, 299–314. [CrossRef]
81. Naing, A.H.; Kim, C.K. Abiotic stress-induced anthocyanins in plants: Their role in tolerance to abiotic stresses. *Physiol. Plant* **2021**, *172*, 1711–1723. [CrossRef]
82. Wu, Z.; Wang, J.; Yan, D.; Yuan, H.; Wang, Y.; He, Y.; Wang, X.; Li, Z.; Mei, J.; Hu, M. Exogenous spermidine improves salt tolerance of pecan-grafted seedlings via activating antioxidant system and inhibiting the enhancement of Na^+/K^+ ratio. *Acta Physiol. Plant* **2020**, *42*, 83. [CrossRef]
83. Banu, M.N.A.; Hoque, M.A.; Watanabe-Sugimoto, M. Proline and glycinebetaine induce antioxidant defense gene expression and suppress cell death in cultured tobacco cells under salt stress. *J. Plant Physiol.* **2009**, *166*, 146–156. [CrossRef]
84. Naing, A.H.; Park, K.I.; Ai, T.N.; Chung, M.Y.; Han, J.S.; Kang, Y.W.; Lim, K.B.; Kim, C.K. Overexpression of snapdragon Delila (Del) gene in tobacco enhances anthocyanin accumulation and abiotic stress tolerance. *BMC Plant Biol.* **2017**, *17*, 65. [CrossRef] [PubMed]
85. Katsuhara, M.; Kawasaki, T. Salt Stress Induced Nuclear and DNA Degradation in Meristematic Cells of Barley Roots. *Plant Cell Physiol.* **1996**, *37*, 169–173. [CrossRef]

86. Ameisen, J. On the origin, evolution, and nature of programmed cell death: A timeline of four billion years. *Cell Death Differ.* **2002**, *9*, 367–393. [CrossRef]
87. Zhao, Y.; Wu, M.; Deng, P.; Zhou, X.W.; Huang, S.Y. Effects of salt stress on growth and chlorophyII fluorescence parameters of *Siraitia grosvenorii* seedlings. *South China Fruits* **2021**, *50*, 103–107. [CrossRef]
88. Athar, H.R.; Ashraf, M. Photosynthesis under Drought Stress. In *Handbook of Photosynthesis*, 2nd ed.; Pessarakli, M., Ed.; CRC Press: Boca Raton, FL, USA, 2005; pp. 793–809.
89. Ehsen, S.; Abideen, Z.; Rizvi, R.F.; Gulzar, S.; Aziz, I.; Gul, B.; Khan, M.A.; Ansari, R. Ecophysiological adaptations and anti-nutritive status of sustainable cattle feed *Haloxylon stocksii* under saline conditions. *Flora* **2019**, *257*, 151425. [CrossRef]
90. Abideen, Z.; Koyro, H.W.; Hussain, T.; Rasheed, A.; Alwahibi, M.S.; Elshikh, M.S.; Hussain, M.I.; Zulfiqar, F.; Mansoor, S.; Abbas, Z. Biomass Production and Predicted Ethanol Yield Are Linked with Optimum Photosynthesis in *Phragmites karka* under Salinity and Drought Conditions. *Plants* **2022**, *11*, 1657. [CrossRef] [PubMed]
91. Ibrahim, W.; Ahmed, I.M.; Chen, X.; Cao, F.; Zhu, S.; Wu, F. Genotypic differences in photosynthetic performance, antioxidant capacity, ultrastructure and nutrients in response to combined stress of salinity and Cd in cotton. *Biometals* **2015**, *28*, 1063–1078. [CrossRef] [PubMed]

International Journal of *Plant Biology*

MDPI

Article

Argemone ochroleuca Phytochemicals and Allelopathic Effect of Their Extracts on Germination of Soybean

Nezelo T. Mlombo [1,*], Zakheleni P. Dube [1], Fikile N. Makhubu [2,3] and Hellen Nxumalo [1]

1 Faculty of Agriculture and Natural Sciences, University of Mpumalanga, Mpumalanga 1200, South Africa; zakheleni.dube@ump.ac.za (Z.P.D.); nxumalohellen0@gmail.com (H.N.)
2 Faculty of Veterinary Science, University of Pretoria, Gauteng 0001, South Africa; fnmakhubu@gmail.com or makhufn@unisa.ac.za
3 College of Agriculture and Environmental Sciences, University of South Africa, Gauteng 0001, South Africa
* Correspondence: nezelomlombo1@gmail.com

Abstract: Soybean is a high-value food source, and the invasive weeds Mexican prickly poppy (*Argemone ochroleuca*) could release allelochemicals that inhibit the growth of this crop. The impact of *A. ochroleuca* on the germination and growth of soybean is not well documented. Therefore, the aim of this study was to evaluate the TLC profiles of different extracts of *A. ochroleuca* and assess the effects of extracts on the germination of soybean seeds. Shoots and roots of *A. ochroleuca* were weighed and 100 g of each was separately extracted with 1000 mL deionized water, hexane or acetone. Ten concentrations of water extracts ranging from 10 to 100 mL per 100 mL of deionized water and three concentrations of acetone and hexane extracts ranging from 2.5 to 7.5 g/L were separately used for seed germination bioassays. Thin-layer chromatography (TLC) analysis was used to compare the chemical profiles in the shoot and root water, and in the hexane and acetone extracts of *A. ochroleuca*. The highest reduction was recorded from the water extract, at 100%. The TLC profiling of *A. ochroleuca* addressed different classes of compounds, including alkaloids, phenolic acids and flavanoids. There is, however, a need to identify the most active phytochemicals in the suppression of germination.

Keywords: inhibition; invasive alien weeds; Mexican prickly poppy; phytochemistry; phytotoxicity

1. Introduction

Invasive alien weeds are a major problem in agriculture, reducing crop yields [1]. *Argemone ochroleuca* (Sweet) is among the most important economically devastating invasive plant species that affect both agricultural and natural ecosystems [2]. This weed has been reported to release allelochemicals that affect crops in agricultural fields [3,4]. Muche et al. [4] observed that *A. ochroleuca* leaf, stem and root, and water extracts inhibited the growth and germination of three sorghum varieties. Abd-ElGawad et al. [5] reported that the chemical composition of *A. ochroleuca* essential oils had highly oxygenated constituents including mono-, sesqui-, di-terpenoids, carotenoids, and hydrocarbons. These compounds have been associated with the inhibition of the germination and growth of crops [5]. Few empirical studies have been conducted to assess the impact of *A. ochroleuca* on crops in South African agricultural fields, leaving the effects speculative due to the limited research on the germination and growth of locally produced crops.

Soybean (*Glycine max* (L.) Merr.) is a high-value leguminous food source and an oil seed crop important for human consumption and animal feeds [6]. Khojely et al. [7] reported that in Africa, the soybean is a non-staple and non-native crop with the prospective to become a commercialized crop due to its multiple applications as food, industrial raw material, and feed. Soybean is also considered a significant source of oil and protein [8] and accounts for about a quarter of global protein and animal feed production [9]. In South Africa, soybean is mostly cultivated in the Free State, KwaZulu-Natal, Mpumalanga, and Gauteng Provinces, but its production is extended to areas where predominantly maize and crops such as groundnut, sunflower, potato and others are traditionally grown [10]. The need for soybean is gradually increasing due to an increase in population and societal development [11]. However, the production of soybean is currently threatened by invasive alien weeds all over the world [12].

Paul and Begum [13] reported that the aqueous extracts of *A. mexicana* root and leaf could reduce the germination of *Lens culinaris*. This conforms to a study by Alagesaboopathi [14], who reported that *A. mexicana*, another *Argemone* species, decreased seed germination of *Sorghum bicolor* by 18% compared to the 89% germination observed in the control. Dar et al. [15] also noted that the shoot extracts of *A. ochroleuca* inhibited the germination of *Farsetia aegyptia* Turra, *Salvia aegyptiaca*, *Hordeum vulgare* and *Medicago sativa* more than their corresponding root extracts. With *A. ochroleuca* spreading across Africa at an alarming rate [2], the understanding of how this invasive species affects economic crops is the starting point in the management decisions [16]. Currently, there is no information on the impact of *A. ochroleuca* on the germination of soybean, one of the most economically important crops belonging to the Leguminosae family, and the information on the phytochemicals present in this plant is inadequate. Therefore, the aim of the current study was to evaluate the TLC profiles of different extracts of *A. ochroleuca* and determine the effects of shoot and root extracts on soybean seed germination.

2. Materials and Methods

2.1. Collection and Preparation of *Argemone ochroleuca* Plant Extracts

Argemone ochroleuca (flowering) plant materials were collected from the University of Mpumalanga farm (25.4365° S, 30.9818° E). The botanical identity of the plant was confirmed by the H.G.W.J. Schweickerdt Herbarium (PRU), Pretoria, South Africa and given a voucher number PRU 130499. The shoots and roots of the weed were separated and cleaned thoroughly with running tap water. The collected shoots and roots were placed inside brown paper bags and dried in an oven (Memmet UN 110, Lasec, Eagle, WI, USA) set at a temperature of 55 °C for 72 h. The dried plant materials were then ground in an electric grinder (BBS1200, Summit Pro Blend, China) and sieved into a conical flask using a 2 mm sieve.

2.2. Preparation of Crude Extracts

In total, 100 g of shoot and root extracts were separately extracted in 1000 mL of water, hexane and acetone.

Aqueous water extract: Extraction was performed using deionized water and the sample was left at room temperature (25 °C) for 24 h. Each mixture was filtered through a 2 µm sieve. The resultant aqueous supernatant was considered as a 100% concentration, and subsequent dilutions of 10, 20, 30, 40, 50, 60, 70, 80, 90 and 100 mL per 100 mL of deionized water were prepared by adding deionized water into 100 mL bottles as previously stated by Namkeleja et al. [17].

Organic solvents: The acetone and hexane mixtures were left at room temperature for 3 days with occasional stirring. The mixture was passed through a Whatman No. 1 filter paper with the resultant filtrates evaporated to dryness using the stream of cold air from the lamina flow. Acetone and hexane extract concentrations of 0, 2.5, 5 and 7.5 g/L were prepared from the roots and shoots. These dilutions were kept at room temperature for 24 h. The extracts were then centrifuged at 4000 rpm for 15 min and the supernatants were used in the germination test [18].

2.3. Germination Bioassay

Germination tests were conducted following the techniques of the International Seed Testing Association (ISTA) [19]. Briefly, soybean seeds (cultivar PAN 1532R-ZXG7) was purchased from Pannar Seeds, South Africa) were separately surface-sterilized for 1 min by soaking them in 3% sodium hypochlorite and then washing with deionized water two times each for 3 min. Ten seeds were placed in a 9 cm diameter Petri dish that was lined with Whatman No. 1 filter paper. The seeds were treated with 10 mL water extracts from *A. ochroleuca* shoot or root concentrations; 0, 10, 20, 30, 40, 50, 60, 70, 80, 90 and 100 mL per 100 mL of deionized water was added using a syringe, whereas for hexane and acetone extract bioassays, seeds were separately treated with the four extract concentrations of 0, 2.5, 5 and 7.5 g/L of deionized water daily. All treatments were replicated three times and trials were duplicated three times for water extracts and two times for hexane and acetone extracts. The experiments were carried out in a growth chamber at room temperature (±25 °C) with 8 h of light for ten days with germination counts taken daily until the tenth day.

Seeds were considered as germinated when both the radicle and plumule were 2 mm long. On the tenth day, the last germination count was made along with the measurement of plumule and radicle lengths. Germination percentage, germination speed, mean germination time, mean daily germination and germination index were computed using formulae previously described by Gairola et al. [20]:

$$\text{Germination \%} = \frac{\text{Number of germinated seeds in each Petri dish}}{\text{Total number of seeds in each Petri dish}} \times 100 \quad (1)$$

$$\text{Germination speed} = \frac{\text{n1}}{\text{d1}} + \frac{\text{n2}}{\text{d2}} + \frac{\text{n3}}{\text{d3}} \pm - - - - - - - - - - \quad (2)$$

$$\text{Mean germination time} = \frac{\text{n1} \times \text{d1} + \text{n2} \times \text{d2} + \text{n3} \times \text{d3} + - - - - -}{\text{Total number of days}} \quad (3)$$

where n = number of germinated seed d = number of days

$$\text{Mean daily germination} = \frac{\text{Total number of seeds germinated}}{\text{Total number of days}} \quad (4)$$

$$\text{Germination index} = \frac{\text{Number of germinated seeds of day of first count}}{\text{Day of first count}} + \cdots \frac{\text{Number of germinated seeds of day of final count}}{\text{Day of final count}} \quad (5)$$

2.4. Thin Layer Chromatography (TLC) Analysis

Here, 2 g of each plant part was separately extracted in 20 mL of acetone, hexane and hot distilled water. The mixture was left at room temperature for 30 min and filtered through Whatman No. 1 filter paper. The procedure was repeated twice to exhaustively extract plant material, and the second collection was done after an hour. Filtrates were concentrated by drying in front of a stream of cold air to obtain crude extracts. The crude extracts were stored in an air-tight container at 25 °C until further use.

Chemical constituents in the acetone, hexane and water extracts were analyzed by thin-layer chromatography (TLC) using aluminum-backed TLC plates (ALIGRAM®SIL g/UV 254 MACHEREY-NAGEL, Merk, Johannesburg, South Africa). Then, 10 mg/mL of the extract redissolved in the respective solvents were loaded in bands of approximately 1 cm in length of the TLC plate. Plates were developed using three eluent systems developed by Kotze et al. [21]:

Ethyl acetate:methanol:water = 40:5.4:4 [EMW] (polar)

Chloroform:ethyl acetate:formic acid = 5:4:1 [CEF] (intermediate polar)

Benzene:ethanol:ammonium hydroxide = 90:10:1 [BEA] (non-polar/basic)

After development in the respective eluent system, the dried plates were visualized under ultraviolet light (254 and 366 nm, Camac Universal UV lamp TL-600). For the detection of compounds not visible under UV light, a vanillin–sulfuric acid spray reagent (1 g vanillin in 28 mL of methanol and 1 mL sulfuric acid) was used.

The retention factor was measured, which is defined as the ratio of the distance traveled by the center of a spot to the distance traveled by the solvent front.

R_f value = distance moved by the component from the origin to the spot center/distance moved from the origin by the solvent front.

2.5. Statistical Analysis

The obtained germination data were subjected to two-way independent (concentration × plant part) analysis of variance (ANOVA) through Statistix 10.0 software [22]. Pre-ANOVA, Shapiro–Wilk normality test were conducted to determine the normality of residual distribution [23]. All the variables that failed the normality test ($p \leq 0.05$) were transformed using $\text{Log}_{10}(x + 1)$, and $\text{arcsine}\sqrt{(x \div 100)}$ was used for percentage variables. The mean separation was achieved using Fisher's least significant difference at a 5% probability level.

3. Results

There were no statistical differences between trials, hence the data were pooled and reanalyzed as one data set. All data on the germination variables of soybean seeds treated with water, acetone, and hexane extracts were not normally distributed, except for germination speed and mean germination time of hexane extract and germination speed under acetone extract, hence data were transformed accordingly.

3.1. Effects of *Argemone ochroleuca* Water Extracts on Soybean Seed Germination

The interactions between water extract concentrations and plant part were highly significant ($p < 0.01$) for all germination variables and seedling length variables. Concentration and plant parts of *A. ochroleuca* as separate factors were also highly significant ($p < 0.01$) for all germination variables and seedling length variables. Increasing the concentrations of both root and shoot extracts significantly decreased all germination variables and seedling length variables (Tables 1 and 2). Root extract concentrations of 90–100, 70–100, 100, 90–100, 100, 30–100 and 20–100 mL per 100 mL of deionized water reduced germination percentage, germination speed, mean germination time, mean daily germination, germination index, plumule and radicle length, respectively. Shoot extract concentrations of 40 and above generally reduced germination variables and seedling length variables (Tables 1 and 2). In relation to the untreated control, *A. ochroleuca* root extracts decreased germination percentage (26–49%), germination speed (15–40%), mean germination time (29%), mean daily germination (24–45%), germination index (33%), plumule length (11–64%) and radicle length (15–94%) (Tables 1 and 2), whereas shoot extracts decreased germination percentage (39–85%), germination speed (32–84%), mean germination time (18–72%), mean daily germination (32–89%), germination index (20–79%), plumule length (15–95%) and radicle length (31–100%).

Table 1. Soybean seeds' germination response to *Argemone ochroleuca* water extracts.

Concentration (%)	Plant Part	Mean Germination Percentage		Mean Germination Speed		Mean Germination Time		Mean Daily Germination		Germination Index	
		Mean [x]	RI [y]	Mean	RI	Mean	RI	Mean	RI	Mean	RI
0	Roots	1.35 (90.00) [abc]	-	1.26 (17.29) [a]	-	1.68 (47.93) [ab]	-	0.28 (0.90) [abc]	-	1.03 (9.90) [ab]	-
0	Shoots	1.38 (90.00) [ab]	-	1.24 (16.76) [abc]	-	1.68 (47.84) [ab]	-	0.28 (0.90) [abc]	-	1.03 (9.90) [ab]	-
10	Roots	1.41 (94.44) [a]	4	1.26 (17.29) [a]	0	1.71 (49.87) [a]	1	0.29 (0.94) [a]	4	1.06 (10.39) [a]	2
10	Shoots	1.45 (95.56) [a]	5	1.25 (16.97) [ab]	1	1.71 (50.20) [a]	2	0.29 (0.96) [a]	5	1.06 (10.51) [a]	3
20	Roots	1.32 (87.50) [abc]	−3	1.19 (14.88) [abcd]	−5	1.66 (45.84) [ab]	−1	0.27 (0.88) [abc]	−2	1.02 (9.62) [abc]	−1
20	Shoots	1.39 (91.11) [a]	1	1.18 (14.43) [abcd]	−5	1.67 (46.76) [ab]	−1	0.28 (0.91) [abc]	1	1.04 (10.02) [ab]	1
30	Roots	1.35 (87.78) [abc]	−0	1.18 (14.57) [abcd]	−6	1.66 (45.49) [ab]	−2	0.27 (0.88) [abc]	−2	1.02 (9.66) [abc]	−1
30	Shoots	1.23 (80.00) [abcd]	−11	1.07 (11.53) [bcde]	−14	1.57 (38.90) [abc]	−7	0.25 (0.80) [abcd]	−9	0.97 (8.80) [abcd]	−6
40	Roots	1.42 (93.33) [a]	5	1.21 (15.52) [abcd]	−3	1.69 (48.10) [ab]	0	0.29 (0.93) [ab]	3	1.05 (10.27) [a]	2
40	Shoots	0.84 (55.56) [fgh]	−39	0.84 (6.64) [fg]	−32	1.38 (25.13) [cd]	−18	0.19 (0.56) [fg]	−32	0.83 (6.11) [de]	−20
50	Roots	1.37 (88.89) [abc]	1	1.15 (13.54) [abcd]	−9	1.64 (44.08) [abc]	−3	0.27 (0.89) [abc]	−1	1.02 (9.78) [abc]	−1
50	Shoots	0.93 (60.00) [efg]	−33	0.91 (7.70) [efg]	−27	1.44 (28.42) [bcd]	−15	0.20 (0.60) [def]	−28	0.85 (6.60) [cde]	−17
60	Roots	1.18 (75.56) [abcde]	−13	1.09 (12.58) [abcde]	−13	1.56 (39.01) [abc]	−8	0.24 (0.76) [abcde]	−15	0.93 (8.31) [abcd]	−10
60	Shoots	0.59 (35.56) [hij]	−57	0.61 (4.36) [hi]	−51	1.06 (16.13) [ef]	−37	0.12 (0.36) [gh]	−55	0.59 (3.91) [fg]	−42
70	Roots	1.10 (71.11) [bcdef]	−19	1.06 (11.43) [cde]	−15	1.53 (35.86) [abc]	−9	0.23 (0.71) [cde]	−18	0.92 (7.82) [abcd]	−11
70	Shoots	0.52 (27.78) [ijk]	−62	0.57 (3.17) [i]	−54	1.02 (12.26) [ef]	−39	0.10 (0.28) [gh]	−63	0.56 (3.06) [fg]	−46
80	Roots	1.08 (72.22) [cdef]	−20	1.05 (10.67) [de]	−16	1.54 (35.44) [abc]	−9	0.23 (0.72) [bcde]	−17	0.93 (7.94) [abcd]	−10
80	Shoots	0.40 (20.00) [jkl]	−71	0.44 (2.27) [i]	−65	0.84 (9.58) [f]	−50	0.08 (0.20) [hi]	−73	0.43 (2.20) [g]	−59
90	Roots	1.00 (65.56) [def]	−26	0.95 (8.96) [ef]	−25	1.45 (31.66) [abc]	−14	0.21 (0.66) [de]	−24	0.87 (7.21) [bcd]	−16
90	Shoots	0.24 (10.00) [kl]	−83	0.19 (0.75) [j]	−84	0.46 (3.76) [g]	−72	0.04 (0.10) [i]	−86	0.25 (1.10) [h]	−76
100	Roots	0.69 (44.44) [ghi]	−49	0.76 (5.96) [gh]	−40	1.19 (20.67) [de]	−29	0.15 (0.44) [fg]	−45	0.69 (4.89) [ef]	−33
100	Shoots	0.21 (7.78) [l]	−85	0.20 (0.82) [j]	−84	0.46 (3.69) [g]	−72	0.03 (0.08) [l]	−89	0.21 (0.86) [h]	−79
p-value		0.0000		0.0000		0.0000		0.0000		0.0000	

Table 1. *Cont.*

Concentration (%)	Plant Part	Mean Germination Percentage		Mean Germination Speed		Mean Germination Time		Mean Daily Germination		Germination Index	
		Mean ^{x}	RI ^{y}	Mean	RI	Mean	RI	Mean	RI	Mean	RI
F-value		4.56		8.26		6.67		5.67		6.54	
$LSD_{0.05}$		0.2890		0.1838		0.2617		0.0547		0.1743	

^{x} Column means followed by the same letter are not significantly different at $p \leq 0.05$ according to Fisher's least significant difference. ^{y} Relative impact (%) = [(treatment/control) − 1] × 100. Values in brackets are untransformed means [arcsine(x/100)]/[Log (x + 1)].

Table 2. Effect of *Argemone ochroleuca* water extracts on soybean plumule and radicle length (mm).

Concentrations (%)	Plant Part							
	Plumule				Radicle			
	Shoots ^{x}	RI ^{y}	Roots	RI	Shoots	RI	Roots	RI
0	1.52 (54.21) ^{abc}	-	1.59 (58.43) ^{ab}	-	1.24 (34.84) ^{ab}	-	1.32 (40.69) ^{a}	-
10	1.67 (59.31) ^{a}	9	1.63 (56.43) ^{a}	2	1.32 (37.57) ^{a}	6	1.31 (30.18) ^{a}	−1
20	1.53 (52.07) ^{abc}	0	1.51 (52.87) ^{abc}	−5	1.21 (26.74) ^{ab}	−3	1.12 (28.28) ^{bc}	−15
30	1.30 (43.16) ^{de}	−15	1.41 (42.98) ^{bcd}	−11	0.86 (19.16) ^{de}	−31	1.00 (20.17) ^{cd}	−24
40	0.86 (25.36) ^{h}	−44	1.56 (50.21) ^{ab}	−2	0.45 (9.09) ^{hi}	−64	1.02 (20.16) ^{cd}	−23
50	0.91 (28.19) ^{gh}	−41	1.35 (38.22) ^{bcd}	−15	0.51 (10.14) ^{gh}	−59	0.75 (13.33) ^{ef}	−43
60	0.49 (11.94) ^{ij}	−68	1.18 (31.02) ^{ef}	−26	0.18 (2.60) ^{jkl}	−85	0.65 (9.72) ^{fg}	−51
70	0.33 (6.48) ^{jk}	−78	1.08 (28.63) ^{fg}	−32	0.06 (0.69) ^{klm}	−95	0.55 (8.03) ^{gh}	−59
80	0.21 (2.89) ^{kl}	−86	0.97 (21.16) ^{gh}	−39	0.02 (0.30) ^{lm}	−98	0.33 (4.41) ^{ij}	−75
90	0.09 (1.26) ^{l}	−94	0.79 (14.97) ^{h}	−50	2.41×10^{-15} (-1.14×10^{-13}) ^{m}	−100	0.21 (2.92) ^{jk}	−84
100	0.08 (0.91) ^{l}	−95	0.56 (9.87) ^{i}	−64	-4.12×10^{-15} (5.71^{-14}) ^{m}	−100	0.08 (0.74) ^{klm}	−94
p-value	0.0000				0.0000			
F-value	12.69				6.47			
$LSD_{0.05}$	0.1811				0.1650			

^{x} Column means followed by the same letter are not significantly different at $p \leq 0.05$ according to Fisher's least significant difference. ^{y} Relative impact (%) = [(treatment/control) − 1] × 100. Values in brackets are untransformed means [Log (x + 1)].

3.2. Effects of *Argemone ochroleuca* Hexane Extracts on Soybean Seed Germination

The interaction between *A. ochroleuca* hexane extract concentrations and plant part were not significant ($p > 0.05$) for germination percentage, mean daily germination, germination index, plumule length and radicle length, whereas the interaction was statistically significant ($p < 0.05$) for germination speed and mean germination time. The concentration and plant parts of *A. ochroleuca* as separate factors were not significant ($p > 0.05$) for germination percentage, mean daily germination and germination index, whereas concentration as a separate factor was statistically significant ($p < 0.05$) for germination speed and mean germination time. Increasing concentrations in shoot extracts significantly decreased germination speed and mean germination time, whereas root extracts had a stimulating effect on the same variables (Table 3). Relative to untreated control, hexane shoot extract decreased germination speed (35–40%) and mean germination time (28–35%), whereas root extracts stimulated germination speed (33–53%) and mean germination time (30–35%). Concentration as a separate factor was statistically significant ($p < 0.05$) for plumule length and radicle length. Increasing concentrations in hexane extracts reduced plumule and radicle lengths by 20% and 25–27%, respectively (Table 4).

Table 3. Effect of *Argemone ochroleuca* hexane extracts on soybean germination speed and mean germination time.

Concentrations (g/L)	Germination Speed				Mean Germination Time			
	Plant Part							
	Shoots ^{x}	RI ^{y}	Roots	RI	Shoots	RI	Roots	RI
0.0	15.44 ^{ab}	-	10.81 ^{cd}	-	40.67 ^{a}	-	28.90 ^{cd}	-
2.5	15.72 ^{a}	2	16.51 ^{a}	53	39.00 ^{ab}	−4	37.50 ^{abc}	30
5.0	9.96 ^{d}	−35	14.37 ^{abc}	33	29.08 ^{bcd}	−28	39.00 ^{ab}	35
7.5	9.19 ^{d}	−40	11.48 ^{bcd}	6	26.48 ^{d}	−35	29.27 ^{bcd}	1
p-value	0.0215				0.0267			
F-value	3.60				3.40			
$LSD_{0.05}$	4.1128				9.9348			

^{x} Column means followed by the same letter are not significantly different at $p \leq 0.05$ according to Fisher's least significant difference. ^{y} Relative impact (%) = [(treatment/control) − 1] × 100. Values in brackets are untransformed means [Log (x + 1)].

Table 4. Effect of *Argemone ochroleuca* hexane extracts on soybean plumule length and radicle length.

Concentrations	Plumule Length ^{x}	RI ^{y}	Radicle Length	RI
0 g/L	1.1294 ^{ab}	-	0.9209 ^{a}	-
2.5 g/L	1.2724 ^{a}	13	0.9987 ^{a}	8
5 g/L	1.0143 ^{bc}	−10	0.6685 ^{b}	−27
7.5 g/L	0.9013 ^{c}	−20	0.6928 ^{b}	−25
p-value	0.0067		0.0024	
F-value	4.12		4.86	
$LSD_{0.05}$	0.2174		0.2074	

^{x} Column means followed by the same letter are not significantly different at $p \leq 0.05$ according to Fisher's least significant difference. ^{y} Relative impact (%) = [(treatment/control) − 1] × 100. Values in brackets are untransformed means [Log (x + 1)].

3.3. Effects of *Argemone ochroleuca* Acetone Extracts on Soybean Seed Germination

The interactions between *A. ochroleuca* acetone extract concentration and plant part were statistically significant ($p < 0.05$) for mean germination time and germination index, highly significant ($p < 0.01$) for plumule length and radicle length, and not significant ($p > 0.05$) for germination percentage, germination speed and mean daily germination. Concentration as a separate factor was not significant ($p > 0.05$) for all germination variables and radicle length, whereas a significant response was observed for plumule length. Plant parts were highly significant ($p < 0.01$) for germination percentage, germination speed, mean daily germination, plumule length and radicle length, and statistically significant ($p < 0.05$) for mean germination time and germination index. Shoot and root extracts significantly decreased germination percentage, germination speed and mean daily germination by 22, 21 and 18%, respectively (Table 5). Increasing concentrations of shoot extracts did not affect the germination variables, whereas the highest concentration in the root extract significantly reduced mean germination time and germination index (Table 6). Seedling plumule and radicle were reduced by the acetone-shoot extract at the concentration of 2.5 g/L, whereas only the plumule was reduced by acetone root extract at concentrations of 7.5 g/L (Table 7). In comparison to the untreated control, shoot extracts reduced only plumule and radicle lengths by 24% and 23%, respectively, whereas *A. ochroleuca* root extracts decreased germination time, germination index, and plumule length by 16, 19 and 33%, respectively.

Table 5. Effect of *Argemone ochroleuca* acetone extracts on soybean seed germination percentage, germination speed and mean daily germination.

Plant Part	Germination Percentage ^{x}	RI ^{y}	Germination Speed	RI	Mean Daily Germination	RI
Shoots	1.27 (85.00) ^{a}	-	17.07 ^{a}	-	0.26 (0.85) ^{a}	-
Roots	0.99 (66.67) ^{b}	−22	13.43 ^{b}	−21	0.22 (0.67) ^{b}	−18
p value	0.0018		0.0076		0.0054	
F value	11.22		7.90		8.66	
$LSD_{0.5}$	0.1722		2.6137		0.0453	

^{x} Column means followed by the same letter are not significantly different at $p \leq 0.05$ according to Fisher's least significant difference. ^{y} Relative impact (%) = [(treatment/control) − 1] × 100. Values in brackets are untransformed means [Arcsine (x/100)] [Log (x + 1)].

Table 6. Effect of *Argemone ochroleuca* acetone extracts on soybean mean germination time and germination index.

Concentration (g/L)	Plant Part							
	Mean Germination Time				Germination Index			
	Shoots ^{x}	RI ^{y}	Roots	RI	Shoots	RI	Roots	RI
0	1.69 (48.40) ^{a}	-	1.58 (37.38) ^{a}	-	1.03 (9.90) ^{a}	-	0.93 (7.70) ^{a}	-
2.5	1.53 (36.78) ^{a}	−10	1.60 (40.10) ^{a}	2	0.90 (7.70) ^{ab}	−13	0.97 (8.43) ^{a}	4
5	1.70 (49.25) ^{a}	0	1.57 (37.43) ^{a}	−0	1.04 (10.08) ^{a}	1	0.93 (7.70) ^{a}	−1
7.5	1.67 (46.03) ^{a}	−1	1.33 (25.50) ^{b}	−16	1.03 (9.72) ^{a}	−1	0.75 (5.50) ^{b}	−19
p value	0.0402				0.0438			
F value	3.03				2.96			
$LSD_{0.5}$	0.1958				0.1633			

^{x} Column means followed by the same letter are not significantly different at $p \leq 0.05$ according to Fisher's least significant difference. ^{y} Relative impact (%) = [(treatment/control) − 1] × 100. Values in brackets are untransformed means [Log (x + 1)].

Table 7. Effects of *Argemone ochroleuca* acetone extracts on soybean plumule and radicle length (mm).

Concetrations (g/L)	Plant Part							
	Plumule				Radicle			
	Shoots [x]	RI [y]	Roots	RI	Shoots	RI	Roots	RI
0	1.71 (82.52) [ab]	-	1.36 (64.57) [c]	-	1.25 (37.82) [ab]	-	0.86 (24.98) [cd]	-
2.5	1.29 (57.45) [c]	−24	1.47 (69.78) [bc]	8	0.96 (36.13) [c]	−23	1.10 (30.92) [bc]	28
5	1.83 (106.55) [a]	7	1.38 (70.03) [c]	2	1.49 (53.33) [a]	19	0.93 (25.15) [cd]	9
7.5	1.71 (82.28) [ab]	−0	0.92 (38.55) [d]	−33	1.30 (47.45) [ab]	4	0.66 (15.62) [d]	−23
p value	0.0001				0.0005			
F value	7.31				6.00			
$LSD_{0.5}$	0.2912				0.2800			

[x] Column means followed by the same letter are not significantly different at $p \leq 0.05$ according to Fisher's least significant difference. [y] Relative impact (%) = [(treatment/control) − 1] × 100. Values in brackets are untransformed means [Log (x + 1)].

The TLC analysis was used to compare the chemical composition in the water, hexane and acetone crude extracts of *A. ochroleuca*. Similar colors with corresponding R_f values are an indication of the same chemical compounds (Tables 8 and 9). The TLC chromatograms were developed in three different solvent systems of different polarities: the CEF (intermediate polarity), EMW (polar), and BEA (non-polar) systems. The TLC chromatograms of *A. ochroleuca* shoots (S) and the root extracts (R) are shown in Figure 1. The CEF and BEA eluted more compounds as compared to EMW. Hexane extracts from shoots and roots indicated the presence of the most compounds when run with the CEF system. The observed compounds based on the color have previously been identified as flavonoids, coumarins, phenolic acids, alkaloids, saponins and anthracene derivatives. Most compounds were eluted from the acetone extracts (shoot and root) and hexane root extract in the BEA system. The fewest compounds were eluted by the EMW system. Similar compounds with an R_f value of 0.15 were identified in the shoots of water, hexane, and acetone extract while in the root, and the same compound was eluted in the acetone extract using the CEF system. A common compound with an R_f value of 0.08 was eluted in the CEF and BEA systems from the root hexane, acetone and water extract, and in the shoot hexane extract.

Table 8. R_f values of compounds separated in CEF, EMW and BEA extracted in water, hexane and acetone from the shoot of *Argemone ochroleuca*.

Solvent System	Extract	Compound Number	R_f Values of Compounds
CEF	SW	1	0.15
	SH	1	0.09
		2	0.15
		3	0.72
		4	0.85
	SA	1	0.15
		2	0.81
		3	0.86

Table 8. *Cont.*

Solvent System	Extract	Compound Number	R_f Values of Compounds
EMW	SW	1	0.07
		2	0.56
	SH	0	-
	SA	1	0.16
		2	0.62
		3	0.79
BEA	SW	0	-
	SH	1	0.08
		2	0.17
		3	0.68
	SA	1	0.08
		2	0.17
		3	0.47
		4	0.68
		5	0.94

CEF—chlorofor:ethyl acetate:formic acid; EMW—ethyl acetate:methanol:water; BEA—benzene:ethanol:ammonium hydroxide; SW—shoot water; SH—shoot hexane; SA—shoot acetone; RA—root acetone; RH—root hexane; RW—root water.

Table 9. R_f values of compounds separated in CEF, EMW and BEA extracted in water, hexane and acetone from the roots of *Argemone ochroleuca*.

Solvent System	Extract	Compound Number	R_f Values of Compounds
CEF	RA	1	0.08
		2	0.15
		3	0.61
		4	0.73
		5	0.80
		6	0.86
	RH	1	0.08
		2	0.13
		3	0.53
		4	0.62
		5	0.72
		6	0.80
		7	0.85
	RW	1	0.08
EMW	RA	1	0.12
		2	0.17
		3	0.31
		4	0.53
		5	0.63

Table 9. *Cont.*

Solvent System	Extract	Compound Number	R_f Values of Compounds
	RH	1	0.14
		2	0.19
		3	0.35
	RW	1	0.19
BEA	RA	1	0.08
		2	0.17
		3	0.23
		4	0.47
		5	0.71
	RH	1	0.08
		2	0.19
		3	0.51
		4	0.67
		5	0.75
		6	0.87
	RW	-	-

CEF—chloroform:ethyl acetate:formic acid; EMW—ethyl acetate:methanol:water; BEA—benzene:ethanol:ammonium hydroxide; SW—shoot water; SH—shoot hexane; SA—shoot acetone; RA—root acetone; RH—root hexane; RW—root water.

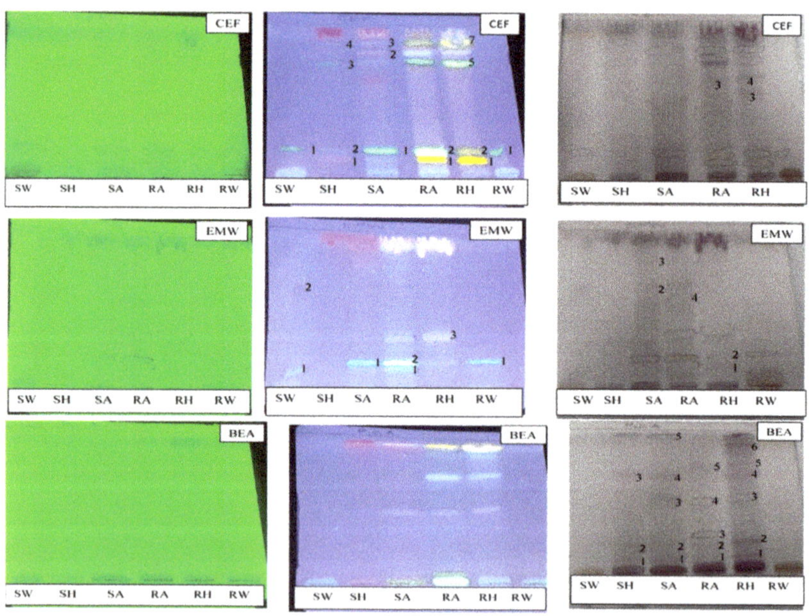

Figure 1. TLC fingerprinting of *Argemone ochroleuca* extracts, only compounds with distinct and identifiable colors are numbered. CEF—chloroform:ethyl acetate:formic acid; EMW—ethyl acetate:methanol:water; BEA—benzene:ethanol:ammonium hydroxide; SW—shoot water; SH—shoot hexane; SA—shoot acetone; RA—root acetone; RH—root hexane; RW—root water. Numbers indicate the compound number.

4. Discussion

The current study indicates that the allelopathic effects of water, hexane and acetone extracts obtained from *A. ochroleuca* shoots and roots inhibited the germination of soybean seeds, with the highest inhibition observed in the germination bioassay of water extracts. The explanation of this could be that most of the allelopathic active compound concentrations were water-soluble, hence the higher concentrations in water extracts [24]. Ashrafi et al. [24] observed that the inhibitory effects of the water-soluble fractions of *Azadirachta indica* were the highest, compared to the n-hexane-soluble and acetone-soluble fractions in all germination bioassays of *Amaranthus rotundus, Cirsium arvense, Digitaria sanguinalis, Sinapis arvensis, Lactuca sativa* and *Lolium ultiforum.* These findings were also reported by Tanveer et al. [25], who observed that there were differences in the inhibitory effects of *Euphorbia dracunculoides* n-hexane, chloroform, ethyl acetate, 1-butanol and aqueous fractions on the germination and seedling growth of maize and chickpea. Tanveer et al. [25] reported that hexane fractions had more suppressive effects on the germination of chickpeas and wheat when compared with chloroform, ethyl acetate and 1-butanol fractions. Sultana et al. [26] attributed the differences in the allelochemical composition of solvents to different compounds in plants with varying polarities and chemical properties affecting their solubility. Lower concentrations of extracts were observed to stimulate germination. This has been attributed to a process called hormesis [27]. Hormesis is an adaptative response where there is an induction of beneficial effects when the organism is exposed to low dosages of harmful chemical or physical agents [28]. In hormesis, after a small stress, special proteins responsible for the removal of damage produced by stressors are over-produced, resulting in not only the removal of damage produced by the current stress, but also the removal of the pre-existing damage, which produces a stimulating effect [29].

Weeds compete with other crops for water, nutrients and space, and they release allelochemicals into the environment that inhibit plant growth [30]. The combined effects of allelochemicals such as fatty acids, fatty acid methyl esters, terpenoids and phenolics that are released have inhibitory effects on plants [31]. Even though this particular trial did not quantify the allelochemicals, the current study has demonstrated similar decreasing trends of *A. ochroleuca* extracts' effects on the germination and seedling length of soybean seeds. These observed trends were concentration- and plant-part-dependent inhibition responses. This phenomenon is very common to many weed extracts used in crop seed germination [4,32,33]. Nxumalo et al. [33] reported a concentration-dependent inhibition response of *A. ochroleuca* extracts on the germination, seedling length and early growth of millet and maize. *Cassia occidentalis* seeds' response effects were found to be more pronounced at lower concentrations of *Psidium guajava* extracts than at higher concentrations [34]. Muche et al. [4] also reported the same effects of *A. ochroleuca* extracts on the germination and seedling length of *Sorghum bicolor* varieties. In the current study, *A. ochroleuca* extracts also affected other germination variables such as germination speed and mean daily germination. M'barek et al. [35] reported the phytotoxic effects of *Tetraclinis articulata* on germination speed, even though in their report the allelochemicals had no effects on the final germination percentage. The delay in seed germination can have some important biological and ecological implications, because it affects the ability of the seedling to establish itself in natural conditions, resulting in uneven plant stands [35].

The current study also found that there were differences in the allelopathic effects of the different plant parts of *A. ochroleuca* on the germination of soybean seeds, with extracts from the shoots inhibiting all measured germination variables more when compared to extracts from the roots. Water and acetone shoot and root extracts and hexane shoot extracts had inhibitory effects on the measured germination and seedling length variables, whereas hexane root extracts stimulated germination and seedling length variables. Generally, the distributions of allelopathic active compounds differ with plant organ, both in quantity and quality [36]. The leaves of *Eucalyptus camaldulensis* were recorded by Nasr et al. [37] to contain the highest total allelochemicals when compared to other plant organs. Ghareib et al. [38] reported that the allelopathic potential induced by low concentrations of the acetone fraction of *Chenopodium murale* stimulated the germination and growth of tomato. According to Muche et al. [4], *A. ochroleuca* weed leaf extracts had more inhibitory effects on *Sorghum bicolor* varieties compared to extracts from the roots and stems. Root, stem and leaf aqueous extracts of *A. philoxeroides* had different effects on the root length, shoot length and fresh weight of *Z. matrella* [39]. Paul and Begum [40] explained the high degree of inhibition of the germination and seedling growth of blackgram, rapeseed and wheat using leaf and root extracts of *A. mexicana* as being due to the fact that *A. mexicana* synthesizes and stores the phytochemicals in leaves and roots.

The TLC analysis identified different classes of compounds from *A. ochroleuca* hexane and acetone extracts, which include flavonoids, lactones, phenolic acids, alkaloids, saponins and anthracene derivatives. Water extract did not elute many compounds as compared to hexane and acetone; their highly suppressive effects could mean that the compounds present in water are phytotoxic. The activities of non-polar compounds and polar ones from hexane and acetone could be attributed to the presence of the compounds shown in TLC plates. Cheng and Cheng [41] reported that the allelochemicals produced by plants exhibiting allelopathy include phenolics, terpenoids, and alkaloids, but water-soluble phenolic compounds have been established to play a major role in the growth suppression of associated plants. Phytochemicals such as phenolics, alkaloids, steroids, terpenes, saponins, and quinones have allelopathic effects on the growth and development of certain plant species [42]. According to Sasikumar et al. [43], the allelopathic effects of *Eucalyptus globulus* can be attributed to volatile terpenes and phenolic acids, and have been reported to be responsible for the inhibitory effects on the germination and seedling growth of various crops. Ghimire et al. [44] reported that alfalfa-derived phenolic compounds and saponins exhibit phytotoxicity effects on the growth of *Digitaria ciliaris, Chenopodium album, Amaranthus lividus, Portulaca oleracea* and *Commelina communis.* Phenolic acids have been identified and isolated from many allelopathic plants, and the role of phenolic compounds in inducing allelopathic abilities is well-established [45]. According to Movafeghi et al. [46], alkaloids, steroids, flavonoids, anthraquinones, amino acids, and polysaccharides were isolated from the seeds, leaves, flowers, stems, and roots of *Peganum harmala,* a weed reported to have phytotoxic effects on plants. Shao et al. [47] reported that alkaloids isolated from the seeds of *Peganum harmala* exerted significant inhibitory activity on lettuce, amaranth, wheat, and ryegrass seed germination. Synowiec et al. [48] reported that the essential oils of *Achillea millefolium, Acorus calamus, Carum carvi, Chamomilla recutita, Foeniculum vulgare, Lavandula angustifolia, Melissa officinalis, Mentha piperita, Salvia officinalis, Solidago canadensis, Tanacetum vulgare* and *Thymus vulgaris* inhibited the germination of *Amaranthus retroflexus, Avena fatua, Bromus secalinus* and *Centaurea cyanus,* which have been reported as notorious weeds that affect the germination of economically important crops (*Avena sativa, Brassica napus* and *Zea mays*). Allelopathic phytochemicals are released from donor plants as volatiles, roots exudates, or foliage leachates, and contain secondary metabolites such as flavonoid phenolics [49], ketones, aldehydes, terpenoids, lactones, cinnamic acid, and quinines [50]. When these compounds are excreted into the rhizosphere, neighboring plants absorb them through the uptake of sap [51], and interfere with the physiologic and biosynthetic machinery of the receiver plant [52,53]. Given the most recent developments in allelochemistry, which provides a physiologically and ecologically solid explanation for plant invasion, it is hypothesized that the phytotoxicity of the applied extract is responsible for the negative response of seeds or seedlings [15].

5. Conclusions

The results show that the there were differences in the allelopathic effects of the three different solvent extracts on soybean germination, with water having the highest effects. The allelopathic effects also differed between the shoot extracts and the root extracts. Lower concentrations had lesser effects on germination, while higher concentrations of *A. ochroleuca* had higher suppressive effects on germination. Flavonoids, lactones, phenolic acids, alkaloids, saponins and anthracene derivatives were observed in the extracts of *A.*

ochroleuca. The results indicate that *A. ochroleuca* has inhibitory effects on the germination of seeds, although the effect should be confirmed further by field experiments. The obtained data might be useful in the management of these weeds within soybean fields. It would be recommended for soybean producers to control this weed at the early growth stage before it causes drastic effects on the crops. This study focused on the allelopathic effects of *A. ochroleua* extracts on the germination of soybean seeds and the chemical profiling of *A. ochroleuca*; however, a similar study evaluating the effects of *A. ochroleuca* on other weeds and the modes of action of these phytochemicals would be recommended. Future research should focus on purifying, identifying, and characterizing active compounds. The obtained data would be useful in the development of bioherbicides.

Author Contributions: Conceptualization, N.T.M., Z.P.D. and F.N.M.; methodology, N.T.M., Z.P.D., F.N.M. and H.N.; validation, N.T.M., Z.P.D. and F.N.M.; formal analysis, N.T.M., Z.P.D., F.N.M. and H.N.; investigation, N.T.M., Z.P.D., F.N.M. and H.N.; resources, N.T.M., Z.P.D. and F.N.M.; data curation, N.T.M., Z.P.D., F.N.M. and H.N.; writing—original draft preparation, N.T.M., Z.P.D. and F.N.M.; writing—review and editing, N.T.M., Z.P.D. and F.N.M.; visualization N.T.M., Z.P.D. and F.N.M.; supervision, Z.P.D. and F.N.M.; sourcing funding, N.T.M. and Z.P.D.; project administration, N.T.M., Z.P.D., F.N.M. and H.N. All authors have read and agreed to the published version of the manuscript.

Funding: This research was funded by the University of Mpumalanga (Vice Chancellor Scholarship) 201743337 and the Nation Research Foundation (NRF) MND210617612476.

Institutional Review Board Statement: Not applicable.

Informed Consent Statement: Not applicable.

Data Availability Statement: Data is contained within the article.

Acknowledgments: The authors duly acknowledge the H.G.W.J. Schweickerdt Herbarium (PRU) at the University of Pretoria.

Conflicts of Interest: The authors declare no conflict of interest.

References

1. Yadav, V.; Singh, N.B.; Singh, H.; Singh, A.; Hussain, I. Allelopathic invasion of alien plant species in India and their management strategies: A review. *Trop. Plant Res.* **2016**, *3*, 87–101.
2. Sanaa, A. Perspectives on the relationship between invisibility, richness, plant size, seed production, seed bank and community productivity of invasive *Argemone ochroleuca* Sweet in Taif, Saudi Arabia. *J. Life Sci.* **2012**, *9*, 953–958.
3. Namkeleja, H.S.; Tarimo, M.T.C.; Ndakidemi, P.A. Allelopathic effects of *Argemone mexicana* to growth of native plant species. *Am. J. Plant Sci.* **2014**, *5*, 1336–1344. [CrossRef]
4. Muche, M.; Molla, E.; Teshome, H. Phytotoxicity of *Argemone ochroleuca* L. on germination and seedling growth of *Sorghum bicolor* L. varieties under in vitro condition. *Am. Eurasian J. Agric. Environ. Sci.* **2018**, *18*, 185–192.
5. Abd-ElGawad, A.M.; El Gendy, A.E.N.G.; Assaeed, A.M.; Al-Rowaily, S.L.; Omer, E.A.; Dar, B.A.; Al-Taisan, W.A.A.; Elshamy, A.I. Essential oil enriched with oxygenated constituents from invasive plant *Argemone ochroleuca* exhibited potent phytotoxic effects. *Plants* **2020**, *9*, 998. [CrossRef]
6. Medic, J.; Atkinson, C.; Hurburgh, C.R. Current knowledge in soybean composition. *J. Am. Oil Chem. Soc.* **2014**, *91*, 363–384. [CrossRef]
7. Khojely, D.M.; Ibrahim, S.E.; Sapey, E.; Han, T. History, current status, and prospects of soybean production and research in sub-Saharan Africa. *Crop J.* **2018**, *6*, 226–235. [CrossRef]
8. Liu, S.; Zhang, M.; Feng, F.; Tian, Z. Toward a "green revolution" for soybean. *Mol. Plant* **2020**, *13*, 688–697. [CrossRef] [PubMed]
9. Graham, P.H.; Vance, C.P. Legumes: Importance and constraints to greater use. *Plant Physiol.* **2003**, *131*, 872–877. [CrossRef]
10. Fourie, H.; De Waele, D.; Mc Donald, A.H.; Mienie, C.; Marais, M.; De Beer, A. Nematode pests threatening soybean production in South Africa, with reference to Meloidogyne. *S. Afr. J. Sci.* **2015**, *111*, 01–09. [CrossRef]
11. Ray, D.K.; Mueller, N.D.; West, P.C.; Foley, J.A. Yield trends are insufficient to double global crop production by 2050. *PLoS ONE* **2013**, *8*, e66428. [CrossRef] [PubMed]
12. Soltani, N.; Dille, J.A.; Burke, I.C.; Everman, W.J.; VanGessel, M.J.; Davis, V.M.; Sikkema, P.H. Perspectives on potential soybean yield losses from weeds in North America. *Weed Technol.* **2017**, *31*, 148–154. [CrossRef]
13. Paul, N.K.; Begum, N. Allelopathic effect of *Argemone mexicana* L. on germination and seedling growth characteristics of Lentil (*Lens culinaris*). *J. Biosci.* **2010**, *18*, 146–147. [CrossRef]

14. Alagesaboopathi, C. Allelopathic effect of different concentration of water extract of *Argemone mexicana* L. on seed germination and seedling growth of *Sorghum bicolor* (L.) Moench. *J. Pharm. Pharm. Sci.* **2013**, *5*, 52–55. [CrossRef]
15. Dar, B.A.; Al-Rowaily, S.L.; Assaeed, A.M.; El-Bana, M.I.; Hegazy, A.K.; Malik, J.A. Allelopathic potential of *Argemone ochroleuca* from different habitats on seed germination of native species and cultivated crops. *Pak. J. Bot.* **2017**, *49*, 1841–1848.
16. Diekmann, M.; Efertz, H.; Baranowski, M.; Dupré, C. Weak effects on plant diversity of two invasive Impatiens species. *Plant Ecol.* **2016**, *217*, 1503–1514. [CrossRef]
17. Namkeleja, H.S.; Tarimo, M.T.; Ndakidemi, P.A. Allelopathic effect of aqueous extract of *Argemone mexicana* L. on germination and growth of *Brachiaria dictyoneura* L. and *Clitoria ternatea* L. *Am. J. Plant Sci.* **2013**, *4*, 2138–2147. [CrossRef]
18. Tarzi, B.G.; Gharachorloo, M.; Baharinia, M.; Mortazavi, S.A. The effect of germination on phenolic content and antioxidant activity of chickpea. *Iran J. Pharm. Res.* **2012**, *11*, 1137–1143.
19. Jones, S. Future seed testing needs for seed analysts and researchers-A personal view. *Seed Technol.* **2013**, *35*, 15–20.
20. Gairola, K.C.; Nautiyal, A.R.; Dwivedi, A.K. Effect of temperatures and germination media on seed germination of Jatropha curcas Linn. *Adv. Biores.* **2011**, *2*, 66–71.
21. Kotze, M.; Eloff, J.N.; Houghton, P.J. Extraction of antibacterial compounds from *Combretum microphyllum* (Combretaceae). *S. Afr. J. Bot.* **2002**, *68*, 62–67. [CrossRef]
22. Analytical Software. *Statistix 10 (Statistix 10.0) Software*; Analytical Software: Tallahassee, FL, USA, 2013.
23. Gomez, A.; Gomez, A.A. *Statistical Procedures for Agricultural Research*, 2nd ed.; John Wiley and Sons: New York, NY, USA, 1984; pp. 1–92.
24. Ashrafi, Z.Y.; Sadeghi, S.; Alizade, H.M.; Mashhadi, H.R.; Mohamadi, E.R. Study of bioassay the allelopathical effect of Neem (*Azadirachta indica*) n-hexane, acetone and water-soluble extracts on six weeds. *Int. J. Biol.* **2009**, *1*, 71–77. [CrossRef]
25. Tanveer, A.; Jabbar, M.K.; Kahliq, A.; Matloob, A.; Abbas, R.N.; Javaid, M.M. Allelopathic effects of aqueous and organic fractions of *Euphorbia dracunculoides* Lam. on germination and seedling growth of chickpea and wheat. *Chil. J. Agric. Res.* **2012**, *72*, 495–501. [CrossRef]
26. Sultana, B.; Anwar, F.; Ashraf, M. Effect of extraction solvent/technique on the antioxidant activity of selected medicinal plant extracts. *Molecules* **2009**, *14*, 2167–2180. [CrossRef] [PubMed]
27. Wang, X.; Wang, J.; Zhang, R.; Huang, Y.; Feng, S.; Ma, X.; Zhang, Y.; Sikdar, A.; Roy, R. Allelopathic effects of aqueous leaf extracts from four shrub species on seed germination and initial growth of *Amygdalus pedunculata* Pall. *Forests* **2018**, *9*, 711. [CrossRef]
28. Zhao, Y.L.; Wang, D.Y. Formation and regulation of adaptive response in nematode *Caenorhabditis elegans*. *Oxid. Med. Cell. Longev.* **2012**, *8*, 38–78.
29. Butov, A.; Johnson, T.; Cypser, J.; Sannikov, I.; Volkov, M.; Sehl, M.; Yashin, A. Hormesis and debilitation effects in stress experiments using the nematode worm *Caenorhabditis elegans*: The model of balance between cell damage and HSP levels. *Exp. Gerontol.* **2001**, *37*, 57–66. [CrossRef] [PubMed]
30. Fahad, S.; Hussain, S.; Chauhan, B.S.; Saud, S.; Wu, C.; Hassan, S.; Tanveer, M.; Jan, A.; Huang, J. Weed growth and crop yield loss in wheat as influenced by row spacing and weed emergence times. *Crop Prot.* **2015**, *71*, 101–108. [CrossRef]
31. Verma, P.; Blaise, D.; Sheeba, J.A.; Manikandan, A. Allelopathic potential and allelochemicals in different intercrops for weed management in rainfed cotton. *Curr. Sci.* **2021**, *120*, 1035–1039. [CrossRef]
32. Mlombo, N.; Dube, Z.P.; Ganyani, L.; Nxumalo, H.; Mnyambo, N.M.; Timana, M. *Argemone ochrolueca* extracts suppression of germination and early growth of bean (*Phaseolus vulgaris*). *Res. Crop.* **2021**, *22*, 508–515.
33. Nxumalo, H.; Dube, Z.P.; Ganyani, L.; Mlombo, N.T.; Timana, M.; Mnyambo, N.M. Potential suppressive effects of Mexican poppy weed residues on germination and early growth of maize and pearl millet crops. *Afr. J. Food Agric. Nutr. Dev.* **2022**, *22*, 19909–19928. [CrossRef]
34. Kawawa, R.C.A.; Muyekho, F.N.; Obiri, J.F.; Agevi, H.; Obiet, L. The allelopathic impact of *Psidium guajava* L., leaf extracts on the germination and growth of *Cassia occidentalis* L., seeds. *J. Agric. Vet. Sci.* **2016**, *9*, 101–105.
35. M'barek, K.; Zribi, I.; Haouala, R. Allelopathic effects of *Tetraclinis articulata* on barley, lettuce, radish and tomato. *Allelopath. J.* **2018**, *43*, 187–202. [CrossRef]
36. Favaretto, A.; Chini, S.; Scheffer-basso, S.; Sobottka, A.; Bertol, C.; Perez, N. Pattern of allelochemical distribution in leaves and roots of tough lovegrass (*Eragrostis plana* Nees.). *Aust. J. Crop Sci.* **2015**, *9*, 1119–1125.
37. Nasr, A.; Zhou, X.; Liu, T.; Yang, J.; Zhu, G. Acetone water mixture is a competent solvent to extract phenolics and antioxidants from four organs of *Eucalyptus camaldulensis*. *Turk. J. Biochem.* **2019**, *44*, 231–239. [CrossRef]
38. Ghareib, H.R.A.; Abdelhamed, M.S.; Ibrahim, O.H. Antioxidative effects of the acetone fraction and vanillic acid from *Chenopodium murale* on tomato plants. *Weed Biol. Manag.* **2010**, *10*, 64–72. [CrossRef]
39. Huang, Y.; Ge, Y.; Wang, Q.; Zhou, H.; Liu, W.; Christie, P. Allelopathic effects of aqueous extracts of *Alternanthera philoxeroides* on the growth of *Zoysia matrella*. *Pol. J. Environ. Stud.* **2017**, *26*, 97–105. [CrossRef]
40. Paul, N.K.; Begum, N. Influence of root and leaf extracts of *Argemone mexicana* on germination and seedling growth of blackgram, rapeseed and wheat. *Bangladesh J. Sci. Ind. Res.* **2007**, *42*, 229–234. [CrossRef]
41. Cheng, F.; Cheng, Z. Research progress on the use of plant Allelopathy in agriculture and physiological and ecological mechanisms of allelopathy. *Front. Plant Sci.* **2015**, *6*, 10–20. [CrossRef]
42. Dayan, F.E.; Cantrell, C.L.; Duke, S.O. Natural products in crop protection. *Bioorg. Med. Chem.* **2009**, *17*, 4022–4034. [CrossRef]

43. Sasikumar, K.; Vijayalakshmi, C.; Parthiban, K.T. Allelopathic effects of *Eucalyptus* on blackgram (*Phaseolus mungo* L.). *Allelopath. J.* **2002**, *9*, 205–214.
44. Ghimire, B.K.; Ghimire, B.; Yu, C.Y.; Chung, I.M. Allelopathic and autotoxic effects of *Medicago sativa*—Derived allelochemicals. *Plants* **2019**, *8*, 233. [CrossRef]
45. Sampietro, D.A.; Soberon, J.R.; Sagariglia, M.A.; Quiroga, E.N.; Vattuone, M.A. Allelopathic plants. 17. Sugarcane (*Saccharum officinarum* L.). *Allelopath. J.* **2007**, *20*, 243–250.
46. Movafeghi, A.; Abedini, M.; Fathiazad, F.; Aliasgharpour, M.; Omidi, Y. Floral nectar composition of *Peganum harmala* L. *Nat. Prod. Res.* **2009**, *23*, 301–308. [CrossRef]
47. Shao, H.; Huang, X.; Zhang, Y.; Zhang, C. Main alkaloids of *Peganum harmala* L. and their different effects on dicot and monocot crops. *Molecules* **2013**, *18*, 2623–2634. [CrossRef]
48. Synowiec, A.; Kalemba, D.; Drozdek, E.; Bocianowski, J. Phytotoxic potential of essential oils from temperate climate plants against the germination of selected weeds and crops. *J. Pest Sci.* **2017**, *90*, 407–419. [CrossRef]
49. Xuan, T.D.; Tsuzuki, E.; Terao, H.; Matsuo, M.; Khanh, T.D. Identification of potential allelochemicals in kava (*Piper methysticum* L.) root. *Allelopath. J.* **2003**, *12*, 197–203.
50. Li, Z.H.; Wang, Q.; Ruan, X.; Pan, C.D.; Jiang, D.A. Phenolics and plant allelopathy. *Molecules* **2010**, *15*, 8933–8952. [CrossRef]
51. Biswas, O.; Paul, K.P.; Ghosh, S.; Karim, S.M.R. Allelopathic effects of parthenium weed debris in soil on the emergence and development of rice. *J. Agrofor. Environ.* **2010**, *4*, 193–196.
52. Bubna, G.A.; Lima, R.B.; Zanardo, D.Y.L.; dos Santos, W.D.; Ferrarese, M.L.L.; Ferrarese-Filho, O. Exogenous cafeic acid inhibits the growth and enhances the lignifcation of the roots of soybean (*Glycine max*). *J. Plant Physiol.* **2011**, *168*, 1627–1633. [CrossRef]
53. Latif, S.; Chiapusio, G.; Weston, L.A. Allelopathy and the role of allelochemicals in plant defence. *Adv. Bot. Res.* **2017**, *82*, 19–54.

forests

MDPI

Review

Fungal Phytopathogens: Their Role in the Spread and Management of Invasive Alien Plants

Katarzyna Patejuk [1,*], Kamil Najberek [2], Paulina Pacek [1], Jan Bocianowski [3] and Wojciech Pusz [1]

1 Division of Plant Pathology and Mycology, Department of Plant Protection, Wrocław University of Environmental and Life Sciences, Pl. Grunwaldzki 24a, 50-363 Wrocław, Poland; wojciech.pusz@upwr.edu.pl (W.P.)
2 Institute of Nature Conservation, Polish Academy of Sciences, Al. Adama Mickiewicza 33, 31-120 Kraków, Poland; najberek@iop.krakow.pl
3 Department of Mathematical and Statistical Methods, Poznań University of Life Sciences, Wojska Polskiego 28, 60-627 Poznań, Poland; jan.bocianowski@up.poznan.pl
* Correspondence: katarzyna.patejuk@upwr.edu.pl

Abstract: Biological invasions pose a major environmental challenge, often facilitating the unregulated dissemination of pathogens and parasites associated with their hosts. These pathogens can severely impact native and cultivated species, with far-reaching ecological and economic consequences. Despite their importance, the mycobiota associated with invasive plant species remains relatively understudied, posing a complex challenge for researchers. The aim of this manuscript is to underscore the most significant threats posed by the uncontrolled transmission of fungal pathogens from invasive alien plants to native environments and agricultural systems, and to identify the factors influencing this phenomenon. We emphasize the role of pathogen spillback and spillover mechanisms in the domestication of invasive alien plants. The influence of environmental, host, and pathogen-related factors on the survival of fungal pathogens were also investigated. Finally, we explore the technical and legal feasibility of using plant pathogens as "green agents" to control invasive alien plants.

Keywords: biological invasion; phytopathogen; pathogens' transfer; invasive plant species monitoring; epidemics

1. Introduction

The global spread of invasive alien species (abb. IAS) is recognized as one of the most significant ecological challenges of the 21st century, with profound implications for biodiversity, agriculture, and ecosystem stability [1]. Human activities, particularly increased global mobility, have accelerated the introduction of these species at an unprecedented rate, far outpacing natural processes. Due to their rapid growth and reproductive advantages, invasive alien plants often outcompete native flora, dominate ecosystems, and disrupt natural habitats [2]. This leads to widespread socio-economic, health, and ecological impacts on a global scale. For example, controlling black cherry (*Prunus serotina* Ehrh.) in Germany alone incurs an estimated annual cost of €25 million [3]. Invasive plants can destabilize national economies by adversely affecting agriculture, forestry, fisheries, and the ecosystems that provide critical livelihoods, particularly in developing regions [3]. Some invasive species, such as water hyacinth and knotweed, obstruct communication routes and infrastructure, while others, like *Ambrosia artemisiifolia* L. (common ragweed) and Sosnowsky's hogweed, pose direct health risks. Additionally, species such as black cherry and *Impatiens parviflora* DC. hinder forest regeneration, while *Solidago canadensis* L. (Canadian goldenrod) and common ragweed contaminate crops and disrupt traditional farming practices [4,5]. However, the challenges posed by invasive plants are not limited to the species themselves. The movement of IAS is often accompanied by the spread of pathogens and parasites associated with these plants. These pathogens play a complex role in biological invasions, acting both as facilitators and inhibitors of the invasive species' success. Once introduced, alien pathogens can pose a serious threat to native and cultivated species, amplifying ecological damage and destabilizing local economies. These intricate interactions between invasive plants and their associated pathogens highlight the need for a deeper understanding of their combined impacts on ecosystems and human livelihoods [6,7].

The dynamics between invasive plants and their associated pathogens can be explained through several ecological theories. The Enemy Release Hypothesis (ERH) [2] posits that invasive species often gain a competitive advantage in new habitats due to the absence or reduced presence of their natural pathogens and herbivores. This enemy absence allows invasive plants to allocate fewer resources toward defense mechanisms and more toward growth and reproduction, enabling their dominance over native species. However, this apparent advantage can be counterbalanced over time as pathogens present in the area of introduction adapt to the invasive host, a phenomenon referred to as the spillback mechanism [7]. In spillback, pathogens from the local environment colonize invasive plants, proliferate, and subsequently increase infection pressures on native co-occurring species, often with greater severity than on the invasive host. On the other hand, the spillover mechanism involves the introduction of alien pathogens alongside invasive plants, which then infect native flora unaccustomed to these new threats [7]. This process can lead to disease-mediated invasions, wherein pathogens become pivotal agents in the displacement of native species. For example, alien pathogens carried by IAS have driven severe epidemics, such as those caused by *Phytophthora infestans* (Mont.) de Bary and *Cronartium ribicola* J.C. Fisch., which have decimated staple crops and forest ecosystems, respectively. These dynamics underscore the potential of pathogens to act as "biological weapons" under the Novel Weapon Hypothesis (NWH), furthering the ecological advantage of invasive plants in their new environments [7]. The role of plant pathogens in invasive plant success is not limited to direct interactions. Indirect effects, such as the facilitation of hybridization between pathogens or the cryptic behavior of endophytic fungi transitioning to pathogenic lifestyles, further complicate this relationship. Additionally, anthropogenic activities such as global trade and urban horticulture have inadvertently facilitated the dissemination of invasive plants and their associated pathogens. It can be inferred that anthropogenic activities contribute to the pathogen spread and adaptation in new environments.

Understanding the interplay between invasive alien plants and their pathogens is critical for addressing the ecological and economic consequences of biological invasions. This manuscript explores the mechanisms through which plant pathogens influence the invasion process, including enemy release, spillover, and spillback. It also highlights the challenges posed by the cryptic lifestyles of pathogens and the genetic shifts driven by environmental changes. By examining these relationships, we aim to inform strategies for monitoring and mitigating the spread of invasive plants and their pathogens, emphasizing the importance of integrated approaches to biodiversity conservation and ecosystem management. We also address the potential risks associated with the intentional introduction of pathogens, used as "green agents" to control IAS plants, and the concerns about this process.

2. The Role of Plant Pathogens in the Spread of Alien Plant Species from the Sight of Ecological Hypotheses

Environmental characteristics, including basal resources, species composition, and top-down control, may determine the success of IAS in novel habitats [8]. Following this establishment, there is often a lag phase during which the population may show little to no expansion. This lag phase can extend from several to numerous years. Notably, the majority of invasive plant species in New Zealand exhibit a lag phase, averaging around 20–30 years, with 4% of species having a lag phase lasting more than 40 years [9]. Historically, in Europe, lag periods for alien woody species average ~150 years, while this extends to around

180 years for terrestrial plants [10]. Often, these lag phases precede a rapid and exponential increase in the population of the alien species [8].

The Enemy Release Hypothesis (abb. ERH) [2,11–13] suggests that the success of invasive plants is influenced by both the characteristics of the alien species itself and the new habitat, where alien species are under low pressure from local enemies, such as pathogens, herbivores, or predators (Figure 1). Furthermore, the new plant often does not co-evolve with natural enemies, and local enemies may not recognize potential hosts or specialize on native hosts. While numerous studies have supported this hypothesis [11,14–16], it remains a subject of controversy, and the findings from tests do not consistently indicate its universal applicability [17–20]. In new areas, the absence of obligatory enemies that controlled the spread of the newly introduced host in its natural area allows invasive plants to reduce their investment into defense and re-allocate the saved resources to improve their competitive abilities. As a result, invasive alien species exploit the environment more efficiently than native plants do. This leads to an increase in the number and spread of alien species, leading to long-term negative impacts on the environment and economy.

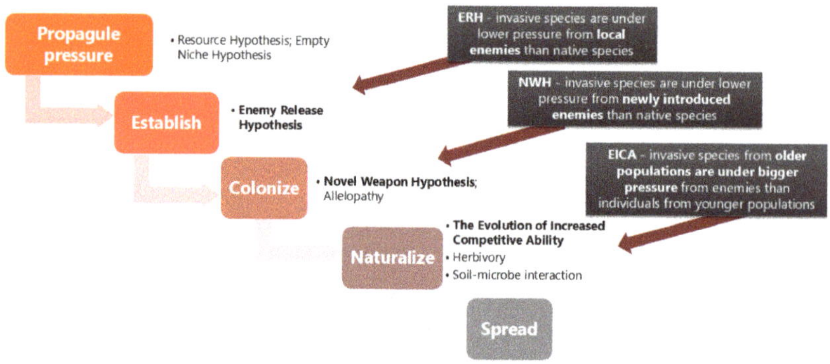

Figure 1. Salient hypothesis applicable at varying invasion stages (following Rai [21]). The graph highlights hypotheses in which plant pathogens play a crucial role.

The success of alien species after their introduction into new areas is explained also by the hypothesis of the evolution of increased competitive ability (abb. EICA) [22] that derived from the ERH hypothesis. The EICA assumes that the pressure from enemies on the species increases over time because local pests and/or pathogens need time to adapt to newly introduced hosts (Figure 1). Consequently, individuals from older populations must invest more into defense against enemies than individuals from younger ones, which results in the decline of characteristics related to competitive abilities. In turn, individuals from the youngest populations at the forefront of the invasion are under the lowest pressure from enemies, which determines their highly competitive abilities [23–26]. The EICA hypothesis has been tested on numerous alien plant species [27–29], and even though not all the results prove the concept of EICA, scientists emphasize the value of this hypothesis in understanding the biology of invasion. Both the ERH and EICA hypotheses together provide a comprehensive understanding of the complexity of the invasion process and the gradual development of environmental resistance to alien species.

Another crucial hypothesis describing the invasion process is the Novel Weapon Hypothesis (NWH). The NWH posits that alien invasive organisms have an advantage over native taxa due to their "biological warfare advantage"; it includes allelopathins and toxins that directly affect the fitness of neighboring plants. Pathogens transmitted from alien to native plants may also serve as such "novel weapons" (Figure 1) [7]. Additionally, in 1999, Simberloff and Von Holle [30] introduced the term *invasional meltdown* to describe how non-native species facilitate one another's invasions, enhancing their survival, spread, and impact. Based on that, invasive alien plants may serve as a wharf to new pathogens and facilitate their arrival.

3. Spillover and Spillback Mechanisms from the Sight of Novel Weapon Hypothesis

Over time, alien species become recognizable to organisms in the environment into which they have been introduced. This is evident, for example, in the gradual establishment of fungi, bacteria and pests. While this phenomenon can slow down the expansion of invasive alien species (IAS), it also may have some negative consequences. Spillback mechanism assumes that local native pathogens colonize alien species, leading to an unnatural increase in the number of colonies of these pathogens. This results in an increase in the infections rate in the environment. Although the IAS plant may still have sufficient resources to sustain itself despite being affected by the new pathogen, the pathogen often returns to its native host, causing stronger effects on native plants than on the alien ones (Figure 2). Examples of this mechanism include the following IAS species: *Avena fatua* L. and *Bromus* sp. affected by barley/cereal yellow dwarf viruses (B/CYDVs) in California; Siam Weed (*Chromolaena odorata* (L.) R.M.King & H.Rob) in India, affected by soil fungi like *Fusarium* spp. and *Fusarium semitectum* Berk. & Ravenel; and Ice plant (*Carpobrotus edulis* (L.) N.E. Br) in Europe, affected by chytrid fungi [7]. As emphasized by Strauss et al. [7], the spillback mechanism occurs more often in the insect–plant relationship than in the fungi–plant relationship.

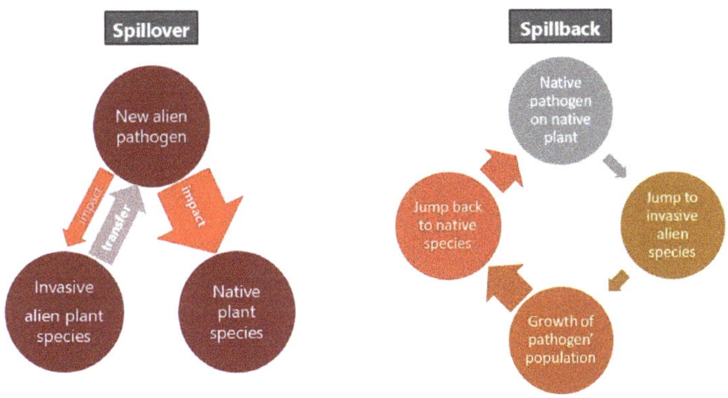

Figure 2. Spillback and spillover mechanisms.

In contrast, spillover is a process wherein a non-native alien pathogen or pest is carried by an invasive alien species to a new area than infect local organisms. The impact on the invasive alien plant, with which it has coexisted for years, is typically smaller than on the new host plant. Native species, with no resistance to the new pathogen, become vulnerable to its attack [7]. Examples of this mechanism include the spread of *C. ribicola*, which causes white pine blister rust in North American pines [31,32], *Cryphonectria parasitica* (Murrill) M.E. Barr, responsible for chestnut blight in North America and Europe [33], and *Austropuccinia psidii* (G. Winter) Beenken, causing myrtle rust that threaten *Myrtaceae* species native to New Zealand and Europe [34]. A recent study by Chen et al. [35] revealed that some plant pathogens, such as fungi from the family *Didymellaceae* associated with the invasive species *Ageratina adenophora* (Spreng.) King & H.Rob., are more virulent toward native plants than their invasive hosts. Notably, many of these pathogens remain unknown to science. This highlights a key challenge in studying spillover mechanisms in interactions between invasive alien species and plant pathogens. Limited plant monitoring in the past and a lack of molecular studies have led to many introductions of plant pathogens being overlooked. Over time, these pathogens spread to new areas, making it difficult to trace their origins [35]. This is why spillover involving insects has historically been better studied, while the lack of accessible and affordable identification methods has limited the number of documented cases for fungal pathogens.

As suggested by Strauss et al. [7], the impact of spillover and spillback on native plants can be so significant that the pathogen may drive the invasion process, promoting the predominance of IAS over closely related native species, leading to disease-mediated invasions (DMI).

3.1. Cryptic Lifestyle

It is crucial to consider that some pathogens, particularly fungi, lead a cryptic, endophytic lifestyle for most of their existence, remaining beyond the reach of researchers [36–39], which can create favorable conditions for the uncontrolled spread (Figure 3). The study of cryptogamous fungi, which requires specialized equipment for detection, also leads to underestimation of the prevalence of this phenomenon [39]. Moreover, research on animals is more prevalent [40–44], and the focus of phytopathology, which deals with plant diseases, is often directed towards crop plants, leaving wild-growing native or non-economically important plants understudied. Additionally, it is challenging to determine whether a particular pathogen colonizing the tissues of an invasive species is an alien, native, cryptogenic, or a cosmopolitan taxon [45]. These factors contribute to invasive plants accumulating undescribed species, including potential pathogens [41]. According to the authors, the main reasons are as follows: (1) limited research on fungi–invasive plant interactions; (2) the amplification of previously rare fungi by introduced invasive alien plants; and (3) the insufficient investigation of fungi co-introduced with IAS plants from their native range. This lack of information creates favorable conditions for the uncontrolled spread of IAS' plant pathogens (Figure 3).

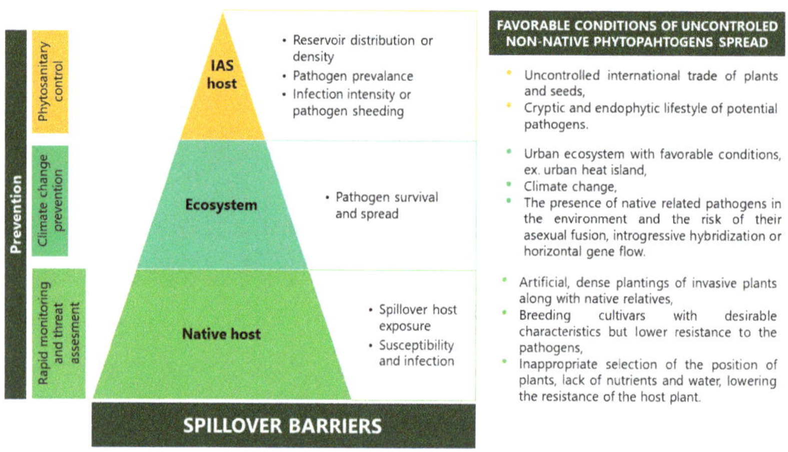

Figure 3. The barriers impacting the spillover mechanism at three following stages: invasive alien species host stage (yellow), ecosystem stage (turquoise), and native host stage (vivid green) (following Sokolow et al. [6]). Actions that can be used to prevent spillover at every stage of its spread are demonstrated on the left, favorable conditions that support transmission are on the right.

The cryptic life of many fungi is one of the most difficult features to monitor, especially in the context of international plant phytosanitary monitoring. Endophytic fungi, settling tissues of plants, can alter their life strategy during their individual lifespan, activating features that enable them to initiate pathogenesis on the host plant [46–48]. They may do this, for instance, in response to changing environmental conditions, increased insolation, temperature, or exposure to other stressors [47,48]. The host plant's reaction to a potentially pathogenic fungal species also varies depending on the species/variety and the microbiome present, where antagonistic relationships between microorganisms can occur [49]. Therefore, potentially cryptic, unnoticed fungal species transferred to a new area may behave differently in relation to the non-native host plant, triggering the parasitic phase in its tissues. This makes the potential prevention of the invasive alien pathogens spread even more challenging and hard to assess. Another factor that can affect the live strategy shifts are changes in environmental conditions caused by climate changes. Increasing temperatures and insolation not only opens the natural passages for pathogens drift, like e.g., melting the glaciers on mountain ranges, which was important in *Biscogniauxia nummularia* (Bull.) Kuntze drift [50,51], but also can lead to initiation of the pathogenesis on the host plant [46–48] (Figure 3).

3.2. Transfer of Genes

The breakage of natural borders and barriers, like oceans and mountain ranges, can lead to the unprecedented meetings of related pathogen species, which had never shared the development or stayed in isolation for a long time, which let them to evolve in separation. These events may potentially lead to uncontrolled gene transfer through introgressive hybridization, asexual fusion, or horizontal transfer (Figure 3) [52], which can resolve in the emergence of virulent and potentially dangerous for our native flora and crops pathogens. A similar was recently observed in the Balkan Peninsula, where *B. nummularia* and *Biscogniauxia anceps* (Sacc.) J.D. Rogers, Y.M. Ju & Cand., underwent introgresive hybrydysation, resulting in the more virulent pathogen *Biscogniauxia destructive* Vujanovic, which causes dieback of beeches in Montenegro [50]. Introgressive hybridization has also substantially contributed to the successful spread of plant pathogens such as *Ophiostoma ulmi* (Buisman) Nannf. and *Ophiostoma novo-ulmi* Brasier, the causal agents of Dutch elm disease, and other tree pathogens such *Melampsora*, the primary cause of rust disease [53]. Also, other authors suggest the emergence of new pathogens that had evolved from interspecific hybridization between endemic and invading fungi, which raise the importance of the geographical barrier even more than before [52–56].

In sight of molecular plant pathology, it is worth concerning the number of "entry points" of the new invasive pathogens and the diversity of the IAS pathogens' population, to assess the possible scenarios of its development in new areas [57]. One of the best-documented example in the history of alien plant pathogens that led to the epidemy outbreak is *P. infestans*, which caused late blight of potatoes. The A1 clonal mating type initially spread from the Americas to Europe in 1845 but the second "entry point" of A2 mating types, which happened in 1980, let the previous clonal genotype to undergo sexual reproduction [58]. This increased the adaptability of the pathogen and provided a new life-cycle stage, oospores, enabling survival in the absence of host plant material between seasons [59]. A very similar situation occurred with myrtle rust in Florida [60,61]. Therefore, in early monitoring of invasive alien pathogens that were transferred with IAS, it is important to recognize the local genetical population diversity. This type of information could aid in decision-making and the establishment of phytosanitary politics.

3.3. Time Matters

When considering a potentially pathogenic fungus introduced with an IAS plant, it is essential to recognize that, upon crossing the geographical barrier, it may adopt an endophytic cryptogamous lifestyle but undergo changes in its characteristics over time. Additionally, the timing of its introduction into a new area should be taken into account. Facultative pathogens, or those with a wide host spectrum, may need time to fully develop, undergoing similar stages of invasion as plants and animals, making them a potential "time bomb". Unfortunately, removing the IAS host transmitting the pathogen might not be sufficient after introducing a new pathogen into the environment. Alien pathogens may survive even if their original hosts do not, as the spill of the pathogen into the external environment may have already occurred [62].

4. Gardening and Urban Areas as Crucial Hubs for the Spread of IAS-Associated Pathogens

4.1. Environmental Chances to Pathogen Spread and Survival

Last but not the least, the environment plays a crucial rule in plant disease development (Figure 3). This triangular relationship (host–pathogen–environment) is unique to phytopathology in comparison to veterinary and medical sciences because terrestrial plants possess little thermal storage capacity, and their immobility precludes escape from an inhospitable environment [63]. During discussions about plant pathogens development, we cannot forget about this factor, especially because of the predominance of fungi in phytopathology, which also are highly dependent on environment. To stress this point, invasive alien species serve as reservoirs for pathogens, with the second most common group of isolated pathogens being fungi belonging to *Ascomycota* (84 different species) [64]. The climatic condition that an introduced IAS pathogen encounters in the new area determines its chances of spreading and surviving (Figure 3). For example, the minimum or maximum temperatures of host and pathogen can differ (as it is in case of coffee rust [65]), limiting the pathogen range of occurrence. However, the same example shows how small fluctuations of temperatures, caused by global rise of temperature, can impact this subtle relation, leading to the expansion of *Hemileia vastatrix* Berk. & Broome in areas that were previously unsuitable for it [65–67]. In sight of pathogens' temperature preferences, there is concern about the impact of urban heat islands in IAS' pathogen establishment and survival, especially considering that urban areas are the first and most invaded areas [68].

4.2. Host Factor: Invasive Alien Plants in Urban Ecosystems

According to Hulme's estimates [69], up to 80% of neophytes in Europe initially arrived due to their decorative values, serving as ornamental plants. Thus, anthropogenic communities, as noted by Tokarska-Guzik et al. [4], harbor the highest number of invasive plants. Thus, urban areas are one of the first areas where the IAS breaks the geographical barrier and enters a new environment. It is estimated that as much as 67.8% of urban areas are invaded by alien plants, making it the environment with the highest occupancy of area by IAS. In comparison, 43.9% of forests and woodlands and 42.9% of croplands are invaded by alien plants [68]. Given that many IAS plants are overrepresented species used in horticulture (387 species, 51% of them being trees) across the world [70,71], this highlights the significant problem of phytopathogen transfer on invasive plants and their implications for global economy and sustainable agriculture development.

An important aspect to consider in sight of the uncontrolled transfer of IAS' pathogens is the botanical families to which alien invasive plants belong. Pathogens with medium specificity to the host often inhabit several plant species closely related to each other, typically within the same botanical family (ex. *Colletotrichum* family) [72]. Consequently, the affinity of invasive alien species with local taxa increases the risk of transmitting potential phytopathogens, which can adapt to new environments, spread to native plants, and pose threats to ornamental, orchard, or horticultural crops. A compelling example is the rose family, with Europe alone hosting 25 alien invasive species belonging to the *Rosaceae* family, including four in the *Cotoneaster* family, one in *Pyracantha*, one in *Rosa*, and four in *Rubus* [70,71]. Similarly, North America is a natural area of 25 species, including two in the *Cotoneaster* family, six in *Rosa*, and eight in *Rubus*. This family includes several economically significant orchard plants in the temperate regions of the northern hemisphere, such as apple trees, pear trees, quinces, chokeberries, wild strawberries, plums, and blackberries. Therefore, any spread of its pathogens can be devastating to local fruit production.

4.3. International Trade Market

The transfer of alien species, along with potentially pathogenic fungi, through seeds and nursery materials is inevitable [73,74] (Figure 3). Global and uncontrolled transfer of seedling caused the spread of many invasive alien pathogens, carried on the hosts tissues to the new areas. A good example is one of the most important pathogens of North America—*C. ribicola*—causing white pine blister rust, which lead to high mortality of western white pine, sugar pine, limber pine and whitebark pine across Canada and the United States [31,32]. In this particular case, the key to the survival of the pathogen was the natural occurrence of its alternative host, representatives of the *Ribes* family (currant and gooseberry), where the pathogen develops part of its lifecycle. It highlights the complexity of the relationships between pathogens, their hosts, and environment, which allows a new plant pathogen to establish itself in a new area. It is also important to note that in gardening, while seeking desirable consumer traits, many of these varieties lack crucial pathogen resistance genes, making them more susceptible to diseases [75] (Figure 3). This, in turn, has implications for the exchange of pathogens and genes between ornamental plants and their invasive relatives, particularly in urban environments, where anthropogenic alterations can weaken the host plant's ability to respond to pathogens [76–79], aiding the potential adaptation of pathogens to new hosts.

4.4. Gardening Strategies That Support Invasive Alien Plants' Pathogen Transfer

It is worth noting that urban centers, another form of anthropogenic environment, may also facilitate the dispersion of pathogens associated with IAS and their subsequent migration beyond metropolitan regions. This is primarily due to plant selection in urban gardening and the unique characteristics of these anthropogenic ecosystems. Unfortunately, IAS are still cultivated in urban centers, primarily due to their ease of cultivation, low environmental requirements, and their resistance to adverse abiotic factors. Notably, in a list of woody plants best adapted to the urban environment, provided by Yang et al. [76], many species with the highest adaptability scores are invasive alien species. This phenomenon can be attributed to the remarkable phenotypic plasticity of invasive alien species, which enables them to maintain productivity and vitality even under unfavorable conditions [80–82]. However, IAS plants are often planted in close proximity to native relatives in dense plantings in small urban areas, which promotes the sharing and drift of pathogens in between closely related species, for example alien and local species of maples, oaks, and coniferous trees (Figure 3). An example of shared pathogens is *Colletotrichum acericola* Patejuk, Piątek, Czachura & Baturo-Cieśniewska, which, in in vitro conditions, has shown the possibility of host jumping to native European maples [74] (unpublished data). The presence of invasive alien ornamental plants (e.g., *Lantana camara* L. and *Ochna serrulate* (Hochst.) Walp.) also facilitate spread and impact of myrtle rust (*A. psidii*) to native Australian and New Zealand *Myrtlaceae* plants, such as *Archirhodomyrtus beckleri* (F.Muell.) A.J.Scott, *Decaspermum humile* (G.Don) A.J.Scott, *Gossia hillii* (Benth.) N.Snow & Guymer, and *Rhodamnia maideniana* C.T.White [83].

Furthermore, urban conditions, characterized by limited water and nutrient availability and air pollution, pose challenges for plant growth. These conditions reduce the resistance of native plants to potentially new pathogens associated with IAS, facilitating their rapid adaptation to new hosts and thereby breaking down the resistance of native plants (Figure 3). Poor plant selection based solely on ornamental features can also expedite the spread of IAS pathogens. As a result, urban areas emerge as pivotal locations for the occurrence of the spillover phenomenon and serve as centers for the adaptation of IAS pathogens to potential new hosts.

5. Counteracting the Spread of Invasive Plants: Pathogens as "Green Agents" and Their Role in Long-Term Monitoring

5.1. Legislation and Programs Against Invasive Alien Plants' Pathogen Spread

The topic of counteracting the invasion of alien species is addressed by numerous legal acts and international agreements, including the Convention on Biological Diversity [1], as well as the EU Strategy for the Conservation of Biodiversity for the period up to 2020 [84]. They assume the identification and categorization of knowledge on the mechanisms that determine spread of IAS organisms, potential methods of IAS elimination, as well as the establishment of a comprehensive legal framework. So far, the mechanical removal of plants, chemical herbicide treatments (offshoot applications, injections, herbicide markers), the removal of the top layer of soil, deep plowing, liming, and limiting seed dispersal are commonly used [4]. Repeated mowing and soil screening (by sifting) are also used. These treatments are associated with huge costs and often require many years of repetition for the treatment to be highly effective [4,85,86]. One alternative method, which does not require such large expenditures, is the use of the IAS' natural enemies, such as fungal or bacterial pathogens, with commercial potential.

Preventing the spread of invasive alien species requires a comprehensive understanding of invasion dynamics and acquiring specialized knowledge about each species and its interactions with the environment [4,87]. In this regard, various international programs have been initiated to collect data and raise awareness about the subject. Examples include the European programs DAISIE (Delivering Alien Invasive Species Inventories for Europe) [88], HARMONIA+, and PANDORA+ [89–91]. Updated databases on invasive species are also accessible, such as NOBANIS [92], containing information about alien species in Central and Northern Europe, and the Polish database "Alien species in Poland" [93].

Alien invasive species vary in factors that determine their invasiveness. To enable an objective assessment of the harmfulness of a given taxon, various metrics have been developed to classify specific characteristics of such species, including the extent of their establishment on a national or regional scale. Tokarska-Guzik et al. [4] listed four categories of invasive species that allowed for a preliminary evaluation of their impact on the environment. The assessment considered features of the taxon, such as the degree of its establishment, population size and dynamics, the type of colonized habitats, and the economic, environmental, and social threats posed by the alien species.

Recognizing the complexity of the invasion issue, recent projects like PANDORA+ have taken into account the significance of pathogen transfer by non-native species and the impact of local pathogens on plant health [90]. Pathogens have been incorporated into the Species Invasiveness Analysis Survey, conducted for each taxon classified as invasive [93–95], with consideration of both economic and sanitary aspects. The European and Mediterranean Plant Protection Organization website [96] provides a partial list of pathogens and pests associated with invasive species.

Developing a global database of pathogens accompanying invasive alien plants and regular monitoring of their health status become crucial in the context of threats posed by the spillover and spillback phenomena. Rapid identification of potential threats and establishing protocols for quick responses can significantly reduce these phenomena and prevent their spread across continents. Historical cases have demonstrated that the spread of foreign pathogens can lead to the near extinction of native or cultivated species, as seen with Dutch elm disease, ash dieback, dogwood anthracnose, chestnut blight, white pine blister rust, potato blight, and banana dieback [97–99]. According to Mułenko et al. [100], these types of parasitic fungi (alien and fast spreading) meet the criteria for invasive organisms, making them a group of organisms that requires constant monitoring of their spread, similar to plants and animals, legislated as obligatory by many countries.

The most important aspects are the prevention and rigorous control of plant material transported across borders. Countries with valuable and vulnerable ecosystems, such as New Zealand or Australia, have implemented strict phytosanitary regulations to prevent the transmission of invasive plants and their pathogens. However, research indicates that many parts of the world do not prioritize this issue, as evidenced by studies showing the scale of transfer of foreign fungi colonizing seeds of non-native plants purchased from commercial stores [73]. There is currently no monitoring or rapid warning system for the transfer of potentially dangerous species on the surface of invasive plants, which could limit their spread. Controlling everything that crosses borders would be extremely difficult if not impossible, but there are some of the most dangerous species that should be investigated. Such species have been identified for Poland and Europe [64]. Similar meta-analyses can be performed for other countries or areas. Perhaps, considering adding pathogens of invasive alien species to international quarantine lists could facilitate data collection, education, and implementation of procedures to prevent the movement of these pathogens to new areas along with their host plants.

5.2. IAS Pathogen on Service: "Green Agents"

Another concerning and contentious topic related to pathogens of invasive alien plants is the concept of creating so-called "green agents," which involves the intentional introduction of natural enemies of invasive alien plants to control them. This approach is often more economically viable than mechanical removal or the use of strong herbicides. Understanding the mycobiota inhabiting neophytes may be particularly useful in the context of developing and utilizing mycoherbicides, which could serve as natural remedies to slow down the expansion of certain invasive species [101–105]. Extensive research on this subject has been carried out in Poland, particularly in the Wigry National Park [57,106], the Drawa National Park [57], and Tatra National Park [107], where long-term monitoring of the health condition of invasive alien plants is combined with mycological analysis. However, only isolated cases of finding pathogenic species associated with invasive plants are known [108,109].

However, creating green agents is a challenging and controversial task. Monitoring invasive plants allows for the identification of suitable pathogens that have already been established in a new area alongside their host plant. This can help in selecting genotypes with satisfactory virulence and effectiveness for developing mycoherbicides. It was tested, for example, on *Puccinia komarovii* var. *glandulifera* R.A. Tanner, C.A. Ellison, L. Kiss & H.C. Evans rust, infecting *Impatiens glandulifera* Royle in England [103]. However, most green agents are foreign to the local environment, raising concerns about the spillover phenomenon, where introduced species can become invasive themselves. Arguments often cited in this discussion include the escape of the Asian ladybug from greenhouses that becomes invasive in Europe [110], or cane toads in Australia that were used to control the native grey-backed cane beetle (*Dermolepida albohirtum* C.O.Waterhouse) [111] and French's beetle (*Lepidiota frenchi* Blackburn) [112]. Nevertheless, current research and procedures accompanying the creation of new green agents are much more refined. CABI (Centre for Agriculture and Bioscience International) is a prominent organization in the world dealing with research and the creation of such biological preparations, and it has achieved numerous successes in this field [112]. Examples of successful green agents include the usage of rust fungus *Maravalia cryptostegiae* (Vestergr.) Y. Ono against invasive alien rubbervine *Cryptostegia grandiflora* R.Br. in Australia [113] and the usage of *Aphthona* flea beetle to successfully control *Euphorbia esula* L., an alien and invasive plant in North America [114]. Currently, ongoing projects are exploring biocontrol methods for the Madagascar rubbervine (*Cryptostegia madagascariensis* Bojer ex Decne.), which threatens Brazil's biodiversity. A promising candidate for this is the rust fungus *Uredo cryptostegiae* Vestergr. [115]. Efforts are also underway to control invasive blackberry on the Galápagos Islands, with promising results from two rust species belonging to the genus *Phragmidium* and two insect species: a leaf-rolling weevil and a sawfly [116]. Additionally, trials are being conducted with three species of weevils to manage garlic mustard in North America [117].

However, green agents face numerous legislative problems. In the light of current European laws [118], the introduction of a pathogen of foreign origin into the environment is considered equivalent to introducing an invasive alien organism and is therefore illegal. This legal stance also hinders research into alien species that are already present in the environment. Despite the procedures accompanying the development of green agents, which prioritize testing in vitro and in vivo infection tests on closely and distantly related native plants, the new legislation has completely halted the progress of these studies in Europe [119,120].

6. Conclusions

The topic of pathogens colonizing the invasive plants is complex, multifaceted, and requires a comprehensive perspective encompassing phytopathology, botany, ecology, and economics to gain a full understanding. Observing the adaptation of plants to new environments and the reciprocal changes in the environment, due to the introduction of invasive species, allows us to be a witness the gradual evolution of both components and its far-reaching consequences. From a nature protection standpoint, the negative impacts of introducing invasive plants of foreign origin are evident. The scientific community, along with activists and educators, has been taking actions to control invasive alien species for years, but still there is still much to be done.

The research supports the notion that while pathogens can act as natural checks on invasive species over time, they also pose significant risks to native ecosystems and agricultural practices. The cryptic lifestyles of many fungal pathogens, combined with the potential for genetic transfer and shifts in life strategy, further complicate the challenge of managing these biological invasions. The role of urban and anthropogenic environments in fostering the establishment and spread of invasive species and their pathogens requires special attention, given their status as hubs for pathogen adaptation and spillover phenomena.

Future management strategies should prioritize an integrated approach that includes rigorous phytosanitary controls, ongoing monitoring, and the development of global pathogen databases. While the use of pathogens as biological control agents presents a promising avenue for mitigating invasions, it requires careful consideration of ecological risks and legislative constraints. Expanding research into the mechanisms underlying plant-pathogen interactions and the impacts of environmental changes, such as climate warming, will be essential for improving our ability to predict, monitor, and mitigate the effects of invasive alien plants and their associated pathogens.

Ultimately, combating the challenges posed by invasive alien plants and their pathogens will demand enhanced international collaboration, comprehensive legislative frameworks, and adaptive management strategies that balance ecological, economic, and societal considerations. By addressing these challenges proactively, we can better safeguard biodiversity, ecosystem services, and human well-being from the multifaceted threats of biological invasions.

Author Contributions: Conceptualization, K.P. and K.N.; writing—original draft preparation, K.P., K.N. and P.P; writing—review and editing, P.P., J.B. and W.P.; visualization K.P. and P.P.; supervision, W.P. and K.N. All authors have read and agreed to the published version of the manuscript.

Funding: This research received no external funding.

Conflicts of Interest: The authors declare no conflicts of interest.

References

1. UNCED. The Convention on Biological Diversity (CBD) from Rio de Janeiro 5 June 1992. In Proceedings of the United Nations Conference on Environment and Development (UNCED), Rio Earth Summit, Rio de Janeiro, Brazil, 3–14 June 1992.
2. Elton, C.S. *The Ecology of Invasions by Animals and Plants*; Methuen: London, UK, 1958.
3. Reinhardt, F.; Herle, M.; Bastiansen, F.; Streit, B. Economic Impact of the Spread of Alien Species in Germany. Available online: https://www.researchgate.net/publication/235702159_Economic_Impact_of_the_Spread_of_Alien_Species_in_Germany (accessed on 1 November 2021).
4. Tokarska-Guzik, B.; Dajdok, Z.; Zając, M.; Zając, A.; Urbisz, A.; Danielewicz, W.; Hołodyński, C. *Rośliny Obcego Pochodzenia w Polsce*; Generalna Dyrekcja Ochrony Środowiska: Warszawa, Poland, 2012; ISBN 9788362940349.
5. Tokarska-Guzik, B. *The Establishment and Spread of Alien Plant Species (Kenophytes) in the Flora of Poland*; Wydawnictwo Uniwersytetu Śląskiego: Katowice, Poland, 2005; ISBN 83-226-1485-3.
6. Sokolow, S.H.; Nova, N.; Pepin, K.M.; Peel, A.J.; Pulliam, J.R.C.; Manlove, K.; Cross, P.C.; Becker, D.J.; Plowright, R.K.; McCallum, H.; et al. Ecological Interventions to Prevent and Manage Zoonotic Pathogen Spillover. *Philos. Trans. R. Soc. Lond. B Biol. Sci.* **2019**, *374*, 20180342. [CrossRef] [PubMed]
7. Strauss, A.; White, A.; Boots, M. Invading with Biological Weapons: The Importance of Disease-Mediated Invasions. *Funct. Ecol.* **2012**, *26*, 1249–1261. [CrossRef]
8. Holzmueller, E.; Jose, S. Invasive Plant Conundrum: What Makes the Aliens so Successful? *J. Trop. Agric.* **2009**, *47*, 18–29.
9. Aikio, S.; Duncan, R.P.; Hulme, P.E. Lag-Phases in Alien Plant Invasions: Separating the Facts from the Artefacts. *Oikos* **2010**, *119*, 370–378. [CrossRef]
10. Kowarik, I. Time Lags in Biological Invasions with Regard to the Success and Failure of Alien Species. In *Plant Invasions—General Aspects and Special Problems*; Pysek, P., Prach, K., Rejmanek, M., Wade, P.M., Eds.; SPB Academic Publishing: Amsterdam, The Netherlands, 1995; pp. 15–38.
11. Maron, J.L.; Vilà, M. When Do Herbivores Affect Plant Invasion? Evidence for the Natural Enemies and Biotic Resistance Hypotheses. *Oikos* **2001**, *95*, 361–373. [CrossRef]

12. Keane, R.M.; Crawley, M.J. Exotic Plant Invasions and the Enemy Release Hypothesis. *Trends Ecol. Evol.* **2002**, *17*, 164–170. [CrossRef]
13. Blackburn, T.M.; Ewen, J.G. Parasites as Drivers and Passengers of Human-Mediated Biological Invasions. *Ecohealth* **2017**, *14*, 61–73. [CrossRef] [PubMed]
14. Mitchell, C.E.; Power, A.O. Release of Invasive Plants from Fungal and Viral Pathogens. *Nature* **2003**, *421*, 625–627. [CrossRef]
15. Cincotta, C.L.; Adams, J.M.; Holzapfel, C. Testing the Enemy Release Hypothesis: A Comparison of Foliar Insect Herbivory of the Exotic Norway Maple (*Acer platanoides* L.) and the Native Sugar Maple (*A. saccharum* L.). *Biol. Invasions* **2009**, *11*, 379–388. [CrossRef]
16. Adams, J.M.; Fang, W.; Callaway, R.M.; Cipollini, D.; Newell, E. A Cross-Continental Test of the Enemy Release Hypothesis: Leaf Herbivory on *Acer platanoides* (L.) Is Three Times Lower in North America than in Its Native Europe. *Biol. Invasions* **2009**, *11*, 1005–1016. [CrossRef]
17. Agrawal, A.A.; Kotanen, P.M. Herbivores and the Success of Exotic Plants: A Phylogenetically Controlled Experiment. *Ecol. Lett.* **2003**, *6*, 712–715. [CrossRef]
18. Parker, J.D.; Hay, M.E. Biotic Resistance to Plant Invasions? Native Herbivores Prefer Non-Native Plants. *Ecol. Lett.* **2005**, *9*, 959–967. [CrossRef] [PubMed]
19. van Kleunen, M.; Fischer, M. Release from Foliar and Floral Fungal Pathogen Species Does Not Explain the Geographic Spread of Naturalized North American Plants in Europe. *J. Ecol.* **2009**, *97*, 385–392. [CrossRef]
20. Schultheis, E.H.; Berardi, A.E.; Lau, J.A. No Release for the Wicked: Enemy Release Is Dynamic and Not Associated with Invasiveness. *Ecology* **2015**, *96*, 2446–2457. [CrossRef] [PubMed]
21. Rai, P.K. What Makes the Plant Invasion Possible? Paradigm of Invasion Mechanisms, Theories and Attributes. *Environ. Skept. Crit.* **2015**, *4*, 36.
22. Blossey, B.; Notzold, R. Evolution of Increased Competitive Ability in Invasive Nonindigenous Plants: A Hypothesis. *J. Ecol.* **1995**, *83*, 887–889. [CrossRef]
23. Siemann, E.; Rogers, W.E.; Dewalt, S.J. Rapid Adaptation of Insect Herbivores to an Invasive Plant. *Proc. R. Soc. B Biol. Sci.* **2006**, *273*, 2763–2769. [CrossRef]
24. Gioria, M.; Osborne, B.A. Resource Competition in Plant Invasions: Emerging Patterns and Research Needs. *Front. Plant Sci.* **2014**, *5*, 501. [CrossRef] [PubMed]
25. Gruntman, M.; Segev, U.; Glauser, G.; Tielbörger, K. Evolution of Plant Defences along an Invasion Chronosequence: Defence Is Lost Due to Enemy Release—But Not Forever. *J. Ecol.* **2017**, *105*, 255–264. [CrossRef]
26. Gioria, M.; Pyšek, P.; Osborne, B.A. Timing Is Everything: Does Early and Late Germination Favor Invasions by Herbaceous Alien Plants? *J. Plant Ecol.* **2018**, *11*, 4–16. [CrossRef]
27. Siemann, E.; Rogers, W.E. Increased Competitive Ability Of An Invasive Tree May Be Limited By An Invasive Beetle. *Ecol. Appl.* **2003**, *13*, 1503–1507. [CrossRef]
28. Siemann, E.; DeWalt, S.J.; Zou, J.; Rogers, W.E. An Experimental Test of the EICA Hypothesis in Multiple Ranges: Invasive Populations Outperform Those from the Native Range Independent of Insect Herbivore Suppression. *AoB Plants* **2016**, *9*, plw087. [CrossRef]
29. Najberek, K.; Solarz, W.; Pusz, W.; Patejuk, K.; Olejniczak, P. Two Sides of the Same Coin: Does Alien Impatiens Balfourii Fall into an Ecological Trap after Releasing from Enemies? *Environ. Exp. Bot.* **2020**, *176*, 104103. [CrossRef]
30. Simberloff, D.; Von Holle, B. Positive Interactions of Nonindigenous Species: Invasional Meltdown? *Biol. Invasions* **1999**, *1*, 21–32. [CrossRef]
31. Maloy, O.C. White Pine Blister Rust. *Plant Health Instr.* **2003**, *3*. [CrossRef]
32. Geils, B.W.; Hummer, K.E.; Hunt, R.S. White Pines, Ribes, and Blister Rust: A Review and Synthesis. *For. Pathol.* **2010**, *40*, 147–185. [CrossRef]
33. Bieniek, P.; Oszako, T.; Pusz, W. Zagrożenie drzewostanów dębowych przez *Cryphonectria parasitica*. *Sylwan* **2020**, *164*, 576–582. [CrossRef]
34. Paap, T.; Santini, A.; Rodas, C.A.; Granados, G.M.; Pecori, F.; Wingfield, M.J. *Myrtus communis* in Europe threatened by the pandemic and South African strains of the myrtle rust pathogen *Austropuccinia psidii* (Sphaerophragmiaceae, Pucciniales). *NeoBiota* **2023**, *84*, 41–46. [CrossRef]
35. Chen, L.; Yang, A.-L.; Li, Y.-X.; Zhang, H.-B. Virulence and Host Range of Fungi Associated with the Invasive Plant *Ageratina adenophora*. *Front. Microbiol.* **2022**, *13*, 857796. [CrossRef]
36. Condé, T.O.; Leão, A.F.; Dutra, Y.L.G.; Rosado, A.W.C.; de Carvalho Neves, S.; Fraga, L.M.S.; Kasuya, M.C.M.; Pereira, O.L. Shedding Light on the Darkness: A New Genus and Four New Species in the Family *Chaetomiaceae* from Brazilian Neotropical Caves Revealed by Multi-Gene Phylogenetic Analyses. *Mycol. Prog.* **2023**, *22*, 49. [CrossRef]
37. Bamisile, B.S.; Dash, C.K.; Akutse, K.S.; Keppanan, R.; Wang, L. Fungal Endophytes: Beyond Herbivore Management. *Front. Microbiol.* **2018**, *9*, 329443. [CrossRef] [PubMed]
38. Hawksworth, D.L. Global Species Numbers of Fungi: Are Tropical Studies and Molecular Approaches Contributing to a More Robust Estimate? *Biodivers. Conserv.* **2012**, *21*, 2425–2433. [CrossRef]
39. Hawksworth, D.L. The Magnitude of Fungal Diversity: The 1.5 Million Species Estimate Revisited. *Mycol. Res.* **2001**, *105*, 1422–1432. [CrossRef]

40. Balding, M.; Williams, K.J.H. Plant Blindness and the Implications for Plant Conservation. *Conserv. Biol.* **2016**, *30*, 1192–1199. [CrossRef] [PubMed]
41. Fukushima, C.S.; Mammola, S.; Cardoso, P. Global Wildlife Trade Permeates the Tree of Life. *Biol. Conserv.* **2020**, *247*, 108503. [CrossRef]
42. Parsley, K.M. Plant Awareness Disparity: A Case for Renaming Plant Blindness. *Plants People Planet* **2020**, *2*, 598–601. [CrossRef]
43. Wandersee, J.H.; Schussler, E.E. Preventing Plant Blindness. *Am. Biol. Teach.* **1999**, *61*, 82–86. [CrossRef]
44. Adamo, M.; Chialva, M.; Calevo, J.; Bertoni, F.; Dixon, K.; Mammola, S. Plant Scientists' Research Attention Is Skewed towards Colourful, Conspicuous and Broadly Distributed Flowers. *Nat. Plants* **2021**, *7*, 574–578. [CrossRef]
45. Pociecha, A.; Solarz, W.; Najberek, K.; Wilk-Woźniak, E. Native, Alien, Cosmopolitan, or Cryptogenic? A Framework for Clarifying the Origin Status of Rotifers. *Aquat. Biol.* **2016**, *24*, 141–149. [CrossRef]
46. Delaye, L.; García-Guzmán, G.; Heil, M. Endophytes versus Biotrophic and Necrotrophic Pathogens—Are Fungal Lifestyles Evolutionarily Stable Traits? *Fungal Divers.* **2013**, *60*, 125–135. [CrossRef]
47. Álvarez-Loayza, P.; White, J.F.; Torres, M.S.; Balslev, H.; Kristiansen, T.; Svenning, J.-C.; Gil, N. Light Converts Endosymbiotic Fungus to Pathogen, Influencing Seedling Survival and Niche-Space Filling of a Common Tropical Tree, *Iriartea deltoidea*. *PLoS ONE* **2011**, *6*, e16386. [CrossRef]
48. Hardoim, P.R.; van Overbeek, L.S.; Berg, G.; Pirttilä, A.M.; Compant, S.; Campisano, A.; Döring, M.; Sessitsch, A. The Hidden World within Plants: Ecological and Evolutionary Considerations for Defining Functioning of Microbial Endophytes. *Microbiol. Mol. Biol. Rev.* **2015**, *79*, 293–320. [CrossRef]
49. Chlebicki, A. Biofilmy Występują Powszechnie w Przyrodzie. *Kosmos* **2015**, *64*, 337–345.
50. Vujanovic, V.; Kim, S.H.; Latinovic, J.; Latinovic, N. Natural Fungicolous Regulators of *Biscogniauxia destructiva* sp. nov. That Causes Beech Bark Tarcrust in Southern European (*Fagus sylvatica*) Forests. *Microorganisms* **2020**, *8*, 1999. [CrossRef] [PubMed]
51. Patejuk, K.; Baturo-Cieśniewska, A.; Pusz, W.; Kaczmarek-Pieńczewska, A. Biscogniauxia Charcoal Canker—A New Potential Threat for Mid-European Forests as an Effect of Climate Change. *Forests* **2022**, *13*, 89. [CrossRef]
52. Drenth, A.; McTaggart, A.R.; Wingfield, B.D. Fungal Clones Win the Battle, but Recombination Wins the War. *IMA Fungus* **2019**, *10*, 18. [CrossRef] [PubMed]
53. Stukenbrock, E.H. The Role of Hybridization in the Evolution and Emergence of New Fungal Plant Pathogens. *Phytopathology* **2016**, *106*, 104–112. [CrossRef]
54. Menardo, F.; Praz, C.R.; Wyder, S.; Ben-David, R.; Bourras, S.; Matsumae, H.; McNally, K.E.; Parlange, F.; Riba, A.; Roffler, S.; et al. Hybridization of Powdery Mildew Strains Gives Rise to Pathogens on Novel Agricultural Crop Species. *Nat. Genet.* **2016**, *48*, 201–205. [CrossRef] [PubMed]
55. Newcombe, G.; Stirling, B.; McDonald, S.; Bradshaw, H.D. *Melampsora × columbiana*, a Natural Hybrid of *M. medusae* and *M. occidentalis*. *Mycol. Res.* **2000**, *104*, 261–274. [CrossRef]
56. Brasier, C.M.; Kirk, S.A. Rapid Emergence of Hybrids between the Two Subspecies of *Ophiostoma novo-ulmi* with a High Level of Pathogenic Fitness. *Plant Pathol.* **2010**, *59*, 186–199. [CrossRef]
57. Baturo-Cieśniewska, A.; Patejuk, K.; Kaczmarek-Pieńczewska, A.; Jurga-Zotow, M. Variability of *Puccinia komarovii* var. *komarovii* Isolates from *Impatiens parviflora* DC. and Its Occurrence in the Drawa and Wigry National Parks in Poland. *Acta Mycol.* **2023**, *58*, 174570. [CrossRef]
58. Fry, W.E.; Goodwin, S.B.; Matuszak, J.M.; Spielman, L.J.; Milgroom, M.G.; Drenth, A. Population Genetics and Intercontinental Migrations of *Phytophthora infestans*. *Annu. Rev. Phytopathol.* **2003**, *30*, 107–129. [CrossRef]
59. Drenth, A.; Tas, I.C.Q.; Govers, F. DNA Fingerprinting Uncovers a New Sexually Reproducing Population of *Phytophthora infestans* in The Netherlands. *Eur. J. Plant. Pathol.* **1994**, *100*, 97–107. [CrossRef]
60. Toome-Heller, M.; Ho, W.W.H.; Ganley, R.J.; Elliott, C.E.A.; Quinn, B.; Pearson, H.G.; Alexander, B.J.R. Chasing myrtle rust in New Zealand: Host range and distribution over the first year after invasion. *Australas. Plant Pathol.* **2020**, *49*, 221–230. [CrossRef]
61. Loope, L.; La Rosa, A.M. *An Analysis of the Risk of Introduction of Additional Strains of the Rust Puccinia psidii Winter ('Ōhi'a Rust) to Hawai'i*; Open File Report 2008–1008; U.S. Geological Survey: Reston, VA, USA, 2008. Available online: http://pubs.usgs.gov/of/2008/1008/ (accessed on 1 December 2024).
62. Solarz, W.; Najberek, K. Alien Parasites May Survive Even If Their Original Hosts Do Not. *Ecohealth* **2017**, *14*, 3–4. [CrossRef] [PubMed]
63. Francl, L.J. The Disease Triangle: A Plant Pathological Paradigm Revisited. *Plant Health Instr.* **2001**, *10*. Available online: https://www.apsnet.org/edcenter/foreducators/TeachingNotes/Pages/DiseaseTriangle.aspx (accessed on 30 September 2024). [CrossRef]
64. Najberek, K.; Olszańska, A.; Tokarska-Guzik, B.; Mazurska, K.; Dajdok, Z.; Solarz, W. Invasive Alien Species as Reservoirs for Pathogens. *Ecol. Indic.* **2022**, *139*, 108879. [CrossRef]
65. Tadesse, Y.; Amare, D.; Kesho, A. Coffee Leaf Rust Disease and Climate Change. *World J. Agric. Sci.* **2021**, *17*, 418–429.
66. Torres Castillo, N.E.; Melchor-Martínez, E.M.; Ochoa Sierra, J.S.; Ramirez-Mendoza, R.A.; Parra-Saldívar, R.; Iqbal, H.M.N. Impact of Climate Change and Early Development of Coffee Rust—An Overview of Control Strategies to Preserve Organic Cultivars in Mexico. *Sci. Total Environ.* **2020**, *738*, 140225. [CrossRef]
67. de Carvalho Alves, M.; Sanches, L. Potential Effects of Spatio-Temporal Temperature Variation for Monitoring Coffee Leaf Rust Progress Under CMIP6 Climate Change Scenarios. *Earth Syst. Environ.* **2022**, *6*, 421–436. [CrossRef]

68. Polce, C.; Cardoso, A.C.; Deriu, I.; Gervasini, E.; Tsiamis, K.; Vigiak, O.; Zulian, G.; Maes, J. Invasive Alien Species of Policy Concerns Show Widespread Patterns of Invasion and Potential Pressure across European Ecosystems. *Sci. Rep.* **2023**, *13*, 8124. [CrossRef] [PubMed]
69. Hulme, P.E. Biological Invasions in Europe: Drivers, Pressures, States, Impacts and Responses. In *Biodiversity Under Threat*; Hester, R., Harrison, R.M., Eds.; Cambridge University Press: Cambridge, UK, 2007.
70. Rejmánek, M.; Richardson, D.M. Trees and Shrubs as Invasive Alien Species—2013 Update of the Global Database. *Divers. Distrib.* **2013**, *19*, 1093–1094. [CrossRef]
71. Richardson, D.M.; Rejmánek, M. Trees and Shrubs as Invasive Alien Species—A Global Review. *Divers. Distrib.* **2011**, *17*, 788–809. [CrossRef]
72. Talhinhas, P.; Baroncelli, R. *Colletotrichum* Species and Complexes: Geographic Distribution, Host Range and Conservation Status. *Fungal Divers.* **2021**, *110*, 109–198. [CrossRef]
73. Franić, I.C.; Prospero, S.; Hartmann, M.; Allan, E.; Auger Rozenberg, M.; Grünwald, N.J.; Kenis, M.; Roques, A.; Schneider, S.; Sniezko, R.; et al. Are Traded Forest Tree Seeds a Potential Source of Nonnative Pests? *Ecol. Appl.* **2019**, *29*, e01971. [CrossRef] [PubMed]
74. Patejuk, K.; Czachura, P.; Owczarek-Kościelniak, M.; Pusz, W.; Najberek, K.; Piątek, M. *Colletotrichum acericola* sp. nov. from Seeds of the Invasive Alien Tree Species *Acer negundo* in Poland. *Plant Pathol.* **2023**, *72*, 1716–1725. [CrossRef]
75. Khoury, C.K.; Brush, S.; Costich, D.E.; Curry, H.A.; de Haan, S.; Engels, J.M.M.; Guarino, L.; Hoban, S.; Mercer, K.L.; Miller, A.J.; et al. Crop Genetic Erosion: Understanding and Responding to Loss of Crop Diversity. *New Phytol.* **2022**, *233*, 84–118. [CrossRef]
76. Yang, J.; Chang, Y.; Yan, P. Ranking the Suitability of Common Urban Tree Species for Controlling $PM_{2.5}$ pollution. *Atmos. Pollut. Res.* **2015**, *6*, 267–277. [CrossRef]
77. Jutras, P.; Prasher, S.O.; Mehuys, G.R. Appraisal of Key Abiotic Parameters Affecting Street Tree Growth. *Arboric. Urban For. (AUF)* **2010**, *36*, 1–10. [CrossRef]
78. Nielsen, C.N.; Bühler, O.; Kristoffersen, P. Soil Water Dynamics and Growth of Street and Park Trees. *Arboric. Urban For. (AUF)* **2007**, *33*, 231–245. [CrossRef]
79. Pauleit, S. Urban Street Tree Plantings: Identifying the Key Requirements. *Proc. Inst. Civ. Eng. Munic. Eng.* **2003**, *156*, 43–50. [CrossRef]
80. Davidson, A.M.; Jennions, M.; Nicotra, A.B. Do Invasive Species Show Higher Phenotypic Plasticity than Native Species and, If so, Is It Adaptive? A Meta-Analysis. *Ecol. Lett.* **2011**, *14*, 419–431. [CrossRef] [PubMed]
81. Hulme, P.E. Phenotypic Plasticity and Plant Invasions: Is It All Jack? *Funct. Ecol.* **2008**, *22*, 3–7. [CrossRef]
82. Richards, C.L.; Bossdorf, O.; Muth, N.Z.; Gurevitch, J.; Pigliucci, M. Jack of All Trades, Master of Some? On the Role of Phenotypic Plasticity in Plant Invasions. *Ecol. Lett.* **2006**, *9*, 981–993. [CrossRef]
83. Buru, J.C.; Diaz Torres, S.; Pegg, G.; Wills, J.; Firn, J. APBSF Project Final Report, Myrtle Rust Impacts on a Wet Sclerophyll Forest in the Tallebudgera Valley: South-East Queensland 2021. Available online: http://www.apbsf.org.au/plant-epidemiology-host-susceptibility-to-myrtle-rust-pbsf004/ (accessed on 6 December 2024).
84. European Commission. *EU Biodiversity Strategy for 2030—Bringing Nature Back into Our Lives*; COM (2020) 380 final; European Commission: Brussels, Belgium, 2020.
85. Najberek, K.; Solarz, W. Gatunki Obce. Przyczyny Inwazyjnych Zachowań i Sposoby Zwalczania. *Kosmos* **2016**, *65*, 81–91.
86. Dajdok, Z.; Pawlaczyk, P. *Inwazyjne Gatunki Roślin Ekosystemów Mokradłowych Polski*; Wydawnictwo Klubu Przyrodników: Świebodzin, Poland, 2009; ISBN 978-83-87846-69-5.
87. Essl, F.; Latombe, G.; Lenzner, B.; Wilson, J.R.U.; Genovesi, P.; Pagad, S.; Seebens, H.; Smith, K. The Convention on Biological Diversity (CBD)'s Post-2020 Target on Invasive Alien Species—What Should It Include and How Should It Be Monitored? *NeoBiota* **2020**, *62*, 99–121. [CrossRef]
88. DAISIE. *Handbook of Alien Species in Europe*; Springer: Dordrecht, The Netherlands, 2009; ISBN 978-1-4020-8279-5.
89. General Directorate for Plant Protection. Available online: https://www.gov.pl/web/gdos/inwazyjne-gatunki-obce-ias (accessed on 29 September 2024).
90. D'hondt, B.; Vanderhoeven, S.; Roelandt, S.; Mayer, F.; Versteirt, V.; Adriaens, T.; Ducheyne, E.; San Martin, G.; Grégoire, J.C.; Stiers, I.; et al. Harmonia + and Pandora +: Risk Screening Tools for Potentially Invasive Plants, Animals and Their Pathogens. *Biol. Invasions* **2015**, *17*, 1869–1883. [CrossRef]
91. NOBANIS. The North European and Baltic Network on Invasive Alien Species. Available online: http://www.nobanis.org (accessed on 29 September 2024).
92. Institute of Nature Conservation of the Polish Academy of Sciences in Kraków. Alien Species in Poland. Available online: http://www.iop.krakow.pl/ias (accessed on 29 September 2024).
93. Halarewicz, A.; Otręba, A.; Danielewicz, W. *Ankieta Oceny Stopnia Inwazyjności Padus serotina Erkh. Borkh. w Polsce, Na Podstawie Protokołu Harmonia+PL—Procedura Oceny Ryzyka Negatywnego Oddziaływania Inwazyjnych i Potencjalnie Inwazyjnych Gatunków Obcych w Polsce*; GDOŚ: Warszawa, Poland, 2017.
94. Wiatrowska, B.; Michalska-Hejduk, D.; Dajdok, Z. *Ankieta Oceny Stopnia Inwazyjności Spiraea tomenstosa L. w Polsce, Na Podstawie Protokołu Harmonia+PL—Procedura Oceny Ryzyka Negatywnego Oddziaływania Inwazyjnych i Potencjalnie Inwazyjnych Gatunków Obcych w Polsce*; GDOŚ: Warszawa, Poland, 2018.

95. Chmura, D.; Chmiel, J.; Danielewicz, W. *Ankieta Oceny Stopnia Inwazyjności Acer negundo L. w Polsce, Na Podstawie Protokołu Harmonia+PL—Procedura Oceny Ryzyka Negatywnego Oddziaływania Inwazyjnych i Potencjalnie Inwazyjnych Gatunków Obcych w Polsce*; GDOŚ: Warszawa, Poland, 2018.
96. EPPO Global Database. Available online: https://gd.eppo.int/ (accessed on 29 September 2024).
97. Kryczyński, S.; Weber, Z. *Fitopatologia Tom 2—Choroby Roślin Uprawnych*; Powszechne Wydawnictwo Rolnicze i Leśne: Poznań, Poland, 2011.
98. Kryczyński, S.; Weber, Z. *Fitopatologia Tom 1. Podstawy Fitopatologii*; Wydawnictwo Rolnicze i Leśne: Poznań, Poland, 2011.
99. Mańka, M.; Grzywacz, A. *Fitopatologia Leśna*; Wydawnictwo Uniwersytetu Przyrodniczego w Poznaniu: Poznań, Poland, 2023.
100. Mułenko, W.; Piątek, M.; Wołczańska, A.; Kozłowska, M.; Ruszkiewicz-Michalska, M. Plant Parasitic Fungi Introduced to Poland in Modern Times. Alien and Invasive Species. *Biol. Invasions Pol.* **2010**, *1*, 49–71.
101. Pusz, W.; Patejuk, K.; Kaczmarek, A. Fungi Colonizing of Small Balsam Seeds (*Impatiens parviflora* DC.) Seeds in Wigry National Park. *Prog. Plant Prot.* **2020**, *60*, 33–40. [CrossRef]
102. Patejuk, K.; Pusz, W. Występowanie *Puccinia komarovii* Tranzschel w Wybranych Lokalizacjach Wigierskiego Parku Narodowego. *Prz. Przyr.* **2018**, *29*, 96–100.
103. Tanner, R.; Ellison, C.; Shaw, R.; Evans, H.; Gange, A. Losing Patience with Impatiens: Are Natural Enemies the Solution? *Outlooks Pest Manag.* **2008**, *19*, 86–91. [CrossRef]
104. Tanner, R.A.; Pollard, K.M.; Varia, S.; Evans, H.C.; Ellison, C.A. First Release of a Fungal Classical Biocontrol Agent against an Invasive Alien Weed in Europe: Biology of the Rust, *Puccinia komarovii* var. *glanduliferae*. *Plant Pathol.* **2015**, *64*, 1130–1139. [CrossRef]
105. Tanner, R.A.; Ellison, C.A.; Seier, M.K.; Kovács, G.M.; Kassai-Jáger, E.; Berecky, Z.; Varia, S.; Djeddour, D.; Singh, M.C.; Csiszár, Á.; et al. *Puccinia komarovii* var. *glanduliferae* var. nov.: A Fungal Agent for the Biological Control of Himalayan Balsam (*Impatiens glandulifera*). *Eur. J. Plant Pathol.* **2014**, *141*, 247–266. [CrossRef]
106. Pusz, W.; Patejuk, K.; Baturo-Cieśniewska, A.; Kaczmarek, A. *Parasitic Fungi occurring on Selected Species of Invasive Plants in Wigry National Park*; Wigry National Park: Krzywe, Poland, 2019.
107. Pusz, W.; Kaczmarek, A.; Baturo-Cieśniewska, A. *Parasitic Fungi Occurring on Selected Species of Invasive Plants in the Tatra National Park*; Tatra National Park: Zakopane, Poland, 2018.
108. Piskorz, R.; Klimko, M. The Effect of *Puccinia Komarovii* Tranzsch. Infection on Characters of *Impatiens parviflora* DC. in Galio Sylvatici-Carpinetum (R. Tx. 1937) Oberd. 1957 Forest Association. *Acta Soc. Bot. Pol.* **2006**, *75*, 51–59. [CrossRef]
109. Ruszkiewicz-Michalska, M.; Mułenko, W. *Padus serotina* (*Rosaceae*), a New Host Plant for Some Species of Parasitic Microfungi. *Acta Mycol.* **2003**, *38*, 51–58. [CrossRef]
110. Masetti, A.; Magagnoli, S.; Lami, F.; Lanzoni, A.; Burgio, G. Long Term Changes in the Communities of Native Ladybirds in Northern Italy: Impact of the Invasive Species *Harmonia axyridis* (Pallas). *BioControl* **2018**, *63*, 665–675. [CrossRef]
111. Randolph, D. Cane Toad: Purposefully Introduced, but Instantly a Regret. Available online: https://ohiostate.pressbooks.pub/sciencebitesvolume2/front-matter/copyright/ (accessed on 1 December 2024).
112. CABI. *Invasive Species Management: A Nature-Based Solution for Climate and Environment*; CABI Briefing: Climate change; CABI: Wallingford, UK, 2021.
113. Shaw, R.; Witt, A.; Cock, M.; Pollard, K.; Thomas, S.; Romney, D. Safeguarding the Environment, Food Security and Livelihoods from Invasive Species Using Biological Controls. In *CABI Study Brief 4*; CABI: Wallingford, UK, 2014.
114. Hinz, H.; Williams, F. Beetles Arrest the Leafy Spurge Scourge in North America. In *CABI Study Brief 18*; CABI: Wallingford, UK, 2016.
115. Available online: https://www.cabi.org/projects/managing-invasive-madagascar-rubbervine-in-brazil/ (accessed on 20 July 2023).
116. Available online: https://www.cabi.org/projects/controlling-the-invasive-blackberry-on-the-galapagos-islands/ (accessed on 20 July 2023).
117. Available online: https://www.cabi.org/projects/biological-control-of-garlic-mustard/ (accessed on 20 July 2023).
118. Kancelaria Sejmu. *Dz.U. 2021 poz. 1718 Ustawa z Dnia 11 Sierpnia 2021 r. o Gatunkach Obcych*; Kancelaria Sejmu: Warsaw, Poland, 2021.
119. Shaw, D. Invasive Plants—The Enemy of Your Enemy Is Your Friend. In *Proceedings of the Our Plants, Our Future*; University of Birmingham: Birmingham, UK, 2021; pp. 6–8.
120. CABI. *Action on Invasives Annual Report 2020*; CABI: Wallingford, UK, 2021.

diversity MDPI

Article

Altitudinal Patterns of Native and Invasive Alien Herbs along Roadsides in the Dayao Mountain National Nature Reserve, Guangxi, China

Bai Li [1,2], Xinying Ni [2] and Caiyun Zhao [2,*]

1 School of Ecology and Environment, Zhengzhou University, Zhengzhou 450001, China
2 State Key Laboratory of Environmental Criteria and Risk Assessment, Chinese Research Academy of Environmental Sciences, Beijing 100012, China
* Correspondence: zhaocy@craes.org.cn; Tel.: +86-138-1114-9270

Abstract: Invasive alien plants have rapidly established and spread in nature reserves via roads and now pose a threat to biodiversity. To understand the mechanism and distribution patterns of invasive alien herbs, we compared the altitude patterns of native and invasive alien herbs based on 105 plots in the Dayao Mountain National Nature Reserve. This study also compared the distribution patterns of new (introduced to China after 1900) and old (introduced to China before 1900) invasive alien herbs. In addition, we examined the effects of climatic factors and human activities on the distribution patterns of species richness. In our study, 151 native herbs species and 18 invasive alien herbs species were observed, of which 12 were new invasive alien herbs. Old invasive alien herbs occurred more frequently and occupied a wider range of altitudes than new invasive alien herbs. The richness of native herbs tended to decrease with increasing altitude, and the altitude patterns of the richness of all invasive herbs and new invasive alien herbs were hump-shaped. Based on an analysis using the linear mixed model, the results indicated that temperature was the main factor limiting the altitude patterns of native herbs, and that temperature and human activities were essential factors in the distribution and spread of all invasive alien herbs and new invasive alien herbs. The intensity of human interference is a crucial driver of the spread of new invasive alien herbs to higher altitudes.

Keywords: plant invasion; mountain; introduction time; climate factors; human activity

1. Introduction

Invasive alien species are one of the major factors threatening global biodiversity and ecosystem functioning [1]. In recent years, with the rapid development of trade and tourism, the rate of introduction of alien species has gradually increased [2–4]. Nature reserves are often considered protected zones for the conservation of biodiversity, ecological landscapes and ecosystems [5–7]. The high diversity of native species in nature reserves can help them compete with alien species and resist their invasion [8]. Harsh climatic conditions also prevent the invasion of alien species in nature reserves located in mountainous regions [9,10]. However, a growing number of studies [11,12] have demonstrated that invasive alien species have already invaded nature reserves and pose an even greater threat to biodiversity in nature reserves located in mountainous regions. Alien plants could invade the grassland areas from roadsides in the Glacier National Park in the United States [13]. Alien species have been observed between 2400 and 3570 m a.s.l. in protected areas of the Central Andes in Argentina [14], and 46 alien species were recorded in five protected areas of south-central Chile [15]. In China, 176 invasive alien plants have been reported in 53 national nature reserves [16]. Therefore, understanding their distribution patterns and impact factors is particularly important for the effective management of invasive alien species.

Roads are also often studied as pathways for invasive alien species because roadside habitats are usually susceptible to disturbance and present a beneficial environment for the establishment and spread of invasive alien species [17,18]. One study compared the distribution patterns of alien species at different altitudes in 65 regions worldwide, and revealed that in most areas roadside alien species decreased with increasing altitude, with a few showing a single peak pattern [19]. In addition, at altitude gradients, roads can help species cross different climatic gradients [20]. Human activities also help invasive alien species to reach higher altitude ranges [21–23]. However, most studies in nature reserves have focused on invasive alien species at low altitudes and in island regions [24,25]. Few studies have examined higher altitudes, particularly the mountainous areas of mainland regions.

In recent years, the acceleration of the infrastructure construction process has facilitated the spread of invasive alien plants [26,27], firstly because disturbance causes damage to roadside habitats and reduces the resistance of native communities to alien plants, and secondly because human activities are gradually shifting to higher altitudes because of the continuous improvement of transportation facilities [18,28]. This phenomenon has provided an opportunity for alien plants to spread to middle and high altitudes. Therefore, human disturbance appears to be more critical than climatic factors in terms of the impact on the altitudinal patterns of alien species [29]. In addition to roads, the climate is also a key factor influencing the distribution patterns of invasive alien plants [30]. Studies have shown that the richness of invasive alien species is lower in mid- and high-altitude regions than at lower altitudes. In the harsh environment at high altitudes, it is difficult for alien species to survive. However, some studies have suggested that the warming climate may facilitate the invasion of alien species into alpine regions [31–33]. Therefore, the climate has been cited as the main factor limiting the invasion of alien species at high altitudes [34]. Compared with invasive alien species, relatively few studies have examined the effects of human disturbances on the altitudinal patterns of native species [18]. In mountainous regions, the distribution patterns of native plants at different altitudes are mainly influenced by climatic conditions and are not related to disturbances [35]. Understanding the factors limiting the distribution patterns of native and invasive alien plants can help to conserve plant diversity in mountain regions.

For invasive alien plants, the time of introduction is also a non-negligible factor [36]. The residence time (time since species introduction) is related to most of the factors that may influence the successful invasion of alien species, including the evolutionary process of the species and the occupation of a more extensive range [37]. The size of the propagule pool and opportunities for dispersal increase as alien species adapt to new environments [38]. Recent studies [39,40] have shown that land use changes caused by human activities have resulted in new alien plants spreading rapidly and shifting their distribution to higher altitudes. However, most studies have only considered the time of introduction as a factor for successful invasion, and few have investigated the differences in the distribution of altitudinal patterns of invasive alien plants in relation to different introduction times.

The Guangxi Dayao Mountain National Nature Reserve, located in central Guangxi, is globally significant for its biodiversity [41]. High mountains and complex topographies in nature reserves provide a variety of microenvironments that promote rich biodiversity [42]. To date, some investigations concerning woodland and fern species have been conducted in the reserve [41,42], but few studies have involved invasive alien species. Our aims were to answer the following questions: (1) Do the altitudinal distributional patterns differ between the native and invasive alien herbs? (2) What are the main factors that determine the distribution patterns of native and invasive herbs at altitude? (3) Are there different factors affecting the distribution patterns of old and new invasive alien herbs at altitude?

2. Materials and Methods

2.1. Study Area

The study was conducted in the Dayao Mountain National Nature Reserve, which is located in the central eastern area of Guangxi, China (110°01′–110°22′ E; 23°52′–24°22′ N). The protected area encompasses an area of 256 km^2 and is divided into seven zones:

Changtan River, Demei Mountain, Hekou, Huaping Mountain, Longjun Mountain, Pingzhu Mountain, and Shengtang Mountain (Figure 1). The center of the nature reserve is at high altitudes, gradually decreasing in all directions, and the geosphere is mainly composed of Cambrian sand, shale, and Devonian clastic rocks [41]. Shengtang Mountain is the central peak, which reaches 1979 m a.s.l. [42]. In this study, several areas of the reserve were surveyed along major roads, and we purposefully selected sites where invasive alien species could be reflected in the reserve. We did not set up sampling sites in areas without major roads and with low human activity levels. The main climatic feature is the significant subtropical mountain climate. The annual hours of sunshine average equal 1268.6 h, the average temperature is 17 °C, the average temperature of the hottest month (July) is 23.9 °C, and the average temperature of the coldest month (January) is 8.3 °C [41]. The annual precipitation range is 1380~2700 mm, with an annual average relative humidity rate of 83%. The forest cover in the reserve is very high, at more than 95% [43]. The main natural vegetation types are deciduous broad-leaved forests and evergreen broad-leaved forests in the low mountainous areas below 1000 m a.s.l. In the higher regions, montane evergreen broad-leaved forests and middle mountain mixed coniferous forests are widely distributed [43].

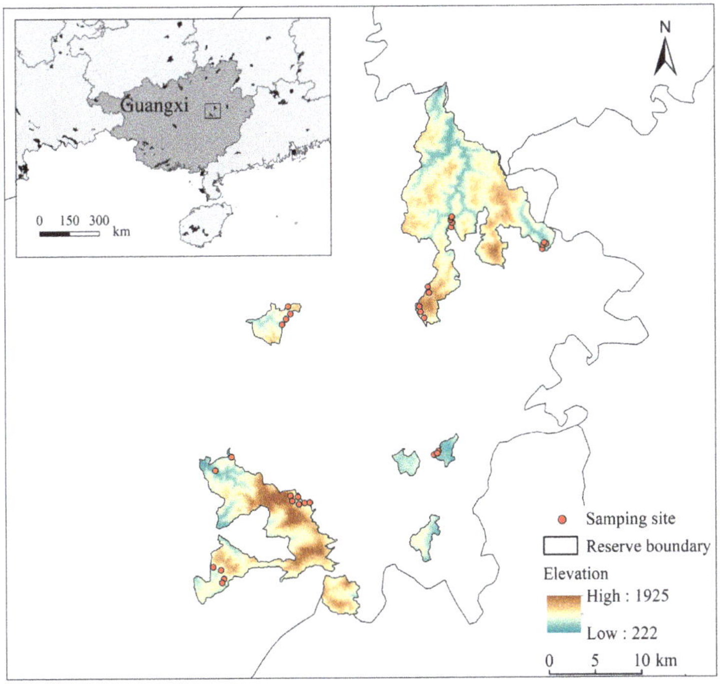

Figure 1. Location of the Dayao Mountain National Nature Reserve and plant survey sample sites.

2.2. Data Collection

The field survey was conducted in July 2021, and the sample sites were distributed across seven parts of the reserve. Thirty-three sampling sites were situated along a road that is intensively used for human activities and by vehicles, covering the altitude range between 200 and 1400 m a.s.l. The sample sites were located in the reserve's experimental area and buffer zone. The road edges were defined as consisting of distinct areas ranging from 0.5 to 4 m adjacent to the road, which are vulnerable to human activities [3]. For each sampling site, a 2 m-wide sample strip was established parallel to the road. Then, three to six 1 m × 1 m plots were randomly marked out in each sample strip, giving a total of

105 plots to be surveyed. For each plot, we recorded all of the native and invasive alien herbs and the individual numbers of each species. The invasive alien herbs and native herbs were distinguished in accordance with Ma [44]. The invasive alien herbs were identified as herbs with harmful effects in accordance with Ma [44]. The location and altitude of each plot were recorded using the Global Positioning System (GPS). Since the 19th century, the rate of introduction of invasive alien plants in China has increased rapidly, with 266 species of invasive plants having been introduced since 1900 [4]. Therefore, we classified invasive alien herbs as either old invasive alien herbs (introduced to China before 1900) or new invasive alien herbs (introduced to China after 1900) (Table S1). The introduction time of each invasive alien herb was obtained from Ma's study [44].

2.3. *Climatic and Human Disturbance Factors*

Three climatic factors were selected, namely the annual precipitation, annual mean temperature, and mean temperature of the warmest quarter, based on previous studies [45, 46]. The climate data for each site were obtained from the WorldClim data website (accessed on 23 December 2021) (https://worldclim.org) by extracting a 30-year (1970–2000) national average of meteorological data. Human disturbance was evaluated based on two indexes: (1) the distance from the plot to the nearest human settlement, which was calculated using ArcGIS; (2) the interference intensity, which was evaluated based on the habitat type around the plot (Table S2) and was estimated using a 4-grade scale: 1 = no significant interference; 2 = weak interference; 3 = moderate interference; 4 = heavy interference [47].

2.4. *Data Analysis*

We examined the potential covariance between the variables using the GGally package (Figure S1). We removed the annual mean temperature because it had strong covariance with other factors (>0.70). The native herbs, all invasive alien herbs, old invasive alien herbs, and new invasive alien herbs were the response variables. The annual precipitation, mean temperature of warmest quarter, interference intensity, and distance from the plot to the nearest human settlement were the explanatory variables. Firstly, four linear models (LM) were built and used to extensively assess the relationship between explanatory and response variables. The normality of the residuals and the homogeneity of the variance were checked using the visual inspection method for the model residuals, and the variables generally conformed to a normal distribution. Secondly, to determine the relative importance of each explanatory variable, four linear mixed models (LMM) were built based on these response variables. Climate and human disturbance factors were treated as fixed effects, and the sampling sites were treated as random effects. These variables in the mixed model were decomposed using the MuMIn and glmm.hp packages, and the contribution of each variable was calculated separately. The model analysis was performed using the lmer function of the lme4 package in R (R Core Team 2022) [48].

3. Results

3.1. *Composition and Distribution of Native and Invasive Alien Herbs*

In this nature reserve, a total of 151 native herb species have been recorded, belonging to 62 families and 123 genera. These species are mainly classified into Poaceae and Asteraceae, with 16 and 11 genera, respectively.

A total of 18 invasive alien herbs were observed, belonging to 8 families and 15 genera. Half of all invasive alien herbs were classified into Asteraceae. When we classified these species as old or new invasive alien herbs, the number of new invasive alien herb species was twice as large as that of old invasive alien herbs. The frequency of occurrence of old invasive alien herbs was higher than that of new invasive alien herbs. For example, the frequency rates of *Bidens pilosa*, *Erigeron canadensis*, and *Ageratum conyzoides* were 40.95%, 34.29%, and 32.38%, respectively. However, the frequency of most new invasive alien herbs was less than 10%, with the exception of *Crassocephalum crepidioides*, *Praxelis clematidea*, and *Erechtites valerianifolius* (Figure 2a). In addition, 18 invasive alien herbs were classified as

12 annual and 6 perennial invasive alien herbs. Annual invasive alien herbs occurred more frequently than perennial invasive alien herbs. The frequency of occurrence of all perennial invasive alien herbs was less than 10% (Figure 2b).

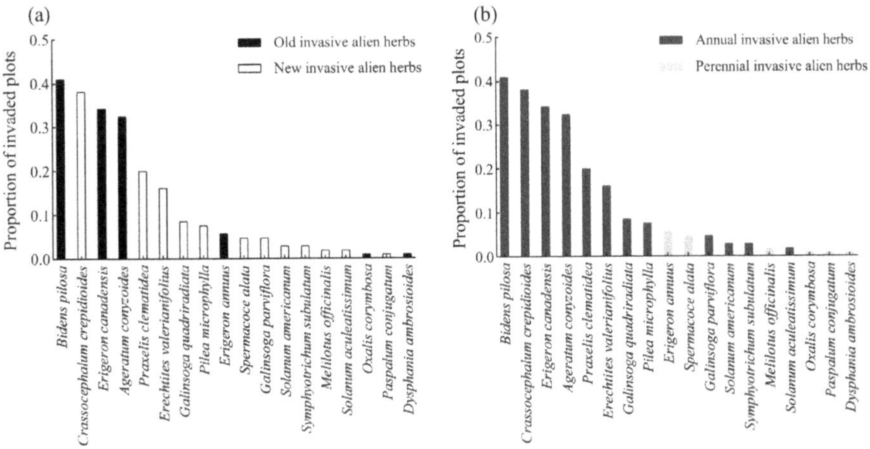

Figure 2. Occurrence frequency rates of (**a**) new and old invasive alien herbs and (**b**) annual and perennial invasive alien herbs in Dayao Mountain Reserve. The y-axis indicates the proportion of samples in which each species was present. Black, white, dark gray, and light gray represent old, new, annual, and perennial invasive alien herbs, respectively.

Old invasive alien herbs were found at higher altitudes and occupied a wider range of altitudes (Figure 3a). For example, *Bidens pilosa, Ageratum conyzoides,* and *Erigeron canadensis* were distributed in the range of 250–1300 m a.s.l. Although of the new invasive alien herbs *Crassocephalum crepidioides* was distributed in the altitude range of 300–1300 m, the other new invasive alien herbs were distributed in a narrow altitude range. Annual invasive alien herbs were distributed in the range of 250–1400 m a.s.l., whereas perennial invasive alien herbs were distributed in a smaller altitude range (Figure 3b).

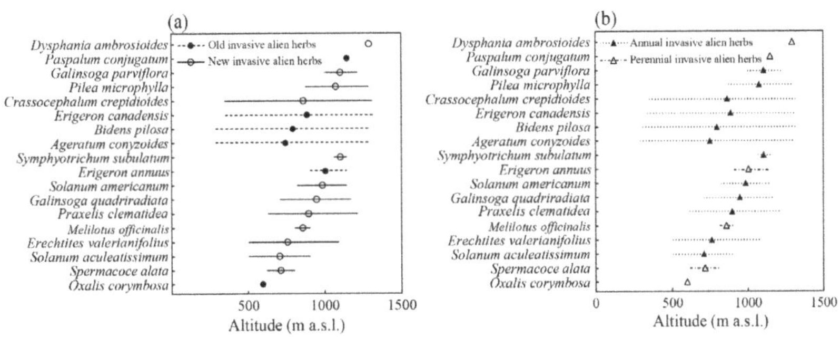

Figure 3. The altitudinal distribution range of (**a**) new and old invasive alien herbs and (**b**) annual and perennial invasive alien herbs. Old and new invasive alien herbs are indicated by dashed and solid lines, respectively. Annual and perennial invasive alien herbs are indicated by dotted lines and dot-dash lines, respectively. The positions of the symbols show the average altitudes at which the species occurs. Solid and hollow circles denote old and new invasive alien herbs, respectively. Solid and hollow triangles indicate annual and perennial invasive alien herbs, respectively.

3.2. *Altitude Distribution Patterns of Native and Invasive Alien Herbs*

The richness of native herbs tends to slowly decrease with increasing altitude (Figure 4a). The richness of invasive alien herbs was represented by a single hump pattern with a peak at 900 m a.s.l. (Figure 4a). The richness of new invasive alien herbs depicted a single hump pattern with a peak at 1000 m a.s.l. (Figure 4b). The richness of old invasive alien herbs was evenly distributed across altitudes (Figure 4b).

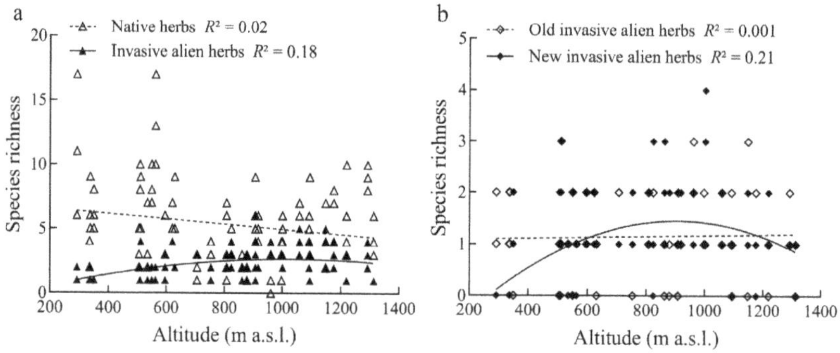

Figure 4. Altitudinal patterns of the richness of (**a**) native and all invasive alien herbs and (**b**) new and old invasive alien herbs in the Dayao Mountain Reserve. Hollow triangles and solid circles represent native herbs and all invasive alien herbs, respectively. Solid and open diamonds represent new and old invasive alien herbs, respectively.

3.3. *Factors Determining the Distribution Patterns of Native and Invasive Herbs*

Based on the results of the linear model analysis (Table 1), the altitude pattern of native herbs was mainly explained by the temperature (Figure 5a) and could not be correlated with human disturbance (Figure 5e).

Table 1. Linear models were constructed for native herbs, all invasive alien herbs, the richness levels of new and old invasive alien herbs, and various environmental factors at the altitude gradient of the reserve.

Species Category	Environment Variables	t	p
Native herbs	Mean temperature of warmest quarter	3.008	0.003
	Annual precipitation	0.415	0.679
	Interference intensity	−1.038	0.302
	Dis. human settlement	−1.060	0.292
All invasive alien herbs	Mean temperature of warmest quarter	−2.469	0.015
	Annual precipitation	0.191	0.848
	Interference intensity	3.913	<0.001
	Dis. human settlement	−0.371	0.712
New invasive alien herbs	Mean temperature of warmest quarter	−4.723	<0.001
	Annual precipitation	−1.733	0.086
	Interference intensity	3.233	0.002
	Dis. human settlement	0.429	0.668
Old invasive alien herbs	Mean temperature of warmest quarter	1.194	0.235
	Annual precipitation	1.767	0.08
	Interference intensity	1.875	0.05
	Dis. human settlement	−0.830	0.408

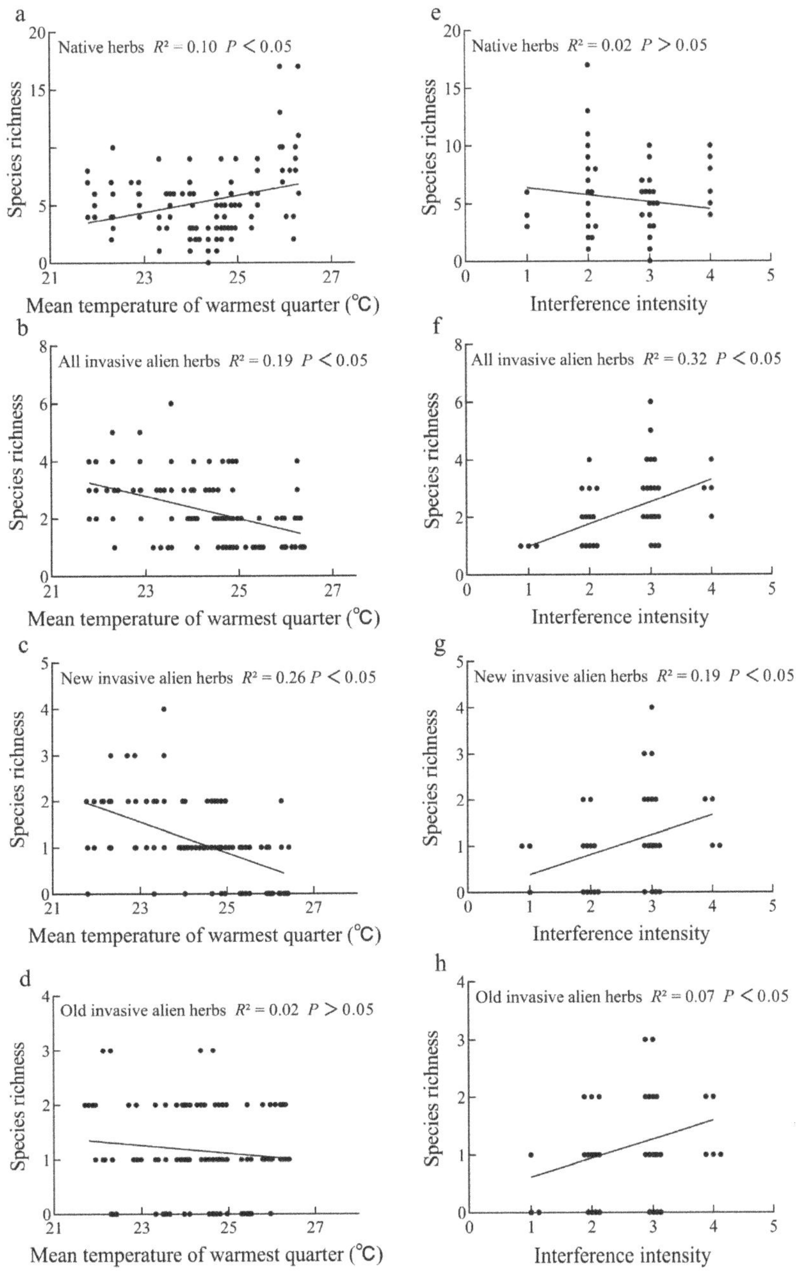

Figure 5. Distribution of richness for (**a**,**e**) native herbs, (**b**,**f**) all invasive alien herbs, and (**c**,**g**) new and (**d**,**h**) old invasive alien herbs with the mean temperature of the warmest quarter and interference intensity in the Dayao Mountain Reserve.

The altitude patterns of all invasive alien herbs were significantly correlated with the temperature and interference intensity (Figure 5b,f). The distribution patterns of the new invasive alien herbs were mainly explained by the temperature and interference intensity

(Figure 5c,g). The altitude patterns of old invasive alien herbs were correlated with the disturbance intensity only (Figure 5h).

Based on the results of the linear mixed model analysis (Table S3), the temperature explained 70.18% of the variation in richness of native herbs and was positively correlated with it ($p < 0.05$, Figure 6a).

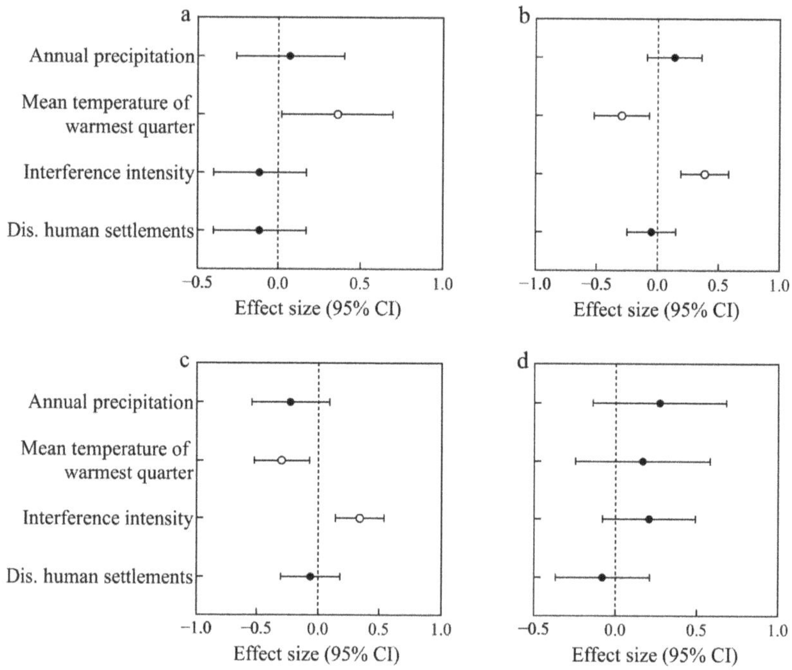

Figure 6. Effect values of human activities and climatic factors on (**a**) native herbs, (**b**) all invasive alien herbs, and (**c**) new and (**d**) old invasive alien herbs. Hollow circles represent species richness significantly correlated with environmental factors, solid circles represent species richness not significantly correlated with environmental factors, and dashed lines indicate an effect value of 0.

The distribution pattern of all invasive alien herbs was significantly negatively correlated with the temperature ($p < 0.05$, Figure 6b) and positively correlated with the interference intensity ($p < 0.05$, Figure 6b). The temperature explained 34.81% of the variation, but the interference intensity explained 39.71% of the variation. The richness pattern of new invasive alien herbs was mainly explained by the temperature and interference intensity (Table S3). The temperature negatively impacted the richness pattern of new invasive alien herbs and explained 53.49% of the variation ($p < 0.05$, Figure 6c), whereas the interference intensity had a positive impact and explained 34.15% of the variation ($p < 0.05$, Figure 6c). None of the environmental factors had a significant effect on the richness of old invasive alien herbs ($p > 0.05$, Figure 6d).

The annual precipitation had no significant effect on the richness of either native or invasive herbs ($p > 0.05$, Figure 6).

4. Discussion

4.1. Altitudinal Patterns of Native and Invasive Alien Herbs

This study found that invasive alien herbs in the Dayao Mountain National Nature Reserve showed a hump-shaped distribution pattern according to altitude. In contrast with our research, invasive alien herbs are currently concentrated at lower altitudes in montane areas [47]. First, this may be in response to the arid environment at low altitudes [49] and the cold climate at high altitudes [50]. Second, our study was based on the altitudinal distribution pattern of roadside species. Invasive alien plants can occupy the remaining ecological niches because of the destruction of the original ecosystem, and then can gradually spread from lower to higher altitudes [51]. In addition, our results also suggested that the factors affecting the altitude patterns of native and invasive alien herbs were different. Native herbs have gradually adapted to the local environment through long-term development and evolution [52]. Under the influence of human activities, ecological damage may also lead to changes in the altitude patterns of native herbs [53].

In this study, differences were identified between the two invasive alien groups. Old invasive alien herbs were evenly distributed across the altitudes, whereas new invasive alien herbs showed a hump-shaped distribution pattern. The invasive alien plants with a longer residence time had a wider distribution range [54]. Becker et al. [55] showed that the altitude of invasive species increases with the time since introduction. Over time, under propagule pressure [56] and long-term genetic evolution [57], the old invasive alien plants have gradually adapted to the colder environment in order to survive at higher altitudes. In addition, human disturbance is an essential factor in the movement of species across different climatic zones [20]. Contrary to the old invasive alien herbs, the number of newly introduced aliens is increasing every year [58], even though they are not adapted to the cold environment of high altitudes [3]. Therefore, the hump-shaped distribution pattern of the new invasive alien herbs also indicated that conditions of drought at low altitudes and cold temperatures at high altitudes are unsuitable for plant survival.

4.2. Effects of Environmental Factors on the Richness of Native and Invasive Herbs

In recent years, the influence of environmental factors on species distribution has received extensive attention in studies of species distribution patterns across altitudinal gradients [25,59,60]. Our study revealed a significant difference in the explanatory variables between native and invasive alien herbs. The richness of native herbs was only positively correlated to temperature. This is the response of plants to climatic constraints, suggesting that higher temperatures allow for greater species richness [61]. Native plants have adapted to the local climate over tens of thousands of years of genetic evolution [62].

In contrast to native herbs, anthropogenic disturbance is an important factor affecting the richness of invasive alien plants. This is because human activities can destroy pristine ecosystems and provide more opportunities for invasive alien species [63,64]. The Dayao Mountain National Nature Reserve is a tourist destination, and tourism is a major human disturbance. With the rise of tourism, road construction has increased in reserves. Roads can expand the distribution range of alien species, acting as invasive pathways for alien species in mountainous areas [65]. The destruction of roadside habitats by pedestrians and vehicles also increases the number of suitable ecological niches for alien species [66]. In addition, we found a significant negative correlation between the richness of invasive alien herbs and temperature. This relationship suggested that mountain microenvironments provide suitable climatic conditions for the survival of invasive alien herbs and that plant invasion may be promoted in such environments [39]. Human activities play a stronger role than climatic factors in explaining the altitudinal pattern of invasive alien herbs in the Dayao Mountain National Nature Reserve.

4.3. Effects of Environmental Factors on the Richness of Invasive Alien Herbs at Different Introduction Times

The responses of old and new invasive alien herbs to temperature were different. The richness of old invasive alien herbs was not related to temperature, and new invasive alien herbs negatively correlated with temperature. Gasso et al. [67] showed that it took about 150 years for invasive alien plants to adapt to the local environment or to reach their maximum range. The old invasive alien herbs were all introduced to the Dayao Mountain National Nature Reserve more than 150 years ago and have gradually adapted to the local climate conditions. Thus, they were observed at higher altitudes. Most of the new invasive alien herbs were introduced to China less than 100 years ago. For example, *Crassocephalum crepidioides* was introduced to China about 90 years ago and *Praxelis clematidea* was introduced about 40 years ago. Therefore, new invasive alien herbs were significantly correlated with temperature. Polce et al. [39] identified a negative correlation between altitudinal distribution patterns and temperature for new alien plants (>1500 years) in Europe. This is consistent with the results of our study. The mesic hypothesis stated that the microclimate is vital for new alien plants and may increase the likelihood of invasion [39,68]. This means that alien species may form a larger pool of potential species in a suitable microclimate.

Studies [69,70] have shown that the impact of human activities on the altitude pattern of invasive alien plants is more critical than climatic factors. Seebens et al. [71] showed that the global invasion of alien species has not slowed in the past few centuries because of rapid economic development and population growth. The development of commercial imports to China was accelerated by war and the ecological damage caused by the prolonged war in the 20th century, causing the introduction rate of new alien plants to peak in this period [4]. Kalwij et al. [3] showed that the movements of both vehicles and livestock are the means by which the propagules of alien species spread along roads, and have led to an annual increase in the upper altitude limit of roadside alien species, as well as in the rate of introduction of new species. Human disturbances along roadsides destroy the local vegetation composition and soil environment, reducing the biological resistance of native plant communities, creating more available resources and empty ecological niches [55], and creating opportunities for the introduction and establishment of new invasive alien species [3]. Lososova et al. [72] also confirmed that habitat changes can affect the distribution of new invasive alien plants. Thus, changes in roadside habitat can increase the establishment of new alien plants and hiking or other human activities can help them to reach higher altitude ranges [15]. Polce et al. [39] found that in Europe, old invasive alien plants were less affected by human activities than newer ones. This is consistent with the results of our study. Old invasive alien herbs generally reach the limit of their potential range after a long period of adaptation in the invaded regions [55]. New invasive alien plants, when introduced into non-native habitats, spread rapidly under the influence of human disturbance and propagule pressure, but their distribution is not uniform [40]. As a result, the number of new alien plants on the roadside has increased significantly. Unlike the new invasive alien herbs, we found that the richness of the old invasive alien herbs responded to human activities similarly to the native herbs. This phenomenon suggested that the old invasive alien herbs have been adapted over a long period and have been distributed in a wide range of altitudes after a long time of adaptation.

5. Conclusions

This study found that the temperature limited the altitudinal distribution patterns of native herbs, human activities facilitated the introduction and spread of alien plants, and suitable temperatures provided the conditions for alien plants to survive. Although the old invasive alien herbs occurred more frequently and occupied a more extensive altitude range than new invasive alien herbs, we found that human activities appeared to be the main driver of new invasive alien herbs in the reserve. The old invasive alien herbs gradually adapted to the local environmental conditions with increasing residence times. The new invasive alien herbs are currently distributed at low and middle altitudes because their distribution has been restricted by climatic conditions. However, new invasive alien herbs have a higher potential spreading risk under the influence of human activities. Therefore, the monitoring and prevention of newly introduced invasive alien herbs should be strengthened to reduce the proliferation risk.

Supplementary Materials: The following supporting information can be downloaded at: https://www.mdpi.com/article/10.3390/d15010105/s1, Figure S1: Results of correlation analysis of explanatory variables. Values above 0.7 are considered to have a strong covariance between variables; Table S1: The checklist of the alien invasive plants in Dayao Mountain National Nature Reserve, Table S2: Plots investigation of herbs in Dayao Mountain National Nature Reserve, Table S3: Linear mixed models constructed for native herbs, all invasive alien herbs, new and old invasive alien herbs richness and various environmental factors at the altitude gradient of the reserve, Table S4: Information on herbs species in plots of Dayao Mountain National Nature Reserve.

Author Contributions: Conceptualization, B.L. and C.Z.; methodology, B.L.; software, B.L. and X.N.; investigation, B.L. and X.N.; data curation, B.L.; writing—original draft preparation, B.L.; writing—review and editing, C.Z.; supervision, C.Z. All authors have read and agreed to the published version of the manuscript.

Funding: This research was supported by grants from the National Key Research and Development Program of China (2020YFC1806300) and the Central Public-Interest Scientific Institution Basal Research Fund (2022YSKY-08).

Institutional Review Board Statement: Not applicable.

Data Availability Statement: The data presented in this study are available on request from the corresponding author.

Conflicts of Interest: The authors declare no conflict of interest. The funders had no role in the design of the study; in the collection, analyses, or interpretation of data; in the writing of the manuscript; or in the decision to publish the results.

References

1. Perrings, C.; Naeem, S.; Ahrestani, F.; Bunker, D.E.; Burkill, P.; Canziani, G.; Elmqvist, T.; Ferrati, R.; Fuhrman, J.A.; Jaksic, F.; et al. Ecosystem Services for 2020. *Science* **2010**, *330*, 323–324. [CrossRef] [PubMed]
2. Hulme, P.E. Trade, transport and trouble: Managing invasive species pathways in an era of globalization. *J. Appl. Ecol.* **2009**, *46*, 10–18. [CrossRef]
3. Kalwij, J.M.; Robertson, M.P.; van Rensburg, B.J. Annual monitoring reveals rapid upward movement of exotic plants in a montane ecosystem. *Biol. Invasions* **2015**, *17*, 3517–3529. [CrossRef]
4. Li, Y.; Shen, Z. Roles of Dispersal Limit and Environmental Filtering in Shaping the Spatiotemporal Patterns of Invasive Alien Plant Diversity in China. *Front. Ecol. Evol.* **2020**, *8*, 544670. [CrossRef]
5. Tittensor, D.P.; Walpole, M.; Hill, S.L.L.; Boyce, D.G.; Britten, G.L.; Burgess, N.D.; Butchart, S.H.M.; Leadley, P.W.; Regan, E.C.; Alkemade, R.; et al. A mid-term analysis of progress toward international biodiversity targets. *Science* **2014**, *346*, 241–244. [CrossRef] [PubMed]
6. Braun, M.; Schindler, S.; Essl, F. Distribution and management of invasive alien plant species in protected areas in Central Europe. *J. Nat. Conserv.* **2016**, *33*, 48–57. [CrossRef]
7. Bhatta, S.; Joshi, L.R.; Shrestha, B.B. Distribution and impact of invasive alien plant species in Bardia National Park, western Nepal. *Environ. Conserv.* **2020**, *47*, 197–205. [CrossRef]
8. Byers, J.E. Impact of non-indigenous species on natives enhanced by anthropogenic alteration of selection regimes. *Oikos* **2002**, *97*, 449–458. [CrossRef]
9. Foxcroft, L.C.; Jarosik, V.; Pysek, P.; Richardson, D.M.; Rouget, M. Protected-Area Boundaries as Filters of Plant Invasions. *Conserv. Biol.* **2011**, *25*, 400–405. [CrossRef]
10. Gonzalez-Moreno, P.; Diez, J.M.; Ibanez, I.; Font, X.; Vila, M. Plant invasions are context-dependent: Multiscale effects of climate, human activity and habitat. *Divers. Distrib.* **2014**, *20*, 720–731. [CrossRef]
11. Thalmann, D.J.K.; Kikodze, D.; Khutsishvili, M.; Kharazishvili, D.; Guisan, A.; Broennimann, O.; Mueller-Schaerer, H. Areas of high conservation value in Georgia: Present and future threats by invasive alien plants. *Biol. Invasions* **2015**, *17*, 1041–1054. [CrossRef]
12. Shackleton, R.T.; Foxcroft, L.C.; Pysek, P.; Wood, L.E.; Richardson, D.M. Assessing biological invasions in protected areas after 30 years: Revisiting nature reserves targeted by the 1980s SCOPE programme. *Biol. Conserv.* **2020**, *243*, 108424. [CrossRef]
13. Tyser, R.W.; Worley, C.A. Alien flora in grasslands adjacent to road and trail corridors in Glacier National-park, Montana (USA). *Conserv. Biol.* **1992**, *6*, 253–262. [CrossRef]
14. McDougall, K.L.; Lembrechts, J.; Rew, L.J.; Haider, S.; Cavieres, L.A.; Kueffer, C.; Milbau, A.; Naylor, B.J.; Nunez, M.A.; Pauchard, A.; et al. Running off the road: Roadside non-native plants invading mountain vegetation. *Biol. Invasions* **2018**, *20*, 3461–3473. [CrossRef]
15. Liedtke, R.; Barros, A.; Essl, F.; Lembrechts, J.J.; Wedegartner, R.E.M.; Pauchard, A.; Dullinger, S. Hiking trails as conduits for the spread of non-native species in mountain areas. *Biol. Invasions* **2020**, *22*, 1121–1134. [CrossRef]

16. Gong, L.; Li, J.; Liu, X.; Zhao, X.; Zhao, C. Analysis of invasive alien species in Chinese national nature reserves. *Ecol. Sci.* **2017**, *36*, 210–216. [CrossRef]
17. Pollnac, F.; Seipel, T.; Repath, C.; Rew, L.J. Plant invasion at landscape and local scales along roadways in the mountainous region of the Greater Yellowstone Ecosystem. *Biol. Invasions* **2012**, *14*, 1753–1763. [CrossRef]
18. Lembrechts, J.J.; Alexander, J.M.; Cavieres, L.A.; Haider, S.; Lenoir, J.; Kueffer, C.; McDougall, K.; Naylor, B.J.; Nunez, M.A.; Pauchard, A.; et al. Mountain roads shift native and non-native plant species' ranges. *Ecography* **2017**, *40*, 353–364. [CrossRef]
19. Guo, Q.; Fei, S.; Shen, Z.; Iannone, B.V., III; Knott, J.; Chown, S.L. A global analysis of elevational distribution of non-native versus native plants. *J. Biogeogr.* **2018**, *45*, 793–803. [CrossRef]
20. Pauchard, A.; Kueffer, C.; Dietz, H.; Daehler, C.C.; Alexander, J.; Edwards, P.J.; Arevalo, J.R.; Cavieres, L.A.; Guisan, A.; Haider, S.; et al. Ain't no mountain high enough: Plant invasions reaching new elevations. Front. *Ecol. Environ.* **2009**, *7*, 479–486. [CrossRef]
21. Barros, A.; Pickering, C.M. Non-native Plant Invasion in Relation to Tourism Use of Aconcagua Park, Argentina, the Highest Protected Area in the Southern Hemisphere. *Mt. Res. Dev.* **2014**, *34*, 13–26. [CrossRef]
22. Yang, M.; Lu, Z.; Fan, Z.; Liu, X.; Hens, L.; De Wulf, R.; Ou, X. Distribution of non-native plant species along elevation gradients in a protected area in the eastern Himalayas, China. *Alp. Bot.* **2018**, *128*, 169–178. [CrossRef]
23. Ibanez, T.; Hart, P.; Ainsworth, A.; Gross, J.; Monello, R. Factors associated with alien plant richness, cover and composition differ in tropical island forests. *Divers. Distrib.* **2019**, *25*, 1910–1923. [CrossRef]
24. Landi, S.; Tordoni, E.; Amici, V.; Bacaro, G.; Carboni, M.; Filibeck, G.; Scoppola, A.; Bagella, S. Contrasting patterns of native and non-native plants in a network of protected areas across spatial scales. *Biodivers. Conserv.* **2020**, *29*, 2035–2053. [CrossRef]
25. Rojas-Sandoval, J.; Ackerman, J.D.; Tremblay, R.L. Island biogeography of native and alien plant species: Contrasting drivers of diversity across the Lesser Antilles. *Divers. Distrib.* **2020**, *26*, 1539–1550. [CrossRef]
26. Pretto, F.; Celesti-Grapow, L.; Carli, E.; Brundu, G.; Blasi, C. Determinants of non-native plant species richness and composition across small Mediterranean islands. *Biol. Invasions* **2012**, *14*, 2559–2572. [CrossRef]
27. Paudel, S.; Benavides, J.C.; MacDonald, B.; Longcore, T.; Wilson, G.W.T.; Loss, S.R. Determinants of native and non-native plant community structure on an oceanic island. *Ecosphere* **2017**, *8*, e01927. [CrossRef]
28. Haider, S.; Kueffer, C.; Bruelheide, H.; Seipel, T.; Alexander, J.M.; Rew, L.J.; Ramon Arevalo, J.; Cavieres, L.A.; McDougall, K.L.; Milbau, A.; et al. Mountain roads and non-native species modify elevational patterns of plant diversity. *Glob. Ecol. Biogeogr.* **2018**, *27*, 667–678. [CrossRef]
29. Marini, L.; Battisti, A.; Bona, E.; Federici, G.; Martini, F.; Pautasso, M.; Hulme, P.E. Alien and native plant life-forms respond differently to human and climate pressures. *Glob. Ecol. Biogeogr.* **2012**, *21*, 534–544. [CrossRef]
30. Sullivan, J.J.; Williams, P.A.; Timmins, S.M.; Smale, M.C. Distribution and spread of environmental weeds along New Zealand roadsides. *N. Z. J. Ecol.* **2009**, *33*, 190–204.
31. Pauchard, A.; Milbau, A.; Albihn, A.; Alexander, J.; Burgess, T.; Daehler, C.; Englund, G.; Essl, F.; Evengard, B.; Greenwood, G.B.; et al. Non-native and native organisms moving into high elevation and high latitude ecosystems in an era of climate change: New challenges for ecology and conservation. *Biol. Invasions* **2016**, *18*, 345–353. [CrossRef]
32. Dainese, M.; Aikio, S.; Hulme, P.E.; Bertolli, A.; Prosser, F.; Marini, L. Human disturbance and upward expansion of plants in a warming climate. *Nat. Clim. Change* **2017**, *7*, 577–580. [CrossRef]
33. Carboni, M.; Gueguen, M.; Barros, C.; Georges, D.; Boulangeat, I.; Douzet, R.; Dullinger, S.; Klonner, G.; van Kleunen, M.; Essl, F.; et al. Simulating plant invasion dynamics in mountain ecosystems under global change scenarios. *Glob. Change Biol.* **2018**, *24*, E289–E302. [CrossRef]
34. Steyn, C.; Greve, M.; Robertson, M.P.; Kalwij, J.M.; le Roux, P.C. Alien plant species that invade high elevations are generalists: Support for the directional ecological filtering hypothesis. *J. Veg. Sci.* **2017**, *28*, 337–346. [CrossRef]
35. Felde, V.A.; Kapfer, J.; Grytnes, J.-A. Upward shift in elevational plant species ranges in Sikkilsdalen, central Norway. *Ecography* **2012**, *35*, 922–932. [CrossRef]
36. Pysek, P.; Manceur, A.M.; Alba, C.; McGregor, K.F.; Pergl, J.; Stajerova, K.; Chytry, M.; Danihelka, J.; Kartesz, J.; Klimesova, J.; et al. Naturalization of central European plants in North America: Species traits, habitats, propagule pressure, residence time. *Ecology* **2015**, *96*, 762–774. [CrossRef]
37. Richardson, D.M.; Pysek, P. Plant invasions: Merging the concepts of species invasiveness and community invasibility. *Prog. Phys. Geogr. -Earth Environ.* **2006**, *30*, 409–431. [CrossRef]
38. Richardson, D.M.; Pysek, P. Naturalization of introduced plants: Ecological drivers of biogeographical patterns. *New Phytol.* **2012**, *196*, 383–396. [CrossRef]
39. Polce, C.; Kunin, W.E.; Biesmeijer, J.C.; Dauber, J.; Phillips, O.L.; Network, A.F.S. Alien and native plants show contrasting responses to climate and land use in Europe. *Glob. Ecol. Biogeogr.* **2011**, *20*, 367–379. [CrossRef]
40. Eichenberg, D.; Bowler, D.E.; Bonn, A.; Bruelheide, H.; Grescho, V.; Harter, D.; Jandt, U.; May, R.; Winter, M.; Jansen, F. Widespread decline in Central European plant diversity across six decades. *Glob. Change Biol.* **2021**, *27*, 1097–1110. [CrossRef]
41. Chen, B.; Wang, L. Investigation and application of wild ornamental ferns in Dayao Mountain National Nature Reserve, Guangxi. *Contemp. Hortic.* **2021**, *44*, 83–86. [CrossRef]
42. Chen, B.; Wang, L.; Sun, C. Investigation of wild woody ornamental plant resources in Dayao Mountain National Nature Reserve. *Chin. Wild Plant Resour.* **2021**, *40*, 61–67. [CrossRef]

43. Wang, L.; Luo, B.; Tan, M.; Chen, B. Study on Flora of Dayao Mountain Nature Reserve in Guangxi. In Proceedings of the Second China Forestry Academic Conference—S8 Wildlife, Wetland and Nature Reserve, Nanning, China, 7 November 2009; pp. 132–136.
44. Ma, J. *Flora of Invasive Aliens in China*, 1st ed.; Shanghai Jiao Tong University Press: Shanghai, China, 2021; pp. 39–378.
45. Xu, Z.; Feng, Z.; Yang, J.; Zheng, J.; Zhang, F. Nowhere to Invade: Rumex crispus and Typha latifolia Projected to Disappear under Future Climate Scenarios. *PLoS ONE* **2013**, *8*, e70728. [CrossRef] [PubMed]
46. Zhang, X.; Wei, H.; Zhao, Z.; Liu, J.; Zhang, Q.; Zhang, X.; Gu, W. The Global Potential Distribution of Invasive Plants: Anredera cordifolia under Climate Change and Human Activity Based on Random Forest Models. *Sustainability* **2020**, *12*, 1491. [CrossRef]
47. Zhang, W.; Yin, D.; Huang, D.; Du, N.; Liu, J.; Guo, W.; Wang, R. Altitudinal patterns illustrate the invasion mechanisms of alien plants in temperate mountain forests of northern China. *For. Ecol. Manag.* **2015**, *351*, 1–8. [CrossRef]
48. R Core Team. *R: A Language and Environment for Statistical Computing*; R Foundation for Statistical Computing: Vienna, Austria, 2022; Available online: https://www.R-project.org/ (accessed on 12 January 2022).
49. Koerner, C. The use of 'altitude' in ecological research. *Trends Ecol. Evol.* **2007**, *22*, 569–574. [CrossRef]
50. Haider, S.; Alexander, J.; Dietz, H.; Trepl, L.; Edwards, P.J.; Kueffer, C. The role of bioclimatic origin, residence time and habitat context in shaping non-native plant distributions along an altitudinal gradient. *Biol. Invasions* **2010**, *12*, 4003–4018. [CrossRef]
51. Applestein, C.; Germino, M.J. Patterns of post-fire invasion of semiarid shrub-steppe reveals a diversity of invasion niches within an exotic annual grass community. *Biol. Invasions* **2022**, *24*, 741–759. [CrossRef]
52. Roy, K.; Goldberg, E.E. Origination, extinction, and dispersal: Integrative models for understanding present-day diversity gradients. *Am. Nat.* **2007**, *170*, S71–S85. [CrossRef]
53. Clark-Tapia, R.; Gonzalez-Adame, G.; Campos, J.E.; Aguirre-Hidalgo, V.; Pacheco-Cruz, N.; Ugalde, J.J.V.T.; Campista-Leon, S.; Peinado-Guevara, L.I.; Alfonso-Corrado, C. Effects of Habitat Loss on the Ecology of Pachyphytum caesium (*Crassulaceae*), a Specialized Cliff-Dwelling Endemic Species in Central Mexico. *Diversity* **2021**, *13*, 421. [CrossRef]
54. Sheppard, C.S.; Brendel, M.R. Competitive ability of native and alien plants: Effects of residence time and invasion status. *Neobiota* **2021**, *65*, 47–69. [CrossRef]
55. Becker, T.; Dietz, H.; Billeter, R.; Buschmann, H.; Edwards, P.J. Altitudinal distribution of alien plant species in the Swiss Alps. *Perspect. Plant Ecol. Evol. Syst.* **2005**, *7*, 173–183. [CrossRef]
56. Ross, L.C.; Lambdon, P.W.; Hulme, P.E. Disentangling the roles of climate, propagule pressure and land use on the current and potential elevational distribution of the invasive weed *Oxalis pes-caprae* L. on Crete. *Perspect. Plant Ecol. Evol. Syst.* **2008**, *10*, 251–258. [CrossRef]
57. Roy, S.; Simon, J.P.; Lapointe, F.J. Determination of the origin of the cold-adapted populations of barnyard grass (*Echinochloa crus-galli*) in eastern North America: A total-evidence approach using RAPD DNA and DNA sequences. *Can. J. Bot. -Rev. Can. De Bot.* **2000**, *78*, 1505–1513. [CrossRef]
58. Essl, F.; Glaser, M.; Schertler, A. New and old invaders in forests in eastern Austria: The role of species attributes and invasion history. *Flora* **2021**, *283*, 151922. [CrossRef]
59. Padalia, H.; Bushra, A.; Singh, G.; Nandy, S.; Rai, I.D.; Rawat, G.S. Spatial patterns of plant functional types and environmental proxies of plant richness in alpine region of Western Himalaya, India. *Biodivers. Conserv.* **2019**, *28*, 2221–2244. [CrossRef]
60. Gao, D.; Fu, L.; Sun, J.; Li, Y.; Cao, Z.; Liu, Y.; Xu, P.; Zhao, J. The mid-domain effect and habitat complexity applied to elevational gradients: Moss species richness in a temperate semihumid monsoon climate mountain of China. *Ecol. Evol.* **2021**, *11*, 7448–7460. [CrossRef]
61. Alexander, J.M.; Kueffer, C.; Daehler, C.C.; Edwards, P.J.; Pauchard, A.; Seipel, T.; Consortium, M. Assembly of nonnative floras along elevational gradients explained by directional ecological filtering. *Proc. Natl. Acad. Sci. USA* **2011**, *108*, 656–661. [CrossRef]
62. Khuroo, A.A.; Weber, E.; Malik, A.H.; Reshi, Z.A.; Dar, G.H. Altitudinal distribution patterns of the native and alien woody flora in Kashmir Himalaya, India. *Environ. Res.* **2011**, *111*, 967–977. [CrossRef]
63. Davis, M.A.; Grime, J.P.; Thompson, K. Fluctuating resources in plant communities: A general theory of invasibility. *J. Ecol.* **2000**, *88*, 528–534. [CrossRef]
64. Rose, M.; Hermanutz, L. Are boreal ecosystems susceptible to alien plant invasion? Evidence from protected areas. *Oecologia* **2004**, *139*, 467–477. [CrossRef] [PubMed]
65. Paiaro, V.; Cabido, M.; Pucheta, E. Altitudinal distribution of native and alien plant species in roadside communities from central Argentina. *Austral Ecol.* **2011**, *36*, 176–184. [CrossRef]
66. Woitke, M.; Dietz, H. Shifts in dominance of native and invasive plants in experimental patches of vegetation. *Perspect. Plant Ecol. Evol. Syst.* **2002**, *5*, 165–184. [CrossRef]
67. Gasso, N.; Pysek, P.; Vila, M.; Williamson, M. Spreading to a limit: The time required for a neophyte to reach its maximum range. *Divers. Distrib.* **2010**, *16*, 310–311. [CrossRef]
68. Huebner, C.D.; Tobin, P.C. Invasibility of mature and 15-year-old deciduous forests by exotic plants. *Plant Ecol.* **2006**, *186*, 57–68. [CrossRef]
69. Tanaka, T.; Sato, T. Contemporary patterns and temporal changes in alien plant species richness along an elevational gradient in central Japan. *Plant Ecol. Evol.* **2016**, *149*, 177–188. [CrossRef]

70. Irl, S.D.H.; Schweiger, A.H.; Steinbauer, M.J.; Ah-Peng, C.; Arevalo, J.R.; Beierkuhnlein, C.; Chiarucci, A.; Daehler, C.C.; Fernandez-Palacios, J.M.; Flores, O.; et al. Human impact, climate and dispersal strategies determine plant invasion on islands. *J. Biogeogr.* **2021**, *48*, 1889–1903. [CrossRef]
71. Seebens, H.; Blackburn, T.M.; Dyer, E.E.; Genovesi, P.; Hulme, P.E.; Jeschke, J.M.; Pagad, S.; Pysek, P.; Winter, M.; Arianoutsou, M.; et al. No saturation in the accumulation of alien species worldwide. *Nat. Commun.* **2017**, *8*, 14435. [CrossRef]
72. Lososova, Z.; Chytry, M.; Tichy, L.; Danihelka, J.; Fajmon, K.; Hajek, O.; Kintrova, K.; Kuehn, I.; Lanikova, D.; Otypkova, Z.; et al. Native and alien floras in urban habitats: A comparison across 32 cities of central Europe. *Glob. Ecol. Biogeogr.* **2012**, *21*, 545–555. [CrossRef]

diversity

MDPI

Article

Niche Characteristics of *Alternanthera philoxeroide*-Invaded Plant Communities in Heterogeneous Habitats and Their Latitudinal Trends

Hao Wu [1,2,3,4,*], Sijin Dong [1], Yanyan Wang [1], Li Wang [1] and Benqiang Rao [1,*]

1 College of Life Sciences, Xinyang Normal University, Xinyang 464000, China; dongsijin6556@163.com (S.D.); 18055073546@163.com (Y.W.); 18736978525@163.com (L.W.)

2 Henan Dabieshan National Field Observation and Research Station of Forest Ecosystem, Xinyang 464000, China

3 Xinyang Academy of Ecological Research, Xinyang 464000, China

4 Wuhan Botanical Garden, Chinese Academy of Sciences, Wuhan 430074, China

* Correspondence: wuhao86868686@163.com (H.W.); rbqxy@163.com (B.R.)

Abstract: Plant invasions are closely related to environmental filtering and biointeractions; however, the variations in invasive plant niches along latitudinal gradients in heterogeneous habitats remain unclear. In this study, we conducted a two-year survey in China spanning 21° N–37° N to explore the niche characteristics of plant species within communities invaded by the amphibious alien weed *Alternanthera philoxeroides* in both terrestrial and aquatic habitats as well as their latitudinal trends. We found that *A. philoxeroides* had the greatest niche breadth in the studied communities. The species pairs with the highest niche similarity were *A. philoxeroides–Digitaria sanguinalis* in terrestrial communities and *Cyperus rotundus–Kyllinga brevifolia* in aquatic communities. The niche similarity between *A. philoxeroides* and its accompanying species in terrestrial habitats was significantly higher than that in aquatic habitats ($t = 5.954$; $p < 0.001$). The niche breadth of *A. philoxeroides* had no obvious latitudinal trend, while the niche breadth of its accompanying species in the terrestrial community significantly decreased with increasing latitude ($F_{7,57} = 4.364$, $p = 0.001$). In the terrestrial communities, the niche similarity between *A. philoxeroides* and its accompanying species significantly decreased with increasing latitude ($F_{7,57} = 3.671$, $p = 0.003$), while the niche overlap significantly increased with increasing latitude ($F_{7,57} = 8.916$, $p < 0.001$). However, the aquatic species' niche characteristics had no obvious latitudinal trends. These findings indicated that habitat heterogeneity significantly affected the species' niche characteristics in *A. philoxeroides*-invaded communities. Environmental filtering at low latitudes allowed the invasive and accompanying species to evolve similar niches, while the cold climate at high latitudes increased the niche overlap between the invader and accompanying species. Our findings are crucial for predicting the dynamics of invasive plant communities under global change and for understanding the mechanisms of species coexistence.

Keywords: *Alternanthera philoxeroides*; habitat heterogeneity; latitudinal gradient; niche; plant invasions

1. Introduction

In the context of global climate change and economic integration, biological invasions are one of the five serious global environmental problems of the 21st century [1,2]. Frequent resource fluctuations and disturbances resulting from human activities have led to increasing plant invasions, causing a dramatic decline in biodiversity and great damage to the structures and functions of ecosystems [3]. Functional differences between native and alien species are crucial factors for accelerating successful plant invasions; invasive plants usually have a higher growth rate and higher photosynthesis, water use efficiency, specific leaf area, nitrogen content and interspecific competitiveness than their native accompanying species [4–7], as well as stronger responses to environmental changes [8–10].

A recent meta-analysis has shown that multiple invasive plants could also have a synergistic invasion effect, resulting in twice the negative effect on neighboring natives than on other alien plants [11]. However, biological invasions and native species diversity often demonstrate positive relationships in large-scale observational studies but are negatively correlated in small-scale experimental studies; this phenomenon is referred to as the invasion paradox [12–14]. Exploring the performance and ecological effect of invasive plants at large spatial scales is helpful for implementing effective invasion prevention and biodiversity protection measures.

A niche represents the utilization degree of various resources by plant species, while niche breadth, niche similarity and niche overlap are important indicators for measuring plant niches; in addition, these factors reflect the role of plant species in communities and their demands for living environments [15,16]. Under conditions with limited resources, species that have similar niches will compete fiercely with each other and cannot stably coexist [17,18]. The expansion of invasive plant populations is usually associated with niche changes, and their niche characteristics are determined by environmental filtering, biointeractions, adaptive evolution and other factors [15,16,19,20]. Many invasive plants have undergone niche shifts and have expanded their climate niches through adaptive evolution and phenotypic plasticity, which is of great benefit for quickly occupying the open niche, and thus facilitates their invasion process [21–23]. In brief, the key to the successful invasion of alien plants is their ability to create 'niche opportunities' through various mechanisms, such as successfully coexisting in native communities through heterogeneous resource use or by increasing their diffusion ability by competing for the same resources with native species [24,25].

Hydrothermal fluctuations caused by latitudinal variations significantly affect the niche changes in invasive plant species. Compared with native plants, invasive plants usually have higher climate tolerance and phenological plasticity, which benefit their ability to form a wide niche breadth to resist climate change [26–29]. Moreover, global warming weakens the biotic resistance of native plant communities at higher latitudes, thus creating favorable niche spaces for invasive plants in those regions [30]. The latitude–niche breadth hypothesis predicts that tropical species evolve into narrower niches due to weak seasonality and environmental variability at low latitudes, while frequent climate changes at high latitudes promote the evolution of wider niches [31]. Rapoport's ecogeographic rules also propose that the distribution ranges of species shrink with decreasing latitude, and intense interspecific competition at low latitudes reduces species niche breadth, which negatively correlates with latitude [32,33]. The germination niche breadth and climate niche breadth of some invasive plants have been found to increase with increasing latitude; however, their water quality niche breadth decreases with increasing latitude, which is due to their long-term adaptations to water eutrophication [27,33]. However, the niche relationships between invasive and native plants along the larger latitudinal gradient are still not well understood.

The alligator weed *Alternanthera philoxeroides*, a type of noxious invasive weed with clonal growth, is native to South America and has extensively invaded the United States, Australia, South Africa, China and other parts of the world [34]. *A. philoxeroides* is an amphibious invader, and in China it has invaded aquatic and terrestrial habitats in more than 20 provinces, causing serious damage to ecological environments and economic development [35]. Previous studies have found that *A. philoxeroides* is more responsive to environmental fluctuations than its native congener *A. sessilis*, and climate warming will promote the climate niche expansion of the terrestrial ecotype of *A. philoxeroides*, intensifying its competitive substitution for native plants at high latitudes [36,37]. In flooded environments, *A. philoxeroides* increases the diameter of its stolon to strengthen its interspecific competitiveness [38]. In addition, with increasing latitudes in China, aquatic *A. philoxeroides* communities have higher β-diversity, while terrestrial *A. philoxeroides* communities have higher similarity [34]. However, the niche characteristics of *A. philoxeroides* and its accompanying species at large spatial scales are unclear.

In this study, we investigated the plant communities invaded by *A. philoxeroides* within the latitudinal range from 21° N to 37° N in mainland China, and we hypothesize that the species niche characteristics of communities invaded by *A. philoxeroides* along a latitudinal gradient have different trends in aquatic and terrestrial habitats. We specifically address the following questions: (1) Do the niche breadth, niche similarity and niche overlap of the *A. philoxeroides* communities in heterogeneous habitats differ? (2) Are the variations in each of the above three indicators different along latitudinal gradients in heterogeneous habitats?

2. Materials and Methods

2.1. Site Selection

From July to August of 2019–2021, the time of year during which the growth period of plant species is vigorous, we selected sites that were invaded by *A. philoxeroides* over an area of more than 100 m^2 to set the sampling plots, and the plots contained low (*A. philoxeroides* coverage < 40%), moderate (*A. philoxeroides* coverage was 40%–60%) and heavy (*A. philoxeroides* coverage > 60%) invaded levels. Starting from 21° N, a total of 8 latitudinal clusters were set, each 2° apart (the low latitudinal gradient containing Cluster 1: 21° N–23° N and Cluster 2: 23° N–25° N; the middle latitudinal gradient containing Cluster 3: 25° N–27° N, Cluster 4: 27° N–29° N, Cluster 5: 29° N–31° N, and Cluster 6: 31° N–33° N; and the high latitudinal gradient containing Cluster 7: 33° N–35° N and Cluster 8: 35° N–37° N) [34,36,39]. Five terrestrial plots were set along each latitudinal cluster, and five aquatic plots were also set in areas with similar habitat conditions along the same latitudinal cluster [35]. The area of each plot was 10 m × 10 m, and plots with the same ecotype were set at intervals of more than 10 km [39]. In total, we set 80 sampling plots (40 terrestrial and 40 aquatic; Figure 1), with the types of terrestrial sites mainly comprising roadsides (13 plots), gardens (8 plots), abandoned land (7 plots), farmland (7 plots) and riparian zones (5 plots), while the types of aquatic sites comprised ponds (12 plots), rivers (10 plots), puddles (9 plots), water channels (6 plots) and paddy fields (3 plots) (Supplementary Table S1). For the sampling sites in which there may be a fluid transition between terrestrial and aquatic habitats or in which there may be intermittent changes in habitat type due to climate fluctuations, we combined the morphological plasticity of *A. philoxeroides* to identify the type of plot in question. The whole study area covered 20 cities within 10 provinces in China and spanned 3 climatic regions, including tropical, subtropical and temperate regions.

2.2. Field Survey

For terrestrial plots and some aquatic plots with shallow water levels, we evenly set three 10 m transects in each plot and then evenly set five quadrats with an area of 0.5 m × 0.5 m along each transect for the plant investigation [35,37,40]. For a few aquatic plots with deep water levels, to facilitate the investigation, we chose a blanket patch of *A. philoxeroides* that was close to the bank and set three continuous 10 m transects along the horizontal direction of the aquatic *A. philoxeroides* patches for investigation. We used the survey method of Wu et al. (2016) to record the names, height and coverage of the plant species in every quadrat [39]. We also recorded the individuals (for nonclonal plants), tillers (for graminaceous plants) or numbers of asexual branches (for clonal plants) as the abundance of plant species in each quadrat [35,39]. The longitude, latitude and altitude of every plot were recorded using a handheld GPS receiver (Garmin Inc., Olathe, KS, USA). Photos or collected samples of unknown plants were taken during the field survey, and we then used the 'Chinese Virtual Herbarium' (https://www.cvh.ac.cn, accessed on 10 March 2023) and the online database 'Flora of China' (http://www.iplant.cn/frps, accessed on 10 March 2023) to conduct species identification.

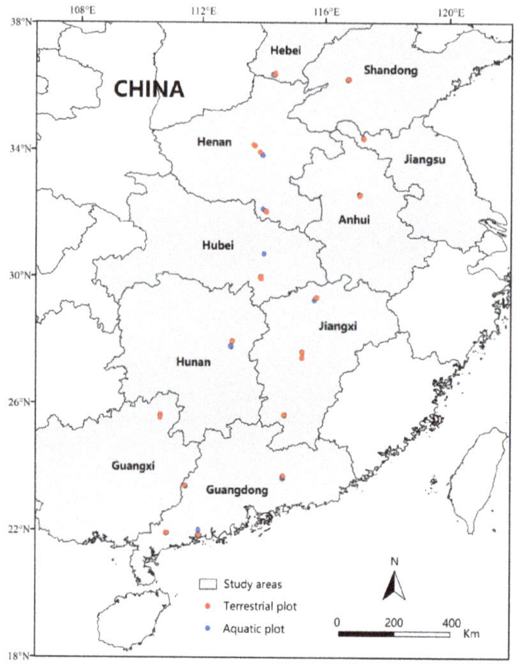

Figure 1. Sampling plots of *A. philoxeroides*-invaded communities in China along latitudinal gradients (21° N to 37° N). Plot size: 10 × 10 m; transect size: 10 m; quadrat size: 0.5 × 0.5 m.

2.3. Data Calculation

2.3.1. Importance Value

We used the importance value (IV) as a comprehensive indicator to measure the dominance of plants in the *A. philoxeroides*-invaded communities. The IV was calculated as follows [39]:

1. IV = (relative abundance + relative height + relative coverage)/3.
2. The total IV was the sum of a plant species' IV in 40 terrestrial or 40 aquatic plots.

2.3.2. Niche Characteristics

Each plot in this study was regarded as a comprehensive resource level. Based on the occurrence frequency of plant species in the plots (Table 1), we used the abundance and IV of 25 main plant species (frequency > 9) in the terrestrial plots vs. 24 main plant species (frequency > 4) in the aquatic plots as comprehensive indicators to measure their niche characteristics. The results calculated using this method better reflected the efficiency of the plant species' utilization of environmental resources and could prevent errors caused by individual size differences in each species.

Niche breadth represents the sum of the multiple environmental resources that are used by organisms. Plants with a higher niche breadth can better utilize resources and tend to be generalized species, while plants with a lower niche breadth tend to be specialized species. We calculated niche breadth values using Levins' method as follows [41]:

$$B_{(L)} = 1/(\mathrm{r} \times \sum_{j=1}^{r} P_{ij}^2) \tag{1}$$

where $B_{(L)}$ is the species' niche breadth, P_{ij} is the relative IV of the *i*th species in the *j*th plot and *r* is the total number of sampling plots.

Niche similarity represents the similarity degree of the species' resource utilization, and a lower niche similarity indicates that there are differences in resource utilization between plant species and that they might coexist in the community for a long time. We calculated the niche similarity value using the Schoener index as follows [42]:

$$C_{ik} = 1 - \frac{1}{2}\sum_{j=1}^{r}|P_{ij} - P_{kj}| \quad (2)$$

where C_{ik} is the niche similarity value between species i and k, and P_{kj} is the relative IV of the kth species in the jth plot.

Niche overlap represents the degree of species with similar niches sharing or competing for common resources when living in the same space; the greater the niche overlap value, the greater the similarity in resource utilization and the more intense the interspecific competition. Niche overlap was calculated as follows [43]:

$$L_{ih} = B_{(L)i}\sum_{j=1}^{r} P_{ij} \times P_{hj} \quad (3)$$

$$L_{hi} = B_{(L)h}\sum_{j=1}^{r} P_{ij} \times P_{hj} \quad (4)$$

where L_{ih} is the niche overlap value of species i overlapping species h, L_{hi} is the niche overlap value of species h overlapping species i, P_{hj} is the relative IV of the hth species in the jth plot, $B_{(L)i}$ is the niche breadth of species i and $B_{(L)h}$ is the niche breadth of species h.

Table 1. Niche breadth ($B_{(L)}$) of plant species with a high frequency in terrestrial and aquatic *A. philoxeroides*-invaded communities.

Code	Plant Species	Frequency	IV	$B_{(L)}$	Code	Plant Species	Frequency	IV	$B_{(L)}$
	Terrestrial Community					Aquatic Community			
1	*Alternanthera philoxeroides*	40	17.360	0.960	1	*Alternanthera philoxeroides*	40	23.752	0.943
2	*Digitaria sanguinalis*	30	2.936	0.414	2	*Paspalum paspaloides*	25	5.427	0.424
3	*Echinochloa crusgalli*	28	1.112	0.314	3	*Polygonum hydropiper*	20	0.727	0.301
4	*Bidens pilosa*	25	0.739	0.354	4	*Echinochloa crusgalli*	18	1.226	0.210
5	*Ipomoea nil*	23	0.599	0.294	5	*Bidens pilosa*	14	0.341	0.232
6	*Eleusine indica*	22	1.064	0.278	6	*Commelina communis*	9	0.534	0.107
7	*Cyperus votundus*	21	0.901	0.204	7	*Cyperus rotundus*	8	0.198	0.126
8	*Setaria viridis*	19	0.972	0.322	8	*Kyllinga brevifolia*	7	0.123	0.146
9	*Commelina communis*	18	1.301	0.285	9	*Eclipta prostrata*	7	0.217	0.086
10	*Cyperus rotundus*	16	0.303	0.269	10	*Humulus scandens*	7	0.444	0.099
11	*Acalypha australis*	16	0.198	0.287	11	*Ipomoea nil*	7	0.232	0.079
12	*Humulus scandens*	15	1.120	0.203	12	*Lemna minor*	6	0.284	0.115
13	*Ageratum conopids*	15	0.680	0.160	13	*Digitaria sanguinalis*	6	0.123	0.120
14	*Erigeron canadensis*	15	0.425	0.138	14	*Acorus calamus*	5	0.146	0.106
15	*Artemisia argyi*	15	0.342	0.278	15	*Glycine soja*	5	0.309	0.075
16	*Eclipta prostrata*	15	0.195	0.214	16	*Alternanthera sessilis*	5	0.184	0.074
17	*Cynodon dactylon*	14	0.695	0.108	17	*Eleusine indica*	5	0.190	0.096
18	*Polygonum hydropiper*	14	0.677	0.129	18	*Trigastrotheca stricta*	5	0.291	0.052
19	*Amaranthus blitum*	13	0.170	0.132	19	*Cyperus votundus*	5	0.169	0.111
20	*Erigeron annuus*	12	0.231	0.123	20	*Polygonum perfoliatum*	4	0.061	0.092
21	*Oxalis corniculata*	10	0.250	0.099	21	*Phragmites australis*	4	0.187	0.083
22	*Paspalum paspaloides*	9	0.435	0.155	22	*Senna tora*	4	0.043	0.060
23	*Broussonetia papyrifera*	9	0.296	0.112	23	*Symphyotrichum subulatum*	4	0.197	0.053
24	*Ludwigia prostrata*	9	0.168	0.113	24	*Microstegium nodosum*	4	0.146	0.041
25	*Portulaca oleracea*	9	0.106	0.122					

2.4. Statistical Analyses

We conducted an independent sample t test (subset for $\alpha = 0.05$) to compare the differences in the plant species' niche characteristics between terrestrial and aquatic communities using SPSS16.0 software (SPSS Inc., Chicago, IL, USA) and conducted a one-way ANOVA and multiple comparisons of least-significant difference (LSD) to examine the variations in niche characteristics along the latitudinal gradient.

3. Results

3.1. Niche Characteristics at the Level of the Whole Community

3.1.1. Niche Breadth

The species niche breadth in the terrestrial communities was 0.099–0.960, and that of the aquatic communities was 0.041–0943. Of the habitat types, the heterogeneous habitats had the largest *A. philoxeroides* niche breadth (terrestrial $B_{(L)} = 0.960$, aquatic $B_{(L)} = 0.943$) (Table 1). In the terrestrial communities, the species with higher niche breadth values were *Digitaria sanguinalis, Bidens pilosa, Setaria viridis* and *Echinochloa crusgalli*, and *Oxalis corniculata* had the lowest niche breadth (Table 1). In the aquatic communities, *Paspalum paspaloides, Polygonum hydropiper, B. pilosa* and *E. crusgalli* also had higher niche breadth values, and *Microstegium nodosum* had the lowest niche breadth (Table 1). The t test showed that there was no significant difference in the niche breadth values of the common species between the terrestrial and the aquatic communities ($t = 1.598$, $p = 0.117$).

3.1.2. Niche Similarity

A total of 25 plant species in the terrestrial community formed 300 species pairs, and the values of their niche similarity were 0–0.547. Among these values, the niche similarity of two species pairs was more than 0.5 (containing *A. philoxeroides–D. sanguinalis, S. viridis–Humulus scandens*), accounting for 0.67% of the total species pairs. The niche similarity values of 11 species pairs were 0.4–0.5, accounting for 3.67% of the total species pairs. The niche similarity values of 46 species pairs were 0.3–0.4, accounting for 15.33% of the total number of species pairs. The niche similarity of 241 species pairs was less than 0.3, accounting for 80.33% of the total number of species pairs. The niche similarity between *A. philoxeroides* and its accompanying species was generally higher than that between each accompanying species (Figure 2a).

There were 276 species pairs formed from 24 plant species in the aquatic communities, with a niche similarity of 0–0.511. The niche similarity of one species pair was more than 0.5 (*C. rotundus–Kyllinga brevifolia*), accounting for 0.36% of the total species pairs. The niche similarity of seven species pairs was 0.4–0.5, accounting for 2.54% of the total species pairs. The niche similarity of 20 species pairs was 0.3–0.4, accounting for 7.25% of the total species pairs. The niche similarity of 248 species pairs was less than 0.3, accounting for 89.85% of the total species pairs (Figure 2b). The t test showed that the niche similarity between *A. philoxeroides* and its accompanying species in the terrestrial communities was significantly higher than that between *A. philoxeroides* and its accompanying species in the aquatic communities ($t = 5.954$, $p < 0.001$).

(a)

	1	2	3	4	5	6	7	8	9	10	11	12	13	14	15	16	17	18	19	20	21	22	23	24
2	0.538																							
3	0.474	0.477																						
4	0.458	0.395	0.353																					
5	0.484	0.345	0.313	0.308																				
6	0.405	0.321	0.250	0.498	0.254																			
7	0.410	0.312	0.331	0.296	0.221	0.214																		
8	0.397	0.246	0.333	0.311	0.298	0.196	0.279																	
9	0.344	0.316	0.239	0.434	0.258	0.329	0.297	0.125																
10	0.368	0.359	0.370	0.338	0.235	0.205	0.196	0.188	0.270															
11	0.383	0.391	0.284	0.345	0.251	0.214	0.288	0.294	0.246	0.466														
12	0.293	0.086	0.334	0.158	0.285	0.175	0.129	0.547	0.092	0.039	0.141													
13	0.283	0.244	0.207	0.344	0.177	0.328	0.279	0.078	0.339	0.209	0.132	0.052												
14	0.343	0.110	0.152	0.129	0.207	0.211	0.080	0.296	0.124	0.135	0.207	0.298	0.077											
15	0.352	0.271	0.212	0.321	0.199	0.231	0.320	0.373	0.341	0.236	0.292	0.258	0.319	0.228										
16	0.355	0.433	0.310	0.296	0.326	0.251	0.195	0.097	0.229	0.376	0.325	0.117	0.150	0.028	0.164									
17	0.285	0.249	0.243	0.205	0.139	0.116	0.176	0.315	0.166	0.062	0.203	0.216	0.012	0.068	0.213	0.120								
18	0.323	0.167	0.216	0.194	0.257	0.209	0.063	0.083	0.212	0.353	0.178	0.159	0.152	0.116	0.064	0.201	0.079							
19	0.304	0.197	0.159	0.182	0.082	0.232	0.173	0.316	0.185	0.152	0.188	0.303	0.052	0.227	0.173	0.111	0.131	0.250						
20	0.289	0.157	0.203	0.114	0.211	0.095	0.117	0.247	0.036	0.196	0.195	0.203	0.050	0.287	0.300	0.178	0.098	0.136	0.105					
21	0.226	0.235	0.133	0.244	0.175	0.164	0.280	0.260	0.207	0.145	0.130	0.216	0.076	0.066	0.210	0.160	0.111	0.074	0.263	0.128				
22	0.204	0.114	0.232	0.107	0.265	0.065	0.132	0.068	0.453	0.267	0.122	0.170	0.030	0.146	0.000	0.112	0.072	0.231	0.224	0.038	0.072			
23	0.211	0.093	0.118	0.166	0.250	0.130	0.154	0.335	0.030	0.093	0.170	0.234	0.087	0.128	0.327	0.123	0.191	0.012	0.109	0.199	0.098	0.000		
24	0.202	0.049	0.227	0.133	0.092	0.212	0.024	0.176	0.085	0.253	0.101	0.158	0.280	0.286	0.076	0.133	0.092	0.143	0.102	0.088	0.029	0.257	0.104	
25	0.209	0.345	0.292	0.157	0.124	0.151	0.091	0.128	0.110	0.106	0.180	0.183	0.018	0.085	0.015	0.249	0.124	0.058	0.146	0.041	0.025	0.096	0.000	0.022

(b)

	1	2	3	4	5	6	7	8	9	10	11	12	13	14	15	16	17	18	19	20	21	22	23
2	0.464																						
3	0.352	0.443																					
4	0.371	0.168	0.228																				
5	0.282	0.303	0.158	0.261																			
6	0.162	0.083	0.229	0.194	0.149																		
7	0.150	0.139	0.207	0.308	0.198	0.467																	
8	0.139	0.139	0.196	0.415	0.326	0.366	0.511																
9	0.154	0.116	0.241	0.125	0.021	0.301	0.132	0.193															
10	0.189	0.092	0.224	0.000	0.176	0.009	0.000	0.000	0.080														
11	0.151	0.164	0.090	0.078	0.114	0.025	0.025	0.008	0.008	0.029													
12	0.154	0.086	0.210	0.128	0.037	0.107	0.134	0.107	0.107	0.186	0.000												
13	0.148	0.034	0.092	0.366	0.074	0.034	0.354	0.325	0.029	0.000	0.290	0.246											
14	0.130	0.127	0.082	0.067	0.163	0.000	0.000	0.000	0.051	0.177	0.374	0.094	0.120										
15	0.128	0.100	0.038	0.033	0.052	0.000	0.016	0.071	0.000	0.029	0.033	0.000	0.016	0.000									
16	0.091	0.055	0.168	0.149	0.050	0.256	0.253	0.315	0.075	0.000	0.025	0.000	0.184	0.000	0.000								
17	0.110	0.055	0.208	0.177	0.037	0.373	0.216	0.164	0.426	0.000	0.016	0.303	0.240	0.000	0.000	0.151							
18	0.084	0.090	0.260	0.147	0.037	0.236	0.384	0.298	0.160	0.000	0.025	0.077	0.083	0.038	0.000	0.226	0.258						
19	0.137	0.006	0.129	0.048	0.103	0.093	0.037	0.000	0.124	0.310	0.016	0.000	0.000	0.177	0.000	0.189	0.137	0.137					
20	0.111	0.060	0.044	0.000	0.103	0.000	0.000	0.000	0.000	0.273	0.033	0.000	0.000	0.177	0.349	0.000	0.000	0.000	0.372				
21	0.117	0.053	0.000	0.041	0.028	0.053	0.000	0.055	0.421	0.000	0.000	0.000	0.000	0.000	0.108	0.000	0.252	0.000	0.321	0.000			
22	0.094	0.054	0.126	0.109	0.078	0.084	0.179	0.142	0.000	0.000	0.231	0.000	0.083	0.215	0.056	0.474	0.182	0.279	0.189	0.056	0.000		
23	0.110	0.000	0.070	0.324	0.046	0.056	0.037	0.000	0.000	0.036	0.016	0.000	0.126	0.000	0.000	0.056	0.056	0.056	0.144	0.088	0.000	0.056	
24	0.108	0.060	0.008	0.000	0.054	0.009	0.000	0.000	0.000	0.167	0.000	0.000	0.000	0.078	0.123	0.000	0.000	0.000	0.184	0.200	0.000	0.056	0.000

Figure 2. Niche similarity of plant species with a high frequency in terrestrial (**a**) and aquatic (**b**) *A. philoxeroides*-invaded communities. Arabic numbers are codes of plant species (as shown in Table 1 for terrestrial and aquatic communities, respectively).

3.1.3. Niche Overlap

The species niche overlap values of the terrestrial communities were 0–0.026. The niche overlap between *A. philoxeroides* and its accompanying species was 0.022–0.026, which was generally higher than that between each accompanying species. The niche overlap of *A. philoxeroides-P. paspaloides* was the highest. The niche overlap of 49 species pairs was more than 0.01, accounting for 16.33% of the total species pairs. The niche overlap values of 64 species pairs were 0.005–0.01, accounting for 21.33% of the total species pairs. The niche overlap of 187 species pairs was less than 0.005, accounting for 62.33% of the total species pairs.

In the aquatic communities, the species niche overlap values were 0–0.029. The niche overlap values between *A. philoxeroides* and its accompanying species were 0.017–0.029, which were generally higher than those between each accompanying species. The niche overlap values of *A. philoxeroides–Symphyotrichum subulatum* and *A. philoxeroides–M. nodosum* were both the highest. The niche overlap values of 58 species pairs were more than 0.01, accounting for 21.01% of the total species pairs. The niche overlap values of 56 species pairs were 0.005–0.01, accounting for 20.29% of the total species pairs. The niche overlap values of 162 species pairs were less than 0.005, accounting for 58.70% of the total species pairs. The t test showed that there was no significant difference in the niche overlap of *A. philoxeroides*–accompanying species between terrestrial and aquatic communities (t = 0.488, p = 0.630).

3.2. Latitudinal Trends of Niche Characteristics

3.2.1. Variations in Niche Breadth along the Latitudinal Gradient

The niche breadth of *A. philoxeroides* had no obvious fluctuations along the latitude gradient. For the accompanying species in the terrestrial communities, compared to the other species, *Commelina communis*, *B. pilosa* and *Eleusine indica* had greater niche breadths at low latitudes; *D. sanguinalis*, *E. indica*, *E. crusgalli* and *S. viridis* had greater niche breadths at middle latitudes; and *S. viridis* and *H. scandens* had greater niche breadths at high latitudes. Conspicuously, *D. sanguinalis* always had greater niche breadths along the whole latitudinal gradient (Figure 3a). The ANOVA showed that the average niche breadth of the accompanying species significantly decreased with increasing latitude in the terrestrial habitats ($F_{7,57} = 4.364$, $p = 0.001$); in particular, the niche breadth of Cluster 2 at low latitudes was significantly higher than that at middle and high latitudes.

In the aquatic communities, *P. hydropiper* and *E. crusgalli* had greater niche breadths at low latitudes, *P. paspaloides* had an extremely large niche breadth at middle latitudes, and *P. paspaloides*, *E. crusgalli* and *P. hydropiper* had greater niche breadths at high latitudes (Figure 3b). The ANOVA showed that the average niche breadth of the accompanying species in the aquatic community had no significant latitudinal trend ($F_{7,45} = 1.170$, $p = 0.343$).

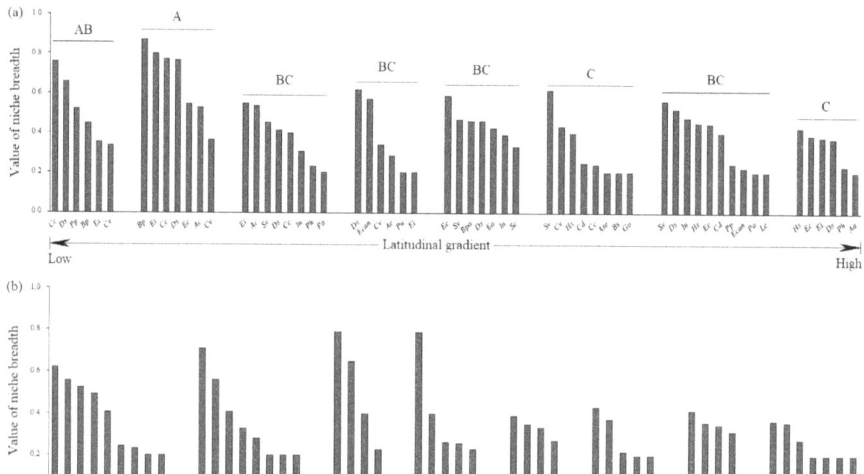

Figure 3. Variations in niche breadth of the accompanying species along a latitudinal gradient in terrestrial (**a**) and aquatic (**b**) *A. philoxeroides*-invaded communities. **Definition of abbreviations:** Aa—*Alopecurus aequalis*; Ac—*Ageratum conyzoides*; As—*Asarum sagittarioides*; Ase—*Alternanthera sessilis*; Bp—*Bidens pilosa*; Bpa—*Broussonetia papyrifera*; Bpl—*Bolboschoenus planiculmis*; Bs—*Beckmannia syzigachne*; Cc—*Commelina communis*; Cd—*Cynodon dactylon*; Cde—*Ceratophyllum demersum*; Cr—*Cyperus rotundus*; Cv—*Cyperus votundus*; Ds—*Digitaria sanguinalis*; Ea—*Erigeron annuus*; Ec—*Echinochloa crusgalli*; Eca—*Echinochloa caudata*; Ecan—*Erigeron canadensis*; Ecr—*Eichhornia crassipes*; Ei—*Eleusine indica*; Ep—*Eclipta prostrata*; Go—*Galium odoratum*; Gs—*Glycine soja*; Hs—*Humulus scandens*; Ia—*Ipomoea aquatica*; In—*Ipomoea nil*; Lc—*Leptochloa chinensis*; Lm—*Lemna minor*; Mn—*Microstegium nodosum*; Mv—*Myriophyllum verticillatum*; Oj—*Oenanthe javanica*; Pa—*Phragmites australis*; Pb—*Panicum bisulcatum*; Ph—*Polygonum hydropiper*; Pp—*Paspalum paspaloides*; Ps—*Pistia stratiotes*; Pu—*Phyllanthus urinaria*; Sc—*Sesbania cannabina*; Ss—*Symphyotrichum subulatum*; Sv—*Setaria viridis*; Ts—*Trigastrotheca stricta*; Vp—*Veronica polita*. A and C, A and BC indicate the significant differences at the $p < 0.01$ level, respectively; while the A and AB, AB and BC, BC and C indicate no significant difference, respectively.

3.2.2. Variations in Niche Similarity along the Latitudinal Gradient

The niche similarity between *A. philoxeroides* and its accompanying species was generally higher than that between each accompanying species, and with increasing latitude, the niche similarity showed an overall downward trend. In the terrestrial community, the species pairs of *A. philoxeroides–C. communis*, *A. philoxeroides–D. sanguinalis*, *A. philoxeroides–B. pilosa*, *C. communis–P. paspaloides* and *B. pilosa–E. indica* had higher niche similarity at low latitudes than at other latitudes; *A. philoxeroides–E. indica*, *A. philoxeroides–D. sanguinalis*, *A. philoxeroides–S. viridis*, *E. crusgalli–S. viridis*, *E. indica–S. viridis* and *H. scandens–S. viridis* had higher niche similarity at middle latitudes; and *A. philoxeroides–H. scandens*, *A. philoxeroides–Ipomoea nil*, *H. scandens–E. crusgalli* and *D. sanguinalis–E. crusgalli* had higher niche similarity at high latitudes (Figure 4a). The ANOVA showed that niche similarity between *A. philoxeroides* and its accompanying species significantly decreased with increasing latitude in the terrestrial habitats ($F_{7,57} = 3.671$, $p = 0.003$), especially the niche similarity of Cluster 2 at low latitudes, which was significantly higher than that at middle and high latitudes (Figure 5a).

Figure 4. Niche similarity between the dominant species of each latitudinal cluster in terrestrial (**a**) and aquatic (**b**) *A. philoxeroides*-invaded communities. **Definition of abbreviations:** Aa—*Alopecurus aequalis*; Ac—*Ageratum conyzoides*; Ap—*Alternanthera philoxeroides*; As—*Asarum sagittarioides*; Ase—*Alternanthera sessilis*; Bp—*Bidens pilosa*; Bpa—*Broussonetia papyrifera*; Bpl—*Bolboschoenus planiculmis*; Bs—*Beckmannia syzigachne*; Cc—*Commelina communis*; Cd—*Cynodon dactylon*; Cde—*Ceratophyllum demersum*; Cr—*Cyperus rotundus*; Cv—*Cyperus votundus*; Ds—*Digitaria sanguinalis*; Ea—*Erigeron annuus*; Ec—*Echinochloa crusgalli*; Eca—*Echinochloa caudata*; Ecan—*Erigeron canadensis*; Ecr—*Eichhornia crassipes*; Ei—*Eleusine indica*; Ep—*Eclipta prostrata*; Go—*Galium odoratum*; Gs—*Glycine soja*; Hs—*Humulus scandens*; Ia—*Ipomoea aquatica*; In—*Ipomoea nil*; Lc—*Leptochloa chinensis*; Lm—*Lemna minor*; Mn—*Microstegium nodosum*; Mv—*Myriophyllum verticillatum*; Oj—*Oenanthe javanica*; Pa—*Phragmites australis*; Pb—*Panicum bisulcatum*; Ph—*Polygonum hydropiper*; Pp—*Paspalum paspaloides*; Ps—*Pistia stratiotes*; Pu—*Phyllanthus urinaria*; Sc—*Sesbania cannabina*; Ss—*Symphyotrichum subulatum*; Sv—*Setaria viridis*; Ts—*Trigastrotheca stricta*; Vp—*Veronica polita*.

In aquatic communities, the species pairs of *A. philoxeroides–Pistia stratiotes*, *A. philoxeroides–Eichhornia crassipes*, *A. philoxeroides–P. paspaloides*, *P. stratiotes–Asarum sagittarioides* and *Bolboschoenus planiculmis–Panicum bisulcatum* had higher niche similarity at low latitudes than at the other latitudes; *A. philoxeroides–P. paspaloides*, *A. philoxeroides–I. nil*, *A. philoxeroides–E. caudata*, *A. philoxeroides–Myriophyllum verticillatum* and *P. paspaloides–P. bisulcatum* had higher niche similarity at middle latitudes; and *A. philoxeroides–Phragmites australis*, *A. philoxeroides–S. subulatum*, *P. paspaloides–Oenanthe javanica* and *Ceratophyllum demersum–E. crusgalli* had higher niche similarity at high latitudes (Figure 4b). The ANOVA showed that the niche similarity between *A. philoxeroides* and its accompanying species of each cluster had no significant latitudinal trend in the aquatic habitat ($F_{7,45} = 0.981$, $p = 0.459$) (Figure 5b).

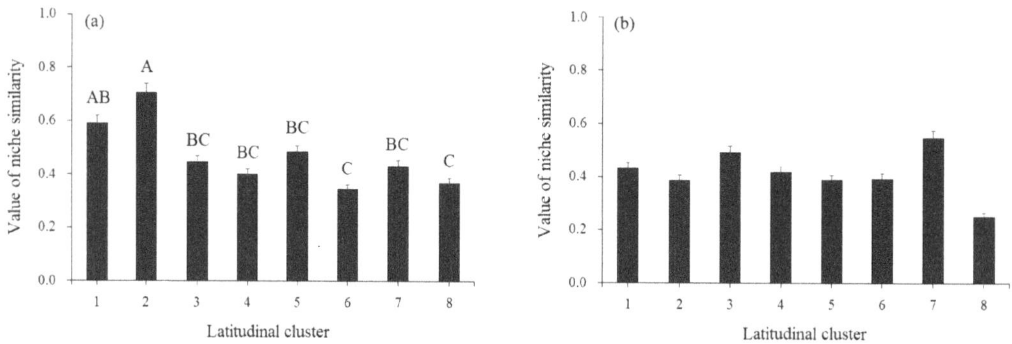

Figure 5. Variations in niche similarity between *A. philoxeroides* and the dominant accompanying species of each latitudinal cluster in terrestrial (**a**) and aquatic (**b**) habitats. A and C, A and BC indicate the significant differences at the $p < 0.01$ level, respectively; while the A and AB, AB and BC, BC and C indicate no significant difference, respectively.

3.2.3. Variations in Niche Overlap along the Latitudinal Gradient

The niche overlap values of *A. philoxeroides* and its accompanying species were generally higher than those of the species accompanying *A. philoxeroides*. With increasing latitude, the niche overlap values between plant species in the *A. philoxeroides*-invaded communities showed an overall downward trend (Figure 6). In the terrestrial communities, the species pairs of *A. philoxeroides–D. sanguinalis*, *A. philoxeroides–B. pilosa*, *B. pilosa–E. indica* and *Ageratum conyzoides–C. votundus* had higher niche overlap values at low latitudes than at other latitudes; *A. philoxeroides–A. conyzoides*, *A. philoxeroides–C. communis*, *A. philoxeroides–C. votundus*, *E. indica–Erigeron canadensis* and *E. crusgalli–Sesbania cannabina* had higher niche overlap values at middle latitudes; and *A. philoxeroides–S. viridis*, *A. philoxeroides–D. sanguinalis*, *A. philoxeroides–I. nil*, *A. philoxeroides–H. scandens*, *A. philoxeroides–E. crusgalli* and *A. philoxeroides–Cynodon dactylon* had higher niche overlap values at high latitudes (Figure 6a). The ANOVA showed that the niche overlap between *A. philoxeroides* and its accompanying species in Cluster 7 at high latitudes was significantly higher than that at low and middle latitudes ($F_{7,57} = 8.916$, $p < 0.001$) (Figure 7a).

In the aquatic communities, *A. philoxeroides–P. stratiotes*, *A. philoxeroides–E. crassipes*, *A. philoxeroides–P. paspaloides*, *P. stratiotes–A. sagittarioides* and *B. planiculmis* had higher niche overlap values at low latitudes than at other latitudes; *A. philoxeroides–P. paspaloides*, *A. philoxeroides–I. nil*, *A. philoxeroides–E. caudata*, *A. philoxeroides–M. verticillatum* and *P. paspaloides–P. bisulcatum* had higher niche overlap at middle latitudes; and *A. philoxeroides–P. australis*, *A. philoxeroides–S. subulatum*, *P. paspaloides–O. javanica*, *C. demersum–E. crusgalli*, *P. paspaloides–P. hydropiper* and *O. javanica–P. hydropiper* had higher niche overlap at high latitudes (Figure 6b). The ANOVA showed that the niche overlap between

A. philoxeroides and its accompanying species of each cluster had no significant latitudinal trend in the aquatic habitats ($F_{7,45} = 1.042$, $p = 0.419$) (Figure 7b).

Figure 6. Niche overlap between the dominant species of each latitudinal cluster in terrestrial (**a**) and aquatic (**b**) *A. philoxeroides*-invaded communities. **Definition of abbreviations:** Aa—*Alopecurus aequalis*; Ac—*Ageratum conyzoides*; Ap—*Alternanthera philoxeroides*; As—*Asarum sagittarioides*; Ase—*Alternanthera sessilis*; Bp—*Bidens pilosa*; Bpa—*Broussonetia papyrifera*; Bpl—*Bolboschoenus planiculmis*; Bs—*Beckmannia syzigachne*; Cc—*Commelina communis*; Cd—*Cynodon dactylon*; Cde—*Ceratophyllum demersum*; Cr—*Cyperus rotundus*; Cv—*Cyperus votundus*; Ds—*Digitaria sanguinalis*; Ea—*Erigeron annuus*; Ec—*Echinochloa crusgalli*; Eca—*Echinochloa caudata*; Ecan—*Erigeron canadensis*; Ecr—*Eichhornia crassipes*; Ei—*Eleusine indica*; Ep—*Eclipta prostrata*; Go—*Galium odoratum*; Gs—*Glycine soja*; Hs—*Humulus scandens*; Ia—*Ipomoea aquatica*; In—*Ipomoea nil*; Lc—*Leptochloa chinensis*; Lm—*Lemna minor*; Mn—*Microstegium nodosum*; Mv—*Myriophyllum verticillatum*; Oj—*Oenanthe javanica*; Pa—*Phragmites australis*; Pb—*Panicum bisulcatum*; Ph—*Polygonum hydropiper*; Pp—*Paspalum paspaloides*; Ps—*Pistia stratiotes*; Pu—*Phyllanthus urinaria*; Sc—*Sesbania cannabina*; Ss—*Symphyotrichum subulatum*; Sv—*Setaria viridis*; Ts—*Trigastrotheca stricta*; Vp—*Veronica polita*.

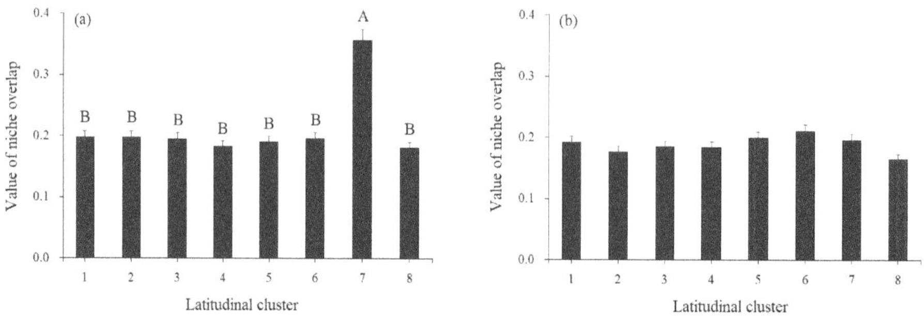

Figure 7. Variations in niche overlap between *A. philoxeroides* and the dominant accompanying species of each latitudinal cluster in terrestrial (**a**) and aquatic (**b**) habitats. A and B indicate the significant differences at the $p < 0.01$ level.

4. Discussion

4.1. Causes of Niche Characteristic Formation in Heterogeneous Habitats

Of the species examined in this study, *A. philoxeroides* had the greatest niche breadth in both aquatic and terrestrial habitats, reflecting that this invader had a strong ecological adaptation capacity and high resource utilization efficiency, which may be attributed to the cold-tolerant genotype and the growth–defense tradeoff of *A. philoxeroides*, which has benefited from adapting to harsh habitat conditions and thus occupies a broader niche space [44–46]. This conclusion also supports the niche breadth hypothesis, which states that there is a positive correlation between the size of a species' geographical range and its niche breadth [17], and that plant populations that continuously match habitat conditions will have greater opportunities for distribution expansion [33,47]. However, *D. sanguinalis*, *B. pilosa* and *S. viridis* in the terrestrial habitats and *P. paspaloides* and *P. hydropiper* in the aquatic habitats also had greater niche breadths, which might be attributed to their bio-properties. *D. sanguinalis* and *S. viridis* had tillering growth and produced massive seeds with strong transmission ability through generative propagation. The tall individuals of *B. pilosa* and *P. hydropiper* would imply that they compete for more light resources, while *P. paspaloides* was found to be a perennial species and had both vegetative and generative propagations (Table S2); it could spread rapidly through the rhizomes and stolons. All of these superior functional traits promoted their stable existence at each stage of community succession. Although the niche breadth of most species was positively correlated with their IV in the *A. philoxeroides* community, some dominant accompanying species, such as *S. viridis* and *C. communis*, were incidentally present, indicating that IV was not the unique factor determining niche breadth, and niche breadth might also be determined by the interactive effect of species distribution frequency and habitat heterogeneity [48].

In our study, niche similarity in the terrestrial communities was generally higher than that in the aquatic communities, which might be due to the higher species diversity of the terrestrial communities [37]. Neutral theory states that species similarity rather than species difference contributes to high community diversity; thus, continuous biointeractions among multiple species in terrestrial *A. philoxeroides* communities promote coexistence groups with high niche similarity [49]. In addition, Wu et al. (2022) found that *A. philoxeroides* invasions caused higher community homogenization in terrestrial habitats than in aquatic habitats, and functional groups of Gramineae and Compositae that have strong invasion resistance widely exist in the community [34], which causes the plant species in terrestrial habitats to have closer geographical affinity and facilitates their convergent evolution. Moreover, environmental filtering caused by strong disturbances (e.g., human disturbances and insect feeding) in terrestrial habitats have led to the repeated colonization of plant species in the same species pool, which have caused similar species to show obvious aggregation and thus facilitated the coexistence of species with high niche similarity [19]. However, the structure of aquatic communities is relatively simple and is more sensitive to environmental changes; thus, aquatic plant species might require a smaller niche similarity to maintain survival [50].

Species niche overlap in terrestrial and aquatic habitats was low, indicating that plant species in the *A. philoxeroides*-invaded communities tended to reduce their niche overlap to coexist in large-scale spaces [18]. However, the niche overlap between *A. philoxeroides* and its accompanying species was higher than that between each accompanying species, indicating that *A. philoxeroides*, as an invader, had stronger competitiveness; it competed for resources with most plant species and even with narrow-niche species. Interestingly, the niche similarity and niche overlap of species pairs such as *A. philoxeroides–D. sanguinalis* and *A. philoxeroides–P. paspaloides* were all high, indicating that they had highly similar living habits and resource requirements and that more intense interspecific competition might thus occur with community succession. Previous studies have found that plant invasions cause a decline in niche overlap among native plants and thus enhance the competition between invasive and native plants [51]. Similarly, we also found that the accompanying species with large niche breadth did not form species pairs with higher niche overlap in the *A. philoxeroides* community, indicating that the resource utilization capacity of the accompanying species was weaker than that of *A. philoxeroides* at large spatial scales and that they might choose to reduce interspecific competition and/or improve mutual assistance for resisting invasion [29].

4.2. Causes of Niche Breadth Variations along the Latitudinal Gradient in Heterogeneous Habitats

Niche theory predicts that with changes in the environmental gradient, the difference in species specificity during the process of adaptive evolution will lead to changes in community structures [16]. Habitat heterogeneity can improve the phenotypic plasticity of plant species, while latitudinal gradients facilitate the continuous evolution of species ecological strategies [45]. In our study, the niche breadths of *A. philoxeroides* in both terrestrial and aquatic habitats were high, and no obvious latitudinal trend was observed, suggesting that the niche breadth of this species might be determined by both adaptive variation and phenotypic plasticity; thus, a larger niche breadth could benefit *A. philoxeroides* in resisting environmental fluctuations [26,45]. Previous studies have found that the phenotypic plasticity of *A. philoxeroides* increased with increasing latitude, reflecting its strong adaptations to changes in temperature, precipitation and herbivory along a latitudinal gradient, thus leading to the large niche breadth of *A. philoxeroides* in each latitudinal cluster [52–54]. This scenario may also have occurred because *A. philoxeroides*, as a pioneer species, has strong environmental tolerance; it can share similar habitat conditions at different latitudes and spread to higher latitudinal regions under climate change [36,55,56].

Temperature is the crucial factor affecting plant growth, and the distributions of plant species in terrestrial ecosystems are more vulnerable to thermal variations than to other variations [10]. Low-latitude regions have sufficient heat and exacerbate the competition between each accompanying species, while the cold climate of high latitudes might mitigate interspecific competition. However, invasive plants can continuously establish populations in new habitats at high latitudes and reduce the niche breadth of the accompanying plants by disturbing their competition and coexistence patterns [47]. In our study, species with larger niche breadths, such as *H. scandens*, which had both vegetative and generative propagations, and *S. viridis*, which had tall individuals (Table S2), also appeared in high-latitude terrestrial communities, which may be due to the continuous evolution of native species to increase their competitive tolerance and thus reduce their sensitivity to bioinvasions [57]. As the community size increased, the proportion of species with extreme specialization also increased, and the difference in their niche breadths was further amplified. Compared to terrestrial communities, aquatic communities that have simpler structures and lower species diversity are more vulnerable to plant invasions [10,32]. *A. philoxeroides* has extensively invaded various aquatic habitats, and tends to form monodominant populations, which results in the niche breadths of the aquatic accompanying species having no obvious latitudinal trend.

4.3. Causes of Niche Similarity and Overlap Variations along the Latitudinal Gradient in Heterogeneous Habitats

In the terrestrial community, the niche similarity between *A. philoxeroides* and its accompanying species at low latitudes was relatively high, indicating that the accompanying species tended to possess climate adaptability similar to that of *A. philoxeroides*, and thermal stability promoted the coevolution of plant species in the low-latitude regions. Each accompanying species also had high niche similarity for preferably adapting to the coexistence environment, which was conducive to the formation of higher species diversity at low latitudes [58,59], and environmental filtering led to obvious aggregation in these regions [19]. *A. philoxeroides* has high phenotypic plasticity and compensatory growth capacity at high latitudes [39], but the cold climate at high latitudes limits the growth of native species. Only a few species with strong environmental tolerance can share resources with *A. philoxeroides* and thus have higher niche similarity, and most of the remaining species pairs have low niche similarity [45]. The niche similarity between *A. philoxeroides* and its accompanying species in the aquatic habitats had no obvious latitudinal trend because the water bodies buffered temperature fluctuations caused by latitude changes; thus, they were more stable than terrestrial habitats [60]. However, *A. philoxeroides* invasion was more significant in aquatic habitats, and the accompanying species thus suffered from stronger competitive exclusion [35].

A. philoxeroides had high niche overlap with its accompanying species in the heterogeneous habitats, indicating that it had a higher competitive advantage than native species. In terrestrial habitats, the niche overlap between *A. philoxeroides* and its accompanying species significantly increased with increasing latitude, particularly in latitudinal Cluster 7, which may imply that *A. philoxeroides* strengthened competition with its accompanying species by increasing the niche overlap under increased environmental stress; in turn, the resource utilization of the accompanying species decreased, reducing their adaptability to harsh environments and thus inhibiting native plant establishment at high latitudes [58,59]. However, the niche overlap between *A. philoxeroides* and its accompanying species in aquatic habitats had no obvious latitudinal trend, possibly due to aquatic plant species experiencing severe *A. philoxeroides* invasion at all latitudinal clusters [61]. In this study, niche overlap between *A. philoxeroides* and *P. stratiotes* and *E. crassipes* was higher at low latitudes; these are all globally invasive weeds with vegetative propagation, indicating that they may have similar survival needs for dissolved oxygen, light and water flow and may also provide interspecific mutual assistance [62,63]. Attention must be paid to this phenomenon of collaborative aquatic plant invasions at low latitudes because with rapid global warming, the superior hydrothermal conditions at low latitudes will accelerate the niche expansion and coevolution of aquatic invasive plants originating from tropical America [61,64]. In addition, there were over 10 forms and varieties of *A. philoxeroides* in the natural ecosystems, which made the taxonomy of *A. philoxeroides* one of the most complicated taxa among *Alternanthera* [65,66]; thus, it is necessary to fully assess the influence of taxonomic differences of *A. philoxeroides* on its niche characteristics in future studies.

5. Conclusions

We found that, of the studied species, *A. philoxeroides* had the greatest niche breadth in both aquatic and terrestrial habitats and that the niche similarity and overlap between *A. philoxeroides* and its accompanying species were all high. The niche similarity of *A. philoxeroides* in the terrestrial communities was significantly higher than that in the aquatic communities, while the niche similarity and overlap between each accompanying species were low. The niche breadth of *A. philoxeroides* had no obvious latitudinal trend, which was attributed to its strong environmental tolerance. In the terrestrial communities, the niche similarity of *A. philoxeroides* and the niche breadth of its accompanying species all significantly decreased with increasing latitude, while the niche overlap between *A. philoxeroides* and its accompanying species significantly increased with increasing latitude. However, the niche characteristics in the aquatic communities had no obvious latitudinal trend. These findings indicate that habitat heterogeneity significantly affects the niche characteristics of the *A. philoxeroides* community and that *A. philoxeroides* invasion might weaken the interspecific competition of its accompanying species. At low latitudes, environmental filtering resulted in species with similar niches in the *A. philoxeroides* communities showing obvious aggregation, and the invasive and native species tended to evolve similar environmental capacities. At high latitudes, the restrictions of the cold climate led to an increase in niche overlap between *A. philoxeroides* and its accompanying species.

Supplementary Materials: The following supporting information can be downloaded at https://www.mdpi.com/article/10.3390/d15050651/s1, Table S1: Sampling plots of *A. philoxeroides*-invaded communities in terrestrial and aquatic habitats; Table S2: The functional traits of species with high-frequency in terrestrial and aquatic *A. philoxeroides*-invaded communities.

Author Contributions: Conceptualization, H.W. and B.R.; methodology, H.W.; formal analysis, H.W. and S.D.; investigation, H.W., S.D., Y.W. and L.W.; data curation, H.W. and S.D.; writing—original draft preparation, H.W. and S.D.; writing—review and editing, H.W., S.D. and B.R.; funding acquisition, H.W. and B.R. All authors have read and agreed to the published version of the manuscript.

Funding: This research was funded by The National Natural Science Foundation of China (31800460), The Xinyang Academy of Ecological Research Open Foundation (2023XYMS16) and The Nanhu Scholars Program for Young Scholars of Xinyang Normal University (XYNU) (2023A017).

Institutional Review Board Statement: Not applicable.

Informed Consent Statement: Not applicable.

Data Availability Statement: Data will be made available upon request.

Acknowledgments: We would like to thank Shaoqi Jia, Junda Wang, Qiubo Ji, Nannan Xiao and Wenhao Wang for their help in field surveys.

Conflicts of Interest: The authors declare no conflict of interest.

References

1. Seebens, H.; Blackburn, T.M.; Dyer, E.E.; Genovesi, P.; Hulme, P.E.; Jeschke, J.M.; Pagad, S.; Pyšek, P.; Winter, M.; Arianoutsou, M.; et al. No saturation in the accumulation of alien species worldwide. *Nat. Commun.* **2017**, *8*, 14435. [CrossRef] [PubMed]
2. Carboni, M.; Livingstone, S.W.; Isaac, M.E.; Cadotte, M.W. Invasion drives plant diversity loss through competition and ecosystem modification. *J. Ecol.* **2021**, *109*, 3587–3601. [CrossRef]
3. Pabst, R.; Dias, F.S.; Borda-de-Água, L.; Rodríguez-González, P.M.; Capinha, C. Assessing and predicting the distribution of riparian invasive plants in continental Portugal. *Front. Ecol. Evol.* **2022**, *10*, 875578. [CrossRef]
4. Lebbink, G.; Dwyer, J.; Fensham, R. An invasive grass species has both local and broad-scale impacts on diversity: Potential mechanisms and implications. *J. Veg. Sci.* **2021**, *32*, 1240–1253. [CrossRef]
5. Yue, M.; Shen, H.; Ye, W.; Li, W.; Chen, J. Winter low temperature disturbance in the southern subtropics of China promotes the competitiveness of an invasive plant. *Biol. Invasions* **2021**, *23*, 2913–2925. [CrossRef]
6. Gross, N.; Liancourt, P.; Butters, R.; Duncan, R.; Hulme, P.E. Functional equivalence, competitive hierarchy and facilitation determine species coexistence in highly invaded grasslands. *New Phytol.* **2015**, *206*, 175–186. [CrossRef]
7. Wang, C.; Yu, Y.; Cheng, H.; Du, D. Which factor contributes most to the invasion resistance of native plant communities under the co-invasion of two invasive plant species? *Sci. Total Environ.* **2022**, *813*, 152628. [CrossRef]
8. Wu, H.; Ding, J.Q. Abiotic and biotic determinants of plant diversity in aquatic communities invaded by water hyacinth (*Eichhornia crassipes* (Mart.) Solms). *Front. Plant Sci.* **2020**, *11*, 1306. [CrossRef]
9. Dubey, D.; Kumar, S.; Dutta, V. In-situ anatomical and elemental response of aquatic macrophytes against nutrient enrichment in freshwater tropical lakes. *Appl. Water Sci.* **2022**, *12*, 163. [CrossRef]
10. Huang, X.; Ke, F.; Li, Q.; Zhao, Y.; Guan, B.; Li, K. Functional traits underlying performance variations in the overwintering of the cosmopolitan invasive plant water hyacinth (*Eichhornia crassipes*) under climate warming and water drawdown. *Ecol. Evol.* **2022**, *12*, e9181. [CrossRef]
11. Kuebbing, S.E.; Nuñez, M.A. Invasive non-native plants have a greater effect on neighbouring natives than other non-natives. *Nat. Plants* **2016**, *2*, 1613. [CrossRef]
12. Byers, J.E.; Noonburg, E.G. Scale dependent effects of biotic resistance. *Ecology* **2003**, *84*, 1428–1433. [CrossRef]

13. Fridley, J.D.; Stachowicz, J.J.; Naeem, S.; Sax, D.F.; Seabloom, E.W.; Smith, M.D.; Stohlgren, T.J.; Tilman, D.; Von Holle, B. The invasion paradox: Reconciling pattern and process in species invasions. *Ecology* **2007**, *88*, 3–17. [CrossRef]
14. Powell, K.I.; Chase, J.M.; Knight, T.M. Invasive plants have scale-dependent effects on diversity by altering species-area relationships. *Science* **2013**, *339*, 316–318. [CrossRef]
15. Tilman, D. Niche tradeoffs, neutrality, and community structure: A stochastic theory of resource competition, invasion, and community assembly. *Proc. Natl. Acad. Sci. USA* **2004**, *101*, 10854–10861. [CrossRef]
16. Guisan, A.; Petitpierre, B.; Brönnimann, O.; Daehler, C.; Kueffer, C. Unifying niche shift studies: Insights from biological invasions. *Trends Ecol. Evol.* **2014**, *29*, 260–269. [CrossRef]
17. Cai, Q.; Welk, E.; Ji, C.; Fang, W.; Sabatini, F.M.; Zhu, J.; Zhu, J.; Tang, Z.; Attorre, F.; Campos, J.A.; et al. The relationship between niche breadth and range size of beech (*Fagus*) species worldwide. *J. Biogeogr.* **2021**, *48*, 1240–1253. [CrossRef]
18. Pastore, A.I.; Barabás, G.; Bimler, M.D.; Mayfield, M.M.; Miller, T.E. The evolution of niche overlap and competitive differences. *Nat. Ecol. Evol.* **2021**, *5*, 330–337. [CrossRef]
19. Cadotte, M.W.; Tucker, C.M. Should Environmental Filtering be Abandoned? *Trends Ecol. Evol.* **2017**, *32*, 429–437. [CrossRef]
20. Liu, C.; Wolter, C.; Courchamp, F.; Roura-Pascual, N.; Jeschke, J.M. Biological invasions reveal how niche change affects the transferability of species distribution models. *Ecology* **2022**, *103*, e3719. [CrossRef]
21. Yuan, Y.; Tang, X.; Liu, M.; Liu, X.; Tao, J. Species distribution models of the *Spartina alterniflora* Loisel in its origin and invasive country reveal an ecological niche shift. *Front. Plant Sci.* **2021**, *12*, 738769. [CrossRef] [PubMed]
22. Fridley, J.D. Extended leaf phenology and the autumn niche in deciduous forest invasions. *Nature* **2012**, *485*, 359–362. [CrossRef] [PubMed]
23. Datta, A.; Schweiger, O.; Ingolf, K. Niche expansion of the invasive plant species *Ageratina adenophora* despite evolutionary constraints. *J. Biogeogr.* **2019**, *46*, 1306–1315.
24. Enders, M.; Havemann, F.; Ruland, F.; Bernard-Verdier, M.; Catford, J.A.; Gómez-Aparicio, L.; Haider, S.; Heger, T.; Kueffer, C.; Kühn, I.; et al. A conceptual map of invasion biology: Integrating hypotheses into a consensus network. *Glob. Ecol. Biogeogr.* **2020**, *29*, 978–991. [CrossRef]
25. Yang, Y.H.; Bao, L.P. Scale-dependent changes in species richness caused by invader competition. *Ecol. Model.* **2022**, *469*, 109996. [CrossRef]
26. Sexton, J.P.; Montiel, J.; Shay, J.E.; Stephens, M.R.; Slatyer, R.A. Evolution of ecological niche breadth. *Annu. Rev. Ecol. Evol. Syst.* **2017**, *48*, 183–206. [CrossRef]
27. Finch, J.; Walck, J.L.; Hidayati, S.N.; Kramer, A.T.; Lason, V.; Havens, K. Germination niche breadth varies inconsistently among three *Asclepias* congeners along a latitudinal gradient. *Plant Biol.* **2019**, *21*, 425–438. [CrossRef]
28. Schuster, M.J.; Wragg, P.D.; Reich, P.B. Phenological niche overlap between invasive buckthorn (*Rhamnus cathartica*) and native woody species. *For. Ecol. Manag.* **2021**, *498*, 11956. [CrossRef]
29. Stotz, G.C.; Pec, G.J.; Cahill, J.F. Is biotic resistance to invaders dependent upon local environmental conditions or primary productivity? A meta-analysis. *Basic Appl. Ecol.* **2016**, *17*, 377–387. [CrossRef]
30. Welshofer, K.B.; Zarnetske, P.L.; Lany, N.K.; Read, Q.D. Short-term responses to warming vary between native vs. exotic species and with latitude in an early successional plant community. *Oecologia* **2018**, *187*, 333–342. [CrossRef]
31. Cirtwill, A.R.; Stouffer, D.B.; Romanuk, T.N. Latitudinal gradients in biotic niche breadth vary across ecosystem types. *Proc. R. Soc. B-Biol. Sci.* **2015**, *282*, 20151589. [CrossRef]
32. Vázquez, D.P.; Stevens, R.D. The latitudinal gradient in niche breadth: Concepts and evidence. *Am. Nat.* **2004**, *164*, E1–E19. [CrossRef]
33. Alahuhta, J.; Virtala, A.; Hjort, J.; Ecke, F.; Johnson, L.B.; Sass, L.; Heino, J. Average niche breadths of species in lake macrophyte communities respond to ecological gradients variably in four regions on two continents. *Oecologia* **2017**, *184*, 219–235. [CrossRef]
34. Wu, H.; Dong, S.J.; Rao, B.Q. Latitudinal trends in the structure, similarity and beta diversity of plant communities invaded by *Alternanthera philoxeroides* in heterogeneous habitats. *Front. Plant Sci.* **2022**, *13*, 1021337. [CrossRef]
35. Wu, H.; Carrillo, J.; Ding, J.Q. Species diversity and environmental determinants of aquatic and terrestrial communities invaded by *Alternanthera philoxeroides*. *Sci. Total Environ.* **2017**, *581*, 666–675. [CrossRef]
36. Lu, X.M.; Siemann, E.; Shao, X.; Wei, H.; Ding, J. Climate warming affects biological invasions by shifting interactions of plants and herbivores. *Glob. Chang. Biol.* **2013**, *19*, 2339–2347. [CrossRef]
37. Wu, H.; Ismail, M.; Ding, J.Q. Global warming increases the interspecific competitiveness of the invasive plant alligator weed, *Alternanthera philoxeroides*. *Sci. Total Environ.* **2017**, *575*, 1415–1422. [CrossRef]
38. Wang, T.; Hu, J.; Wang, R.; Liu, C.; Yu, D. Tolerance and resistance facilitate the invasion success of *Alternanthera philoxeroides* in disturbed habitats: A reconsideration of the disturbance hypothesis in the light of phenotypic variation. *Environ. Exp. Bot.* **2018**, *153*, 135–142. [CrossRef]
39. Wu, H.; Carrillo, J.; Ding, J.Q. Invasion by alligator weed, *Alternanthera philoxeroides*, is associated with decreased species diversity across the latitudinal gradient in China. *J. Plant Ecol.* **2016**, *93*, 311–319. [CrossRef]
40. Lu, X.M.; Siemann, E.; He, M.Y.; Wei, H.; Shao, X.; Ding, J. Climate warming increases biological control agent impact on a non-target species. *Ecol. Lett.* **2015**, *18*, 48–56. [CrossRef]
41. Ma, Y.; Li, Q.; Pan, S.; Liu, C.; Han, M.; Brancelj, A. Niche and interspecific associations of *Pseudoanabaena limnetica*-Exploring the influencing factors of its succession stage. *Ecol. Indic.* **2022**, *138*, 108806. [CrossRef]

42. Broennimann, O.; Fitzpatrick, M.C.; Pearman, P.B.; Petitpierre, B.; Pellissier, L.; Yoccoz, N.G.; Thuiller, W.; Fortin, M.-J.; Randin, C.; Zimmermann, N.E.; et al. Measuring ecological niche overlap from occurrence and spatial environmental data. *Glob. Ecol. Biogeogr.* **2012**, *21*, 481–497. [CrossRef]
43. Martorell, C.; Almanza-Celis, C.A.; Pérez-García, E.A.; Sánchez-Ken, J.G. Co-existence in a species-rich grassland: Competition, facilitation and niche structure over a soil depth gradient. *J. Veg. Sci.* **2015**, *26*, 674–685. [CrossRef]
44. Luo, L.; Kong, X.; Gao, Z.; Zheng, Y.; Yang, Y.; Li, X.; Yang, D.; Geng, Y.; Yang, Y. Comparative transcriptome analysis reveals ecological adaption of cold tolerance in northward invasion of *Alternanthera philoxeroides*. *BMC Genom.* **2020**, *21*, 532. [CrossRef] [PubMed]
45. Stotz, G.C.; Salgado-Luarte, C.; Escobedo, V.M.; Valladares, F.; Gianoli, E. Global trends in phenotypic plasticity of plants. *Ecol. Lett.* **2021**, *24*, 2267–2281. [CrossRef]
46. Müller, C.; Junker, R.R. Chemical phenotype as important and dynamic niche dimension of plants. *New Phytol.* **2022**, *234*, 1168–1174. [CrossRef]
47. Lancaster, L.T. On the macroecological significance of eco-evolutionary dynamics: The range shift-niche breadth hypothesis. *Philos. Trans. R. Soc. B-Biol. Sci.* **2022**, *377*, 20210013. [CrossRef]
48. Bar-Massada, A. Complex relationships between species niches and environmental heterogeneity affect species co-occurrence patterns in modelled and real communities. *Proc. R. Soc. B-Biol. Sci.* **2015**, *282*, 20150927. [CrossRef]
49. Segura, A.M.; Calliari, D.; Kruk, C.; Conde, D.; Bonilla, S.; Fort, H. Emergent neutrality drives phytoplankton species coexistence. *Proc. R. Soc. B-Biol. Sci.* **2011**, *278*, 2355–2361. [CrossRef]
50. Pennington, R.T.; Richardson, J.E.; Lavin, M. Insights into the historical construction of species-rich biomes from dated plant phylogenies, neutral ecological theory and phylogenetic community structure. *New Phytol.* **2006**, *172*, 605–616. [CrossRef]
51. dos Santos Ribas, L.G.; de Cassia-Silva, C.; Petsch, D.K.; Silveira, M.L.; Lima-Ribeiro, M.S. The potential invasiveness of an aquatic macrophyte reflects founder effects from native niche. *Biol. Invasions* **2018**, *20*, 3347–3355. [CrossRef]
52. He, D.; Chen, Y.; Zhao, K.; Cornelissen, J.H.C.; Chu, C. Intra- and interspecific trait variations reveal functional relationships between specific leaf area and soil niche within a subtropical forest. *Ann. Bot.* **2018**, *121*, 1173–1182. [CrossRef]
53. van Boheemen, L.A.; Atwater, D.Z.; Hodgins, K.A. Rapid and repeated local adaptation to climate in an invasive plant. *New Phytol.* **2019**, *222*, 614–627. [CrossRef]
54. Yang, Y.; Liu, M.; Pan, Y.; Huang, H.; Pan, X.; Sosa, A.; Hou, Y.; Zhu, Z.; Li, B. Rapid evolution of latitudinal clines in growth and defence of an invasive weed. *New Phytol.* **2021**, *230*, 845–856. [CrossRef]
55. Morin, X.; Chuine, I. Niche breadth, competitive strength and range size of tree species: A trade-off based framework to understand species distribution. *Ecol. Lett.* **2006**, *9*, 185–195. [CrossRef]
56. Foster, S.L.; Kharouba, H.M.; Kharouba, H.M. Testing the assumption of environmental equilibrium in an invasive plant species over a 130 year history. *Ecography* **2022**, *2022*, e06284. [CrossRef]
57. Golivets, M.; Wallin, K.F. Neighbour tolerance, not suppression, provides competitive advantage to non-native plants. *Ecol. Lett.* **2018**, *21*, 745–759. [CrossRef]
58. MacDougall, A.S.; Gilbert, B.; Levine, J.M. Plant invasions and the niche. *J. Ecol.* **2009**, *97*, 609–615. [CrossRef]
59. Carscadden, K.A.; Emery, N.C.; Arnillas, C.A.; Cadotte, M.W.; Afkhami, M.E.; Gravel, D.; Livingstone, S.W.; Wiens, J.J. Niche breadth: Causes and consequences for ecology, evolution, and conservation. *Q. Rev. Biol.* **2020**, *95*, 179–214. [CrossRef]
60. Anufriieva, E.V.; Shadrin, N.V. Extreme hydrological events destabilize aquatic ecosystems and open doors for alien species. *Quat. Int.* **2018**, *475*, 11–15. [CrossRef]
61. Wu, H.; Ding, J.Q. Global change sharpens the double-edged sword of aquatic alien plants in China and beyond. *Front. Plant Sci.* **2019**, *10*, 787. [CrossRef]
62. Kriticos, D.J.; Brunel, S. Assessing and managing the current and future pest risk from water hyacinth, (*Eichhornia crassipes*), an invasive aquatic plant threatening the environment and water security. *PLoS ONE* **2016**, *11*, e0120054. [CrossRef] [PubMed]
63. Jawed, A.; Kar, P.; Verma, R.; Shukla, K.; Hemanth, P.; Thakur, V.K.; Pandey, L.M.; Gupta, R.K. Integration of biological control with engineered heterojunction nano-photocatalysts for sustainable and effective management of water hyacinth weed. *J. Environ. Chem. Eng.* **2022**, *10*, 106976. [CrossRef]
64. Banerjee, A.K.; Liang, X.R.; Harms, N.E.; Tan, F.; Lin, Y.; Feng, H.; Wang, J.; Li, Q.; Jia, Y.; Lu, X.; et al. Spatio-temporal pattern of cross-continental invasion: Evidence of climatic niche shift and predicted range expansion provide management insights for smooth cordgrass. *Ecol. Indic.* **2022**, *140*, 109052. [CrossRef]
65. Jia, X.; Pan, X.-Y.; Sosa, A.; Li, B.; Chen, J. Differentiation in growth and biomass allocation among three native *Alternanthera philoxeroides* varieties from Argentina. *Plant Species Biol.* **2010**, *25*, 85–92. [CrossRef]
66. Jiang, P.; Xu, G.; He, Y.; Sun, T.; Liu, C.; Chen, C.; Zuo, R.; Sun, C. Complete chloroplast genome of *Alternanthera philoxeroides* by de novo sequencing. *Mitochondrial DNA Part B-Resour.* **2021**, *6*, 1826–1828. [CrossRef]

plants

MDPI

Article

Reconstructed Global Invasion and Spatio-Temporal Distribution Pattern Dynamics of *Sorghum halepense* under Climate and Land-Use Change

Ming Yang [1,2], Haoxiang Zhao [1], Xiaoqing Xian [1], Yuhan Qi [1], Qiao Li [1], Jianying Guo [1], Li Chen [2] and Wanxue Liu [1,*]

[1] State Key Laboratory for Biology of Plant Diseases and Insect Pests, Institute of Plant Protection, Chinese Academy of Agricultural Sciences, Beijing 100193, China

[2] School of Life Sciences, Hebei University, Baoding 071000, China

* Correspondence: liuwanxue@caas.cn

Abstract: *Sorghum halepense* competes with crops and grass species in cropland, grassland, and urban environments, increasing invasion risk. However, the invasive historical dynamics and distribution patterns of *S. halepense* associated with current and future climate change and land-use change (LUC) remain unknown. We first analyzed the invasive historical dynamics of *S. halepense* to explore its invasion status and expansion trends. We then used a species distribution model to examine how future climate change and LUC will facilitate the invasion of *S. halepense*. We reconstructed the countries that have historically been invaded by *S. halepense* based on databases with detailed records of countries and occurrences. We ran biomod2 based on climate data and land-use data at 5′ resolution, assessing the significance of environmental variables and LUC. *Sorghum halepense* was widely distributed worldwide through grain trade and forage introduction, except in Africa. Europe and North America provided most potential global suitable habitats (PGSHs) for *S. halepense* in cropland, grassland, and urban environments, representing 48.69%, 20.79%, and 84.82%, respectively. The future PGSHs of *S. halepense* increased continuously in the Northern Hemisphere, transferring to higher latitudes. Environmental variables were more significant than LUC in predicting the PGSHs of *S. halepense*. Future PGSHs of *S. halepense* are expected to increase, exacerbating the invasion risk through agricultural LUC. These results provide a basis for the early warning and prevention of *S. halepense* worldwide.

Keywords: climate change; *Sorghum halepense*; land-use change; potential global suitable habitats; invasive alien plants

1. Introduction

Global changes, including international trade, climate change, and land-use change (LUC), are direct drivers of biological invasion [1,2]. The continuous increase in invasive alien plants (IAPs) is closely related to the globalization of trade, in which landscape gardening and the long-distance carriage of grass have been introduced in grassland and agriculture [3–5]. More than 500 species of IAPs are traded daily worldwide [6]. In the United States, 61% of the 1285 IAPs, including half of the state-regulated IAPs, are available via the plant trade [7]. Therefore, analyzing the invasive historical dynamics of IAPs is beneficial for understanding their invasion status and expansion trends. With global warming, IAPs could create new habitats at high elevations because of the suitable thermal conditions of climate change [8]. In addition, individuals first settle at low elevations and subsequently transfer to higher-elevation regions, causing LUCs in plant communities and increasing the possibility of IAPs [9,10]. Meanwhile, IAPs at low elevations are more likely to be affected by LUC [8]. These factors have a profound influence on native ecological communities, species interactions, and suitable ranges based on the ecology of invasive species [11]. The emergence and expansion of IAPs are notably increasing with climate warming and are expected to become more pronounced in association with LUC [12,13]. Therefore, to decrease the impact of IAPs, predicting the spatial dynamics of invasive species through an invasive risk assessment with species distribution models is beneficial for management to formulate relevant policies and measures for IAPs.

Sorghum halepense (Fabaceae) is among the ten most noxious and destructive global annual weeds [14]. Its exact native range is western Asia, the Mediterranean, and North Africa. However, broad distributions worldwide have recently been introduced as a consequence of crop seeds and forage cultivation, increasing invasion risk and threatening agriculture, economic development, and human health. The high density of *S. halepense* resulting from faster seed germination has led to the loss of plant diversity, drastically affecting cotton, soybean, and wheat production in Asia and North America [15–17]. *Sorghum halepense* seedlings, which have significant similarities with annual grains, when mixed with sorghum and maize plants have decrease fresh weight and increased crop yield losses in North America [15,18]. Mature *S. halepense* with a higher leaf area is a host of numerous pests, nematodes, and fungal pathogens for annual crops, including the sorghum midge (*Contarinia sorghicola*), *Meloidogyne incognita*, and leaf spot diseases (*Cercospora sorghi*) [19,20]. In addition, it can develop wider subterranean root systems that produce secondary metabolites, threatening medical plant development in Africa, such as *Plantago ovate* and *Ocimum basilicum* [21]. Furthermore, the pollen of *S. halepense* probably contributes to hay fever in North America, impacting human health. Additionally, *S. halepense* seeds are easily established in rangeland and agricultural fields, where they grow more effectively than C_4 prairie plants, reducing native flora diversity [14]. *Sorghum halepense* extension has dramatically affected soil nutrition, soil temperature, and water content, leading to a C_3 species decrease as well as a C_4 group biomass increase [14]. Therefore, predicting the potential global suitable habitats (PGSHs) of *S. halepense* in ecological systems associated with future climate change and LUC has considerable practical value for agriculture and livelihoods.

Species distribution models (SDMs) are typically used to forecast the impacts of climate change on the potential spatial distribution of invasive species in ecology and biogeography [22]. SDMs combine species presence–absence data and related environmental conditions for distribution prediction, including generalized linear models (GLMs), maximum entropy (MaxEnt), random forest (RF), and CLIMEX [23–26]. Biomod2 is the most popular and well-constructed platform in the SDM community, composed of ten individual models (GLM, gradient boosting models (GBMs), RF, artificial neural network (ANN), generalized additive models (GAMs), flexible discriminant analysis (FDA), MaxEnt, classification tree analysis (CTA), surface range envelope (SRE), and multivariate adaptive regression splines (MARSs)) that can be randomly selected to produce ensemble models (EMs) [27,28]. With the high prediction accuracy achieved by minimizing the disadvantages of each individual model, EMs are more robust than individual models in the risk assessment of IAPs [29]. For instance, predicting the potential worldwide distribution of *Ambrosia* L. and the suitable areas of *Ageratum houstonianum*, *Chromolaena odorata*, *Hyptis suaveolens*, *Lantana camara*, *Mikania micrantha*, and *Parthenium hysterophorus* in Nepal has been widely attempted in biomod2 [30,31].

In the present study, we (a) reconstructed the historically invaded countries, analyzing the movement means and dispersal dynamics of *S. halepense* by collecting geographical information that detailed the longitude, latitude, and invasive countries as time elapsed; (b) predicted the PGSHs of *S. halepense* under climate change and LUC in the 2030s and 2050s, mainly extracting three land-use types (cropland, grassland, and urban areas); (c) analyzed the probability of suitable habitats for *S. halepense* among the three land-use types; and finally (d) screened the significant environmental variables for the PGSHs of *S. halepense*.

2. Results

2.1. Reconstruction of Historically Invaded Countries

The native regions of *S. halepense* were mainly distributed in southern Asia, eastern Africa, and southern Europe, following CABI and WOS (Figure 1; Excel S1). In Asia, the earliest introduction of Nepal was in 1805, and it was subsequently transferred from the native region of southern Asia to eastern Asia. As a result, it is widely distributed in western, southern, and eastern Asia. In 1900, *S. halepense* was introduced into the Mediterranean Sea in Europe and transferred to western Europe in 1880 [32]. *Sorghum halepense* reached northern Europe in the subsequent two decades and is now distributed throughout almost all of Europe. In the 1800s, *S. halepense* was discovered in the United States and spread to southern North America between 1900 and 2022 [33]. However, in South America, although it was first discovered in Argentina in the 1800s [34], it spread from southern North America and was introduced into new countries until the 1980s. In Africa, it spread to northeastern and northern Africa between the 1900s and 2022 [35]. It has been widely distributed throughout Oceania since *S. halepense* was introduced to Australia in 1871 [36]. In summary, *S. halepense* is now widespread in most countries worldwide, except Africa, which has a potential invasion risk.

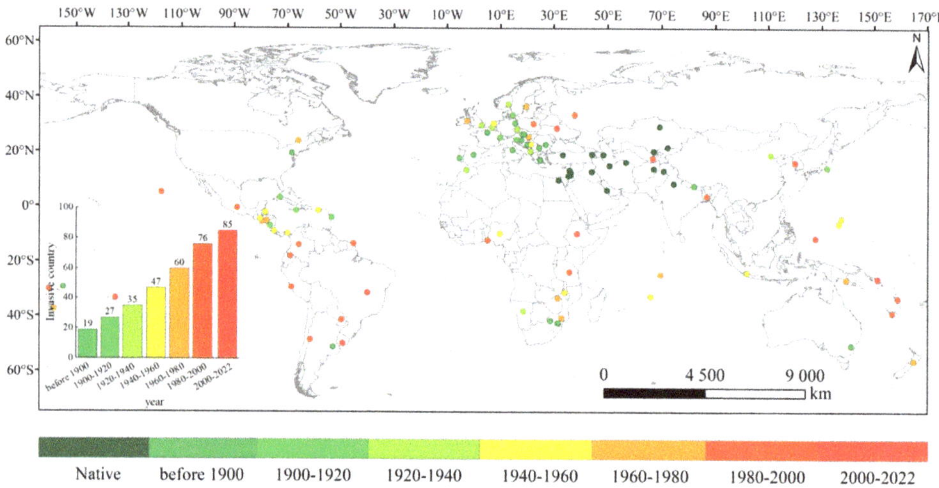

Figure 1. Reconstructed global countries invaded by *Sorghum halepense*. Eight time intervals are shown until 2022 from green to red, except for native countries. Temporal accumulation histograms of invaded countries are displayed in the bottom left corner.

2.2. Model Performance and Significant Environmental Variables

We calculated the model accuracy of ANN, CTA, FDA, GBM, GLM, MARS, MaxEnt, RF, and EM using ROC, TSS, and KAPPA values (Figure S1). The mean ROC of the eight models was 0.885, 0.943, 0.948, 0.960, 0.947, 0.952, 0.951, and 0.983, respectively (Table S1). The mean TSS of the eight individual models was 0.666, 0.818, 0.773, 0.809, 0.775, 0.778, 0.785, and 0.880, respectively. The mean KAPPA of the eight models was 0.685, 0.810, 0.768, 0.808, 0.777, 0.782, 0.778, and 0.880, respectively. The mean ROC, TSS, and KAPPA of the EM were 0.984, 0.851, and 0.856, respectively, which were higher than those of the individual models, indicating that the EM-predicted PGSHs of *S. halepense* were reliable.

We analyzed the significant environmental variables to predict the PGSHs of *S. halepense* using the EM, and the mean contributions are listed in Table 1. The precipitation of the coldest quarter (bio19, 0.317) was the most significant environmental variable, followed by the precipitation of the driest quarter (bio17, 0.108); the minimum temperature of the coldest month (bio6, 0.091); the annual precipitation (bio12, 0.085); the mean diurnal range (the mean of the monthly max temp-min temp) (bio2, 0.078); the max temperature of the warmest month (bio5, 0.067); precipitation seasonality (the coefficient of variation) (bio15, 0.048); and LUC (0.019). The response curves of the significant variables are shown in Figure S2. When the suitability probability achieved the maximum values, the survival probability of *S. halepense* was more reliable. Bio19 and bio17 were the most significant environmental variables for the predicted PGSHs of *S. halepense* in the three land-use types, achieving the approximate maximum values at 200 mm and 140 mm, respectively.

Table 1. Significant screened environmental variables and contribution importance.

Variable	Description	Unit	Importance
bio2	Mean diurnal range (mean of monthly max temp-min temp)	°C	0.078
bio5	Max temperature of warmest month	°C	0.067
bio6	Min temperature of coldest month	°C	0.091
bio12	Annual precipitation	mm	0.085
bio15	Precipitation seasonality (coefficient of variation × 1)	-	0.048
bio17	Precipitation of driest quarter	mm	0.108
bio19	Precipitation of coldest quarter	mm	0.317
LUC	Land-use change		0.019

2.3. The PGSHs of S. halepense under Current and Future Climate Scenarios

The PGSHs of *S. halepense* extracted from the three land-use types were mainly distributed in eastern, southeastern, central, and western Asia; western Europe; southern North America; southeastern South America; east-central, southwestern, and northern Africa; and southeastern and southwestern Oceania, as projected for near-current climate and future climate scenarios (Figures 2 and 3). Compared with the near-current climate, the predicted global total cropland and urban areas would increase under future scenarios, yet the predicted total grassland area would decrease to a certain extent (Table 2). As a result, the PGSHs of *S. halepense* notably increased in the three land-use types, achieving a maximum under SSP5-8.5 in the 2030s and the 2050s.

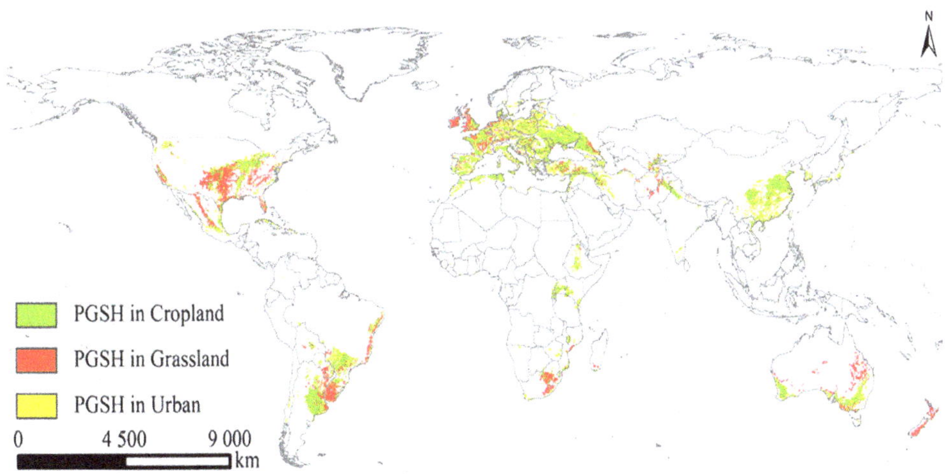

Figure 2. Potential global suitable habitats (PGSHs) of *Sorghum halepense* in three land-use types (cropland, grassland, and urban land) under the near-current climate scenario.

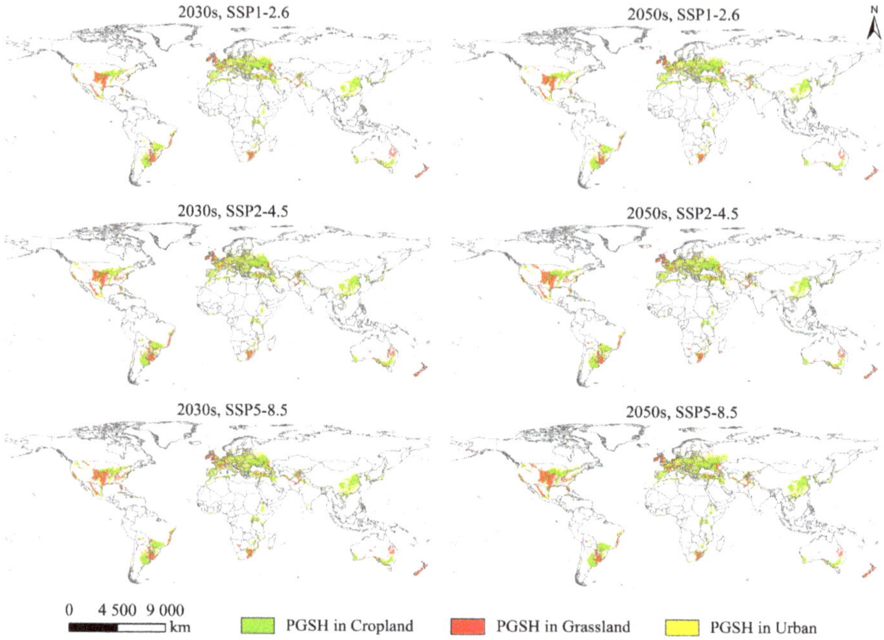

Figure 3. Potential global suitable habitats (PGSHs) of *Sorghum halepense* in three land-use types (cropland, grassland, and urban land) under SSP1-2.6, SSP2-4.5, and SSP5-8.5 in the 2030s and 2050s.

Table 2. Potential global suitable habitats (PGSHs) of *Sorghum halepense* in three land-use types (cropland, grassland, and urban land) for the near-current climate and different future scenarios.

Land Use (×104 km)	Cropland			Grassland			Urban		
	Suitable Area	Total Area		Suitable Area	Total Area		Suitable Area	Total Area	
Near-current	921.99	2177.42	42.34%	280.41	1675.80	16.73%	50.42	61.89	81.47%
2030s, SSP1-2.6	1026.27	2150.03	47.73%	297.00	1555.53	19.09%	75.24	89.92	83.67%
2030s, SSP2-4.5	1073.00	2279.08	47.08%	317.87	1638.99	19.39%	72.92	87.09	83.73%
2030s, SSP5-8.5	1079.24	2340.77	46.11%	315.38	1613.94	19.54%	78.81	93.78	84.04%
2050s, SSP1-2.6	1060.59	2178.33	48.69%	281.28	1448.65	19.42%	85.00	101.57	83.69%
2050s, SSP2-4.5	1091.27	2345.28	46.53%	315.51	1577.10	20.01%	82.24	99.20	82.90%
2050s, SSP5-8.5	1113.91	2377.42	46.85%	334.58	1609.21	20.79%	95.63	112.74	84.82%

In cropland, the PGSHs of *S. halepense* were located in eastern Asia (eastern Japan and South Korea); southeastern Asia (southern and eastern China); central Asia (southern and northern India, northeastern Pakistan, Afghanistan, southeastern Uzbekistan, western Tajikistan, and southern Kazakhstan); and western Asia (Iran, northern Iraq, northern Syria, and Turkey); nearly all western countries in Europe; southern North America (central and western United States, Mexico, Cuba, and the Dominican Republic); southeastern South America (southern and eastern Brazil, central Bolivia, Paraguay, eastern Argentina, and Chile); northern Africa (northern Morocco, northern Algeria, northern Tunisia); east-central Africa (Ethiopia and Kenya); southwestern Africa (Uganda, Malawi, Mozambique, Zimbabwe, and South Africa); and southeastern and southwestern Oceania (southeastern and southwestern Australia).

In grassland, the PGSHs of *S. halepense* were mainly located in central Asia (Pakistan, Afghanistan, Tajikistan, Uzbekistan, and Kazakhstan); western Asia (Turkey); western

Europe (northern Mediterranean Sea and west coast); southern North America (the United States and Mexico); southeastern South America (central Bolivia, eastern and southwestern Brazil, Paraguay, Uruguay, and northeastern Argentina); southern Africa (Ethiopia, eastern Mozambique, Lesotho, southern Madagascar, and South Africa); and southeastern and southwestern Oceania (Australia and New Zealand). However, in urban areas, the PGSHs of *S. halepense* were broadly distributed worldwide.

The suitable areas for *S. halepense* in cropland were mainly located in Europe under the near-current climate scenario, spread over approximately 311.95×10^4 km, accounting for 33.83% of the PGSHs of *S. halepense*, followed by Asia (228.21×10^4, 24.75%); South America (127.91×10^4, 13.87%); North America (117.19×10^4, 12.71%); Africa (79.29×10^4, 8.60%); and Oceania (54.21×10^4, 5.88%). In the 2030s, future suitable cropland areas achieved the maximum extent under SSP2-4.5 in Europe, at approximately 379.84×10^4 km, accounting for 33.83% of the PGSHs of *S. halepense*, followed by Asia (249.93×10^4, 24.35%); South America (151.74×10^4, 14.38%); North America (147.60×10^4, 14.38%); Africa (85.07×10^4, 8.29%); and Oceania (55.69×10^4, 5.43%). In the 2050s, future suitable cropland areas for *S. halepense* were mainly distributed in Europe, spread over approximately 407.34×10^4 km, accounting for 36.57% of the PGSHs of *S. halepense*, followed by Asia (254.32×10^4, 22.83%); North America (160.65×10^4, 14.42%); South America (148.04×10^4, 13.29%); Africa (87.94×10^4, 7.89%); and Oceania (52.28×10^4, 4.69%).

Suitable areas for *S. halepense* in grassland were mainly located in North America under the near-current climate scenario, spread over approximately 95.63×10^4 km, accounting for 34.10% of the PGSHs of *S. halepense*, followed by Europe (55.47×10^4, 19.78%); South America (46.70×10^4, 16.65%); Oceania (43.16×10^4, 15.39%); Africa (19.51×10^4, 6.96%); and Asia (18.2×10^4, 6.49%). In the 2030s, future suitable grassland areas achieved the maximum extent under SSP5-8.5 in North America at approximately 116.18×10^4 km, accounting for 36.84% of the PGSHs of *S. halepense*, followed by Europe (66.83×10^4, 21.19%); South America (43.55×10^4, 13.81%); Oceania (39.18×10^4, 12.42%); Asia (25.53×10^4, 8.09%); and Africa (22.30×10^4, 7.07%). In the 2050s, future suitable grassland areas for *S. halepense* were mainly distributed in North America, spread over approximately 131.39×10^4 km, accounting for 39.27% of the PGSHs of *S. halepense*, followed by Europe (81.90×10^4, 24.48%); South America (39.69×10^4, 11.86%); Oceania (33.83×10^4, 10.11%); Asia (26.96×10^4, 8.06%); and Africa (18.82×10^4, 5.62%).

Suitable areas for *S. halepense* in urban environments were mainly located in Europe under the near-current climate scenario, spread over approximately 14.98×10^4 km, accounting for 29.71% of the PGSHs of *S. halepense*, followed by North America (14.18×10^4, 28.12%); Asia (13.00×10^4, 25.78%); South America (3.02×10^4, 5.99%); Africa (2.50×10^4, 4.96%); and Oceania (1.24×10^4, 2.46%). In the 2030s, future suitable urban areas achieved the maximum extent under SSP5-8.5 in North America at approximately 25.01×10^4 km, accounting for 31.73% of the PGSHs of *S. halepense*, followed by Europe (20.47×10^4, 25.97%); Asia (20.05×10^4, 25.44%); Africa (4.81×10^4, 6.10%); South America (4.06×10^4, 5.15%); and Oceania (2.46×10^4, 3.12%). In the 2050s, future suitable urban areas for *S. halepense* were mainly distributed in North America, spread over approximately 34.10×10^4 km, accounting for 35.62% of the PGSHs of *S. halepense*, followed by Europe (24.43×10^4, 25.52%); Asia (21.42×10^4, 22.37%); Africa (5.65×10^4, 5.90%); South America (4.17×10^4, 4.36%); and Oceania (3.67×10^4, 3.83%).

2.4. Changes in PGSHs of S. halepense

Future increased cropland habitats were virtually all distributed in the North Hemisphere, including eastern, central, and western Asia (China, Iran, Kazakhstan, and Turkey); southwestern Europe (Russia and Belarus); and central North America (Canada and the United States), as well as parts of South America (Brazil) and Africa (Cote d' Ivoire) (Figure 4). The suitable area achieved its maximum extent under SSP5-8.5 in the 2050s, at approximately 172.09×10^4 km^2. Future decreased cropland habitats were mainly distributed in central Asia (Pakistan and India), as well as sporadically in southern North America (Mexico); central South America (Argentina, Bolivia, Brazil, and Paraguay); and eastern and central Africa (Ethiopia, Congo, and Zimbabwe). The suitable area achieved its minimum extent under SSP5-8.5 in the 2050s, at approximately 29.21 × 10^4 km^2.

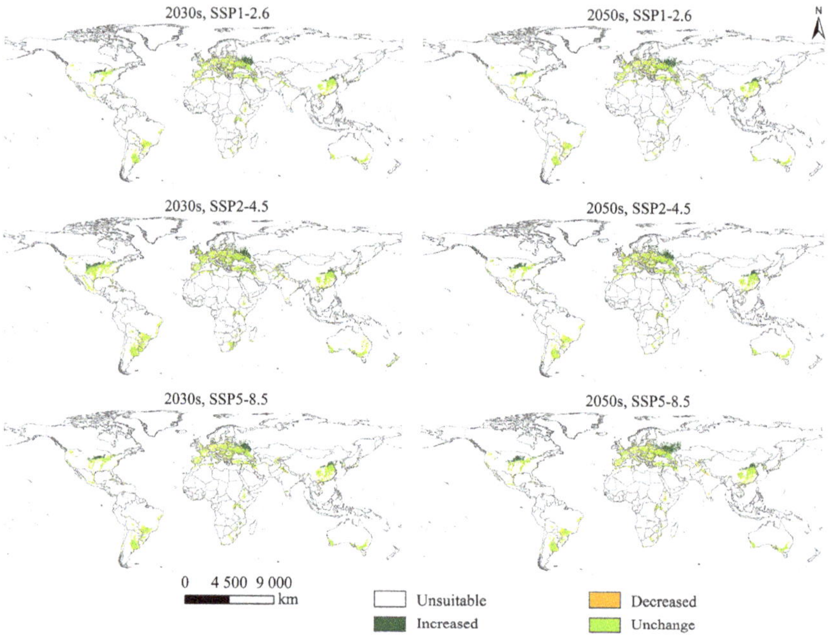

Figure 4. Changes in suitable cropland area for *Sorghum halepense* from different future scenarios (SSP1-2.6, SSP2-4.5, and SSP5-8.5) in the 2030s and 2050s compared to the near-current climate scenario.

Future increased grassland habitats were virtually all distributed in the Northern Hemisphere, including central and western Asia (Afghanistan, Kyrgyzstan, Tajikistan, and Turkey); southern and southwestern Europe (Austria, Norway, Russia, San Marino, Slovenia, Switzerland, and the United Kingdom); and central North America (the United States), as well as parts of southern Africa (Namibia and South Africa) (Figure S3). The suitable area achieved the maximum extent under SSP5-8.5 in the 2050s at approximately 72.58 × 10^4 km^2. Future decreased grassland habitats were virtually all distributed in eastern, central, and western Oceania (Australia). The suitable area achieved the minimum extent under SSP5-8.5 in the 2050s at approximately 12.64 × 10^4 km^2.

In the future scenarios, there was no decrease in suitable urban habitats, which were distributed in the Northern Hemisphere, including eastern Asia (China), southern Europe (Russia), and central North America (the United States) (Figure S4). The suitable area achieved its maximum extent under SSP5-8.5 in the 2050s, at approximately 3.62 × 10^4 km^2.

2.5. Trend of Suitabillity Probability for *S. halepense* according to Latitudinal Gradient

Compared with the near-current climate, the PGSHs of *S. halepense* in croplands tended to have high latitudinal gradients with a higher suitability probability under SSP1-2.6, SSP2-4.5, and SSP5-8.5 in the 2030s and 2050s (Figure 5). In the Northern Hemisphere, the suitable area for *S. halepense* in croplands was positioned at 28° N–39° N under a near-current climate, while it increased to 31° N–42° N, 35° N–50° N, and 58° N–61° N under future scenarios. In the Southern Hemisphere, the suitable area for *S. halepense* in croplands was positioned at 22° S–43° S under a near-current climate, while remaining firm at 22° S–42° S in future scenarios.

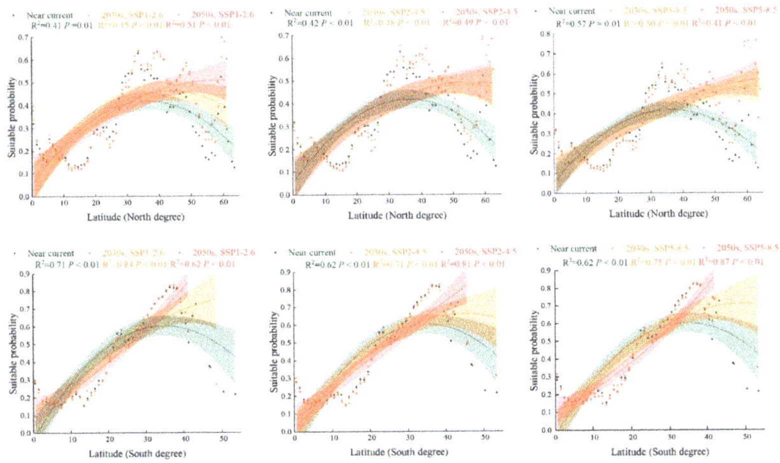

Figure 5. The PGSHs of *Sorghum halepense* in cropland for the Southern and Northern Hemispheres in the near-current climate scenario and under SSP1-2.6, SSP2-4.5, and SSP5-8.5 in the 2030s and 2050s. Green color means the near near-current climate scenario, orange color means the different climate scenario in the 2030s, and red is in the 2050s. For each group, we added 95% confidence bands via binomial fitting.

The global suitability probability of *S. halepense* in grasslands increased slightly for higher latitudinal gradients under SSP1-2.6, SSP2-4.5, and SSP5-8.5 in the 2030s and 2050s compared with the near-current climate scenario (Figure 6). In the Northern Hemisphere, the suitable area for *S. halepense* in grasslands was positioned at 20° N under the near-current climate, while it increased to 20° N–26° N under future scenarios. In the Southern Hemisphere, the suitable area for *S. halepense* in grasslands was positioned at 30° S–41° S under the near-current climate, while it increased to 31° S–46° S under future scenarios.

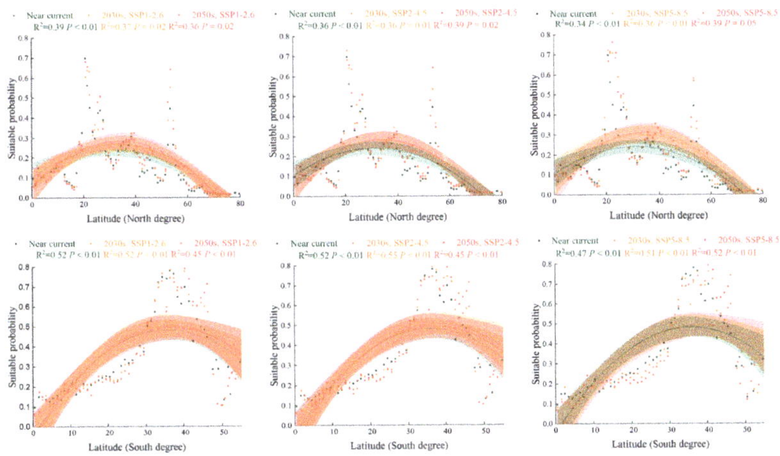

Figure 6. The PGSHs of *Sorghum halepense* in grassland for the Southern and Northern Hemispheres in the near-current climate scenario and under SSP1-2.6, SSP2-4.5, and SSP5-8.5 in the 2030s and 2050s. Green color means the near near-current climate scenario, orange color means the different climate scenario in the 2030s, and red is in the 2050s. For each group, we added 95% confidence bands via binomial fitting.

The global suitability probability of *S. halepense* in urban areas increased dramatically for higher latitudinal gradients under SSP1-2.6, SSP2-4.5, and SSP5-8.5 in the 2030s and the 2050s compared with the near-current climate scenario (Figure 7). In the Northern Hemisphere, the suitable area for *S. halepense* in urban land was positioned at 20° N–52° N under a near-current climate, while it increased to 19° N–61° N under future scenarios. In the Southern Hemisphere, the suitable area for *S. halepense* in urban land was positioned at 22° S–33° S under the near-current climate, while it increased to 26° S–41° S under future scenarios.

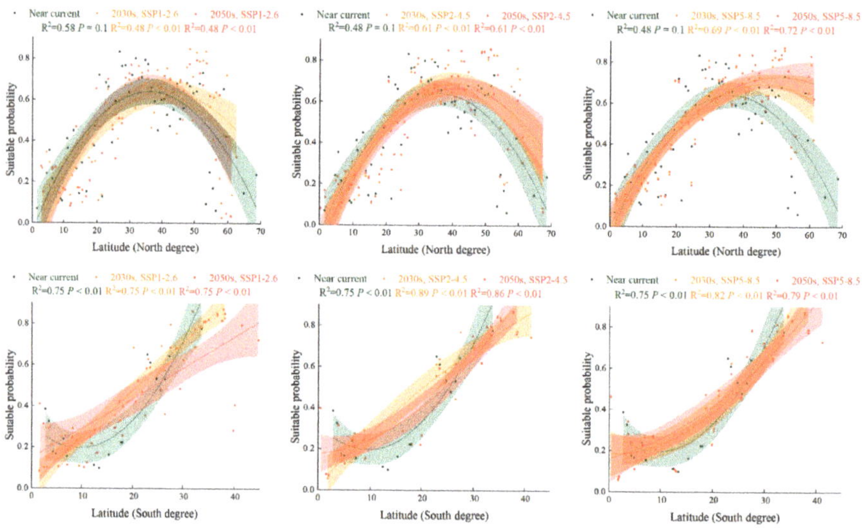

Figure 7. The PGSHs of *Sorghum halepense* in urban land for the Southern and Northern Hemispheres in the near-current climate scenario and under SSP1-2.6, SSP2-4.5, and SSP5-8.5 in the 2030s and 2050s. Green color means the near near-current climate scenario, orange color means the different climate scenario in the 2030s, and red is in the 2050s. For each group, we added 95% confidence bands via binomial fitting.

3. Discussion

Sorghum halepense, with its high reproduction rate and quick dispersal, is a well-known IAP worldwide that has a negative influence on crops, such as cotton, soybean, and wheat, in croplands and grasslands [15–17]. This study analyzed the dispersal modes and distribution dynamics of *S. halepense* according to reconstructions of the global countries historically invaded by *S. halepense* mixed in with crop seeds and introduced as forage based on CABI, GBIF, and WOS [5]. The expansion of IAPs is driven by climate change and possibly associated with LUC [13]. The impacts of climate change, LUC, and their interactions on habitat suitability for IAPs have been broadly investigated; however, few studies have examined how future LUC-associated climate change might affect IAPs [37,38]. This is the first study to explore the joint effect of LUC and climate change on the PGSHs of *S. halepense* under the near-current climate and future climate scenarios using the biomod2 model, which could be a vital premise for the management and prevention of IAPs.

3.1. Historical Invasion Reconstruction

The globalization of trade has facilitated the invasion of IAPs by intentionally moving them away from their native range for commercial purposes or accidentally introducing them into new environments [39]. The future dynamics of IAPs remain unclear and will be associated with socioeconomic changes and increased human populations, which are difficult to predict [40]. *Sorghum halepense* has been either introduced as a potential fodder or imported and exported mixed with grains, which are vital factors for rapid invasion via global trade [41]. Early in the 1800s, as a potential fodder plant and pasture grass, it was probably introduced from Turkey to South Carolina and Argentina and subsequently introduced to Australia and the southeastern United States, California, and New Mexico by farmers in the middle of the 1800s [39]. Our results showed that the countries invaded by *S. halepense* are broadly distributed worldwide, except for central Asia (Myanmar and Mongolia); northern Europe (Byelarus, Latvia, Lithuania, and Estonia); northern North America (Greenland); northern South America (Venezuela, Guyana, and Surname); and central, western, and northern Africa (Chad, Mali, and Libya). Furthermore, global crop reproduction and pasturing livestock, which are vulnerable to population growth and climate change, are likely to affect the invasion of *S. halepense* [42]. How socioeconomic drivers will impact the future distribution of IAPs is unknown, but the influence of socioeconomic dynamics could facilitate the invasion and establishment of IAPs in new environments depending on climate change and LUC [40].

3.2. Impact of Climate Change and LUC on Suitable Habitats

A comparison of the effects of climate change and LUC on IAPs has shown that the impact of LUC on the suitable ranges for invasive plants is weaker than that of climate change [43]. Our results showed that the contribution of LUC to predicting the distribution of *S. halepense* was smaller than that of climate change (Table 2). Climate change plays an important role in the distribution of IAPs at larger scales, whereas LUC has a stronger influence on variables at smaller scales [44]. Considering the invasion of *S. halepense* in croplands and grasslands, we explored the combined effects of climate change and LUC on the distribution of invasive plants at the global scale. With rising temperatures and elevated CO_2 emissions, the PGSHs of *S. halepense* will increase in the future and transfer to higher altitudes in cropland, grassland, and urban areas. Climate change may facilitate plant transfer along elevational gradients. IAPs respond faster to climate change than native plants [45,46]. Warming occurs in mid-to-high latitude regions; therefore, more adaptable plants are expected to exhibit elevated shifts [39].

Climate change may expand the range of IAPs by changing their traits, such as greater high-temperature tolerance, better adaptable growth with elevated CO_2 emissions, and latitude dependence [47,48]. Temperature is a vital factor that restricts plant growth and reproduction and is closely correlated with latitude [2]. Therefore, IAPs are expected to be transferred to higher latitudes with climate warming [49]. Our results showed that the future PGSHs of *S. halepense* in the three land-use types transferred to higher latitudes under the future scenarios. In contrast, future changes in precipitation are more variable than those in temperature under elevated emission scenarios [50]. *Sorghum halepense*, with its high adaptability in new environments, could tolerate a maximum temperature of 35 °C and temperatures as low as −26 °C at the coldest time of year. In addition, the mean annual precipitation that fits with *S. halepense* growth ranges widely from 300 to 2000 mm. The odds of future climate change that will cause extreme climates are significantly increasing. Our results showed that the future PGSHs of *S. halepense* are expected to increase in the Northern Hemisphere, achieving their maximum extent under SSP5-8.5 in the 2050s (Table 2). This is similar to other IAPs, whose potentially suitable habitats will change in response to climate change. For instance, the suitable global habitats of three ragweeds (*Ambrosia artemisiifolia*, *A. psilostachya*, and *A. trifida*) are expected to increase by the 2050s [30].

With future climate warming, many regions will become increasingly hot, leading to frequent droughts that will cause physiological stress and mortality in invasive and native plants, which is expected to cause grassland degradation [51]. However, IAPs' responses to temperature, precipitation, and landscape changes outcompete those of native plants; therefore, invasive plants will become dominant [52]. Previous studies have shown that *S. halepense* has a climate suitability of 50–90% in plains and grasslands in the United States, which poses a potential risk of transforming native grasslands into invaded prairies [53].

Our results showed that the global grassland area projected for future LUC scenarios decreased to some extent, yet the invasive area of *S. halepense* progressively increased (Table 2; Figure 3). In addition, cropland expansion has been gradually increasing in tropical areas since 2000 and is closely correlated with urban expansion [54]. To satisfy the housing demands of an increasing urban population, urban land is expected to increase dramatically in the future. More than 60% of global croplands are distributed near urban land; therefore, urban expansion is expected to be transferred from croplands [55,56]. Hence, the future PGSHs of *S. halepense* in cropland and urban areas will increase northward, and more suitable areas will be located in eastern China, western Europe, and northern North America under future climate scenarios (Table 2; Figure 3). Similarly, future increases in the PGSHs of *S. halepense* in grasslands will be located in western Europe and northern North America.

3.3. Prevention and Control

The unintentional introduction of *S. halepense* via the global grain trade and its intentional introduction as a forage crop have been key to *S. halepense*'s spread [5]. Our results showed that future increases in the PGSHs of *S. halepense* in the three land-use types will mainly be located in eastern China, Turkey, Iran, Afghanistan, Tajikistan, and southern Kazakhstan in Asia; Austria, Norway, San Marino, Slovenia, Switzerland, the United Kingdom, western Russia, and Belarus in Europe; the northern United States and southern Canada in North America; southern Brazil in South America; and southern Cote d' Ivoire, South Africa, Ghana, and Ethiopia in Africa. For newly invaded countries such as San Marino, Belarus, Cote d' Ivoire, and Ghana, identifying early warnings of *S. halepense* through risk assessment is important. However, for the newly invaded regions, customs quarantine controls in each country will play a crucial role in resisting the expansion and spread of *S. halepense*. Furthermore, regarding the invasion and establishment of *S. halepense*, there are a few measures to control population outbreaks, such as chemical control and integrated pest management (IPM). Selective herbicides may be effective against *S. halepense*, including quizalofop, nicosulfuron, and glyphosate, achieving 88–97% effectiveness [57]. In addition, the pre-sowing treatment of soybean and sorghum with glyphosate, Merlin, and Stromp is a vital method for controlling *S. halepense* [58–60].

4. Materials and Methods

4.1. Occurrence Data and Reconstructed Historically Invaded Countries

The global occurrences of *S. halepense* were collected from the Global Biodiversity Information Facility (GBIF: accessed on 13 December 2022, https://www.gbif.org/); aBarcode of Life Data Systems (BOLD: accessed on 19 December 2022, http://boldsystems.org/); the Southwest Environmental Information Network (SNIet: accessed on 6 December 2022, http://swbiodiversity.org); the Ministry of Agriculture and Rural Affairs of the People's Republic of China (accessed on 7 December 2022, http://www.moa.gov.cn/); some published literature on Web of Science (WOS: accessed on 6 December 2022, https://www.webofscience.com/); and field sampling in China. A total of 32,972 occurrences of *S. halepense* were found in 124 countries (including regions, as stated below) (Supplementary Excel). To avoid a sampling bias in constructing the species distribution model, occurrences of *S. halepense* were filtered using ENMTools, retaining only one occurrence in a 10 × 10 km raster [61]. The final occurrence data comprised 21,443 occurrences of *S. halepense* (Figure S5).

To reconstruct the countries historically invaded by *S. halepense*, we first removed 21 countries, including 18 countries that had no detailed geographical information on the occurrence of *S. halepense* and three countries without recorded databases and literature, compared with countries recorded by the Centre for Agriculture and Biosciences International (CABI: accessed on 21 February 2023, https://www.cabi.org/) and European and Mediterranean Plant Protection Organization (EPPO: accessed on 21 February 2023, https://www.eppo.int/) based on GBIF and WOS. The earliest recorded time for field sampling and the literature review was assumed to be time of invasion in each country. We confirmed 18 countries to which *S. halepense* was native based on CABI and EPPO. The invaded countries recorded before 1900 were regarded as one category, and other invaded countries from 1900 to 2022 with an interval of 20 years were used to reconstruct the global spatio-temporal dynamics of *S. halepense* invasion. In addition, we calculated the variation dynamics of the number of invasive countries over time and analyzed the invasive trends of *S. halepense*. Geographical observations were set to the World Geodetic System 1984 geographic coordinate system and imported to ArcGIS 10.8. We presented the capital of each country invaded by *S. halepense* in ArcGIS.

4.2. Land-Use Harmonization Data

Land-use harmonization data were used to examine how changes in the PGSHs of *S. halepense* may be affected by LUC. The annual datasets for 2015–2100 with a 1 km resolution estimated the fractional land-use patterns, underlying land-use transitions, and key agricultural management information [62]. In addition, we obtained historical reconstructions of land-use in the annual fraction state layers for 2015, representing near-current climate data. The future harmonized land-use dataset covered 2040 and 2060 to match future periods of the baseline, including SSP1-2.6, built with IMAGE; SSP2-4.5, built with MESSAGE-GLOBIOM; and SSP5-8.5, built with REMIND-MAGPIE [63]. The harmonized land classifications considered were mainly cropland, grassland, urban areas, and others, which could be transferred from one land-use type to another (Figure S6).

We predicted the potentially suitable and unsuitable habitats for *S. halepense* during the near-current and future periods, quantifying the proportion of land use for each classification in every grid cell. In addition, the predicted PGSHs of *S. halepense* under the near-current and future climate scenarios were mainly located in three land-use types: cropland, grassland, and urban areas. Therefore, by analyzing the decreased, increased, or unchanged PGSHs for *S. halepense* towards the middle of the century, we could examine the response to climate change and LUC from the near-current climate to future scenarios.

4.3. Climate Data

We downloaded 19 historical environmental variables from 1970 to 2000 and the future period (2021–2040 and 2041–2060) at a 5 min spatial resolution from WorldClim v2.1 [64]. Future data were based on BCC-CSM2-MR global climate models (GCMs) for four shared socio-economic pathways (SSPs): SSP1-2.6, SSP2-4.5, and SSP5-8.5. SSP1-2.6 is regarded as the future ideal scenario, in that 1.5 °C global warming is avoided, while the extreme scenario SSP5-8.5 is more likely to occur [65,66]. Hence, climate change and LUC analyses mainly focused on the emission scenarios SSP2-4.5 and SSP5-8.5 to estimate the spatial dynamics and PGSHs of *S. halepense*. In addition, strong correlations between environmental variables could lead to multicollinearity [67]. We used ENMTools to examine the variable correlations ($|r| \geq 0.8$) and removed variables that were strongly correlated (Figure S7). Finally, we retained seven significant environmental variables for *S. halepense* (Table 2).

4.4. Model Construction and Evaluation

We predicted the PGSHs of *S. halepense* under the near-current climate and future scenarios based on occurrence records and environmental variables that included climate data and harmonized land-use data using eight individual models, including GLM, GBM, RF, Maxnet, CTA, ANN, FDA, and MARS in the biomod2 4.2.3 package in Rstudio [68]. The training data randomly used 75% of the occurrence data, and the remaining 25% was selected as the testing dataset, with five model replicates. Ten thousand selected global pseudoabsence points were randomly replicated once to run the models. Finally, 40 models were reconstructed, and their performances were evaluated using the test dataset. In the present study, we selected a single model in which the true skill statistic (TSS) was higher than 0.8 and the area under the receiver operating characteristic (ROC) curve (AUC) was higher than 0.9 from 40 models to construct an EM predicting the PGSHs of *S. halepense* under the near-current climate and future scenarios.

TSS, AUC, and KAPPA were used to estimate the model performance [69]. TSS is an independent threshold calculated as sensitivity and specificity-1 ranging from −1 to +1 [70]. A value closer to +1 indicates excellent performance, and zero or less indicates worse performance than random [70]. The AUC is a metric that varies in true-positive and -negative rates and ranges from 0 to 1, with values closer to 1 indicating more perfect discrimination. KAPPA ranges from −1 to +1, and values closer to +1 indicate perfect performance [70]. The probability (P) of the presence of *S. halepense* was generated in the ASCII raster and ranged from 0 to 1 in the model results. We classified the PGSHs of *S. halepense* into two categories based on the maximum KAPPA value: unsuitable habitats ($0 < p \leq 0.37$) and suitable habitats ($0.37 < p \leq 1$).

5. Conclusions

Our study was the first to reconstruct the countries historically invaded by *S. halepense* and explore how climate change and LUC will influence the PGSHs of *S. halepense* in the three land-use types (croplands, grasslands, and urban areas). *Sorghum halepense* is widely found worldwide except in Africa due to the import and export trade and the introduction of fodder grass species. Our results found that environmental variables were more significant than LUC in predicting the PGSHs of *S. halepense*. In addition, future croplands and urban areas are expected to increase continuously, while grassland areas are expected to decrease. The future PGSHs of *S. halepense* in the three land-use types are expected to increase continuously in the Northern Hemisphere and transfer to higher latitudes with climate warming. Furthermore, new invaded countries, such as San Marino, Belarus, Cote d'Ivoire, and Ghana, need risk assessments of *S. halepense* to prevent its introduction. The threatened regions in the six continents need to increase plant quarantines at customs facilities in each country. The reconstruction of the countries historically invaded by *S. halepense* and the predicted PGSHs of *S. halepense* will help provide a more reliable risk assessment.

Supplementary Materials: The following supporting information can be downloaded at: https://www.mdpi.com/article/10.3390/plants12173128/s1, Figure S1: Single model performance with ROC and TSS values, Figure S2: Significant environmental variables for predicting the potential global suitable habitats (PGSHs) of *Sorghum halepense*, Figure S3: Changes in suitable grassland areas for *Sorghum halepense* under different future scenarios (SSP1-2.6, SSP2-4.5, and SSP5-8.5) in the 2030s and 2050s compared to the near-current climate, Figure S4: Changes in suitable urban areas for *Sorghum halepense* under different future scenarios (SSP1-2.6, SSP2-4.5, and SSP5-8.5) in the 2030s and 2050s compared to the near-current climate, Figure S5: Global distribution occurrences of *Sorghum halepense*, Figure S6: Global land-use transitions under future scenarios. Each arrow represents a change from a specific land-use type to another land-use type. (A) the transitions from SSP1-2.6 in the 2030s to the near-current climate, (B) the transitions from SSP1-2.6 in the 2050s to SSP1-2.6 in the 2030s, (C) the transitions from SSP2-4.5 in the 2030s to the near-current climate, (D) the transitions from SSP2-4.5 in the 2050s to SSP2-4.5 in the 2030s, (E) the transitions from SSP5-8.5 in the 2030s to the near-current climate, (F) the transitions from SSP5-8.5 in the 2050s to SSP5-8.5 in the 2030s, Figure S7: Correlation of twenty environmental variables via ENMTools, Table S1: The mean ROC, TSS, and KAPPA values of eight individual models and the EM, Table S2: Environmental variables projected to predict the potential global suitable habitats (PGSHs) of *Sorghum halepense*, Excel S1: Countries historically invaded by *Sorghum halepense*.

Author Contributions: All authors contributed to the study conception and design. Data acquisition, H.Z.; data analysis, M.Y. and Y.Q.; writing—original draft preparation, M.Y.; supervision, X.X.; writing—review and editing, Q.L., J.G., L.C. and W.L.; funding acquisition, W.L. All authors have read and agreed to the published version of the manuscript.

Funding: This work was supported by the National Key R&D Program of China (grant no. 2022YFC2601100, grant no. 2021YFC2600400); the Key R&D Program of Yunnan Province (grant no. 202103AF140007); and Technology Innovation Program of the Chinese Academy of Agricultural Sciences (grant no. caascx-2022-2025-IAS).

Data Availability Statement: The data presented in this study are available in this article.

Conflicts of Interest: The authors declare no conflict of interest.

References

1. Pyšek, P.; Hulme, P.E.; Simberloff, D.; Bacher, S.; Blackburn, T.M.; Carlton, J.T.; Dawson, W.; Essl, F.; Foxcroft, L.C.; Genovesi, P.; et al. Scientists' warning on invasive alien species. *Biol. Rev.* **2020**, *95*, 1511–1534. [CrossRef]
2. Walther, G.-R.; Roques, A.; Hulme, P.; Sykes, M.; Pyšek, P.; Kühn, I.; Zobel, M.; Bacher, S.; Botta-Dukat, Z.; Bugmann, H.; et al. Alien species in a warmer world: Risks and opportunities. *Trends Ecol. Evol.* **2009**, *24*, 686–693. [CrossRef] [PubMed]
3. Ribeiro, J.; Bingre, P.; Strubbe, D.; Santana, J.; Capinha, C.; Araújo, M.B.; Reino, L. Exploring the effects of geopolitical shifts on global wildlife trade. *Bioscience* **2022**, *72*, 560–572. [CrossRef] [PubMed]
4. Wu, Z.-Y.; Milne, R.I.; Liu, J.; Nathan, R.; Corlett, R.T.; Li, D.-Z. The establishment of plants following long-distance dispersal. *Trends Ecol. Evol.* **2023**, *38*, 289–300. [CrossRef] [PubMed]
5. Wilson, J.R.U.; Dormontt, E.E.; Prentis, P.J.; Lowe, A.J.; Richardson, D.M. Something in the way you move: Dispersal pathways affect invasion success. *Trends Ecol. Evol.* **2009**, *24*, 136–144. [CrossRef] [PubMed]
6. Humair, F.; Humair, L.; Kuhn, F.; Kueffer, C. E-commerce trade in invasive plants. *Conserv. Biol.* **2015**, *29*, 1658–1665. [CrossRef] [PubMed]
7. Beaury, E.M.; Patrick, M.; Bradley, B.A. Invaders for sale: The ongoing spread of invasive species by the plant trade industry. *Front. Ecol. Environ.* **2021**, *19*, 550–556. [CrossRef]
8. Freeman, B.G.; Lee-Yaw, J.A.; Sunday, J.M.; Hargreaves, A.L. Expanding, shifting and shrinking: The impact of global warming on species' elevational distributions. *Glob. Ecol. Biogeogr.* **2018**, *27*, 1268–1276. [CrossRef]
9. Elsen, P.R.; Monahan, W.B.; Merenlender, A.M. Topography and human pressure in mountain ranges alter expected species responses to climate change. *Nat. Commun.* **2020**, *11*, 1974. [CrossRef]
10. Nogués-Bravo, D.; Araújo, M.B.; Romdal, T.; Rahbek, C. Scale effects and human impact on the elevational species richness gradients. *Nature* **2008**, *453*, 216–219. [CrossRef]
11. Pecl, G.T.; Araújo, M.B.; Bell, J.D.; Blanchard, J.; Bonebrake, T.C.; Chen, I.C.; Clark, T.D.; Colwell, R.K.; Danielsen, F.; Evengård, B.; et al. Biodiversity redistribution under climate change: Impacts on ecosystems and human well-being. *Science* **2017**, *355*, eaai9214. [CrossRef] [PubMed]
12. MacDonald, A.J.; McComb, S.; O'Neill, C.; Padgett, K.A.; Larsen, A.E. Projected climate and land use change alter western blacklegged tick phenology, seasonal host-seeking suitability and human encounter risk in California. *Glob. Chang. Biol.* **2020**, *26*, 5459–5474. [CrossRef] [PubMed]
13. Taheri, S.; García-Callejas, D.; Araújo, M.B. Discriminating climate, land-cover and random effects on species range dynamics. *Glob. Chang. Biol.* **2021**, *27*, 1309–1317. [CrossRef] [PubMed]
14. Klein, P.; Smith, C. Invasive Johnsongrass, a threat to native grasslands and agriculture. *Biologia* **2020**, *76*, 413–420. [CrossRef]
15. Nouri, H.R.; Talab, Z.A.; Tavassoli, A. Effect of weed allelopathic of sorghum (*Sorghum halepense*) on germination and seedling growth of wheat, Alvand cultivar. *Ann. Biol. Res.* **2012**, 3, 1283–1293.
16. Bridges, D.C.; Chandler, J.M. Influence of johnsongrass (*Sorghum halepense*) density and period of competition on cotton yield. *Weed Sci.* **1987**, *35*, 63–67. [CrossRef]
17. Williams, C.S.; Hayes, R.M. Johnsongrass (*Sorghum halepense*) competition in soybeans (*Glycine max*). *Weed Sci.* **1984**, *32*, 498–501. [CrossRef]
18. Mitskas, M.B.; Tsolis, C.E.; Eleftherohorinos, I.G.; Damalas, C.A. Interference between corn and johnsongrass (*Sorghum halepense*) from seed or rhizomes. *Weed Sci.* **2003**, *51*, 540–545. [CrossRef]
19. Holm, L.R.G.; Plucknett, D.L.; Pancho, J.V.; Herberger, J.P. *The World's Worst Weeds. Distribution and Biology*; University Press of Hawaii: Honolulu, HI, USA, 1977; p. 621.
20. Warwick, S.I.W.I.; Black, L.D. The biology of Canadian weeds.: 61. *Sorghum halepense (L.) PERS. Can. J. Plant Sci.* **1983**, *63*, 997–1014. [CrossRef]
21. Asgharipour, M. Inhibitory effects of *Sorghum Halepens* root and leaf extracts on germination and early seedling growth of widely used medicinal plants. *Adv. Environ. Biol.* **2010**, *4*, 316–324.
22. Guillera-Arroita, G.; Lahoz-Monfort, J.J.; Elith, J.; Gordon, A.; Kujala, H.; Lentini, P.E.; McCarthy, M.A.; Tingley, R.; Wintle, B.A. Is my species distribution model fit for purpose? Matching data and models to applications. *Glob. Ecol. Biogeogr.* **2015**, *24*, 276–292. [CrossRef]
23. Lopatin, J.; Dolos, K.; Hernández, H.J.; Galleguillos, M.; Fassnacht, F.E. Comparing generalized linear models and random forest to model vascular plant species richness using LiDAR data in a natural forest in central Chile. *Remote Sens. Environ.* **2016**, *173*, 200–210. [CrossRef]

24. Guisan, A.; Thuiller, W. Predicting species distribution: Offering more than simple habitat models. *Ecol. Lett.* **2005**, *8*, 993–1009. [CrossRef] [PubMed]
25. Favretti, M. Remarks on the maximum entropy principle with application to the maximum entropy theory of ecology. *Entropy* **2018**, *20*, 11. [CrossRef] [PubMed]
26. Kriticos, D.; Maywald, G.; Yonow, T.; Zurcher, E.; Herrmann, N.; Sutherst, R. *CLIMEX. Version 4. Exploring the Effects of Climate on Plants, Animals and Diseases*; CSIRO: Canberra, Australia, 2015.
27. Thuiller, W. BIOMOD—Optimizing predictions of species distributions and projecting potential future shifts under global change. *Glob. Chang. Biol.* **2003**, *9*, 1353–1362. [CrossRef]
28. Hao, T.; Elith, J.; Guillera-Arroita, G.; Lahoz-Monfort, J.J. A review of evidence about use and performance of species distribution modelling ensembles like BIOMOD. *Divers. Distrib.* **2019**, *25*, 839–852. [CrossRef]
29. Stohlgren, T.J.; Ma, P.; Kumar, S.; Rocca, M.; Morisette, J.T.; Jarnevich, C.S.; Benson, N. Ensemble habitat mapping of invasive plant species. *Risk Anal.* **2010**, *30*, 224–235. [CrossRef]
30. Xian, X.; Zhao, H.; Wang, R.; Huang, H.; Chen, B.; Zhang, G.; Liu, W.; Wan, F. Climate change has increased the global threats posed by three ragweeds (*Ambrosia* L.) in the Anthropocene. *Sci. Total Environ.* **2023**, *859*, 160252. [CrossRef]
31. Shrestha, U.B.; Sharma, K.P.; Devkota, A.; Siwakoti, M.; Shrestha, B.B. Potential impact of climate change on the distribution of six invasive alien plants in Nepal. *Ecol. Indic.* **2018**, *95*, 99–107. [CrossRef]
32. De Almeida, J.; Freitas, H. Exotic naturalized flora of continental Portugal. A reassessment. *Bot. Complut.* **2006**, *30*, 117.
33. McWhorter, C.G. Introduction and Spread of Johnsongrass in the United States. *Weed Sci.* **1971**, *19*, 496–500. [CrossRef]
34. Holm, L.G.; Pancho, J.V.; Herberger, J.P.; Plucknett, D.L. A Geographic Atlas of World Weeds. Krieger Publishing Co.: Malabar, FL, USA, 1991; p. 391.
35. Jacques-félix, H. *The Gramineae of Tropical Africa. I. Generalities, Classification and Description of Genera*; Institut de Recherches Agronomiques Tropicales: Paris, France, 1962.
36. Parsons, W.T.; Cuthbertson, E.G. *Noxious Weeds of Australia*; Inkarta Press: Melbourne, Australia, 1992.
37. Puchalka, R.; Dyderski, M.K.; Vitkova, M.; Sadlo, J.; Klisz, M.; Netsvetov, M.; Prokopuk, Y.; Matisons, R.; Mionskowski, M.; Wojda, T.; et al. Black locust (*Robinia pseudoacacia* L.) range contraction and expansion in Europe under changing climate. *Glob. Chang. Biol.* **2021**, *27*, 1587–1600. [CrossRef] [PubMed]
38. Murray, J.V.; Stokes, K.E.; van Klinken, R.D. Predicting the potential distribution of a riparian invasive plant: The effects of changing climate, flood regimes and land-use patterns. *Glob. Chang. Biol.* **2012**, *18*, 1738–1753. [CrossRef]
39. Gottfried, M.; Pauli, H.; Futschik, A.; Akhalkatsi, M.; Barančok, P.; Benito Alonso, J.L.; Coldea, G.; Dick, J.; Erschbamer, B.; Fernández Calzado, M.a.R.; et al. Continent-wide response of mountain vegetation to climate change. *Nat. Clim. Chang.* **2012**, *2*, 111–115. [CrossRef]
40. Bonnamour, A.; Gippet, J.M.W.; Bertelsmeier, C. Insect and plant invasions follow two waves of globalisation. *Ecol. Lett.* **2021**, *24*, 2418–2426. [CrossRef] [PubMed]
41. Ikeda, M.; Nishi, T.; Asai, M.; Muranaka, T.; Konuma, A.; Tominaga, T.; Shimono, Y. The role of weed seed contamination in grain commodities as propagule pressure. *Biol. Invasions* **2022**, *24*, 1707–1723. [CrossRef]
42. Godber, O.F.; Wall, R. Livestock and food security: Vulnerability to population growth and climate change. *Glob. Chang. Biol.* **2014**, *20*, 3092–3102. [CrossRef] [PubMed]
43. Gong, X.; Chen, Y.; Wang, T.; Jiang, X.; Hu, X.; Feng, J. Double-edged effects of climate change on plant invasions: Ecological niche modeling global distributions of two invasive alien plants. *Sci. Total Environ.* **2020**, *740*, 139933. [CrossRef]
44. Sirami, C.; Caplat, P.; Popy, S.; Clamens, A.; Arlettaz, R.; Jiguet, F.; Brotons, L.; Martin, J.-L. Impacts of global change on species distributions: Obstacles and solutions to integrate climate and land use. *Glob. Ecol. Biogeogr.* **2017**, *26*, 385–394. [CrossRef]
45. Liu, Y.; Oduor, A.M.O.; Zhang, Z.; Manea, A.; Tooth, I.M.; Leishman, M.R.; Xu, X.; van Kleunen, M. Do invasive alien plants benefit more from global environmental change than native plants? *Glob. Chang. Biol.* **2017**, *23*, 3363–3370. [CrossRef]
46. Lenoir, J.; Gégout, J.C.; Marquet, P.A.; de Ruffray, P.; Brisse, H. A Significant upward shift in plant species optimum elevation during the 20th century. *Science* **2008**, *320*, 1768–1771. [CrossRef] [PubMed]
47. Molina-Montenegro, M.; Naya, D. Latitudinal patterns in phenotypic plasticity and fitness-related traits: Assessing the climatic variability hypothesis (CVH) with an invasive plant species. *PLoS ONE* **2012**, *7*, e47620. [CrossRef] [PubMed]
48. Bongaarts, J. IPBES, 2019. Summary for policymakers of the global assessment report on biodiversity and ecosystem services of the Intergovernmental Science-Policy Platform on Biodiversity and Ecosystem Services. *Popul. Dev. Rev.* **2019**, *45*, 680–681. [CrossRef]
49. Walther, G.-R.; Post, E.; Convey, P.; Menzel, A.; Parmesan, C.; Beebee, T.J.C.; Fromentin, J.-M.; Hoegh-Guldberg, O.; Bairlein, F. Ecological responses to recent climate change. *Nature* **2002**, *416*, 389–395. [CrossRef] [PubMed]
50. Finch, D.M.; Butler, J.L.; Runyon, J.B.; Fettig, C.J.; Kilkenny, F.F.; Jose, S.; Frankel, S.J.; Cushman, S.A.; Cobb, R.C.; Dukes, J.S.; et al. Effects of Climate Change on Invasive Species. In *Invasive Species in Forests and Rangelands of the United States: A Comprehensive Science Synthesis for the United States Forest Sector*; Poland, T.M., Patel-Weynand, T., Finch, D.M., Miniat, C.F., Hayes, D.C., Lopez, V.M., Eds.; Springer International Publishing: Cham, Switzerland, 2021; pp. 57–83.
51. Ravi, S.; Law, D.J.; Caplan, J.S.; Barron-Gafford, G.A.; Dontsova, K.M.; Espeleta, J.F.; Villegas, J.C.; Okin, G.S.; Breshears, D.D.; Huxman, T.E. Biological invasions and climate change amplify each other's effects on dryland degradation. *Glob. Chang. Biol.* **2022**, *28*, 285–295. [CrossRef] [PubMed]

52. Yu, K.; Okin, G.S.; Ravi, S.; D'Odorico, P. Potential of grass invasions in desert shrublands to create novel ecosystem states under variable climate. *Ecohydrology* **2016**, *9*, 1496–1506. [CrossRef]
53. Barney, J.N.; DiTomaso, J.M. Global climate niche estimates for bioenergy crops and invasive species of agronomic origin: Potential problems and opportunities. *PLoS ONE* **2011**, *6*, e17222. [CrossRef]
54. Grassini, P.; Eskridge, K.M.; Cassman, K.G. Distinguishing between yield advances and yield plateaus in historical crop production trends. *Nat. Commun.* **2013**, *4*, 2918. [CrossRef]
55. Bren d'Amour, C.; Reitsma, F.; Baiocchi, G.; Barthel, S.; Güneralp, B.; Erb, K.-H.; Haberl, H.; Creutzig, F.; Seto, K.C. Future urban land expansion and implications for global croplands. *Proc. Natl. Acad. Sci. USA* **2017**, *114*, 8939–8944. [CrossRef]
56. Thebo, A.; Drechsel, P.; Lambin, E. Global assessment of urban and peri-urban agriculture: Irrigated and rainfed croplands. *Environ. Res. Lett.* **2014**, *9*, 114002. [CrossRef]
57. Johnson, W.G.; Li, J.; Wait, J.D. Johnsongrass control, total nonstructural carbohydrates in rhizomes, and regrowth after application of herbicides used in herbicide-resistant corn (*Zea mays*). *Weed Technol.* **2003**, *17*, 36–41. [CrossRef]
58. Defelice, M.S.; Witt, W.W.; Martin, J.R. Johnsongrass (*Sorghum halepense*) control and soil moisture relationships in no-tillage, doublecropped soybeans (*Glycine max*). *Weed Sci.* **1987**, *35*, 108–114. [CrossRef]
59. Brown, S.M.; Chandler, J.M.; Morrison, J.E. Glyphosate for johnsongrass (*Sorghum halepense*) control in no-till Sorghum (*Sorghum bicolor*). *Weed Sci.* **1988**, *36*, 510–513. [CrossRef]
60. Tóth, V.; Lehoczky, É. The analysis of effect of different herbicides on Johnson-grass in maize. *Commun. Agric. Appl. Biol. Sci.* **2007**, *72*, 279–282. [PubMed]
61. Warren, D.L.; Matzke, N.J.; Cardillo, M.; Baumgartner, J.B.; Beaumont, L.J.; Turelli, M.; Glor, R.E.; Huron, N.A.; Simões, M.; Iglesias, T.L.; et al. ENMTools 1.0: An R package for comparative ecological biogeography. *Ecography* **2021**, *44*, 504–511. [CrossRef]
62. Chen, G.; Li, X.; Liu, X. Global land projection based on plant functional types with a 1-km resolution under socio-climatic scenarios. *Sci. Data* **2022**, *9*, 125. [CrossRef] [PubMed]
63. Stehfest, E.; Vuuren, D.; Kram, T.; Bouwman, A.; Alkemade, R.; Bakkenes, M.; Biemans, H.; Bouwman, A.; Elzen, M.; Janse, J.; et al. *Integrated Assessment of Global Environmental Change with IMAGE 3.0. Model Description and Policy Applications*; Netherlands Environmental Assessment Agency (PBL): Zuid-Holland, The Netherlands, 2014.
64. Fick, S.E.; Hijmans, R.J. WorldClim 2: New 1-km spatial resolution climate surfaces for global land areas. *Int. J. Climatol.* **2017**, *37*, 4302–4315. [CrossRef]
65. Schwalm, C.R.; Glendon, S.; Duffy, P.B. RCP8.5 tracks cumulative CO_2 emissions. *Proc. Natl. Acad. Sci. USA* **2020**, *117*, 19656–19657. [CrossRef]
66. Lee, J.-Y.; Marotzke, J.; Bala, G.; Cao, L.; Corti, S.; Dunne, J.P.; Engelbrecht, F.; Fischer, E.; Fyfe, J.C.; Jones, C.; et al. *Climate Change 2021: The Physical Science Basis. Contribution of Working Group I to the Sixth Assessment Report of the Intergovernmental Panel on Climate Change*; Masson-Delmotte, V., Zhai, P., Pirani, A., Connors, S.L., Péan, C., Berger, S., Caud, N., Chen, Y., Goldfarb, L., Gomis, M.I., et al., Eds.; Cambridge University: Cambridge, UK; New York, NY, USA, 2021.
67. Dormann, C.F.; Elith, J.; Bacher, S.; Buchmann, C.; Carl, G.; Carré, G.; Marquéz, J.R.G.; Gruber, B.; Lafourcade, B.; Leitão, P.J.; et al. Collinearity: A review of methods to deal with it and a simulation study evaluating their performance. *Ecography* **2013**, *36*, 27–46. [CrossRef]
68. Thuiller, W.; Georges, D.; Engler, R.; Breiner, F. Ensemble Platform for Species Distribution Modeling; R Package, 2023.
69. Fielding, A.H.; Bell, J.F. A review of methods for the assessment of prediction errors in conservation presence/absence models. *Environ. Conserv.* **1997**, *24*, 38–49. [CrossRef]
70. Allouche, O.; Tsoar, A.; Kadmon, R. Assessing the accuracy of species distribution models: Prevalence, kappa and the true skill statistic (TSS). *J. Appl. Ecol.* **2006**, *43*, 1223–1232. [CrossRef]

plants

MDPI

Article

Modeling the Potential Distribution Patterns of the Invasive Plant Species *Phytolacca americana* in China in Response to Climate Change

Qianru Nan [1], Chunhui Li [2], Xinghao Li [1], Danni Zheng [1], Zhaohua Li [1] and Liya Zhao [1,*]

1 School of Resources and Environmental Science, Hubei University, Wuhan 430062, China; 202121108012639@stu.hubu.edu.cn (Q.N.); li_xhao@163.com (X.L.); 202221108012260@stu.hubu.edu.cn (D.Z.); zli@hubu.edu.cn (Z.L.)

2 Agricultural Development Service Centre of Enshi Tujia and Miao Autonomous Prefecture, Enshi 44500, China; esfgk@163.com

* Correspondence: zhaoly0128@hubu.edu.cn

Abstract: *Phytolacca americana*, introduced to China in the 20th century for its medicinal properties, has posed a significant ecological and agricultural challenge. Its prolific fruit production, high reproductive coefficient, adaptability, and toxic roots and fruits have led to the formation of monoculture communities, reducing native species diversity and posing threats to agriculture, human and animal health, and local ecosystems. Understanding its potential distribution patterns at a regional scale and its response to climate change is essential for effective monitoring, management, and control. In this study, we utilized the Maxent model to simulate potential habitat areas of *P. americana* across three timeframes (current, 2050s, and 2070s) under three climate change scenarios (SSP126, SSP245, and SSP585). Leveraging data from 556 *P. americana* sites across China, we employed ROC curves to assess the prediction accuracy. Our findings highlight key environmental factors influencing *P. americana*'s geographical distribution, including the driest month's precipitation, the coldest month's minimum temperature, the wettest month's precipitation, isothermality, and temperature annual range. Under current climate conditions, *P. americana* potentially inhabits 280.26×10^4 km^2 in China, with a concentration in 27 provinces and cities within the Yangtze River basin and its southern regions. While future climate change scenarios do not drastically alter the total suitable area, the proportions of high and low-suitability areas decrease over time, shifting towards moderate suitability. Specifically, in the SSP126 scenario, the centroid of the predicted suitable area shifts northeastward and then southwestward. In contrast, in the SSP245 and SSP585 scenarios, the centroid shifts northward.

Keywords: *Phytolacca americana*; biological invasion; Maxent; climate change; potentially suitable area

1. Introduction

Recently, intensified global environmental changes have significantly increased the threat of biological invasions to ecosystems and biodiversity [1]. Given the backdrop of significant global climate change, understanding the spatial distribution of biological populations and their interplay with the environment has become pivotal for unraveling biodiversity patterns and devising strategies for its preservation [2]. Extensive research demonstrates that climate change profoundly impacts and will continue to influence species distribution patterns [3,4]. Consequently, predicting species' potential habitats and their migration trends in response to climate change is vital for informing future species management, ecological conservation, and cultivation strategies [5,6].

The emergence of advanced computational statistical techniques and the development of global information systems have facilitated the exploration of direct correlations between environmental factors, such as climate and topography, and species data, revolutionizing the field of ecology [7,8]. Among these innovations, species distribution models (SDMs) have emerged as a crucial tool for deciphering species distribution patterns [9]. SDMs leverage known distribution data and corresponding environmental variables to simulate species' geographical distribution and their responses to climate change, employing various algorithms across temporal and spatial scales [2,10].

The Maxent model, based on the maximum entropy principle developed by Phillips, is widely recognized for predicting geographical distributions of species [11–14]. Its notable advantage lies in its capacity to maintain a high accuracy and stability, even in cases with incomplete species data and limited sample sizes [15]. Furthermore, Maxent stands out for its speedy computations and flexible operation [14–16], with its predictions easily visualized using ArcGIS [17]. The Maxent model has also proven valuable in forecasting the potential distribution of invasive and endangered species, including *Xanthium spinosum* L. (Asteraceae) [18], *Lonicera japonica* Thunb. (Caprifoliaceae) [19], *Ziziphus jujuba* var. *spinosa* (Rhamnaceae) [20], *Pinus massoniana* Lamb. (Pinaceae) [21], and *Litsea cubeba* (Lour.) Pers. (Lauraceae) [22].

Native to North America, *Phytolacca americana*, a perennial herb from the Phytolaccaceae family, was introduced to China for its medicinal properties. Since its discovery in Hangzhou in 1935, it has rapidly invaded numerous regions of the country, colonizing forest margins, roadsides, village lands, wastelands, and disrupting tea plantations, orchards, and bamboo groves [23]. This invasion poses significant threats to local agricultural and forestry production and biodiversity [24,25]. *P. americana* was included in the "List of Invasive Species in China's Natural Ecosystems (Fourth Batch)" in 2016, mandating strict control and regulation of its cultivation [26]. Consequently, conducting a comprehensive study of *P. americana*'s suitable and potential distribution areas assumes a paramount importance for the effective monitoring, early warning, and prevention of its invasion. It is also critical for sustaining agroforestry production, conserving diversity, maintaining productivity, and restoring ecological environments. Presently, research on the invasive species *P. americana* primarily focuses on its invasion mechanisms, biological characteristics, and ecological hazards [27–29]. However, there is limited domestic and international research concerning its suitable distribution areas and potential sites.

In this study, leveraging distribution data and relevant environmental variables, we employed the Maxent model and ArcGIS to forecast potential suitable areas for *P. americana* across various climate scenarios, both current and future. Our objectives encompassed (1) predicting potentially suitable habitats for *P. americana* under current climate scenarios, (2) identifying primary environmental variables constraining its geographic distribution, and (3) forecasting potentially suitable habitats under future climate scenarios. Our aim was to unravel shifts in *P. americana*'s present and future geographical distribution patterns. This research strives to provide a scientific foundation for the monitoring and management of this highly invasive species.

2. Materials and Methods

2.1. Collection and Screening of Species Occurrence Data

The distribution data for *P. americana* in China were primarily sourced from two databases: the Global Biodiversity Information Facility (GBIF) (https://www.gbif.org, accessed on 1 May 2023) and the Chinese National Plant Specimen Resource Center (http://www.cvh.ac.cn/, accessed on 1 May 2023). Additionally, we augmented our dataset with field survey data collected using GPS terminals. The data collection period spanned from 1935 to 2023.

To ensure data accuracy, we performed a thorough screening of the occurrence data, eliminating duplicate entries and addressing missing latitude and longitude information. This meticulous data curation process yielded a comprehensive dataset consisting of 556 unique distribution points for *P. americana* across China. The resulting distribution map is depicted in Figure 1.

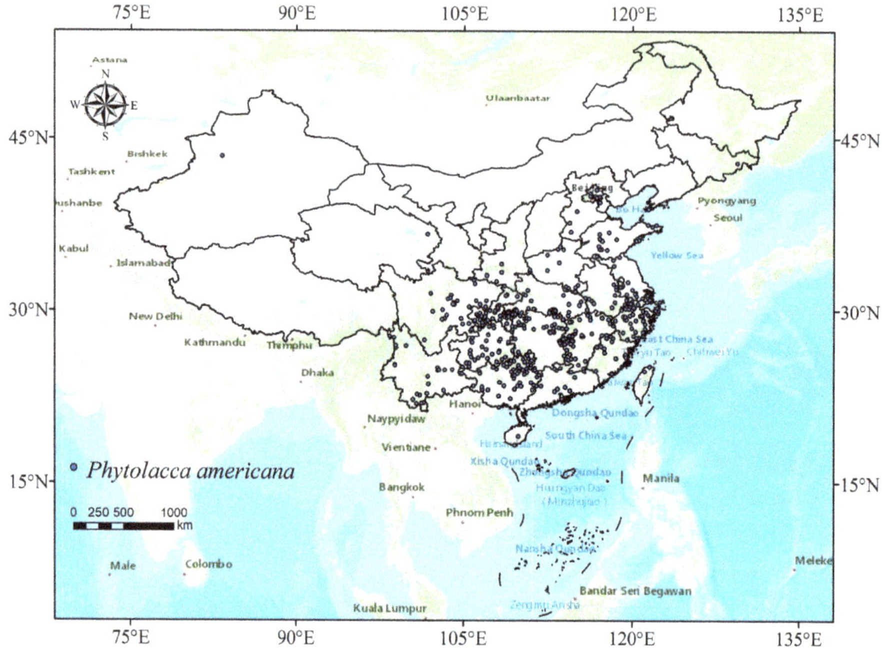

Figure 1. Current distribution of *Phytolacca americana* L. across China, based on data compiled up to May 2023.

2.2. Data Acquisition and Variables

We sourced historical and projected climate data from the World Climate Database (https://www.worldclim.org, accessed on 1 May 2023), with a spatial resolution of 2.5 arc minutes (approximately 5 km × 5 km). The selected periods for analysis encompassed current climate conditions (1970–2000), as well as two future projections, the 2050s (2041–2060) and the 2070s (2061–2080), as outlined in Table 1.

Table 1. Environmental variables considered for modeling the potentially suitable habitat of *Phytolacca americana* L.

Environmental Variables	Names
2.5m-bio1	Annual mean temperature
2.5m-bio2	Mean diurnal range
2.5m-bio3	Isothermality (2/7) (×100)
2.5m-bio4	Temperature seasonality (standard deviation × 100)
2.5m-bio5	Max temperature of warmest month
2.5m-bio6	Min temperature of coldest month
2.5m-bio7	Temperature annual range
2.5m-bio8	Mean temperature of wettest quarter
2.5m-bio9	Mean temperature of driest quarter
2.5m-bio10	Mean temperature of warmest quarter
2.5m-bio11	Mean temperature of coldest quarter
2.5m-bio12	Annual precipitation
2.5m-bio13	Precipitation of wettest month
2.5m-bio14	Precipitation of driest month
2.5m-bio15	Precipitation seasonality (variation coefficient)
2.5m-bio16	Precipitation of wettest quarter
2.5m-bio17	Precipitation of driest quarter

Table 1. *Cont.*

Environmental Variables	Names
2.5m-bio18	Precipitation of warmest quarter
2.5m-bio19	Precipitation of coldest quarter
Alt	Altitude

To model the future scenarios, we utilized data from the Beijing Climate Center (BCC) Climate System Model 2 Medium Resolution, within the framework of the Sixth International Coupled Model Intercomparison Program (CMIP6). This dataset included projections corresponding to three distinct Shared Socioeconomic Pathways (SSPs): SSP1_2.6 (sustainable development), SSP2_4.5 (moderate development), and SSP5_8.5 (normal development) [30].

2.3. Variable Selection

To mitigate the impact of spatial covariance among environmental variables on the accuracy of the Maxent model and enhance its predictive performance, we employed a systematic variable selection process guided by the following principles:

A Maxent pre-model was constructed using 20 environmental variables, allocating 25% of the data for testing and 75% for training. Default parameters were utilized, and the calculations were repeated ten times with the Bootstrap method. A total of 500 iterations and 10,000 background points were specified. We employed the jackknife test to assess the contribution of each environmental variable [30–33].

A Principal Component Analysis (PCA) was conducted, using SPSS 25.0 software, on the initial set of 20 environmental variables. Principal components with eigenvalues exceeding 1 were retained. From each principal component, environmental variables with loading coefficients exceeding 0.5 were selected for further analysis. Spearman correlation tables were generated [34].

Exclusion Criteria: We excluded environmental variables with a high correlation ($|r| \geq 0.8$), low percentage contribution, and limited biological significance [35]. When two environmental variables exhibited a correlation coefficient exceeding 0.8, the less influential variable was eliminated using the jackknife test in Maxent. Only variables with greater contributions to the model were retained [36,37]. Ultimately, we identified nine environmental variables for the Maxent modeling process: bio2, bio3, bio6, bio7, bio8, bio10, bio12, bio13, and bio14.

2.4. Parameter Selection and Model Accuracy Evaluation

We employed Maxent 3.4.1, following the methodology proposed by Phillips et al. (2017) [38], to construct the maximum entropy model for our study. To ensure the probability of *P. americana* distribution approached normality, we allocated 75% of the data for model training, reserving the remaining 25% for accuracy validation [39]. The primary parameters were configured as follows: Maximum Iterations: 500, Replicated Run Type: Bootstrap, No. of Replicates: 10, and Output Format: logistic.

We incorporated *P. americana* distribution data and nine environmental variables into the Maxent model. The "Do jackknife" function within the software was employed to assess the contribution of each environmental factor to the model construction [40,41]. Response curves were generated to visualize the relationship between distribution probabilities and climate factors (see Figure S1). The model accuracy was evaluated using the receiver operating characteristic (ROC) curve [17]. We conducted 10 model replicates, and the average result was considered the prediction of potential species distribution [42].

The model performance was assessed using the ROC area under the curve (AUC) [43]. The AUC is independent of the model threshold settings and serves as a reliable measure of accuracy [43]. AUC values range from 0 to 1, with interpretations as follows: an AUC < 0.7 indicates a poor predictive performance, from 0.7 to 0.8 suggests moderate performance, from 0.8 to 0.9 signifies good performance, and from 0.9 to 1.0 indicates excellent performance [44]). Upon running the model with the specified parameter settings in ten replicates, the average AUC from these iterations was calculated and found to be 0.907 (see Figure S2). This high AUC value indicates an excellent predictive performance, underscoring the accuracy and credibility of our prediction results.

2.5. Classification of Suitable Areas

We took the output ASC file from Maxent and imported it into ArcGIS, converting it into grid data. It was then overlaid onto a map of China's administrative districts for visualization. Using ArcGIS, we employed the reclassification function to classify different areas and utilized grid calculation tools to determine their corresponding areas. For this study, we employed the Natural Breaks (Jenks) classification method to categorize suitable areas into four grades: highly suitable areas ($p \geq 0.48$), moderately suitable areas ($0.30 \leq p < 0.48$), low-suitability areas ($0.10 \leq p < 0.30$), and unsuitable areas ($p < 0.10$).

3. Results

3.1. Analysis of the Contribution of Environmental Variables

Following model optimization, the final model included nine bioclimatic factors:

Mean diurnal range (bio2), isothermality (bio3), minimum temperature of coldest month (bio6), temperature annual range (bio7), mean temperature of wettest quarter (bio8), mean temperature of warmest quarter (bio10), annual precipitation (bio12), precipitation of wettest month (bio13), precipitation of driest month (bio14). Their respective contributions to the model construction are as follows: bio14: 68.7%, bio6: 13.6%, bio13: 6.7%, bio3: 3.4%, bio7: 2.3%, bio12: 1.7%, bio2: 1.6%, bio8: 1.1%, and bio10: 0.9% (Table 2).

Table 2. Contribution rate of environmental variables.

Variable	Bioclimatic Factors	Percent Contribution
bio14	Precipitation of driest month	68.7
bio6	Min temperature of coldest month	13.6
bio13	Precipitation of wettest month	6.7
bio3	Isothermality (2/7) (×100)	3.4
bio7	Temperature annual range	2.3
bio12	Annual precipitation	1.7
bio2	Mean diurnal range	1.6
bio8	Mean temperature of wettest quarter	1.1
bio10	Mean temperature of warmest quarter	0.9

3.2. Current Potentially Suitable Areas of *P. americana* in China

Figure 2 illustrates the division of *P. americana* distribution based on the Maxent model predictions, visualizing its potential suitable areas. Under current climate conditions, potentially suitable areas for *P. americana* are dispersed across 27 provinces and municipalities, including the Hong Kong and Macao Special Administrative Regions. These areas are predominantly concentrated in the Yangtze River basin and its southern vicinity.

Highly suitable areas are primarily situated in southeastern Sichuan, eastern Guizhou, southern Hubei, southern Anhui, southern Jiangsu, northern Guangdong, the northern Guangxi Zhuang Autonomous Region, northwestern and southeastern Fujian, Zhejiang, Shanghai, Jiangxi, Hunan, and Chongqing, with a sporadic presence in eastern Shandong and northern Taiwan. These highly suitable areas cover a total area of 98.68×10^4 km^2, constituting 10.26% of China's total area.

Moderately suitable areas border the highly suitable regions, encompassing a total area of 69.10×10^4 km^2, accounting for 24.66% of the total suitable area.

Low-suitability areas predominantly include northeastern Sichuan, western Guizhou, central Hubei, central Anhui, northern Jiangsu, central Guangdong, the central Guangxi

Zhuang Autonomous Region, most of Fujian, southern Henan, and southern Shaanxi, covering a total area of 112.48×10^4 km^2, which is 40.13% of the total suitable area.

The combined area of low-suitability and unsuitable areas reaches 793.99×10^4 km^2, representing 82.55% of China's total area.

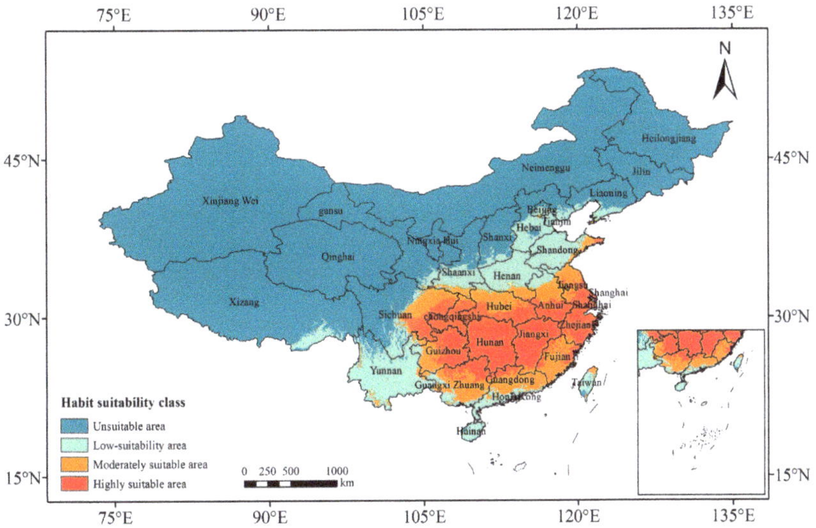

Figure 2. Predictions of the potentially suitable area of *Phytolacca americana* L. under current climate conditions based on the Maxent model. The color-coded areas denote different levels of suitability: red, highly suitable areas; orange, moderately suitable areas; light blue, low-suitability areas; and blue, unsuitable areas.

3.3. Future Suitable Areas of *P. americana* under Different Climate Scenarios

We employed the Maxent model, following the same criteria as mentioned earlier, to predict the potential suitability areas for *P. americana* in the 2050s and 2070s under three climate scenarios: SSP126, SSP245, and SSP585. The results included a spatial distribution map of future predicted potential suitability areas for *P. americana* (Figure 3), a dynamic change map of each suitability class (Figure 4), and changes in area for each suitability category (Table 3).

Across the three climate scenarios in the 2050s and 2070s, the location and area of each suitability class for *P. americana* exhibited varying degrees of change (Table 3). Notably, there was a significant increase in moderately suitable areas and an overall decrease in high- and low-suitability areas. There was a modest increase in highly suitable areas in the 2050s under the SSP245 and SSP585 scenarios and in low-suitability areas in the 2070s under the SSP126 scenario. The percentage change in the total area of suitable areas increased over time, with varying trends under different climate scenarios.

Looking specifically at moderately suitable areas, they increased in size over time under all three climate scenarios, with reduced fragmentation (Figures 3 and 4). In the 2050s, the SSP126 scenario exhibited the largest area of moderately suitable regions (82.6×10^4 km^2), marking a 19.53% increase compared to current climatic conditions, representing the most significant change (Table 3). The expansion of moderately suitable areas was primarily attributed to transformations from highly suitable areas at multiple junctions and from low-suitability areas in various regions. In the 2070s, the SSP126 scenario continued to lead, with the largest moderately suitable area (76.09×10^4 km^2), reflecting a 10.10% increase (Table 3).

In contrast, the total area of highly suitable and low-suitability areas declined, primarily converting into moderately suitable areas. The most substantial decrease in highly suitable areas occurred in the 2050s under the SSP126 scenario, with a 15.86% reduction (Table 3). The most significant decline in low-suitability areas was observed in the 2050s under the SSP245 scenario, with a 9.14% decrease (Table 3). Many of the areas transitioning from low-suitability to moderately suitable became unsuitable, notably in northern Yunnan, southern Liaoning, and the junction of Shaanxi, Shanxi, and Henan (Figure 4).

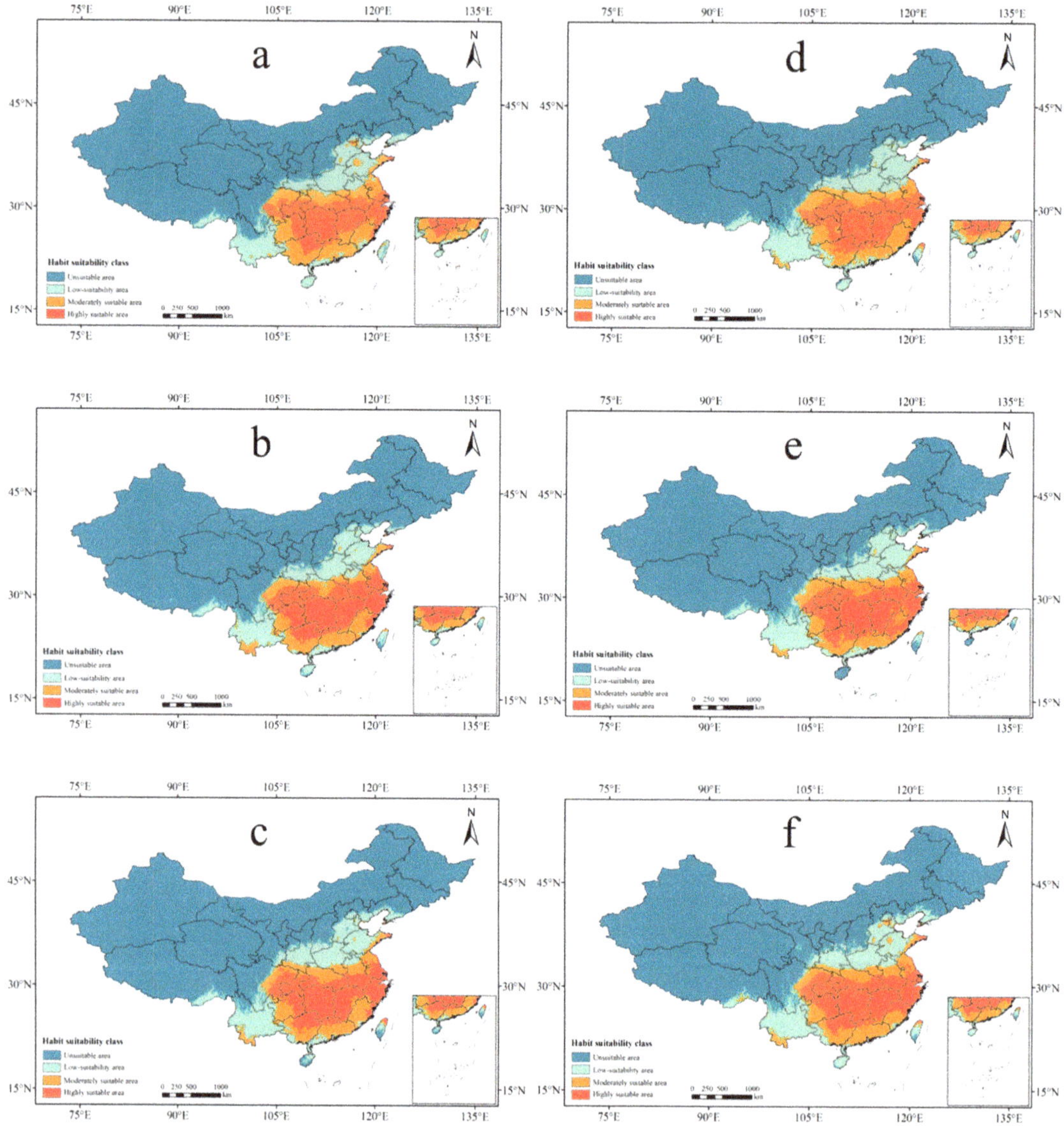

Figure 3. Dynamic change map of the predicted potentially suitable area of *Phytolacca americana* L. based on different climate scenarios in the future. ((**a**) 2050s-SSP126; (**b**) 2050s-SSP245; (**c**) 2050s-SSP585; (**d**) 2070s-SSP126; (**e**) 2070s-SSP245; (**f**) 2070s-SSP585).

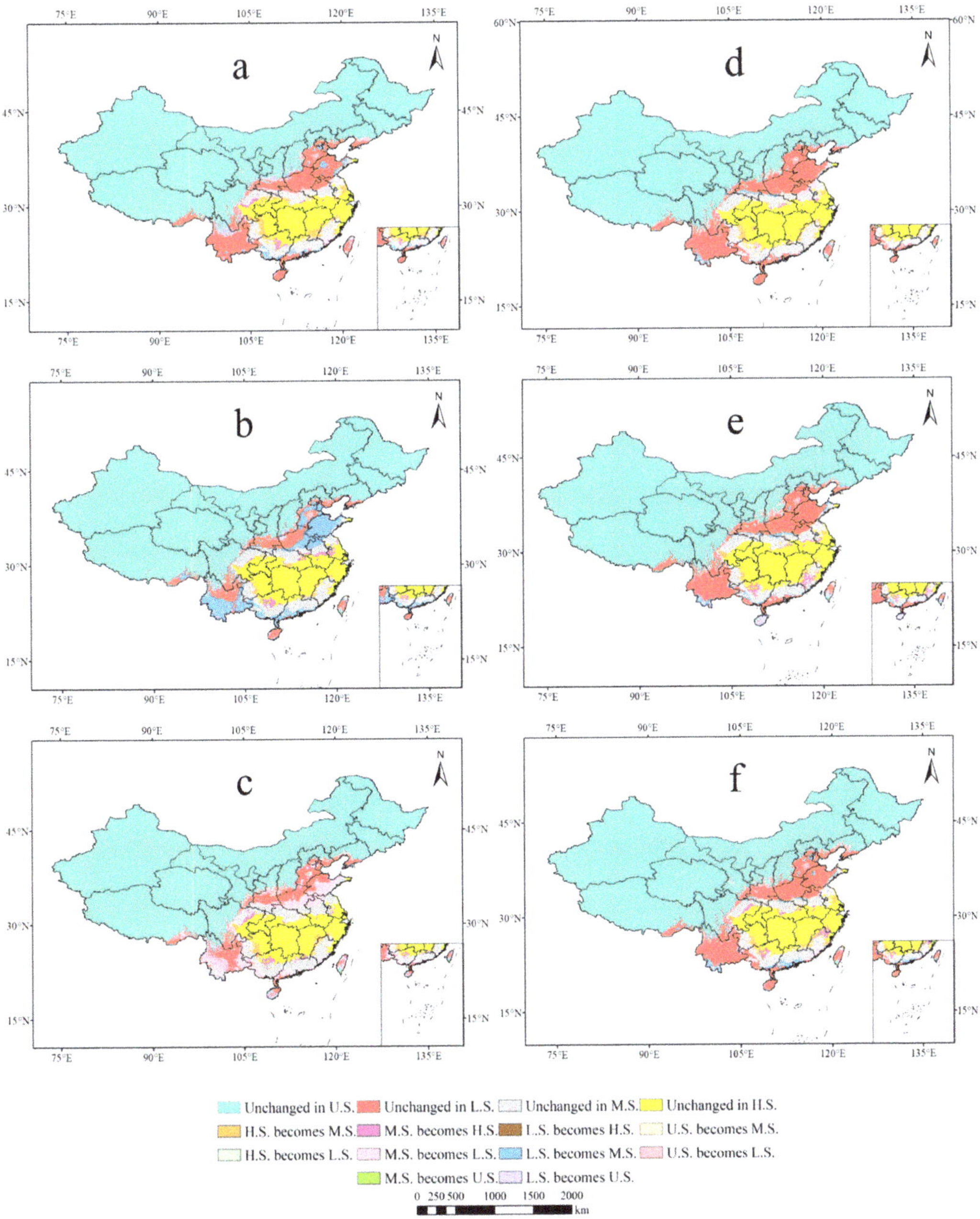

Figure 4. Predictions of the potentially suitable area of *Phytolacca americana* L. under different future climate scenarios based on the Maxent model. ((**a**) 2050s-SSP126; (**b**) 2050s-SSP245; (**c**) 2050s-SSP585; (**d**) 2070s-SSP126; (**e**) 2070s-SSP245; (**f**) 2070s-SSP585; U.S.: unsuitable; L.S.: low-suitability; M.S.: moderately suitable; H.S.: highly suitable).

Table 3. Potential suitable area of *Phytolacca americana* L. under different climate change scenarios.

Scenario	Period	Highly Suitable Area	Change	Moderately Suitable Area	Change	Low-Suitability Area	Change	Total Area	Total Change
		Area ($\times 10^4$ km^2)		Area ($\times 10^4$ km^2)		Area ($\times 10^4$ km^2)			
Current		98.68		69.10		112.48		280.26	
SSP126	2050s	83.02	−15.86%	82.60	19.53%	104.94	−6.70%	270.56	−3.46%
	2070s	87.51	−11.32%	76.09	10.10%	115.99	3.12%	279.59	−0.24%
SSP245	2050s	101.00	2.35%	70.19	1.58%	102.20	−9.14%	273.40	−2.45%
	2070s	97.46	−1.24%	72.50	4.91%	107.49	−4.43%	277.45	−1.00%
SSP585	2050s	94.42	−4.32%	79.07	14.43%	111.99	−0.44%	285.48	1.86%
	2070s	102.50	3.88%	74.93	8.42%	109.01	−3.08%	286.44	2.20%

3.4. Centroid Distributional Shifts under Future Climate Conditions

Figure 5 illustrates the calculation of the centroid location, direction, and distance using ArcGIS under various climate scenarios and time periods.

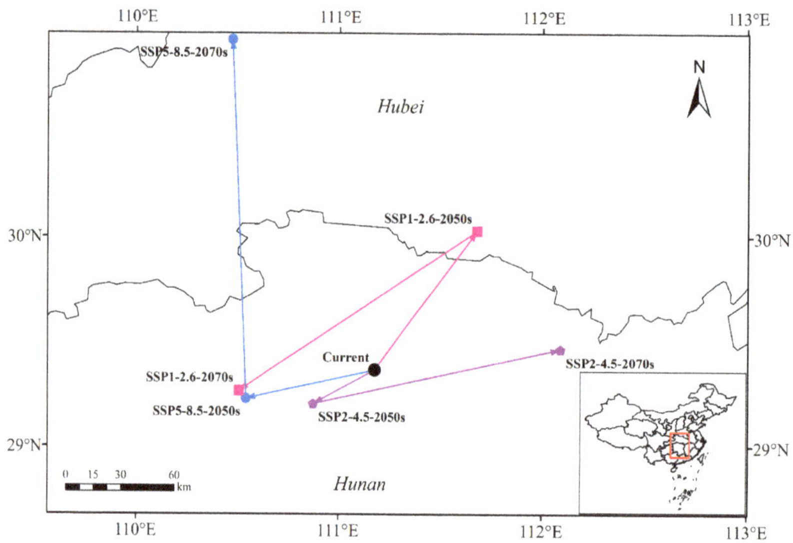

Figure 5. Total suitable habitat centroid distribution shifts under climate change for *Phytolacca americana* L. (The red square represents the location in China of the main migration areas of the centroids.).

Under current climate conditions, the centroid of suitable habitat was at coordinates 111°10′36″ E, 29°21′51″ N.

For the SSP126 climate scenario:

In the 2050s, the centroid was projected to shift 88.54 km to the northeast (111°40′47″ E, 30°1′42″ N). In the 2070s, it was expected to move 142.04 km to the southwest (111°30′32″ E, 29°15′51″ N). Overall, it would have shifted 66.09 km southwest.

For the SSP245 climate scenario:

In the 2050s, the centroid was projected to move 34.33 km southwest (110°52′33″ E, 29°12′15″ N). In the 2070s, it was expected to shift 121.75 km northeast (112°5′22″ E, 29°27′54″ N). Overall, it would have moved 89.54 km to the northeast.

For the SSP585 climate scenario:

In the 2050s, the centroid was projected to move 63.59 km southwest (110°32′37″ E, 29°13′42″ N). In the 2070s, it was expected to shift 190.61 km north (110°28′19″ E, 30°56′21″ N). Overall, it would have moved 188.08 km to the northwest.

4. Discussion

4.1. Main Environmental Variables Affecting the Occurrence of *P. americana*

The Maxent model, considering percentage contribution, ranked importance, and jackknife tests, identifies the primary factors influencing the distribution of *P. americana*. Precipitation of the driest month, minimum temperature of the coldest month, precipitation of the wettest month, isothermality, and temperature annual range emerge as the dominant variables shaping its current distribution.

These findings highlight the pivotal roles of temperature and precipitation in determining the species' distribution. Research has shown that *P. americana* thrives in warm and humid climates [45]. Temperature affects plant photosynthesis, while precipitation plays a critical role in regulating carbon sequestration and transpiration by influencing soil moisture. Together, these factors impact various physiological processes throughout the plant's life cycle, from seed germination to growth and development [46]. Hydrothermal conditions are key variables that restrict the geographical distribution of plants. Terrestrial plants exhibit high sensitivity to temperature and precipitation [47,48]. Consequently, areas with warm and humid climatic conditions are more conducive to the growth and proliferation of *P. americana*. These environmental insights provide valuable information for understanding the current distribution patterns of *P. americana* and can be instrumental in managing and controlling its spread.

The present study reveals that *P. americana* primarily inhabits coastal regions rich in hydrothermal resources and the subtropical monsoon climate zone south of the Yangtze River in China. Several factors contribute to this distribution pattern.

Climatic Suitability: The species thrives in areas characterized by humid and rainy summers, seasonal high temperatures, and ample sunshine—conditions conducive to its growth and reproduction [49,50]. These climatic characteristics align with the observed distribution.

Human Activities: Human interventions, including agricultural production, trade, and transportation, often serve as significant drivers in the spread of invasive species like *P. americana* [51,52]. Frequent economic and trade exchanges in the distribution areas may contribute to its expansion.

Unsuitable Regions: Conversely, regions such as the Tibetan Plateau, inland Northwest China, and much of Northeast China are unsuitable for *P. americana*. These areas feature temperate continental and temperate monsoon climates, characterized by low annual rainfall and cold, dry winters. As such, they pose a low risk of invasion or are entirely unsuitable for the species' growth.

In summary, the current distribution of *P. americana* can be attributed to a combination of favorable climatic conditions, human activities, and the unsuitability of certain regions. These insights are valuable for understanding the species' distribution patterns and can inform management and control efforts.

4.2. Future Changes in *P. americana* Distribution

The study indicates that the potential suitable areas for *P. americana* will undergo changes over time and under various climate scenarios. While specific patterns may vary, a consistent trend emerges: no significant alteration in the total suitable area, expansion of moderately suitable areas, and reductions in highly suitable and low-suitability areas.

Climate-Induced Shifts in Centroid Location:

Depending on future climatic conditions, the centroid of *P. americana* suitability exhibits directional movement. Under the SSP245 and SSP585 scenarios, it generally migrates to higher latitudes over time. SSP585 shows more pronounced shifts, especially by the 2070s.

Conversely, under the SSP126 scenario, the centroid moves northeastward in the 2050s but significantly shifts southwestward by the 2070s.

Implications of Climate Change:

Centroid movements in the present study (Figure 5) align with previous findings that climate change can drive shifts in species distributions. Examples include *P. massoniana* extending northward [21], *Vaccinium membranaceum* Douglas ex Torr. (Ericaceae) transitioning to high-elevation areas [53], and *Z. spinosa*'s suitable area gradually moving northward [20].

Factors Influencing Shifts:

The SSP126 scenario, projecting reduced carbon dioxide emissions and temperature stabilization, leads to a relatively slow rate of temperature increase. Consequently, *P. americana* suitability does not exhibit a significant northward shift, reflecting the influence of sustainable development and a slower temperature rise.

Hydrothermal Considerations:

The subtropical monsoon climate of high-suitability areas plays a crucial role. Changes in precipitation patterns under the SSP126 scenario significantly impact suitability. Precipitation of the driest month (bio14), a critical contributing factor, tends to fluctuate, influencing the centroid's movement (Figure 6).

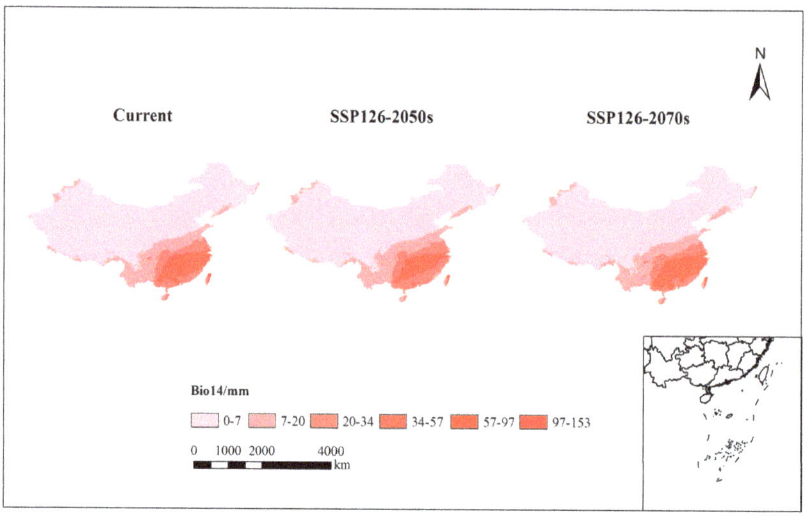

Figure 6. Changes in the highest-contributing climate variable, precipitation of the driest month (bio14), under the SSP126 future climate scenario versus the current climate scenario for *Phytolacca americana* L.

In summary, the study highlights the potential impact of climate scenarios on *P. americana* distribution. The findings underscore the importance of climate-induced shifts in species habitats and the role of precipitation patterns in shaping these changes.

4.3. Prevention and Control Strategies

China's location in the East Asian monsoon zone, with its proximity to the Pacific Ocean, exposes it to significant climate change impacts. Studies indicate that rising temperatures, strengthened summer winds, increased rainfall, and a warmer, wetter climate are expected in coastal areas due to global climate change. These changes align with *P. americana*'s preference for warm and humid conditions, potentially expanding its suitable habitat in the future [54–56].

The modeled potential distribution maps in this study serve as crucial references for effective prevention and control strategies for *P. americana*. Three key recommendations are proposed:

Strict Regulation of Highly Suitable Areas: Regions such as Zhejiang, Shanghai, Jiangxi, Hunan, and Chongqing are identified as highly suitable for *P. americana* under the current climate scenario. Consequently, strict regulations should be enforced to control the introduction and planting of this invasive species in these areas.

Enhanced Monitoring for New Suitable Habitats: Proactive monitoring of potential new suitable habitats should be intensified to prevent *P. americana* invasion driven by human activities. Identifying and addressing emerging suitable areas can help curb its spread.

Protection of Agricultural Areas: Agricultural production zones, including croplands and orchards, require robust prevention and control measures against *P. americana* intrusion to safeguard native species. Early intervention methods, such as chemical and biological controls, as well as the timely removal of fruit-bearing plants and aerial parts after fruiting, can effectively mitigate its impact.

These recommendations underscore the importance of proactive and adaptive management strategies to address the potential expansion of *P. americana* in response to climate change, safeguarding native ecosystems and agricultural productivity.

4.4. Limitations of the Study

While our study provides valuable insights into the potential distribution of *P. americana*, several limitations should be acknowledged:

Limited Environmental Factors: Our analysis focused solely on bioclimatic variables as predictors of species distribution. Species distributions are influenced by a myriad of biotic factors (e.g., interspecific competition, predation, and disease) and abiotic factors (e.g., soil, topography, and human activities). Future studies should consider a broader range of factors, such as soil type, land use, and anthropogenic impacts, to provide a more comprehensive understanding of *P. americana*'s distribution dynamics.

Incomplete Response Curve Data: Although we provided response curves for dominant environmental factors, some secondary factors, such as elevation, annual mean temperature, and temperature seasonality, were not included due to data limitations. Incorporating these factors into future research could enhance the precision of our models.

Limited Geographic Scope: Our dataset only included occurrences within China; it does not fully capture the global distribution of *P. americana*. The response curves derived from this dataset may not entirely represent the species' environmental responses across its entire range. Future studies should aim to gather data from a broader geographic range to improve model accuracy.

Large-Scale Analysis: Our study primarily operates at a large geographic scale. Future research could benefit from conducting more fine-scale analyses in specific geographic regions. Excluding unsurveyed and inaccessible areas in these smaller-scale studies can enhance precision and accuracy.

Model Validation: While the Maxent model is widely accepted for species distribution modeling, employing alternative models, such as BioCLIM and DOMAIN, in future research could provide additional validation and strengthen our findings.

Acknowledging these limitations helps refine the scope of future research and ensures a more comprehensive understanding of *P. americana*'s distribution patterns and ecological interactions.

5. Conclusions

In this study, we employed the Maxent model to predict the potential suitable areas for *P. americana* in China. Examining future climate scenarios, our projections indicate that the total area of potential suitable habitat for *P. americana* is unlikely to undergo significant changes. However, the centroid of suitable areas is expected to shift northeastward in the 2050s under the SSP126 climate scenario, followed by a substantial southwestward movement in the 2070s. Under the SSP245 and SSP585 climate scenarios, the centroid generally migrates toward higher latitudes over time.

Considering these findings, we propose rigorous monitoring of the currently highly suitable areas for *P. americana*, especially agricultural production areas such as croplands and gardens. Meanwhile, we recommend intensified surveillance of potential new suitable habitats, based on projected future distribution patterns of the species. Additionally, to address the limitations of this study, future research can further improve the results of this study by considering a broader range of environmental factors (e.g., soil type, land use, and anthropogenic impacts), a wider range of geographic factors (e.g., global scale), and more model validation (e.g., BioCLIM, DOMAIN, etc.). In conclusion, the outcomes of this study offer crucial insights for future monitoring, early warning systems, and management strategies concerning *P. americana*.

Supplementary Materials: The following supporting information can be downloaded at: https://www.mdpi.com/article/10.3390/plants13081082/s1, Figure S1: Response curve of environmental factors. Figure S2: Reliability test of the distribution model created for *Phytolacca americana* L.

Author Contributions: Conceptualization, Q.N. and L.Z.; Methodology, Q.N.; Validation, Q.N.; Formal analysis, Q.N.; Investigation, Q.N., C.L., D.Z. and L.Z.; Resources, C.L. and Z.L.; Data curation, Q.N. and D.Z.; Writing—original draft, Q.N.; Writing—review & editing, X.L. and L.Z.; Visualization, Q.N.; Supervision, C.L. and L.Z.; Project administration, C.L., Z.L. and L.Z.; Funding acquisition, X.L. and L.Z. All authors have read and agreed to the published version of the manuscript.

Funding: This research was funded by National Natural Science Foundation of China, grant number 31801984.

Data Availability Statement: Data are contained within the article and Supplementary Materials.

Conflicts of Interest: The authors declare no conflict of interest.

References

1. Ostad-Ali-Askar, K.; Su, R.; Liu, L. Water resources and climate change. *J. Water Clim. Chang.* **2018**, *9*, 239. [CrossRef]
2. Huang, R.; Yu, T.; Zhao, H.; Zhang, S.; Jing, Y.; Li, Y. Prediction of suitable distribution area of the endangered plant *Acer catalpifolium* under the background of climate change in China. *J. Beijing For. Univ.* **2021**, *43*, 33. [CrossRef]
3. Gao, H.; Qian, Q.; Deng, X.; Peng, Y.; Xu, D. Predicting the Distributions of *Morus notabilis* C. K. Schneid under Climate Change in China. *Forests* **2024**, *15*, 352. [CrossRef]
4. Thakur, S.; Rai, I.D.; Singh, B.; Dutt, H.C.; Musarella, C.M. Predicting the suitable habitats of *Elwendia persica* (Boiss.) in the Indian Himalayan Region (IHR). *Plant Biosyst.* **2023**, *157*, 769–778. [CrossRef]
5. Ali, H.; Jaffar, U.D.; Luciano, B.; Shoaib, H.; Muhammad, K.; Muhammad, Y.; Muhammad, A.N. Expanding or shrinking? range shifts in wild ungulates under climate change in Pamir-Karakoram mountains, Pakistan. *PLoS ONE* **2022**, *16*, e0260031. [CrossRef] [PubMed]
6. Wei, B.; Wang, R.; Hou, K.; Wang, X.; Wu, W. Predicting the current and future cultivation regions of *Carthamus tinctorius* L. using MaxEnt model under climate change in China. *Glob. Ecol. Conserv.* **2018**, *16*, e00477. [CrossRef]
7. Ye, X.; Zhao, G.; Zhang, M.; Cui, X.; Fan, H.; Liu, B. Distribution Pattern of Endangered Plant *Semiliquidambar cathayensis* (Hamamelidaceae) in Response to Climate Change after the Last Interglacial Period. *Forests* **2020**, *11*, 434. [CrossRef]
8. Araújo, M.B.; Antoine, G. Five (or so) challenges for species distribution modelling. *J. Biogeogr.* **2006**, *33*, 1677–1688. [CrossRef]
9. Brown, J.L.; Bennett, J.R.; French, C.M. SDMtoolbox 2.0: The next generation Python-based GIS toolkit for landscape genetic, biogeographic and species distribution model analyses. *PeerJ* **2017**, *5*, e4095. [CrossRef]
10. Sun, S.; Zhang, Y.; Huang, D.; Wang, H.; Cao, Q.; Fan, P.; Yang, N.; Zheng, P.; Wang, R. The effect of climate change on the richness distribution pattern of oaks (*Quercus* L.) in China. *Sci. Total Environ.* **2020**, *744*, 140786. [CrossRef]
11. Ab Lah, N.Z.; Yusop, Z.; Hashim, M.; Mohd Salim, J.; Numata, S. Predicting the Habitat Suitability of *Melaleuca cajuputi* Based on the MaxEnt Species Distribution Model. *Forests* **2021**, *12*, 1449. [CrossRef]
12. Ancillotto, L.; Mori, E.; Bosso, L.; Agnelli, L.; Russo, D. The Balkan long-eared bat (*Plecotus kolombatovici*) occurs in Italy—First confirmed record and potential distribution. *Mamm. Biol.* **2019**, *96*, 61–67. [CrossRef]
13. Conolly, J.; Manning, K.; Colledge, S.; Dobney, K.; Shennan, S. Species distribution modelling of ancient cattle from early Neolithic sites in SW Asia and Europe. *Holocene* **2012**, *22*, 997–1010. [CrossRef]
14. Phillips, S.J.; Robert, P.A.; Robert, E.S. Maximum entropy modeling of species geographic distributions. *Ecol. Model.* **2006**, *190*, 231–259. [CrossRef]
15. Ye, X.; Zhang, M.; Lai, W.; Yang, M.; Fan, H.; Zhang, G.; Chen, S.; Liu, B. Prediction of potential suitable distribution of *Phoebe bournei* based on MaxEnt optimization model. *Acta Ecol. Sin.* **2021**, *41*, 8135–8144. [CrossRef]

16. Hu, W.; Wang, Y.; Dong, P.; Zhang, D.; Yu, W.; Ma, Z.; Chen, G.; Liu, Z.; Du, J.; Chen, B.; et al. Predicting potential mangrove distributions at the global northern distribution margin using an ecological niche model: Determining conservation and reforestation involvement. *For. Ecol. Manag.* **2020**, *478*, 118517. [CrossRef]
17. Moreno, R.; Ricardo, Z.; Molina, J.R.; Vasquez, A.; Herrera, M. Predictive modeling of microhabitats for endemic birds in South Chilean temperate forests using Maximum entropy (Maxent). *Ecol. Inform.* **2011**, *6*, 364–370. [CrossRef]
18. Hamit, S.; Abdushalih, N.; Xu, Z.; Jiesisi, A.; Shao, H.; Yilihar, V. Simulation of potential distribution patterns of the invasive plant species *Xanthium spinosum* L. (Bathurst burr) in Xinjiang under climate change. *Acta Ecol. Sin.* **2019**, *39*, 1551–1559. [CrossRef]
19. Lemke, D.; Hulme, P.E.; Brown, J.A.; Tadesse, W. Distribution modelling of Japanese honeysuckle (*Lonicera japonica*) invasion in the Cumberland Plateau and Mountain Region, USA. *For. Ecol. Manag.* **2011**, *262*, 139–149. [CrossRef]
20. Qian, Z.; Mi, Z.; Lu, C.; Zhang, X.; Chen, L.; Wang, S.; Niu, J.; Wang, Z. Predicting potential distribution of *Ziziphus spinosa* (Bunge) H.H. Hu ex F.H. Chen in China under climate change scenarios. *Ecol. Evol.* **2022**, *12*, e8629. [CrossRef]
21. Chi, Y.; Wang, G.; Zhu, M.; Jin, P.; Hu, Y.; Shu, P.; Wang, Z.; Fan, A.; Qian, P.; Han, Y.; et al. Potentially suitable habitat prediction of *Pinus massoniana* Lamb. in China under climate change using Maxent model. *Front. For. Glob. Chang.* **2023**, *6*, 1144401. [CrossRef]
22. Shi, X.; Wang, J.; Zhang, L.; Chen, S.; Zhao, A.; Ning, X.; Fan, G.; Wu, N.; Zhang, L.; Wang, Z. Prediction of the potentially suitable areas of *Litsea cubeba* in China based on future climate change using the optimized MaxEnt model. *Ecol. Indic.* **2023**, *148*, 110093. [CrossRef]
23. Zhou, B.; Yan, X.; Xiao, Y.; Zhang, Z.; Li, X.; Yang, J. Traits of reproductive biology associated with invasiveness in alien invasive plant *Phytolacca americana*. *Ecol. Environ. Sci.* **2013**, *22*, 567–574.
24. Follak, S.; Schwarz, M.; Essl, F. Notes on the occurrence of *Phytolacca americana* L. in crop fields and its potential agricultural impact. *BioInvasions Rec.* **2022**, *11*, 620–630. [CrossRef]
25. Woolum, J.A.; Akpunonu, P.; Johnson, M.; Webb, A.N. Human exposures to *Phytolacca americana* in Kentucky: Characterization of toxicities, treatments, and outcomes. *Clin. Toxicol.* **2022**, *220*, 106962. [CrossRef]
26. Ministry of Ecology and Environment of the People's Republic of China, Ministry of Environmental Protection. Announcement on the Publication of the List of Alien Invasive Species in Natural Ecosystems of China (Fourth Batch). 2016. Available online: https://www.mee.gov.cn/gkml/hbb/bgg/201612/W020161226561612030090.pdf (accessed on 1 May 2023).
27. Strgulc Krajšek, S.; Kladnik, A.; Skočir, S.; Bačič, M. Seed Germination of Invasive *Phytolacca americana* and Potentially Invasive *P. acinosa*. *Plants* **2023**, *12*, 1052. [CrossRef] [PubMed]
28. Xu, Y.; Ye, X.; Yang, Q.; Weng, H.; Liu, Y.; Ahmad, S.; Zhang, G.; Huang, Q.; Zhang, T.; Liu, B. Ecological niche shifts affect the potential invasive risk of *Phytolacca americana* (Phytolaccaceae) in China. *Ecol. Process.* **2023**, *12*, 1. [CrossRef]
29. Marinas, I.; Baicea, C.; Oprea, E.; Geana, I.; Gird, C.E.; Chifiriuc, M. Chemical composition, antimicrobial and antioxidant activity of *Phytolacca americana* L. fruits and leaves extracts. *Farmacia* **2021**, *69*, 883–889. [CrossRef]
30. Riahi, K.; Vuuren, D.V.; Kriegler, E.; Edmonds, J.; O'Neill, B.C.; Fujimori, S.; Bauer, N.; Calvin, K.V.; Dellink, R.; Fricko, O.; et al. The Shared Socioeconomic Pathways and their energy, land use, and greenhouse gas emissions implications: An overview. *Glob. Environ. Chang.* **2017**, *42*, 153–168. [CrossRef]
31. Zeng, Y.; Low, B.W.; Yeo, D.C. Novel methods to select environmental variables in MaxEnt: A case study using invasive crayfish. *Ecol. Model.* **2016**, *341*, 5–13. [CrossRef]
32. Václavík, T.; Meentemeyer, R.K. Equilibrium or not? Modelling potential distribution of invasive species in different stages of invasion. *Divers. Distrib.* **2012**, *18*, 73–83. [CrossRef]
33. Warren, B.H.; Bakker, F.T.; Bellstedt, D.U.; Bytebier, B.; Claßen-Bockhoff, R.; Dreyer, L.L.; Edwards, D.; Forest, F.; Galley, C.; Hardy, C.R.; et al. Consistent phenological shifts in the making of a biodiversity hotspot: The Cape flora. *BMC Evol. Biol.* **2011**, *11*, 39. [CrossRef]
34. Graham, M.H. Confronting multicollinearity in ecological multiple regression. *Ecology* **2003**, *84*, 2809–2815. [CrossRef]
35. Che, L.; Cao, B.; Bai, C.; Wang, J.; Zhang, L. Predictive distribution and habitat suitability assessment of *Notholirion bulbuliferum* based on MaxEnt and arcGIS. *Chin. J. Ecol.* **2014**, *33*, 1623–1628.
36. Yang, X.; Kushwaha, S.P.; Saran, S.; Xu, J.; Roy, P.S. Maxent modeling for predicting the potential distribution of medicinal plant, *Justicia adhatoda* L. in Lesser Himalayan foothills. *Ecol. Eng.* **2013**, *51*, 83–87. [CrossRef]
37. Wisz, M.S.; Pottier, J.; Kissling, W.D.; Pellissier, L.; Lenoir, J.; Damgaard, C.F.; Dormann, C.F.; Forchhammer, M.C.; Grytnes, J.-A.; Guisan, A.; et al. The role of biotic interactions in shaping distributions and realised assemblages of species: Implications for species distribution modelling. *Biol. Rev.* **2013**, *88*, 15–30. [CrossRef]
38. Phillips, S.J.; Anderson, R.P.; Dudík, M.; Schapire, R.E.; Blair, M.E. Opening the black box: An open-source release of Maxent. *Ecography* **2017**, *40*, 887–893. [CrossRef]
39. Phillips, S.J.; Dudík, M. Modeling of species distributions with Maxent: New extensions and a comprehensive evaluation. *Ecography* **2008**, *31*, 161–175. [CrossRef]
40. Merow, C.; Smith, M.J.; Silander, J.A., Jr. A practical guide to MaxEnt for modeling species' distributions: What it does, and why inputs and settings matter. *Ecography* **2013**, *36*, 1058–1069. [CrossRef]
41. Peterson, A.T.; Papeş, M.; Eaton, M.D. Transferability and model evaluation in ecological niche modeling: A comparison of GARP and Maxent. *Ecography* **2007**, *30*, 550–560. [CrossRef]
42. Flory, A.R.; Kumar, S.; Stohlgren, T.J.; Cryan, P.M. Environmental conditions associated with bat white-nose syndrome mortality in the north-eastern United States. *J. Appl. Ecol.* **2012**, *49*, 680–689. [CrossRef]

43. Wan, G.; Wang, L.; Jin, L.; Chen, J. Evaluation of environmental factors affecting the quality of *Codonopsis pilosula* based on chromatographic fingerprint and MaxEnt model. *Ind. Crops Prod.* **2021**, *170*, 113783. [CrossRef]
44. Buckman-Sewald, J.; Whorton, C.R.; Root, K.V. Developing macrohabitat models for bats in parks using maxent and testing them with data collected by citizen scientists. *Int. J. Biodivers. Conserv.* **2014**, *6*, 171–183. [CrossRef]
45. Wang, S.; Jia, H.; Zhang, Z.; Hu, L.; Chen, G. Competitive Effects between Invasive Plant *Phytolacca americana* and three Forage Species. *Acta Agrestia Sin.* **2021**, *29*, 95–102. [CrossRef]
46. Punyasena, S.W.; Eshel, G.; McElwain, J.C. The influence of climate on the spatial patterning of Neotropical plant families. *J. Biogeogr.* **2007**, *35*, 117–130. [CrossRef]
47. Bradie, J.; Leung, B. A quantitative synthesis of the importance of variables used in MaxEnt species distribution models. *J. Biogeogr.* **2017**, *44*, 1344–1361. [CrossRef]
48. Li, H.; Zhang, C.; Wang, S.; Ma, W.; Liu, F.; Chen, Q.; Zhou, Q.; Xia, X.; Niu, B. Response of vegetation dynamics to hydrothermal conditions on the Qinghai-Tibet Plateau in the last 40 years. *Acta Ecol. Sin.* **2022**, *42*, 4770–4783. [CrossRef]
49. Damalas, C.A. Distribution, biology, and agricultural importance of *Galinsoga parviflora* (Asteraceae). *Weed Biol. Manag.* **2008**, *8*, 147–153. [CrossRef]
50. Lei, M.; Lu, Y.; Zhou, C.; Liu, Z.; Xu, Y.; Xia, J. High slope stability remote real-time monitoring system used in subtropical monsoon climate zone in western Guangdong. *Zhongshan Daxue Xuebao/Acta Sci. Natralium Univ. Sunyatseni* **2015**, *54*, 30–36. [CrossRef]
51. Chen, G.; Qiang, S. Human activities are the principle cause of biotic homogenization. *Acta Ecol. Sin.* **2011**, *31*, 4016–4107.
52. Vibrans, H. Epianthropochory in Mexican weed communities. *Am. J. Bot.* **1999**, *86*, 476–481. [CrossRef]
53. Prevéy, J.S.; Parker, L.E.; Harrington, C.A.; Lamb, C.T.; Proctor, M.F. Climate change shifts in habitat suitability and phenology of huckleberry (*Vaccinium membranaceum*). *Agric. For. Meteorol.* **2020**, *280*, 107803. [CrossRef]
54. Jueterbock, A.; Smolina, I.; Coyer, J.A.; Hoarau, G. The fate of the Arctic seaweed *Fucus distichus* under climate change: An ecological niche modeling approach. *Ecol. Evol.* **2016**, *6*, 1712–1724. [CrossRef]
55. Kuang, X.; Han, Y. Response of China Climate to Vegetation Change under Mid-Holocene Background Based on Dynamic Downscaling Simulation. *Adv. Earth Sci.* **2021**, *36*, 1301–1312. [CrossRef]
56. Voosen, P. Global temperatures in 2020 tied record highs. *Science* **2021**, *371*, 334–335. [CrossRef]

plants

MDPI

Article

Invasion Status, Mechanisms, and Future Distribution Prediction of *Solidago canadensis* in the Trade Port Region: A Case Study of Ningbo Port, China

Xu Luo [1,2,*], Sixiao Shen [1,2], Ke Liao [1], Saiqiang Li [1], Qinqin Pan [1], Jiahao Ma [1], Weiqiang Li [1] and Xiaodong Yang [1,2,*]

[1] Ningbo University Donghai Academy, Zhejiang Ocean Development Think Tank Alliance, Ningbo 315211, China; 2211420021@nbu.edu.cn (S.S.); 2111073012@nbu.edu.cn (K.L.); 2211420010@nbu.edu.cn (S.L.); 2211420052@nbu.edu.cn (Q.P.); 2311110180@nbu.edu.cn (J.M.); 2311110174@nbu.edu.cn (W.L.)

[2] Department of Geography & Spatial Information Techniques, Ningbo University, Ningbo 315211, China

* Correspondence: luoxu@nbu.edu.cn (X.L.); yangxiaodong@nbu.edu.cn (X.Y.); Tel.: +86-574-8760-9522 (X.Y.)

Abstract: Trade ports are the first places where alien species invade and the source of their spread to other areas. Controlling invasions in these regions can effectively reduce invasion pressure and disrupt the spread pathways of invasive species, thereby significantly reducing their threat to local ecosystems and biodiversity loss. Based on 595 field survey plots, the Generalized Linear Model (GLM) and Species Distribution Model (MaxEnt) were employed to analyze and predict the invasion mechanisms and future possible distribution of *Solidago canadensis* in the Ningbo Port, China. The results indicate that the invasion of *S. canadensis* in the Ningbo Port was particularly severe, with a 67.7% occurrence rate of all sampling plots in the field survey, and a risk level classified as Grade 1. Biodiversity ($p < 0.001$) and the minimum temperature of the coldest month ($p < 0.01$) significantly affect the invasiveness. Highly diverse communities could resist the invasion of alien species, which support Elton's diversity–invasibility hypothesis. Low temperatures had a restrictive effect on the invasion of *S. canadensis*. The total suitable area continued to expand under three different climate change scenarios compared to current conditions (increased by 3.73%, 5.67%, and 3.74% by the 2070s). The total potential habitat area of *S. canadensis* reached its maximum extent (89.77%) under the medium greenhouse gas emission scenario in the 2050s. Meanwhile, the medium suitable area exhibited the greatest fluctuation among the three climate scenarios. Under the low emission condition, the medium suitable area of *S. canadensis* diminished by 63.10 km^2, but in the medium and high emission condition, its area expanded by 91.13 km^2 and 16.20 km^2, respectively. Under future climate warming scenarios, the invasion risk of *S. canadensis* in Ningbo Port will continue to increase. The results of our study reveal the diffusion mechanisms of invasive plants at the colonization source, providing important theoretical support for invasive alien species' initial prevention and control.

Keywords: invasive alien species; distribution patterns; ecological risk; species distribution model; future climate scenarios

1. Introduction

Influenced by global trade and climate change, invasive alien species (IAS) have become a significant threat to global ecological security [1,2]. Taking China as an example, the number of IAS increased from 126 in 2005 to 403 in 2020 [3], causing economic losses of approximately USD 14.45 billion, which severely threatens ecological security, biodiversity, and food security [4]. As the core nodes of global trade, ports handle over 80% of the world's cargo volume and serve as major hubs for trade and maritime traffic [5]. The frequent ship transportation and human activities in the port region make these regions the first to be colonized by invasive species, and the richness of alien plants in these regions is obviously higher than that in near-natural ecosystems [6]. The research of 54 waterway ports in Central Europe over the past 40 years has shown that alien species account for 41% of all species and increase with the decrease in the distance from the ocean [7]. Moreover, studies shown a positive correlation between the first record of alien species and international import volume [8]. It was projected that, by 2050, the global shipping volume will further increase by 240% to 1209%, leading to significant changes in the risk of alien species invasion in port areas [9]. Thus, effectively controlling IAS in the port region plays a crucial role in preventing their wide spread and in formulating control measures.

The successful establishment of IAS is influenced by factors such as species characteristics, community diversity, climate, and soil nutrition [10–13]. Studies have demonstrated that variations in species traits play a crucial role in determining the success of invasions [12,14]. For example, *Conyza canadensis* successfully invaded eastern China due to its superior light acquisition ability compared to local species [15]. *Ailanthus altissima* successfully invaded all administrative regions in Italy through competition [16]. Moreover, some studies using comparative methods have found that many IAS, such as *Ageratina adenophora* and *Jacobaea vulgaris*, enhance the probability of their successful invasion by adjusting the nutrient distribution and functional traits of leaves [17–19]. The Elton's diversity–invasibility hypothesis posits that a higher species diversity in local communities reduces the likelihood of successful invasion by alien plants [20]. The successful establishment of IAS is also closely related to environmental conditions. For instance, Dawson et al. [13] found that abundant resources facilitate biological invasions. However, in contrast, Funk et al. [21] concluded that areas with resource scarcity can also be successfully invaded because IAS have a higher resource utilization efficiency than local species in resource-poor environments. Most existing studies primarily focus on analyzing community survey data post-invasion, aiming to uncover the relationship among invasion intensity, diversity, and environmental factors to explain invasion mechanisms [22–24]. They relatively lack analyses of the colonization source, specifically in the trade port region where alien species first arrive. This limitation hinders the implementation of control measures at the source.

Solidago canadensis is a perennial herb that belongs to the *Asteraceae* family, indigenous to the United States and Canada [25]. Initially imported to China in 1935 as an ornamental species, it subsequently escaped cultivation and became an invasive problem [26]. *S. canadensis* has a strong competitive ability due to its tall, rapid growth and high reproductive capacity [15,27], and can quickly form monodominant stands in invaded areas, severely threatening local biodiversity and ecological balance [28]. Moreover, it exhibits strong allelopathic effects, as its root exudates can inhibit the growth and development of various crops [29], resulting in reduced crop yields and quality decline. Thus, understanding the invasion mechanisms of *S. canadensis* and seeking control methods have become a hot issue for local governments and the academic community.

Ningbo Port is the largest foreign trade port in the world by cargo throughput, with approximately 25% of its containers originating from North America, which is the native region of *S. canadensis* [30]. The reproductive bodies of this species are spread through containers, ballast water, or transport vehicles along roads, rivers, and other pathways [31,32]. Among them, ballast water alone can transfer 3000 to 4000 different species every day [33]. Furthermore, a large number of artificial structures in port regions (such as docks and breakwaters) provide important spatial resources for the settlement and spread of IAS [34]. Thus, port regions are highly vulnerable to invasion and the establishment of IAS. Our study investigated the current invasion status of *S. canadensis* in the Ningbo Port region, analyzed the influencing factors, and aimed to reveal the invasion mechanisms and predict its invasion trends under different future climate scenarios. The specific objectives of this study were: (1) to analyze the current geographic distribution and invasion risk level of *S. canadensis* in Ningbo Port; (2) to reveal its invasion mechanisms in port regions; and (3) to predict the trend of its potential distribution area in the port region under different climate scenarios in the future.

2. Results

2.1. Invasion Status and Risk Level

We surveyed a total of 205 species in 595 sample plots; these species belong to 53 families, with the *Asteraceae* family having the highest proportion (19.2%) (Figure 1b). The next were *Poaceae and Fabaceae*, accounting for 9.6% and 6.1%, respectively. *S. canadensis* was widely distributed in the study area, exhibiting a scattered multi-point distribution pattern. *S. canadensis* was present in 403 of the total plots, accounting for 67.7% (Figure 1a). In addition, *S. canadensis* held a competitive advantage in the plots, with mean values of relative height, relative cover, relative abundance, and ecological importance of 0.459, 0.379, 0.381, and 0.457, respectively. Other species had a more dispersed distribution of height, coverage, abundance, and importance value. All values of *S. canadensis* were significantly higher than those of other species for relative height ($p < 0.001$), relative cover ($p < 0.001$), relative abundance ($p < 0.01$), and importance ($p < 0.001$) (Figure 1c). Among them, the plant height showed the most significant difference.

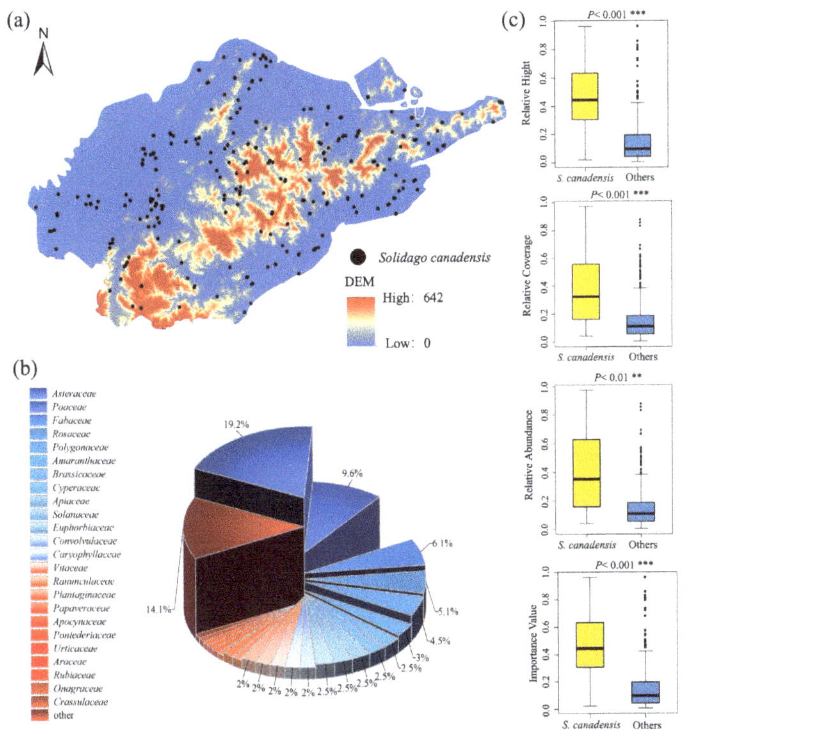

Figure 1. Invasion status of *S. canadensis* in Ningbo Port. (**a**) The distribution map of *S. canadensis*; (**b**) showed the proportion of plant families, with the lower part of the family (<1.0%) being merged into the others; and (**c**) presented the T test results for *S. canadensis* compared to other species.

Based on the formulas mentioned in the Section 4.2 and Table S1 (list in the Supplementary Materials), the risk assessment of *S. canadensis* indicated an evaluation score of 2.92, classifying it as a level 1 invasion risk. This was consistent with the classification in the Chinese Alien Invasive Species Information System (https://www.iplant.cn/ias/, accessed on 13 May 2025), where it was categorized as a malicious invader.

2.2. The Relationship Between Invasion Intensity, Plant Diversity, and Environmental Factors

We analyzed the selected variables using the GLM. The R^2 of the model was 0.45, indicated a relatively good explanatory power. The results of GLM show that the Shannon–Wiener Diversity Index was the most important factor on the invasiveness of *S. canadensis*, and there was an extremely significant negative correlation between them ($p < 0.001$). Among the variables, bio6 and bio18 showed significant negative correlations with the invasiveness of *S. canadensis* ($p < 0.01$). Notably, bio6 contributes more than bio18 to this relationship. Oppositely, it had a significant positive correlation with bio1 ($p < 0.01$) and bio17 ($p < 0.01$). Although a negative correlation between invasiveness and bio8 was observed, it was not statistically significant (Figure 2).

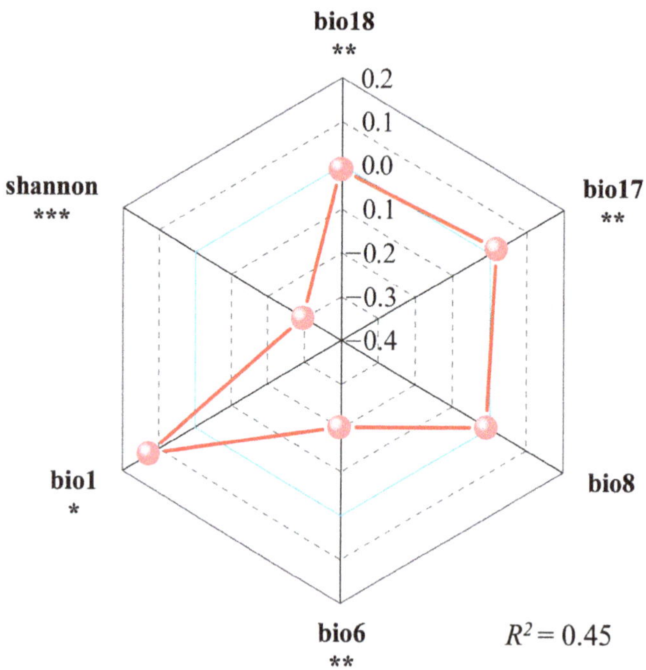

Figure 2. The relationship of IAS ecological importance to species diversity and environmental factors. *** $p < 0.001$, ** $p < 0.01$, and * $p < 0.05$.

2.3. Potential Distribution of S. canadensis Under Climate Scenarios

The AUC value of this model was 0.845, which indicates that the performance of the model had relatively good predictive accuracy (Figure S1). The percentage contribution of each environmental factors is shown in Figure 3a, with bio16 having the highest contribution (26.1%), followed by land cover (9.7%), bio19 (9.6%), bio1 (9.2%), bio18 (8.6%), bio8 (7.3%), bio15 (6.4%), and bio3 (5.9%). The total contribution of these factors was 82.7%. The permutation importance of the environmental factors is shown in Figure 3b, with bio18 having the highest importance (12.1%), followed by bio1 (11.2%), bio18 (9.2%), bio3 (8.6%), bio19 (7.5%), bio14 (7.5%), bio13 (7.0%), and bio6 (6.0%). After reclassification, the total suitable area of *S. canadensis* under current climate conditions was estimated: 244.24 km^2 in the low suitable areas, 458.88 km^2 in the medium suitable areas, and 240.55 km^2 in the highly suitable areas, with the total suitable habitat area accounting for 84.10%.

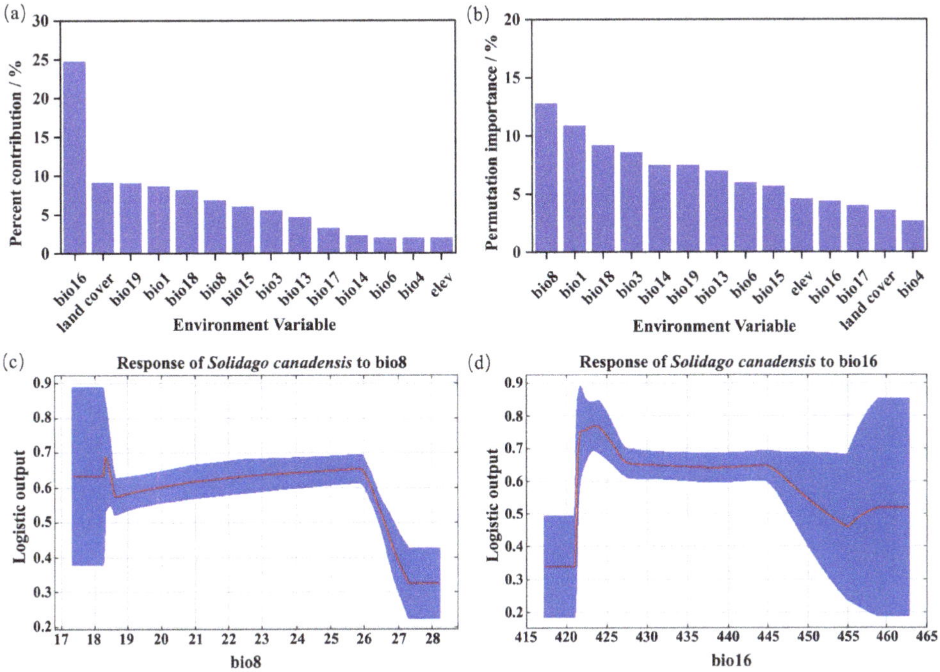

Figure 3. Response curves of high-weight influencing factors for *S. canadensis* under current climate conditions. (**a**,**b**) represent the percent contribution and permutation importance of environmental factors to *S. canadensis*, respectively. (**c**,**d**) respectively represent the response curves of the most influential environmental variables for *S. canadensis*.

Single-factor modeling was performed for bio8 and bio16. The results show that the probability of *S. canadensis* distribution fluctuated significantly with changes in the environmental variables. When bio8 was between 17 and 18 °C, the probability of *S. canadensis* presence remained stable at 0.64 (Figure 3c). As the temperature increased to 18.2~18.5 °C, the probability gradually increased, reaching a peak value of 0.69. It was stable within the range of 18.6 °C~25.9 °C. When the temperature exceeded 25.9 °C, the distribution probability decreased rapidly and stabilized at 0.33 within the range of 27.3~28.1 °C. When bio16 was between 417 and 421 mm, the probability of *S. canadensis* distribution was 0.33 (Figure 3d). As the precipitation increased to 421~424 mm, the probability gradually rose, peaking at 0.76, stably existing in the range of 427~445 mm. When the precipitation exceeded 424 mm, the distribution probability showed a decreasing trend, stabilizing at 0.51 between 458 and 464 mm.

Under all SSP scenarios, the total suitable habitat area for *S. canadensis* was higher than that under current climate conditions, although the increase was not substantial (Figure 4). The highly suitable habitat area had a relatively wide distribution, mainly concentrated in the five major port regions of Ningbo Port, extending inland. Under low greenhouse gas emission conditions and medium greenhouse gas emission conditions (from 2050s to 2070s), the percentage of total suitable habitat area decreased from 88.12% to 87.83% and from 89.77% to 89.46%, respectively. Under high greenhouse gas emission conditions, the proportion of total suitable habitat area increased from 86.39% to 87.84%. Specifically, the highly suitable habitat area increased by 0.66 km^2, the medium suitable area decreased by 63.10 km^2, and the low suitable area increased by 59.09 km^2 under low emission greenhouse gas conditions. Under medium greenhouse gas emission conditions, the highly suitable area decreased by 84.03 km^2, the medium suitable area increased by 91.13 km^2, and the low suitable habitat area decreased by 10.55 km^2. Under high greenhouse gas emission conditions, the highly suitable area decreased by 25.17 km^2, the medium suitable area increased by 16.20 km^2, and the low suitable area increased by 25.25 km^2. In summary, under scenarios of low greenhouse gas emissions and medium greenhouse gas emissions, the total suitable habitat area for *S. canadensis* reached its maximum proportion in the 2050s and then declined slowly. However, under high greenhouse gas emission conditions, the total suitable habitat area continued to increase.

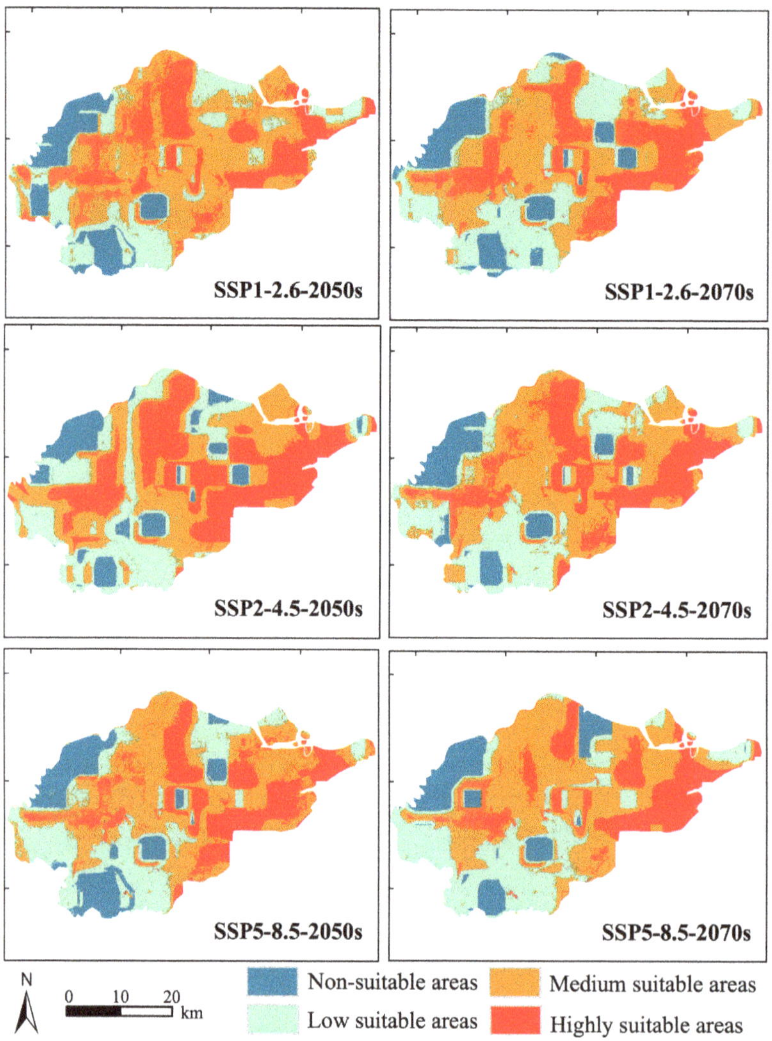

Figure 4. Potential geographic distribution of *S. canadensis* under different future climate change scenarios.

3. Discussion

3.1. The Invasion of S. canadensis Was Severe in the Port Region

Our field survey of *S. canadensis* in Ningbo Port showed that *S. canadensis* was present in 403 out of 595 plots, accounting for 67.73%, indicating a severe invasion. This was due to the long history of *S. canadensis* invasion in the area. *S. canadensis* first escaped to Zhejiang and Jiangsu provinces after entering Shanghai city [35]. Furthermore, ports were the first areas where *S. canadensis* established itself through transportation routes. Ningbo Port, with its numerous inner and outer ports and frequent human activity, facilitated the efficient spread of the species via rail, road, and water transport [2]. *S. canadensis* reproduces sexually and can be wind-dispersed, with a single plant producing 15,000–25,000 seeds annually. The higher average wind speed in coastal cities accelerated seed dispersal [36]. Our results show that *S. canadensis* had significant advantages in its traits. Its height usually ranges from 1 to 3 m, which is quite different from that of other local species. This made it more competitive in the struggle for resources and enabled it to become an advantageous species. Meanwhile, the competitive advantage in abundance and cover enhanced shading within the community, led to vegetation homogenization, and reduced community diversity [37,38]. The study conducted by Wang et al. further demonstrated that the height of *S. canadensis* was a critical factor at all stages of invasion, as it occupied distinct spatial niches compared to local species [15,39].

The risk assessment of the invasion of *S. canadensis* showed it as level 1, and it is regarded as a malicious invader. This classification indicated that *S. canadensis* poses a high ecological threat and had a significant dispersal potential in the Ningbo Port region. As a malicious IAS, its high reproductive capacity, broad adaptability, and allelopathic effects enabled it to rapidly occupy local ecological niches, substantially inhibiting the establishment and development of native plant community, which resulted in biodiversity decline and impairment in ecosystem functions [29,40]. Our results indicate the severity of the invasion risk posed by *S. canadensis*. Therefore, regulatory authorities should enhance monitoring and management in port trade areas to reduce the spread of invasive alien plants.

3.2. The Major Underlying Drivers of S. canadensis Invasion

Our results show a significant negative relationship between the invasiveness of *S. canadensis* and species diversity within the community, which was highly consistent with Elton's diversity–invasibility hypothesis [20]. Native species occupy multidimensional ecological niches, reducing resource availability to resist alien species. This mechanism was also confirmed in experimental studies on IAS such as *Phragmites australis*, *Senecio*, and *Cytisus scoparius* [41,42]. Furthermore, the mixed allelopathic substances in high-diversity communities can counteract the singular allelopathic effects of *S. canadensis*, weakening its suppression of the growth and development of local species [35].

Climate change influences multiple mechanisms underlying the success of biological invasions [43], with temperature playing a particularly significant role [44]. Our results show that the invasion intensity of *S. canadensis* in Ningbo Port decreased with a lower minimum temperature of the coldest month. The possible reason is that *S. canadensis* is a temperate plant, and changes in climate niches are relatively rare among terrestrial invasive plants [45]. Thus, the low-temperature limitation threshold of *S. canadensis* suppressed its invasion, while native plants were unaffected [44]. Another major negative influencing factor was bio18. It may be that *S. canadensis* is better adapted to drier or moderately humid environments, while excessive rainfall could potentially affect root respiration and inhibit its invasion. These two negative influencing factors may suggest that the success of *S. canadensis* invasion depends on specific climatic windows. The increase in bio1 shortens the plant growth cycle, which may improve the overwinter survival rate of *S. canadensis* and its seeds [46,47], while bio17 may suppress local plants (such as annual herbs), releasing their ecological niche. The effects of these two factors alleviate the competitive pressure on *S. canadensis*, enhancing its invasion intensity within the community.

3.3. The Potential Area of *S. canadensis* Under Different Climate Scenarios

Under the current climate scenario, *S. canadensis* exhibited strong adaptability in the Ningbo Port region. The total area of suitable habitat accounted for 84.1%, primarily consisting of medium and highly suitable zones, indicating elevated invasion pressure. The response curve under the current climate conditions showed that *S. canadensis* had a relatively long suitable growth range in the Ningbo Port region. Under the context of global warming, the total suitable habitat area for *S. canadensis* was higher than that under the current climate scenario, which may be due to its preference for moderately warmer and wetter climatic conditions. The increasing trend under future high emission climate scenarios was evident. This indicates that the invasion of *S. canadensis* in the Ningbo Port has not yet reached saturation, and its invasion risk is rising, suggesting a potential outbreak. This result aligns with the findings of Seebens et al. [48]. Spatially, the potential highly suitable habitat was relatively dispersed in all three climate scenarios and was mainly located near five major ports. Among them, Meishan Port and its surrounding areas in the southeastern part showed the most concentration. This was because ports, as sources of invasion, were the initial areas where *S. canadensis* established itself and spread rapidly. Ports acted as "stepping stones" for alien species to invade inland areas, and studies in Taiwan and Hainan have also highlighted the serious issue of alien species invasions in ports [49,50]. The changes in various suitable habitat zones within the port region under the three climatic scenarios differed. The reason may be related to the sensitivity of *S. canadensis* to dynamic changes in temperature and precipitation, with Ningbo Port being mostly coastal lowlands, with a humid climate and significant influence from monsoons and typhoons. Medium suitable habitat areas may be concentrated in the transition zones from the ports to the inland (such as farmland and wetland edges), where *S. canadensis* is more sensitive to changes in temperature and precipitation. The response curve of high–weight factors also reflected that *S. canadensis* prefers warm conditions but not excessive heat. Therefore, regulatory authorities should strengthen the prevention, monitoring, and management of IAS in the port region. Control measures should be prioritized based on the suitability of habitat areas, particularly for highly suitable and medium suitable areas. A multi-scale, cross-regional monitoring network should be established using satellite remote sensing and drone patrols. Additionally, physical removal should be carried out during the growth window of *S. canadensis*, such as before flowering and seeding or during the coldest month. Its spread could also be mitigated by increasing local species diversity.

Our study mainly focused on exploring the impact of climate change on the distribution of IAS, with less consideration of human activities. Meanwhile, the optimal model we adopted only considered the land cover factor, which might be inadequate. Because of the frequent human activities in port regions, the propagules can be spread through cargo transportation, vehicle tires, personnel clothing, and other ways. Therefore, future studies should consider more factors related to human activities (such as traffic density, land use intensity etc.) to better understand their impact on IAS distribution patterns. In addition to Elton's diversity–invasibility hypothesis, the current theories about the invasion mechanism of alien species also include the Darwin puzzle (naturalization hypothesis and preadaptation hypothesis). However, we did not analyze genetic diversity or niche overlap to validate these theories, which may limit the generality and scalability of our study. Furthermore, the invasion mechanisms at trade ports in different regions also vary. Future research should expand the scale of study, implement long-term monitoring, and comprehensively consider various influencing factors. These future studies are conducive to a more comprehensive understanding of the contributions and applicability of different invasion mechanisms. This will provide stronger theoretical support for managing IAS, protecting ecosystems, and preserving biodiversity.

4. Materials and Methods

4.1. Study Area

Ningbo Port is the port with the highest number of large and extra-large deep-water berths in mainland China. As of the end of 2023, it has 205 large deep-water berths of over 1×10^4 tons and 129 extra-large deep-water berths of over 5×10^4 tons (data from the Ningbo Port's official website (www.nbport.com.cn, accessed on 8 April 2025)). It consists of several port regions, including Beilun, Zhenhai, Ningbo, Daxie, Chuanshan North, and Meishan. Ningbo Port is located in Ningbo city, in southeastern China, covering a total land area of 9816 km^2 (Figure 5) [51]. The climate type is characterized by subtropical monsoon climate, with 22 days of high temperatures, 18 days of frost, an average annual temperature of 17.4 °C, 1850 h of annual sunshine, and annual precipitation of 1480 mm [52,53]. The local vegetation belongs to the northern subzone of the central subtropical evergreen broadleaf forest, with dominant woody plants such as *Fagaceae* and *Lauraceae* [54].

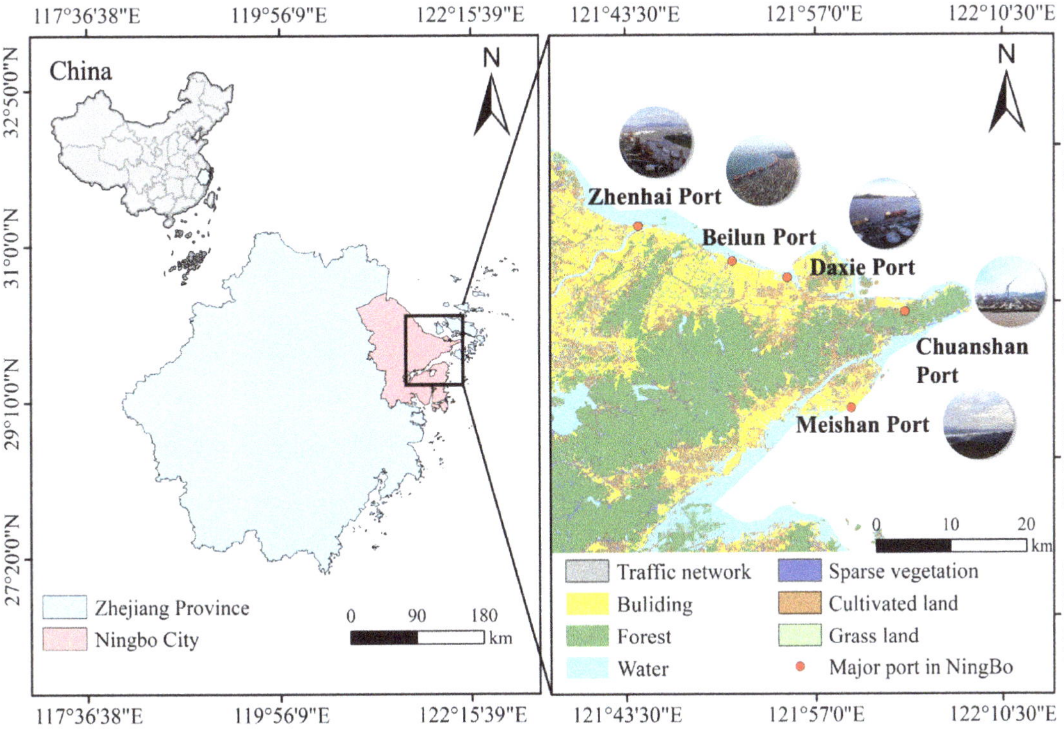

Figure 5. Map of the study area.

4.2. Methods

4.2.1. Field Community Surveys

Through the reviewing of the literature, the distribution status of IAS in the Ningbo Port and its surrounding areas was initially understood, and potential distribution areas were identified (Figure 5). Comprehensive field surveys were conducted in these potential areas. Surveys were conducted during the plant growing seasons from 2022 to 2023. The survey habitats included forests, wetlands, and grasslands. A simple random sampling method was used to establish 1 m × 1 m herbaceous plots along different survey routes within the study area, totaling 595 plots. We recorded the coordinates (latitude and longitude) and elevation and conducted a community survey for all plants within the plots. The survey indicators included species names, abundance, heights, and coverage. The types of communities in the study area are shown in Table 1.

Table 1. Vegetation community types in the study area.

	Community Type	Companion Species	Number of Plots	Coverage (Mean ± SD %) *	Height (Mean ± SD cm) *
1	*Solidago canadensis*	*Setaria viridis, Lactuca indica, Vicia sepium, Artemisia indica*	166	34.7 ± 19	64.5 ± 50.4
2	*Veronica persica*	*Geranium carolinianum, Vicia sepium, Stellaria media*	81	20.2 ± 16.9	7.1 ± 5.1
3	*Galium spurium*	*Veronica persica, Solidago canadensis, Artemisia indica*	60	12.50 ± 8.9	9.9 ± 17.3
4	*Alternanthera philoxeroides*	*Solidago canadensis, Erigeron canadensis*	45	32.50 ± 20.6	25.1 ± 13.3
5	*Artemisia indica*	*Veronica persica, Erigeron sumatrensis, Trifolium repens*	45	12.70 ± 5.9	6.2 ± 14.1
6	*Setaria viridis*	*Symphyotrichum subulatum, Solidago canadensis, Anthriscus sylvestris*	41	15.0 ± 5.9	72.4 ± 30.3
7	*Oxalis corniculata*	*Solidago canadensis, Stellaria media*	36	8.20 ± 3.8	9.4 ± 6.2
8	*Symphyotrichum subulatum*	*Artemisia argyi, Erigeron canadensis*	29	15.10 ± 8.6	67.1 ± 41.3
9	*Spartina alterniflora*	*Phragmites australis*	26	50.80 ± 28.2	75.7 ± 42.2
10	*Digitaria sanguinalis*	*Solidago canadensis, Symphyotrichum subulatum, Humulus scandens*	24	20.80 ± 13.2	41.2 ± 19.7
11	*Geranium carolinianum*	*Stellaria media, Cerastium glomeratum*	22	16.90 ± 7.8	8.4 ± 8.6
12	*Erigeron canadensis*	*Oxalis corniculata, Pseudognaphalium affine*	20	28.80 ± 12.5	55.6 ± 50.5
		Total number of plots	595		

* The coverage and height refer to those of the dominant species.

4.2.2. Environmental Data

The environmental data used in this study included climate data, soil data, and topographic data. Current climate condition data were obtained from 6 meteorological observation stations provided by the local meteorological department, consisting of a total of 19 variables (Table 2). We selected future climate data from three climate change scenarios (SSP1-2.6, SSP2-4.5, SSP5-8.5), with low greenhouse gas emission, medium greenhouse gas emissions, and high greenhouse gas emissions under the BCC-CSM2-MR model for prediction, with a resolution of 2.5 arc-minutes for all scenarios. The soil data included soil salinity, moisture content, and pH. These were determined through laboratory experiments. Specifically, we collected 0–20 cm of topsoil in each quadrat center and conducted laboratory experiments after completing community surveys. The topographic data were sourced from the Geospatial Data Cloud, with a resolution of 30 m × 30 m.

Table 2. Climate data.

Type	Factors	Description
Climate data	bio1	Average temperature for the year
	bio2	Mean daily temperature variation
	bio3	The consistency of temperature (bio2/bio7) (×100)
	bio4	Seasonal temperature variation
	bio5	Highest temperature in the warmest month
	bio6	Minimum temperature of the coldest month
	bio7	Annual temperature variation (bio5-bio6)
	bio8	Average temperature of the wettest season
	bio9	Temperature average of the driest season
	bio10	Temperature average of the warmest season
	bio11	Temperature average of the coldest season
	bio12	Total precipitation for the year
	bio13	Total precipitation in the wettest month
	bio14	Total precipitation in the driest month
	bio15	Seasonal variation in precipitation (coefficient of variation)
	bio16	Total precipitation in the wettest season
	bio17	Precipitation during the driest season
	bio18	Total precipitation in the warmest season
	bio19	Total precipitation in the coldest season

4.2.3. Risk Assessment of Biological Invasions

According to Helen et al. [55] and Jian et al. [56], we developed a risk assessment system for IAS to evaluate the invasion risk of *S. canadensis* (Table S1). The evaluation formulas are provided in Equations (1)–(5). Risk assessment values ranging from 2.8 to 4.0 correspond to level 1 (high risk), 1.2 to 2.8 to level 2 (medium risk), and 0 to 1.2 to level 3 (low risk). The specific calculation formula is as follows:

$$P = \sqrt[4]{P_1 \times P_2 \times P_3 \times P_4} \quad (1)$$

P_1 represents the introduction and colonization risk, P_2 represents the spread risk, P_3 represents the potential hazards and impacts, and P_4 represents the control of hazards.

The calculation formula for P_1, the introduction and colonization risk, is as follows:

$$P_1 = 0.3 \times P_{11} + 0.3 \times P_{12} + 0.2 \times P_{13} + 0.2 \times P_{14} \quad (2)$$

P_{11} represents environmental factor suitability, P_{12} represents food factor suitability, P_{13} represents growth and reproduction characteristics, and P_{14} represents the situation of natural enemies.

The formula of P_2, the spread risk, is as follows:

$$P_2 = \sqrt[4]{P_{21} \times P_{22} \times P_{23} \times P_{24}} \quad (3)$$

P_{21} represents distribution, P_{22} represents existing management measures, P_{23} represents spread capacity, and P_{24} represents the suitable habitat range.

The calculation formula for P_3, the potential hazards and impacts, is as follows:

$$P_3 = max(P_{31}, P_{32}, P_{33}) \quad (4)$$

P_{31} represents the impact on socio-economic factors, P_{32} represents the impact on the ecological environment, and P_{33} represents the importance of the affected targets.

P_4, the hazard control, is calculated as follows:

$$P_4 = (P_{41} + P_{42} + P_{43})/3 \quad (5)$$

P_{41} represents the difficulty of identification and verification, P_{42} represents the difficulty of monitoring and surveying, and P_{43} represents the difficulty of control and management.

4.2.4. Invasion Mechanism Analysis

To reveal the invasion mechanisms of *S. canadensis*, our study followed the method of Kreft et al. [57], and first calculated the ecological importance value of *S. canadensis* and the Shannon–Wiener Index. Subsequently, the T test method was used to compare the growth dominance of *S. canadensis* with that of other species [58]. We then used the Generalized Linear Model (GLM) to analyze the effects of species diversity and environmental factors on *S. canadensis*. To avoid the impact of multicollinearity, we chose environmental variables and excluded those with high collinearity; we finally selected 6 variables, which were bio1, bio6, bio8, bio17, bio18, and the Shannon–Wiener Diversity Index.

4.2.5. Prediction of Suitable Habitat Area for *S. canadensis*

We used the Maximum Entropy Model (MaxEnt) in Species Distribution Models (SDMs) to predict the growth areas of *S. canadensis* under different future climate scenarios [59]. This model is a quantitative approach based on niche theory, utilizing species distribution data and relevant environmental data, combined with Global Climate Models (GCMs), to study species' environmental tolerance and map their potential distribution. Prior to prediction, climate and topographic factors were resampled using the ArcMap 10.8 software. Additionally, considering the presence of autocorrelation and multicollinearity among environmental factors, we employed a Spearman correlation analysis to identify the environmental variables (Figure S2). Factors with a correlation coefficient higher than 0.75 were modeled using the Jackknife method for single-factor modeling, determining the contribution of each factor, and the final prediction variables were selected. These variables were categorized into three types. The first category was climate factors, which included biol, bio2, bio3, bio4, bio6, bio7, bio8, bio12, bio13, biol4, bio15, bio16, bio17, bio18, and bio19. The second category was topographic factors, with only one variable: elevation (elev). The third category was land cover type.

To reduce overfitting issues in the niche model analysis, the "*kuenm*" package in the R 4.3.3 software was used to evaluate different regularization factors, and feature types were adjusted based on distribution points to obtain the optimal combination for running the MaxEnt model. We used 75% of the dataset to parameterize the model and the remaining 25% for validation. Model accuracy was evaluated using the area under receiver operator curve (AUC), which ranges from 0 to 1; higher AUC values mean a better performance of this model [60]. Finally, the probability of species presence was estimated based on presence records, and background points were randomly generated by identifying the maximum entropy distribution. Using the natural breaks method in the ArcMap software, the habitat suitability was classified into four levels: non-, low, medium, and highly suitable level. Furthermore, the total area of *S. canadensis* distribution under current and future climate conditions was estimated by multiplying the number of "presence" grid cells by their spatial resolution values.

5. Conclusions

Our study was based on field survey data and involved analyzing and simulating the current invasion drivers of *S. canadensis* and its suitable habitat distribution under climate change scenarios. The results show that the problem of invasive alien species in the port area was serious, and due to high-intensity human interference, the level of invasion risk was relatively high. Biodiversity and the lowest temperature (especially that of the coldest month) were the key drivers of the invasion of *S. canadensis*. A highly diverse plant community could inhibit the invasion of *S. canadensis*. Meanwhile, *S. canadensis* was sensitive to low temperatures and had a distinct low-temperature limiting threshold. Under three global climate warming scenarios, the suitable habitat area of *S. canadensis* continued to increase, reaching its maximum under the medium greenhouse gas emission scenario in the 2050s. Our results provide scientific evidence for assessing and managing the risk of invasive plants, offering significant practical theoretical value for guiding source control of invasive species.

Supplementary Materials: The following supporting information can be downloaded at: https://www.mdpi.com/article/10.3390/plants14101546/s1, Table S1: Table of risk assessment index system of invasive alien species; Figure S1: ROC curve validation of the potential distribution prediction results for *Solidago canadensis*.; Figure S2: The results of correlation analysis of climate factors.

Author Contributions: Conceptualization, X.L. and X.Y.; methodology, X.L.; software, S.S. and K.L.; validation, X.L., S.S. and S.L.; formal analysis, S.S. and K.L.; investigation, X.L., S.L., Q.P., J.M. and W.L.; writing—original draft preparation, S.S., K.L. and X.L.; writing—review and editing, X.Y. and S.S.; visualization, Q.P.; supervision, X.L.; project administration, X.L.; funding acquisition, X.Y. All authors have read and agreed to the published version of the manuscript.

Funding: This work was supported by the Key Technology Breakthrough Plan Project of Science and Innovation Yongjiang 2035 (Grant No. 2023Z146, 2024Z249, 2024Z262), the National Natural Science Foundation of China (Grant No. 42371027) and Ningbo Natural Science Foundation (Grant No. 2021J084).

Data Availability Statement: The data presented in this study are available upon request from the corresponding authors.

Acknowledgments: We thank our lab members for their helpful discussions.

Conflicts of Interest: The authors declare no conflicts of interest.

References

1. Early, R.; Bradley, B.A.; Dukes, J.S.; Lawler, J.J.; Olden, J.D.; Blumenthal, D.M.; Gonzalez, P.; Grosholz, E.D.; Ibañez, I.; Miller, L.P.; et al. Global Threats from Invasive Alien Species in the Twenty-First Century and National Response Capacities. *Nat. Commun.* **2016**, *7*, 12485. [CrossRef] [PubMed]
2. Hulme, P.E. Unwelcome Exchange: International Trade as a Direct and Indirect Driver of Biological Invasions Worldwide. *One Earth* **2021**, *4*, 666–679. [CrossRef]
3. Hao, Q.; Ma, J.-S. Invasive Alien Plants in China: An Update. *Plant Divers* **2022**, *45*, 117–121. [CrossRef] [PubMed]
4. Xu, H.; Ding, H.; Li, M.; Qiang, S.; Guo, J.; Han, Z.; Huang, Z.; Sun, H.; He, S.; Wu, H.; et al. The Distribution and Economic Losses of Alien Species Invasion to China. *Biol. Invasions* **2006**, *8*, 1495–1500. [CrossRef]
5. Kang, L.; Wu, W.; Yu, H.; Su, F. Global Container Port Network Linkages and Topology in 2021. *Sensors* **2022**, *22*, 5889. [CrossRef]
6. Potgieter, L.J.; Li, D.; Baiser, B.; Kühn, I.; Aronson, M.F.J.; Carboni, M.; Celesti-Grapow, L.; de Matos, A.C.L.; Lososová, Z.; Montaño-Centellas, F.A.; et al. Cities Shape the Diversity and Spread of Nonnative Species. *Annu. Rev. Ecol. Evol. Syst.* **2024**, *55*, 157–180. [CrossRef]
7. Jehlík, V.; Dostálek, J.; Frantík, T. Alien Plants in Central European River Ports. *Neobiota* **2019**, *45*, 93–115. [CrossRef]
8. Seebens, H.; Essl, F.; Dawson, W.; Fuentes, N.; Moser, D.; Pergl, J.; Pyšek, P.; Van Kleunen, M.; Weber, E.; Winter, M.; et al. Global Trade Will Accelerate Plant Invasions in Emerging Economies under Climate Change. *Glob. Change Biol.* **2015**, *21*, 4128–4140. [CrossRef] [PubMed]
9. Sardain, A.; Sardain, E.; Leung, B. Global Forecasts of Shipping Traffic and Biological Invasions to 2050. *Nat. Sustain.* **2019**, *2*, 274–282. [CrossRef]
10. Meyerson, L.A.; Mooney, H.A. Invasive Alien Species in an Era of Globalization. *Front. Ecol. Environ.* **2007**, *5*, 199–208. [CrossRef]

11. Pyšek, P.; Hulme, P.E.; Simberloff, D.; Bacher, S.; Blackburn, T.M.; Carlton, J.T.; Dawson, W.; Essl, F.; Foxcroft, L.C.; Genovesi, P.; et al. Scientists' Warning on Invasive Alien Species. *Biol. Rev.* **2020**, *95*, 1511–1534. [CrossRef] [PubMed]
12. Gioria, M.; Hulme, P.E.; Richardson, D.M.; Pyšek, P. Why Are Invasive Plants Successful? *Annu. Rev. Plant Biol.* **2023**, *74*, 635–670. [CrossRef] [PubMed]
13. Dawson, W.; Rohr, R.P.; Van Kleunen, M.; Fischer, M. Alien Plant Species with a Wider Global Distribution Are Better Able to Capitalize on Increased Resource Availability. *New Phytol.* **2012**, *194*, 859–867. [CrossRef]
14. Huenneke, L.F.; Hamburg, S.P.; Koide, R.; Mooney, H.A.; Vitousek, P.M. Effects of Soil Resources on Plant Invasion and Community Structure in Californian Serpentine Grassland. *Ecology* **1990**, *71*, 478–491. [CrossRef]
15. Wang, C.; Cheng, H.; Wei, M.; Wang, S.; Wu, B.; Du, D. Plant Height and Leaf Size: Which One Is More Important in Affecting the Successful Invasion of Solidago Canadensis and Conyza Canadensis in Urban Ecosystems? *Urban For. Urban Green.* **2021**, *59*, 127033. [CrossRef]
16. Lazzaro, L.; Bolpagni, R.; Buffa, G.; Gentili, R.; Lonati, M.; Stinca, A.; Acosta, A.T.R.; Adorni, M.; Aleffi, M.; Allegrezza, M.; et al. Impact of Invasive Alien Plants on Native Plant Communities and Natura 2000 Habitats: State of the Art, Gap Analysis and Perspectives in Italy. *J. Environ. Manag.* **2020**, *274*, 111140. [CrossRef]
17. Feng, Y.-L.; Lei, Y.-B.; Wang, R.-F.; Callaway, R.M.; Valiente-Banuet, A.; Inderjit; Li, Y.-P.; Zheng, Y.-L. Evolutionary Tradeoffs for Nitrogen Allocation to Photosynthesis versus Cell Walls in an Invasive Plant. *Proc. Natl. Acad. Sci. USA* **2009**, *106*, 1853–1856. [CrossRef]
18. Huang, K.; Kong, D.-L.; Lu, X.-R.; Feng, W.-W.; Liu, M.-C.; Feng, Y.-L. Lesser Leaf Herbivore Damage and Structural Defense and Greater Nutrient Concentrations for Invasive Alien Plants: Evidence from 47 Pairs of Invasive and Non-Invasive Plants. *Sci. Total Environ.* **2020**, *723*, 137829. [CrossRef]
19. Lin, T.; Doorduin, L.; Temme, A.; Pons, T.L.; Lamers, G.E.M.; Anten, N.P.R.; Vrieling, K. Enemies Lost: Parallel Evolution in Structural Defense and Tolerance to Herbivory of Invasive Jacobaea Vulgaris. *Biol. Invasions* **2015**, *17*, 2339–2355. [CrossRef]
20. Elton, C.S. *The Ecology of Invasions by Animals and Plants*; Springer Nature: Berlin, Germany, 2020; pp. 23–30.
21. Funk, J.L.; Vitousek, P.M. Resource-Use Efficiency and Plant Invasion in Low-Resource Systems. *Nature* **2007**, *446*, 1079–1081. [CrossRef]
22. Raheem, A.; Yohanna, P.; Li, G.; Noh, N.J.; Iqbal, B.; Tang, J.; Du, D.; Alahmadi, T.A.; Ansari, M.J.; Zhan, A.; et al. Unraveling the Ecological Threads: How Invasive Alien Plants Influence Soil Carbon Dynamics. *J. Environ. Manag.* **2024**, *356*, 120556. [CrossRef] [PubMed]
23. Wang, C.; Cheng, H.; Wu, B.; Jiang, K.; Wang, S.; Wei, M.; Du, D. The Functional Diversity of Native Ecosystems Increases during the Major Invasion by the Invasive Alien Species, *Conyza canadensis*. *Ecol. Eng.* **2021**, *159*, 106093. [CrossRef]
24. Hejda, M.; Sádlo, J.; Kutlvašr, J.; Petřík, P.; Vítková, M.; Vojík, M.; Pyšek, P.; Pergl, J. Impact of Invasive and Native Dominants on Species Richness and Diversity of Plant Communities. *Preslia* **2021**, *93*, 181–201. [CrossRef]
25. Benelli, G.; Pavela, R.; Cianfaglione, K.; Nagy, D.U.; Canale, A.; Maggi, F. Evaluation of Two Invasive Plant Invaders in Europe (*Solidago canadensis* and *Solidago gigantea*) as Possible Sources of Botanical Insecticides. *J. Pest Sci.* **2019**, *92*, 805–821. [CrossRef]
26. Wang, C.; Jiang, K.; Zhou, J.; Wu, B. *Solidago canadensis* Invasion Affects Soil N-Fixing Bacterial Communities in Heterogeneous Landscapes in Urban Ecosystems in East China. *Sci. Total Environ.* **2018**, *631–632*, 702–713. [CrossRef]
27. Wang, W.; Zhu, Q.; Dai, S.; Meng, L.; He, M.; Chen, S.; Zhao, C.; Dan, X.; Cai, Z.; Zhang, J.; et al. Effects of *Solidago canadensis* L. on Mineralization-Immobilization Turnover Enhance Its Nitrogen Competitiveness and Invasiveness. *Sci. Total Environ.* **2023**, *882*, 163641. [CrossRef]
28. Wang, C.; Jiang, K.; Liu, J.; Zhou, J.; Wu, B. Moderate and Heavy *Solidago canadensis* L. Invasion Are Associated with Decreased Taxonomic Diversity but Increased Functional Diversity of Plant Communities in East China. *Ecol. Eng.* **2018**, *112*, 55–64. [CrossRef]
29. Abhilasha, D.; Quintana, N.; Vivanco, J.; Joshi, J. Do Allelopathic Compounds in Invasive *Solidago canadensis* s.l. Restrain the Native European Flora? *J. Ecol.* **2008**, *96*, 993–1001. [CrossRef]
30. Ningbo Municipal Bureau of Statistics. *Ningbo Statistical Yearbook*; China Statistics Press: Beijing, China, 2022; pp. 258–271.
31. Wilson, J.R.U.; Dormontt, E.E.; Prentis, P.J.; Lowe, A.J.; Richardson, D.M. Something in the Way You Move: Dispersal Pathways Affect Invasion Success. *Trends Ecol. Evol.* **2009**, *24*, 136–144. [CrossRef]
32. Hulme, P.E. Trade, Transport and Trouble: Managing Invasive Species Pathways in an Era of Globalization. *J. Appl. Ecol.* **2009**, *46*, 10–18. [CrossRef]
33. Lan, R.; Qiao, B. Study on the Invasion Mechanism of Alien Species through Ship Activities. *IOP Conf. Ser. Earth Environ. Sci.* **2020**, *446*, 052104. [CrossRef]
34. Dafforn, K. Future-Proofing Our Ports against Biological Invasion. *Curr. Opin. Environ. Sustain.* **2025**, *73*, 101518. [CrossRef]
35. Adomako, M.O.; Ning, L.; Tang, M.; Du, D.-L.; Van Kleunen, M.; Yu, F.-H. Diversity- and Density-Mediated Allelopathic Effects of Resident Plant Communities on Invasion by an Exotic Plant. *Plant Soil* **2019**, *440*, 581–592. [CrossRef]

36. Zhang, Z.; Wen, G.; Bu, D.; Sun, G.; Qiang, S. Long−Distance Wind Dispersal Drives Population Range Expansion of Solidago Canadensis. *Plants* **2022**, *11*, 2734. [CrossRef]
37. Vojík, M.; Boublík, K. Fear of the Dark: Decline in Plant Diversity and Invasion of Alien Species Due to Increased Tree Canopy Density and Eutrophication in Lowland Woodlands. *Plant Ecol.* **2018**, *219*, 749–758. [CrossRef]
38. Biswas, S.R.; He, D.; Li, J.; Gong, L.; Biswas, P.L.; Zhuo, Z.; Xu, M.; Yang, X.; Yan, E. Putting Space into Trait Ecology: Trait, Environment and Biodiversity Relationships at Multiple Spatial Scales. *J. Ecol.* **2024**, *112*, 613–628. [CrossRef]
39. Blossey, B.; Notzold, R. Evolution of Increased Competitive Ability in Invasive Nonindigenous Plants: A Hypothesis. *J. Ecol.* **1995**, *83*, 887–889. [CrossRef]
40. Mei, D.; Jian-Zhong, L.; Wen-Ju, Z.; Jia-Kuan, C.; Bo, L. Canada Goldenrod (*Solidago canadensis*): An Invasive Alien Weed Rapidly Spreading in China. *J. Syst. Evol.* **2006**, *44*, 72–85.
41. Levine, J.M.; Adler, P.B.; Yelenik, S.G. A Meta-analysis of Biotic Resistance to Exotic Plant Invasions. *Ecol. Lett.* **2004**, *7*, 975–989. [CrossRef]
42. Byun, C.; De Blois, S.; Brisson, J. Plant Functional Group Identity and Diversity Determine Biotic Resistance to Invasion by an Exotic Grass. *J. Ecol.* **2013**, *101*, 128–139. [CrossRef]
43. Hulme, P.E. Climate Change and Biological Invasions: Evidence, Expectations, and Response Options. *Biol. Rev.* **2017**, *92*, 1297–1313. [CrossRef] [PubMed]
44. Yang, W.; Sun, S.; Wang, N.; Fan, P.; You, C.; Wang, R.; Zheng, P.; Wang, H. Dynamics of the Distribution of Invasive Alien Plants (Asteraceae) in China under Climate Change. *Sci. Total Environ.* **2023**, *903*, 166260. [CrossRef] [PubMed]
45. Petitpierre, B.; Kueffer, C.; Broennimann, O.; Randin, C.; Daehler, C.; Guisan, A. Climatic Niche Shifts Are Rare Among Terrestrial Plant Invaders. *Science* **2012**, *335*, 1344–1348. [CrossRef]
46. Zhou, X.-H.; He, W.-M. Climate Warming Facilitates Seed Germination in Native but Not Invasive *Solidago canadensis* Populations. *Front. Ecol. Evol.* **2020**, *8*, 595214. [CrossRef]
47. Li, H.; Ali, A.; Luo, X.; Liao, K.; Sun, N.; Xu, M.; Sha, L.; He, D.; Du, Y.; Sun, W.; et al. China's Subtropical Deciduous Plants Are More Sensitive to Climate Change than Evergreen Plants by Flowering Phenology. *Glob. Change Biol.* **2024**, *30*, e17168. [CrossRef]
48. Seebens, H.; Blackburn, T.M.; Dyer, E.E.; Genovesi, P.; Hulme, P.E.; Jeschke, J.M.; Pagad, S.; Pyšek, P.; Winter, M.; Arianoutsou, M.; et al. No Saturation in the Accumulation of Alien Species Worldwide. *Nat. Commun.* **2017**, *8*, 14435. [CrossRef]
49. Dawson, W.; Moser, D.; Van Kleunen, M.; Kreft, H.; Pergl, J.; Pyšek, P.; Weigelt, P.; Winter, M.; Lenzner, B.; Blackburn, T.M.; et al. Global Hotspots and Correlates of Alien Species Richness across Taxonomic Groups. *Nat. Ecol. Evol.* **2017**, *1*, 0186. [CrossRef]
50. Lu, J.; Li, S.; Wu, Y.; Jiang, L. Are Hong Kong and Taiwan Stepping-Stones for Invasive Species to the Mainland of China? *Ecol. Evol.* **2018**, *8*, 1966–1973. [CrossRef]
51. Li, X.; Fu, J.; Jiang, D.; Lin, G.; Cao, C. Land Use Optimization in Ningbo City with a Coupled GA and PLUS Model. *J. Clean. Prod.* **2022**, *375*, 134004. [CrossRef]
52. Li, S.; Pan, Q.; Yang, X.; Gao, Y.; Xu, M. Differential Influences of Island Biogeographic Effects on the Diversity among Three Plant Life-Forms. *Glob. Ecol. Conserv.* **2025**, *58*, e03499. [CrossRef]
53. Tang, Y.T.; Chan, F.K.S.; Griffiths, J. City Profile: Ningbo. *Cities* **2015**, *42*, 97–108. [CrossRef]
54. Wu, Z. *Chinese Vegetation*, 1st ed.; Science Press: Beijing, China, 1980; pp. 823–836.
55. Roy, H.E.; Rabitsch, W.; Scalera, R.; Stewart, A.; Gallardo, B.; Genovesi, P.; Essl, F.; Adriaens, T.; Bacher, S.; Booy, O.; et al. Developing a Framework of Minimum Standards for the Risk Assessment of Alien Species. *J. Appl. Ecol.* **2018**, *55*, 526–538. [CrossRef]
56. Ou, J.; Lu, C.; O'Toole, D.K. A Risk Assessment System for Alien Plant Bio-Invasion in Xiamen, China. *J. Environ. Sci.* **2008**, *20*, 989–997. [CrossRef] [PubMed]
57. Kreft, H.; Jetz, W. Global Patterns and Determinants of Vascular Plant Diversity. *Proc. Natl. Acad. Sci. USA* **2007**, *104*, 5925–5930. [CrossRef]
58. Tello, R.; Crewson, P.E. Hypothesis Testing II: Means. *Radiology* **2003**, *227*, 1–4. [CrossRef] [PubMed]
59. Anibaba, Q.A.; Dyderski, M.K.; Jagodziński, A.M. Predicted Range Shifts of Invasive Giant Hogweed (*Heracleum mantegazzianum*) in Europe. *Sci. Total Environ.* **2022**, *825*, 154053. [CrossRef]
60. Shi, X.; Wang, J.; Zhang, L.; Chen, S.; Zhao, A.; Ning, X.; Fan, G.; Wu, N.; Zhang, L.; Wang, Z. Prediction of the Potentially Suitable Areas of *Litsea cubeba* in China Based on Future Climate Change Using the Optimized MaxEnt Model. *Ecol. Indic.* **2023**, *148*, 110093. [CrossRef]

forests

MDPI

Article

Evaluating the Impact of Climate Change and Human Activities on the Potential Distribution of Pine Wood Nematode (*Bursaphelenchus xylophilus*) in China

Liang Zhang [1], Ping Wang [1,2,*], Guanglin Xie [1,2] and Wenkai Wang [1,2]

1 Institute of Entomology, College of Agriculture, Yangtze University, Jingzhou 434025, China; 13349832924@163.com (L.Z.); xieguanglin@yangtzeu.edu.cn (G.X.); wwk@yangtzeu.edu.cn (W.W.)
2 MARA Key Laboratory of Sustainable Crop Production in the Middle Reaches of the Yangtze River (Co-Construction by Ministry and Province), College of Agriculture, Yangtze University, Jingzhou 434025, China
* Correspondence: wangping1992@yangtzeu.edu.cn

Abstract: *Bursaphelenchus xylophilus* is a pest that interferes with the health of forests and hinders the development of the forestry industry, and its spread is influenced by changes in abiotic factors and human activities. The potential distribution areas of *B. xylophilus* in China under four shared-economic pathways were predicted using the optimized MaxEnt model (version 3.4.3), combining data from a variety of environmental variables: (1) prediction of natural environmental variables predicted under current climate models; (2) prediction of natural environmental variables + human activities under current climate models; and (3) prediction of natural environmental variables under the future climate models (2050s and 2070s). Meanwhile, whether the niche of *B. xylophilus* has changed over time is analyzed. The results showed that human activities, precipitation in the driest month, annual precipitation, and elevation had significant effects on the distribution of *B. xylophilus*. In the current conditions, human activities greatly reduced the survival area of *B. xylophilus*, and its suitable distribution area was mainly concentrated in the southwestern and central regions of China. Under the influence of climate change in the future, the habitat of *B. xylophilus* will gradually spread to the northeast. In addition, the ecological niche overlap analysis showed that *B. xylophilus* in future climate was greater than 0.74. This study provides important information for understanding the ecological adaptation and potential risk of *B. xylophilus*, which can help guide the decision making of pest control and forest protection.

Keywords: *Bursaphelenchus xylophilus*; climate change; human activity; MaxEnt model; habitat suitability

1. Introduction

The spatial distribution of organisms is a fundamental spatial characteristic influenced by prolonged interactions between species and their environment [1]. Currently, the impacts of global climate change and human interference on biodiversity and ecosystems are important research priorities in the fields of ecology, environmental science, and conservation biology [2]. As societies evolve, human activities alter land use patterns and create novel landscape interfaces, which in turn influence the spatial distribution and diversity of species. Numerous examples can be given to illustrate the above points: activities such as urbanization, agricultural expansion, and deforestation can result in the loss and fragmentation of habitats, thereby impeding the ability of numerous species to locate suitable living environments, which in turn affects their survival and reproduction [3]. Furthermore, climate change can affect temperature and precipitation patterns, as well as increase the frequency of climatic extremes. These changes will inevitably alter the geographical distribution of species and the stability of ecosystems [4]. Since the advent of the Industrial Revolution, the global climate has been warming due to the greenhouse effect caused by human activities [5]. According to the Fifth Assessment Report (AR5) of the Intergovernmental Panel on Climate Change (IPCC), the average temperature of the Earth will have risen by 0.3 to 4.5 °C by the end of the 21st century [6]. In order to respond to these challenges, scientists are engaged in research aimed at elucidating the impact of climate change and human activities on the geographic distribution patterns of species [7]. These studies help elucidate species' responses to environmental changes and potential future distribution.

Forests are the largest terrestrial ecosystems on Earth and play an irreplaceable role as carbon sinks and in combating climate change [8]. Pine trees, as an important silvicultural species, are widely planted globally, and they not only provide timber and other forest products but also play key roles in ecosystem services, such as fixing soil and reducing erosion and wind and sand control, which have important economic, cultural, and ecological values [9]. Nevertheless, pine trees are susceptible to pine wilt disease (PWD), a significant global quarantine forest disease caused by *Bursaphelenchus xylophilus*, which is highly detrimental to pines; once infected, pines typically succumb within a few months [10]. The species was initially identified in North America and subsequently disseminated to Japan, China, Korea, Portugal, and other countries [11]. A recent report by the National Forestry and Grassland Administration (NFGA, 2024) indicates that PWD has spread northward and westward to Jilin and Gansu Provinces; this epidemic has shown a rapid proliferation and spread, posing a serious threat to the ecological security, biosecurity, and economic development of forest ecosystems [12].

Under the background of intensified global climate change, the area of China's forests infested by PWD has been gradually increasing, and the scope of transmission has been expanding [13]. Analyzing the characteristics of the evolution of the geographical distribution of *B. xylophilus* and the prediction of the future development trend under the influence of climate change and human interference can provide the theoretical basis and technical support for the government and the forestry management department to effectively monitor, predict and control PWD in a timely manner.

Species distribution modeling is an important tool for studying the geographical distribution of species and their response to climate change and human activities [14,15]. Models predict the potential distribution of species under different environmental conditions by analyzing their current distribution data and environmental variables (e.g., climate, topography, vegetation, etc.). The MaxEnt model is widely used; it is based on the principle of maximum entropy as conceptual information, which means that in the absence of more information, the model will tend to choose the most uniform probability distribution. It is able to infer the potential distribution of a species from limited data on the presence points of the species and still maintains a high predictive accuracy when dealing with incomplete data [16–18]. In addition to MaxEnt, other models such as CLIMEX, GARP, and ENFA have their own characteristics and applications [19–21]. These models have applications in several fields, including conservation planning for endangered species, habitat assessment for wildlife, estimation of potential acreage for crops, and monitoring and control of quarantine pests [22]. In particular, in the context of pest management, ecological niche modeling can help to predict the potential distribution areas of pests, thus providing a scientific basis for monitoring and controlling their spread [23,24].

In this study, we compare predictions based on different environmental factors using the MaxEnt model, which can reveal how human activities and climate change individually or jointly affect the distribution of *B. xylophilus*. We aimed to (1) identify the most critical factors affecting the distribution of *B. xylophilus* within China; (2) compare the differences in the distribution and size of *B. xylophilus* habitats under scenarios with and without anthropogenic disturbance in the current period; (3) explore changes in the size of potential geographic distribution areas of *B. xylophilus* under different future climate scenarios; and (4) clarify the pattern of spatial evolution and development trend of *B. xylophilus*. The results of this study can provide a scientific basis and reference for localities and governments to formulate reasonable prevention and control strategies, which can help protect forestry resources, maintain ecological balance, and promote ecological civilization.

2. Materials and Methods

2.1. Data

To generate the occurrence records of *B. xylophilus* used in the modeling, we collected data from multiple sources, specifically the following: (1) information regarding endemic regions affected by *B. xylophilus* was obtained from the National Forestry and Grassland Administration (http://www.forestry.gov.cn, accessed on 2 June 2024); (2) book materials and online references (CNKI, https://www.cnki.net, accessed on 20 May 2024; WOS, https://www.webofscience.com/wos, accessed on 19 May 2024) [25,26]; and (3) two online public databases, the Global Biodiversity Information Facility (GBIF) (https://doi.org/10.15468/dl.uvs3yu, accessed on 18 May 2024) and an iNaturalist (https://www.inaturalist.org, accessed on 20 May 2024). For available record locations without specific latitude and longitude coordinates, we instead used Google Earth software (http://ditu.google.cn, accessed on 3 June 2024) to obtain this information. Ultimately, through these sources, a total of 1385 occurrence points were collected for this study (Figure 1). These records provided the necessary database for modeling.

In order to prevent model overfitting, the "ENMTools" package was employed to guarantee the quality of the dataset utilized for model construction and to minimize potential biases in spatial clustering due to data duplication, thereby enhancing the accuracy and reliability of the model [27]. The tool automatically identifies the cell size of the environmental factor grid and eliminates redundant data within the same grid, which in this study was set to 2.5 arcmin (~21 km^2 at the equator). This approach is rapid and effective, and the results of the analysis are more meaningful. Consequently, the resulting occurrence records are necessarily smaller than the actual distribution area. Ultimately, 963 occurrence data points for *B. xylophilus* were retained for the construction of the MaxEnt model.

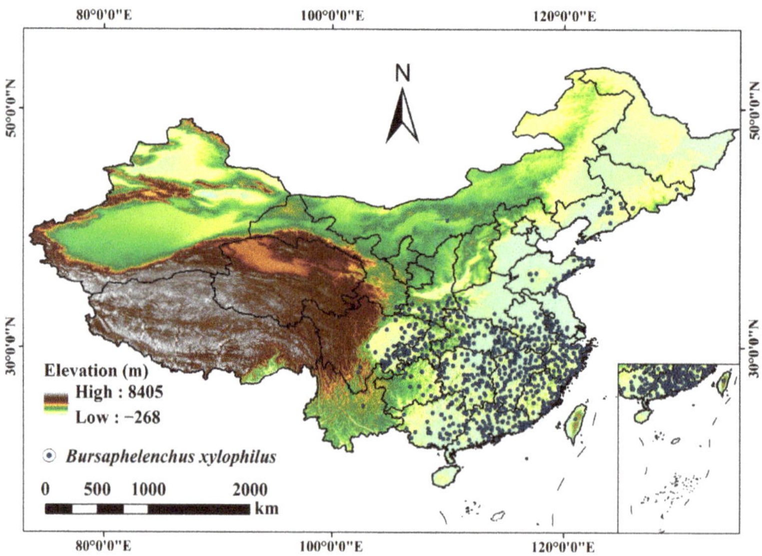

Figure 1. Occurrence records of *B. xylophilus* within China.

In this study, in order to analyze and predict the potential distribution area of *B. xylophilus*, we initially selected 31 environmental variables that may affect the distribution of *B. xylophilus* (Table 1). These variables cover a wide range of aspects, including bioclimatic, topographic, normalized difference vegetation index, ultraviolet radiation, and human activities [28]. Climate data at a resolution of 2.5 arcmin from the 1970s to 2000s were downloaded from WorldClim version 2.1 for the current period. Future climate data for the 2050s and 2070s were simulated under four shared socioeconomic pathways (SSPs) using the BCC-CSM2-MR global circulation model. Elevation data at 2.5 arcmin resolution were downloaded from WorldClim, and slope and aspect were extracted using ArcGIS Map software (version 10.8.1). We downloaded the normalized difference vegetation index (NDVI) data from the Center for Resource and Environmental Sciences and data from the Chinese Academy of Sciences (https://www.resdc.cn/, accessed on 19 January 2024). In addition, we have downloaded ultraviolet radiation data (https://www.ufz.de/gluv/index.php?en=32367, accessed on 19 January 2024). Human activity data were downloaded from the Socioeconomic Data and Applications Center (https://sedac.ciesin.columbia.edu, accessed on 15 January 2024), including the global human influence index and global human footprint. Finally, we used the "Resample" and "Extract" tools in the ArcGIS Map software to standardize the 31 environmental variables into a consistent format for subsequent analysis and modeling [29]. These data provide a comprehensive environmental background for studying the distribution of *B. xylophilus* in China, helping to understand the relationship between the species and its environment and to predict the possible effects of future climate change on the distribution of the species.

The selection and treatment of environmental variables is a crucial step when constructing a species distribution model, as the interconnections between variables may lead to autocorrelation and multicollinearity problems, which can reduce the predictive accuracy of the model [30]. To address these issues, we first assessed the contribution of the 31 environmental variables in the MaxEnt model using the Jackknife method, which identifies the variables that contribute most to the model predictions. Subsequently, a Pearson correlation analysis was conducted utilizing the "ENMTools" package (version 1.1.3) within the R platform (Figure S1). Environmental variables with correlation coefficients of $|r| \leq 0.9$ were selected for modeling to avoid multicollinearity issues. When the correlation coefficient $|r| > 0.9$ between two environmental variables was observed, the variable with the higher contribution was chosen for inclusion in the model. This implies that in the case of highly correlated variables, only those that contribute the most to the model are retained. Through the aforementioned steps, 13 bioclimatic factors were ultimately incorporated into the MaxEnt model (Figure 2). These variables were deemed to be the most crucial and mutually independent environmental factors in the model, thereby enhancing the predictive accuracy and reliability of the model.

Table 1. Screening of 31 environmental variables.

Abbreviation	Environmental Variables	Operation ($|r| > 0.9$)
Bio1	Annual mean temperature (°C)	Retain
Bio2	Mean diurnal range (°C)	Retain
Bio3	Isothermality	Retain
Bio4	Temperature seasonality	Retain
Bio5	Maximum temp of warmest month (°C)	Eliminate
Bio6	Minimum temp of coldest month (°C)	Eliminate
Bio7	Temperature annual range (°C)	Eliminate
Bio8	Mean temp of wettest quarter (°C)	Eliminate
Bio9	Mean temp of driest quarter (°C)	Eliminate
Bio10	Mean temp of warmest quarter (°C)	Eliminate
Bio11	Mean temp of coldest quarter (°C)	Eliminate
Bio12	Annual precipitation (mm)	Retain
Bio13	Precipitation of wettest month (mm)	Eliminate
Bio14	Precipitation of driest month (mm)	Retain
Bio15	Precipitation seasonality (mm)	Retain
Bio16	Precipitation of wettest quarter (mm)	Eliminate
Bio17	Precipitation of driest quarter (mm)	Eliminate

Table 1. *Cont.*

Abbreviation	Environmental Variables	Operation (|r| > 0.9)
Bio18	Precipitation of warmest quarter (mm)	Eliminate
Bio19	Precipitation of coldest quarter (mm)	Eliminate
Bio20	Elevation (m)	Retain
Bio21	NDVI	Retain
Bio22	Slope	Retain
Bio23	Aspect	Retain
Bio24	Annual_mean_UV-B	Eliminate
Bio25	UV-B_seasonality	Eliminate
Bio26	Mean_UV-B_of_highest_month	Eliminate
Bio27	Mean_UV-B_of_lowest_month	Eliminate
Bio28	Sum_of_UV-B_radiation_of_highest_quarter	Eliminate
Bio29	Sum_of_UV-B_radiation_of_lowest_quarter	Eliminate
Bio30	Global human footprint	Retain
Bio31	Global human influence index	Retain

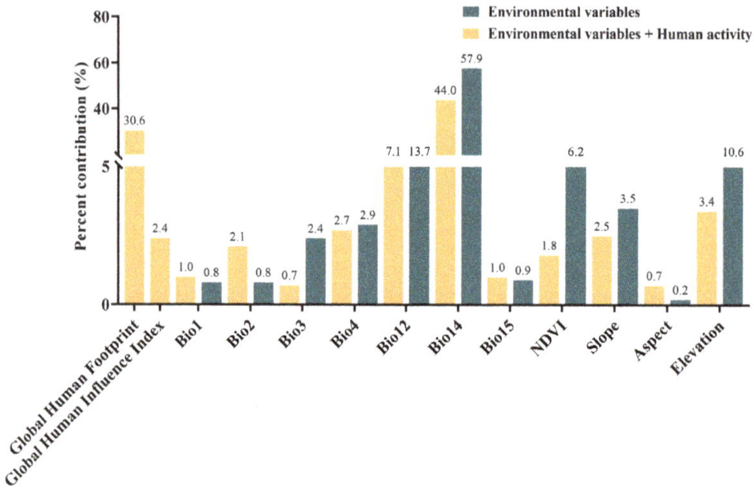

Figure 2. Contribution of 13 environmental variables after screening.

2.2. Evaluation and Validation of MaxEnt Model

Regularized multipliers (RMs) and feature combinations (FCs) serve as crucial parameters within the MaxEnt model, and optimizing these parameters can substantially enhance the model's accuracy [31]. The RMs and FCs parameters were optimally tuned using the "ENMeval" package in the R software (version 4.3.3). Feature combinations encompass 5 feature types, namely L (linear features), Q (quadratic features), P (product features), T (threshold features), and H (hinge features). Initially, RMs were configured from 0 to 4 at intervals of 0.5 based on the preserved distribution data and environmental variables. Additionally, 6 FCs were designated to identify the optimal parameter combinations: L, LQ, H, LQH, LQHP, and LQHPT. Subsequently, 48 models incorporating various RM and FC combinations were constructed, and the model with the lowest delta AICc value among the feature combinations was selected as the optimal model (Figure S2). Finally, under natural environmental variable conditions, RMs were set to 0.5 with the FCs designated as "LQHP" (Figure S2A). Under conditions incorporating both natural environment variables and human activities, RMs were set to 1 with the FCs specified as "LQHPT" (Figure S2B). The additional parameter settings for the optimal model were as follows: 25% of the distribution points for each species were selected as the test set, while 75% were used for training. The maximum number of iterations was set to 5000, the maximum number of background points was limited to 10,000, and the repetitions were carried out 10 times. The closer the test omission rate is to the theoretical omission rate, the more accurate the model construction [32].

In ecological and environmental science research, the area under the curve (AUC) of the receiver operating characteristic curve (ROC) is an important metric for assessing the accuracy and predictive performance of modeling results [33]. The generalization ability of the model can be assessed by comparing the AUC values of the training and test sets and ensures that the model performs well on the training data and accurately predicts new and unseen data [34]. The AUC value determines the accuracy of the model, which ranges from 0 to 1. The larger the AUC value, the higher the accuracy of the model prediction. AUC values below 0.6 indicate a failure of the model performance, rendering the predictions unreliable. Values between 0.6 and 0.7 are considered poor, between 0.7 and 0.8 are considered fair, between 0.8 and 0.9 are considered good, and those above 0.9 are considered excellent. These values help to ensure the reliability and applicability of the findings.

2.3. Changes in the Potential Distribution Areas of *B. xylophilus*

The impact of natural environmental factors and human activities on the geospatial distribution pattern of *B. xylophilus* was investigated by developing three distinct models: (1) prediction of natural environmental factors (including bioclimate + topography + ultraviolet radiation + the normalized difference vegetation index) under the current climate model; (2) prediction of natural environmental factors (including bioclimate + topography + ultraviolet radiation + the normalized difference vegetation index) + human activity (including global human footprint + global human influence index) under the current climate model; and (3) prediction of natural environmental factors (including future bioclimate + topography + ultraviolet radiation + the normalized difference vegetation index) under the future climate model. Among them, model (1) and model (2) are based on current climate model projections, while model (3) is based on future climate model projections. Models (1) and (2) were employed to assess the influence of anthropogenic factors on the suitability of habitat for *B. xylophilus*, while models (1) and (3) were utilized to examine the potential effect of climate change on the suitability of habitat for *B. xylophilus*.

The average of the optimal MaxEnt model run through 10 repetitions was utilized as the final result, which was then evaluated based on the logistic value of the probability of species presence (P). The final results were converted to raster form and visualized using ArcGIS Map software, and the proportion of area in each suitable habitat was calculated. For comparison, the same thresholds were used for categorization in both the current and future climate models [35]. The suitability results for *B. xylophilus* were classified into two categories based on the natural breaks (Jenks) method in the ArcGIS Map software: unsuitable habitat ($P < 0.172121$) and suitable habitat ($P \geq 0.172121$).

2.4. Change in Potential Distribution Center Shift under Future Climate Scenarios

We applied the "Centroid Changes (Lines)" tool in SDMToolbox v2.6 to identify potential geographic distribution centers in areas of high probability of species presence in each time period. The locations of potential distribution centers were compared across time and under different carbon emission scenarios. This helps to understand how the distribution of species changes in response to different climate change scenarios [36]. In addition, connecting the calculated potential distribution center points for different time periods represents the route of spatial change in the main suitable growing areas for *B. xylophilus* under future climate conditions.

2.5. Dynamics of Potential Distribution Areas of *B. xylophilus* under Future Climate Scenarios

We applied the "Distribution Changes between Species Distribution Models (SDMs)" function in SDMToolbox v2.6 to create binary maps with a threshold value of 0.172121, and the binary maps of each category in different future climate change scenarios were compared with the current climate scenarios to assess the future changes in the distribution of *B. xylophilus* [37]. Four results were obtained: "Expansion" (species range increases), "No occupancy" (species has no distribution under the new scenario), "Unchanged" (species range remains unchanged), and "Contraction" (species range decreases). In addition, we analyzed the extent of future climate change impacts on *B. xylophilus* in China using multivariate environmental similarity surfaces (MESSs) [38]. Environmental variables from contemporary potentially suitable areas were used as the reference layer to calculate the similarity between future climate conditions and the current one. The S value reflects the degree of similarity between the climate conditions at a point and the climate conditions in the reference layer during a given period. Negative values indicate that at least one of the environmental variables at the point has a value outside the range of the corresponding value in the reference layer, which is referred to as a climate anomaly, while a maximum value of 100 indicates that the climate at the point is completely normal. This operation is realized by running the "density.tools.Novel" tool in the "maxent.jar" file in the command window, and the ASCII file exported from the model was imported into ArcGIS Map 10.8.1 software for drawing.

2.6. Niche Dynamics Analyses

We conducted ecological niche quantification and overlap analyses using the "ecospat" software package (version 4.0.0) to explore climatic ecological niche changes between the ranges of *B. xylophilus* under contemporary and future climate scenarios [39]. Firstly, the presence and non-presence areas of *B. xylophilus* were binarised using ArcGIS Map software, with presence marked as 1 and non-presence marked as 0. Then, relevant environmental variables were extracted from each occurrence record. Subsequently, Schoener's D index was employed to quantify the degree of overlap between ecological niches across climate scenarios, with values ranging from 0 (indicating no overlap) to 1 (representing complete overlap).

3. Results

3.1. Model Accuracy and Evaluation

The AUC values were applied to assess the accuracy of the optimized MaxEnt model in the prediction of species distributions. The results of this study showed that under contemporary climatic conditions, the average training AUC value of 10 repetitions of natural environmental variables was 0.9338, and the average testing AUC was 0.9285, while the average training AUC value of the model for the coexistence of natural environmental variables and human activities was 0.9500 and the average testing AUC value was 0.9465, which suggests that human activities may play an important role in predicting the results of the potential distributional areas of *B. xylophilus* potential distribution areas. In addition, the mean training AUC values and mean test AUC values for the different future carbon emission scenarios under the four climate change scenarios were greater than 0.9 (Table 2). This indicates that the optimized MaxEnt model is more accurate and precise and confirms the stability of the model's robustness and generalization ability.

Table 2. Model accuracy evaluation.

Shared Socioeconomic Pathways	Train AUC (Avg)	Test AUC (Avg)
Current-Environmental variables	0.9338	0.9285
Current-Environmental variables + Human activity	0.9500	0.9465
Future-SSP1.0-2.6 2040–2060	0.9345	0.9277
Future-SSP1.0-2.6 2060–2080	0.9338	0.9284
Future-SSP2.0-4.5 2040–2060	0.9349	0.9287
Future-SSP2.0-4.5 2060–2080	0.9342	0.9291
Future-SSP3.0-7.0 2040–2060	0.9357	0.9302
Future-SSP3.0-7.0 2060–2080	0.9324	0.9266
Future-SSP5.0-8.5 2040–2060	0.9351	0.9301
Future-SSP5.0-8.5 2060–2080	0.9351	0.9286

3.2. Main Environmental Variables Affecting Distribution of *B. xylophilus*

In this study, the effect of each variable on the geographic distribution of *B. xylophilus* was assessed using MaxEnt model outputs and Jackknife methods. This study found that the potential geographic distribution of *B. xylophilus* in the presence of only natural environmental factors was influenced by precipitation in the driest month (Bio14, 57.9%), annual precipitation (Bio12, 13.7%), elevation (Bio20, 10.6%), and the normalized difference vegetation index (Bio21, 6.2%), with a cumulative contribution of 88.4% (Figure 2). The main variables affecting the potential geographic distribution of *B. xylophilus* under a combination of natural environmental and human activities were precipitation in the driest month (Bio14, 44%), global human footprint (Bio30, 30.6%), and annual precipitation (Bio12, 7.1%), with a cumulative contribution of 81.7%. The variables with the largest changes in contribution were Bio14 (13.9% decrease), Bio12 (6.6% decrease), Bio20 (7.2% decrease), and Bio21 (4.4% increase) (Figure 2). This indicates that precipitation, the normalized difference vegetation index, elevation, and human activities are the main factors influencing the distribution of *B. xylophilus*.

3.3. Prediction of the Potential Distribution of *B. xylophilus* under Climate and Human Interference in the Current

The MaxEnt model was employed to predict the spatial extent and habitat of *B. xylophilus* with and without anthropogenic disturbance (Figure 3), and the results showed that it was mainly distributed in central and southeastern China, which was similar to the actual occurrence area, indicating that the optimized MaxEnt model was able to simulate the potential distribution area of *B. xylophilus* well. Under the influence of natural environmental conditions, the occurrence areas of *B. xylophilus* were mainly concentrated in Liaoning, Shandong, Zhejiang, Anhui, Jiangsu, Shaanxi, Henan, Hubei, Hunan, Guangdong, Fujian, Chongqing, Sichuan, and Guangxi (Figure 3A). The suitable habitat covered an area of 138.58×10^4 km^2, representing 14.59% of China's total land area (Table S1).

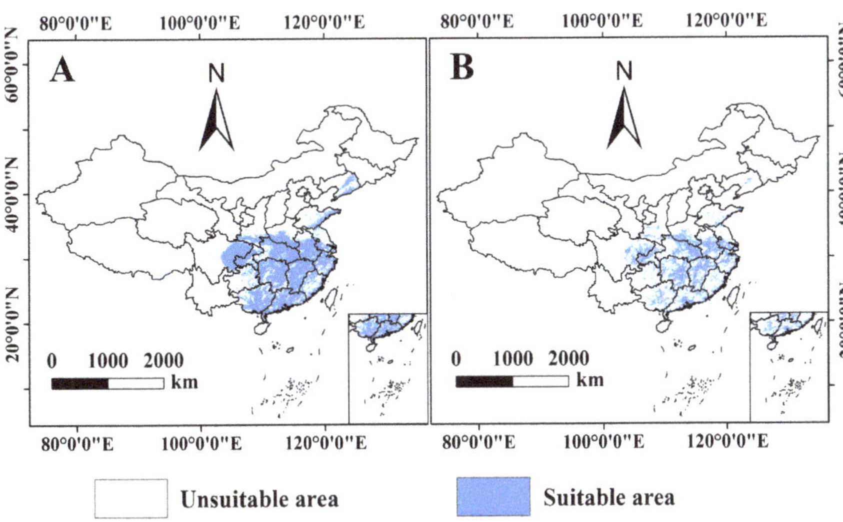

Figure 3. Distribution of suitable habitat for *B. xylophilus* without (**A**) and with (**B**) human activity under current climate models in China.

Taking into account human activities, the combined suitable habitat area for *B. xylophilus* in China was 80.13×10^4 km^2, making up 8.44% of China's total land area. Conversely, unsuitable habitat covered 869.23×10^4 km^2, comprising 91.56% of China's total area (Figure 3B). Compared with the area affected by natural environmental factors alone, there was a decrease of 42.18%, and the total area of suitable habitat for *B. xylophilus* decreased by 58.45 × 10^4 km^2, which represents 6.16% of China's total land area (Table S1), indicating that human activities have reduced the area of suitable habitat for *B. xylophilus*.

3.4. Prediction of the Potential Distribution of B. xylophilus under Different Climate Scenarios in the Future

In the context of climate change, the projected suitable ranges of *B. xylophilus* were based on future climate change scenarios with different common socioeconomic pathways and representative concentration pathways. The results showed that under different future climate scenarios, the range of suitable habitats for *B. xylophilus* was generally consistent with the current climate scenario and was mainly distributed in northeast, central and southeastern China, including Liaoning, Shandong, Zhejiang, Anhui, Jiangsu, Shaanxi, Henan, Hubei, Hunan, Guangdong, Fujian, Chongqing, Sichuan and Guangxi (Figure 4). Under the future climate scenarios, the estimated area of suitable habitat for *B. xylophilus* ranges from 137.87 × 10^4 km^2 to 141.55 × 10^4 km^2, accounting for 14.52% to 14.92% of China's total area (Table S1). Among them, SSP3.0-7.0-2070s predicted the largest area of suitable habitat, followed by SSP5.0-8.5-2070s, while SSP1.0-2.6-2050s predicted the smallest area of suitable habitat. Over time, the area of suitable habitat for *B. xylophilus* showed a decreasing and then increasing trend in the low greenhouse gas emission scenario (SSP1.0-2.6, SSP2.0-4.5) and the medium greenhouse gas emission scenario (SSP3.0-7.0), i.e., the area of suitable habitat decreased in the 2050s and increased in the 2070s. However, only the SSP5.0-8.5 scenario shows a continuous increase (Table S1).

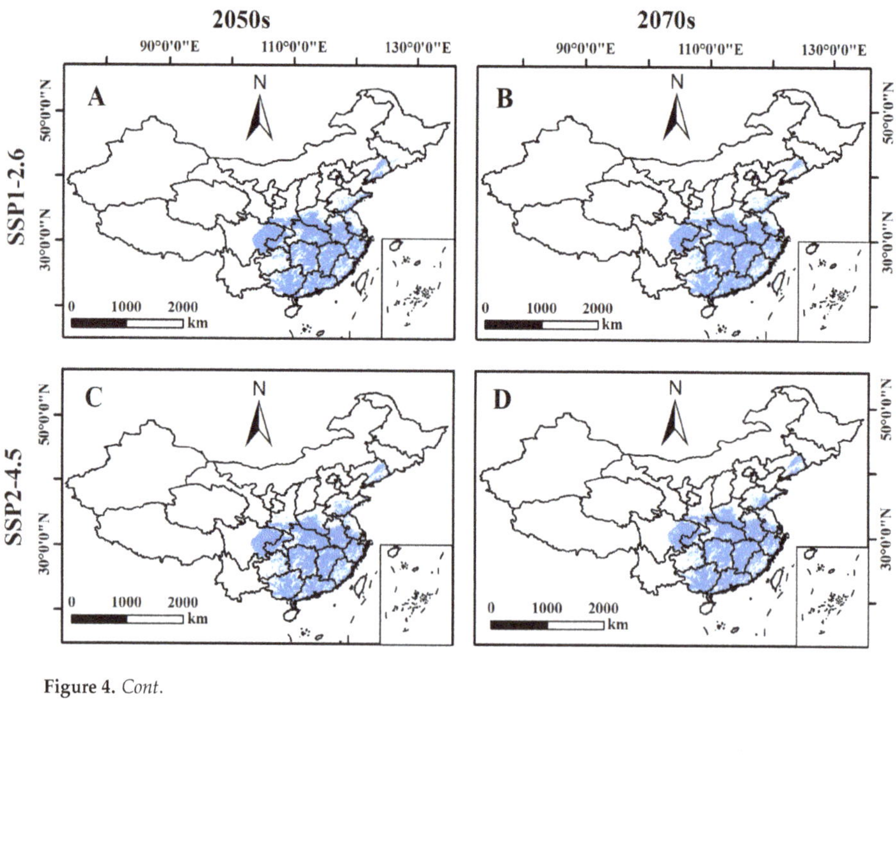

Figure 4. *Cont.*

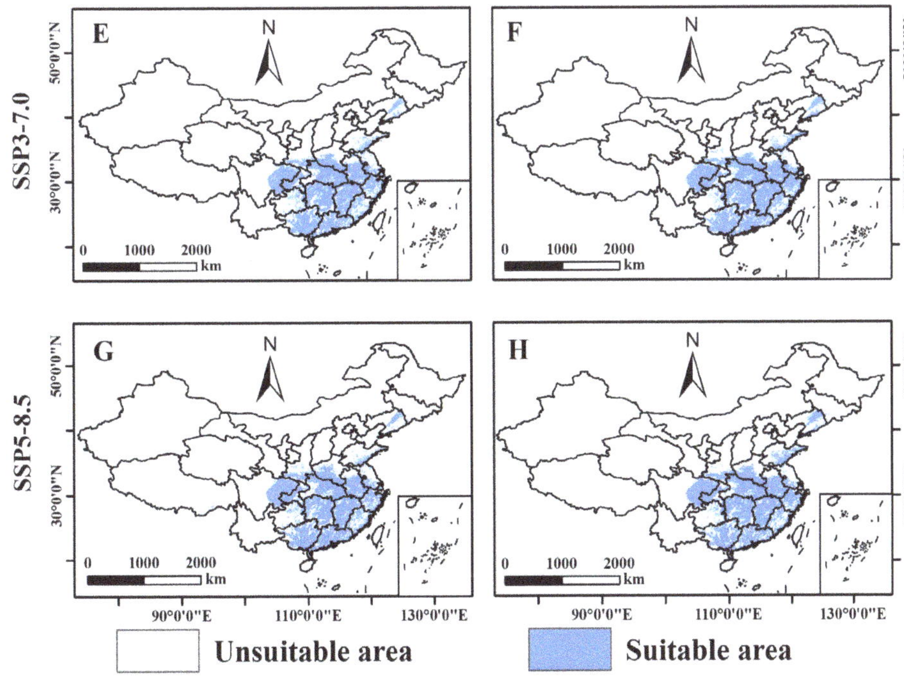

Figure 4. Suitable areas for *B. xylophilus* in China under different future climate scenarios. (**A**) SSP1.0-2.6-2050s; (**B**) SSP1.0-2.6-2070s; (**C**) SSP2.0-4.5-2050s; (**D**) SSP2.0-4.5-2070s; (**E**) SSP3.0-7.0-2050s; (**F**) SSP3.0-7.0-2070s; (**G**) SSP5.0-8.5-2050s; (**H**) SSP5.0-8.5-2070s.

3.5. Relative Changes in the Potential Distribution Area of B. xylophilus under Future Climate Scenarios

Relative changes in the potential distribution of *B. xylophilus* were derived by comparing differences in current and future distribution areas (Figures 5 and 6). The results showed that under future climate scenarios, the expansion area ranged from 18.16×10^4 km^2 to 29.82×10^4 km^2, and the contraction area ranged from 6.71×10^4 km^2 to 14.39×10^4 km^2 (Table S2). The expansion is primarily observed in Liaoning, Jilin, Shandong, Henan, Hubei, Guangxi, Guangdong, Guizhou, Jiangxi, Jiangsu, Anhui, Shaanxi, and Fujian. Conversely, the contraction area is predominantly found in Guangxi, Guangdong, Guizhou, Jiangxi, Fujian, Jiangsu, Shandong, Jilin, Xizang, and Taiwan Island (Figure 5). Among them, the largest expansion area was predicted for the high greenhouse gas (GHG) emission scenario SSP5.0-8.5-2070s, followed by SSP3.0-7.0-2070s, while the smallest expansion area was predicted for SSP1.0-2.6-2070s, which was development-pattern- and climate-change-dependent. Whereas the largest contraction area is in the low greenhouse gas (GHG) emission scenario (SSP1.0-2.6-2050s), the smallest is in the medium greenhouse gas (GHG) emission scenario (SSP3.0-7.0-2070s). Furthermore, under future climate scenarios, Inner Mongolia, Shanxi, Ningxia Hui Autonomous Region, Gansu, Qinghai, and Shaanxi exhibit a high degree of similarity with the climatic conditions of their origins (Figure 6), indicating that the climatic zones of the potential distribution area of *B. xylophilus* differ significantly from the current climate reference layer (Table S3).

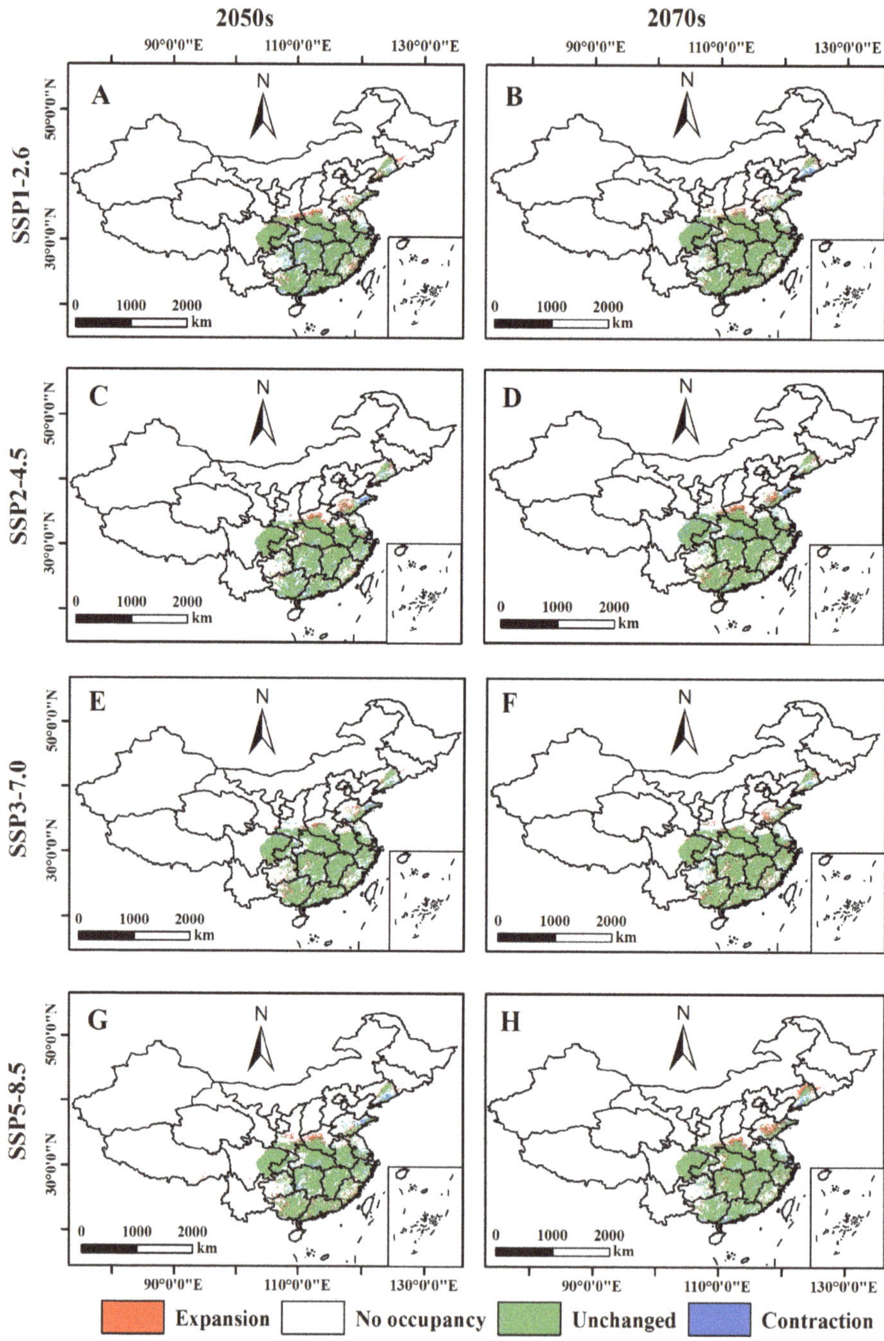

Figure 5. Relative changes in *B. xylophilus* within China under different future climate scenarios. (**A**) SSP1.0-2.6-2050s; (**B**) SSP1.0-2.6-2070s; (**C**) SSP2.0-4.5-2050s; (**D**) SSP2.0-4.5-2070s; (**E**) SSP3.0-7.0-2050s; (**F**) SSP3.0-7.0-2070s; (**G**) SSP5.0-8.5-2050s; (**H**) SSP5.0-8.5-2070s.

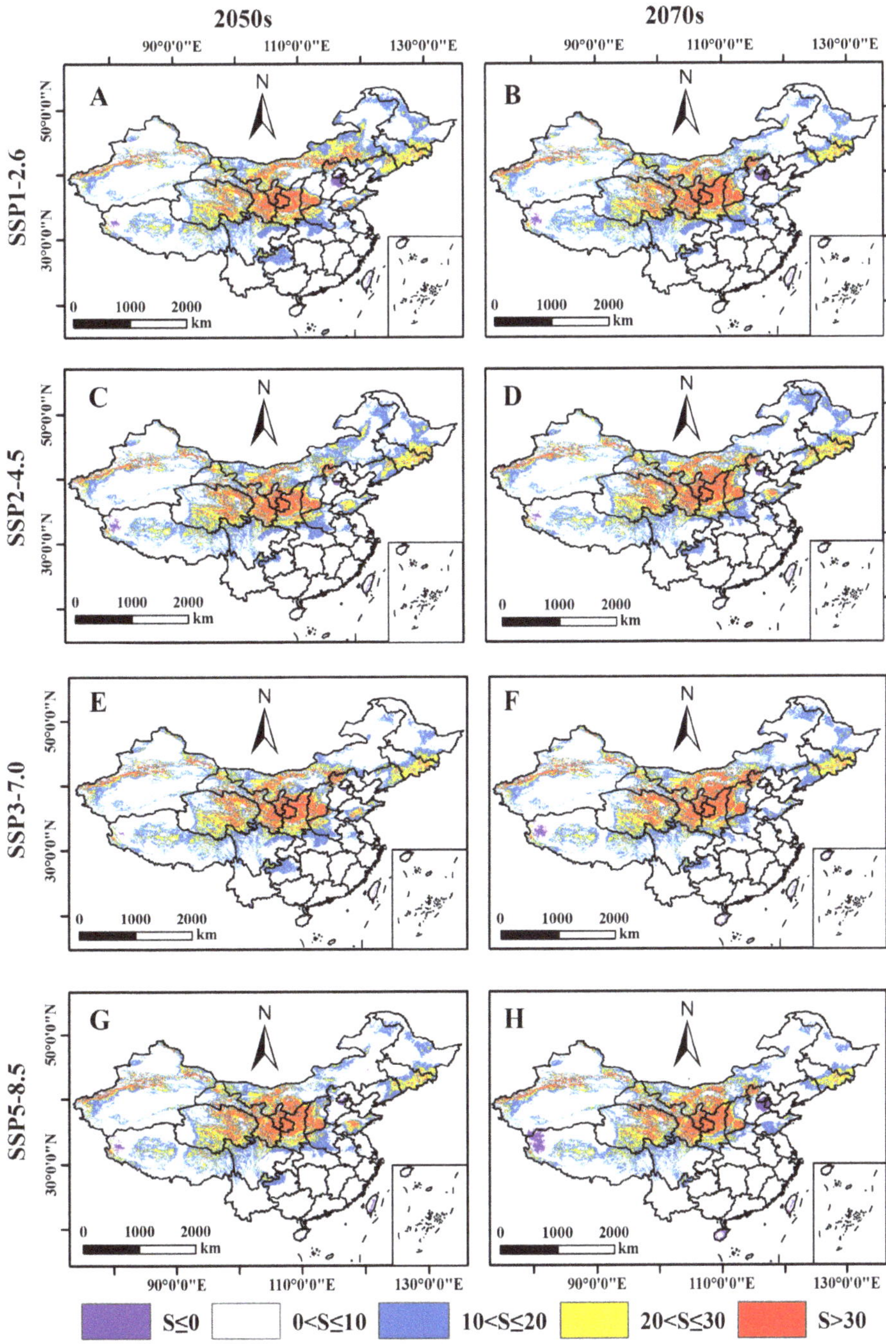

Figure 6. MESS analysis for *B. xylophilus* under future climate scenarios in China. (**A**) SSP1.0-2.6-2050s; (**B**) SSP1.0-2.6-2070s; (**C**) SSP2.0-4.5-2050s; (**D**) SSP2.0-4.5-2070s; (**E**) SSP3.0-7.0-2050s; (**F**) SSP3.0-7.0-2070s; (**G**) SSP5.0-8.5-2050s; (**H**) SSP5.0-8.5-2070s.

3.6. Potential Distribution Center Shifts of *B. xylophilus* under Different Scenarios in the Future

We applied the "Centroid Changes (Lines)" tool to compare changes in the center of distribution of *B. xylophilus* under different carbon emission scenarios, which allowed us to assess the potential impacts of different emission reduction strategies on the movement routes and ecosystem stability of *B. xylophilus*. The current distribution center of *B. xylophilus* is located in Hubei Province at 113.18° N, 31.60° E (Figure 7 and Table 3). In the SSP1.0-2.6 pathway, the distribution center in the 2050s is located at 31.53° N, 113.49° E, and that in the 2070s is located at 32.35° N, 114.13° E, with the distribution center shifting 122.58 km to the northeast. In the SSP2.0-4.5 pathway, the distribution center was located at 31.71° N, 113.63° E in the 2050s and 31.75° N, 113.05° E in the 2070s, shifting the center of distribution by 20.86 km to the northwest. The distribution center in the SSP3.0-7.0 pathway was located at 31.63° N, 113.30° E in the 2050s and 31.55° N, 113.13° E in the 2070s and moved to the southwest by 7.20 km. In the SSP5.0-8.5 pathway, the distribution center for *B. xylophilus* was located at 31.87° N, 113.36° E in the 2050s and at 31.55° N, 113.40° E in the 2070s, with a southeast move of 21.02 km. As time goes on, the distribution center of *B. xylophilus* remained relatively stable between time points, suggesting that the MaxEnt model performs well in predicting the stability of the potential future range of the species, implying that the model's selection and trade-offs of environmental variables were accurate (Figure 7 and Table 3).

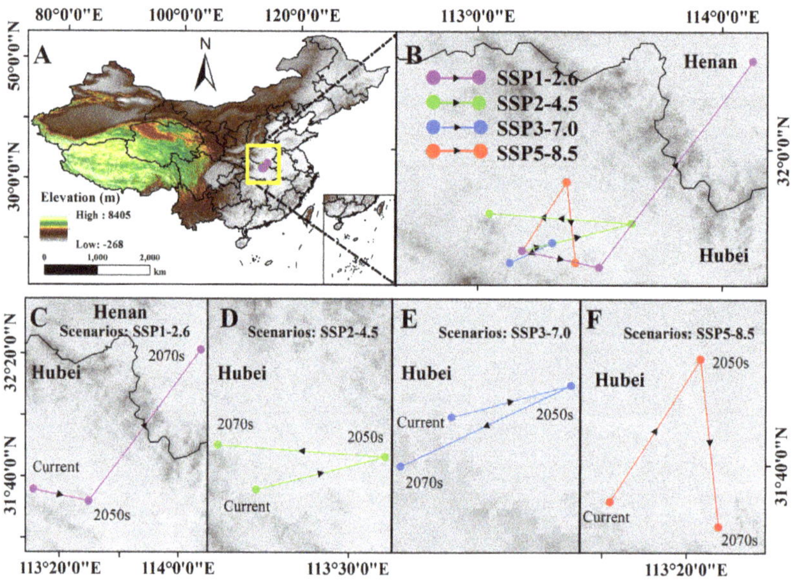

Figure 7. (**A**) Moving trajectories of potential distribution center routes for *B. xylophilus* in China; (**B**) The *B. xylophilus* movement routes under different shared socio-economic path models; (**C**) SSP1.0-2.6; (**D**) SSP2.0-4.5; (**E**) SSP3.0-7.0; (**F**) SSP5.0-8.5.

Table 3. Trends in longitude, latitude, and migration distance of *B. xylophilus* in different periods.

Shared Socioeconomic Pathways	Longitude (°E)	Latitude (°N)	Center Migration Distance (km)
Current	113.18	31.60	-
Future-SSP1.0-2.6 2040–2060	113.49	31.53	30.52
Future-SSP1.0-2.6 2060–2080	114.13	32.35	122.58
Future-SSP2.0-4.5 2040–2060	113.63	31.71	43.98
Future-SSP2.0-4.5 2060–2080	113.05	31.75	20.86
Future-SSP3.0-7.0 2040–2060	113.30	31.63	11.93
Future-SSP3.0-7.0 2060–2080	113.13	31.55	7.20
Future-SSP5.0-8.5 2040–2060	113.36	31.87	34.76
Future-SSP5.0-8.5 2060–2080	113.40	31.55	21.02

3.7. Niche Dynamics of *B. xylophilus*

The ecological niche similarity of *B. xylophilus* was analyzed under different climatic scenarios using ecological niche overlap (Figure 8). The Schoene's D index ranged from 0.74 (current vs. SSP5.0-8.5-2070s) to 0.89 (current vs. SSP1.0-2.6-2070s). The higher overlap values indicate that the ecological niche of *B. xylophilus* is highly similar under different climate scenarios. This implies that despite future changes in climatic conditions, the ecological needs and adaptations of *B. xylophilus* remain largely consistent with current climatic variables and may not change significantly in its future range because it is able to adapt to these changes.

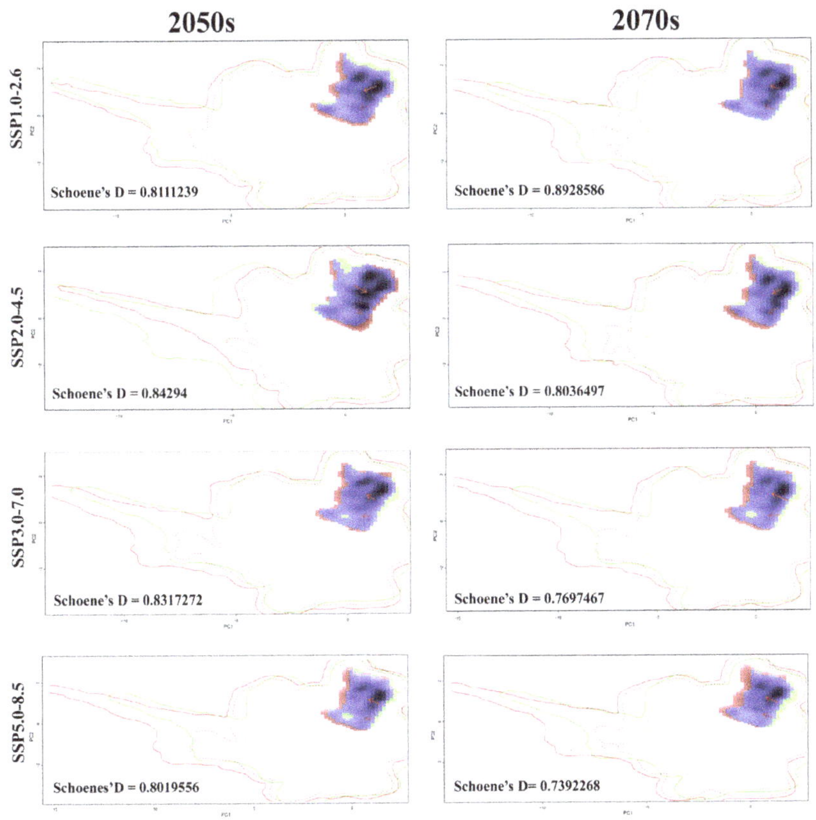

Figure 8. Future analysis of *B. xylophilus* ecological niche dynamics under different climate scenarios. The blue areas show the stable habitats of *B. xylophilus*, the green areas show the habitats it will lose, and the pink areas show the habitats that are shifting. The red arrow shows the direction of niche migration for *B. xylophilus* in future climates. Solid and dashed lines connect the current and future distribution centers, with the dashed lines outlining the potential range of distribution changes.

4. Discussion

The accuracy of ecological niche model prediction results depends on the selection of input data and adjustment of model parameters, both of which directly affect the output of the results [40,41]. In this experiment, we used more distribution point data than online books and public online databases and obtained the most recent distribution point data through the National Forestry and Grassland Administration (2024, NFGA) and the literature [25,26], which improved the accuracy and timeliness of the study. We also considered a variety of environmental factors, including climate change, terrain, ultraviolet radiation, NDVI, elevation, and human activities, which are considered to be key factors influencing the distribution and dynamics of *B. xylostella*. In addition, in order to prevent model overfitting, the MaxEnt model parameter was adjusted in this study, which helped to improve the accuracy and precision of prediction. The results of this study pointed out that all average AUC values exceeded 0.92, indicating the high predictive reliability of the model and the reliability of the findings, which is crucial for the development of science-based management strategies.

In this study, the potentially suitable habitats of *B. xylophilus* were found to be limited by Bio14, Bio12, Bio20, and Bio21 under the influence of no anthropogenic disturbance through the combined analysis of the contribution rate, the sequence importance value, and the Jackknife method, which indicates that the combined effect of humid and thermal conditions plays an important role in the potential distribution of *B. xylophilus* [42,43]. With climate change, there will be a gradual decrease in precipitation in the future and a decrease in water content in the soil, thus exacerbating drought stress. Changes in temperature and precipitation patterns due to climate change may have significant impacts on the species. Global warming may cause species to expand their ranges to higher latitudes or higher altitudes, while changes in precipitation patterns may affect species' food chains and habitats [44]. Thus, instability in precipitation and temperature may threaten the distribution of *B. xylophilus*, leading to changes in their range and thus spreading to form new distribution patterns [45]. With the involvement of human activities, the global human footprint (Bio30) and the global human influence index (Bio31) are important human factors influencing the distribution of *B. xylophilus*. Expression of agricultural work, road construction, and urbanization can all lead to fragmentation of natural habitats, which in turn reduces areas suitable for *B. xylophilus* survival.

The actual distribution of *B. xylophilus* and the potential distribution area predicted by the model are clustered in northern and central China. These areas have suitable geo-climatic conditions for *B. xylophilus* growth, providing a favorable environment for the survival, reproduction, and spread of *B. xylophilus* [46]. In addition, the transmission of *B. xylophilus* is mainly dependent on its insect vectors, *Monochamus alternatus* and *Monochamus saltuarius*. The larvae of these insects bore holes within the trunks of trees to form channels where *B. xylophilus* is usually found in the woody parts of pine trees. As the larvae of the family Cerambycidae develop within the pine tree, they may carry *B. xylophilus*, and when they develop into adults, they burrow out of the pine tree. Cerambycidae insects may transmit *B. xylophilus* to healthy pines as they search for new pines in which to lay their eggs [47,48]. Therefore, the range and number of these insect vectors play a key role in the spread of pine nematode dispersal. The climatic conditions in central and northern provide favorable conditions for the growth of pine trees, as well as suitable environments for the reproduction and activities of Cerambycidae [49]. These factors work together to make southern, central, and northeastern China high-risk areas for the spread and propagation of the PWD.

Climate change has altered the phenology and geographical distribution of the *B. xylophilus*, resulting in expansion and contraction of its range [50]. The results show that the expansion areas occur mainly in southern, central, and northeastern China, especially in Liaoning, Jilin, Shandong, Henan, Fujian, and Guangdong. These areas may become more suitable for the survival and reproduction of *B. xylophilus* due to changes in temperature and precipitation patterns caused by climate change. On the contrary, the contraction areas of *B. xylophilus* mainly occurred in Guizhou, Jiangsu, and Liaoning, which may be due to environmental conditions that are no longer suitable for *B. xylophilus* survival due to climate change [51]. Under the low–medium carbon greenhouse gas (GHG) emission scenarios (SSP1.0-2.6 and SSP2.0-4.5), the contracted area of *B. xylophilus* was larger than the expanded area, suggesting that suitable habitats for *B. xylophilus* will be continuously lost with climate change. However, under the medium-to-high carbon greenhouse gas (GHG) emission scenarios (SSP3.0-7.0 and SSP5.0-8.5), the contracted area of *B. xylophilus* was smaller than the expanded area, which may indicate that *B. xylophilus* is less dependent on its current habitat and is able to adapt to a wider range of environmental conditions [52]. Similarly, it has been suggested that slight changes in the potential distribution of species as a result of climate change may be a consequence of the occurrence of their ecological amplitude and ecological niche width [53]. In this study, the ecological niche overlap analysis revealed that the ecological niche overlap of *B. xylophilus* was greater than 0.74 under different climatic scenarios, which indicates that *B. xylophilus* has a high ecological niche width and adaptability to maintain its ecological niche under different climatic conditions. In addition, the suitability areas of *B. xylophilus* did not change significantly under future climate change, which may imply that *B. xylophilus* has a strong adaptive capacity and can maintain its distribution range normally under the background of climate change.

Changes in temperature and precipitation patterns due to climate change can affect the geographic distribution of *B. xylophilus*, leading to a shift in its range center. This shift may occur towards higher or lower latitudes, depending on the different shared socioeconomic pathways (SSP) scenarios, which model the extent and rate of climate change [54]. Under the SSP1.0-2.6 and SSP2.0-4.5 scenarios, the *B. xylophilus* moves to higher latitudes, possibly because projected climate change under the low greenhouse gas (GHG) emissions scenario results in more favorable temperature and precipitation conditions at higher latitudes. In contrast, *B. xylophilus* migration to lower latitudes under the SSP3.0-7.0 and SSP5.0-8.5 scenarios may be due to the degradation of suitable areas at lower latitudes as a result of projected climate change under the medium–high greenhouse gas (GHG) emissions scenarios, forcing the species to search for new suitable habitats [55]. It is worth noting that migration to lower latitudes may be relative and does not necessarily mean that climatic conditions at lower latitudes become more favorable, but may simply be due to the degradation of suitable areas at higher latitudes [56]. In addition, geographical anomalies may also cause species to migrate to lower latitudes, suggesting that changes in species' distributions may not only be due to climate change but may also be influenced by other factors, such as human activities that help certain species to overcome natural barriers and thus expand their ranges.

From the results of this study, we found that *B. xylophilus* is at risk of spreading to the northeast (Liaoning and Jilin) and central regions (Henan and Shandong) in the future, and that failure to prevent and control it in time may lead to the death of a large number of pine trees, which constitutes an ecological disaster. To address this challenge, China has taken a series of comprehensive control measures [57], which include the following: 1. timely detection of outbreaks through quarantine and monitoring to prevent the spread of the disease; 2. removal of infected trees to reduce the source of transmission of the pathogen; 3. control of insect vectors that transmit the *B. xylophilus*; 4. utilization of trunk injection techniques to inject insecticides into the tree, which distributes the drug throughout the tree by transpiration to kill the nematode; and 5. adoption of the natural enemies of the nematode, such as predatory nematodes or other microorganisms, to control pine nematode populations. A combination of the above measures can effectively control the spread of *B. xylophilus* and protect forest resources [58]. Nevertheless, due to the complexity and rapid spread of *B. xylophilus*, control requires sustained efforts and interdisciplinary cooperation to meet the challenges posed by *B. xylophilus* and thus protect China's forest resources and ecological balance [59,60].

5. Conclusions

In this study, an optimized MaxEnt model was used to assess the potential distribution area and dynamics of *B. xylophilus* in China based on references and the latest distribution data published by the National Forestry and Grassland Administration, and a variety of environmental factors, including climate change, terrain, vegetation, ultraviolet radiation, and human interference. The prediction results showed that temperature, precipitation, NDVI, elevation, and human activities were the key factors affecting the distribution of *B. xylophilus*. Under current climatic conditions, suitable habitat areas are mainly distributed in Liaoning, Shandong, Zhejiang, Anhui, Jiangsu, Shaanxi, Henan, Hubei, Hunan, Guangdong, Fujian, Chongqing, Sichuan and Guangxi. Between 2050 and 2070, the potential range area of *B. xylophilus* is at risk of spreading to the northeast, suggesting that focused monitoring in the northeast is needed to prevent *B. xylophilus* from damaging pines in the region. Furthermore, the high ecological niche similarity indicates that *B. xylophilus* will be able to locate similar habitats under future climatic conditions, which may facilitate its spread and ability to adapt to new environments. The results of this study will provide valuable references for the development of sustainable management strategies for agriculture and forestry in China under the challenge of climate change, which will help to reduce economic losses and promote the healthy and stable development of ecosystems.

Supplementary Materials: The following supporting information can be downloaded at https://www.mdpi.com/article/10.3390/f15071253/s1, Figure S1: Correlation among the 31 bioclimatic variables; Figure S2: Tuning parameters for *Bursaphelenchus xylophilus* in predicting potential distribution regions using the MaxEnt model. Table S1: Area of *B. xylophilus* under current (with and without human activity disturbance) and future climate scenarios; Table S2: Relative change in potential range size of *B. xylophilus* under different future climate scenarios; Table S3: Multivariate environmental similarity surfaces (MESS) affecting the distribution of *B. xylophilus* under future climate conditions.

Author Contributions: Conceptualization, L.Z.; methodology, L.Z.; software, L.Z.; validation, L.Z. and P.W.; formal analysis, L.Z.; investigation, P.W. and G.X.; resources, W.W.; data curation, P.W. and G.X.; writing—original draft preparation, L.Z.; writing—review and editing, L.Z.; visualization, L.Z. and P.W.; supervision, W.W.; project administration, W.W.; funding acquisition, P.W. and W.W. All authors have read and agreed to the published version of the manuscript.

Funding: This work was funded by the Natural Science Foundation of Hubei Province (No. 2024AFB254) and the National Natural Science Foundation of China (No. 31672327).

Data Availability Statement: The data in this study are available from the corresponding author.

Acknowledgments: We appreciate the constructive comments of the anonymous reviewers and processing editors and their help in revising the manuscript.

Conflicts of Interest: The authors declare no conflicts of interest.

References

1. Sun, S.X.; Zhang, Y.; Huang, D.Z.; Wang, H.; Cao, Q.; Fan, P.X.; Yang, N.; Zheng, P.M.; Wang, R.Q. The effect of climate change on the richness distribution pattern of oaks (*Quercus* L.) in China. *Sci. Total Environ.* **2020**, *744*, e140786. [CrossRef] [PubMed]
2. Wei, X.J.; Xu, D.P.; Zhuo, Z.H. Predicting the impact of climate change on the geographical distribution of Leafhopper, *Cicadella viridis* in China through the MaxEnt model. *Insects* **2023**, *14*, 586. [CrossRef] [PubMed]
3. Dawson, T.P.; Jackson, S.T.; House, J.I.; Prentice, I.C.; Mace, G.M. Beyond predictions: Biodiversity conservation in a changing climate. *Science* **2011**, *332*, 53–58. [CrossRef] [PubMed]
4. Urban, M.C. Accelerating extinction risk from climate change. *Science* **2015**, *348*, 571–573. [CrossRef] [PubMed]
5. Pound, K.L.; Larson, C.A.; Passy, S.I.; Webb, T. Current distributions and future climate-driven changes in diatoms, insects and fish in U.S. streams. *Global Ecol. Biogeogr.* **2021**, *30*, 63–78. [CrossRef]
6. Liu, T.; Liu, H.Y.; Wang, Y.J.; Yang, Y.X. Climate change impacts on the potential distribution pattern of *Osphya* (Coleoptera: Melandryidae), an old but small beetle group distributed in the Northern Hemisphere. *Insects* **2023**, *14*, 476. [CrossRef] [PubMed]
7. Hardy, P.B.; Kinder, P.M.; Sparks, T.H.; Dennis, R.L.H. Elevation and habitats: The potential of sites at different altitudes to provide refuges for phytophagous insects during climatic fluctuations. *J. Insect. Conserv.* **2010**, *14*, 297–303. [CrossRef]
8. Wu, Z.; Li, M.; Wang, B.; Tian, Y.; Quan, Y.; Liu, J. Analysis of factors related to forest fires in different forest ecosystems in China. *Forests* **2022**, *13*, 1021. [CrossRef]
9. Sharma, A.; Jose, S.; Bohn, K.K.; Andreu, M.G. Effects of reproduction methods and overstory species composition on understory light availability in longleaf pine-slash pine ecosystems. *For. Ecol. Manag.* **2012**, *284*, 23–33. [CrossRef]
10. Tuomola, J.; Gruffudd, H.; Ruosteenoja, K.; Hannunen, S. Could pine wood nematode (*Bursaphelenchus xylophilus*) cause pine wilt disease or even establish inside healthy trees in finland Now-Or ever? *Forests* **2021**, *12*, 1679. [CrossRef]
11. Shi, L.; Wang, L.; Shi, X.; Luo, L.; Ye, J. Negative effects of free-living nematodes on the populations of *Bursaphelenchus xylophilus* in dead pine trees. *Biol. Control* **2022**, *168*, 104858. [CrossRef]

12. Takai, K.; Soejima, T.; Suzuki, T.; Kawazu, K. Development of a water-soluble preparation of emamectin benzoate and its preventative effect against the wilting of pot-grown pine trees inoculated with the pine wood nematode, *Bursaphelenchus xylophilus*. *Pest. Manag. Sci.* **2001**, *57*, 463–466. [CrossRef] [PubMed]
13. Hu, L.; Wu, X.; Ding, X.; Ye, J. Comparative transcriptomic analysis of candidate effectors to explore the infection and survival strategy of *Bursaphelenchus xylophilus* during different interaction stages with pine trees. *BMC Plant Biol.* **2021**, *21*, 224. [CrossRef] [PubMed]
14. Zhang, X.L.; Qin, X.D.; Alvarez, F.; Chen, Z.N.; Wu, Z.J. Potential impact of land–use change on habitat quality in the distribution range of crocodile lizards in China. *Ecol. Evol.* **2022**, *12*, 9390. [CrossRef]
15. Wang, D.; Cui, B.C.; Duan, S.S.; Chen, J.J.; Fan, H.; Lu, B.B.; Zheng, J.H. Moving north in China: The habitat of *Pedicularis kansuensis* in the context of climate change. *Sci. Total Environ.* **2019**, *697*, 133979. [CrossRef] [PubMed]
16. Zhang, H.; Song, J.Y.; Zhao, H.X.; Li, M.; Han, W.H. Predicting the distribution of the invasive species *Leptocybe invasa*: Combining MaxEnt and geodetector models. *Insects* **2021**, *12*, 92. [CrossRef] [PubMed]
17. Iannella, M.; Console, G.; Cerasoli, F.; De Simone, W.; D'Alessandro, P.; Biondi, M. A step towards SDMs: A "couple-and-weigh" framework based on accessible data for biodiversity conservation and landscape planning. *Divers. Distrib.* **2021**, *27*, 2412–2427. [CrossRef]
18. Hadi Ahmad, M.; Abubakar, A.; Yusoff Ishak, M.; Shehu Danhassan, S.; Zhang, J.H.; Aiatalo, J.M. Modeling the influence of daily temperature and precipitation extreme indices on vegetation dynamics in Katsina State using statistical downscaling model (SDM). *Ecol. Indic.* **2023**, *155*, 110979. [CrossRef]
19. Wang, Z.L.; Xu, D.P.; Liao, W.K.; Xu, Y.; Zhuo, Z.H. Predicting the current and future distributions of *Frankliniella occidentalis* (Pergande) based on the MaxEnt species distribution model. *Insects* **2023**, *14*, 458. [CrossRef]
20. Harte, J.; Umemura, K.; Brush, M.; Chase, J. DynaMETE: A hybrid MaxEnt–plus–mechanism theory of dynamic macroecology. *Ecol. Lett.* **2021**, *24*, 935–949. [CrossRef]
21. Guga, S.; Xu, J.; Riao, D.; Li, K.W.; Han, A.; Zhang, J.Q. Combining MaxEnt model and landscape pattern theory for analyzing interdecadal variation of sugarcane climate suitability in Guangxi, China. *Ecol. Indic.* **2021**, *131*, 108152. [CrossRef]
22. Soliman, M.M.; Al-Khalaf, A.A.; El-Hawagry, M.S.A. Effects of climatic change on potential distribution of *Spogostylum ocyale* (Diptera: Bombyliidae) in the middle east using Maxent modelling. *Insects* **2023**, *14*, 120. [CrossRef] [PubMed]
23. Lee, C.M.; Lee, D.S.; Kwon, T.S.; Athar, M.; Park, Y.S. Predicting the global distribution of *Solenopsis geminata* (Hymenoptera: Formicidae) under climate change using the MaxEnt model. *Insects* **2021**, *12*, 229. [CrossRef] [PubMed]
24. Zheng, Y.X.; Chao, Y.; Matsushita, N.; Lian, C.L.; Geng, Q.F. Analysis of the distribution pattern of the ectomycorrhizal fungus *Cenococcum geophilum* under climate change using the optimized MaxEnt model. *Ecol. Evol.* **2023**, *13*, 10565. [CrossRef] [PubMed]
25. Ouyang, X.; Chen, A.; Li, Y.; Han, X.; Lin, H. Predicting the potential distribution of pine wilt disease in china under climate change. *Insects* **2022**, *13*, 1147. [CrossRef] [PubMed]
26. Xiao, Y.; Guo, Q.; Xie, N.; Yuan, G.; Liao, M.; Gui, Q.; Ding, G. Predicting the global potential distribution of *Bursaphelenchus xylophilus* using an ecological niche model: Expansion trend and the main driving factors. *BMC Ecol Evol.* **2024**, *24*, 48. [CrossRef] [PubMed]
27. Warren, D.L.; Matzke, N.J.; Cardillo, M.; Baumgartner, J.B.; Beaumont, L.J.; Turelli, M.; Glor, R.E.; Huron, N.A.; Simoes, M.; Iglesias, T.L.; et al. ENMTools 1.0: An R package for comparative ecological biogeography. *Ecography* **2021**, *44*, 504–511. [CrossRef]
28. Wu, T.W.; Lu, Y.X.; Fang, Y.J.; Xin, X.G.; Li, L.; Li, W.P.; Jie, W.H.; Zhang, J.; Liu, Y.M.; Zhang, L.; et al. The Beijing climate center climate system model (BCC-CSM): The main progress from CMIP5 to CMIP6. *Geosci. Model Dev.* **2019**, *12*, 1573–1600. [CrossRef]
29. Li, D.X.; Li, Z.X.; Liu, Z.W.; Yang, Y.J.; Khoso, A.G.; Wang, L.; Liu, D.G. Climate change simulations revealed potentially drastic shifts in insect community structure and crop yields in China's farmland. *J. Pest Sci.* **2022**, *96*, 1–15. [CrossRef]
30. Phillips, S.J.; Anderson, R.P.; Schapire, R.E. Maximum entropy modeling of species geographic distributions. *Ecol. Model.* **2006**, *190*, 231–259. [CrossRef]
31. Kaky, E.; Nolan, V.; Alatawi, A.; Gilbert, F. A comparison between ensemble and MaxEnt species distribution modelling approaches for conservation: A case study with Egyptian medicinal plants. *Ecol. Inf.* **2020**, *60*, 101150. [CrossRef]
32. Maruthadurai, R.; Das, B.; Ramesh, R. Predicting the invasion risk of rugose spiraling Whitefly, *Aleurodicus rugioperculatus*, in India based on CMIP6 projections by MaxEnt. *Pest Manag. Sci.* **2023**, *79*, 295–305. [CrossRef] [PubMed]
33. Yoon, S.; Lee, W.H. Assessing potential European areas of Pierce's disease mediated by insect vectors by using spatial ensemble model. *Front. Plant Sci.* **2023**, *14*, 1209694. [CrossRef]
34. Aidoo, O.F.; Souza, P.G.C.; Silva, R.S.; Santana, P.A.; Picanço, M.C.; Kyerematen, R.; Sètamou, M.; Ekesi, S.; Borgemeister, C. Climate-induced range shifts of invasive species (*Diaphorina citri* Kuwayama). *Pest Manag. Sci.* **2022**, *78*, 2534–2549. [CrossRef]
35. Yang, S.L.; Wang, H.M.; Tong, J.P.; Bai, Y.; Alatalo, J.M.; Liu, G.; Fang, Z.; Zhang, F. Impacts of environment and human activity on grid-scale land cropping suitability and optimization of planting structure, measured based on the MaxEnt model. *Sci. Total Environ.* **2022**, *836*, 155356. [CrossRef] [PubMed]
36. Gao, R.H.; Liu, L.; Zhao, L.J.; Cui, S.P. Potentially suitable geographical area for *Monochamus alternatus* under current and future climatic scenarios based on optimized MaxEnt model. *Insects* **2023**, *14*, 182. [CrossRef]
37. Huang, M.Y.; Ge, X.Z.; Shi, H.L.; Tong, Y.G.; Shi, J. Prediction of current and future potential distributions of the Eucalyptus pest *Leptocybe invasa* (Hymenoptera: Eulophidae) in China using the CLIMEX model. *Pest Manag. Sci.* **2019**, *75*, 2958–2968. [CrossRef] [PubMed]

38. Wang, B.X.; Hof, A.R.; Matson, K.D.; Langevelde, F.; Ma, C.S. Climate change, host plant availability, and irrigation shape future region–specific distributions of the *Sitobion* grain aphid complex. *Pest Manag. Sci.* **2023**, *79*, 2311–2324. [CrossRef]
39. Liu, T.; Liu, H.; Li, Y.; Yang, Y. Staying on the current niche: Consensus model reveals the habitat loss of a critically endangered dragonfly *Libellula angelina* under climate changes. *J Insect Conserv.* **2024**, *28*, 483–499. [CrossRef]
40. Ge, X.Z.; He, S.Y.; Zhu, C.Y.; Wang, T.; Xu, Z.C.; Zong, S.X. Projecting the current and future potential global distribution of *Hyphantria cunea* (Lepidoptera: Arctiidae) using CLIMEX. *Pest Manag. Sci.* **2019**, *75*, 160–169. [CrossRef]
41. Gao, H.; Qian, Q.Q.; Liu, L.J.; Xu, D.P. Predicting the distribution of *Sclerodermus sichuanensis* (Hymenoptera: Bethylidae) under climate change in China. *Insects* **2023**, *14*, 475. [CrossRef] [PubMed]
42. Zhao, M.Z.; Duan, Q.; Shen, X.Y.; Zhang, S.Y. Climate change influences the population density and suitable area of *Hippotiscus dorsalis* (Hemiptera: Pentatomidae) in China. *Insects* **2023**, *14*, 135. [CrossRef] [PubMed]
43. Santana, P.A.; Kumar, L.; Da Silva, R.S.; Picanço, M.C. Global geographic distribution of *Tuta absoluta* as affected by climate change. *J. Pest Sci.* **2019**, *92*, 1373–1385. [CrossRef]
44. Kumar, A.V.; Zimova, M.; Martin, T.E.; Mills, L.S. Contrasting seasonal effects of climate change influence density in a cold–adapted species. *Global Change Biol.* **2022**, *28*, 6228–6238. [CrossRef] [PubMed]
45. Santana, P.A.; Kumar, L.; Da Silva, R.S.; Pereira, J.L.; Picanço, M.C. Assessing the impact of climate change on the worldwide distribution of *Dalbulus maidis* (DeLong) using MaxEnt. *Pest Manag. Sci.* **2019**, *10*, 2706–2715. [CrossRef] [PubMed]
46. Mazziotta, A.; Lundström, J.; Forsell, N.; Moor, H.; Eggers, J.; Subramanian, N.; Aquilué, N.; Morán–Ordóñez, A.; Brotons, L.; Snäll, T. More future synergies and less trade–offs between forest ecosystem services with natural climate solutions instead of bioeconomy solutions. *Global Change Biol.* **2022**, *28*, 6333–6348. [CrossRef] [PubMed]
47. Lee, B.; Yang, J.; Beckett, S.; Ren, Y. Preliminary trials of the ethanedinitrile fumigation of logs for eradication of *Bursaphelenchus xylophilus* and its vector insect *Monochamus alternatus*: C 2 N 2 fumigation of logs to eradicate *B. xylophilus* and its vector insect *M. alternatus. Pest. Manag. Sci.* **2017**, *73*, 1446–1452. [CrossRef]
48. Wang, J.; Zhang, S.; Zheng, Y. Feeding preferences and responses of *Monochamus saltuarius* to volatile components of host pine trees. *Insects* **2022**, *13*, 888. [CrossRef] [PubMed]
49. Meng, F.; Liu, Z.; Li, Y.; Zhang, X. Genes encoding potential molecular mimicry proteins as the specific targets for detecting *Bursaphelenchus xylophilus* in PCR and loop-mediated isothermal amplification assays. *Front. Plant Sci.* **2022**, *13*, 890949. [CrossRef]
50. Wedegärtner, R.E.M.; Lembrechts, J.J.; Van der Wal, R.; Barros, A.; Chauvin, A.; Janssens, I.; Graae, B.J. Hiking trails shift plant species' realized climatic niches and locally increase species richness. *Divers. Distrib.* **2022**, *28*, 1416–1429. [CrossRef]
51. Haran, J.; Roques, A.; Bernard, A.; Robinet, C.; Roux, G. Altitudinal barrier to the spread of an invasive species: Could the pyrenean chain slow the natural spread of the pinewood nematode? *PLoS ONE.* **2015**, *10*, e0134126. [CrossRef] [PubMed]
52. Spaak, J.W.; Carpentier, C.; De Laender, F.; Mayfield, M. Species richness increases fitness differences, but does not affect niche differences. *Ecol. Lett.* **2021**, *24*, 2611–2623. [CrossRef] [PubMed]
53. Yin, X.; Jarvie, S.; Guo, W.; Deng, T.; Mao, L.; Zhang, M.; Chu, C.; Qian, H.; Svenning, J.; He, F.; et al. Niche overlap and divergence times support niche conservatism in eastern Asia-eastern north America disjunct plants. *Global Ecol. Biogeogr.* **2021**, *30*, 1990–2003. [CrossRef]
54. Ling, P.Y.; Aguilar-Amuchastegui, N.; Baldwin-Cantello, W.; Rayden, T.; Gordon, J.; Dainton, S.; Bagwill, A.L.; Pacheco, P. Mapping global forest regeneration–an untapped potential to mitigate climate change and biodiversity loss. *Environ. Res. Lett.* **2023**, *18*, 54025. [CrossRef]
55. Zhang, Z.X.; Kass, J.M.; Mammola, S.; Koizumi, I.; Li, X.C.; Tanaka, K.; Ikeda, K.; Suzuki, T.; Yokota, M.; Usio, N. Lineage-level distribution models lead to more realistic climate change predictions for a threatened crayfish. *Divers. Distrib.* **2021**, *27*, 684–695. [CrossRef]
56. Jiang, K.; Teuling, A.J.; Chen, X.; Huang, N.; Wang, J.; Zhang, Z.; Gao, R.; Men, J.; Zhang, Z.; Wu, Y.; et al. Global land degradation hotspots based on multiple methods and indicators. *Ecol. Indic.* **2024**, *158*, 111462. [CrossRef]
57. Zhou, X.; Chen, S.; Lu, F.; Guo, K.; Huang, L.; Su, X.; Chen, Y. Nematotoxicity of a Cyt–like protein toxin from *Conidiobolus obscurus* (entomophthoromycotina) on the pine wood nematode *Bursaphelenchus xylophilus. Pest. Manag. Sci.* **2021**, *77*, 686–692. [CrossRef]
58. Shi, F.; Ge, S.; Hou, Z.; Xu, Y.; Tao, J.; Wu, H.; Zong, S. Species–specific primers for rapid detection of *Monochamus saltuarius*, an effective vector of *Bursaphelenchus xylophilus* in China. *J. Appl. Entomol.* **2022**, *146*, 636–647. [CrossRef]
59. Park, C.G.; Son, J.; Lee, B.; Cho, J.H.; Ren, Y. Comparison of ethanedinitrile (C_2N_2) and metam sodium for control of *Bursaphelenchus xylophilus* (Nematoda: Aphelenchidae) and *Monochamus alternatus* (Coleoptera: Cerambycidae) in naturally infested logs at low temperatures. *J. Econ. Entomol.* **2014**, *107*, 2055–2060. [CrossRef]
60. Gruffudd, H.R.; Schröder, T.; Jenkins, T.A.R.; Evans, H.F. Modelling pine wilt disease (PWD) for current and future climate scenarios as part of a pest risk analysis for pine wood nematode *Bursaphelenchus xylophilus* (Steiner and Buhrer) Nickle in Germany. *J. Plant Dis. Prot.* **2019**, *126*, 129–144. [CrossRef]

plants

MDPI

Article

Contrasting Responses of Alien and Ancient Forest Indicator Plant Species to Fragmentation Process in the Temperate Lowland Forests

Mirjana Šipek [1,*], Lado Kutnar [2], Aleksander Marinšek [2] and Nina Šajna [1]

1 Faculty of Natural Sciences and Mathematics, University of Maribor, 2000 Maribor, Slovenia
2 Slovenian Forestry Institute, 1000 Ljubljana, Slovenia
* Correspondence: mirjana.sipek1@um.si

Abstract: Fragmentation is one of the major threats to biodiversity. In a fragmented landscape, forest specialists are losing suitable forest habitats with specific site and microclimate conditions, which results in their local extinction. Conversely, the invasion of alien species is facilitated by open forest areas and increased boundaries between forest fragments and adjacent land. We studied the effect of fragmentation in terms of fragment size impact on overall plant species richness and on selected ecologically important groups' richness, composition, and diversity. We surveyed vegetation in the interior of 47 fragments of various sizes and one unfragmented reference forest. Our results reveal that the effect of fragmentation is complex and differs for studied plant groups. Decreasing fragment size negatively affects the overall plant richness and richness of native and ancient forest indicator plants as well as their diversity, while the effect is positive for alien plants. The highest proportion of ancient forest indicator plant species and the lowest proportion of alien plants in the unfragmented forest underline the great conservation value of forest fragments. At the same time, our results reveal that large and diverse forest ecosystems are susceptible to biological invasions as well.

Keywords: forest fragment; indicator plants; invasion; diversity; habitat modification; habitat conservation

1. Introduction

Fragmentation has a complex effect on biological communities by altering the biotic and abiotic components of an ecosystem [1]. The most evident negative effects of forest fragmentation are habitat loss, wide forest edge formation, declining connectivity, and consequent isolation of populations [2,3]. The effect of fragmentation on biota is either negative, neutral, or positive [4,5]. The abundance of certain species may decline sharply, or they may even become locally extinct, and other species may remain stable, while the abundance of some may increase dramatically [3].

To understand the effect of habitat fragmentation on species richness, island biogeography theory provides a theoretical framework [6,7]. A fragmented landscape is considered a dynamic system where species are disappearing and appearing on a local scale (fragment level) because of extinction and immigration processes, respectively. The intensity of each process depends on fragment size, isolation, and connectivity. Larger fragments tend to have higher species richness, species extinction is slower, and they do not suffer as much from the negative consequences of isolation (e.g., lost connectivity within a metapopulation) in comparison to smaller fragments [6].

Considering the island biogeography theory, some forest species are expected to disappear in smaller forest fragments and an influx of species from the surrounding landscape, which is a source of colonizers, is expected [8,9]. Forest specialist species in small and isolated forest fragments are subjected to decreased fitness because of a lack of pollinators [10] or inbreeding following population isolation [11–13]. This may result in lower seed set and abundance [13] and eventually cause local extinction of these species. Ancient forest indicator plant species (AFI) is a group of forest specialist plants that differ ecologically from other forest plants. They are poor colonizers, characterized by greater shade and stress tolerance with predominant geophyte and hemicryptophyte life-form [14]. Therefore, Hermy et al. [14] suggested that the presence of AFI may serve as an important biodiversity species indicator for forests.

Moreover, the fragmented landscape is highly susceptible to biological invasions because of the increased boundary between forest and adjacent non-forest land [15–17]. Therefore, the entry of non-forest and alien species into forest fragments is among the most commonly reported observations [2,3,18,19]. In smaller and more isolated forest fragments, species turnover should appear to a greater extent than in larger, more connected fragments [3,17]. This process is especially important in light of nature conservation because, in an anthropogenically modified landscape, fragmentation and biological invasions are one of the main causes of forest ecosystem degradation [20,21]. However, the effect of fragment size on plant species richness is highly variable, showing a lack of consistency [1,9,22,23].

Slovenia is among European countries with the highest share of forests, comprising around 60% of the country's area [24,25]. Larger, continuous forested areas prevail in hilly areas, while lowland forests are more fragmented, consisting of various spatial configurations with the most prominent difference among them—the fragment size. Forest fragments embedded in the agricultural matrix and intertwined with urbanized areas host alien plant species, among them several invasive ones [19]. Lowlands in NE Slovenia are an appropriate landscape to study fragmentation effects because forest fragments of various sizes are placed in a relatively homogenous area with similar abiotic conditions regarding soil and climate. Current research has shown that the invasion level of lowland forests differs among various forest habitat types [26], while the effect of the forest fragments' size on invasion level is yet to be studied. Moreover, Marinšek and Kutnar [26] reported that invasive alien plants have a negative effect on the regeneration of native forest species. Therefore, a better understanding of the spatial characteristics of forest fragments in the invasion process is needed for appropriate management planning to prevent new invasions and to enable the conservation of forest ecosystems.

In our study, we evaluated the fragmentation process with two descriptors: i) forest fragment size and ii) the proportion of forest in a buffer zone of 500 m in the surroundings of the forest fragment as a surrogate for isolation/connectivity. The effect of fragment size on plant richness per unit area is less predictable [15] than per the whole fragment, where the correlation is consistently positive because of increasing habitat heterogeneity [27]. We expect that the effect of forest fragment size among different ecological plant groups varies. The study aimed to test whether the size of the lowland forest fragment affects (1) plant species richness of overall vascular plants, with special consideration to native, alien, alien invasive, and ancient forest indicator species, (2) diversity per unit area, and (3) plant species composition expressed as a ratio between different ecological groups.

2. Results

2.1. Floristic Richness and Composition

In the forest fragments at Apaško polje and Dravsko polje, we found 124 native plant species (37 AFI) and 17 alien plants, 10 of which are considered invasive (Table 1).

In the unfragmented forest, 87 native (30 AFI) and 9 alien plants were listed. Among aliens, 6 were invasive, while the remaining 3 alien tree species (*Juglans nigra, Fraxinus americana* and *Populus x canadensis*) were planted in the Murska šuma forest for forestry purposes in the past and, in Slovenia, are considered as naturalized (Table 1).

Table 1. List of alien plant species and ancient forest indicator plants (AFI) present in forest fragments of Apaško polje and Dravsko polje (N = 47) and unfragmented forest Murska šuma (N = 29). Invasive alien plants (according to Jogan et al. [28], *P. americana* was later recognized as invasive) are in bold.

Alien Plant	% in Forest Fragments	% in Unfragmented Forest	AFI	% in Forest Fragments	% in Unfragmented Forest
***Acer negundo* L.**	2.13	3.45	*Acer campestre* L.	17.02	100
***Ailanthus altissima* (Mill.) Swingle**	4.26	0	*Asarum europaeum* L.	12.77	37.93
Allium sp.	2.13	0	*Athyrium filix-femina* (L.) Roth	0	37.93
***Ambrosia artemisiifolia* L.**	0	3.45	*Berberis vulgaris* L.	2.13	0
***Duchesnea indica* (Andrews) Focke**	4.26	0	*Brachypodium sylvaticum* (Huds.) Beauv.	19.15	51.72
***Erigeron annuus* (L.) Pers.**	4.26	6.9	*Carex pallescens* L.	2.13	0
Fraxinus americana L.	0.00	6.9	*Carex pendula* Huds.	0	44.83
***Impatiens glandulifera* Royle**	8.51	20.69	*Carex remota* L.	2.13	44.83
***Impatiens parviflora* DC.**	36.17	3.44	*Carex sylvatica* Huds.	12.77	89.66
Juglans nigra L.	0	13.79	*Circaea lutetiana* L.	4.26	86.21
Oxalis fontana Bunge	4.26	0	*Clematis vitalba* L.	2.13	0
***Parthenocissus quinquefolia* (L.) Planch.**	6.38	0	*Convallaria majalis* L.	2.13	0
***Phytolacca americana* L.**	17.02	0	*Cornus mas* L.	0	3.45
Populus x canadensis Moench	0	6.9	*Cornus sanguinea* L.	19.15	65.52
Prunus laurocerasus L.	4.26	0	*Corylus avellana* L.	19.15	24.14
Prunus serotine Ehrh.	12.77	0	*Crataegus* sp.	14.89	44.83
Quercus rubra L.	2.13	0	*Dactylis glomerata subsp. lobata* (Drejer) Lindb. F.	2.13	0
***Robinia pseudoacacia* L.**	34.04	0	*Dentaria bulbifera* L.	0	17.24
***Rudbeckia laciniata* L.**	4.26	0	*Dryopteris carthusiana* (Vill.) H.P. Fuchs	14.89	0
***Solidago gigantea* Aiton**	25.53	34.48	*Dryopteris filix-mas* (L.) Schott	4.26	6.90
Taxus baccata L.	2.13	0	*Euonymus europaeus* L.	31.91	31.03
			Festuca gigantea (L.) Vill.	2.13	0
			Festuca heterophylla Lam.	6.38	0
			Galeobdolon montanum (Pers.) Rchb.	14.89	89.66
			Luzula luzuloides (Lam.) Dandy and Wilmott	12.77	0
			Luzula pilosa (L.) Willd.	4.26	0
			Maianthemum bifolium (L.) F.W.Schmidt	0	3.45
			Melampyrum sp.	2.13	0
			Milium effusum L.	0	44.83
			Oxalis acetosella L.	6.38	31.03
			Paris quadrifolia L.	0	58.62
			Polygonatum multiflorum (L.) All.	2.13	34.48
			Pteridium aquilinum (L.) Kuhn	17.02	0
			Pulmonaria officinalis L.	6.38	100
			Ranunculus auricomus L.	2.13	0
			Rhamnus catharticus L.	4.26	0
			Sanicula europaea L.	0	6.90
			Scrophularia nodosa L.	2.13	10.34
			Stachys sylvatica L.	0	79.31
			Stellaria holostea L.	0	6.90
			Stellaria nemorum L.	0	13.79
			Tilia cordata Mill.	19.15	0
			Tilia platyphyllos Scop.	2.13	0
			Ulmus laevis Pall.	0	79.31
			Ulmus minor Mill.	14.89	0
			Vaccinium myrtillus L.	8.51	0
			Viburnum opulus L.	14.89	10.34
			Vinca minor L.	2.13	0
			Viola reichenbachiana Jord. ex Boreau	4.26	55.17

Of the overall identified alien plants (21), 5 were common in unfragmented forest and smaller fragments across Dravsko and Apaško polje. Those are *Acer negundo*, *Erigeron annuus*, *Impatiens glandulifera*, *I. parviflora*, and *Solidago gigantea*, all of them invasive.

2.2. Effect of Fragment Size on Plant Species Richness

We found a significant effect of fragment size on the richness and percentage of alien plants, percentage of native, and percentage of AFI (Table 2, Figure 1). Small fragments had a high number of AFI (Figure 2), which interfered with the models for native plants and their subgroup AFI (Figure 1a,b,g,h). With the increasing size of large fragments, the richness, proportion, and cover of alien plants start to fall toward the value of the unfragmented forest. This is also true for IAS (Tables 2 and 3, Figure 1).

Table 2. Summary of linear (y = a *q + r), quadratic (y = a *q + b *q^2 + r), and polynomial (y = a *q + b *q^2 + c *q^3 + r) regression analyses of forest fragment size effect on plant species richness, composition, and diversity of plant communities. Significant coefficient estimates a, b, and c are in bold and marked as ** <0.01, * <0.05, $^{+}$ <0.1. Grey color represents statistically non-significant models. Murska šuma is not included in the analysis.

	Adjusted R^2	F-Statistic	Coefficient Estimates			Standard Error
			a	b	c	
Plant Species Richness						
Overall	0.07	**2.63** $^{+}$	10.66 $^{+}$	6.52		5.45
Native	0.05	2.21	6.34	8.02		4.87
Alien	0.19	**4.58** **	**4.32** **	−1.50	−2.65 $^{+}$	1.43
IAS	0.03	1.43	1.73	−0.67	−1.53	1.16
AFI	0.08	3.09 $^{+}$	1.27	**5.43** *		2.24
% Native	0.17	**5.79** **	**−22.87** **	15.79 $^{+}$		8.17
% Alien	0.20	**4.79** **	**22.87** **	−15.79 $^{+}$	−12.52	8.04
% IAS	0.03	1.50	9.14	−7.91	−6.44	6.47
% AFI	0.19	**6.45** **	−13.80	**38.92** **		11.49
Composition						
Alien/Native	0.19	**4.63** **	**0.30** *	−0.22 $^{+}$	−0.20 $^{+}$	0.11
IAS/Native	0.05	1.87	0.13	−0.11	−0.11	0.09
AFI/Other native	0.31	**8.02** **	−0.24	**0.95** **	**−0.67** **	0.24
Diversity						
Shannon	0.13	**7.64** **	**1.24** **			0.45
Simpson's	0.13	**7.95** **	**0.40** **			0.14
Evenness	0.09	**5.67** *	**0.31** *			0.13

Table 3. Species richness, proportion, and cover of native, alien, invasive alien, and ancient forest indicator plants in the unfragmented forest, large, medium, and small forest fragments. Average values with standard deviation are given. Different uppercase and lowercase letters represent significant differences at $p < 0.05$ for Tukey HSD or Mann–Whitney pairwise test, respectively.

Species Richness	Native		Alien		IAS		AFI	
	Average	SD	Average	SD	Average	SD	Average	SD
Unfragmented	27.90 ^{A}	5.27	1.24 ^{a}	1.41	0.72 ^{a}	0.92	12.38 ^{A}	2.31
Large	16.88 ^{B}	4.12	3.00 ^{b}	1.85	1.75 ^{a}	1.49	4.50 ^{B}	2.73
Medium	11.91 ^{B}	4.29	1.61 ^{a,b}	1.53	0.83 ^{a}	1.15	2.52 ^{B}	1.75
Small	14.25 ^{B}	5.57	0.81 ^{a}	0.98	0.75 ^{a}	0.93	3.88 ^{B}	2.60
Proportion (%)								
Unfragmented	96.99 ^{a}	3.43	3.76 ^{a}	4.04	2.10 ^{a}	2.60	43.03 ^{a}	5.99
Large	84.12 ^{b}	11.23	15.88 ^{b}	11.23	9.15 ^{a}	8.20	21.68 ^{b}	13.66
Medium	88.95 ^{b}	8.52	11.05 ^{b}	8.52	5.69 ^{a}	6.76	17.55 ^{b}	9.65
Small	95.58 ^{a}	5.37	4.42 ^{a}	5.37	4.03 ^{a}	4.96	25.71 ^{b}	15.34
Cover (%)								
Unfragmented	93.15 ^{a}	5.40	4.23 ^{a}	8.27	1.89 ^{a}	2.86	59.37 ^{A}	17.02
Large	65.84 ^{b}	15.91	11.95 ^{b}	7.94	7.34 ^{a}	8.15	18.88 ^{B}	19.37
Medium	51.00 ^{b}	21.33	5.74 ^{a}	13.70	2.38 ^{a}	4.46	8.33 ^{B}	15.01
Small	55.49 ^{b}	18.78	6.36 ^{a,c}	17.03	5.88 ^{a}	15.86	16.41 ^{B}	24.43

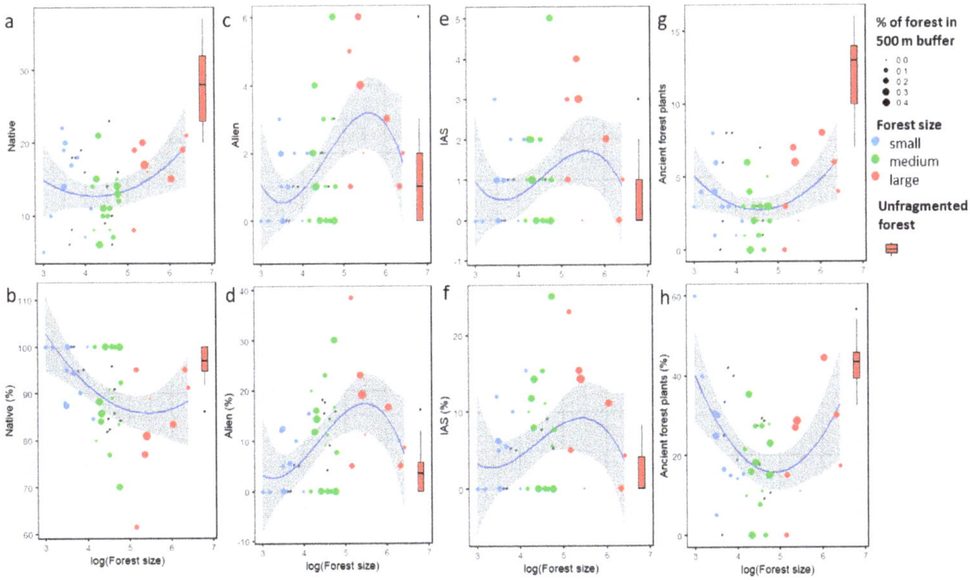

Figure 1. Effect of forest fragment size on richness (**a**) and proportion (**b**) of native plant species, alien (**c**,**d**), invasive alien (**e**,**f**), and ancient forest plants (**g**,**h**) in a plot. The color and size of the dots correspond to the forest fragment size group and % of forest cover in 500 m buffer, respectively. Control unfragmented forest Murska šuma is shown by a bar (minimum, 1st quartile, median, 3rd quartile, maximum).

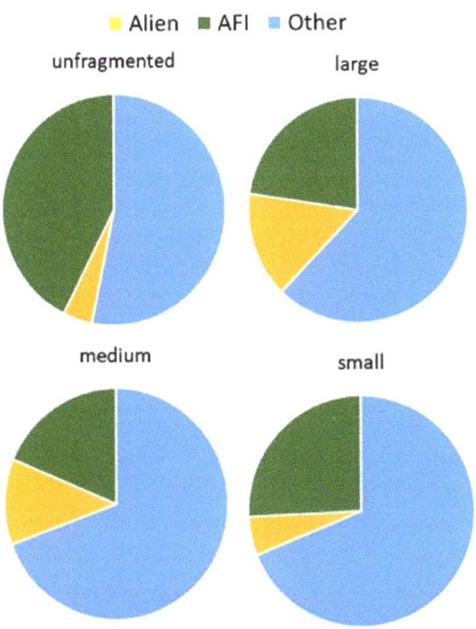

Figure 2. Plant composition quality assessed by an alien, AFI, and other plants in forest groups according to their size: unfragmented forest, large, medium, and small forest fragments.

2.3. Effect of Forest Fragment Size on Plant Species Composition Quality

The ratio of alien and native as well as IAS and native plant species increases with increasing forest size. However, for large forest fragments larger than 30 ha, the curve bands toward the values for the unfragmented forest. The opposite trend was found for the ratio of AFI and other native plants. According to regression analysis, the significance was shown only for the composition of alien/native and AFI/other native plants (Table 2, Figures 2 and 3).

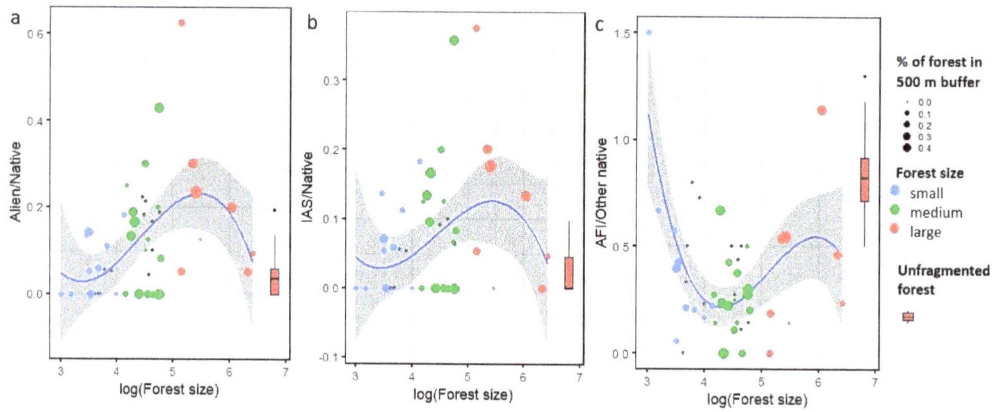

Figure 3. Effect of forest fragment size on plant composition assessed by the ratio between (**a**) alien and native plants, (**b**) invasive alien and native plants, and (**c**) ancient forest indicator plants and the rest of native plants. The color and size of the dots correspond to the forest fragment size group and % of forest cover in 500 m buffer. Control unfragmented forest Murska šuma is shown by a bar (minimum, 1st quartile, median, 3rd quartile, maximum).

Considering plot level, significantly higher richness, proportion, and cover of AFI was found in the unfragmented forest. In the same forest, the lowest proportion and cover of alien plants and IAS was detected (Figure 1, Table 3).

2.4. Effect of Forest Fragment Size on Diversity

We found a significant positive effect of forest fragment size on overall plant species richness and plant diversity assessed by Shannon diversity, Simpson's diversity, and evenness (Tables 2 and 4, Figure 4).

Table 4. Diversity indices in an unfragmented forest, large, medium, and small forest fragments. Different uppercase and lowercase letters represent significant differences at $p < 0.05$ for Tukey HSD or Mann–Whitney pairwise test, respectively.

	Species Richness		Shannon		Simpson's		Evenness	
	Average	SD	Average	SD	Average	SD	Average	SD
Unfragmented	29.14 ^{A}	6.15	2.54 ^{a}	0.26	0.87 ^{a}	0.03	0.76 ^{a}	0.04
Large	19.88 ^{B}	3.59	1.99 ^{b}	0.29	0.8 ^{b}	0.06	0.67 ^{b}	0.07
Medium	13.52 ^{C}	5.06	1.47 ^{c}	0.32	0.68 ^{c}	0.11	0.58 ^{b}	0.12
Small	15.06 ^{B,C}	6.1	1.5 ^{b,c}	0.63	0.65 ^{c,b}	0.21	0.55 ^{b}	0.17

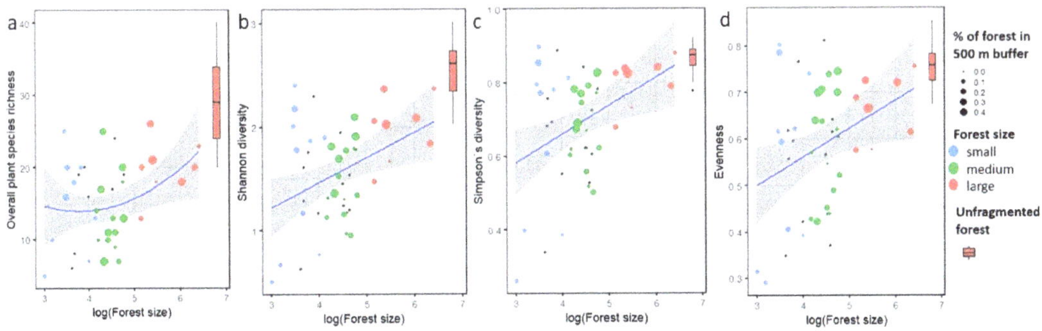

Figure 4. Effect of forest fragment size on plant diversity evaluated by (**a**) plant richness, (**b**) Shannon diversity, (**c**) Simpson's diversity, and (**d**) evenness. The color and size of the dots correspond to the forest fragment size group and % of forest cover in 500 m buffer. Control unfragmented forest Murska šuma is shown by a bar (minimum, 1st quartile, median, 3rd quartile, maximum).

Species richness increased by 32–54%, Shannon diversity increased by 22–32%, Simpson's index by 8–25%, and evenness by 12–28% if we compare different-sized forest fragments to the unfragmented forest (Table 4).

3. Discussion

Fragmentation of the European forests is ongoing nowadays. After 2015, an abrupt increase in forest harvest was recorded in the majority of European countries [25], including Slovenia [29]. Given this trend, it is even more important to understand better the impact of fragmentation and reduction of forest fragment size on its biodiversity. In this study, we focused on responses of alien and ancient forest indicator plant species according to forest fragment size, because these two plant groups may well indicate ecosystem degradation and preservation, respectively. Moreover, the study area is also the habitat of several endangered plant species such as *Gagea spathacea* (Hayne) Salisb., *Carex acutiformis* Ehrh., *C. riparia* Curt., and *Chimaphila umbellata* (L.) W. Barton (our unpublished records). This further supports the effort to study changes in species composition in such a landscape.

If we compare the total sampled area of all forest fragments to the total sampled area of a large unfragmented forest, higher species richness confirmed high cumulative species number for the fragmented sampled area. A similar observation was found by Herrera and Laterra [30] when assessing the effects of fragmentation on grasslands. This biodiversity and biogeography phenomenon is well known as distance decay similarity [30]. In a larger spatial extent, differences in species composition increase due to larger environmental heterogeneity and dispersal limitation [27,31].

Our results confirm that forest fragment size affects plant community composition besides other taxa such as beetles [32], birds [33,34], small mammals [35], and others. Larger forest fragments contain higher richness and a higher proportion of AFI per unit area, compared to smaller forest fragments, while there are fewer alien plants. Higher plant diversity in larger forest fragments and plant community composition compared to smaller fragments support the importance of forest fragment size in a fragmented landscape.

The pan-European survey revealed the vulnerability of forest habitats to alien plant invasions and predicted the further spread of alien species, especially under certain circumstances such as fragmentation [36]. The invasion of alien species affects biodiversity by reducing the richness and abundance of native species and alters several ecosystem functions [37]. Invasive alien plants overcome biogeographic barriers and cause plant communities' homogenization [38]. Moreover, they can form dense stands in forest gaps and aggravate the natural rejuvenation of the native forest tree species [26]. Therefore, it is critical to understand how forest fragment size, as the most obvious change in fragmented landscapes, affects alien plant species, including invasive species. We would expect smaller forest fragments to be more susceptible to alien plant invasion because the forest edge has proportionally more contact with the surrounding non-forest matrix, which is a donor for alien plants. However, our results show that in the case of small forest fragments, species richness was either overestimated or underestimated. If extinction rates are slow and species from the matrix penetrate forest fragments, species richness can be in smaller forest fragments for at least temporary enhancement [1]. For example, species richness of native plant species in small forest fragments in our study was inflated because of high light availability, allowing the establishment of native non-forest specialist plants. Additionally, we have shown a high richness of AFI species in small forest fragments because AFI species are long-lived plants. On the other hand, these temporary higher richness values for native plants give the impression that in small forest fragments, alien species have low abundance and richness, particularly if we compare the native vs. alien species ratio.

Our results confirm the general presence of different alien plant species in the lowland forest ecosystems regardless of their size. Forest fragment size did affect alien, but not invasive alien plant, richness. The highest richness, proportion, and a cover of aliens the invasive plants among them were recorded in larger forest fragments, while their presence was the lowest in unfragmented and small fragments. The level of invasion does not solely depend on forest fragment characteristics but is a subset of many variables including matrix quality [3,16], which is the most likely reason for such a result in our study. The most common alien plant in sampled forest fragments was *Solidago gigantea*, which is recognized as one of the invasive plants with the highest impact in Europe [39]. It was present in 26% of plots in fragmented and in 34% of the unfragmented forest.

Ancient forest indicator plant richness, relative abundance, and cover significantly increase with increasing forest fragment size. According to our dataset, the AFIs have confirmed their indicative value, but only for fragments larger than 6 ha due to the long lifetime of the AFIs. Unfragmented forest, serving as a control site, had the highest share of native and ancient forest indicator plant species in comparison to smaller forest fragments of various sizes. Our results indicate species turnover in sampled forest fragments. Even if a proportion of native species was similar in all plots, their composition differed at the expense of AFI, whereby once more the indicative value of this plant group is evident. This is consistent with findings from the mixed-oak forest in the northern Iberian Peninsula, where fragmentation mainly negatively affected the diversity of herbaceous forest specialist species, while other growth forms were less affected [40]. The study of Naaf et al. [13] provides a great insight into the population biology of three ancient forest indicator plants as a response to habitat fragmentation, all of them also present at our study site: *Oxalis acetosella*, *Anemone nemorosa* L., and *Polygonatum multiflorum*. All three herbs showed sensitivity to habitat fragmentation and isolation, which was species-specific due to different life-history traits. Habitat fragmentation resulted in a smaller population size and lower genetic diversity of these herbs. However, because of vegetative reproduction and long lifespan, the negative effects of habitat loss for some understory plants may be observed decades after forest fragmentation [41]. We can expect this to be true in our study for *Adoxa moschatellina* L., *Anemone nemorosa*, and *Paris quadrifolia*, as all three herbs also persisted in smaller forest fragments.

The high cumulative numbers of native plants, including AFI, illustrate the conservation value of small forest fragments in a predominantly biodiversity-depleted agricultural matrix. Our results have shown that they must have a critical size to maintain their species conservation area in the future. In our study area, the critical size was 13 ha. Unfragmented forest serving as a control site had the highest share of native and ancient forest indicator plant species in comparison to smaller forest fragments of various sizes. Moreover, in the unfragmented forest, although not entirely without alien plants, their overall richness, proportion, and cover per plot were lower than in smaller forest fragments. To summarize these findings, we must attribute higher conservation value to the unfragmented forest and maintain forest fragments of at least 13 ha in the studied agricultural landscape.

4. Materials and Methods

4.1. Study Area

The study area consists of forest fragments of various sizes (Illyrian *Quercus-Carpinus betulus* forest by EUNIS habitat classification) [42] located on the planes of Apaško polje and Dravsko polje near rivers Mura and Drava, respectively, in NE Slovenia (Figure 5). Apaško polje has an average elevation of 220 m a.s.l., while the average elevation of Dravsko polje is 250 m a.s.l. Both plains are part of the Danube River basin in the sub-Pannonian biogeographical region with a temperate continental climate. The 10-year average precipitation is 837 mm and 938 mm at Apaško polje and Dravsko polje, respectively, and the average temperature in both plains is 11.2 °C in the period from 2011 to 2021 [43]. Most of the study area is covered with shallow and sandy soils on deposited gravel or sand, suitable for agriculture, but is sensitive to summer droughts. Prevailing agricultural land use at both plains intertwines with forest fragments and settlements. Apaško polje area has a rural character, while the Dravsko polje area is more urbanized, with larger towns and long roadside settlements [44,45].

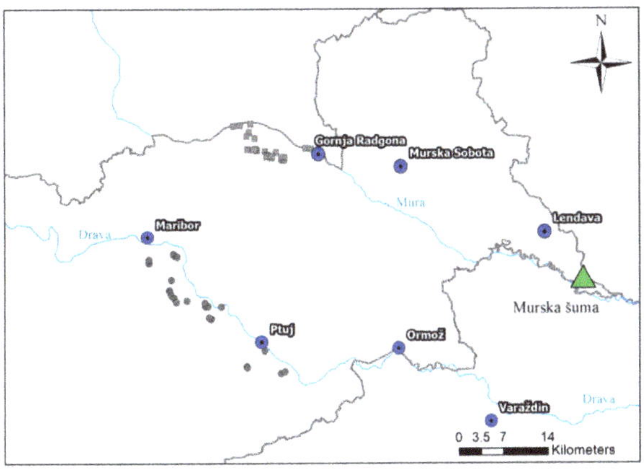

Figure 5. Map indicating the study area in the NE part of Slovenia, Europe. In detail, the distribution of forest fragments in studied areas is indicated: Apaško polje (grey squares) and Dravsko polje (grey dots). The control of the unfragmented forest area of Murska šuma is indicated by the green triangle.

Our study also included the extensive forest "Murska šuma" located near the Mura River in the NE Slovenia (Figure 5). The elevation of the area is around 190 m a.s.l., the 10-year average precipitation is 812 mm, and the average temperature is 11.7 °C in the period from 2011 to 2021 [30]. More details about Murska šuma are in Marinšek and Kutnar [26].

4.2. Data Collection

According to the satellite imagery (Google Earth Pro) [46] and later field inspection, we selected 23 and 24 forest fragments at Apaško polje and Dravsko polje, respectively. Forest fragments are defined as smaller areas (in this study between 0.1 and 260 ha) with typical forest vegetation surrounded along the entire perimeter with other land use types (e.g., arable land, settlements, roads). We drew polygons of sampled forest fragments using Google Earth Pro to determine the perimeter (forest edge), area, and centroid of each fragment. In Arc GIS [47] we overlaid polygons of forest fragments with layer Corine Land Cover [48] and extracted the area of the forest cover in a 500 m buffer zone surrounding fragments of interest.

We evaluated fragmentation with two descriptors for each studied forest fragment: forest fragment area and connectivity in terms of forest cover in the 500 m buffer zone surrounding each fragment.

Approximately in the centroid of each forest fragment, we surveyed vegetation according to the Standard European method [49] in the summer of 2019. Cover estimation was assessed using the Braun-Blanquet cover-abundance scale. All plots were of 100 m^2 (10 m × 10 m) to ensure comparability. The size of the plots was determined to avoid the edge and to ensure that even in the smallest fragments forest interior was sampled.

Forest fragments were classified according to their size: large (>13 ha; 8 plots), medium (1.5–13 ha; 23 plots), and small (0.1–1.5 ha; 16 plots) fragments.

To increase the reliability of the study of fragment size effect, we included 29 plots from the extensive forest area of Murska šuma (<900 ha) which was treated as a control site in this study—an unfragmented forest area to serve as "negative control". Plots were of 200 m^2 with vegetation belonging to the Illyrian oak-hornbeam habitat type (for more details see Marinšek and Kutnar [26]). Because differences in plot size may affect species richness, we compared species richness as well as the proportions of predetermined plant groups (Table 5). To avoid the influence of the plot size, all comparisons were repeated by weighting the number of plant species by the plot size logarithm (2.3 for 200 m^2 plots and 2 for 100 m^2 plots). All trends for the plant species richness remained the same, even after the weight of the plot size. However, we must emphasize that the samples from forest Mura šuma were not included in the regression analysis but served only as a control of the reliability of the direction of obtained trends.

Table 5. Classification of forest vegetation based on three species' characteristics: origin, invasiveness, and ancient forest indicator plants *sensu* Hermy et al. [14].

Classification	Expected Forest Fragment Size Effect	Expected Effect of Forest Cover in Buffer
Origin (native vs. alien) plant species	Higher alien plant invasion potential in smaller fragments	The higher richness of alien plant species at lower forest cover
Invasive alien species (IAS)	Higher richness and proportion of IAS in smaller fragments	Higher IAS richness with decreasing forest cover
Ancient forest indicator plant species (AFI)	Higher richness and proportion in unfragmented forest	The increasing richness and proportion with increasing forest cover

A total area of 4700 m^2 (representing 5.99% of fragmented forest area) and 5800 m^2 (representing 6.44% of unfragmented forest area) were sampled in fragmented forests at Apaško polje and Dravsko polje and the unfragmented forest area of Murska šuma, respectively.

4.3. Plant Species Classification

We classified plants according to their origin, invasiveness, and fidelity to (old) forest habitats. The origin of the plants was based on Martinčič et al. [50]. We considered only neophytes (plants that were introduced in Europe after 1500) as alien plant species. Plant species that do not naturally occur in the study area but are native in other parts of Slovenia were also considered alien (e.g., *Taxus baccata* which is frequently planted in urban hedgerows). The invasiveness status of alien plants followed expert judgment, according to which plants with naturalization status 5 were considered invasive species [50]. As forest specialist species, we considered slow-colonizing species unable to colonize new forests quickly. Those plants are recognized in Central Europe for having an indication potential and exhibiting high fidelity to old forest stands (ancient forest indicator plants—AFI) [14]. Plant species nomenclature was followed by Martinčič et al. [50].

4.4. Data Analysis

Different vegetation layers in the forest were merged into a single one, considering the independence assumption in JUICE software [51]. Plant diversity per plot was quantified by calculating species richness, Shannon index, Simpson index, and evenness in JUICE software.

Descriptive statistics were used to summarize species richness, proportion, and cover of different functional groups of plants per plot in the reference forest and three groups of fragmented forests. Differences among groups were assessed by ANOVA and post-hoc Tukey HSD. If assumptions for parametric tests were violated, we used Kruskal-Wallis and post-hoc Mann–Whitney tests.

Fragment size and % cover of forest in 500 m buffer were significantly correlated (linear regression, $p < 0.0001$). Therefore, only fragment size was included in the models as an explanatory variable.

We built linear ($y = a *q + r$), quadratic ($y = a *q + b *q^2 + r$), and polynomial ($y = a *q + b *q^2 + c *q^3 + r$) regression models (*lm()*) to evaluate the effect of fragment size on plant species richness (overall plant richness, native, alien, IAS, and AFI), composition (alien/native, IAS/native and AFI/other natives), and diversity (Shannon, Simpson's, and evenness). The final models were selected according to the criterion of the best fit. The distribution of response variables was visualized using functions *ggdensity()* and *hist()*. To test the normality of model residuals, we used a function *plot()*. The level of the significance was at $p < 0.05$. All statistical analyses were performed in R [52].

5. Conclusions

Our study addressed the effect of lowland forest fragment size on plant species richness and diversity with a focus on ecologically important groups (native, alien, alien invasive, and ancient forest indicator species). Our results revealed that the effects of fragmentation are complex and differ among studied plant groups. By decreasing forest fragment size, overall plant richness, and the richness of native and ancient forest indicators, plants per plot, and their diversity, decrease. On the other hand, the effects of forest fragment size on alien plants and, among them, invasive ones, were more complex. Their richness, proportion, and cover declined in forest fragments larger than 23 ha to the value recorded for the unfragmented forest. Our results confirm the high conservation value of unfragmented habitats because the highest proportion of AFI was listed there. However, the presence of alien plants in the same plots indicates that even large and diverse forest ecosystems are susceptible to biological invasions. According to our results, we suggest sustaining forest fragments of at least 13 ha in the studied agricultural landscape to enable their species conservation role in the future. Cumulative plant species richness was higher in smaller forest fragments than in unfragmented forest Murska šuma. Even though small forest fragments have high importance for biodiversity in otherwise biologically depleted landscapes, higher nature conservation value in terms of native plant richness, diversity, and compositional quality should be attributed to large unfragmented forest habitats. Our results serve as a case study that can help determine critical sizes of temperate lowland forest fragments which should be maintained in a landscape if we want to preserve their nature conservation value.

Author Contributions: All authors contributed to the study's conception and design. Data collection was performed by M.Š., L.K. and A.M. Material preparation and analysis were performed by M.Š. The first draft of the manuscript was written by M.Š. and N.Š. All authors commented on previous versions of the manuscript. All authors have read and agreed to the published version of the manuscript.

Funding: The study was partly supported by the Slovene Ministry of Higher Education, Science and Technology within research programs P1-0403 and J1-2457. The survey of Murska šuma was done in the framework of the project GoForMura, funded by the EEA Financial Mechanism 2009–2014 (SI02) and Government Office for Development and European Cohesion Policy of the Republic of Slovenia. The finalization of the study was funded by the Slovenian Ministry of Agriculture, Forestry and Food (JGS task 1/2) and by the Slovenian Research Agency (research core funding No. P4-0107).

Informed Consent Statement: Not applicable.

Data Availability Statement: The datasets generated during and/or analyzed during the current study are available from the corresponding author upon reasonable request.

Acknowledgments: We acknowledge Andrej Kosednar for his help with the fieldwork.

Conflicts of Interest: The authors have no conflict of interest to declare. All co-authors have seen and agree with the contents of the manuscript and there is no financial interest to report.

References

1. Debinski, D.M.; Holt, R.D. A survey and overview of habitat fragmentation experiments. *Conserv. Biol.* **2022**, *14*, 342–355. [CrossRef]
2. Fahrig, L. Effects of habitat fragmentation on biodiversity. *Annu. Rev. Ecol. Evol. Syst.* **2003**, *34*, 487–515. [CrossRef]
3. Laurance, W.F. Theory meets reality: How habitat fragmentation research has transcended island biogeographic theory. *Biol. Conserv.* **2008**, *141*, 1731–1744. [CrossRef]
4. Matthews, T.J.; Cottee-Jones, H.E.; Whittaker, R.J. Habitat Fragmentation and the Species–Area Relationship: A Focus on Total Species Richness Obscures the Impact of Habitat Loss on Habitat Specialists. *Divers. Distrib.* **2014**, *20*, 1136–1146. [CrossRef]
5. Deák, B.; Valkó, O.; Török, P.; Kelemen, A.; Bede, Á.; Csathó, Á.I.; Tóthmérész, B. Landscape and habitat and filters jointly drive richness and abundance of grassland specialist plants in terrestrial habitat islands. *Landsc. Ecol.* **2018**, *33*, 1117–1132. [CrossRef]
6. MacArthur, R.H.; Wilson, E.O. *The Theory of Island Biogeography*; Princeton University Press: Princeton, NJ, USA, 1967; ISBN 9780691088365.
7. Matthews, T. On the Biogeography of Habitat Islands: The Importance of Matrix Effects, Noncore Species, and Source-Sink Dynamics. *Q. Rev. Biol.* **2021**, *96*, 73–104. [CrossRef]
8. Nascimento, H.; Andrade, A.; Camargo, J.; Laurance, W.F.; Laurance, S.G.; Ribeiro, J. Effects of the surrounding matrix on tree recruitment in Amazonian forest fragments. *Conserv. Biol.* **2006**, *20*, 853–860. [CrossRef]
9. Zulka, K.P.; Abensperg-Traun, M.; Milasowszky, N.; Bieringer, G.; Gereben-Krenn, B.-A.; Holzinger, W.; Hölzler, G.; Rabitsch, W.; Reischütz, A.; Querner, P.; et al. Species Richness in Dry Grassland Patches of Eastern Austria: A Multi-Taxon Study on the Role of Local, Landscape and Habitat Quality Variables. *Agric. Ecosyst. Environ.* **2014**, *182*, 25–36. [CrossRef]
10. Valdivia, C.E.; Simonetti, J.A.; Henriquez, C.A. Depressed pollination of Lapageria rosea Ruiz et Pav. (Philesiaceae) in the fragmented temperate rainforest of southern South America. *Biodivers. Conserv.* **2005**, *15*, 1845–1856. [CrossRef]
11. Hauser, T.P.; Damgaard, C.; Loeschcke, V. Effects of inbreeding in small plant populations: Expectations and implications for conservation. In *Conservation Genetics*; EXS, Loeschcke, V., Jain, S.K., Tomiuk, J., Eds.; Birkhäuser: Basel, Switzerland, 1994; p. 68. ISBN 978-3-0348-9657-3.
12. Schlaepfer, D.R.; Braschler, B.; Rusterholz, H.P.; Baur, B. Genetic effects of anthropogenic habitat fragmentation on remnant animal and plant populations: A meta-analysis. *Ecosphere* **2018**, *9*, 17. [CrossRef]
13. Naaf, T.; Feigs, J.T.; Huang, S.; Brunet, J.; Cousins, S.A.O.; Decocq, G.; De Frenne, P.; Diekmann, M.; Govaert, S.; Hedwall, P.-O.; et al. Sensitivity to habitat fragmentation across European landscapes in three temperate forest herbs. *Landsc. Ecol.* **2021**, *36*, 2831–2848. [CrossRef]
14. Hermy, M.; Honnay, O.; Firbank, L.; Grashof-Bokdam, C.; Lawesson, J.E. An ecological comparison between ancient and other forest plant species of Europe, and the implications for forest conservation. *Biol. Conserv.* **1999**, *91*, 9–22. [CrossRef]
15. Ohlemüller, R.; Walker, S.; Wilson, J.B. Local vs. Regional factors as determinants of the invasibility of indigenous forest fragments by alien plant species. *Oikos* **2006**, *112*, 493–501. [CrossRef]
16. Vilà, M.; Ibáñez, I. Plant invasions in the landscape. *Landsc. Ecol.* **2011**, *26*, 461–472. [CrossRef]
17. Mavimbela, L.Z.; Sieben, E.J.J.; Procheş, Ş. Invasive alien plant species, fragmentation and scale effects on urban forest community composition in Durban, South Africa. *N. Z. J. For.* **2018**, *48*, 19. [CrossRef]
18. Gignac, L.D.; Dale, M.R.T. Effects of size, shape, and edge on vegetation in remnants of the upland boreal mixed-wood forest in agro-environments of Alberta, Canada. *Can. J. Bot.* **2007**, *85*, 273–284. [CrossRef]
19. Šipek, M.; Šajna, N. Alien plant diversity in urban forest «Miklavški gozd». *Hladnikia* **2020**, *46*, 53–65.
20. Decocq, G.; Andrieu, E.; Brunet, J.; Chabrerie, O.; De Frenne, P.; De Smedt, P.; Deconchat, M.; Diekmann, M.; Ehrmann, S.; Giffard, B.; et al. Ecosystem Services from Small Forest Patches in Agricultural Landscapes. *Curr. For.* **2016**, *2*, 30–44. [CrossRef]
21. Zahawi, R.A.; Oviedo-Brenes, F.; Peterson, C.J. A degradation debt? Large-scale shifts in community composition and loss of biomass in a tropical forest fragment after 40 years of isolation. *PLoS ONE* **2017**, *12*, e0183133. [CrossRef] [PubMed]
22. Possley, J.; Woodmansee, S.W.; Maschinski, J. Patterns of Plant Composition in Fragments of Globally Imperiled Pine Rockland Forest: Effects of Soil Type, Recent Fire Frequency, and Fragment Size. *Nat. Areas J.* **2008**, *28*, 379–394. [CrossRef]
23. Gallé, R.; Tölgyesi, C.; Császár, P.; Bátori, Z.; Gallé-Szpisjak, N.; Kaur, H.; Maák, I.; Torma, A.; Batáry, P. Landscape Structure is a Major Driver of Plant and Arthropod Diversity in Natural European Forest Fragments. *Ecosphere* **2022**, *13*, e3905. [CrossRef]
24. Pisek, R. Forest Area. Available online: http://kazalci.arso.gov.si/sl/content/clone-povrsina-gozda-0 (accessed on 1 September 2019).
25. Ceccherini, G.; Duveiller, G.; Grassi, G.; Lemoine, G.; Avitabile, V.; Pilli, R.; Cescatti, A. Abrupt increase in harvested forest area over Europe after 2015. *Nature* **2020**, *583*, 72–77. [CrossRef]
26. Marinšek, A.; Kutnar, L. Occurrence of invasive alien plant species in the floodplain forests along the Mura River in Slovenia. *Period. Biol.* **2017**, *119*, 251–260. [CrossRef]
27. Kohn, D.D.; Walsh, D.M. Plant species richness-the effect of island size and habitat diversity. *J. Ecol.* **1994**, *82*, 367–377. [CrossRef]

28. Jogan, N.; Bačič, M.; Strgulc Krajšek, S. Neobiota of Slovenia: Invasive Alien Species in SLOVENIA and the Impact on the Conservation of Biodiversity and Sustainable Use of Resources. (In Slovenian: Neobiota Slovenije: Invazivne Tujerodne Vrste v Sloveniji ter Vpliv na Ohranjanje Biotske Raznovrstnosti in Trajnostno Rabo Virov). In *Final Report, CRP «Konkurenčnost Slovenije 2006–2013»*; University of Ljubljana (Biotechnical faculty): Ljubljana, Slovenia, 2012.
29. Kutnar, L.; Kermavnar, J.; Pintar, A.M. Climate change and disturbances will shape future temperate forests in the transition zone between Central and SE Europe. *Ann. For. Res.* **2021**, *64*, 67–87. [CrossRef]
30. Herrera, L.P.; Laterra, P. Relative influence of size, connectivity and disturbance history on plant species richness and assemblages in fragmented grasslands. *Appl. Veg. Sci.* **2011**, *14*, 181–188. [CrossRef]
31. Nekola, J.C.; White, P.S. The distance decay of similarity in biogeography and ecology. *J. Biogeogr.* **1999**, *26*, 867–878. [CrossRef]
32. Gaublomme, E.; Eggermont, H.; Hendrickx, F. Local extinction processes rather than edge effects affect ground beetle assemblages from fragmented and urbanised old beech forests. *Insect. Conserv. Divers.* **2013**, *7*, 82–90. [CrossRef]
33. McIntyre, N. Effects of forest patch size on avian diversity. *Landsc. Ecol.* **1995**, *10*, 85–99. [CrossRef]
34. dos Anjos, L.; Bochio, G.M.; Medeiros, H.R.; de Arruda Almeida, B.; Lindsey, B.R.A.; Calsavara, L.C.; Ribeiro, M.C.; Domingues, T.; José, M. Insights on the functional composition of specialist and generalist birds throughout continuous and fragmented forests. *Ecol. Evol.* **2019**, *9*, 6318–6328. [CrossRef]
35. Pardini, R.; de Souza, S.M.; Braga-Neto, R.; Metzger, J.P. The role of forest structure, fragment size and corridors in maintaining small mammal abundance and diversity in an Atlantic forest landscape. *Biol. Conserv.* **2005**, *124*, 253–266. [CrossRef]
36. Wagner, V.; Chytrý, M.; Jiménez-Alfaro, B.; Pergl, J.; Hennekens, S.; Biurrun, I.; Knollová, I.; Berg, C.; Vassilev, K.; Rodwell, J.S.; et al. Alien plant invasions in European woodlands. *Divers. Distrib.* **2017**, *23*, 969–981. [CrossRef]
37. Pyšek, P.; Hulme, P.E.; Simberloff, D.; Bacher, S.; Blackburn, T.M.; Carlton, J.M.; Dawson, W.; Essl, F.; Foxcroft, L.C.; Genovesi, P.; et al. Scientists' warning on invasive alien species. *Biol. Rev.* **2020**, *95*, 1511–1534. [CrossRef] [PubMed]
38. Daru, B.H.; Davies, T.J.; Willis, C.G.; Meineke, E.K.; Ronk, A.; Zobel, M.; Pärtel, M.; Antonelli, A.; Davis, C.C. Widespread homogenization of plant communities in the Anthropocene. *Nat. Commun.* **2021**, *12*, 698. [CrossRef]
39. Nentwig, W.; Bacher, S.; Kumschick, S.; Pyšek, P.; Vilà, M. More than "100 worst" alien species in Europe. *Biol. Invasions* **2018**, *20*, 1611–1621. [CrossRef]
40. Rodríguez-Loinaz, G.; Amezaga, I.; Onaindia, M. Does forest fragmentation affect the same way all growth-forms? *J. Environ. Manag.* **2012**, *94*, 125–131. [CrossRef] [PubMed]
41. Honnay, O.; Jacquemyn, H.; Bossuyt, B.; Hermy, M. Forest fragmentation effects on patch occupancy and population viability of herbaceous plant species. *New Phytol.* **2005**, *166*, 723–736. [CrossRef]
42. Davies, C.E.; Moss, D.; Hill, M.O. *EUNIS Habitat Classification Revised 2004*; European Topic Centre on Nature Protection and Biodiversity: Paris, France, 2004.
43. ARSO. ARCHIVE—Observed and Measured Meteorological Data in Slovenia (In Slovenian: ARHIV—Opazovani in Merjeni meteorološki Podatki po Sloveniji). Available online: https://meteo.arso.gov.si/met/sl/archive/ (accessed on 29 January 2022).
44. Ščap, D. *Apaško Polje on the Path of Tolerance through History (In Slovenian: Apaško Polje po Poti Strpnosti Skozi Zgodovino*; Občina Apače: Apače, Slovenia, 2018.
45. Vodeb, V.; Mladenovič, L.; Nikšič, M.; Cotič, B.; Šuklje Erjavec, I.; Praper, S.; Gulič, A.; Sendi, R.; Urban Planning Institute of the Republic of Slovenia; Miklavčič, T.; et al. *National Report on Urban Development—Habitat III*; Ministry of the Environment and Spatial Planning, Spatial Planning, Construction and Housing Directorate: Ljubljana, Slovenia, 2016. Available online: https://www.gov.si/assets/ministrstva/MOP/Publikacije/9d3d8bd710/porocilo_urbani_razvoj_HabitatIII_en.pdf (accessed on 10 January 2022).
46. Google Earth Pro. Available online: https://www.google.com/intl/sl/earth/versions/#earth-pro (accessed on 20 September 2020).
47. ESRI. *ArcGIS 10.2.2 for Desktop*; Environmental Systems Research Institute: Redlands, CA, USA, 2014.
48. EEA. Data and Maps: Copernicus Land Monitoring Service—Corine Land Cover. Available online: https://www.eea.europa.eu/data-and-maps/data/copernicus-land-monitoring-service-corine (accessed on 5 September 2018).
49. Braun-Blanquet, J. *Pflanzensoziologie. Grundzüge der Vegetationskunde*; Springer: Wien, Austria, 1964.
50. Martinčič, A.; Wraber, T.; Jogan, N.; Podobnik, A.; Turk, B.; Vreš, B.; Ravnik, V.; Frajman, B.; Strgulc Krajšek, S.; Trčak, B.; et al. *Small flora of Slovenia. Identification Key of Vascular Plants, 4th ed., Supplemented and Amended Edition (In Slovenian: Mala Flora Slovenije. Ključ za Določanje Praprotnic in Semenk, Četrta, Dopolnjena in Spremenjena Izdaja)*; Tehniška založba Slovenije: Ljubljana, Slovenia, 2007; ISBN 978-961-251-026-8.
51. Tichý, L. JUICE, software for vegetation classification. *J. Veg. Sci.* **2002**, *13*, 451–453. [CrossRef]
52. R Core Team. *R: A Language and Environment for Statistical Computing*; R Foundation for Statistical Computing: Wien, Austria, 2021. Available online: https://www.R-project.org/ (accessed on 5 September 2018).

plants

MDPI

Article

Non-Native Flora of the Mediterranean Lesvos Island (East Aegean, Greece): Floristic Analysis, Traits, and Assessment

Alexandros Galanidis [1,*], Ioannis Bazos [2] and Panayiotis G. Dimitrakopoulos [1]

[1] Biodiversity Conservation Laboratory, Department of Environment, University of the Aegean, 81100 Mytilene, Greece; pdimi@env.aegean.gr

[2] Department of Ecology and Systematics, Faculty of Biology, School of Sciences, National and Kapodistrian University of Athens, 15784 Athens, Greece; ibazos@biol.uoa.gr

* Correspondence: agal@env.aegean.gr

Abstract: A checklist of Lesvos Island's non-native vascular flora is presented. Through the literature and a roadside survey, we recorded 187 non-native plant taxa, representing 12% of the total regional flora. A total of 37 taxa were reported for the first time for Lesvos, including three taxa that are also new to the Greek non-native flora. The dominant families were Fabaceae, Asteraceae, Poaceae, Amaranthaceae, and Solanaceae. *Amaranthus*, *Oxalis*, *Solanum*, *Opuntia*, and *Prunus* were the most species rich genera. Most taxa were neophytes, naturalized and perennial, while phanerophytes and therophytes were the predominant life forms. Animal (insect) pollination was the principal pollination mode, whereas the main dispersal mechanism was zoochory, with most taxa forming a capsule. The majority of the taxa preferred agricultural and ruderal habitats and originated from the Americas and Asia. The primary introduction pathway was escape from confinement, concerning taxa introduced for ornamental and agricultural purposes.

Keywords: alien species; invasive species; survey; checklists; islands; life-history traits; origin; pathways; species–area curves

1. Introduction

Biological invasions are an expanding component of global environmental change [1]. They can cause serious ecological, socio-economic, and health problems and affect an ecosystem's functioning, structure, composition, and biodiversity [2]. Driven by the globalization of trade and transport [3,4], land use changes [5,6], climate alteration [7], and access to new source pools [8], non-native species are increasing in number, population size, and spread. Future projections predict an increase in alien species numbers worldwide, especially in Europe [9].

In contrast to other Mediterranean-type ecosystems of the world, the Mediterranean Basin is regarded as less susceptible to invasion due to its long and gradual history of human intervention [10] and the fact that more European species were intentionally introduced by European settlers to the colonies than the opposite [11]. Moreover, compared to temperate ecosystems, the summer drought stress limitations in the area appear to prevent the establishment of non-adapted plants, a remark that is consistent with recent studies indicating lower numbers of alien species in the Mediterranean than in temperate Europe [12].

Islands are defined by distinct boundaries, where most biotic assemblages and ecological processes in them are fundamentally restricted by obvious geographical constraints [13]. Excluding the tropics, the Mediterranean Basin is the richest area in islands than anywhere else, with more than 10,000 islands and islets. Mediterranean islands are considered biodiversity hotspots that are rich in rare habitats and endemic and range-restricted plant species [11,14]. Compared to continental Europe, Mediterranean islands have been considered less invaded, as revealed by crosschecks between their floras [12]. However, Mediterranean islands tend to be more vulnerable to impacts by alien taxa than other comparable in size mainland areas [11,15]. Biogeographical (e.g., unsaturated island communities, smaller regional species pool, lower habitat diversity), ecological (e.g., less competitive native species, lack of key plant functional groups), and socio-economic factors (e.g., abandonment of agricultural land, development of tourism, intense anthropogenic activity particularly in coastal and riparian habitats) contribute to this pattern [11,12,15–17].

Furthermore, recent studies showed a drastic rise in the number of non-native species in both Europe and the Mediterranean Basin [3,18]. This trend has been considerably accelerated for Mediterranean islands regarding the establishment success and distribution range of their alien species [15]. Although the levels of invasion in Mediterranean insular ecosystems are considered context-dependent [17] and habitat-specific [19], various factors linked to human presence are also known to facilitate alien species introduction and establishment in the area. Human population density [16,20], road networks [21–23], transport infrastructures [16,23], tourism development [14,21,23], or the percentage of agricultural areas and artificial surfaces [23] are eventually connected to increased propagule pressure that is considered to promote the colonization and persistence of alien species [24].

Being geographically isolated, islands miss opportunities to be naturally colonized by non-native species, and this fact reveals, in greater strength, the role of humans as a key dispersal vector. The deliberate use of alien species in agriculture, forestry, horticulture, or even as ornamental plants has been intensified in the Mediterranean since the middle of the 20th century [11]. Along with various unintentional pathways, such as transport contaminants on seed supplies or transport stowaways on vehicles or machinery, Mediterranean islands are facing increased introduction pressure, making their ecosystems some of the most threatened by biological invasions globally.

Checklists are an essential management tool for the protection and conservation of biodiversity and ecosystem functions, constituting a structural part of species richness and population trend monitoring systems. Such lists link taxonomic, spatial, and temporal information and provide a benchmark for detecting the most vulnerable areas and species, support decision-making, and ensure the efficient allocation of resources aimed at avoiding or mitigating the negative effects of biological invasions [25,26]. In this work, along with the compilation of the non-native flora checklist for Lesvos, one of the largest Mediterranean islands, we aim to analyze various life-history, reproductive, and dispersal traits, habitat preferences, chorological data, and introduction pathways. Furthermore, we explore possible correlations or dependencies of these parameters with the current position of non-native taxa on the introduction–naturalization–invasion continuum (INIC).

2. Results

2.1. Species Richness and Taxonomy

A total of 187 taxa (species and subspecies) were recorded. Most of them (118; 63%) were observed in the field, while 69 taxa records (37%) were obtained from the literature (Table S1). The alien plant species richness of Lesvos Island corresponds to 11.6% of the total flora on the island; when excluding casual aliens, this percentage falls to 7.5%. Among the recorded taxa, there were 37 new entries for Lesvos Island (e.g., with no records previously reported), including three records that were new for Greece as well (Table S1). The majority of these new taxa were perennial, woody, casual neophytes.

The taxa belong to 60 different families with the most abundant ones being Fabaceae (17 taxa; 9.1%), Asteraceae (15 taxa; 8%), Poaceae (14 taxa; 7.5%), Amaranthaceae (13 taxa; 7%), Solanaceae (12 taxa; 6.4%), Rosaceae (8 taxa; 4.3%), and Cactaceae and Plantaginaceae (6 taxa; 3.2% each). The top five families were also the most abundant in naturalized taxa, while the highest number of invasive taxa were found for Amaranthaceae (seven taxa; 54% of the family's taxa) and Asteraceae (seven taxa; 46.7% of the family's taxa) (Figure 1). Thirty families (50%) were represented by one taxon only.

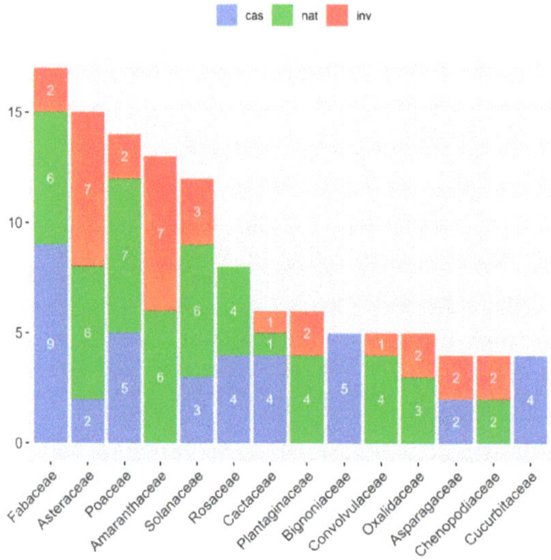

Figure 1. Number of non-native taxa per family across invasion status categories. cas: casual; nat: naturalized; inv: invasive taxa.

The total number of different genera was 137, and the families with the highest numbers of genera were Fabaceae (15 genera), Poaceae (13 genera), Asteraceae (9 genera), Solanaceae (6 genera), and Bignoniaceae and Rosaceae (5 genera each). Thirty-three families (55%) were represented by one genus only. The most abundant genus was *Amaranthus*, with 13 taxa, followed by *Oxalis* and *Solanum* (five taxa each) and *Opuntia* and *Prunus* (four taxa each). About 60% of the genera (112 taxa) were represented by one taxon only.

2.2. Residence and Invasion Status

2.2.1. Residence Status

Most of the taxa were neophytes (152; 81%). Archaeophytes were mainly naturalized (69%), while neophytes were more equally distributed across INIC (38% casuals, 36% naturalized, and 26% invasive). Established taxa (naturalized and invasive) accounted for 74% of the archaeophytes and 63% of the neophytes. There were significant differences in the counts of taxa per residence status between invasion status categories (Fisher's exact test, $p < 0.001$). Invasives were over-represented among neophytes and under-represented among archaeophytes, while for naturalized taxa, the outcome was reversed, as more taxa than expected were observed for archaeophytes and fewer for neophytes (Figure 2).

2.2.2. Invasion Status

Of the 187 taxa, 35% were casual, 42% naturalized, and 23% invasives, resulting in a 65% majority of the taxa being already established on the island. The families with the highest number of invasive taxa were Amaranthaceae and Asteraceae (seven taxa each), while most of the naturalized taxa were found in Poaceae (seven taxa each) and Amaranthaceae, Asteraceae, Fabaceae, and Solanaceae (six taxa each).

Comparing the top five richest families between the whole non-native flora and the invasion status categories (established taxa included), similar patterns were found, apart from the invasive taxa where Aizoaceae and Euphorbiaceae, although low in the whole alien flora ranking, emerged in the top five families (with three taxa each).

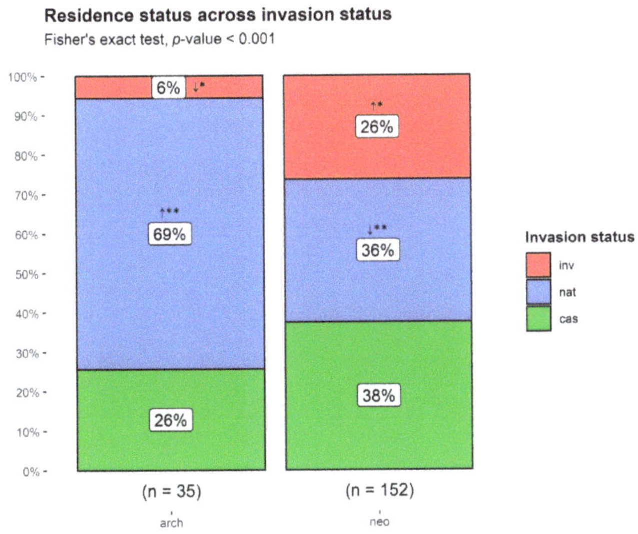

Figure 2. Proportion of non-native taxa by residence status across invasion status categories. arch: archaeophyte; neo: neophyte. cas: casual; nat: naturalized; inv: invasive taxa. ↑ observed counts higher than expected by chance; ↓ observed counts lower than expected by chance. ** $p < 0.01$; * $p < 0.05$.

2.3. *Life-History Traits*

2.3.1. Life Span

Most of Lesvos' alien plants were perennials (64%), while 36% were annual, and one taxon was biennial. Most perennial and annual alien taxa were naturalized (42% and 43%, respectively).

2.3.2. Growth Form

The most common growth form was herbaceous (58%), followed by shrub (21%), tree (20%), and subshrub (6%). There were significant differences in the counts of taxa per growth form between the invasion status categories (Fisher's exact test, $p < 0.001$). These differences were due to herbs, where the observed counts were lower than expected by chance for casual species, and due to trees, where casuals were more than expected (Figure 3A).

2.3.3. Life Form

In total, phanerophytes (40%) and therophytes (36%) were the most frequent life forms, followed by hemicryptophytes (16%), chamaephytes (9%), and geophytes (6%). Woody plants (phanerophytes and chamaephytes) represented 48% of the taxa. Among the established taxa, therophytes were dominant (39%), followed by phanerophytes (29%) and hemicryptophytes (19%). There were significant differences in the counts of taxa per life form between the invasion status categories (Fisher's exact test, $p = 0.005$). These differences were due to phanerophytes, where the observed counts for casual species were higher than expected by chance (Figure 3B).

2.3.4. Dispersal Mechanisms

The main dispersal mechanism for Lesvos aliens was zoochory (49%), followed by barochory (25%) and anemochory (24%), with no significant differences among the invasion status groups. Zoochory was also the main dispersal mechanism for invasive

(52%) and naturalized (47%) taxa. The majority of the taxa (89%) were dispersed by a single mechanism.

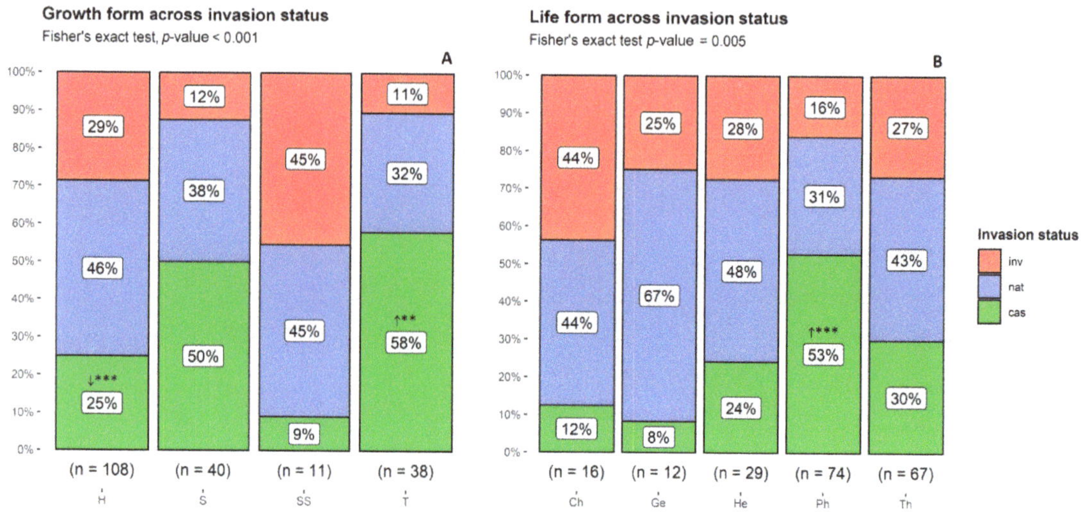

Figure 3. Proportion of non-native taxa by (**A**) growth form and (**B**) life form across invasion status categories. H: herb; S: shrub; SS: subshrub; T: tree. Ch: chamaephyte; Ge: geophyte; He: hemicryptophyte; Ph: phanerophyte; Th: therophyte. cas: casual; nat: naturalized; inv: invasive taxa. ↑: observed counts higher than expected by chance; ↓ observed counts lower than expected by chance. *** $p < 0.001$; ** $p < 0.01$.

2.3.5. Pollination

Most of the taxa (57%) relied on animal pollinators, predominantly insects (105 taxa). Self-pollination was followed by 42% of the taxa, while wind pollination occurred in 12% of the taxa. Although there were no significant differences between categories, more casuals than naturalized or invasives and more archaeophytes than neophytes were pollinated by animals.

2.3.6. Flowering

The taxa mainly started to flower in May (29%), followed by April (18%), June (17%), and July (12%), whereas fewer taxa were found to enter their flowering phase in October, November, and December (one, one, and two taxa, respectively) (Figure 4A). The flowering response data showed that most of the taxa (67%) flowered in June, followed by July (61%), May (58%), and August (55%). The lowest response was observed in December and January (3% each) (Figure 4B). For nearly half of the taxa (46%), the flowering span lasted two to three months.

2.3.7. Fruit Types

Capsule was the predominant fruit type (35%), followed by taxa that produced berries (17%), achenes (16%), and legumes (10%). Most of the invasive (45%) and naturalized (35%) taxa likewise produced capsules as their fruit type. Significant differences for fruit type were found between the invasion status categories (Fisher's exact test, $p = 0.0015$), as fewer casual taxa than expected by chance were found forming an achene.

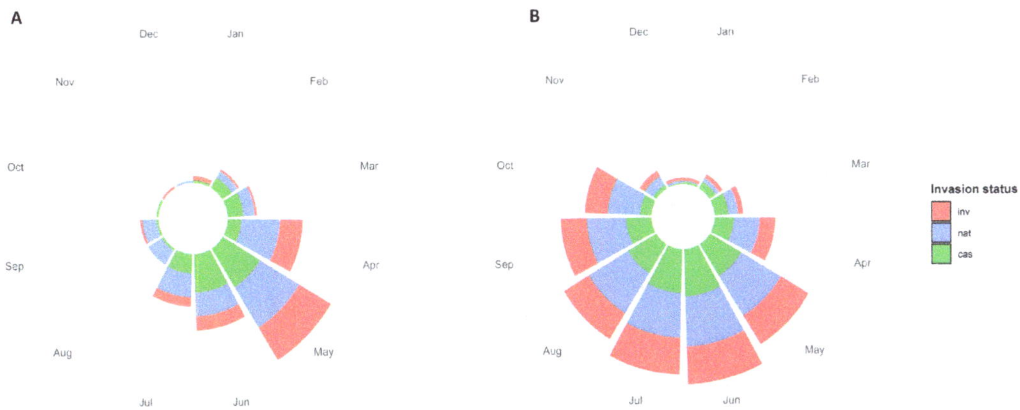

Figure 4. Circular histograms for (**A**) the start of flowering time and (**B**) the number of taxa flowering per month across invasion categories. Each triangle represents the number of non-native taxa that flower in that month. cas: casual; nat: naturalized; inv: invasive taxa.

2.4. Habitats

Nearly all of the alien plants recorded on Lesvos Island showed a preference for ruderal and agricultural habitats (94%), with the remaining taxa found in rather low proportions to other habitats; 10 taxa were found in rocky, 9 taxa in freshwater, 6 taxa in coastal, 5 taxa in woodland, and 2 taxa in phryganic habitats. Established taxa were found in all habitats. There were significant differences in the counts of taxa per habitat between the invasion status categories (Fischer exact test $p < 0.001$). More casuals than expected by chance were found in ruderal and agricultural habitats, while invasive taxa were over-represented in coastal habitats (Figure 5).

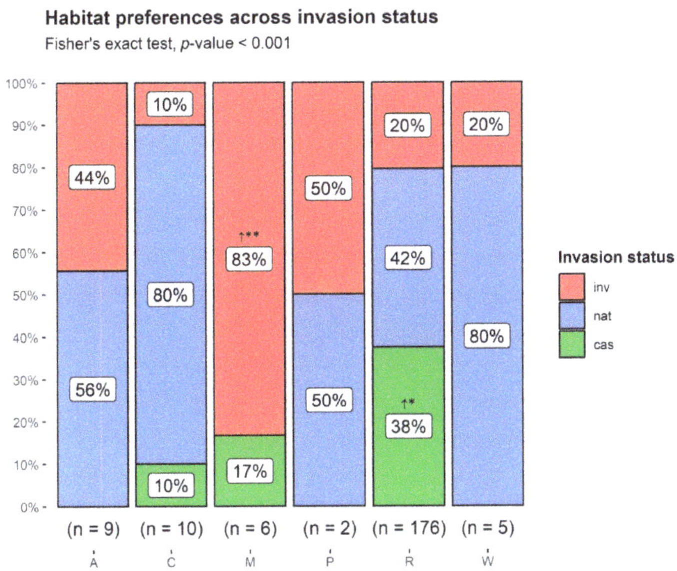

Figure 5. Proportion of non-native taxa by habitat type preferences across invasion status categories. A: freshwater; C: cliffs, rocks, walls, ravines, boulders; M: coastal; P: xeric Mediterranean phrygana and grasslands; R: ruderal and agricultural; W: woodlands and scrub. cas: casual; nat: naturalized; inv: invasive taxa. ↑ observed counts higher than expected by chance. ** $p < 0.01$; * $p < 0.05$.

2.5. Origin

Most of the taxa were of Asian (32%) and S. American (30%) origin, followed by N. American (25%), African (13%), European (12%), and Australian (5%) (Figure 6). The taxa of Asian origin were less frequently observed as invasives than expected by chance (Fisher's exact test, $p = 0.001$) (Figure 7). South African taxa predominated among the taxa with African origin, covering 40% of them and primarily being invasive (60%). Concerning the European taxa, most of them were specifically of Mediterranean origin (59%) and mainly naturalized, comprising, however, only 7% of the whole alien flora.

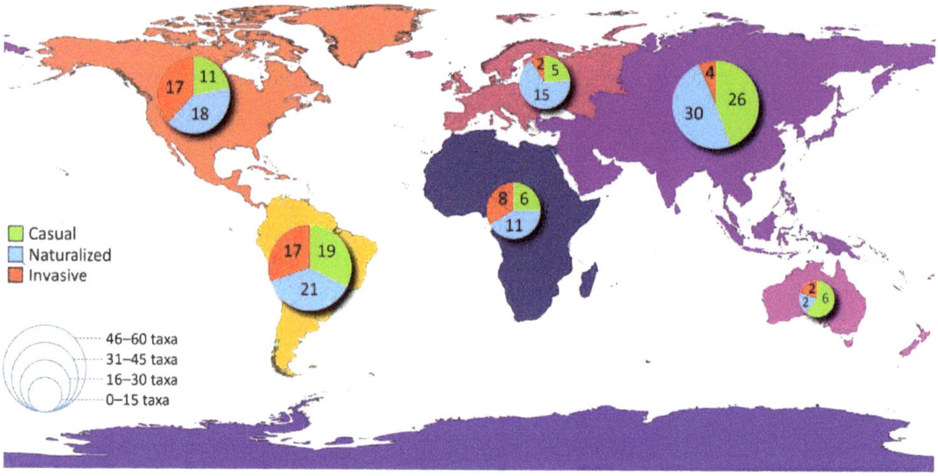

Figure 6. Global map with the continent of origin for non-native plant taxa on Lesvos Island. The size of the pie indicates the number of taxa originating from each continent. Pie sector size indicates the number of taxa per invasion status category. Map created with QGIS 3.34 "Prizren".

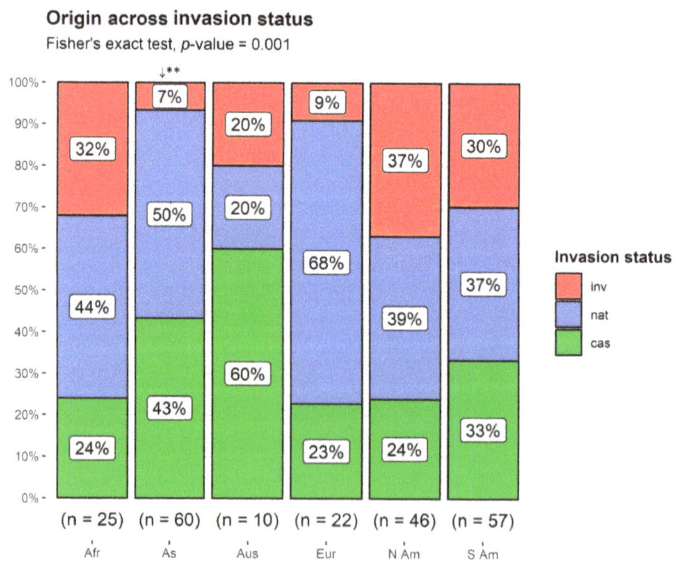

Figure 7. Proportion of non-native taxa by native origin across invasion status categories. Afr: Africa; As: Asia; Aus: Australia; Eur: Europe; N Am: North America: S Am: South America. cas: casual; nat: naturalized; inv: invasive taxa. ↓ observed counts lower than expected by chance. ** $p < 0.01$.

The established taxa have their native range predominantly in the Americas (60%), followed by Asian taxa (28%).

2.6. Introduction Pathways

A total of 300 introduction pathways were defined for the 187 non-native taxa of Lesvos Island, corresponding on average to 1.6 pathways per taxon ($SD = 0.65$). Most of the taxa (52%) were polyvectic, i.e., utilized more than one introduction pathway. On average, casual, naturalized, and invasive taxa were introduced via analogous numbers of known pathways per taxon ($M = 1.58$, $SD = 0.68$, $M = 1.62$, $SD = 0.67$, and $M = 1.62$, $SD = 0.58$, respectively).

Deliberate introductions were responsible for the entry of 93% of the alien plants on Lesvos Island, while 24% were unintentionally introduced. The ratio of unintentional to intentional introductions was lower in archaeophytes, where one out of seven taxa was unintentionally introduced. Regarding invasion status categories, there were significant differences (Fisher's exact test, $p < 0.001$) in the counts of taxa per pathway type (deliberate vs. unintentional introductions). More casuals than expected by chance were introduced via intentional and fewer via unintentional pathways, while for invasives, the pattern was reversed, as more taxa than expected by chance were introduced via unintentional and fewer via deliberate pathways (Figure 8A).

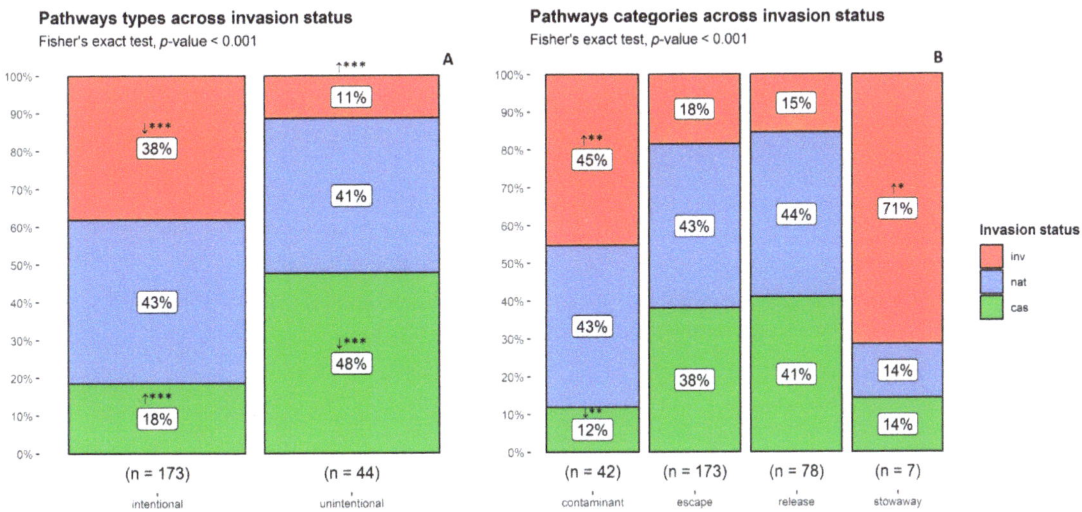

Figure 8. Proportion of non-native taxa by (**A**) pathway types and (**B**) pathway categories across invasion status categories. cas: casual; nat: naturalized; inv: invasive taxa. ↑ observed counts higher than expected by chance; ↓ observed counts lower than expected by chance. *** $p < 0.001$; ** $p < 0.01$; * $p < 0.05$.

The escape from confinement pathway was involved in nearly all cases (93%) and in all deliberate introductions, with most of them resulting from ornamental, horticultural, or agricultural plantings (122, 97, and 70 taxa. respectively). The release in nature pathway was responsible for 42% of the deliberate taxa introductions, with most of them concerning landscape/flora "improvement" in the wild. For the unintentional introductions, the transport-contaminant pathway was responsible for 23% of the introductions, with most of them regarding seed contamination, while the transport-stowaway pathway was responsible for only 4% of the taxa, mostly attached to people and their luggage/equipment and machinery/equipment.

The invasion status categories displayed significant differences in the distribution of pathway categories (Fisher's exact test, $p < 0.001$). Casuals followed the transport-contaminant pathway less often than expected by chance, while invasives introduced as transport contaminants and transport stowaways were more frequent than expected (Figure 8B).

The distribution of pathway categories displayed significant differences across life forms (Fisher's exact test, $p < 0.001$) due to therophytes. Therophytes followed the transport-contaminant pathway more often than expected by chance and were less frequent than expected in the escape from confinement (Figure 9).

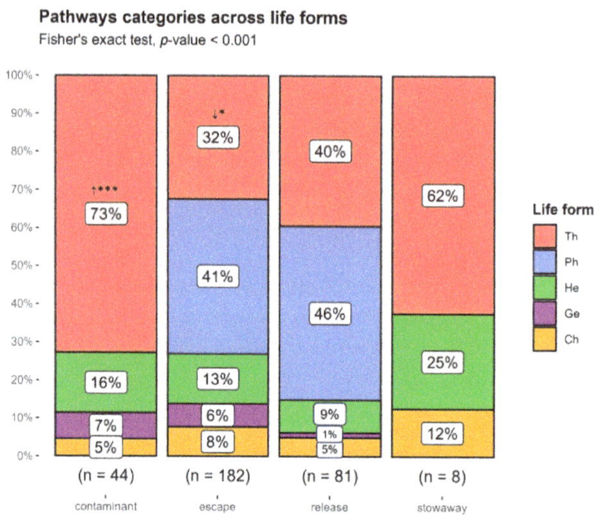

Figure 9. Proportion of non-native taxa by pathway category across life form types. Ch: chamaephyte; Ge: geophyte; He: hemicryptophyte; Ph: phanerophyte; Th: therophyte. ↑ observed counts higher than expected by chance; ↓ observed counts lower than expected by chance. *** $p < 0.001$; * $p < 0.05$.

3. Discussion

3.1. Species-Area Curves

The present study reports 187 non-native plant taxa (including casuals) which correspond to ca. 12% of the total flora on Lesvos Island. This percentage is comparable with data reported for Stromboli, Salina, Crete, Panarea, the uninhabited island of Cabrera, and Filicudi. It is considerably higher than the corresponding values reported for the Tuscan Archipelago islands, the much smaller and uninhabited island of Dragonera, Rhodes, and the whole of Greece. However, regardless of size, it is lower than other Mediterranean islands, such as Mallorca, Menorca, Linosa, Formentera, Sardinia, Sicily, or Vulcano (Table S2).

The 187 non-native plant taxa of Lesvos Island fall short of the alien species numbers reported for larger Mediterranean islands such as Sicily, Sardinia, Corsica, Crete, or Mallorca; the number is also lower than that on some smaller islands such as Menorca or Ibiza (Table S2). Despite the lack of a mechanistic understanding, it is well known that species richness tends to increase with area, and alien plants appear to conform to this general pattern. In particular, reported SAR intercepts (c-values) are lower for non-native species compared to native species, while SAR exponents (z-values) have been found to be similar between them, though higher across islands compared to the mainland [27]. Conclusively, it seems that alien species richness increases with area at the same rate as native species richness (similar slopes) but accumulates faster on islands than on the mainland (steeper slopes).

In our study, the species–area curves of total and alien species richness for twenty Mediterranean islands (Table S2, Figure 10) confirmed a lower intercept for non-natives compared to natives (3.5 vs. 5.5), which indicates relatively higher alpha diversity for the native flora on these islands [28]. However, with a z-value = 0.27, within the range of 0.2–0.45 proposed by [29] for insular systems, the slope was steeper for non-native flora. Area explained a greater proportion of the variance in total species richness (Adjusted R^2 = 0.89) compared to that for aliens (Adjusted R^2 = 0.81), which aligns with expectations, likely reflecting the influence of heavily colonized and possibly disturbed anthropogenic zones, along with poorly invaded and possibly less disturbed natural areas [30]. This outcome, besides reflecting an alerting faster accumulation rate for alien species, reveals higher β-diversity for non-native flora among the Mediterranean islands. In other words, it indicates low levels of similarity between islands and a more aggregative distribution pattern [28] as a result of the spatial environmental and anthropogenic mosaics that these Mediterranean islands form.

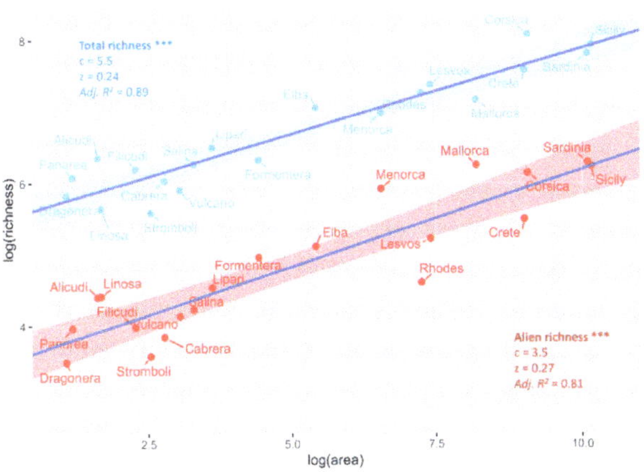

Figure 10. Species–area curves (log–log) of total vs. non-native plant species richness for twenty Mediterranean islands. Shaded areas represent the 95% confidence intervals around the lines. Blue area: total richness; red area: non-native richness. *** $p < 0.001$.

Still, the number of alien species in an area depends on the combined effect of macroecological, biogeographical, and anthropogenic determinants at local, regional, and global scales. An area can be related to a broad range of the above determinants, such as climate, resource heterogeneity, or habitat diversity. Moreover, it can be linked to human population size and its consequent land use intensity, trade volume, or disturbances that increase propagule pressure in that area [31]. On Lesvos Island, anthropogenic determinants associated with introduction pressure and the establishment potential of non-native plant species have been found to regulate their spatial distribution [16].

3.2. *Taxonomy*

The richest families of Lesvos Island's alien flora, i.e., Fabaceae, Asteraceae, Poaceae, Amarantaceae, and Solanaceae, are likewise dominant in the alien flora of Greece [32] and also in other alien floras in Europe [33,34], China [35], and globally [36]. Fabaceae, Asteraceae, and Poaceae are also the most abundant families in the native flora of Lesvos, in the Mediterranean Biome [10], and in the Mediterranean basin [37]. These families include many cultivated species and additionally contain species with invasiveness-favoring traits such as a high reproductive rate and grazing resistance (Asteraceae), resilience (Poaceae), or increased dispersal ability (Poaceae and Fabaceae) [10]. In our results, Fabaceae and Poaceae account for the highest number of casual taxa (nine and five, respectively).

Compared to the global naturalized alien flora [36], Lesvos aliens include 17 taxa from the top 50 and 40 taxa from the top 200 most widely distributed naturalized aliens of the world, although in various invasion statuses. Furthermore, of the 30 most widely distributed invasive aliens worldwide, seven taxa (*Lantana camara*, *Leucaena leucocephala*, *Lonicera japonica*, *Eleusine indica*, *Sorghum halepense*, *Arundo donax*, and *Xanthium spinosum*) also occur as aliens on Lesvos; three of them (*L. camara*, *L. leucocephala*, and *L. japonica*) are in the top 10, as naturalized, casual, and naturalized aliens respectively, while *E. indica* and *X. spinosum* are also considered invasive. Counting the Mediterranean zonobiome alone [36], from the top 10 naturalized species with the highest frequency of occurrence, two taxa (*Erigeron bonariensis* and *Amaranthus hybridus*) also occur on Lesvos as invasives.

Fifteen taxa of the Lesvos alien flora are also included in the National List of Alien Invasive Species (HELLAS-ALIENS) [38], while two taxa (*Acacia saligna* and *Ailanthus altissima*) are included in the third update of the list of Invasive Alien Species (IAS) of Union Concern.

3.3. Life Forms

Phanerophytes and therophytes are the prevailing life forms for non-native plants on Lesvos Island. This pattern, common to Mediterranean islands, may reflect two different plant strategies for coping with the stressful typical long dry seasons of the Mediterranean climate: tolerance for phanerophytes and avoidance for therophytes [39]. The biological spectrum of Lesvos Island is quite similar to the one found in the Mediterranean basin, for all life forms but phanerophytes [10], and to those found for Mallorca, and Ibiza, particularly for phanerophytes, hemicryptophytes, and chamaephytes [40].

Non-native phanerophytes are generally associated with direct anthropogenic introductions for forestry, horticulture, and ornamental/gardening purposes [10,40] and are expected to expand their range due to land use changes, both in agricultural and urban environments [41]. Therophytes constitute a high proportion of weeds that are accidentally introduced to ruderal and agricultural habitats. On Lesvos, non-native phanerophytes have been exclusively introduced via deliberate pathways, particularly via escape from confinement, for ornamental (85% of taxa) and horticulture (78% of taxa) purposes, and via release in nature, for landscape/flora/fauna improvement (97% of taxa). Likewise, nearly half of therophytes have been introduced via the transport-contaminant pathway, with most of them (94% of the taxa) as seed contaminants.

Major differences have been observed in the biological spectrum between the non-native and native flora of Lesvos. Aliens display a notably higher proportion of phanerophytes (40% vs. 7%) and lower proportions of therophytes (36% vs. 49%), hemicryptophytes (16% vs. 21%), and geophytes (6% vs. 15%) [42]. The biological spectrum of Lesvos aliens also differs from that of the alien flora of Greece [32], which reveals fewer phanerophytes (22%), and more therophytes (44%) and geophytes (9%).

The proportion of woody alien taxa (48%) is analogically higher than that in the local native flora (14%; [42]) and in the alien flora of Greece (28%; [32]). Naturalized woody taxa (16%) constitute a smaller proportion of the island's alien flora than that reported for the total and insular global alien taxa (32% and 34%, respectively [36]) and for the alien flora of Greece (32%; [32]).

3.4. Density and Naturalization Index

Alien flora density, defined as species per unit area (e.g., taxa/km^2), can be regarded as a proxy for invasion intensity, reflecting the extent or severity of alien invasion within ecosystems [43]. It is assumed that the higher the density of IAS in an area, the higher their ability to change the structure and diversity of native species communities and thus decrease the function and services of the ecosystem. Alien flora density assumes a constant, proportional relationship between richness and area. This linearity oversimplifies ecological complexity and invasion dynamics, overlooking factors like habitat heterogeneity or disturbance regime history. However, this metric allows for straightforward and standardized comparisons across different-sized regions, highlighting areas under ecological stress and human pressures that should be prioritized for management. Therefore, despite its limitations, this approach, alongside traditional SAR analysis, could be still useful in revealing local invasion pressures. Notably, Lesvos, with a value of 0.11, has one of the lowest alien flora densities among Mediterranean islands and exceeds only the values for Sicily, Sardinia, Crete, Corsica, and Rhodes, while the corresponding density for Greece is very low (Table S2).

The naturalization rate, defined as the proportion of naturalized to all alien taxa (including casuals), for Lesvos is 0.42 or 42%. This is considerably lower than that of Rhodes, Elba, Linosa, Sicily, and Greece as a whole, comparable to Cabrera, Tuscan archipelago islands, Menorca, and Formentera, and larger than that of Dragonera, Crete, Balearics, Ibiza, Mallorca, Sardinia, and Corsica (Table S2). Since naturalized species lay one step in the INIC before becoming invaders, the proportion of naturalized taxa in an alien flora requires special attention for the efficient management of invasive species as it provides a better understanding of invasion processes, species invasiveness, and invasibility at community or regional scale [44]. In addition, introduction pathways should be synchronously considered since fluctuations in naturalization rates among groups of species connected to different aspects of human activities and thus to different pathways have been reported [45].

As invasions are often prompted by stochastic events, residence time is positively connected to invasion success [46]. The longer a species is present in a region, its chances of overcoming various barriers to invasion and shifting its position to the INIC or expanding its range increase, with propagule pressure playing a key role in that [47]. In our data, there is a clear prevalence of neophytes on archaeophytes both for invasive (21% vs. 1%) and naturalized (29% vs. 13%) taxa. Similar trends for invasive neophytes, though in lower proportions, have been reported for Sardinia (12% vs. 1%; [48]) and the Balearics (13% vs. 1%; [49]), while analogous proportions have been shown for Corsica (20% vs. 1%; [48]) and the Tuscan archipelago (20% vs. 1%; [50]).

3.5. Reproductive and Dispersal Traits

While pollinator limitation is not considered a significant constraint for the spread of non-native plants, pollination syndrome can be regarded as an invasion stage-dependent trait. A study of Czech Republic flora revealed that in the early stages of invasion, pollination by insects prevails, while in the subsequent stages, wind- and self-pollinated species become more common [51]. A likely interpretation for that shift could be introduction bias, as species with showy flowers, and thus probably insect-pollinated, are frequently preferred for ornamental and horticultural purposes. In our study, more casuals were insect-pollinated (59%) compared to naturalized (57%) and invasive taxa (50%), while garden and horticultural escapes, which would be expected to be most likely insect-pollinated, decreased from 53% among casuals to 51% and 35% among naturalized and invasive taxa, respectively.

It has been suggested that among other factors, the chances of an alien reaching a suitable site depend upon its mobility; thus, successful alien species are usually better dispersers than non-successful ones [52]. Moreover, alien species that can spread rapidly in an area are frequently associated with both natural and anthropogenic dispersal modes [53]. Most taxa in our data relied on natural dispersal strategies, but zoochory, which was the top dispersal mechanism, embraced, in our case, anthropochory as well. Thus, a rapid expansion of range is something we could expect, especially since 52% of invasive and 47% of naturalized taxa used zoochory.

3.6. Origin

Most of Lesvos alien plants (55%) originate from the Americas, which is a common pattern for many Mediterranean islands [e.g., Tuscan (43%; [50]), Sardinia (34%; [48]), Corsica (29%; [48]) or Rhodes (21%; [54])], but also for Greece (36%; [32]). Regarding the naturalized taxa only, fewer taxa of American origin were observed on Lesvos compared to the Mediterranean Basin [10] (39% vs. 55%, respectively).

3.7. Introduction Pathways

Shrubs and trees exclusively, along with 95% of the perennial taxa, were introduced via intentional pathway categories (escape from confinement and release in nature), exposing their primary use for horticulture or ornamental purposes. In general, direct human involvement in the introduction of aliens is believed to promote their invasiveness, even though unintentional introductions refer to equally widespread species [55]. Our findings, although missing a temporal aspect, do not support this pattern as in our dataset, there were fewer invasive taxa intentionally introduced than expected. Moreover, concerning unintentional introductions, invasive taxa were over-represented among both contaminants and stowaways, which stand out for their superior dispersal ability.

4. Materials and Methods

4.1. Study Area

Lesvos Island, located in the NE Aegean Sea, is the 3rd largest island in Greece and the 8th largest in the Mediterranean Basin (Figure 11). With an area of 1636 km^2 and a coastline of ca. 400 km, it hosts about 90,000 inhabitants. Lesvos is a semi-mountainous island with two high peaks, Lepetymnos (968 m a.s.l.) in the north and Olympos (967 m a.s.l.) in the south, and no major rivers or lakes. The climate is typical Mediterranean (Csa subtype of the Köppen–Geiger classification), with large spatial and seasonal fluctuations due to the regional effect of mountains and atmospheric circulation patterns [56]. The average annual rainfall ranges from 415 mm in the west to 725 mm in the east, while the average annual temperature reaches 17.7 °C. These patterns form three climatic zones: one semi-arid in the west, one dry sub-humid in the central and east, and a transitional semi-arid zone in between them [57]. Lesvos is dominated by volcanic rocks and crystalline schists [58] and used to be joined to the Anatolian mainland until late Palaeolithic/Mesolithic times. It presents an exceptional geological heterogeneity that is reflected in its vegetation [59,60], which includes pine, oak, and chestnut forests, scattered and widespread Mediterranean formations (phrygana and maquis), and important wetlands and riparian zones, as well as extended olive groves and arable lands at lowlands [59,60]. The native flora of Lesvos Island is estimated at 1611 taxa [61]. Recent archaeological excavations on Lesvos revealed traces of human presence since the Paleolithic era [62]. Human activity appears to have been continuous since then, shaping the landscape and vegetation of the island. Over the last decades, there has been a steady decline in the agricultural sector in favor of tourism.

Figure 11. Map of the Mediterranean Basin with Lesvos Island indicated in red. The map was created in RStudio (v.2023.6.0.421) using the *ggplot2* (v.3.4.2) [63] and *sf* (v.1.0.19) [64] packages.

4.2. Data Sources and Terminology

This is the first attempt to synthesize the non-native flora of Lesvos Island in order to compile a checklist of Lesvos alien plants. Data were gathered in two ways: (i) by a systematic literature review and (ii) by a thorough roadside survey, a technique that is considered both time- and cost-effective [65–67].

The literature review included various sources including peer-reviewed journals (e.g., Phytologia Balcanica, Wildenowia, Bocconea, etc.), dedicated floras and plant checklists (e.g., [68,69]), and the relational database "Alien Plants in Greece: A web-based platform" (https://www.alienplants.gr, accessed on 18 September 2024; [32]). The roadside survey included all main and secondary roads, as well as several mountainous, agricultural, riparian, and coastal footpaths, thus covering most of the island's surface. Following species' phenological phases [16], the survey took place in four sampling rounds between 2020 and 2021, while sporadic records were added through 2022. Species presence along with descriptive and spatial data were entered in the open-source QField mobile GIS app (OPENGIS.ch) during the fieldwork [16].

For the compilation of the database, a taxon was included in the checklist if there was any record (bibliographic and/or roadside survey) affirming at least one wild locality of it on Lesvos Island. Moreover, taxa were included when they were commonly cultivated in human-mediated sites such as fields, roadsides, gardens, parks, etc. [70]. Species names were standardized and attributed to families according to World Flora Online (v.2023.03) [71], with records being checked by using the R package *WorldFlora* (v.1.13.2) [72].

For "alien", we followed the definitions of Pyšek et al., 2004 [46], i.e., plants present in any part of Greece that have arrived with or without human intervention from an area in which they are alien. Translocated species were not considered. Concerning the taxa's invasion status, i.e., the stage they have reached along the introduction–naturalization–invasion continuum, the taxa were classified into "casual", "naturalized", and "invasive" as in [73]. For the residence time status, i.e., information on how long an alien species has been introduced in the region, the taxa were distinguished as "archaeophytes" and "neophytes" (presumably introduced before or after 1492, respectively). Regarding chorology, the taxa were grouped by their geographical origin (native range) at the continent level, including Europe (other than Greece), Africa, Asia, America, and Australia. More precise biogeographical information is discussed, when necessary, in the text. For the analyses of taxa introduction pathways, the classification of the Convention on Biological Diversity [74] was adopted. The specific scheme implements six main categories that are further divided into various subcategories.

Values for plant life-history traits were primarily obtained from [32]. For life forms, we used Raunkiaer's system classification, which is regarded as the most accepted and utilized system in the world due to its simplicity and efficiency, with the exception of warmer climates such as subtropics or tropics [75]. Residence and invasion status information mainly followed [32] and refers to the whole of Greece. Chorological information was mainly gathered from [32,76], while introduction pathway data were mostly obtained from [77]. Additional sources included the literature, online databases, and field observations.

Some of the taxa were linked with more than one life-history trait, origin, habitat, or pathway category; in such cases, the taxa were considered representative of each category. The Lesvos alien plants checklist was archived on the Global Biodiversity Information Facility (GBIF) [78].

4.3. Data Analyses

To test for significant differences in the representation of life-history traits, residence status, origin, habitat preferences, and introduction pathways of the taxa among the invasion status categories, their counts were analyzed by row × column contingency tables (i.e., life form × status) using Fisher's exact test. To determine for which taxa the counts were lower or higher than expected by chance, adjusted standardized residuals of Chi-square tests were compared with critical values of the normal distribution. All analyses were performed in RStudio (v.2023.6.0.421) [79].

5. Conclusions

This publication provides the first comprehensive survey of non-native plants for Lesvos Island in Greece, using a combined method of a literature review and a roadside survey. We anticipate that this work will serve as a valuable reference for comparative studies across Mediterranean islands while paving the way for further investigation of the critical issue of island invasion dynamics. Our results highlight that Lesvos, although hosting several of the most widely distributed naturalized and invasive aliens of the world, compared to other Mediterranean islands, displays a relatively low density of alien species with an intermediate rate of naturalization. Most of the taxa prefer disturbed ruderal and agricultural habitats, whereas the frequently prone to invasion freshwater and coastal habitats exhibit low numbers of aliens. Deliberate introduction pathways are the major source of alien species, with unintentional introductions favoring particular invasive taxa.

As land abandonment and tourism development increase, more alien species are likely to enter the island. Furthermore, already present non-native taxa are expected to establish or increase their spread and consequential impacts on natural ecosystems, human or other animal health, and the economy. Although still less invaded, Mediterranean islands are more prone to invasions than analogous mainland areas. This is critical for conservation efforts and indicates the importance of such studies and the need for a comprehensive management strategy that will include monitoring, prevention, early detection, and control actions towards non-native invasive species.

Supplementary Materials: The following supporting information can be downloaded at https://www.mdpi.com/article/10.3390/plants13233375/s1: Table S1: Alien taxa checklist of Lesvos Island with information on record type; Table S2: Area, total and non-native plant species richness, and naturalized non-native taxa for twenty-three Mediterranean islands and Greece; Table S3: Contingency tables of the counts of taxa per categorical trait between invasion status categories. Refs. [80–84] are cited in the Supplementary Materials.

Author Contributions: Conceptualization, A.G. and P.G.D.; methodology, A.G. and P.G.D.; validation, A.G., I.B., and P.G.D.; formal analysis, A.G.; investigation, A.G. and P.G.D.; resources, P.G.D.; data curation, A.G.; writing—original draft preparation, A.G.; writing—review and editing, A.G., I.B., and P.G.D.; visualization, A.G.; project administration, P.G.D.; funding acquisition, A.G. and P.G.D. All authors have read and agreed to the published version of the manuscript.

Funding: This research was co-financed by Greece and the European Union (European Social Fund-ESF) through the Operational Program "Human Resources Development, Education and Lifelong Learning 2014–2020" in the context of the project "An Integrative Framework for the Study of Alien Flora" (MIS 5049419).

Data Availability Statement: The original data presented in this study are available in GBIF—the Global Biodiversity Information Facility—at https://doi.org/10.15468/wj8syb.

Conflicts of Interest: The authors declare no conflicts of interest. The funders had no role in the design of this study; in the collection, analyses, or interpretation of data; in the writing of this manuscript; or in the decision to publish the results.

References

1. Pyšek, P.; Hulme, P.E.; Simberloff, D.; Bacher, S.; Blackburn, T.M.; Carlton, J.T.; Dawson, W.; Essl, F.; Foxcroft, L.C.; Genovesi, P.; et al. Scientists' Warning on Invasive Alien Species. *Biol. Rev.* **2020**, *95*, 1511–1534. [CrossRef] [PubMed]
2. Feng, Y.-L.; Du, D.; van Kleunen, M. Global Change and Biological Invasions. *J. Plant Ecol.* **2022**, *15*, 425–428. [CrossRef]
3. Hulme, P.E.; Pyšek, P.; Nentwig, W.; Vila, M. Will Threat of Biological Invasions Unite the European Union. *Science* **2009**, *324*, 40–41. [CrossRef] [PubMed]
4. Seebens, H.; Blackburn, T.M.; Dyer, E.E.; Genovesi, P.; Hulme, P.E.; Jeschke, J.M.; Pagad, S.; Pyšek, P.; Winter, M.; Arianoutsou, M.; et al. No Saturation in the Accumulation of Alien Species Worldwide. *Nat. Commun.* **2017**, *8*, 14435. [CrossRef]
5. Pauchard, A.; Alaback, P.B. Influence of Elevation, Land Use, and Landscape Context on Patterns of Alien Plant Invasions along Roadsides in Protected Areas of South-Central Chile. *Conserv. Biol.* **2004**, *18*, 238–248. [CrossRef]

6. Sánchez-Ortiz, K.; Taylor, K.J.M.; Palma, A.D.; Essl, F.; Dawson, W.; Kreft, H.; Pergl, J.; Pyšek, P.; van Kleunen, M.; Weigelt, P.; et al. Effects of Land-Use Change and Related Pressures on Alien and Native Subsets of Island Communities. *PLoS ONE* **2020**, *15*, e0227169. [CrossRef]
7. Dainese, M.; Aikio, S.; Hulme, P.E.; Bertolli, A.; Prosser, F.; Marini, L. Human Disturbance and Upward Expansion of Plants in a Warming Climate. *Nat. Clim. Change* **2017**, *7*, 577–580. [CrossRef]
8. Seebens, H.; Blackburn, T.M.; Dyer, E.E.; Genovesi, P.; Hulme, P.E.; Jeschke, J.M.; Pagad, S.; Pyšek, P.; van Kleunen, M.; Winter, M.; et al. Global Rise in Emerging Alien Species Results from Increased Accessibility of New Source Pools. *Proc. Natl. Acad. Sci. USA* **2018**, *115*, E2264–E2273. [CrossRef]
9. Seebens, H.; Bacher, S.; Blackburn, T.M.; Capinha, C.; Dawson, W.; Dullinger, S.; Genovesi, P.; Hulme, P.E.; van Kleunen, M.; Kühn, I.; et al. Projecting the Continental Accumulation of Alien Species through to 2050. *Glob. Change Biol.* **2020**, *27*, 970–982. [CrossRef]
10. Arianoutsou, M.; Delipetrou, P.; Vilà, M.; Dimitrakopoulos, P.G.; Celesti-Grapow, L.; Wardell-Johnson, G.; Henderson, L.; Fuentes, N.; Ugarte-Mendes, E.; Rundel, P.W. Comparative Patterns of Plant Invasions in the Mediterranean Biome. *PLoS ONE* **2013**, *8*, e79174. [CrossRef]
11. Hulme, P.; Brundu, G.; Camarda, I.; Dalias, P.; Lambdon, P.; Lloret, F.; Médail, F.; Moragues, E.; Suehs, C.; Traveset, A.; et al. Assessing the Risks to Mediterranean Islands Ecosystems from Alien Plant Introductions. In *Plant Invasions: Human Perception, Ecological Impacts and Management*; Tokarska-Guzik, B., Brock, J.H., Brundu, G., Child, L., Daehler, C.C., Pyšek, P., Eds.; Backhuys Publishers: Leiden, The Netherlands, 2008; pp. 39–56. ISBN 978-3-8236-1528-6.
12. Guarino, R.; Chytrý, M.; Attorre, F.; Landucci, F.; Marcenò, C. Alien Plant Invasions in Mediterranean Habitats: An Assessment for Sicily. *Biol. Invasions* **2021**, *23*, 3091–3107. [CrossRef]
13. Chiarucci, A.; Fattorini, S.; Foggi, B.; Landi, S.; Lazzaro, L.; Podani, J.; Simberloff, D. Plant Recording across Two Centuries Reveals Dramatic Changes in Species Diversity of a Mediterranean Archipelago. *Sci. Rep.* **2017**, *7*, 5415. [CrossRef] [PubMed]
14. Fois, M.; Podda, L.; Médail, F.; Bacchetta, G. Endemic and Alien Vascular Plant Diversity in the Small Mediterranean Islands of Sardinia: Drivers and Implications for Their Conservation. *Biol. Conserv.* **2020**, *244*, 108519. [CrossRef]
15. Celesti-Grapow, L.; Bassi, L.; Brundu, G.; Camarda, I.; Carli, E.; D'Auria, G.; Del Guacchio, E.; Domina, G.; Ferretti, G.; Foggi, B.; et al. Plant Invasions on Small Mediterranean Islands: An Overview. *Plant Biosyst. Int. J. Deal. All Asp. Plant Biol.* **2016**, *150*, 1119–1133. [CrossRef]
16. Dimitrakopoulos, P.G.; Koukoulas, S.; Michelaki, C.; Galanidis, A. Anthropogenic and Environmental Determinants of Alien Plant Species Spatial Distribution on an Island Scale. *Sci. Total Environ.* **2022**, *805*, 150314. [CrossRef]
17. Vilà, M.; Pino, J.; Montero, A.; Font, X. Are Island Plant Communities More Invaded than Their Mainland Counterparts? *J. Veg. Sci.* **2010**, *21*, 438–446. [CrossRef]
18. Butchart, S.H.M.; Walpole, M.; Collen, B.; van Strien, A.; Scharlemann, J.P.W.; Almond, R.E.A.; Baillie, J.E.M.; Bomhard, B.; Brown, C.; Bruno, J.; et al. Global Biodiversity: Indicators of Recent Declines. *Science* **2010**, *328*, 1164–1168. [CrossRef]
19. Traveset, A.; Brundu, G.; Carta, L.; Mprezetou, I.; Lambdon, P.; Manca, M.; Médail, F.; Moragues, E.; Rodríguez-Pérez, J.; Siamantziouras, A.-S.D.; et al. Consistent Performance of Invasive Plant Species within and among Islands of the Mediterranean Basin. *Biol. Invasions* **2008**, *10*, 847–858. [CrossRef]
20. Malavasi, M.; Santoro, R.; Cutini, M.; Acosta, A.T.R.; Carranza, M.L. The Impact of Human Pressure on Landscape Patterns and Plant Species Richness in Mediterranean Coastal Dunes. *Plant Biosyst. Int. J. Deal. All Asp. Plant Biol.* **2016**, *150*, 73–82. [CrossRef]
21. Lazzaro, L.; Viciani, D.; Dell'Olmo, L.; Foggi, B. Predicting Risk of Invasion in a Mediterranean Island Using Niche Modelling and Valuable Biota. *Plant Biosyst. Int. J. Deal. All Asp. Plant Biol.* **2017**, *151*, 361–370. [CrossRef]
22. Lozano, V.; Marzialetti, F.; Carranza, M.L.; Chapman, D.; Branquart, E.; Dološ, K.; Große-Stoltenberg, A.; Fiori, M.; Capece, P.; Brundu, G. Modelling Acacia Saligna Invasion in a Large Mediterranean Island Using PAB Factors: A Tool for Implementing the European Legislation on Invasive Species. *Ecol. Indic.* **2020**, *116*, 106516. [CrossRef]
23. Pretto, F.; Celesti-Grapow, L.; Carli, E.; Brundu, G.; Blasi, C. Determinants of Non-Native Plant Species Richness and Composition across Small Mediterranean Islands. *Biol. Invasions* **2012**, *14*, 2559–2572. [CrossRef]
24. Dimitrakopoulos, P.G.; Koukoulas, S.; Galanidis, A.; Delipetrou, P.; Gounaridis, D.; Touloumi, K.; Arianoutsou, M. Factors Shaping Alien Plant Species Richness Spatial Patterns across Natura 2000 Special Areas of Conservation of Greece. *Sci. Total Environ.* **2017**, *601–602*, 461–468. [CrossRef] [PubMed]
25. Reyserhove, L.; Desmet, P.; Oldoni, D.; Adriaens, T.; Strubbe, D.; Davis, A.J.S.; Vanderhoeven, S.; Verloove, F.; Groom, Q. A Checklist Recipe: Making Species Data Open and FAIR. *Database* **2020**, *2020*, baaa084. [CrossRef]
26. D'Antraccoli, M.; Bedini, G.; Peruzzi, L. Next Generation Floristics: A Workflow to Integrate Novel Methods in Traditional Floristic Research. *Plant Biosyst. Int. J. Deal. All Asp. Plant Biol.* **2022**, *156*, 594–597. [CrossRef]
27. Baiser, B.; Li, D. Comparing Species–Area Relationships of Native and Exotic Species. *Biol. Invasions* **2018**, *20*, 3647–3658. [CrossRef]
28. He, C.; Fang, L.; Xiong, X.; Fan, F.; Li, Y.; He, L.; Shen, X.; Li, S.; Ji, C.; Zhu, J. Environmental Heterogeneity Regulates Species-Area Relationships through the Spatial Distribution of Species. *For. Ecosyst.* **2022**, *9*, 100033. [CrossRef]
29. Whittaker, R.J.; Fernandez-Palacios, J.M. *Island Biogeography: Ecology, Evolution, and Conservation*; OUP: Oxford, UK, 2007; ISBN 978-0-19-856611-3.

30. D'Antraccoli, M.; Peruzzi, L.; Conti, F.; Galasso, G.; Roma-Marzio, F.; Bartolucci, F. Floristic Richness in a Mediterranean Hotspot: A Journey across Italy. *Plants* **2024**, *13*, 12. [CrossRef]
31. Blackburn, T.M.; Cassey, P.; Pyšek, P. Species–Area Relationships in Alien Species: Pattern and Process. In *The Species–Area Relationship: Theory and Application*; Cambridge University Press: Cambridge, UK, 2021; pp. 133–154.
32. Dimopoulos, P.; Bazos, I.; Kokkoris, I.P.; Zografidis, A.; Karadimou, E.; Kallimanis, A.S.; Raus, T.; Strid, A. *A Guide to the Alien Plants of Greece with Reference to the Natura 2000 Protected Area Network*; Natural Environment and Climate Change Agency (NECCA): Athens, Greece, 2020; ISBN 978-618-85104-1-8.
33. Galasso, G.; Conti, F.; Peruzzi, L.; Alessandrini, A.; Ardenghi, N.M.G.; Bacchetta, G.; Banfi, E.; Barberis, G.; Bernardo, L.; Bouvet, D.; et al. A Second Update to the Checklist of the Vascular Flora Alien to Italy. *Plant Biosyst. Int. J. Deal. All Asp. Plant Biol.* **2024**, *158*, 297–340. [CrossRef]
34. Spampinato, G.; Laface, V.L.A.; Posillipo, G.; Cano Ortiz, A.; Quinto Canas, R.; Musarella, C.M. Alien Flora in Calabria (Southern Italy): An Updated Checklist. *Biol. Invasions* **2022**, *24*, 2323–2334. [CrossRef]
35. Weber, E.; Sun, S.-G.; Li, B. Invasive Alien Plants in China: Diversity and Ecological Insights. *Biol. Invasions* **2008**, *10*, 1411–1429. [CrossRef]
36. Pyšek, P.; Pergl, J.; Essl, F.; Lenzner, B.; Dawson, W.; Kreft, H.; Weigelt, P.; Winter, M.; Kartesz, J.; Nishino, M.; et al. Naturalized Alien Flora of the World: Species Diversity, Taxonomic and Phylogenetic Patterns, Geographic Distribution and Global Hotspots of Plant Invasion. *Preslia* **2017**, *89*, 203–274. [CrossRef]
37. Arianoutsou, M.; Bazos, I.; Delipetrou, P.; Kokkoris, Y. The Alien Flora of Greece: Taxonomy, Life Traits and Habitat Preferences. *Biol. Invasions* **2010**, *12*, 3525–3549. [CrossRef]
38. Arianoutsou, M.; Adamopoulou, C.; Andriopoulos, P.; Bazos, I.; Christopoulou, A.; Galanidis, A.; Kalogianni, E.; Karachle, P.K.; Kokkoris, Y.; Martinou, A.F.; et al. HELLAS-ALIENS. The Invasive Alien Species of Greece: Time Trends, Origin and Pathways. *NeoBiota* **2023**, *86*, 45–79. [CrossRef]
39. Allen, H.D. Plants and Animals. In *Mediterranean Ecogeography*; Routledge: London, UK, 2001; ISBN 978-1-315-83852-6.
40. Cerrato, M.D.; Cortés-Fernández, I.; Ribas-Serra, A.; Mir-Rosselló, P.M.; Cardona, C.; Gil, L. Time Pattern Variation of Alien Plant Introductions in an Insular Biodiversity Hotspot: The Balearic Islands as a Case Study for the Mediterranean Region. *Biodivers Conserv.* **2023**, *32*, 2585–2605. [CrossRef]
41. Wagner, V.; Chytrý, M.; Jiménez-Alfaro, B.; Pergl, J.; Hennekens, S.; Biurrun, I.; Knollová, I.; Berg, C.; Vassilev, K.; Rodwell, J.S.; et al. Alien Plant Invasions in European Woodlands. *Divers. Distrib.* **2017**, *23*, 969–981. [CrossRef]
42. Bazos, I. Study of the Flora and Vegetation of Lesvos Island (East Aegean Islands, Greece). PhD Thesis, National and Kapodistrian University of Athens, Athens, Greece, 2005.
43. Sharma, G.P.; Esler, K.J.; Blignaut, J.N. Determining the Relationship between Invasive Alien Species Density and a Country's Socio-Economic Status. *S. Afr. J. Sci.* **2010**, *106*, 1–6. [CrossRef]
44. Pyšek, P.; Richardson, D.M. The Biogeography of Naturalization in Alien Plants. *J. Biogeogr.* **2006**, *33*, 2040–2050. [CrossRef]
45. Brock, K.C.; Daehler, C.C. Plant Naturalization Trends Reflect Socioeconomic History and Show a High Likelihood of Inter-Island Spread in Hawai'i. *Invasive Plant Sci. Manag.* **2021**, *14*, 135–146. [CrossRef]
46. Pyšek, P.; Richardson, D.M.; Rejmánek, M.; Webster, G.L.; Williamson, M.; Kirschner, J. Alien Plants in Checklists and Floras: Towards Better Communication between Taxonomists and Ecologists. *Taxon* **2004**, *53*, 131–143. [CrossRef]
47. Pyšek, P.; Jarošík, V. Residence Time Determines the Distribution of Alien Plants. In *Invasive Plants: Ecological and Agricultural Aspects*; Inderjit, Ed.; Birkhäuser-Verlag: Basel, Switzerland, 2005; pp. 77–96. ISBN 978-3-7643-7137-1.
48. Puddu, S.; Podda, L.; Mayoral, O.; Delage, A.; Hugot, L.; Petit, Y.; Bacchetta, G. Comparative Analysis of the Alien Vascular Flora of Sardinia and Corsica. *Not. Bot. Horti Agrobot. Cluj Napoca* **2016**, *44*, 337–346. [CrossRef]
49. Podda, L.; Arguimbau, P.F.; García-Berlanga, O.M.; Mascia, F.; Bacchetta, G. Comparison of the Vascular Exotic Flora in Continental Islands: Sardinia (Italy) and Balearic Islands (Spain). *An. Del JardÍN BotÁNico De Madr.* **2010**, *67*, 157–176. [CrossRef]
50. Lazzaro, L.; Ferretti, G.; Giuliani, C.; Foggi, B. A Checklist of the Alien Flora of the Tuscan Archipelago (Italy). *Webbia* **2014**, *69*, 157–176. [CrossRef]
51. Pyšek, P.; Jarošík, V.; Chytrý, M.; Danihelka, J.; Kühn, I.; Pergl, J.; Tichý, L.; Biesmeijer, J.C.; Ellis, W.N.; Kunin, W.E.; et al. Successful Invaders Co-Opt Pollinators of Native Flora and Accumulate Insect Pollinators with Increasing Residence Time. *Ecol. Monogr.* **2011**, *81*, 277–293. [CrossRef]
52. van Kleunen, M.; Dawson, W.; Maurel, N. Characteristics of Successful Alien Plants. *Mol. Ecol.* **2015**, *24*, 1954–1968. [CrossRef]
53. Zhou, Q.; Wu, J.; Cui, X.; Li, X.; Liu, Z.; Musa, A.; Ma, Q.; Yu, H.; Liang, W.; Jiang, S.; et al. Geographical Distribution of the Dispersal Ability of Alien Plant Species in China and Its Socio-Climatic Control Factors. *Sci. Rep.* **2021**, *11*, 7187. [CrossRef]
54. Galanos, C.J. The Alien Flora of Terrestrial and Marine Ecosystems of Rodos Island (SE Aegean), Greece. *Willdenowia* **2015**, *45*, 261–278. [CrossRef]
55. Pyšek, P.; Jarošík, V.; Pergl, J. Alien Plants Introduced by Different Pathways Differ in Invasion Success: Unintentional Introductions as a Threat to Natural Areas. *PLoS ONE* **2011**, *6*, e24890. [CrossRef]
56. Kosmas, C.; Danalatos, N.G.; Gerontidis, S. The Effect of Land Parameters on Vegetation Performance and Degree of Erosion under Mediterranean Conditions. *CATENA* **2000**, *40*, 3–17. [CrossRef]
57. Kosmas, C.; Gerontidis, S.; Marathianou, M. The Effect of Land Use Change on Soils and Vegetation over Various Lithological Formations on Lesvos (Greece). *CATENA* **2000**, *40*, 51–68. [CrossRef]

58. Margari, V.; Gibbard, P.L.; Bryant, C.L.; Tzedakis, P.C. Character of Vegetational and Environmental Changes in Southern Europe during the Last Glacial Period; Evidence from Lesvos Island, Greece. *Quat. Sci. Rev.* **2009**, *28*, 1317–1339. [CrossRef]
59. Bazos, I.; Yannitsaros, A. Pteridophyte Flora of Lesvos (East Aegean Islands. Greece). *Edinb. J. Bot.* **1999**, *56*, 421–448. [CrossRef]
60. Biel, B. Contributions to the Flora of the Aegean Islands of Lesvos and Limnos, Greece. *Willdenowia* **2002**, *32*, 209–219. [CrossRef]
61. Strid, A. The Botanical Exploration of Greece. *Plant Syst. Evol.* **2020**, *306*, 27. [CrossRef]
62. Galanidou, N.; Athanassas, C.; Cole, J.; Iliopoulos, G.; Katerinopoulos, A.; Magganas, A.; McNabb, J. The Acheulian Site at Rodafnidia, Lisvori, on Lesbos, Greece: 2010–2012. In *Vertebrate Paleobiology and Paleoanthropology*; Springer: Berlin/Heidelberg, Germany, 2016; pp. 119–138. ISBN 978-94-024-0873-7.
63. Wickham, H.; Seidel, D. *Ggplot2: Elegant Graphics for Data Analysis*; Springer-Verlag: New York, NY, USA, 2023; ISBN 978-3-319-24277-4.
64. Pebesma, E.; Bivand, R. *Spatial Data Science: With Applications in R*; Chapman and Hall/CRC: Boca Raton, FL, USA, 2023.
65. Brundu, G.; Aksoy, N.; Brunel, S.; Eliáš, P.; Fried, G. Rapid Surveys for Inventorying Alien Plants in the Black Sea Region of Turkey. *EPPO Bull.* **2011**, *41*, 208–216. [CrossRef]
66. Catry, F.X.; Moreira, F.; Deus, E.; Silva, J.S.; Águas, A. Assessing the Extent and the Environmental Drivers of Eucalyptus Globulus Wildling Establishment in Portugal: Results from a Countrywide Survey. *Biol. Invasions* **2015**, *17*, 3163–3181. [CrossRef]
67. McAvoy, T.J.; Snyder, A.L.; Johnson, N.; Salom, S.M.; Kok, L.T. Road Survey of the Invasive Tree-of-Heaven (*Ailanthus altissima*) in Virginia. *Invasive Plant Sci. Manag.* **2012**, *5*, 506–512. [CrossRef]
68. Chilton, L. *Plant List for Lesvos*; Revised edition 1998–2009; Marengo Publications: Marengo, IA, USA, 2009; ISBN 978-1-900802-50-5.
69. Strid, A. *Atlas of the Aegean Flora. Part 1: Text & Plates. Part 2: Maps.*; Botanic Garden and Botanical Museum Berlin, Freie Universität Berlin: Berlin, Germany, 2006; Volume 33, ISBN 978-3-921800-97-3.
70. Dal Cin D'Agata, C.; Skoula, M.; Brundu, G. A Preliminary Inventory of the Alien Flora of Crete (Greece). *Bocconea* **2009**, *23*, 301–315.
71. WFO. World Flora Online. Version 2023.03. Available online: https://www.worldfloraonline.org/ (accessed on 11 October 2023).
72. Kindt, R. WorldFlora: An R Package for Exact and Fuzzy Matching of Plant Names against the World Flora Online Taxonomic Backbone Data. *Appl. Plant Sci.* **2020**, *8*, e11388. [CrossRef]
73. Richardson, D.M.; Pysek, P.; Rejmanek, M.; Barbour, M.G.; Panetta, F.D.; West, C.J. Naturalization and Invasion of Alien Plants: Concepts and Definitions. *Divers. Distrib.* **2000**, *6*, 93–107. [CrossRef]
74. Convention on Biological Diversity (CBD). *Pathways of Introduction of Invasive Species, Their Prioritization and Management*; Convention on Biological Diversity (CBD): Montreal QC, Canada, 2014.
75. Irl, S.D.H.; Obermeier, A.; Beierkuhnlein, C.; Steinbauer, M.J. Climate Controls Plant Life-Form Patterns on a High-Elevation Oceanic Island. *J. Biogeogr.* **2020**, *47*, 2261–2273. [CrossRef]
76. POWO. Plants of the World Online. Facilitated by the Royal Botanic Gardens, Kew. Available online: http://www.plantsoftheworldonline.org/ (accessed on 24 January 2023).
77. Saul, W.-C.; Roy, H.E.; Booy, O.; Carnevali, L.; Chen, H.-J.; Genovesi, P.; Harrower, C.A.; Hulme, P.E.; Pagad, S.; Pergl, J.; et al. Assessing Patterns in Introduction Pathways of Alien Species by Linking Major Invasion Data Bases. *J. Appl. Ecol.* **2017**, *54*, 657–669. [CrossRef]
78. Galanidis, A.; Dimitrakopoulos, P.G. *Alien Flora of the Mediterranean Lesvos Island (East Aegean, Greece). Checklist dataset Version 1.1*; Department of Environment, University of the Aegean: Mytilene, Greece, 2024. [CrossRef]
79. *Posit team RStudio: Integrated Development Environment for R 2023*; R Studio Team: Boston, MA, USA, 2023.
80. Bartolucci, F.; Peruzzi, L.; Galasso, G.; Alessandrini, A.; Ardenghi, N.M.G.; Bacchetta, G.; Banfi, E.; Barberis, G.; Bernardo, L.; Bouvet, D.; et al. A Second Update to the Checklist of the Vascular Flora Native to Italy. *Plant Biosyst. Int. J. Deal. All Asp. Plant Biol.* **2024**, *158*, 219–296. [CrossRef]
81. Carta, A.; Forbicioni, L.; Frangini, G.; Pierini, B.; Peruzzi, L. An Updated Inventory of the Vascular Flora of Elba Island (Tuscan Archipelago, Italy). *Ital. Bot.* **2018**, *6*, 1–22. [CrossRef]
82. Chiarucci, A.; Guarino, R.; Pasta, S.; Rosa, A.L.; Cascio, P.L.; Médail, F.; Pavon, D.; Fernández-Palacios, J.M.; Zannini, P. Species–Area Relationship and Small-Island Effect of Vascular Plant Diversity in a Young Volcanic Archipelago. *J. Biogeogr.* **2021**, *48*, 2919–2931. [CrossRef]
83. Pasta, S.; Ardenghi, N.M.G.; Badalamenti, E.; Mantia, T.L.; Console, S.L.; Parolo, G. The Alien Vascular Flora of Linosa (Pelagie Islands, Strait of Sicily): Update and Management Proposals. *Willdenowia* **2017**, *47*, 135–144. [CrossRef]
84. Dimopoulos, P.; Raus, T.; Bergmeier, E.; Constantinidis, T.; Iatrou, G.; Kokkini, S.; Strid, A.; Tzanoudakis, D. *Vascular Plants of Greece: An Annotated Checklist*; Englera; Botanic Garden and Botanical Museum Berlin—dahlem: Berlin, Germany; Hellenic Botanical Society: Athens, Greece, 2013; ISBN 978-3-921800-88-1.

diversity

MDPI

Article

Effects of Invasive Plant Diversity on Soil Microbial Communities

Xiaoyan Wang †, Xue Wang †, Wei Wang [1], Jiang Wang * and Feihai Yu *

School of Life Science, Taizhou University, Taizhou 318000, China
* Correspondence: wangjiang@tzc.edu.cn (J.W.); feihaiyu@126.com (F.Y.)
† These authors contributed equally to this work.

Abstract: Native plant communities can be invaded by different numbers of alien plant species or by the same number of alien plant species with different levels of evenness. However, little is known about how alien invasive plant species richness and evenness affect soil microbial communities. We constructed native herbaceous plant communities invaded by exotic plants with different richness (1, 2, 4 and 8 species) and evenness (high and low) and analyzed soil physico-chemical properties and the diversity and composition of soil fungal and bacterial communities by high-throughput Illumina sequencing. Overall, the species richness and evenness of invasive plants had no significant effect on bacterial and fungal alpha diversity (OTUs, Shannon, Simpson, Chao1 and ACE) or the soil physico-chemical properties. However, invasive species richness had a significant impact on the relative abundance of the most dominant fungi, Ascomycota and *Bipolaris*, and the dominant bacteria, Actinobacteriota, which increased with increasing invasive species richness. The relative abundance of the dominant microbial groups was significantly correlated with the relative abundance of some specific invasive plants in the community. This study sheds new light on the effects of plant co-invasion on soil microbial communities, which may help us understand the underlying mechanisms of multiple alien plant invasion processes from the perspective of soil microorganisms.

Keywords: bacterial community; co-invasion; diversity effect; fungal community; invasive species evenness; invasive species richness

1. Introduction

Invasions by alien plants are a world-wide problem that has received great attention [1–4]. Alien plant invasions may not only displace native plant species and damage local flora and biodiversity but may also negatively influence many other ecosystem functions and services [5–9] in relation to the regional context [10]. Thus, there is a great need to develop measures to efficiently manage and control alien plant invasions [11,12]. To this end, we need to first assess the impacts of alien plant invasions on ecosystem functions such as soil microbial communities [13–16]. Soil microbial communities can alleviate the effects of invasive plants on local plant communities by degrading allelopathic substances [17], promoting further invasion via symbiosis with arbuscular mycorrhizal fungi or inhibiting further invasion via the accumulation of pathogens [18,19].

Previous studies have shown that alien plant invasions can have profound effects on soil microbial communities [14,20,21]. For instance, alien plant invasions can affect the soil properties [22–25], species richness and composition of soil microbial communities [26,27], thus destroying the long-term balance between native plants and soil microorganisms [15,16,27]. Some studies have shown that plant invasions can improve soil microbial diversity by providing more energy sources from secondary metabolites or a greater quantity and quality of litter [26,28–30]. Other studies have found that plant invasions reduce soil microbial diversity (e.g., AM fungi) [31–33] or have no significant effect [33,34].

One overlooked phenomenon in invasion ecology is that a native plant community can be invaded by different numbers of alien plant species [35]. Thus, while many studies have examined the impacts of alien plant invasions on soil microbial communities, the impacts of alien invasive species richness have never been considered. Recent studies found that increasing invasive plant species richness significantly increased the productivity of alien invasive plants and decreased that of native plant communities [35–38]. This suggests that increasing alien plant species can promote the invasion success of alien plants into native plant communities [35]. Since previous studies have shown that changes in both alien plant productivity and native plant productivity can impact soil microbial communities [39–42], we hypothesize that alien plant richness can influence soil microbial communities.

Another overlooked phenomenon is that a native plant community can be invaded by the same number of alien plant species with different levels of evenness [35]. As an important aspect of species diversity, evenness is often used to explain inconsistencies in species richness effects [43–46]. In several cases, evenness explained more biodiversity effects than species richness [47,48]. For instance, it has been shown that species richness had impacts on bacterial communities through the modulation of plant evenness in native grassland communities [46,49]. To our best knowledge, little is known about the effect of invasive plant evenness on soil microbial communities. We speculate that a high level of invasive plant evenness can result in high microbial diversity due to the different contributions of all invasive plants to soil micro-environment changes. By contrast, when the evenness of alien invasive plants is low, only one or two alien species are dominant, so alien plant diversity has similar effects to a monoculture of the dominant species [49,50].

We constructed native herbaceous plant communities invaded by alien plants with different richness (1, 2, 4 and 8 species) and evenness (high and low) to explore the impacts of invasive plant species diversity on soil fungal and bacterial communities. Specifically, we addressed the following questions: (1) Do alien invasive plant richness and evenness affect the diversity and composition of soil fungal communities? (2) Do alien invasive plant richness and evenness affect the diversity and composition of soil bacterial communities? (3) Do alien invasive plant richness and evenness affect the abundance of some specific microbial taxa?

2. Materials and Methods

2.1. Native and Invasive Species Preparation

The plant species pool used to construct experimental plant communities consisted of 18 native plant species and 10 alien invasive plant species that commonly grow in grasslands around Taizhou city (121.43° E, 28.68° N), Zhejiang Province, China [35]. Eight of the native species were perennials (*Achyranthes bidentata* Blume, *Persicaria filiformis* (Thunb.) Nakai, *Aster trinervius* subsp. *ageratoides* (Turczaninow) Grierson, *Bellis perennis* L., *Carpesium abrotanoides* L., *Patrinia scabiosifolia* Link, *Penthorum chinense* Pursh and *Plantago asiatica* L.), and ten were annual/biennial. The ten invasive species include four perennials (*Alternanthera philoxeroides* (Mart.) Griseb., *Mirabilis jalapa* L., *Solidago canadensis* L., and *Talinum paniculatum* (Jacq.) Gaertn.) and six annuals (*Ageratum conyzoides* L., *Symphyotrichum subulatum* (Michx.) G.L.Nesom, *Bidens frondosa* L., *Bidens pilosa* L., *Celosia argentea* L., and *Sesbania cannabina* (Retz.) Poir.). Detailed information on these species was described by Wang et al. (2022) [35].

Seeds (except *Alternanthera philoxeroides*) of all native and invasive species were collected from field sites around Taizhou city and stored at 4 °C till use. In April 2020, these seeds were sown into plastic containers (30 cm long × 30 cm wide × 30 cm high) filled with a mixture of river sand and peat. The plastic containers were placed in a greenhouse at Taizhou University, Zhejiang Province, China. For *A. philoxeroides*, ramets (asexual individuals) were propagated from stem cuttings. One month later, seedlings with similar sizes were used to construct plant communities. Height differences within the same species were limited to 2 cm.

2.2. Experimental Design

On 27 May 2020, 96 native plant communities were constructed in 96 containers (each 120 cm in diameter × 74 cm in height) with 18 native plant species. In each container, each native species was initially planted with three seedlings. The containers were filled with a mixture of soil (0.62 ± 0.17 g kg^{-1} total N, 0.13 ± 0.03 g kg^{-1} total P, *mean* ± *SE*, *n* = 5), peat and sand at a volume ratio of 2:1:1. Slow release fertilizer (N:P:K = 14:14:14, Osmocote exact standard 3–4 M) was added to the soil at a dose of 3 g L^{-1}. After one week, individuals of invasive plants with four levels of species richness (1, 2, 4 and 8 species) and two levels of evenness (high and low) were introduced to 90 of the 96 native plant communities (Table A1). For the 1-species treatment, individuals of each of the 10 invasive species were planted in three containers, making 30 communities. For each of the 2-species, 4-species and 8-species treatments, 10 different species mixtures were constructed by randomly selecting them from the species pool in such a way that each species had an equal chance of being selected. Each of these mixtures was replicated twice: one was used for high evenness (2-species, 12:12; 4-species, 6:6:6:6; 8-species, 3:3:3:3:3:3:3:3) and the other for low evenness (2-species, 22:2; 4-species, 18:3:2:1; 8-species, 13:3:2:2:1:1:1:1). For communities with invasive plants, the number of individuals of invasive species in each community was equal (i.e., 24 individuals). In the remaining six native plant communities, invasive species were not introduced. For a more detailed description of the experiment, see Wang et al. (2022) [35].

2.3. Soil Sample Collection and Physico-Chemical Analysis

On 11 October 2020, we took five soil samples (0–10 cm deep) under each of the 90 communities invaded by alien plant species [51,52]. The five soil samples from each container were thoroughly homogenized to form a composite sample. Then, a sub-sample of 10 mL was preserved at −80 °C until DNA extraction for soil microbial analysis. Another sub-sample was air-dried for analysis of soil physico-chemical properties. Soil pH was measured in a 1:2.5 soil:water (w:w) mixture using a digital pH meter (Mettler Toledo FE20, Shanghai, China). Soil organic matter was determined by the Walkley-Black acid digestion method. Soil N-NH_4^+, N-NO_3^- (extracted from a 1:5 suspension of soil in 2 M KCl solution) and total N (digested with concentrated H_2SO_4 and $HClO_4$) were measured using a continuous flow analyzer (Auto Analyzer 3, Bran and Luebbe, Norderstedt, Germany). Total P (digested with concentrated H_2SO_4 and $HClO_4$) and available P (extracted with 0.5 M $NaHCO_3$) were determined by an inductively coupled plasma emission spectrometer (Optima 2100DV, PerkinElemer, Waltham, MA, USA).

2.4. Microbial Sample Collection and Sequencing

Soil DNA was extracted using a PowerSoil™ DNA Isolation Kit (MoBio, Carlsbad, CA, USA) following the manufacturer's instructions and then sent to Novogene Co., Ltd., Beijing, China for amplification, library preparation and sequencing. The V4 region of the bacterial 16S rDNA gene was amplified using primers 515F (GTGCCAGCMGCCGCG-GTAA) and 806R (GGACTACNNGGGTATCTAAT) [53]. The fungal ITS region rDNA was amplified using primers ITS5-1737F (GGAAGTAAAAGTCGTAACAAGG) and ITS2-2043R (GCTGCGTTCTTCATCGATGC) [54]. The sequencing was done on an Illumina Novaseq 6000 platform, with 250 bp paired-end reads generated. The raw sequencing data have been deposited into the NCBI Sequence Read Archive (SRA) database (Accession Number: PRJNA870235 and PRJNA861831).

2.5. Bioinformatic Analysis

Barcode extraction, paired read assembly and quality filtering of the raw data were processed using the Software QIIME v.1.9.1 [55]. Chimeric sequences were identified and removed using UCHIME [56]. Operational taxonomic units (OTUs) were identified at the 97% similarity level with the UPARSE v.7.1b [57]. The taxonomy of each OTU of bacteria and fungi was assigned by the Uclust method using SILVA release 132 (https://www.

arb-silva.de/documentation/release-132, accessed on 5 July 2021) and UNITE database (http://qiime.org/scripts/assign_taxonomy.html, accessed on 5 July 2021), respectively.

2.6. Statistical Analysis

Alpha diversity (Shannon, Simpson, Chao1 and ACE indices) and beta diversity were calculated using the QIIME software (Version 1.9.1) [55]. The effects of invasive plant species richness (log_2 scale), evenness and their interaction on alpha diversity index, soil physico-chemical properties and the relative abundance of the most abundant phyla and genera of soil fungi and bacteria were tested by linear mixed models. In these models, invasive species richness was treated as a fixed continuous term, evenness as a fixed categorical term and species composition as a random term [35]. The treatment with a single invasive species was considered a low evenness treatment because monocultures were on the regression lines of uneven rather than even treatments [35,58]. We performed Principal Co-ordinates Analysis (PCoA) and Adonis to test whether species composition of soil microbial communities differed between different treatments based on Weighted Unifrac distance. Linear regressions were used to assess the relationships of invasive species richness with the alpha diversity index, and relative abundances of phyla and genera of fungi and bacteria. Residuals of all variables were checked for homoscedasticity and normality. All statistical analyses were conducted using R 4.1.2 (https://www.r-project.org/, accessed on 10 June 2022).

3. Results

3.1. Effects of Invasive Species Diversity on Alpha Diversity of Soil Microbial Communities

Alien invasive species richness, evenness or their interaction did not significantly affect alpha diversity (OTUs, Shannon, Simpson, Chao1 and ACE) of either soil fungal communities (Table 1A, Figure 1A–E) or soil bacterial communities (Table 1B, Figure 1F–J).

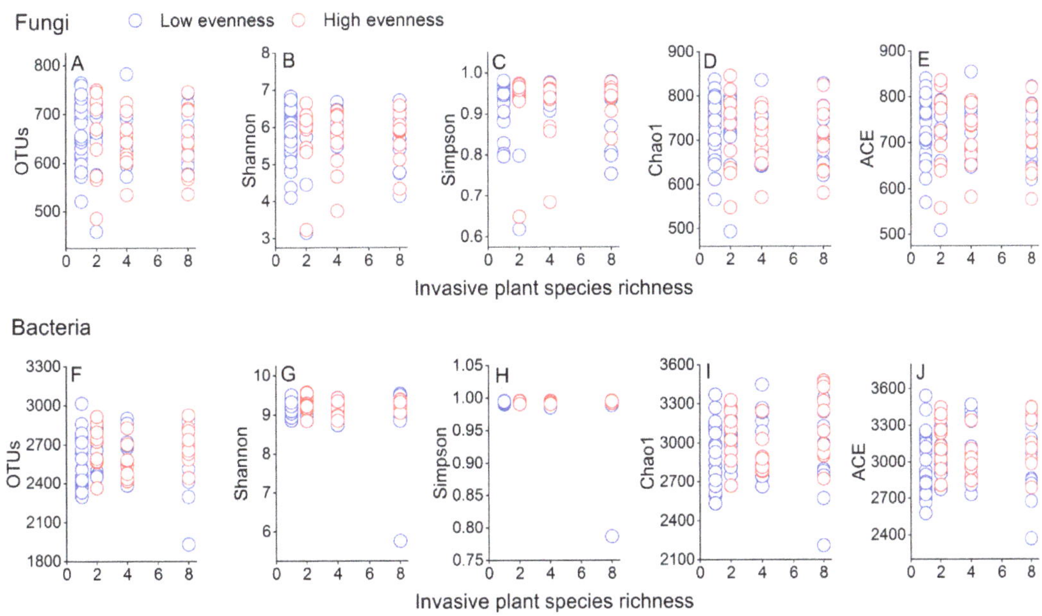

Figure 1. Effects of invasive species richness and evenness on the OTUs, Shannon, Simpson, Chao1 and ACE of soil fungal communities (**A–E**) and soil bacterial communities (**F–J**). Raw data points are shown. See Table 1A,B for the results of the linear mixed models.

Table 1. Summary of linear mixed models testing the effects of invasive species richness, evenness and their interaction on alpha diversity of soil fungal communities (**A**), soil bacterial communities (**B**) and soil physico-chemical properties (**C**). Species composition was included as a random term. Degrees of freedom are (1, 38), (1, 48) and (1, 48) for the effects of richness, evenness and their interaction, respectively.

Variable	Richness (R)		Evenness (E)		R′ E	
	F	*P*	*F*	*P*	*F*	*P*
(A) Soil fungal communities						
OTUs	0.515	0.477	0.337	0.564	0.298	0.588
Shannon	0.555	0.461	0.007	0.933	0.292	0.592
Simpson	0.286	0.596	0.033	0.857	0.781	0.381
Chao1	0.284	0.597	0.170	0.682	0.357	0.553
ACE	0.341	0.563	0.128	0.722	0.316	0.577
(B) Soil bacterial communities						
OTUs	0.585	0.449	1.344	0.252	0.008	0.931
Shannon	0.568	0.456	1.856	0.180	0.768	0.385
Simpson	2.155	0.150	2.204	0.144	1.433	0.237
Chao1	1.668	0.204	2.688	0.108	0.587	0.447
ACE	1.232	0.274	2.529	0.118	0.150	0.701
(C) Soil physico-chemical properties						
pH	0.012	0.913	0.004	0.951	0.448	0.506
Organic matter	0.019	0.891	0.724	0.399	1.317	0.257
Total N	1.508	0.227	1.257	0.268	2.303	0.136
NH_4_N	0.021	0.886	1.835	0.182	0.843	0.363
NO_3_N	0.005	0.944	2.427	0.126	2.755	0.104
Total P	0.475	0.495	0.277	0.602	0.037	0.849
Available P	0.058	0.811	2.977	0.091	0.801	0.375
Available K	0.023	0.882	0.884	0.352	3.272	0.077

3.2. *Effects of Invasive Species Diversity on Soil Physico-Chemical Properties*

Alien invasive species richness, evenness or their interaction had no significant effect on any of the soil physico-chemical properties (soil pH, organic matter, total N, NH_4_N, NO_3_N, total P, available P and available K; Table 1C, Table 2).

Table 2. Soil physico-chemical properties under communities with different levels of invasive plant species richness and evenness (mean ± SE). See Table 1C for the results of the linear mixed models.

Richness	Evenness	pH	Organic matter (g/kg)	Total N (mg/g)	NH_4_N (mg/kg)	NO_3_N (mg/kg)	Total P (mg/g)	Available P (mg/kg)	Available K (mg/kg)
1	Low	7.27 ± 0.04	285.05 ± 17.48	0.77 ± 0.07	10.45 ± 1.20	33.76 ± 3.24	3.45 ± 0.06	122.05 ± 7.49	163.27 ± 8.90
2	Low	7.35 ± 0.05	303.00 ± 37.82	0.78 ± 0.08	11.90 ± 2.29	37.29 ± 6.68	3.43 ± 0.14	125.64 ± 17.72	182.28 ± 31.35
	High	7.34 ± 0.06	328.04 ± 27.28	0.58 ± 0.06	8.74 ± 1.33	35.72 ± 7.62	3.48 ± 0.11	122.13 ± 15.25	189.38 ± 24.27
4	Low	7.34 ± 0.05	275.48 ± 37.52	0.67 ± 0.09	9.96 ± 1.69	36.20 ± 6.91	3.32 ± 0.10	132.27 ± 12.46	180.98 ± 27.74
	High	7.26 ± 0.09	303.12 ± 18.88	0.65 ± 0.06	8.21 ± 0.61	31.32 ± 3.71	3.47 ± 0.12	105.33 ± 11.21	161.78 ± 13.50
8	Low	7.27 ± 0.04	296.57 ± 27.73	0.66 ± 0.05	11.12 ± 2.22	43.31 ± 6.00	3.39 ± 0.20	139.36 ± 13.88	191.89 ± 18.65
	High	7.29 ± 0.05	278.40 ± 11.35	0.70 ± 0.05	11.11 ± 1.48	26.67 ± 4.40	3.37 ± 0.20	115.02 ± 16.12	150.62 ± 16.54

3.3. *Effects of Invasive Species Diversity on Soil Microbial Community Composition*

Soil fungal communities in the 1-, 2- and 4-species treatments were separated from those in the 8-species treatment along the first axis of PCoA (Figure 2A). Results of the Adonis test also confirmed that species composition of soil fungal communities differed between the lower (1-, 2- and 4-species) and the higher (8-species) species richness treatments (Table 3A). For soil bacterial communities, the 1- and 8-species treatments were distinguished from the 2- and 4-species treatments along the first axis of PCoA (Figure 2B), and the results of the Adonis test also confirmed that the differences were significant (Table 3A). However, the Adonis test showed that neither the species compositions of soil fungal communities nor soil bacterial communities differ significantly between the high and the low evenness treatment within each level of species richness (Table 3B).

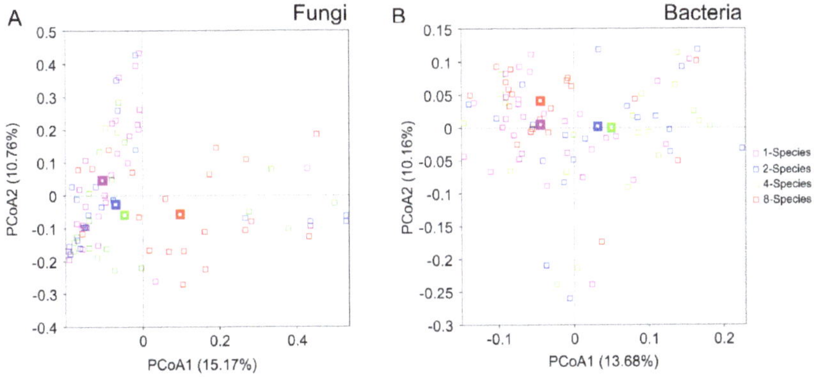

Figure 2. Results of PCoA for soil fungal (**A**) and bacterial (**B**) communities. See Table 3A for the results of Adonis test. The bold boxes indicate the central location of the sample points under different richness treatments.

Table 3. Adonis test for differences in community composition of fungi and bacteria between different richness and evenness treatments of invasive species based on Bray-Curtis distance.

		df	*F*	R^2	*P*
	(A) Comparison among different richness of alien invasive plant species				
Fungi					
	1 vs. 2 species	1 (48)	0.704	0.014	0.872
	1 vs. 4 species	1 (48)	1.135	0.023	0.291
	1 vs. 8 species	1 (48)	3.048	0.060	0.001
	2 vs. 4 species	1 (38)	0.747	0.019	0.807
	2 vs. 8 species	1 (38)	1.803	0.045	0.030
	4 vs. 8 species	1 (38)	1.967	0.049	0.016
Bacteria					
	1 vs. 2 species	1 (48)	1.333	0.027	0.113
	1 vs. 4 species	1 (48)	1.767	0.036	0.017
	1 vs. 8 species	1 (48)	1.149	0.023	0.242
	2 vs. 4 species	1 (38)	0.763	0.020	0.812
	2 vs. 8 species	1 (38)	1.595	0.040	0.026
	4 vs. 8 species	1 (38)	1.791	0.045	0.021
	(B) Comparison among different evenness of alien invasive plant species				
Fungi					
	Low vs. high evenness of the 2 species	1 (18)	0.910	0.048	0.517
	Low vs. high evenness of the 4 species	1 (18)	0.637	0.034	0.929
	Low vs. high evenness of the 8 species	1 (18)	0.737	0.039	0.844
Bacteria					
	Low vs. high evenness of the 2 species	1 (18)	0.694	0.037	0.930
	Low vs. high evenness of the 4 species	1 (18)	0.814	0.043	0.593
	Low vs. high evenness of the 8 species	1 (18)	0.958	0.051	0.501

3.4. Effects of Invasive Species Diversity on Relative Abundances of Soil Microbial Taxa

Regarding soil fungal taxa, Ascomycota, Basidiomycota and Rozellomycota were the three most abundant phyla (Figure 3A). Among them, invasive plant species richness had a significant effect on Ascomycota and Rozellomycota (Table 4); with increasing invasive plant species richness, the relative abundance of Ascomycota increased significantly, while that of Rozellomycota decreased (Figures 3A and 4A). Among the 20 most abundant fungal genera, invasive plant species richness had a significant effect on *Bipolaris*, unidentified_*Rozellomycota*_sp and *Clitopilus* (Table 4A); with increasing invasive plant species richness, the relative abundance of *Bipolaris*, the most abundant genus, increased significantly, while that of unidentified_*Rozellomycota*_sp and *Clitopilus* decreased (Figures 3C and 4A).

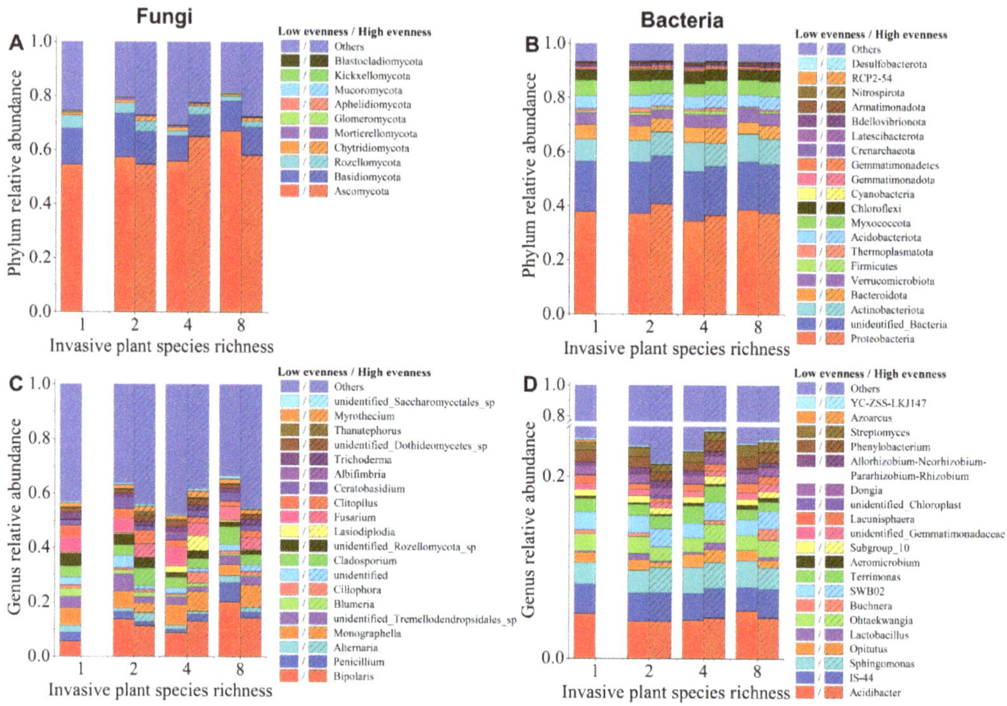

Figure 3. Relative abundance of the 10 most abundant phyla of soil fungi (**A**), the 20 most abundant phyla of soil bacteria (**B**) and the 20 most abundant genera of soil fungi (**C**) and bacteria (**D**) under the communities with different invasive plant species richness and evenness.

Table 4. Summary of linear mixed models testing the effects of invasive species richness, evenness and their interaction on the relative abundance of (**A**) soil fungal and (**B**) soil bacterial taxa. Values are in bold when $p < 0.05$. Species composition was included as a random term. Degrees of freedom are (1, 38), (1, 48) and (1, 48) for the effects of richness, evenness and their interaction, respectively.

Variable	Richness (R)		Evenness (E)		R × E	
	F	*P*	*F*	*P*	*F*	*P*
(A) Fungi						
Phylum						
Ascomycota	4.896	0.033	0.119	0.732	0.894	0.349
Rozellomycota	7.365	0.010	0.093	0.762	0.180	0.673
Genus						
Bipolaris	5.203	0.028	0.022	0.883	0.275	0.602
unidentified_*Rozellomycota*_sp	7.835	0.008	0.043	0.836	0.011	0.918
Clitopilus	6.076	0.018	0.606	0.440	0.222	0.639

Table 4. *Cont.*

Variable	Richness (R)		Evenness (E)		R × E	
	F	*P*	*F*	*P*	*F*	*P*
(B) Bacteria						
Phylum						
Actinobacteriota	4.920	0.033	1.838	0.182	0.844	0.363
Bdellovibrionota	0.252	0.619	0.210	0.649	7.782	0.008
Armatimonadota	1.065	0.309	4.358	0.042	0.018	0.894
Genus						
Sphingomonas	1.465	0.234	0.067	0.796	4.576	0.038
Buchnera	0.314	0.579	0.918	0.343	7.560	0.008
Streptomyces	8.822	0.005	6.024	0.018	0.126	0.724

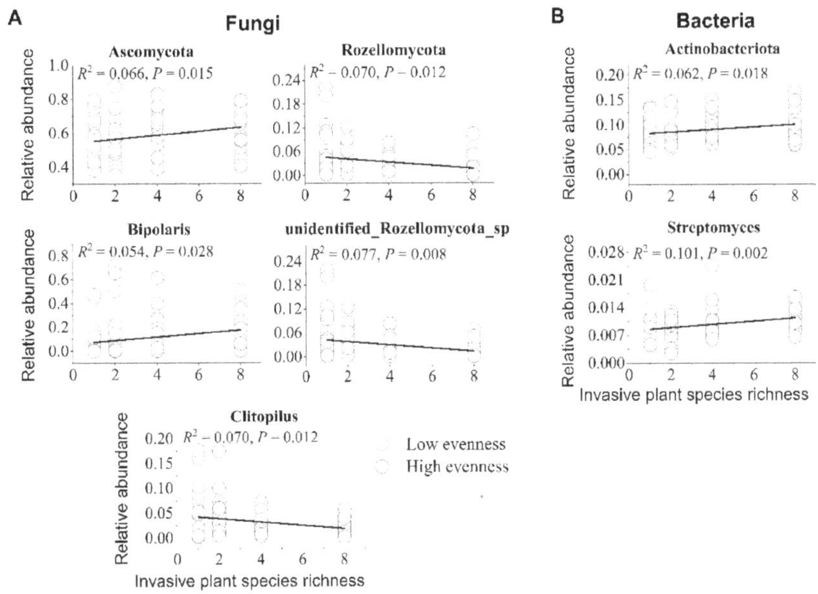

Figure 4. Relationships of the relative abundance of the fungal phylum of Ascomycota and Rozellomycota, the fungal genus of *Bipolaris*, unidentified_*Rozellomycota*_sp and *Clitopilus* (**A**), the bacterial phylum of Actinobacteriota and the bacterial genus of *Streptomyces* (**B**) with invasive plant richness. Raw data points are shown. See Table 4 for the results of the linear mixed models.

Regarding soil bacterial taxa, Proteobacteria, unidentified_Bacteria and Actinobacteriota were the three most abundant phyla (Figure 3B). Invasive plant richness has a significant influence on Actinobacteriota, which increased significantly with the increase of invasive plant species richness (Table 4 and Figures 3B and 4B). Invasive plant evenness had a significant effect on Armatimonadota, and invasive plant richness and evenness had a significant interactive effect on Bdellovibrionota (Table 4B, Figure 3B). Among the 20 most abundant genera of bacteria, invasive plant species richness and evenness had a significant effect on *Streptomyces*, which increased significantly with the increase of invasive species richness (Table 4B, Figures 3D and 4B). In addition, the interaction of invasive plant richness and evenness had a significant effect on *Sphingomonas* and *Buchnera* (Table 4B, Figure 3D).

4. Discussion

A large body of studies has investigated the effects of alien plant invasions on soil microbial communities [14–16,20,21], but few have considered the effects of invasive species diversity (richness and evenness). While we found no significant effect of alien invasive plant species richness and evenness on the alpha diversity of soil fungal and bacterial communities, there was a significant effect of invasive species richness on the community composition of both soil fungi and bacteria and on the relative abundances of some specific microbial taxa.

4.1. *Effects of Invasive Species Diversity on Alpha Diversity of Soil Microbial Communities*

Although, in the same experimental setup, increasing invasive species richness was found to significantly change the species compositions of the plant communities by increasing the productivity of alien invasive plant species and decreasing that of native plant species [35], it had no impact on alpha diversity of either soil fungal or bacterial communities. Previous studies testing the role of native plant richness showed that alpha diversity of soil fungal communities can have a positive [16,59,60], negative [61,62] or no relationship [63–67] with native plant species richness. The positive relationship could be because of more diverse microhabitats and/or resources created by the higher richness of plant species [16,63,68]. For example, soil fungal richness increased with native plant richness through the increase of soil total carbon [16]. The negative relationship results from the greater impact of context (e.g., plant biomass, soil pH and plant species identity) [61,62]. For example, soil resources and plant species had a stronger effect on soil biota than plant species richness [69].

The neutral effect of invasive plant species richness on soil microbial diversity in this study may be related to the antagonistic effects among different invasive plants on soil microorganisms [20]. This can be explained from two aspects: different invasive plants have opposite effects on the same microbial group, and the same invasive plant communities have opposite effects on different microbial groups. In this study, the first aspect can be confirmed by the opposite effects of the relative abundance of different invasive plants on specific microbial groups (Table A2). For example, the relative abundance of the invasive plants *Alternanthera philoxeroides*, *Sesbania cannabina* and *Talinum paniculatum* was significantly negatively correlated with the relative abundance of *Bipolaris*, the genus with the highest relative abundance of fungi, while the relative abundance of *Bidens frondosa* showed the opposite trend (Table A2).

For the second aspect, when multiple invasive plants co-exist, some species favor some particular microbial taxa and inhibit others, with the result that there is no significant change, as reported before [23]. In this study, this explanation can be evidenced by the opposite trend of the effect of invasive species diversity on the relative abundance of some fungal taxa (i.e., Ascomycota vs. Rozellomycota and *Bipolaris* vs. unidentified_*Rozellomycota*_sp and *Clitopilus*; Figure 4A). The same situation is present in soil bacterial communities, i.e., increasing invasive species richness increased the relative abundance of Actinobacteriota and decreased that of *Streptomyces* (Figure 4B).

Soil is an important environmental factor for microbial life. Many studies on the effects of plant richness on microbial life have found that this relationship is ultimately explained not only by plant species richness but also by soil physico-chemical properties, such as soil pH [70,71]. In this study, although soil microbial diversity was significantly negatively correlated with some soil physico-chemical properties (Table A3), invasive species richness and evenness had no significant effect on soil physico-chemical properties. Therefore, invasive species diversity may not change soil microbial diversity via altering soil physico-chemical properties.

4.2. Effects of Invasive Species Diversity on Soil Microbial Taxa and Community Composition

Specific taxa of microorganisms play an important role in maintaining microbial community structure and affecting ecosystem functions [72,73] (In this study, invasive species richness and evenness have significant effects on some specific taxa of fungi and bacteria (Table 4; Figure 4). Our results revealed that high invasive plant richness promoted the most abundant fungal phylum of Ascomycota and the most abundant fungal genus of *Bipolaris* and suppressed the fungal phylum of Rozellomycota and fungal genus of unidentified_*Rozellomycota*_sp and *Clitopilus*.

Ascomycota is one of the most common phyla among soil fungi and is the most abundant phyla in the soil of many invasive sites [16,29]. Similarly, Ascomycota possessed more than 50% relative abundance in each treatment of this study and increased significantly with increasing invasive species richness. As Ascomycota contained a wide diversity of fungal taxa, ranging from pathogens to mutualists, its ecological function depends on the comprehensive effect of each member. The increasing tendency of relative abundance of Ascomycota with invasive species richness in this study may be partly explained by its member *Bipolaris*, the most abundant genus of fungi, which also increased significantly with increasing invasive species richness. The genus of *Bipolaris* includes more than 20 species of pathogens that can infect a broad range of grasses [74–76], including invasive plants (e.g., *Microstegium*) [77]. Our results suggest that increasing invasive species richness may facilitate pathogen emergence and amplification, and these effects may inhibit native plant growth and promote further invasions. On the other hand, the accumulation of pathogens may also inhibit the invasion of alien plants [77].

For soil bacterial communities, Actinobacteriota, one of the largest taxonomic units of bacteria, increased significantly with increasing invasive species richness. As plant growth-promoting rhizobacteria, Actinobacteriota constitute important drivers of rhizosphere nutrient cycling and play a role as biocontrol agents against a range of pathogenic fungi and promote plant growth by phosphate solubilization, secondary metabolite production and antimicrobial synthesis [78–80]. Our results agreed with the findings of a previous study, which showed that increasing native crop diversity increased the relative abundance of Actinobacteriota [81]. In this study, the increased relative abundance of Actinobacteriota in more diverse invasive plant communities has the same trend with its member *Streptomyces*, which is the most abundant and arguably the most important actinomycetes, and is a good source of bioactive compounds, antibiotics, and extracellular enzymes [79,80,82]. Although studies have shown that plant invasion (e.g., *Ageratina adenophora*) increases the relative abundance of *Streptomyces* [83], the relative abundances of Actinobacteriota and *Streptomyces* were all negatively related to the relative abundance of the invasive plants *Sesbania cannabina* and *Talinum paniculatum* (Table A2). Therefore, invasive species richness may affect this microbe (Actinobacteriota and *Streptomyces*) through species-species interactions and species-environment interactions, and this effect may be beneficial to the growth and competitiveness of invasive plants themselves [83].

5. Conclusions

We conclude that alien invasive plant species diversity had no effect on the alpha diversity of both soil fungal and bacterial communities, likely because the dominant microbial groups showed opposite responses to alien invasive species diversity. One caveat is that the soil samples were collected only after one growing season of invasion, so the impact of invasive species diversity on the alpha diversity of soil microbial communities may not be big enough. Thus, future studies could test the longer-term effect of alien invasive plant species diversity on the alpha diversity of soil microbes. Given the important effects of soil microbial communities on plant growth, especially pathogens, it is possible that soil microorganisms may have important effects on native plants or invasive plants to alter community composition. Therefore, it is worth further researching if invasive plant diversity can affect the plant community composition through the influence of soil pathogens.

Author Contributions: Conceptualization, X.W. (Xiaoyan Wang), X.W. (Xue Wang), J.W. and F.Y.; methodology, X.W. (Xiaoyan Wang), X.W. (Xue Wang) and W.W.; formal analysis, X.W. (Xiaoyan Wang) and W.W.; writing—original draft preparation, X.W. (Xiaoyan Wang) and J.W.; writing—review and editing, X.W. (Xue Wang), J.W. and F.Y.; funding acquisition, X.W. (Xiaoyan Wang) and J.W. All authors have read and agreed to the published version of the manuscript.

Funding: This research was funded by Basic Public Welfare Research Projects of Zhejiang Province (No. LGN19C150004); Natural Science Foundation of Zhejiang Province (Nos. LY22C030001, LTY22C030004); National Natural Science Foundation of China (No. 31870504); 2021 Taizhou College Students Science and Technology Innovation Project (No. 2).

Institutional Review Board Statement: There are no humans or vertebrates involved in the current study.

Informed Consent Statement: No human subjects were involved or recruited in the current study.

Data Availability Statement: Data will be available upon request.

Conflicts of Interest: The authors declare no conflict of interest.

Appendix A

Table A1. Species composition ratio in different richness levels and evenness levels.

	1-species									
Invasive species	1	2	3	4	5	6	7	8	9	10
Bidens pilosa	24									
Bidens frondosa		24								
Ageratum conyzoides			24							
Sesbania cannabina				24						
Talinum paniculatum					24					
Celosia argentea						24				
Solidago canadensis							24			
Mirabilis jalapa								24		
Alternanthera philoxeroides									24	
Aster subulatus										24

	2-species																			
	High evenness										Low evenness									
Invasive species	1	2	3	4	5	6	7	8	9	10	1	2	3	4	5	6	7	8	9	10
Bidens pilosa	12					12					22					2				
Bidens frondosa				12					12					22					22	
Ageratum conyzoides	12							12			2							22		
Sesbania cannabina					12	12									2	22				
Talinum paniculatum				12						12				2						2
Celosia argentea		12					12					2					22			
Solidago canadensis					12			12							22			2		
Mirabilis jalapa		12								12		22								22
Alternanthera philoxeroides			12				12						22				2			
Aster subulatus			12						12				2						22	

	4-species																			
	High evenness										Low evenness									
Invasive species	1	2	3	4	5	6	7	8	9	10	1	2	3	4	5	6	7	8	9	10
Bidens pilosa	6	6	6	6							18	3	2	1						
Bidens frondosa	6				6	6	6				3				3	18	1			
Ageratum conyzoides				6		6	6			6				18		1	2			3
Sesbania cannabina		6	6	6					6			2	3	2					18	
Talinum paniculatum	6	6						6		6	2	1						2		18
Celosia argentea					6	6		6	6						1	3	18	18	2	
Solidago canadensis			6		6	6		6					18		2	2	3	3		
Mirabilis jalapa							6	6	6	6								1	3	2
Alternanthera philoxeroides	6				6		6			6	1				18					1
Aster subulatus		6	6	6					6			18	1	3					1	

	8-species																			
	High evenness										Low evenness									
Invasive species	1	2	3	4	5	6	7	8	9	10	1	2	3	4	5	6	7	8	9	10
Bidens pilosa			3	3	3	3	3	3	3	3			1	3	2	1	13	1	1	2
Bidens frondosa	3	3			3	3	3	3	3	3	13	1			1	2	1	1	2	3
Ageratum conyzoides		3	3		3	3	3	3	3	3		13	3		1	1	1	2	2	1
Sesbania cannabina	3	3		3	3	3		3	3	3	3	1		1	1	13		3	1	1
Talinum paniculatum	3		3	3	3		3	3	3	3	2		1	13	2		2	2	3	2
Celosia argentea	3	3	3	3	3	3	3	3			2	1	2	2	13	3	1	1		
Solidago canadensis	3	3	3	3	3	3	3		3		1	1	2	1	1	1	1		13	
Mirabilis jalapa	3	3	3	3		3	3	3		3	1	3	1	1		1	2	13		1
Alternanthera philoxeroides	3	3	3	3			3	3	3	3	1	2	1	1			3	1	1	13
Aster subulatus	3	3	3	3	3	3			3	3	1	2	13	2	3	2			1	1

The numbers in the third row represent the number of experiment replicates. For 1-species, there are three replicates for each species, which was not shown in the table. Other numbers represent the seedling number of different species planted in each community.

Table A2. Pearson correlation analysis between top 20 predominant phyla and genus and percentage of initial plant number for 10 invasive plants.

	Bidens pilosa	*Bidens frondosa*	*Ageratum conyzoides*	*Sesbania cannabina*	*Talinum paniculatum*	*Celosia argentea*	*Solidago canadensis*	*Mirabilis jalapa*	*Alternanthera philoxeroides*	*Aster subulatus*
Fungi										
Phylum										
Ascomycota	**−0.375 ***	0.265	***−0.328***	**−0.465 ****	−0.248	0.052	−0.241	***0.337***	**−0.474 ****	***−0.340***
Basidiomycota	0.195	0.124	0.080	**0.531 ****	0.179	−0.114	0.202	−0.219	**0.367 ***	−0.182
Rozellomycota	0.265	−0.225	**0.541 ****	**0.391 ***	0.066	0.065	0.099	0.186	**0.481 ****	0.174
Chytridiomycota	−0.178	−0.281	***0.314***	−0.166	−0.007	0.243	−0.106	−0.152	0.229	0.044
Mortierellomycota	−0.151	−0.037	0.293	***0.352***	−0.102	−0.278	***0.331***	−0.256	**0.370 ***	**0.598 ****
Glomeromycota	0.126	0.206	0.145	−0.220	0.237	***−0.312***	0.066	**0.382 ***	−0.173	−0.045
Mucoromycota	−0.001	−0.078	−0.126	**0.492 ****	0.174	0.069	0.102	−0.100	0.018	0.148
Monoblepharomycota	0.255	−0.122	0.052	−0.189	**0.549 ****	0.063	−0.005	−0.077	−0.100	−0.079
Genus										
Bipolaris	***−0.309***	**0.419 ***	***−0.352***	**−0.389 ***	**−0.426 ***	−0.350	−0.139	0.197	**−0.429 ***	−0.282
Alternaria	0.026	**−0.366 ***	−0.225	−0.090	0.094	**0.380 ***	−0.043	0.161	−0.281	0.068
Monographella	−0.181	−0.242	**0.514 ****	−0.155	0.174	0.297	−0.045	−0.083	−0.213	−0.155
unidentified_*Tremellodendropsidales*_*sp*	0.038	−0.007	−0.162	***0.323***	0.101	−0.076	0.128	***−0.340***	**0.398 ***	−0.196
Blumeria	0.028	−0.182	−0.142	**0.388 ***	0.270	0.153	−0.106	**0.516 ****	**0.417 ***	0.160
unidentified	−0.026	−0.058	−0.148	0.288	0.075	−0.089	0.156	***−0.304***	**0.366 ***	−0.191
Cladosporium	−0.020	***−0.317***	−0.230	0.051	0.006	0.101	−0.116	0.280	−0.187	−0.078
unidentified_*Rozellomycota*_*sp*	0.279	−0.237	**0.588 ****	**0.445 ***	0.048	0.097	0.056	0.298	**0.476 ****	0.124
Clitopilus	**0.668 ****	0.195	0.293	**0.570 ****	0.157	−0.035	0.187	0.113	0.256	0.137
Albifimbria	−0.124	−0.164	−0.114	−0.153	0.037	0.266	0.090	−0.096	−0.123	***0.346***
Trichoderma	−0.050	0.093	**0.440 ***	0.112	***0.329***	−0.164	−0.034	−0.178	0.137	0.121
unidentified_*Dothideomycetes*_*sp*	***0.335***	−0.038	0.049	−0.051	0.289	−0.089	−0.181	0.104	−0.058	−0.167
Bacteria										
Phylum										
Proteobacteria	−0.013	0.207	−0.139	0.197	0.267	−0.062	0.225	0.058	0.060	**−0.498 ****
unidentified_Bacteria	0.242	−0.147	0.221	0.229	−0.186	−0.024	0.027	0.062	**0.488 ****	−0.113
Actinobacteriota	**−0.431 ***	0.061	−0.180	**−0.478 ****	**−0.358 ***	−0.117	−0.271	0.152	−0.277	0.169
Verrucomicrobiota	−0.001	−0.100	0.212	−0.209	0.142	0.056	−0.115	−0.295	**−0.357 ***	−0.055
Firmicutes	0.006	−0.019	−0.008	−0.185	−0.133	−0.055	−0.176	−0.064	−0.182	**0.506 ****
Thermoplasmatota	0.149	−0.269	**0.394 ***	0.068	0.013	0.294	−0.203	−0.066	0.159	**0.527 ****
Acidobacteriota	0.132	**−0.400 ***	−0.139	0.109	0.055	0.158	0.233	−0.186	−0.060	0.172
Myxococcota	−0.029	−0.169	−0.056	**0.490 ****	0.244	**0.421 ***	−0.021	−0.028	−0.165	−0.228
Gemmatimonadota	−0.005	−0.192	***0.349***	***−0.353***	−0.219	−0.022	0.055	−0.190	0.106	0.065
Gemmatimonadetes	−0.024	**0.363 ***	−0.019	**−0.438 ***	−0.241	−0.270	***0.339***	−0.178	0.007	0.076
Latescibacterota	0.228	−0.286	−0.155	***0.332***	**0.406 ***	0.218	0.152	−0.081	−0.017	0.023
Bdellovibrionota	0.045	−0.092	0.249	0.221	0.188	0.191	0.020	−0.038	−0.047	**−0.377 ***
Nitrospirota	−0.106	−0.036	0.042	−0.123	−0.063	−0.065	**0.381 ***	−0.287	0.005	**−0.383 ***
RCP2-54	−0.005	−0.025	0.121	***0.333***	−0.034	0.107	0.118	0.172	**0.374 ***	−0.164
Desulfobacterota	0.072	0.097	0.146	−0.221	−0.056	−0.025	−0.092	−0.039	−0.138	**0.723 ****
Genus										
Acidibacter	−0.086	−0.271	0.227	0.010	**0.386 ***	0.113	0.037	−0.116	***−0.328***	−0.067
IS-44	***0.341***	**−0.424 ***	−0.105	**0.376 ***	***0.354***	0.116	−0.180	0.125	0.035	0.094
Sphingomonas	−0.284	***0.313***	−0.165	***−0.302***	−0.239	***−0.341***	**0.373 ***	−0.099	0.102	−0.296
Opitutus	0.089	0.071	***0.320***	***−0.309***	0.048	−0.014	−0.237	−0.189	***−0.325***	0.171
Lactobacillus	−0.059	−0.068	−0.010	−0.033	−0.065	−0.097	−0.045	−0.130	−0.111	**0.469 ****
Ohtaekwangia	0.146	−0.278	−0.272	**0.430 ***	0.084	0.142	−0.021	0.037	0.088	−0.062

Table A2. *Cont.*

	Bidens pilosa	*Bidens frondosa*	*Ageratum conyzoides*	*Sesbania cannabina*	*Talinum paniculatum*	*Celosia argentea*	*Solidago canadensis*	*Mirabilis jalapa*	*Alternanthera philoxeroides*	*Aster subulatus*
SWB02	0.184	−0.178	−0.121	***0.337***	0.066	0.214	0.158	−0.025	−0.057	0.063
Terrimonas	0.192	−0.019	**0.368 ***	0.007	0.190	−0.172	0.100	0.235	0.044	0.165
Aeromicrobium	−0.237	***0.331***	−0.242	−0.237	−0.240	−0.144	***−0.343***	−0.199	−0.283	−0.202
Subgroup_10	−0.053	−0.186	0.018	−0.188	−0.071	−0.131	0.252	−0.076	−0.010	**0.445 ***
Lacunisphaera	−0.109	−0.113	0.225	−0.065	0.194	0.078	−0.061	***−0.307***	−0.241	**−0.380 ***
Dongia	−0.073	−0.065	−0.052	0.075	0.244	−0.118	0.229	−0.044	0.216	***−0.342***
Allorhizobium-Neorhizobium-Pararhizobium-Rhizobium	***−0.324***	**0.401 ***	−0.002	0.080	***−0.343***	−0.288	−0.062	0.154	0.239	**−0.358 ***
Streptomyces	−0.268	−0.017	−0.089	**−0.524 ****	**−0.363 ***	−0.199	**−0.410 ***	−0.160	**−0.372 ***	−0.070
Azoarcus	**0.383 ***	0.007	−0.210	0.247	**0.359 ***	***0.344***	−0.224	0.077	0.288	***−0.307***
YC-ZSS-LKJ147	0.059	−0.290	***0.315***	***−0.314***	−0.232	−0.010	−0.034	−0.252	0.039	−0.112

Only significant or critical significant results are shown here. Values in the table are correlation coefficients, with bold font indicating a significant correlation and bold and italic font indicating critical significance. Asterisk indicates a significant correlation. "*" represents $0.01 \leq p < 0.05$; "**" represents $p < 0.01$.

Table A3. Pearson correlation analysis between soil physico-chemical properties and fungal and bacterial diversity indices.

	Fungi					Bacteria				
	OTUs	Shannon	Simpson	Chao1	ACE	OTUs	Shannon	Simpson	Chao1	ACE
pH	0.069	0.092	0.101	0.070	0.065	−0.009	0.060	0.044	0.030	−0.005
Organic matter	0.004	0.066	0.009	−0.013	−0.004	0.054	−0.154	***−0.191***	0.044	0.063
Total N	0.012	0.002	−0.012	−0.009	−0.010	−0.008	−0.063	−0.046	−0.011	0.028
NH_4_N	−0.080	−0.043	−0.066	−0.082	−0.079	0.006	0.073	0.052	−0.047	−0.052
NO_3_N	**−0.275 ****	−0.106	−0.033	**−0.239 ***	**−0.241 ***	−0.067	0.042	0.065	−0.140	−0.124
Total P	**−0.252 ***	***−0.191***	−0.123	**−0.265 ****	**−0.240 ***	0.039	0.074	0.073	0.029	0.051
Available P	**−0.225 ***	**−0.207 ***	***−0.182***	***−0.188***	***−0.178***	**−0.225 ***	−0.068	0.017	**−0.261 ***	**−0.250 ***
Available K	−0.039	−0.097	−0.139	−0.025	−0.015	−0.158	−0.034	−0.020	**−0.208 ***	**−0.203 ***

Values in the table are correlation coefficients, with bold font indicating a significant correlation and bold and italic font indicating critical significance. Asterisk indicates a significant correlation. "*" represents $0.01 \leq p < 0.05$; "**" represents $p < 0.01$.

References

1. van Kleunen, M.; Dawson, W.; Essl, F.; Pergl, J.; Winter, M.; Weber, E.; Kreft, H.; Weigelt, P.; Kartesz, J.; Nishino, M.; et al. Global exchange and accumulation of non-native plants. *Nature* **2015**, *525*, 100–103. [CrossRef] [PubMed]
2. Lambertini, M.; Leape, J.; Marton-Lefèvre, J.; Mittermeier, R.A.; Rose, M.; Robinson, J.G.; Stuart, S.N.; Waldman, B.; Genovesi, P. Invasives: A major conservation threat. *Science* **2011**, *333*, 404–405. [CrossRef] [PubMed]
3. Atwater, D.Z.; Callaway, R.M. Extended consequences of selection by exotic invaders on natives. *New Phytol.* **2022**, *233*, 588–591. [CrossRef] [PubMed]
4. Wang, W.; Shi, J.; Xie, Q.; Jiang, Y.; Yu, N.; Wang, E. Nutrient exchange and regulation in arbuscular mycorrhizal symbiosis. *Mol. Plant* **2017**, *10*, 1147–1158. [CrossRef] [PubMed]
5. Pysek, P.; Hulme, P.E.; Simberloff, D.; Bacher, S.; Blackburn, T.M.; Carlton, J.T.; Dawson, W.; Essl, F.; Foxcroft, L.C.; Genovesi, P.; et al. Scientists' warning on invasive alien species. *Biol. Rev.* **2020**, *95*, 1511–1534. [CrossRef]
6. Simberloff, D.; Martin, J.L.; Genovesi, P.; Maris, V.; Wardle, D.A.; Aronson, J.; Courchamp, F.; Galil, B.; Garcia-Berthou, E.; Pascal, M.; et al. Impacts of biological invasions: What's what and the way forward. *Trends Ecol. Evol.* **2013**, *28*, 58–66. [CrossRef]
7. Vila, M.; Espinar, J.L.; Hejda, M.; Hulme, P.E.; Jarosik, V.; Maron, J.L.; Pergl, J.; Schaffner, U.; Sun, Y.; Pysek, P. Ecological impacts of invasive alien plants: A meta-analysis of their effects on species, communities and ecosystems. *Ecol. Lett.* **2011**, *14*, 702–708. [CrossRef]
8. Xu, H.; Liu, Q.; Wang, S.; Yang, G.; Xue, S. A global meta-analysis of the impacts of exotic plant species invasion on plant diversity and soil properties. *Sci. Total Environ.* **2022**, *810*, 152286. [CrossRef]
9. Pisani, D.; Pazienza, P.; Perrino, E.V.; Caporale, D.; De Lucia, C. The economic valuation of ecosystem services of biodiversity components in protected areas: A review for a framework of analysis for the Gargano National Park. *Sustainability* **2021**, *13*, 11726. [CrossRef]
10. Stinca, A.; Musarella, C.M.; Rosati, L.; Laface, V.L.A.; Licht, W.; Fanfarillo, E.; Wagensommer, R.P.; Galasso, G.; Fascetti, S.; Esposito, A.; et al. Italian vascular flora: New findings, updates and exploration of floristic similarities between regions. *Diversity* **2021**, *13*, 600. [CrossRef]
11. Byun, C.; Lee, E.J. Ecological application of biotic resistance to control the invasion of an invasive plant, *Ageratina altissima*. *Ecol. Evol.* **2017**, *7*, 2181–2192. [CrossRef] [PubMed]
12. Hoffberg, S.L.; Mauricio, R.; Hall, R.J. Control or re-treat? Model-based guidelines for managing established plant invasions. *Biol. Invasions* **2018**, *20*, 1387–1402. [CrossRef]
13. Zhang, Z.; Liu, Y.; Brunel, C.; van Kleunen, M. Soil-microorganism-mediated invasional meltdown in plants. *Nat. Ecol. Evol.* **2020**, *4*, 1612–1621. [CrossRef] [PubMed]
14. Zhang, G.L.; Bai, J.H.; Tebbe, C.C.; Huang, L.B.; Jia, J.; Wang, W.; Wang, X.; Yu, L.; Zhao, Q.Q. Plant invasion reconstructs soil microbial assembly and functionality in coastal salt marshes. *Mol. Ecol.* **2022**, *31*, 4478–4494. [CrossRef] [PubMed]
15. Torres, N.; Herrera, I.; Fajardo, L.; Bustamante, R.O. Meta-analysis of the impact of plant invasions on soil microbial communities. *BMC Evol. Biol.* **2021**, *21*, 172. [CrossRef] [PubMed]
16. Shen, C.; Wang, J.; He, J.Z. Plant diversity enhances soil fungal diversity and microbial resistance to plant invasion. *Appl. Environ. Microbiol.* **2021**, *87*, e00251-00221. [CrossRef]
17. Li, Y.-P.; Feng, Y.-L.; Chen, Y.-J.; Tian, Y.-H. Soil microbes alleviate allelopathy of invasive plants. *Sci. Bull.* **2015**, *60*, 1083–1091. [CrossRef]
18. Zhang, F.-J.; Li, Q.; Chen, F.-X.; Xu, H.-Y.; Inderjit; Wan, F.-H. Arbuscular mycorrhizal fungi facilitate growth and competitive ability of an exotic species Flaveria bidentis. *Soil Biol. Biochem.* **2017**, *115*, 275–284. [CrossRef]
19. van der Putten, W.H.; Klironomos, J.N.; Wardle, D.A. Microbial ecology of biological invasions. *ISME J.* **2007**, *1*, 28–37. [CrossRef]
20. Stefanowicz, A.M.; Stanek, M.; Nobis, M.; Zubek, S. Species-specific effects of plant invasions on activity, biomass, and composition of soil microbial communities. *Biol. Fertil. Soils* **2016**, *52*, 841–852. [CrossRef]
21. Jacobs, K.; Conradie, T.; Jacobs, S. Microbial communities in the Fynbos region of South Africa: What happens during woody alien plant invasions. *Diversity* **2020**, *12*, 254. [CrossRef]
22. Kuebbing, S.E.; Classen, A.T.; Simberloff, D. Two co-occurring invasive woody shrubs alter soil properties and promote subdominant invasive species. *J. Appl. Ecol.* **2014**, *51*, 124–133. [CrossRef]
23. Gibbons, S.M.; Lekberg, Y.; Mummey, D.L.; Sangwan, N.; Ramsey, P.W.; Gilbert, J.A. Invasive plants rapidly reshape soil properties in a grassland ecosystem. *mSystems* **2017**, *2*, e00178-00116. [CrossRef] [PubMed]
24. Qu, T.; Du, X.; Peng, Y.; Guo, W.; Zhao, C.; Losapio, G. Invasive species allelopathy decreases plant growth and soil microbial activity. *PLoS ONE* **2021**, *16*, e0246685. [CrossRef] [PubMed]
25. Ehrenfeld, J.G. Effects of exotic plant invasions on soil nutrient cycling processes. *Ecosystems* **2003**, *6*, 503–523. [CrossRef]
26. Gaggini, L.; Rusterholz, H.-P.; Baur, B. The invasive plant Impatiens glandulifera affects soil fungal diversity and the bacterial community in forests. *Appl. Soil Ecol.* **2018**, *124*, 335–343. [CrossRef]
27. Bell, J.K.; Siciliano, S.D.; Lamb, E.G. A survey of invasive plants on grassland soil microbial communities and ecosystem services. *Sci. Data* **2020**, *7*, 86. [CrossRef] [PubMed]
28. Zhang, P.; Li, B.; Wu, J.; Hu, S. Invasive plants differentially affect soil biota through litter and rhizosphere pathways: A meta-analysis. *Ecol. Lett.* **2019**, *22*, 200–210. [CrossRef]

29. Wang, C.; Zhou, J.; Liu, J.; Jiang, K.; Xiao, H.; Du, D. Responses of the soil fungal communities to the co-invasion of two invasive species with different cover classes. *Plant Biol.* **2018**, *20*, 151–159. [CrossRef]
30. Liao, C.; Peng, R.; Luo, Y.; Zhou, X.; Wu, X.; Fang, C.; Chen, J.; Li, B. Altered ecosystem carbon and nitrogen cycles by plant invasion: A meta-analysis. *New Phytol.* **2008**, *177*, 706–714. [CrossRef]
31. Vogelsang, K.M.; Bever, J.D. Mycorrhizal densities decline in association with nonnative plants and contribute to plant invasion. *Ecology* **2009**, *90*, 399–407. [CrossRef] [PubMed]
32. Shah, M.A.; Beaulieu, M.-E.; Reshi, Z.A.; Qureshi, S.; Khasa, D.P. A cross-city molecular biogeographic investigation of arbuscular mycorrhizas in *Conyza canadensis* rhizosphere across native and non-native regions. *Ecol. Process.* **2015**, *4*, 7. [CrossRef]
33. Řezáčová, V.; Řezáč, M.; Gryndler, M.; Hršelová, H.; Gryndlerová, H.; Michalová, T. Plant invasion alters community structure and decreases diversity of arbuscular mycorrhizal fungal communities. *Appl. Soil Ecol.* **2021**, *167*, 104039. [CrossRef]
34. Custer, G.F.; van Diepen, L.T.A. Plant invasion has limited impact on soil microbial α-diversity: A meta-analysis. *Diversity* **2020**, *12*, 112. [CrossRef]
35. Wang, X.; Wang, J.; Hu, B.; Zheng, W.-L.; Li, M.; Shen, Z.-X.; Yu, F.-H.; Schmid, B.; Li, M.-H. Richness, not evenness, of invasive plant species promotes invasion success into native plant communities via selection effects. *Oikos* **2022**, *2022*, e08966. [CrossRef]
36. Wilsey, B.J.; Teaschner, T.B.; Daneshgar, P.P.; Isbell, F.I.; Polley, H.W. Biodiversity maintenance mechanisms differ between native and novel exotic-dominated communities. *Ecol. Lett.* **2009**, *12*, 432–442. [CrossRef]
37. Kuebbing, S.E.; Souza, L.; Sanders, N.J. Effects of co-occurring non-native invasive plant species on old-field succession. *For. Ecol. Manag.* **2014**, *324*, 196–204. [CrossRef]
38. Kuebbing, S.E.; Classen, A.T.; Sanders, N.J.; Simberloff, D. Above- and below-ground effects of plant diversity depend on species origin: An experimental test with multiple invaders. *New Phytol.* **2015**, *208*, 727–735. [CrossRef]
39. Zak, D.R.; Tilman, D.; Parmenter, R.R.; Rice, C.W.; Fisher, F.M.; Vose, J.; Milchunas, D.; Martin, C.W. Plant production and soil microorganisms in late-successional ecosystems: A continental-scale study. *Ecology* **1994**, *75*, 2333–2347. [CrossRef]
40. Waldrop, M.P.; Holloway, J.M.; Smith, D.B.; Goldhaber, M.B.; Drenovsky, R.E.; Scow, K.M.; Dick, R.; Howard, D.; Wylie, B.; Grace, J.B. The interacting roles of climate, soils, and plant production on soil microbial communities at a continental scale. *Ecology* **2017**, *98*, 1957–1967. [CrossRef]
41. McTee, M.R.; Lekberg, Y.; Mummey, D.; Rummel, A.; Ramsey, P.W. Do invasive plants structure microbial communities to accelerate decomposition in intermountain grasslands? *Ecol. Evol.* **2017**, *7*, 11227–11235. [CrossRef] [PubMed]
42. Broughton, L.C.; Gross, K.L. Patterns of diversity in plant and soil microbial communities along a productivity gradient in a Michigan old-field. *Oecologia* **2000**, *125*, 420–427. [CrossRef] [PubMed]
43. Wilsey, B.J.; Chalcraft, D.R.; Bowles, C.M.; Willig, M.R. Relationships among indices suggest that richness is an incomplete surrogate for grassland biodiversity. *Ecology* **2005**, *86*, 1178–1184. [CrossRef]
44. Wang, X.-Y.; Ge, Y.; Gao, S.; Chen, T.; Wang, J.; Yu, F.-H. Evenness alters the positive effect of species richness on community drought resistance via changing complementarity. *Ecol. Indic.* **2021**, *133*, 108464. [CrossRef]
45. Wang, J.; Yu, F.-H. Effects of functional diversity and functional dominance on complementary light use depend on evenness. *J. Veg. Sci.* **2018**, *29*, 726–736. [CrossRef]
46. Sun, Y.-Q.; Wang, J.; Shen, C.; He, J.-Z.; Ge, Y. Plant evenness modulates the effect of plant richness on soil bacterial diversity. *Sci. Total Environ.* **2019**, *662*, 8–14. [CrossRef]
47. Zhang, Y.; Chen, H.Y.H.; Reich, P.B. Forest productivity increases with evenness, species richness and trait variation: A global meta-analysis. *J. Ecol.* **2012**, *100*, 742–749. [CrossRef]
48. Filstrup, C.T.; King, K.B.S.; McCullough, I.M. Evenness effects mask richness effects on ecosystem functioning at macro-scales in lakes. *Ecol. Lett.* **2019**, *22*, 2120–2129. [CrossRef]
49. Lamb, E.G.; Kennedy, N.; Siciliano, S.D. Effects of plant species richness and evenness on soil microbial community diversity and function. *Plant Soil* **2011**, *338*, 483–495. [CrossRef]
50. Kirwan, L.; LÜScher, A.; SebastiÀ, M.T.; Finn, J.A.; Collins, R.P.; Porqueddu, C.; Helgadottir, A.; Baadshaug, O.H.; Brophy, C.; Coran, C.; et al. Evenness drives consistent diversity effects in intensive grassland systems across 28 European sites. *J. Ecol.* **2007**, *95*, 530–539. [CrossRef]
51. Sadiqi, S.; Hamza, M.; Ali, F.; Alam, S.; Shakeela, Q.; Ahmed, S.; Ayaz, A.; Ali, S.; Saqib, S.; Ullah, F.; et al. Molecular characterization of bacterial isolates from soil samples and evaluation of their antibacterial potential against MDRS. *Molecules* **2022**, *27*, 6281. [CrossRef]
52. Porazinska, D.L.; Farrer, E.C.; Spasojevic, M.J.; Bueno de Mesquita, C.P.; Sartwell, S.A.; Smith, J.G.; White, C.T.; King, A.J.; Suding, K.N.; Schmidt, S.K. Plant diversity and density predict belowground diversity and function in an early successional alpine ecosystem. *Ecology* **2018**, *99*, 1942–1952. [CrossRef] [PubMed]
53. Bergmann, G.T.; Bates, S.T.; Eilers, K.G.; Lauber, C.L.; Caporaso, J.G.; Walters, W.A.; Knight, R.; Fierer, N. The under-recognized dominance of Verrucomicrobia in soil bacterial communities. *Soil Biol. Biochem.* **2011**, *43*, 1450–1455. [CrossRef] [PubMed]
54. Wei, W.; Yang, M.; Liu, Y.; Huang, H.; Ye, C.; Zheng, J.; Guo, C.; Hao, M.; He, X.; Zhu, S. Fertilizer N application rate impacts plant-soil feedback in a sanqi production system. *Sci. Total Environ.* **2018**, *633*, 796–807. [CrossRef]
55. Caporaso, J.G.; Kuczynski, J.; Stombaugh, J.; Bittinger, K.; Bushman, F.D.; Costello, E.K.; Fierer, N.; Peña, A.G.; Goodrich, J.K.; Gordon, J.I.; et al. QIIME allows analysis of high-throughput community sequencing data. *Nat. Methods* **2010**, *7*, 335–336. [CrossRef] [PubMed]

56. Edgar, R.C.; Haas, B.J.; Clemente, J.C.; Quince, C.; Knight, R. UCHIME improves sensitivity and speed of chimera detection. *Bioinformatics* **2011**, *27*, 2194–2200. [CrossRef]
57. Edgar, R.C. UPARSE: Highly accurate OTU sequences from microbial amplicon reads. *Nat. Methods* **2013**, *10*, 996–998. [CrossRef]
58. Schmitz, M.; Flynn, D.F.B.; Mwangi, P.N.; Schmid, R.; Scherer-Lorenzen, M.; Weisser, W.W.; Schmid, B. Consistent effects of biodiversity on ecosystem functioning under varying density and evenness. *Folia Geobot.* **2013**, *48*, 335–353. [CrossRef]
59. Peay, K.G.; Baraloto, C.; Fine, P.V.A. Strong coupling of plant and fungal community structure across western Amazonian rainforests. *ISME J.* **2013**, *7*, 1852–1861. [CrossRef]
60. Dassen, S.; Cortois, R.; Martens, H.; de Hollander, M.; Kowalchuk, G.A.; van der Putten, W.H.; de Deyn, G.B. Differential responses of soil bacteria, fungi, archaea and protists to plant species richness and plant functional group identity. *Mol. Ecol.* **2017**, *26*, 4085–4098. [CrossRef]
61. Yang, N.; Zhang, Y.; Li, J.; Li, X.; Ruan, H.; Bhople, P.; Keiblinger, K.; Mao, L.; Liu, D. Interaction among soil nutrients, plant diversity and hypogeal fungal trophic guild modifies root-associated fungal diversity in coniferous forests of Chinese Southern Himalayas. *Plant Soil* **2022**. [CrossRef]
62. Tedersoo, L.; Bahram, M.; Cajthaml, T.; Põlme, S.; Hiiesalu, I.; Anslan, S.; Harend, H.; Buegger, F.; Pritsch, K.; Koricheva, J.; et al. Tree diversity and species identity effects on soil fungi, protists and animals are context dependent. *ISME J.* **2016**, *10*, 346–362. [CrossRef] [PubMed]
63. Waldrop, M.P.; Zak, D.R.; Blackwood, C.B.; Curtis, C.D.; Tilman, D. Resource availability controls fungal diversity across a plant diversity gradient. *Ecol. Lett.* **2006**, *9*, 1127–1135. [CrossRef] [PubMed]
64. Prober, S.M.; Leff, J.W.; Bates, S.T.; Borer, E.T.; Firn, J.; Harpole, W.S.; Lind, E.M.; Seabloom, E.W.; Adler, P.B.; Bakker, J.D.; et al. Plant diversity predicts beta but not alpha diversity of soil microbes across grasslands worldwide. *Ecol. Lett.* **2015**, *18*, 85–95. [CrossRef] [PubMed]
65. Nguyen, N.H.; Williams, L.J.; Vincent, J.B.; Stefanski, A.; Cavender-Bares, J.; Messier, C.; Paquette, A.; Gravel, D.; Reich, P.B.; Kennedy, P.G. Ectomycorrhizal fungal diversity and saprotrophic fungal diversity are linked to different tree community attributes in a field-based tree experiment. *Mol. Ecol.* **2016**, *25*, 4032–4046. [CrossRef]
66. Fierer, N.; McCain, C.M.; Meir, P.; Zimmermann, M.; Rapp, J.M.; Silman, M.R.; Knight, R. Microbes do not follow the elevational diversity patterns of plants and animals. *Ecology* **2011**, *92*, 797–804. [CrossRef]
67. Chen, Y.; Xu, T.; Fu, W.; Hu, Y.; Hu, H.; You, L.; Chen, B. Soil organic carbon and total nitrogen predict large-scale distribution of soil fungal communities in temperate and alpine shrub ecosystems. *Eur. J. Soil Biol.* **2021**, *102*, 103270. [CrossRef]
68. Hooper, D.U.; Bignell, D.E.; Brown, V.K.; Brussard, L.; Dangerfield, J.M.; Wall, D.H.; Wardle, D.A.; Coleman, D.C.; Giller, K.E.; Lavelle, P.; et al. Interactions between aboveground and belowground biodiversity in terrestrial ecosystems: Patterns, mechanisms, and feedbacks. *BioScience* **2000**, *50*, 1049–1061. [CrossRef]
69. Tedersoo, L.; Bahram, M.; Põlme, S.; Kõljalg, U.; Yorou, N.S.; Wijesundera, R.; Villarreal Ruiz, L.; Vasco-Palacios, A.M.; Thu, P.Q.; Suija, A.; et al. Fungal biogeography. Global diversity and geography of soil fungi. *Science* **2014**, *346*, 1256688. [CrossRef]
70. Fierer, N. Embracing the unknown: Disentangling the complexities of the soil microbiome. *Nat. Rev. Microbiol.* **2017**, *15*, 579–590. [CrossRef]
71. de Vries, F.T.; Manning, P.; Tallowin, J.R.B.; Mortimer, S.R.; Pilgrim, E.S.; Harrison, K.A.; Hobbs, P.J.; Quirk, H.; Shipley, B.; Cornelissen, J.H.C.; et al. Abiotic drivers and plant traits explain landscape-scale patterns in soil microbial communities. *Ecol. Lett.* **2012**, *15*, 1230–1239. [CrossRef]
72. Liu, S.; Yu, H.; Yu, Y.; Huang, J.; Zhou, Z.; Zeng, J.; Chen, P.; Xiao, F.; He, Z.; Yan, Q. Ecological stability of microbial communities in Lake Donghu regulated by keystone taxa. *Ecol. Indic.* **2022**, *136*, 108695. [CrossRef]
73. Banerjee, S.; Schlaeppi, K.; van der Heijden, M.G.A. Keystone taxa as drivers of microbiome structure and functioning. *Nat. Rev. Microbiol.* **2018**, *16*, 567–576. [CrossRef] [PubMed]
74. Pratt, R.G. Johnsongrass, yellow foxtail, and broadleaf signalgrass as new hosts for six species of *Bipolaris*, *Curvularia*, and *Exserohilum* pathogenic to bermudagrass. *Plant Dis.* **2006**, *90*, 528. [CrossRef] [PubMed]
75. Kumar, J.; Schäfer, P.; Hückelhoven, R.; Langen, G.; Baltruschat, H.; Stein, E.; Nagarajan, S.; Kogel, K.-H. *Bipolaris sorokiniana*, a cereal pathogen of global concern: Cytological and molecular approaches towards better control. *Mol. Plant Pathol.* **2002**, *3*, 185–195. [CrossRef] [PubMed]
76. Flory, S.L.; Kleczewski, N.; Clay, K. Ecological consequences of pathogen accumulation on an invasive grass. *Ecosphere* **2011**, *2*, 120. [CrossRef]
77. Stricker, K.B.; Harmon, P.F.; Goss, E.M.; Clay, K.; Luke Flory, S. Emergence and accumulation of novel pathogens suppress an invasive species. *Ecol. Lett.* **2016**, *19*, 469–477. [CrossRef] [PubMed]
78. Merzaeva, O.V.; Shirokikh, I.G. Colonization of plant rhizosphere by actinomycetes of different genera. *Microbiology* **2006**, *75*, 226–230. [CrossRef]
79. Kaushik, R.; Pandit, M.K.; Meyerson, L.A.; Chaudhari, D.S.; Sharma, M.; Dhotre, D.; Shouche, Y.S. Contrasting composition, diversity and predictive metabolic potential of the rhizobacterial microbiomes associated with native and invasive *Prosopis* congeners. *Curr. Microbiol.* **2021**, *78*, 2051–2060. [CrossRef]
80. Franco-Correa, M.; Quintana, A.; Duque, C.; Suarez, C.; Rodríguez, M.X.; Barea, J.-M. Evaluation of actinomycete strains for key traits related with plant growth promotion and mycorrhiza helping activities. *Appl. Soil Ecol.* **2010**, *45*, 209–217. [CrossRef]

81. Stefan, L.; Hartmann, M.; Engbersen, N.; Six, J.; Schöb, C. Positive effects of crop diversity on productivity driven by changes in soil microbial composition. *Front. Microbiol.* **2021**, *12*, 660749. [CrossRef] [PubMed]
82. Olanrewaju, O.S.; Babalola, O.O. Streptomyces: Implications and interactions in plant growth promotion. *Appl. Microbiol. Biotechnology* **2019**, *103*, 1179–1188. [CrossRef] [PubMed]
83. Li, Q.; Wan, F.; Zhao, M. Distinct soil microbial communities under Ageratina adenophora invasions. *Plant Biol.* **2022**, *24*, 430–439. [CrossRef] [PubMed]

plants

MDPI

Article

Four Invasive Plant Species in Southwest Saudi Arabia Have Variable Effects on Soil Dynamics

Ahmed M. Abbas [1,2,*], Wagdi S. Soliman [3], Maryam M. Alomran [4], Nahaa M. Alotaibi [4] and Stephen J. Novak [5,*]

1 Department of Biology, College of Science, King Khalid University, P.O. Box 9004, Abha 61413, Saudi Arabia
2 Department of Botany and Microbiology, Faculty of Science, South Valley University, Qena 83523, Egypt
3 Horticulture Department, Faculty of Agriculture and Natural Resources, Aswan University, Aswan 81528, Egypt
4 Department of Biology, College of Science, Princess Nourah bint Abdulrahman University, P.O. Box 84428, Riyadh 11671, Saudi Arabia; mmalomran@pnu.edu.sa (M.M.A.); namialotaibi@pnu.edu.sa (N.M.A.)
5 Department of Biological Sciences, Boise State University, Boise, ID 83725, USA
* Correspondence: ahhassan@kku.edu.sa (A.M.A.); snovak@boisestate.edu (S.J.N.); Tel.: +96-65-4072-1385 (A.M.A.); +1-208-426-3548 (S.J.N.)

Abstract: Predicting the direction and magnitude of change in soil dynamics caused by invasive plant species has proven to be difficult because these changes are often reported to be species- and habitat-specific. This study was conducted to determine changes in three soil properties, eight soil ions, and seven soil microelements under established stands of four invasive plants, *Prosopis juliflora*, *Ipomoea carnea*, *Leucaena leucocephala*, and *Opuntia ficus-indica*. Soil properties, ions, and microelements were measured in sites invaded by these four species in southwest Saudi Arabia, and these values were compared to the results for the same 18 parameters from adjacent sites with native vegetation. Because this study was conducted in an arid ecosystem, we predict that these four invasive plants will significantly alter the soil properties, ions, and microelements in the areas they invaded. While the soils of sites with the four invasive plant species generally had higher values for soil properties and ions compared to sites with native vegetation, in most instances these differences were not statistically significant. However, the soils within sites invaded by *I. carnea*, *L. leucocephala*, and *P. juliflora* had statistically significant differences for some soil parameters. For sites invaded by *O. puntia ficus-indica*, no soil properties, ions, or microelements were significantly different compared to adjacent sites with native vegetation. Sites invaded by the four plant species generally exhibited differences in the 11 soil properties, but in no instance were these differences statistically significant. All three soil properties and one soil ion (Ca) were significantly different across the four stands of native vegetation. For the seven soil microelements, significantly different values were detected for Co and Ni, but only among stands of the four invasive plant species. These results indicate that the four invasive plant species altered soil properties, ions, and microelements, but for most of the parameters we assessed, not significantly. Our results do not support our initial prediction, but are in general agreement with previous published findings, which indicate that the effects of invasive plants on soil dynamics vary idiosyncratically among invasive species and among invaded habitats.

Keywords: *Prosopis juliflora*; *Ipomoea carnea*; *Leucaena leucocephala*; *Opuntia ficus-indica*; soil properties; soil microelements; invasive plant species; ecosystem modification

1. Introduction

In regions where native plants and animals did not meet the needs of early colonists, these colonists began moving plants and animals around the globe, with some of these non-native species becoming invasive [1,2]. Invasive plant species are often described as having negative impacts on ecosystem processes [3]. They not only pose a threat to global biodiversity hotspots, but also to all the habitats (i.e., forest, grasslands, etc.) they invade.

There are many ways that invasive species damage native communities and ecosystems, especially in remote locations and on islands, where native species may be more vulnerable to these non-natives [4,5]. The extent to which invasive plants alter ecosystem processes in their new range compared to the ecosystem processes of native species in these areas has been shown to vary among invasive species and invaded habitats [6–8]. Many accounts of the impacts of invasive species are available for certain habitats and regions around the world [9,10], but details on the impacts of invasive species in other habitats and regions (e.g., in Saudi Arabia) are largely unknown, but see Assaeed et al. [11].

Invasive plant species can cause rapid changes in soil properties relative to native communities, and these differences may hinder efforts to restore native plant communities [8,12]. Invasive plants modify soil conditions either directly, by depositing leaf litter of varying quality and quantity [13,14], or indirectly, by altering the native microbial communities and the biotic interactions they normally engage in [14,15]. Invasive plants may alter soil properties in ways that increase their competitive advantage. For example, the soils of sites infested with invasive plants have lower levels of accessible phosphorus, total nitrogen, and carbon than sites with native vegetation [16,17]. However, the soil carbon/nitrogen ratio of soils under invasive plants has been reported to increase [8]. Additionally, the capacity of soils in invaded wetlands to store and release carbon was influenced by plant biomass input in soils, the sediment deposition rate of the tidal salt marsh, carbon turnover, and organic carbon stability, which are generally reported to be associated with the accumulation and decomposition of organic matter [18]. Invasive species in tropical dry forests negatively alter soil characteristics [19]. In contrast, the presence of invasive species in degraded habitats can improve soil moisture conditions and can facilitate the reclamation of such degraded areas [20].

Invasive species that displace native plants create unique plant communities that alter or create new plant–soil nutrient cycles and carbon cycles and change the capacity of soil to accumulate N and P [21]. In fact, the ability of invasive plants to alter the concentration and stoichiometry of soil nutrients has been widely studied [22–25]. Some invasive plants may alter the soil properties (and other ecological processes) to such an extent that the ecosystem services associated with native communities are threatened or are no longer provided [19,26–28]. Such invasive plant species are referred to as "transformer species" [29]. However, recent assessments of invasive plant species highlight the need to distinguish between non-native (invasive) species that have minimal ecological consequences and non-native species that have major ecological consequences in their new habitats [30,31]. There is ample evidence that invasive species can alter ecological processes and reduce native biodiversity, yet the ecological consequences of invasive tree species and how they modify soil properties and conditions are still not well understood and not adequately documented [32].

Other studies have reported no statistically significant difference among the properties and microelements of soils from habitats infested with invasive plant species compared to the soils of equivalent habitats occupied by native plant species [6,8,12,32,33]. Moreover, several studies that have compared the capacity of multiple invasive species to alter soil properties within the same experiments design have revealed contrasting results among the species examined [6,8,32]. Other studies have reported that the same invasive species alters soil dynamics in different ways in different habitats (communities) or under different experimental conditions [6,8,12]. These results indicate that the effects of invasive plants on soil dynamics vary in an idiosyncratic manner among invasive species and among invaded habitats (see [6,8]). Thus, the alteration of soil properties by invasive plants is often species- and habitat–specific.

Invasive plant species in Saudi Arabia have negative ecological consequences and high economic costs in natural ecosystems and agriculture areas, respectively [34]. These invasive plants have displaced many native plants from species-rich areas, especially in high-elevation habitats, wadis, and meadows. Some non-native plant species are believed to have been introduced into Saudi Arabia during past Hajj seasons, while others were introduced by early settlers for use as food, medicine, ornamental uses, livestock consumption, shade, and other reasons [34]. The introduction of non-native plants continues and is even increasing due to increased international trade, commerce, travel, and accidental means [35]. In Saudi Arabia, the establishment of non-native plant species is also facilitated by human disturbance of natural habitats, including the removal of native plants for road-building activities. This process explains the presence of invasive species along roads in mountainous regions of southwest Saudi Arabia [34]. Additionally, many other regions of southwest Saudi Arabia, plant diversity hotspots such as wadis and Red Sea islands, now contain many invasive plant species [34]. Unfortunately, scientific research focused on the invasive plant species in this region of Saudi Arabia is lacking (but see Assaeed et al., [11] and Alfarhan et al., [36]). Thus, there is an urgent need to conduct research on the invasive plants in southwest Saudi Arabia and to better understand their ecological consequences.

Six non-native plant species have been identified as the most important invasive species in Saudi Arabia [34]: *Prosopis juliflora*, *Cuscuta campestris*, *Cenchrus echinatus*, *Verbesina encelioides*, *Leucaena leucocephala*, and *Prosopis koelziana*. Among these six species, *P. juliflora* is the most damaging invasive plant species. It was reported as a ruderal plant, mostly found in urban and suburban areas, and is rarely observed in native plant communities of the central region of Saudi Arabia [34]. Unlike other tropical and subtropical countries, the invasion of *L. leucocephala* in Saudi Arabia does not have severe impacts on native communities because it mostly occupies urban areas. *Opuntia ficus-indica* and *Opuntia dillenii* are widely distributed across Saudi Arabia. *Ipomoea carnea* ssp. *fistulosa*, a pantropical invasive plant, has wide ecological amplitude. According to Al-Sodany [37], the growth of *I. carnea* follows a seasonal pattern in which its growth is greatest from September and October. This study reported significant differences for growth parameters (e.g., height, crown diameter, crown volume, and size index), demographic characteristics (e.g., birth, mortality, and survival rates), and reproductive output (e.g., number of flowers and fruits) among populations of *I. carnea* in different habitats [37]. *I. carnea* subsp. *fistulosa* reproduces clonally, through the production of underground rhizomes that grow laterally through the soil [38].

This research was designed to assess the direction and magnitude of change in soil properties, soil ions, and soil microelements caused by the invasion of four invasive plant species, *Prosopis juliflora*, *Ipomoea carnea*, *Leucaena leucocephala*, and *Opuntia ficus-indica*, in southwest Saudi Arabia. This was achieved by sampling and analyzing soils in areas invaded by the four species and comparing these results to those obtained for soils from adjacent areas where native vegetation occurred. A meta-analysis by Waring et al. [39] showed that the ability of plants to alter soil dynamics is significantly stronger in arid ecosystems; therefore, we predict that all four invasive plants included in this study will significantly alter soil dynamics in the areas they have invaded. The results of this study will be used to test this prediction.

2. Results

2.1. Comparison between Invaded and Native Soil Samples

The soil properties, ion concentrations, and microelement concentrations of sites in southwest Saudi Arabia invaded by *Prosopis juliflora*, *Ipomoea carnea*, *Leucaena leucocephala*, and *Opuntia ficus-indica* were compared to the soils from sites with native vegetation. Most soil properties, ion concentrations, and soil microelements concentrations did not significantly differ between the *P. juliflora* stands and native stands, except for several microelements, including Cr, Co, Cu, and Ni (Table 1). The concentration of these four microelements was significantly higher for the soil under native vegetation compared to the soil of sites invaded by the tree *P. juliflora*. Except for pH, organic matter (OM), Ca, and SO_4, no soil properties, ions, or microelements were significantly different for soils from stands invaded by the shrub *I. carnea* and soils from sites with native vegetation (Table 2).

Table 1. Soil properties, ion concentrations, and microelement concentrations of soils in stands invaded by *Prosopis juliflora* and adjacent stands with native vegetation in southwest Saudi Arabia.

	Invasive	Native	*t*-Test	*p*-Values
Soil properties				
pH	7.35 ± 0.15	7.13 ± 0.03	1.960	0.234
EC μS	3837.67 ± 1712.79	1797.67 ± 357.34	1.359	0.308
OM %	8.03 ± 1.99	5.39 ± 0.31	1.726	0.259
Ions (cations and anions)				
Na^+ ppm	88.09 ± 33.55	87.95 ± 12.63	0.001	0.982
K^+ ppm	454.52 ± 77.13	435.58 ± 20.53	0.056	0.824
Ca^{2+} ppm	0.92 ± 0.28	0.72 ± 0.06	0.477	0.528
Mg^{2+} ppm	1.03 ± 0.49	0.62 ± 0.25	0.551	0.499
SO_4^{2-} ppm	46.68 ± 6.28	54.89 ± 11.58	0.389	0.567
NO_3^- ppm	666.67 ± 425.57	60.00 ± 20.00	2.028	0.228
PO_4^{3-} ppm	4.08 ± 0.60	3.59 ± 1.66	0.078	0.794
Ammonia (NH_4^+) ppm	233.33 ± 88.19	100.00 ± 23.09	2.139	0.217
Microelements				
Mn ppm	172.92 ± 122.92	270.83 ± 90.60	0.411	0.556
Zn mg/L	43.54 ± 3.15	67.92 ± 8.90	6.665	0.061
Fe ppm	10,445.83 ± 3732.17	22,385.42 ± 7700.57	1.947	0.235
Cr ppm	4.99 ± 0.47	12.05 ± 0.90	48.472	0.002
Co ppm	4.25 ± 0.48	8.91 ± 0.42	53.902	0.002
Cu ppm	5.42 ± 1.63	12.50 ± 0.63	16.514	0.015
Ni ppm	9.69 ± 1.31	79.58 ± 9.42	54.015	0.002

Data represented as means ± SE, *n* = 6.

Table 2. Soil properties, ion concentrations, and microelement concentrations of soils in stands invaded by *Ipomoea carnea* and adjacent stands with native vegetation in southwest Saudi Arabia.

	Invasive	Native	*t*-Test	*p*-Values
Soil properties				
pH	7.11 ± 0.09	7.73 ± 0.07	28.782	0.006
EC μS	1736.67 ± 777.63	357.33 ± 63.34	3.126	0.152
OM %	8.44 ± 1.17	4.99 ± 0.20	8.393	0.044
Ions (cations and anions)				
Na^+ ppm	70.61 ± 28.87	58.42 ± 21.24	0.116	0.751
K^+ ppm	334.65 ± 90.64	372.31 ± 79.34	0.098	0.770
Ca^{2+} ppm	0.77 ± 0.15	0.27 ± 0.04	10.843	0.030
Mg^{2+} ppm	0.49 ± 0.16	0.21 ± 0.03	2.770	0.171
SO_4^{2-} ppm	49.25 ± 0.52	31.27 ± 0.61	511.249	0.001
NO_3^- ppm	80.00 ± 11.55	73.33 ± 24.04	0.063	0.815
PO_4^{3-} ppm	4.42 ± 2.04	2.67 ± 0.73	0.651	0.465
Ammonia (NH_4^+) ppm	133.33 ± 40.55	40.00 ± 11.55	4.900	0.091
Microelements				
Mn ppm	229.17 ± 182.30	269.58 ± 126.90	0.033	0.864
Zn mg/L	55.42 ± 11.59	45.00 ± 17.14	0.253	0.641
Fe ppm	10,564.58 ± 2597.91	16,856.25 ± 9276.55	0.427	0.549
Cr ppm	9.96 ± 6.17	12.97 ± 7.86	0.091	0.778
Co ppm	7.52 ± 2.82	9.76 ± 4.41	0.183	0.691
Cu ppm	10.21 ± 6.01	19.38 ± 9.57	0.658	0.463
Ni ppm	35.69 ± 25.95	66.25 ± 32.37	0.543	0.502

Data represented as means ± SE, *n* = 6.

The pH was significantly higher for the soil under native vegetation, while OM, CA, and SO_4 were significantly higher for the soils from stands with an *I. carnea* overstory (Table 2). Soil properties, ion concentrations, and soil microelements did not significantly differ between stands in which the tree *L. leucocephala* had invaded and stands with native vegetation, except for pH, which was significantly higher in soil from areas with native vegetation (Table 3). No soil properties, ion concentrations, or soil microelements were significantly different for soils from stands invaded by the cactus *O. ficus-indica* and stands of native vegetation (Table 4).

Table 3. Soil properties, ion concentrations, and microelement concentrations of soils in stands invaded by *Leucaena leucocephala* and adjacent stands with native vegetation in southwest Saudi Arabia.

	Invasive	Native	t-Test	p-Values
Soil properties				
pH	6.97 ± 0.22	7.64 ± 0.07	8.894	0.041
EC μS	7250.00 ± 4145.68	324.00 ± 12.22	2.791	0.170
OM %	12.22 ± 4.06	4.19 ± 0.09	3.901	0.119
Ions (cations and anions)				
Na^+ ppm	297.22 ± 221.53	77.89 ± 31.77	0.960	0.383
K^+ ppm	468.29 ± 41.84	536.73 ± 139.35	0.221	0.663
Ca^{2+} ppm	1.52 ± 1.07	0.38 ± 0.16	1.096	0.354
Mg^{2+} ppm	1.17 ± 0.44	0.48 ± 0.20	2.029	0.227
$SO_4{}^{2-}$ ppm	56.42 ± 12.45	41.07 ± 8.84	1.010	0.372
$NO_3{}^-$ ppm	173.33 ± 75.13	43.33 ± 8.82	2.953	0.161
$PO_4{}^{3-}$ ppm	2.33 ± 0.30	15.20 ± 10.33	1.551	0.281
Ammonia ($NH_4{}^+$) ppm	73.33 ± 13.33	53.33 ± 24.04	0.529	0.507
Microelements				
Mn ppm	73.33 ± 13.33	53.33 ± 24.04	0.529	0.507
Zn mg/L	57.71 ± 10.28	40.63 ± 15.66	0.831	0.413
Fe ppm	23,525.00 ± 9341.35	17,068.75 ± 10,154.34	0.219	0.664
Cr ppm	9.60 ± 1.21	12.24 ± 8.09	0.104	0.763
Co ppm	7.52 ± 0.34	8.09 ± 4.66	0.015	0.908
Cu ppm	12.29 ± 3.61	16.25 ± 10.23	0.133	0.734
Ni ppm	18.52 ± 2.13	45.50 ± 37.03	0.529	0.507

Data represented as means ± SE, $n = 6$.

Table 4. Soil properties, ion concentrations, and microelement concentrations of soils in stands invaded by *Opuntia ficus-indica* and adjacent stands with native vegetation in southwest Saudi Arabia.

	Invasive	Native	t-Test	p-Values
Soil properties				
pH	7.08 ± 0.04	7.09 ± 0.03	0.084	0.786
EC μS	288.60 ± 68.97	191.23 ± 20.50	1.831	0.247
OM %	5.68 ± 2.14	3.36 ± 0.80	1.026	0.368
Ions (cations and anions)				
Na^+ ppm	51.36 ± 6.46	35.95 ± 6.49	2.828	0.168
K^+ ppm	545.34 ± 98.00	491.97 ± 167.64	0.076	0.797
Ca^{2+} ppm	0.27 ± 0.04	0.25 ± 0.00	0.143	0.725
Mg^{2+} ppm	0.20 ±0.02	0.17 ± 0.05	1.690	0.263
$SO_4{}^{2-}$ ppm	37.47 ± 0.96	31.50 ± 3.56	2.616	0.181
$NO_3{}^-$ ppm	126.67 ± 86.67	46.67 ± 6.67	0.847	0.409
$PO_4{}^{3-}$ ppm	17.70 ± 9.75	1.17 ± 0.92	2.848	0.167
Ammonia ($NH_4{}^+$) ppm	86.67 ± 13.33	73.33 ± 13.33	0.500	0.519
Microelements				
Mn ppm	312.50 ± 160.20	181.25 ± 76.38	0.547	0.501
Zn mg/L	81.25 ± 29.05	32.50 ± 6.68	2.674	0.177
Fe ppm	17,970.83 ± 5340.86	9552.08 ± 2189.11	2.127	0.218
Cr ppm	14.43 ± 3.84	33.13 ± 24.06	0.589	0.486
Co ppm	12.40 ± 0.74	10.10 ± 2.36	0.869	0.404
Cu ppm	18.96 ± 4.71	17.71 ± 4.65	0.036	0.859
Ni ppm	133.13 ± 17.51	52.78 ± 33.61	4.493	0.101

Data represented as means ± SE, $n = 6$.

2.2. Soil Properties, Ion Concentrations, and Microelement Concentrations for the Four Invasive Plant Species and Adjacent Stands of Native Vegetation

There were large differences in soil properties and ion concentrations in the soils from stands with native vegetation and the soil of stands invaded by the four invasive plant species. We detected significant differences in soil pH, EC, OM, and Ca among the four stands of native vegetation, but not for the stands of the four invasive species (Figure 1). In soils from the native sites, pH ranged from 7.1 to 7.7, and in the invaded site pH ranged from 6.95 to 7.35. pH values were higher in the areas invaded by *P. juliflora* and *O. ficus-indica* (compared to adjacent native sites) and lower in the areas invaded by *I. carnea* and *L. leucocephala* (compared to adjacent native sites). Conversely, EC and OM values were higher in the soils of invaded stands compared to the soil of native sites. The largest increase in the values of EC and OM occurred for sites invaded by *L. leucocephala* (Figure 1). Similarly, the values of Na, Ca, and Mg were highest in the soils of stands that were invaded by all four invasive plants, and the largest values were observed in areas invaded by *L. leucocephala* (Figure 1). The concentrations of K in soils of invaded stands of all four invasive species and adjacent native stands were not significantly different (Figure 1). Variation among soil SO_4 values for the four invasive species in both invaded and native stands were not significantly different (Figure 1). However, the NO_3 and ammonia values were generally higher in the soils of stands invaded by the four species compared to the soils of sites with native vegetation. The largest increases in NO_3 values were observed for areas invaded by *P. juliflora*, followed by areas invaded by *L. leucocephala*, with the largest increases in ammonia values occurring in areas invaded with *I. carnea*, followed by areas infested with *P. juliflora* (Figure 1). The soil PO_4 content exhibited variable results (Figure 1). For instance, the soil PO_4 content increased in the stands invaded by *O. ficus-indica* compared to the soil of native stands and decreased in areas invaded by *P. juliflora* (Figure 1).

There were also differences in the concentration of microelements in the soils from stands invaded by the four invasive plant species and adjacent stands of native vegetation (Figure 2). The concentration of Mn, Zn, and Fe showed higher values in areas invaded by *O. ficus-indica* and *L. leucocephala*, while these microelements had lower concentrations in areas invaded by *P. juliflora* and *I. carnea*. These differences, however, were not statistically significant. In addition, variation in Cu and Cr in the four stands of native vegetation and the four sites with invasive plants were not significantly different, but values for both microelements were lowest for the soil of sites invaded by *P. juliflora* (Figure 2). The concentrations of Co and Ni significantly differed among stands of the four invasive plant species and was significantly lower for the soils of sites invaded by *P. juliflora*, *I. carnea*, and *L. leucocephala*. The four stands of native vegetation exhibited differences for all seven microelements, but none of these differences were statistically significant (Figure 2).

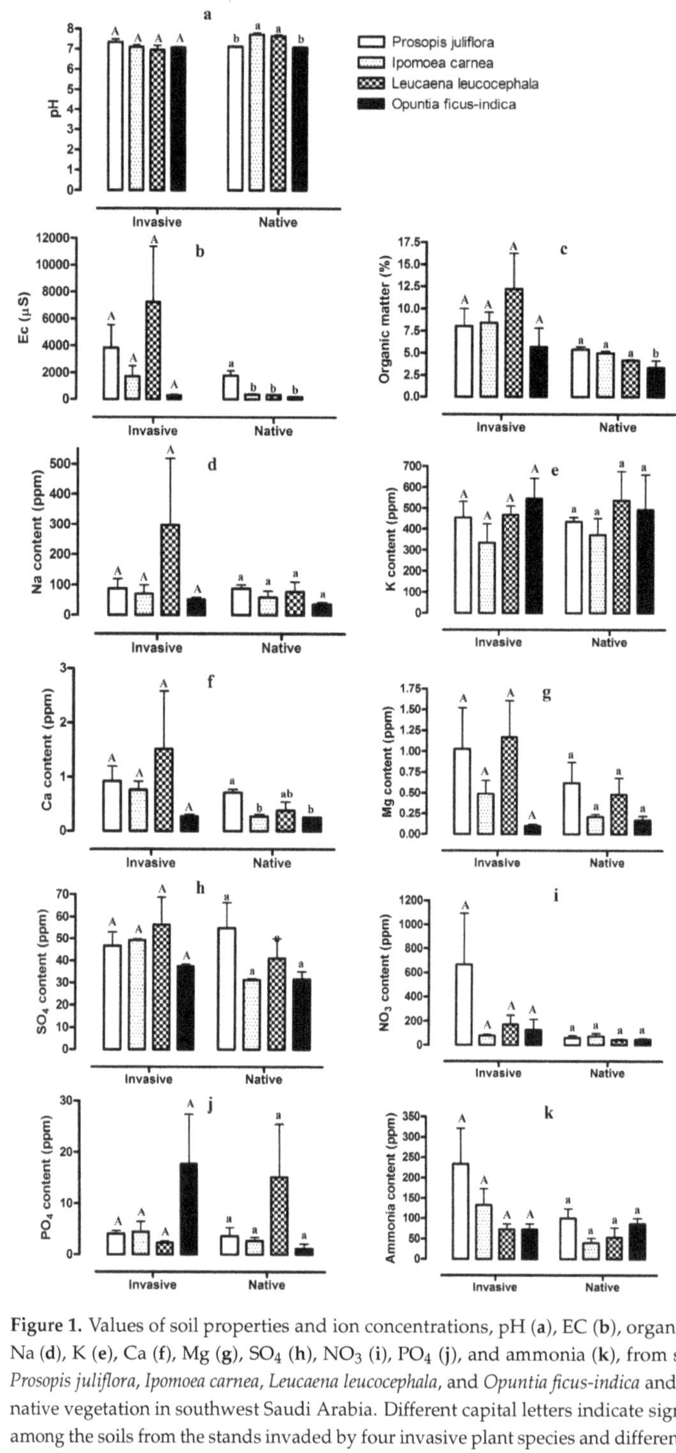

Figure 1. Values of soil properties and ion concentrations, pH (**a**), EC (**b**), organic matter [OM] (**c**), Na (**d**), K (**e**), Ca (**f**), Mg (**g**), SO_4 (**h**), NO_3 (**i**), PO_4 (**j**), and ammonia (**k**), from stands invaded by *Prosopis juliflora*, *Ipomoea carnea*, *Leucaena leucocephala*, and *Opuntia ficus-indica* and adjacent stands of native vegetation in southwest Saudi Arabia. Different capital letters indicate significant differences among the soils from the stands invaded by four invasive plant species and different lower-case letters indicate significant differences among the soils from stands of native vegetation ($n = 6$, $p < 0.05$).

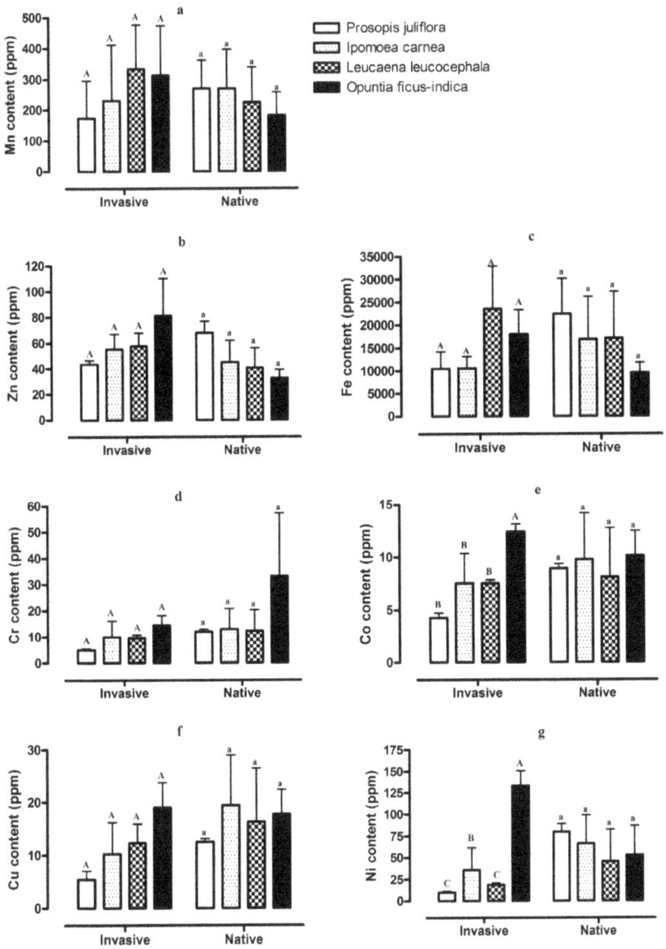

Figure 2. Values of soil microelements, Mn (**a**), Zn (**b**), Fe (**c**), Cr (**d**), Co (**e**), Cu (**f**), and Ni (**g**), from stands invaded by *Prosopis juliflora*, *Ipomoea carnea*, *Leucaena leucocephala*, and *Opuntia ficus-indica* and adjacent stands of native vegetation in southwest Saudi Arabia. Different capital letters indicate significant differences among the soils from stands invaded by four invasive plant species and different lower-case letters indicate significant differences among the soils from stands of native vegetation ($n = 6$, $p < 0.05$).

3. Discussion

Invasive species are those species (taxa) that are introduced outside of their native range and persist, reproduce, and spread beyond their original points of introduction in their new range [40,41]. Invasive plant species can outcompete and reduce the abundance of native plant species [42], alter soil properties [26], and homogenize the biodiversity of invaded communities and ecosystems [43]. Invasive plant species may alter soil properties in their new ranges, but the magnitude and direction of these changes are sometimes unique to each invasive species and often vary among habitats for the same non-native species [6,8,12,32,33,44,45]. To the best of our knowledge, this is the first study to investigate the biogeochemical consequences of the invasion of *Prosopis juliflora*, *Ipomoea carnea*, *Leucaena leucocephala*, and *Opuntia ficus-indica* in Saudi Arabia. At the outset of this study we predicted that all four invasive plant species would significantly alter soil properties and microelements in the areas they invaded. Our results do not support this prediction.

Soil carbon, nitrogen, and phosphorus have a large influence on many ecosystem processes, and the direction of change for these three elements was widespread across the species and sites studied. Our study revealed higher levels for soil carbon, nitrogen, and phosphorus levels in soils of invaded habitats compared to the soils of sites with native vegetation. A recent meta-analysis discovered that invasive plant species were associated with significantly higher litter deposition and decomposition rates as well as increases in soil nitrogen mineralization and nitrification [46,47]. Ehrenfeld [48] and Ehrenfeld et al. [13] proposed that the characteristics of many invasive plant species, in comparison to native plants, may influence the specific way in which invasive plants alter ecosystem properties. These characteristics include larger size, faster growth, higher photosynthetic rates, higher nutrient concentrations in living tissue and litter, and extremely efficient nitrogen-fixing symbioses. Rout and Callaway [49] state that if invasive species increase net primary production and nitrogen cycling in their invasive ranges but not in their native ranges, their inherent traits are unlikely to drive these processes because these same ecosystem effects should also occur in their native ranges. This statement also highlights the context-specific nature of traits on invasion outcomes.

Another possibility is that invasive plants, once established, will undergo rapid evolution through natural selection [50,51], especially for key traits related to nutrient cycling. For example, Feng et al. [47] provided evidence that the invasive plant *Ageratina adenophora*, which is native to Mexico but has spread throughout subtropical regions around the globe, appears to have developed leaves that decompose more easily in its invaded range. Determining whether and how invasive species have more negative ecosystem consequences than native species can answer critical questions in ecology, evolution, and biogeography [52].

We detected higher NO_3 levels in the soil from sites infested with all four invasive species compared to soils with native vegetation, even though these differences were not statistically significant. This trend is consistent with previous research indicating a positive relationship between the levels of soil NO_3 and the abundance of invasive plant species [53]. Previous studies have also reported that plant invasions occur due to several mechanisms that help them uptake and use nutrients more efficiently [23,54]. The mechanisms associated with invasiveness appear to be different in nutrient-poor soils compared to nutrient-rich soils. Many studies indicate that the success of invasive plants in nutrient-poor soils is dependent on strategies such as increased nutrient-use efficiency [24, 55], especially at shorter time scales [22]; long nutrient residence times [56]; high resistance to low levels of nutrients [57]; and high plasticity of stoichiometric ratios [23].

We also found that the soil concentrations of Na, Ca^{2+}, and Mg^{2+} increase in the soils of invaded areas compared to soils from areas with native plants. According to Duda et al. [58], as *Halogeton glomeratus* invades native *Krascheninnikovia lanata*-dominated vegetation, soil concentrations of nitrate, phosphorus, potassium, and sodium increases with invasion. Rodgers et al. [59] reported that the availability of nitrogen, phosphorus, calcium, and magnesium in the soil, as well as the soil pH, were consistently and considerably higher in soils in North American temperate deciduous forests invaded by the European forb *Alliaria petiolata*. Similarly, our results reveal higher values for most soil properties and soil ions in sites invaded by the four invasive plant species included in this study, even if many of these values were not statistically significant (Tables 1–4, Figure 1).

Invasive plants frequently play a role in modifying microelement (heavy metal) biogeochemical cycles in ecosystems [60]. Invading plants can increase soil microelement bioavailability in different ways: microelement secretion from leaves and release via humus breakdown and desorption. For instance, most rooted macrophytes can operate as a nutrient pump to absorb the nutrients contained in sediment pools, such as N and P, and subsequently remove them by tissue leaching and secretion and litter decomposition above the soil surface [61]. Invasive plants that perform a similar function can increase the bioavailability of soil microelements [62]. Additionally, microelements contained in plant tissues can be released into the environment by degradation of litter, other organic matter, and diagenesis [62,63]. During the decomposition of invasive plant litter, an increase in soil pH might indirectly result in the mobilization of microelements in the soil. Microelements can be released as sulfides from insoluble sulfide compounds in sediments. The microelements are then oxidized by sediment oxidation and enzymes in the roots of invasive plants [62]. In the same way, plant growth can cause microelements to be released from sediment particles into the pore water of wetlands and concentrate microelements in the rhizosphere of plants [64,65]. This can also be carried out by some native plants.

Increased mobilization can lead to higher rates of leaching and faster uptake of these microelements by plants, which can lower their total concentration in soil by a large amount [66], even though this would be expected to increase bioavailability. In addition to their role in the biogeochemical cycling of microelements, some invasive plants can reduce microelement mobility by adsorption, absorption, immobilization, and "returning elements to belowground pools". *Spartina alterniflora*, the most pervasive invasive plant in China's coastal wetland, has caused a considerable decrease in Cs^{137} and Pb^{210} adsorption from beach deposition of silt–fine sand and mud–sand [60]. The roots of hydrophytes, especially invasive hydrophytes, frequently show signs of iron plaque. As a major mechanism for tolerating hazardous metals, it adsorbs and immobilizes heavy metals efficiently, hence reducing their mobility [67]. Furthermore, invasive plants with interactions with symbiotic microorganisms may be able to sequester microelement toxicity through microbial biosynthesis of antioxidants and enzymes [68]. Therefore, the appearance of invasive species in areas contaminated with toxic microelements may not increase the potential enrichment of these toxic elements in the food chain and would in fact contribute to phytostabilization [69].

The results of this study reveal that the microelement values of soils sampled from three of the invasive species analyzed are generally lower than that of the soils from adjacent stands of native vegetation (Tables 1–3). The only exception to this pattern occurred for soils from sites invaded by *O. ficus-indica* (Table 4). Among the four invasive plant species, there were no significant differences in the levels of Mn, Zn, Fe, Cr, and Cu in the soils where they invaded; however, the levels of Co and Ni were significantly lower for soils invaded by *P. jliflora*, *I. carnea*, and *L. leucocephala* compared to those invaded by *O. ficus-indica* (Figure 2).

4. Materials and Methods

Saudi Arabia contains a variety of ecosystems, ranging from high-altitude mountains (Jabals) (up to 3050 m asl), valleys or river drainages that are seasonally dry (wadis), meadows (Raudhas), salt pans (Sabkhas), lava flows (Harrats), deep sands (Nafud), and irrigation canals. Winters are generally cool with occasional frost in the mountains of the northwestern region. The summer months are hot, with temperatures sometimes reaching above 50 °C. Rainfall is typically unpredictable and erratic in most parts of the country and mainly falls in winter and spring (100–150 mm), except for the southwestern region of the country, which can receive much more precipitation (>600 mm from September to November). Humidity levels are low in the interior of the country (15–20%) and high along the coast (55–75%).

4.1. Species Selection

Four species were selected for study: *Prosopis juliflora*, *Ipomoea carnea*, *Leucaena leucocephala* and *Opuntia ficus-indica*. All these species form mono-specific stands in invaded localities and are recognized as being among the most invasive plant species in Saudi Arabia [34].

4.2. Field Sampling

Within each habitat, large stands of the four invasive plant species were selected to serve as "invaded" sampling locations. The size of the mono-specific stands varied and was usually 10–12 ha for *P. juliflora*, 2–6 ha for *I. carnea*, 0.6–1 ha for *L. leucocephala*, and 0.5–2 ha for *O. ficus-indica*. The invaded study sites were heavily infested by one of the four invasive species and were located on one side of a distinct ecotone. On the other side of the ecotone were areas consisting of native vegetation; these were the "native" sampling locations. Our experimental design therefore consisted of paired invaded and native sampling locations for each of the four invasive species. Fieldwork was conducted on 3 March 2022 for *P. juliflora*, 15 March 2022 for *I. carnea*, 3 April 2022 for *L. leucocephala*, and 23 April 2022 for *O. ficus-indica*.

Soil samples from the paired invaded and native sampling locations were generally located 20 m from each other. Soil samples were collected at a depth of 0–15 cm at each sampling location. At each invasive and native sampling location, a 20 m transect line was positioned perpendicular to the axis of the ecotone. Ten subsamples were collected at 2 m intervals along the transects established in invaded and native stands at each sampling location and these 10 subsamples were combined into one homogenized soil sample. Six sampling locations were selected for collecting paired soil samples for each of the four invasive plant species, yielding a total of six replicates per species. This approach means that we collected 120 soil subsamples for each species, and a total of 480 soil subsamples for the entire study.

4.3. Soil Analyses

Each soil sample replicate was separated into two subsamples. One subsample was dried at 105 °C for standard chemical assays. The soil pH and electrical conductivity (EC) were measured electrometrically in two different soil–H_2O solutions (1:1, vol/vol and 1:2, vol/vol, respectively) [70]. The total carbon and nitrogen content were determined using dry combustion (FLASH EA1112 CHN analyzer; Thermo Fisher, Waltham, MA, USA). The available P was determined with the Olsen method [71]. Extractable microelements were quantified following digestion with diethylene-triamine-penta-acetic acid [72]. We determined the concentration of seven microelements: manganese (Mn), zinc (Zn), iron (Fe), chromium (Cr), cobalt (Co), copper (Cu), and nickel (Ni). The total microelement content was quantified following digestion with nitric acid [73]. The analysis was conducted using inductively coupled plasma optical emission spectrometry (Spectro Genesis; SPECTRO Analytical Instruments, Kleve, Germany). The second soil subsample was used to determine nitrate and ammonium colorimetrically, following 2 M potassium chloride extraction [74]. Nitrate and ammonium were quantified at the end of the incubation period.

4.4. Statistical Analysis

The data generated in this study were tested for normality using the Kolmogorov–Smirnov test and for homogeneity of variance using the Brown–Forsythe test. For each of the four invasive plant species, mean values were calculated based on the results of the six replicates for each species and variation in the data was calculated as the standard error (SE) of the mean. Paired *t*-tests were used to determine differences between invaded and native soil samples for each of the four invasive species. Statistically significant differences in the soil properties, ion concentrations, and microelement concentrations for the stands of the four invasive plants and the native stands were tested using one-way analysis of variance (ANOVA, *F* test) [75]. The Tukey's honestly significant difference (HSD) test was used to determine the significance among means for treatments with a significant *F*-test. A significance level of $p < 0.05$ was applied for all analyses. All statistical analyses were performed using JMP (version 4.0 SAS Institute, Cary, NC, USA).

5. Conclusions

The results reported here do not support our prediction that we would detect significant changes in soil properties and microelements in areas infested by the four invasive plant species compared to the soils from stands of native vegetation. However, our results do indicate that the alteration of soil parameters varies among the four invasive plant species and varies for the different soil parameters we examined. For instance, the soils of sites with the four invasive plant species generally had higher values for soil properties compared to sites with native vegetation, but in most cases these differences were not statistically significant. The soils of stands invaded by *P. juliflora* and *I. carnea* had the largest number of statistically significant differences: four microelements and four soil properties, respectively. While the soils of sites invaded by *O. ficus-indica* and sites with native vegetation exhibited no significant differences for the 18 soil parameters. The soil of areas invaded by the four invasive species showed no significant differences for the three soil properties and the eight soil ions we measured, and no significant differences for five of the seven soil microelements. The levels of Co and Ni were significantly higher for the sites invaded by *O. ficus-indica* compared to the levels of the other three invasive species. Our results are in general agreement with previously published findings that indicate that the effects of invasive plants on soil dynamics vary idiosyncratically among invasive species and among invaded habitats (i.e., these effects are species- and habitat specific). These four invasive plants may simply be incapable of significantly altering soil properties and soil microelements. Conversely, our results do suggest a trend that might signal significant alteration of soil dynamics in the future, especially if the abundance of these invasive plant species increases. Thus, the invasion of these four plant species in southwest Saudi Arabia should be closely monitored and studied to determine their ecological consequences and their impacts on the native vegetation in this region.

Author Contributions: Conceptualization, A.M.A. and S.J.N.; methodology, A.M.A.; software, W.S.S.; validation, A.M.A., W.S.S., M.M.A., N.M.A. and S.J.N.; formal analysis, W.S.S.; investigation, A.M.A.; resources, A.M.A.; data curation, S.J.N.; writing—original draft preparation, W.S.S. and S.J.N.; writing—review and editing, A.M.A. and S.J.N.; visualization, W.S.S.; supervision, A.M.A., W.S.S. and S.J.N.; project administration, A.M.A., M.M.A. and N.M.A.; funding acquisition, A.M.A. and N.M.A. All authors have read and agreed to the published version of the manuscript.

Funding: The authors received funding for this study from the Deanship of Scientific Research at the King Khalid University through the Research Group Project under grant number (RGP. 2/92/43) and Princess Nourah bint Abdulrahman University Researchers Supporting Project number (PNURSP2023R356), Princess Nourah bint Abdulrahman University, Riyadh, Saudi Arabia.

Data Availability Statement: Not applicable.

Acknowledgments: The authors extend their appreciation to the Deanship of Scientific Research at the King Khalid University for funding through the Research Group Project (RGP. 2/92/43) and Princess Nourah bint Abdulrahman University Researchers Supporting Project number (PNURSP2023R356), Princess Nourah bint Abdulrahman University, Riyadh, Saudi Arabia. SJN acknowledges financial support from the Department of Biological Sciences at Boise State University to work on this project and manuscript.

Conflicts of Interest: The authors declare no conflict of interest.

References

1. Mack, R.N. The motivation for importing potentially invasive plant species: A primal urge. In Proceedings of the VI International Rangeland Congress, International Rangeland Congress, Aitkenvale, QLD, Australia, 19–23 July 1999; Volume 2, pp. 557–562.
2. Lockwood, J.L.; Hoopes, M.F.; Marchetti, M.P. *Invasion Ecology*; John Wiley & Sons: New York, NY, USA, 2013; p. 444.
3. Czortek, P.; Królak, E.; Borkowska, L.; Bielecka, A. Impacts of soil properties and functional diversity on the performance of invasive plant species *Solidago canadensis* L. on post-agricultural wastelands. *Sci. Total Environ.* **2020**, *729*, 139077. [CrossRef]
4. Mack, R.N. *Assessing the Extent, Status, and Dynamism of Plant Invasions: Current and Emerging Approaches. Invasive Species in a Changing World*; Island Press: Washington, WA, USA, 2000; pp. 141–168.
5. Davis, M.A. Biotic globalization: Does competition from introduced species threaten biodiversity? *Bioscience* **2003**, *53*, 481–489. [CrossRef]
6. Dassonville, N.; Vanderhoeven, S.; Vanparys, V.; Hayez, M.; Gruber, W.; Meerts, P. Impacts of alien plants on soil nutrients are correlated with initial site conditions in NW Europe. *Oecologia* **2008**, *157*, 131–140. [CrossRef]
7. Hejda, M.; Pyšek, P.; Jarošík, V. Impact of invasive plants on the species richness, diversity and composition of invaded communities. *J. Ecol.* **2009**, *97*, 393–403. [CrossRef]

8. Gibbons, S.M.; Lekberg, Y.; Mummey, D.L.; Sangwan, N.; Ramsey, P.W.; Gilbert, J.A. Invasive plants rapidly reshape soil properties in a grassland ecosystem. *MSystems* **2017**, *2*, e00178-16. [CrossRef]
9. Jamil, M.D.; Waheed, M.; Akhtar, S.; Bangash, N.; Chaudhari, S.K.; Majeed, M.; Hussain, M.; Ali, K.; Jones, D.A. Invasive plants diversity, ecological status, and distribution pattern in relation to edaphic factors in different habitat types of district Mandi Bahauddin, Punjab, Pakistan. *Sustainability* **2022**, *14*, 13312. [CrossRef]
10. Clements, D.R.; Upadhyaya, M.K.; Joshi, S.; Shrestha, A. *Global Plant Invasions*; Springer: Cham, Switzerland, 2022; p. 381.
11. Assaeed, A.M.; Alharthi, A.S.; Abd-ElGawad, A.M. Impacts of *Nicotiana glauca* Graham invasion on the vegetation composition and soil: A case study of Taif, western Saudi Arabia. *Plants* **2021**, *10*, 2587. [CrossRef] [PubMed]
12. Unger, I.M.; Kremer, R.J.; Veum, K.S.; Goyne, K.W. Immediate and long-term effects of invasive plant species on soil characteristics. *Soil Ecol. Lett.* **2022**, *4*, 276–288. [CrossRef]
13. Ehrenfeld, J.G.; Kourtev, P.; Huang, W. Changes in soil functions following invasions of exotic understory plants in deciduous forests. *Ecol. Appl.* **2001**, *11*, 1287–1300. [CrossRef]
14. Weidenhammer, J.D.; Callaway, R.M. Direct and indirect effects of invasive plants on soil chemistry and ecosystem function. *J. Chem. Ecol.* **2010**, *36*, 59–69. [CrossRef]
15. Qu, T.; Du, X.; Peng, Y.; Guo, W.; Zhao, C.; Losapio, G. Invasive species allelopathy decreases plant growth and soil microbial activity. *PLoS ONE* **2021**, *16*, e0246685. [CrossRef] [PubMed]
16. Coykendall, K.E.; Houseman, G.R. *Lespedeza cuneata* invasion alters soils facilitating its own growth. *Biol. Invasions* **2014**, *16*, 1735–1742. [CrossRef]
17. Anning, A.K.; Gyamfi, B.; Effah, A.T. *Broussonetia papyrifera* controls nutrient return to soil to facilitate its invasion in a tropical forest of Ghana. *J. Plant Ecol.* **2018**, *11*, 909–918. [CrossRef]
18. Wang, W.; Sardans, J.; Wang, C.; Asensio, D.; Bartrons, M.; Peñuelas, J. Species-specific impacts of invasive plant success on vertical profiles of soil carbon accumulation and nutrient retention in the Minjiang River tidal estuarine wetlands of China. *Soil Syst.* **2018**, *2*, 5. [CrossRef]
19. Maass, J.M.; Balvanera, P.; Castillo, A.; Daily, G.C.; Mooney, H.A.; Ehrlich, P.; Quesada, M.; Miranda, A.; Jaramillo, V.J.; García-Oliva, F.; et al. Ecosystem services of tropical dry forests: Insights from long-term ecological and social research on the Pacific Coast of Mexico. *Ecol. Soc.* **2005**, *10*. [CrossRef]
20. Vasquez-Valderrama, M.; González-M, R.; López-Camacho, R.; Baptiste, M.P.; Salgado-Negret, B. Impact of invasive species on soil hydraulic properties: Importance of functional traits. *Biol. Invasions* **2020**, *22*, 1849–1863. [CrossRef]
21. Liao, C.; Peng, R.; Luo, Y.; Zhou, X.; Wu, X.; Fang, C.; Chen, J.; Li, B. Altered ecosystem carbon and nitrogen cycles by plant invasion: A meta-analysis. *New Phytol.* **2008**, *177*, 706–714. [CrossRef] [PubMed]
22. Funk, J.L.; Vitousek, P.M. Resource-use efficiency and plant invasion in low-resource systems. *Nature* **2007**, *446*, 1079–1081. [CrossRef]
23. González, A.L.; Kominoski, J.S.; Danger, M.; Ishida, S.; Iwai, N.; Rubach, A. Can ecological stoichiometry help explain patterns of biological invasions? *Oikos* **2010**, *119*, 779–790. [CrossRef]
24. Matzek, V. Superior performance and nutrient-use efficiency of invasive plants over non-invasive congeners in a resource-limited environment. *Biol. Invasions* **2011**, *13*, 3005–3014. [CrossRef]
25. Sardans, J.; Peñuelas, J. The role of plants in the effects of global change on nutrient availability and stoichiometry in the plant-soil system. *Plant Physiol.* **2012**, *160*, 1741–1761. [CrossRef] [PubMed]
26. Pejchar, L.; Mooney, H.A. Invasive species, ecosystem services and human well-being. *Trends Ecol. Evol.* **2009**, *24*, 497–504. [CrossRef] [PubMed]
27. Ehrenfeld, J.G.; Stander, E.K. Habitat function in urban riparian zones. *Urban Ecosyst. Ecol.* **2010**, *55*, 103–118.
28. Elgersma, K.J.; Ehrenfeld, J.G.; Yu, S.; Vor, T. Legacy effects overwhelm the short-term effects of exotic plant invasion and restoration on soil microbial community structure, enzyme activities, and nitrogen cycling. *Oecologia* **2011**, *167*, 733–745. [CrossRef] [PubMed]
29. Richardson, D.M.; Pyšek, P.; Rejmanek, M.; Barbour, M.G.; Panetta, F.D.; West, C.J. Naturalization and invasion of alien plants: Concepts and definitions. *Divers. Distrib.* **2000**, *6*, 93–107. [CrossRef]
30. Davis, M.A.; Chew, M.K.; Hobbs, R.J.; Lugo, A.E.; Ewel, J.J.; Vermeij, G.J.; Brown, J.H.; Rosenzweig, M.L.; Gardener, M.R.; Carroll, S.P.; et al. Don't judge species on their origins. *Nature* **2011**, *474*, 153–154. [CrossRef]
31. Reaser, J.K.; Simpson, A.; Guala, G.F.; Morisette, J.T.; Fuller, P. Envisioning a national invasive species information framework. *Biol. Invasions* **2020**, *22*, 21–36. [CrossRef]
32. Vujanović, D.; Losapio, G.; Milić, S.; Milić, D. The Impact of multiple species invasion on soil and plant communities increases with invasive species co-occurrence. Front. *Plant Sci.* **2022**, *13*, 875824. [CrossRef]
33. Jamin, J.; Diehl, D.; Meyer, M.; Davis, J.; Schaumann, G.E. Physio-chemical soil properties affected by invasive plants in southwest Germany (Rhineland-Palatinate)—A case study. *Soil Syst.* **2022**, *6*, 93. [CrossRef]
34. Thomas, J.; El-Sheikh, M.A.; Alfarhan, A.H.; Alatar, A.A.; Sivadasan, M.; Basahi, M.; Al-Obaid, S.; Rajakrishnan, R. Impact of alien invasive species on habitats and species richness in Saudi Arabia. *J. Arid Environ.* **2016**, *127*, 53–65. [CrossRef]
35. Lehan, N.E.; Murphy, J.R.; Thorburn, L.P.; Bradley, B.A. Accidental introductions are an important source of invasive plants in the continental United States. *Am. J. Bot.* **2013**, *100*, 1287–1293. [CrossRef]

36. Alfaran, A.H.; Thomas, J.; Rajakrishnan, R. Impact assessment and management of invasive species in plant diversity centers and agricultural fields in Saudi Arabia. In *Invasive Alien Species: Observations and Issues from Around the World. Volume 2: Issues and Invasions in Asia and the Pacific Region*; Pullaiah, T., Ielmini, M.R., Eds.; John Wiley & Sons: New York, NY, USA, 2021; pp. 207–225.
37. Al-Sodany, Y.M. A new record to the flora of Saudi Arabia: *Ipomoea carnea* Jacq., Convolvulaceae. *World J. Res. Rev.* **2016**, 3, 25–30.
38. Shaltout, K.H.; Al-Sodany, Y.M.; Eid, E.M. The biology of Egyptian woody perennials-2. *Ipomoea Carnea Jacq. Ass. Univ. Bull. Environ. Res.* **2006**, *9*, 75–91.
39. Waring, B.G.; Alvarez-Cansino, L.; Barry, K.E.; Becklund, K.K.; Dale, S.; Gei, M.G.; Keller, A.B.; Lopez, O.R.; Markesteijn, L.; Mangan, S.; et al. Pervasive and strong effects of plants on soil chemistry: A meta-analysis of individual plant 'Zinke' effects. *Proc. R. Soc. B.* **2015**, *282*, 20151001. [CrossRef]
40. Mack, R.N.; Simberloff, D.; Lonsdale, W.M.; Evans, H.; Clout, M.; Bazzaz, F.A. Biotic invasions: Causes, epidemiology, global consequences, and control. *Ecol. Appl.* **2000**, *10*, 689–710. [CrossRef]
41. Skočajić, D.; Nešić, M.L. *Invasive Species: Routes of Introduction, Establishment, and Expansion*; Springer: Berlin/Heidelberg, Germany, 2021; pp. 571–582.
42. Gaertner, M.; Den Breeyen, A.; Hui, C.; Richardson, D.M. Impacts of alien plant invasions on species richness in Mediterranean-type ecosystems: A meta-analysis. *Prog. Phys. Geogr.* **2009**, *33*, 319–338. [CrossRef]
43. Joshi, C.; De Leeuw, J.; van Duren, I.C. Remote sensing and GIS applications for mapping and spatial modelling of invasive species. *Proc. ISPRS* **2004**, *35*, B7.
44. Pyšek, P.; Jarošík, V.; Hulme, P.E.; Pergl, J.; Hejda, M.; Schaffner, U.; Vilà, M. A global assessment of invasive plant impacts on resident species, communities and ecosystems: The interaction of impact measures, invading species' traits and environment. *Glob. Chang. Biol.* **2012**, *18*, 1725–1737. [CrossRef]
45. Castro-Díez, P.; Godoy, O.; Alonso, A.; Gallardo, A.; Saldaña, A. What explains variation in the impacts of exotic plant invasions on the nitrogen cycle? A meta-analysis. *Ecol. Lett.* **2014**, *17*, 1–12. [CrossRef]
46. Ashton, I.W.; Hyatt, L.A.; Howe, K.M.; Gurevitch, J.; Lerdau, M.T. Invasive species accelerate decomposition and litter nitrogen loss in a mixed deciduous forest. *Ecol. Appl.* **2005**, *15*, 1263–1272. [CrossRef]
47. Feng, Y.L.; Lei, Y.B.; Wang, R.F.; Callaway, R.M.; Valiente-Banuet, A.; Inderjit; Li, Y.P.; Zheng, Y.L. Evolutionary tradeoffs for nitrogen allocation to photosynthesis versus cell walls in an invasive plant. *Proc. Nat. Acad. Sci. USA* **2009**, *106*, 1853–1856. [CrossRef] [PubMed]
48. Ehrenfeld, J.G. Implications of invasive species for belowground community and nutrient processes. *Weed Technol.* **2004**, *18*, 1232–1235. [CrossRef]
49. Rout, M.E.; Callaway, R.M. An invasive plant paradox. *Science* **2009**, *324*, 734. [CrossRef] [PubMed]
50. Novak, S.J. The role of evolution in the invasion process. *Proc. Nat. Acad. Sci. USA* **2007**, *104*, 3671–3672. [CrossRef] [PubMed]
51. Xu, C.-Y.; Julien, M.H.; Fatemi, M.; Girod, C.; van Klinken, R.; Gross, C.L.; Novak, S.J. Phenotypic divergence during the invasion of Pyla canescens in Australia and France: Evidence for selection-driven evolution. *Ecol. Lett.* **2010**, *13*, 32–34. [CrossRef] [PubMed]
52. Sax, D.F.; Stachowicz, J.J.; Gaines, S.D. (Eds.) *Species Invasions: Insights into Ecology, Evolution and Biogeography*; Sinauer: Sunderland, MA, USA, 2005.
53. Gilliam, F.S. Response of the herbaceous layer of forest ecosystems to excess nitrogen deposition. *J. Ecol.* **2006**, *94*, 1176–1191. [CrossRef]
54. Daehler, C.C. Performance comparisons of co-occurring native and alien invasive plants: Implications for conservation and restoration. *Annu. Rev. Ecol. Evol. Syst.* **2003**, *34*, 183–211. [CrossRef]
55. Ostertag, R.; Verville, J.H. Fertilization with nitrogen and phosphorus increases abundance of non-native species in Hawaiian montane forests. *Plant Ecol.* **2002**, *162*, 77–90. [CrossRef]
56. Laungani, R.; Knops, J.M. Species-driven changes in nitrogen cycling can provide a mechanism for plant invasions. *Proc. Natl. Acad. Sci. USA* **2009**, *106*, 12400–12405. [CrossRef]
57. Kueffer, C. Reduced risk for positive soil-feedback on seedling regeneration by invasive trees on a very nutrient-poor soil in Seychelles. *Biol. Invasions* **2010**, *12*, 97–102. [CrossRef]
58. Duda, J.J.; Freeman, D.C.; Emlen, J.M.; Belnap, J.; Kitchen, S.G.; Zak, J.C.; Sobek, E.; Tracy, M.; Montante, J. Differences in native soil ecology associated with invasion of the exotic annual chenopod, *Halogeton glomeratus*. *Biol. Fertil. Soils* **2003**, *38*, 72–77. [CrossRef]
59. Rodgers, V.L.; Wolfe, B.E.; Werden, L.K.; Finzi, A.C. The invasive species *Alliaria petiolata* (garlic mustard) increases soil nutrient availability in northern hardwood-conifer forests. *Oecologia* **2008**, *157*, 459–471. [CrossRef]
60. Li, H.; Meng, Z.; Dang, X.; Yang, P. Soil Properties under artificial mixed forests in the desert-yellow river coastal transition zone, China. *Forests* **2022**, *13*, 1174. [CrossRef]
61. Odum, W.E. Comparative ecology of tidal freshwater and salt marshes. *Annu. Rev. Ecol. Syst.* **1988**, *19*, 147–176. [CrossRef]
62. Windham, L.; Weis, J.S.; Weis, P. Patterns and processes of mercury release from leaves of two dominant salt marsh macrophytes, *Phragmites australis* and *Spartina alterniflora*. *Estuaries* **2001**, *24*, 787–795. [CrossRef]
63. Ermakov, V.V.; Korobova, E.M.; Degtyarev, A.P.; Tyutikov, S.F.; Karpova, E.A.; Petrunina, N.S. Impact of natural and man-made factors on migration of heavy metals in the Ardon River basin (North Ossetia). *J. Soils Sediments* **2016**, *16*, 1253–1266. [CrossRef]

64. Chai, M.; Shi, F.; Li, R.; Shen, X. Heavy metal contamination and ecological risk in *Spartina alterniflora* marsh in intertidal sediments of Bohai Bay, China. *Mar. Pollut. Bull.* **2014**, *84*, 115–124. [CrossRef]
65. Li, J.; Lu, H.; Liu, J.; Hong, H.; Yan, C. The influence of flavonoid amendment on the absorption of cadmium in *Avicennia marina* roots. *Ecotoxicol. Environ. Saf.* **2015**, *120*, 1–6. [CrossRef]
66. Selvi, F.; Carrari, E.; Colzi, I.; Coppi, A.; Gonnelli, C. Responses of serpentine plants to pine invasion: Vegetation diversity and nickel accumulation in species with contrasting adaptive strategies. *Sci. Total Environ.* **2017**, *595*, 72–80. [CrossRef]
67. Li, J.; Du, Z.; Zou, C.B.; Dai, Z.; Du, D.; Yan, C. The mutual restraint effect between the expansion of *Alternanthera philoxeroides* (Mart.) Griseb and cadmium mobility in aquatic environment. *Ecotoxicol. Environ. Saf.* **2018**, *148*, 237–243. [CrossRef]
68. El-Bakatoushi, R.; Alframawy, A.M.; Tammam, A.; Youssef, D.; El-Sadek, L. Molecular and physiological mechanisms of heavy metal tolerance in *Atriplex halimus*. *Int. J. Phytoremed.* **2015**, *17*, 789–800. [CrossRef]
69. Ellili, A.; Rabier, J.; Prudent, P.; Salducci, M.D.; Heckenroth, A.; Lachaâl, M.; Laffont-Schwob, I. Decision-making criteria for plant-species selection for phytostabilization: Issues of biodiversity and functionality. *J. Environ. Manag.* **2017**, *201*, 215–226. [CrossRef] [PubMed]
70. Nelson, D.A.; Sommers, L. Total carbon, organic carbon, and organic matter. *Methods Soil Anal. Part 2 Chem. Microbiol. Prop.* **1983**, *9*, 539–579.
71. Olsen, S.R. *Estimation of Available Phosphorus in Soils by Extraction with Sodium Bicarbonate (No. 939)*; US Department of Agriculture: Washington, DC, USA, 1954.
72. Berghage, R.D.; Krauskopf, D.M.; Warncke, D.D.; Widders, I. Micronutrient testing of plant growth media: Extractant identification and evaluation. *Commun. Soil Sci. Plant Anal.* **1987**, *18*, 1089–1109. [CrossRef]
73. Zheljazkov, V.D.; Warman, P.R. Comparison of three digestion methods for the recovery of 17 plant essential nutrients and trace elements from six composts. *Compos. Sci. Util.* **2002**, *10*, 197–203. [CrossRef]
74. Cataldo, D.A.; Maroon, M.; Schrader, L.E.; Youngs, V.L. Rapid colorimetric determination of nitrate in plant tissue by nitration of salicylic acid. *Commun. Soil Sci. Plant Anal.* **1975**, *6*, 71–80. [CrossRef]
75. Snedecor, G.W.; Cochran, W.G. *Statistical Methods*, 8th ed.; Iowa State University Press: Ames, IA, USA, 1989; p. 1191.

diversity

MDPI

Article

Invasive Glossy Buckthorn (*Frangula alnus*) Has Weak Impact on Native Understory Plant and Saprophagous Macroarthropod Communities

Gabrielle Roberge [1,*], Ira Tanya Handa [1,2], Tristan Juette [2,3] and Daniel Kneeshaw [1,2,*]

1 Institut des Sciences de l'Environnement, Université du Québec à Montréal, 141 Avenue du Président-Kennedy, Montréal, QC H2X 2Y4, Canada; handa.ira_tanya@uqam.ca
2 Département des Sciences Biologiques, Université du Québec à Montréal, 141 Avenue du Président-Kennedy, Montréal, QC H2X 2Y4, Canada; tristan.juette@umontreal.ca
3 Faculté de Médecine Vétérinaire, Université de Montréal, 3200 Rue Sicotte, Saint-Hyacinthe, QC J2S 2M2, Canada
* Correspondence: roberge.gabrielle.2@uqam.ca (G.R.); kneeshaw.daniel@uqam.ca (D.K.)

Abstract: Glossy buckthorn (*Frangula alnus* Miller) is an invasive alien plant species (IAPS) rapidly expanding in North America but is largely understudied compared to the common buckthorn (*Rhamnus cathartica*). Our study investigated the effects of a 27-year-old *F. alnus* invasion on native understory plant and saprophagous macroarthropod communities in a wet deciduous woodland in Southern Québec, Canada. We hypothesized a decreased taxonomic diversity and a change in community composition of both indicator taxa with increasing *F. alnus* density. The understory plant and saprophagous macroarthropod communities were characterized, respectively, through vegetation surveys and pitfall trapping across a density gradient of 43 plots invaded by *F. alnus*. Our results demonstrated that *F. alnus* did not exert a strong influence on species community composition, although the homogenization of understory plant communities was observed. Despite several decades of *F. alnus* invasion at our study site, the consequences on the selected indicator taxa were overall relatively small, suggesting that the magnitude of effects is variable. We suggest that further investigation at a larger scale should be performed to evaluate the effect of *F. alnus* on a broad diversity of indicators and understand any context dependency.

Keywords: invasive plant species; Diplopoda; biodiversity; community ecology; millipede; Isopoda; woodlice

1. Introduction

Invasive alien plants species (IAPS) are defined as exotic plant species that are successfully established in a novel environment and are causing negative impacts on biodiversity, local ecosystems, human health, and ecosystem services [1]. Several have been recognized as being able to cause great and potentially irreversible biological disturbance to the ecosystems they invade [1–4]. IAPSs have strong competitive effects including advantageous reproductive mechanisms, earlier and later growth phenology, no (or few) natural predators, and the capacity to alter biotic and abiotic conditions in invaded ecosystems [1,5,6]. For instance, in Eastern North America, species such as *Phragmites australis* ssp. *australis* (Cav.) Steud. (common reed) propagate very rapidly by abundant seed and vegetative propagation and can form dense monocultures [7]. Some IAPS, like *Ailanthus altissima* (Mill.) Swingle, (tree of heaven) and *Rhamnus cathartica* L. (common buckthorn) have been reported to produce allelopathic substances, potentially inhibiting the survival of nearby plants [8–11]. Other IAPSs like *Rosa multiflora* Thunb. (Japanese rose) and *Berberis thunbergii* DC. (Japanese barberry) have dense foliage and form thickets, which greatly limit access to light for indigenous understory plants [12]. Given that the presence and impacts of IAPSs are projected to increase worldwide due to global trade and habitat modification [1], it is important to understand which species pose the greatest threat by understanding their interactions with local species [13].

The competitive ability of many IAPSs is postulated to negatively impact the diversity, relative abundance, and total abundance of companion species thus modifying community structure at multiple trophic levels as well as modifying food web interactions [14–17]. Changes in the plant community composition can lead to cascading effects through altered soil chemistry, leaf litter biomass, and leaf litter quality for detrital consumers thus affecting, in turn, microbial activity and community composition [18–20]. Earlier work has shown that ground-dwelling arthropod abundance, community structure, and diversity may be at risk due to IAPSs [21–23]. A decrease in resource diversity can lead to a decrease in consumer diversity [24], also referred to as the resource specialization hypothesis, whether taxonomically or functionally [25]. In their meta-analysis on the effect of plant invasions on arthropods, Litt et al. observed that total arthropod abundance decreased in 62% of the studies, but evidence varied depending on the functional groups that were evaluated [26]. Schuh and Larsen [27] documented a decrease in the abundance and diversity of crawling predatory insects, caused by the lack of prey in areas invaded by *R. cathartica*. Brousseau et al. [28] reported that the presence of *Reynoutria* spp. strongly decreased ground-dwelling arthropod richness and reduced correlations between feeding traits of predators and palatability traits of prey, suggesting that *Reynoutria* spp. disrupts functional linkages in the food web. In short, a loss of plant diversity due to IAPSs could lead to a diversity loss in ground-dwelling arthropods of both green (grazing) and brown (detrital) food webs [16,24,29].

The glossy buckthorn (*Frangula alnus* Miller) is a shrub in the buckthorn family (Rhamnaceae) and considered to be invasive in Eastern North America. Native to Asia and Europe, it was introduced in the late 1800s and was imported for its medicinal and ornamental properties [7,30]. Like other IAPS, the density of its foliage and thickets, as well as the late senescence of leaves in the fall and its rapid and effective reproduction mechanisms make it a strong competitor of understory plant species [7,31]. Its distribution has been increasing at a fast rate in Eastern North America since the mid-1900s [30,32,33]. However, studies on the impact of *F. alnus* on floral diversity are equivocal. Some research shows no effect on understory plant communities [34,35] or even an increase in shrub species richness [36]. Other studies have revealed negative relationships with certain groups of plants, such as a decrease in the density of native tree seedlings, herbaceous plant cover and richness, and changes in plant community structure in favor of shade-tolerant species [37,38]. In addition, Stokdyk and Hermann [39,40] observed that leaf extracts of *F. alnus* altered soil dynamics and total extractable nitrogen mineralization, causing changes in the microbial community functions and litter decomposition. Finally, the only known study assessing *F. alnus* and arthropod diversity found an increase in the abundance and diversity of generalist pollinating insects followed the removal of *F. alnus* [41].

Our goal was to study the impact of *F. alnus* invasion on two indicator communities, understory plants and saprophagous macroarthropods, to better assess the biodiversity risks associated with an *F. alnus* invasion in a wet deciduous woodland. Understory plants were selected as they are most likely to suffer from shading by *Frangula alnus* [3,7,37,38,42,43]. Moreover, saprophagous macroarthropods (detritus eaters larger than 5 mm) such as millipedes and woodlice were selected because they are known to be sensitive to disturbance [44,45] and are the first line of organic matter shredders in the brown food web playing an essential role in the decomposition of organic matter [46–49]. Despite the importance of saprophagous macroarthropods to ecosystem function [50,51], no study has assessed their relationship with *F. alnus*.

Using a density gradient of the *F. alnus* invaded plots, we hypothesized that the increase in the density of *F. alnus* will result in the following: (1) simplify the structure of plant species communities in the understory layer and (2) reduce species diversity. Additionally, we hypothesized that the increase in *F. alnus* density, as well as the diversity loss of native understory plant species would achieve the following: (3) alter species composition and (4) decrease species diversity of saprophagous macroarthropod communities.

2. Materials and Methods

The study site was located in the Boisé des Terres Noires nature reserve that covers 12 ha of deciduous forest in a 600 ha area, which comprised a diversity of habitats including wetlands, early succession woodlands, and peatlands [52]. These habitats are surrounded by agricultural land in L'Assomption, QC, Canada (45.81863, −73.4742) (Figure 1). Since 2016, a significant increase in *F. alnus* has been documented [53], but the invasion likely began following the commercial exploitation of peatland soil in 1996 [54].

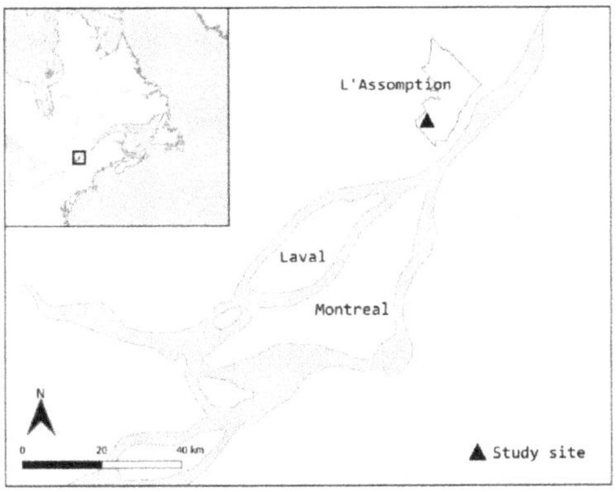

Figure 1. Location of the study site, in L'Assomption, Québec, Eastern Canada.

This invasion estimate was confirmed by coring the biggest individuals which were found to be around 27 years old (personal communication). The soil is acidic (average pH of 4.00 across the study site), which favors glossy buckthorn over common or European buckthorn [55]. Woodlands on the site were mainly dominated by pioneer trees, such as *Acer rubrum* L. (red maple), *Populus tremuloides* (Michx.) Á. Löve and D. Löve (quaking aspen), and *Betula populifolia* M. (gray birch). Several species of *Rubus* and ferns were also found in great abundance.

To assess the influence of *F. alnus* on native understory plant and saprophagous macroarthropod communities, we first visually estimated the percent cover of *F. alnus* (height > 1 m) along the transects perpendicular to an access trail. To prevent a border effect, all transects were started 10 m away from the trail. Based on this percent cover evaluation ranging from less than 5% to 100% cover of *F. alnus*, we established 43 sample plots, located at a minimum distance of 40 m from each other along the *F. alnus* density gradient. Stem density was then evaluated by counting individuals of *F. alnus* (height > 1 m) in 5 m^2 plots (high density) or in 50 m^2 plots (low density) and subsequently estimated per m^2.

2.1. Understory Plant Characterization

From 15 July to 25 August 2020, all vascular plants in the understory (height < 1 m) were identified and cover abundance was evaluated in 4 m^2 (four 1 m^2 circular plots) to the nearest 5% and then estimated per m^2. Unidentifiable plants in the field were sampled and photographed for further identification in the laboratory using identification keys [56]. A ratio of the cover abundance of each understory species to the total understory plant cover was calculated. Authorities were selected based on their use in North American and the most recent published work from the International Plant Name Index (IPNI) database [57].

2.2. Saprophagous Macroarthropod Sampling

Multi-Pher® pitfall traps (20.5 cm height and 10 cm in diameter) were used to collect saprophagous macroarthropods and were installed in the center of each plot in early June 2020, two weeks prior to activating traps, to minimize any disturbance related to digging [58]. The traps contained 100 mL of 50% ethanol for preservation, and the collected arthropods were then preserved in 80% ethanol in the laboratory. The traps were active for three periods during the summer of 2020—two 15-day periods and one 14-day period on 1 and 29 July and 27 August 2020, respectively, for a total trapping effort of 44 days. Data from five traps in the first sampling and three traps in the third sampling period were excluded from the analyses due to wildlife disturbance or technical issues that compromised the traps. Despite the shortcomings of this method [59–61], capture with the pitfall traps remains an effective and economical way to make a comparative assessment of edaphic arthropods [22,25,27,58,62]. All millipedes and woodlice were counted and identified to species [50,63–66], except for some juvenile Parajulidae (n = 5) whose species could not be identified because their secondary sexual characteristics (gonopods) were not developed. The average abundance of saprophagous macroarthropods was used to calculate diversity indices and analyze community structure.

2.3. Environmental Characterization

Environmental variables (soil temperature at ground level, soil humidity, soil pH, tree canopy cover, cover of woody debris, leaf litter biomass and composition) were measured given their potential impact on soil arthropod communities, understory plant communities, and potential relationship with *F. alnus* [27,42,67–69]. A HOBO data logger (Onset Computer Corporation, Bourne, MA, USA) was placed on the ground, under the leaf litter layer, in the center of 24 plots to record the temperature every 2 h from 11 June to 28 August 2020. Soil humidity was measured three times over the summer in each plot (at less than 1 m from the plot center) with a Field Scout TDR 300 moisture meter (Spectrum Technologies, Yellow Springs, OH, USA) with 7.6 cm rods. Care was taken to avoid rainfall events in the 24 h preceding soil humidity measurements. Soil pH was measured once (3 replicates) in the summer with a pHA10 Ecosens probe (YSI Inc., Bridgend, UK). Tree canopy cover (to the nearest 5%) and cover of woody debris (to the nearest 5%) were estimated for each plot. On 30 September 2020, three samples (15 cm × 15 cm × 5 cm) of the litter in the center of each plot were collected and dried at 35 °C. The leaf litter was sorted by species, and the total litter dry mass for each plot was determined. The average ambient air temperature for the summer of 2020 was 20.6 °C (±4.6 °C) [70]. Heavy rainfall (Max = 27 mm) occurred during the first arthropod sampling period from 23 to 24 June 2020. But generally, the trapping periods had little rain and were mostly sunny or cloudy [71].

2.4. Statistical Analysis

To obtain a general portrait of the environmental conditions at our site that corresponded to the gradient of *F. alnus* (range from 0 to >4 individuals m^{-2}), linear regression tests were carried out to test whether the density of *F. alnus* was correlated with soil temperature, soil humidity, leaf litter biomass, and the percentage of woody debris cover. We also used linear regression models to test whether canopy cover and soil pH influenced the density of *F. alnus*. Redundancy analyses (RDA) were used to explore community structures across plots for both understory plants and saprophagous macroarthropod communities. To analyze plant communities, rare species (<5% average cover and found in less than 5% of plots) were removed. Plant cover ratios were transformed using Hellinger's distance to minimize the effect of double zeros [72–74] and soil humidity, soil pH, and canopy cover were used as constraint variables. To assess saprophagous macroarthropod communities, the mean abundance of saprophagous macroarthropods was log-transformed to avoid the horseshoe effect [72]. Species with fewer than five individuals and unidentified juveniles were removed. Along with the density of *F. alnus*, other environmental factors that were likely to influence community structure, such as understory plant species, woody debris cover, and litter biomass were used as constraint variables and were standardized and centered on 0.

Community metrics such as species richness (*S*), species abundance, the effective number of species, Shannon–Wiener index (*H′*), and the evenness index (*J′*) of the native understory plants (including rare plants) and saprophagous macroarthropod communities were calculated and tested as response variables in the linear regression models assessing the potential effect of *F. alnus* density. The Shannon–Wiener index was obtained from the following formula:

$$H' = -\sum_{i=1}^{S} Pi ln Pi$$

where *Pi* corresponded to the proportion of species cover *i* on the total cover (for understory plants) or the proportion of species *i* individuals on the total number of individuals (for saprophagous macroarthropods). Evenness was calculated as follows:

$$J' = H'/\ln(S)$$

The effective number of species (e*H′*) is an index based on the Shannon–Wiener index and was used to provide a more intuitive interpretation of diversity [75].

For native understory plants, the abundance and evenness index were square root transformed to meet linear model conditions. One extreme data point was removed for the evenness (plot L4) to fit conditions of normality. For saprophagous macroarthropods, the percentage of *F. alnus* leaves in the litter and the effective number of understory plant species were also used as explanatory variables, in addition to litter biomass (mg m^{-2}), pH, soil humidity (%) and woody debris (%)). All analyses and data processing were carried out with the R program version 4.0.4 [76] using the vegan library [77].

3. Results

3.1. Site Characteristics

In general, the environmental characteristics were stable across all 43 plots with no significant differences observed in soil humidity, leaf litter biomass, and soil temperature as a function of the density gradient of *F. alnus* (Table 1). However, woody debris cover increased significantly with *F. alnus* density (Table 1). Moreover, we observed that the density of *F. alnus* was not predicted significantly by tree canopy cover, but soil pH was a significant predictor (Table 1).

Table 1. Linear regression model output of the environmental variables associated with density gradient of *F. alnus* at the Boisé des Terres Noires study site. Variables that were most likely to influence the density of *F. alnus* were used as independent variables, such as tree canopy cover (%) and pH. Others that might be influenced by *F. alnus*, such as soil humidity (%), leaf litter biomass (g m^{-2}), ground-level temperature (°C), and woody debris cover (%) were used as dependent variables to test if they varied across the gradient of *F. alnus*.

Dependent Variable	Independent Variable	Mean ± SD	Estimate	±SE	T-Value	R^2	*p*-Value
F. alnus density (nb m^{-2})	Canopy cover (%)	58.95 ± 13.30	−0.009	0.006	−1.360	0.043	0.181
	pH	4.00 ± 0.40	−0.482	0.206	−2.344	0.118	<0.05 *
Humidity (%) [1]	*F. alnus* density (nb m^{-2})	20.46 ± 10.80	−0.034	0.092	−0.364	0.003	0.718
Leaf litter biomass (g m^{-2}) [1]		269.26 ± 65.90	0.238	0.267	0.893	0.019	0.377
Ground temperature (°C) [1]		20.15 ± 0.48	0.086	0.077	1.122	0.054	0.274
Woody debris cover (%) [1]		13.80 ± 6.50	0.510	0.083	6.16	0.481	<0.001 ***

[1] Square root transformation to fit the linear regression model conditions; * $p < 0.05$; *** $p < 0.001$.

3.2. Native Understory Plants

A total of 54 species of understory plants were found throughout the study site (Table 2). Among the most common species, saplings *A. rubrum* accounted for around 11%

of the total vegetation cover across all study plots, making it the second most abundant understory plant after *F. alnus*. The most abundant shrubs were ericaceous species such as *Rhododendron canadense* (L.) Torr., *Vaccinium angustifolium* A., and *Kalmia angustifolia* L., as well as *Aronia melanocarpa* (Willd.) Torr. (Rosaceae). Each of these species represented, on average, between 3% and 8% of total vegetation cover (Table 2). The most abundant herbaceous plant species was *Maianthemum canadense*, with an average cover of around 4%, as well as four species of pteridophytes, with a combined average cover of around 12% (*Dryopteris spinulosa* (Mull.) Watt., *Onoclea sensibilis* L., *Osmunda claytoniana* L., *Osmunda cinnamomea* L.). The rare species (n = 27) altogether accounted for, on average, less than 5% of the total vegetation cover. Other exotic plant species were also reported including *Phragmites australis* ssp. *australis* and *Galeopsis ladanum* L. These represented less than 3% of the vegetation cover and were found in less than 5% of the sampling stations. We also observed a general decrease in total native plant abundance across the gradient of *F. alnus*, where the highest abundance was observed when *F. alnus* was around two individuals m^{-2} ($R^2 = 0.11$, $p < 0.05$). On the site, with an average of 20.5%, Frangula alnus was the species with the largest percentage of total understory plant cover (Table 2).

Table 2. Dominant understory plant species: trees, shrubs, and herbaceous plants (>5% presence in plots and >5% vegetation cover) in all our 43 plots on our site. Shade tolerance, frequency (percentage of presence in plot) and average cover (percentage) of total vegetation for each of understory plant species.

Family	Genre	Species	Shade Tolerance	Frequency (%)	Average Cover ± SE
		Native and cosmopolite species			
		Trees			
Rosaceae	*Prunus*	*virginiana* L.	Tolerant	16.3	0.6 ± 0.3
Sapindaceae	*Acer*	*rubrum* L.	Medium	100	11.2 ± 1.4
Salicaceae	*Populus*	*tremuloides* M.	Intolerant	30.2	0.5 ± 0.1
		Shrubs			
Aquifoliaceae	*Ilex*	*mucronata* L.	Tolerant	34.9	1.4 ± 0.3
Caprifoliaceae	*Viburnum*	*cassinoides* L.	Tolerant	41.9	2.7 ± 0.7
Cornaceae	*Cornus*	*alternifolia* L.	Tolerant	7	1.1 ± 0.7
Ericaceae	*Kalmia*	*angustifolia* L.	Intolerant	48.8	3.4 ± 0.8
	Rhododendron	*canadense* L.	Tolerant	55.8	6.2 ± 1.4
	Vaccinium	*angustifolium* A.	Intolerant	58.1	6.2 ± 1.2
	Chamaedaphne	*calyculata* L.	Intolerant	9.3	0.3 ± 0.2
Rosaceae	*Aronia*	*melanocarpa* W.	Intolerant	88.4	8.3 ± 1.0
	Rubus	*occidentalis* L.	Tolerant	14	1.8 ± 1.2
	Rubus	*allegheniensis* P.	Intolerant	9.3	0.6 ± 0.3
	Rubus	*Pubescence* R.	Tolerant	44.1	5.1 ± 1.4
	Rubus	*hispidus* L.	Tolerant	16.3	1.8 ± 0.9
	Rubus	*idaeus* L.	Intolerant	16.3	2.3 ± 1.0
	Spiraea	*latifolia* A.	Medium	76.7	3.5 ± 0.6
		Herbs			
Asparagaceae	*Maianthemum*	*Canadense* D.	Medium	69.8	4.3 ± 0.9
Osmundaceae	*Osmunda*	*claytoniana* L.	Tolerant	7	0.7 ± 0.5
Polypodiaceae	*Dryopteris*	*spinulosa* M.	Tolerant	20.9	4.3 ± 1.9
	Onoclea	*sensibilis* L.	Medium	7	1.1 ± 1.0
	Osmunda	*cinnamomea* L.	Tolerant	11.6	1.2 ± 0.8
Primulaceae	*Lysimachia*	*borealis* R.	Tolerant	62.8	3.0 ± 0.8
Vitaceae	*Parthenocissus*	*quinquefolia* L.	Tolerant	23.3	1.8 ± 0.7

Table 2. *Cont.*

Family	Genre	Species	Shade Tolerance	Frequency (%)	Average Cover ± SE
		Exotic species			
Lamiaceae	*Galeopsis*	*ladanum* L.	Medium	2.3	<0.1
Poaceae	*Phragmites*	*australis* ssp. *australis* C.	Intolerant	4.7	<0.1
Rosaceae	***Frangula***	***alnus* M.**	**Tolerant**	**93**	**20.5 ± 3.0**
		***Rare species*: 27**			<5%
Richness of native species: 51					
Total richness: 54					

3.3. Native Understory Plant Communities

The redundancy analysis model for understory plant community composition only explained 14.3% of the variation along the two first axes (Figure 2). In plots with a high density of *F. alnus*, we observed shrub species such as *Cornus alternifolia* L., *Ilex mucronata* (L.) M.Powell, Savol. & S. Andrews, *Rubus occidentalis* L. the vine *Parthenocissus quinquefolia* (L.) Planch., herbaceous species like *M. canadense* and *Circaea lutetiana* L. and tree seedlings of *A. rubrum*. However, as the density of *F. alnus* only explained a small percentage of variation (4.4%) on the second axis, plant species composition was driven more strongly by other environmental factors, such as soil humidity and tree canopy cover, which were both associated with axis 1 and explained more than twice as much variation (9.9%).

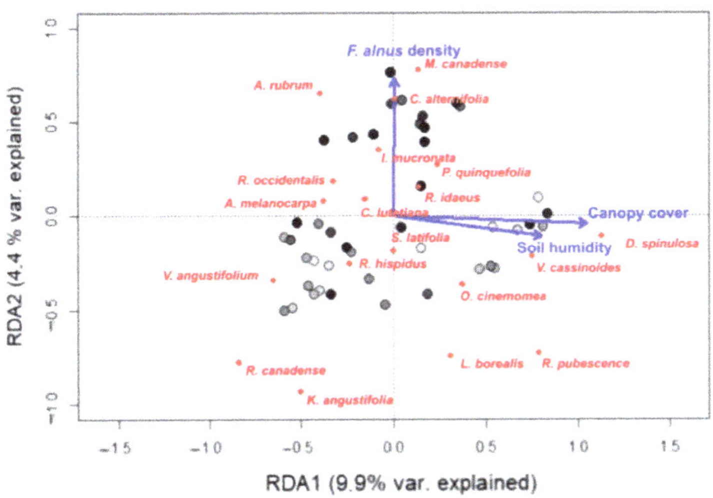

Figure 2. Redundancy analysis (see statistical output in Table S1) of percent cover of native understory plant species and environmental variables, according to each sampling plot (represented by dots). For each plot, density of *F. alnus* is represented by a gray gradient to visualize the degree of invasion, whereby pale plots have a lower density (0 individual m^{-2}) and gradually increase to dark for high density plots (>4 individual m^{-2}). Plant species, in red, have been centered and transformed (Hellinger). Significant constraint variables (canopy cover, soil humidity, soil pH and density of *F. alnus*), in blue, have also been standardized, centered on 0.

Surprisingly, shade-tolerant plant species were present on both ends (low and high density) of the *F. alnus* density gradient [56,78–80]. Conversely, peatland species like *K. angustifolia*, *R. canadense*, *V. angustifolium*, *Rubus hispidus* L., and *Spiraea latifolia* (A.) F.

Seym. were associated with plots with lower tree canopy cover and lower humidity. These species are mostly shade-intolerant, and they tend to prefer higher luminosity and soil humidity and more acidic organic soils [79,81].

Except for species richness ($R^2 = 0.002$, $p = 0.78$), we found a consistent negative, albeit weak, response of plant understory abundance and diversity measured with multiple diversity metrics along the gradient of *F. alnus* invasion (Table 3). The effective number of species ($R^2 = 0.177$, $p < 0.005$) and the Shannon–Wiener diversity index ($R^2 = 0.21$, $p < 0.005$) both decreased as *F. alnus* density increased, but the variation caused by *F. alnus* was not very strong [82]. The effective number of understory plant species variations was explained mostly by *F. alnus* density ($p < 0.01$) and leaf litter biomass ($p < 0.05$, $R^2 = 0.2219$, Table S3). Pielou's evenness in plant communities ($R^2 = 0.26$, $p < 0.001$) still showed that when the density of *F. alnus* increased, understory plant communities tended to homogenize. Although the models showed a noticeable effect, the variation caused by the increasing density of *F. alnus* varied throughout the models.

Table 3. Linear regressions of density of *F. alnus* (individual m^{-2}) as a predictive variable on five diversity indices of understory native plant species (height < 1 m). Evenness index and total abundance were square root transformed to fit linear model conditions. All plant diversity indices were based on the ratio of mean species abundance (to the closest 5%, on 1 m^2) to total plant abundance (% cover).

	Descriptive Statistics		Linear Model Output				
Dependent Variable	**Mean ± SE**	**Median (Min–Max)**	**Estimate**	**SE**	**T-Value**	**R^2**	***p*-Value**
Specific richness (*S*)	11.4 ± 2.5	11 (7–17)	−0.093	0.331	−0.281	0.002	0.78
Effective number of species	7.1 ± 2.2	6.97 (2.60–12.9)	−0.783	0.264	−2.967	0.177	**<0.005** **
Shannon-Weiner index (*H'*)	1.9 ± 0.3	1.94 (0.95–2.56)	−0.129	0.039	−3.316	0.211	**<0.005** **
Evenness (*J'*) [1]	0.8 ± 0.11	0.80 (0.39–0.91)	−0.030	0.008	−3.731	0.272	**<0.001** ***
Total abundance of native understory plant species (%) [1]	38.3 ± 16.7	36.90 (9.5–79.45)	−0.392	0.172	−2.276	0.112	**<0.05** *

[1] Square root transformation to fit the linear regression model conditions. * $p < 0.05$; ** $p < 0.005$; *** $p < 0.001$.

3.4. Saprophagous Macroarthropods

A total of 2566 saprophagous macroarthropods (millipedes and woodlice) were collected across all sampling periods, totaling nine species, and one which could only be identified to the genus level (*Cleidogona* sp.). Only one species of woodlice was captured, an exotic generalist species *Trachelipus rathkii* Brandt, 1833, that was many times more abundant than any other sampled species on the site (n = 1728) and known to be among one of the ten most abundant woodlouse species in North America [67]. All other eight species collected were diplopods; three of which were exotic (*Ophyiulus pilosus* Newport, 1842), *Choneiulus palmatus* Němec, 1895, and *Polydesmus inconstans* Latzel, 1884). *O. pilosus* abundance slightly increased along the gradient of *F. alnus*, whereas *C. palmatus* and *P. inconstans* were relatively stable (no significant relation to *F. alnus* density). Moreover, some species were only found in very low abundance, such as *Cleidogona* sp. (n = 2) and *Oriulus venustrus* Wood, 1864 (n = 6) (Table 4).

Table 4. Abundance, habitat, and origin (native/exotic) for each species of saprophagous macroarthropods (Diplopoda and Isopoda) captured in the 43 pitfall traps during the summer of 2020 on the Boisé des Terres Noires study site, in L'Assomption, Qc.

Class	Order	Family	Genre and Species	Total Abundance	Origin (Native/Exotic)	Habitat
Diplopoda	Chordeumatida	Cleidogonidae	*Cleidogona* sp.	2	Native	NA
	Julida	Blaniulidae	*Choneiulus palmatus* N.	56	Exotic	Urban, disturbed habitats
	Julida	Julidae	*Ophyiulus pilosus* N.	138	Exotic	Wetlands, Urban
	Julida	Parajulidae	*Uroblaniulus canadensis* N.	99	Native	Open woodland, mesic
	Julida	Parajulidae	*Oriulus venustus* W.	6	Native	Open woodland
	Julida	Parajulidae	*Parajulidae* sp.	5	Native	NA
	Polydesmida	Polydesmidae	*Polydesmus inconstans* L.	178	Exotic	Urban
	Polydesmida	Polydesmidae	*Pseudopolydesmus serratus* S.	266	Native	Mixed forests
	Spirobolida	Spirobolidae	*Narceus annularis* R.	88	Native	Woodland, Urban
Malacostraca	Isopoda	Trachelipodidae	*Trachelipus rathkii* B.	1728	Exotic	Terrestrial, generalist

3.5. Saprophagous Macroarthropod Community Composition

Overall, the selected variables generally predicted the composition of the saprophagous macroarthropod communities (Figure 3, 25.2% total var. explained). The density of *F. alnus* had a weak influence on the variation in community composition as it was associated mostly with axis 2 of the RDA (5.1%), compared to other vectors that were mostly associated with axis 1, such as native plant cover, woody debris cover, and soil humidity (20.1% var. explained). Nonetheless, some native species (*Uroblaniulus canadensis* Newport, 1844, *N. annularis* Raf. 1820, and *P. serratus* Say, 1821) [49] were associated with plots with higher native understory plant diversity and a lower *F. alnus* density. Species such as *P. inconstans* and *U. canadensis* were influenced by humidity and woody debris cover. Indeed, *U. canadensis* was associated with plots with lower humidity levels and more understory native plant diversity, whereas *P. inconstans* was mostly found where woody debris cover, and soil humidity were high. Exotic *Trachelipus rathkii* and *Choneiulus palmatus* [83] were also associated with these environmental vectors but to a lesser degree.

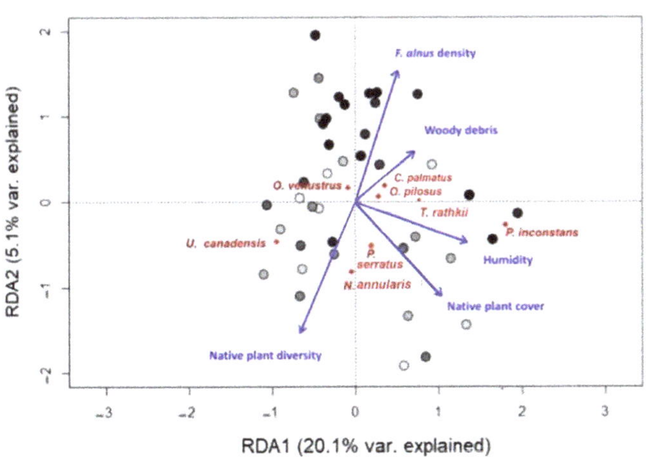

Figure 3. Redundancy analysis (see statistical output in Table S2) of average abundance of saprophagous macroarthropod communities and environmental variables across sampling plots

(represented by dots). For each plot, density of *F. alnus* is represented by a gray gradient to visualize the degree of invasion, whereby pale dots have a lower density (0 individual m^{-2}) and gradually increase to darker dots for high density plots (>4 individual m^{-2}). The species, in red, have been log-transformed to minimize the effect of very abundant species and the horseshoe effect. Significant constraint variables (soil humidity (%), woody debris cover (%), plant diversity (effective number of native plant species), and native understory plant cover (%)), in blue, were standardized, centered on 0, and vectors were scaled (×1.2) for better visualization.

No species were strongly associated with high *F. alnus* density as we observed that most of the species were present in plots where *F. alnus* density was lower. Similar to our observations with understory plants, these results suggest only a small influence of *F. alnus* density and a stronger influence of other measured environmental variables.

3.6. Saprophagous Macroarthropod Abundance and Diversity

Overall, the total abundance of the saprophagous macroarthropods did not vary across the gradient of *F. alnus* invasion, but the diversity metrics did vary significantly (Table 5). As with our observations on understory plants, an increase in *F. alnus* was associated with a decrease in all diversity indices of saprophagous macroarthropod communities (Table 5). Low diversity plots were homogenized by the dominance of *T. rathkii*. Plots with a higher density of *F. alnus* showed a decrease in the effective number of detritivore species ($R^2 = 0.10$, $p < 0.05$), the Shannon–Wiener index ($R^2 = 0.15$, $p < 0.02$), and evenness ($R^2 = 0.14$, $p < 0.02$). Similarly to understory plants, we observed a decrease in the diversity indices, but the variation was weakly explained by the density of *F. alnus* [82]. Species richness also was negatively impacted ($R^2 = 0.14$, $p < 0.02$), but since the total number of species per plot varied only from two to six, the data could not be normalized to fit model assumptions and therefore should be interpreted with caution.

Table 5. Descriptive statistics and linear regression outputs of density of *F. alnus* (individual m^{-2}) as a predictive variable on five community indices (effective number of species, Shannon-Wiener index, and evenness index) for saprophagous macroarthropod communities. Mean abundance of millipedes and woodlice were calculated across three trapping periods in summer 2020 and square root transformed to fit linear model conditions.

	Descriptive Statistics		Linear Models Outputs				
Dependent Variable	**Mean ± SE**	**Median (Min–Max)**	**Estimate**	**Std Error**	**T-Value**	**R^2**	***p*-Value**
Species richness (*S*)	4.47 ± 0.13	5 (2–6)	−0.27995	0.1070	−2.616	0.1431	**<0.02 ***
Effective number of species	2.59 ± 0.11	2.49 (1.24–4.14)	−0.19505	0.09306	−2.096	0.100	**<0.05 ***
Shannon-Wiener index (*H′*)	0.91 ± 0.04	0.914 (0.26–1.42)	−0.09680	0.03677	−2.633 7	0.1446	**<0.02 ***
Evenness index (*J′*)	0.60 ± 0.02	0.612 (0.216–0.881)	−0.05082	0.02005	−2.535	0.1355	**<0.02 ***
Mean abundance [1]	20.39 ± 1.8	20.3 (4.3–52.6)	0.2541	0.1696	1.498	0.05191	0.142

[1] Square root transformation to fit the linear regression model conditions. * $p < 0.05$.

Since the increase in density of *F. alnus* was associated with a decrease in saprophagous macroarthropod diversity, other environmental factors, such as the *F. alnus* ratio in the leaf litter and the understory plant diversity (effective number of species), were tested as possible predictors affecting saprophagous macroarthropod communities. Neither a reduction in the understory native plant diversity nor an increase in the *F. alnus* ratio in the leaf litter influenced saprophagous macroarthropod diversity.

4. Discussion

Contrary to our hypothesis that an increase in the density of *F. alnus* would alter community composition and lower the diversity of both indicator taxa, we did not find a strong alteration of the community composition due to increasing *F. alnus*, but our results did show a tendency towards homogenization indicated through negative, but generally weak, effects on the multiple diversity metrics for the native plant understory (Table 3) and saprophagous macroarthropod communities (Table 5).

Invasive alien species are one of the biggest causes of biodiversity loss worldwide [1,2], but results are equivocal as to the magnitude of the impacts on native ecosystems [35,38,84–89]. Our results demonstrate that the invasion by *F. alnus*, like other IAPSs, was associated with plant community changes and the loss of native plant diversity [36–38]. However, species richness and variation in other community indices were not strongly explained by an increasing density of *F. alnus* (Table 3), suggesting that glossy buckthorn may be less competitive than other IAPSs or that its effect may be scale-dependent [90].

Exceptional competitors are not ubiquitous among IAPSs. Hedja et al. [15], who assessed the potential effects of thirteen IAPSs on native diversity, showed that only a few caused an obvious adverse effect on native plant species (i.e., *Reynoutria* spp. and *Heracleum mantegazzianum*). Indeed, *Reynoutria* spp., have allelopathic properties and can reduce native species ten-fold [91]. Invasion by another strong competitor, *Centaurea maculosa*, can also lead to strong decreases in native plant abundance and diversity [92]. However, the case of *F. alnus* is not clearcut; previous studies have suggested that an increase in the density of *F. alnus* can lead to an increase in native plant species richness [34,36] or no significant effect on herbaceous plant communities [42,93]. It is thus not surprising that we only observed a weak, albeit significant, decrease in our measured diversity indices of native understory plants as the density of *F. alnus* increased. This pattern was driven in part by a decrease in shade-intolerant peatland species such as *V. angustifolium* and *K. angustifolia*.

Our data support the observation that not all groups of organisms nor ecosystems respond similarly to invasion by IAPSs [5,26,28,89]. We hypothesized that the loss of diversity and homogenization of plant species communities should, as a corollary, lower the diversity of the primary consumers [16,24–27]. Perhaps, because of the only slight decrease in understory plant diversity, we observed only a weak decrease in saprophagous macroarthropod diversity indices (Table 5). Moreover, we also observed that all exotic saprophagous macroarthropod species either increased in abundance or were unaffected when *F. alnus* increased in density. These exotic species are known to be generalist consumers and generally resilient to disturbance caused by changing plant communities and edaphic conditions [49,94]. We also observed that native species, such as *U. canadensis* and *N. annularis* were associated with plots that had a lower *F. alnus* density and a higher understory plant diversity, implying that these species were more sensitive to invasion by *F. alnus*. Soil humidity and woody debris were associated with saprophagous macroarthropod community structure, but no mechanisms responsible for the decrease in diversity could be identified within our selected variables, as neither the *F. alnus* ratio in the sampled leaf litter nor the native understory plant diversity explained saprophagous macroarthropod diversity loss in our study system (Table S4). Schuh and Larsen [27] also found an adverse impact on ground-dwelling arthropods following *R. cathartica* invasion but could not identify the causal mechanisms. While we understand that *F. alnus* can alter the pH, nitrogen mineralization, and microbial activity of soils [39,40], leaf litter traits were not assessed in detail in our study, which makes it difficult to speculate about the effect of changes to leaf litter quality on saprophagous macroarthropod consumers. Future investigations to understand cascading effects in ecosystems invaded by *F. alnus* and other IAPSs on saprophagous arthropods could benefit from a functional trait-based approach, given that there is frequently a covariance between detritus and detritivore traits [95–97].

Our results suggest that although *F. alnus* had a general moderate effect on the diversity of plant understory and a weak effect on saprophagous macroarthropod communities after

27 years, it may be a smaller threat for native biodiversity than other IAPSs. Its negative influence was stronger for understory plants than for saprophagous macroarthropods, as plots with a density of *F. alnus* higher than 1.5 individuals per m^2 decreased in the average effective number of plant species. Because older stems tend to occur in dense even-aged thickets, it may be that more significant decreases in biodiversity will occur at a higher density threshold or later in the invasion process, as observed for other invasive species [98]. Since Mills et al. [35] only observed a weak impact of *F. alnus* after 15 years in a similar ecosystem, we suspect that the threat to biodiversity posed by *F. alnus* occurs slowly. Despite success in colonizing and expanding its range, it appears that the invasiveness of *F. alnus*, although it seemed to interact with the species evenness, does not necessarily correlate with the severity of its impacts [86,87]. Nonetheless, we recommend pursuing research on the impacts of *F. alnus* on other indicator taxa and in different contexts, as some homogenization of plant and macroarthropod communities at our invaded site was observed.

Although the environmental variables that we monitored did not vary along our *F. alnus* gradient (Table 1), other relevant abiotic and biotic environmental variables may be influenced by *F. alnus*, making it difficult to understand the mechanisms at play following invasion [85]. Having multiple study sites could give us a better general understanding of the interaction of *F. alnus* with different indicator taxa as different sites and ecosystems can respond differently when invaded by the same IAPS [28,88], accounting in part for the lack of consistency in the scientific literature when it comes to assessing IAPSs [87,99,100]. A uniformized impact metric, as proposed by Barney et al. [99], could help to facilitate comparisons between the studies.

5. Conclusions

Our study demonstrated that *F. alnus* negatively influenced understory plant diversity and caused understory plant communities to homogenize, but the effect was not as strong as other IAPS. Consequently, the trophic level effects on saprophagous macroarthropods were not observed as much as expected. Nonetheless, our study supports the need for an ongoing re-evaluation and monitoring of IAPS as their mere presence may not be synonymous with strong negative effects on native biodiversity.

Supplementary Materials: The following supporting information can be downloaded at: https://www.mdpi.com/article/10.3390/d16090584/s1, Table S1. Statistical output of redundancy analysis (RDA) of environmental variables (standardized) associated with understory plants specific community (Hellinger transformation) between all 43 plots in our study site, at L'Assomption; Table S2. Statistical outputs of redundancy analysis (RDA) of constraint variables (standardized) that may be associated with specific variation in shredder detritivores communities (logarithmic transformation) between 43 sampling plots in study site, at L'Assomption; Table S3. The most explanatory multiple linear regressions models of the variation in the effective number of species of understory plants with several explanatory variables (*F. alnus* density, Litter biomass, pH). These models are derived by comparing several models and choosing those whose AICc are the lowest based on the MuMin package in R [101]; Table S4. Linear regressions models using *Frangula alnus* ratio in leaf litter and effective number of native understory plant species as predictive variables on different diversity index of saprophagous macroarthropods; Table S5. Linear regressions of density of *F. alnus* (per m^2), *F. alnus* in leaf litter (%) and effective number of plant species as predictive variables on five diversity indices of saprophagous macroarthropods species. Abundance of detritivores have been transformed (square roots) to fit linear model conditions. All arthropods' data are based on the average of three collects within summer 2020.

Author Contributions: G.R., D.K. and I.T.H. contributed to the study conception and design. G.R. prepared material and acquired data. Data analyses were carried out by G.R. and T.J. The first draft was written by G.R., with editorial support from D.K., I.T.H. and T.J. for the subsequent versions. All authors have read and agreed to the published version of the manuscript.

Funding: Financial support was provided by a Mitacs Accelerate partnership grant, as well as NSERC Discovery grants to D.K. and I.T.H.

Institutional Review Board Statement: Not applicable for this study.

Data Availability Statement: All data and codes are available on https://doi.org/10.17632/v7vjwph992.2.

Acknowledgments: We thank the members of Kneeshaw lab (C. Gabillot, M. Meyer, G. Bergen, P. McNie) and Handa lab members (R. Sepulveda-Mina, L. J. Raymond-Léonard, J. St-Pierre, E. Gagnon Koudji, F. Jaeger), for assisting in the data sampling, identification, and meaningful exchanges as well as the FRQNT-funded Center of Forest Research for their support. We are particularly grateful J. Campeau-Poirier for his assistance in the field work and sampling. Moreover, P.-M. Brousseau kindly helped confirm the identification of arthropods and provided advice on statistical analysis and data interpretation. We are also grateful to R. Dumas, M. Leboeuf and all members at the Fiducie des écosystèmes de Lanaudière for permission and support to work at the Boisé des Terres Noires field site. We thank C. Garceau and all members of the non-profit organization CANOPÉE for their partnership in the Mitacs Acceleration project. C. Garceau was a great contributor in realization of the methodology protocol. Finally, we thank S. Pellerin and C. Lavoie for their relevant feedback on an earlier draft.

Conflicts of Interest: First author, Gabrielle Roberge, and co-authors declare no financial and competing interest to disclose.

References

1. IPBES (Intergovernmental Science-Policy Platform on Biodiversity and Ecosystem Services). *Thematic Assessment Report on Invasive Alien Species and Their Control*; IPBES Secretariat: Bonn, Germany, 2023.
2. Ruckelshaus, M.H.; Jackson, S.T.; Mooney, H.A.; Jacobs, K.L.; Kassam, K.-A.S.; Arroyo, M.T.K.; Báldi, A.; Bartuska, A.M.; Boyd, J.; Joppa, L.N.; et al. The IPBES global assessment: Pathways to action. *Trends Ecol. Evol.* **2020**, *35*, 407–414. [CrossRef] [PubMed]
3. Kumar Rai, P.; Singh, J.S. Invasive alien plant species: Their impact on environment, ecosystem services and human health. *Ecol. Indic.* **2020**, *111*, 106020. [CrossRef]
4. Lowry, E.; Rollinson, E.J.; Laybourn, A.J.; Scott, T.E.; Aiello-Lammens, M.E.; Gray, S.M.; Mickley, J.; Gurevitch, J. Biological invasions: A field synopsis, systematic review, and database of the literature. *Ecol. Evol.* **2013**, *3*, 182–196. [CrossRef]
5. Pyšek, P.; Jarošík, V.; Hulme, P.E.; Pergl, J.; Hejda, M.; Schaffner, U.; Vilà, M. A global assessment of invasive plant impacts on resident species, communities and ecosystems: The interaction of impact measures, invading species' traits and environment. *Glob. Change Biol.* **2012**, *18*, 1725–1737. [CrossRef]
6. Vilà, M.; Espinar, J.L.; Hejda, M.; Hulme, P.E.; Jarošík, V.; Maron, J.L.; Pergl, J.; Schaffner, U.; Sun, Y.; Pyšek, P. Ecological impacts of invasive alien plants: A meta-analysis of their effects on species, communities and ecosystems. *Ecol. Lett.* **2011**, *14*, 702–708. [CrossRef] [PubMed]
7. Lavoie, C. *50 Plantes Envahissantes: Protéger la Nature et L'agriculture*; Les Publications du Québec: Québec, QC, Canada, 2019; ISBN 978-2-551-26390-5.
8. Callaway, R.M.; Ridenour, W.M. Novel weapons: Invasive success and the evolution of increased competitive ability. *Front. Ecol. Environ.* **2004**, *2*, 436–443. [CrossRef]
9. Ehrenfeld, J.G. Effects of exotic plant invasions on soil nutrient cycling processes. *Ecosystems* **2003**, *6*, 503–523. [CrossRef]
10. Pisula, N.L.; Meiners, S.J. Relative allelopathic potential of invasive plant species in a young disturbed woodland. *J. Torrey Bot. Soc.* **2010**, *137*, 81–87. [CrossRef]
11. Warren, R.J.; Labatore, A.; Candeias, M. Allelopathic invasive tree (*Rhamnus cathartica*) alters native plant communities. *Plant Ecol.* **2017**, *218*, 1233–1241. [CrossRef]
12. Yates, E.D.; Levia, D.F., Jr.; Williams, C.L. Recruitment of three non-native invasive plants into a fragmented forest in southern Illinois. *For. Ecol. Manag.* **2004**, *190*, 119–130. [CrossRef]
13. Rawat, Y.S.; Singh, G.S.; Tekleyohannes, A.T. Impacts of invasive plant management on forest biodiversity and ecosystem services. *Front. For. Glob. Chang.* **2024**, *7*, 1403746. [CrossRef]
14. Burkle, L.A.; Mihaljevic, J.R.; Smith, K.G. Effects of an invasive plant transcend ecosystem boundaries through a dragonfly-mediated trophic pathway. *Oecologia* **2012**, *170*, 1045–1052. [CrossRef] [PubMed]
15. Hejda, M.; Pyšek, P.; Jarošík, V.C. Impact of invasive plants on the species richness, diversity and composition of invaded communities. *J. Ecol.* **2009**, *97*, 393–403. [CrossRef]
16. McCary, M.A.; Mores, R.; Farfan, M.A.; Wise, D.H. Invasive plants have different effects on trophic structure of green and brown food webs in terrestrial ecosystems: A meta-analysis. *Ecol. Lett.* **2016**, *19*, 328–335. [CrossRef]
17. Salamon, J.A.; Wissuwa, J.; Jagos, S.; Koblmüller, M.; Ozinger, O.; Winkler, C.; Frank, T. Plant species effects on soil macrofauna density in grassy arable fallows of different age. *Eur. J. Soil Biol.* **2011**, *47*, 129–137. [CrossRef] [PubMed]
18. Gooden, B.; French, K. Non-interactive effects of plant invasion and landscape modification on native communities. *Divers. Distrib.* **2014**, *20*, 626–639. [CrossRef]

19. Heneghan, L.; Clay, C.; Brundage, C. Rapid Decomposition of Buckthorn Litter May Change Soil Nutrient Levels. *Ecol. Restor.* **2002**, *20*, 108–111. [CrossRef]
20. McLeod, M.L.; Cleveland, C.C.; Lekberg, Y.; Maron, J.L.; Philippot, L.; Bru, D.; Callaway, R.M. Exotic invasive plants increase productivity, abundance of ammonia-oxidizing bacteria and nitrogen availability in intermountain grasslands. *J. Ecol.* **2016**, *104*, 994–1002. [CrossRef]
21. Garcia-Tejero, S.; Taboada, A. Microhabitat heterogeneity promotes soil fertility and ground-dwelling arthropod diversity in Mediterranean wood-pastures. *Agric. Ecosyst. Environ.* **2016**, *233*, 192–201. [CrossRef]
22. Gerber, E.; Krebs, C.; Murrell, C.; Moretti, M.; Rocklin, R.; Schaffner, U. Exotic invasive knotweeds (*Fallopia* spp.) negatively affect native plant and invertebrate assemblages in European riparian habitats. *Biol. Conserv.* **2008**, *141*, 646–654. [CrossRef]
23. Lalk, S.; Hartshorn, J.; Coyle, D.R. Invasive Woody Plants and Their Effects on Arthropods in the United States: Challenges and Opportunities. *Ann. Entomol. Soc. Am.* **2021**, *114*, 192–205. [CrossRef]
24. Southwood, T.R.; Brown, V.; Reader, P. The relationships of plant and insect diversities in succession. *Biol. J. Linn. Soc.* **1979**, *12*, 327–348. [CrossRef]
25. Brousseau, P.M.; Gravel, D.; Handa, I.T. Traits of litter-dwelling forest arthropod predators and detritivores covary spatially with traits of their resources. *Ecology* **2019**, *100*, e02815. [CrossRef]
26. Litt, A.R.; Cord, E.E.; Fulbright, T.E.; Schuster, G.L. Effects of invasive plants on arthropods. *Conserv. Biol.* **2014**, *28*, 1532–1549. [CrossRef]
27. Schuh, M.; Larsen, K.J. *Rhamnus cathartica* (Rosales: Rhamnaceae) invasion reduces ground-dwelling insect abundance and diversity in Northeast Iowa forests. *Plant-Insect Interact.* **2015**, *44*, 647–657. [CrossRef]
28. Brousseau, P.M.; Chauvat, M.; De Almeida, T.; Forey, E. Invasive knotweed modifies predator–prey interactions in the soil food web. *Biol. Invasions* **2021**, *23*, 1987–2002. [CrossRef]
29. Siemann, E. Experimental tests of effects of plant productivity and diversity on grassland arthropod diversity. *Ecology* **1998**, *79*, 2057–2070. [CrossRef]
30. Aiello-Lammens, M.E. Observations of extended lag phase of nonnative invasive *Frangula alnus* (Rhamnaceae) may be spatial-scale dependent. *J. Torrey Bot. Soc.* **2020**, *147*, 125–139. [CrossRef]
31. Godwin, H. *Frangula alnus* Miller. *J. Ecol.* **1943**, *31*, 77–92. [CrossRef]
32. Pellerin, S.; Mercure, M.; Desaulniers, A.S.; Lavoie, C. Changes in plant communities over three decades on two disturbed bogs in southeastern Québec. *Appl. Veg. Sci.* **2009**, *12*, 107–118. [CrossRef]
33. De Kort, H.; Mergeay, J.; Jacquemyn, H.; Honnay, O. Transatlantic invasion routes and adaptive potential in North American populations of the invasive glossy buckthorn, *Frangula alnus*. *Ann. Bot.* **2016**, *118*, 1089–1099. [CrossRef] [PubMed]
34. Houlahan, J.E.; Findlay, C.S. Effect of invasive plant species on temperate wetland plant diversity. *J. Torrey Bot. Soc.* **2004**, *18*, 1132–1138. [CrossRef]
35. Mills, J.E.; Reinartz, J.A.; Meyer, G.A.; Young, E.B. Exotic shrub invasion in an undisturbed wetland has little community-level effect over a 15-year period. *Biol. Invasions* **2009**, *11*, 1803–1820. [CrossRef]
36. Possessky, S.L.; Williams, C.E.; Moriarity, W.J. Glossy buckthorn, *Rhamnus frangula* L.: A threat to riparian plant communities of the northern Allegheny Plateau (USA). *Nat. Areas J.* **2000**, *20*, 290–292.
37. Fagan, M.; Peart, D. Impact of the invasive shrub glossy buckthorn (*Rhamnus frangula* L.) on juvenile recruitment by canopy trees. *For. Ecol. Manag.* **2004**, *194*, 95–107. [CrossRef]
38. Frappier, B.; Lee, T.D.; Olson, K.F.; Eckert, R.T. Small-scale invasion pattern, spread rate, and lag-phase behavior of *Rhamnus frangula* L. *For. Ecol. Manag.* **2003**, *186*, 1–6. [CrossRef]
39. Stokdyk, J.P.; Herrman, K.S. Short-term impacts of frangula alnus litter on forest soil properties. *Water Air Soil Pollut.* **2014**, *225*, 2000. [CrossRef]
40. Stokdyk, J.P.; Herrman, K.S. Effects of *Frangula alnus* on soil microbial communities and biogeochemical processes in Wisconsin forests. *Plant Soil* **2016**, *409*, 65–75. [CrossRef]
41. Fiedler, A.K.; Landis, D.A.; Arduser, M. Rapid Shift in Pollinator Communities Following Invasive Species Removal. *Restor. Ecol.* **2012**, *20*, 593–602. [CrossRef]
42. Hamelin, C.; Gagnon, D.; Truax, B. Exotic invasive shrub glossy buckthorn reduces restoration potential for native forest herbs. *Sustainability* **2017**, *9*, 249. [CrossRef]
43. Kaul, A.D.; Wilsey, B.J. Exotic species drive patterns of plant species diversity in 93 restored tallgrass prairies. *Ecol. Appl.* **2021**, *31*, e2252. [CrossRef]
44. Wenk, E.S.; Callaham, M.A.; O'Brien, J.J.; Hanson, P.J. Soil macroinvertebrate communities across a productivity gradient in deciduous forests of Eastern North America. *Northeast. Nat.* **2016**, *23*, 25–44. [CrossRef]
45. Paoletti, M.G.; Hassall, M. Woodlice (Isopoda: Oniscidea): Their potential for assessing sustainability and use as bioindicators. *Agric. Ecosyst. Environ.* **1999**, *74*, 157–165. [CrossRef]
46. David, J.F.; Handa, I.T. The ecology of saprophagous macroarthropods (millipedes, woodlice) in the context of global change. *Biol. Rev.-Camb.* **2010**, *85*, 881–895. [CrossRef] [PubMed]
47. Hopkin, S.P.; Read, H.J. *Biology of Millipedes*; Oxford University Press: Oxford, UK, 1992; 233p.
48. Langor, D.W.; deWaard, J.R.; Snyder, B.A. Myriapoda of Canada. *ZooKeys* **2019**, *819*, 169–186. [CrossRef]
49. Shelley, R. The millipeds of eastern Canada (Arthropoda: Diplopoda). *Can. J. Zool.* **1988**, *66*, 1638–1663. [CrossRef]

50. Handa, I.T.; Aerts, R.; Berendse, F.; Berg, M.P.; Bruder, A.; Butenschoen, O.; Chauvet, E.; Gessner, M.O.; Jabiol, J.; Makkonen, M.; et al. Consequences of biodiversity loss for litter decomposition across biomes. *Nature* **2014**, *509*, 218–221. [CrossRef]
51. Coyle, D.R.; Nagendra, U.J.; Taylor, M.K.; Campbell, J.H.; Cunard, C.E.; Joslin, A.H.; Mundepi, A.; Phillips, C.A.; Callaham, M.A., Jr. Soil fauna responses to natural disturbances, invasive species, and global climate change: Current state of the science and a call to action. *Soil Biol. Biochem.* **2017**, *110*, 116–133. [CrossRef]
52. UPA (Union des Producteurs Agricoles). *Consultation Publique du Projet Oléoduc Énergie est de Transcanada—Section Québécoise*; Lanaudière: Joliette, QC, Canada, 2016.
53. Gabillot, C.; Meyer, J.; Handa, I.T.; Kneeshaw, D. *Le nerprun au Boisé des Terres Noires*; Université du Québec à Montréal: Montréal, QC, Canada, 2019.
54. CPTAQ (Commission de Protection du Territoire Agricole du Québec). *Décision*; CPTAQ, Ed.; L'Assomption: Montréal, QC, Canada, 1996.
55. Pansu, M.; Gautheyrou, J. *Handbook of Soil Analysis: Mineralogical, Organic and Inorganic Methods*; Springer: Berlin/Heidelberg, Germany; New York, NY, USA, 2006.
56. Brouillet, L.; Marie-Victorin, F.; Rouleau, E. *Flore Laurentienne*, 3rd ed.; Morin, G., Ed.; Chenelière Éducation: Montréal, QC, Canada, 2002; ISBN 9782891058179.
57. International Plant Names Index. The Royal Botanic Gardens, Kew, Harvard University Herbaria & Libraries and Australian National Herbarium. Available online: https://www.ipni.org/ (accessed on 13 August 2024).
58. Schirmel, J.; Lenze, S.; Katzmann, D.; Buchholz, S. Capture efficiency of pitfall traps is highly affected by sampling interval. *Entomol. Exp. Appl.* **2010**, *136*, 206–210. [CrossRef]
59. Brown, G.R.; Matthews, I.M. A review of extensive variation in the design of pitfall traps and a proposal for a standard pitfall trap design for monitoring ground-active arthropod biodiversity. *Ecol. Evol.* **2016**, *6*, 3953–3964. [CrossRef]
60. Duelli, P.; Obrist, M.K.; Schmatz, D.R. Biodiversity evaluation in agricultural landscapes: Above-ground insects. *Agric. Ecosyst. Environ.* **1999**, *74*, 33–64. [CrossRef]
61. Topping, C.J.; Sunderland, K.D. Limitations to the use of pitfall traps in ecological studies exemplified by a study of spiders in a field of winter wheat. *J. Appl. Ecol.* **1992**, *29*, 485–491. [CrossRef]
62. Sabu, T.K.; Shiju, R.T.; Vinod, K.V.; Nithya, S. A comparison of the pitfall trap, winkler extractor and berlese Funnel for sampling ground-dwelling arthropods in tropical montane cloud forests. *J. Insect Sci.* **2011**, *11*, 28. [CrossRef] [PubMed]
63. Borror, D.; Triplehorn, C.; Johnson, N. *An Introduction to the Study of Insects*; Saunders College Publishing: Brooks Cole, PA, USA, 1992; ISBN 0030253977.
64. Brousseau, P.-M. *Clé D'identification des Mille-Pattes des Milieux Forestiers du Mont St-Bruno. n.p*; Université du Québec à Montréal: Montréal, QC, Canada.
65. Shelley, R. The millipeds of central Canada (Arthropoda: Diplopoda), with reviews of the Canadian fauna and diplopod faunistic studies. *Can. J. Zool.* **2002**, *80*, 1863–1875. [CrossRef]
66. Shultz, J.W. A guide to the identification of the terrestrial Isopoda of Maryland, USA (Crustacea). *ZooKeys* **2018**, *801*, 207–228. [CrossRef]
67. Jass, J.; Klausmeier, B. Endemics and immigrants: North American terrestrial isopods (Isopoda, Oniscidea) north of Mexico. *Crustaceana* **2000**, *73*, 771–799. [CrossRef]
68. Noël, F.; Séchet, E. Crustacés Isopodes terrestres du Nord-Ouest de la France (Crustacea, Isopoda, Oniscidea). *Invertébr. Armor.* **2007**, *2*, 1–48.
69. Shakir, M.M.; Ahmed, S. Seasonal abundance of soil arthropods in relation to meteorological and edaphic factors in the agroecosystems of Faisalabad, Punjab, Pakistan. *Int. J. Biometeorol.* **2015**, *59*, 605–616. [CrossRef]
70. EnvironnementCanada, Historical Data. Gouvernement du Canada. 2021. Available online: https://climate.weather.gc.ca/ (accessed on 1 October 2021).
71. WeatherSpark. Daily Precipitation in the Summer of 2020 at L'Assomption. 2021. Available online: https://weatherspark.com/h/s/147212/2020/1/Historical-Weather-Summer-2020-at-L%E2%80%99Assomption-Quebec-Canada#Figures-PrecipitationProbability (accessed on 1 October 2021).
72. Anderson, M.J.; Ellingsen, K.E.; McArdle, B.H. Multivariate dispersion as a measure of beta diversity. *Ecol. Lett.* **2006**, *9*, 683–693. [CrossRef]
73. Legendre, P.; Gallagher, E.D. Ecologically meaningful transformations for ordination of species data. *Oecologia* **2001**, *129*, 271–280. [CrossRef]
74. Zeleny, D. Analysis of Community Ecology Data in R. 2021. Available online: https://www.davidzeleny.net/anadat-r/doku.php/en:rda_cca (accessed on 14 June 2022).
75. Jost, L. Entropy and diversity. *Oikos* **2006**, *113*, 363–375. [CrossRef]
76. R Core Team. *R: A Language and Environment for Statistical Computing*; R Foundation for Statistical Computing: Vienna, Austria, 2021.
77. Oksanen, J.; Kindt, R.; Legendre, P.; O'Hara, B.; Stevens, M.H.H.; Oksanen, M.J.; Suggests, M.A.S.S. The vegan package. *Community Ecol. Package* **2007**, *10*, 631–637.
78. Carter, G.A.; Teramura, A.H. Vine photosynthesis and relationships to climbing mechanics in a forest understory. *Am. J. Bot.* **1988**, *75*, 1011–1018. [CrossRef]

79. Humbert, L. *Tolérance Et Largeur De Niche Des Espèces Herbacées Des Forêts Québécoises Vis-À-Vis Du Ph Et De La Lumière*; Université du Québec à Montréal: Montréal, QC, USA, 2006.
80. Petrulaitis, L.; Gudžinskas, Z. The first records of two alien woody species, *Cornus alternifolia* and *Cornus amomum*, in Lithuania. *Bioinvasions Rec.* **2020**, *9*, 384–392. [CrossRef]
81. Stanton, K.M.; Weeks, S.S.; Dana, M.N.; Mickelbart, M.V. Light exposure and shade effects on growth, flowering, and leaf morphology of *Spiraea alba* Du Roi and *Spiraea tomentosa* L. *HortScience* **2010**, *45*, 1912–1916. [CrossRef]
82. Cohen, J. *Statistical Power Analysis for the Behavioral Sciences*; Lawrence Erlbaum Associates: Routledge, NY, USA, 2013; ISBN 0-8058-0283-5.
83. Jędryczkowski, W.B. Millipedes (diplopoda) of Warsaw and Mazovia. *Memorab. Zool.* **1982**, *36*, 253–261.
84. Davis, M.A.; Thompson, K. Eight ways to be a colonizer; two ways to be an invader: A proposed nomenclature scheme for invasion ecology. *Bull. Ecol. Soc. Am.* **2000**, *81*, 226–230.
85. Diekmann, M.; Effertz, H.; Baranowski, M.; Dupré, C. Weak effects on plant diversity of two invasive Impatiens species. *Plant Ecol.* **2016**, *217*, 1503–1514. [CrossRef]
86. Lavoie, C. *40 Autres Plantes Envahissantes*; Les Publication du Québec: Québec, QC, Canada, 2022; ISBN 978-2-551-26773-6.
87. Ricciardi, A.; Cohen, J. The invasiveness of an introduced species does not predict its impact. *Biol. Invasions* **2007**, *9*, 309–315. [CrossRef]
88. Qureshi, H.; Anwar, T.; Habib, N.; Ali, Q.; Haider, M.Z.; Yasmin, S.; Munazir, M.; Basit, Z.; Waseem, M. Multiple comparisons of diversity indices invaded by Lantana camara. *Braz. J. Biol.* **2021**, *81*, 83–91. [CrossRef]
89. Robertson, M.P.; Harris, K.R.; Coetzee, J.A.; Foxcroft, L.C.; Dippenaar-Schoeman, A.S.; Van Rensburg, B.J. Assessing local scale impacts of *Opuntia stricta* (Cactaceae) invasion on beetle and spider diversity in Kruger National Park, South Africa. *Afr. Zool.* **2011**, *46*, 205–223. [CrossRef]
90. McMillan, N.A.; Fuhlendorf, S.D.; Davis, C.A.; Hamilton, R.G.; Neumann, L.K.; Cady, S.M. A plea for scale, and why it matters for invasive species management, biodiversity and conservation. *J. Appl. Ecol.* **2023**, *60*, 1468–1480. [CrossRef]
91. Aguilera, A.G.; Alpert, P.; Dukes, J.S.; Harrington, R. Impacts of the invasive plant *Fallopia japonica* (Houtt.) on plant communities and ecosystem processes. *Biol. Invasions* **2010**, *12*, 1243–1252. [CrossRef]
92. Kedzie-Webb, S.A.; Sheley, R.L.; Borkowski, J.J.; Jacobs, J.S. Relationships between Centaurea maculosa and indigenous plant assemblages. *West. N. Am. Nat.* **2001**, *61*, 43–49.
93. Frappier, B.; Eckert, R.T.; Lee, T.D. Experimental removal of the non-indigenous shrub *Rhamnus frangula* (glossy buckthorn): Effects on native herbs and woody seedlings. *Northeast. Nat.* **2004**, *11*, 313–322. [CrossRef]
94. Richmond, C.E.; Breitburg, D.L.; Rose, K.A. The role of environmental generalist species in ecosystem function. *Ecol. Model.* **2005**, *188*, 279–295. [CrossRef]
95. Brousseau, P.M.; Gravel, D.; Handa, I.T. On the development of a predictive functional trait approach for studying terrestrial arthropods. *J. Anim. Ecol.* **2018**, *87*, 1209–1220. [CrossRef]
96. Raymond-Léonard, L.J.; Gravel, D.; Handa, I.T. A novel set of traits to describe Collembola mouthparts: Taking a bite out of the broad chewing mandible classification. *Soil Biol. Biochem.* **2019**, *138*, 107608. [CrossRef]
97. Marchand, T.; Lecerf, A.; Brousseau, P.-M.; Chauvat, M.; Danger, M.; Forey, E.; Handa, I.T.; Hedde, M.; Maunoury-Danger, F.; Santonja, M.; et al. The Detri2match conceptual framework: Matching detritivore and detritus traits to unravel consumption rules in a context of decomposition. *Funct. Ecol.* 2024, *in press*.
98. Panetta, F.D.; Gooden, B. Managing for biodiversity: Impact and action thresholds for invasive plants in natural ecosystems. *NeoBiota* **2017**, *34*, 53–66. [CrossRef]
99. Barney, J.N.; Tekiela, D.R.; Dollete, E.S.; Tomasek, B.J. What is the "real" impact of invasive plant species? *Front. Ecol. Environ.* **2013**, *11*, 322–329. [CrossRef]
100. Matthews, J.; van der Velde, G.; Collas, F.P.; de Hoop, L.; Koopman, K.R.; Hendriks, A.J.; Leuven, R.S. Inconsistencies in the risk classification of alien species and implications for risk assessment In the European Union. *Ecosphere* **2017**, *8*, e01832. [CrossRef]
101. Barton, K. MuMIn: Multi-Model Inference. R Package Version 1.43.17. CRAN. R Package Website. 2020. Available online: https://CRAN.R-project.org/package=MuMIn (accessed on 12 January 2022).

forests

Article

Effects of Herbaceous Plant Encroachment on the Soil Carbon Pool in the Shrub Tundra of the Changbai Mountains

Xiaoyun Xu [1], Yinghua Jin [1], Jiawei Xu [1,*], Yingjie Zhang [2] and Jiaxing Yang [1]

[1] Key Laboratory of Geographical Processes and Ecological Security in Changbai Mountains, Ministry of Education, School of Geographical Sciences, Northeast Normal University, No. 5268, Renmin Street, Changchun 130024, China; xuxy930@nenu.edu.cn (X.X.); jinyh796@nenu.edu.cn (Y.J.); yangjx007@nenu.edu.cn (J.Y.)

[2] School of Teacher Education, Hebei Normal University, Shijiazhuang 050024, China; zhangyingjie@hebtu.edu.cn

* Correspondence: xujw634@nenu.edu.cn; Tel.: +86-431-85099550

Abstract: Under global warming, vegetation composition changes induced by plant encroachment have a significant impact on the carbon balance of tundra ecosystems. The encroachment of herbaceous plants into indigenous shrub communities has changed the aboveground and belowground litter carbon input and the characteristics in the shrub tundra of the Changbai Mountains. However, the impact of variations in litter characteristics and litter carbon input on the dynamics of soil organic carbon (SOC) pool concentrations and SOC stability remains ambiguous. In this study, aboveground and belowground litter and soil samples were collected for lab experiments. Our results showed that the increase in aboveground litter and belowground litter due to *Deyeuxia purpurea* encroachment increased the SOC concentration. Simultaneously, *D. purpurea* encroachment decreased the soil C/N by decreasing the components of both aboveground and belowground litter that were resistant to decomposition (C/N and lignin/N) and increased the soil mineralization ability and available N concentrations, increased the CO_2 release rate, and ultimately decreased the SOC concentration. *D. purpurea* encroachment enhanced soil decomposition capacity by increasing the concentration of organic carbon molecular structures, such as carbohydrates, in the aboveground and belowground litter, thereby increasing the concentration of decomposable organic carbon molecular structures and active organic carbon in the soil, while simultaneously reducing the concentration of recalcitrant organic carbon. Even more, *D. purpurea* encroachment reduced the recalcitrant components of the aboveground and belowground litter enhanced soil mineralization capability and increased soil nitrogen concentration, which collectively increased the carbon oxidation state (COX) and decreased SOC stability. In general, global warming has led to herbaceous plant encroachment, which changes the aboveground and belowground litter carbon inputs and properties in the tundra, in turn reducing the SOC concentration and soil carbon pool stability, enhancing soil carbon emission capacity, and increasing atmospheric CO_2 concentration, forming a vicious cycle.

Keywords: soil organic carbon; *Deyeuxia purpurea*; soil active organic carbon; molecular structure of soil organic carbon; litter decomposition

1. Introduction

The soil organic carbon (SOC) pool, containing an estimated 1462–1548 Pg of carbon, surpasses the combined carbon content of both the vegetation and atmospheric pools within terrestrial ecosystems [1]. In alpine environments, the decomposition of organic matter is notably slower due to lower temperatures, resulting in a higher concentration of SOC compared to other ecological zones [2]. Notably, the northern coniferous forests and the Arctic tundra serve as significant carbon sinks, collectively holding nearly one-third of the total terrestrial carbon [3]. Climate change significantly influences both the concentration and stability of SOC, as well as the composition of vegetation. A critical area of investigation within the global carbon cycle pertains to how changes in vegetation, driven by climate warming, affect the SOC pool [4–6]. Consequently, international research on carbon dynamics has increasingly concentrated on understanding the variability and primary influences on SOC concentration and composition in cold biomes. This focus is essential for elucidating the impacts of global climate change, thereby providing foundational knowledge for predicting future carbon cycle dynamics.

Plant encroachment can lead to alterations in local vegetation types and diversity, which in turn affect the nature and volume of litter produced. This shift in litter characteristics directly influences soil carbon storage and its dynamics [7]. Litter not only represents the principal input for soil organic carbon (SOC) but also provides essential material and energy for decomposers. Consequently, any modification in litter quality and quantity impacts soil microbial communities and enzyme activities, thereby modulating the decomposition rates of both litter and SOC, which ultimately govern the flux of carbon into and out of the soil system [8]. There exists a dual perspective on the effects of increased litter input on SOC dynamics. On one hand, it is proposed that an increase in litter can elevate SOC concentrations and potentially slow the decomposition of pre-existing SOC, suggesting a stabilization effect [9,10]. Conversely, other studies indicate that augmented litter inputs might initially enhance SOC mineralization, thus reducing SOC concentrations and augmenting carbon emissions [11]. Furthermore, the quality of litter plays a pivotal role in SOC formation. Cotrufo [12] argues that litter rich in easily decomposable compounds can be rapidly assimilated by soil microbes, leading to the production of microbial by-products and metabolites that contribute to stable SOC through aggregation and mineral association. However, high-quality litter can also induce a positive priming effect, accelerating the decomposition of native soil carbon and potentially decreasing overall SOC concentrations due to increased CO_2 emissions [13]. This dichotomy underscores the complex interplay between litter quality, microbial activity, and SOC stability, highlighting the need for further research to fully understand these dynamics.

Research concerning soil organic carbon (SOC) stability suggests that an increase in plant diversity leads to enhanced litter species richness, thereby providing more substrates for soil hydrolases and promoting extensive microbial proliferation. This dynamic results in elevated levels of SOC, dissolved organic carbon (DOC), easily oxidizable organic carbon (EOC), and microbial biomass carbon (MBC) while concurrently diminishing the stability of the soil carbon pool [14–16]. However, Zhang [17] observed that the increase in aboveground and belowground litter species diversity would significantly increase SOC stability. The chemical composition of litter also plays a crucial role in affecting SOC stability by modifying microbial biomass and substrate utilization [12]. Cotrufo [12] posits that the presence of high-quality litter components, such as proteins, amino acids, and monosaccharides, which are easily decomposed by microbes, can lead to the formation of stable mineral-bound SOC, thereby enhancing SOC stability. Conversely, other research suggests that litter rich in high-quality components might stimulate microbial activity, accelerating SOC mineralization and potentially reducing SOC stability [18–21]. This indicates a complex relationship between plant encroachment, litter characteristics, and SOC dynamics that warrants further investigation.

In the tundra zone of the Changbai Mountains, the predominant vegetation type is shrub tundra. Over the past three decades, with rising temperatures and an extended growing season, significant vegetative changes have been observed. Notably, the encroachment of *Deyeuxia purpurea*, a low-altitude herbaceous species, has led to a decline in native tundra shrubs like *Rhododendron aureum* and *Vaccinium uliginosum* [22]. This encroachment has resulted in the emergence of five distinct vegetation groups, including areas dominated by *Rhododendron aureum* (*Rho*), with varying degrees of *D. purpurea* encroachment classified as mild (*Mil*), moderate (*Mod*), severe (*Sev*), and complete dominance by *D. purpurea* (*Dey*) (Figure 1b). The encroachment of *D. purpurea* is anticipated to increase both the quantity and diversity of litter, alongside accelerating the decomposition rate of aboveground litter [23]. Despite these changes, the precise impact of *D. purpurea*'s encroachment on the SOC concentration and characteristics within the shrub tundra of Changbai Mountains remains unclear, necessitating further research to clarify these ecological transformations.

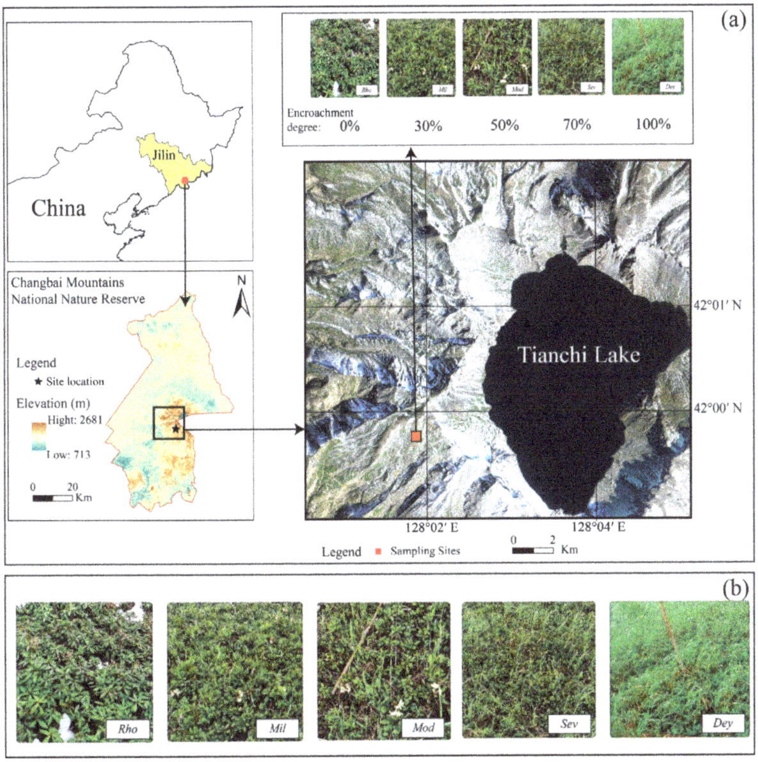

Figure 1. Overview of the study area (**a**) and five types of tundra vegetation created by *Deyeuxia purpurea* encroachment (**b**).

In this study, field samples from the Changbai Mountains tundra were evaluated to elucidate dynamic changes in the SOC concentration and components following herbaceous plant encroachment. Concentrations of various components (DOC, EOC, MBC, POC [particle organic carbon], MAOC [mineral-associated organic carbon]), ^{13}C NMR spectra, aboveground and underground litter quality, total carbon input, lignin, cellulose, and litter components were compared under different degrees of encroachment. Soil physical and chemical properties (pH, total carbon (TC) concentration, total nitrogen (TN) concentration, total phosphorus (TP) concentration, nitrate nitrogen (NO_3^-) concentration, and ammonium nitrogen (NH_4^+) concentration) were also analyzed. The aims of this study were as follows: (a) to measure changes in litter properties and contents caused by the encroachment of *D. purpurea*, (b) to clarify how the encroachment of *D. purpurea* affects soil C emissions, and (c) to clarify how the properties of the SOC pool respond to the encroachment of *D. purpurea*.

2. Materials and Methods

2.1. Study Area

The tundra zone is situated at the highest area of the Changbai Mountains volcanic cone. (Soil in the tundra zone can freeze and become snow-covered in winter.) The climate is characterized by extremely cold winters, cool and brief summers, low temperatures throughout the year, and substantial precipitation. However, the seasonal distribution is unbalanced, with the highest temperatures in the summer, high humidity levels throughout the year, and more than 260 days of wind and foggy conditions annually. The average temperature during the growing season (June to September) is 6.37 °C, while the average precipitation is 153.20 mm (Figure A1) [24]. The predominant soil type in the Changbai Mountains alpine tundra zone is Entisols, which is influenced by volcanic eruption parent material comprising volcanic ejecta. Through intense physical weathering and water erosion, three subclasses are formed: primitive (stony) tundra soil, tundra soil, and scrub tundra soil. The extent of soil development is minimal, with a thickness of typically 10–20 cm, attributable to the brief soil formation period characteristic of tundra soil [24]. Simultaneously, the significant variation in water and thermal conditions across different altitudes in the Changbai Mountains results in a characteristic vertical zonation of vegetation, which includes, from highest to lowest elevation, the polar frigid alpine tundra belt, the subfrigid birch forest belt, the subfrigid coniferous forest belt, and the frigid temperate red pine broadleaf mixed forest belt [25].

The sample plot (43°59′35″ N, 128°1′18″ E, elevation 2200 m) was on the tundra within the area of *D. purpurea* encroachment on the western slope of the Changbai Mountains (Figure 1a). The area was 20 m × 20 m with a 5–8° slope and mostly level ground. The plot contained five representative plant communities: *Mil* (*D. purpurea* coverage was 30%), *Mod* (*D. purpurea* coverage was 50%), *Sev* (*D. purpurea* coverage was 70%), *Dey* (all *D. purpurea*), and *Rho* (no *D. purpurea* encroachment, all *Rhododendron aureum*) (Figure 1b). In every plant community, three 1 m × 1 m quadrats were erected (for a total of 15).

2.2. Sample Collection

The aboveground litter collection methods varied for *D. purpurea*, a perennial bud herb, and *R. aureum*, a perennial evergreen shrub. Nylon nets were used to cover the entire *R. aureum* plant in September 2021, and the litter within the nets, representing the *R. aureum* aboveground litter throughout the year, was collected in September 2022. At the end of the growing season, the aboveground portion of *D. purpurea* was collected in close proximity to the ground (roughly representing the aboveground litter). Using the root drilling method, fine roots (diameter less than 2 mm) were examined at the end of the growing season. The fine roots of *D. purpurea* and *R. aureum* from the current year were selected after the samples were sieved out of the sand and washed with water. Every year, there were approximately equal amounts of new and dead roots. Drying and weighing the aboveground and underground litter of *D. purpurea* and *R. aureum* enabled the quantification of the overall volume of the litter and determination of its biochemical characteristics.

In September 2022, a five-point sampling method was used to collect soil from each quadrat. Samples were brought to the laboratory and passed through a 2 mm sieve to remove visible roots and gravel. Part of the soil was stored in a refrigerator at 4 °C for the measurement of soil moisture content and MBC and DOC concentration, while the rest was air-dried for the measurement of other physical and chemical properties.

2.3. Litter Chemical Properties

Aboveground and belowground litter TN and TC concentrations were analyzed via the dry combustion method using an elemental analyzer (EA3000; Euro Vector, Milan, Italy). Aboveground and belowground litter lignin and cellulose concentrations were analyzed via the acid-washing method using a lignin analyzer (FIWE, VELP, Milan, Italy).

2.4. Soil Physico-Chemical Properties

Plant growth and soil microbial decomposition in the Changbai Mountains tundra zone are primarily concentrated between June and September [23]. Therefore, soil CO_2 respiration rate was measured via the EGM-5 portable soil carbon flux measurement system (PP Systems, Haverhill, MA, USA) from June to September 2022. Soil moisture content was measured by oven-drying the soil samples to a constant weight at 105 °C. Soil pH was measured in 1/5 soil and distilled water suspensions using a pH meter (Time Power, Beijing, China). Soil TN and TC concentrations were analyzed via the dry combustion method using an elemental analyzer (EA3000; Euro Vector, Milan, Italy). SOC concentration was consistent with the TC concentration because the pre-experimental data indicated that the tundra on Changbai Mountains soil had a low inorganic carbon concentration. Soil TP concentration was analyzed via the molybdenum-antimony anti-colorimetric method using a spectrophotometer. Before colorimetry, the soil samples were subjected to a reaction with 5 mL of nitric acid and 0.8 mL of perchloric acid solution in a microwave digestion apparatus (CEM MARS6, CEM, Charlotte, NC, USA) to yield the TP digestion solution. Soil NO_3^- and NH_4^+ concentrations were analyzed via the potassium chloride (30 mL, 2 mol/L) extraction method using a continuous flow analyzer (SAN++; Skalar, Breda, Holland). DOC concentration was measured via the ultrapure water extraction method using a total organic carbon analyzer (1030S; OI, Houston, TX, USA). EOC concentration was measured via potassium permanganate oxidation using a full-wavelength microplate reader (SynergyH4, Bio Tek, South Burlington, VT, USA). MBC concentration was measured via chloroform fumigation-potassium sulfate extraction using a total organic carbon analyzer (1030S; OI, USA). MAOC and POC concentrations were determined via the sodium hexametaphosphate—wet sieve method using an elemental analyzer (EA3000; Euro Vector, Italy).

2.5. Quantification of Litter- and Soil-Derived Organic Carbon Molecular Structure

Measurements were made on aboveground and belowground litter and soil organic carbon molecular structure using a superconducting nuclear magnetic resonance spectrometer (AVANCE NEO 400 WB, Bruker, Bill, Switzerland), including seven functional groups [26] (Table A1). Calculations were then performed using the method described by Nelson [27] and Baldock [26] to yield six components: carbohydrate component, protein component, lignin component, lipid component, carbonyl component, and char component. Among them, carbohydrate and protein were considered easily decomposed organic carbon molecular structures. Lignin, lipid, carbonyl, and char were considered recalcitrant decomposed organic carbon molecular structures [26,27].

The oxidation state of carbon (COX) could be used to indicate the stability of aboveground litter, belowground litter, and soil samples. COX was inversely proportional to sample stability. COX was calculated from the normalized molecular formula of CXHYOZNW as described by Masiello [28].

2.6. Statistical Analysis

Before statistical analysis, the data were assessed for normality and homogeneity of variance, then log-transformed when necessary. Since the compositional data were calculated using a combination of model formulas and the actual conditions of the study area, some values of 0 remain untreated by log-ratio transformed. Aboveground and belowground litter carbon input, aboveground and belowground litter properties, composition of the soil carbon pool, and characteristics under various degrees of encroachment were analyzed using single-factor analysis (one-way ANOVA) and LSD testing. SPSS 2019 was used for ANOVA, and Origin 2019 was used to generate histograms. Correlation analyses were utilized to describe the relationship between aboveground and belowground litter properties, soil properties and SOC concentration, and COX by Origin 2019. Structural equation modeling (SEM) was conducted using the SPSS AU to analyze the direct and indirect consequences of total aboveground and belowground litter carbon input, aboveground and belowground litter properties, soil properties, soil organic carbon pools, and molecular structure on SOC concentration and COX. The SEM fit was evaluated using p-values > 0.05, chi-square values/degree values (χ^2/df), the goodness-of-fit index (GFI), and the root mean square error of approximation (RMSEA).

3. Results

3.1. Effects of Herbaceous Plant Encroachment on Soil Carbon Input

The total litter carbon input of aboveground and belowground litter increased significantly as a result of *D. purpurea* encroachment in the *R. aureum* community (Figure 2a). The total carbon input of the aboveground litter was significantly lower (p = 0.02) than that of the belowground litter (Figure 2a). D. purpurea encroachment caused significant reductions in C/N and lignin/N (p < 0.001). On average, C/N and lignin/N were higher in belowground litter than in aboveground litter (Figure 2b,c).

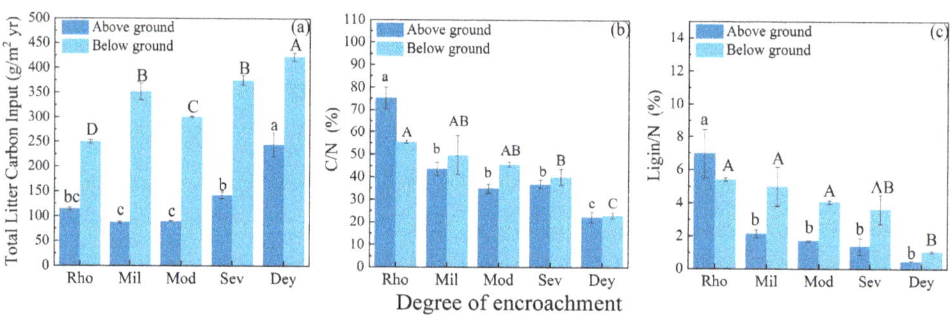

Figure 2. Histogram of the total carbon input, C/N and lignin/N of belowground and aboveground litter for various degrees of *Deyeuxia purpurea* encroachment. (**a**) Shows the total carbon input, (**b**) shows C/N, (**c**) shows lignin/N. Different lowercase letters indicate significant differences in the total carbon input, C/N, and lignin/N of aboveground litter (p < 0.05), while distinct uppercase letters indicated differences in the total carbon input, C/N, and lignin/N of belowground litter (p < 0.05). *Mil* (*D. purpurea* coverage was 30%), *Mod* (*D. purpurea* coverage was 50%), *Sev* (*D. purpurea* coverage was 70%), *Dey* (all *D. purpurea*), *Rho* (no *D. purpurea* encroachment, all *Rhododendron aureum*).

In the aboveground and underground litter, *D. purpurea* encroachment resulted in increases in the protein and carbohydrate components and decreases in lignin, lipid, and char compounds. However, carbonyl components were not significantly affected by encroachment (Figure A2a,b). On average, the carbohydrate, protein, lipid, and char concentrations of belowground litter were higher than those of aboveground litter, and the lignin concentration of belowground litter was lower than that of aboveground litter (Figure A2a,b).

The COX in aboveground litter and belowground litter gradually increased with *D. purpurea* encroachment, indicating a decrease in aboveground and belowground litter stability (Figure A2c). At the same time, the COX of belowground litter was lower than that of aboveground litter (Figure A2c).

3.2. Effects of Herbaceous Encroachment on Soil Carbon Emissions

During the sampling period, the soil respiration rate in each of the different degrees of encroachment first increased and then decreased. Simultaneously, the rate of soil respiration increased as the degree of *D. purpurea* encroachment increased throughout the sampling period (Figure 3). As the degree of *D. purpurea* encroachment increased, the average soil respiration rate also increased over the course of the growing season. This might be due to the fact that August temperature and precipitation were higher than September, which was suitable for microbial activity and would accelerate soil respiration. In addition, soil respiration was accelerated due to the fact that more organic matter was imported in August as a result of strong plant growth.

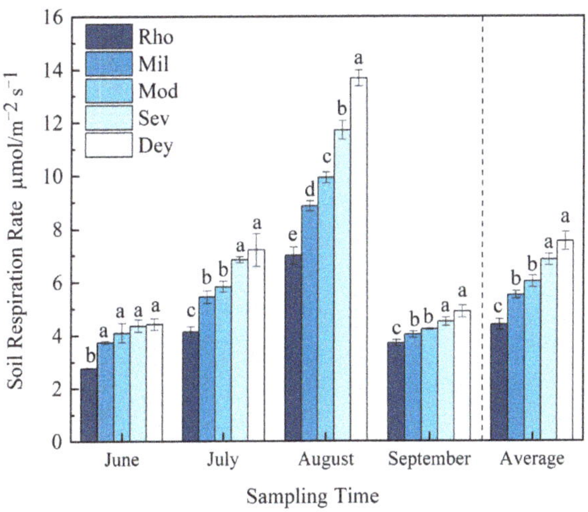

Figure 3. Soil respiration rates with respect to the extent of *D. purpurea* encroachment in various months. Significant differences between sites at the same sampling time are indicated by distinct lowercase letters ($p < 0.05$). The X-axis represents soil respiration rate from June to September, and "Average" represents the average soil respiration rate from June to September.

3.3. Effects of Herbaceous Plant Encroachment on the Soil Organic Carbon Pool

The SOC concentration decreased significantly in response to *D. purpurea* encroachment (Figure 4a). There were no significant differences between estimates for the other two communities, except the SOC concentration in the *Dey* community was much lower than those in the other communities ($p = 0.03$). At the same time, COX increased significantly with different degrees of *D. purpurea* encroachment increase, indicating a decrease in soil stability (Figure 4b). Among them, the COX values in *Dey* and *Sev* were significantly different from each other and significantly higher than in the other communities ($p < 0.001$).

The active organic carbon concentrations and molecular structure in SOC were altered significantly by *D. purpurea* encroachment ($p = 0.01$, Figure 5). On average, the DOC concentration and MBC concentration exhibited an increase trend as the degree of *D. purpurea* encroachment increased (Figure 5a,c). The DOC and MBC concentrations in the *Dey* community were significantly higher than those of the other communities. The EOC concentration and MAOC concentration (the soil grain-size distribution in Figure A4) exhibited a declining trend as the degree of *D. purpurea* encroachment increased (Figure 5b,e). The EOC and MAOC concentrations in the *Dey* community were significantly lower than the *Rho* community, and the POC concentration exhibited an increasing and then decreasing trend as the degree of *D. purpurea* encroachment increased (Figure 5d). At the same time, the POC concentration in the *Dey* community was significantly lower than in the *Rho* community.

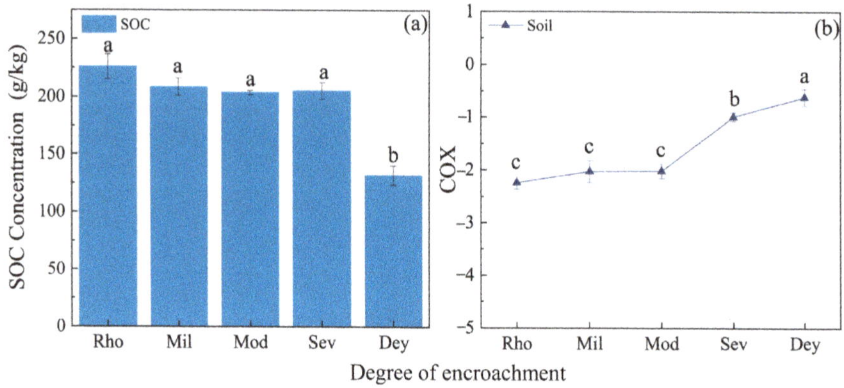

Figure 4. Effects of different degrees of *Deyeuxia purpurea* encroachment on SOC concentration and COX. (**a**) SOC concentration, (**b**) COX. Significant differences between sites are indicated by different lowercase letters ($p < 0.05$). *Mil* (*D. purpurea* coverage was 30%), *Mod* (*D. purpurea* coverage was 50%), *Sev* (*D. purpurea* coverage was 70%), *Dey* (all *D. purpurea*), *Rho* (no *D. purpurea* encroachment, all *Rhododendron aureum*).

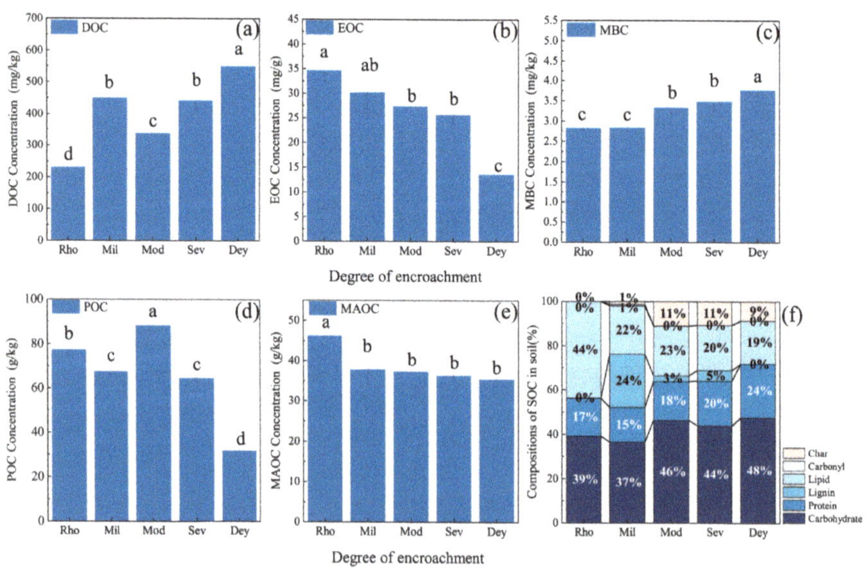

Figure 5. Changes in active organic carbons and molecular structure in the soil organic carbon according to the extent of *Deyeuxia purpurea* encroachment. Different letters denote significant differences between sites ($p < 0.05$). (**a**) DOC concentration, (**b**) EOC concentration, (**c**) MBC concentration, (**d**) POC concentration, (**e**) MAOC concentration, (**f**) Molecular structure.

Overall, the *D. purpurea* encroachment increased the concentrations of carbohydrates, proteins, and char components in the SOC and decreased the lipid component concentration. No significant differences were observed in the lignin and the carbonyl components concentrations with respect to encroachment ($p < 0.001$, Figure 5f).

3.4. Main Influencing Factors of Soil Organic Carbon Concentration and COX

Based on Pearson correlation analysis (Figure A3), the characteristics of soil, aboveground litter, and belowground litter were significantly related to SOC concentration and COX.

Structural equation modeling (SEM) was used to determine the key factors affecting the SOC concentration and COX (Figure 6). The primary factors influencing the SOC concentration were the total aboveground and belowground litter carbon input, aboveground and belowground litter properties (C/N and lignin/N), soil properties (C/N, NO_3^-, and NH_4^+), and soil carbon emissions (CO_2, Figure 6a). Among these, an increase in total aboveground and belowground litter carbon input directly promoted an increase in the SOC concentration, explaining 35% of the variance in the SOC concentration. A decrease in recalcitrant decomposed components (C/N and lignin/N) in aboveground and belowground litter directly promoted an increase in the SOC concentration, explaining 30% of the variance in the SOC concentration. The increase in total aboveground and belowground litter carbon input and decrease in refractory components in aboveground and belowground litter also reduced the SOC concentration by increasing the soil mineralization rate and soil CO_2 release rate, respectively, which indirectly explained the changes in the SOC concentration by 20% and 11%.

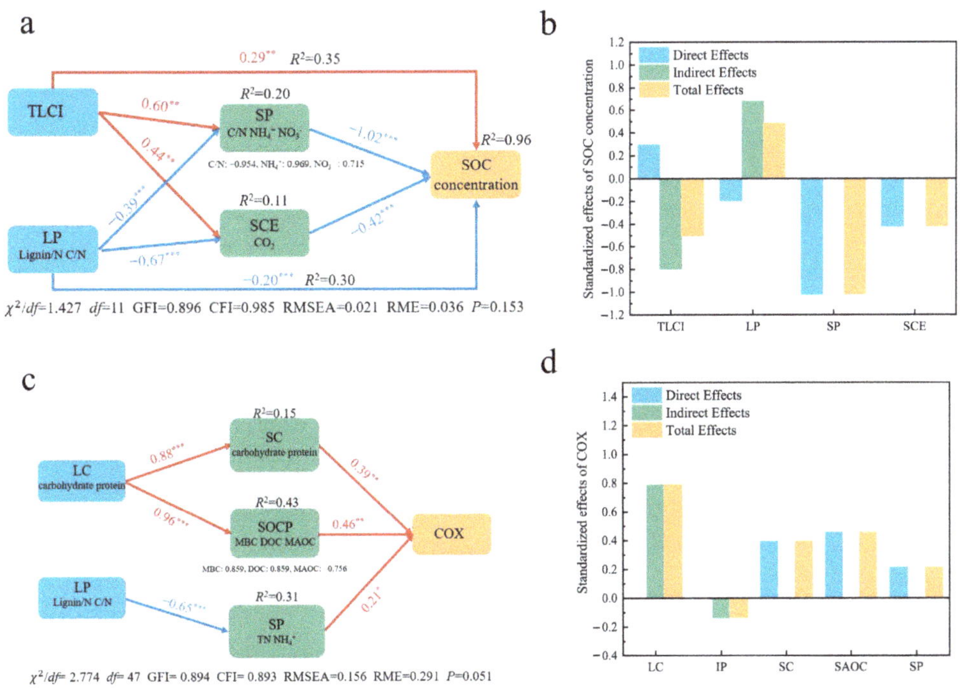

Figure 6. Structural equation model (SEM) of the total aboveground and belowground litter carbon input (TLCI), aboveground and belowground litter properties (LP), soil properties (SP), and soil carbon emission (SCE) with SOC concentration (**a**). Standardized effects of driving factors on SOC

concentration (**b**). SEM of aboveground and belowground litter components (LC), aboveground and belowground litter properties (LP), soil properties (SP), soil components (SC), and soil organic carbon pool (SOCP) with COX (**c**). Standardized effects of determinants of COX (**d**). Solid red lines and solid blue lines represent significantly positive and negative relationships, respectively. * $0.01 < p < 0.05$; ** $0.001 < p < 0.01$; *** $p < 0.001$.

The primary factors influencing the change in COX were the aboveground and belowground litter organic carbon molecular components (carbohydrates and proteins), aboveground and belowground litter properties (C/N and lignin/N), soil properties (TN and NH_4^+), soil organic carbon pool (MBC, DOC, and MAOC), and soil organic carbon components (carbohydrates and proteins) (Figure 6c). Among these, the increase in easily degradable organic carbon (carbohydrates and proteins) in aboveground and belowground litter increased the COX and reduced SOC stability via increases in active organic carbon (DOC and MBC) in soil and a decrease in recalcitrant organic carbon (MAOC), explaining 15% and 43% of the variance in SOC stability, respectively. The decrease in refractory components (C/N and lignin/N) in aboveground and belowground litter indirectly explained 3% of the variance in SOC stability by increasing the soil mineralization rate, soil N concentration, and COX and reducing SOC stability.

4. Discussion

4.1. Effects of Herbaceous Plant Encroachment on SOC Concentration

Total aboveground and belowground litter carbon input increased significantly as *D. purpurea* encroachment continued, consistent with the findings of other studies. This may be because *D. purpurea* has a large specific leaf area and a high net photosynthetic rate [29,30]. This enabled *D. purpurea* and *R. aureum* to obtain an advantage in the competition for resources, such as light and nutrients [31], and increased both aboveground and belowground biomass in *D. purpurea* [29]. Concurrently, as all *D. purpurea* were deciduous toward the end of the growing season, and a segment of *R. aureum* were also deciduous at this time [32], the total aboveground and belowground litter carbon input increased with the ongoing encroachment of *D. purpurea*. The SOC concentration in the community eventually decreased with an increase in the degree of encroachment in this study, despite the fact that the increase in total litter carbon input directly promoted an increase in the SOC concentration and explained 35% of the variance in the SOC concentration. This contradicted the results of most studies showing that the SOC concentration increases with an increase in the amount of litter. This occurred because SOC concentration resulted from a dynamic balance between total aboveground and belowground litter carbon input and soil mineralization and decomposition. [33]. This study revealed that the increase in total aboveground and belowground litter carbon input, alongside the reduction of recalcitrant components in both litter types, diminished SOC concentration by enhancing soil mineralization and CO_2 release rates, thereby accounting for 20% and 11% of the variation in SOC concentration, respectively (Figure 6a). *D. purpurea* encroachment increased soil respiration rate dramatically, indicating that *D. purpurea* encroachment enhanced the soil mineralization capacity. *D. purpurea* encroachment increased substantial nutrient input to the soil from both aboveground and belowground litter, hence enhancing microbial activity and accelerating nutrient turnover rate. Furthermore, the increase in the concentration of easily decomposed organic carbon compounds, such as carbohydrates, in both aboveground and belowground litter resulting from *D. purpurea* encroachment enhanced the metabolic capability of microorganisms [34]. This led to enhanced microbial capacity to decompose and utilize soil aged carbon, as well as an enhanced capacity for soil mineralization. In addition, soil characteristics had a direct negative effect on the SOC concentration

(Figure 6a). In this study, the soil C/N ratio was significantly reduced and was lower than the optimal ratio for microbial growth (25/1); additionally, the pH was significantly reduced (Table A2). The concentrations of NO_3^- and NH_4^+ increased, which promoted the growth and reproduction of microorganisms, increased the microbial decomposition ability, increased SOC consumption, and reduced the SOC concentration in the soil [34]. Therefore, in this study, SOC concentration consumed by soil mineralization generated by microbial utilization might be higher than SOC concentration input directly into the soil by aboveground and belowground litter decomposition, resulting in a significant decrease in SOC concentration in the soil with *D. purpurea* encroachment.

4.2. Effects of Herbaceous Plant Encroachment on SOC Stability

The ongoing *D. purpurea* encroachment resulted in an increase in COX in the soil, signifying a decline in soil stability, which was associated with alterations in the molecular structure of organic carbon and characteristics in both aboveground and belowground litter. *D. purpurea* encroachment led to a reduction in the growth and metabolic cycle of surface vegetation, altering the vegetation type. This resulted in a diminished cycle of aboveground and belowground litter input to the soil, as well as an expedited decomposition of both aboveground and belowground litter. The faster degradation of aboveground and belowground litter was accompanied by an increase in the concentration of easily decomposed organic carbon compounds, including carbohydrates and proteins, within the aboveground and belowground litter. (Figure A2a,b). This study revealed that an augmentation in the molecular structure of easily decomposed organic carbon in both aboveground and belowground litter indirectly diminished soil stability by 43% through the increase in active organic carbon concentration and the decrease in recalcitrant organic carbon concentration in the soil. (Figure 6c). This was because the increase in easily decomposed organic carbon, such as carbohydrates, in aboveground and belowground litter would provide a large number of decomposition substrates for soil hydrolases. This would increase the organic carbon sources that microorganisms can absorb and utilize [35,36], improving the efficiency of the "microbial carbon pump" [37,38], which promotes the accumulation of MBC and DOC concentration and reduces soil stability. In addition, a decrease in soil pH caused an increase in the fungal concentration, which in turn promoted the production of carbon decomposition enzymes [13,39]. In addition, it also accelerated the process by which fungi and enzymes broke down the easily decomposed organic carbon components in aboveground and belowground litter, increased active organic carbon concentration in the soil [40], accelerated the process by which microorganisms mineralize plant carbon sources, decreased the adsorption of plant residues by minerals, decreased recalcitrant organic carbon (MAOC) concentration [41], and decreased soil stability. Furthermore, an increase in the concentration of easily decomposed organic carbon molecular structures in the aboveground and belowground litter would indirectly decrease soil stability by 15% due to an increase in organic carbon molecular structures in the soil, such as carbohydrates and proteins (Figure 6c). Increased easily decomposable organic carbon components concentrations, such as carbohydrates, could provide adequate nutrition for soil microbes [42]. This could cause an increase in extracellular enzyme activity, which in turn speeds up the breakdown of plant carbon sources and soil aged carbon [43] and decreases soil stability.

In addition to the effect of the easily decomposed organic carbon molecular structure in aboveground and belowground litter on soil stability, the aboveground and belowground litter properties also had an effect on soil stability [12]. This study showed that the decrease in the recalcitrant decomposed components of aboveground and belowground litter (Lignin/N, C/N) following *D. purpurea* encroachment indirectly accounted for 31% of the reduction in soil stability by increasing soil TN and NH_4^+ concentrations (Figure 6c). On the one hand, high TN concentration in the soil alleviated nitrogen limitations on microbial activity, thereby increasing decomposition rates, enhancing microbial substrate utilization efficiency [21,44], and ultimately reducing soil stability. On the other hand, high-quality aboveground and belowground litter, with lower C/N ratios, were more efficiently utilized by microorganisms, stimulating the production of hydrolytic enzymes that accelerated the recalcitrant decomposed soil components, such as lignin [12], further reducing soil stability. Consequently, the decrease in SOC stability observed in this study was primarily attributed to an increase in the molecular structure of easily decomposed organic carbon in aboveground and belowground litter, leading to its greater input into the soil. This resulted in higher concentrations of decomposable and active organic carbon in the soil, a decline in recalcitrant organic carbon, and accelerated soil mineralization due to enhanced microbial activity stimulated by the reduction in recalcitrant components of the aboveground and belowground litter.

5. Limitations of the Study

There might be some possible limitations in this study. First, a total of 15 sample plots were only established in this study. There were limitations due to the location of the study within a national nature reserve, which restricted the number of sample plots that could be established. Second, this study focused exclusively on changes in CO_2 among greenhouse gases, while changes in other greenhouse gases, such as CH_4 and N_2O, were not considered. In the future, we can further measure other greenhouse gases to make the study more comprehensive. Third, the Fe and Al concentrations were not determined in this study, which might have influenced the variation in MAOC concentration. However, the volcanic ash and weathering materials of the Changbai Mountains are rich in potassium, resulting in a distinct tundra ecosystem that differs from other tundras. This ecosystem exhibits relatively low acidity, with pH values generally ranging from 5 to 6. The minimum pH recorded at the sampling sites in this study was 4.19, indicating that most soils do not qualify as strongly acidic. Therefore, the Fe and Al concentrations might have little influence on the variation in MAOC concentration.

6. Conclusions

D. purpurea encroachment in the Changbai Mountains tundra zone directly increased SOC concentration within the community by enhancing aboveground and belowground litter inputs by 113.25% and 68.7%, respectively, while reducing the recalcitrant components of the litter. However, the low C/N ratio in aboveground and belowground litter caused by *D. purpurea* encroachment could create a favorable environment for microbial growth and activity by reducing the soil C/N ratio and accelerating microbial decomposition, leading to increased SOC concentration consumption. Simultaneously, the increase in total aboveground and belowground litter carbon inputs and easily decomposed components also indirectly reduced SOC concentration within the community by increasing soil mineralization capacity and increasing CO_2 release by 71.20%. Ultimately, the *D. purpurea* encroachment resulted in a decrease of 41.63% in SOC concentration due to the dynamic balance between the total aboveground and belowground litter carbon inputs and soil mineralization and decomposition. Simultaneously, with the *D. purpurea* encroachment, there was an increase in easily decomposed organic carbon concentrations (such as carbohydrates and proteins) of more than 8% in the aboveground and belowground litter, which led to an increase in active organic carbon concentrations (such as MBC) of more than 33% in the soil and a decrease in recalcitrant organic carbon concentrations (such as MAOC) of more than 23%. At the same time, it lowered recalcitrant deposition components (such as lipids) in aboveground and belowground litter by more than 6%, accelerated soil mineralization, and decreased SOC stability by 74.90%. Therefore, *D. purpurea* encroachment would decrease the recalcitrant decomposed components concentrations in aboveground and belowground litter, increase easily decomposed organic carbon concentrations, and accelerate the decomposed of carbon pool in the tundra soil. We posited that this would progressively convert the tundra zone from a carbon sink to a carbon source, thereby expediting the release of nutrients and further facilitating the encroachment of herbaceous plants such as *D. purpurea*, which would in turn accelerate CO_2 emissions.

Author Contributions: Conceptualization, X.X. and J.X.; methodology, X.X. and Y.Z.; software, X.X. and Y.Z.; validation, Y.J.; formal analysis, X.X.; investigation, X.X. and J.Y.; resources, J.X. and Y.J.; data curation, X.X. and J.Y.; writing—original draft preparation, X.X.; writing—review and editing, X.X., J.X., Y.J. and Y.Z.; visualization, X.X. and Y.Z.; supervision, J.X.; project administration, Y.J.; funding acquisition, J.X. and Y.J. All authors have read and agreed to the published version of the manuscript.

Funding: This research was funded by the Natural Science Foundation of Jilin, China, grant number 20220101151JC, the National Natural Science Foundation of China, grant number 41571078 and Science Research Project of Hebei Education Department, grant number SQ2023237.

Data Availability Statement: Data are available upon a reasonable request.

Conflicts of Interest: The author declares no conflict of interest.

Abbreviations

The following abbreviations are used in this manuscript:

ANOVA	Analysis of variance
CO2	Carbon dioxide
COX	Carbon oxidation state
DOC	Dissolved organic carbon
EOC	Easily oxidizable organic carbon
LSD	Least significant difference
MAOC	Mineral-associated organic carbon
MBC	Microbial biomass carbon
NH4+	Ammonium nitrogen
NO3−	Nitrate nitrogen
POC	Particle organic carbon
SEM	Structural equation model
SOC	Soil organic carbon
TC	Total carbon
TN	Total nitrogen
TP	Total phosphorus

Appendix A

Table A1. Definition of particular regions in the ^{13}C NMR spectra.

Functional Groups	Chemical Shifts	Carbon Types
Alkyl	0–45 ppm	Alkyl carbon
N-alkyl/methoxy	45–60 ppm	N-alkyl methoxy carbon
O-alkyl	60–95 ppm	Alkoxy carbon
O_2-alkyl	95–110 ppm	Di-alkoxy carbon
Aromatic	110–145 ppm	Aromatic carbon
O-aromatic	145–165 ppm	Phenolic aromatic carbon
Carbonyl	165–210 ppm	Carbonyl carbon

Table A2. Physical and chemical properties of surface soil. Different lowercase letters represent significant differences in the physical and chemical properties of surface soil at different sampling points.

Sampling Site	pH	TC (g/kg)	TN (g/kg)	C/N (%)	TP (g/kg)	NH_4^+ (mg/kg)	NO_3^- (mg/kg)
Rho	(4.28 + 0.01) ^{a}	(225.87 + 10.88) ^{a}	(10.47 + 0.22) ^{bc}	(21.60 + 1.17) ^{a}	(1.10 + 0.03) ^{c}	(10.56 + 0.23) ^{c}	(0.18 +0.07) ^{b}
Mil	(4.27 + 0.02) ^{a}	(208.59 + 7.39) ^{a}	(9.27 + 0.47) ^{c}	(22.65 + 1.61) ^{a}	(0.98 + 0.08) ^{c}	(10.76 + 0.06) ^{c}	(0.06 + 0.02) ^{b}
Mod	(4.23 + 0.02) ^{ab}	(203.82 + 1.87) ^{a}	(9.56 + 0.20) ^{c}	(21.34 + 0.61) ^{a}	(1.21 + 0.06) ^{bc}	(11.55 + 0.15) ^{b}	(0.06 + 0.02) ^{b}
Sev	(4.25 + 0.04) ^{ab}	(205.25 + 7.15) ^{a}	(11.42 + 0.01) ^{b}	(17.97 + 0.62) ^{b}	(1.30 + 0.06) ^{b}	(11.38 + 0.06) ^{b}	(0.27 + 0.19) ^{b}
Dey	(4.19 + 0.02) ^{b}	(131.85 + 8.45) ^{b}	(13.57 + 0.92) ^{a}	(9.83 + 0.98) ^{b}	(1.62 + 0.13) ^{a}	(16.78 + 0.14) ^{a}	(4.92 + 3.24) ^{a}

Different lowercase letters correspond to significant differences in the chemical characteristics of soil ($p < 0.05$).

Appendix B

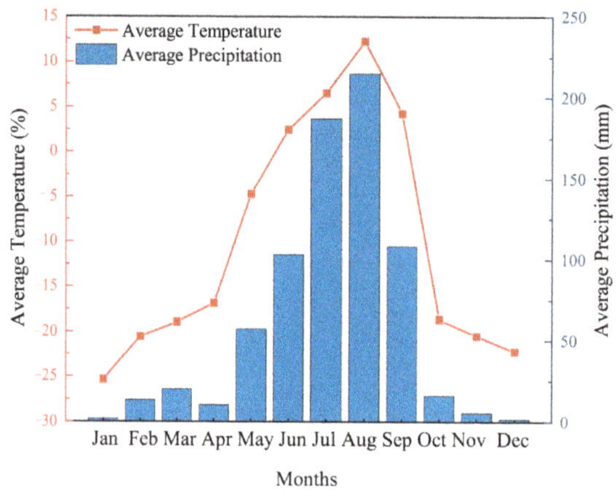

Figure A1. Monthly variations in average temperature and precipitation in the Changbai Mountains tundra, 2022.

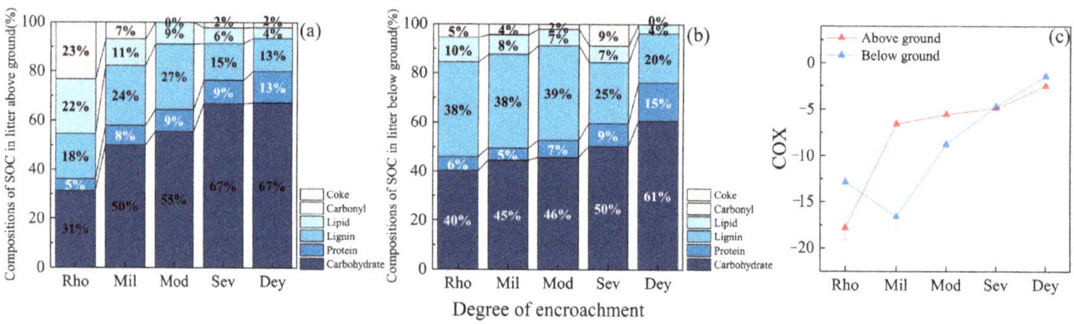

Figure A2. Changes in molecular structure of organic carbon and COX in aboveground and belowground litters under different degrees of *Deyeuxia purpurea* encroachment. (**a**) Aboveground litter, (**b**) belowground litter, (**c**) COX.

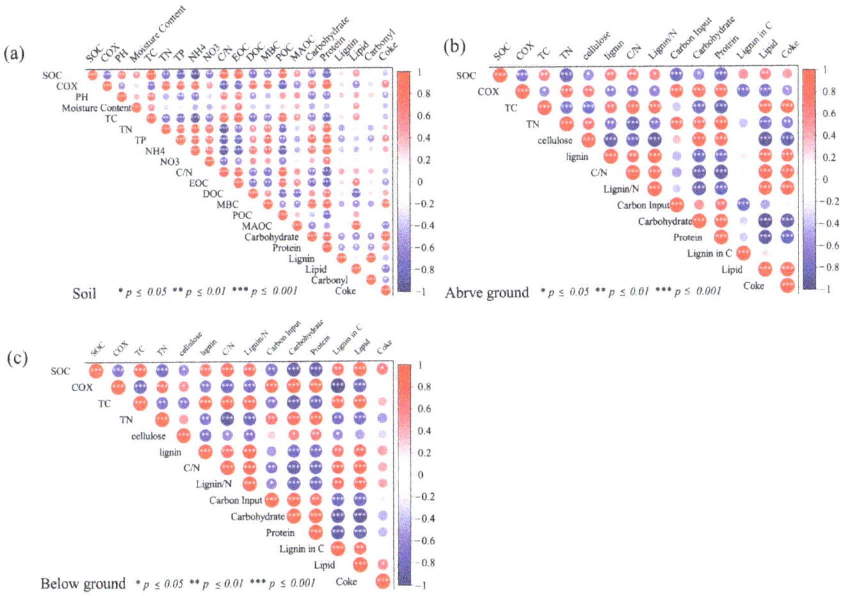

Figure A3. Correlation analysis of SOC and COX with various indicators in soil, aboveground litter, and belowground litter. (**a**) Soil, (**b**) aboveground litter, (**c**) belowground litter. * $0.01 < p < 0.05$; ** $0.001 < p < 0.01$; *** $p < 0.001$.

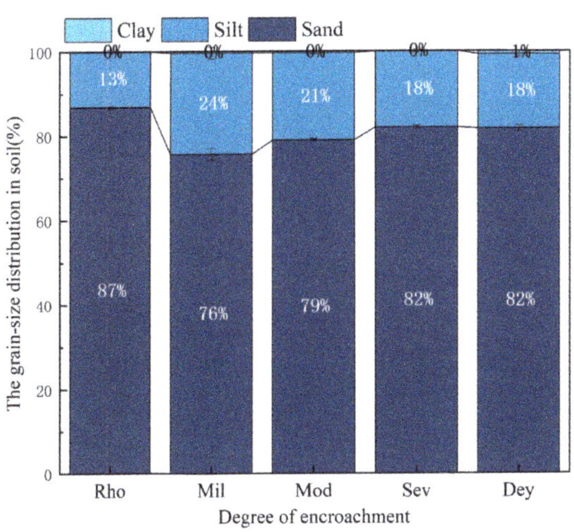

Figure A4. Variation in soil grain-size distribution.

References

1. Batjes, N.H. Projected Changes in Soil Organic Carbon Stocks Upon Adoption of Recommended Soil and Water Conservation Practices in the Upper Tana River Catchment, Kenya. *Land Degrad. Dev.* **2014**, *25*, 278–287. [CrossRef]
2. Schulze, E.D.; Freibauer, A. Environmental science—Carbon unlocked from soils. *Nature* **2005**, *437*, 205–206. [CrossRef]
3. Pan, Y.; Chen, J.M.; Birdsey, R.; McCullough, K.; He, L.; Deng, F. Age structure and disturbance legacy of North American forests. *Biogeosciences* **2011**, *8*, 715–732. [CrossRef]
4. Zhang, Y.; An, C.B.; Zhang, W.S.; Zheng, L.Y.; Zhang, Y.Z.; Lu, C.; Liu, L.Y. Drivers of mountain soil organic carbon stock dynamics: A review. *J. Soils Sediments* **2023**, *23*, 64–76. [CrossRef]

5. Zhao, G.Y.; Li, S.Y.; Zhou, W.Z.; He, L.; Zou, R.; Yu, J.; Chen, Z.; Bai, X.; Zhang, J.; Sun, X. Comparative Analysis of Soil Organic Carbon and Soil Oxide Minerals across Different Climates and Forest Types. *Austrian J. For. Sci.* **2024**, *141*, 53–78.
6. Ofiti, N.O.E.; Schmidt, M.W.; Abiven, S.; Hanson, P.J.; Iversen, C.M.; Wilson, R.M.; Kostka, J.E.; Wiesenberg, G.L.B.; Malhotra, A. Climate warming and elevated CO2 alter peatland soil carbon sources and stability. *Nat. Commun.* **2023**, *14*, 7533. [CrossRef]
7. Xu, M.P.; Zhi, R.C.; Jian, J.N.; Feng, Y.Z.; Han, X.H.; Zhang, W. Changes in Soil Organic C Fractions and C Pool Stability Are Mediated by C-Degrading Enzymes in Litter Decomposition of *Robinia pseudoacacia* Plantations. *Microb. Ecol.* **2023**, *86*, 1189–1199. [CrossRef]
8. Abiven, S.; Menasseri, S.; Chenu, C. The effects of organic inputs over time on soil aggregate stability—A literature analysis. *Soil Biol. Biochem.* **2009**, *41*, 1–12. [CrossRef]
9. Mitchell, E.; Scheer, C.; Rowlings, D.; Cotrufo, M.F.; Conant, R.T.; Friedl, J.; Grace, P. Trade-off between 'new' SOC stabilisation from above-ground inputs and priming of native C as determined by soil type and residue placement. *Biogeochemistry* **2020**, *149*, 221–236. [CrossRef]
10. Zhong, Y.; Yan, W.; Wang, R.; Shangguan, Z. Differential responses of litter decomposition to nutrient addition and soil water availability with long-term vegetation recovery. *Biol. Fertil. Soils* **2017**, *53*, 939–949. [CrossRef]
11. Gao, J.; Feng, J.; Zhang, X.; Yu, F.H.; Xu, X.; Kuzyakov, Y. Drying-rewetting cycles alter carbon and nitrogen mineralization in litter-amended alpine wetland soil. *Catena* **2016**, *145*, 285–290. [CrossRef]
12. Cotrufo, M.F.; Wallenstein, M.D.; Boot, C.M.; Denef, K.; Paul, E. The Microbial Efficiency-Matrix Stabilization (MEMS) framework integrates plant litter decomposition with soil organic matter stabilization: Do labile plant inputs form stable soil organic matter? *Glob. Change Biol.* **2013**, *19*, 988–995. [CrossRef] [PubMed]
13. Tian, J.; Zong, N.; Hartley, I.P.; He, N.; Zhang, J.; Powlson, D.; Zhou, J.; Kuzyakov, Y.; Zhang, F.; Yu, G.; et al. Microbial metabolic response to winter warming stabilizes soil carbon. *Glob. Change Biol.* **2021**, *27*, 2011–2028. [CrossRef]
14. Jin, Z.; Shah, T.; Zhang, L.; Liu, H.; Peng, S.; Nie, L. Effect of straw returning on soil organic carbon in rice-wheat rotation system: A review. *Food Energy Secur.* **2020**, *9*, e200. [CrossRef]
15. Yan, W.; Zhong, Y.; Zhu, G.; Liu, H.; Peng, S.; Nie, L. Nutrient limitation of litter decomposition with long-term secondary succession: Evidence from controlled laboratory experiments. *J. Soils Sediments* **2020**, *20*, 1858–1868. [CrossRef]
16. Wang, Q.; Yu, Y.; He, T.; Wang, Y. Aboveground and belowground litter have equal contributions to soil CO_2 emission: An evidence from a 4-year measurement in a subtropical forest. *Plant Soil* **2017**, *421*, 7–17. [CrossRef]
17. Zhang, W.P.; Fornara, D.; Yang, H.; Yu, R.P.; Callaway, R.M.; Li, L. Plant litter strengthens positive biodiversity-ecosystem functioning relationships over time. *Trends Ecol. Evol.* **2023**, *38*, 473–484. [CrossRef]
18. Pérez-Harguindeguy, N.; Díaz, S.; Cornelissen, J.H.C.; Vendramini, F.; Cabido, M.; Castellanos, A. Chemistry and toughness predict leaf litter decomposition rates over a wide spectrum of functional types and taxa in central Argentina. *Plant Soil* **2000**, *218*, 21–30. [CrossRef]
19. Shaw, M.R.; Harte, J. Control of litter decomposition in a subalpine meadow-sagebrush steppe ecotone under climate change. *Ecol. Appl.* **2001**, *11*, 1206–1223. [CrossRef]
20. Mcculley, R.L.; Archer, S.R.; Boutton, T.W.; Hons, F.M.; Zuberer, D.A. Soil respiration and nutrient cycling in wooded communities developing in grassland. *Ecology* **2004**, *85*, 2804–2817. [CrossRef]
21. Liao, J.D.; Boutton, T.W. Soil microbial biomass response to woody plant invasion of grassland. *Soil Biol. Biochem.* **2008**, *40*, 1207–1216. [CrossRef]
22. Zong, S.W.; Xu, J.W.; Wu, Z.F. Investigation and Mechanism Analysis on the Invasion of *Deyeuxia angustifolia* to Tundra Zone in Western Slope of Changbai Mountain. *J. Mt. Sci.* **2013**, *31*, 448–455.
23. Zhang, Y.J. Effects of *Deyeuxia purpurea* Expansion on the Changes in the Litter Decomposition of the Changbai Mountain Tundra and Soil Nutrients. Ph.D. Thesis, Northeast Normal University, Changchun, China, 2022.
24. Zong, S.W.; Wu, Z.F.; Du, H.B. Study on Climate Change in Alpine Tundra of the Changbai Mountain in Growing Season in Recent 52 Years (in China). *Arid Zone Res.* **2013**, *30*, 41–49.
25. Zong, S.W. Mechanism Research on the Vegetation Changes of the Sub-Alpine Tundra, Changbai Mountains. Ph.D. Thesis, Northeast Normal University, Changchun, China, 2014.
26. Baldock, J.A.; Masiello, C.A.; Gélinas, Y.; Hedges, J.I. Cycling and composition of organic matter in terrestrial and marine ecosystems. *Mar. Chem.* **2004**, *92*, 39–64. [CrossRef]
27. Nelson, P.N.; Baldock, J.A. Estimating the molecular composition of a diverse range of natural organic materials from solid-state ^{13}C NMR and elemental analyses. *Biogeochemistry* **2005**, *72*, 1–34. [CrossRef]
28. Masiello, C.A.; Gallagher, M.E.; Randerson, J.T.; Deco, R.M.; Chadwick, O.A. Evaluating two experimental approaches for measuring ecosystem carbon oxidation state and oxidative ratio. *J. Geophys. Res. Biogeosciences* **2008**, *113*, G03010. [CrossRef]
29. Ehrenfeld, J.G. Effects of exotic plant invasions on soil nutrient cycling processes. *Ecosystems* **2003**, *6*, 503–523. [CrossRef]
30. Nie, M.; Shang, L.; Liao, C.; Li, B. Changes in Primary Production and Carbon Sequestration After Plant Invasions. In *Impact of Biological Invasions on Ecosystem Services*; Vilà, M., Hulme, P., Eds.; Springer: Cham, Switzerland, 2017. [CrossRef]

31. Lambers, H.; Chapin, F.S.; Pons, T.L. *Plant Physiological Ecology*; Springer: New York, NY, USA, 1998; 540p.
32. Biederbeck, V.O.; Janzen, H.H.; Campbell, C.A.; Zentner, R.P. Lsbile soil organic-matter as influenced by cropping practices in an arid environmentent. *Soil Biol. Biochem.* **1994**, *26*, 1647–1656. [CrossRef]
33. Fearnside, P.M.; Barbosa, R.I. Soil carbon changes from conversion of forest to pasture in Brazilian Amazonia. *For. Ecol. Manag.* **1998**, *108*, 147–166. [CrossRef]
34. De Deyn, G.B.; Cornelissen JH, C.; Bardgett, R.D. Plant functional traits and soil carbon sequestration in contrasting biomes. *Ecol. Lett.* **2008**, *11*, 516–531. [CrossRef]
35. Feng, X.; Wang, S. Plant influences on soil microbial carbon pump efficiency. *Glob. Change Biol.* **2023**, *29*, 3854–3856. [CrossRef] [PubMed]
36. Huang, Y.; Liang, C.; Duan, X.; Chen, H.; Li, D. Variation of microbial residue contribution to soil organic carbon sequestration following land use change in a subtropical karst region. *Geoderma* **2019**, *353*, 340–346. [CrossRef]
37. Shao, P.; Liang, C.; Lynch, L.; Xie, H.; Bao, X. Reforestation accelerates soil organic carbon accumulation: Evidence from microbial biomarkers. *Soil Biol. Biochem.* **2019**, *131*, 182–190. [CrossRef]
38. Lauer, F.; Koesters, R.; Du Preez, C.C.; Amelung, W. Microbial residues as indicators of soil restoration in South African secondary pastures. *Soil Biol. Biochem.* **2011**, *43*, 787–794. [CrossRef]
39. Huang, R.; Crowther, T.W.; Sui, Y.; Sun, B.; Liang, Y. High stability and metabolic capacity of bacterial community promote the rapid reduction of easily decomposing carbon in soil. *Commun. Biol.* **2021**, *4*, 1376. [CrossRef]
40. Geethanjali, P.A.; Jayashankar, P.M. A Review on Litter Decomposition by Soil Fungal Community. *IOSR J. Pharm. Biol. Sci.* **2016**, *11*, 1–3. [CrossRef]
41. Angst, G.; Mueller, K.E.; Castellano, M.J.; Vogel, C.; Wiesmeier, M.; Mueller, C.W. Unlocking complex soil systems as carbon sinks: Multi-pool management as the key. *Nat. Commun.* **2023**, *14*, 2967. [CrossRef]
42. Cragg, S.M.; Friess, D.A.; Gillis, L.G.; Trevathan-Tackett, S.M.; Terrett, O.M.; Watts, J.E.; Distel, D.L.; Dupree, P. Vascular Plants Are Globally Significant Contributors to Marine Carbon Fluxes and Sinks. *Annu. Rev. Mar. Sci.* **2020**, *12*, 469–497. [CrossRef]
43. Ding, Y.; Wang, D.; Zhao, G.; Chen, S.; Sun, T.; Sun, H.; Wu, C.; Li, Y.; Yu, Z.; Li, Y.; et al. The contribution of wetland plant litter to soil carbon pool: Decomposition rates and priming effects. *Environ. Res.* **2023**, *224*, 115575. [CrossRef]
44. Yang, J.; Wu, F.; Wei, X.; Zhang, X.; Wu, Q.; Yue, K.; Ni, X. Global Positive Effects of Litter Inputs on Soil Nitrogen Pools and Fluxes. *Ecosystems* **2023**, *26*, 860–872. [CrossRef]

forests

MDPI

Article

Characterization of Chinese Tallow Invasion in the Southern United States

Mohammad M. Bataineh [1,*], Jacob S. Fraser [2] and Lauren S. Pile Knapp [2]

[1] USDA Forest Service, Southern Research Station 2500 Shreveport Hwy, Pineville, LA 71360, USA
[2] USDA Forest Service, Northern Research Station, 202 Anheuser Busch Natural Resources Building, Columbia, MO 65211, USA; jacob.fraser@usda.gov (J.S.F.); lauren.pile@usda.gov (L.S.P.K.)
* Correspondence: mohammad.bataineh@usda.gov

Abstract: Chinese tallow is a non-native invasive tree expanding in range and abundance throughout the southern United States. Several biogeographical studies mapping tallow distribution and examining key underlying environmental factors relied on the U.S. Forest Service Forest Inventory and Analysis (FIA) data, representing forestlands at scales of ~2400 ha. However, given that most invasive trees, like tallow, are cosmopolitan and dynamic in nature, FIA data fails to capture the extent and severity of the invasion especially outside areas classified as forestlands. To develop tallow maps that more adequately depict its distribution at finer spatial scales and to capture observations in non-forestlands, we combined verified citizen science observations with FIA data. Further, we described spatiotemporal patterns and compared citizen science to FIA and other previously published distribution maps. From our work, although tallow is prevalent in the south, Louisiana, Texas, and Mississippi were the most invaded states. Tallow was associated with flatwoods and prairie grasslands of the Gulf Coast. Annual extreme minimum temperatures of less than −12.2 °C (10 °F) represented the northern limit of naturalized tallow populations. Tallow's northward and inland expansion was captured in citizen science and FIA data, indicating a tallow spread rate ranging from 5 to 20 km annually over the last decade. Systematic sampling, such as FIA, and citizen science data both have their own unique pitfalls. However, the use of citizen science data can complement invasive plant distribution mapping, especially when combined with data from established systematic monitoring networks. This approach provides for a more complete understanding of invasive tree extent and spatiotemporal dynamics across large landscapes.

Keywords: *Triadica sebifera*; popcorn tree; citizen science; invasive plants

1. Introduction

Chinese tallow [*Triadica sebifera* (L.) Small; Euphorbiaceae] is a non-native invasive tree rapidly expanding in range and abundance throughout the southeastern United States [1]. Tallow is native to China, Taiwan, and northern Vietnam with naturalized populations in Japan, India, Pakistan, Bangladesh, Sudan, Europe, Australia, Panama, and Peru [2–4]. In its native range, tallow is cultivated for a variety of uses including food, medicine, and industrial applications [5]. It was introduced in the Gulf Coastal Plain to establish local soap and oil industries and is currently favored by beekeepers as forage for honey production [5,6]. Earlier accounts of tallow spatial distribution and abundance indicate occurrence throughout the Gulf Coastal Plain and along the Atlantic Coast as far as North Carolina [7,8]. Three-, five-, and six-fold increments in tallow tree density were reported for Texas, Mississippi, and Louisiana, respectively, over an average period of 14 years [8]. An eight-fold increase in occupied area in southeast Texas was reported from the 1970s to the 1990s [7]. Tallow occurrence is associated with low-lying areas, especially along waterways and riparian zones, in prairies and forests of the Gulf Coastal Plain [7,9]. Tallow encroachment into uplands of the Upper Coastal Prairie of Texas occurred circa the

1970s [7]. In forested areas, tallow occurred more frequently within private ownerships and in recently disturbed or younger stands [9,10] and was less likely to occur in mature forest stands and plantations [11]. Northward expansion of tallow invasion range was predicted in relation to a warming climate [3,9,11], and the observed spread appears to at least partially support the general predicted expansion pattern [12]. The expansion of tallow's current invasion range appears to be constrained by propagule supply and dispersal, and birds in particular are important dispersers [13,14], but perennial water sources may provide unbroken pathways for spread, especially when outlying populations are established [3].

Several factors aid in the current extensive distribution of tallow. The high fecundity of large seeds, coupled with relatively persistent soil seed banks, and the short time to maturity provide tallow with the propagule pressure needed to enhance establishment likelihood in recipient communities [1]. Sheer propagule pressure was suggested as one of the mechanisms allowing tallow to outcompete native trees [15]. Substantial recruitment potential into small-diameter tree sizes [7,8], however, indicates a lack of effective site filters on tallow establishment and early-regeneration processes. Rapid early-growth rates and the ability to establish itself across a wide variety of sites and soil conditions allow tallow to outcompete native vegetation [1,16,17]. Phenotypic plasticity, especially to shade, appears to provide tallow with an advantage over native trees in terms of survival and growth [17,18]. Tallow's shorter lifespan and smaller sizes at maturity compared to fast-growing native tree species [16], however, may limit its persistence and dominance when recruitment of native trees is not limited or natives' recruitment is facilitated through release treatments [7,15,16].

Although several biogeographical studies have mapped the distribution and examined key factors and mechanisms underlying the current tallow invasion [8,9,19–21], data for those studies were sourced from the U.S. Forest Service Forest Inventory and Analysis (FIA) program. FIA sampling plots represent forestlands at scales of ~2400 ha, not capturing observations in non-forestland such as urban settings and open lands, which can be important sources of propagule supply and dispersal for invasive trees. In some cases, FIA data were aggregated at county [9] or state levels [8] masking tallow's presence and dynamics at management-relevant spatial scales. Citizen science data from a variety of platforms (e.g., iNaturalist and EDDMapS) can contribute to species distribution mapping and complement existing occurrence records even at lower identification quality [22,23]. In this study, we use citizen science data from two platforms, iNaturalist and EDDMapS, to: (1) provide a contemporary Chinese tallow distribution map, (2) describe ecological associations of tallow occurrence with ecological regions and climatic hardiness zones, (3) describe associated spatiotemporal occurrence patterns based on observational and systematic data, (4) compare measured compositional and structural conditions between invaded and non-invaded forestland plots, and (5) compare citizen science distribution maps to FIA-based maps and other maps and projections reported elsewhere.

2. Materials and Methods

2.1. Study Area

The study area encompassed all 13 southern states, from Virginia to Florida along the Atlantic Coast and from Florida to Texas along the Gulf of Mexico, including the northern tier of the southern states of Oklahoma, Arkansas, Tennessee, and Kentucky (Figure 1). The 13 southern states encompass a wide variety of habitats and ecosystems, including nine ecological provinces stretching from the Everglades through the Outer Coastal Plain and Southeastern Mixed Forest provinces to the Central Appalachian and Central Interior Forests, including the Lower Mississippi, Ouachita, Ozark, and Prairie Parkland provinces [24]. This 13-state area represents the current maximum invasive extent of tallow. The population of people within the study area exceeded 112 million by 1 April 2020 [25]. Climate varies with latitude and topography but is mainly humid temperate to humid subtropical, with tropical influence at the southern extent. The annual mean temperature range is from 13 °C to 21 °C [26]. The annual mean minimum temperature range is from 0 °C at the northern extent to 21 °C along the southern coast lines [27]. The annual mean maximum temperatures range from 16 °C to 32 °C [27]. The region receives from 102 to over 178 cm of precipitation annually [27]. The annual frost-free period is 180–315 days, spanning 6 of the 13 USDA plant hardiness zones [28]. Low-lying, flat areas stretch from the Coastal Plains poleward, gradating into gently rolling topography ranging in elevation from 300 to 400 m above sea level.

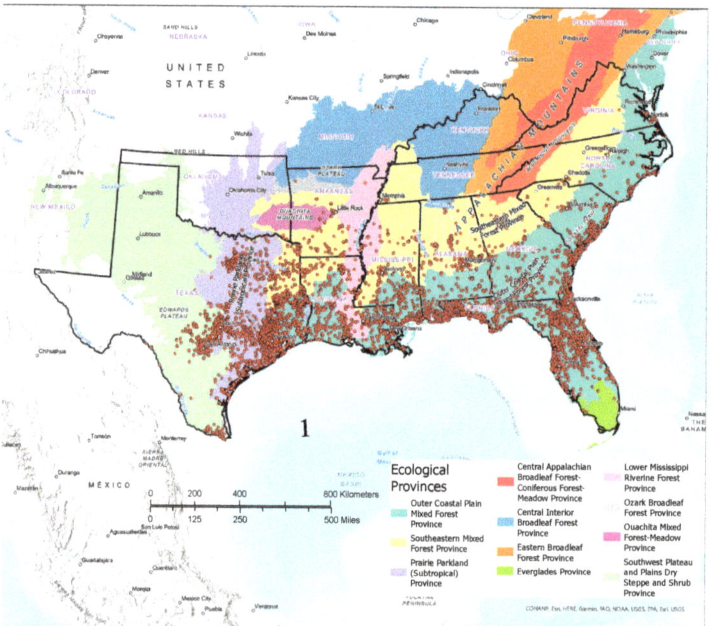

Figure 1. Study area extent with overlapping ecological provinces and tallow observations from1991 to 2023 from the citizen science databases iNaturalist and EDDMapS.

2.2. Data Sources

Data from two online citizen science platforms were downloaded. Data reflected observations collected and compiled through these platforms as of 1 May 2023. The two citizen science data platforms were iNaturalist and the Early Detection and Distribution Mapping System (EDDMapS). iNaturalist began in 2008 as a final master's project developed by graduate students at UC Berkeley. EDDMapS was established in 2005 by the Center for Invasive Species and Ecosystem Health at the University of Georgia. iNaturalist is an online social network connecting people to nature with the aim of advancing and informing biodiversity and conservation science. iNaturalist provides a consensus-based organism identification and recording tool with the aid of image recognition technology. EDDMapS is an online mapping and tracking system designed to simplify invasive and certain native pest species recordings without the aid of a Geographic Information System. EDDMapS is more focused on invasive species, and the platform aggregates invasive species data from other sources, such as herbaria records, in addition to iNaturalist.

Only verified observations from both platforms were used. The iNaturalist consensus-based verification characterizes observations as "Research Grade" when two or more observers agree on a species-level or lower taxon. The iNaturalist verifiable observations must have a date, a photo, and a location to qualify as research grade status; otherwise, observations are labeled "Casual". EDDMapS observations are reviewed and verified by local and state verifiers rather than social network members or consensus-based verification.

iNaturalist had 13,008 verified tallow observations, whereas EDDMapS had 18,917 verified observations. EDDMapS included polygon-delineated observations, mainly associated with the U.S. Forest Service National Forest System. To merge and maintain consistency among datasets, polygon records were converted to points at the centroid of each polygon. For EDDMapS data, observations without a date, expert verification, or positive detection status were excluded from the verified records. EDDMapS observations with high uncertainty in recorded date (e.g., within 183 days) were also excluded. EDDMapS duplicate records (i.e., observations imported from iNaturalist) were also excluded. The subset of verified observations encompassed within the 13 southern states was used as the verified and compiled dataset, with a total of 27,540 records.

FIA data for the most recent inventories for each of the 13 southern states were downloaded using the EVALIDator 2.1.0 application [29]. State-level retrievals for forestland, at least 10% tree canopy cover and a minimum area of 0.4 ha (one acre), were selected. Number of live trees (at least 2.54 cm in diameter) and seedlings (less than 2.54 cm in diameter) was used as the tallow presence metric. The FIA records were queried for living tallow trees and seedlings by county for each of the 13 states to create county-level maps, a total of 243 counties. For the FIA plot-level observations, state-level raw FIA data in comma-delimited tables were downloaded from FIA DataMart 2.0 [30]. After downloading, tallow cover was queried from the raw FIA data via the package rFIA (v1.0.0) in R v4.3.1 (R Foundation for Statistical Computing, Vienna, Austria) [31]. The tallow cover proportion, regardless of height, on each of the four FIA subplots was averaged to obtain the FIA plot-level cover estimates used in mapping. The rFIA package was used to calculate plot-level forest structural (basal area, stem density), species composition (Shannon's diversity index, Shannon's equitability index, and species richness), annual removals, and annual mortality metrics. The plots were filtered to only include the most recent inventory cycle across all states, and plots with new detections of tallow were assigned to the "invaded" category. Non-invaded plots were the remaining forested plots from the same time-period, and the FIA plot data spanned the period from 2001 to 2022 across the 13 states.

2.3. Analytical Approach

Spatially explicit ecosystem classification data from two sources were used to provide the ecological context for tallow invasion patterns. The Ecoregions of the Conterminous United States and the National Hierarchical Framework of Ecological Units were used as the two ecosystem classification sources [24,32]. Both classifications delineate ecological units based on similarities in physiography, climate, edaphic factors, and potential natural vegetation and provide a spatial framework of nested hierarchies from continental to subregional scales. Level III and IV hierarchies of the Ecoregions classification [32] and Sections and Subsections of Ecological Units classification [24] were used. These hierarchical levels provide mapping of ecological units at regional to subregional scales and use similar factors in the recognition of variability among ecological units.

As minimum temperature is cited as a key climatic constraint for tallow spread and distribution [3], our contemporary tallow distribution was overlaid on the 2012 USDA plant hardiness zone map [33]. Plant hardiness zones represent geographical regions with relatively homogeneous mean annual minimum winter temperature in 2.8 °C (5 °F) half zones [33]. Hardiness zones were mapped by interpolating the mean minimum daily temperature of the coldest month for the period 1976–2005, considering the effect of elevation and air mass movement on extreme daily minimum temperatures [33].

To assess spatiotemporal patterns of tallow invasion, observation dates were used to map tallow spatial trends over time. Although observation dates do not necessarily reflect effective tallow establishment dates nor do they faithfully reflect tallow's range expansion, observation dates provide an overview of the timing during which tallow was observed and recorded by citizen scientists. The earliest EDDMapS tallow record was reported on 6 May 1991, with the following observations being recorded on 6 September 1995. These early records were compiled by EDDMapS from the Florida Exotic Pest Plant Council

Database. The earliest records were reported for Florida, followed by Louisiana, in 1999 and Texas in 2002. iNaturalist's earliest tallow record was reported on 22 May 2004 in Texas, with observations made in the same year in Florida. Similarly, the tallow FIA plot observation spatial trends were mapped using the inventory year, regardless of the FIA inventory cycle.

To compare forest conditions between plots invaded by tallow and non-invaded plots, we used the Wilcoxon rank sum test in the base stats package using R v4.3.1 (R Foundation for Statistical Computing, Vienna, Austria) to test for statistically significant differences in the medians of forest condition metrics between groups [34].

Overlay analysis, point pattern analysis, and spatial joins were all conducted in ArcGIS Pro 3.1.2 (Esri Inc., Redlands, CA, USA). The mean center tool from the spatial statistics toolbar was used to identify the geographic centers of tallow observations. Tallow observations were compiled cumulatively by year, and the mean center was calculated at each time step to identify spatial shifts in the mean center over time. To compare the citizen science distribution maps to the FIA-based maps, heat map symbology was used in overlays of tallow citizen science observations with FIA-based maps. Merged FIA plot-level data and citizen science observations were aggregated into polygon bins to provide an overall, pooled data sources, tallow distribution map.

3. Results

3.1. Contemporary Tallow Distribution by Political Boundary and Ecological Province

Verified citizen science data from two platforms, iNaturalist and EDDMapS, were used to provide a contemporary Chinese tallow distribution map (Figure 1). Based on this map, tallow was most frequent in Texas, followed by Florida and Louisiana, with these three states accounting for 87% of the total verified observations. Harris County, TX, had the highest number of verified citizen science observations, followed by Okaloosa County, FL, Santa Rosa County, FL, Trinity County, TX, Angelina County, TX, and Sabine County, TX. These counties represented 30% of the total verified citizen science observations from 1991 to 2023.

The Outer Coastal Plain Mixed Forest Province of the National Hierarchical Framework of Ecological Units [24] contained the majority of tallow observations (Table 1 and Figure 2). Most FIA plot observations (96%) were within forested provinces, whereas 22% of the citizen science observations were within prairie and steppe provinces, highlighting the forestland inventory focus of FIA.

Table 1. Proportion (%) of total Forest Inventory and Analysis plots and citizen science observations within provinces of the National Hierarchical Framework of Ecological Units.

Province	FIA [1]	Citizen Science
Outer Coastal Plain Mixed Forest	74	62
Southeastern Mixed Forest	13	9
Lower Mississippi Riverine Forest	9	3
Prairie Parkland (Subtropical)	4	17
Southwest Plateau and Plains Dry Steppe and Shrub	0.02	5

[1] Forest Inventory and Analysis plots.

Within the Outer Coastal Plain Mixed Forest Province, the Coastal Plains and Flatwoods-Western Gulf Section had the highest number of tallow observations, regardless of data source (29% and 21% for FIA and citizen science, respectively). The Gulf Coastal Plains and Flatwoods Section and the Louisiana Coastal Prairie and Marshes Section combined contained 27% of the FIA observations but only 8% of the citizen science observations (Figure 2). The Atlantic Coastal Flatwoods had 7% and 10% of the FIA plot and citizen science observations, respectively. The Florida Coastal Lowlands-Atlantic Section had 1% of the FIA and 8% of the citizen science observations. The Central Gulf Prairie and Marshes Section had 3% of the FIA and 7% of the citizen science observations.

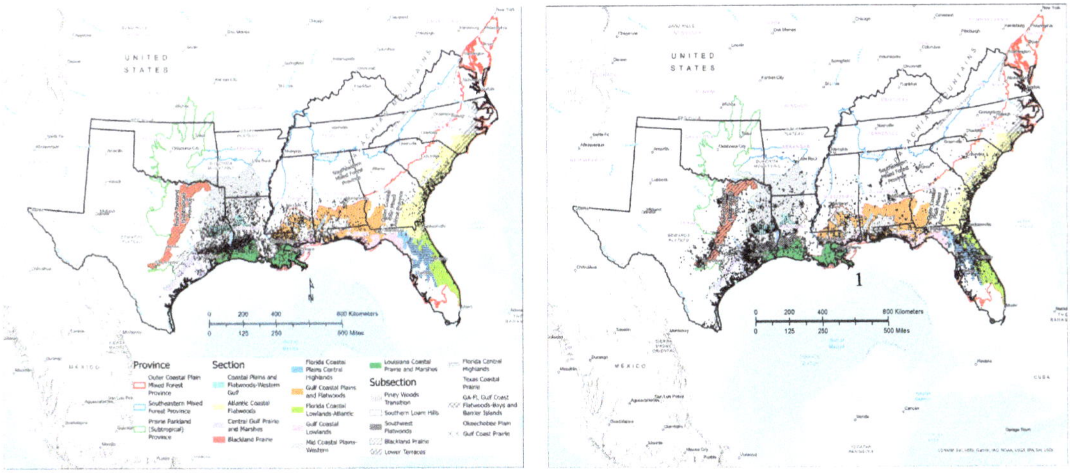

(**a**) (**b**)

Figure 2. Sections and Subsections of the National Hierarchical Framework of Ecological Units with tallow: (**a**) Forest Inventory and Analysis (FIA) plot observations for the period from 2001 to 2022, and (**b**) citizen science observations for the period from 1991 to 2023, as reported in the iNaturalist and EDDMapS databases.

Within the Coastal Plains and Flatwoods-Western Gulf Section, the Piney Woods Transition Subsection had 4% and 10% of the FIA plot and citizen science observations, respectively (Figure 2). The Southwest Flatwoods Subsection had 16% of the FIA and 6% of the citizen science total observations. The Southern Loam Hills Subsection of the Coastal Plains and Flatwoods-Western Gulf and the Gulf Coastal Plains and Flatwoods Sections had 15% of the FIA but 8% of the citizen science observations. While 6% of the total verified citizen science observations were within the Blackland Prairie Subsection, only ten FIA plots had tallow within this subsection. The Gulf Coast Prairie Subsection of the Louisiana Coastal Prairie and Marshes Section contained 6% of the FIA plot and 1% of the citizen science observations.

Using the Ecoregions of the Conterminous United States classification [32], 91% of the total verified citizen science observations were within seven Level III Ecoregions (Figure 3). On the other hand, 90% of the FIA plot observations were within five of the seven Level III Ecoregions containing the majority of citizen science observations. The Texas Blackland Prairies and Edwards Plateau had few to no FIA plot observations, respectively (Figure 3a). The Southern Coastal Plain and South Central Plains ecoregions had the most frequent citizen science tallow observations with 35% and 25% of the total, respectively (Figure 3b). For FIA plots, the Southern Coastal Plain and South Central Plains ecoregions had 24% and 40% of the total observations, respectively. At Level IV of the Ecoregions hierarchy, 55% and 54% of the total observations for FIA and citizen science sources, respectively, were within six ecoregions (Figure 4). Flatwoods, Southern Tertiary Uplands, and Southern Pine Plains and Hills had the highest number of FIA plot observations (Figure 4a). The Southern Tertiary Uplands, Eastern Florida Flatwoods, and Northern Humid Gulf Coastal Prairies had the highest number of citizen science observations (Figure 4b).

3.2. Tallow Distribution within Plant Hardiness Zones

The USDA plant hardiness zones 8a, 8b, and 9a encompassed 99% and 92% of the total tallow observations for the FIA plot and citizen science observations, respectively (Figure 5). These three hardiness zones represent annual extreme minimum temperatures ranging from −12.2 °C (10 °F) to −3.9 °C (25 °F) over a 30-year period from 1976 to 2005 [33].

Hardiness zone 7b represented the northern limit of naturalized populations of tallow, with few observations reported in zones 7a, 6a, and 6b.

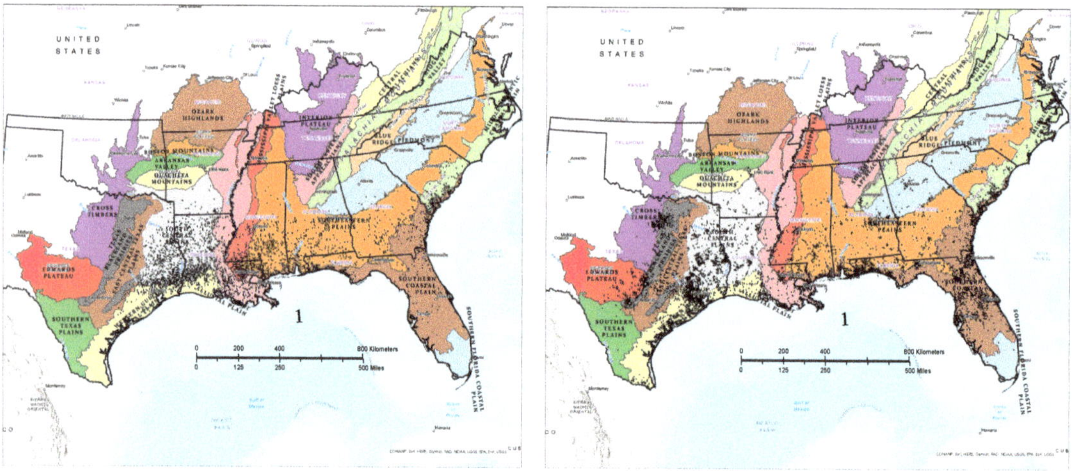

(a) (b)

Figure 3. Level III Ecoregions of the Conterminous United States classification with tallow: (**a**) Forest Inventory and Analysis (FIA) plot observations for the period from 2001 to 2022, and (**b**) citizen science observations for the period from 1991 to 2023, as reported in the iNaturalist and EDDMapS databases.

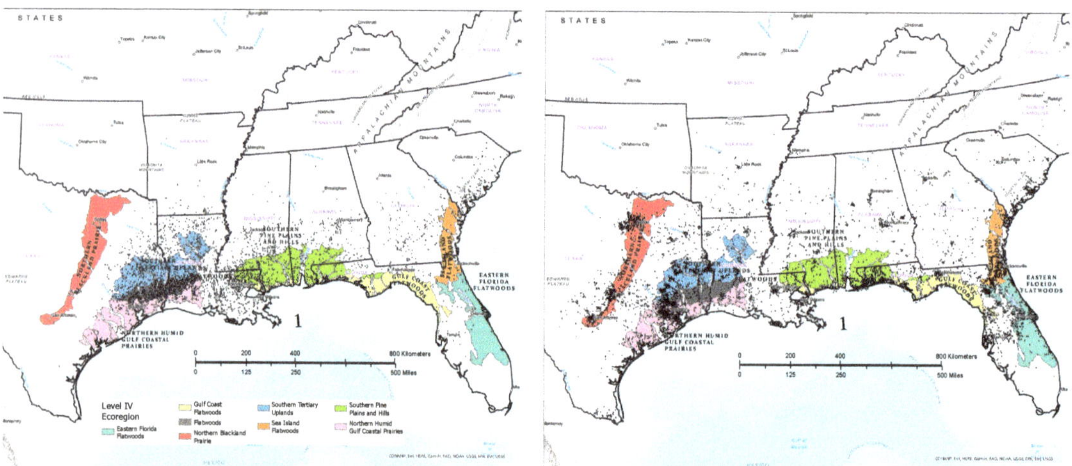

(a) (b)

Figure 4. Level IV Ecoregions of the Conterminous United States classification with tallow: (**a**) Forest Inventory and Analysis (FIA) plot observations for the period from 2001 to 2022, and (**b**) citizen science observations for the period from 1991 to 2023, as reported in the iNaturalist and EDDMapS databases.

3.3. *Structural and Compositional Characteristics of Tallow-Invaded Forestlands*

Based on the FIA data, in the 13-state region, there were 1225 newly invaded plots and 12,018 non-invaded plots included in the analysis. Comparisons between invaded and non-invaded plots showed that there was significantly lower basal area and Shannon's diversity index and significantly higher mean annual mortality and mean annual removals in the invaded plots (Table 2).

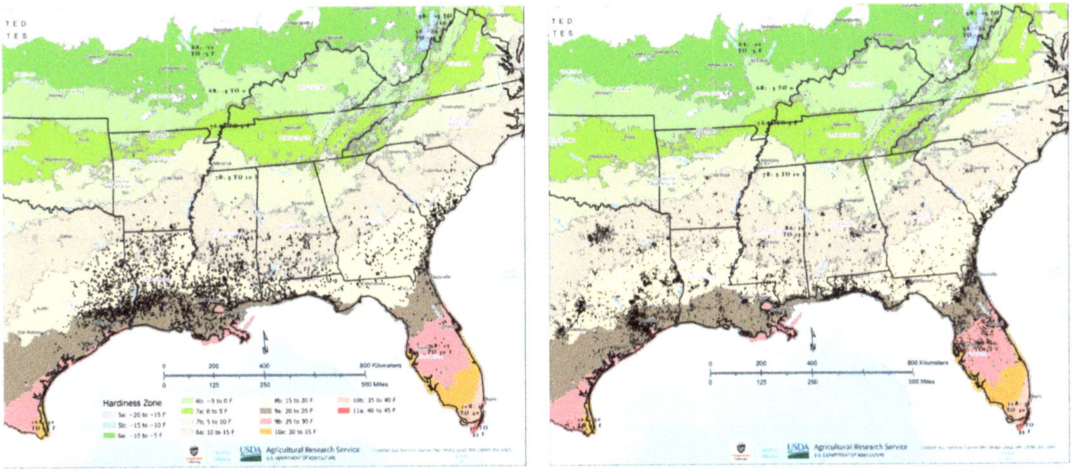

(a) (b)

Figure 5. USDA 2012 plant hardiness zones with tallow: (**a**) Forest Inventory and Analysis (FIA) plot observations for the period from 2001 to 2022, and (**b**) citizen science observations for the period from 1991 to 2023, as reported in the iNaturalist and EDDMapS databases.

Table 2. Mean and standard error summary statistics and statistical test results of eight response variables based on forest structural metrics, species composition, and disturbance rates for USDA Forest Service Forest Inventory and Analysis (FIA) plots grouped by tallow invasion status.

Variable	Non-Invaded	Invaded	*p* Value
Sample size	12,018	1225	
Basal area (m^2 ha^{-1})	20.88 (0.10)	19.81 (0.31)	0.0004
Density (trees ha^{-1})	1381.60 (10.61)	1401.96 (34.13)	0.5262
Shannon's diversity index	1.15 (0.01)	1.12 (0.01)	0.0204
Shannon's equitability index	0.19 (0.00)	0.19 (0.00)	0.9279
Species richness	5.98 (0.03)	5.82 (0.08)	0.4566
Mean annual mortality (% BA)	1.24 (0.02)	1.61 (0.07)	0.0037
Mean annual removals (% BA)	2.00 (0.04)	2.37 (0.14)	0.0018

3.4. Temporal Shifts in Tallow Detection

Shifts in the geographic mean centers indicated a westward and outward tallow spread along the Gulf Coast (Figure 6). For the FIA plots, the mean centers shifted more than 130 km westward early in the observation period, from 2001 to 2003 (Figure 6a). From 2003 to 2005, the FIA mean centers shifted westward farther into Louisiana by 19 km annually. From 2005 to 2007, the FIA mean center shifts were in a northeasterly direction at 20 km and 7 km, respectively. The FIA mean center shifted west by 10 km in 2008 and farther west by 5 km in 2009. From 2009 to 2022, the FIA mean centers continued shifting in a northeast direction at an average annual rate of 5 km. From 1991 to 2023, the mean centers of citizen science observations shifted consistently westward and outward along the Gulf Coast as new observations were added across the study area (Figure 6b). The earliest citizen science observations were recorded for Florida in 1991 in EDDMapS. As more observations were added from 1995 to 1998, the mean center shifted from Central Florida westward along the Gulf Coast (Figure 6b). By 1999, tallow observations in the Florida Panhandle and around Alexandria, Louisiana, resulted in further mean center shifts, 17 to 19 km annually, westward along the Gulf Coast. By 2002, observations around Lufkin, Texas, shifted the mean center even farther west, 363 km, to the southwestern corner of Stone County, Mississippi. In 2005, observations near Greenville, Mississippi, Montgomery,

Alabama, and Lake Village, Arkansas, shifted the mean center east. However, observations in East Texas in 2006 and 2007 shifted the mean center farther west by 196 km and 81 km, respectively. In 2010 and 2011, observations along the Atlantic Coast of Georgia and South Carolina, northern Mississippi, and Central Arkansas stationed the mean center around Mobile, Alabama. By 2017, more observations were recorded for Texas pulling the mean center westward, 20 km annually, into Louisiana through 2023.

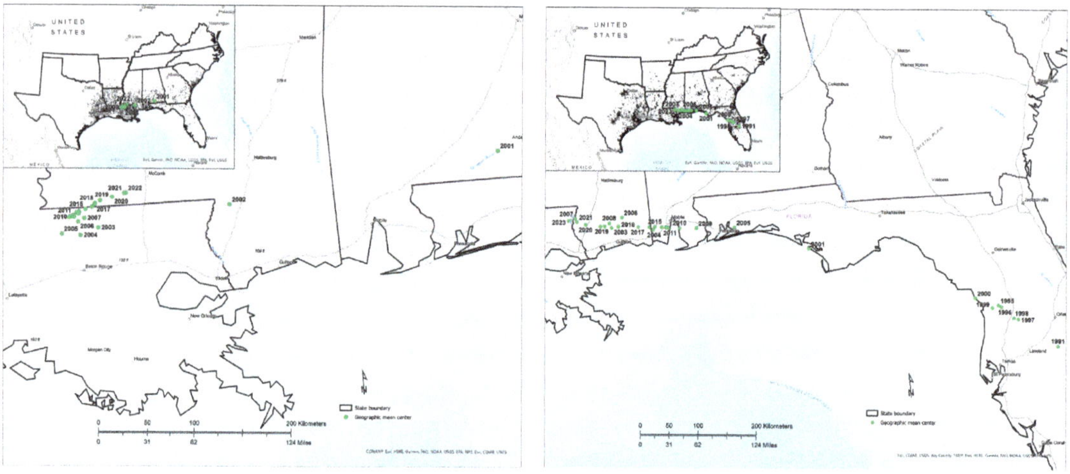

(**a**) (**b**)

Figure 6. Tallow observations geographic mean centers by year showing spread as reported for (**a**) Forest Inventory and Analysis (FIA) plots for the period from 2001 to 2022 and (**b**) citizen science data for the period from 1991 to 2023, as recorded in the iNaturalist and EDDMapS databases.

3.5. Comparison of Citizen Science and FIA Occurrence and Abundance

The FIA-based county-level map showed a preponderance of tallow on forestlands in Louisiana and Texas (Figure 7a). Liberty County, TX, had the highest density of live tallow trees and seedlings, followed by Calcasieu Parish, LA, Hardin County, TX, Allen Parish, LA, and Beauregard Parish, LA. Harris County, TX, which had the highest number of verified citizen science observations, ranked the 11th highest county using the FIA-based density of live tallow trees and seedlings. Superimposing verified tallow citizen science observations as a heat map on the FIA-based county map showed an agreement but also indicated concentrations of observations that are not currently depicted in the FIA-based county map (Figure 7b). Comparing the FIA-based county-level map (Figure 7a) to that reported in [9] for 2007, southeastern Texas as well as southwestern Louisiana and Mississippi continue to be areas of high tallow abundance. Similar to that reported in [10], Harris, Liberty, Hardin, and Jasper counties of Texas were some of the most heavily infested.

Tallow cover, regardless of plant height, on FIA plots was the highest in Texas, Louisiana, and Mississippi across the inventory years from 2001 to 2022 (Figure 8a). Virginia, Kentucky, and Oklahoma had no tallow observations on the FIA plots across those years. Heat map overlays of verified citizen science observations for the period from 1991 to 2023 showed a general agreement for hot spots of tallow occurrence but also highlighted tallow presence on non-forestland, especially around urban areas such as Dallas, TX, and Jacksonville, FL (Figure 8b). Using FIA plot cover data, tallow was first recorded in Alabama in 2001, with records in 2002 for Texas, Louisiana, Florida, Georgia, and South Carolina. Tallow cover spread northward and outward along the Gulf Coast and was recorded as far north as the Ozark St. Francis National Forest in 2019 (Figure 8a). The majority of FIA plots, 84%, across inventory years had less than 20% tallow cover, which is in agreement with that previously reported [12].

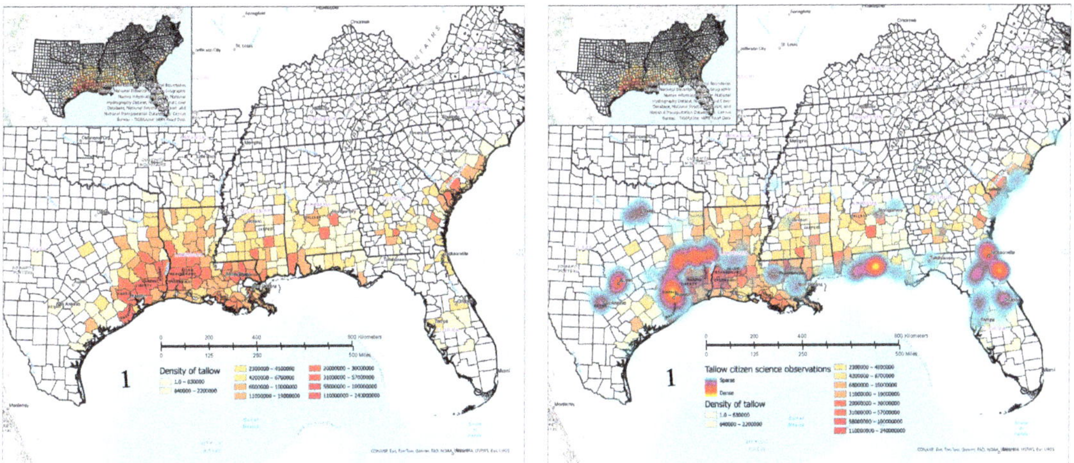

(**a**) (**b**)

Figure 7. Tallow distribution across the 13 southern states based on (**a**) density of live trees and seedlings on forestlands by county using Forest Inventory and Analysis (FIA) most recent inventories and (**b**) heat map of citizen science tallow observations for the period from 1991 to 2023 superimposed on FIA county-level map.

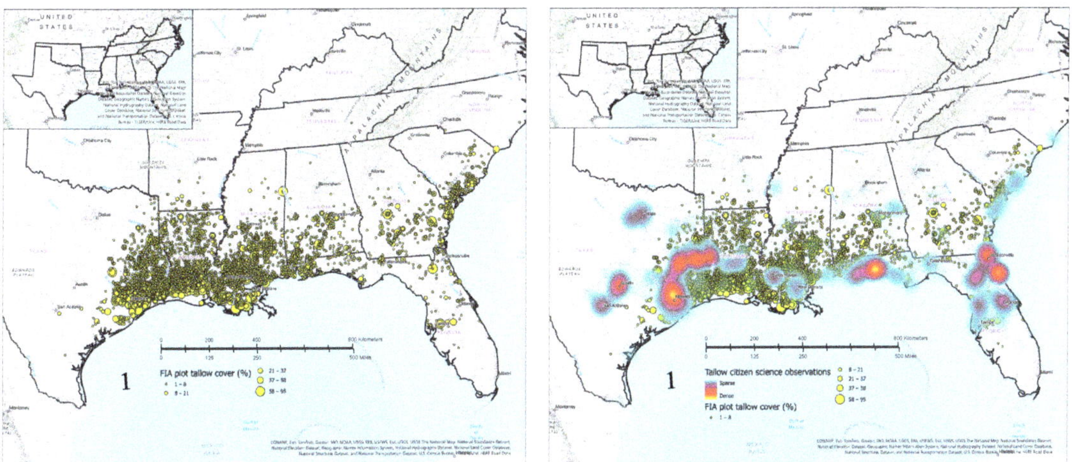

(**a**) (**b**)

Figure 8. Tallow distribution across the 13 southern states based on (**a**) cover (%) on Forest Inventory and Analysis (FIA) plots for the period from 2001 to 2022; (**b**) heat map of citizen science tallow observations for the period from 1991 to 2023 superimposed on FIA plot-level map.

Recently published tallow abundance maps using only FIA [21] reported that only a few East Texas counties and Louisiana parishes had low-to-minimal tallow abundance. However, compared to our citizen science (Figure 1) and pooled FIA plot-level/citizen science maps (Figure 9), tallow observations were concentrated east of the San Bernard/Navasota rivers and south of U.S. Highway 84 in Texas, with additional hotspots around Dallas, Austin, and San Antonio. Our maps also indicated that tallow in Louisiana is concentrated in three hotspots centered around Allen Parrish, Tangipahoa Parrish, and Jefferson Davis Parish. All counties south of Jackson, MS, were reported to have moderate-to-high tallow abundance [21]. Our maps also indicate that tallow is most frequent along the Mississippi Gulf Coast, with hotspots around Jackson and Laurel, MS.

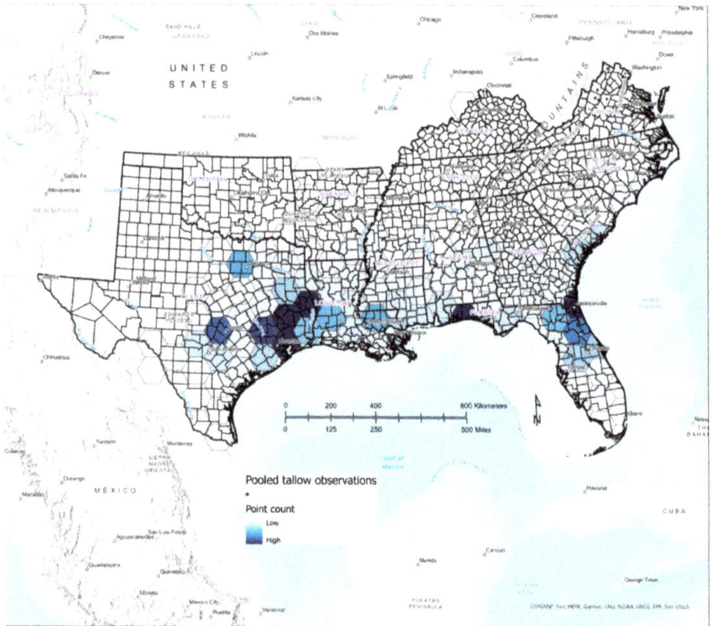

Figure 9. Polygon bins of tallow observations using pooled Forest Inventory and Analysis (FIA) plot data for the period from 2001 to 2022 and citizen science data, as reported in the iNaturalist and EDDMapS, for the period from 1991 to 2023.

4. Discussion

Chinese tallow is highly prevalent in the southern United States, ranking as one of the most abundant tree species in several states [8]. Observations from two online citizen science platforms and FIA data underscored the widespread distribution of tallow throughout the South and highlighted invasion hotspots. In agreement with previous reports [8,20], Louisiana, Texas, and Mississippi were the most impacted states, with tallow extending over most of Louisiana, the eastern half of Texas, and the southern portion of Mississippi. Invasion hotspots were centered around the southeastern corner of Texas, extending into Louisiana along the Interstate 10 Corridor into Mississippi, and covering the majority of the De Soto National Forest (Figure 9).

Tallow occurrences were associated with prairies and forests of the Gulf Coastal Plain. In agreement with previous reports [7,9], tallow observations compiled in this study were mainly associated with mixtures of pine and hardwood forests as well as prairie grasslands of the Coastal Plains. Flat, alluvial plains and terraces of poor to well-drained soils, generally known as flatwoods, were associated with most tallow observations, thus indicating a higher likelihood of invasion potential for these ecosystems. Prairies and marshes along the Gulf coasts of Texas and Louisiana were also vulnerable. In invaded forestlands, tallow was associated with lower basal areas, lower diversity, and higher mortality and removals. This is similar to previously published work associating tallow occurrence and abundance with areas of overstory disturbance and lower species richness [35–37].

The minimum daily temperature appears to be a key climatic constraint for tallow invasion. In our maps, USDA plant hardiness zone 7b represented the northern limit of naturalized tallow populations, which is in line with observations reported in [2]. Thus, indicating that annual extreme minimum temperatures of less than −12.2 °C (10 °F) may be limiting to tallow growth and development [9]. Minimum temperature, or cold stress, was also reported to be a critical factor in predicting tallow distribution under current climate and tallow expansion under future climate scenarios [3,9]. For example, acute extreme temperatures as low as −4.6 to −7.8 °C can impact the physiological performance and morphological traits of tallow, but subsequent growth may not be impacted [38]. Under current climate conditions, tallow was predicted to have a suitable climate in southern Missouri, Oklahoma, Tennessee, and Virginia [3]. Our maps indicated a limited number of tallow observations for these states, with no records in Missouri. Although climatic conditions may be favorable for tallow's invasion, the invasion range may also be constrained by propagule supply and dispersal [3]. It is worth mentioning, however, that states outside the advancing front of tallow's range may not have citizen science campaigns in place to allow for early detection, relying on a small number of plant experts for detection.

Tallow's invasion range is advancing northward and inland along the Gulf Coast. Tallow's general advancement was captured in citizen science and FIA data (Figure 6). The earliest tallow observations were recorded in the eastern portions of the southern United States, Florida, for citizen science, and Alabama and Georgia for FIA. As observation records became more robust, a consistent inland shift in geographic mean centers in a westerly and northeasterly directions was observed (Figure 6). Although distance associated with shifts in geographic mean centers may be viewed as a surrogate for tallow spread rate, we caution against this use for early observation records. Reliable estimates of the spread rate require effective detection and monitoring, which may not have been possible at the beginning of the data record for all sources. Effective delineation of tallow extent at the initiation of records for those data sources would also not have been possible. For instance, tallow introductions occurred in the late 1700s for Savannah, GA [39], and Charleston, NC [40], and plantings occurred in the early 1900s in Texas, Louisiana, and near Jacksonville, FL [6,40,41]. In the Texas prairies, tallow was well established by the 1940s, and remarkable population expansion into upland habitats occurred by the 1970s [7]. These temporal patterns of tallow establishment and spread were not captured in contemporary citizen science or FIA data. However, as tallow extent was captured more effectively by FIA and citizen science data over time, shifts in mean centers were more indicative of tallow spread rate, especially when records from remeasured FIA monitoring plots were incorporated, ranging from 5 to 20 km annually over the last decade. Lower tallow rates of spread, ranging from 1 to 3.3 km per year, were previously reported [11–13]. Our results indicate that tallow has yet to reach the range extent (Figure 9) predicted based on climatic factors [3]. Nonetheless, our results support previously reported trends of tallow spread northward and inland [3,9–12].

5. Conclusions

Citizen science data generally mirrored FIA observations. Tallow observations within prairies were more frequently reported by citizen scientists. FIA, however, included tallow observations along the Louisiana Coastal Prairies and Marshes. This may reflect the FIA forestland inventory focus as well as the association of urban development with coastal and prairie systems, where citizen science observations were more concentrated. As expected, the FIA-based tallow distribution did not capture observations in non-forestland such as urban settings, with concentrations around Dallas, Travis, and Bexar counties of Texas not being depicted in the FIA-based maps (Figures 7a and 8a). Despite the general agreement between FIA and citizen science data in depicting tallow distribution in the southern United States, each data source has limitations for depicting the spatiotemporal dynamics of tallow invasion at this scale. The systematic grid and remeasurements of FIA plots provide better estimates of change detection, spread rates, and invasibility level determination while also allowing for the characterization of demographic, structural, and compositional attributes associated with invasion. Citizen science observations, however, better reflect the extent of tallow distribution, especially in non-forestland settings, and provide more efficient early detection by delineating important sources of propagule supply and dispersal at finer spatial scales. The use of citizen science data can enhance species distribution mapping, especially when combined with data from established monitoring networks such as FIA. Integrating the use of citizen science data complements existing occurrence records and provides more comprehensive mapping and monitoring of invasive species extent and spatiotemporal dynamics.

Author Contributions: Conceptualization, M.M.B.; methodology, M.M.B., J.S.F. and L.S.P.K.; validation, M.M.B. and J.S.F.; formal analysis, M.M.B. and J.S.F.; investigation, M.M.B., J.S.F. and L.S.P.K.; resources, M.M.B., J.S.F. and L.S.P.K.; data curation, M.M.B. and J.S.F.; writing—original draft preparation, M.M.B.; writing—review and editing, M.M.B., J.S.F. and L.S.P.K.; visualization, M.M.B.; supervision, M.M.B.; project administration, M.M.B. All authors have read and agreed to the published version of the manuscript.

Funding: Support for the Article Processing Charge was provided by the U.S. Forest Service Southern Research Station Center for Forest Health and Disturbance—Research Work Unit 4552 Insects, Diseases, and Invasive Plants.

Data Availability Statement: Publicly available datasets from iNaturalist (www.inaturalist.org) and the Early Detection and Distribution Mapping System (www.eddmaps.org) were used in this study. All analyses and data generated during this study are included in this article.

Acknowledgments: This work would not have been possible without the contributions of citizen scientists and the two online citizen science platforms, iNaturalist (www.inaturalist.org) and the Early Detection and Distribution Mapping System (www.eddmaps.org). We thank all data contributors to these platforms. Forest Inventory and Analysis Program data contributions from field personnel, data analysts, and quality assurance personnel are acknowledged and greatly appreciated. We thank the administrative staff at the Center for Forest Health and Disturbance at the Southern Research Station for their support.

Conflicts of Interest: The authors declare no conflicts of interest.

References

1. Pile, L.S.; Wang, G.G.; Stovall, J.P.; Siemann, E.; Wheeler, G.S.; Gabler, C.A. Mechanisms of Chinese tallow (*Triadica sebifera*) invasion and their management implications—A review. *For. Ecol. Manag.* **2017**, *404*, 1–13. [CrossRef]
2. Bruce, K.A.; Cameron, G.N.; Harcombe, P.A.; Jubinsky, G. Introduction, Impact on Native Habitats, and Management of a Woody Invader, the Chinese Tallow Tree, *Sapium sebiferum* (L.) Roxb. *Nat. Areas J.* **1997**, *17*, 255–260.
3. Pattison, R.R.; Mack, R.N. Potential distribution of the invasive tree *Triadica sebifera* (Euphorbiaceae) in the United States: Evaluating climex predictions with field trials. *Glob. Chang. Biol.* **2008**, *14*, 813–826. [CrossRef]
4. DeWalt, S.J.; Siemann, E.; Rogers, W.E. Geographic distribution of genetic variation among native and introduced populations of Chinese tallow tree, *Triadica sebifera* (Euphorbiaceae). *Am. J. Bot.* **2011**, *98*, 1128–1138. [CrossRef] [PubMed]
5. Vogt, J.T.; Olatinwo, R.; Ulyshen, M.D.; Lucardi, R.D.; Saenz, D.; McKenney, J.L. An overview of *Triadica sebifera* (Chinese Tallowtree) in the Southern United States, emphasizing pollinator impacts and classical biological control. *Southeast Nat.* **2021**, *20*, 536–559. [CrossRef]
6. Jamieson, G.; McKinney, R. Stillingia oil. *Oil Soap* **1938**, *15*, 295–296. [CrossRef]
7. Bruce, K.A.; Cameron, G.N.; Harcombe, P.A. Initiation of a new woodland type on the Texas Coastal Prairie by the Chinese tallow tree (*Sapium sebiferum* (L.) Roxb.). *Torrey Bot. Soc.* **1995**, *122*, 215–225. [CrossRef]
8. Oswalt, S.N. *Chinese Tallow (Triadica sebifera (L.) Small) Population Expansion in Louisiana, East Texas, and Mississippi*; U.S. Department of Agriculture, Forest Service, Southern Research Station: Asheville, NC, USA, 2010.
9. Gan, J.; Miller, J.H.; Wang, H.; Taylor, J.W. Invasion of tallow tree into southern US forests: Influencing factors and implications for mitigation. *Can. J. For. Res.* **2009**, *39*, 1346–1356. [CrossRef]
10. Wang, H.-H.; Buchhorn, J.L.; Grant, W.E. Effects of Management on Range Expansion by Chinese Tallow in the Forestlands of Eastern Texas. *J. For.* **2014**, *112*, 346–353. [CrossRef]
11. Wang, H.-H.; Grant, W.E.; Swannack, T.M.; Gan, J.; Rogers, W.E.; Koralewski, T.E.; Miller, J.H.; Taylor, J.W. Predicted range expansion of Chinese tallow tree (*Triadica sebifera*) in forestlands of the southern United States. *Divers. Distrib.* **2011**, *17*, 552–565. [CrossRef]
12. Suriyamongkol, T.; McGrew, E.; Culpepper, L.; Beck, K.; Wang, H.-H.; Grant, W.E. Recent Range Expansions by Chinese Tallow (*Triadica sebifera* (L.) Small), the Most Prevalent Invasive Tree in the Forestlands of Eastern Texas. *Southeast Nat.* **2016**, *15*, 68–75. [CrossRef]
13. Renne, I.J.; Gauthreaux, S.A., Jr.; Gresham, C.A. Seed dispersal of the Chinese tallow tree (*Sapium sebiferum* (L.) Roxb.) by birds in coastal South Carolina. *Am. Midl. Nat.* **2000**, *144*, 202–215. [CrossRef]

14. Renne, I.J.; Spira, T.P.; Bridges, W.C. Effects of habitat, burial, age and passage through birds on germination and establishment of Chinese tallow tree in Coastal South Carolina. *J. Torrey Bot. Soc.* **2001**, *128*, 109–119. [CrossRef]
15. Siemann, E.; Rogers, W.E. Recruitment limitation, seedling performance and persistence of exotic tree monocultures. *Biol. Invasions* **2006**, *8*, 979–991. [CrossRef]
16. Pile, L.S.; Vickers, L.; Stambaugh, M.; Norman, C.; Wang, G.G. The tortoise and the hare: A race between native tree species and the invasive Chinese tallow. *For. Ecol. Manag.* **2019**, *445*, 110–121. [CrossRef]
17. Jones, R.H.; McLeod, K.W. Shade tolerance in seedlings of Chinese tallow tree, American sycamore, and cherrybark oak. *Bull. Torrey Bot. Club* **1989**, *116*, 371–377. [CrossRef]
18. Lin, J.; Harcombe, P.A.; Fulton, M.R.; Hall, R.W. Sapling growth and survivorship as affected by light and flooding in a river floodplain forest of southeast Texas. *Oecologia* **2004**, *139*, 399–407. [CrossRef]
19. Fan, Z.; Tan, Y.; Crosby, M.K. Factors associated with the spread of Chinese Tallow in East Texas forestlands. *Open J. Ecol.* **2012**, *2*, 121–130. [CrossRef]
20. Yang, S.; Fan, Z.; Liu, X.; Ezell, A.W. Predicting the spread of Chinese tallow (*Triadica sebifera*) in the southeastern United States forestland: Mechanism and risk factors at the regional scale. *For. Ecol. Manag.* **2021**, *482*, 118892. [CrossRef]
21. Nepal, S.; Moser, W.K.; Fan, Z. Spatiotemporal Invasion Severity of Chinese Tallow (*Triadica sebifera*) and Invasibility of Forest Types in Southern US Forestlands. *For. Sci.* **2021**, *67*, 491–500. [CrossRef]
22. Hochmair, H.H.; Scheffrahn, R.H.; Basille, M.; Boone, M. Evaluating the data quality of iNaturalist termite records. *PLoS ONE* **2020**, *15*, e0226534. [CrossRef]
23. Reinhardt, J.R.; Russell, M.B.; Senay, S.; Lazarus, W. Assessing the current and potential future distribution of four invasive forest plants in Minnesota, U.S.A., using mixed sources of data. *Sci. Rep.* **2020**, *10*, 12738. [CrossRef]
24. Cleland, D.T.; Avers, P.E.; McNab, W.H.; Jensen, M.E.; Bailey, R.G.; King, T.; Russell, W.E. *National Hierarchical Framework of Ecological Units*; Yale University Press: New Haven, CT, USA, 1997.
25. United States Census Bureau. 2020 Population Census Quick Facts. Available online: https://www.census.gov/quickfacts/geo/chart/US/POP010220 (accessed on 7 December 2023).
26. National Oceanic and Atmospheric Adminstration, SCIPP. Historical Climate Trends Tool. Available online: https://www.charts.climate.lsu.edu/trends (accessed on 7 December 2023).
27. National Oceanic and Atmospheric Adminstration, NCfEI. U.S. Climate Atlas. Available online: https://www.ncei.noaa.gov/access/climateatlas (accessed on 7 December 2023).
28. U.S. Department of Agriculture, Agricultural Research Service. USDA Plant Hardiness Zone Map. Available online: https://www.planthardiness.ars.usda.gov (accessed on 7 December 2023).
29. USDA Forest Service. Forest Inventory and Analysis National Program. Forest Inventory EVALIDator Web-Application Version 2.1.0. Available online: https://apps.fs.usda.gov/fiadb-api/evalidator (accessed on 18 October 2023).
30. USDA Forest Service. Forest Inventory and Analysis National Program. Forest Inventory DataMart Web-Application Version 2.0. Available online: https://apps.fs.usda.gov/fia/datamart/datamart.html (accessed on 18 October 2023).
31. Stanke, H.; Finley, A.O.; Weed, A.S.; Walters, B.F.; Domke, G.M. rFIA: An R package for estimation of forest attributes with the US Forest Inventory and Analysis database. *Environ. Model. Softw.* **2020**, *127*, 104664. [CrossRef]
32. Omernik, J.M.; Griffith, G.E. Ecoregions of the Conterminous United States: Evolution of a Hierarchical Spatial Framework. *Environ. Manag.* **2014**, *54*, 1249–1266. [CrossRef]
33. Daly, C.; Widrlechner, M.P.; Halbleib, M.D.; Smith, J.I.; Gibson, W.P. Development of a new USDA plant hardiness zone map for the United States. *J. Appl. Meteorol. Climatol.* **2012**, *51*, 242–264. [CrossRef]
34. R Core Team. *R: A Language and Environment for Statistical Computing*; R Foundation for Statistical Computing: Vienna, Austria, 2023.
35. Henkel, T.K.; Chambers, J.Q.; Baker, D.A. Delayed tree mortality and Chinese tallow (*Triadica sebifera*) population explosion in a Louisiana bottomland hardwood forest following Hurricane Katrina. *For. Ecol. Manag.* **2016**, *378*, 222–232. [CrossRef]
36. Johns, D.T.; Williams, B.; Williams, H.M.; Stroupe, M. Seedling survival and natural regeneration for a bottomland hardwood planting on sites differing in site preparation. In Proceedings of the 10th Biennial Southern Silvicultural Research Conference, USDA For. Serv., Gen. Tech. Rep. SRS-30, Southern Research Station, Asheville, NC, USA, 16–18 February 1999; pp. 145–150.
37. Pile, L.S.; Wang, G.G.; Knapp, B.O.; Walker, J.L.; Stambaugh, M.C. Chinese tallow (*Triadica sebifera*) invasion in maritime forests: The role of anthropogenic disturbance and its management implication. *For. Ecol. Manag.* **2017**, *398*, 10–24. [CrossRef]
38. Xie, H.; Pile Knapp, L.S.; Yu, M.; Wang, G.G. Morphological and physiological response of Chinese tallow (*Triadica sebifera*) to an extreme cold spell in subtropical, coastal forests of China. *Front. Ecol. Evol.* **2022**, *10*, 945886. [CrossRef]
39. Bell, M.I. *Some Notes and Refl Ections upon a Letter from Benjamin Franklin to Noble Wimberly Jones, 7 October 1772*; Ashantilly Press: Darien, GA, USA, 1966.
40. Elliott, S. *A Sketch of the Botany of South-Carolina and Georgia*; JR Schenck: Charleston, SC, USA, 1821; Volume 1.
41. Howes, F. The Chinese tallow tree (*Sapium sebiferum* Roxb.): A source of drying oil. *Kew Bull.* **1949**, *4*, 573–580. [CrossRef]

forests

MDPI

Article

Pest Risk Assessment of *Aeolesthes sarta* (Coleoptera: Cerambycidae) in Pakistan under Climate Change Scenario

Umer Hayat [1,†], Muhammad Akram [1,†], Sumeet Kour [2,†], Tahreem Arif [3] and Juan Shi [1,*]

[1] Sino-France Joint Laboratory for Invasive Forest Pests in Eurasia, Beijing Forestry University, Beijing 100083, China
[2] School of Agriculture, Geography, Environment, Ocean and Natural Sciences, The University of the South Pacific, Laucala Campus, Suva 999210, Fiji
[3] Department of Silviculture, School of Forestry, Beijing Forestry University, Beijing 100083, China
* Correspondence: shi_juan@263.net
† These authors contributed equally to this work.

Abstract: *Aeolesthes sarta* (Solsky 1871) (Coleoptera: Cerambycidae) is a polyphagous longhorned beetle species that primarily damages broadleaved tree species. This pest is distributed in the western and northern regions of Pakistan, where it caused serious damage to *Populus* spp. plantations. However, the growth and dispersal patterns of insects and pests are changing due to climate change. Modeling the range expansion or contraction of *A. sarta* development regions in Pakistan was the goal of the current study, assuming climate change might influence the geographical distribution of *A. sarta* in Pakistan. Under historical and future climatic conditions, *A. sarta* distribution areas were estimated using the CLIMEX model. Three time periods, 2030 (early century), 2070 (late century), and 2100 (end century), were forecasted for habitat suitability using the two climate change scenarios (CCSs) A1B and A2. Under the historic climatic condition (HCC), *A. sarta* was distributed in most areas of Pakistan, and its optimum habitat accounted for 71.67% of its total potential distribution. In the early-century period, optimum habitat dropped to 50.60% and 52.22% under A1B and A2 scenarios in the suitable condition. In the late-century period, optimum habitat further reduced to 31.76% and 30.60% under A1B and A2 scenarios. Moreover, at the end-century period, severe range shrinkage was predicted in the optimum habitat (19.99% under both CSSs). The model predicted a shift in the suitable habitat areas for *A. sarta* to the west and north. Furthermore, most climatically suitable areas under historic conditions became unsuitable during the end-century period. These projected results will assist in identifying the impacts of global warming on the possible distribution of *A. sarta*, thereby offering vital information for developing early forecasting and pest-prevention techniques to prevent further loss of forest and woodland trees.

Keywords: *Aeolesthes sarta*; city longhorned beetle (CLB); Coleoptera; Cerambycidae; forest pest; climate change; A1B–A2 scenarios; ecological niche model; CLIMEX; invasive species

1. Introduction

Approximately 30% of the Earth's land area is covered by forests [1]. Natural forests comprise about 5% of Pakistan's total land area, located mainly in the country's north and northwest mountainous regions [2]. Forests are essential to mitigate climate change as they serve as a carbon sink [2–4]. However, pathogens and pests each year kill millions of trees across commercial woodlands and natural forests [5]. The loss of trees resulting from major pest invasions can be catastrophic for total net productivity and carbon sequestration.

A hostile biotic potential threat to trees is pest infestations. The Coleoptera and Lepidoptera orders are home to the most frequently reported pest species [5]. In the Coleoptera order, Cerambycidae is one of the world's most prolific, diverse, ecologically and commercially significant families, which includes longhorn beetles [6]. *Aeolesthes sarta* (City Longhorned Beetle) is one of the dominant polyphagous pest species of the

Cerambycidae family, chiefly invading broadleaved tree species such as *Acer* spp., *Juglans* spp., *Malus* spp., *Platanus* spp., *Populus* spp., *Salix* spp., and *Ulmus* spp. [7]. Larvae of this family are internal feeders of living or dead tree tissues [8]. This behavior of internal feeding causes structural damage, which interjects the flow of water and nutrition, resulting in the loss of branches and, ultimately, whole tree death [9–12].

With a significant spread into Afghanistan, Iran, and other countries in central Asia, it is believed that *A. sarta* emerged in Pakistan and the western part of India [13,14]. Invasive pests thrive in areas with warm temperatures and their preferred host tree species [15]. This species may be highly hazardous in regions with hot and dry weather [16]. The longhorn beetle is a common pest in central Asia that attacks most broadleaved tree species. *A. sarta* can attack and proliferate on the main stem and major branches [17]. Infestations are particularly evident in highland forests, which may contribute to the decrease in poplar tree forests, which are a dynamic source of wood for the wood industry/market [18]. A severe infestation (high population) may cause the afflicted trees' canopy to decline and their leaves to dry off within two to three years [19].

Aeolesthes sarta has gained notoriety in Quetta and throughout Balochistan because it has harmed poplar, willow, and elm trees; for example, in the region (1900–1907), over 3000 trees were damaged by *A. sarta* [20]. *A. sarta* is one of the most destructive poplar borers [21] and has seriously harmed the number of *Populus* plantations across Pakistan [18,21,22].

The scientific foundation for assessing the hazards that plant pests cause to both cultivated and wild flora is supplied by pest risk assessments [23]. By calculating the potential for the introduction (entrance, transfer, and development) and spreading of plant pests and the ensuing effects on crops and plants in the larger environment, they include the rigorous synthesis of information and characterization of hazards [24].

Over the past century, global warming has been a major climate change concern [25]. The patterns of species dispersion have changed, affecting the areas where various species may survive and decreasing biodiversity [26]. Due to their ectothermic nature and wide variation in population size across time and geography, insects can acclimate to climate change well [27]. Many pests that widely disperse will also be able to expand their range more quickly than more stationary or habitat-specific species [28]. We can acquire vital information to control the spread of invasive pests by simulating how climate change impacts these pests [29].

Ecological niche models, including CLIMEX, CLIMATE, MaxEnt, DOMAIN, GARP, HABITAT, and ANUCLIM/BIOCLIM have all proven effective in evaluating the possible geographic dispersion of various invasive species [30]. Among these models, CLIMEX is a specialized toolset designed to investigate how climate affects invasive species and assess how such species might proliferate in the future [31]. Biological data can be frequently adjusted and simulated in CLIMEX, improving the findings. The key benefit of CLIMEX is that it can be theoretically tuned using knowledge of population demographics and phenological evidence [32].

Research on the precise potential distribution of *A. sarta* in Pakistan under projected climatic conditions is lacking. We modified parameter values following the literature published [7,18,21,22] to better match the known distribution in Pakistan. We hypothesized that a changing climate would affect the *A. sarta* range and the region in Pakistan where it thrives. Using the CLIMEX model, we looked into the potential distribution of *A. sarta* in Pakistan and assessed high-risk areas for its invasion. We assessed the vulnerability of different regions in Pakistan for *A. sarta* establishment under historical and future climates (A1B and A2 Climate Change Scenarios (CCSs)).

2. Materials and Methods

2.1. Study Species

Pest species known as *Aeolesthes sarta* (City Longhorned Beetle) grow in temperate locations with warm, dry weather [7]. This pest species is considered harmful since it can harm and destroy healthy and weak trees [21]. On a regional level, climate change is anticipated to modify the region's ideal for growth and development. Numerous studies in Pakistan have described the damage [18,21,22] and mitigation measures [33]; however, ecological niche models have never been used to predict the state of *A. sarta* development areas in a semi-arid nation such as Pakistan under a future climate change scenario. Therefore, this species was chosen for the current modeling research. We divided Pakistan into six administrative parts (Punjab (PB), Khyber Pakhtunkhwa (KPK), Azad Jammu and Kashmir (AJK), Balochistan (BLC), Sindh (SD), and Gilgit Baltistan (GB)) so that we could illustrate the *A. sarta* distribution well (Figure 1).

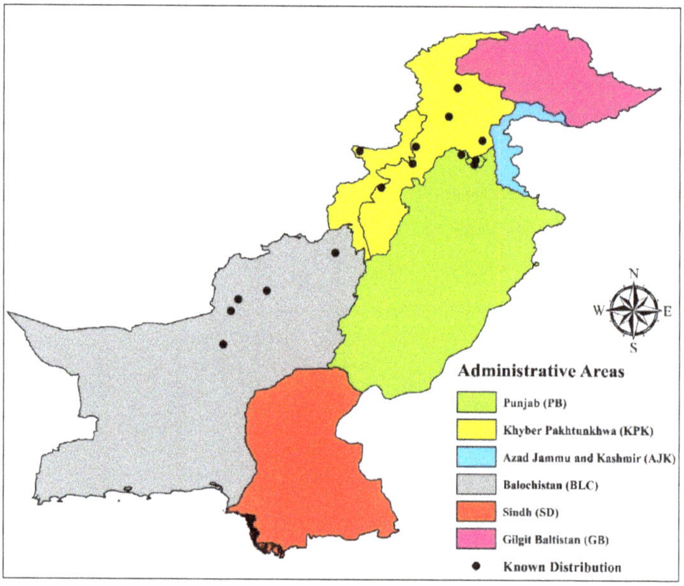

Figure 1. Map of Pakistan with six administrative areas (highlighted in six colors). Black dots represent the present known distribution of *Aeolesthes sarta* in Pakistan.

2.2. CLIMEX Model

In order to provide a set of yearly indices, the CLIMEX model incorporates generic morphological and physiological data and weekly interactions of the species to climate. The CLIMEX model takes on species distribution increases and decreases in response to suitable and unsuitable climatic conditions. The species' annual growth (GA) rate is used to compute the temperature index (TI) and moisture index (MI) and is further used to calculate the species' weekly growth [34]. Four different stresses (cold, dry, hot, and wet) impact the GA and add to stress indices (SI). Based on all these variables, CLIMEX generates an ecoclimatic index (EI) range of 0 to 100 [35]. EI = 0 represents that no growth is possible and EI = 100 indicates ideal growth. Geographical areas in this study that have an EI > 15, 0 < EI = 15, and EI = 0 indicate climatic conditions that are optimum, suitable, and unsuitable habitats for the establishment, survival, and proliferation of *A. sarta*. The CLIMEX uses a set of variables to determine the EI for a specific species [36]. The species' responsiveness determines these attributed to the environmental stresses stated above and its preferred temperature and moisture levels. These parameters' values might be inferred from the current distribution range or determined empirically. Therefore, the parameter fitting operation requires a thorough understanding of species biology and the current distribution spectrum [15]. The variables are modified until a good fit is attained. Following the fitting of a baseline model (under historical/current climatic conditions), predictions for future climatic conditions under various climate change scenarios could be established [37].

2.3. Climatic Data

Historical and future climatic data downloaded from CliMond (https://www.climond.org/ (accessed on 16 August 2022)). The 10′ gridded data had 26,323 location points for Pakistan and monthly averaged data on maximum and minimum temperature, rainfall, and daily relative humidity at 0900 h and 1500 h for each location. Historical climatic data averaged from 1961 to 1990 and finally averaged in 1975 [38].

Future climatic data are available based on two global circulation models (GCMs) and two separate climate change scenarios. These two GCMs were chosen for data generation because they performed better than the other available GCMs [38]. Data were acquired and utilized from the CliMond website for the 2030 (early-century), 2070 (late-century), and 2100 (end-century) time periods. Only data from the A1B and A2 SRES scenarios from CSIRO-Mk3.0 were employed to model the distribution of *A. sarta* in future climate change scenarios [39].

2.4. Collection of *A. sarta* Distribution Data in Pakistan

Aeolesthes sarta known distribution data in Pakistan collected from the Centre of Agriculture and Bioscience International (CABI →https://www.cabi.org/ (accessed on 15 November 2022)), Global Biodiversity Information Facility (GBIF →https://www.gbif.org/ (accessed on 15 November 2022)), European and Mediterranean Plant Protection Organization (EPPO →https://gd.eppo.int/ (accessed on 15 November 2022)), and from literature review [7,18,21,22,31] were used. Except in challenging environments, *A. sarta* has disseminated throughout Pakistan's western and northern mountainous regions.

2.5. Parameters Fitting

2.5.1. Temperature Index (TI) and Population Degree Days (PDD)

Aeolesthessarta is more common in temperate climates with dry and cold weather [7]; thus, while generating the "species parameter file," the Temperate template included with the CLIMEX software was used. Physiological tolerance criteria specific to each species were used to calibrate the model. According to Ahmad et al. [21], oviposition does not occur at <11 °C or >35 °C, whereas at 40 °C, species survive without laying viable eggs. However, Vanhanen et al. [40] used 20 °C and 37 °C as DV1 and DV2 values. Initially, we ran the model following Vanhanen et al. [40] and Ahmad et al. [21] using 11 °C, 20 °C, 35 °C, and 40 °C as DV0, DV1, DV2, and DV3 values, but we did not get all the known occurrence points covered, i.e., Kalam. Then, we adjusted the values to 10 °C (DV0), 15 °C (DV1), 37 °C (DV2), and 40 °C (DV3) and reran the model; it was able to predict the currently known occurrence. Therefore, we finalized these values as temperature indices to forecast the potential global distribution.

A. sarta completes its life cycle in almost two years [9,13,21]; therefore, we set the population day degree (PDD) value as 700, according to Vanhanen et al. [40].

2.5.2. Moisture Index (MI)

Another variable of the CLIMEX model is the soil moisture index (MI), which provides information about the amount of precipitation. The lower soil moisture threshold (SM0), lower optimal soil moisture (SM1), upper optimum soil moisture (SM2), and upper soil moisture threshold (SM3) are the four additional characteristics that the CLIMEX model uses to classify the MI [34]. SM0 was adjusted to 0 to allow for normal species development [40]. Soil moisture values for optimal growth (SM1 and SM2) were established at 0.001 and 1.5 to suit the observed occurrence of species in known distribution. Following a similar procedure, an upper soil moisture level (SM3) of 2.5 was adjusted [40].

2.5.3. Diapause Index

Diapause duration was reported to be around 90 days [9,13], so we set the DPD (diapause development days) as 90 days. DPD0 (diapause induction day length), DPT0 (diapause induction temperature), DPT1 (diapause termination temperature), and DPSW (summer or winter diapause) were set to be 12 days, 13 °C, 10 °C, and 0 days, respectively, according to [40].

2.5.4. Stress Index (SI)

Cold stress (CS), heat stress (HS), dry stress (DS), and wet stress (WS) are the four environmental stress indices in the CLIMEX modeling system that are reflective of hostile circumstances that restrict the population development of a species [34]. We only took CS and HS into consideration for this research study. Stress starts to restrict a species' development when the temperature falls below that species' cold stress threshold temperature (TTCS) at a particular rate (THCS). In the current study, 9 °C was considered the TTCS for *A. sarta* since *A. sarta* development is inhibited below this temperature [21].

Based on the TTCS result, the THCS was set to −0.00001. Similarly, heat stress affects a species as soon as the temperature surpasses its heat stress temperature threshold (TTHS) at a specific rate (THHS). In the present research, we set the TTHS to 41 °C and the THHS to 0.005 since *A. sarta* did not survive over 40 °C [21]. The parametric values used to run the CLIMEX model are shown in Table 1.

Table 1. *Aeolesthes sarta* parameter values used in the CLIMEX model.

Parameters	Code	Settled Values
Temperature		
Limiting low temperature (°C)	DV0	10
Lower optimal temperature (°C)	DV1	15
Upper optimal temperature (°C)	DV2	37
Limiting high temperature (°C)	DV3	40
Population degree day	PDD	700
Moisture Index		
Limiting low soil moisture	SM0	0
Lower optimal soil moisture	SM1	0.001
Upper optimal soil moisture	SM2	1.5
Limiting high soil moisture	SM3	2.5
Diapause Index		
Diapause induction day length	DPD0	12
Diapause induction temperature (°C)	DPT0	13
Diapause termination temperature (°C)	DPT1	10
Diapause development days	DPD	90
Summer or winter diapause	DPSW	0
Cold Stress		
CS temperature threshold (°C)	TTCS	9
CS temperature rate	THCS	−0.00001
Heat Stress		
HS temperature threshold (°C)	TTHS	41
HS temperature rate	THHS	0.005

3. Results

3.1. Potential Distribution of A. sarta in Pakistan under Historic Climatic Conditions (HCC)

The model calculation results illustrate that the known distributions gathered from CABI, EPPO, GBIF, and the literature review were within the potential distribution area projected by CLIMEX under historic climate conditions (HCC) (Figure 2). This demonstrates that the finalized parameters can accurately predict Pakistan's *A. sarta* distribution pattern. Using the map developed by ArcMap, we estimated the potential area of *A. sarta* distribution in Pakistan to be 7.32 HT-Km^2 (100,000 square kilometers), or 83.43% of Pakistan's total area (Figure 3).

The area covering optimum habitat (EI > 15) was 6.26 HT-Km^2, which is 71.67% of the total area of Pakistan, accounting for 85.91% of the total potential distribution area. This potential distribution comprises 100% of the area of Punjab (PB), 78.23% of Khyber Pakhtunkhwa (KPK), 67.48% of Azad Jammu and Kashmir (AJK), 80.58% of Balochistan (BLC), 37.45% of Sindh (SD), and 3.53% of Gilgit Baltistan (GB).

The area covering suitable habitat (0 < EI = 15) was 1.03HT-Km^2, 11.76% of the total area of Pakistan, accounting for 14.09% of the total potential distribution area. This covers 15.60% of KPK, 13.35% of AJK, 12.45% of BLC, 19.47% of SD, and 22.48% of GB. However, only 1.45 HT-Km^2, which is 16.57% of the total area of Pakistan, covers unsuitable habitats (EI = 0).

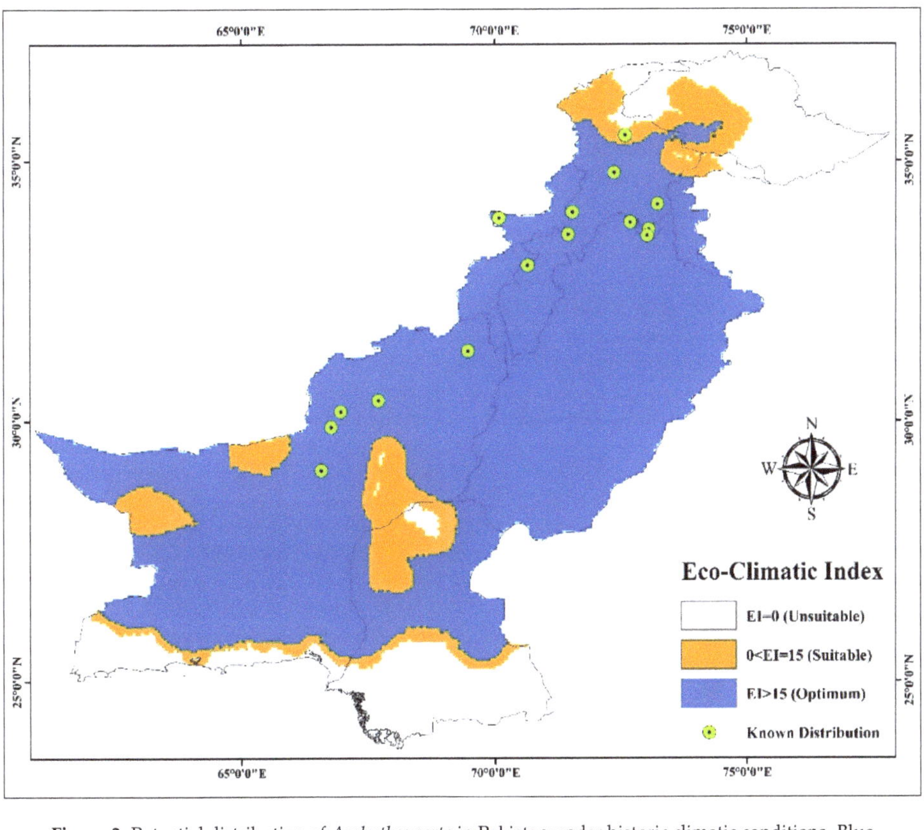

Figure 2. Potential distribution of *Aeolesthes sarta* in Pakistan under historic climatic conditions. Blue color indicates optimum habitat (EI > 15), orange color indicates suitable habitat (0 < EI = 15), white color indicates unsuitable habitat (EI = 0), and green circles indicate the current known distribution of *A. sarta* in Pakistan.

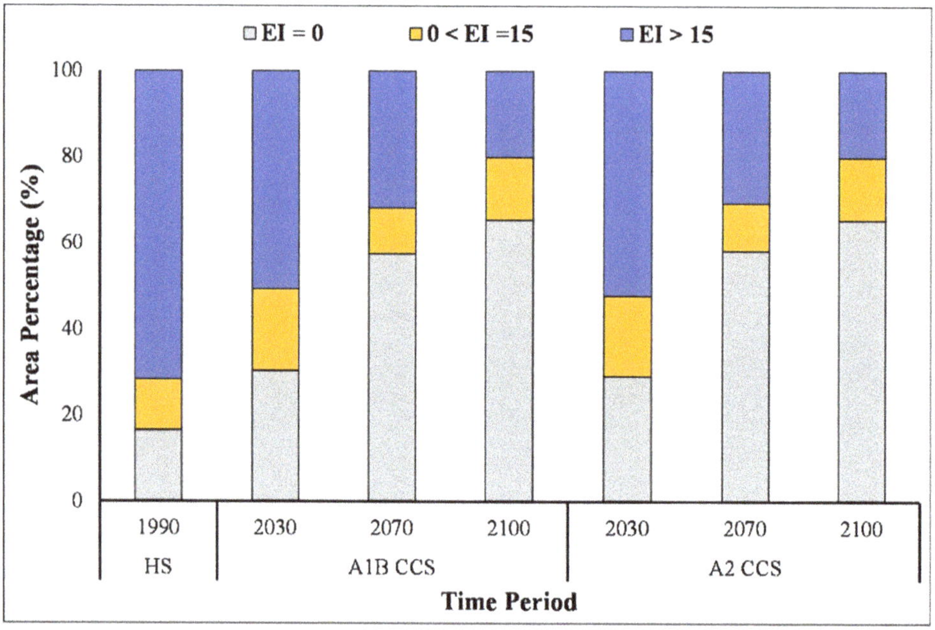

Figure 3. The global land area percentage of *Aelosthes sarta* within the three ecoclimatic index ranges under the historic and future climatic conditions under A1B and A2 climate change scenarios. Optimum habitat (EI > 15); suitable habitat (0 < EI = 15); and unsuitable habitat (EI = 0). HS = historic; CSS = climate change scenario.

3.2. Potential Distribution of A. sarta in Pakistan under Future Climatic Condition (FCC)

Under the future climatic condition (FCC), the model projected a gradual upsurge in the suitable habitat for *A. sarta* development areas for the early-century period (Figures 3 and 4). The unsuitable areas' percentage increased compared with HCC. Under A1B and A2 CCSs for the early-century period, the optimum habitat area for *A. sarta* development decreased to 50.60% (4.44 HT Km^2) and 52.22% (4.58 HT Km^2) compared with HCC 71.67% (6.28 HT Km^2), while the suitable habitat area increased to 19.02% (1.67 HT Km^2) and 18.71% (1.64 HT Km^2) compared with HCC 11.76% (1.03 HT Km^2). The unsuitable habitat area increased to 30.38% (2.66 HT Km^2) and 29.08% (2.55 HT Km^2) compared with historic climatic conditions and 16.57% (1.45 HT Km^2), respectively.

For the late-century period under A1B and A2 CCSs, the optimum habitat area for *A. sarta* development further decreased to 31.76% (2.78 HT Km^2) and 30.60% (2.68 HT Km^2) compared with historic climatic conditions, 71.67% (6.28 HT Km^2). In comparison, the suitable habitat area also decreased to 10.73% (0.94 HT Km^2) and 11.13% (0.98 HT Km^2) compared with HCC 11.76% (1.03 HT Km^2), and the unsuitable habitat area further increased to 57.51% (5.04 HT Km^2) and 58.26% (5.11 HT Km^2) compared with the historic climatic conditions of 16.57% (1.45 HT Km^2), respectively.

However, during the end-century period, the optimum habitat area for *A. sarta* development further decreased to 19.99% (1.75 HT Km^2) under A1B and A2 CCSs compared with HCC 71.67% (6.28 HT Km^2); the suitable habitat area increased to 14.60% (1.28 HT Km^2) under A1B and A2 CCSs compared with the late-century period but decreased compared with HCC 11.76% (1.03 HT Km^2); and the unsuitable habitat area further increased to 65.41% (5.74 HT Km^2) under A1B and A2 CCSs compared with HCC 16.57% (1.45 HT Km^2), respectively.

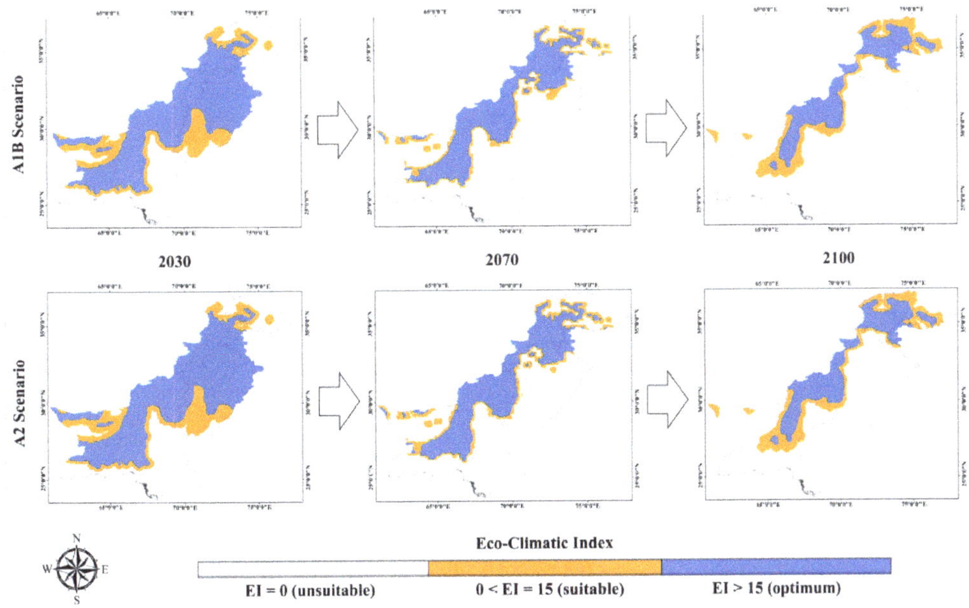

Figure 4. Forecast potential distribution of *Aelosthes sarta* in Pakistan under future climatic conditions for time periods of 2030, 2070, and 2100 under A1B and A2 climate change scenarios. The projected global distributions (three colors) narrate the ecoclimatic index (EI) from the CLIMEX model: blue →optimum (EI > 15); orange →suitable (0 < EI = 15), and white →unsuitable (EI = 0).

3.3. Administrative Area-Wise Distribution under Historic and Future Climatic Conditions

The optimum habitat area distribution was forecasted to decrease in five administrative areas except for GB (Table 2; Figure 5).

In Punjab (PB), 100% (20.63 TT Km^2) of the area under HCC was optimum for *A. sarta* development, but this trend decreased under FCC. In the early-century period, optimum habitat area decreased to 64.69% (13.34 TT Km^2) and 68.78% (14.18 TT Km^2) under A1B and A2 CCSs. In the late-century period, optimum habitat area decreased to 14.71% (3.03 TT Km^2) and 13.07% (2.69 TT Km^2) under both CCSs, and in the end-century period, the optimum habitat area further decreased to 1.29% (0.27 TT Km^2), whereas the unsuitable habitat area increased to 95.88% (19.77 TT Km^2) under both CCSs.

In Khyber Pakhtunkhwa (KPK) under HCC, 78.23% (7.85 TT Km^2) of the area comprised optimum habitat, and suitable habitat area accounted for 15.60% (1.57 TT Km^2). In the early-century period, the optimum habitat area shrunk to 78.79% (8.01 TT Km^2) and 79.99% (8.03 TT Km^2); in the late-century period, the optimum habitat area further reduced to 67.57% (6.78 TT Km^2) and 66.18% (6.64 TT Km^2) and in the end-century period, the optimum habitat area further reduced to 51.18% (5.14 TT Km^2) under both CCSs. However, suitable habitats decreased in the early and late centuries but increased in the end-century period to 27.73% (2.28 TT Km^2) under both CSSs.

In Azad Jammu and Kashmir (AJK), under HCC, optimum and suitable habitat areas comprise 67.48% (0.93 TT Km^2) and 13.35% (0.18 TT Km^2). In the early-century period, the optimum habitat area increased to 68.93% (0.95 TT Km^2) and 68.45% (0.94 TT Km^2), and suitable habitat decreased to 12.62% (0.17 TT Km^2) and 12.38% (0.17 TT Km^2) under both CCSs. However, in the late- and end-century periods, the optimum habitat decreased, and suitable habitat increased under both CCSs.

In Balochistan (BLC), for the early-century period, optimum habitat area reduced to 60.94% (20.91 TT Km^2) and 62.58% (21.47 TT Km^2), while the suitable habitat area increased to 21.07% (7.23 TT Km^2) and 20.99% (7.20 TT Km^2) under both CCSs compared with HCC 80.58% (27.65 TT Km^2) and 12.45% (4.27 TT Km^2). However, in the late- and end-century periods, optimum and suitable habitats decreased under both CCSs. In the end-century period, optimum habitat covers 25.55% (8.77 TT Km^2) while suitable habitat covers 18.65% (6.4 TT Km^2) tunder both CCSs.

In Sindh (SD), under HCC, the optimum and suitable habitat areas covered 37.35% (5.21 TT Km^2) and 19.47% (2.71 TT Km^2). In the early, late, and end centuries, optimum and suitable habitat areas were reduced to the minimum, while unsuitable habitat areas increased to the maximum under both CCSs.

In Gilgit Baltistan (GB) under HCC, optimum and suitable habitat areas covered 3.53% (0.24 TT Km^2) and 22.48% (1.53 TT Km^2). In the early-century period, optimum habitat area increased to 7.95% (0.54 TT Km^2) and 7.51% (0.51 TT Km^2), while suitable habitat also increased to 23.47% (1.59 TT Km^2) and 23.66% (1.61 TT Km^2) under both CCSs. However, in the late and end centuries, optimum and suitable habitat areas gradually increased with a decrease in unsuitable areas. At the end of the century, the optimum habitat area increased to 36.03% (2.45 TT Km^2) under both CCSs.

Based on the assessment of the model's findings, predictive maps on a regional basis of Pakistan show that the potential climate zone for *A. sarta* is confined in different areas by cold and heat stress (Figure 6). The northern region, such as the GB potential climatic zone boundaries, is restricted by cold stress under historic and early-century climatic conditions, which reduce under FCC at the end of the century. In contrast, the potential climatic zone limits of the southern, central, and eastern regions such as SD, PB, and south BLC are constrained by heat stress. However, the cold stress severity decreased under A1B and A2 CCSs, whereas heat stress increased over time.

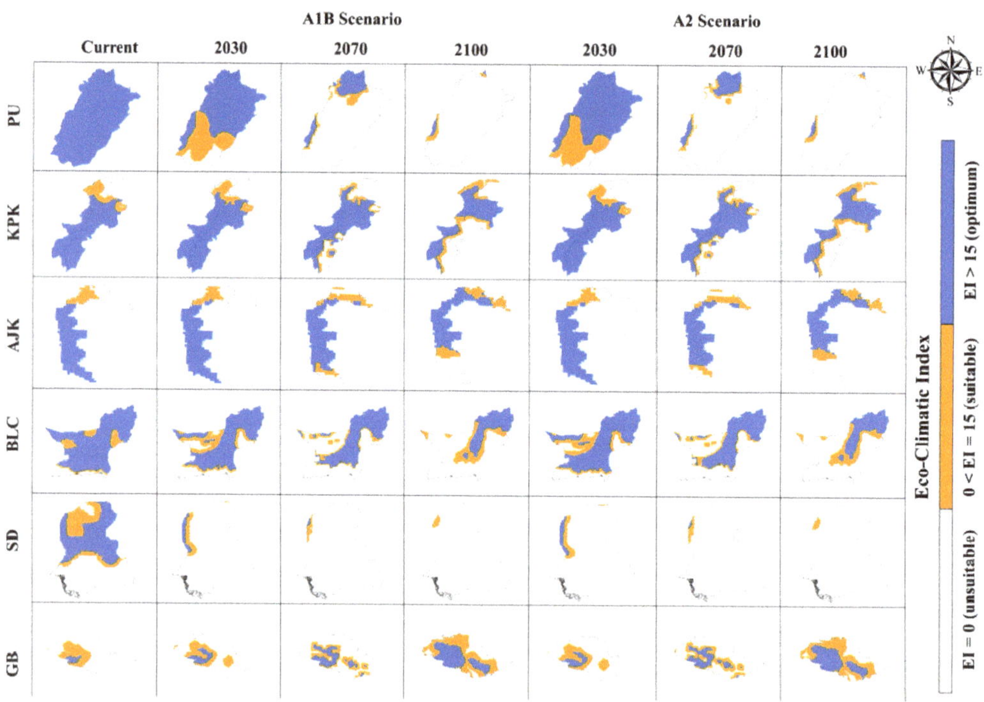

Figure 5. Predicted administrative area-wise potential distribution of *Aelosthes sarta* in Pakistan under historic and future climatic conditions in the time periods of 2030, 2070, and 2100 under A1B and A2 scenarios. Predicted regional distributions (three colors) narrate the ecoclimatic index (EI) from the CLIMEX model: blue = optimum habitat (EI > 15); orange = suitable habitat (0 < EI = 15); white = unsuitable habitat (EI = 0). PB = Punjab; KPK = Khyber Pakhtunkhwa; AJK = Azad Jammu and Kashmir; BLC = Balochistan; SD = Sindh; GB = Gilgit Baltistan.

Table 2. Administrative area-wise land area percentage of *Aelosthes sarta* within the three ecoclimatic index ranges under the historic and future climatic conditions under A1B and A2 scenarios.

Location	Scenario	Area (10,000 Km^2)											
		EI = 0				0 < EI = 15				EI > 15			
		1990	2030	2070	2100	1990	2030	2070	2100	1990	2030	2070	2100
PB	A1B scenario	0	1.70	15.98	19.77	0	5.59	1.61	0.58	20.62	13.34	3.03	0.27
KPK	A1B scenario	0.62	0.58	1.76	2.61	1.57	1.45	1.50	2.28	7.85	8.01	6.78	5.14
AJK	A1B scenario	0.26	0.25	0.25	0.28	0.18	0.17	0.28	0.30	0.93	0.95	0.85	0.78
BLC	A1B scenario	2.39	6.18	14.41	19.15	4.27	7.23	4.07	6.39	27.65	20.91	15.83	8.77
SD	A1B scenario	5.99	12.95	13.63	13.74	2.71	0.61	0.24	0.17	5.21	0.35	0.04	0
GB	A1B scenario	5.02	4.65	3.96	1.37	1.53	1.59	1.69	2.97	0.24	0.54	1.14	2.45
PB	A2 scenario	0	0.98	16.56	19.77	0	5.46	1.37	0.58	20.62	14.18	2.69	0.27
KPK	A2 scenario	0.62	0.69	1.78	2.62	1.57	1.32	1.62	2.28	7.85	8.03	6.64	5.14
AJK	A2 scenario	0.26	0.26	0.30	0.28	0.18	0.17	0.28	0.31	0.93	0.94	0.80	0.78
BLC	A2 scenario	2.39	5.64	14.64	19.15	4.27	7.20	4.36	6.40	27.65	21.47	15.32	8.77
SD	A2 scenario	5.99	12.94	13.61	13.74	2.71	0.60	0.27	0.17	5.21	0.37	0.04	0
GB	A2 scenario	5.02	4.67	3.77	1.37	1.53	1.61	1.84	2.97	0.24	0.51	1.18	2.45

EI = ecoclimatic index; PB = Punjab; KPK = Khyber Pakhtunkhwa; AJK = Azad Jammu and Kashmir; BLC = Balochistan; SD = Sindh; GB = Gilgit Baltistan.

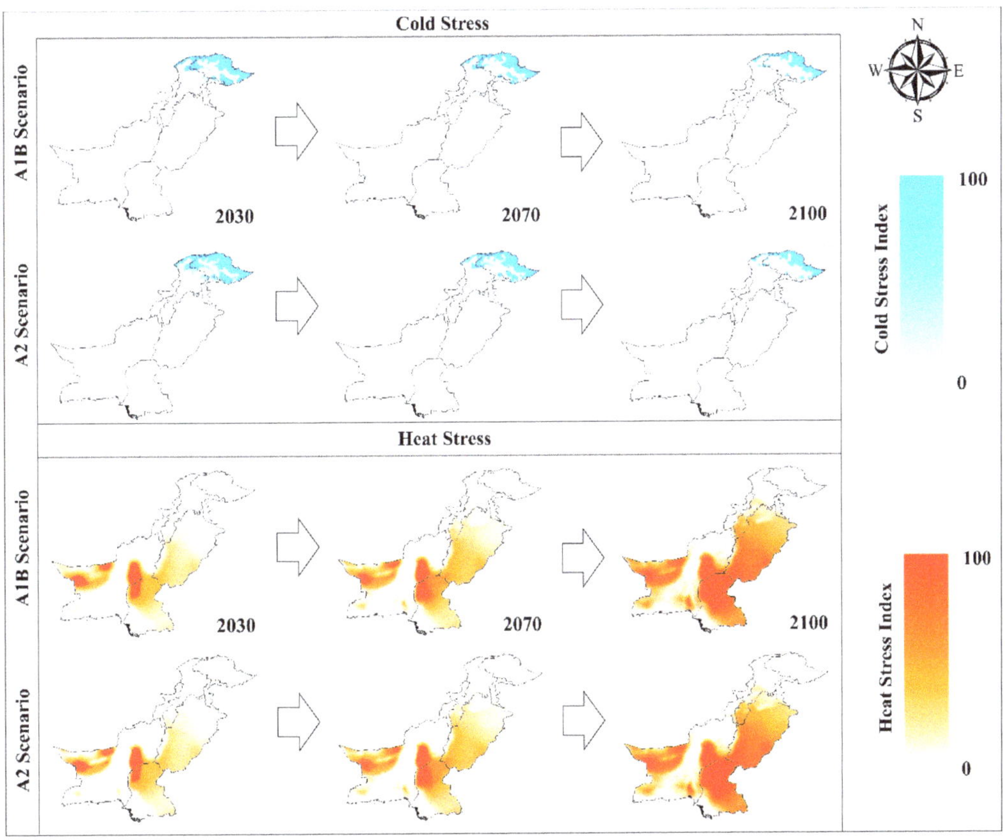

Figure 6. The predicted region in Pakistan where *Aelosthes sarta* survival is affected by cold and heat stress for different periods under the scenarios of A1B and A2 climate change.

4. Discussion

We aimed to provide a robust method for investigating the geographical distribution of *A. sarta* in Pakistan under the historic and future climates using CLIMEX. The results of this study demonstrated that, compared with the historic climate, future climatic conditions under A1B and A2 CCSs strongly influence the distribution pattern of *A. sarta* in Pakistan.

The habitat range of many species is significantly impacted by climate change, particularly cold-blooded species such as insects [41]. With the improvement of tools and techniques for climate modeling, it is now convenient to forecast the geographical range of a specified pest species under projected climate conditions, making timely monitoring and management strategies even more feasible [32,42]. CLIMEX can simulate a species' seasonal environmental adaptability [43] and numerous insect species' habitats have been predicted using it [15,43,44]. The notion that *A. sarta* exhibits polyphagous traits and has wreaked havoc on the plantations and forests in its native environment [7] suggests that it has the potential to damage trees seriously. As a result, we forecasted its possible distribution in Pakistan under the historic and future A1B and A2 climatic scenarios. All the cities in Pakistan where the distribution of *A. sarta* is documented fall within our projected range, demonstrating our model's accuracy.

Our CLIMEX models predicted that most of the regions in Pakistan, such as PB (EI > 15 = 100%), KPK (EI > 15 = 78.23), AJK (EI > 15 = 67.48%), BLC (EI > 15 = 80.58%), and SD (EI > 15 = 37.58%) under HCC were optimum for *A. sarta* development; the models also forecasted reduced development of the species at the end-century period under A1B and A2 CCSs. With some variations in climate change scenarios, the suitable regions predicted to be suitable under HCC would become unsuitable by the end of the century. The variations in the expected regions are caused by their various GHG emission scenarios [45]. Numerous researchers have stated that global warming will affect the possible insect/pest dispersion pattern [15,42–44].

Model predictions revealed that at the end-century period, the optimum habitat area reduced to northwest BLC, northwest KPK, central AJK, and eastern GB areas. Areas within the current known distribution of *A. sarta* are likely to remain in optimum habitat range for its development under FCC.

CLB is a diapause species [7]; adults undergo a suspended development phase during cold winter and require optimum temperatures such as 20 °C [46] to break dormancy and come out from the pupal chamber in spring [9,21]; therefore, its growth and mobility, as well as its dispersal, can be impacted by temperature.

We observed the shift of EI values from eastern to western regions and southern to northern regions of Pakistan. Central, eastern, and southern regions of Pakistan mainly comprise plain areas with dominancy of agricultural lands, while western and northern regions comprise hills and mountain ranges reaching up to 6600 m msl [15]. Moreover, due to the diversity of elevation, the surface temperatures of the central, eastern, and southern regions are higher than the western and northern mountain ranges. Almost 86% of natural forests in Pakistan are present in the country's northern regions (AJK, GB, and KPK) [47], with a diverse distribution of *A. sarta* preferred host tree species such as *Juglans* spp., *Malus* spp., *Platanus* spp., *Populus* spp., *Prunus* spp., *Pyrus* spp., *Salix* spp., *Ulmus* spp. [7,18,21,22,31], and *Quercus* spp. [7,48]. Therefore, the availability of preferred host species and suitable environmental conditions (temperature and elevation) makes these regions optimum for *A. sarta* development.

Pakistan is a region that is highly susceptible to climate change [49]. The optimum temperature range for *A. sarta* survival is between 10 and 40 °C. As a result, the temperature changes and generally higher temperatures in central and southern Pakistan would not be conducive to *A. sarta* habitation, which would probably cause a shift in the *A. sarta* distribution under FCC. Furthermore, the development of *A. sarta* in the PB, SD, and south BLC was observed to be constrained by increasing heat stress at the end of the 21st century period, but the GB region became more suitable due to a potential decline in cold stress (Figure 6). In addition, these regions have semi-arid to dry climates, making them more sensitive to climate change than Pakistan's northern region because even slight temperature changes can result in high heat in such climates. Over the past 50 years, extreme temperature variations and increased frequency of heavy rainstorms have been seen globally [39]. The frequency of cold days, cold nights, and frost occurrences has also decreased owing to changes in the climate, while the frequency of hot days/nights and heat waves has climbed [50,51].

The CLIMEX model predictions are mainly based on climatic variables, although non-climatic variables such as biological linkages, genetic evolution, visual barriers, and human activities might impact studies of species dispersion [52]. Recent research suggests invasive pest range changes under future climates may be predicted using genetic variation and climatic models [53]. The impact of non-climatic variables on the potential spread of *A. sarta* will be further investigated with the conception and advancement of species distribution models. In the following research, we will continue investigating *A. sarta* distribution in Pakistan, analyze the precise impacts of changes in the *A. sarta* external environment on its biological characteristics, and set up a method for evaluating model performance.

According to the study's model, in the next 78 years, most of Pakistan's western and northern regions will be a potentially suitable distribution area for *A. sarta*. Since the commencement of the 21st century, cultivation of trees on farmland and woodland in northern Pakistan has amplified considerably and has been utilized in furniture, wood, and sawmill industries and shipped to other parts of the country to fulfil the demand for timber and fuelwood. Additionally, *A. sarta* spends its egg, larvae, pupae, and young adult stages inside a tree trunk [7], so it easily spreads in the host for a long time. If efficient control measures are not implemented, *A. sarta* will continue to expand in Pakistan. *A. sarta* advanced surveillance and early warning technology can precisely predict its temporal dynamics and provide a firm assurance for halting it from spreading and causing more damage.

Once a tree is detected to be infected by *A. sarta*, it is recommended to fall and burn the contaminated tree [18,54]. Furthermore, it is recommended to follow sanitary/phytosanitary procedures before using the wood for trading, such as stem debarking and kiln drying [48]. Biological control techniques can be more economical, practical, ecologically safe, and long lasting [55]; therefore, we highly recommend the use of biological control agents such as the use of click beetle *Alaus* larvae (a predator of *A. sarta* grubs) [22], *Proctolaelaps* spp. (Acarina) (a predator of *A. sarta* grubs) [22], parasitoid *Sclerodermus turkmenicus* (a parasite of *A. sarta* larvae) [56], and *Beauveria bassiana* white muscardine fungus (a parasite of *A. sarta* adults) [18,56,57]. All these are very effective biological agents in stem borer control. We highly advise avoiding excessive chemical use for *A. sarta* control as these are not environmentally friendly and costly [5,58].

5. Conclusions

In this model-based research study, the CLIMEX model was used, which is the first study of its kind, leveraging accurate parametric variables to forecast the probable distribution of *A. sarta* in Pakistan under historic and future climatic conditions (A1B and A2 climate change scenarios). This work illustrated a helpful technique for mapping *A. sarta* habitat adaptability. The results show that in the next 78 years, most of Pakistan's western and northern regions will be potential distribution areas for *A. sarta*. With the warming of the climate, the optimum habitats are mainly located in the western and northern parts of Pakistan, while the central, eastern, and southern parts of Pakistan appear likely to become unsuitable areas. Four of the six administrative areas will remain the optimum habitat under FCC, such as BLC, KPK, AJK, and GB. To avoid the further spread of *A. sarta* in Pakistan and to guarantee the forest's security and the forest industry, Pakistan should continue to undertake stringent monitoring, preventive, and control measures.

Author Contributions: Conceptualization, U.H., M.A., S.K. and J.S.; investigation and methodology, U.H., M.A. and S.K.; formal analysis, U.H., M.A. and S.K.; investigation, U.H.; resources, U.H., M.A., S.K. and T.A.; data curation, UH; writing—original draft preparation, U.H. and T.A.; writing—review and editing, U.H., M.A., S.K., T.A. and J.S.; supervision and project administration, J.S.; funding acquisition, J.S. All authors have read and agreed to the published version of the manuscript.

Funding: This research was funded by (1) National Key R&D Program of China (Grant No. 2021YFC2600400); (2) National Natural Science Foundation of China (NSFC) (Grant No. 32171794); and (3) Forestry Science and Technology Innovation Special of Jiangxi Forestry Department (Grant No. 201912).

Data Availability Statement: The original contributions presented in the study are included in the article; further inquiries can be directed to the corresponding author.

Acknowledgments: We would like to thank our lab mates at the Sino-France Joint Laboratory for Invasive Forest Pests in Eurasia, Beijing Forestry University, for administrative and technical support.

Conflicts of Interest: The authors declare no conflict of interest.

Abbreviations

CCS	Climate change scenario
HCC	Historic climatic condition
FCC	Future climatic condition

References

1. Keenan, R.J.; Reams, G.A.; Achard, F.; de Freitas, J.V.; Grainger, A.; Lindquist, E. Dynamics of global forest area: Results from the FAO Global Forest Resources Assessment 2015. *Forest Ecol. Manag.* **2015**, *352*, 9–20. [CrossRef]
2. Khan, I.; Hayat, U.; Mujahid, A.; Majid, A.; Chaudhary, A.; Badshah, M.T.; Huang, J. Evaluation of growing stock, biomass and soil carbons and their association with a diameter: A case study from planted chir pine (*Pinus roxburghii*) forest. *Appl. Ecol. Environ. Res.* **2021**, *19*, 1457–1472. [CrossRef]
3. Hayat, U.; Khan, I.; Roman, M. A response of tropical tree *Terminalia arjuna* (Arjun) [Combretaceae-Myrtales] seeds and seedlings towards the presowing seed treatment in forest nursery of Sargodha, Pakistan. *Int. J. Biosci.* **2020**, *16*, 495–505. [CrossRef]
4. Farooqi, T.J.A.; Hayat, U.; Roman, M.; Abbas, H.; Hussain, S. Comparative study determining the impacts of broadleaved and Needle leaved forest harvesting on hydrology and water yield: State of knowledge and research outlook. *Int. J. Biosci.* **2020**, *16*, 231–240. [CrossRef]
5. Balla, A.; Silini, A.; Cherif-Silini, H.; ChenariBouket, A.; Moser, W.K.; Nowakowska, J.A.; Oszako, T.; Benia, F.; Belbahri, L. TheThreat of Pests and Pathogens andthe Potential for Biological Control in Forest Ecosystems. *Forests* **2021**, *12*, 1579. [CrossRef]
6. Rossa, R.; Goczał, J. Global diversity and distribution of longhorn beetles (Coleoptera: Cerambycidae). *Eur. Zool. J.* **2021**, *88*, 289–302. [CrossRef]
7. Hayat, U. City longhorn beetle (*Aeolesthes sarta*): A review of the species, its morphology, distribution, damage, prevention and control. *J. For. Sci.* **2022**, *68*, 199–212. [CrossRef]
8. Ślipiński, A.; Escalona, H. *Australian Longhorn Beetles (Coleoptera: Cerambycidae). Vol. 2, Subfamily Cerambycinae*; CSIRO Publishing: Clayton, CA, USA, 2016; p. 640. [CrossRef]
9. Khan, S.A.; Bhatia, S.; Tripathi, N. Entomological investigation on *Aeolesthes sarta* (Solsky), a major pest on walnut trees (*Juglans regia* L.) in Kashmir valley. *J. Acad. Ind. Res.* **2013**, *2*, 325.
10. Mazaheri, A.; Khajehali, J.; Marzieh, K.; Hatami, B. Laboratory and field evaluation of insecticides for the control of *Aeolesthes sarta* Solsky (Col.: Cerambycidae). *J. Crop Prot.* **2015**, *4*, 257–266.
11. Morewood, W.D.; Hoover, K.; Neiner, P.R.; McNeil, J.R.; Sellmer, J.C. Host tree resistance against the polyphagous wood-boring beetle *Anoplophora glabripennis*. *Entomol. Et Appl.* **2004**, *110*, 79–86. [CrossRef]
12. Poland, T.M.; Haack, R.A.; Petrice, T.R.; Miller, D.L.; Bauer, L.S.; Gao, R. Field evaluations of systemic insecticides for control of *Anoplophora glabripennis* (Coleoptera: Cerambycidae) in China. *J. Econ. Entomol.* **2006**, *99*, 383–392. [CrossRef]
13. Farashiani, M.E.; Shamohammadi, D.; Sadeghi, S.E. Biological study of Sart long horn beetle, *Aeolesthes sarta* Solsky (Coleoptera: Cerambycidae) in the laboratory. *J. Entomol. Soc. Iran* **2000**, *20*, 77–90.
14. Orlinskii, A.D. Outcomes of the EPPO project on quarantine pests for forestry 1. *Eppo. Bull.* **2006**, *36*, 497–511. [CrossRef]
15. Hayat, U.; Qin, H.; Zhao, J.; Akram, M.; Shi, J.; Ya, Z. Variation in the potential distribution of *Agrotis ipsilon* (Hufnagel) globally and in Pakistan under current and future climatic conditions. *Plant Prot. Sci.* **2021**, *57*, 148–158. [CrossRef]
16. CABI. Forestry Compendium. 2006. Available online: https://www.cabi.org/fc (accessed on 15 November 2022).
17. Farashiani, M.E.; Sadeghi, S.E.; Abaii, M. Geographic distribution and hosts of Sart longhorn beetle, *Aeolesthes sarta* Solsky (Col.: Cerambycidae) in Iran. *J. Entomol. Soc. Iran* **2001**, *20*, 81–96.

18. Arshad, M.; Hafiz, I.A. Microbial trials of a pathogenic fungus, *Beauveria bassiana* (Bals.) Vuill. against the adults of *Aeolesthes sartus* Solsky (Cerambycidae: Coleoptera). *Pak. J. Zool.* **1983**, *15*, 213–215.
19. Krivosheina, N.P.; Tokgaev, T.B. The formation of trunk-insect complexes on irrigated areas in the Kopet-dag foothills. *Izv. Akad. Nauk. Turkm. SSR Biol. Nauk.* **1985**, *5*, 34–40.
20. Stebbing, E.P. *Indian Forest Insects of Economic Importance: Coleoptera*; Eyre & Spottiswoode: London, UK, 1914; p. 648.
21. Ahmad, M.I.; Hafiz, I.A.; Chaudhry, M.I. Biological studies on *Aeolesthes sarta* Solsky attacking poplars in Pakistan. *Pak. J. For.* **1977**, *27*, 123–129.
22. Gul, H.; Chaudhry, M.I. Some observations on natural enemies of poplar borers in Pakistan. *Pak. J. For.* **1992**, *42*, 214–222.
23. EFSA Panel on Plant Health (PLH); Jeger, M.; Bragard, C.; Caffier, D.; Candresse, T.; Chatzivassiliou, E.; Dehnen-Schmutz, K.; Grégoire, J.C.; Jaques Miret, J.A.; MacLeod, A.; et al. Guidance on quantitative pest risk assessment. *Efsa J.* **2018**, *16*, e05350. [CrossRef] [PubMed]
24. FAO (Food and Agriculture Organization of the United Nations). *ISPM (International Standards for Phytosanitary Measures) No. 11. Pest Risk Analysis for Quarantine Pests*; FAO: Rome, Italy, 2017; 36p. Available online: https://www.ippc.int/en/publications/639/ (accessed on 28 November 2022).
25. Stocker, T.F.; Qin, D.; Plattner, G.K.; Tignor, M.M.; Allen, S.K.; Boschung, J.; Nauels, A.; Xia, Y.; Bex, V.; Midgley, P.M. *Climate Change 2013: The Physical Science Basis. Contribution of Working Group I to the Fifth Assessment Report of IPCC the Intergovernmental Panel on Climate Change*; Cambridge University Press: Cambridge, UK, 2014. [CrossRef]
26. Gao, X.; Zhao, Q.; Wei, J.; Zhang, H. Study on the Potential Distribution of *Leptinotarsa decemlineata* and Its Natural Enemy *Picromerusbidens* Under Climate Change. *Front. Ecol. Evol.* **2022**, *9*, 786436. [CrossRef]
27. Bale, J.S.; Masters, G.J.; Hodkinson, I.D.; Awmack, C.; Bezemer, T.M.; Brown, V.K.; Butterfield, J.; Buse, A.; Coulson, J.C.; Farrar, J.; et al. Herbivory in global climate change research: Direct effects of rising temperature on insect herbivores. *Glob. Change Biol.* **2002**, *8*, 1–16. [CrossRef]
28. Pöyry, J.; Luoto, M.; Heikkinen, R.K.; Kuussaari, M.; Saarinen, K. Species traits explain recent range shifts of Finnish butterflies. *Glob. Change Biol.* **2009**, *15*, 732–743. [CrossRef]
29. Wei, J.; Peng, L.; He, Z.; Lu, Y.; Wang, F. Potential distribution of two invasive pineapple pests under climate change. *Pest Manag. Sci.* **2020**, *76*, 1652–1663. [CrossRef]
30. Ge, X.Z.; He, S.Y.; Zhu, C.Y.; Wang, T.; Xu, Z.C.; Zong, S.X. Projecting the current and future potential global distribution of *Hyphantriacunea* (Lepidoptera: Arctiidae) using CLIMEX. *Pest Manag. Sci.* **2019**, *75*, 160–169. [CrossRef] [PubMed]
31. Kriticos, D.J.; Maywald, G.F.; Yonow, T.; Zurcher, E.J.; Herrmann, N.I.; Sutherst, R. Exploring the effects of climate on plants, animals and diseases. *CLIMEX Version* **2015**, *4*, 184.
32. Early, R.; Rwomushana, I.; Chipabika, G.; Day, R. Comparing, evaluating and combining statistical species distribution models and CLIMEX to forecast the distributions of emerging crop pests. *Pest Manag. Sci.* **2022**, *78*, 671–683. [CrossRef]
33. Kamran, K.; Kakar, A.; Arif, S.; Iqbal, A. Evaluation of insect repellent and insecticide implantation techniques against *Aeolesthes sarta* Solsky in Quetta district of Baluchistan province, Pakistan. *Pak. J. Entomol. Zool. Stud.* **2017**, *5*, 273–276.
34. Sutherst, R.W.; Maywald, G.F.; Kriticos, D.J. *CLIMEX Version 3: User's Guide*; Hearne Scientific Software: South Yarra, Australia, 2007; p. 47.
35. Sutherst, R.W.; Maywald, G.F.; Bourne, A.S. Including species interactions in risk assessments for global change. *Glob. Change Biol.* **2007**, *13*, 1843–1859. [CrossRef]
36. Shabani, F.; Kumar, L.; Solhjouy-fard, S. Variances in the projections, resultingfrom CLIMEX, Boosted Regression Trees and Random Forests techniques. *Theor. Appl. Climatol.* **2017**, *129*, 801–814. [CrossRef]
37. Kriticos, D.J.; Watt, M.S.; Potter, K.J.B.; Manning, L.K.; Alexander, N.S.; Tallent-Halsell, N. Managing invasive weeds under climate change: Considering the current and potential future distribution of Buddlejadavidii. *Weed Res.* **2011**, *51*, 85–96. [CrossRef]
38. Kriticos, D.J.; Webber, B.L.; Leriche, A.; Ota, N.; Macadam, I.; Bathols, J.; Scott, J.K. CliMond: Global high-resolution historical and future scenario climate surfaces for bioclimatic modelling. *Methods Ecol. Evol.* **2012**, *3*, 53–64. [CrossRef]
39. IPCC. Climate Change 2007: Synthesis Report. In *Contribution of Working Groups I, II and III to the Fourth Assessment Report of the Intergovernmental Panel on Climate Change*; Core Writing Team, Pachauri, R.K., Reisinger, A., Eds.; IPCC: Geneva, Switzerland, 2007; p. 104. Available online: https://www.ipcc.ch/site/assets/uploads/2018/02/ar4_syr_full_report.pdf (accessed on 10 August 2022).
40. Vanhanen, H.; Veteli, T.O.; Niemelä, P. Potential distribution ranges in Europe for *Aeolesthes sarta*, Tetropiumgracilicorne and Xylotrechusaltaicus, a CLIMEX analysis. *EPPO Bull.* **2008**, *38*, 239–248. [CrossRef]
41. Aljaryian, R.; Kumar, L.; Taylor, S. Modelling the current and potential future distributions of the sunn pest *Eurygasterintegriceps* (Hemiptera: Scutelleridae) using CLIMEX. *Pest Manag. Sci.* **2016**, *72*, 1989–2000. [CrossRef] [PubMed]
42. Wei, J.F.; Zhao, Q.; Zhao, W.Q.; Zhang, H.F. Predicting the potential distributions of the invasive cycad scale *Aulacaspisyasumatsui* (Hemiptera: Diaspididae) under different climate change scenarios and the implications for management. *PeerJ* **2018**, *6*, e4832. [CrossRef]
43. Zhao, J.; Song, C.; Ma, L.; Yan, X.; Shi, J.; Hao, C. The Impacts of Climate Change on the Potential Distribution of *Plodia interpunctella* (Hübner)(Lepidoptera: Pyralidae) in China. *Insects* **2022**, *13*, 636. [CrossRef]
44. Ding, W.C.; Li, H.Y.; Wen, J.B. Climate Change Impacts on the Potential Distribution of *Apocheimacinerarius* (Erschoff) (Lepidoptera:Geometridae). *Insects* **2022**, *13*, 59. [CrossRef]

45. Farooq, S.; Maqbool, M.M.; Bashir, M.A.; Ullah, M.I.; Shah, R.U.; Ali, H.M.; Al Farraj, D.A.; Elshikh, M.S.; Hatamleh, A.A.; Bashir, S.; et al. Production suitability of date palm under changing climate in a semi-arid region predicted by CLIMEX model. *J. King Saud Univ. Sci.* **2021**, *33*, 101394. [CrossRef]
46. Mazaheri, A.; Khajehali, J.; Hatami, B. Oviposition preference and larval performance of *Aeolesthes sarta* (Coleoptera: Cerambycidae) in six hardwood tree species. *J. Pest Sci.* **2011**, *84*, 355–361. [CrossRef]
47. Bhatti, S.I. An over View of Forests in Pakistan. World Environmental Day 16th June 2011. pp. 88–97. Available online: https://pecongress.org.pk/images/upload/books/(9)%20An%20Overview%20of%20Forests%20in%20Pakistan%20By%20Engr.%20Saeed%20Iqbal%20.pdf (accessed on 18 August 2022).
48. EPPO. Aeolesthessarta. Datasheets on quarantine pests. *EPPO Bull.* **2005**, *35*, 387–389.
49. Chaudhry, Q.; Rasul, G.; Kamal, A.; Ahmad, M.; Mahmood, S. Technical Report on Karachi Heat Wave June 2015. Government of Pakistan Ministry of Climate Change 2015. Available online: http://www.ndma.gov.pk/files/heatwave.pdf (accessed on 18 August 2022).
50. Caesar, J.; Alexander, L.; Vose, R. Large-scale changes in observed daily maximum and minimum temperatures: Creation and analysis of a new gridded data set. *J. Geophys. Res. Atmos.* **2006**, *111*, D05101. [CrossRef]
51. Tebaldi, C.; Hayhoe, K.; Arblaster, J.M.; Meehl, G.A. Going to the extremes. *Clim. Change* **2006**, *79*, 185–211. [CrossRef]
52. Guisan, A.; Thuiller, W. Predicting species distribution: Offering more than simple habitat models. *Ecol. Lett.* **2005**, *8*, 993–1009. [CrossRef] [PubMed]
53. Chen, Y.; Liu, Z.; Régnière, J.; Vasseur, L.; Lin, J.; Huang, S.; Ke, F.; Chen, S.; Li, J.; Huang, J.; et al. Large-scale genome-wide study reveals climate adaptive variability in a cosmopolitan pest. *Nat. Commun.* **2021**, *12*, 7206. [CrossRef]
54. Naves, P.; de Sousa, E. Threshold temperatures and degree-day estimates for development of post-dormancy larvae of Monochamusgalloprovincialis (Coleoptera: Cerambycidae). *J. Pest Sci.* **2009**, *82*, 1–6. [CrossRef]
55. Trumbore, S.; Brando, P.; Hartmann, H. Forest health and global change. *For. Health* **2015**, *349*, 814–818. [CrossRef] [PubMed]
56. CABI. Trirachyssartus—Invasive Species Compendium 2022. Available online: https://www.cabi.org/isc/datasheet/3430 (accessed on 15 August 2022).
57. Khan, A.A.; Kundoo, A.A. Pests of walnut. In *Pests and Their Management*; Springer: Singapore, 2018; pp. 605–647. [CrossRef]
58. Hussain, A.; Muhammad, A.; Hayat, U.; Ahmad, B.; Murtaza, A.M.; Khalid, K.M.B.; Ullah, S. Response of rice cultivars and insecticides against Rice stem borer (*Scirpophagaincertulus*) in Pakistan (Swat). *J. Biodivers. Environ. Sci.* **2019**, *15*, 88–94.

forests

Article

Assessing the Global Pest Risk of *Aeolesthes sarta* with Regards to the Host Specie *Populus alba* under Climate Change Scenarios

Umer Hayat [1,†], Sumeet Kour [2,†], Muhammad Akram [1], Juan Shi [1,*] and Rinto Wiarta [3]

1 Sino-France Joint Laboratory for Invasive Forest Pests in Eurasia, Beijing Forestry University, Beijing 100083, China; oomarcassi6116@gmail.com (U.H.); akramforestr11223@gmail.com (M.A.)
2 School of Agriculture, Geography, Environment, Ocean and Natural Sciences, The University of the South Pacific, Laucala Campus, Suva 999210, Fiji; sumeetkour@gmail.com
3 Department of Forest Management, School of Forestry, Beijing Forestry University, Beijing 100083, China; rwiarta@gmail.com
* Correspondence: shi_juan@263.net
† These authors contributed equally to this work.

Abstract: *Aeolesthes sarta* or *Trirachys sarta* is a polyphagous long-horned beetle that has caused severe damage to the *Populus alba* forests/plantations in its regions of origin. Climate change could accelerate the introduction and spread of invasive pest species, potentially causing ecological damage and economic losses. Furthermore, globalization and increased trade can inadvertently transport pests across borders into regions where they do not already occur. Hence, it is crucial to identify areas where the climate is most suitable for the establishment of *A. sarta*'s and which areas of the world are suitable for the growth of *P. alba* under climate change scenarios. This study employed the CLIMEX model to estimate the potential global distribution of *A. sarta* and its correlation with its dominant host, *P. alba,* under current climatic conditions and potential future scenarios, namely the A1B and A2 climate change scenarios (CCSs). Under current climatic conditions, the model indicates that the establishment of a climatically suitable habitat for *A. sarta* extends beyond its current known range. The model estimated that, under the world's current climatic conditions, 41.06% of the world can provide suitable areas (EI > 0) for the survival of *A. sarta.* For *P. alba*, under the current climatic conditions, suitable regions for the growth of *P. alba* are present in all continents (excluding Antarctica); under the world's current climatic conditions, 53.52% of the world can provide suitable areas for the growth of *P. alba* (EI > 0). Climate change will significantly alter the number of suitable habitats for *A. sarta* development and *P. alba* growth globally. In future climatic conditions, the number areas capable of supplying suitable habitats (EI > 0) for *A. sarta* will slightly decrease to 40.14% (under A1B and A2 CCSs), while, for *P. alba*, the number areas capable of supplying suitable habitats will also marginally decrease to 50.39% (under A1B scenario), and this figure is estimated to drop to 48.41% (under A2 scenario) by the end century (2100). Asia, Europe, North America, South America, and Oceania have a high percentage of highly suitable areas for *A. sarta* development and *P. alba* growth under current climatic conditions; however, according to estimates of future climatic conditions, by the end century, only Asia, Europe, North America, and Oceania will have a high percentage of highly suitable areas for *A. sarta* development and *P. alba* growth. The range of highly suitable habitats is likely to increase in the northern hemisphere; however, this range is expected to shrink with regards to the southern hemisphere. The range contraction was higher under the A2 climate change scenario due to a higher warming trend than in the A1B scenario. Due to climate change, the range of *A. sarta* development shifted, as did the *P. alba* growth range, which, thanks to the suitable environmental conditions for the growth of *P. alba*, makes all those regions vulnerable to the introduction and development of *A. sarta.* Strict monitoring, prevention, and control measures at borders, airports, and seaports before the trade of *P. alba* and other suitable host species wood (logs/billets) are highly recommended to prevent the spread of *A. sarta* and ensure biodiversity security. It is expected that the *A. sarta* and *P. alba* climate models presented here will be useful for management purposes since both can be adapted to guide decisions about imparting resources to regions where the threat of pest invasion remains and away from regions where climate suitability is predicted to decrease in the future.

Keywords: *Aeolesthes sarta; Trirachys sarta; Populus alba;* CLIMEX; climate change; A1B–A2 scenarios; ecological niche model; invasive species; forest pest

1. Introduction

Over the past century, global warming has become a significant climate change challenge [1]. The dispersal patterns of species have been altered, changing the habitats that different species may thrive in and reducing biodiversity [2]. By modeling the effects of climate change on invasive pests and their predators, we can gain insight into the tools necessary to control their spread [3]. Climate change is an important topic because it can have direct and indirect effects on insect populations (by, for example, altering the frequency of species and hosts) [2].

Aeolesthes sarta or *Trirachys sarta*, commonly known as the City or Sarta Longhorned Beetle (belonging to the *Trirachys* genus), is one of the foremost species of the Cerambycidae family [4]. It is polyphagous, and mainly attacks broadleaved tree species, including *Populus* spp., *Juglans* spp., *Acer* spp., *Salix* spp., *Malus* spp., *Platanus* spp., and *Ulmus* spp. [4]. Larvae of this family primarily feed internally on live or dead plant tissues [5]. Boring by larvae can cause structural damage to host trees and block the movement of nutrients and water, resulting in the death of multiple branches and the tree as a whole [6–9].

A. sarta is thought to have originated in Pakistan and the western part of India, with significant distribution into Afghanistan, Iran, and other nations in Central Asia [10,11]. It may thrive in forests up to 2000 m above sea level [12]. The region's warm temperature and the fact that it hosts the pest's favored host tree species provides an ideal environment for it to thrive in [13]. The species is highly problematic in regions with hot, dry climates [14]. In central Asia, the longhorn beetle is a major pest affecting most broadleaved trees [4].

Poplar wood is broadly utilized to produce a wide range of goods, including pulp, paper, and other fiber-based products [15]. Poplars are also helpful for nontimber purposes, such as aesthetic planting, windbreaks, and soil binders [16]. Poplars are grown in over 70 countries worldwide, both in natural forest mixes and artificial forests [17]. However, the vast majority of poplars still thrive in natural forests [15]. *Populus alba* is among the dominant host species severely and repeatedly attacked and damaged by *A. sarta* in its countries of origin [4,18–21]. *A. sarta* can attack and reproduce on the main stem and large branches distinctively [18]. Infestations are especially visible in highland forests, where there may be declines in poplar tree forests. Despite being a key supplier of wood for the industry [20], *Populus* trees are entirely destroyed by the *A. sarta* borer, rendering it ill-suited for industrial use [19]. There were 200 trees of *P. alba* inspected for infestation in the Mustang, Balochistan area; 100% were found to be severely infested, and 34% were destroyed entirely [19]. *Populus alba* in the *Populus* genus has been reported to be among the dominant host species infested by *A. sarta* in the Mashhad and Zahedan regions of Iran, where the infestation rate was deemed to be 100% and the mortality rates was deemed to be 49 and 11%, respectively [18]. Because of climate change and changing ecological constraints, the breeding of drought-tolerant species such as *Populus alba* is becoming increasingly appealing [22].

The polyphagy is a major factor in the possibility of colonization. Compared to oligophagous and monophagous species, polyphagous species tend to have a larger potential for colonization [23]. Longhorn beetles, or Cerambycinae, appear more prone to introduction. Most species thrive on living plants, while numerous Cerambycinae complete their life cycle in live or dead wood, such as *A. sarta* [4]. Due to globalization, cerambycids have followed trade routes [23]. The documented established pathway includes importing timber for building homes or furniture, importing plants for planting in nurseries [24], and importing timber for pallets, pulp and wood packaging, and other wood-derived goods [23]. The danger of invasion is expected to rise due to escalating trade volumes, faster potential vector imports, and a wider variety of sources and introduction locations [24].

Regional changes in climate, most notably those associated with increases in temperature, impact several ecological systems [25]. Despite morphological and physiological alterations in one or both species, climate change may sabotage relationships between species [26]. Climate patterns, notably temperature, directly and significantly impact the phenology of poikilotherms, primarily invertebrates like insects [27].

Global warming alters pest distribution patterns and raises the possibility of invasive species causing ecological damage. The accelerated exchange of breeding and planting materials may introduce more insect pests and pathogens into new habitats as poplar cultivation develops more intensely and becomes ubiquitous. In order to track this species more effectively, it is crucial to investigate the shifting patterns of potentially favorable habitats for *A. sarta* under different climate change scenarios. However, in this study, we examine climate change's effect on *Populus alba,* one of the dominant host species of *A. sarta,* and correlate it with the distribution pattern of *A. sarta* globally. Until now, no detailed research studies have estimated the global distribution pattern of *A. sarta* and its preferred host, *P. alba,* for the future under different climate change scenarios. The present study is the first to investigate the pest's distribution pattern concerning its host species. Only a single study has been available till now, which mainly focused on the European continent and used very basic biological parameters to forecast the potential distribution of *A. sarta* [12]. Therefore, employing accurate and extensive information about the pest's known habitat, ecology, and biology, an attempt has been made to forecast the possible distribution of *A. sarta* globally under climate change scenarios using the CLIMEX model concerning its preferred host, *P. alba* (Figure 1). We assumed that climate change would alter the distribution pattern of insect pests and the host tree species globally. This study aimed to determine the possible global distribution of *A. sarta* and its preferred host, *P. alba,* under current climatic conditions and forecast the potential global distribution of *A. sarta* and its preferred host, *P. alba,* for the end of the century (2100) under different climate change scenarios (A1B and A2, respectively).

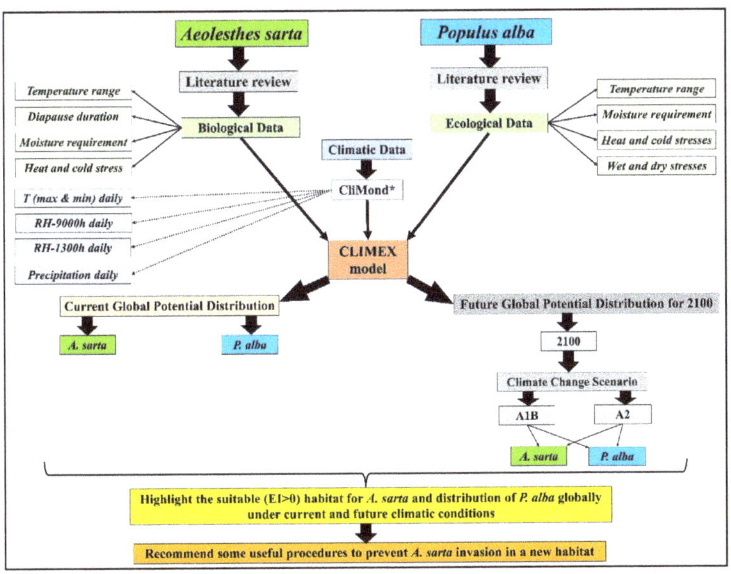

Figure 1. Experimental layout of the present study. T = Temperature; RH = Relative Humidity; * https://www.climond.org/CLIMEX.aspx (accessed on 2 March 2023).

2. Materials and Methods

2.1. CLIMEX Model

CLIMEX software version 4.0 (Hearne Scientific Software, Auckland, New Zealand) was used to model the potential distribution of *A. sarta* and *P. alba* under current and future climatic conditions using meteorological data and the physiological characteristics of the pest and host species. A "Compare Locations model" for a single species was utilized for this study. The annual growth index (GI) was determined as a function of soil moisture and temperature, and the possibility of the population sustaining unfavorable conditions was assessed using stress indices (cold, hot, dry, and wet) [28]. The "Ecoclimatic Index" (EI), an annual index of climatic readiness, was created by combining weekly calculations of growth indices and stress indices to provide an overall assessment of a location's favorability for the species to survive there permanently [29,30]. In this study, geographical areas with a $15 < EI \geq 30$ reflect climatic conditions ideal for the establishment, survival, and proliferation of *A. sarta* and *P. alba* (Table 1).

Table 1. Eco-Climatic index values used in the study and their representation.

Eco-Climatic Index	Representation
EI = 0	Unsuitable habitat—The climate and environment are not conducive to the growth and establishment of *A. sarta* and *P. alba*.
0 < EI = 15	Marginally suitable habitat—The climate and environment are marginally conducive to the growth and establishment of *A. sarta* and *P. alba*.
15 < EI = 30	Suitable habitat—The climate and environment are conducive to the growth and establishment of *A. sarta* and *P. alba*.
EI > 30	Highly suitable habitat—The climate and environment are highly conducive to the growth and establishment of *A. sarta* and *P. alba*.

Known species distribution data along with laboratory-derived ecological data were used to parameterize the model [6,7,10,19,31–36]. The parameters were iteratively adjusted according to the satisfactory agreement between the potential and known global distribution of *A. sarta* and *P. alba* [31,37]. The parameters were then verified by comparing the physiological and ecological data of *A. sarta* and *P. alba* to ensure they were biologically and ecologically reasonable.

2.2. Climatic Data

Climatic data were downloaded from CliMond [https://www.climond.org/CLIMEX.aspx (accessed on 20 February 2023)] and used to help model the potential distribution of *A. Sarta* and *P. alba* under current and future climatic conditions (Table 2). CliMond provides a fine-scale 10′ dataset with long-term monthly climate means for maximum temperature (T max), minimum temperature (T min), precipitation (P total), and relative humidity at 0900 (9 AM) and 1500 h (3 PM) [38]. For forecasting, we used A1B (2.8 °C rise in global temperature) and A2 (3.4 °C rise in global temperature) as per the Special Report on climate change scenarios (SRES) [39].

Table 2. Climatic data were used to run the CLIMEX model.

Climate Change Scenarios	Period	Data Set	Source
	Current	**CM10_1975H_CX_MM_V1.2**	**Climond**
A1B *	2100	CM10_2100_A1B_CS_Cx_MM_V1.2	Climond
A2 **	2100	CM10_2100_A2_CS_Cx_MM_V1.2	Climond

* 2.8 °C rise in global temperature at the end of the 21st century (a future characterized by very rapid economic growth, low population growth, and the rapid introduction of new and more efficient technology). ** 3.4 °C rise in global temperature at the end of the 21st century (A future characterized by very high population growth, moderate economic growth, high demand for agricultural products, loss of grasslands to food crops, and reduced regulations).

2.3. *Aeolesthes sarta and Populus alba Global Distribution Data*

Known occurrence record data of *A. sarta* and *P. alba* were gathered from the Centre of Agriculture and Bioscience International [CABI → https://www.cabi.org/ (accessed on 8 February 2023)], Global Biodiversity Information Facility [GBIF → https://www.gbif.org/ (accessed on 8 February 2023)], European and Mediterranean Plant Protection Organization [EPPO → https://gd.eppo.int/ (accessed on 8 February 2023)], and from a literature review [4,6,7,10,19,31–36]. When published research papers did not provide precise geographic coordinates for positive sites, the coordinates for the site were pinpointed using Google Earth (https://www.google.com/earth/ accessed on 10 February 2023) and Google Maps (https://www.google.com/maps/preview accessed on 10 February 2023). All occurrences that were already recorded twice have been removed.

2.4. *Pest (Aelosthes sarta) Biological Data*

Due to its preference for dry, cold weather, *A. sarta* is most often found in temperate zones [4]; thus, the Temperate template included with CLIMEX software was used to create a "species parameter file." Physiological tolerance thresholds for *A. sarta* were used to calibrate the model. The existing occurrence data of *A. sarta* was used to determine the species climatic requirements. Parameter values for temperature, moisture, and diapause index (Table 3) were defined using inferred data from the published literature associated with developmental threshold temperatures, moisture, and diapause [4,6,15]. By repeatedly running the parameter file with different parameter values; the stress index values were fine-tuned until the predicted climate suitability patterns were consistent with *A. sarta's* observed distribution.

Table 3. Parameter values used in the CLIMEX model for *Aelosthes sarta* and *Populus alba*.

Parameters	Code	Values	
		Aelosthes sarta	*Populus alba*
Temperature			
Limiting low temperature (°C)	DV0	10	−5
Lower optimal temperature (°C)	DV1	15	4
Upper optimal temperature (°C)	DV2	37	20
Limiting high temperature (°C)	DV3	40	35
Moisture Index			
Limiting low soil moisture	SM0	0	0.5
Lower optimal soil moisture	SM1	0.001	1
Upper optimal soil moisture	SM2	1.5	1.5
Limiting high soil moisture	SM3	2.5	2
Diapause Index			
Diapause induction day length	DPD0	12	-
Diapause induction temperature (°C)	DPT0	13	-
Diapause termination temperature (°C)	DPT1	10	-
Diapause development days	DPD	90	-
Summer or winter Diapause	DPSW	0	-
Cold Stress			
CS temperature threshold (°C)	TTCS	9	−15
CS temperature rate	THCS	−0.00001	−0.0001
Heat Stress			
HS temperature threshold (°C)	TTHS	41	35
HS temperature rate	THHS	0.005	0.005
Wet Stress			
Wet stress threshold	SMWS	-	2.5
Wet stress accumulation rate	HWS	-	0.002
Dry Stress			
Dry stress threshold	SMDS	-	0.2
Dry stress accumulation rate	HDS	-	−0.005
Population degree day	PDD	700	-

2.4.1. Temperature Index

In the beginning, the model was run using the following temperature values: DV0 = 11 °C, DV1 = 20 °C, DV1 = 35 °C, and DV4 = 40 °C [12,19]. However, these values did not predict *A. sarta's* occurrence in northern India, China, and Afghanistan. Hence, DV0, DV1, and DV2 were adjusted to 10 °C, 15 °C, and 37 °C [31] to include positive sites in Northern India, China, and Afghanistan and were finalized to forecast the potential distribution globally. The population day degree (PDD) value for *A. sarta* was set to be 700 [12,31], as it completes its lifecycle in almost two years [4,6,10,19].

2.4.2. Diapause Index

For *A. sarta*, Diapause Development Days (DPD) were set at 90 days [31] as, according to the literature, *A. sarta* diapause duration is approximately 90 days [4,6,10]. Other diapause parameters, such as Diapause induction day length (DPD0), Diapause induction temperature (DPT0), Diapause termination temperature (DPT1), and Summer or winter Diapause (DPSW) were set at 12 days, 13 °C, 10 °C, and 0 days, respectively [12,31].

2.4.3. Moisture Index

Another parameter the CLIMEX model uses is the soil moisture index (MI). SM0, SM1, SM2, and SM3 (for detail names, please see Table 3) are the four additional characteristics that the CLIMEX model uses to further categorize the MI [30]. For normal species development, SM0 was attuned to 0 [12,31]. SM1 and SM2 were set at 0.001 and 1.5, respectively [31]. Following a similar practice, the value for SM3 was set at 2.5 [12,31].

2.4.4. Stress Index

Dry stress (DS), wet stress (WS), cold stress (CS), and heat stress (HS) are four environmental stress indices in the CLIMEX that depict unfavorable conditions that limit a species' population development [30]. For *A. sarta,* only HS and CS were considered. The cold stress restricts species' development when the temperature falls below the cold stress threshold temperature (TTCS) of that species at a particular rate (THCS). In the present study, TTCS was set at 9 °C as, below this temperature, *A. sarta* development is impeded [19]. Following the TTCS value, the THCS value was set to −0.00001. This rate provided an appropriate fit to the observed occurrence of *A. sarta*. However, when temperatures rise above a species' heat stress temperature threshold (TTHS), the growth of that species is hindered. As *A. sarta* cannot survive above temperatures of 40 °C [19], the TTHS value was set at 41 °C. The weekly rate (THHS) was set at 0.005 [31]. The observed global distribution of *A. sarta* was seemingly in agreement with these values.

2.5. *Host (Populus alba) Ecological Data*

Populus alba is primarily found in temperate climates with moist and relatively cold weather [36]; thus, a Temperate template included with CLIMEX software was used to create a "species parameter file." Parameter values for temperature and moisture index (Table 3) were determined using data inferred from the literature [32–36,40,41]. By running the parameter file with varying parameter values, we were able to fine-tune the stress index values so that the predicted climate suitability patterns were consistent with *P. alba's* observed global distribution.

2.5.1. Temperature Index

Populus alba cannot tolerate frigid temperatures, and its growth is inhibited at low temperate (e.g., −5 °C) [36,40], whereas temperatures higher than 35 °C might be problematic for the growth of *P. alba* in temperate regions [41,42]. Therefore, lower (DV0) and upper (DV3) temperature thresholds for the temperature after which species stopped growing were set to −5 and 35 °C. The DV0, DV1, DV2, and DV3 values used to run the model were −5 °C, 4 °C, 20 °C, and 35 °C, respectively.

2.5.2. Moisture Index

Populus alba is a tree species that grows well in both moist and dry soils [43]. However, areas with moist soil conditions and moderate to heavy precipitation rates favor the growth rate of *P. alba* [36,40–42]. As a result, restricting low soil moisture (SM0) was set at 0.5 to enable normal species growth, and SM1 and SM2 soil moisture values were set up for optimal growth. In addition, the SM3 soil moisture upper limit was set at 2.5.

2.5.3. Stress Index

For *P. alba*, all four stress indices (CS, HS, WS, and DS) were considered. The cold stress restricts species' development when the temperature falls below the cold stress threshold temperature (TTCS) of that species at a particular rate (THCS). Three different cold temperature values (−5 °C, −15 °C, and −45 °C) were used to run the model. At −5 °C, many known distribution points were missing; at −15 °C and −45 °C, all known distributions were covered, and no notable difference was found. Therefore, in the current study, TTCS was set at −15 °C as *P. alba* growth is inhibited at extremely low temperatures [36,41]. A THCS value of −0.0001 was determined from the TTCS data. The same corresponds to heat stress; once temperatures rise above a species' heat stress threshold (TTHS) at a given rate (THHS), the growth and development of that species will be impeded. In the current study, TTHS was established at 35 °C because *P. alba* does not thrive above this temperature [41]. The weekly rate (THHS) was set at 0.005. Likewise, dry stress also affects a species' development. In the present research, SMDS was set at 0.2 as *P. alba* cannot grow well under low soil moisture conditions [42]. The dry stress accumulation rate (HDS) was set at −0.005. Through an algorithm of iteration, the wet stress value was mapped to the known distribution of *P. alba*. Therefore, the wet stress threshold (SMWS) was set at 2.5 with an accumulation rate (HWS) of 0.002, respectively.

2.6. Model Verification

The CLIMEX model was verified by assessing its ability to correctly predict the current known occurrence of *A. sarta* and *P. alba*. EI > 0 were viewed as correctly predicted occurrences. After verification, the parameter file was run with the "Compare Locations" function to look at the predicted global occurrence.

2.7. Map Generation

ArcMap version 10.7 was used to map CLIMEX output. To create a continuous surface, we processed the variance of the potential *A. sarta* and *P. alba* distributions using ArcMap's Spatial Analyst module. The data were then reclassified using the Kriging interpolation method, and the global potential *A. sarta* and *P. alba* distribution under current and future climatic conditions was visualized.

3. Results

3.1. Model Fitting/Verification and Current Global Distribution

Under current climate conditions, the predicted global distribution of *A. sarta* and *P. alba* is in very close agreement with the species' known geographic range. The current distribution of *A. sarta* and *P. alba* agrees well with the modeled climatic suitability for the location (Figure 2—yellow dots; Figure 3—red dots).

Under current climatic conditions, 41.06% of the world (excluding Antarctica) can provide habitats suitable for the survival of *A. sarta* (EI > 0), except for the Scandinavian and Nordic countries, Alaska, Canada, northern, eastern, and central Russia, and wet tropical countries (except the southern part of Brazil). Similarly, 80% of Southeast Asia is suitable for the survival of *A. sarta*. In total, 52 countries, including Ukraine, Moldova, Romania, Hungary, Croatia, Serbia, Bulgaria, Greece, Italy, France, Spain, Portugal, Cyprus, Morocco, Algeria, Tunisia, Libya, Egypt, Syria, Iraq, Lebanon, Israel, Jordan, Palestine, Saudi Arabia, Angola, Zambia, Zimbabwe, Namibia, Botswana, South Africa, Uruguay, Argentina, Brazil, Mexico, USA, Turkey, Azerbaijan, Iran, Kazakhstan, Uzbekistan, Turkmenistan,

China, Pakistan, India, Afghanistan, Nepal, Myanmar, Japan, South Korea, Australia, and New Zealand can provide highly suitable habitats (EI > 30), accounting for 27.07% (36.41 million km^2) of the global area under the current climatic conditions (Figures 2 and 4). Whereas potentially suitable habitats (15 < EI = 30) and marginal habitats (0 < EI = 15) cover 9.15% (12.30 million km^2) and 4.84% (6.51 million km^2) of the globe, respectively.

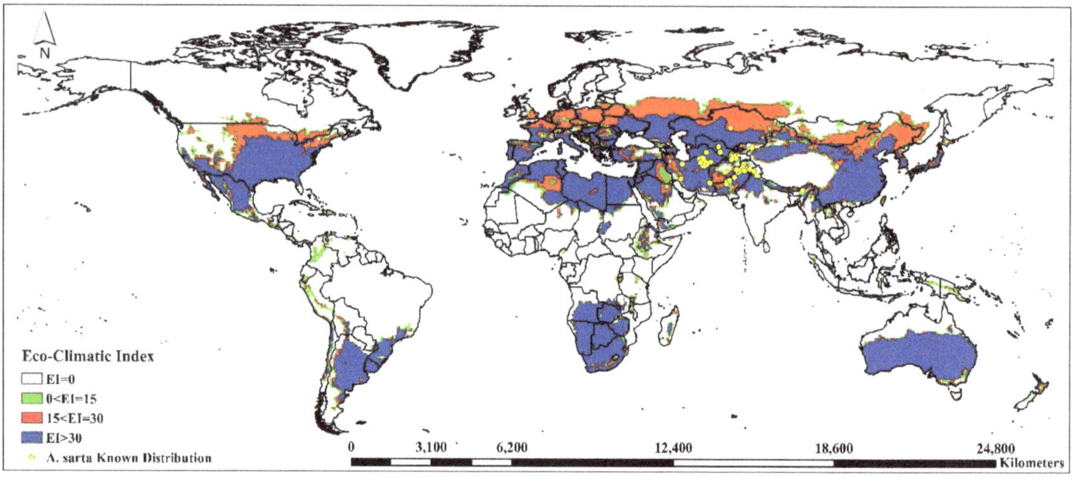

Figure 2. Predicted potential global distribution of *A. sarta* under current climatic conditions. The predicted global distributions (four colors) relate to the ecoclimatic index (EI) from the CLIMEX model: highly suitable (EI > 30), suitable (15 < EI = 30), marginally suitable (0 < EI = 15), unsuitable (EI = 0); yellow dots represent the known occurrence record of *A. sarta* (CABI, GBIF, EPPO, and Hayat 2022).

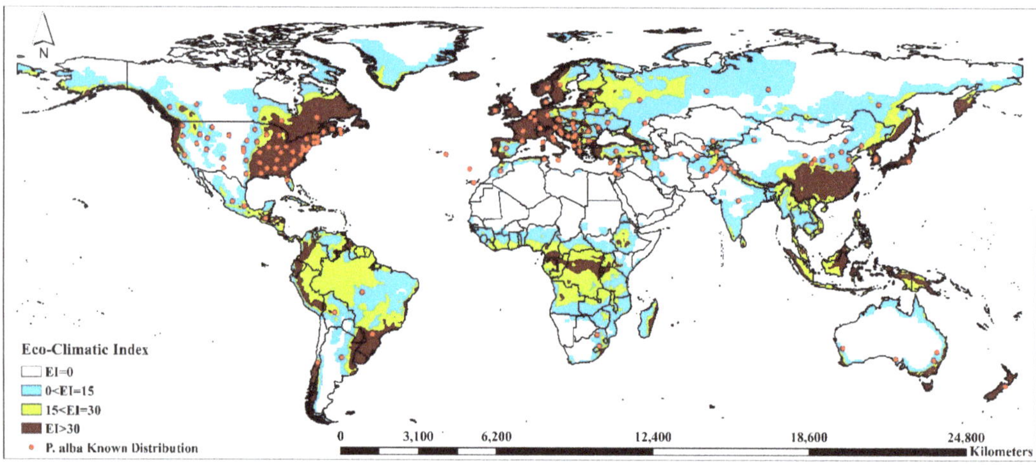

Figure 3. Predicted potential climatic favorability of *P. alba* under current climatic conditions. The predicted global climatic favorability (four colors) relate to the ecoclimatic index (EI) from the CLIMEX model: highly suitable (EI > 30), suitable (15 < EI = 30), marginally suitable (0 < EI = 15), unsuitable (EI = 0); red dots represent the known occurrence record of *P. alba* (CABI, GBIF, EPPO).

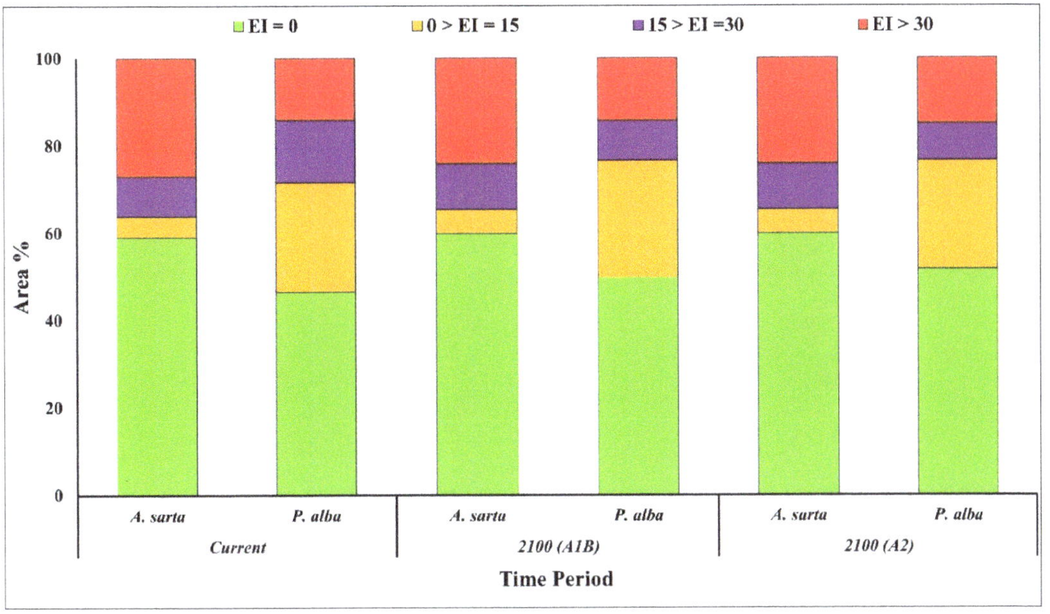

Figure 4. Percentage of the land area distribution of *Aelosthes sarta* and *Populus alba* within the four ranges of the ecoclimatic index under current and future climatic conditions on a global scale under A1B and A2 climate change scenarios. Highly suitable (EI > 30); suitable (15 < EI = 30); marginally suitable (0 < EI = 15); unsuitable (EI = 0).

However, around 85 countries under the current climatic conditions have suitable environmental conditions (EI > 0) for the growth of *P. alba*, which covers 53.52% of the globe's land area (excluding Antarctica), except for northern Canada, northern African states, and central Australia. Similarly, almost the entirety of Europe, South America, and the southern and eastern North American regions are suitable for *P. alba* growth. Regions that have highly suitable environmental conditions (EI > 30) cover 14.24% (20.16 million km^2) of the globe's land area (Figures 3 and 4). Whereas regions with suitable environmental conditions (15 < EI = 30) and marginal environmental conditions (0 < EI = 15) account for 14.31% (20.26 million km^2) and 24.97% (35.35 million km^2) of the globe's land area, respectively.

3.2. Potential Global Distribution under Future Climate Conditions

The model forecasted a decreasing trend in the number of areas capable of providing highly suitable habitats for *A. sarta* at the end of century 2100 under A1B and A2 CCSs (Figures 4 and 5). No noticeable differences were observed for both scenarios. This decrease was mainly observed in the southern hemisphere and central part of the world. Overall, the highly suitable area (EI > 30) decreased to 24.20% (32.55 million km^2) from 27.07% (36.41 million km^2) compared to the current climatic condition. However, a shift in number of areas capable of providing highly suitable habitats in the northern hemisphere was observed. Furthermore, the model forecasted an increase in total countries with highly suitable habitats (EI > 30), with up to 68 countries being capable of providing highly suitable habitats by the end century under both CSSs, compared to 52 countries under current climatic conditions. The countries that are most likely going to be capable of providing highly suitable habitats (EI > 30) at the end of the century (2100) are Australia, New Zealand, Afghanistan, Pakistan, India, China, Kazakhstan, Iran, Azerbaijan, Turkey, Kyrgyzstan, Russia, Mongolia, Tajikistan, Bhutan, Democratic People's Republic of Korea, Republic of Korea, Japan, Nepal, Canada, USA, Mexico, Argentina, Uruguay, South Africa,

Namibia, Jordan, Lebanon, Tunisia, Algeria, Morocco, Germany, UK, Spain, Portugal, France, Italy, Greece, Bulgaria, Serbia, Croatia, Hungary, Romania, Moldova, Ukraine, Cyprus, Belgium, Poland, Bosnia and Herzegovina, Albania, Macedonia, Estonia, Lithuania, Belarus, Slovenia, Slovakia, Czech Republic, Luxembourg, Latvia, Montenegro, Armenia, Georgia, The Netherlands, Sweden, Finland, Switzerland, Austria, Denmark. Compared to current climatic conditions, suitable and marginally suitable habitat areas will likely increase to 10.46% and 5.74% by 2100 under both CCSs.

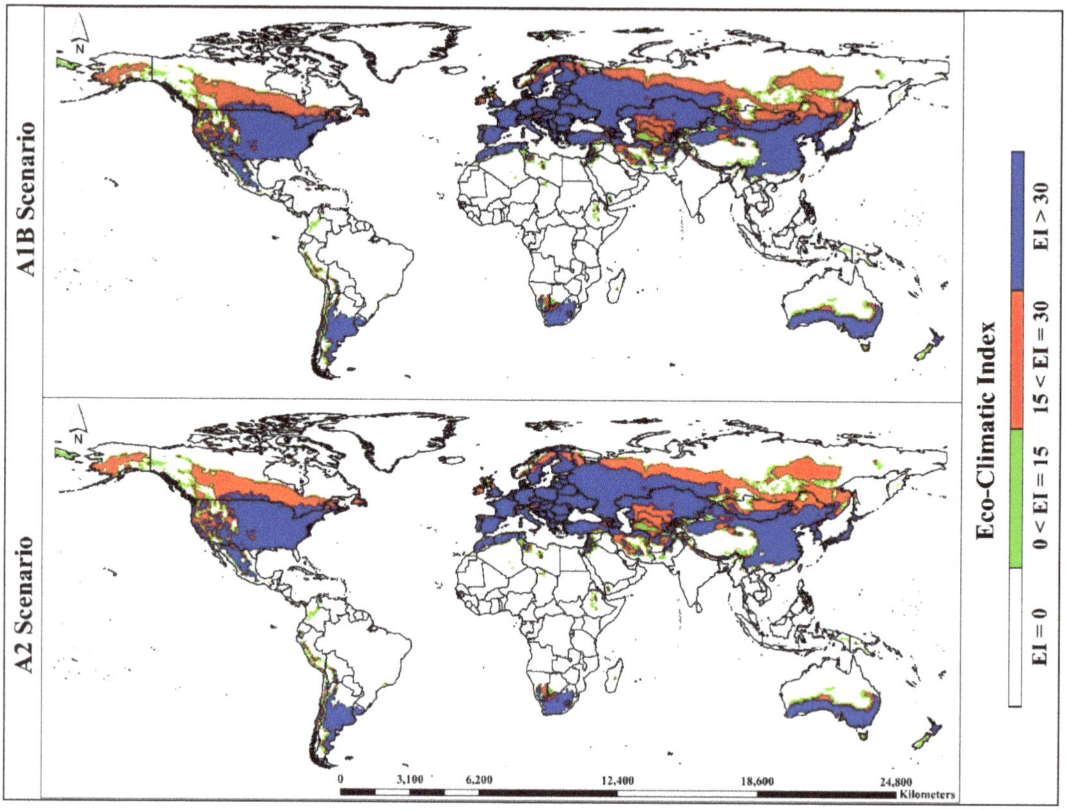

Figure 5. Predicted potential global distribution of *A. sarta* under future climatic conditions (2100) under A1B and A2 climate change scenarios (CCSs). The predicted global distributions (four colors) relate to the ecoclimatic index (EI) from the CLIMEX model: highly suitable (EI > 30), suitable (15 < EI = 30), marginally suitable (0 < EI = 15), unsuitable (EI = 0).

On a global level, the model forecasted a gradual increase in the number of highly suitable habitat (EI > 30) areas for the growth of *P. alba* till the end century under both CCSs (Figures 4 and 6). Under A1B CCS, highly suitable habitat areas increased to 14.41% (20.40 million km^2), and under A2 CCS, this increase is forecasted to be 15.10% (21.38 million km^2). Under current climatic conditions, this figure is estimated to be 14.24% (20.16 million km^2). This increase was mainly observed in the northern hemisphere and central part of the world. The countries that are likely to be capable of providing highly suitable habitat areas under future climatic conditions (EI > 30) (at the end of the century; 2100) are the USA, Canada, New Zealand, Afghanistan, Pakistan, India, China, Azerbaijan, Turkey, Russia, Bhutan, North Korea, South Korea, Japan, Nepal, Malaysia, Indonesia, Papua New Guinea, Fiji, Canada, Argentina, Uruguay, Germany, UK, Ireland, Spain, Portugal, France,

Italy, Greece, Bulgaria, Serbia, Croatia, Hungary, Romania, Moldova, Cyprus, Belgium, Poland, Estonia, Lithuania, Belarus, Slovenia, Slovakia, Luxembourg, Latvia, Armenia, Georgia, The Netherlands, Sweden, Finland, Switzerland, Austria, Denmark, Australia.

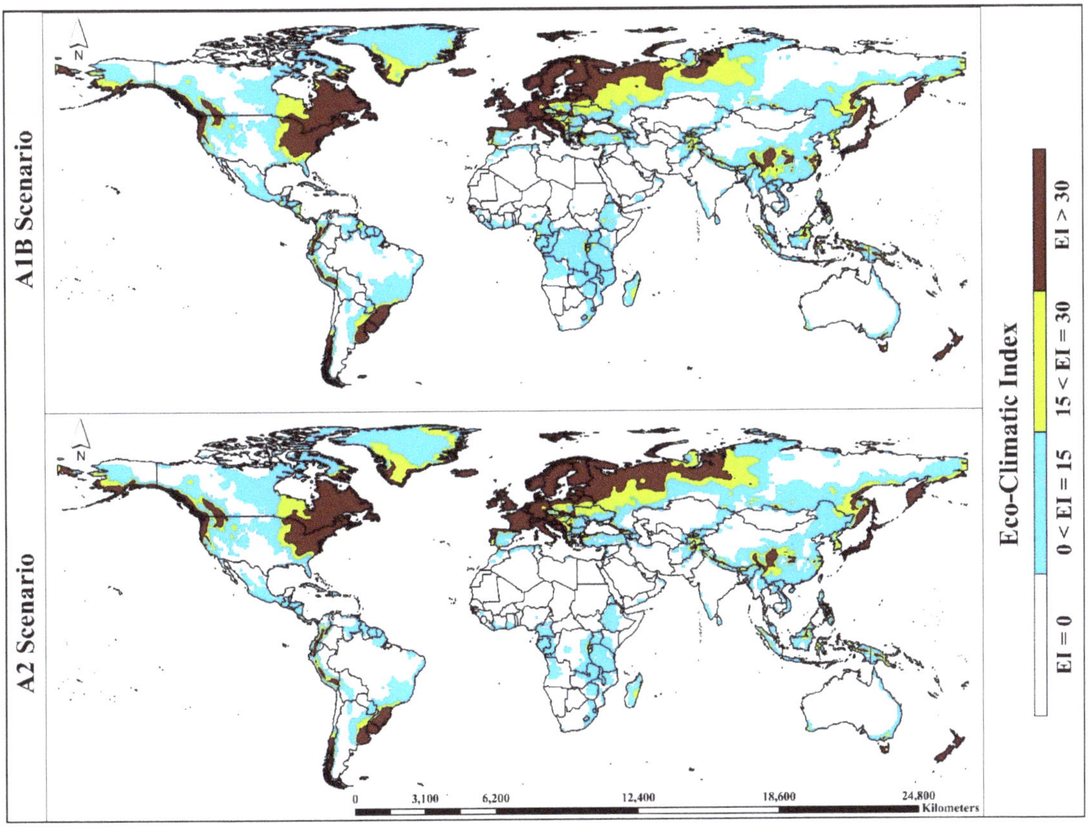

Figure 6. Predicted potential climatic favorability of *P. alba* under future climatic conditions 2100 under A1B and A2 climate change scenarios (CCSs). The predicted global climatic favorability (four colors) relates to the ecoclimatic index (EI) from the CLIMEX model: highly suitable (EI > 30), suitable (15 < EI = 30), marginally suitable (0 < EI = 15), unsuitable (EI = 0).

3.3. Continent-Wise Distribution under Future Climate Conditions

Under current climate conditions, out of the 48 countries in Asia, 20 are capable of providing highly suitable habitats (EI > 30) for the development of *A. sarta*, accounting 26.79% (11.68 million km^2) of the globe's land area. According to our predictions, this will reduce to 14 countries by the end of the century (2100) under A1B and A2 CCSs. Under future climatic conditions, highly suitable, suitable, and marginally suitable habitat areas will increase up to global area figures of 27.11% (11.82 million km^2), 16.45% (7.17 million km^2), and 7.35% (3.20 million km^2) in 2100 under both CSSs (Table 4). While, under current climatic conditions, areas with highly suitable habitats for the growth of *P. alba* account for 10.16% (5.04 million km^2) of the globe's land area, and this is predicted to increase to 10.53% (5.23 million km^2) under A1B CCS and decrease to 7.64% (3.79 million km^2) under A2 CCS at the end century.

Table 4. Continent-wise percentage of the land area distribution of *Aelosthes sarta* and *Populus alba* within the four ranges of the ecoclimatic index under current and future climatic conditions on a global scale under A1B and A2 climate change scenarios.

	Eco-Climatic Index		Asia		Europe		Africa		North America		South America		Oceania	
			%	Mi. km^2	%	Mi. km^2	%	Mi. km^2	%	Mi. km^2	%	Mi. km^2	%	Mi. km^2
Current	EI = 0	*A. sarta*	53.81	23.46	37.59	3.60	63.70	18.84	65.06	15.30	77.62	13.53	29.60	2.32
		P. alba	53.21	26.41	5.44	0.72	56.58	13.32	53.10	18.17	15.19	2.15	74.10	5.04
	0 < EI = 15	*A. sarta*	5.96	2.60	5.61	0.54	3.74	1.11	4.36	1.03	5.05	0.88	3.53	0.28
		P. alba	27.39	13.59	39.70	5.25	20.12	4.74	20.21	6.91	28.21	3.99	12.66	0.86
	15 < EI = 30	*A. sarta*	13.44	5.86	28.35	2.72	3.28	0.97	8.27	1.94	2.78	0.48	2.92	0.23
		P. alba	9.25	4.59	24.26	3.21	17.73	4.17	7.85	2.69	38.53	5.45	2.23	0.15
	EI > 30	*A. sarta*	26.79	11.68	28.44	2.73	29.28	8.66	22.31	5.25	14.55	2.54	63.95	5.01
		P. alba	10.16	5.04	30.60	4.05	5.57	1.31	18.84	6.44	18.07	2.56	11.00	0.75
2100 (A1B)	EI = 0	*A. sarta*	49.09	21.40	7.29	0.70	91.06	26.93	43.38	10.20	82.99	14.46	60.34	4.73
		P. alba	52.60	26.11	3.48	0.46	72.90	17.16	36.04	12.33	59.43	8.41	84.93	5.78
	0 < EI = 15	*A. sarta*	7.35	3.20	3.38	0.32	2.18	0.65	7.07	1.66	4.96	0.87	7.21	0.56
		P. alba	29.68	14.73	13.84	1.83	25.23	5.94	34.18	11.69	24.24	3.43	7.03	0.48
	15 < EI = 30	*A. sarta*	16.45	7.17	10.39	1.00	1.64	0.48	18.22	4.28	2.52	0.44	6.03	0.47
		P. alba	7.19	3.57	23.23	3.07	1.63	0.38	9.68	3.31	4.92	0.70	2.01	0.14
	EI > 30	*A. sarta*	27.11	11.82	78.94	7.57	5.12	1.51	31.34	7.37	9.52	1.66	26.41	2.07
		P. alba	10.53	5.23	59.45	7.87	0.24	0.06	20.10	6.88	11.41	1.61	6.03	0.41
2100 (A2)	EI = 0	*A. sarta*	49.09	21.40	7.29	0.70	91.06	26.93	43.38	10.20	82.99	14.46	60.34	4.73
		P. alba	53.23	26.42	3.77	0.50	82.61	19.44	33.48	11.45	66.34	9.39	85.85	5.84
	0 < EI = 15	*A. sarta*	7.35	3.20	3.38	0.32	2.18	0.65	7.07	1.66	4.96	0.87	7.21	0.56
		P. alba	29.52	14.65	12.40	1.64	16.42	3.86	34.72	11.88	19.03	2.69	7.34	0.50
	15 < EI = 30	*A. sarta*	16.45	7.17	10.39	1.00	1.64	0.48	18.22	4.28	2.52	0.44	6.03	0.47
		P. alba	9.61	4.77	19.30	2.55	0.90	0.21	10.85	3.71	4.16	0.59	1.44	0.10
	EI > 30	*A. sarta*	27.11	11.82	78.94	7.57	5.12	1.51	31.34	7.37	9.52	1.66	26.41	2.07
		P. alba	7.64	3.79	64.52	8.54	0.08	0.02	20.96	7.17	10.46	1.48	5.37	0.37

Note: Highly suitable (EI > 30); suitable (15 < EI = 30); marginally suitable (0 < EI = 15); unsuitable (EI = 0).

Under current climatic conditions, out of the 44 countries in Europe, 15 provide highly suitable habitats for *A. sarta* development, accounting for 28.44% (2.73 million km^2) of the globe's total land area. Under future climatic conditions and A1B and A2 CCSs, 32 countries are likely to have highly suitable habitat areas (EI > 30), accounting for 78.94% (7.57 million km^2) of the globe's total land area. Comparatively, under current climatic conditions, regions offering highly suitable habitat areas for *P. alba* growth account for 30.60% (4.05 million km^2) of the globe's total land area, and by the end of the century, this figure is forecasted to gradually increase to 59.54% (7.87 million km^2) and 64.52% (8.54 million km^2) under A1B and A2 CCSs, respectively. In our predictions, unsuitable habitat (EI = 0) areas for *P. alba* growth reduced compared to current climatic conditions under future climatic conditions under both CCSs at the end century.

In Africa, only 11 countries out of 54 offer highly suitable habitats (EI > 30) for *A. sarta* development, accounting for 29.29% (8.66 million km^2) of the globe's total land area under current climate conditions, and this will supposedly reduce to only four countries covering 5.12% (1.51 million km^2) of the globe's total land area by the end of the century (2100) under A1B and A2 CCSs. However, highly suitable habitat areas for the growth of *P. alba* under future climatic conditions are predicted to reduce to 0.24% (0.06 million km^2) and 0.08% (0.02 million km^2) under A1B and A2 CCSs at the end of century compared to current climatic conditions, for which the same value is estimated to be 5.57% (1.31 million km^2).

In the North American continent, only two countries, Mexico and America, offer highly suitable habitat areas for the development of *A. sarta*, accounting for 22.31% (5.25 million km^2) of the globe's total land area under current climatic conditions; however, this figure is estimated to increase to three countries, including South Canada, covering 31.34% (7.37 million km^2) at the end of the century under both CCSs. At the same time, the unsuitable habitat (EI = 0) areas will reduce to 43.38% (10.20 million km^2) of the globe's total land area in 2100 compared to current climatic conditions, the corresponding value of which is 65.06% (15.30 million km^2). Highly suitable habitats for the growth of *P. alba* under current climatic conditions account for 18.84% (6.44 million km^2) of the globe's total land area, which is predicted to increase to 20.10% (6.88 million km^2) and 20.96% (7.17 million km^2) under A1B and A2 CCSs at the end century. Unsuitable habitat areas also experienced respective decreases to 36.04 (12.33 million km^2) and 33.48% (11.45 million km^2) under A1B and A2 CCSs in our predictions for 2100. Under current climatic conditions, this figure is estimated to be 53.10% (18.17 million km^2).

Out of the 12 countries in South America, under current climatic conditions, 5 offer highly suitable habitats (EI > 30) for *A. sarta* development, accounting for 14.55% (2.54 million km^2) of the globe's total land area; whereas, at the end of the century (2100), this will reduce to 4 countries, accounting for 9.52% (1.66 million km^2) of the globe's total land area under A1B and A2 CCSs. However, under future climatic conditions, the land area comprising unsuitable habitats increases up to 82.99% (14.66 million km^2). For current climatic conditions, this figure was estimated to be 77.62% (13.43 million km^2). Highly suitable habitat areas for *P. alba* growth account for 18.07% (2.56 million km^2) of the globe's total land area under current climatic conditions, and this is predicted to gradually decrease to 11.41% (1.61 million km^2) and 10.46% (1.48 million km^2) under A1B and A2 CCSs, respectively, by the end century.

In Oceania, under current climatic conditions, the global land area comprising highly suitable habitats for *A. sarta* development is 63.95% (5.01 million km^2), while this area continues to reduce up to 26.41% (2.07 million km^2) at the end of the century under A1B and A2 CCSs. Under future climatic conditions, unsuitable areas decreased by up to 60.34%. (4.73 million km^2). For the current climatic conditions, this figure was estimated to be 29.60% (2.32 million km^2). However, highly suitable habitat areas for the growth of *P. alba* under future climatic conditions are predicted to reduce to 6.03% (0.41 million km^2) and 5.37% (0.37 million km^2) under A1B and A2 CCSs at the end century. For the current climatic conditions, this figure was estimated to be 11% (0.75 million km^2).

Global predictive maps based on the model's results demonstrate that cold stress and heat stress limit the potential climatic zone for *A. sarta* to specific regions (Figure 7). Cold stress limits the potential climatic zone boundaries in the northern hemisphere, but heat stress limits those in the southern hemisphere and the central region of the world. However, the intensity of cold and heat stress increased with time under A1B and A2 CCSs.

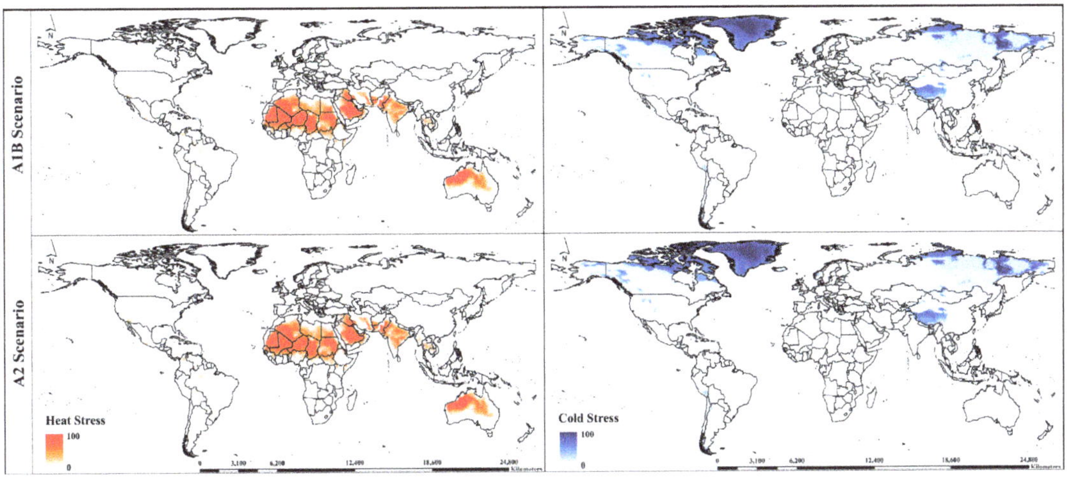

Figure 7. The predicted region where *A. sarta* survival is affected globally by cold stress and heat stress under A1B and A2 climate change scenarios at the end of this century (2100).

Globally, cold, heat, and dry stress tend to limit the potential climatic zone for *P. alba* on predicted maps (Figure 8), but in the northern hemisphere, only cold stress limits the potential climatic zone. However, heat stress and dry stress limit the potential climate zone range in the southern hemisphere and the central region of the earth. In contrast, heat stress and dry stress intensified over time under A1B and A2 CCSs, but the intensity of cold stress decreased.

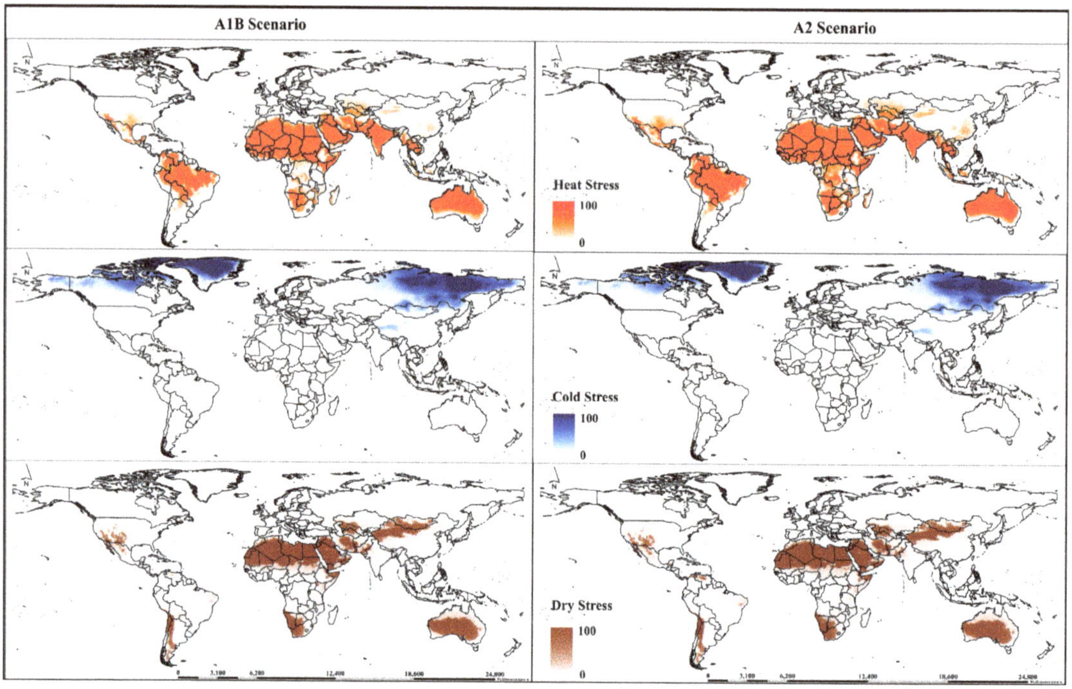

Figure 8. The predicted region where *P. alba* survival is affected globally by cold, heat, and dry stresses under A1B and A2 climate change scenarios at the end of this century.

4. Discussion

Our goal was to use a modeling technique (CLIMEX) to demonstrate a comprehensive approach to analyzing the distribution pattern of *A. sarta* and its dominant host, *P. alba*, under current and future climatic conditions. The modeled climate suitability for *A. sarta* and *P. alba* under current climatic conditions extends well beyond the existing recorded occurrences and provides considerable insight into the likelihood of the pest outbreaks in the region having host distribution. The findings of this study confirmed that, in comparison to the current climate, the distribution of *A. sarta* and growth pattern of *P. alba* are significantly altered by future climatic conditions under A1B and A2 CCSs.

The climate is among the most crucial factors that significantly impact species' dispersion globally [37]. The habitat range of many species, especially cold-blooded species like insects, is significantly impacted by climate change [44] (not only insects but also the forest species growth/distribution pattern) [45]. Estimating the geographic ranges of a targeted pest species under future climatic scenarios has been made simpler due to advancements in climate modeling methodologies and technology, which has further improved the monitoring of pests and pest management strategies [46]. An earlier attempt to characterize climatic influences on *A. sarta* distributions using CLIMEX was explored in a study by Vanhanen et al., wherein they explored its distribution in Europe only [12].

The ecological niche of *A. sarta* concerning its dominant host species, *P. alba*, was modeled globally for the first time in the present study, which also evaluates the significance of environmental factors in relation to modeling ranges and spatial patterns. Additionally, the present study is the very first to forecast the global distribution of *A. sarta* and its dominant host species, *P. alba*, under A1B and A2 climate change scenarios. The only prior use of this method, regarding *A. sarta* generally, was outlined in a study by Vanhanen et al. [12]. Notably, we observe that the maps produced by this approach vary markedly from those seen in the study of Vanhanen et al. (*A. sarta* distribution only restricted to southern Europe) [12] (Figures 2 and 5). In particular, we noted that, under both CCSs, by 2100, highly suitable habitat areas (EI > 30) are estimated to stretch from south to north, covering entire central regions in Europe, North America, and NE and NW Asia. Likewise, by 2100, habitat areas highly suitable for *P. alba* are estimated to expand towards the northern hemisphere, covering almost the entirety of Europe, Eastern USA, SE Canada, and the entirety of New Zealand.

According to an ecological niche model for *A. sarta* and *P. alba*, under current climatic conditions, the range for the distribution of suitable habitats for *A. sarta* is primarily restricted to the southern and central regions of the world due to cold stress, while *P. alba*'s distribution range stretches from the southern to the northern hemisphere, excluding central African regions, where heat and dry stresses restrict the distribution as warming induces the dry stress by raising the demand for atmospheric water [47,48]. Moreover, in northern regions where cold stress restricted the distribution under cold climatic conditions, *P. alba* finds less suitable conditions under cold climatic conditions, suffering from high moisture, frigid temperature, and frost [36]. However, it is predicted that under future climatic conditions (A1B and A2 scenarios), the distribution of suitable habitats for *A. sarta* will expand to the Northern Hemisphere but shrink in the Southern Hemisphere, while warm temperate and tropical coastlines remain beyond the *A. sarta* suitable habitat areas (EI = 0).

Furthermore, *P. alba's* suitable growth range under future climatic conditions will expand to the northern hemisphere, covering Europe and Eastern and Southern North America, but shrink in the southern hemisphere. As heat and dry stresses are predicted to increase at the end century, the central part of the world becomes unsuitable for the growth of *P. alba*. Climate change will cause species' distribution range to shift or expand or decrease, significantly impacting species occurrence [3,46].

No extensive datasets or maps have been developed for the global distribution and growth pattern of *A. sarta* and *P. alba*, respectively. The present study implies that *A. sarta* can establish itself in areas broader than initially expected, and *P. alba* forests in these areas are at high risk. Understanding the scope of this species' distribution can help us take the necessary measures to limit the spread of this pest. The global distribution of *A. sarta* throughout the warm and dry temperate zones demonstrated its vast climate niche and extensive climatic adaptability. The projection of *P. alba's* suitable growth regions also exhibits the availability of the favorable host species of *A. sarta*. In this climate zone lies a variety of phenotypic traits that *A. sarta* exhibits and the availability of preferred host species [4].

The climates of Asia, Europe, North America, and Oceania are the most suitable for introducing and growing *P. alba*. These continents support the growth of *P. alba* under current and future climatic conditions. However, the likelihood of *A. sarta* occurring in any of these continents is mainly unknown (except in some Asian countries). The same may be stated about New Zealand, a nation in Oceania projected to be capable of providing moderate to highly suitable habitats for *A. sarta*, though this has not been thoroughly investigated. If *A. sarta* were introduced in any of these areas, *P. alba* plantations would be highly affected, as has already been the case in Asian countries Pakistan, Iran, India, Turkmenistan, and Tajikistan [4], where infested trees become unsuitable for pulp and wood exploitation.

Several studies have reported both latitudinal and altitudinal temperature-related range alterations in various insects [49–51]. These demonstrate how strongly climate influences species distribution and offer a solid framework for climatic modeling. Distribution models are an initial step in assessing the danger of a pest, together with the distributions of its possible hosts. However, it has been demonstrated that the absence of congeneric or native host plants hinders the establishment of alien phytophagy insects [52,53]. However, our results revealed that, in many Asian, European, North American, and Oceania forests, the growth of *P. alba*, commercial stands, and urban areas facilitates the establishment of the *A. sarta*.

Over the past 50 years, a continuous increase in the trade of wood-based products has been observed globally [54]. Around thirty items are made from poplar wood, making it substantial for trade [55]. As commerce in commodities with related wood packaging, such as forest products, expands globally, there is an ongoing issue with the movement of wood and tree pests [54]. *A. sarta* invades trees inside a domain of 500 m from its nursery tree [12,56]. Although capable of long-distance dispersion, human-aided dispersal through *P. alba* wood shipment would still be the major route for long-distance dispersion if a species wished to establish itself in a new continent. In the USA, through human-aided dispersal, spongy moth (*Lymantria dispar*) has been observed to populate new remote areas outside of its continuous dispersion span; however, its constant dispersion span in the Eastern USA only expands westward by natural dispersal at a rate of 2 km per year [57], primarily due to strong flight [58]. Similar to the Spongy moth in North America, *A. sarta* has had an invasion sequence in Asia comparable to its original distribution zone [4,56]. The proliferation of invasive species, i.e., the global distribution of *A. sarta*, poses profound economic implications [59,60].

Generally, models are calibrated using data onto the native range or a composite of the native and invaded area [31,37]. We analyzed worldwide occurrence data To configure the model used in the present study and capture the range of suitable environmental conditions for the development of *A. sarta* and the growth of *P. alba*. As a result, the model is more responsive [61]. The model would understate the target species' potential distribution if the parameters were fitted within a narrow range. The rationality of the model relies on how precisely the predictions match reality [31,37]. Map comparisons reveal good agreement regarding simulated and actual *A. sarta* and *P. alba* distributions, demonstrating that climate-based models may produce helpful forecasts with little input data [13,31,37]. The geographical distribution of *A. sarta* is influenced by various biotic and abiotic variables, such as the availability of the host tree species [4]. Our *P. alba* distribution model demonstrated that, out of many preferred host species, *P. alba* is mainly available in the regions highly suitable for developing *A. sarta* under current and future climate conditions.

Since the egg, larva, and pupal life stages of *A. sarta* are within the host tree, it is simple for it to disperse with the host for a very long time [4]. *A. sarta* will continue to spread extensively over the world if robust control measures are not implemented. Therefore, intelligent detection systems should be installed in airports and seaports to precisely detect and intercept *A. sarta* eggs, larvae, or pupa. Additionally, before trading *P. alba* wood/logs/billets, phytosanitary measures, such as stem debarking and kiln drying, are recommended. In the case of invasion, the use of Click beetle Alaus larvae, which feed on *A. sarta* larvae; *Proctolaelaps* spp, which feeds on *A. sarta* grubs [20]; *Sclerodermus turkmenicus*, which are parasites of *A. sarta* larvae [14]; and *Beauveria bassiana*, which infects the *A. sarta* adults [62], are recommended as biological control agents. We recommend avoiding the use of pesticides as they are not environmentally friendly [63].

In this work, the aforementioned modeling software was used for the first time on a global scale to simulate the ecological niche of *A. sarta* insight with its dominant host, *P. alba*, using precision parametric variables. The habitat suitability forecasts may pinpoint areas where the species may establish itself as a potential invasive pest because the distribution is projected at the species level. The CLIMEX forecasting of known distribution and the impact of environmental conditions on *A. sarta* development have a high degree of accuracy, which suggests that this modeling strategy has great potential for assisting in the creation of pest control plans.

5. Conclusions

Globally, invasive insect species like *A. sarta* seriously threaten native biodiversity and ecological processes. Economic globalization and climate change have exacerbated their invasion. This paper is the first study of its kind to use the CLIMEX model to predict the global habitat distribution of *A. sarta* regarding its dominant host species, *P. alba*, under current and future climatic conditions (using A1B and A2 climate change scenarios). This study demonstrated a practical method for creating a global habitat suitability map for *A. sarta* development and *P. alba* growth. Overall, under future climatic conditions, the global land area comprising highly suitable habitats for the development of *A. sarta* is estimated to decrease globally, and the same can be said for global land area comprising highly suitable habitats for *P. alba* growth range areas; however, regions like North America, Europe, Oceania, and most parts of Asia will remain highly suitable for the growth of *P. alba* and the development of *A. sarta*. Under climate change scenarios (A1B and A2), due to changes in temperature intensity, areas of cold stress decreasing, and areas of heat stress increasing, a shift will occur toward the northern hemisphere regarding the global land area of suitable habitats for *A. sarta*, while the global land area of suitable habitats will shrink at the southern hemisphere; whereas, under climate change scenarios, *P. alba* growth regions are estimated to shrink with the same pattern as that of *A. sarta* due to the heat stress, cold stress, and dry stress. For *A. sarta*, the expanding areas are mainly located in Asia, Europe, and North America, while for *P. alba*, the continents where the suitability of climatic conditions will increase are Europe, North America, Oceania, and Asia. Regions where the eco-climatic suitability index decreases for *A. sarta* development and *P. alba* growth under future climatic conditions are Africa (central states of Africa), South America (states with hot and dry stresses), Oceania (excluding New Zealand), some parts of Australia (where hot and dry stresses predicted to be minimum), and some regions in Asia. Strict monitoring, prevention, and control measures at borders, airports, and seaports before the trade of *P. alba* and other suitable host species wood (logs/billets) are highly recommended to prevent the spread of *A. sarta* and ensure biodiversity security. The *A. sarta* and *P. alba* climate models presented here should be useful for quarantine pest management purposes because they can be used to direct resources toward locations where the threat of pest invasion is high and away from places where climate suitability is expected to decline in the future.

Author Contributions: Conceptualization, U.H., S.K., M.A. and J.S.; investigation and methodology, U.H. and S.K.; formal analysis, U.H., S.K. and M.A.; investigation, U.H.; resources, U.H. and S.K.; data curation, U.H. and R.W.; writing—original draft preparation, U.H.; writing—review and editing, U.H., S.K., M.A., J.S. and R.W.; supervision and project administration, J.S.; funding acquisition, J.S. All authors have read and agreed to the published version of the manuscript.

Funding: This study was supported by the Forestry Science and Technology Innovation Special of Jiangxi Forestry Department (201912) and the National Natural Science Foundation of China (32171794).

Data Availability Statement: Not applicable.

Conflicts of Interest: The authors declare no conflict of interest.

Abbreviations

CCS	Climate Change Scenario
EI	Ecoclimatic index

References

1. Stocker, T.F.; Qin, D.; Plattner, G.K.; Tignor, M.M.; Allen, S.K.; Boschung, J.; Nauels, A.; Xia, Y.; Bex, V.; Midgley, P.M. *Climate Change 2013: The Physical Science Basis. Contribution of Working Group I to the Fifth Assessment Report of IPCC the Intergovernmental Panel on Climate Change*; Cambridge University Press: Cambridge, UK, 2014. [CrossRef]
2. Gao, X.; Zhao, Q.; Wei, J.; Zhang, H. Study on the Potential Distribution of Leptinotarsa decemlineata and Its Natural Enemy Picromerusbidens under Climate Change. *Front. Ecol. Evol.* **2022**, *9*, 786436. [CrossRef]
3. Wei, J.; Peng, L.; He, Z.; Lu, Y.; Wang, F. Potential distribution of two invasive pineapple pests under climate change. *Pest Manag. Sci.* **2020**, *76*, 1652–1663. [CrossRef] [PubMed]

4. Hayat, U. City longhorn beetle (*Aeolesthes sarta*): A review of the species, its morphology, distribution, da-mage, prevention and control. *J. For. Sci.* **2022**, *68*, 199–212. [CrossRef]
5. Ślipiński, A.; Escalona, H. Australian Longhorn Beetles (Coleoptera: Cerambycidae). In *Subfamily Cerambycinae*; CSIRO Publishing: Clayton, Australia, 2016; Volume 2, p. 640. [CrossRef]
6. Khan, S.A.; Bhatia, S.; Tripathi, N. Entomological investigation on *Aeolesthes sarta* (Solsky), a major pest on walnut trees (*Juglans regia* L.) in Kashmir valley. *J. Acad. Ind. Res.* **2013**, *2*, 325.
7. Mazaheri, A.; Khajehali, J.; Marzieh, K.; Hatami, B. Laboratory and field evaluation of insecticides for the control of *Aeolesthes sarta* Solsky (Col.: Cerambycidae). *J. Crop Prot.* **2015**, *4*, 257–266.
8. Morewood, W.D.; Hoover, K.; Neiner, P.R.; McNeil, J.R.; Sellmer, J.C. Host tree resistance against the polyphagous wood-boring beetle *Anoplophora glabripennis*. *Entomol. Exp. Appl.* **2004**, *110*, 79–86. [CrossRef]
9. Poland, T.M.; Haack, R.A.; Petrice, T.R.; Miller, D.L.; Bauer, L.S.; Gao, R. Field evaluations of systemic insecticides for control of *Anoplophora glabripennis* (Coleoptera: Cerambycidae) in China. *J. Econ. Entomol.* **2006**, *99*, 383–392. [CrossRef]
10. Farashiani, M.E.; Shamohammadi, D.; Sadeghi, S.E. Biological study of Sart long horn beetle, *Aeolesthes sarta* Solsky (Coleoptera: Cerambycidae) in the laboratory. *J. Entomol. Soc. Iran* **2000**, *20*, 77–90.
11. Orlinskii, A.D. Outcomes of the EPPO project on quarantine pests for forestry 1. *EPPO Bull.* **2006**, *36*, 497–511. [CrossRef]
12. Vanhanen, H.; Veteli, T.O.; Niemelä, P. Potential distribution ranges in Europe for *Aeolesthes sarta*, *Tetropium gracilicorne* and *Xylotrechus altaicus*, a CLIMEX analysis. *EPPO Bull.* **2008**, *38*, 239–248. [CrossRef]
13. Hayat, U.; Qin, H.; Zhao, J.; Akram, M.; Shi, J.; Ya, Z. Variation in the potential distribution of *Agrotis ipsilon* (Hufnagel) globally and in Pakistan under current and future climatic conditions. *Plant Prot. Sci.* **2021**, *57*, 148–158. [CrossRef]
14. CABI (2022): Trirachys Sartus—Invasive Species Compendium. Available online: https://www.cabi.org/isc/datasheet/3430 (accessed on 8 February 2023).
15. Ištok, I.; Šefc, B.; Hasan, M.; Popović, G.; Sedlar, T. Fiber characteristics of white poplar (*Populus alba* L.) juvenile wood along the Drava river. *Drv. Ind.* **2017**, *68*, 241–247. [CrossRef]
16. Ostry, M.E. *A Guide to Insect, Disease, and Animal Pests of Poplars (No. 677)*; US Department of Agriculture, Forest Service: Washington, DC, USA, 1989; pp. 1–120.
17. Ball, J.; Carle, J.; Del Lungo, A. Contribution of poplars and willows to sustainable forestry and rural development. *UNASYLVA-FAO*. 2005, 56, pp. 3–9. Available online: https://www.fao.org/3/a0026e/a0026e03.pdf (accessed on 15 March 2023).
18. Farashiani, M.E.; Sadeghi, S.E.; Abaii, M. Geographic distribution and hosts of Sart longhorn beetle, Aeolesthes sarta Solsky (Col.: Cerambycidae) in Iran. *J. Entomol. Soc. Iran* **2001**, *20*, 81–96.
19. Ahmad, M.I.; Hafiz, I.A.; Chaudhry, M.I. Biological studies on *Aeolesthes sartus* Solsky attacking poplars in Pakistan. *Pak. J. For.* **1977**, *27*, 122–129.
20. Gul, H.; Chaudhry, M.I. Some observations on natural enemies of poplar borers in Pakistan. *Pak. J. For.* **1992**, *42*, 214–222.
21. Arshad, M.; Hafiz, I.A. Microbial trials of a pathogenic fungus, *Beauveria bassiana* (Bals.) Vuill. against the adults of *Aeolesthes sartus* Solsky (Cerambycidae: Coleoptera). *Pak. J. Zool.* **1983**, *15*, 213–215.
22. Isebrands, J.G.; Richardson, J. *Poplars and Willows: Trees for Society and the Environment*; CABI International: Wallingford, UK; FAO: Roma, Italy, 2014; pp. 1–7. [CrossRef]
23. Cocquempot, C.; Lindelöw, Å. Longhorn beetles (Coleoptera, Cerambycidae). In *Alien Terrestrial Arthropods of Europe*; Roques, A., Ed.; BioRisk: Singapore, 2010; Chapter 8.1; Volume 4, pp. 193–218. [CrossRef]
24. Cocquempot, C. Alien longhorned beetles (Coleoptera Gerambycidae): Original interceptions and introductions in Europe, mainly in France, and notes about recently imported species. *Redia* **2006**, *89*, 35–50.
25. IPCC. *Climate Change 2014: Synthesis Report. Contribution of Working Groups I, II and III to the Fifth Assessment Report of the Intergovernmental Panel on Climate Change*; Core Writing Team, Pachauri, R., Meyer, L., Eds.; IPCC: Geneva, Switzerland, 2014; p. 151. Available online: https://www.ipcc.ch/site/assets/uploads/2018/02/AR5_SYR_FINAL_Front_matters.pdf (accessed on 10 February 2023).
26. Tamura, Y.; Osawa, T.; Tabuchi, K.; Yamasaki, K.; Niiyama, T.; Sudo, S.; Ishigooka, Y.; Yoshioka, A.; Takada, M.B. Estimating plant–insect interactions under climate change with limited data. *Sci. Rep.* **2022**, *12*, 10554. [CrossRef] [PubMed]
27. Damos, P.; Savopoulou-Soultani, M. Temperature-driven models for insect development and vital thermal requirements. *Psyche* **2012**, *2012*, 123405. [CrossRef]
28. Sutherst, R.W.; Maywald, G.F.; Russell, B.L. Estimating vulnerability under global change: Modular modelling of pests. *Agric. Ecosyst. Environ.* **2000**, *82*, 303–319. [CrossRef]
29. Sutherst, R.W.; Maywald, G.F.; Kriticos, D.J. *CLIMEX Version 3: User's Guide*; Hearne Scientific Software Pty Ltd.: Victoria, Australia, 2007.
30. Sutherst, R.W.; Maywald, G.F.; Bourne, A.S. Including species interactions in risk assessments for global change. *Glob. Chang. Biol.* **2007**, *13*, 1843–1859. [CrossRef]
31. Hayat, U.; Akram, M.; Kour, S.; Arif, T.; Shi, J. Pest Risk Assessment of *Aeolesthes sarta* (Coleoptera: Cerambycidae) in Pakistan under Climate Change Scenario. *Forests* **2023**, *14*, 253. [CrossRef]
32. Nevidomov, A.M. Ecophytocoenotic patterns in the distribution of poplar forests in the floodplains of southeastern European Russia. *Bot. Zhurnal* **1994**, *79*, 47–58.
33. Modir-Rahmati, A. Cultivation of poplar in Iran. *Holzzucht* **1997**, *51*, 41–43.

34. Pasiecznik, N.M.; Smith, I.M.; Watson, G.W.; Brunt, A.A.; Ritchie, B.; Charles, L.M.F. CABI/EPPO distribution maps of plant pests and plant diseases and their important role in plant quarantine. *EPPO Bull.* **2005**, *35*, 1–7. [CrossRef]
35. Afshan, N.S.; Iqbal, S.H.; Khalid, A.N.; Niazi, A.R. Some additions to the uredinales of Azad Jammu and Kashmir (AJ & K), Pakistan. *Pak. J. Bot.* 2011, 43, pp. 1373–1379. Available online: http://www.pakbs.org/pjbot/PDFs/43(2)/PJB43(2)1373.pdf (accessed on 28 March 2023).
36. Pasiecznik, N. *'Populus alba (Silver-Leaf Poplar)', CABI Compendium*; CABI International: Wallingford, UK, 2022. [CrossRef]
37. Kour, S.; Khurma, U.; Brodie, G.; Singh, S. Modeling the potential global distribution of suitable habitat for the biological control agent *Heterorhabditis indica*. *Ecol. Evol.* **2022**, *12*, e8997. [CrossRef]
38. Kriticos, D.J.; Webber, B.L.; Leriche, A.; Ota, N.; Macadam, I.; Bathols, J.; Scott, J.K. CliMond: Global high-resolution historical and future scenario climate surfaces for bioclimatic modelling. *Methods Ecol. Evol.* **2012**, *3*, 53–64. [CrossRef]
39. IPCC. *Climate Change 2007: Synthesis Report. Contribution of Working Groups I, II and III to the Fourth Assessment Report of the Intergovernmental Panel on Climate Change*; Core Writing Team, Pachauri, R.K., Reisinger, A., Eds.; IPCC: Geneva, Switzerland, 2007; p. 104. Available online: https://www.ipcc.ch/site/assets/uploads/2018/02/ar4_syr_full_report.pdf (accessed on 25 February 2022).
40. Sekawin, M. Genetics of Populus alba. *Ann. For.* **1975**, *6*, 159–189.
41. Gucker, C.L. Populus alba and hybrids. In *Fire Effects Information System*; U.S. Department of Agriculture, Forest Service, Rocky Mountain Research Station, Fire Sciences Laboratory (Producer): Missoula, MT, USA, 2010. Available online: https://www.fs.usda.gov/database/feis/plants/tree/popspp/all.html (accessed on 15 March 2023).
42. Domingo, I.L.; Gordon, J.C. Physiological responses of an aspen-poplar hybrid to air temperature and soil moisture. *Bot. Gaz.* 1974, 135, pp. 184–192. Available online: https://www.journals.uchicago.edu/doi/abs/10.1086/336750 (accessed on 16 March 2023).
43. Rédei, K. Early evaluation of promising white poplar (*Populus alba* L.) clones in sandy ridges between the rivers Danube and Tisza in Hungary. *Silva Lusit.* **1998**, *6*, 63–71.
44. Aljaryian, R.; Kumar, L.; Taylor, S. Modelling the current and potential future distributions of the sunn pest *Eurygaster integriceps* (Hemiptera: Scutelleridae) using CLIMEX. *Pest Manag. Sci.* **2016**, *72*, 1989–2000. [CrossRef]
45. Hamann, A.; Wang, T. Potential effects of climate change on ecosystem and tree species distribution in British Columbia. *Ecology* **2006**, *87*, 2773–2786. [CrossRef] [PubMed]
46. Wei, J.; Zhao, Q.; Zhao, W.; Zhang, H. Predicting the potential distributions of the invasive cycad scale *Aulacaspis yasumatsui* (Hemiptera: Diaspididae) under different climate change scenarios and the implications for management. *PeerJ* **2018**, *6*, e4832. [CrossRef] [PubMed]
47. Mattson, W.J.; Haack, R.A. The role of drought in outbreaks of plant-eating insects. *Bioscience* **1987**, *37*, 110–118. [CrossRef]
48. McIntyre, N.E. Ecology of urban arthropods: A review and a call to action. *Ann. Entomol. Soc. Am.* **2000**, *93*, 825–835. [CrossRef]
49. Walther, G.R.; Post, E.; Convey, P.; Menzel, A.; Parmesan, C.; Beebee, T.J.; Fromentin, J.M.; Hoegh-Guldberg, O.; Bairlein, F. Ecological responses to recent climate change. *Nature* **2022**, *416*, 389–395. [CrossRef]
50. Battisti, A. Forests and climate change—Lessons from insects. *Forest* **2004**, *1*, 17–24. [CrossRef]
51. Battisti, A.; Stastny, M.; Netherer, S.; Robinet, C.; Schopf, A.; Roques, A.; Larsson, S. Expansion of geographic range in the pine processionary moth caused by increased winter temperatures. *Ecol. Appl.* **2005**, *15*, 2084–2096. [CrossRef]
52. Niemelä, P.; Mattson, W.J. Invasion of North American forests by European phytophagous insects. *BioScience* **1996**, *46*, 741–753. [CrossRef]
53. Roques, A.; Auger-Rozenberg, M.A.; Boivin, S. A lack of native congeners may limit colonization of introduced conifers by indigenous insects in Europe. *Can. J. For. Res.* **2006**, *36*, 299–313. [CrossRef]
54. Allen, E.; Noseworthy, M.; Ormsby, M. Phytosanitary measures to reduce the movement of forest pests with the international trade of wood products. *Biol. Invasions* **2017**, *19*, 3365–3376. [CrossRef]
55. Dhiman, R.C. Availability of poplar wood-a boon for plywood industry. *Plywood Gaz.* **2004**, 64–72.
56. EPPO. *Trirachys sartus*. EPPO Datasheets on Pests Recommended for Regulation. 2023. Available online: https://gd.eppo.int (accessed on 8 June 2023).
57. Liebhold, A.M. Alien species as agents of global change ecology and management of the gypsy moth in north america as a case history. In *Proceedings: International Symposium of the Kanazawa University 21st Century COE Program*; Kamata, N., Ed.; Kanazawa University: Kanazawa, Japan, 2003; Volume 1, pp. 71–75.
58. Akram, M.; Hayat, U.; Shi, J.; Anees, S.A. Association of the Female Flight Ability of Asian Spongy Moths (*Lymantria dispar asiatica*) with Locality, Age and Mating: A Case Study from China. *Forests* **2022**, *13*, 1158. [CrossRef]
59. Paini, D.R.; Sheppard, A.W.; Cook, D.C.; De Barro, P.J.; Worner, S.P.; Thomas, M.B. Global threat to agriculture from invasive species. *Proc. Natl. Acad. Sci. USA* **2016**, *113*, 7575–7579. [CrossRef]
60. Pimentel, D. *Biological Invasions: Economic and Environmental Costs of Alien Plant, Animal, and Microbe Species*, 2nd ed.; CRC Press: Boca Raton, FL, USA, 2011; ISBN 978-1-4398-2007-0.
61. Webber, B.L.; Yates, C.J.; Le Maitre, D.C.; Scott, J.K.; Kriticos, D.J.; Ota, N.; McNeill, A.; Le Roux, J.J.; Midgley, G.F. Modelling horses for novel climate courses: Insights from projecting potential distributions of native and alien *Australian acacias* with correlative and mechanistic models. *Divers. Distrib.* **2011**, *17*, 978–1000. [CrossRef]

62. Khan, A.A.; Kundoo, A.A. Pests of walnut. In *Pests and Their Management*; Springer: Singapore, 2018; pp. 605–647. [CrossRef]
63. Hussain, A.; Muhammad, A.; Hayat, U.; Ahmad, B.; Murtaza, A.M.; Khalid, K.M.B.; Ullah, S. Response of rice cultivars and insecticides against Rice stem borer (*Scirpophaga incertulus*) in Pakistan (Swat). *J. Biodivers. Environ. Sci.* 2019, 15, pp. 88–94. Available online: https://www.innspub.net/wp-content/uploads/2022/04/JBES-V15-No2-p88-94.pdf (accessed on 28 March 2023).

plants

MDPI

Article

Vascular Plant Species Inventory of Mexico's Revillagigedo National Park: Awareness of Alien Invaders as a *Sine Qua Non* Prerequisite for Island Conservation

Alejandra Domínguez-Meneses [1], Juan Esteban Martínez-Gómez [1,*], Teresa Mejía-Saulés [2], Israel Acosta-Rosado [3] and Stefan Stadler [4]

1 Red de Interacciones Multitróficas, Instituto de Ecología A.C. (INECOL), Xalapa 91073, Veracruz, Mexico; alejandra.dominguez@posgrado.ecologia.edu.mx
2 Red de Biología Evolutiva, Instituto de Ecología A.C. (INECOL), Xalapa 91073, Veracruz, Mexico; teresa.mejia@inecol.mx
3 XAL Herbarium, Instituto de Ecología A.C. (INECOL), Xalapa 91073, Veracruz, Mexico; israel.acosta@inecol.mx
4 Frankfurt Zoo, 60316 Frankfurt am Main, Germany; stefan.stadler@stadt-frankfurt.de
* Correspondence: juan.martinez@inecol.mx

Abstract: The Revillagigedo Archipelago, located in the Eastern Pacific Ocean, stands out for its unique biological richness and endemism. These islands remained uninhabited until the second half of the twentieth century, allowing a better conservation status than on other oceanic islands. However, the continuous introduction of potentially invasive alien plant species, and the lack of adequate control or eradication actions, jeopardize the conservation and restoration of these islands' fragile ecosystems. We present the most complete vascular plant species inventory and an updated list of alien plant species of the Revillagigedo Archipelago, which was compiled through an extensive review of national and international plant collections and other sources. Our 272 species list includes 106 alien plant species (39.3%; 104 in Socorro, and 16 in Clarion): 67 (24.8%) are naturalized, 14 (5.2%) are casual aliens, and 25 (9.3%) subsist under cultivation. The documented alien species belong to 73 families. Annual and perennial herbs are the prevailing life forms in the alien flora, while naturalized species are primarily native to North America. The number of introduced species has increased significantly since the islands became inhabited. Many of the recently introduced species pose a major invasion risk like on other islands of the world.

Keywords: Eastern Pacific; exotic plants; invasive species; invasiveness risk; oceanic islands

1. Introduction

The Revillagigedo Archipelago, located in the eastern North Pacific, is a group of four oceanic islands: Socorro, Clarion, San Benedicto, and Roca Partida [1]. These islands are famous for their invaluable biological richness, which, thanks to their isolation from the continent, led to the evolution of a high number of endemic species [2], as well as many protected and endangered species [3]. Declared as a Protected Natural Area in the category of Biosphere Reserve in 1994 [4], the Revillagigedo Archipelago entered the World Heritage List of the United Nations Educational, Scientific, and Cultural Organization in November 2016 [5]. In November 2017, it was declared a National Park [6], and with a surface of 148,087 km^2, it is the largest marine sanctuary in North America free of fishing activities [7].

Herbivores were introduced to these islands in different periods. Domestic sheep (*Ovis aries*) were brought to Socorro in 1869 [8]. Overgrazing by sheep facilitated the transformation into open and eroded habitats of a significant percentage of the forests and scrublands in the southeast half of the island [9]. The drastic impact on the island´s vegetation probably played a major role in the extinction of the Socorro Elf Owl (*Micrathene whitneyi graysoni*) in the 1930s [10] and the extinction of the Socorro Dove (*Zenaida graysoni*) in the wild in the mid-1970s [9]. After the removal of sheep in 2012, passive recovery of the vegetation cover has been observed in the most affected areas in the south of the island. However, a considerable amount of this recovery does not involve native species [11].

Clarion was also subjected to herbivore introductions, domestic pigs (*Sus scrofa*) in 1980, domestic rabbits (*Oryctolagus cuniculus*) in 1983, and sheep taken from Socorro between 1990 and 1991 [12]. The introduction of these mammals affected the island's vegetation drastically, and some native plant species, like *Opuntia* sp., have diminished drastically [13]. Although pigs and sheep have been eradicated [14], the transformed landscape is more susceptible to the establishment of alien plant species.

As on other islands, many alien plant species were most likely introduced unintentionally, with their seeds traveling in the animals' fodder, wool, fur, and excrement, as well as crop contaminants [15]. The absence of permanent human settlements, until the establishment of a naval garrison on Socorro in 1957 [1] and a naval station on Clarion in 1979 [12], prevented the intentional introduction of alien plants. Afterwards, more alien plant species arrived due to human intervention, even though introductions were strictly prohibited since the publication of the management program of the Biosphere Reserve in 2007 [16] and, more recently, with the establishment of the National Park [17].

Several botanical expeditions have been conducted in the Revillagigedo Archipelago since the XIX century [18]. Since then, relevant work has been carried out to compile and describe the archipelago's flora, for instance, Johnston [19] and references therein, Levin and Moran [1], and Rebman et al. [20]. Additionally, other valuable contributions have been made, such as the revision of pteridophyte species by Mickel and Smith [21], or punctual additions by Dávila et al. [22] and Benavides et al. [23]. Despite these studies, the Revillagigedo Archipelago still lacks a comprehensive plant inventory that integrates all of the confirmed available information and an up-to-date list of alien plant species.

Island ecosystems are particularly vulnerable to invasions by alien plant species; these species bring negative impacts to the native flora and, subsequently, to the native fauna, particularly specialist and endemic species [24]. The Revillagigedo Archipelago hosts several exotic plant species known to be extremely invasive in other islands. For instance, the guava tree (*Psidium guajava*) is one of the most harmful invasive species in the Galapagos Archipelago [25]; however, its potential impacts on Socorro remain unassessed.

The successful management of invasive plant species requires active work to prevent new introductions, such as surveillance to detect emerging populations and persistent efforts to eradicate or at least control the worst invaders [26]. Even though it is understood that invasions are context-specific, and invasiveness will only materialize when certain environmental conditions are met [27], invasive patterns of species elsewhere can help anticipate how they will behave when introduced into a new territory [26].

In this article, we present the most complete and up-to-date list of native and introduced flora of the Revillagigedo Archipelago. Furthermore, this study aims to categorize alien plant species through their known invasiveness risk to forecast their impact on the native vegetation in the archipelago and to prevent possible negative outcomes. We studied and compared the invasion patterns on Socorro and Clarion across time to propose a way to establish management priorities while taking notice of limited logistical and financial resources. Also, the status of native and alien plant species of the Revillagigedo Archipelago is compared with other insular regions of the world.

2. Materials and Methods

2.1. Study Area

The Revillagigedo Archipelago is composed of four volcanic islands: Socorro, Clarion, San Benedicto, and Roca Partida (Figure 1). Socorro, the largest and most diverse island, is located 700 km west of Manzanillo, Colima [28]. It has an area of 131.32 km^2 [29]. The climate from 0 to 400 m is tropical semi-arid, while from 400 to 1150 m, it has a sub-humid subtropical climate [30]. The annual average temperature is 24.7 °C. The average annual accumulated precipitation is 313.8 mm; most precipitation occurs between July and October and the dry season is observed between February and June. During the rainy season, tropical storms and hurricanes are frequent [28].

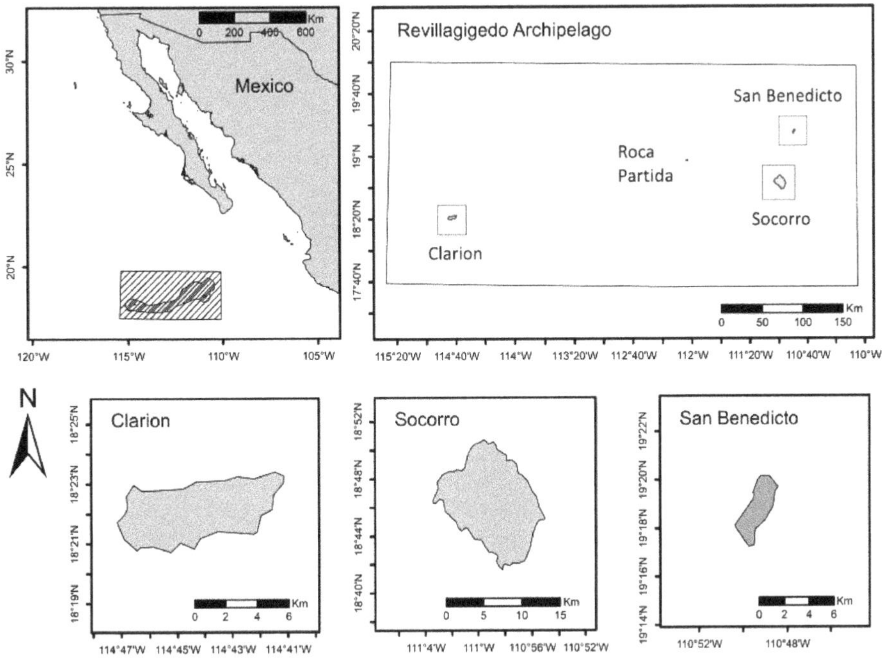

Figure 1. Location of the Revillagigedo Archipelago National Park in the Mexican Pacific Ocean, in relation to the mainland, and location of the three main islands.

The vegetation of Socorro has been classified into nine types [28]: coastal halophiles, *Conocarpus erectus* scrub, grasslands (*Cenchrus ciliaris, Cynodon dactylon, Melinis repens* [31]), *Croton masonii* scrub, *Pteridium caudatum–Dodonaea viscosa* scrub, mountain mesophilic forest (*Ilex socorroensis* and *Guettarda insularis*), deciduous tropical forest (*Ficus cotinifolia*), evergreen tropical forest (*Psidium-Sideroxylon* and *Hippomane mancinella*), and prairie (*Coreocarpus insularis, Lepechinia hastata socorrensis* and *Brickellia peninsularis*) [28].

Socorro's fauna consists of seabirds, migrating shorebirds, seasonal transients, eight endemic terrestrial birds [32], and one reptile, the endemic Socorro Lizard (*Urosaurus auriculatus*), as well as numerous species of invertebrates, among which the land crab (*Johngarthia planata*) is the most representative. The exotic fauna consists of two bird species, the Mourning Dove (*Zenaida macroura*) and the Northern Mockingbird (*Mimus polyglottos*), two mammals, namely, feral cats (*Felis catus*) and the domestic mouse (*Mus musculus*), and one reptile: the Tropical House Gecko (*Hemidactylus frenatus*).

The human population on Socorro is comprised of personnel from the Mexican Navy (between 40 and 70 people), who live in the naval garrison in the island´s southeast; additionally, there is an intermittent presence of civilian personnel and researchers. Food consumed by this population is transported from the continent, with very few exceptions, like fruits of introduced species (guava, coconut) and some species of fish. In the past, drinking water was transported to the island by ship; today, it is obtained from a desalination plant located at the naval garrison (Martínez-Gómez, unpublished field notes).

Clarion is the furthest island from the mainland and second in both size (19.58 km^2) [29] and biodiversity [12]. The highest point on Clarion reaches 308 m, while the elevation on the rest of the island averages around 150 m. The vegetation on this island is primarily herbaceous (*Brickellia peninsularis*), with only a few shrubs like *Rhamnus humboldtiana* with a height of no more than 3 m [12]. Its fauna includes eight bird species, which breed on the island, four of them being Clarion endemics, and several bird migrants [14], the endemic lizard (*U. clarionensis*), and two endemic snakes (Clarion Island Racer, *Masticophis anthonyi* and the Clarion Nightsnake, *Hypsiglena unaocularus*) [33] as well as insects, spiders, land snails, land crabs and introduced European rabbits [13], and the Western Spiny-tailed Iguana (*Ctenosaura pectinata*) [34]. Clarion is inhabited primarily by Mexican Navy personnel, whose number fluctuates between 10 and 20 people, and as on Socorro, their food is transported by ship (Martínez-Gómez, unpublished field notes).

The remaining two islands lack introduced plants. San Benedicto is a small island of 6.09 km^2 [29], with a maximum altitude of 335 m [35]. One of the volcanoes of this island, the Barcena, erupted between 1952 and 1953, destroying most of its original vegetation [35]. Since this event, the vegetation has recovered [35]. Unlike Socorro and Clarion, it does not seem to have been colonized by alien plant species [20]. San Benedicto hosts large numbers of seabirds, occasional bird migrants, and visitors from Socorro, as well as land crabs, insects, and spiders [13]. It does not have a permanent human population and it is not visited as frequently as Socorro or Clarion. Roca Partida is an islet of only 0.02 km^2 [29] with a maximum height of 50 m. It lacks vascular vegetation and its fauna consists only of resting seabirds [13].

2.2. Inventory of Plant Species Present on the Revillagigedo Archipelago

We built an inventory of the vascular plant species of the Revillagigedo Archipelago, compiled from published lists [1,19–23,36,37]. Each published species was corroborated with herbarium specimens or photographic records. We collected plant specimens during expeditions conducted on the archipelago by members of our team over an extended period. Initially, we visited Socorro during July 1993 and March–May 1994, and visited Socorro and Clarion during March–April 1995. After a pause, we resumed our collecting efforts on Socorro during December 2013–January 2014, December 2017–January 2018 [31], and November–December 2018. We visited San Benedicto in December 2014–Jan 2015 and January 2016 to obtain photographic records of the vegetation. Then, we carried out a comprehensive review of herbarium specimens from the archipelago in the Enciclovida.mx website [38] of the National Information System on Biodiversity (SNIB), and consulted multiple herbaria from Mexico and the United States (Supplementary Materials S1). Current names, basonyms, or synonyms were validated with the following global floristic databases: The Plant List website [39], World Flora Online website [40], the International Plant Names Index website [41], and the Plants of the World Online website [42].

The introduced plant species compiled in our inventory for the Revillagigedo Archipelago were assigned a category adapted from Richardson et al. [43] and Pyšek et al. [44] to designate their invasion status, whereas *alien* species are those whose presence is assumed to be the result of human-mediated transport. We decided to include those alien species that were introduced for horticulture, forestry, or ornamental purposes and have not escaped from cultivation, under the category of *Cultivated* [45], so we lay a precedent over the presence of potentially invasive species, although for now, they are only represented by a few individuals living under cultivation. *Casual* alien species are alien plants that may flourish and even reproduce occasionally outside cultivation, but do not form self-sustaining populations and may rely on repeated introductions for their persistence.

Taking into consideration the isolation of the archipelago, we reduced the minimum residency time required to consider a plant as *Naturalized*. In this category, we included species that have self-replacing populations for at least five years, as in the case of recognized highly invasive species with already large numbers or formed groups at considerable distance from their parental plants. Also, they have the potential to spread over a larger area, without direct human intervention, by recruitment from seeds or ramets capable of independent growth [27]. When a species was present on more than one island with different invasion stages, the highest status was applied [46]. Although invasive plant species are certainly present on the Revillagigedo Archipelago, we did not assign the category *Invasive*, because there is not enough information available on the spread rates and impacts of naturalized species for this region. Instead, we performed an invasiveness risk assessment based on the available information for each species from other insular regions (see below).

2.3. Species Traits: Origin, Life History, and Date of Introduction

The putative origin of alien plants in this inventory, as well as their life form and habit, was determined using the following global floristic databases: World Flora Online [40], Plants of the World Online [42], and the International Plant Name Index [41]. The native range of the species was grouped into seven regions resulting from the classification of major biogeographically defined areas of the Biodiversity Information Standards Organization (originally Taxonomic Databases Working Group [47]): North America (including Mexico), South America (including Central America), Africa, Europe, Tempered Asia, Tropical Asia, and Australasia. Since many species are native to more than one region, they were assigned to all regions where they are considered native [46]. Plants whose native range is unknown because they are a result of recent human-mediated hybridization were not included in any category. Since cultivated species are not yet found in the wild, they were excluded from the analyses of habit, life form, and date of introduction [48], but were considered in the origin determination and invasiveness risk assessment. Habit was classified as *Grass, Herb, Vine, Shrub*, or *Tree* and life form as *Annual* (for annuals and biennials) or *Perennial* [49].

In order to define an approximate date of introduction, we searched for the year of the first published record, or the first collected specimen available at herbaria [45], both for native and introduced species. Then, we built native and introduced species accumulation curves across years, for both Socorro and Clarion, and compared the distributions of these curves with Kolmogorov–Smirnov two-sample tests [50,51] using the R "fBasics" package [52]. These tests, which provide the cumulative probability that two samples have the same distributions [53], were used to assess whether native and introduced species have been registered in a similar way over time, as the number of scientific expeditions increased, or if introduced species records respond to other factors, such as the establishment of human settlements on the islands or species introduction restrictions. The proportions of native and introduced species in 30-year periods for both islands were estimated from the total number of species known for each period. A multiple sample test for equality of proportions was performed using the R "stats" package version 4.0.3 [54]. We ran *post hoc* tests to identify which proportions were different using the routine binom_test with a Holm–Bonferroni adjustment for multiple comparisons, with an experiment-wise significance level set at α = 0.05.

2.4. Invasiveness Risk Assessment for Alien Plant Species

Weed Risk Assessment (WRA) systems, originally created to evaluate the invasiveness risk of plant species before their introduction into certain island countries [55–57], can also be used to estimate the invasiveness risk of recently introduced alien species in a new environment. An example of this approach is the categorization of plants for their invasiveness risk as presented in the Global Compendium of Weeds [58]. This compendium considers a list of all invasive plant species anywhere in the world, along with their characteristics, such as the means of entry of plants into new environments, the routes of dispersal and the impact they have had on ecosystems, and the relationship the plant has had with humans [59]. This scoring system calculates a plant's risk potential as a ranking that can be compared to other weed species [58].

In addition to classifying alien plants by their invasion status in the archipelago, we included an invasiveness risk assessment for every species using the rating proposed in the latest edition of the Global Compendium of Weeds [58]. This rating is obtained by calculating a score, which grades each species in terms of (1) its ability to enter new environments, (2) to spread and establish over wider areas by its dispersal options, and

(3) by its impacts in these environments. This score allows one to rate invasiveness risk as *Extreme*, *High*, *Medium*, and *Low*. Since this system owes its strength to the number of times that plant has been historically reported as a weed in different regions of the world, there is a possibility that some species are included in the compendium, but are not rated because there are not enough records to assign a score.

2.5. Comparison of the Status of Native and Alien Plant Species of Different Insular Regions

To compare trends among different insular regions of the world, we evaluated the relationship between the number of species versus island area for both native and alien plants. We fitted a generalized linear model (GLM) using the quasi-Poisson family with a log link function [60]; the Poisson regression is appropriate when modeling an overdispersed count variable. The number of species was used as the response variable and island area as the independent variable. Analyses were performed separately for native and invasive species. Island size was obtained from INEGI [29] and other sources [61–69].

3. Results

3.1. Inventory Overview

Our work allowed us to assemble the most accurate list of vascular plant species for the Revillagigedo Archipelago. We eliminated spelling and taxonomic errors from existing lists, like double or even triple records of the same species that were, in fact, synonyms or basonyms, as well as non-existent species. Our final compilation is based on a thorough revision of published listings and herbarium specimens and resulted in 272 species for the entire archipelago belonging to 73 families (the complete and substantiated species list, by island, is available in Supplementary Material S2). Socorro has 237 species, Clarion has 62, and San Benedicto has 11 (of which four endemic and one native species are believed to be absent from the island since the eruption of Barcena Volcano in 1952) [35]. Thirty-three species are strictly endemic to Socorro, eight to Clarion, three to Socorro and Clarion, four to San Benedicto, and four to the Revillagigedo Archipelago. We found 106 alien species, four of which were collected on Socorro, but have not been observed since 2010 (Martínez-Gómez, unpublished field notes; Table 1). Six species were not included in our inventory because they have only been mentioned in the species catalog compiled by Captain Ignacio Cabrera [16]; these species lack herbarium or photographic support and other reports did not confirm their presence. Although some authors suggest treating insular forms and island varieties as a single species or as equivalents of their continental counterparts (e.g., González-Gallegos et al. [70] and Wood et al. [71], we consider them distinct forms when noticeable morphological differences have been observed previously (sensu Levin and Moran [1]).

Table 1. Revillagigedo Archipelago vascular plant species inventory. S = Socorro, C = Clarion, SB = San Benedicto. (*) Symbol marks whether the species is Endemic or Introduced and/or Probably Not Present; according to each column.

Family	Scientific Name	Island	Endemic	Introduced	Probably Not Present
Acanthaceae	*Elytraria imbricata* (Vahl.) Pers.	S			
Aizoaceae	*Sesuvium portulacastrum* (L.) L.	C			
Amaranthaceae	*Amaranthus palmeri* S. Watson	S		*	
Amaranthaceae	*Gomphrena nitida* Rothr.	S		*	
Amaranthaceae	*Gomphrena serrata* L.	S		*	
Amaranthaceae	*Iresine diffusa* Humb. & Bonpl. ex Willd.	C			
Anacardiaceae	*Mangifera indica* L.	S		*	
Annonaceae	*Annona muricata* L.	S		*	
Apiaceae	*Daucus montanus* Humb. & Bonpl. ex Schult.	S			
Apocynaceae	*Cascabela thevetia* (L.) Lippold	S		*	

Table 1. *Cont.*

Family	Scientific Name	Island	Endemic	Introduced	Probably Not Present
Apocynaceae	*Metastelma californicum* Benth.	S			
Apocynaceae	*Metastelma cuneatum* Brandegee	S			
Apocynaceae	*Metastelma minutiflorum* Wiggins	S			
Aquifoliaceae	*Ilex socorroensis* Brandegee	S	*		
Araliaceae	*Oreopanax xalapensis* (Kunth) Decne. & Planch.	S			
Arecaceae	*Cocos nucifera* L.	S, C		*	
Arecaceae	*Phoenix dactylifera* L.	S		*	
Aristolochiaceae	*Aristolochia islandica* Pfeiffer	SB	*		*
Aristolochiaceae	*Aristolochia socorroensis* Pfeifer	S	*		
Aristolochiaceae	*Aristolochia* sp.	C	*		
Asparagaceae	*Agave americana* L.	S		*	
Asphodelaceae	*Aloe vera* (L.) Burm.f.	S		*	
Aspleniaceae	*Asplenium formosum* Willd.	S			
Aspleniaceae	*Asplenium sessilifolium* Desv.	S			
Asteraceae	*Ageratina pacifica* (B.L. Rob. ex B.L. Rob.) R.M. King & H. Rob.	S			
Asteraceae	*Ageratina pichinchensis* (Kunth) R.M. King & H. Rob.	S			
Asteraceae	*Ambrosia confertiflora* DC.	S		*	
Asteraceae	*Bahiopsis chenopodina* (E. Greene) E. E. Schill. & Panero	S			
Asteraceae	*Bidens socorrensis* Moran & G. A. Levin	S	*		
Asteraceae	*Brickellia peninsularis* Brandegee	S	*		
Asteraceae	*Brickellia peninsularis* Brandegee var. *amphithalassa* I. M. Johnst.	C	*		
Asteraceae	*Coreocarpus insularis* (Brandegee) E. B. Smith	S	*		
Asteraceae	*Eclipta prostrata* (L.) L.	S			
Asteraceae	*Erigeron crenatus* Eastwood ex I. M. Johnst.	SB	*		*
Asteraceae	*Erigeron socorrensis* Brandegee	S	*		
Asteraceae	*Gamochaeta sphacelata* (Kunth) Cabrera	S			
Asteraceae	*Gnaphalium attenuatum* DC.	S			
Asteraceae	*Laennecia confusa* (Cronq.) G. L. Nesom	S			
Asteraceae	*Perityle socorrosensis* Rose	S, C, SB	*		
Asteraceae	*Blumea viscosa* (Miller) D'Arcy	S		*	
Asteraceae	*Sonchus asper* (L.) Hill	S		*	
Asteraceae	*Sonchus tenerrimus* L.	S		*	
Asteraceae	*Tridax procumbens* L.	S		*	
Asteraceae	*Eremosis littoralis* Gleason	S	*		
Boraginaceae	*Cordia curassavica* (Jacq.) Roemer & Schultes	S			
Boraginaceae	*Cordia cylindrostachya* (Ruiz & Pav.) Roem. & Schult.	S		*	
Boraginaceae	*Heliotropium curassavicum* L.	S			
Boraginaceae	*Heliotropium curassavicum* L. var. *oculatum* (Heller) Thorne	C			
Boraginaceae	*Heliotropium procumbens* Miller	S		*	
Boraginaceae	*Tournefortia capitata* M. Martens & Galeotti	S			
Boraginaceae	*Tournefortia hartwegiana* Steud.	S			
Brassicaceae	*Lepidium lasiocarpum* Torr. & A. Gray var. *latifolium* C. L. Hitchc.	C			
Brassicaceae	*Lepidium virginicum* L. ssp. *pubescens* (Greene) C. L. Hitchc.	S			
Burseraceae	*Bursera epinnata* (Rose) Engl.	S			
Cactaceae	*Opuntia* sp. 1	C			
Cactaceae	*Opuntia* sp. 2	S			
Campanulaceae	*Calcaratolobelia cordifolia* (Hook. & Arn.) Wilbur	S			
Campanulaceae	*Triodanis perfoliata* subsp. *biflora* (Ruiz & Pav.) Lammers	S			

Table 1. *Cont.*

Family	Scientific Name	Island	Endemic	Introduced	Probably Not Present
Caricaceae	*Carica papaya* L.	S		*	*
Caryophyllaceae	*Arenaria lanuginosa* (Michx.) Rohrb.	S			
Chenopodiaceae	*Dysphania ambrosioides* (L.) Mosyakin & Clemants	S		*	
Cleomaceae	*Arivela viscosa* (L.) Raf.	S, C		*	
Combretaceae	*Conocarpus erectus* L.	S			
Combretaceae	*Terminalia catappa* L.	S		*	
Commelinaceae	*Commelina erecta* L.	C			
Convolvulaceae	*Cressa truxillensis* Kunth	S, C			
Convolvulaceae	*Ipomoea carnea* subsp. *fistulosa* (Mart. ex Choisy) D.F. Austin	S		*	
Convolvulaceae	*Ipomoea halierca* I. M. Johnst.	C	*		
Convolvulaceae	*Ipomoea imperati* (Vahl) Griseb.	S			
Convolvulaceae	*Ipomoea indica* (Burm. f.) Merrill	C			
Convolvulaceae	*Ipomoea pes-caprae* (L.) R. Br.	S, C, SB			
Convolvulaceae	*Ipomoea purpurea* (L.) Roth	S		*	
Convolvulaceae	*Ipomoea triloba* L.	S			
Convolvulaceae	*Distimake quinquefolius* (L.) A.R Simões & Staples	S		*	
Cucurbitaceae	*Citrullus lanatus* (Thunb.) Matsum. & Nakai.	S		*	*
Cucurbitaceae	*Lagenaria siceraria* (Mol.) Standl.	S		*	*
Cucurbitaceae	*Luffa cylindrica* (L.) M. Roem.	S, C		*	*
Cycadaceae	*Cycas revoluta* Thunb.	S		*	
Cyperaceae	*Bolboschoenus maritimus* (L.) Palla ssp. *paludosus* (A. Nelson) T. Koyama	C			
Cyperaceae	*Bulbostylis nesiotica* (I. M. Johnst.) Fern.	S, C, SB	*		SB *
Cyperaceae	*Cyperus duripes* I. M. Johnst.	S, C, SB	*		
Cyperaceae	*Cyperus ligularis* Hemsl.	S			
Cyperaceae	*Cyperus rotundus* L.	S		*	
Cyperaceae	*Cyperus sordidus* J. Presl & C. Presl	S			
Cyperaceae	*Eleocharis mutata* (L.) Roem. & Schult.	C			
Dennstaedtiaceae	*Pteridium caudatum* (L.) Maxon	S			
Dryopteridaceae	*Ctenitis equestris* (Kunze) Ching	S			
Dryopteridaceae	*Dryopteris knoblochii* A. R. Smith	S			
Dryopteridaceae	*Polystichum hartwegii* (Klotzsch) Hieron	S			
Euphorbiaceae	*Acalypha umbrosa* Brandegee	S	*		
Euphorbiaceae	*Croton masonii* I. M. Johnst.	S	*		
Euphorbiaceae	*Euphorbia anthonyi* (Brandegee) G. A. Levin var. *anthonyi*	S	*		
Euphorbiaceae	*Euphorbia anthonyi* (T.S. Brandegee) G. Levin	SB	*		
Euphorbiaceae	*Euphorbia anthonyi* Brandegee var. *clarionensis* (Brandegee) I. M. Johnst.	C	*		
Euphorbiaceae	*Euphorbia californica* Benth.	C			
Euphorbiaceae	*Euphorbia heterophylla* L.	S		*	
Euphorbiaceae	*Euphorbia hirta* L. var. *hirta*	S		*	
Euphorbiaceae	*Euphorbia hyssopifolia* L.	S		*	
Euphorbiaceae	*Euphorbia incerta* Brandegee	S			
Euphorbiaceae	*Euphorbia thymifolia* L.	S, C		*	
Euphorbiaceae	*Hippomane mancinella* L.	S			
Euphorbiaceae	*Ricinus communis* L.	S		*	
Fabaceae	*Guilandina bonduc* L.	S, C			
Fabaceae	*Canavalia rosea* (Sw.) DC.	S, C			
Fabaceae	*Cassia fistula* L.	S		*	
Fabaceae	*Crotalaria incana* L.	S		*	
Fabaceae	*Delonix regia* (Hook.) Raf.	S		*	
Fabaceae	*Desmanthus bicornutus* S. Watson	S		*	
Fabaceae	*Desmodium procumbens* (Mill.) Hitchc.	S		*	
Fabaceae	*Desmodium scorpiurus* (Sw.) Desf.	S		*	

Table 1. *Cont.*

Family	Scientific Name	Island	Endemic	Introduced	Probably Not Present
Fabaceae	*Galactia striata* (Jacq.) Urban	C			
Fabaceae	*Indigofera suffruticosa* Miller	S		*	
Fabaceae	*Leucaena leucocephala* (Lam.) de Wit	S		*	
Fabaceae	*Macroptilium atropurpureum* (DC.) Urban	S, C		*	
Fabaceae	*Neptunia plena* (L.) Benth.	S, C		*	
Fabaceae	*Phaseolus lunatus* L.	S			
Fabaceae	*Pithecellobium dulce* (Roxb.) Benth.	S		*	
Fabaceae	*Prosopis juliflora* (Sw.) DC.	S		*	
Fabaceae	*Rhynchosia minima* (L.) DC.	S, C		*	
Fabaceae	*Senna alata* (L.) Roxb.	S		*	
Fabaceae	*Senna obtusifolia* (L.) Irwin & Barneby	S		*	
Fabaceae	*Sophora tomentosa* L.	C			
Fabaceae	*Tamarindus indica* L.	S		*	
Fabaceae	*Vachellia campechiana* (Mill.) Seigler & Ebinger	S		*	
Fabaceae	*Vachellia farnesiana* (L.) Wight & Arn. var. *farnesiana*	S		*	
Fabaceae	*Zapoteca formosa* subsp. *rosei* (Wiggins) H.M. Hern.	S, C	*		
Gentianaceae	*Centaurium capense* C. R. Broome	S			
Gentianaceae	*Zeltnera wigginsii* (Broome) Mansion	S			
Goodeniaceae	*Scaevola plumieri* (L.) Vahl	S			
Hypericaceae	*Hypericum eastwoodianum* I. M. Johnst.	S	*		
Lamiaceae	*Lepechinia hastata* (A. Gray) Epling ssp. *socorrensis* Moran	S	*		
Lamiaceae	*Hyptis pectinata* (L.) Poit.	S		*	
Lamiaceae	*Salvia pseudomisella* Moran & G. A. Levin	S	*		
Lamiaceae	*Salvia misella* Kunth	S		*	
Lamiaceae	*Teucrium townsendii* Vasey & Rose var. *affine* (Brandegee) Moran	S	*		
Lamiaceae	*Teucrium townsendii* Vasey & Rose var. *townsendii*	C	*		
Lamiaceae	*Teucrium affine* Vasey & Rose var. *dentosum* I.M. Johnst.	SB	*		*
Lauraceae	*Persea americana* Mill.	S		*	
Lycopodiaceae	*Huperzia dichotoma* (Jacq.) Trevis	S			
Malvaceae	*Abutilon californicum* Benth.	S			
Malvaceae	*Corchorus aestuans* L.	S		*	
Malvaceae	*Corchorus orinocensis* Kunth	S		*	
Malvaceae	*Gossypium hirsutum* L.	S		*	
Malvaceae	*Hibiscus diversifolius* Jacq.	S		*	
Malvaceae	*Hibiscus furcellatus* Desr.	S			
Malvaceae	*Talipariti tiliaceum* var. *pernambucense* (Arruda) Fryxell	S		*	
Malvaceae	*Malvastrum americanum* (L.) Torrey	S		*	
Malvaceae	*Malvastrum coromandelianum* (L.) Garcke	S		*	
Malvaceae	*Malvella leprosa* (Ortega) Krapov.	C			
Malvaceae	*Melochia pyramidata* L.	S, C		S *	
Malvaceae	*Pavonia hastata* Cav.	S			
Malvaceae	*Sida acuta* Burm.f.	S		*	
Malvaceae	*Sida barclayi* Baker f.	S		*	
Malvaceae	*Sida ciliaris* L.	S		*	
Malvaceae	*Sida nesogena* I. M. Johnst.	S	*		
Malvaceae	*Sida rhombifolia* L.	S		*	
Malvaceae	*Sida salviifolia* C. Presl	S		*	
Malvaceae	*Triumfetta socorrensis* Brandegee	S	*		
Malvaceae	*Waltheria indica* L.	S, C			
Meliaceae	*Cedrela odorata* L.	S		*	
Meliaceae	*Melia azedarach* L.	S		*	

Table 1. *Cont.*

Family	Scientific Name	Island	Endemic	Introduced	Probably Not Present
Meliaceae	*Swietenia humilis* Zucc.	S		*	
Meliaceae	*Swietenia macrophylla* King	S		*	
Molluginaceae	*Glinus radiatus* (Ruíz López & Pavón) Rohrb.	S		*	
Molluginaceae	*Mollugo verticillata* L.	S		*	
Moraceae	*Ficus benjamina* L.	S		*	
Moraceae	*Ficus cotinifolia* Kunth	S			
Myrtaceae	*Psidium guajava* L.	S		*	
Myrtaceae	*Psidium socorrense* I. M. Johnst.	S	*		
Nyctaginaceae	*Boerhavia coccinea* Miller	S, C			
Nyctaginaceae	*Boerhavia erecta* L.	S		*	
Nyctaginaceae	*Boerhavia scandens* L.	C			
Nyctaginaceae	*Bougainvillea glabra* Choisy	S		*	
Oleaceae	*Forestiera rhamnifolia* Griseb.	S			
Onagraceae	*Oenothera resicum* Benavides, Kuethe, Ortiz-Alcaráz & León de la Luz	C	*		
Ophioglossaceae	*Botrychium socorrense* W.H. Wagner	S	*		
Ophioglossaceae	*Ophioglossum reticulatum* L.	S			
Orchidaceae	*Acianthera unguicallosa* (Ames & C. Schweinf.) Solano	S	*		
Orchidaceae	*Dendrophylax porrectus* (Rchb. f.) Carlsward & Whitten	S			
Orchidaceae	*Epidendrum nitens* Rchb. f.	S			
Orchidaceae	*Epidendrum rigidum* Jacq.	S			
Orchidaceae	*Guarianthe aurantiaca* (Bateman ex Lindl.) Dressler & W.E. Higgins	S			
Orchidaceae	*Sarcoglottis schaffneri* (Rchb. f.) Ames	S			
Orobanchaceae	*Castilleja bryantii* Brandegee var. *socorrensis* (Moran) J. M. Egger	S	*		
Papaveraceae	*Argemone ochroleuca* Sweet	S		*	
Passifloraceae	*Passiflora suberosa* L.	S			
Piperaceae	*Peperomia socorronis* Trel.	S	*		
Piperaceae	*Peperomia tetraphylla* (G. Forster) Hook. & Arn.	S			
Plantaginaceae	*Nuttallanthus texanus* (Scheele) D. A. Sutton	S			
Plantaginaceae	*Scoparia dulcis* L.	S		*	
Poaceae	*Aristida adscensionis* L.	S			
Poaceae	*Aristida tenuifolia* Hitchc.	C	*		
Poaceae	*Aristida vaginata* A. Hitchc.	S	*		
Poaceae	*Bothriochloa pertusa* (L.) A. Camus	S		*	
Poaceae	*Cenchrus ciliaris* L.	S, C		*	
Poaceae	*Cenchrus echinatus* L.	S, C		*	
Poaceae	*Cenchrus myosuroides* Kunth	S, SB			SB *
Poaceae	*Chloris barbata* Swartz	S, C		*	
Poaceae	*Cynodon dactylon* (L.) Pers.	S		*	
Poaceae	*Dactyloctenium aegyptium* (L.) Willd.	S, C		*	
Poaceae	*Digitaria bicornis* (Lam.) Roem. & Schult.	S		*	
Poaceae	*Echinochloa colona* (L.) Link	S		*	
Poaceae	*Eragrostis tenella* (L.) P. Beauv. ex Roem. & Schult.	S		*	
Poaceae	*Eragrostis ciliaris* (L.) Link.	S		*	
Poaceae	*Eragrostis prolifera* Vasey	SB			
Poaceae	*Eragrostis pectinacea* (Michx.) Nees	S		*	
Poaceae	*Eriochloa acuminata* (Presl). Kunth.	C			
Poaceae	*Heteropogon contortus* (L.) P. Beauv. ex Roem. & Schult.	S, C			
Poaceae	*Jouvea pilosa* (Presl) Scribn.	S			
Poaceae	*Leptochloa panicea* (Retz.) Ohwi	S		*	
Poaceae	*Melinis repens* (Willdenow) Zizka	S, C		*	

Table 1. *Cont.*

Family	Scientific Name	Island	Endemic	Introduced	Probably Not Present
Poaceae	*Muhlenbergia solisii* (G. A. Levin) P. M. Peterson	S	*		
Poaceae	*Oplismenus hirtellus* (L.) Beauv.	S			
Poaceae	*Paspalum longum* Chase	S	*		
Poaceae	*Schizachyrium mexicanum* (Hitchc.) A. Camus	S			
Poaceae	*Schizachyrium sanguineum* (Retz.) Alston	S			
Poaceae	*Setaria geminata* (Forssk) Veldkamp	S			
Poaceae	*Setaria parviflora* (Poir.) Kerguelen	S			
Poaceae	*Sorghastrum pohlianum* Dávila, L.I. Cabrera & R. Lira	S	*		
Poaceae	*Sporobolus purpurascens* (Swartz) Hamilton	S			
Poaceae	*Sporobolus pyramidatus* (Lam.) Hitchc.	C			
Poaceae	*Urochloa maxima* (Jacq.) R.D. Webster	S		*	
Poaceae	*Urochloa reptans* (L.) Stapf	C		*	
Polygonaceae	*Antigonon leptopus* Hook. & Arn.	S		*	
Polypodiaceae	*Pecluma alfredii* (Rosenst.) M. G. Price var. *occidentalis* A. R. Smith	S			
Polypodiaceae	*Polypodium californicum* Kaulfuss	S			
Polypodiaceae	*Polypodium polypodioides* (L.) Watt. var. *aciculare* Weath.	S			
Portulacaceae	*Portulaca oleracea* L.	S, C		*	
Portulacaceae	*Portulaca pilosa* L	S, C			
Potamogetonaceae	*Potamogeton nodosus* Poiret	S			
Psilotaceae	*Psilotum nudum* (L.) P. Beauv.	S			
Pteridaceae	*Hemionitis radiata* (L.) Christenh.	S			
Pteridaceae	*Cheilanthes peninsularis* Maxon var. *insularis* Weath.	S, C	*		
Pteridaceae	*Pityrogramma ebenea* (L.) Proctor	S			
Rhamnaceae	*Frangula discolor* (Donn.Sm.) Grubov	S			
Rhamnaceae	*Rhamnus humboldtiana* Willd. ex Schult.	C			
Rosaceae	*Prunus serotina* Ehrh.	S			
Rosaceae	*Rubus* sp.	S			
Rubiaceae	*Chiococca alba* (L.) C. L. Hitchc.	S			
Rubiaceae	*Galium mexicanum* Kunth	S			
Rubiaceae	*Guettarda insularis* Brandegee	S	*		
Rubiaceae	*Houstonia mucronata* (Benth.) B.L. Rob.	S			
Rubiaceae	*Mitracarpus hirtus* (L.) DC.	S		*	
Rubiaceae	*Spermacoce nesiotica* (Robinson) G. A. Levin	S, C	*		
Rutaceae	*Citrus* × *limon* (L.) Osbeck	S		*	
Rutaceae	*Citrus x× aurantium* L.	S		*	
Rutaceae	*Zanthoxylum fagara* (L.) Sarg.	C			
Rutaceae	*Zanthoxylum insulare* Rose	S			
Sabiaceae	*Meliosma nesites* I. M. Johnst.	S	*		
Santalaceae	*Phoradendron quadrangulare* (Kunth) Griseb.	S			
Santalaceae	*Phoradendron tonduzii* Trel.	S			
Sapindaceae	*Cardiospermum halicacabum* L.	S, C			
Sapindaceae	*Dodonaea viscosa* (L.) Jacq.	S, C			
Sapindaceae	*Sapindus saponaria* L.	C			
Sapindaceae	*Sapindus* sp.	S		*	
Sapotaceae	*Manilkara zapota* (L.) P. Royen	S		*	
Sapotaceae	*Sideroxylon socorrense* (Brandegee) T. D. Penn.	S	*		
Solanaceae	*Cestrum pacificum* Brandegee	S	*		
Solanaceae	*Datura aff. discolor* Bernh.	C			
Solanaceae	*Nicotiana stocktonii* T. S. Brandegee	S, C, SB	*		
Solanaceae	*Physalis clarionensis* Waterf.	C	*		
Solanaceae	*Physalis mimulus* Waterf.	S	*		
Solanaceae	*Physalis philadelphica* Lam.	C		*	
Solanaceae	*Physalis pruinosa* L.	S		*	

Table 1. *Cont.*

Family	Scientific Name	Island	Endemic	Introduced	Probably Not Present
Solanaceae	*Solanum americanum* Miller	S		*	
Solanaceae	*Solanum ferrugineum* Jacq.	S			
Thelypteridaceae	*Thelypteris oligocarpa* (Willd.) Ching	S			
Verbenaceae	*Citharexylum danirae* León de la Luz & F. Chiang	S	*		
Verbenaceae	*Lantana involucrata* L. var. *socorrensis* Moldenke	S	*		
Verbenaceae	*Lantana velutina* M. Martens & Galeotti	S			
Verbenaceae	*Verbena sphaerocarpa* L. M. Perry	S	*		
Zygophyllaceae	*Tribulus cistoides* L.	S, C		*	

The 106 introduced species included in our inventory represent 39% of the total species present on the Revillagigedo Archipelago. When considering only naturalized species, they account for 29% of the total species present (0.41 naturalized/native ratio). Regarding Socorro, naturalized species represent 28% of the total species richness (0.5 naturalized/native ratio), while for Clarion, they represent 21% (0.28 naturalized/native ratio).

Socorro accounted for 104 alien species belonging to 39 families, among which the most represented families are Fabaceae (18 species), Poaceae (14 species), Malvaceae (13 species), and Euphorbiaceae (five species). In total, 66 alien species were classified as naturalized, 14 as casuals, and 24 as cultivated. Clarion presented 16 alien species, belonging to nine families, among which the most represented families are Poaceae (six species) and Fabaceae (three species) (Figure 2). Thirteen species were classified as naturalized and three as cultivated. San Benedicto did not have any alien species.

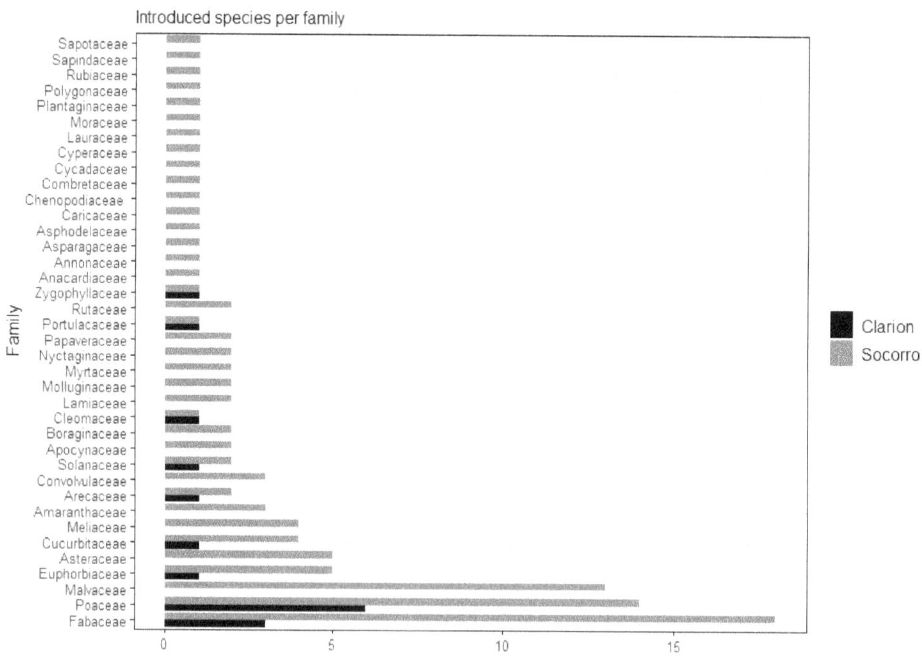

Figure 2. Number of alien plant species by family present in the Revillagigedo Archipelago.

3.2. *Origin and Life History*

The majority of alien species in the Revillagigedo Archipelago have their native range in the Americas (Table 2). North America contributes 66 species and South America 58 species. Africa and Asia follow; Africa contributed 32 species, Temperate Asia 32 species, and Tropical Asia 27 species. Australasia with 13 species and Europe with 5 species were the least represented regions. North America, Africa, and Asia were the regions with most naturalized species.

Table 2. Contribution of the world regions where alien species introduced into the Revillagigedo Archipelago are native, broken down by invasion status. Note that the total number exceeds the 104 species recorded in the archipelago because some species originate from more than one continent. Delimitation of regions followed the Biodiversity Information Standards Organization [47]. (-) Symbol indicates the absence of species in that category.

Region	Cultivated	Casual	Naturalized	Total
Africa	5	1	26	32
Asia Tem.	7	-	25	32
Asia Trop.	7	-	19	26
Australasia	4	-	9	13
Europe	-	-	5	5
N. Am.	11	12	43	66
S. Am.	9	11	38	58

The most common life history among the naturalized and casual species of Socorro are annual herbs (24 species; 30.0%) and perennial herbs (18 species, 22.4%; Figure 3). Perennial woody species account for ten trees (12.5%) and seven shrubs (8.7%), while there is only one annual shrub (1.3%). Grasses are represented by nine annual species (11.3%) and six perennial species (7.5%). Vines are represented by four perennial species (5%) and one annual species (1.3%). On Clarion, grasses are the most common naturalized species, with four annual (30.7%) and two perennial species (15.4%). Herbs account for three annual species (23.1%) and two perennial herbs (15.4%), while vines are represented by two perennial species (15.4%).

3.3. *Year of Introduction*

The dates of first collection or first report available for each species were used a proxy of the year of introduction. On Socorro, only 11 of the 80 casual and naturalized species were reported before the naval garrison was established in 1957. Fifty-three species were first reported between the establishment of the naval garrison and the publication of the Biosphere Reserve management program in 2007 [16], which prohibited the introduction of exotic species. Twelve more species have been reported since (Figure 4a). On Clarion, two introduced species were reported before the establishment of the naval garrison in 1979, another five species were reported between 1979 and the publication of the management program in 2007, and six species have been reported since its publication (Figure 4b).

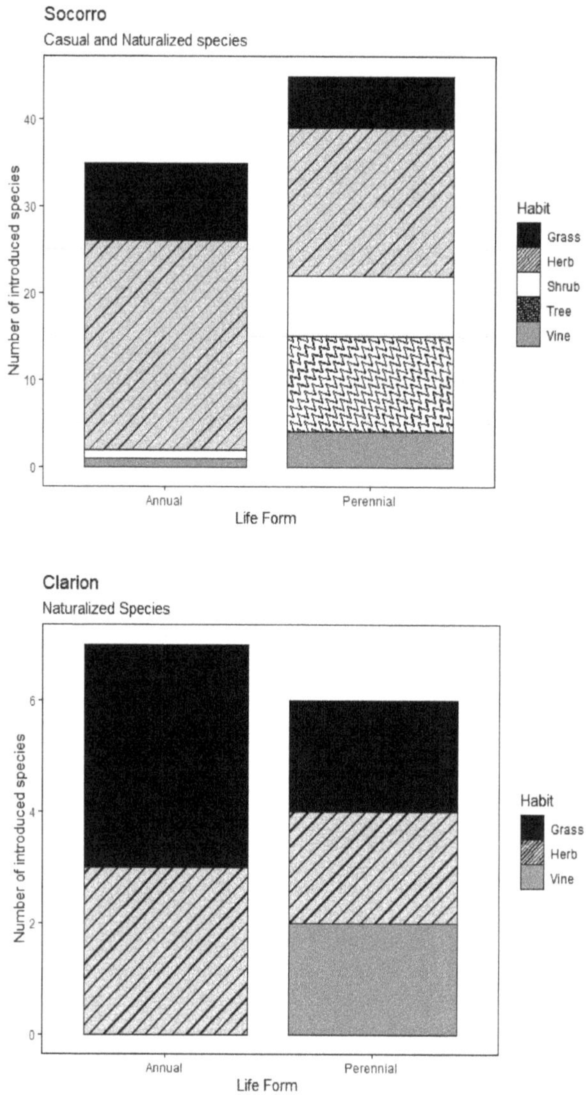

Figure 3. Habit and life form of casual and naturalized plant species on Socorro and naturalized plant species on Clarion.

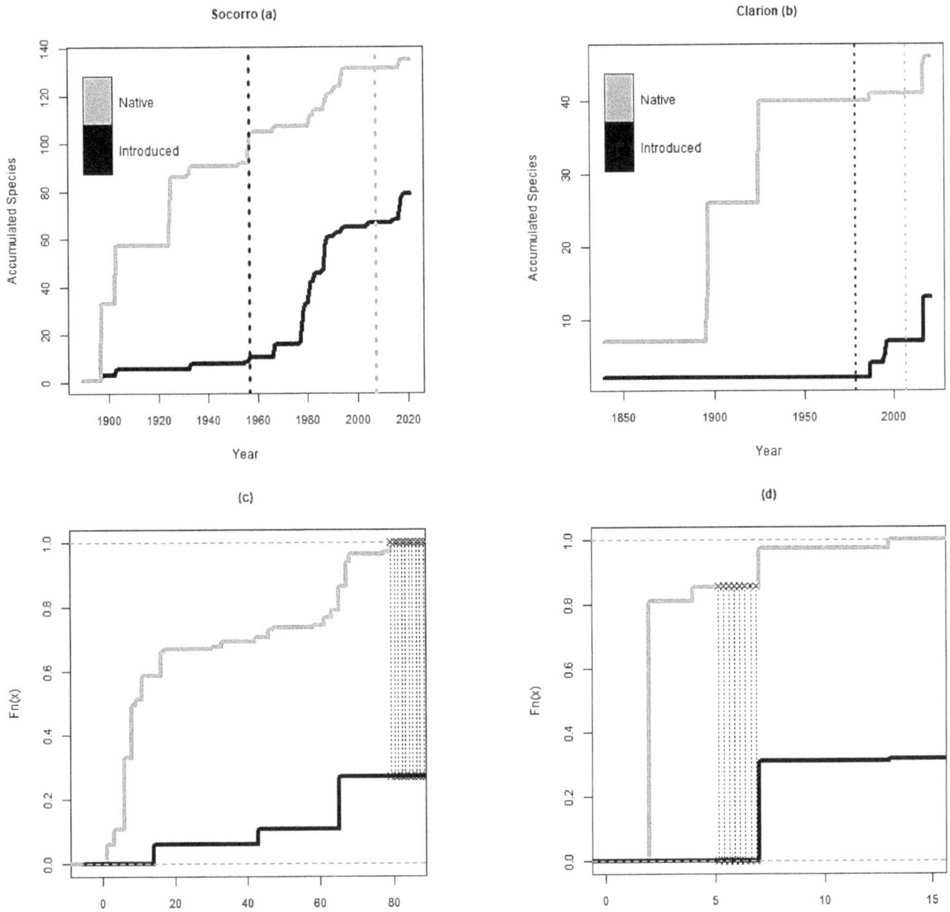

Figure 4. Species accumulation curves of native and introduced plant species for (**a**) Socorro and (**b**) Clarion. Black dotted lines mark the year in which the naval garrison was established on each island and gray dotted lines mark the year when the introduction of alien species was prohibited by the management program of the Biosphere Reserve [36]. Kolmogorov–Smirnov two-sample test plots: thin black dotted lines show the maximum difference between the two cumulative distributions for (**c**) Socorro and (**d**) Clarion.

Kolmogorov–Smirnov tests showed that the species accumulation curves of native and introduced species on Socorro and Clarion belong to statistically different distributions; native and introduced species have not been recorded in a similar way over time (D = 0.7293, $p \leq 0.001$ for Socorro, D = 0.8579, $p \leq 0.001$ for Clarion; Figure 4 c,d). The test for equality of proportions yielded significant differences between the introduced/native species over five 30–year periods on Socorro ($\chi^2 = 61.41$, df = 4, $p \leq 0.001$; Figure 5). The Holm–Bonferroni-adjusted p-values yielded a significant difference between the periods spanning from 1881 to 1970 and from 1971 to 2021 ($p \leq 0.001$; Figure 6). For Clarion, although the test for equality of proportions showed a significant difference ($\chi^2 = 12.048$, df = 5, $p = 0.034$; Figure 5), the Holm–Bonferroni-adjusted p-values did not show significant differences between the proportion of introduced/native species over the 30–year periods (Figure 6).

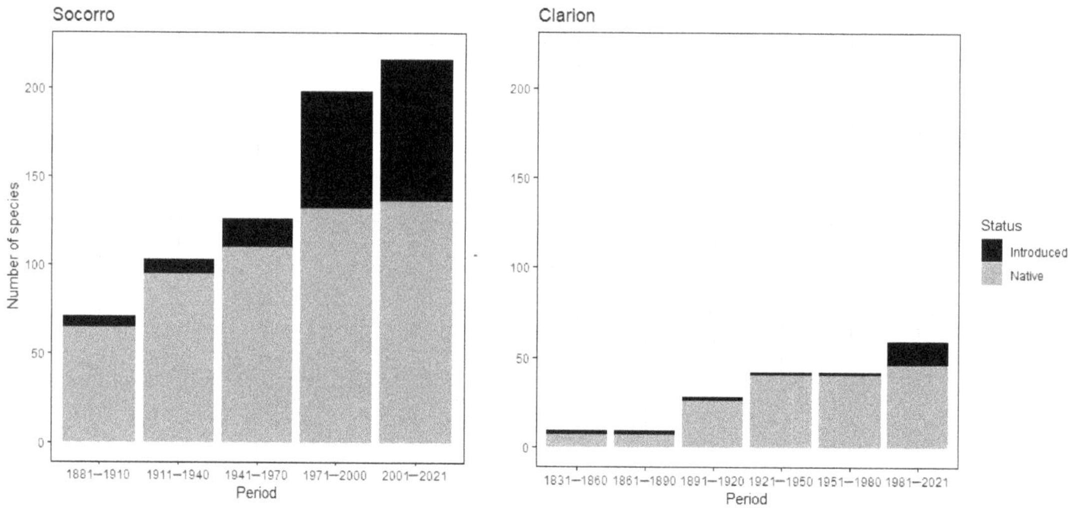

Figure 5. Total number of native and introduced (casual and naturalized) plant species recorded on Socorro and Clarion, for 30-year periods, from the year of the first species reported to 2021.

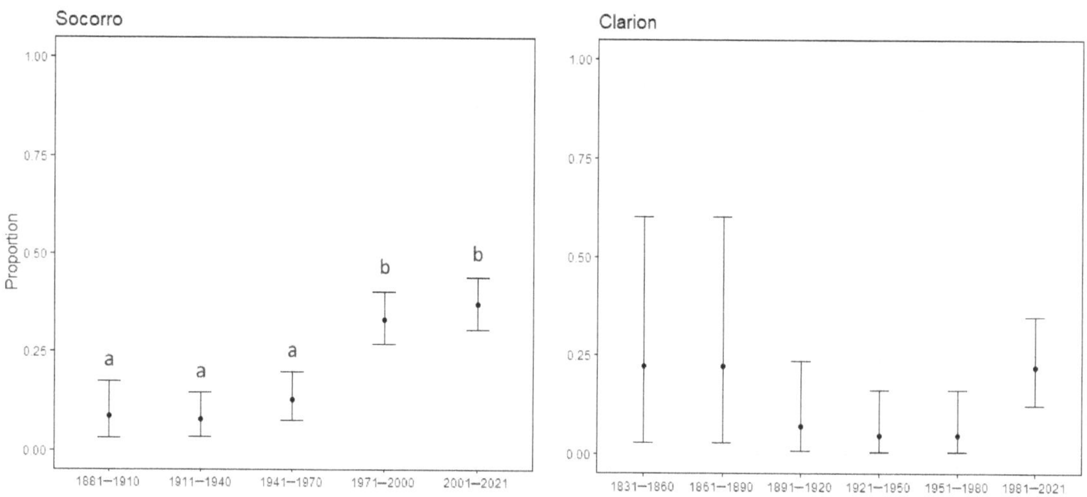

Figure 6. Proportion of introduced species by total number of species known at each 30-year period on Socorro and Clarion, from the year when the first species was reported on each island to 2021. Ninety-five percent confidence intervals were derived from a binomial distribution. In Socorro, the Holm–Bonferroni adjustment was able to separate the 30-year periods in two significantly different groups, marked with a and b in the figure. Although at least one significant pairwise difference between proportions on Clarion was detected by the multiple sample test for equality of proportions, the Holm–Bonferroni adjustment failed to identify it.

3.4. Invasiveness Risk Assessment for Alien Plant Species

Following the GCW 2017 risk classification [58] allowed the invasiveness risk for the 106 alien plants found in the archipelago to be assigned as follows: 18 species are considered to present an extremely high risk, 18 species represent a high risk, 32 species a medium risk, and 22 species present a low risk of invasiveness. Although eleven species found in the Revillagigedo are included in the GCW 2017, the information or number of reports necessary to determine their risk of invasiveness is not available; five more species are not included in the compendium (Table 3, Figure 7).

Table 3. Alien plant species in the Revillagigedo Archipelago with Extreme and High invasiveness risk according to the rating and score assigned by the Global Compendium of Weeds [58]. S = Socorro, C = Clarion, N = naturalized, Cas = casual, Cul = cultivated.

Family	Species	Island	Invasion Status	GCW Score	GCW Rating
Fabaceae	*Leucaena leucocephala* (Lam.) de Wit	S	N	64	Extreme
Fabaceae	*Pithecellobium dulce* (Roxb.) Benth.	S	Cas	51.2	Extreme
Fabaceae	*Vachellia farnesiana* (L.) Wight & Arn. var. *farnesiana*	S	N	51.2	Extreme
Cyperaceae	*Cyperus rotundus* L.	S	N	44.8	Extreme
Fabaceae	*Macroptilium atropurpureum* (DC.) Urban	S, C	N	44.8	Extreme
Fabaceae	*Senna obtusifolia* (L.) Irwin & Barneby	S	N	44.8	Extreme
Poaceae	*Cenchrus echinatus* L.	S, C	N	44.8	Extreme
Poaceae	*Cynodon dactylon* (L.) Pers.	S	N	44.8	Extreme
Portulacaceae	*Portulaca oleracea* L.	S, C	N	44.8	Extreme
Euphorbiaceae	*Ricinus communis* L.	S	N	43.2	Extreme
Arecaceae	*Phoenix dactylifera* L.	S	Cul	38.4	Extreme
Meliaceae	*Swietenia macrophylla* King	S	Cul	38.4	Extreme
Asteraceae	*Sonchus asper* (L.) Hill ssp. *asper*	S	N	35.84	Extreme
Fabaceae	*Crotalaria incana* L.	S	N	35.84	Extreme
Myrtaceae	*Psidium guajava* L.	S	N	34.56	Extreme
Poaceae	*Cenchrus ciliaris* L.	S, C	N	33.6	Extreme
Poaceae	*Dactyloctenium aegyptium* (L.) Willd.	S, C	N	33.6	Extreme
Poaceae	*Echinochloa colona* (L.) Link	S	N	33.6	Extreme
Arecaceae	*Cocos nucifera* L.	S, C	Cul	28.8	High
Asteraceae	*Tridax procumbens* L.	S	N	26.88	High
Malvaceae	*Sida acuta* Burm.f.	S	N	26.88	High
Chenopodiaceae	*Dysphania ambrosioides* (L.) Mosyakin & Clemants	S	N	26.4	High
Anacardiaceae	*Mangifera indica* L.	S	Cul	25.92	High
Fabaceae	*Cassia fistula* L.	S	Cul	25.92	High
Meliaceae	*Cedrela odorata* L.	S	N	25.92	High
Meliaceae	*Melia azedarach* L.	S	Cul	25.92	High
Convolvulaceae	*Ipomoea purpurea* (L.) Roth	S	Cul	24	High
Polygonaceae	*Antigonon leptopus* Hook. & Arn.	S	N	24	High
Solanaceae	*Solanum americanum* Miller	S	N	24	High
Poaceae	*Melinis repens* (Willdenow) Zizka	S, C	N	20.16	High
Caricaceae	*Carica papaya* L.	S	Cul	19.44	High
Combretaceae	*Terminalia catappa* L.	S	Cul	19.44	High
Euphorbiaceae	*Euphorbia heterophylla* L.	S	Cas	19.2	High
Zygophyllaceae	*Tribulus cistoides* L.	S, C	N	19.2	High
Poaceae	*Eragrostis amabilis* (L.) Wight & Arn.	S	N	17.92	High
Fabaceae	*Tamarindus indica* L.	S	Cul	17.28	High

3.5. Comparison of the Status of Native and Alien Plant Species of Different Insular Regions

The GLM model derived from the quasi-Poisson family shows that smaller insular regions, on average, show a 1:1 ratio between native and introduced species, while in larger insular regions, native species exceed introduced ones, on average (Figure 8). According to this model, area is a good predictor of the number of introduced and native species on insular regions; however, the analyses of residual deviance suggest that other variables should be considered (χ^2 = 400.9, df = 9, $p \leq 0.001$, for introduced species; χ^2 = 1482.2, df = 9, $p \leq 0.001$, for native species) (Table 4).

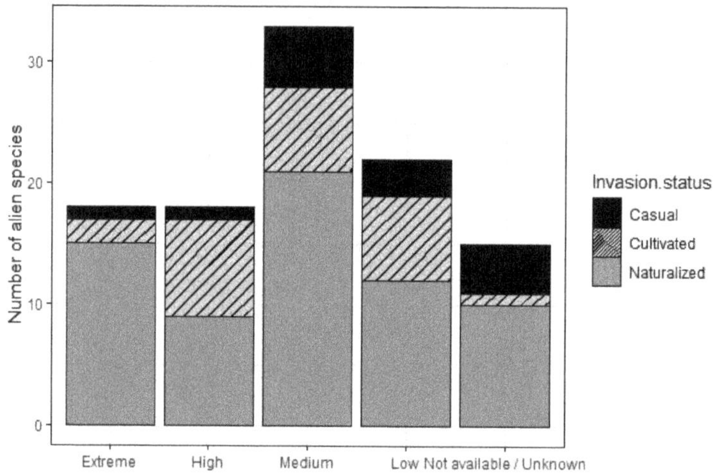

Figure 7. Total number of alien plant species by invasion status and GCW 2017 invasiveness risk rating categories in the Revillagigedo Archipelago.

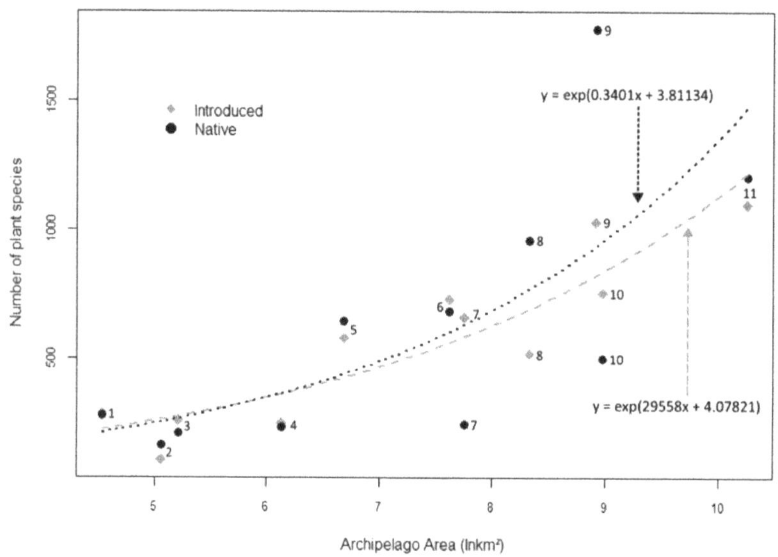

Figure 8. Number of introduced and native plant species from eleven insular regions around the world, as follows: (1) Cook, (2) Revillagigedo, (3) Juan Fernandez, (4) Seychelles, (5) Madeira, (6) Mauritius, (7) Azores, (8) French Polynesia, (9) Canaries, (10) Galapagos, (11) Hawaii. GLM based on the quasi-Poisson family are depicted: native species by black dots and invasive species by the gray broken line.

Table 4. Generalized linear model (GLM) for the number of introduced and native plant species versus archipelago or island surface using a quasi-Poisson model with a log link function.

	Estimate	SE	*t*-Value	*p*
Introduced				
Intercept	4.07821	0.41238	9.889	3.92×10^{-6}
ln area (km^2)	0.29558	0.04989	5.925	0.000222
Dispersion parameter: 42.16917				
Null deviance: 1989.76 (10 df)				
Residual deviance: 400.98 (9 df)				
Native				
Intercept	3.81134	0.81587	4.672	0.00117
ln area (km^2)	0.3401	0.09741	3.491	0.00682
Dispersion parameter: 169.7447				
Null deviance: 3750.7 (10 df)				
Residual deviance: 1484.2 (9 df)				

4. Discussion

4.1. Inventory Overview

The impact that invasive plant species may have on the native flora and fauna of the Revillagigedo Archipelago requires a complete assessment. Invasive plants represent a potential threat to the survival of endangered bird species, such as the endemic Socorro Mockingbird (*Mimus graysoni*) and Socorro Parakeet (*Psittacara brevipes*), and other endemic and native species, because these plants can take over key areas required for their breeding success. Invasive plants may also jeopardize reintroduction efforts and the establishment of the Socorro Dove. There are numerous examples of the risks posed by invasive plants to island ecosystems. For instance, in the Galapagos Islands, where introduced plant species outnumber native species [25], many of the invasive species, particularly trees, vines, and grasses, compete aggressively with native plants, displacing them and altering a globally unique habitat, home to numerous endemic species [72]. The Hawaiian Archipelago, rich in endemic animals and plants, is also one of the ecosystems most threatened by invasive plants [73]. On these islands, once-abundant plant species are now threatened, and native habitats are dominated by highly invasive exotic plants that are capable of dramatically altering community structure and ecosystem processes [73]. In other archipelagos and islands, invasive plants are one of the main threats to endemic species and plant diversity, such as the Azores Islands, Madeira and the Canary Islands in the Atlantic Ocean, the Juan Fernandez Archipelago in the south of the Pacific Ocean, and the Mascarene Islands in the southwest of the Indian Ocean [74].

Knowledge on the flora of the Revillagigedo Archipelago has grown significantly over time. Johnston [19] provided the first account that included a list for the Revillagigedo Archipelago; he derived this list from collecting trips carried out between 1839 and 1925. More than five decades later, Levin and Moran [1] published a more complete inventory, which included specimens collected between 1957 and 1988. They listed 117 native and 47 naturalized species for Socorro, while 41 native and one naturalized species were listed for Clarion. Several inventories have been published since, for instance, the management programs for the Revillagigedo Archipelago Biosphere Reserve [36] and the Revillagigedo National Park [37]. These inventories used scientific publications, internal reports, and databases such as EncicloVida [38].

More recently, Rebman et al. [20] reported 201 plant species for Socorro, of which 67 were considered non-native. On Clarion, they reported 58 species, of which 12 were considered non-native. On San Benedicto, they reported eleven species and no alien plants. Rebman et al. [20] considered 12 alien species as native, while other authors considered them introductions because of the date of first collection and closeness to human settlements (e.g., Levin and Moran [1]). Also, they claimed to report 38 "new" records (21 for Socorro, 16 for Clarion, and 1 for San Benedicto); however, 11 of these records were reported or collected during previous expeditions. This highlights the need for a thorough revision of Mexican and international herbaria when attempting to build a species inventory.

Since Levin and Moran's flora, 57 additional species have been introduced to Socorro and 14 to Clarion. We report three new records for Socorro: *Cassia fistula*, which was assigned a *Cultivated* status; *Senna alata*, with a *Casual* status, and *Urochloa maxima*, which was found to be already *Naturalized*. It should be noted that these three species belong to families known for their high invasive potential (Fabaceae and Poaceae) [49], and it is imperative to establish immediate control actions to prevent their spread throughout the island. We also report, for the first time, four ornamental introduced species that have been kept under human care in the naval garrison: *Agave americana*, *Aloe vera*, *Cycas revoluta*, and *Ficus benjamina*. Although, for the moment, these plants do not pose a significant invasiveness risk, we recommend that they be monitored and considered examples of the continuous introduction of alien plant species into the island.

Leucaena leucocephala is the most common invasive species worldwide. This legume grows in the form of a tree or shrub, with high tolerance to drought and nitrogen-fixing root nodules, and thrives in arid and nutrient-poor environments [49]. The impact of *L. leucocephala* as a weed around the world is well known [58]. In the Hawaiian Archipelago, this species has spread quickly, invading lowland plant communities and hindering the proliferation of native species. By 2014, this species had already fully covered the lowlands of O'ahu and Kaua'i, as well as large lowland areas in the rest of that archipelago [64]. In Galapagos, the dissemination and impact of *L. leucocephala* has been less dramatic, but its presence has been reported in agricultural areas of Floreana, Isabela, San Cristobal, and Santa Cruz Islands, and plants have already been detected in the National Park area [75]. In the South Pacific, the Vanuatu islands have large extensions of monodominant shrubs of this invasive species [76]. In their global analysis of naturalized alien flora, Pyšek et al. [77] also found that *L. leucocephala* was present with an invasive status in 29.5% of the 349 regions (mainland and islands) considered in the study, of which 55 were Pacific Islands. On Socorro, this species was first registered by our team in 2015 and its continuous presence confirmed by Martínez-Gómez [31] and Rebman et al. [20]. *Leucaena leucocephala* has the highest GCW score in the Revillagigedo Archipelago.

Another invasive species of tree posing a high risk to Socorro native plants and endemic birds is the guava tree (*Psidium guajava*). This species, whose presence on Socorro was first reported by Cruz-Cisneros in 1967 (Supplementary Material S2), has numerous reports on its invasiveness around the world [58,77]. This species is considered one of the most damaging invasive plants in the Galapagos Archipelago [25], where it has become the most abundant invasive species in rural areas [62], and has displaced entire endemic tree forests such as *Scalesia pedunculata* [76].

Some herb species also stand out for their extreme invasiveness risk. *Senna obtusifolia* is a Fabaceae species with almost 200 reports of its invasiveness from around the world, including islands in the Caribbean, Pacific, and Indian Oceans [58]. This species threatens the ecosystems by competing with vulnerable native species and forming monospecific stands [78]. In Socorro, it can be observed covering large patches between 300 and 350 m in the southeastern region [31]. However, it is the grasses that have most successfully colonized the areas devoid of their original vegetation cover. *Melinis repens* is a fast-growing pioneer grass with the potential to colonize highly disturbed sites [79,80] and that has taken advantage of the bare and eroded soil areas on Socorro. On Desecheo Island, Puerto Rico, García-Cancel et al. [81] found that from 2000 to 2016, this grass went from being nearly absent to being present in almost all exposed areas of the island. In Socorro, Walter and Levin [9] described the rapid spread of *M. repens* in the south end of the island since its establishment in the late 1970s. Recently, Martínez-Gómez [31] confirmed that this species, along with *Cenchrus ciliaris* [20], had become the dominant cover in the most disturbed open areas. *C. ciliaris* poses an extreme invasiveness risk [58] and it is considered a threat to native plant communities in many islands of the world because it forms extensive, dense, and low-diversity stands, which inhibit the growth of native plant species; it also promotes frequent and intense fires [81–83], which have already occurred on occasion near the naval base on Socorro (Martínez-Gómez, unpublished field notes).

Worldwide, the families Asteraceae, Poaceae, and Fabaceae are the most represented in naturalized floras [77]. Insular ecosystems also show this pattern. Kueffer et al. [49] found, in 30 insular regions in the Atlantic, Caribbean, Pacific, and Western Indian Oceans, that Fabaceae, Poaceae, and Asteraceae were the best represented. In the Eastern Pacific, Guézou et al. [62] found that Fabaceae, Asteraceae, and Poaceae were also the most common in the Galapagos. On Socorro and Clarion, we observed that Fabaceae, Poaceae, and Malvaceae were the most common families.

4.2. Origin and Life History

The percentage of naturalized species native to North America (65%) found in Revillagigedo is higher than the 17% found among alien species with this origin in mainland and island regions around the world reported by Pyšek et al. [77]. This could be the result of the proximity of this continent to the archipelago. Furthermore, naval personnel and their families most likely introduced, intentionally or unintentionally, species that were familiar and closer to them [15]. Thus, only 11 of the 43 naturalized species native to North America were collected before the first naval garrison was established on Socorro in 1957 and only one was reported for Clarion before the establishment of the naval garrison in 1979. Species native to Temperate and Tropical Asia in the Revillagigedo Archipelago are present in higher percentages than those for the global naturalized flora, reported by Pyšek et al. [77], with 37.9% and 28.8% compared with 20.3% and 11.7%, respectively. We also found a greater representation of African species, with 35%, compared with the global proportion of 14.3%; this is most likely a result of the high number of African grasses [84] and other ruderal species, which, once introduced, took advantage of the overgrazed and eroded areas, rapidly colonizing the bare soil [81,85].

Regarding species' life history (life form and habit), herbs were the best represented in the Revillagigedo Archipelago, where 53% of all naturalized species are herbs (30.3% annuals and 22.7% perennials). This finding is coincident with the trend observed in Galapagos by Guézou et al. [62], where herbs accounted for almost half of the alien species found in the archipelago. At a global level, the best-represented life histories among naturalized flora on island regions are perennial herbs, composing more than one third of all species (33%), while annual herbs represent 19%. Annual and perennial grasses in the Revillagigedo Archipelago are present in a higher percentage than the global trend at 15% and 9%, against 2.5% and 4.6%, respectively. It is noticeable that Revillagigedo's percentage of annuals (47%) doubles the 22% found in other island regions, while the proportion of perennial woody species is lower at 16.7% compared to the 34.4% found on islands worldwide [77]. When compared with the percentages of oceanic islands only, Revillagigedo's perennial woody species percentage is also lower than the 57% reported by Kueffer et al. [49], as well as the percentage of perennials (53%), which is lower than the 93% reported by these authors.

4.3. Year of Introduction

Th results from the comparisons between native/introduced species proportions over time showed that on Socorro, the proportion of introduced species became significantly higher in the years after the establishment of naval facilities. However, Clarion did not present a significant difference in these proportions after the establishment of the naval garrison, possibly as an effect of the small total number of species present on the island. Nevertheless, these results emphasize the need to reconsider and reinforce the biosecurity practices carried out in the archipelago.

It is necessary to highlight that, among the 106 introduced species included in our list (including species under cultivation), there are 19 species whose first reports are subsequent to the publication of the management program of the Biosphere Reserve in 2007, despite the explicit prohibition to introduce plants. This finding is particularly alarming since, among the species that have entered and become established after this prohibition, there are some that represent an extreme risk to the ecosystem due to their high potential for invasiveness, and, therefore, require immediate actions of control and eradication; these species include *Leucaena leucocephala, Ricinus communis* [86], and *Tridax procumbens* [20]. The extreme invasiveness of *R. communis* has also been reported for numerous countries around the world [58]. In fact, Pyšek et al. [77] found that this species is the second most widely distributed naturalized species on the planet, being recorded as an invader in 259 of 483 mainland regions and 161 of 361 island regions considered in their study, with 45 of the latter being located in the Pacific Ocean. On the Galapagos Archipelago, this species is considered one of the worst invaders [87], while *Tridax procumbens*, being one of the 200 most invasive plant species in the world, was found to be reported as invasive on 28 Pacific islands.

4.4. *Invasiveness Risk Assessment for Alien Plant Species*

Even when the minimum residency time for the invasion status classification was adapted so it would be relevant to the fragility of these islands and the characteristics of the alien plant species present on the archipelago, the mere fact of being able to classify the invasion status for some species such as *Leucaena leucocephala* or *Ricinus communis* may not be enough to carry out timely control actions. This is the case because, as soon as naturalization is achieved, the impacts may be both immediate and costly [58]. Taking this into account, we believe that by using weed risk assessment systems, like the Global Compendium of Weeds, that allow us to extrapolate knowledge about the invasion success of species in other areas of the world, we can advance to a prediction of the possible effects of the species in these new areas [26]. Nevertheless, we agree that this approach must not be taken at a distance without addressing the situation in the field, because the invasion history of each species on the Revillagigedo Archipelago itself should also be considered as a determinant for management actions. For example, *Melia azedarach* and *Psidium guajava* present the same score and risk category; however, the history of the presence of *P. guajava* on Socorro is older, its distribution much broader, and the risk of dispersal to less disturbed areas greater. Complementing the categorization of the invasiveness risk of the species with records of their invasion success on the island in question will thus provide a more complete framework for decision making. Therefore, we implemented a pilot control program based on mechanical methods for *Leucaena leucocephala, Psidium guajava, Acacia farnesiana, Citrus aurantium, Pithecellobium dulce, Cedrela odorata, Swietenia humilis* and *Ricinus communis*; preliminary results suggest that chemical methods should be incorporated to outcompete species with strong vegetative growth [86].

4.5. *Comparison of the Status of Native and Alien Plant Species of Different Insular Regions*

Less than 40% of the flora in the Revillagigedo Archipelago is alien, which is a percentage lower than those reported for other islands or archipelagos. For example, Pyšek et al. [77] found that, for 41% of the island regions included in their global study, naturalized alien plants accounted for more than 40% of their flora. Nevertheless, Socorro, which was the first island in Revillagigedo to be inhabited, is slowly approximating the trend presented by most inhabited oceanic islands, where there is practically a one-to-one ratio between the number of native and naturalized species [88]. There are several examples of this phenomenon in the Pacific Ocean. For the Galapagos Islands, 754 introduced species are registered (60% of an approximate total of 1254 species) [62]; of the total of these introduced species, about 45% are considered naturalized, and, of these species, 7% are considered invasive and an additional 20% potentially invasive [25]. In the Hawaiian archipelago, the number of naturalized species is around 1100 (48% of an approximate total of 2307 species) [64]. French Polynesia and the Cook Islands are also close to this 1:1 introduced/native ratio at 0.75 and 0.97, respectively, while the Juan Fernández Archipelago has already surpassed it with a ratio of 1.1. Examples from the Indian Ocean show a similar scene. La Reunion presents a 0.93 ratio, while the Seychelles and Mauritius surpass the 1:1 ratio at 1.06 and 1.07, respectively [24].

In our exploration of the relationship between the size of islands or island groups and the number of native and introduced plant species, we found that insular regions with smaller areas were the ones closer to this 1:1 ratio. However, there are other variables that may be involved in the native and introduced species richness. Other studies have identified size of the native flora, human population size [65], human development (measured by the gross domestic product (GDP) per capita) [49,57,77], habitat heterogeneity [61,89], geographic region and altitude [77] as significant contributors to the number of alien species. Among the reasons why the Revillagigedo Archipelago percentages of introduced species are lower than the ones presented in other tropical oceanic islands, despite human occupancy, may be the late colonization and the lack of a permanent human population relying on the islands' resources.

5. Conclusions

This is the first work to curate and gather information from botanical material collected more than a century ago, located at Mexican and international institutions. This resulted in a complete and accurate list of the vascular plant species present in the Revillagigedo National Park, substantiated by an extensive review of herbaria. Certainly, additional curatorial work and taxonomic revisions will change the inventory composition. An awareness of the presence and risk posed by the steadily increasing number of alien species is crucial to propose and carry out timely control or eradication actions [89], thereby optimizing costs and resources as part of a restoration program on these Mexican islands. *Leucaena leucocephala*, *Psidium guajava*, *Ricinus communis*, and *Cenchrus ciliaris* are among the species that require immediate control actions. Constant monitoring, control efforts, and regular botanical surveys are also necessary to maintain the number of alien species in the Revillagigedo Archipelago. Control and eradication measures for alien plants will not only benefit engendered species such as the Socorro Mockingbird and Socorro Parakeet, but will also facilitate the reintroduction of the Socorro Dove.

Supplementary Materials: The following supporting information can be downloaded at: https://www.mdpi.com/article/10.3390/plants12193455/s1, Supplementary Materials S1: Consulted Herbaria; Supplementary Materials S2: Complete species list, by island.

Author Contributions: Conceptualization, A.D.-M. and J.E.M.-G.; methodology, A.D.-M. and J.E.M.-G.; formal analysis, A.D.-M. and J.E.M.-G.; investigation, A.D.-M., J.E.M.-G., T.M.-S., and I.A.-R.; data curation, A.D.-M., J.E.M.-G., T.M.-S., and I.A.-R.; writing—original draft preparation, A.D.-M., J.E.M.-G., and S.S. writing—review and editing, J.E.M.-G. and S.S.; funding acquisition, J.E.M.-G. All authors have read and agreed to the published version of the manuscript.

Funding: This work was supported by Consejo Nacional de Ciencia y Tecnología (CONACyT), grant number 2019-000037-02NACF-18646 (CVU: 712072), the National Commission of Protected Areas program CONANP/PROMANP/MB/037/2017, the Island Endemics Foundation, Mohamed bin Zayed Species Conservation Fund, and INECOL. The APC was funded by the corresponding author.

Informed Consent Statement: Not applicable.

Data Availability Statement: Data are available upon request.

Acknowledgments: We are sincerely grateful for the long-term support provided by the Mexican Navy (SEMAR) to conduct scientific research in the Revillagigedo Archipelago; M. Ramírez and A. Paizanni (December 2013–January 2014 Socorro expedition) and Aguirre-Fey (November–December 2018 Socorro expedition) for their hard work collecting plant specimens in Socorro Island; C.M. Durán-Espinosa, E. Pérez-Calix, O.M. Palacios-Wassenar, A. Flores-Palacios, and M. Farrera for their invaluable assistance in the identification of plant species deposited at INECOL´s plant collections, as well as for making their research available for valuable databases and literature. This manuscript benefited from the valuable comments provided by A. Moerenschlager, H.M. Horblit, B.Y. Bedolla-García, J.G. Sánchez-Ken and three anonymous reviewers.

Conflicts of Interest: The authors declare no conflict of interest. Funders had no role in the design of the study; in the collection, analyses, or interpretation of data; in the writing of the manuscript; or in the decision to publish the results.

References

1. Levin, G.; Moran, R. *The Vascular Flora of Isla Socorro, México. Memoir*; San Diego Society of Natural History: San Diego, CA, USA, 1989; pp. 1–71.
2. Brattstrom, B.H. Biogeography of the Islas Revillagigedo, Mexico. *J. Biogeogr.* **1990**, *17*, 177–183. [CrossRef]
3. Martínez-Gómez, J.E. La avifauna del archipiélago de Revillagigedo. Amenazas, estrategias e instrumentos para la conservación. In *Conservación de Aves: Experiencias en México*; Gómez Silva, H., Oliveras de Ita, A., Eds.; CIPAMEX: Mexico City, Mexico, 2002.
4. SEDESOL. Decreto por el que se Declara Como Área Natural Protegida con el Carácter de Reserva de la Biosfera, la Región Conocida Como Archipiélago de Revillagigedo, Integrada por Cuatro Áreas: Isla San Benedicto, Isla Clarión o Santa Rosa, Isla Socorro o Santo Tomás e Isla Roca Partida. Diario Oficial de la Federación. 1994. Available online: https://faolex.fao.org/docs/pdf/mex179968.pdf (accessed on 16 November 2022).
5. UNESCO. Decision: 40 COM 8B.14. In Proceedings of the World Heritage Committee: 40th Session. WHC/16/40.COM/19, Istanbul, Turkey, 10–20 July 2016; pp. 196–199.
6. SEMARNAT. Decreto por el que se Declara Como Área Natural Protegida, con el Carácter de Parque Nacional, la Región Conocida Como Revillagigedo, Localizada en el Pacífico Mexicano. Diario Oficial de la Federación. 2017. Available online: https://www.dof.gob.mx/nota_detalle_popup.php?codigo=5505736 (accessed on 16 November 2022).
7. March Mifsut, I.J.; González Moreno, J. *Revillagigedo: Refugio de Gigantes*; MAPorrúa: Mexico City, Mexico, 2018; p. 282.
8. Jehl, J.R., Jr.; Parkes, K.C. "Replacements" of Landbird Species on Socorro Island, Mexico. *Auk* **1983**, *100*, 551–559. [CrossRef]
9. Walter, H.S.; Levin, G.A. Feral sheep on Socorro Island: Facilitators of alien plant colonization and ecosystem decay. *Divers. Distrib.* **2008**, *14*, 422–431. [CrossRef]
10. Jehl, J.; Parkes, K. The status of the avifauna of the Revillagigedo Islands, Mexico. *Wilson Bull.* **1982**, *94*, 1–19.
11. Ortíz-Alcaraz, A.; Maya-Delgado, Y.; Cortés-Calva, P.; Aguirre-Muñoz, A.; Rojas-Mayoral, E.; Cordoba-Matson, M.V.; Ortega-Rubio, A. Recovery of Vegetation Cover and Soil after the Removal of Sheep in Socorro Island, Mexico. *Forests* **2016**, *7*, 91. [CrossRef]
12. Mendez-Guardado, P. Analysis of the Environmental Impact Caused by Introduced Animals in the Clarion Island, Archipelago of Revillagigedo, Colima, Mexico. In *Global Change and Protected Areas*; Visconti, G., Beniston, M., Iannorelli, E.D., Barba, D., Eds.; Springer: Dordrecht, The Netherlands, 2001; pp. 323–329.
13. Brattstrom, B.H. Food Webs and Feeding Habits on the Revillagigedo Islands, Mexico. *Pac. Sci.* **2015**, *69*, 181–195. [CrossRef]
14. Wanless, R.M.; Aguirre-Muñoz, A.; Angel, A.; Jacobsen, J.K.; Keitt, B.S.; McCann, J. Birds of Clarion Island, Revillagigedo Archipelago, Mexico. *Wilson J. Ornithol.* **2009**, *121*, 745–751. [CrossRef]
15. Mack, R. Motivations and consequences of the human dispersal of plants. In *The Great Reshuffling: Human Dimensions in Invasive Alien Species, IUCN*; McNeely, J.A., Ed.; World Conservation Union: Gland, Switzerland, 2001; pp. 23–34.
16. SEMARNAT. Aviso por el que se Informa al Público en General que la Secretaría de Medio Ambiente y Recursos Naturales ha Concluido la Elaboración del Programa de Manejo del Área Natural Protegida con el Carácter de Reserva de la Biosfera Archipiélago de Revillagigedo, Ubicada en el Océano Pacífico, Establecida Mediante Decreto Presidencial Publicado el 6 de Junio de 1994. Diario Oficial de la Federación. 2007. Available online: https://www.dof.gob.mx/nota_detalle_popup.php?codigo=5008088 (accessed on 16 November 2022).
17. SEMARNAT. Acuerdo por el que se da a Conocer el Resumen del Programa de Manejo del Área Natural Protegida con la Categoría de Parque Nacional Revillagigedo. Diario Oficial de la Federación. 2018. Available online: https://dof.gob.mx/nota_detalle_popup.php?codigo=5545530 (accessed on 16 November 2022).
18. McLellan, M.E. Expedition of the California Academy of Sciences to the Revillagigedo Islands. *Science* **1925**, *62*, 171–173. [CrossRef]
19. Johnston, I. The flora of the Revillagigedo Islands. In *Proceedings of the California Academy of Sciences*; 4th Series; The Academy: San Francisco, CA, USA, 1931; Volume 20, pp. 9–104.
20. Rebman, J.; Vanderplank, S.E.; Ezcurra, E. Vascular plants of the Revillagigedo Islands expedition in February 2017. In *Report on the Terrestrial Expedition to the Revillagigedo Archipelago, 2020*; Vanderplank, S.E., Wall, M.A., Ezcurra, E., Eds.; San Diego Society of Natural History: San Diego, CA, USA, 2020; Volume 48, pp. 53–76.
21. Mickel, J.; Smith, A.R. *The Pteridophytes of Mexico*; The New York Botanical Garden Press: New York, NY, USA, 2004; Volume 88.
22. Dávila, P.; Cabrera, L.I.; Lira, R. A new species of Sorghastrum (Poaceae) from Isla Socorro, Colima, Mexico. *Brittonia* **1998**, *50*, 301–308. [CrossRef]
23. Benavides, E.; Kuethe, J.R.; Ortiz-Alcaraz, A.; De la Luz, J.L. *Oenothera resicum* (Onagraceae), a new species and the first record of the family from the Revillagigedo Archipelago, Mexico. *Phytotaxa* **2019**, *416*, 59–66. [CrossRef]
24. Kiehn, M. Invasive alien species and islands. In *The Biology of Island Floras*; Bramwell, D., Caujapé-Castells, J., Eds.; Cambridge University Press: Cambridge, UK, 2011; pp. 365–384.
25. Magee, J.F.; McMullen, C.K.; Reaser, J.K.; Spitzer, E.; Struve, S.; Tufts, C.; Tye, A.; Woodruff, G. Green invaders of the Galapagos Islands. *Science* **2001**, *294*, 1279. [CrossRef]

26. Rejmánek, M. Invasive plants: Approaches and predictions. *Austral. Ecol.* **2000**, *25*, 497–506. [CrossRef]
27. Richardson, D.; Pyšek, P. Plant invasions: Merging the concepts of species invasiveness and community invasibility. *Prog. Phys. Geogr. Earth Environ.* **2006**, *30*, 409–431. [CrossRef]
28. Flores-Palacios, A.; Martínez-Gómez, J.E.; Curry, R.L. La Vegetación de Isla Socorro, Archipiélago de Revillagigedo, México. *Bol. Soc. Bot. Mex.* **2009**, *84*, 13–23. [CrossRef]
29. INEGI. *Catálogo del Territorio Insular Mexicano*; Instituto Nacional de Estadística y Geografía: Aguascalientes, Mexico, 2015; p. 243.
30. Valero, A.; Schipper, J.; Allnutt, T. Islands of Socorro, Clarion, San Benedicto, and Roca Partida in the Pacific Ocean off the Coast of Mexico. Available online: https://www.worldwildlife.org/ecoregions/nt0216 (accessed on 2 September 2022).
31. Martínez-Gómez, J. Informe Final: Programa de Manejo de Áreas Naturales Protegidas (PROMANP): Coordinación del Programa Nacional para la Conservación de la Paloma de Socorro. Convenio de concertación CONANP/PROMANP/MB/037/2017. 2017. Available upon request.
32. Rodriguez-Estrella, R.; Leon de la Luz, J.L.; Breceda, A.; Castellanos, A.; Cancino, J.; Llinas, J. Status, density and habitat relationships of the endemic terrestrial birds of Socorro Island, Revillagigedo Islands, Mexico. *Biol. Conserv.* **1996**, *76*, 195–202. [CrossRef]
33. Mulcahy, D.G.; Martínez-Gómez, J.E.; Aguirre-León, G.; Cervantes-Pasqualli, J.A.; Zug, G.R. Rediscovery of an Endemic Vertebrate from the Remote Islas Revillagigedo in the Eastern Pacific Ocean: The Clarión Nightsnake Lost and Found. *PLoS ONE* **2014**, *9*, e97682. [CrossRef]
34. González-Sánchez, V.H.; Johnson, J.D.; González-Solís, D.; Fucsko, L.A.; Wilson, L.D. A review of the introduced herpetofauna of Mexico and Central America, with comments on the effects of invasive species and biosecurity methodology. *Zookeys* **2021**, *1022*, 79–154. [CrossRef]
35. Brattstrom, B.H. Bárcena Volcano, 1952: A 60-year report on the repopulation of San Benedicto Island, Mexico, with a review of the ecological impacts of disastrous events. *Pac. Conserv. Biol.* **2015**, *21*, 38–59. [CrossRef]
36. SEMARNAT-CONANP. Programa de Conservación y Manejo de la Reserva de la Biosfera Archipiélago de Revillagigedo, México. 2004. Available online: https://www.conanp.gob.mx/que_hacemos/pdf/programas_manejo/revillagigedo.pdf (accessed on 22 December 2021).
37. SEMARNAT-CONANP. Programa de Manejo del Parque Nacional Archipiélago de Revillagigedo, México. 2018. Available online: https://simec.conanp.gob.mx/pdf_libro_pm/82_libro_pm.pdf (accessed on 22 December 2021).
38. CONABIO. 2021. Available online: https://enciclovida.mx/ (accessed on 28 September 2023).
39. The Plant List. Version 1.1. 2013. Available online: http://www.theplantlist.org/ (accessed on 10 August 2021).
40. World Flora Online (WFO). 2021. Available online: https://www.worldfloraonline.org/ (accessed on 28 September 2023).
41. International Plant Names Index (IPNI). 2020. Available online: https://www.ipni.org/ (accessed on 28 September 2023).
42. Plants of the World Online (POWO). 2019. Available online: https://powo.science.kew.org/ (accessed on 28 September 2023).
43. Richardson, D.M.; Pyšek, P.; Rejmánek, M.; Barbour, M.G.; Panetta, F.D.; West, C.J. Naturalization and invasion of alien plants: Concepts and definitions. *Divers. Distrib.* **2000**, *6*, 93–107. [CrossRef]
44. Pyšek, P.; Richardson, D.M.; Rejmánek, M.; Webster, G.L.; Williamson, M.; Kirschner, J. Alien plants in checklists and floras: Towards better communication between taxonomists and ecologists. *TAXON* **2004**, *53*, 131–143. [CrossRef]
45. Dorjee, J.; Barry, S.; Buckmaster, A.J.; Downey, P.O. Weeds in the land of Gross National Happiness: Knowing what to manage by creating a baseline alien plant inventory for Bhutan. *Biol. Invasions* **2020**, *22*, 2899–2914. [CrossRef]
46. Ansong, M.; Pergl, J.; Essl, F.; Hejda, M.; van Kleunen, M.; Randall, R.; Pyšek, P. Naturalized and invasive alien flora of Ghana. *Biol. Invasions* **2019**, *21*, 669–683. [CrossRef]
47. Brummitt, R.K.; Pando, F.; Hollis, S.; Brummitt, N. *World Geographical Scheme for Recording Plant Distributions*, 2nd ed.; International Working Group on Taxonomic Databases for Plant Sciences (TDWG): Geneva, Switzerland, 2001.
48. Omer, A.; Kordofani, M.; Gibreel, H.H.; Pyšek, P.; van Kleunen, M. The alien flora of Sudan and South Sudan: Taxonomic and biogeographical composition. *Biol. Invasions* **2021**, *23*, 2033–2045. [CrossRef]
49. Kueffer, C.; Daehler, C.C.; Torres-Santana, C.W.; Lavergne, C.; Meyer, J.-Y.; Otto, R.; Silva, L. A global comparison of plant invasions on oceanic islands. *Perspect. Plant Ecol. Evol. Syst.* **2010**, *12*, 145–161. [CrossRef]
50. Tokeshi, M. Species Abundance Patterns and Community Structure. In *Advances in Ecological Research*; Begon, M., Fitter, A.H., Eds.; Academic Press: Cambridge, MA, USA, 1993; Volume 24, pp. 111–186.
51. Stürmer, S.L.; Stürmer, R.; Pasqualini, D. Taxonomic diversity and community structure of arbuscular mycorrhizal fungi (Phylum Glomeromycota) in three maritime sand dunes in Santa Catarina state, south Brazil. *Fungal Ecol.* **2013**, *6*, 27–36. [CrossRef]
52. Wuertz, D.; Setz, T.; Chalabi, Y.; Maechler, M.; Setz, M. fBasics: Rmetrics-Markets and Basic Statistics, R Package Version 3042.89.1. 2020. Available online: https://packages.debian.org/bullseye/r-cran-fbasics (accessed on 30 April 2022).
53. Crawley, M. *The R Book*, 2nd ed.; John Wiley & Sons: Hoboken, NJ, USA, 2013.
54. R_Core_Team. *R: A Language and Environment for Statistical Computing*; R Foundation for Statistical Computing: Vienna, Austria, 2020.
55. Pheloung, P. Determining the Weed Potential of New Plant Introductions to Australia. 1995. Available online: https://nre.tas.gov.au/Documents/WA-Weed-Risk-Assessment-REPORT.pdf (accessed on 15 May 2022).
56. Pheloung, P.C.; Williams, P.A.; Halloy, S.R. A weed risk assessment model for use as a biosecurity tool evaluating plant introductions. *J. Environ. Manag.* **1999**, *57*, 239–251. [CrossRef]

57. Daehler, C.C.; Denslow, J.S.; Ansari, S.; Kuo, H.-C. A Risk-Assessment System for Screening Out Invasive Pest Plants from Hawaii and Other Pacific Islands. *Conserv. Biol.* **2004**, *18*, 360–368. [CrossRef]
58. Randall, R.P. *A Global Compendium of Weeds*, 3rd ed.; RP Randall: Perth, WA, Australia, 2017.
59. Randall, R.P. Can a plant's cultural status and weed history provide a generalised weed risk score? In Proceedings of the 20th Australasian Weeds Conference, Perth, WA, Australia, 11–15 September 2016; pp. 5–12.
60. Martínez-Gómez, J.E.; Matías-Ferrer, N. Island Biogeography of Avian Haemosporidians in the Neotropical Region. In *Avian Malaria and Related Parasites in the Tropics: Ecology, Evolution and Systematics*; Santiago-Alarcon, D., Marzal, A., Eds.; Springer International Publishing: Cham, Switzerland, 2020; pp. 281–329.
61. Beierkuhnlein, C.; Walentowitz, A.; Welss, W. FloCan—A Revised Checklist for the Flora of the Canary Islands. *Diversity* **2021**, *13*, 480. [CrossRef]
62. Guézou, A.; Trueman, M.; Buddenhagen, C.E.; Chamorro, S.; Guerrero, A.M.; Pozo, P.; Atkinson, R. An Extensive Alien Plant Inventory from the Inhabited Areas of Galapagos. *PLoS ONE* **2010**, *5*, e10276. [CrossRef]
63. Castro, S.A.; Daehler, C.C.; Silva, L.; Torres-Santana, C.W.; Reyes-Betancort, J.A.; Atkinson, R.; Jaramillo, P.; Guezou, A.; Jaksic, F.M. Floristic homogenization as a teleconnected trend in oceanic islands. *Divers. Distrib.* **2010**, *16*, 902–910. [CrossRef]
64. Gustafson, R.J.; Herbst, D.R.; Rundel, P.W. *Hawaiian Plant Life: Vegetation and Flora*; University of Hawaii Press: Honolulu, HI, USA, 2014.
65. Blackburn, T.M.; Delean, S.; Pyšek, P.; Cassey, P. On the island biogeography of aliens: A global analysis of the richness of plant and bird species on oceanic islands. *Glob. Ecol. Biogeogr.* **2016**, *25*, 859–868. [CrossRef]
66. Cuevas, J.G.; Marticorena, A.; Cavieres, L.A. New additions to the introduced flora of the Juan Fernández Islands: Origin, distribution, life history traits, and potential of invasion. *Rev. Chil. Hist. Nat.* **2004**, *77*, 523–538. [CrossRef]
67. Meyer, J.-Y.; Lavergne, C. Beautés fatales: Acanthaceae species as invasive alien plants on tropical Indo-Pacific Islands. *Divers. Distrib.* **2004**, *10*, 333–347. [CrossRef]
68. Meyer, J.-Y. Threat of invasive alien plants to native flora and forest vegetation of eastern Polynesia (1). *Pac. Sci.* **2004**, *58*, 357–375. [CrossRef]
69. Lavergne, C.; Rameau, J.-C.; Figier, J. The Invasive Woody Weed Ligustrum robustum subsp. walkeri Threatens Native Forests on La Réunion. *Biol. Invasions* **1999**, *1*, 377–392. [CrossRef]
70. González-Gallegos, J.G.; Bedolla-García, B.Y.; Cornejo-Tenorio, G.; Fernández-Alonso, J.L.; Fragoso-Martínez, I.; García-Peña, M.d.R.; Harley, R.M.; Klitgaard, B.; Martínez-Gordillo, M.J.; Wood, J.R.I.; et al. Richness and Distribution of Salvia Subg. Calosphace (Lamiaceae). *Int. J. Plant Sci.* **2020**, *181*, 831–856. [CrossRef]
71. Wood, J.R.I.; Muñoz-Rodríguez, P.; Williams, B.R.M.; Scotland, R.W. A foundation monograph of Ipomoea (Convolvulaceae) in the New World. *PhytoKeys* **2020**, *143*, 1–823. [CrossRef] [PubMed]
72. Tye, A.; Soria, M.; Gardener, M. A strategy for Galapagos weeds. In *Turning the Tide: The Eradication of Invasive Species*; Veitch, C.R., Clout, M.N., Eds.; Invasive Species Specialist Group, IUCN: Gland, Switzerland; Cambridge, UK, 2002; pp. 336–341.
73. Vorsino, A.E.; Fortini, L.B.; Amidon, F.A.; Miller, S.E.; Jacobi, J.D.; Price, J.P.; Gon, S.O., III; Koob, G.A. Modeling Hawaiian Ecosystem Degradation due to Invasive Plants under Current and Future Climates. *PLoS ONE* **2014**, *9*, e95427. [CrossRef] [PubMed]
74. Caujapé-Castells, J.; Tye, A.; Crawford, D.J.; Santos-Guerra, A.; Sakai, A.; Beaver, K.; Lobin, W.; Vincent Florens, F.B.; Moura, M.; Jardim, R.; et al. Conservation of oceanic island floras: Present and future global challenges. *Perspect. Plant Ecol. Evol. Syst.* **2010**, *12*, 107–129. [CrossRef]
75. Rentería, J.; Atkinson, R.; Buddenhage, C. *Estrategias para la Erradicación de 21 Especies de Plantas Potencialmente Invasoras en Galápagos*; Fundación Charles Darwin, Departamento de Botánica, Programa de Especies Invasoras en Galápagos: Puerto Ayora, Ecuador, 2007; p. 108.
76. Mueller-Dombois, D.; Fosberg, F.R. *Vegetation of the Tropical Pacific Islands*; Springer Science & Business Media: Berlin/Heidelberg, Germany, 1998; p. 733.
77. Pyšek, P.; Pergl, J.; Essl, F.; Lenzner, B.; Dawson, W.; Kreft, H.; Weigelt, P.; Winter, M.; Kartesz, J.; Nishino, M.; et al. Naturalized alien flora of the world. *Preslia* **2017**, *89*, 203–274. [CrossRef]
78. Dunlop, E.A.; Wilson, J.C.; Mackey, A.P. The potential geographic distribution of the invasive weed Senna obtusifolia in Australia. *Weed Res.* **2006**, *46*, 404–413. [CrossRef]
79. David, A.S.; Menges, E.S. Microhabitat preference constrains invasive spread of non-native natal grass (*Melinis repens*). *Biol. Invasions* **2011**, *13*, 2309–2322. [CrossRef]
80. Stokes, C.A.; MacDonald, G.E.; Adams, C.R.; Langeland, K.A.; Miller, D.L. Seed Biology and Ecology of Natalgrass (*Melinis repens*). *Weed Sci.* **2011**, *59*, 527–532. [CrossRef]
81. García-Cancel, J.G.; Swinnerton, K.; Albarracín, R.; Feliciano, A.; Figuerola-Hernández, C.E.; Silander, S. Update on the Desecheo Island Flora. *Caribb. J. Sci.* **2019**, *49*, 295–304. [CrossRef]
82. Daehler, C.C.; Goergen, E.M. Experimental Restoration of an Indigenous Hawaiian Grassland after Invasion by Buffel Grass (*Cenchrus ciliaris*). *Restor. Ecol.* **2005**, *13*, 380–389. [CrossRef]
83. Jackson, J. Is there a relationship between herbaceous species richness and buffel grass (*Cenchrus ciliaris*)? *Austral Ecol.* **2005**, *30*, 505–517. [CrossRef]

84. Barbosa, F.G. The future of invasive African grasses in South America under climate change. *Ecol. Inform.* **2016**, *36*, 114–117. [CrossRef]
85. Visser, V.; Wilson, J.R.U.; Fish, L.; Brown, C.; Cook, G.D.; Richardson, D.M. Much more give than take: South Africa as a major donor but infrequent recipient of invasive non-native grasses. *Glob. Ecol. Biogeogr.* **2016**, *25*, 679–692. [CrossRef]
86. Martínez-Gómez, J.E. Proyecto para remover a las especies de árboles exóticos introducidos en Isla Socorro. Oficio de Autorización SGPA/DGVS/0792/16. Solicitud Complementaria: Oficio de Autorización SGPA/DGVS/1152/19. Dirección General de Vida Silvestre, Secretaría de Medio Ambiente y Recursos Naturales. 2015. Available upon request.
87. Tye, A. Invasive plant problems and requirements for weed risk assessment in the Galápagos islands. In *Weed Risk Assessment*; Groves, R.H., Panetta, F.D., Virtue, J.G., Eds.; CSIRO Publishing: Melbourne, VIC, Australia, 2001; Volume 153, pp. 154–175.
88. Sax, D.F.; Gaines, S.D. Species invasions and extinction: The future of native biodiversity on islands. *Proc. Natl. Acad. Sci. USA* **2008**, *105*, 11490–11497. [CrossRef]
89. Denslow, J.S.; Space, J.C.; Thomas, P.A. Invasive Exotic Plants in the Tropical Pacific Islands: Patterns of Diversity. *Biotropica* **2009**, *41*, 162–170. [CrossRef]

International Journal of
Plant Biology

MDPI

Article

Preliminary Evidence of Foliar Spray Effectiveness Against the Invasive Cactus *Cylindropuntia pallida* (Rose), F.M. Knuth in South Africa

Keletso Makaota [1], Thabiso Michael Mokotjomela [2,*], Caswell Munyai [3], Thembelihle Joyce Mbele [4] and Nontembeko Dube [1]

1 Center for Invasion Biology, Department of Zoology and Entomology, University of Free State, Bloemfontein 9310, South Africa; duben@ufs.ac.za (N.D.)
2 Centre for Invasion Biology, School of Life Sciences, University of KwaZulu-Natal, Pietermaritzburg 3200, South Africa
3 School of Life Sciences, University of KwaZulu-Natal, Pietermaritzburg 3200, South Africa; munyaic2@ukzn.ac.za
4 South African National Biodiversity Institute, Free State National Botanical Garden, Bloemfontein 9310, South Africa
* Correspondence: mokotjomelat@yahoo.co.uk or mokotjomelat@ukzn.ac.za; Tel.: +27-733246118

Abstract: Using the biometric differences (i.e., plant physical characteristics) between the *C. pallida* populations previously treated with herbicide and those of the untreated populations, we tested the prediction that herbicidal treatment suppresses *C. pallida* plant growth in South Africa, where invasive cacti display high resilience to herbicide treatment. We also determined whether the surrounding communities knew of *C. pallida* invasion and whether they experienced any negative impacts. Overall, biometric analyses supported the study's prediction because *C. pallida* plant height in treated populations was significantly lower than in untreated populations and before treatment. The average plant height of populations not treated with herbicide (i.e., experimental control) and those measured before treatment were not significantly different, but both were significantly greater than the heights of populations previously treated with herbicide. Similarly, the mean number of fruits, cladodes and juveniles per *C. pallida* plant treated with herbicide was significantly lower than in the untreated populations. We found a significant positive correlation between all measured parameters and *C. pallida* plant height. Out of 39 participants, 16 (41.0%) confirmed *C. pallida* presence in and around their properties, with 21 (53.9%) participants having experienced negative impacts directly and indirectly through fatal injuries on pets and livestock in the sampled sites. We confirmed the effectiveness of herbicide in suppressing the invasion of *C. pallida* and that people experience substantial negative impacts in areas where the species has established.

Keywords: society livelihoods; invasive species; natural resources; herbicides

1. Introduction

Cactus invasion is a significant problem in South Africa [1–4]. Cacti species are among the most widespread species globally, and South Africa has approximately 35 taxa [5]. Historically, more than 200 cacti species have been introduced into South Africa for various socio-economic purposes [4,6]. In particular, the multiple introductions through activities such as the horticulture trade and via commercial and subsistence agriculture could have partly increased the propagule pressure and invasion success of cacti species [1,7,8]. In addition, cacti species generally have a high tolerance to extreme conditions of temperatures ranging from −6 to 55 °C [9], and have prolific reproduction rates, which increase the chances of population growth and range expansion [8,10,11]. Consequently, in combination with the high diversity of local environmental conditions [12], this has led to most cacti species forming large, dense infestations in South Africa that reduce the quality of grazing land [11,13].

Because many alien plant species are associated with human life as ornamentals, food, and medicine [8,14–17], many studies have argued that creating an awareness of their negative impacts could have positive repercussions for their effective management [18–21]. Specifically, investigating the local perceptions of the affected area could ensure greater awareness and a systematic understanding of the local community's needs and how they relate to and interact with target alien species [18,20,22]. The awareness of alien species is important for detecting new populations of emerging species, and so encouraging a rapid management response [16,21,23,24]. Community awareness has contributed positively to many alien control projects through positive participation, planning, and decision-making [21]. However, creating an effective level of awareness may be challenging due to the limited understanding of the scientific content and of species identification by citizen scientists, resulting in delayed response action because of lengthy verification processes.

Since 1995, the management of plant invasions, including those of cacti species, has been prioritised to reduce their negative impacts in South Africa, and this has led to the establishment of clearing programmes, such as Working for Water [5,25–27]. Also, a new unit was developed in 2009 that is dedicated to clearing the emerging alien populations in order to avoid the high costs and the accumulated negative impacts associated with established large populations [16,23]. Furthermore, South Africa has developed and enacted legislation and policy in the form of the National Environmental Management: Biodiversity Act (NEM:BA, Act 10 of 2004) and the Alien and Invasive Species Regulations (hereafter called NEM:BA-A&IS Regulations; Department of Environmental Affairs, 2014; revised in 2020). According to the NEM:BA-A&IS Regulations of 2020, ten cacti species are listed under category 1a of alien species, including *Cylindropuntia pallida* (Rose), F.M. Knuth, and they are earmarked as targets in the national eradication programme [27,28].

Cylindropuntia pallida (Rose), F.M. Knuth, commonly known as the pink-flowered thistle cholla, is an emerging alien invader in the cactus family [29,30]. *Cylindropuntia pallida* reproduces vegetatively, with easily detachable spiny segments (cladodes) and fruits that can root within 20 days after being in contact with the soil, even without being watered [31]. The main dispersal strategy is through the movement of animals and humans by attaching to fur, skin, clothing, and even vehicle tyres [11,31]. Other dispersal agents are flood waters and heavy rain [11,32]. *Cylindropuntia pallida* is currently in the spreading phase of the invasion process in South Africa [33,34]. It can grow in a great variety of habitats, including grasslands, shrublands, and arid regions [32,35], and in South Africa, it prefers to invade the dry savanna and karoo biomes [29]. It can also resist drought and survive short periods of frost [32].

Different control methods have been considered to manage the invasion, including physical removal [31], chemical control [30], and the most effective biological control [30,36,37]. Although expensive when applied over vast areas, a foliar spray method is currently used to eradicate *C. pallida* populations in South Africa, with a success rate of over 90% [34,35]. The herbicide consists of triclopyr as the active ingredient, which is mixed with a dye to ensure that no plant parts are missed [38]. Triclopyr is a systemic herbicide that acts by disrupting the cell growth of a plant after absorption, leading to the death of the *C. pallida* plant within three months after application [35,39]. Since 1995 to this end, the South African government has spent about ZAR 15 billion on alien plant control operations.

The budget has risen exponentially since 2010, with the annual spending in 2019 being around ZAR 2 billion [4]. However, a piece of highly desirable information—especially given the lack of an account of adequate progress and field evidence of success and what would be the return on investment for so many millions of rands spent on the containment and control of biological invasions in South Africa [21,27,35,40,41].

This study seeks to contribute knowledge on the effectiveness of the management approach in eradicating *C. pallida* in South Africa. Although the cactus invasions are being effectively cleared with a foliar spray and followed with post-treatment monitoring [35], there are recovering populations in some of the sites [35]. Although less understood, *C. pallida* dispersal mainly occurs through landscaping human activities [15,16], and domestic animal interaction [11,32,34]. There are ongoing studies on reproductive and dispersal strategies both in the wild and captivity to determine the feasibility of eradication and the role of humans in driving the invasion processes for *C. pallida* in South Africa [42].

Against this background, we aimed to provide evidence of effective management by comparing the biometric differences between the *C. pallida* populations previously treated with herbicide and those of the untreated populations using plant height as a proxy for growth. Since cactus reproduces both sexually and asexually, we compared the numbers of cladodes, fruits and juveniles in the *C. pallida* populations as a reproduction proxy between the two populations. We predicted that herbicidal treatment would suppress the growth and reproduction of the *C. pallida* plants. We also determined whether the surrounding communities knew of *C. pallida*'s presence and whether they had experienced any negative impacts.

2. Materials and Methods

2.1. Study Species

The study species (i.e., *C. pallida*) is a typical deciduous succulent with white spines and produces pink flowers in summer (Figure 1). These features also distinguish it from other cactus species [11,31].

2.2. Study Sites

Overall, the sites were located in the Northern Cape (NC), Free State (FS) and Eastern Cape (EC) provinces and were named according to the nearest town as a reference. The nine sites (i.e., Maphiniki, Barkley East, Edenburg, Philippolis, Bethulie, Colesberg, Richmond, Cookhouse and Cradock) located in areas in which *C. pallida* occurs and were identified as targets for clearing with the collaboration of the South African National Biodiversity Institute's regional scientists (Figure 2). All the sites selected for this study similar in that the primary land use was livestock farming and grass was the dominant vegetation [43]. Because of the erratic distribution of the populations of *C. pallida* as an emerging alien cactus species in South Africa [35], the sites were far from each other and did not allow full replication within the site. Each site comprised a reasonable *C. pallida* population (i.e., >50 mature plants in a continuous patch; [35]).

The sites were classified into two groups. The first group comprised all nine sites, which had measurements of height as the proxy for *C. pallida* plant size taken independently, where there was no herbicide treatment (i.e., experimental control), before and after treatment. The height of *C. pallida* plants was measured between November 2018 and March 2023 in different sites with established populations (see Table S1).

Figure 1. *Cylindropuntia pallida* plants/stands showing their pink flowers, which are typically used for identification, and spiny cladodes in Katu, Northern Cape (Photo credits: Dr Thabiso M Mokotjomela).

Another group entailed sites with measurements of quantifying attributes indicating growth and reproduction of *C. pallida* plants (i.e., number of cladodes, fruits and juveniles), where there is no prior and after herbicide treatment. The site grouping was mainly dependent on the consistency of the variables measured. For example, Cookhouse (EC) and Richmond (NC) had *C. pallida* populations that had not been previously cleared or treated with herbicides, while Colesberg (FS) and Edenburg (FS) were sites where *C. pallida* populations had been treated once with a herbicidal foliar spray in 2018. However, the number of cladodes, fruits and juveniles was sampled from November to March 2023 in the four sites above.

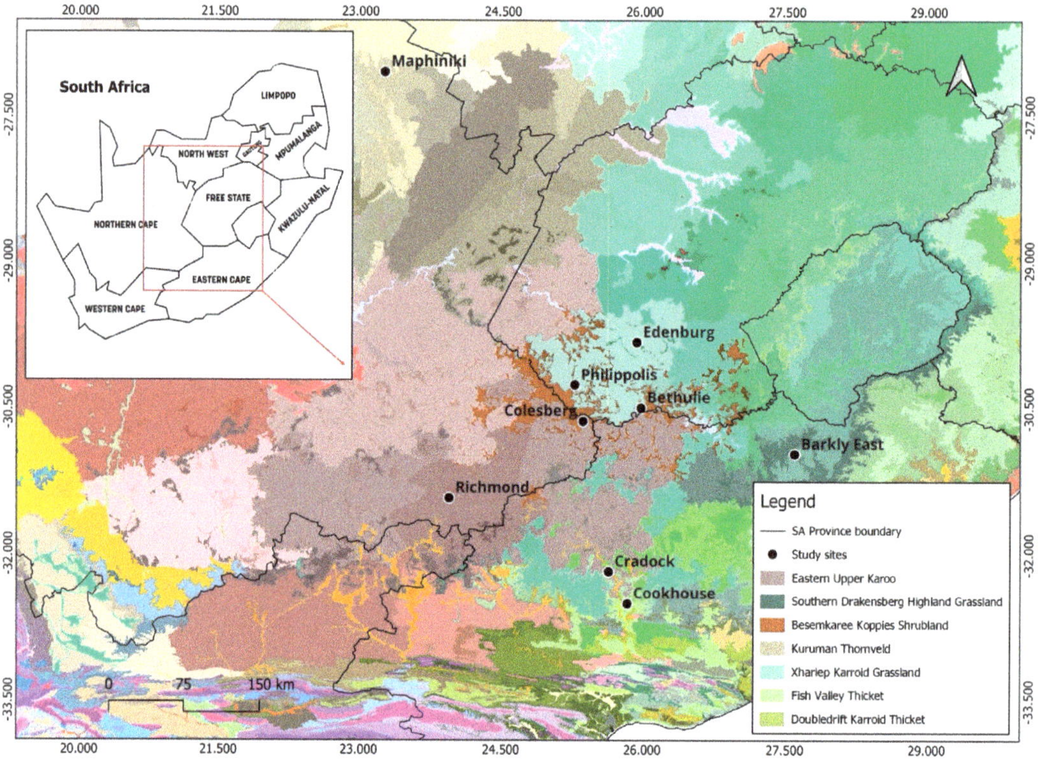

Figure 2. The nine study sites (i.e., Maphiniki, Barkly East, Edenburg, Philippolis, Bethulie, Colesberg, Richmond, Cookhouse and Cradock) overlaid on the national vegetation map [44].

2.3. Management of Cactus in South Africa

In South Africa, the alien cactus is treated with the foliar spray method with a 2% concentration of herbicide active ingredients fluroxypyr (Pyridyloxy compound 320 g/L); and a triclopyr (Pyridyloxy 480 g/L) following Mokotjomela et al. [35] and this study. The application rate was 50–75 mL Garlon/50 mL mineral oil/10 litres of water by a certified applicator. The herbicide was mixed with spray oil, and blue dye was applied using the 20 L knapsack sprayer with calibrated nozzles to allow the coarse spraying of the whole shoots of each plant until drip-off occurred [31]. The spraying was performed during cool times when air temperatures were less than 30 °C to avoid extreme conditions leading to plant dormancy and stomatal closure [45], and during active growing seasons: November–March 2017 and 2018.

2.4. Plant Biometry Data Collection

Data collection was performed concurrently with eradication during the summer season of 2018 to 2022 (in February and March). To determine the effect of herbicide treatment, the height of each *C. pallida* plant was measured with a measuring tape held against a person standing near each stand. The plants' height was measured vertically, starting at the ground surface. The measurements were taken as the baseline before herbicide application, and sites were revisited for monitoring only after three years for this study. The height measurements were taken in *C. pallida* populations being cleared and the newly identified and targeted for future clearing but not included in the annual operations plan. Such height measurements were patchy and restricted to only nine sites. We included every plant taller than 0.25 m.

To determine the growth and reproduction of *C. pallida*, in a 2 m^2 plot marked at every 10 m interval of each transect, we counted and recorded the number of cladodes, and fruits in each mature plant. We marked four transects in each site as replicates, and each transect was at least 100 m long and 2 m wide. This method was repeated across four sites [46]. The locations of all sampled *C. pallida* plants were recorded using a Garmin Oregon 750 model device. The number of cladodes was estimated by counting the exact number of cladodes in a portion of the canopy and then multiplying it by the total number of equal portions for each plant canopy [47]. The same approach was used to determine the number of fruits, and adjustments were made based on the density of fruits per portion.

To determine the possibility of the further invasion and population expansion of *C. pallida* populations in each site, the number of juvenile plants within a 2 m^2 quadrat was counted around each mature plant in each transect per site at every 10 m interval of each transect following the procedure for choice of quadrat sizes based on vegetation cover type and size as outlined in Hill [46].

The transects were used as within-site replicates because there were not enough sites with large populations (i.e., >50 plants) and sites for replication. Also, since Mokotjomela et al. [35] found no correlation between *C. pallida* plant size and the toxicity of the herbicide, the assumption is that the *C. pallida* plants at the sprayed and unsprayed sites had the same size/density before treatment in two of the four sites.

2.5. Interviews with Communities Affected by C. pallida

Interviews were conducted with participants residing in the Cradock (EC) and Richmond (NC) towns that were invaded by *C. pallida*. Cradock (now Nxuba), a town in Eastern Cape Province, was an additional site where interviews were conducted because the two towns had human populations around them, which could give better insight and supply reliable samples to reveal people's perceptions about *C. pallida*. The other field sites (Cookhouse, Edenburg, and Colesberg) had *C. pallida* populations located on private farms with a limited number of human participants in the vicinity.

The snowballing method was used to select the interviewees after those living closest to *C. pallida* stands had initially been identified (see [20,48]). We investigated the local context in which engagement was sought to ensure awareness following Shackleton et al. [20] and systematically asked about the needs of the community affected by *C. pallida*. For ethical considerations, face-to-face interviews were conducted with informed consent, and questions were translated into each of the interviewees' home languages (including Afrikaans, isiXhosa, Sesotho and Setswana when necessary [20]. The questions were structured to seek information about the presence of *C. pallida* and when the communities had started noticing it and its negative impacts on people and animals, and to find out whether the species was abundant, dispersing, and expanding the population, and whether any action was being taken to control the species. The questionnaire comprised a set of twenty questions, which were a mixture of open-ended and closed questions. However, only eight main questions relevant to understanding the awareness and presence of *C. pallida* and its negative impacts were considered for analyses in the current study.

The interviews were conducted in July 2023; each was about 15 min, and 39 participants were interviewed. The study's ethical clearance was obtained from the University of Free State, with Ethics Clearance number UFS-HSD2023/1208.

3. Data Analysis

3.1. *Comparisons of* C. pallida *Biometrics Between Sites*

To determine the herbicidal effect on the *C. pallida* plant height between the populations treated with herbicide and after treatment, a univariate general linear model analysis of variance was applied to analyse the data in SPSS version 29. The *C. pallida* plant height data for nine sites was classified into three site conditions namely: "No Treat"—independent measurements of height where no herbicide treatment is applied; "Before Treat"—measurements taken in each site prior to application of herbicide, and "After Treat"—the measurements taken after at least three years of site monitoring. The *C. pallida* plant height was fitted as the dependent variable; the site conditions (i.e., "No Treat", "Before Treat" and "After Treat") were independent variables, while site was treated as a random effect. Dunnett's post hoc test determined significant differences between the mean *C. pallida* plant height in different site conditions, with "No Treat" considered as the experimental control at a significance of $p < 0.05$.

The number of cladodes and the fruits and the juvenile plants were compared to determine if plant growth and reproduction, respectively, are similar in populations previously treated with herbicide and those not treated. We applied a generalised linear model to analyse the counts' data using SPSS version 29. To build a model, the counts of the cladodes per *C. pallida* plant, its fruits, and the juveniles were specified as the response variables and their classifications (i.e., "After Treat" and "No Treat") as well as the individual study sites were defined as the predictor variables. We compared the variance coefficients for the Generalised Linear Models fitted with Poisson and negative binomial error distributions to select the best-fitting regression model. Model selection relied on the Akaike Information Criterion (AIC) and checks of the data dispersion residuals.

Finally, the non-parametric Partial correlation was performed to investigate whether there was a relationship between the *C. pallida* plant height and its cladodes, fruits, and juveniles to infer if large and uncontrolled plants can encourage further spread and invasions. Additionally, height was used as a proxy for environmental gradients and was included as a control variable.

3.2. *Interviews: Knowledge of and Community Interactions with* C. pallida

The data, in the form of responses from the different participants, were pooled before analysis because we had a limited sample, partly due to the local people's reluctance to be interviewed, and there were very few human settlements near the *C. pallida* populations. Overall, 39 individuals were interviewed on eight key questions that focused on community awareness and known interactions with *C. pallida*; 11 were from Cradock (Eastern Cape), and 28 were from Richmond (Northern Cape).

The responses to the questions were captured verbatim in an Excel spreadsheet and later were transcribed, and each response was coded to generate countable data. Then, we applied the generalised linear model regressions to compare the grouped responses per question between towns, using the Pearson chi-square analysis at the significance of $p < 0.05$ in SPSS version 29. The questions were fitted into the model as response variables, while the towns were considered predictor variables. The questions and their responses were then classified into two groups, each containing four questions: the presence of *C. pallida* in the local area, the period of presence and negative impacts on people and animals, the behaviour of *C. pallida*—its abundance, spread/dispersal, and population—and actions taken against *C. pallida*.

4. Results

4.1. Cylindropuntia pallida *Biometry Comparisons*

Overall, biometric analyses supported the study's prediction because *C. pallida* plant height in treated populations was significantly lower than in untreated populations and before treatment ($F = 152.15$; $df = 1$; $p < 0.001$). The treatments were also significantly different ($F = 38.68$; $df = 2$; $p < 0.001$). Dunnett's post hoc revealed that the average plant height for *C. pallida* populations not treated with herbicide (Mean ± SE: 56.8 ± 2.2; N = 105) and that one measured before treatment (Mean ± SE: 52.9 ± 1.7; N = 170) were not significantly different ($p = 0.168$), but both were significantly greater height ($p < 0.001$) than populations previously treated with herbicide (Mean ± SE: 40.2 ± 1.7; N = 163; Figure 3). Similarly, the sites displayed significantly different average height of *C. pallida* plants ($F = 21.01$; $df = 2$; $p < 0.001$).

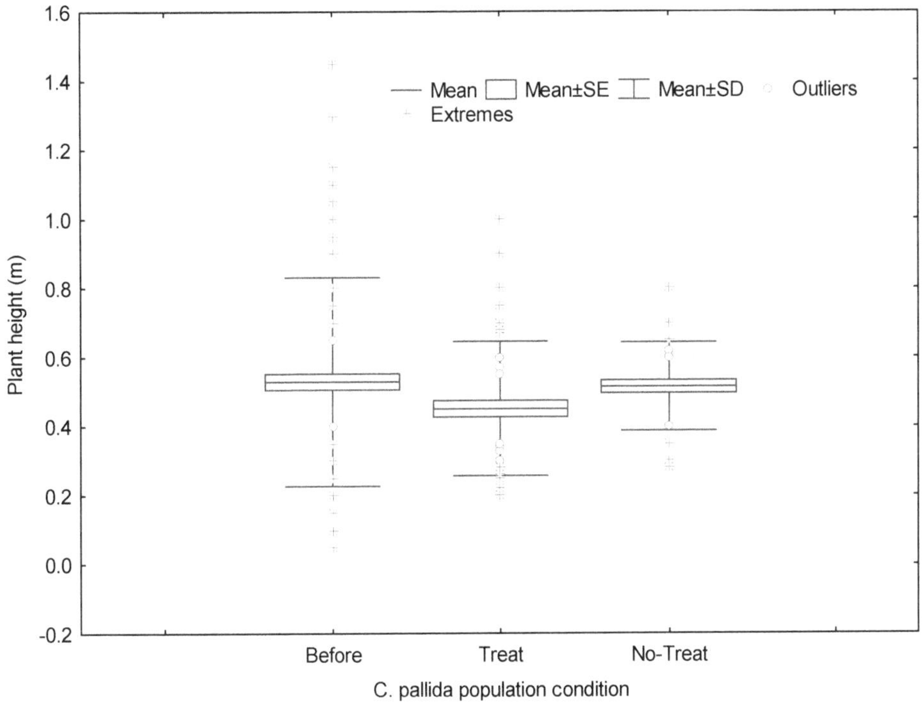

Figure 3. Variation in *C. pallida* plant height (m) in populations not treated with herbicide (No Treat), before treatment (Before Treat), and after treatment (After Treat) across nine sites.

Regarding the number of cladodes, the populations treated with foliar spray generally had significantly lower numbers than those not treated with herbicides (Wald $\chi^2 = 73.3$; $df = 1$; $p < 0.001$; Table 1). Similarly, the mean number of fruits per *C. pallida* plant in populations treated with herbicide was significantly lower (Wald $\chi^2 = 624.4$; $df = 1$; $p < 0.001$; Figure 3), and average number of juveniles around adult plants than the untreated populations (Wald $\chi^2 = 102.4$; $df = 1$; $p < 0.001$; Figure 3) than the populations not treated with herbicide.

Table 1. Variation in mean number of cladodes, fruits and juveniles in four sites, and the GLM statistical comparisons. 'a' is the parameter used as a reference during multiple comparisons. Superscript letters indicate the significant difference among the average number of cladodes, fruits and juveniles.

	Number of Cladodes per plant		GLM comparison		
Site	Mean (N)	SE	χ^2	Df	Sig.
	Overall (Treated vs. No Treatment)		73.29	1	<0.001
Colesberg	76.3 (53) [a]	15.3	15.228	1	<0.001
Cookhouse	209.1 (57) [b]	14.7	1.328	1	0.249
Edenburg	33.4 (45) [c]	16.6	59.006	1	<0.001
Richmond	166.7 (48) [d]	16.1	a	-	-
	Number of fruits per plant		GLM comparison		
Site	Mean (N)	SE	χ^2	Df	Sig.
	Overall (Treated vs. No Treatment)		624.45	1	<0.001
Colesberg	0.6 (53) [a]	18.5	1009.19	1	<0.001
Cookhouse	100.3 (57) [b]	19.4	365.952	1	<0.001
Edenburg	0.6 (45) [ac]	20	0.083	1	0.773
Richmond	94.8 (48) [bd]	17.8	a	-	-
	Number of juveniles		GLM comparison		
Site	Mean (N)	SE	χ^2	Df	Sig.
	Overall (Treated vs. No Treatment)		27.28	1	<0.001
Colesberg	2.5 (53) [a]	0.3	15.73	1	<0.001
Cookhouse	6.1 (57) [b]	0.3	6.47	1	0.011
Edenburg	1.4 (45) [ac]	0.3	34.612	1	<0.001
Richmond	3.5 (48) [bd]	0.3	a	-	-

The various sites also showed significantly different variation in the numbers of cladodes (Wald $\chi^2 = 99.7$; df = 3; $p < 0.001$; Table 1), fruits (Wald $\chi^2 = 428.5$; df = 3; $p < 0.001$; Table 1), and juveniles (Wald $\chi^2 = 37.7$; df = 3; $p < 0.001$; Table 1). The Bonferroni post hoc test showed that while Richmond had significantly higher numbers of cladodes, fruits, and juveniles than the other sites, the numbers of cladodes and fruits were significantly lower than those in Cookhouse ($p < 0.05$; Table 1).

Partial correlation analysis revealed that there was a significant relationship between all the variables measured and the height of the plants (Table 2). The strongest correlation was observed between plant height and the number of cladodes, followed by plant height and the number of fruits present. However, when height was used as a control variable for environmental gradients, there was no significant correlation to the number of fruits recorded in *C. pallida* plants, suggesting the strong influence of the environment on plant growth (see Table 2).

Table 2. Partial correlation analysis between the height of the plants and the number of cladodes, number of fruits, and number of juveniles measured in four sites/populations for *C. pallida* and uses of plant height to control for environmental gradients (Correlation is significant at $p < 0.001$ level).

Control Variables			Gladodes	Fruits	Juveniles	Height
-none-	Gladodes	Correlation	1.000	0.698	0.609	0.817
		Significance (2-tailed)		<0.001	<0.001	<0.001
		Df	0	201	201	201
	Fruits	Correlation	0.698	1.000	0.578	0.757
		Significance (2-tailed)	<0.001		<0.001	<0.001
		Df	201	0	201	201
	Juveniles	Correlation	0.609	0.578	1.000	0.479
		Significance (2-tailed)	<0.001	<0.001		<0.001
		Df	201	201	0	201
	Height	Correlation	0.817	0.757	0.479	1.000
		Significance (2-tailed)	<0.001	<0.001	<0.001	
		Df	201	201	201	0
Height	Gladodes	Correlation	1.000	0.211	0.430	
		Significance (2-tailed)		0.003	<0.001	
		Df	0	200	200	
	Fruits	Correlation	0.211	1.000	0.377	
		Significance (2-tailed)	0.003		<0.001	
		Df	200	0	200	
	Juveniles	Correlation	0.430	0.377	1.000	
		Significance (2-tailed)	<0.001	<0.001		
		Df	200	200	0	

4.2. Community Interviews on Their Knowledge of and Interactions with C. pallida

A total of 39 participants were included in the study; 28 (71.8%) lived in Richmond and 11 (28.2%) in Cradock. *C. pallida* was present in both towns. When the participants were questioned about whether the plant was present on their private properties, 16 (41.0%) confirmed that the plant was present on their properties. The other 23 participants (59.0%) responded that no *C. pallida* plants were present on their properties. When asked how long the plant had been present in the local area and towns, the Cradock participants had been aware of the plant for only three years or less. However, the participants in Richmond had been aware of the presence of the plant for as long as five years (Figure 4a). No significant difference was observed between the responses in the two towns (Pearson $\chi^2 = 2.5$; df = 5; $p = 0.776$).

Regarding the abundance of *C. pallida* plants on the participants' properties, only two of the participants in Cradock had *C. pallida* plants, and they estimated them to be only a few plants (Figure 4a,b). In Richmond, an equal ratio of participants had plants either present or absent on their properties, and of the 14 participants who had plants present on their properties, 12 defined their number as relatively few (1–25 individual plants); two small-scale farmers reported having a moderate number of plants (25–50 individual plants) on their properties. No significant difference was observed between the towns with respect to the presence (Pearson $\chi^2 = 3.3$; df = 1; $p = 0.059$) and abundance (Pearson $\chi^2 = 3.5$; df = 1; $p = 0.175$; Figure 4a) of *C. pallida* plants on their private properties.

Regarding the negative impacts of *C. pallida* on human livelihoods, 21 (53.9%, N = 39) participants claimed that they or their household members had previously been pricked and injured by the plants (Figure 5). Of the 39 participants, 21 had experienced the direct negative impacts of *C. pallida* (Figure 5c). Some participants had pets and livestock that had experienced those negative impacts (Figure 5d) including six who stated that their pets had been injured. The other three said that their livestock (goats/sheep) had been wounded by *C. pallida* plants (Figure 5e). The participants who owned livestock said that the sheep would not stay still or in place after being pricked and that the injured goats could not go out to graze while they were limping.

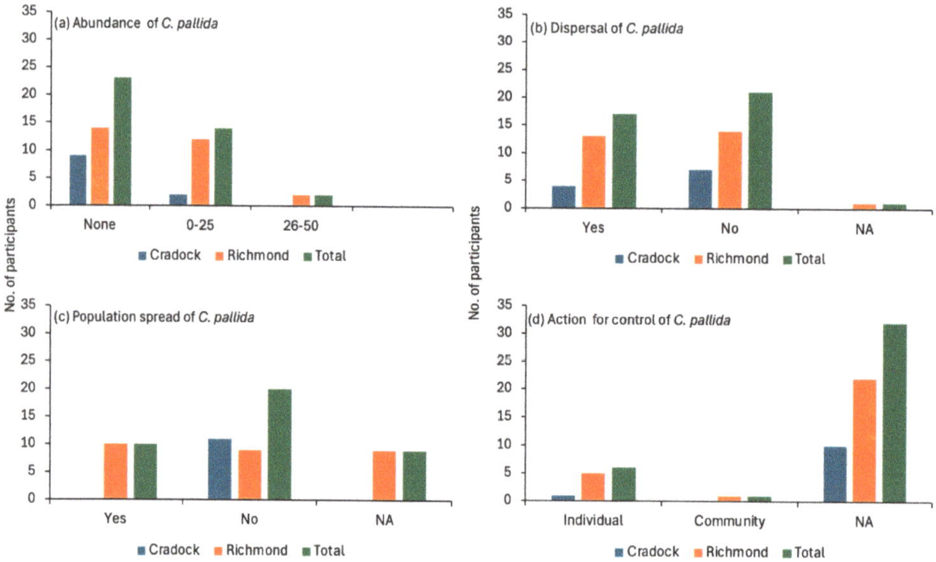

Figure 4. Interview responses of participants in Cradock and Richmond. Questions: (**a**) Presence—Is the plant present on your property? (**b**) Time—How long has the participant been aware of the plant's presence in their respective towns? (**c**) Impacts on people—Have the participants experienced any negative impacts from the cactus plants? (**d**) Impacts on animals—Have any of the pets/livestock of the participants experienced negative impacts?

Figure 5. Digital field evidence of the impacts of *C. pallida*: (**a–g**) show direct impacts on people and domestic livestock; (**h–j,m**) show birds that died after being trapped in *C. pallida*; (**k,l,n,o**) (apparently a mongoose)and (**q**) (apparently an aardwolf) show unidentified mammal skeletons with *C. pallida* spines on them; (**p**) shows a striped mouse with *C. pallida* spine; and (**r**) shows a dead lizard.

When questioned about the dispersal of the plant (whether the propagules attached to their apparel, tools, or animals), most of the participants answered 'No' and added that they generally stayed away from the plant; however, 17 of the participants (44%, N = 39) answered 'Yes' (Figure 5b), and some added that propagules had sometimes attached to their clothing without their noticing. One of the farmers also said that he had seen propagules attached to some of his animals, posing a risk of further spread. No significant difference was observed between the two towns regarding the dispersal of *C. pallida* (Pearson $\chi^2 = 0.9$; df = 1; $p = 0.654$; Figure 5).

Regarding the management of *C. pallida* plants, participants were asked whether and, if so, how the plants in their hometown were being cleared. Most of the participants (82.1%; N = 39) in both towns said 'No' because they were not aware of any clearing being performed, while the 18.0% who answered 'Yes' indicated that individuals cleared *C. pallida* on their properties (Figure 6c). When asked whether the plant was spreading further in their properties, all the participants in Cradock answered 'No'; therefore, 11 (82%) local participants did not have *C. pallida* in their properties (see Figure 6a). In Richmond, 28 (36.0%) participants responded 'Yes' to the question (Figure 6d); this was significantly greater than the observation in Cradock (Pearson $\chi^2 = 14.6$; df = 2; $p < 0.001$; Figure 6).

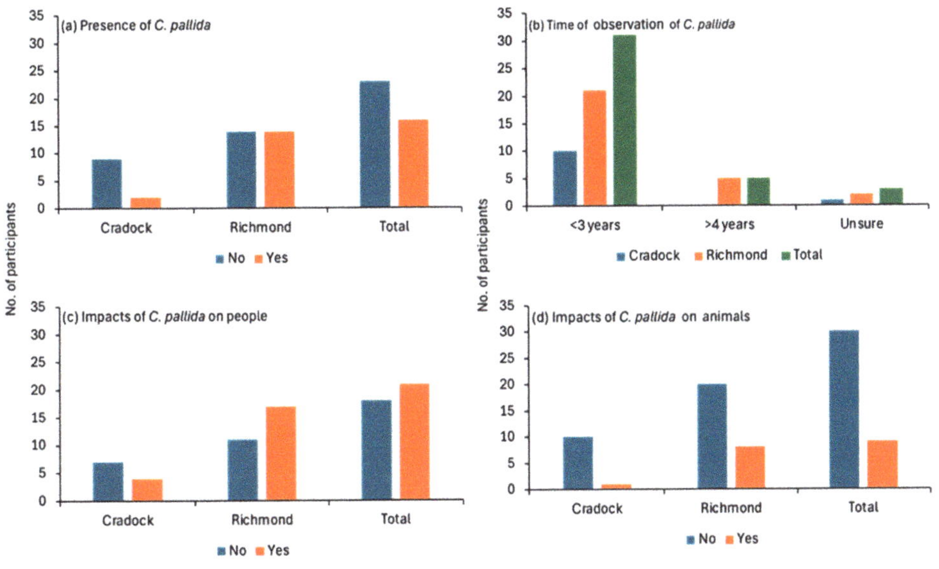

Figure 6. Interview responses for participants in Cradock and Richmond. Questions: (**a**) Please estimate the abundance of the plant on your property. (none = no *C. pallida*; plants; few = 1–25 individual plants; moderate = 25–50 individual plants); (**b**) Dispersal—Is the plant further dispersing by attaching to apparel, tools, and animals? (**c**) Population spreading—Is the plant spreading further on your property? (NA = Not applicable; plant is not present on the property); (**d**) Control action—Are the plants being cleared and, if so, who is clearing the plants? (NA = Not applicable; not aware of any clearing). Individual = Individual members clearing their properties; Community = a small group of community members clearing the plants.

5. Discussion

Our study provides evidence supporting the study's prediction that herbicidal foliar spray substantially suppresses *C. pallida* growth—this was attested by a significantly lower plant height, number of cladodes, fruits and juveniles in the population treated with herbicide compared to the untreated populations. It is worthwhile to note that, fewer plants observed in the treated sites were individuals that recovered from cleared populations following herbicide application. Corroborating the previous studies elsewhere [31], we show that thwarted growth and few fruits, as well as the cladodes, suggest that herbicidal treatment may reduce *C. pallida*'s future invasion, as shown for South Africa. Using structured interviews and field observations, we also showed that the local communities and animals experienced the negative impacts of *C. pallida* in selected areas where it has established.

It is contended that the observed growth suppression is a likely effect of the herbicidal treatment administered. It has been shown that the efficiency of herbicides differs on the physiological principles involved during the herbicidal action on the target plant, with systemic herbicides killing plants after their absorption by either accelerating or retarding the metabolic activities of weed plants [49,50]. Systemic herbicides are known to create oxidative stress and damage critical physiological processes [50], and thus, we argue that triclopyr might have a similar effect in *C. pallida* plants. It has been shown that the triclopyr effects are identical to excessive auxins, leading to increased uncontrolled plant growth [39], and ester formulations against root and stem sprouting species while remaining persistent in plants until they die [39], i.e., 1–5 months depending on the *C. pallida* plant size [31]. Noting that the impact of herbicides may not always manifest in visible injuries [50,51], we ascribe the retarded *C. pallida* plant growth to the effect of the herbicide that may have inhibited the critical photosynthetic and other metabolic processes, thus restricting the availability of resources for plant growth [50]. Similarly, effective foliar spray control using triclopyr herbicide has been reported in Spain and Australia, where *C. pallida* is invasive, and this was only observed where the whole plant was sprayed [31]. Further comparison of our results with other cactus management studies is limited because other countries have different environmental regulations regarding the use of herbicides. For example, some South African and Australian herbicides to control *C. pallida* are not allowed in some parts of Europe [45]. Thus, since *C. pallida* plants observed in all sites not previously treated with herbicides thrived, long-term monitoring is also required to observe further changes in plant growth, other attributable factors driving its invasion, and the timing of the changes to inform adjustments in the management strategy.

Our finding that the *C. pallida* plant height in populations treated with herbicide confirmed the effectiveness of the foliar spray method in containing its invasion. Consistently, the earlier studies [42–55], argued that plant characteristics can be used as reliable indicators of the impact that management has on alien plant species. In this study, the taller *C. pallida* plants had many cladodes, fruits and juveniles (in Richmond and Cookhouse), which supports the observation by previous studies (e.g., [42,56–58]) reporting the positive correlation between plant size, i.e., height and fecundity in alien plants. High fecundity is an essential driver of the spread of alien species and successful invasion by many species [54,59,60]. This observation suggests that if management intervention for *C. pallida* is stopped, further invasion may occur since we recorded many fruits and cladodes correlated with the number of juveniles. We also argue that the significantly greater number of *C. pallida* juveniles observed in Richmond was likely an effect of the presence of a larger human population around the study sites, which reportedly facilitates dispersal and invasions because of people's multiple uses of the alien species [2,61,62]. Furthermore, *C. pallida*'s dispersal is stimulated by attachment to clothing, shoes, vehicle tyres, and domestic animals in South Africa [11,34,35]. It is possible that many plants in Cookhouse could be ascribed to the biophysical disturbance created by farm livestock and other human activity on the land [17], and this combined with the greater plant height, points out to higher chances of invasion than in populations treated with herbicide (i.e., Edenburg and Colesberg), which had fewer fruits (i.e., low fecundity). The suppression of growth and reproduction in *C. pallida* plants can also limit their competitive advantage over native plant species in consuming natural resources [60], reduce available dispersal propagules, i.e., cladodes and seeds in this study [31,34,59], and their overall density-based impacts [35,63,64]. Taller plants have also been associated with more significant negative impacts [18,65]; therefore, we suggest that *C. pallida* populations with taller plants be prioritised and targeted for management to reduce the risk of further invasion. Indeed, monitoring the effectiveness of implemented management efforts could guide decision-making for management's investment in effective control methods [66], financial costs, and time [67].

Previous studies have shown that people's perceptions of invasive alien species are influenced by factors such as the species' history, the extent of invasion, benefits, and negative impacts [20,68–70]. For example, in our study, two participants consistently indicated that *C. pallida* was beneficial in keeping burglars away and had medicinal properties, as Balandran-Quintana et al. [71] reported. Our analysis also suggests that about half of the participants were aware of the *C. pallida* plants both in the wilderness and on their properties as ornamental plants [11,20]. This, taken together with (for example) other invasive alien cactus such as *Opuntia fiscus indica*, which has been declared to support people's livelihoods [72], results in a conflict of interest in their management [20,72–75] reported that the community of eastern Free State typically opposed the clearing of *Acacia mearnsii*, arguing that it was a critical source of firewood. However, the finding that they were unaware that it was prohibited corroborates the report by Mhlongo et al. [70] on *Lantana camara*, which pointed to the need for more awareness campaigns about alien species and their negative impacts on the environment [16,21,75]. Jubase et al. [21] found that a lack of awareness about the negative impacts of invasive alien plants in small towns could result from limited public awareness and engagement activities. Formal education about invasive alien species could greatly influence people's perceptions of the negative impacts of invasive alien species [76,77].

Our mixed (i.e., non-significant differences) results indicated that a proportion of the participants were still unaware of the negative impacts of *C. pallida* in the study areas, which could lead to the further unintentional and intentional spread of the species as ornamental plants [21]. Another possible explanation for the mixed results is the different extent of invasion between Richmond and Cradock, with most Cradock participants having few plants on their properties. However, that could change as the invasion intensifies because the negative impacts are thought to be directly proportional to the extent of the invasion [42,69,78]. The reported fatal injuries to people and animals in the study sites can be ascribed to the sharp and hooked spines of *C. pallida*. Previous studies also reported reduced grazing space for domestic livestock such as sheep and goats [11,31,34,35,37,79,80]. The negative impacts of alien cacti species on people and domestic livestock are well-known for *O. stricta*, which is widespread in South Africa [81] and in parts of Kenya [78]. A study in Spain has associated the manual clearing of *C. pallida* plants with the risk of injury and its further spread [31]. Our study reported fatalities to animals and injuries to humans caused by *C. pallida*, and such observation is also consistent with the report of Deltoro et al. [31] from Spain, highlighting the unreported loss of biodiversity. The observed fatal animal injuries may disrupt ecological services because it has been shown that different organisms, especially vertebrates, have distinct roles in maintaining habitat functionality [82]. Thus, more research would help explain the negative impacts of *C. pallida*. Our results also provide field evidence that could help management to reprioritise the limited resources to clear areas with heavy *C. pallida* infestations in South Africa.

About half of the participants (N = 39) in the current study indicated that *C. pallida* was present in low abundance on their properties and their mixed perceptions of whether the species was dispersing in the wilderness to a low level of awareness [20,21], and a lack of scientific knowledge of ecological processes in invasion biology [22,83,84]. It has been argued that the effective management of alien and invasive species is partly hindered by limited knowledge about the species, leading to wrong responsive action and planning [22,24,85,86]. For example, the use of wrong control methods was supported by the responses showing that the plant was spreading further, even after attempting to remove it. Although Korperlainen and Pietilainen [8] found that human activities cause the dispersal of most invasive alien species, it was apparent that citizens in Richmond and Cradock were unaware of how communities could be facilitating the dispersal of *C. pallida*. However, the participants pointed out that *C. pallida* disperses by attaching to human apparel, vehicle tyres, tools, and livestock, which is consistent with previous reports elsewhere in South Africa [11,34] and in Spain [31,32]. A raised awareness could guide people on how to interact with the plant in a way that prevents dispersal.

Regarding the management efforts that were in place, the few people in the current study who had *C. pallida* on their properties responded by attempting to clear it manually, using different methods such as mechanical means (with gardening tools) or burning, although the efforts were futile, possibly because of a lack of adequate knowledge of the species [22,83]. A recent study by Mhlongo et al. [70] pointed out that communities in Limpopo had attempted to clear invasive alien plants without the proper knowledge, leading to resprouting and further spread. This partly indicates that some people who have *C. pallida* in their gardens do not want too much of it, probably because of its negative impacts; however, they do not know about using suitable methods or that manual removal has a high risk of injury [31] and unintentional dispersal [32]. Therefore, we propose that a basic knowledge of clearing methods should be shared with community members during awareness campaigns to minimise the chances of unintentional spread during operations, as Novoa et al. [2] suggested. Several studies have consistently shown positive returns from regular and well-planned awareness campaigns in managing biological invasions [22].

6. Conclusions

Overall, we found that the foliar spray management used to treat *C. pallida* in South Africa effectively suppressed plant growth, reproduction, and possibly further invasions. Treated populations had shorter plants with fewer cladodes and fruits than untreated sites. We also found that taller plants are highly likely to reproduce, which may allow the invasion to thrive. We recommend continuing the herbicide treatment, prioritising taller *C. pallida* plants to prevent further spread in South Africa, as *C. pallida* spreads mainly through human activities and animal movements. However, this recommendation must apply to small populations since foliar spray methods have dire financial implications for large populations. Additional comparative studies on the vegetation and soil characteristics are required to improve the knowledge of the impacts of *C. pallida*. We also noted that some local people were familiar with *C. pallida* plants in the study sites, but their short history with the invasion of the species and the limited extent of the invasion could be the reason for their mixed perceptions of *C. pallida*'s impacts, dispersal, and further spread on their properties. It is important to note that the recorded negative consequences of *C. pallida* and the pertinent recommendations apply only to the sampled sites. Therefore, we recommend increasing awareness campaigns in communities affected by *C. pallida* to educate residents about its impacts, dispersal pathways, and strategies for preventing further invasions. Given the ability of this species to recover, management of dispersal pathways is highly recommended to effectively contain further invasion [87], and future studies should focus on integrating foliar spray with other management strategies such as biological control.

Supplementary Materials: The following supporting information can be downloaded at https://www.mdpi.com/article/10.3390/ijpb16040113/s1. Supplementary Material Table S1: R-C-E-C sites.

Author Contributions: T.M.M., N.D., K.M. and T.J.M. conceptualised the study, developed the field methods, and collected the data; T.M.M. and K.M. analysed the field data, drew the study map, and wrote the manuscript; C.M., N.D. and T.J.M. read and edited the draft of the manuscript and provided important insights into how to improve the study. All authors have read and agreed to the published version of the manuscript.

Funding: This research was funded by the Centre for Invasion Biology, and the APC was funded by the University of Free State.

Institutional Review Board Statement: The study was conducted in accordance with the Declaration of Helsinki. The study interviews' component was ethically cleared by the University of Free State under Ethics Clearance number UFS-HSD2023/1208.

Informed Consent Statement: Informed consent was obtained from all subjects involved in the study.

Data Availability Statement: Available on request because the nationwide study is ongoing.

Acknowledgments: We thank the Centre for Invasion Biology (housed in Stellenbosch University) for funding the running costs of the study. We also thank the University of the Free State for providing transport for the fieldwork and Patrick Mohasi for assisting with the interviews. Loyd R. Vukeya drew the maps showing the study sites and vegetation types. We thank all the community members of Cradock and Richmond who agreed to participate in the study, as well as the property owners who allowed us access to collect data.

Conflicts of Interest: The authors declare no conflicts of interest.

References

1. Kaplan, H.; Wilson, J.R.U.; Klein, H.; Henderson, L.; Zimmermann, H.G.; Manyama, P.; Ivey, P.; Richardson, D.M.; Novoa, A. A proposed national strategic framework for the management of Cactaceae in South Africa. *Bothalia* **2017**, *47*, a2149. [CrossRef]
2. Novoa, A.; Dehnen-Schmutz, K.; Fried, J.; Vimercati, G. Does public awareness increase support for invasive species management? Promising evidence across taxa and landscape types. *Biol. Invasions* **2017**, *19*, 3691–3705. [CrossRef]
3. Masocha, M.; Dube, T. Global terrestrial biomes at risk of cacti invasion identified for four species using consensual modelling. *J. Arid Environ.* **2018**, *156*, 77–86. [CrossRef]
4. Van Wilgen, B.W.; Van Measey, J.; Richardson, D.M.; Wilson, J.R. *Biological Invasions in South Africa*; Springer: Cham, Switzerland, 2020. [CrossRef]
5. Novoa, A.; Le Roux, J.J.; Robertson, M.P.; Wilson, J.R.; Richardson, D.M. Introduced and invasive cactus species: A global review. *AoB Plants* **2015**, *7*, plu078. [CrossRef] [PubMed]
6. Novoa, A.; Le Roux, J.J.; Richardson, D.M.; Wilson, J.R. Level of environmental threat posed by horticultural trade in Cactaceae. *Conserv. Biol.* **2017**, *31*, 1066–1075. [CrossRef] [PubMed]
7. Lockwood, J.L.; Cassey, P.; Blackburn, T.M. The more you introduce the more you get: The role of colonization pressure and propagule pressure in invasion ecology. *Divers. Distrib.* **2009**, *15*, 904–910. [CrossRef]
8. Korpelainen, H.; Pietiläinen, M. What Makes a Good Plant Invader? *Life* **2023**, *13*, 1596. [CrossRef] [PubMed]
9. Nobel, P.S.; Geller, G.N.; Kee, S.C.; Zimmerman, A.D. Temperatures and thermal tolerances for cacti exposed to high temperatures near the soil surface. *Plant Cell Environ.* **1986**, *9*, 279–287. [CrossRef]
10. Williamson, M.H.; Fitter, A. The characters of successful invaders. *Biol. Conserv.* **1996**, *78*, 163–170. [CrossRef]
11. Walters, M.; Figwueiredo, E.; Zimmermann, H.G.; Mashope, B.K. *Naturalised and Invasive Succulents of Southern Africa*; ABC Taxa: Brussels, Belgium, 2011.
12. Richardson, D.M.; Thuiller, W. Home away from home—Objective mapping of high-risk source areas for plant introductions. *Divers. Distrib.* **2007**, *13*, 299–312. [CrossRef]
13. Milton, S.J.; Dean, W.R.J. Disturbance, drought and dynamics of desert dune grassland, South Africa. *Plant Ecol.* **2000**, *150*, 37–51. [CrossRef]

14. Kalwij, J.M.; Robertson, M.P.; van Rensburg, B.J. Annual monitoring reveals rapid upward movement of exotic plants in a montane ecosystem. *Biol. Invasions* **2015**, *17*, 3517–3529. [CrossRef]
15. Howard, P.L. Human adaptation to invasive species: A conceptual framework based on a case study metasynthesis. *Ambio* **2019**, *48*, 1401–1430. [CrossRef]
16. Mokotjomela, T.M.; Nemurangoni, T.; Mundalamo, T.; Jaca, T.P.; Kuhudzai, A.G. The value of dump sites for monitoring biological invasions in South Africa. *Biol. Invasions* **2022**, *24*, 971–986. [CrossRef]
17. Pellegrini, A.F.; Reich, P.B.; Hobbie, S.E.; Coetsee, C.; Wigley, B.; February, E.; Georgiou, K.; Terrer, C.; Brookshire, E.N.J.; Ahlström, A.; et al. Soil carbon storage capacity of drylands under altered fire regimes. *Nat. Clim. Change* **2023**, *13*, 1089–1094. [CrossRef]
18. Novoa, A.; Kumschick, S.; Richardson, D.M.; Rouget, M.; Wilson, J.R. Native range size and growth form in Cactaceae predict invasiveness and impact. *NeoBiota* **2016**, *30*, 75–90. [CrossRef]
19. Novoa, A.; Shackleton, R.; Canavan, S.; Cybele, C.; Davies, S.J.; Dehnen-Schmutz, K.; Fried, J.; Gaertner, M.; Geerts, S.; Griffiths, C.L.; et al. A framework for engaging stakeholders on the management of alien species. *J. Environ. Manag.* **2018**, *205*, 286–297. [CrossRef] [PubMed]
20. Shackleton, R.T.; Richardson, D.M.; Shackleton, C.M.; Bennett, B.; Crowley, S.L.; Dehnen-Schmutz, K.; Estévez, R.A.; Fischer, A.; Kueffer, C.; Kull, C.A.; et al. Explaining people's perceptions of invasive alien species: A conceptual framework. *J. Environ. Manag.* **2019**, *229*, 10–26. [CrossRef] [PubMed]
21. Jubase, N.; Shackleton, R.T.; Measey, J. Public awareness and perceptions of invasive alien species in small towns. *Biology* **2021**, *10*, 1322. [CrossRef] [PubMed]
22. Pluess, T.; Jarošík, V.; Pyšek, P.; Cannon, R.; Pergl, J.; Breukers, A.; Bacher, S. Which factors affect the success or failure of eradication campaigns against alien species? *PLoS ONE* **2012**, *7*, e48157. [CrossRef] [PubMed]
23. Wilson, J.R.; Ivey, P.; Manyama, P.; Nanni, I. A new national unit for invasive species detection, assessment and eradication planning. *S. Afr. J. Sci.* **2013**, *109*, 1–13. [CrossRef]
24. Mokotjomela, T.M.; Vukeya, L.R.; Guo, D.; Mbele, T.J.; Xivuri, T.; Kuhudzai, A.G. Assessing success in attempts to eradicate an emerging invader plant: *Tephrocactus articulatus* (Pfeiff.) Backeb, in arid areas of South Africa. *J. Arid Environ.* **2024**, *220*, 105101. [CrossRef]
25. Cheek, M.D.; Crouch, N.R. Assessment of the invasive status of newly recorded cactus species in the central Tugela River basin. *Bothalia-Afr. Biodivers. Conserv.* **2015**, *45*, a1953. [CrossRef]
26. Henderson, L.; Wilson, J.R. Changes in the composition and distribution of alien plants in South Africa: An update from the Southern African Plant Invaders Atlas. *Bothalia-Afr. Biodivers. Conserv.* **2017**, *47*, a2172. [CrossRef]
27. Van Wilgen, B.W.; Wilson, J.R. (Eds.) *The Status of Biological Invasions and Their Management in South Africa in 2017*; South African National Biodiversity Institute, Kirstenbosch and DST-NRF Centre of Excellence for Invasion Biology: Stellenbosch, South Africa, 2018.
28. Department of Environmental Affairs. *National Environmental Management: Biodiversity Act 2004 (Act No. 10 of 2004) Alien and Invasive Species Lists*; Government Gazette of South Africa: Pretoria, South Africa, 2020.
29. Henderson, L. *Invasive Alien Plants in South Africa*; Plant Protection Research Institute Handbook No. 21; Agricultural Research Council: Pretoria, South Africa, 2020; ISBN 978-0-620-86146-5.
30. Paterson, I.D.; Klein, H.; Muskett, P.C.; Griffith, T.C.; Mayonde, S.; Mofokeng, K.; Mnqeta, Z.; Venter, N. Biological control of Cactaceae in South Africa. *Afr. Entomol.* **2021**, *29*, 713–734. [CrossRef]
31. Deltoro, V.I.; Ballester, G.A.; Oltra, J.E.; Perez-Botella, J.O.; Perez-Rovira, P.A.; Gómez-Serrano, M.A.; Juan, J.I. The practicalities of eradicating an extremely invasive cactus: Hudson pear *Cylindropuntia rosea* in the Valencia region (East Spain). *Aliens Invasive Spec. Bull.* **2013**, *33*, 23–27.
32. CABI. Cylindropuntia pallida. In *Invasive Species Compendium*; CAB International: Wallingford, UK, 2024.
33. Richardson, D.M.; Pyšek, P.; Rejmanek, M.; Barbour, M.G.; Panetta, F.D.; West, C.J. Naturalization and invasion of alien plants: Concepts and definitions. *Divers. Distrib.* **2000**, *6*, 93–107. [CrossRef]
34. Mokotjomela, T.M.; Xivuri, T.M.; Manyama, P. Eradication of the emerging alien cactus species, *Cylindropuntia pallida* F.M. Knuth, and active restoration in arid areas of South Africa. In Proceedings of the 8th World Conference: Ecological Restoration, Cape Town, South Africa, 24–28 September 2019; pp. 24–28.
35. Mokotjomela, T.M.; Nelufule, T.; Scott, Z.; Vukeya, L.R.; Xivuri, T.; Matsokane, K.; Mweli, N.; Magqabi, F.L.; Jaca, T. The invasion threat of the emerging alien cactus *Cylindropuntia pallida* (Rosa), FM Knuth in South Africa and the potential for control using herbicides. *Environ. Monit. Assess.* **2024**, *196*, 673. [CrossRef]
36. Moran, V.C.; Hoffmann, J.H.; Hill, M.P. A context for the 2011 compilation of reviews on the biological control of invasive alien plants in South Africa. *Afr. Entomol.* **2011**, *19*, 177–185. [CrossRef]
37. Zozo, E. The Use of the Cochineal Insect, *Dactylopius tomentosus* Lamarck, as a Biological Control Agent for the Invasive Alien Thistle Cholla, *Cylindropuntia pallida* (Rose) FM Knuth in South Africa. Master's Thesis, Rhodes University, Makhanda, South Africa, 2022.

38. Sheehan, M.R.; Potter, S. *Managing Opuntioid Cacti in Australia: Best Practice Control Manual for Austrocylindropuntia, Cylindropuntia and Opuntia Species*; Department of Primary Industries and Regional Development: Perth, Australia, 2017.
39. Tu, M.; Hurd, C.; Randall, J.M. Weed Control Methods Handbook: Tools & Techniques for Use in Natural Areas. Available via TNC. 2001. Available online: http://www.invasive.org/gist/products/handbook/methods-handbook.pdf (accessed on 3 December 2023).
40. Roura-Pascual, N.; Richardson, D.M.; Krug, R.M.; Brown, A.; Chapman, R.A.; Forsyth, G.G.; Le Maitre, D.C.; Robertson, M.P.; Stafford, L.; Van Wilgen, B.W.; et al. Ecology and management of alien plant invasions in South African fynbos: Accommodating key complexities in objective decision making. *Biol. Conserv.* **2009**, *142*, 1595–1604. [CrossRef]
41. Van Wilgen, B.W. A brief, selective history of researchers and research initiatives related to biological invasions in South Africa. In *Biological Invasions in South Africa*; Van Wilgen, B.W., Measey, J., Richardson, D.M., Wilson, J.R., Zengeya, T.A., Eds.; Springer International Publishing: Cham, Switzerland, 2020; pp. 33–64.
42. Pyšek, P.; Richardson, D.M. Traits associated with invasiveness in alien plants: Where do we stand? In *Biological Invasions, Ecological Studies*; Nentwig, W., Ed.; Springer: Berlin/Heidelberg, Germany, 2007; pp. 97–125.
43. Mucina, L.; Rutherford, M.C. *The Vegetation of South Africa, Lesotho and Swaziland. Strelitzia 19*; South African National Biodiversity Institute: Pretoria, South Africa, 2006.
44. Dayaram, A.; Harris, L.R.; Grobler, B.A.; Van der Merwe, S.; Rebelo, A.G.; Ward Powrie, L.; Vlok, J.H.; Desmet, P.G.; Qabaqaba, M.; Hlahane, K.M.; et al. Vegetation map of South Africa, Lesotho and Swaziland 2018: A description of changes since 2006. *Bothalia-Afr. Biodivers. Conserv.* **2019**, *49*, a2452. [CrossRef]
45. Novoa, A.; Brundu, G.; Day, M.D.; Deltoro, V.; Essl, F.; Foxcroft, L.C.; Fried, G.; Kaplan, H.; Kumschick, S.; Lloyd, S.; et al. Global actions for managing cactus invasions. *Plants* **2019**, *8*, 421. [CrossRef] [PubMed]
46. Hill, D.A. (Ed.) *Handbook of Biodiversity Methods: Survey, Evaluation and Monitoring*; Cambridge University Press: Cambridge, UK, 2005.
47. Mokotjomela, T.M. A Comparison of Bird Foraging Preferences for Fruits of Indigenous and Alien Shrubs and Seed Dispersal Potentials in the Cape Floristic Region. Ph.D. Thesis, Stellenbosch University, Stellenbosch, South Africa, 2012.
48. Nenungwi, L.; Mokotjomela, T.M.; Vukeya, L.R.; Slabbert, M.M.; Prinsloo, K.L. A human–wildlife conflict: Potential impacts of fatal harvesting approaches on medicinal plants in Free State Province, South Africa. *S. Afr. Geogr. J.* **2024**, *107*, 20–38. [CrossRef]
49. Koli, P.; Bhardwaj, N.R.; Mahawer, S.K. Agrochemicals: Harmful and beneficial effects of climate changing scenarios. In *Climate Change and Agricultural Ecosystems*; Woodhead Publishing: Sawston, UK, 2019; pp. 65–94.
50. Singh, S.; Tiwari, S. Responses of plants to herbicides: Recent advances and future prospectives. *Plant Life Under Chang. Environ.* **2020**, *2020*, 237–250.
51. Garnier, E.; Stahl, U.; Laporte, M.A.; Kattge, J.; Mougenot, I.; Kühn, I.; Laporte, B.; Amiaud, B.; Ahrestani, F.S.; Bönisch, G.; et al. Towards a thesaurus of plant characteristics: An ecological contribution. *J. Ecol.* **2017**, *105*, 298–309. [CrossRef]
52. Kwinda, M.; Siebert, S.J.; Van Coller, H.; Masehela, T.S. Identifying Plant Functional Traits of Weeds in Fields Planted with Glyphosate-Tolerant Maize for Preferable Weed Management Practices. *Agriculture* **2024**, *14*, 223. [CrossRef]
53. Herron, P.M.; Martine, C.T.; Latimer, A.M.; Leicht-Young, S.A. Invasive plants and their ecological strategies: Prediction and explanation of woody plant invasion in New England. *Divers. Distrib.* **2007**, *13*, 633–644. [CrossRef]
54. Coutts, S.R.; Van Klinken, R.D.; Yokomizo, H.; Buckley, Y.M. What are the key drivers of spread in invasive plants: Dispersal, demography or landscape: And how can we use this knowledge to aid management? *Biol. Invasions* **2011**, *13*, 1649–1661. [CrossRef]
55. Starzomski, B. Indicators of ecosystem change. In *Encyclopedia of Quality of Life and Well-Being Research*; Michalos, A.C., Ed.; Springer: Dordrecht, The Netherlands, 2014. [CrossRef]
56. Godinez-Alvarez, H.; Valverde, T.; Ortega-Baes, P. Demographic trends in the Cactaceae. *Bot. Rev.* **2003**, *69*, 173–201. [CrossRef]
57. Van Kleunen, M.; Weber, E.; Fischer, M. A meta-analysis of trait differences between invasive and non-invasive plant species. *Ecol. Lett.* **2010**, *13*, 235–245. [CrossRef] [PubMed]
58. Gilbert, M.; Guichard, S.; Freise, J.; Grégoire, J.C.; Heitland, W.; Straw, N.; Tilbury, C.; Augustin, S. Forecasting Cameraria ohridella invasion dynamics in recently invaded countries: From validation to prediction. *J. Appl. Ecol.* **2005**, *42*, 805–813. [CrossRef]
59. Rejmánek, M. A theory of seed plant invasiveness: The first sketch. *Biol. Conserv.* **1996**, *78*, 171–181. [CrossRef]
60. Le Roux, J.J.; Clusella-Trullas, S.; Mokotjomela, T.M.; Mairal, M.; Richardson, D.M.; Skein, L.; Wilson, J.R.; Weyl, O.L.; Geerts, S. Biotic interactions as mediators of biological invasions: Insights from South Africa. In *Biological Invasions in South Africa*; Van Wilgen, B., Measey, J., Richardson, D., Wilson, J., Zengeya, T., Eds.; Springer International Publishing: Cham, Switzerland, 2020; pp. 387–427.
61. Essl, F.; Latombe, G.; Lenzner, B.; Pagad, S.; Seebens, H.; Smith, K.; Wilson, J.R.; Genovesi, P. The Convention on Biological Diversity (CBD)'s Post-2020 target on invasive alien species–what should it include and how should it be monitored? *NeoBiota* **2020**, *62*, 99–121. [CrossRef]

62. Van Kleunen, M.; Essl, F.; Pergl, J.; Brundu, G.; Carboni, M.; Dullinger, S.; Early, R.; González-Moreno, P.; Groom, Q.J.; Hulme, P.E.; et al. The changing role of ornamental horticulture in alien plant invasions. *Biol. Rev.* **2018**, *93*, 1421–1437. [CrossRef] [PubMed]
63. Pachepsky, E.; Levine, J.M. Density dependence slows invader spread in fragmented landscapes. *Am. Nat.* **2011**, *177*, 18–28. [CrossRef]
64. Zhu, J.; Lukić, N.; Pagel, J.; Schurr, F.M. Density dependence of seed dispersal and fecundity profoundly alters the spread dynamics of plant populations. *J. Ecol.* **2023**, *111*, 1735–1748. [CrossRef]
65. Pyšek, P.; Jarošík, V.; Hulme, P.E.; Pergl, J.; Hejda, M.; Schaffner, U.; Vilà, M. A global assessment of invasive plant impacts on resident species, communities and ecosystems: The interaction of impact measures, invading species' traits and environment. *Glob. Change Biol.* **2012**, *18*, 1725–1737. [CrossRef]
66. Kraaij, T.; Baard, J.A.; Rikhotso, D.R.; Cole, N.S.; Van Wilgen, B.W. Assessing the effectiveness of invasive alien plant management in a large fynbos protected area. *Bothalia* **2017**, *47*, a2105. [CrossRef]
67. McConnachie, M.M.; Cowling, R.M.; Van Wilgen, B.W.; McConnachie, D.A. Evaluating the cost-effectiveness of invasive alien plant clearing: A case study from South Africa. *Biol. Conserv.* **2012**, *155*, 128–135. [CrossRef]
68. Ngorima, A.; Shackleton, C.M. Livelihood benefits and costs from an invasive alien tree (*Acacia dealbata*) to rural communities in the Eastern Cape, South Africa. *J. Environ. Manag.* **2019**, *229*, 158–165. [CrossRef]
69. Ruwanza, S.; Thondhlana, G. People's perceptions and uses of invasive plant Psidium guajava in Vhembe Biosphere Reserve, Limpopo Province of South Africa. *Ecosyst. People* **2022**, *18*, 64–75. [CrossRef]
70. Mhlongo, E.S.; Ruwanza, S.; Dalu, T. Perceptions, Knowledge, and Invasion Extent of Lantana camara on Household Yards in Rural Communities in Limpopo Province, South Africa. *Soc. Nat. Resour.* **2024**, *37*, 1218–1239. [CrossRef]
71. Balandrán-Quintana, R.R.; González-León, A.; Islas-Rubio, A.R.; Madera-Santana, T.J.; Soto-Valdez, H.; Mercado-Ruiz, J.N.; Peralta, E.; Robles-Osuna, L.E.; Vásquez-Lara, F.; Carvallo-Ruiz, T.; et al. An overview of Cholla (*Cylindropuntia* spp.) from Sonora, Mexico. *J. Prof. Assoc. Cactus Dev.* **2018**, *20*, 162–176. [CrossRef]
72. Mdweshu, L.; Maroyi, A. Local perceptions about utilization of invasive alien species Opuntia ficus-indica in three Local Municipalities in the Eastern Cape Province, South Africa. *Biodiversitas* **2020**, *21*. [CrossRef]
73. Zengeya, T.; Ivey, P.; Woodford, D.J.; Weyl, O.; Novoa, A.; Shackleton, R.; Richardson, D.; Van Wilgen, B. Managing conflict-generating invasive species in South Africa: Challenges and trade-offs. *Bothalia* **2017**, *47*, 1e11. [CrossRef]
74. Höbart, R.; Schindler, S.; Essl, F. Perceptions of alien plants and animals and acceptance of control methods among different societal groups. *NeoBiota* **2020**, *58*, 33–54. [CrossRef]
75. Mukwada, G.; Chingombe, W.; Taru, P. Strifes of the frontier: An assessment of Acacia mearnsii related park-community conflicts in the Golden Gate Highlands National Park, South Africa. *J. Integr. Environ. Sci.* **2016**, *13*, 37–54. [CrossRef]
76. Potgieter, L.J.; Gaertner, M.; O'Farrell, P.J.; Richardson, D.M. Perceptions of impact: Invasive alien plants in the urban environment. *J. Environ. Manag.* **2019**, *229*, 76–87. [CrossRef]
77. Cordeiro, B.; Marchante, H.; Castro, P.; Marchante, E. Does public awareness about invasive plants pays off? An analysis of knowledge and perceptions of environmentally aware citizens in Portugal. *Biol. Invasions* **2020**, *22*, 2267–2281. [CrossRef]
78. Shackleton, R.T.; Witt, A.B.; Aool, W.; Pratt, C.F. Distribution of the invasive alien weed, Lantana camara, and its ecological and livelihood impacts in eastern Africa. *Afr. J. Range Forage Sci.* **2017**, *34*, 1–11. [CrossRef]
79. Johnson, S.B.; Hosking, J.R.; Chinnock, R.J.; Holtkamp, R.H. The Biology of Australian Weeds 53: *Cylindropuntia rosea* (DC.) Backeb.; *Cylindropuntia tunicata* (Lehm.) FM Knuth. *Plant Prot. Q.* **2009**, *24*, 42–49.
80. Al-Robai, S.A.; Howladar, S.M.; Mohamed, H.A.; Ahmed, A.A. *Cylindropuntia rosea* (DC.) Backeb, (Cactaceae): A new generic alien record in the flora of Saudi Arabia. *J. Asia-Pac. Biodivers.* **2018**, *11*, 320–323. [CrossRef]
81. Mokotjomela, T.M.; Thabethe, V.; Downs, C. Comparing germination metrics of *Opuntia ficus-indica* and *O. robusta* between two sets of bird species (Pied Crows and two smaller species). *Acta Oecol.* **2021**, *110*, 103676. [CrossRef]
82. Mokotjomela, T.M.; Vukeya, L.R.; Pamla, L.; Scott, Z. The critical role of coastal protected areas in buffering impacts of extreme climatic conditions on bird diversity and their ecosystem services' provisioning in the Eastern Cape Province, South Africa. *Ecol. Evol.* **2023**, *13*, e10452. [CrossRef] [PubMed]
83. Foxcroft, L.C.; Van Wilgen, B.W.; Abrahams, B.; Esler, K.J.; Wannenburgh, A. Knowing-doing continuum or knowing-doing gap? Information flow between researchers and managers of biological invasions in South Africa. In *Biological Invasions in South Africa*; Van Wilgen, B., Measey, J., Richardson, D., Wilson, J., Zengeya, T., Eds.; Springer International Publishing: Cham, Switzerland, 2020; pp. 831–853.
84. Mokotjomela, T.M.; Rahlao, S.J.; Vukeya, L.R.; Baltzinger, C.; Mangane, L.V.; Willis, C.K.; Mutshinyalo, T.M. The Diversity of Alien Plant Species in South Africa's National Botanical and Zoological Gardens. *Diversity* **2023**, *15*, 407. [CrossRef]
85. Le Maitre, D.C.; Richardson, D.M.; Chapman, R.A. Alien plant invasions in South Africa: Driving forces and the human dimension: Working for water. *S. Afr. J. Sci.* **2004**, *100*, 103–112.

86. Marchante, E.; Marchante, H.; Morais, M.; Freitas, H. Combining methodologies to increase public awareness about invasive alien plants in Portugal. In *2nd International Workshop on Invasive Plants in the Mediterranean Type Regions of the World*; European Environment Agency: Trabzon, Turkey, 2010; pp. 227–239.
87. Faulkner, K.T.; Burness, A.; Byrne, M.J.; Kumschick, S.; Peters, K.; Roberston, M.P.; Saccaggi, D.L.; Weyl, O.L.; Williams, V.L. South Africa's Pathways of Introduction and Dispersal and How They Have Changed Over Time. In *Biological Invasions in South Africa*; Van Wilgen, B., Measey, J., Richardson, D., Wilson, J., Zengeya, T., Eds.; Springer International Publishing: Cham, Switzerland, 2020; pp. 313–354.

plants

MDPI

Article

Organic Waste from the Management of the Invasive *Oxalis pes-caprae* as a Source of Nutrients for Small Horticultural Crops

Paula Lorenzo [1,*], Cristina Galhano [2] and Maria Celeste Dias [1]

[1] Associate Laboratory TERRA, Centre for Functional Ecology (CFE)—Science for People & the Planet, Department of Life Sciences, University of Coimbra, 3000-456 Coimbra, Portugal; celeste.dias@uc.pt
[2] Polytechnic Institute of Coimbra, Coimbra Agriculture School, Bencanta, 3045-601 Coimbra, Portugal; cicgalhano@esac.pt
* Correspondence: paulalorenzo@uc.pt

Abstract: The management of invasive plants is a challenge when using traditional control methods, which are ineffective for large areas, leading to the abandonment of invaded areas and the subsequent worsening of the situation. Finding potential uses for waste resulting from invaders' management could motivate their control in the long-term, concurrently providing new bio-based resources with different applications. *Oxalis pes-caprae* is an invasive plant, widely distributed worldwide, which spreads aggressively through bulbils, creating a dense ground cover. This study was designed to assess the potential of *Oxalis* aboveground waste for use as fertilizer and in ameliorating deficit irrigation effects in growing crops. *Diplotaxis tenuifolia* (wild rocket) seedlings were planted in pots with soil mixed with *Oxalis* waste at 0, 2.2 and 4.3 kg m^{-2} or with commercial fertilizer, left to grow for 27 days and then irrigated at 100% or 50% field capacity for 14 days. The incorporation of the *Oxalis* waste improved the biomass, photosynthesis, sugars, total phenols and total antioxidant capacity in the crop, achieving commercial fertilization values, as well as increasing the phosphorus in soils. However, *Oxalis* waste seems not to directly affect plants' relative water contents. Our results support the use of *Oxalis* waste as fertilizer, which can encourage the long-term control of this invasive species.

Keywords: Bermuda buttercup; invasion control; organic fertilizer; photosynthesis; water deficit amelioration

1. Introduction

Invasive plants are widely known to negatively impact ecosystem biodiversity, functioning and services in areas where they are not native [1–4]. The traditional management of invasive plants generally produces relatively good results in the short term but, in the long term, for large invaded areas, it often fails due to a lack of costly follow-up measures after the end of management projects [5]. Thus, invasive propagules that escape initial control efforts are able to recolonize cleared areas. To overcome this situation, several invasive-plant ecologists have recently suggested a more integrated and sustainable strategy to control the worst invasive plants, i.e., to find potential uses for the waste generated during control operations as a complementary way to reduce costs related to the management of these invasive plants [4–6].

Oxalis pes-caprae L. (Oxalidaceae) is a South African geophyte that has become invasive in several Mediterranean climatic regions of the world (e.g., [7–9]). This geophyte can grow up to 40 cm high and annually produces a basal rosette of leaves and yellow flowers organized in umbellate cymes [10]. In the European Mediterranean invader range, *O. pes-caprae* has a prolific level of vegetative reproduction, rapidly spreading through bulbils (e.g., [7,11–13]), and it is listed among the 10 worst invasive plants in terms of ecological and economic impacts [14]. It can expand over different types of habitat, such as disturbed, ruderal and agricultural areas, orchards, grasslands and abandoned fields [15]. The lifespan of this species in Europe ranges from early autumn to late spring [16,17], resulting in a dense and homogeneous mat that outcompetes native plants. In fact, introduced *O. pes-caprae* tetraploids collected in the western Mediterranean basin can produce more aboveground biomass than *O. pes-caprae* tetraploids collected in their native range, in South Africa, when grown in competition with European species, suggesting an enhanced competitive ability [18]. In Spain, for example, *O. pes-caprae* can produce relatively high quantities of fresh aboveground biomass, at an average of 20.95 Mg ha^{-1} [13].

Although *O. pes-caprae* is considered a noxious weed, active management to control its invasion has been scarcely implemented in semi-natural ecosystems. Chemical formulations gave good results in controlling *O. pes-caprae* invasion in Australian agricultural fields (chlorsulfuron, [19]) and on a Mediterranean island (glyphosate [16]). However, the application of synthetic herbicides should be reduced due to toxic side effects [20]. Alternatively, the post-emergence application of the bioherbicide pelargonic acid showed good results in controlling *O. pes-caprae* in Mediterranean vineyards [21]. Other methods, such as the consumption of bulbs and tubers by pigs and turkeys, or the grazing of leaves by the noctuid moth, *Klugeana philoxalis*, have been suggested for Australian invaded areas [19,22]. However, no further information on the effectiveness of these methods has been reported. The control of invasions while feeding livestock could be useful for greatly reducing *O. pes-caprae* biomass, but this approach is discouraged, since the oxalate present in leaves produces poisoning if eaten in high quantities [23,24]. Recently, the management of olive orchards with a cover crop mixture (vetch/pea/barley) resulted in a reduction in *O. pes-caprae* density for three consecutive years [25]. Nevertheless, the authors of that study did not indicate whether the cover crop mixture also reduced the *O. pes-caprae* biomass.

On the other hand, the repeated mowing of biomass can contribute to bulbil exhaustion [26], without ecological, environmental or livestock health issues. The use of *O. pes-caprae* biomass resulting from control operations as a basis for creating new products may represent a complementary tool to make its management more attractive [5]. Several ecological studies revealed that *O. pes-caprae* has some fertilizing properties. For example, fresh biomass (leaves and flowers) of *O. pes-caprae* incorporated into soil, after a decomposition period, increased *Lactuca sativa* L. growth [13]. In addition, *O. pes-caprae* invasion can influence phosphorous (P) dynamics in soils, with invaded soils having higher contents of available P than non-invaded soils [27]. This is related to the release of oxalic acid from leaves after plant death and decomposition, as oxalate is a chelating agent that improves P availability is soils [27,28].

The main objective of this study was to assess whether the untreated above-ground biomass of *O. pes-caprae* in soil has potential value when it is applied as an organic fertilizer and/or whether it confers stress tolerance upon horticultural crops. To this end, we designed a greenhouse experiment in order to test the effects of different doses of *O. pes-caprae* biomass on the biometric and physiological biomarkers of a horticultural crop, *Diplotaxis tenuifolia* (wild rocket), under well or deficit irrigation conditions. We hypothesize that waste from *O. pes-caprae* biomass will improve the physiological performance of wild rocket by providing soil nutrients and some tolerance to water-deficit stress. The data provided by the present study can contribute to the creation of new products from *O. pes-caprae* waste, partially recovering funds allocated to its management, to reduce commercial fertilization inputs, to improve crops' resilience to adverse environmental conditions, and to sequester carbon by incorporating waste biomass into soils.

2. Results

2.1. Plant Height, Biomass and Relative Water Content of Wild Rocket

The height increment before different levels of water irrigation was significantly affected by *Oxalis* waste dose and the covariate (Figure 1A). The wild rocket plants growing with *Oxalis* waste and F were taller than the control plants (C). Additionally, *Oxalis* plants grown with waste at a dose of 4.3 kg m^{-2} (O4.3) were 1.6-fold taller than the plants grown with the commercial organic fertilizer (F) (Figure 1A). The height increment after the application of different water regimes was significantly affected by the water irrigation factor and the interaction between this factor and the *Oxalis* waste dose (Figure 1B). The O4.3 soil produced taller plants than the *Oxalis* waste at a dose of 2.2 kg m^{-2} (O2.2) under 50% field capacity irrigation (Figure 1B).

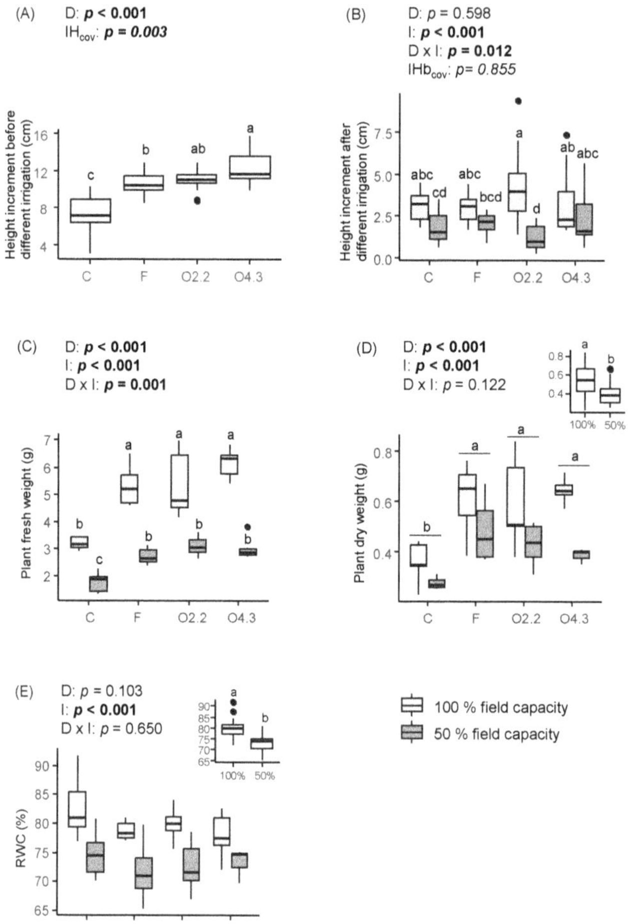

Figure 1. Effects of *Oxalis pes-caprae* waste dose (D), water irrigation (I) and the interaction between these two factors (D × I) on height increments before and after different irrigation (**A**,**B**), plant fresh and dry weight (**C**,**D**), and relative water content (RWC) (**E**) of the horticultural species *Diplotaxis tenuifolia*. Boxplots show the median values inside the boxes that bound the 25th and 75th percentile values. The dots represent values outside of the 10th and 90th percentile (represented by whiskers). IH_{cov}: initial height covariable, IHb_{cov}: initial height before different irrigation levels covariable, C: control, F: commercial fertilizer, O2.2: *Oxalis* waste at a dose of 2.2 kg m^{-2}, O4.3: *Oxalis* waste at a dose of 4.3 kg m^{-2}. Different letters indicate statistically significant differences at $p \leq 0.05$ (in bold) level using general/generalized linear models followed by Sidak's post hoc test, n = 7. When the interaction factor (D × I) is significant, statistical letters refer first to differences among C100%, C50%, F100%, F50%, O100%, and O50% treatments. Otherwise, statistical letters refer to differences between C, F, and O treatments when factor D is significant and/or to differences between 100% and 50% treatments (small figure inside the figure) when factor I is significant.

The *Oxalis* waste dose, water irrigation and *Oxalis* waste dose × water irrigation interaction had a significant effect on the fresh weights of wild rocket plants (Figure 1C). The plants growing in soils with *Oxalis* waste and commercial fertilization showed a higher increase in fresh weight (about 1.7-fold on average) than those in the control soils in both the 100% and the 50% water irrigation level (Figure 1C). However, the dry weight of wild rocket plants was only affected by the *Oxalis* waste dose and water irrigation (Figure 1D). This parameter was increased by 59–68% in O4.3, O2.2, and F compared with the control, and by 27% more at the 100% compared with the 50% irrigation level (Figure 1D).

Water irrigation significantly affected the relative water content (RWC) of the wild rocket plants, as it was higher in the well-watered plants (Figure 1E).

2.2. Gas Exchange, Chlorophyll a Fluorescence, and Pigment Contents of Wild Rocket

The *Oxalis* waste dose, water irrigation, and the interaction between these two factors significantly affected the net CO_2 assimilation rate, stomatal conductance, transpiration rate, and water use efficiency (WUE) (Figures 2A, 2B, 2C and 2D, respectively). However, the intercellular CO_2 concentration was only affected by the *Oxalis* waste dose and water irrigation (Figure 2E). The net CO_2 assimilation rate was 1.6- and 2-fold higher in the O4.3 treatment than in the O2.2 and F treatments, respectively, under 100% water irrigation, and 1.8- and 2.1-fold higher in the O4.3 and F treatments, respectively, than in the C under 50% field capacity (Figure 2A). Similarly, the wild rocket plants growing in the O4.3 soils showed increased stomatal conductance compared with those in the F and C soils under 100% irrigation (Figure 2B). This parameter was also increased by 1.5–1.89-fold in the plants from the F, O2.2, and O4.3 treatments compared to the control with 50% irrigation (Figure 2B). The F, O2.2, and O4.3 wild rocket plants also had 1.3–1.5-fold higher transpiration rate than the C plants under 50% irrigation (Figure 2C). Commercial fertilization improved the WUE of the plants compared to the O2.2 and C treatments under 50% field capacity (Figure 2D).

The F_v/F_m, Φ_{PSII}, and NPQ parameters were significantly affected by the *Oxalis* waste dose and water irrigation (Figures 3A, 3C and 3E, respectively). However, F_v'/F_m' and qP were affected by the *Oxalis* waste dose, water irrigation, and the interaction between these two factors (Figures 3B and 3D, respectively). The F_v/F_m was highest in wild rocket plants growing in the O4.3 treatment and lowest in the plants growing in the F (Figure 3A). One hundred percent field capacity increased the F_v/F_m (Figure 3A). By contrast, the wild rocket plants from the *Oxalis* waste treatments had higher F_v'/F_m' than those from the C under 50% irrigation (Figure 3B). Additionally, the F_v'/F_m' in the O4.3 plants was higher compared to that in the F plants (Figure 3B). The wild rocket plants growing in soils with *Oxalis* waste had Φ_{PSII} values that were 17–28% higher than those in the plants in the F and C soils (Figure 3C). The plants gown under 100% water irrigation also showed higher Φ_{PSII} values (Figure 3C). The *Oxalis* waste increased the qP by 15–23% in the wild rocket plants grown under 100% field capacity (Figure 3D). However, under 50% field capacity, only the O4.3 plants had higher qP compared to the F plants (Figure 3D). The application of *Oxalis* waste at 2.2 kg m^{-2} reduced the NPQ parameter and the 50% water regime increased this parameter (Figure 3E).

The *Oxalis* waste dose, water irrigation, and the interaction between these two factors had a significant effect on the contents of chlorophyll *a* and *b* (Figures 4A and 4B, respectively). The content of chlorophyll *a* was reduced in the wild rocket plants from the F treatment under 100% irrigation and in the plants from the F and O4.3 treatments under 50% irrigation (Figure 4A). Similarly, chlorophyll *b* was reduced in the plants growing in soils with F under 100% field capacity (Figure 4B). The content of carotenoids was affected by the *Oxalis* waste dose and water irrigation, as it was higher in the plants from the C and O2.2 treatments and in the well-watered plants (Figure 4C).

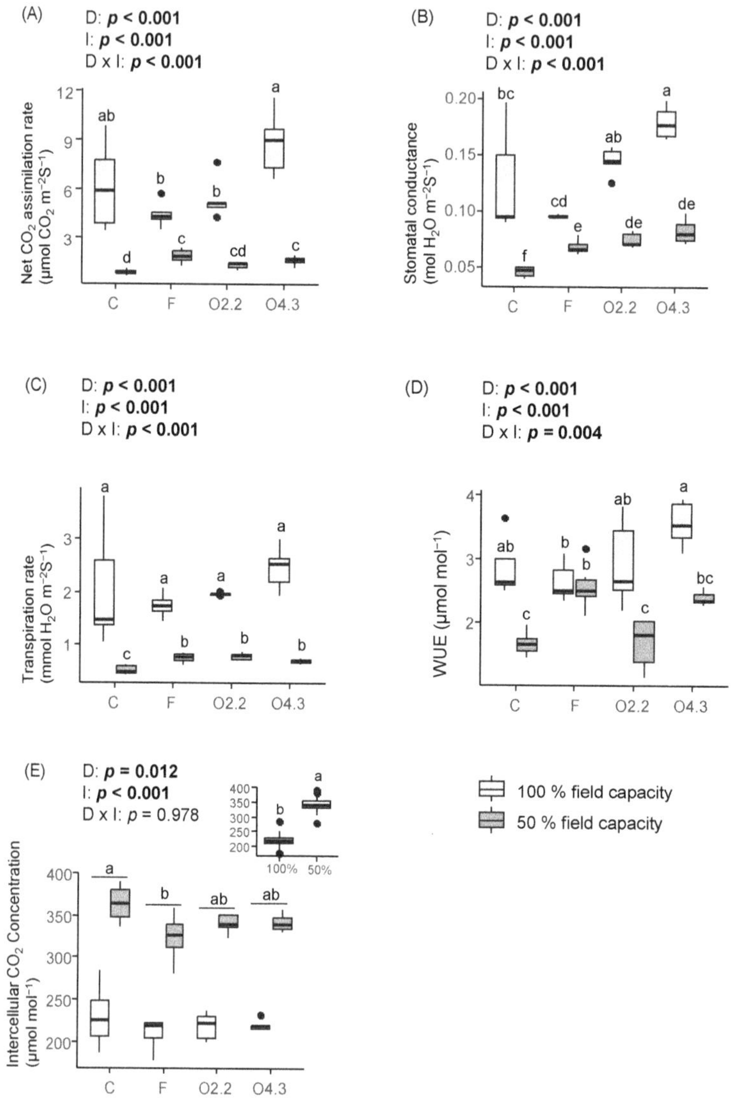

Figure 2. Effects of *Oxalis pes-caprae* waste dose (D), water irrigation (I), and the interaction between these two factors (D × I) on net CO_2 assimilation rate (**A**), stomatal conductance (**B**), transpiration rate (**C**), water use efficiency (WUE) (**D**), and intercellular CO_2 concentration (**E**) of the horticultural species *Diplotaxis tenuifolia*. Boxplots show the median values inside the boxes that bound the 25th and 75th percentile values. The dots represent values outside of the 10th and 90th percentile (represented by whiskers). C: control, F: commercial fertilizer, O2.2: *Oxalis* waste at a dose of 2.2 kg m^{-2}, O4.3: *Oxalis* waste at a dose of 4.3 kg m^{-2}. Different letters indicate statistically significant differences at $p \leq 0.05$ (in bold) level using general/generalized linear models followed by Sidak's post hoc test, $n = 7$. When the interaction factor (D × I) is significant, statistical letters refer first to differences among the C100%, C50%, F100%, F50%, O100%, and O50% treatments. Otherwise, statistical letters refer to differences between the C, F, and O treatments when factor D is significant and/or to differences between the 100% and 50% treatments (small figure inside the figure) when factor I is significant.

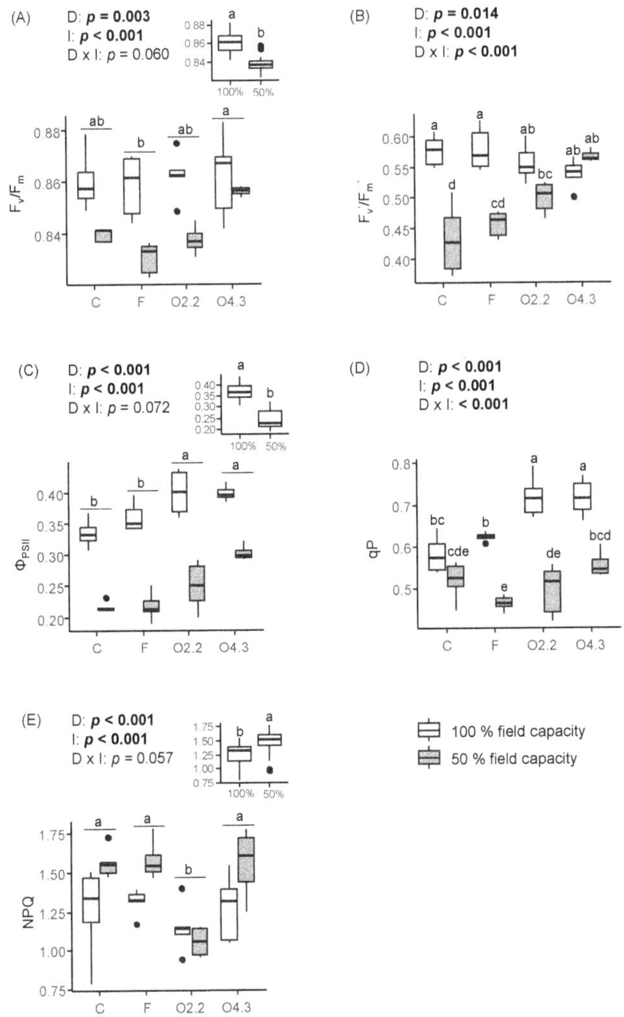

Figure 3. Effects of *Oxalis pes-caprae* waste dose (D), water irrigation (I), and the interaction between these two factors (D × I) on the maximum quantum efficiency of photosystem II (F_v/F_m) (**A**), efficiency of excitation energy capture by open PSII reaction centers (F_v'/F_m') (**B**), effective efficiency of photosystem II (Φ_{PSII}) (**C**), photochemical quenching (qP) (**D**), and non-photochemical quenching (NPQ) (**E**) of the horticultural species *Diplotaxis tenuifolia*. Boxplots show the median values inside the boxes that bound the 25th and 75th percentile values. The dots represent values outside of the 10th and 90th percentile (represented by whiskers). C: control, F: commercial fertilizer, O2.2: *Oxalis* waste at a dose of 2.2 kg m^{-2}, O4.3: *Oxalis* waste at a dose of 4.3 kg m^{-2}. Different letters indicate statistically significant differences at $p \leq 0.05$ (in bold) level using general/generalized linear models followed by Sidak's post hoc test, $n = 7$. When the interaction factor (D × I) is significant, statistical letters refer first to differences among the C100%, C50%, F100%, F50%, O100%, and O50% treatments. Otherwise, statistical letters refer to differences between the C, F, and O treatments when factor D is significant and/or to differences between the 100% and 50% treatments (small figure inside the figure) when factor I is significant.

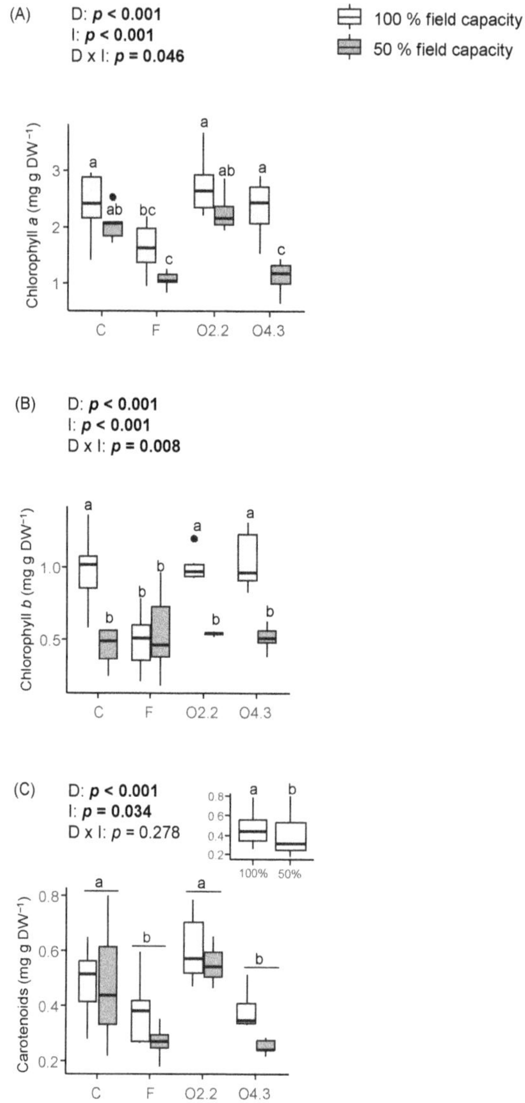

Figure 4. Effects of *Oxalis pes-caprae* waste dose (D), water irrigation (I), and the interaction between these two factors (D × I) on the contents of chlorophylls *a* and *b* (**A**,**B**) and carotenoids (**C**) in the horticultural species *Diplotaxis tenuifolia*. Boxplots show the median values inside the boxes that bound the 25th and 75th percentile values. The dots represent values outside of the 10th and 90th percentile (represented by whiskers). C: control, F: commercial fertilizer, O2.2: *Oxalis* waste at a dose of 2.2 kg m^{-2}, O4.3: *Oxalis* waste at a dose of 4.3 kg m^{-2}, DW: dry weight. Different letters indicate statistically significant differences at $p \leq 0.05$ (in bold) level using general/generalized linear models followed by Sidak's post hoc test, $n = 7$. When the interaction factor (D × I) is significant, statistical letters refer first to differences among the C100%, C50%, F100%, F50%, O100%, and O50% treatments. Otherwise, statistical letters refer to differences between the C, F, and O treatments when factor D is significant and/or to differences between the 100% and 50% treatments (small figure inside the figure) when factor I is significant.

2.3. Total Soluble Sugars and Starch of Wild Rocket

The content of total soluble sugars was significantly affected by the *Oxalis* waste dose, water irrigation, and *Oxalis* waste dose × water irrigation interaction (Figure 5A). The wild rocket plants growing in the O4.3, O2.2, and F soils accumulated 4.4, 2.8, and 3.7 times more total soluble sugars than those plants growing in the control soils under 100% irrigation conditions (Figure 5A). Similarly, the O4.3 and O2.2 plants produced 1.6–1.7 times more soluble sugars than the F and C plants under deficit water irrigation (Figure 5A). On the other hand, the content of starch was affected by the *Oxalis* waste dose and water irrigation factors, as it was higher in the plants growing in soils mixed with *Oxalis* waste at 2.2 kg m^{-2} and in the plants grown under 100% water irrigation (Figure 5B).

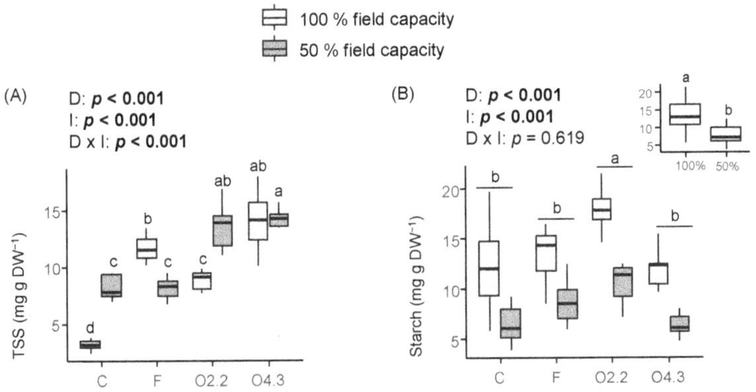

Figure 5. Effects of *Oxalis pes-caprae* waste dose (D), water irrigation (I), and the interaction between these two factors (D × I) on the contents of total soluble sugars (TSS) (**A**) and starch (**B**) in the horticultural species Diplotaxis tenuifolia. Boxplots show the median values inside the boxes that bound the 25th and 75th percentile values. C: control, F: commercial fertilizer, O2.2: *Oxalis* waste at a dose of 2.2 kg m^{-2}, O4.3: *Oxalis* waste at a dose of 4.3 kg m^{-2}, DW: dry weight. Different letters indicate statistically significant differences at $p \leq 0.05$ (in bold) level using general/generalized linear models followed by Sidak's post hoc test, $n = 7$. When the interaction factor (D × I) is significant, statistical letters refer first to differences among the C100%, C50%, F100%, F50%, O100%, and O50% treatments. Otherwise, statistical letters refer to differences between the C, F, and O treatments when factor D is significant and/or to differences between the 100% and 50% treatments (small figure inside the figure) when factor I is significant.

The *Oxalis* waste dose, water irrigation, and the interaction between these two factors had a significant effect on the total antioxidant capacity (TAA), total phenols, and orthodiphenols (Figures 6A, 6B and 6C, respectively). However, the flavonoids content was only affected by the *Oxalis* waste dose and the interaction between this factor and water irrigation (Figure 6D). The wild rocket plants from the F and *Oxalis* waste treatments had a 16–43% higher TAA and total phenols contents than those from the control when a 100% irrigation level was applied (Figures 6A and 6B, respectively). In addition, the TAA and total phenols of the wild rocket plants treated with the *Oxalis* waste was 26–28% higher than that in the F and C plants under reduced irrigation (Figures 6A and 6B, respectively). The F and *Oxalis* waste treatments increased the of orthodiphenols contents in the well-watered plants (Figure 6C). However, this parameter was only increased by the O2.2 treatment under 50% irrigation (Figure 6C). By contrast, the wild rocket plants growing in soils with commercial fertilization and *Oxalis* waste at 4.3 kg m^{-2} had a 58–203% higher content of flavonoids compared to the C and O2.2 treatments when 100% water irrigation was applied, but no differences were found for this parameter under 50% irrigation (Figure 6D).

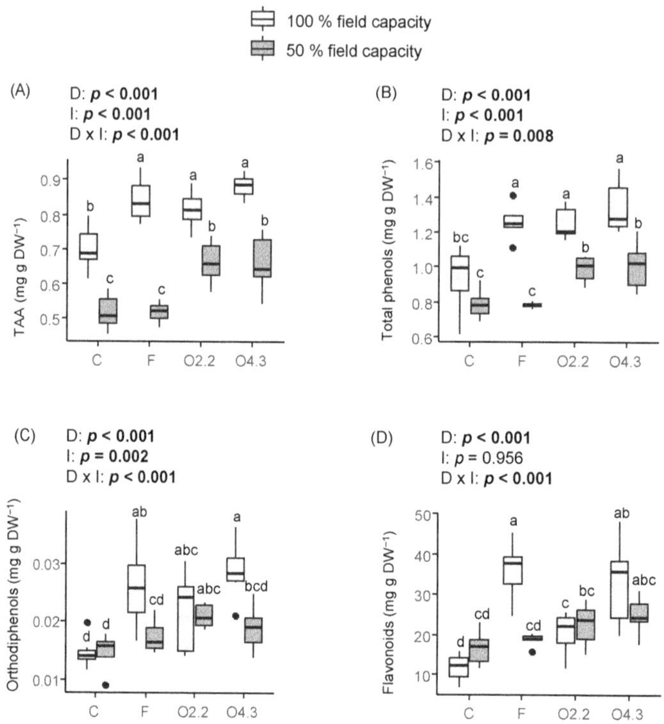

Figure 6. Effects of *Oxalis pes-caprae* waste dose (D), water irrigation (I), and the interaction between these two factors (D × I) on the total antioxidant activity (TAA) (**A**), total phenols (**B**), orthodiphenols (**C**), and flavonoids (**D**) in the horticultural species Diplotaxis tenuifolia. Boxplots show the median values inside the boxes that bound the 25th and 75th percentile values. The dots represent values outside of the 10th and 90th percentile (represented by whiskers). C: control, F: commercial fertilizer, O2.2: *Oxalis* waste at a dose of 2.2 kg m^{-2}, O4.3: *Oxalis* waste at a dose of 4.3 kg m^{-2}, DW: dry weight. Different letters indicate statistically significant differences at $p \leq 0.05$ (in bold) level using general/generalized linear models followed by Sidak's post hoc test, $n = 7$. When the interaction factor (D × I) is significant, statistical letters refer first to differences among the C100%, C50%, F100%, F50%, O100%, and O50% treatments.

2.4. Soil Nutrients

The *Oxalis* waste, the sampling date, and the interaction between them did not affect the total carbon and nitrogen in the soil (Figure 7A,B). However, the content of extractable soil PO_4–P was significantly affected by the *Oxalis* waste, the sampling date, and the *Oxalis* waste × sampling date interaction (Figure 7C). The soil mixed with the *Oxalis* waste and left to decompose for eight weeks almost doubled the amount of PO_4–P compared to the other treatments (Figure 7C).

2.5. Multivariate Analysis

The multivariate analysis (PCA) showed a clear separation between the well-watered (scores in the right quadrant) and water deficit (scores in the left quadrant) treatments, and among the *Oxalis* waste dose (C, O2.2, O4.3, F) treatments, suggesting a different physiological response of the wild rocket plants to the irrigation levels and *Oxalis* waste doses (Figure 8). Under well-watered conditions, the scores were more spread in the right quadrant, and the C scores were more separated from the F, O2.2, and O4.3. However, the water deficit treatment reduced the score spread, reducing the separation between the O2.2,

O4.3, C, and F, and separated the O4.3 scores more from the F and O2.2 (which became closer). Under well-watered conditions, the O2.2 and O4.3 were better characterized by the high levels of Φ_{PSII} (Figure 8).

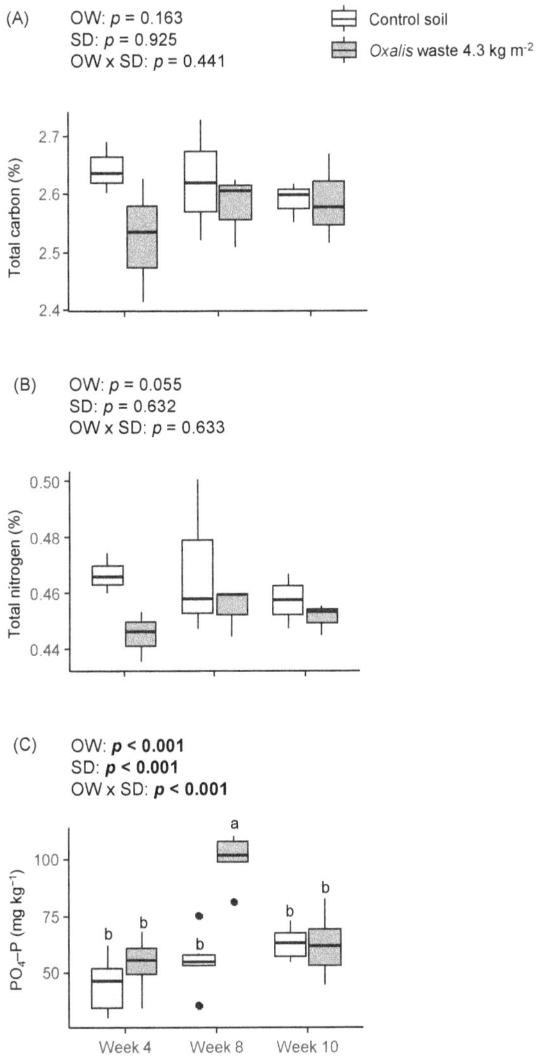

Figure 7. Effects of *Oxalis pes-caprae* waste (OW), the sampling date (SD), and the interaction between these two factors (OW × SD) on the total carbon (**A**), total nitrogen (**B**), and extractable phosphorous (**C**) in the soils during the *Oxalis* waste decomposition. Boxplots show the median values inside the boxes that bound the 25th and 75th percentile values. The dots represent values outside of the 10th and 90th percentile (represented by whiskers). Different letters indicate statistically significant differences at $p \leq 0.05$ level (in bold) using repeated measured models followed by Sidak's post hoc test, $n = 6$. When the interaction factor (OW × SD) is significant, statistical letters refer first to differences between the Week 4 control, Week 4 *Oxalis* waste, Week 8 control, Week 8 *Oxalis* waste, Week 10 control, and Week 10 *Oxalis* waste treatments. The absence of letters on a figure indicates that there are no significant statistical differences for any analysis factor.

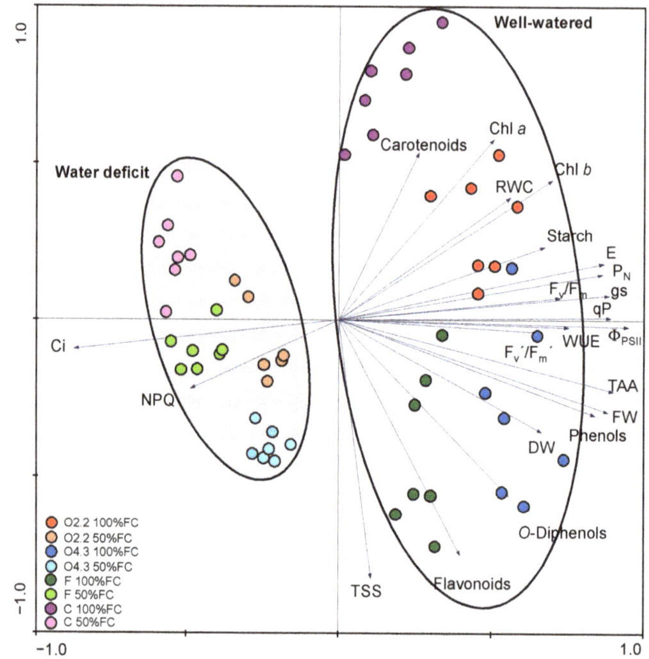

Figure 8. Principal component analysis (PCA) (x—first component and y—second component) of the physiological parameters in the wild rocket under well-watered and water deficit conditions. The first axis (PC1) explains 52% of the total variance and the second axis (PC2) explains 14% of the total variance. Colored circles represent sample scores and arrows show gradients resulting from water status, weight, pigments, carbohydrates, photosynthesis, oxidative stress, and antioxidant profiles. C: control, F: commercial fertilizer, O2.2: *Oxalis* waste at a dose of 2.2 kg m^{-2}, O4.3: *Oxalis* waste at a dose of 4.3 kg m^{-2}, FC: field capacity, Chl *a*: chlorophyll *a*, Chl *b*: chlorophyll *b*, Ci: intercellular CO_2 concentration, DW: dry weight, E: transpiration rate, F_v/F_m: maximum quantum efficiency of photosystem II, F_v'/F_m': efficiency of excitation energy capture by open PSII reaction centers, gs: stomatal conductance, FW: fresh weight, NPQ: non-photochemical quenching, P_N: net CO_2 assimilation rate, qP: photochemical quenching, RWC: relative water content, TAA: total antioxidant activity, TSS: total soluble sugars, WUE: water use efficiency, Φ_{PSII}: effective efficiency of photosystem II.

3. Discussion

In the present study, the application of *Oxalis* waste in soil affected the wild rocket's physiology, modulating several biochemical and physiological processes, which led to a biomass production similar to that recorded in the treatment with commercial organic fertilizer. However, the *Oxalis* dose and the irrigation level influenced this response.

When water availability deviates from the needed quantity, it induces several relevant impairments, leading to weaker development and productivity losses in agriculture [29]. The leaf relative water content (RWC) is a physiological biomarker widely used to assess plants' water status. In the present study, the reduction in irrigation of 50% induced a decrease in the stomata aperture (lower transpiration rate), but this protective mechanism was not effective in preventing leaf water loss, leading to a decline in the leaf RWC in the plants under all the treatments, indicating the impairment of the leaf water status. This reduction was accompanied by a general loss of plant fresh biomass, together with other physiological and biochemical alterations, suggesting a water deficit stress condition. However, although the F_v/F_m, an indicator of plant performance, was reduced by the 50%

irrigation level, the values obtained were above 0.75, suggesting no significant reduction in the photosystem II efficiency [30]. This difference in the response profiles of the plants at 50% irrigation was notable in the multivariate analysis (Figure 8). The PCA analysis also highlighted that the O2.2 and O4.3 physiological responses were different from the C, and closer to the F, or even distinct, as in the case of the O4.3 under water deficit stress (Figure 8). The application of *Oxalis* waste to the wild rocket plants seems not to have influenced the RWC levels, but they affected other plant physiological and morphological traits according to the irrigation level and *Oxalis* waste dose. For instance, the beneficial application of O2.2 and O4.3 was noticeable in the photosynthesis, photosynthetic pigment contents, and carbohydrate levels. Under 100% irrigation, both *Oxalis* doses were effective in maintaining a net CO_2 assimilation rate similar to or higher than those of the F and C plants, and the dose of O4.3 favored stomatal conductance more, despite the lack of changes registered in the intercellular CO_2 availability of the mesophyll cells. Moreover, important differences in the light dependent reactions of photosynthesis were also observed. *Oxalis* waste seems to have enhanced the effective efficiency of photosystem II (possibly due to the maintenance of high levels of chlorophylls) and the photochemical quenching (Figure 8). Additionally, the O2.2 dose seems to have induced the accumulation of reserve carbohydrates (starch) and maintain the levels of protective pigments and carotenoids, similar to C plants, while the O4.3 dose acted similarly to the F plants, in terms of total soluble sugars accumulation and photoprotective processes related to NPQ.

In turn, under 50% irrigation, the *Oxalis* application enabled the plants to maintain a transpiration rate and a stomal conductance similar to those of the F plants, and higher than those of the C. Additionally, at the photochemistry level, the *Oxalis* waste acted mostly in the improvement of the efficiency of the excitation energy capture by the open PSII reaction centers and in photochemical quenching. These adjustments possibly enabled the plants to maintain a net CO_2 assimilation rate (and WUE in O4.3 doses) similar to or higher than those of the F and C plants. Interestingly, both *Oxalis* doses promoted the accumulation of TSS, but only the O2.2 dose induced a higher investment in energy reserves (starch) and a reduction in the photoprotective mechanism related to NPQ.

Regarding the antioxidant status of the plants, under 100% irrigation, the O2.2 and O4.3 maintained TAA, total phenols, orthodiphenols, and flavonoids levels similar to those of the F plants, and higher than those of the C plants. However, under 50% irrigation, the *Oxalis* application in the soil favored the TAA and total phenols, indicating that these plants invested more in antioxidant protection.

All these physiological responses induced by the *Oxalis* waste application resulted in the production of biomass (plant fresh and dry mass) similar to that obtained when commercial fertilizer was used, and higher than the biomass produced in the C plants, under both irrigation levels. This effect was less obvious in terms of plant height investment (despite the differences noticed before the water deficit irrigation). This clearly indicates a beneficial effect of *Oxalis* waste application in soil, providing a growth and development effect similar to the commercial organic fertilizer. In line with our results, Lorenzo et al. [13] reported previously that the *Oxalis* waste present in soils promoted plant physiological performance (stem and root biomass accumulation). Our results also showed that the application of *Oxalis* waste increased the content of phosphorous (P) after eight weeks. This could be related to the fact that *Oxalis* biomass in soil decomposes rapidly, releasing high amounts of oxalic acid and increasing soil acidity [13,28]. In turn, oxalates in soils are associated with a high P available content, which could the improve P nutritional status of plants [28]. Phosphorus is the main element of several compounds, such as ATP, NADPH, sugar phosphates, phospholipids, and nucleic acids, and it is strictly related to photosynthesis, carbohydrate metabolism, and protein synthesis [31]. High P availability (P fertilizers) has been correlated with improved net CO_2 assimilation rates, transpiration rates, and chlorophylls contents, and with the quantum efficiency of photosystem II [32–35]. In the present work, the increase in P availability in the O2.2 and O4.3 soils may explain the good performance of the wild rocket plants, inducing similar effects to the application of the commercial fertilizer.

The fact that the *Oxalis* waste biomass incorporation in the soil ultimately produced crops with similar characteristics to those obtained with the horticultural commercial fertilizer strongly suggests a use as organic fertilizer for the aboveground *Oxalis* waste. In invaded ranges, *O. pes-caprae* mainly reproduces asexually, through bulbils [11,12]. In addition, *O. pes-caprae* individuals from invaded ranges produce higher numbers of bulbils compared to individuals from the native South African range [18]. Each new bulbil will produce new offspring bulbils in the following season. The frequent movement of aboveground biomass during the growing season can lead to an exhaustion of bulbils [26] and reduce the number of new bulbils for the next generation (Dr. Mariana Castro, personal communication). Therefore, using *Oxalis* biomass as a source of nutrients for agricultural purposes can encourage active management to control *Oxalis* populations, which is urgently needed for persistent widely spread invasive plants [5]. However, despite not being the aim of the present study, this approach should be preferably completed with an economic analysis to evaluate the viability of the waste's use in reducing *Oxalis* control costs [5,36].

4. Materials and Methods

4.1. Plant Material and Soil Collection and Preparation

Fresh aboveground waste (including leaves and flowers) from the invasive *Oxalis pes-caprae* management was collected at the Botanical Garden of the University of Coimbra, Portugal, in February 2023, cut into 2–5 cm long pieces and immediately used.

Seedlings of the horticultural species *Diplotaxis tenuifolia* (L.) DC (wild rocket) were obtained from a local supplier (Pombalverde Lda., Leiria, Portugal).

Agricultural soil (top soil, 20 cm layer) devoted to organic agriculture was collected at the Polytechnic Institute of Coimbra, Coimbra Agriculture School, Coimbra, Portugal, at the end of January 2023 and sieved through a 1 cm sieve to remove large debris and stones. Soil characteristics were described by Lorenzo et al. [37]. Soil was mixed with silica sand (0.7 mm diameter, AXTON SILICE) in a proportion of 3:1 v/v for the experiment.

4.2. Experimental Set-Up, Stress Implementation, and Biometric Parameters Analysis

We established a pot experiment with two factors: *Oxalis* waste at doses of 0 (negative control), 2.2, and 4.3 kg m^{-2} (C, O2.2, O4.3, respectively) [13] and irrigation level (100% and 50% soil water capacity) under greenhouse conditions with natural light at the University of Coimbra, Coimbra, Portugal. Commercial organic fertilizer (F, Adubo orgânico Siro BIO N-P-K 6-7-8, Portugal) at a recommended dose of 0.075 kg m^{-2} was used as a positive control. Each treatment had 14 replicates.

On 9 February 2023, 600 mL pots (10 cm in diameter) were filled with ~800 g of sand–soil mixed with *Oxalis* waste at abovementioned doses or only sand-soil. *Oxalis* waste was totally incorporated into the sand–soil mix. Next, pots were gently watered with tap water and left to decompose for four weeks to avoid phytotoxic effects [13]. Two seedlings of wild rocket were transplanted to each pot on 9 March 2023, thinned to one after six days and then allowed to grow for twenty-one days. Commercial fertilizer was added on sand–soil surface to F treatment at transplant. Pots were watered with tap water as needed. On 5 April 2023, all pots from all treatments were watered to 100% soil water capacity. Subsequently, well-watered pots continued to receive irrigation to 100% soil water capacity, while water deficit pots received only irrigation to 50% soil water capacity. Soil water content was assessed by weighting pots twice per week until the end of experiment, on 19 April 2023.

Maximum plant height was registered weekly from transplant to the end of the experiment and used to calculate the plant height increment as the difference between the last and the first records before and after different water irrigation regimes. At the end of experiment, half of the plants (n = 7) from each treatment were randomly selected and cut at aboveground level to determine fresh and dry (60 °C until constant weight) weights. The other half of plants per treatment were used to assess photosynthesis, leaf relative water content (RWC), pigment content, total soluble sugars, starch, total phenols, flavonoids, orthodiphenols, and antioxidant activity. Photosynthesis, RWC, and leaf nutrients were determined on fresh leaves. The remaining parameters were determined after collecting leaf samples, which were immediately frozen in liquid nitrogen and kept at −80 °C.

4.3. Determination of Wild Rocket Water Status, Photosynthesis, and Pigment Levels

Fresh leaves were weighted and, after incubation of the leaves in water for 48 h at 4 °C, the turgid weight was also determined. Leaf dry weight was determined after 10 days at 80 °C. Leaf relative water content was calculated as follows: (fresh weight − dry weight)/(turgid weight − dry weight) × 100.

Photosynthesis, gas exchange, and chlorophyll *a* fluorescence were determined in situ with a portable photosynthesis system LI-6400XT (LI-COR, Lincoln, NE, USA) coupled with a leaf chamber 6400-40 LCF with a light source (LED-based fluorescence and light source accessory for the LI-6400). Net CO_2 assimilation rate (*Pn*), stomatal conductance (g_s), transpiration rate (*E*), and the ratio of intercellular to extracellular CO_2 concentration (C_i/C_a) were determined between 9 a.m. and 11 a.m., under 400 µmol mol^{-1} of CO_2, light intensity of 300 µmol (photon) m^{-2} s^{-1}, and temperature of ~24–25 °C. Water use efficiency (WUE) was calculated as WUE = Pn/E. Simultaneously, chlorophyll *a* fluorescence was also measured. Under light conditions, the steady-state fluorescence was averaged at 2.5 s, and the maximum fluorescence was established after the application of a saturating light pulse (0.8 s). Next, leaves were shaded and the F0′ was determined. Leaves were adapted to dark conditions for at least 40 min. A light saturating pulse (>5000 µmol photons m^{-2} s^{-1} for 0.8 s) was applied and the maximum quantum efficiency of photosystem II ($F_v/F_m = (F_m - F_0)/F_m$) was determined. Additionally, the effective efficiency of photosystem II [$\Phi_{PSII} = (F_m{'} - F_s)/F_m{'}$], efficiency of excitation energy capture by open PSII reaction centers [$F_v{'}/F_m{'} = (F_m{'} - F_0)/F_m{'}$], photochemical quenching [$qP = (F_m{'} - F_s)/(F_m{'} - F_0{'})$], and non-photochemical quenching [$NPQ = (F_m - F_m{'})/F_m{'}$) were also calculated.

Chlorophyll *a* and *b* and carotenoid levels were determined according to Sims and Gamon [38]. Frozen leaves (approximately 50 mg) were ground with cold acetone/50 mM Tris pH 7.8 (80:20, *v*/*v*) and after centrifugation for 5 min at 5000× *g*, the absorbance of the supernatant was read at 470, 537, 647, and 663 nm with a spectrophotometer (Jenway®, model 7305, Cole-Parmer Ltd., Stone, UK). The pigment content was expressed as mg g dry weight $(DW)^{-1}$.

4.4. Wild Rocket Carbohydrate Contents

Total soluble sugars were determined according to Irigoyen et al. [39]. Frozen leaf samples (approximately 80 mg) were ground with 5 mL of ethanol at 80% (*v*/*v*). After incubation for 1 h at 80 °C, an aliquot (10 µL) was homogenized with 250 µL of anthrone + water + H_2SO_4 and incubated at 100 °C for 10 min. After 10 min, the homogenates were cooled on ice and centrifuged for 10 min at 5000× *g*. The absorbance of the mixture was read at 625 nm in a HEALES MB-580 microplate reader (Shenzhen Huisong Technology Development Co., Ltd, Shenzhen, China). The content of total soluble sugars was determined using a calibration curve of glucose (y = 12.527x + 0.1174, r^2 = 0.986) and expressed as mg g dry weight $(DW)^{-1}$.

Starch content was determined according to Osaki et al. [40]. Frozen leaf samples (approximately 80 mg) were ground with 3 mL of perchloric acid (30%, *v*/*v*). After 1 h at 60 °C, an aliquot (10 µL) was homogenized with 250 µL of anthrone + water + H_2SO_4. After 10 min at 100 °C, the homogenate was cooled on ice and centrifuged for 10 min at 5000× *g*. The absorbance of the supernatant was read at 625 nm in a HEALES MB-580 microplate reader. The starch content was determined using a calibration curve of glucose (y = 15.237x + 0.1582, r^2 = 0.989) and expressed as mg g dry weight $(DW)^{-1}$.

4.5. Antioxidant Contents in Wild Rocket Leaves

Approximately 100 mg of frozen leaves were ground with 1.5 mL of methanol. After incubation for 30 min at 40 °C, the homogenate was centrifuged for 15 min at 5000× *g*. The supernatant (leaf methanolic extract) was collected and used for analysis of total antioxidant activity, total phenols, total flavonoids, and orthodiphenols. Total antioxidant capacity was determined using the $ABTS^{+\bullet}$ free cation radical scavenging activity method, according to Re et al. [41]. An aliquot of the leaf methanolic extract (10 μL) was homogenized with 400 μL of ABTS solution [2,20-azinobis(3-ethylbenzothiazoline-6-sulphonic acid)] and incubated for 10 min at 30 °C. The absorbance was read at 734 nm in a Thermo Scientific Multiskan FC spectrophotometer and the total antioxidant capacity was calculated using a calibration curve of gallic acid ($y = 0.0011x + 0.00003$, $r^2 = 0.994$) and expressed as mg g dry weight $(DW)^{-1}$. Total phenols were determined according to López-Orenes et al. [42]. An aliquot of the leaf methanolic extract (20 μL) was homogenized with 405 μL of Folin–Ciocalteu and 75 μL Na_2CO_3 (20%) and after incubation for 30 min at 37 °C, the absorbance of the mixture was measured at 765 nm in a Thermo Scientific Multiskan FC spectrophotometer (Waltham, MA, USA). Total phenols were calculated using a gallic acid calibration curve ($y = 102.91x + 0.049$, $r^2 = 0.984$) and expressed as mg g dry weight $(DW)^{-1}$. Total flavonoids were assessed based on López-Orenes et al. [42]. An aliquot of the leaf methanolic extract (37.5 μL) was homogenized with 37.5 μL of methanol. In total, 75 μL of $NaNO_2$ (5%) was added and after mixing, 75 μL of $AlCl_3$ (10%) was added. The mixture was agitated and after 6 min in dark, 125 μL of NaOH (1M) was added. The absorbance of the mixture was read at 510 nm in a Thermo Scientific Multiskan FC spectrophotometer. Total flavonoid content was calculated using a rutin calibration curve ($y = 0.4186x + 0.1447$, $r^2 = 0.998$) and expressed as mg g dry weight $(DW)^{-1}$. Orthodiphenols were determined according to the molybdate assay [43]. An aliquot of the leaf methanolic extract (80 μL) was mixed with 40 μL of sodium molybdate (5% w/v in 50% methanol) and 80 μL of methanol. After incubation for 15 min at 20 °C, the absorbance of the mixture was read at 370 nm in a Thermo Scientific Multiskan FC spectrophotometer. Orthodiphenol content was calculated using a gallic acid calibration curve ($y = 3203.9x + 0.0396$, $r^2 = 0.999$) and expressed as mg g dry weight $(DW)^{-1}$.

4.6. Evaluation of Soil Nutrients during Oxalis waste Decomposition

Additionally, we conducted a pot experiment to determine the evolution of the main nutrients (nitrogen, N; carbon, C; and phosphorous, P) in soil during *Oxalis* waste decomposition without plant growth interference. On 9 February, 600 mL pots (10 cm in diameter) were filled with sand–soil mixed with *Oxalis* waste at a dose of 4.3 kg m^{-2} (O) or only sand–soil (C), with six replicates per treatment. Pots were left to decompose for 10 weeks (until 19 April 2023) and received full water as needed. Soil samples were collected at four, eight, and ten weeks after *Oxalis* waste incorporation, air-dried at room temperature, sieved through a 1 mm sieve and stored until soil analysis. Soil samples from the same treatment were pooled in pairs and an aliquot of each pair was used to estimate N, C, and P. Total soil N and C were estimated by dry combustion of ~100 mg soil with a CN 802 Elemental Analyzer (VELP, Scientifica, Usmate Velate, Italy; protein factor: none, O_2 flow rate: 300 mL min^{-1}, O_2 factor: 0.7 mL mg^{-1}). N and C results were obtained using standard aspartic acid and $CaCO_3$ curves, respectively. We determined extractable soil PO_4–P by extracting 2 g of soil with 40 mL of 2% acetic acid for 1 h of shaking, and then filtered it through a 0.45 μm filter, analyzed it for PO_4–P following the molybdenum blue method [44], and read the absorbance of extracts at 700 nm. PO_4–P results were obtained using a standard H_2PO_4 curve.

4.7. Statistical Analyses

Statistical analyses were conducted in R v4.3.3 [45]. In previous statistical analyses, we checked for normality of each dependent variable using the Shapiro–Wilk test ('stats' package). Next, we ran two-way general linear models (LMs, lm function, 'stats' package) for normal variables or generalized linear models (GLMs, glm function, 'stats' package) with Gaussian, Gamma or Inverse Gaussian error (link: log, identity or $1/mu^2$, respectively) in case of non-normal variables, to test for the effects of *Oxalis* waste dose (C, O2.2, O4.3, F), water irrigation (100 or 50% of field capacity), and the interaction between these two factors on fresh and dry aboveground biomass, photosynthetic parameters, RWC, pigments content, total soluble sugars, starch, total phenols, flavonoids, orthodiphenols, and antioxidant activity. We conducted a one-way GLM to explore the effect of *Oxalis* waste dose on height increment before different water regime with initial plant height as a covariable. Similarly, we ran a two-way GLM to evaluate the effect of *Oxalis* waste dose, water irrigation, and *Oxalis* waste dose × water irrigation interaction on height increment after different water regime with plant height before water irrigation treatment as a covariable. To test for the effect of *Oxalis* waste decomposition on the contents of total N and C and P over 10 weeks, we ran a two-way repeated measured model (lme function, 'nlme' package) with *Oxalis* waste (control and *Oxalis*), sampling date (week 4, week 8, and week 10), and the interaction between them as fixed factors and pot as random factor. The normality and homoscedasticity of residuals were checked after each model using the Shapiro–Wilk test and 'DHARMa' package. Post hoc mean comparisons were tested pairwise using the emmeans function ('emmeans' package) by comparing the emmeans obtained in each model. Additionally, a principal component analysis (PCA) was conducted using the program "CANOCO for Windows" v4.02 [46]. The level of significance was set at $p \leq 0.05$.

5. Conclusions

In conclusion, the results of this work show that the application of *Oxalis* waste in soil improves wild rocket's physiological performance according to the dose. *Oxalis* doses seem to act in several processes, such as photosynthesis, carbohydrate metabolism, and protective mechanism (antioxidants, NPQ, and carotenoids), modulating plant responses differently, but helping to respond similarly to the commercial organic fertilizer used. The obtained data support the use of this waste as a fertilizer under different irrigation levels, conferring some tolerance to water deficit stress. In turn, *Oxalis* application in soil seems not to influence plants' relative water contents directly. The use of aboveground *Oxalis* biomass mowed during control operations as fertilizer can ultimately reboot the management of this broadly spread invasive species. Further studies are needed to perform a cost–benefit analysis of the process in order to reduce control costs.

Author Contributions: Conceptualization, P.L. and M.C.D.; methodology, P.L., C.G. and M.C.D.; validation, P.L. and M.C.D.; formal analysis, P.L.; investigation, P.L. and M.C.D.; resources, P.L., C.G. and M.C.D.; data curation, P.L. and M.C.D.; writing—original draft preparation, P.L. and M.C.D.; writing—review and editing, P.L., C.G. and M.C.D.; supervision, P.L. and M.C.D.; funding acquisition, P.L. and M.C.D. All authors have read and agreed to the published version of the manuscript.

Funding: This work was supported by FCT—Fundacão para a Ciência e Tecnologia, I.P.—in the framework of the project UIDP/04004/2020 and with DOI identifier 10.54499/UIDP/04004/2020 (https://doi.org/10.54499/UIDP/04004/2020). P.L. and M.C.D. were supported by the Portuguese FCT (P.L.: SFRH/BPD/88504/2012, IT057-18-7248; M.C.D.: SFRH/BPD/100865/2014).

Data Availability Statement: The data that support the findings of this study are available from the corresponding author, P.L., upon request.

Acknowledgments: The authors thank Matilde Lamy, Sílvia Costa, and Francisco Farias, for the field, laboratory and technical support.

Conflicts of Interest: The authors declare no conflicts of interest.

References

1. Le Maitre, D.C.; Gaertner, M.; Marchante, E.; Ens, E.J.; Holmes, P.M.; Pauchard, A.; O'Farrell, P.J.; Rogers, A.M.; Blanchard, R.; Blignaut, J.; et al. Impacts of invasive Australian acacias: Implications for management and restoration. *Diver. Distrib.* **2011**, *17*, 1015–1029. [CrossRef]
2. Pyšek, P.; Jarošik, V.; Hulme, P.E.; Pergl, J.; Hejda, M.; Schaffner, U.; Vila, M. A global assessment of invasive plant impacts on resident species, communities and ecosystems: The interaction of impact measures, invading species' traits and environment. *Glob. Chang. Biol.* **2012**, *18*, 1725–1737. [CrossRef]
3. Pyšek, P.; Hulme, P.E.; Simberloff, D.; Bacher, S.; Blackburn, T.M.; Carlton, J.T.; Dawson, W.; Essl, F.; Foxcroft, L.C.; Genovesi, P.; et al. Scientists' warning on invasive alien species. *Biol. Rev.* **2020**, *95*, 1511–1534. [CrossRef]
4. Souza-Alonso, P.; Rodriguez, J.; Gonzalez, L.; Lorenzo, P. Here to stay. Recent advances and perspectives about *Acacia* invasion in Mediterranean areas. *Ann. For. Sci.* **2017**, *74*, 55–75. [CrossRef]
5. Lorenzo, P.; Morais, M.C. Strategies for the management of aggressive invasive plant species. *Plants* **2023**, *12*, 2482. [CrossRef]
6. Ulm, F.; Estorninho, M.; de Jesus, J.G.; de Sousa Prado, M.G.; Cruz, C.; Máguas, C. From a lose–lose to a win–win situation: User-friendly biomass models for *Acacia longifolia* to aid research, management and valorisation. *Plants* **2022**, *11*, 2865. [CrossRef]
7. Vilà, M.; Bartomeus, I.; Gimeno, I.; Traveset, A.; Moragues, E.V.A. Demography of the invasive geophyte *Oxalis pes-caprae* across a Mediterranean island. *Ann. Bot.* **2006**, *97*, 1055–1062. [CrossRef]
8. Costa, J.; Ferrero, V.; Castro, M.; Jorge, A.; Afonso, A.; Loureiro, J.; Castro, S. Pollen flow between flowers of the same morph in invasive populations of *Oxalis pes-caprae* L. in the western Mediterranean region. *Plant Biosyst. Int. J. Deal. Asp. Plant Biol.* **2016**, *150*, 923–931. [CrossRef]
9. Ferrero, V.; Navarro, L.; Castro, S.; Loureiro, J.; Sánchez, J.M.; Carvallo, G.O.; Barrett, S.C. Global patterns of reproductive and cytotype diversity in an invasive clonal plant. *Biol. Invasions* **2020**, *22*, 1691–1703. [CrossRef]
10. Pedraja, O.S. *Oxalis* L. In *Flora Iberica IX*; Castroviejo, S., Muñoz Garmendia, F., Navarro, C., Quintanar, A., Buira, A., Eds.; Real Jardín Botánico, CSIC: Madrid, Spain, 2015; Volume 9, p. 564.
11. Castro, S.; Ferrero, V.; Costa, J.; Sousa, A.J.; Castro, M.; Navarro, L.; Loureiro, J. Reproductive strategy of the invasive *Oxalis pes-caprae*: Distribution patterns of floral morphs, ploidy levels and sexual reproduction. *Biol. Invasions* **2013**, *15*, 1863–1875. [CrossRef]
12. Ferrero, V.; Barrett, S.C.H.; Castro, S.; Caldeirinha, P.; Navarro, L.; Loureiro, J.; Rodríguez-Echeverría, S. Invasion genetics of the Bermuda buttercup (*Oxalis pes-caprae*): Complex intercontinental patterns of genetic diversity, polyploidy and heterostyly characterize both native and introduced populations. *Mol. Ecol.* **2015**, *24*, 1927–1941. [CrossRef]
13. Lorenzo, P.; González, L.; Ferrero, V. Effect of plant origin and phenological stage on the allelopathic activity of the invasive species *Oxalis pes-caprae*. *Am. J. Bot.* **2021**, *108*, 971–979. [CrossRef]
14. Gallardo, B. Europe's top 10 invasive species: Relative importance of climatic, habitat and socio-economic factors. *Ethol. Ecol. Evol.* **2014**, *26*, 130–151. [CrossRef]
15. Gimeno, I.; Vila, M.; Hulme, P.E. Are islands more susceptible to plant invasion than continents? A test using *Oxalis pes-caprae* L. in the western Mediterranean. *J. Biogeogr.* **2006**, *33*, 1559–1565. [CrossRef]
16. Lazzaro, L.; Ferretti, G.; Bianchi, E.; Benesperi, R. Treatment by glyphosate-based herbicide allowed recovering native species after *Oxalis pes-caprae* L. invasion: Indications from a Mediterranean island. *Plant Biosyst. Int. J. Deal. Asp. Plant Biol.* **2019**, *153*, 651–659. [CrossRef]
17. BioDiversity4All. Available online: https://www.biodiversity4all.org/taxa/53169-Oxalis-pes-caprae (accessed on 20 March 2024).
18. Tavares, D.; Loureiro, J.; Martins, A.; Castro, M.; Roiloa, S.; Castro, S. Genetically based phenotypic differentiation between native and introduced tetraploids of *Oxalis pes-caprae*. *Biol. Invasions* **2019**, *21*, 229–243. [CrossRef]
19. Pierce, J.R. The biology of Australian weeds 31 *Oxalis pes-caprae* L. *Plant Prot. Q.* **1997**, *12*, 110–119.
20. Torretta, V.; Katsoyiannis, I.A.; Viotti, P.; Rada, E.C. Critical review of the effects of glyphosate exposure to the environment and humans through the food supply chain. *Sustainability* **2018**, *10*, 950. [CrossRef]
21. Muñoz, M.; Torres-Pagán, N.; Jouini, A.; Araniti, F.; Sánchez-Moreiras, A.M.; Verdeguer, M. Control of problematic weeds in Mediterranean vineyards with the bioherbicide pelargonic acid. *Agronomy* **2022**, *12*, 2476. [CrossRef]
22. Kluge, R.L.; Claassens, M. *Klugeana philoxalis* Geertsema (Noctuidae: Cuculliinae), the first potential biological control agent for the weed *Oxalis pes-caprae* L. *J. Entomol. Soc. South. Afr.* **1990**, *53*, 191–198.
23. Hulme, P.E. Invasions, islands and impacts: A Mediterranean perspective. In *Island Ecology*; Fernández-Palacios, J.M., Morici, C., Eds.; Asociación Española de Ecología Terrestre: Madrid, Spain, 2024; pp. 359–383.
24. Herbert, E.W.; Dittmer, K.E. Acute and chronic oxalate toxicity in Miniature Horses associated with soursob (*Oxalis pes-caprae*) ingestion. *Equine Vet. Educ.* **2017**, *29*, 549–557. [CrossRef]
25. Volakakis, N.; Kabourakis, E.M.; Rempelos, L.; Kiritsakis, A.; Leifert, C. Effect of different cover crops on suppression of the weed *Oxalis pes-caprae* L., soil nutrient availability, and the performance of table olive trees 'Kalamon' cv. in Crete, Greece. *Agronomy* **2022**, *12*, 2523. [CrossRef]
26. MITECO 2013. Catálogo Español de Especies Exóticas Invasoras. *Oxalis pes-caprae* L. OXAPES/EEI/FL06X. Available online: https://www.miteco.gob.es/content/dam/miteco/es/biodiversidad/temas/conservacion-de-especies/Oxalis_pes-caprae_2013_tcm30-69848.pdf (accessed on 20 March 2024).

27. Sala, A.; Verdaguer, D.; Vilà, M. Sensitivity of the invasive geophyte *Oxalis pes-caprae* to nutrient availability and competition. *Ann. Bot.* **2007**, *99*, 637–645. [CrossRef] [PubMed]
28. Cannon, J.P.; Allen, E.B.; Allen, M.F.; Dudley, L.M.; Jurinak, J.J. The effects of oxalates produced by *Salsola tragus* on the phosphorous nutrition of *Stipa pulchra*. *Oecologia* **1995**, *102*, 265–272. [CrossRef] [PubMed]
29. Xue, J.; San Zhou, S.; Wang, W.; Huo, L.; Zhang, L.; Fang, X.; Zhihong Yang, Z. Water availability effects on plant growth, seed yield, seed quality in *Cassia obtusifolia* L., a medicinal plant. *Agric. Water Manag.* **2018**, *195*, 104–113. [CrossRef]
30. Dias, M.C.; Azevedo, C.; Costa, M.; Pinto, G.; Santos, C. *Melia azedarach* plants show tolerance properties to water shortage treatment: An ecophysiological study. *Plant Physiol. Biochem.* **2014**, *75*, 123–127. [CrossRef]
31. Carstensen, A.; Herdean, A.; Schmidt, S.B.; Sharma, A.; Spetea, C.; Pribil, M.; Husted, S. The Impacts of phosphorus deficiency on the photosynthetic electron transport chain. *Plant Physiol.* **2018**, *177*, 271–284. [CrossRef]
32. Tariq, A.; Pan, K.; Olatunji, O.A.; Graciano, C.; Li, Z.; Sun, F.; Sun, X.; Song, D.; Chen, W.; Zhang, A.; et al. Phosphorous application improves drought tolerance of *Phoebe zhennan*. *Front. Plant Sci.* **2017**, *8*, 663–676. [CrossRef]
33. Attarzadeh, M.; Balouchi, H.; Rajaie, M.; Movahhedi Dehnavi, M.; Salehi, A. Growth and nutrient content of *Echinacea purpurea* as affected by the combination of phosphorus with arbuscular mycorrhizal fungus and pseudomonas florescent bacterium under different irrigation regimes. *J. Environ. Manag.* **2019**, *231*, 182–188. [CrossRef]
34. Khan, F.; Siddique, A.B.; Shabala, S.; Zhou, M.; Zhao, C. Phosphorus plays key roles in regulating plants' physiological responses to abiotic stresses. *Plants* **2023**, *12*, 2861. [CrossRef]
35. Kayoumu, M.; Iqbal, A.; Muhammad, N.; Li, X.; Li, L.; Wang, X.; Gui, H.; Qi, Q.; Ruan, S.; Guo, R.; et al. Phosphorus availability affects the photosynthesis and antioxidant system of contrasting low-P-tolerant cotton genotypes. *Antioxidants* **2023**, *12*, 466. [CrossRef] [PubMed]
36. Ortega, Z.; Romero, F.; Paz, R.; Suárez, L.; Benítez, A.N.; Marrero, M.D. Valorization of invasive plants from Macaronesia as filler materials in the production of natural fiber composites by rotational molding. *Polymers* **2021**, *13*, 2220. [CrossRef]
37. Lorenzo, P.; Guilherme, R.; Barbosa, S.; Ferreira, A.J.; Galhano, C. Agri-food waste as a method for weed control and soil amendment in crops. *Agronomy* **2022**, *12*, 1184. [CrossRef]
38. Sims, D.; Gamon, J. Relationships between leaf pigment content and spectral reflectance across a wide range of species, leaf structures and developmental stages. *Remote Sens. Environ.* **2002**, *81*, 337–354. [CrossRef]
39. Irigoyen, J.J.; Emerich, D.W.; Sanchezdiaz, M. Water stress induced changes in concentrations of proline and total soluble sugars in nodulated alfalfa (*Medicago sativa*) plants. *Physiol. Plant.* **1992**, *84*, 55–60. [CrossRef]
40. Osaki, M.; Shinano, T.; Tadano, T. Redistribution of carbon and nitrogen compounds from the shoot to the harvesting organs during maturation in field crops. *Soil Sci. Plant Nutr.* **1991**, *37*, 117–128. [CrossRef]
41. Re, R.; Pellegrini, N.; Proteggente, A.; Pannala, A.; Yang, M.; Rice-Evans, C. Antioxidant activity applying an improved ABTS radical cation decolorization assay. *Free Radic. Biol. Med.* **1999**, *26*, 1231–1237. [CrossRef]
42. López-Orenes, A.; Dias, M.C.; Ferrer, M.A.; Calderón, A.; Moutinho-Pereira, J.; Correia, C.; Santos, C. Different mechanisms of the metalliferous *Zygophyllum fabago* shoots and roots to cope with Pb toxicity. *Environ. Sci. Pollut. Res.* **2018**, *25*, 1319–1330. [CrossRef]
43. Giertych, M.J.; Karolewski, P.; de Temmerman, L.O. Foliage age and pollution alter content of phenolic compounds and chemical elements in *Pinus nigra needles*. *Water Air Soil Pol.* **1999**, *110*, 363–377. [CrossRef]
44. Allen, S.E.; Grimshaw, H.M.; Rowland, A.P. Chemical Analysis. In *Methods of Plant Ecology*; Moore, P.D., Chapman, S.B., Eds.; Blackwell: Oxford, UK, 1986; pp. 285–344.
45. R Development Core Team. *R: A Language and Environment for Statistical Computing*; R Foundation for Statistical Computing: Vienna, Austria, 2015.
46. ter Braak, C.J.F.; Smilauer, P. *CANOCO Reference Manual and User's Guide to Canoco for Windows—Software for Canonical Community Ordination (Version 4)*; Microcomputer Power: Ithaca, NY, USA, 1998.

plants

MDPI

Article

Emerging Invasive Weeds in Iran: Occurrence, Ecological Impacts, and Sustainable Management

Ali Reza Yousefi [1], Sirwan Babaei [2,3,*], Iraj Nosratti [4], Ehsan Zeidali [5], Masoumeh Babaei [6], Ebrahim Asadi Oskouei [7], Hesan Saberi [1], Mandeep Redhu [3] and Amir Sadeghpour [3,*]

1 Department of Plant Production and Genetics, University of Zanjan, Zanjan 45371-38791, Iran
2 Department of Plant Production and Genetics, University of Kurdistan, Sanandaj 66177-15175, Iran
3 Department of Plant, Soil and Agricultural System, Southern Illinois University, Carbondale, IL 62901, USA
4 Department of Plant Production & Genetics, The Campus of Agriculture and Natural Resources, Razi University, Kermanshah 67144-14971, Iran
5 Department of Agronomy and Plant Breeding, Faculty of Agriculture, Ilam University, Ilam 77111-69391, Iran
6 Department of Biological Sciences, University of Kurdistan, Sanandaj 66177-15175, Iran
7 Research Institute of Meteorology and Atmospheric Science, Tehran 13611-14977, Iran
* Correspondence: sirwan.babaei@siu.edu or s.babaei@uok.ac.ir (S.B.); amir.sadeghpour@siu.edu (A.S.)

Abstract: Invasive weeds pose a growing threat to global biodiversity, ecosystem stability, and agricultural productivity with significant ecological and economic consequences. In Iran, the rapid spread of invasive species such as *Boreava orientalis*, *Azolla* spp., *Ibicella lutea*, *Physalis divaricata*, *Picnomon acarna*, *Cynanchum acutum*, *Vicia hyrcanica*, *Eichhornia crassipes*, and *Ambrosia psilostachya* has severely affected native ecosystems, disrupted ecological processes, and threatened food security. These species exhibit aggressive traits such as rapid maturity, high reproductive rates, seed dormancy, and allelopathy that enable them to outcompete native species and successfully invade and dominate delicate habitats. Despite their documented impacts, a critical gap remains in understanding their biology, ecology, and management, particularly in understudied regions like Iran. This review synthesizes current knowledge on major invasive weeds affecting Iranian agroecosystems, with a focus on their ecological impacts and the urgent need for sustainable management strategies. It presents an integrated framework that combines ecological, biological, and management perspectives to address invasiveness, particularly in highly adaptable species like *B. orientalis* and *A. psilostachya*. This review highlights the critical role of interdisciplinary collaboration, advanced technology, and community involvement in developing effective strategies. It offers practical guidance for researchers, policymakers, and agricultural stakeholders, serving as a model for managing invasive species in other vulnerable regions. Ultimately, it supports global efforts to safeguard biodiversity, improve crop productivity, and strengthen ecological resilience against the growing threat of invasive species.

Keywords: invasive weeds; biological invasions; global biodiversity; ecological impact; agricultural weeds; weed management

1. Introduction

Invasive plants and weeds pose serious ecological and economic challenges [1]; however, invasive species are particularly concerning due to their ability to spread rapidly. These species can establish in new regions, sometimes even without human intervention, often leading to substantial environmental, economic, and health impacts [2,3]. In recent years, their rapid and unregulated expansion has posed a critical threat to global ecosystem diversity, structure, and functionality [4,5]. Recent projections suggest that climate change, global transport, and socio-economic shifts will be key drivers of future alien plant invasions, intensifying their impact on native biodiversity even in the most suitable environments [6,7]. From 1980 to 2019, economic losses due to biological invasions were found to be similar to those from natural hazards, for instance, USD 1208.0 billion compared to USD 1913.6 billion from storms and USD 1139.4 billion from earthquakes [8].

Iran's unique geographical position and diverse topography give rise to a wide range of climatic conditions, from arid deserts to lush forests, making it a hotspot for biodiversity. This rich ecological variation supports over 8000 plant species. According to recent estimates, 311 alien plant species have been recorded in Iran, among which 13 are currently classified as invasive [9]. While some of these species are fully naturalized, others possess invasive traits and may pose future ecological risks. This underscores the urgent need for systematic monitoring and proactive management. Additionally, only about 200 to 250 plant species (less than 0.3%) are considered weeds in Iranian agroecosystems, reflecting both the selectivity of weed classification and the growing concern over emerging invasives with harmful ecological and economic impacts [10,11]. The ecosystems are increasingly threatened by the introduction and spread of invasive plant species, which can outcompete native flora and disrupt ecological balances [12]. The lack of stringent biosecurity measures has further exacerbated this issue, allowing invasive species to establish and spread across various regions. Considering Iran's distinct agricultural, environmental, and economic conditions, understanding the traits driving the success of invasive weeds is crucial. Therefore, this review article aims to fill existing knowledge gaps by identifying key traits that promote invasiveness in Iran's ecosystems, evaluating the ecological and economic impacts on biodiversity and agriculture, and exploring sustainable management strategies to mitigate their establishment and spread.

Rather than attempting to catalog all invasive alien plant species in Iran, this review focuses on the nine most problematic and emerging species identified through national surveys, peer-reviewed literature, and expert consultations, allowing for a more detailed analysis of their classification, origin, life forms, ecological and economic impacts, and management strategies. This review is organized thematically to reflect the ecological complexity and management relevance of emerging invasive weeds in Iran. We first provide a general overview of invasive weed threats and their introduction pathways. Subsequent sections explore the ecological, economic, and agricultural impacts of these species, followed by detailed profiles of specific species. Management challenges and current control strategies are then discussed, culminating in a section on sustainable and integrated solutions. This structure is designed to move from a broad ecological context to specific species and ultimately toward practical recommendations.

2. Concept and Criteria of Weed Invasiveness

Weed invasiveness refers to the ability of certain plant species to establish, spread, and persist in new environments, often resulting in significant ecological or economic consequences. Unlike general weediness, which typically denotes nuisance plants in crop production, invasiveness specifically involves aggressive colonization beyond a species' native range. Invasive plants characteristically possess traits such as rapid growth, high reproductive capacity, efficient dispersal, broad ecological tolerance, and competitive ability, often aided by reduced biotic pressure due to the absence of natural enemies like herbivores and pathogens. They often exhibit high adaptability and competitive ability, enabling them to establish, thrive, and outcompete native flora. This advantage is often linked to evolutionary trade-offs, where reduced herbivore pressure enables plants to invest less in defense mechanisms and more in growth and competition [13].

Most invasive plant species belong to a few families such as *Fabaceae*, *Asteraceae*, *Cyperaceae*, *Poaceae*, and *Mimosaceae*. For example, Mitchell and Power (2003) examined 473 European plant species that had naturalized in the United States; they found that these species faced 84% fewer fungal pathogens and 24% fewer viral pathogens in their introduced ranges compared to their native habitats, illustrating how reduced biotic stress can favor invasiveness [14]. Westbrooks (1998) identified 12 key traits commonly found in invasive weeds that enable them to thrive in new ecosystems, many of which overlap with the general attributes described above [9]. These include the rapid maturity, long-term viability, and dormancy of propagules, compatibility with dispersal vectors such as contaminated crop seed, allelopathy, physical deterrents like thorns, parasitism, mimicry of crop seeds, nutrient storage in vegetative tissues for stress tolerance, and high photosynthetic efficiency even in resource-limited environments [9].

Despite these insights, predicting invasiveness remains challenging due to an incomplete understanding of the ecological and evolutionary mechanisms underlying plant invasions [15]. While a species' behavior in its native range is often considered, it is not a reliable predictor of its invasive potential in new regions [16]. Thus, a consistent and predictive framework for assessing invasiveness risk is still lacking.

Invasive species disrupt biogeographic boundaries and negatively impact ecosystems by reducing native species richness and abundance, increasing extinction risks, altering the genetic structure of native flora, and affecting animal behavior and trophic interactions [4]. Their ecological consequences include a drastic decline in biodiversity, loss of endangered species and their habitats, and reduced availability of suitable habitats for native insects, birds, and wildlife. Invasive plants also diminish food resources for fauna and interfere with essential ecological processes such as plant succession and disturbance regimes, including the intensity and frequency of wildfires. Furthermore, they disrupt key animal–plant interactions, including pollination, seed dispersal, and host-specific relationships, ultimately threatening ecological balance and agricultural sustainability [17].

Additionally, invasive species can alter ecosystem functioning by impacting nutrient cycling, contaminant flow, hydrological patterns, and habitat structure, thereby shifting natural ecosystem dynamics [4]. Despite these impacts, predicting which species will become invasive, and under what environmental or management conditions, remains a major challenge in invasion ecology, particularly in rapidly changing agroecosystems.

In summary, the invasive weed species discussed in this review share a suite of biological and ecological characteristics that enhance their success in new environments. These include rapid growth, early maturity, high reproductive capacity (both sexual and vegetative), seed dormancy and persistence in soil seed banks, phenotypic plasticity, allelopathy, and efficient dispersal mechanisms. Many of these species also benefit from reduced biotic pressure in their introduced ranges, such as escaping natural enemies.

Taxonomically, they span a range of families known for invasiveness, particularly Asteraceae (e.g., *Ambrosia psilostachya*), Brassicaceae (e.g., *Boreava orientalis*), Solanaceae (e.g., *Physalis divaricata*), Fabaceae (e.g., *Vicia hyrcanica*), as well as aquatic groups like Pontederiaceae and Salviniaceae (e.g., *Eichhornia crassipes*, *Azolla* spp.), and Martyniaceae (*Ibicella lutea*). These families are globally recognized for producing species that are highly adaptable, competitive, and ecologically disruptive, particularly in disturbed or resource-rich habitats.

3. Detailed Species Profiles

To enable a targeted and comparative assessment of major invasive weed species in Iran, taxa have been categorized into two groups according to their dominant ecological habitats and principal impact domains (Table 1). This classification facilitates a habitat-specific evaluation of the invasion dynamics and associated ecological consequences.

Table 1. Major invasive weed species in Iran categorized by ecological habitat and origin status (alien or native invasive).

Ecological Habitats	Scientific Name	Family	Common Name	Synonym	Origin	Life Cycle	Weed Status in Iran
Agricultural field weeds	*Ambrosia psilostachya* DC.	*Asteraceae*	Cuman ragweed	*Ambrosia californica* Rydb.; *Ambrosia coronopifolia* Torr. & A. Gray; *Ambrosia cumanensis* auct. non Kunth; *Ambrosia rugelii* Rydb.	Mexico	Annual Perennial	Serious/Principal Weed
	Boreava orientalis Jaub. & Spach	*Brassicaceae*	Waxy leaved mustard or yellow weed	*Isatis quadrivalvis*	Western Asia	Annual	Serious/Principal Weed
	Cynanchum acutum L.	*Asclepiadaceae*	Swallowwort or climbing milkweed	*Solenostemma acutum* (L.) Wehmer; *Vincetoxicum acutum* (L.) Kuntze	Mediterranean region	Perennial	Common Weed
	Ibicella lutea (Lindl.) Van Eselt.	*Martyniaceae*	Devil's Claw	*Martynia lutea* Lindl.; *Proboscidea lutea* (Lindl.) Stapf.	Argentina and South America	Annual	Occasional/Emerging Weed
	Picnomon acarna (L.) Cass.	*Asteraceae*	Soldier thistle	*Cirsium acarna* (L.) Moench. *Carduus acarna* L.	Mediterranean region	Annual	Common Weed
	Physalis divaricata D. Don	*Solanaceae*	Annual Groundcherry	*Physalis halicacabum*	Latin America and South America	Annual	Common Weed
	Vicia hyrcanica Fisch. & C.A. Mey.	*Fabaceae*	Hyrcan vetch	*Hypechusa hyrcanica* (Fisch. & C.A.Mey.) Alef.	Mediterranean region	Annual	Occasional/Emerging Weed
Aquatic Invasive weeds	*Azolla filiculoides* Lam.	*Salviniaceae*	Pacific Mosquitofern	-	Africa, Asia, and America	Annual	Serious/Principal Weed
	Eichhornia crassipes (Mart.) Solms	*Pontederiaceae*	Water hyacinth	*Eichhornia speciosa* Kunth; *Piaropus crassipes* (Mart.) Raf.	Amazon Basin in South America	Perennial	Serious/Principal Weed

3.1. *Agricultural Field Invasive Weeds*

3.1.1. *Ambrosia psilostachya* DC

- Origin and global invasiveness: The genus *Ambrosia* includes approximately 40 species and numerous subspecies, many of which are known for their allergenic properties [7]. *A. psilostachya* DC. was first recorded in Europe in the 1800s [18] and has since expanded its distribution to include North America, Europe, Asia, and other parts of the Americas [19]. It is an herbaceous perennial plant native to Mexico, where it plays a role in local ecosystems. It disrupts native biodiversity and agricultural productivity across the United States, Canada, Australia, Russia, and parts of Europe. Its presence has also been documented in China, France, Germany, and South Africa, reflecting its widespread introduction across Europe, Asia, and Africa (Figure 1) [7]. In Iran, the species was initially identified in 1991 in Anzali, Guilan Province (Figure 1). Subsequent investigations by the Guilan Research Institute confirmed its establishment in the region [20]. Since its first detection, the species has spread to several surrounding areas, including Rezvanshahr, Talesh, Astara counties, and the Rasht Trench (Figure 1).
- Morphological description, biology, and ecology: *A. psilostachya* is an erect herbaceous perennial characterized by horizontal running rootstocks, allowing it to propagate both vegetatively via rhizomes and sexually through seed production. The stems are rigidly upright and may be unbranched or heavily branched in the upper parts of the plant. Leaves are thick, hairy, and oval-lanceolate in outline, typically reaching lengths of up to 5 inches and widths of 2 inches. They are deeply dissected into narrow lobes, often with secondary lobing. The leaf surface, covered in fine hairs,

tends to accumulate dust, which can hinder herbicide adhesion and reduce control efficacy [21,22]. The inflorescence is composed of unisexual flower heads, with male (staminate) and female flowers located on different parts of the same plant. Staminate flowers are small, bead-like, yellow to greenish, and short-stalked, initially densely clustered and later spreading during maturation. Pistillate (female) flowers are solitary and sparse, typically found at the base of the floral cluster, along the stem, or in axils, accompanied by large leaf-like bracts [23].

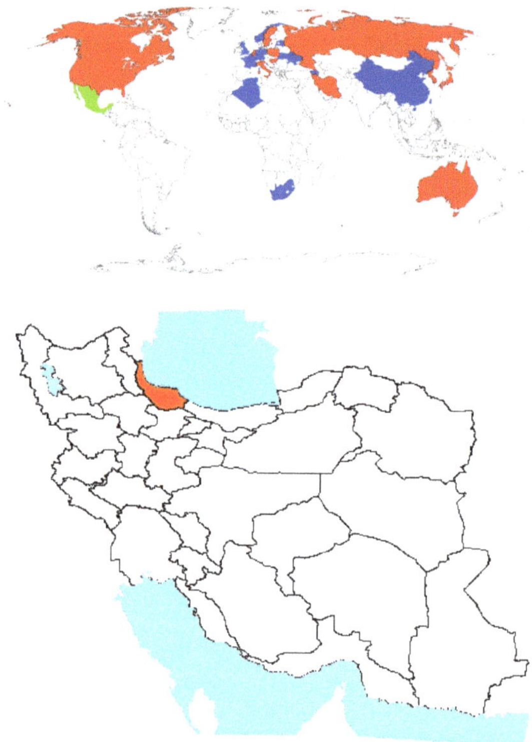

Figure 1. Global and national distribution of *Ambrosia psilostachya* DC. **Top panel**: Global distribution map showing native (green), invasive (red), and introduced (blue) ranges based on the literature and GBIF records (GBIF Backbone Taxonomy. Checklist dataset https://doi.org/10.15468/39omei accessed via GBIF.org on 20 August 2025). **Bottom panel**: Occurrence invasive (red) in Iran based on herbarium records, published studies, and recent field observations by the authors. Note: Absence of occurrence in some regions may reflect data gaps or lack of confirmed records, not necessarily ecological unsuitability.

A. psilostachya exhibits ecological plasticity, readily colonizing a wide range of habitats, including coastal areas, dunes, sandy soils [24], riversides, and disturbed ruderal environments [25]. Although detailed climatic data for its spread are limited [23], this species thrives well under macroclimatic conditions that are conducive to coastal zones, characterized by relatively narrow diurnal and seasonal temperature fluctuations, increased precipitation during the wettest months, and consistently warmer ambient temperatures [24].

- Agricultural impact and management strategies: *A. psilostachya* poses considerable ecological and agricultural challenges due to its allelopathic properties and prolific spread along roadsides and non-agricultural lands [26]. The species releases toxic compounds that inhibit the growth of surrounding vegetation, adversely impacting crops such as peanuts, sunflowers, corn, soybeans, and wheat [27]. In agricultural systems, high densities of *A. psilostachya* have been linked to notable yield reductions. Additionally, the species produces highly allergenic pollen, ranking as the second most significant allergen in the United States, affecting nearly 25% of the population [28]. The impact of pollen on public health has also been highlighted in earlier work [29]. In New Zealand, risk assessments have identified *A. psilostachya* as a high-risk invasive species, emphasizing the need for robust control strategies [30].

The management of *A. psilostachya* incorporates a combination of physical, chemical, and biological methods. Cutting and herbicide application are the most commonly used strategies in agroecosystems [31]. Chemical control has been explored using herbicides such as florasulam, prosulfuron, oxadiargyl, metribuzin, bifenox, and clopyralid; however, these have demonstrated susceptibility to dicamba and 2,4-D [32]. The use of herbicides with multiple modes of action is recommended to mitigate the development of resistance [33,34]. Previous studies revealed that *A. trifida* responds strongly to dicamba and 2,4-D, with saflufenacil causing substantial reductions in plant biomass within eight weeks of application [35]. Biological control using the beetle *Ophraella communa* (Coleoptera: *Chrysomelidae*) has shown promise against *A. artemisiifolia*, reducing plant biomass by defoliating before seed formation [36], although concerns remain regarding flower damage and reduced pollen production [37].

3.1.2. *Boreava orientalis*: Jaub. & Spach

- Origin and global invasiveness: It is native to parts of western Asia, including Iran, Pakistan, Syria, and Turkey (Figure 2). It is widespread from Kütahya to Istanbul in Turkey and from Armenia to the Euphrates [38]. In Iran, it was initially reported from Khorasan and Chaharmahal Bakhtiari (not as a weed), but it has been more recently identified in Kurdistan croplands [39]. According to the Manual of Alien Plants of Belgium records also document its presence in Belgium, Germany, France, the United Kingdom, and Pakistan [40].
- Morphological description, biology, and ecology: *B. orientalis* Jaub. & Spach is an annual glabrous herb that reaches 15–30 cm in height and typically branches above the middle of the plant into a loose, ebracteate inflorescence [41]. It exhibits a glaucous appearance with yellow flowers and simple leaves. Basal leaves are oblong-lanceolate, clasping, auriculate or sagittate, with entire margins and acute apices, while cauline leaves are also sagittate. The petals are approximately 5 mm long, and the calyx is open, not or scarcely saccate. Petals are spathulate to oblong–lanceolate, filaments are slightly dilated, and the ovary is 1–2-ovulate with a capitate stigma. The slender pedicels, 6–8 mm long in fruit, are erect or spreading. The fruit is indehiscent, one-seeded, beaked, ovate, and may be winged or wingless; it is 8–10 mm long, somewhat tubercled, and plicate between the wings. The radicle is incumbent, and the species has a chromosome number of 14n with small-sized chromosomes [41].

Specific climatic requirements for *B. orientalis* remain largely undocumented; however, this species is considered synonymous with *Isatis* sp., and thus likely shares similar ecological preferences. It requires vernalization for flowering, with crown and seedling rosettes exhibiting different responses to temperature variations [42]. The species appears to be adapted to environments with cold winters and warm to hot summers, consistent with habitats occupied by *Isatis* species across the Eurasian steppes, Morocco, and even Arizona. For example, *Isatis tinctoria* commonly coexists with *Artemisia tridentata*, a species capable of tolerating field temperatures ranging from −18 °C to 38 °C [43], with leaf tissue tolerances reaching up to 46 °C [44]. The dispersal of *B. orientalis* fruits typically occurs near the parent plant—within 50 cm—although greater distances can be achieved via insects, water, human and animal activity, or contaminated seed lots [45]. Seed viability in soil is relatively short-lived, with most seeds losing viability within 12 months, likely due to factors such as soil moisture content [42].

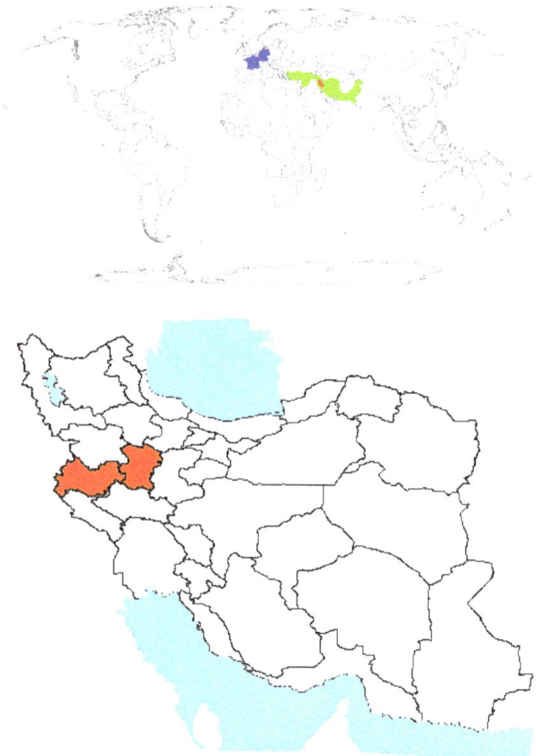

Figure 2. Global and national distribution of *Boreava orientalis* Jaub. & Spach. **Top panel**: Global distribution map showing native (green), invasive (red), and introduced (blue) ranges based on the literature and GBIF records. **Bottom panel:** Occurrence invasive (red) in Iran based on herbarium records, published studies, and recent field observations by the authors. Note: The absence of occurrence in some regions may reflect data gaps or lack of confirmed records and not necessarily ecological unsuitability.

- Agricultural impact and management challenges: *B. orientalis* is a winter annual weed that commonly infests irrigated and rain-fed fields of wheat, barley, chickpeas, and rapeseed [39,46]. It is rarely found in non-arable lands such as pastures and gardens, likely due to the absence of annual tillage. This species has also been reported in winter wheat fields on the Anatolian Plateau of Turkey [47]. Notably, *B. orientalis* exhibits allelopathic activity through the release of isothiocyanates derived from glucosinolate breakdown, which may contribute to its competitiveness and render it unpalatable or toxic to livestock [48].

Despite its growing presence in crop systems, there are currently no published studies detailing specific control measures for *B. orientalis*. While it has not yet caused widespread crop damage, field evidence indicates that commonly used herbicides, such as tribenuron, have spread to livestock [48], and 2,4-D are ineffective against this species in western Iran [49,50], highlighting the need for targeted management strategies.

3.1.3. *Cynanchum acutum* L.

- Origin and global invasiveness: The genus name "*Cynanchum*" is derived from the Greek meaning "dog strangler", likely referencing its twining growth habit, while "acutum" alludes to its sharply pointed leaves [51]. Its origin lies within the Mediterranean Basin, encompassing Southern Europe, North Africa, and parts of Western Asia (Figure 3) [52]. The perennial vine thrives in warm, arid climates and is frequently found in coastal regions, on sandy dunes, and in disturbed habitats [53]. In addition to its invasive potential, *Cynanchum* L. has been used in traditional medicinal practices in some cultures.

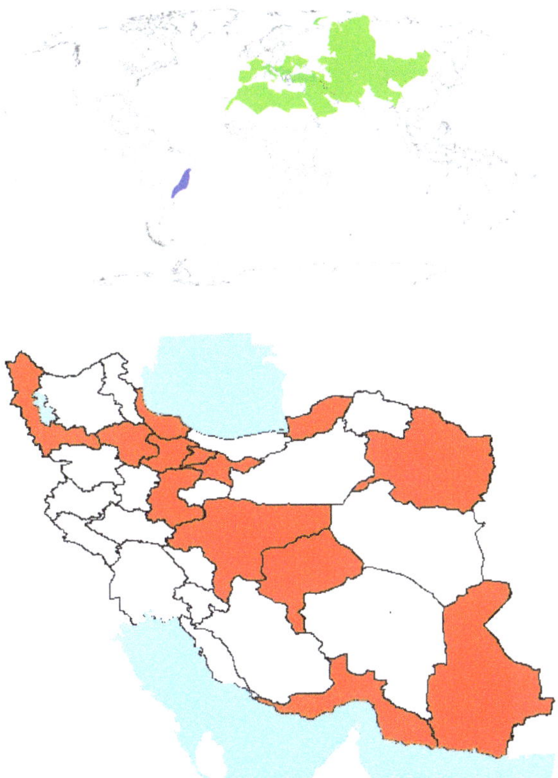

Figure 3. Global and national distribution of *Cynanchum acutum* L. **Top panel**: Global distribution map showing native (green), invasive (red), and introduced (blue) ranges based on the literature and GBIF records. **Bottom panel:** Occurrence invasive (red) in Iran based on herbarium records, published studies, and recent field observations by the authors. Note: The absence of occurrence in some regions may reflect data gaps or lack of confirmed records and not necessarily ecological unsuitability.

Native to Ukraine and Russia [54], *C. acutum* demonstrates a broad geographic range extending from southern Europe through southwest Asia and into North Africa (Figure 3) [55]. In Iran, the species is widespread across diverse climatic regions and is commonly found in the Iranoturani phytogeographical zone. It occurs extensively in the northern provinces (Gorgan and Guilan), as well as the northwestern, central, eastern, and southern provinces throughout the country (Figure 3). Additionally, the species has been introduced to southeastern Brazil, further illustrating its adaptive capacity to varied environmental conditions (Figure 3).

- Morphological description, biology, and ecology: *C. acutum* is a perennial species distinguished by either glabrous surfaces or multicellular hairs, exhibiting fibrous, fleshy, or woody roots with greenish stems. Its leaves are opposite, petiolate, and typically cordate, elliptical, ovate, or obovate, with entire margins and acute apices. Occasionally, small leafy stipules are present. The plant produces extra-axillary inflorescences that can be raceme-like, corymbose, or umbel-like. Flowers range in diameter from 3 to 15 mm, displaying imbricate or contorted aestivation with dextrorse orientation. They are nectariferous and have a divided corolla that is contorted in the bud stage, appearing rotate, sub-rotate, or tubular in shape. Flower coloration varies from white to green and yellow, and occasionally reddish. The calyx is free to the base, composed of erect sepals often containing basal glands. The seeds are oval, flat, and brown, well-adapted for wind dispersal, facilitating long-distance propagation [41,56].

C. acutum is commonly found across diverse habitats, including beaches, riverbeds, hedges, fields, gardens, roadsides, and relatively saline soils, showing a preference for moist environments [55]. It is highly adaptable to various soil types due to its phenotypic plasticity, which enables it to survive under a broad range of environmental conditions. Seed germination initiates at approximately 17 °C and reaches optimal levels at 25 °C. The species has a robust root system that penetrates at least 10 cm deep into the soil, complicating eradication efforts.

- Ecological impact and management strategies: *C. acutum* possesses a robust root system and an ascending, twining stem that wraps around tree trunks, extending into branches and obstructing light penetration to the canopy. This growth behavior damages tree buds and inhibits photosynthesis. It produces flowers and seeds on exposure to light [57]. Control methods include mechanical removal and herbicide application. This deep-rooting behavior enhances its ability to persist and regenerate, making it a challenging invasive species. Effective herbicides include glufosinate ammonium [58], paraquat, and glyphosate [56], while picloram and nicosulfuron have shown limited efficacy [59]. As such, the development of effective, long-term management strategies is essential for mitigating its ecological and agricultural impact.

3.1.4. *Ibicella lutea* (Lindl.) Van Eselt.

- Origin and global invasiveness: The family *Martyniaceae* comprises five genera and 16 species globally, with the greatest concentration in Argentina and other parts of South America (Figure 4) [60]. A distinguishing morphological trait of this family is the presence of fruits with a curved beak, which may be longer or shorter than the fruit body [61]. The genus *Ibicella*, part of *Martyniaceae*, includes eight species that are naturally distributed from South America to Mexico [62]. *I. lutea* (Lindl.) Van Eselt., commonly known as the yellow unicorn plant, has previously been misidentified in various studies. For instance, a study on Lesos Island in the eastern Aegean region misclassified a population of *I. lutea*, later clarified to be aligned with *Proboscidea louisianica* (Mill.) Thell. ssp. *louisianica* based on Bretting's interspecific classification [63,64]. While *I. lutea* has been recorded in neighboring countries such as Turkey, no members of the genus *Proboscidea* have yet been documented there [65].

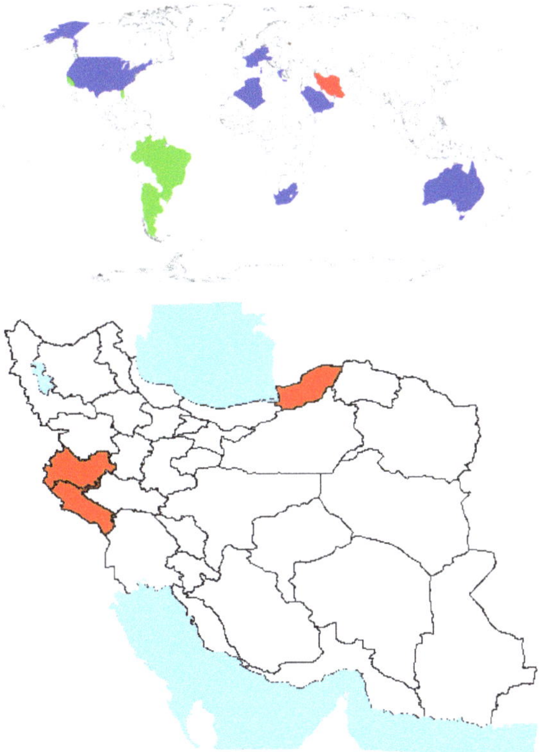

Figure 4. Global and national distribution of *Ibicella lutea* (Lindl.) Van Eselt. **Top panel**: Global distribution map showing native (green), invasive (red), and introduced (blue) ranges based on the literature and GBIF records. **Bottom panel:** Occurrence invasive (red) in Iran based on herbarium records, published studies, and recent field observations by the authors. Note: The absence of occurrence in some regions may reflect data gaps or lack of confirmed records and not necessarily ecological unsuitability.

Species within the *Martyniaceae* family demonstrate a strong ecological resilience, particularly in degraded habitats and arid to semi-arid regions. These plants are commonly found along roadsides, as well as in sandy deserts, meadows, beaches, pastures, and sand dunes, thriving up to elevations of 1800 m [63]. Native to the arid regions of Texas and Mexico, members of *Ibicella* have dispersed throughout South and North America, with introductions from the United States to other parts of the world [66]. The species was first reported in Greece, which also makes it the first record for the Eastern Mediterranean region (Figure 4) [64]. In Iran, *I. lutea* was first documented as a weed in cotton fields in the Gorgan and Gonbad regions, with its introduction speculated to have originated from Saudi Arabia [67]. Recent surveys conducted during 2018–2019 confirmed its presence in western Iran, particularly in the Ilam and Kermanshah Provinces. These included sites such as the landscape of Ilam University and a sugar beet field in Badreh Gard Salimi Village in Homeil, Eslamabad Gharb City (Figure 4). It exhibits a strong capacity to rapidly colonize disturbed habitats, particularly during warm seasons, due to its prolific seed production, efficient seed and fruit dispersal, and climatic adaptability. The absence of natural predators or competitors in newly invaded areas further facilitates its establishment. This ecological advantage contributes to its aggressive spread, posing serious challenges for summer crops and altering landscape dynamics [68].

- Morphological description, biology, and ecology: *I. lutea*, a member of the *Martyniaceae* family in the order Lamiales, is an annual plant with tuberous roots, native to the New World. The family includes both annual and perennial herbaceous plants, noted for their glandular hairs and strong fragrance [69]. *I. lutea* typically reaches heights of 20–120 cm and is characterized by hairy, nearly circular leaves and clusters of bright yellow flowers with red spots internally. The inflorescences bear long petioles that may either terminate or split, with broad, scaly bracts situated just below the flowers [64]. The flowers are bisexual, composed of five unequal sepals and a bilabiate corolla, with stamens arranged epipetally—two long, two short, and one vestigial. The superior ovary exhibits parietal placentation and terminates in a two-lobed stigma that often closes upon contact. The fruit, initially fleshy, matures into a woody capsule with a prominent beaked endocarp, culminating in two curved, horn-like projections longer than the fruit body itself [69]. The seeds are dark, flattened, and uneven in texture, while the fruit morphology varies from smooth to rough, resembling the spiny capsules of *Datura stramonium* L.

Ecologically, *Martyniaceae* plants demonstrate notable resilience, commonly found in arid and semi-arid environments. Their claw-shaped pods are well-adapted for zoochorous dispersal, attaching to the fur and hooves of large animals, which facilitates the spread of seeds across diverse habitats [70,71]. This adaptation plays a significant role in the invasive potential of *I. lutea*. Flowering typically begins by late October, and while immature fruits are present, it remains uncertain whether they can reach full maturity under agricultural conditions. The distinctive morphology of *I. lutea* fruit has also contribued to its anthropogenic dissemination, as humans have used the horn-shaped seed pods ornamentally, contributing to its unintentional spread [68]. Known commonly as "Devil's Claw" or "Yellow Unicorn Plant", the fruit's bifurcated, claw-like form gives the species its colloquial name and enhances its ecological and cultural significance.

- Agricultural impact and management strategies: Field observations and research indicate that *I. lutea* poses a growing threat to both agricultural and natural ecosystems, particularly in arid and semi-arid regions across the southern United States. Its large size and vigorous growth habit have demonstrated significant competitive ability against both pasture and row crops. Studies show that *I. lutea* can reduce cotton (*Gossypium hirsutum* L.) yields by 60 to 74% [72], with weekly interference resulting in incremental yield losses of up to 5% [73]. Furthermore, its competitive effect has been observed to extend as far as 0.5 m from neighboring plants by the season's end [72]. Although previously undocumented in Iran, recent reports confirm its invasion in the Ilam and Kermanshah Provinces, indicating a recent introduction [68].

Effective management of *I. lutea* begins with preventive strategies. Due to its invasive potential, it should be incorporated into regional floristic records. It is crucial to understand the mechanism underlying its rapid spread in *I. lutea* in the Ilam and Kermanshah Provinces during warm seasons. Immediate implementation of plant quarantine regulations and ongoing surveillance of farms and orchards are essential to contain its spread and limit its ecological and agricultural impact [68]. Moreover, this includes avoiding the use of unpacked or contaminated seed lots and refraining from applying fresh manure, which may harbor viable weed seeds. Physical exclusion methods, such as restricting livestock access to infested sites and the proper cleaning of tools, footwear, and machinery, are critical in minimizing seed dispersal. Enhancing field sanitation, such as by cementing irrigation channels and adopting pressurized irrigation systems, can also create conditions less favorable to weed establishment [74]. Mechanical control methods, including early-season cutting of plants before fruit formation, combined with crop rotation, can help suppress weed populations. Chemical control has also proven effective, with registered herbicides such as 2,4-D showing good efficacy [75]. Additionally, pre-plant herbicides like trifluralin have been successfully used in bromoxynil- and glyphosate-resistant cotton fields [76]. Eco-physiological studies are needed to assess germination response to local climatic factors, particularly its optimal germination temperature of 20 °C and its salinity tolerance threshold (~80 mmol). Field experiments must evaluate their competitive behavior in summer crops and assess landscape-level ecological impacts. These integrated strategies offer a multifaceted approach to managing *I. lutea* and safeguarding crop productivity.

3.1.5. *Picnomon acarna* (L.) Cass.

- Origin and global invasiveness: *P. acarna* (L.) Cass. is a spiny, annual, and herbaceous plant native to the Mediterranean region (Figure 5). Initially prevalent in neglected areas and rangelands, *P. acarna* has become a major invasive weed in rainfed agricultural fields, especially in western Iran [77]. It has also been reported in the western regions of Victoria and the southeastern parts of South Australia (Figure 5). The species' non-palatability and resilience to soil compaction and erosion have enabled its proliferation in Iranian forests and pastures (Figure 5) [78]. The increasing adoption of conservation tillage and reduced soil disturbance in rainfed systems [79] has further facilitated the spread of *P. acarna*.

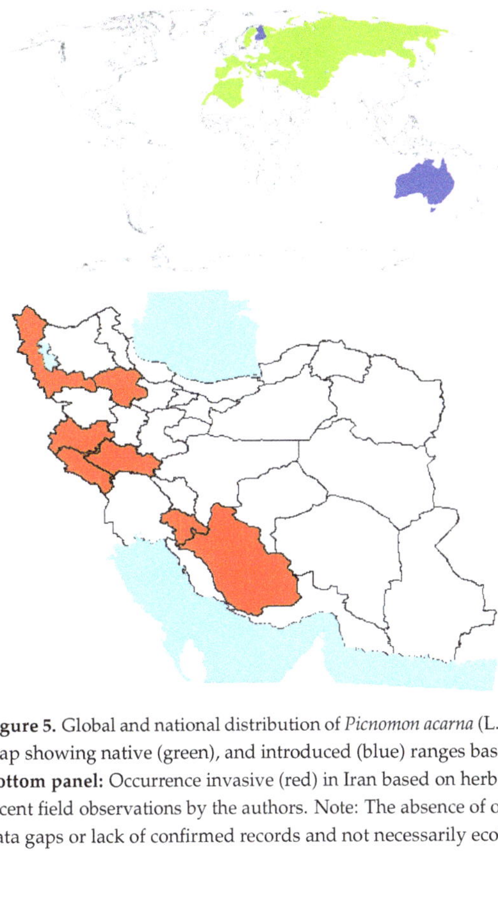

Figure 5. Global and national distribution of *Picnomon acarna* (L.) Cass. **Top panel**: Global distribution map showing native (green), and introduced (blue) ranges based on the literature and GBIF records. **Bottom panel:** Occurrence invasive (red) in Iran based on herbarium records, published studies, and recent field observations by the authors. Note: The absence of occurrence in some regions may reflect data gaps or lack of confirmed records and not necessarily ecological unsuitability.

- Morphological description, biology, and ecology: Belonging to the Asteraceae family, it exhibits typical morphological traits such as spiny, hairy leaves that can accumulate dust and reduce herbicide absorption [22]. A notable characteristic of this plant is its height, reaching approximately 50 cm, with leaves bearing spines that range from 10 to 15 cm in length [80]. The species produces seeds with a pappus measuring 1 to 2 cm in length, aiding in wind dispersal over long distances [81]. Notably, its seeds are primarily photoblastic, meaning that they require light for germination [82]. This germination strategy contributes to its success in open and disturbed habitats. Its effective seed dispersal and environmental adaptability enhance its invasive potential across diverse habitats [82].
- Agricultural impact and management strategies: The proliferation of *P. acarna* has intensified in rainfed agricultural systems due to favorable environmental and management conditions [83]. This invasive species poses a significant threat to key rainfed crops, such as wheat (*Triticum aestivum* L.) and chickpeas (*Cicer arietinum* L.). Its rigid, spinous structure severely complicates harvest operations, particularly for crops like chickpeas and lentils, which are traditionally harvested by hand. In fields heavily infested with *P. acarna*, farmers often abandon cultivation altogether due to the high risk of injury posed by the weed's long spines [84]. Up to 25% yield reduction can occur at a density of 16 plants of *P. acarna* per square meter [82].

The seeds of *P. acarna* are photoblastic, requiring light for germination, which presents challenges for tillage-based management strategies. Tillage operations conducted in early spring, when soil moisture is sufficient, may inadvertently promote emergence by exposing buried seeds to light [82]. An effective cultural strategy involves inter-row cultivation during the warmer parts of the growing season, particularly in legume crops. This approach helps bury seeds deeper in the soil, reducing light exposure and thereby inhibiting germination [85]. In cereal crops such as wheat and barley, increasing crop density has also shown promise in suppressing *P. acarna* by shading the soil surface and limiting the light availability necessary for seed germination, ultimately reducing weed establishment and competition. Although herbicides have proven effective in managing *P. acarna*, their application remains limited in Iranian rainfed crops, including wheat, barley, chickpeas, and lentils [83].

3.1.6. *Physalis divaricata* D. Don

- Origin and global invasiveness: *P. divaricata* D. Don, commonly known as Annual Groundcherry, is primarily native to Latin America and South America (Figure 6). Vargas et al. (2001) classify it as native to Mexico, while Ramírez and Davenport (2016) recognize Colombia as a major centre of its natural distribution [86,87].

This species has spread widely across parts of the Middle East and South Asia, including Afghanistan, Pakistan, Turkey, Iraq, Syria, Saudi Arabia, Kuwait, Oman, and Yemen [52]. It was introduced into Iran in 1984, with its first recorded presence in Fars Province [67]. The annual species *P. divaricata* D. Don is the most prevalent among *Physalis* species in Iran and has become well-established in crop fields, particularly in western regions. It is also recognized as toxic [88], posing potential risks to agricultural and ecological systems (Figure 6). Similar to *I. lutea*, its invasive behavior is also driven by its rapid colonization of disturbed habitats during warm seasons. Its success is attributed to prolific seed production, efficient fruit and seed dispersal, and climatic adaptability. The absence of natural enemies and lack of evolutionary coexistence in newly invaded regions further enhances its establishment across diverse ecosystems, promoting persistent weed infestation.

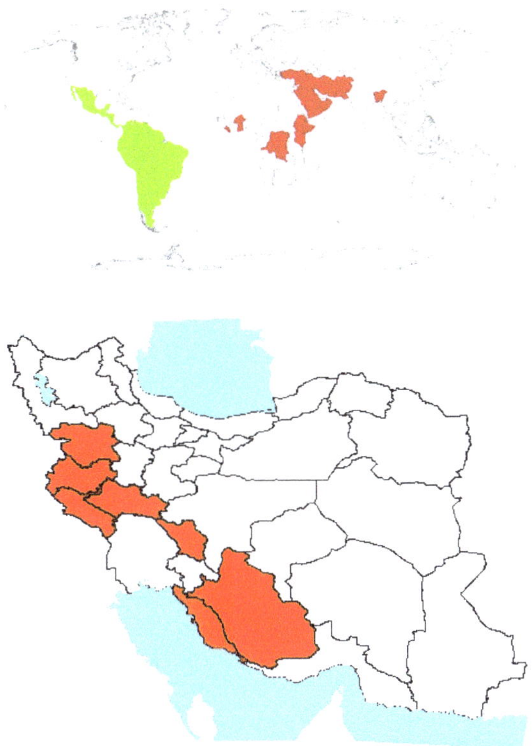

Figure 6. Global and national distribution of *Physalis divaricata* D. Don. **Top panel**: Global distribution map showing native (green) and invasive (red) ranges based on the literature and GBIF records. **Bottom panel:** Occurrence invasive (red) in Iran based on herbarium records, published studies, and recent field observations by the authors. Note: The absence of occurrence in some regions may reflect data gaps or lack of confirmed records and not necessarily ecological unsuitability.

- Morphological description, biology and ecology: The *Solanaceae* family includes approximately 2500 species across 90 genera, predominantly found in warm regions such as Central and South America, and temperate climates worldwide. This diverse plant family ranges from herbaceous annuals to woody perennials. Their leaves are typically alternate with variable shapes and sizes, while the flowers are mostly unisexual and occasionally hermaphroditic. Floral structures usually exhibit pentamerous symmetry, although variability exists. Fruits are commonly berries or capsules [67]. As a member of this family, it is an annual plant that produces berry-type fruits which transition in color from green to purple black upon ripening. The immature berries are known to be toxic [89]. Fruiting generally occurs by late June, with seed germination peaking approximately four weeks later. Seed viability remains high when buried at depths of 10–30 cm, whereas surface-level seeds are more susceptible to desiccation and mortality. Germination typically begins in March following winter moisture absorption in January, with full fruit development occurring by April [90]. The seeds can exhibit a germination rate of 93% under alternating temperatures of 10/20 °C (night/day), although germination drops by 41% under dark conditions at the same temperature. Salinity also affects germination: a concentration of 22.85 mM reduces the maximum germination by 50%. Additionally, soil pH levels between 6 and 7 are optimal for germination, indicating the species' sensitivity to environmental factors such as temperature, light, salinity, and acidity [91].

Ecologically, *P. divaricata* demonstrates morphological plasticity and thrives under varying environmental conditions. Despite its widespread occurrence, it remains under-researched. The extent of its impact is influenced by factors such as the timing of weed emergence, crop type, competition from other weeds, and crop density [92].

- Agricultural impact and management strategies: *P. divaricata* is recognized as a problematic weed across various agricultural systems in Iran due to its entangling growth habit, which complicates mechanical harvesting, particularly in fine grain and sugar beet crops [90,93]. In potato fields and sugar beet systems, its synchronous growth cycle exacerbates competition with crops [94,95]. Its competitive presence is most pronounced during the summer months in summer crops across the Lorestan and Chaharmahal Bakhtiari Provinces [94]. The weed has also been reported in orchards in Lorestan [94] and Dare Shahr in Ilam Province [68]. Notably, high-density infestations of *P. divaricata* can lead to crop yield reductions of 50–60% [96]. Additionally, *P. divaricata* has been observed in tomato fields in southeastern Iran [97], highlighting its broad ecological range and significant agricultural impact.

The effective management of *P. divaricata* requires an integrated approach that combines preventive, chemical, cultural, and biological strategies to reduce weed density, limit its spread, and minimize control costs. By implementing these measures, farmers can effectively manage weeds, reduce control costs, and maintain weed-free fields, ultimately enhancing agricultural productivity [74]. Regular field monitoring is essential for assessing weed populations and guiding timely interventions. Preventive measures such as avoiding the use of unpacked or contaminated seeds and refraining from applying fresh or decomposed manure are critical in reducing initial seed introductions into agricultural fields [74]. Weed emergence around irrigation streams and field margins can be suppressed by cementing main water channels and utilizing pressurized irrigation systems, which create less favorable environments for weed establishment.

Cultural control practices, particularly crop rotation, have also proven effective. Rotating with wheat significantly reduced the emergence and biomass of *P. divaricata*, offering a practical approach to long-term weed suppression [90]. Biological control is another emerging area of interest. Isolate 6A of *Alternaria alternata* has demonstrated pathogenic effects on *P. divaricata*, particularly at the six-leaf stage, when applied at a concentration of 10^8 spores mL^{-1}. This highlights the potential of *A. alternata* as a biological agent in integrated weed management programs [98]. Further research is required to optimize application methods, dosages, and assess non-target impacts on other flora and soil biota.

Chemical control remains an essential component of *P. divaricata* management. The application of contact herbicides, like paraquat, or systemic options, such as glyphosate, around canals and ditches has proven effective in reducing weed infestations. Among pre-emergent herbicides, chloridazone has demonstrated significant efficacy, inhibiting over 50% of seed germination in treated soils compared to the untreated control at a 5% significance level [99]. Three weeks after application, all seedlings exposed to chloridazone were destroyed, while those in the control plots remained unaffected. In contrast, herbicides such as trifluralin and ethalfluralin, whether applied alone or in combination, did not significantly reduce germination rates [99]. Additionally, foramsulfuron and nicosulfuron at 56 g ai ha^{-1} in corn fields, and phenmedipham at 547 g ai ha^{-1} in sugar beet fields, have shown promising results in reducing weed density and growth [100].

3.1.7. *Vicia hyrcanica* Fisch. & C.A. Mey.

- Origin and global invasiveness: *V. hyrcanica* Fisch. & C.A. Mey. is a climbing annual winter legume characterized by weak stems and seed-based dispersal. It is native to northern Iran and has progressively extended its range into western parts of the country [80]. The genus *Vicia* is made up of herbaceous legumes, primarily distributed across temperate regions, particularly in Mediterranean areas (Figure 7) [101,102]. Despite their wide distribution, comprehensive information on the *Vicia* genus's natural distribution, taxonomy, and agricultural potential remains underexplored. In Iran, *V. hyrcanica* is commonly found in rainfed fields, pastures, and perennial horticultural systems. In recent years, the species has rapidly spread into both disturbed and undisturbed habitats across the country (Figure 7).

Figure 7. Global and national distribution of *Vicia hyrcanica* Fisch. & C.A. Mey. **Top panel**: Global distribution map showing native (green) and invasive (red), ranges based on the literature and GBIF records. **Bottom panel:** Occurrence invasive (red) in Iran based on herbarium records, published studies, and recent field observations by the authors. Note: The absence of occurrence in some regions may reflect data gaps or lack of confirmed records and not necessarily ecological unsuitability.

- Morphological description, biology, and ecology: *V. hyrcanica* typically germinates in late winter, aligning its life cycle with that of cultivated legume crops in Iran. The stem is erect and can reach a length of up to 80 cm. Its leaves consist of three to eight pairs of linear to narrowly elliptic leaflets, each measuring between 1.5 and 3.0 cm. The species produces solitary, yellow flowers that range from 1.0 to 1.8 cm in length. The mature legume is circular in cross-section, approximately 4 cm long, and dark brown to black in color. Upon ripening, the pod dehisces into two twisted halves, facilitating seed dispersal [103].

A notable trait of *V. hyrcanica* is its relatively large seeds, which possess a hard seed coat. This physical dormancy often delays germination following seed dispersal, contributing to its persistence and invasive behavior [104]. The scarification of its hard seed coat, promoted by cold temperatures and soil tillage, enhances germination rates. Therefore, adopting no-tillage practices may serve as a viable strategy to suppress its emergence. The species is particularly common in low-input agricultural systems and demonstrates strong tolerance to water stress. This trait enables *V. hyrcanica* to outcompete crops in drought-prone conditions [105].

- Agricultural impact and management strategies: *V. hyrcanica* is widely recognized as an important cover crop species in Iran [106]. Despite efforts in domestication and breeding, many cover crops, including *V. hyrcanica*, retain seed dormancy mechanisms inherited from their wild ancestors [107,108]. These dormancy traits can limit their effective integration into cropping systems and raise concerns about their potential weediness. The presence of seed dormancy allows *V. hyrcanica* to persist in the soil seed bank and germinate unpredictably, potentially disrupting future crop cycles. Studies have shown that both genotype and environmental factors during seed development, as well as postharvest storage conditions, influence dormancy levels and germination timing [108–110]. Additionally, climatic conditions during the parental plant's development significantly impact seed germination potential [111], increasing the species' risk of becoming invasive in new areas.

This increasing prevalence has raised environmental concerns and emphasized the need for management interventions. In cereal crops such as wheat and barley, a broad spectrum of herbicides has been reported to effectively control *V. hyrcanica*. Species such as *V. villosa* (hairy vetch), which share morphological seed characteristics with *V. hyrcanica*, are particularly difficult to manage in rainfed legumes like chickpeas and lentils due to the difficulty in distinguishing their seeds during cleaning and processing. This similarity complicates weed control and contributes to the proliferation of *V. hyrcanica* populations in legume-based cropping systems [112]. Effective management requires both chemical and cultural strategies tailored to crop type and environmental conditions.

3.2. *Aquatic Invasive Weeds*

3.2.1. *Azolla filiculoides* Lam. and *Azolla pinnata*

- Origin and global invasiveness: *Azolla* spp., commonly referred to as mosquitofern, represent a genus of small aquatic ferns with a rich evolutionary history. The first recorded collection of *Azolla* was made by Baptiste-Jean Lamarck in South America, although genetic differentiation among species indicates varied origins [113]. Fossil evidence suggests that *Azolla* has existed for over 70 million years, tracing its origins back to the early Cenozoic era [114]. Historically believed to have originated in the Americas, subsequent genetic studies have revealed a more complex and widespread origin. Today, *Azolla* species are recognized as having diverse evolutionary roots across tropical, subtropical, and warm temperate regions of Africa, Asia, and the Americas (Figure 8) [115], with some species naturally occurring in parts of Europe and Asia [116]. Due to its rapid vegetative growth and nitrogen-fixing ability, *Azolla* has been introduced to several regions, including Australia and parts of Africa, for agricultural and ecological purposes. However, its widespread introduction has led to invasiveness in multiple countries, such as the United States, the United Kingdom, France, Iran, and South Africa (Figure 8).

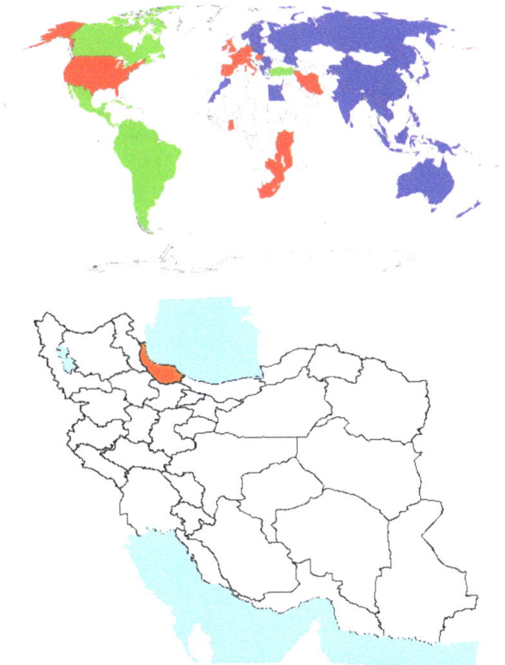

Figure 8. Global and national distribution of *Azolla* spp. **Top panel**: Global distribution map showing native (green), invasive (red), and introduced (blue) ranges based on the literature and GBIF records. **Bottom panel**: Occurrence invasive (red) in Iran based on herbarium records, published studies, and recent field observations by the authors. Note: The absence of occurrence in some regions may reflect data gaps or lack of confirmed records and not necessarily ecological unsuitability.

- Morphological description, biology, and ecology: *Azolla* spp. (*Salviniaceae*) are small, floating aquatic ferns characterized by polygonal or triangular shapes [117]. Most species measure between 3 and 4 cm, except for *Azolla nilotica*, which may exceed this size. The stems are covered with minute, alternate, scaly leaves arranged in two overlapping ranks. Roots are unbranched but bear fine lateral rootlets that give a feathery appearance when submerged [118]. Color and size can vary significantly depending on environmental conditions, such as light and nutrient availability [119]. *Azolla* spp. reproduce both vegetatively and sexually. While vegetative propagation is the predominant method [120], sexual reproduction occurs, involving the formation of megaspores and microspores; however, it is considered a rare phenomenon in natural populations [113].

This broad phylogeographic distribution likely reflects the genus's ability to disperse naturally over millennia, facilitated by its capacity to adapt to a range of aquatic environments. *Azolla* spp. are commonly found in freshwater bodies, such as pits, pools, lakes, and rivers, in temperate and tropical regions. Their growth is closely linked to factors including air temperature, humidity, light intensity, and the availability of nitrates and phosphates [121]. Optimal growth occurs at 20–30 °C [122]. Humidity levels below 60% cause desiccation [123], and the preferred pH range is between 4.5 and 7 [124]. The genus plays a significant ecological role, particularly in nitrogen fixation, through its symbiotic relationship with the cyanobacterium *Anabaena azollae*, contributing to soil fertility in rice paddies and wetland systems. It forms dense mats on water surfaces, which can inhibit light penetration, reduce oxygen levels, and outcompete native aquatic vegetation, thereby disrupting aquatic ecosystems. Despite its utility in biofertilization and wastewater treatment, the unregulated proliferation of *Azolla* spp. necessitates targeted management strategies to mitigate its environmental impacts [120].

- Ecological impact and management challenges: *Azolla* spp. can severely disrupt aquatic ecosystems by reducing oxygen levels, altering water acidity, limiting light penetration, and disturbing aquatic food webs [125]. In northern Iran, particularly in the Guilan and Mazandaran Provinces, the dense proliferation of *Azolla* in Anzali Wetland has led to increased toxicity and mortality of fish and other aquatic organisms, pushing the ecosystem toward collapse [126] (Figure 8). In rice paddies, the rapid surface growth of *Azolla* forms a thick mat that interferes with rice transplanting by bending and submerging seedlings, thus impeding early crop establishment [126].

Management strategies for *Azolla* spp. include mechanical removal and chemical herbicides such as paraquat, Dicot [127], and glyphosate [128]. However, due to their environmental risks, chemical methods are generally not recommended. Also, conventional methods like chemical and mechanical control are often costly and impractical in such ecosystems. Among biological approaches, the use of the beetle *Stenopelmus* (Coleoptera: *Erirhinidae*) has proven to be the most effective and environmentally sustainable method for controlling *Azolla* spp. populations globally [129]. Research should prioritize the optimization of biological control strategies, especially in sensitive and economically constrained regions such as the Anzali Wetland in Guilan Province, Iran. Investigations into potential shifts in host specificity are also critical to ensure the safety of non-target species. Comprehensive risk assessments and long-term ecological monitoring must accompany the implementation of biological control to ensure environmental compatibility and long-term sustainability.

3.2.2. *Eichhornia crassipes* (Mart.) Solms

- Origin and global invasiveness: *E. crassipes* (Mart.) Solms (Family: Pontederiaceae), commonly known as water hyacinth, is a floating aquatic plant native to the Amazon Basin in South America, where it naturally inhabits rivers, lakes, and wetlands (Figure 9) [130]. It is well adapted to tropical and subtropical climates, thriving in nutrient-rich freshwater environments. The species was first described in 1842 by the German botanist Kunth. Initially introduced outside its native range as an ornamental plant in the late 19th and early 20th centuries, *E. crassipes* quickly gained popularity due to its attractive appearance. However, its ability to reproduce rapidly through both vegetative and sexual means has made it one of the most invasive aquatic plants globally [131].

Following its introduction, *E. crassipes* spread aggressively across multiple continents, particularly impacting Asia by the 1940s, with notable infestations in Japan and Indonesia [132]. Today, it is reported in over 50 countries, including regions in Africa, Asia, Australia, and North America (Figure 9) [133]. Severe infestations have been documented in Africa, India, and Spain, where the plant has disrupted aquatic systems and local economies [134]. In Iran, *E. crassipes* was first observed in the Ainak Wetland near Rasht, Guilan Province. Its rapid growth led to a significant expansion from one hectare in 2011 to 600 hectares by 2015 [135], and to 800 hectares by 2019, according to the Iran Environment Center. Although high salinity along the southern Caspian coast limits its spread, the favorable climate and abundant freshwater in northern Iran continue to support its proliferation (Figure 9). Prompt management efforts are essential to mitigate their growing environmental and economic impacts.

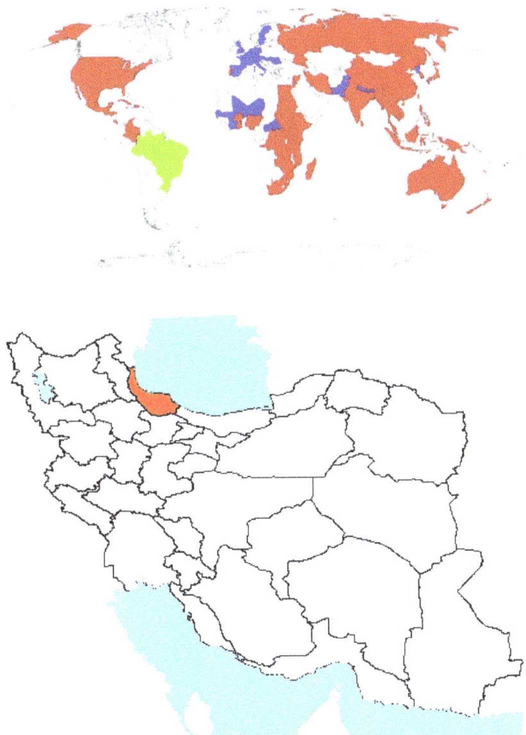

Figure 9. Global and national distribution of *Eichhornia crassipes* (Mart.) Solms. **Top panel**: Global distribution map showing native (green), invasive (red), and introduced (blue) ranges based on the literature and GBIF records. **Bottom panel**: Occurrence invasive (red) in Iran based on herbarium records, published studies, and recent field observations by the authors. Note: The absence of occurrence in some regions may reflect data gaps or lack of confirmed records and not necessarily ecological unsuitability.

- Morphological description, biology, and ecology: The plant is characterized by its circular to oval-shaped leaves with flexible, spongy-covered petioles and lilac to blue flowers. The roots are fibrous and unbranched [136], facilitating nutrient absorption while minimizing resistance in aquatic conditions. Mature plants exhibit pendant roots, rhizomes, stolons, leaves, inflorescences, and fruit. The air-filled sacs in their leaves and stems allow it to remain on the water's surface [137].

E. crassipes demonstrates rapid growth, with populations capable of doubling within two weeks in favorable aquatic environments. Its development is influenced by nutrient levels, temperature, salinity, herbivory, and pathogens [138]. The growth of *E. crassipes* ceases when water temperatures exceed 40 °C or drop below 10 °C, with leaves particularly susceptible to damage in frosty conditions [139]. A decline in its spread during 2016 was linked to a preceding cold winter [140]. Elevated salinity and electrical conductivity (EC) have also been shown to limit habitat suitability, restricting the plant's distribution and survival [141,142].

- Ecological impact and management challenges: *E. crassipes* poses severe threats to aquatic ecosystems by reducing water pH and oxygen levels, obstructing waterways, accelerating sedimentation [143], and disrupting native biodiversity and activities such as fishing, irrigation, and transportation [142]. Its dense mats significantly hinder light penetration, degrading water quality and aquatic health, which, in turn, threatens food security [144]. Furthermore, it depletes essential nutrients in water bodies [145], contributing to the ecological decline of wetlands. The unchecked proliferation of *E. crassipes* often leads to habitat monopolization, outcompeting native flora and fauna [146]. In some wetlands, the increased biomass of this invasive species has resulted in ecosystem collapse. The economic burden of managing *E. crassipes* is substantial, with global control efforts costing an estimated USD 124 million annually [147]. Despite its invasive nature, it has also been investigated for beneficial applications such as phytoremediation and bioenergy production [148].

Due to its aggressive growth and resilience, *E. crassipes* remains difficult to control, necessitating a combination of management techniques [149]. Physical control, including manual and mechanical removal, is effective for small-scale infestations in shallow or confined water bodies. Biological control options, using specific insects and fungi, offer environmentally sustainable alternatives. Chemical control remains one of the most common large-scale strategies; however, concerns over environmental toxicity have driven research into safer systemic herbicides [150,151]. Herbicides like 2,4-D and bispyribac-sodium have demonstrated effectiveness in controlling *E. crassipes* [152–154]. In Iran, glyphosate and 2,4-D have been successfully applied to reduce its spread and mitigate associated ecological and economic damages [155]. Integrated management combining these approaches is essential for long-term control.

4. Policy and Stakeholder Engagement

The effective management of invasive weed species in Iran requires a collaborative framework that involves local farmers, researchers, and policymakers. Farmers, as frontline observers, play a critical role in the early detection of invasive species; however, their ability to respond is often constrained by inadequate awareness and limited access to control resources [156,157]. Enhancing public education through targeted awareness campaigns and integrating traditional ecological knowledge with scientific practices can strengthen community participation in management strategies [158,159]. Policymakers are responsible for formulating and enforcing regulations. Yet, challenges such as regulatory fragmentation and insufficient inter-agency coordination impede the comprehensive management of invasive species [157,160]. The absence of specific legislation targeting invasive weeds becomes a significant policy gap. Moreover, public involvement in policy design is minimal, further limiting efficacy. Research efforts, while essential, are hindered by insufficient baseline data and restricted funding.

5. Conclusions

This review underscores the alarming proliferation of invasive weed species across Iran's agroecosystems, rangelands, and aquatic environments. Key invaders such as *B. orientalis*, *I. lutea*, *P. divaricata*, *Azolla* spp., *E. crassipes*, and *A. psilostachya* exhibit high phenotypic plasticity, robust reproductive strategies, and resistance to common control measures. Their ecological traits, including allelopathy, dormancy, and aggressive vegetative spread, enable them to displace native flora, alter ecosystem dynamics, and reduce agricultural productivity. The review highlights a critical gap in ecological data for many species, hindering the development of effective, species-specific management protocols. Current management strategies remain fragmented and emphasize the urgent need for multi-tiered approaches that combine various control methods, technological innovation, and stakeholder engagement. Invasive weeds in Iran are not merely a national concern but reflect broader global trends driven by climate change and globalization. This synthesis serves as a foundation for targeted research and coordinated management, reinforcing the necessity of proactive, interdisciplinary efforts to mitigate further degradation of Iran's ecosystems and agricultural resilience.

6. Future Research Direction

The future management of invasive plant species requires the integration of advanced technologies, ecological understanding, and collaborative frameworks. Effective mitigation requires integrated approaches that consider the species-specific and ecosystem-specific dynamics. Continued scientific investigation, policy development, and community engagement are essential for minimizing the ecological and agricultural impacts of invasive plants and preserving biodiversity and ecosystem resilience.

1. Research should prioritize the development of herbicide formulations specifically tailored to the physiological and ecological traits of invasive species.
2. The introduction of non-native plants such as *Azolla* spp., *I. lutea*, and *P. acarna* into new habitats, including Iran, underscores the urgent need for comprehensive risk assessments and monitoring systems. Remote sensing and GIS technologies can significantly enhance early detection, spatial analysis, and targeted control strategies.
3. For managing *E. crassipes*, studies should investigate the biotic and abiotic factors in its native range that naturally regulate its growth to inform new management tactics.
4. Research should focus on developing tillage methods, particularly nocturnal tillage, to inhibit *P. acarna* seed germination. As its photoblastic seeds require light, nighttime soil disturbance may prevent emergence by keeping seeds in darkness. Studies must identify optimal tillage depth, timing, and frequency, while assessing impacts on soil health, biodiversity, and crop yield.
5. Future research should focus on the weed's germination biology under various environmental conditions. Germination rates of 93% of *P. divaricata* at 10/20 °C (night/day) and the 50% reduction under salinity levels of 22.85 mM indicate that environmental manipulation could be leveraged for control.
6. For *B. orientalis*, there is a need to assess the effectiveness of shifting from post-emergence to pre-emergence herbicide strategies. Field trials across diverse agro-ecological zones are required to evaluate the efficacy of pre-emergent applications in suppressing germination and reducing the risk of resistance.
7. Managing invasive species like *B. orientalis* and *A. psilostachya* calls for coordinated action involving adaptive management, public awareness, and continuous monitoring. While chemical control remains common, biological methods, such as the use of *O. communa* against *A. psilostachya*, offer sustainable alternatives.

7. Methodology

For this review, we gathered information on invasive plant species from a comprehensive selection of scientific sources, including peer-reviewed journal articles, books, and reputable websites. This review does not attempt to provide a complete inventory of all invasive alien plant species in Iran. Instead, it concentrates on the nine most problematic and emerging invasive weeds identified through national surveys, the peer-reviewed literature, and expert consultations. This targeted selection enables detailed coverage of each species' classification, origin, life form, ecological and economic impacts, and management strategies.

In this study the nine most problematic weeds species were selected based on (i) documented invasive traits, such as rapid growth, prolific reproduction, seed dormancy, allelopathy, and broad ecological tolerance; (ii) evidence of rapid spread in Iranian agroecosystems or potential to become major threats; (iii) significant ecological or economic impacts reported in national weed surveys and peer-reviewed studies; and (iv) expert consultations with Iranian weed scientists to validate species relevance and priority.

Data collection focused on ecology, biology, distribution, and control strategies, ensuring that only reliable and scientifically validated references were used. Sources for global and national distribution included the Global Biodiversity Information Facility (GBIF), herbarium records from national and regional collections, published distribution reports in the scientific literature, data from national plant protection organizations, and recent field surveys conducted by the authors between 2018 and 2024.

All occurrence records were cross-checked to remove duplicates or questionable entries. The final validated dataset was mapped using ArcMap, Version 10.5.1, a geographic information system (GIS) software, to visualize the native, invasive, and introduced ranges of each species. This methodological approach ensured an accurate synthesis of existing knowledge and provided a robust foundation for assessing the spread and impact of these invasive plants.

Author Contributions: Conceptualization, S.B. and A.S.; methodology, S.B. and A.S.; software, S.B.; validation, S.B., A.S., and A.R.Y.; formal analysis, S.B.; investigation, S.B. and A.S.; resources, A.S.; data curation, S.B.; writing—original draft preparation, S.B.; writing—review and editing, A.S., A.R.Y., I.N., E.Z., M.B., E.A.O., H.S., and M.R.; visualization, A.S.; supervision, S.B.; project administration, S.B. All authors have read and agreed to the published version of the manuscript.

Funding: This research received no external funding.

Data Availability Statement: The original contributions presented in this study are included in the article. Further inquiries can be directed to the corresponding author.

Acknowledgments: During the preparation of this manuscript, the authors used ChatGPT, version 4 (OpenAI, GPT-4) for language editing. The authors have reviewed and edited the output and take full responsibility for the content of this publication.

Conflicts of Interest: The authors declare no conflicts of interest.

References

1. Chandrasena, N.R. 'Seeing Weeds with New Eyes' Part II–Some Historical Perspectives and 'Proto Weeds'. *Weeds—J. Asian-Pac. Weed Sci. Soc.* **2020**, *2*, 1–16.
2. MacLaren, C.; Storkey, J.; Menegat, A.; Metcalfe, H.; Dehnen-Schmutz, K. An Ecological Future for Weed Science to Sustain Crop Production and the Environment. A Review. *Agron. Sustain. Dev.* **2020**, *40*, 28. [CrossRef]
3. Montagnani, C.; Gentili, R.; Brundu, G.; Caronni, S.; Citterio, S. Accidental Introduction and Spread of Top Invasive Alien Plants in the European Union through Human-Mediated Agricultural Pathways: What Should We Expect? *Agronomy* **2022**, *12*, 423. [CrossRef]
4. Pyšek, P.; Hulme, P.E.; Simberloff, D.; Bacher, S.; Blackburn, T.M.; Carlton, J.T.; Dawson, W.; Essl, F.; Foxcroft, L.C.; Genovesi, P.; et al. Scientists' Warning on Invasive Alien Species. *Biol. Rev.* **2020**, *95*, 1511–1534. [CrossRef]
5. Panzavolta, T.; Bracalini, M.; Benigno, A.; Moricca, S. Alien Invasive Pathogens and Pests Harming Trees, Forests, and Plantations: Pathways, Global Consequences and Management. *Forests* **2021**, *12*, 1364. [CrossRef]
6. Enders, M.; Havemann, F.; Ruland, F.; Bernard-Verdier, M.; Catford, J.A.; Gómez-Aparicio, L.; Haider, S.; Heger, T.; Kueffer, C.; Kühn, I.; et al. A Conceptual Map of Invasion Biology: Integrating Hypotheses into a Consensus Network. *Glob. Ecol. Biogeogr.* **2020**, *29*, 978–991. [CrossRef]
7. Essl, F.; Biró, K.; Brandes, D.; Broennimann, O.; Bullock, J.M.; Chapman, D.S.; Chauvel, B.; Dullinger, S.; Fumanal, B.; Guisan, A.; et al. Biological Flora of the British Isles: *Ambrosia artemisiifolia*. *J. Ecol.* **2015**, *103*, 1069–1098. [CrossRef]
8. Turbelin, A.J.; Cuthbert, R.N.; Essl, F.; Haubrock, P.J.; Ricciardi, A.; Courchamp, F. Biological Invasions Are as Costly as Natural Hazards. *Perspect. Ecol. Conserv.* **2023**, *21*, 143–150. [CrossRef]
9. Westbrooks, R.G. *Invasive Plants: Changing the Landscape of America: Fact Book*; Federal Interagency Committee for the Management of Noxious and Exotic Weeds: Washington, DC, USA, 1998.
10. Holm, L.G.; Plucknett, D.L.; Pancho, J.V.; Herberger, J.P. *The World's Worst Weeds: Distribution and Biology*; University Press of Hawaii: Honolulu, HI, USA, 1977.

11. Sohrabi, S.; Vilà, M.; Zand, E.; Gherekhloo, J.; Hassanpour-bourkheili, S. Alien Plants of Iran: Impacts, Distribution and Managements. *Biol. Invasions* **2023**, *25*, 97–114. [CrossRef]
12. Heshmati, I.; Khorasani, N.; Shams-Esfandabad, B.; Riazi, B. Forthcoming Risk of Prosopis Juliflora Global Invasion Triggered by Climate Change: Implications for Environmental Monitoring and Risk Assessment. *Environ. Monit. Assess.* **2019**, *191*, 72. [CrossRef]
13. Stastny, M.; Schaffner, U.; Elle, E. Do Vigour of Introduced Populations and Escape from Specialist Herbivores Contribute to Invasiveness? *J. Ecol.* **2005**, *93*, 27–37. [CrossRef]
14. Mitchell, C.E.; Power, A.G. Release of Invasive Plants from Fungal and Viral Pathogens. *Nature* **2003**, *421*, 625–627. [CrossRef]
15. Weber, E. *Invasive Plant Species of the World: A Reference Guide to Environmental Weeds*; CABI Publishing: Wallingford, UK, 2003.
16. Hallett, S.G. Dislocation from Coevolved Relationships: A Unifying Theory for Plant Invasion and Naturalization? *Weed Sci.* **2006**, *54*, 282–290. [CrossRef]
17. Zimdahl, R.L.; Basinger, N.T. *Fundamentals of Weed Science*; Elsevier: Amsterdam, The Netherlands, 2024.
18. Chauvel, B.; Dessaint, F.; Cardinal-Legrand, C.; Bretagnolle, F. The Historical Spread of *Ambrosia artemisiifolia* L. in France from Herbarium Records. *J. Biogeogr.* **2006**, *33*, 665–673. [CrossRef]
19. Cheraghian, A. *A Guide for Diagnosis and Detection of Quarantine Pests: Perennial Ragweed. Ambrosia Psilostachya DC. Asterales: Asteraceae*; Bureau of Plant Pest Surveillance and Pest Risk Analysis; Plant Protection Organization: Tehran, Iran, 2016.
20. Montagnani, C.; Gentili, R.; Smith, M.; Guarino, M.F.; Citterio, S. The Worldwide Spread, Success, and Impact of Ragweed (*Ambrosia* Spp.). *CRC Crit. Rev. Plant Sci.* **2017**, *36*, 139–178. [CrossRef]
21. Sharifi Kaliani, F.; Babaei, S.; ZafarSohrabpour, Y. Study of the Effects of Dusts on the Morphological and Physiological Traits of Some Crops. *J. Plant Prod. Res.* **2021**, *28*, 205–220. [CrossRef]
22. Sharifi Kalyani, F.; Babaei, S.; Zafarsohrabpour, Y.; Nosratti, I.; Gage, K.; Sadeghpour, A. Investigating the Impacts of Airborne Dust on Herbicide Performance on *Amaranthus retroflexus*. *Sci. Rep.* **2024**, *14*, 3785. [CrossRef]
23. Bassett, I.J.; Crompton, C.W. The Biology of Canadian Weeds: 11. *Ambrosia artemisiifolia* L. and *A. Psilostachya* DC. *Can. J. Plant Sci.* **1975**, *55*, 463–476. [CrossRef]
24. Fried, G.; Belaud, A.; Chauvel, B. Ecology and Impact of an Emerging Invasive Species in France: Western Ragweed (*Ambrosia psilostachya* DC.). *Rev. d'Écologie (Terre Vie)* **2015**, *70*, 53–67. [CrossRef]
25. Ardenghi, N.; Polani, F. The Flora of the Province of Pavia (Lombardy, Northern Italy). 1. The Oltrepò Pavese. *Nat. Hist. Sci.* **2016**, *3*, 51–79. [CrossRef]
26. Béres, I.; Kazinczi, G.; Narwal, S. Allelopathic Plants. 4. Common Ragweed (*Ambrosia elatior* L. Syn *A. Artemisiifolia*). *Allelopath. J.* **2002**, *9*, 27–34.
27. Leather, G.R. Weed Control Using Allelopathic Crop Plants. *J. Chem. Ecol.* **1983**, *9*, 983–989. [CrossRef] [PubMed]
28. Katz, D.S.W.; Batterman, S.A. Allergenic Pollen Production across a Large City for Common Ragweed (*Ambrosia artemisiifolia*). *Landsc. Urban. Plan.* **2019**, *190*, 103615. [CrossRef] [PubMed]
29. Wodehouse, R.P. Hayfever Plants. Their Appearance, Distribution, Time of Flowering, and Their Role in Hayfever, with Special Reference to North America. *Madroño* **1946**, *8*, 201–204.
30. Gerber, E.; Schaffner, U.; Gassmann, A.; Hinz, H.; Seier, M.; Müller-Schärer, H. Prospects for Biological Control of *Ambrosia artemisiifolia* in Europe: Learning from the Past. *Weed Res.* **2011**, *51*, 559–573. [CrossRef]
31. Milakovic, I.; Fiedler, K.; Karrer, G. Management of roadside populations of invasive A mbrosia artemisiifolia by mowing. *Weed Res.* **2014**, *54*, 256–264. [CrossRef]
32. Ganie, Z.A.; Sandell, L.D.; Jugulam, M.; Kruger, G.R.; Marx, D.B.; Jhala, A.J. Integrated Management of Glyphosate-Resistant Giant Ragweed (*Ambrosia trifida*) with Tillage and Herbicides in Soybean. *Weed Technol.* **2016**, *30*, 45–56. [CrossRef]
33. Moradi, M.; Babaei, S.; Tahmasebi, I.; Nosratti, I. Assessing Resistance to Clodinafop-Propargyl Herbicide in Wild Oat *Avena Ludoviciana* Populations in Wheat Fields of Some Western and Southwestern Regions of Iran. *J. Appl. Res. Plant Prot.* **2023**, *12*, 461–473.
34. Moradi, M.; Babaei, S.; Nosratti, I.; Tahmasebi, I. Evaluation of the Resistance of *Galium aparine* L. to Common Herbicides in Wheat Fields of Western and Southwestern Iran. *J. Appl. Res. Plant Prot.* **2024**, *13*, 373–385.
35. Soltani, N.; Shropshire, C.; Sikkema, P. Giant Ragweed (*Ambrosia trifida* L.) Control in Corn. *Can. J. Plant Sci.* **2011**, *91*, 577–581. [CrossRef]
36. Zhou, Z.S.; Guo, J.Y.; Zheng, X.W.; Luo, M.; Chen, H.S.; Wan, F.H. Reevaluation of Biosecurity of *Ophraella communa* against Sunflower (*Helianthus annuus*). *Biocontrol Sci. Technol.* **2011**, *21*, 1147–1160. [CrossRef]
37. Bonini, M.; Gentili, R.; Müller-Schärer, H. Ragweed Management and the Potential Benefits and Risks of *Ophraella communa* in Northern Italy: Researchers Meet Their Stakeholders. *Not. Soc. Bot. Ital.* **2017**, *26*, 1–4.
38. Moazzeni, H.; Zarre, S.; Al-Shehbaz, I.A.; Mummenhoff, K. Phylogeny of Isatis (Brassicaceae) and Allied Genera Based on Its Sequences of Nuclear Ribosomal DNA and Morphological Characters. *Flora—Morphol. Distrib. Funct. Ecol. Plants* **2010**, *205*, 337–343. [CrossRef]

39. Babaei, S.; Adeli, T.; Tahmasebi, I.; Mozafarian, V. Introducing Waxy Leaved Mustard (*Boreava orientalis* Jaub. and Spach.) as a Problematic Weed in Wheat Fields of Kurdistan Province. *Cereal Res.* **2019**, *9*, 261–269. [CrossRef]
40. Verloove, F.; Groom, Q.; Brosens, D.; Desmet, P.; Reyserhove, L. Manual of the Alien Plants of Belgium. 2020. Available online: https://alienplantsbelgium.myspecies.info/ (accessed on 2 April 2024).
41. Davis, P.H.; Cullen, J.; Coode, M.J.E. *Flora of Turkey and the East Aegean Islands*; Edinburgh University Press: Edinburgh, UK, 1965.
42. McConnell, E.G.; Evans, J.O.; Dewey, S.A. Dyer's Woad. In *Biology and Management of Noxious Rangeland Weeds*; Sheley, R.L., Petroff, J.K., Eds.; OSU Press: Corvallis, OR, USA, 1999; pp. 231–237.
43. Loik, M.E.; Redar, S.P. Microclimate, Freezing Tolerance, and Cold Acclimation along an Elevation Gradient for Seedlings of the Great Basin Desert Shrub, *Artemisia tridentata*. *J. Arid. Environ.* **2003**, *54*, 769–782. [CrossRef]
44. Loik, M.E.; Harte, J. High-Temperature Tolerance of *Artemisia tridentata* and *Potentilla gracilis* under a Climate Change Manipulation. *Oecologia* **1996**, *108*, 224–231. [CrossRef]
45. Young, J.A.; Evans, R.A. Germination of Dyers Woad. *Weed Sci.* **1971**, *19*, 76–78. [CrossRef]
46. Adeli, T.; Tahmasebi, I.; Babaei, S.; Sadeghpour, A. Assessing Allelopathic Potential: *Boreava orientalis* Impact on Triticum Aestivum. *J. Crop Prot.* **2023**, *12*, 389–401.
47. Tezel, C. Competitive Effects and Control of Weeds in Winter Wheat on the Anatolian Plateau, Turkey. Master's Thesis, Oregon State University, Corvallis, OR, USA, 1975.
48. Oleszek, W. Allelopathic Effects of Volatiles from Some Cruciferae Species on Lettuce, Barnyard Grass and Wheat Growth. *Plant Soil.* **1987**, *102*, 271–273. [CrossRef]
49. Naghib Alsadati, M.; Babaei, S.; Tahmasebi, I.; Kiani, H. Evaluation of Airborne Dust Effect on the Efficiency of Atlantis OD, Clodinafop Propargyl and 2,4-D+MCPA Herbicides on Weed Control in Wheat. *Iran. J. Field Crop Sci.* **2020**, *50*, 4. [CrossRef]
50. Babaei, S.; Ahmadi, S.; Tahmasebi, I.; Sarseyfi, M. Evaluation the Efficacy of Chemical and Non-Chemical Weed Control Approaches in Strawberry (*Fragaria Ananassa*) Fields in Kurdistan. *Iran. J. Weed Sci.* **2022**, *18*, 11–22. [CrossRef]
51. CABI Compendium Invasive Species | CABI Digital Library. Available online: https://www.cabidigitallibrary.org/product/QI (accessed on 9 June 2025).
52. Plants of the World Online | Kew Science. Available online: https://powo.science.kew.org/ (accessed on 9 June 2025).
53. Wilson, B.G. *Mediterranean Wild Flowers*; Harpercollins Publishers Ltd.: London, UK, 2001.
54. Lawlor, F.M. Herbicidal Treatment of the Invasive Plant *Cynanchum rossicum* and Experimental Post-Control Restoration of Infested Sites. Master's Thesis, State University of New York College of Environmental Science and Forestry, New York, NY, USA, 2000.
55. Ar, B.; Tuttu, G.; Gülçin, D.; Özcan, A.U.; Kara, E.; Sürmen, M.; Çiçek, K.; Velázquez, J. Response of an invasive plant species (*Cynanchum acutum* L.) to changing climate conditions and its impact on agricultural land-scapes. *Land* **2022**, *11*, 1438. [CrossRef]
56. Zaifie, M. *Flora of Iran (No. 28 of Estabarq Genus)*; Forest and Rangeland Research Institute: Tehran, Iran, 1999.
57. Soteres, J.K.; Murray, D.S. Root Distribution and Reproductive Biology of Honeyvine Milkweed (*Cynanchum Laeve*). *Weed Sci.* **1982**, *30*, 158–163. [CrossRef]
58. Berti, A.; Sattin, M.; Baldoni, G.; Del Pino, A.M.; Ferrero, A.; Montemurro, P.; Tei, F.; Viggiani, P.; Zanin, G. Relationships between Crop Yield and Weed Time of Emergence/Removal: Modelling and Parameter Stability across Environments. *Weed Res.* **2008**, *48*, 378–388. [CrossRef]
59. Badli, K. *Investigation of Sulfonylurea Group Herbicides in Grain Maize*; Moghan Pest and Diseases Research Department: Moghan, Iran, 1997.
60. Ihlenfeldt, H.D. Martyniaceae. In *Flowering Plants·Dicotyledons: Lamiales (except Acanthaceae including Avicenniaceae)*; Springer: Berlin/Heidelberg, Germany, 2004; pp. 283–288.
61. Thieret, J.W. The Martyniaceae in the Southeastern United States. *J. Arnold Arbor.* **1977**, *58*, 25–39. [CrossRef]
62. Gutierrez, R. A Phylogenetic Study of the Plant Family Martyniaceae (Order Lamiales). Ph.D. Thesis, Arizona State University, Tempe, AZ, USA, 2011.
63. Bretting, P. The Taxonomic Relationship between *Proboscidea louisianica* and *Proboscidea Fragrans* (Martyniaceae). *Southwest. Nat.* **1983**, *28*, 445–449. [CrossRef]
64. Yannitsardos, A.; Bazos, I. Ibicella (Stapf) Van Eseltine: A Genus of the American Family Martyniaceae New for Greece. *Ann. Musei Goulandris* **2006**, *11*, 271–279.
65. Sevgi, E.; Hançer, Ç.K.; Yılmaz, H.; Akkaya, M. A New Alien Species Record for the Flora of Turkey: *Proboscidea louisianica* (Miller) Thell. *Eurasian J. For. Sci.* **2017**, *5*, 19–25. [CrossRef]
66. Small, E. *North American Cornucopia: Top 100 Indigenous Food Plants*; CRC Press: Boca Raton, FL, USA, 2013.
67. Mozaffarian, V. *Plant Classification: Taxonomic Morphology (Vol. 1)*; Amirkabir Publishing Institute: Tehran, Iran, 2010.

68. Saboori, A.; Rezapanah, M.; Shirazi, J.; Rahnama, K.; Mounes; Bakhshi; Jahanshahi Afshar, F.; Najafi, H. (Eds.) Introducing the New Invasive Weed of Devils Claw (Ibicella Lutea) in the Landscape of Ilam City and Sugar Sugar Beet Field in Homail-Kermanshah Province. In Proceedings of the 24th Iranian Plant Protection Congress, Tehran, Iran, 3–6 September 2022; Iranian Research Institute of Plant Protection: Tehran, Iran, 2022.
69. Wetherwax, M.; Heckard, L.R. Martyniaceae. 2012. Available online: https://ucjeps.berkeley.edu/eflora/eflora_display.php?tid=79174 (accessed on 12 May 2025).
70. Morgan, J.W. Effects of Population Size on Seed Production and Germinability in an Endangered, Fragmented Grassland Plant. *Conserv. Biol.* **1999**, *13*, 266–273. [CrossRef]
71. Almazán-Núñez, R.C.; Alvarez-Alvarez, E.A.; Sierra-Morales, P.; Rodríguez-Godínez, R. Fruit Size and Structure of Zoochorous Trees: Identifying Drivers for the Foraging Preferences of Fruit-Eating Birds in a Mexican Successional Dry Forest. *Animals* **2021**, *11*, 3343. [CrossRef]
72. Mercer, K.L.; Pawlak, J.A.; Murray, D.S.; Verhalen, L.M.; Riffle, M.S.; McNew, R.W. Distance-of-Influence of Devil's-Claw (*Proboscidea louisianica*) on Cotton (*Gossypium hirsutum*). *Weed Technol.* **1990**, *4*, 87–91. [CrossRef]
73. Riffle, M.S.; Thilsted, W.E.; Murray, D.S.; Ahring, R.M.; Waller, G.R. Germination and Seed Production of Unicorn-Plant (*Proboscidea louisianica*). *Weed Sci.* **1988**, *36*, 787–791. [CrossRef]
74. Monaco, T.J.; Weller, S.C.; Ashton, F.M. *Weed Science: Principles and Practices*; John Wiley & Sons: Hoboken, NJ, USA, 2002.
75. Prather, T.; Miller, T.; Hulting, A. Control of Problem Weeds. In *Pacific Northwest Weed Management Handbook*; Oregon State University Agricultural: Corvallis, OR, USA, 2011.
76. Lyon, L.L.; Keeling, J.W.; Dotray, P.A. Evaluation and Adaptation of the HADSS® Computer Program in Texas Southern High Plains Cotton. *Weed Technol.* **2004**, *18*, 315–324. [CrossRef]
77. Bagheri, A.; Sohrabi, N.; Mondani, F.; Nosratti, I. Weed Infestation Is Affected by Chickpea Farmer Demographics and Agronomic Practices. *Weed Res.* **2021**, *61*, 45–54. [CrossRef]
78. Mirdavoodi, H.; Marvi Mohadjer, M.R.; Zahedi Amiri, G.; Etemad, V. Disturbance Effects on Plant Diversity and Invasive Species in Western Oak Communities of Iran (Case Study: Dalab Forest, Ilam). *Iran. J. For. Poplar Res.* **2013**, *21*, 1–15. [CrossRef]
79. Hemmat, A.; Eskandari, I. Tillage System Effects upon Productivity of a Dryland Winter Wheat–Chickpea Rotation in the Northwest Region of Iran. *Soil. Tillage Res.* **2004**, *78*, 69–81. [CrossRef]
80. Assadi, M.; Ramak Maassoumi, A.; Khatamsaz, M. *Flora of Iran*; Ministry of Agriculture, Islamic Republic of Iran: Tehran, Iran, 1989.
81. Parsons, W.T.; Parsons, W.T.; Cuthbertson, E. *Noxious Weeds of Australia*; CSIRO Publishing: Clayton South, Australia, 2001.
82. Nosratti, I.; Almaleki, S.; Chauhan, B.S. Seed Germination Ecology of Soldier Thistle (*Picnomon acarna*): An Invasive Weed of Rainfed Crops in Iran. *Weed Sci.* **2019**, *67*, 261–266. [CrossRef]
83. Nosratti, I.; Sabeti, P.; Chaghamirzaee, G.; Heidari, H. Weed Problems, Challenges, and Opportunities in Iran. *Crop Prot.* **2020**, *134*, 104371. [CrossRef]
84. Bagheri, A.; Zargarian, N.; Mondani, F.; Nosratti, I. Artificial Neural Network Potential in Yield Prediction of Lentil (*Lens culinaris* L.) Influenced by Weed Interference. *J. Plant Prot. Res.* **2020**, *60*, 284–295. [CrossRef]
85. Stanley, K.A.; Shirtliffe, S.J.; Benaragama, D.; Syrovy, L.D.; Duddu, H.S.N. Field Pea and Lentil Tolerance to Interrow Cultivation. *Weed Technol.* **2018**, *32*, 205–210. [CrossRef]
86. Vargas, O.; Martínez, M.; Dávila, P. Two New Species of Physalis (Solanaceae) Endemic to Jalisco, Mexico. *Brittonia* **2001**, *53*, 505–510. [CrossRef]
87. Ramírez, F.; Davenport, T.L. The Phenology of the Capuli Cherry [*Prunus Serotina* Subsp. *Apuli* (Cav.) McVaugh] Characterized by the BBCH Scale, Landmark Stages and Implications for Urban Forestry in Bogotá, Colombia. *Urban. For. Urban. Green.* **2016**, *19*, 202–211. [CrossRef]
88. Singh, H.; Singh, T.; Singh, A.P.; Kaur, S.; Arora, S.; Singh, B. Hepatoprotective Effect of *Physalis divaricata* in Paracetamol Induced Hepatotoxicity: In Vitro, in Silico and in Vivo Analysis. *J. Ethnopharmacol.* **2022**, *290*, 115024. [CrossRef]
89. Defelice, M.S. The Black Nightshades, *Solanum nigrum* L. et al.—Poison, Poultice, and Pie1. *Weed Technol.* **2003**, *17*, 421–427. [CrossRef]
90. Nazarialam, J.; Alizadeh, H.; Rahimian Mashhadi, H. Aspects of Management and Biology of Annual Ground Cherry (*Physalis divaricata*) in Sugar Beet Fields. *Iran. Crop Sci.* **2010**, *41*, 569–576.
91. Ghorbani, R.; Zeidali, E.; Hosseini, M. The Effect of Some Environmental Factors on Germination and Emergence of Weeds of Hyoscyamus Niger, *Physalis divaricata* and *Rumex crispus*. *Crop Ecophysiol.* **2016**, *36*, 661–674.
92. Harker, K.N.; Blackshaw, R.E.; Clayton, G.W. Wild Oat (*Avena fatua*) vs. Redstem Filaree (*Erodium cicutarium*) Interference in Dry Pea. *Weed Technol.* **2007**, *21*, 235–240. [CrossRef]
93. Rashed Mohassel, M.H.; Nadjafi, H. *Biology and Control of Weeds*; Ferdowsi University of Mashhad: Mashhad, Iran, 2001.
94. Mousavi, S.K.; Ahmadi, A. Influence of Environmental Factors on Germination of Ground Cherry, *Physalis divaricata*. *Appl. Entomol. Phytopathol.* **2008**, *76*, 59–78.

95. Amjadi, H.; Heidari, G.; Babaei, S.; Sharifi, Z. Evaluation of Yield, Yield Components and Some Quality Traits of Tuber of Potato (*Solanum tuberosum* L.) under Different Weed and Nutritional Management Practices. *Front. Plant Sci.* **2025**, *15*, 1495541. [CrossRef]
96. Hoagland, R.E. *Microbes and Microbial Products as Herbicides: An Overview*; American Chemical Society: Washington, DC, USA, 1990.
97. Shamshiri, M.; Heydarnejad, J.; Kamali, M.; Pouramini, N.; Massumi, H. Identification of Wild Hosts of Tomato Yellow Leaf Curl Virus in South-Eastern Iran. *Arch. Phytopathol. Plant Prot.* **2019**, *52*, 917–929. [CrossRef]
98. Darvishnia, M.; Ziedali, E.; Azadbakht, N.; Darvishnia, F.; Golzardi, F. Biocontrol of *Physalis Divericata* with Antagonistic Plant Pathogens in Lorestan Province. *Weed Res.* **2011**, *2*, 43–59.
99. Ahmadkhani, E.; Nosratti, I.; Sabeti, P. Evaluation of Various Herbicides for Controlling Annual Ground Cherry *Physalis divaricata* in Sugar Beet *Beta Vulgaris* Fields. *J. Appl. Res. Plant Prot.* **2023**, *11*, 97–109. [CrossRef]
100. Zand, A.; Nizamabadi, N.; Baghestani, M.A.; Shimi, P.; Mousavi, S.K. *Guidelines for Chemical Control of Weeds in Iran*, 6th ed.; Mashhad University Jihad Publications: Mashhad, Iran, 2019.
101. Frediani, M.; Maggini, F.; Gelati, M.; Cremonini, R. Repetitive DNA Sequences as Probes for Phylogenetic Analysis in Vicia Genus. *Caryologia* **2004**, *57*, 379–386. [CrossRef]
102. Jaaska, V. Isozyme Variation and Phylogenetic Relationships in Vicia Subgenus Cracca (Fabaceae). *Ann. Bot.* **2005**, *96*, 1085–1096. [CrossRef]
103. Parsa, A. *Flora of Iran*; Forest and Rangeland Research Institute Publication: Tehran, Iran, 2000; Volume 33.
104. Honarmand, S.J.; Nosratti, I.; Nazari, K.; Heidari, H. Factors Affecting the Seed Germination and Seedling Emergence of Muskweed (*Myagrum perfoliatum*). *Weed Biol. Manag.* **2016**, *16*, 186–193. [CrossRef]
105. Amiri Deh Ahmadi, S.; Parsa, M.; Bannayan, M.; Nassiri Mahallati, M.; Deihimfard, R. Yield Gap Analysis of Chickpea under Semi-Arid Conditions: A Simulation Study. *Int. J. Plant Prod.* **2014**, *8*, 531–548.
106. Javanmard, A.; Nikdel, H.; Amani Machiani, M. Evaluation of Forage Quantity and Quality in Domestic Populations of Hairy Vetch (*Vicia villosa* L.), Vetch (*Vicia sativa* L.) and Caspian Vetch (*Vicia hyrcanica*) under Rainfed Condition. *J. Agric. Sci. Sustain. Prod.* **2019**, *29*, 15–31.
107. Bewley, J.D. Seed Germination and Dormancy. *Plant Cell* **1997**, *9*, 1055. [CrossRef] [PubMed]
108. Kissing Kucek, L.; Riday, H.; Rufener, B.P.; Burke, A.N.; Eagen, S.S.; Ehlke, N.; Krogman, S.; Mirsky, S.B.; Reberg-Horton, C.; Ryan, M.R. Pod Dehiscence in Hairy Vetch (*Vicia villosa* Roth). *Front. Plant Sci.* **2020**, *11*, 82. [CrossRef]
109. Nosratti, I.; Soltanabadi, S.; Honarmand, S.J.; Chauhan, B.S. Environmental Factors Affect Seed Germination and Seedling Emergence of Invasive *Centaurea Balsamita*. *Crop Pasture Sci.* **2017**, *68*, 583–589. [CrossRef]
110. Parker, T.A.; y Teran, J.C.; Palkovic, A.; Jernstedt, J.; Gepts, P. Genetic Control of Pod Dehiscence in Domesticated Common Bean: Associations with Range Expansion and Local Aridity Conditions. *bioRxiv* **2019**. [CrossRef]
111. Dürr, C.; Dickie, J.B.; Yang, X.Y.; Pritchard, H.W. Ranges of Critical Temperature and Water Potential Values for the Germination of Species Worldwide: Contribution to a Seed Trait Database. *Agric. For. Meteorol.* **2015**, *200*, 222–232. [CrossRef]
112. Nosratti, I.; Mahdavi-Rad, S.; Heidari, H.; Saeidi, M. Differential Tolerance of Pumpkin Species to Bentazon, Metribuzin, Trifluralin, and Oxyfluorfen. *Planta Daninha* **2017**, *35*, e017165650. [CrossRef]
113. Carrapiço, F. Azolla as a Superorganism. Its Implication in Symbiotic Studies. In *Symbioses and Stress: Joint Ventures in Biology*; Seckbach, J., Grube, M., Eds.; Springer: Dordrecht, The Netherlands, 2010; pp. 225–241.
114. Kenrick, P. *A History of Plants in Fifty Fossils*; Smithsonian Institution: Washington, DC, USA, 2020.
115. Heckman, C.W. Ecoclimatological Survey of the Wetland Biota in the Tropical Wet-and-Dry Climatic Zone. *Glob. Ecol. Biogeogr. Lett.* **1997**, *6*, 97–114. [CrossRef]
116. Costa, N.; Martins, D.; Rodella, R.; Rodrigues-Costa, A. Anatomical Leaf Changes in *Eichhornia crassipes* Due to Herbicides Application. *Planta Daninha* **2011**, *29*, 17–23. [CrossRef]
117. Lumpkin, T.A.; Plucknett, D.L. Azolla: Botany, Physiology, and Use as a Green Manure. *Econ. Bot.* **1980**, *34*, 111–153. [CrossRef]
118. Lejeune, A.; Peng, J.; Le Boulengé, E.; Larondelle, Y.; Van Hove, C. Carotene Content of Azolla and Its Variations during Drying and Storage Treatments. *Anim. Feed. Sci. Technol.* **2000**, *84*, 295–301. [CrossRef]
119. Hussner, A. Alien Aquatic Plant Species in European Countries. *Weed Res.* **2012**, *52*, 297–306. [CrossRef]
120. Farahpour-Haghani, A.; Jalaeian, M.; Alinia, F.; Majidi-Shilsar, F. First Report of Rush Veneer Nomophila Noctuella (Lepidoptera: Crambidae: Spilomelinae) on Water Fern in Wetlands and Rice Fields. In Proceedings of the 22nd Iranian Plant Protection Congress, Karaj, Iran, 27–30 August 2016. [CrossRef]
121. Sadeghi, R.; Zarkami, R.; Sabetraftar, K.; Van Damme, P. Application of Classification Trees to Model the Distribution Pattern of a New Exotic Species *Azolla filiculoides* (Lam.) at Selkeh Wildlife Refuge, Anzali Wetland, Iran. *Ecol. Modell.* **2012**, *243*, 8–17. [CrossRef]
122. Watanabe, I.; Berja, N.S. The Growth of Four Species of Azolla as Affected by Temperature. *Aquat. Bot.* **1983**, *15*, 175–185. [CrossRef]

123. Pirasteh, M. *The Protection and Restoration of Anzali Wetland*; Department of Environmental Protection: Gilan, Iran, 1995.
124. Sabet Raftar, K. Azolla Analysis of the Environmental Impacts on Aquatic Ecosystems of the Wetlands. Ph.D. Thesis, University of Tehran, Tehran, Iran, 1994.
125. Cronk, J.K.; Fennessy, M.S. *Wetland Plants: Biology and Ecology*; CRC Press: Boca Raton, FL, USA, 2016.
126. Padasht–Dehkaii, F.; Tajadoditalab, K.; Hosseini–Chaleshtare, M.; Rabiei, M.; Sharafi, N.; Shokrivahed, H.; Alizadeh, M.; Alinia, F.; Omrani, M.; Kavosi, M. *Paddy Rice Cultivation Guide*; Agricultural Research, Education and Extension Organization (AREEO): Tehran, Iran, 2015.
127. Axelsen, S.; Julian, C. Weed Control in Small Dams. Part II Control of Salvinia, Azolla and of Water Hyacinth. *Qld. Agric. J.* **1988**, *114*, 291–298.
128. Ashton, P.J. *Azolla Infestations in South Africa: History of the Introduction, Scope of the Problem and Prospects for Management*; Department of Water Affairs and Forestry: Pretoria, South Africa, 1992.
129. Winston, R.; Schwarzländer, M.; Hinz, H.L.; Day, M.D.; Cock, M.J.; Julien, M. *Biological Control of Weeds: A World Catalogue of Agents and Their Target Weeds*, 5th ed.; CABI Databases: Wallingford, UK, 2014.
130. Lock, J.M.; Gopal, B. Water Hyacinth. (Aquatic Plant Studies 1). *J. Trop. Ecol.* **1988**, *4*, 92–93. [CrossRef]
131. Villamagna, A.M.; Murphy, B.R. Ecological and socio-economic impacts of invasive water hyacinth (*Eichhornia crassipes*): A review. *Freshw. Biol.* **2011**, *55*, 282–298. [CrossRef]
132. Coetzee, J.A.; Hill, M.P.; Ruiz-Téllez, T.; Starfinger, U.; Brunel, S. Monographs on Invasive Plants in Europe N° 2: *Eichhornia crassipes* (Mart.) Solms. *Bot. Lett.* **2017**, *164*, 303–326. [CrossRef]
133. Dagno, K.; Lahlali, R.; Diourté, M.; Jijakli, H. Fungi Occurring on Waterhyacinth (*Eichhornia crassipes* (Martius) Solms-Laubach) in Niger River in Mali and Their Evaluation as Mycoherbicides. *J. Aquat. Plant Manag.* **2012**, *50*, 25–32.
134. Gunnarsson, C.; Mattsson, C. Water Hyacinth-Trying to Turn an Environmental Problem into an Agricultural Resource. *Waste Manag.* **1997**, *27*, 117–129. [CrossRef] [PubMed]
135. Rohi, A.M.; Naderi Jelodar, M.; Roshan Tabari, M.A.; Afraei, V.; Parakande, F. Ecology of Aquatic Plant Control Plant (*Eichhoornia Crassipes*) in Aquatic Ecosystems of Mazandaran Province. *J. Aquat. Casp. Sea* **2017**, *2*, 25–36.
136. Ben Bakrim, W.; Ezzariai, A.; Karouach, F.; Sobeh, M.; Kibret, M.; Hafidi, M.; Kouisni, L.; Yasri, A. *Eichhornia crassipes* (Mart.) Solms: A Comprehensive Review of Its Chemical Composition, Traditional Use, and Value-Added Products. *Front. Pharmacol.* **2022**, *13*, 842511. [CrossRef]
137. Mujere, N. Water Hyacinth: Characteristics, Problems, Control Options, and Beneficial Uses. In *Impact of Water Pollution on Human Health and Environmental Sustainability*; IGI Global: Hershey, PA, USA, 2016; pp. 343–361.
138. Moran, P.J. Leaf Scarring by the Weevils *Neochetina Eichhorniae* and *N. bruchi* Enhances Infection by the Fungus *Cercospora piaropi* on Water Hyacinth, *Eichhornia crassipes*. *BioControl* **2005**, *50*, 511–524. [CrossRef]
139. Duke, J.A. Handbook of Energy Crops. 1983. Available online: http://www.rain-tree.com/papaya.htm (accessed on 22 October 2024).
140. Mirzajani, A.; Naderi, S.; Parvaneh Moghadam, D. Distribution Survey and Some Biological Aspects of Water Hyacinth in Anzali Wetland, Guilan Province. *Iran. J. Plant Biol.* **2019**, *11*, 51–62.
141. Olivares, E.; Colonnello, G. Salinity Gradient in the Mánamo River, a Dammed Distributary of the Orinoco Delta, and Its Influence on the Presence of *Eichhornia crassipes* and *Paspalum repens*. *Interciencia* **2000**, *25*, 242–248.
142. Malik, A. Environmental Challenge Vis a Vis Opportunity: The Case of Water Hyacinth. *Environ. Int.* **2007**, *33*, 122–138. [CrossRef]
143. Troutman, J.P.; Rutherford, D.A.; Kelso, W. Patterns of Habitat Use among Vegetation-Dwelling Littoral Fishes in the Atchafalaya River Basin, Louisiana. *Trans. Am. Fish. Soc.* **2007**, *136*, 1063–1075. [CrossRef]
144. Schmitz, D.C.; Schardt, J.D.; Leslie, A.J.; Dray, F.J.; Osborne, J.A.; Nelson, B.V. The Ecological Impact and Management History of Three Invasive Alien Aquatic Plant Species in Florida. In *Biological Pollution: The Control and Impact of Invasive Exotic Species, Proceedings of a Symposium held at Indianapolis, IN, USA, 25–26 October 1991*; Indiana Academy of Science: Indianapolisna, IN, USA, 1993; pp. 173–194.
145. Brendonck, L.; Maes, J.; Rommens, W.; Dekeza, N.; Nhiwatiwa, T.; Barson, M.; Callebaut, V.; Phiri, C.; Moreau, K.; Gratwicke, B. The Impact of Water Hyacinth (*Eichhornia crassipes*) in a Eutrophic Subtropical Impoundment (Lake Chivero, Zimbabwe). II. Species Diversity. *Arch. Hydrobiol.* **2003**, *158*, 389–405. [CrossRef]
146. Inderjit, S. *Invasive Plants: Ecological and Agricultural Aspects*; Birkhäuser: Basel, Switzerland, 2005; ISBN 3764371374.
147. Wainger, L.A.; Harms, N.E.; Magen, C.; Liang, D.; Nesslage, G.M.; McMurray, A.M.; Cofrancesco, A.F. Evidence-Based Economic Analysis Demonstrates That Ecosystem Service Benefits of Water Hyacinth Management Greatly Exceed Research and Control Costs. *PeerJ* **2018**, *6*, e4824. [CrossRef] [PubMed]
148. Gunnarsson, C.C.; Petersen, C.M. Water Hyacinths as a Resource in Agriculture and Energy Production: A Literature Review. *Waste Manag.* **2007**, *27*, 117–129. [CrossRef] [PubMed]
149. Khotsa, M.D.; Mkolo, N.M.; Motshudi, M.C.; Mphephu, M.M.; Makhafola, M.A.; Naidoo, C.M. A Comprehensive Review of the Biology, Ecological Impacts, and Control Strategies of Eichhornia crassipes. *Diversity* **2025**, *17*, 564. [CrossRef]

150. Mudge, C.R.; Netherland, M. Response of Giant Bulrush, Water Hyacinth, and Water Lettuce to Foliar Herbicide Applications. *J. Aquat. Plant Manag.* **2014**, *52*, 75–80.
151. FAO. Progress on Water Hyacinth (*Eichhornia crassipes*) Management. Available online: https://www.fao.org/4/y5031e/y5031e0c.htm (accessed on 9 June 2025).
152. *ICID Aquatic Weeds and Their Management*; International Commission on Irrigation and Drainage: New Delhi, India, 2002.
153. Siemering, G.S.; Hayworth, J.D.; Greenfield, B.K. Assessment of Potential Aquatic Herbicide Impacts to California Aquatic Ecosystems. *Arch. Environ. Contam. Toxicol.* **2008**, *55*, 415–431. [CrossRef]
154. Chauhan, B.S.; Mahajan, G. *Recent Advances in Weed Management*; Springer: New York, NY, USA, 2014; ISBN 978-1-4939-1018-2.
155. Mirzajani, A.; Nezamabadi, N.; Salavatian, S.; Bagheri, S.; Salahi, M.; Rufchaei, R. Chemical Control of Water hyacinth by Some Herbicides and Their Effect on Some Aquatic Invertebrate. 2022. Available online: https://www.researchsquare.com/article/rs-1242651/v1 (accessed on 12 April 2023).
156. Martinez, B.; Reaser, J.K.; Dehgan, A.; Zamft, B.; Baisch, D.; McCormick, C.; Giordano, A.J.; Aicher, R.; Selbe, S. Technology Innovation: Advancing Capacities for the Early Detection of and Rapid Response to Invasive Species. *Biol. Invasions* **2019**, *22*, 75–100. [CrossRef]
157. Williams, F.; Constantine, K.L.; Ali, A.A.; Karanja, T.W.; Kibet, S.; Lingeera, E.K.; Muthike, G.; Rwomushana, I.; Godwin, J.; Day, R. An Assessment of the Capacity and Responsiveness of a National System to Address the Threat of Invasive Species: A Systems Approach. *CABI Agric. Biosci.* **2021**, *2*, 42. [CrossRef]
158. Ballard, H.L.; Fernandez-Gimenez, M.E.; Sturtevant, V.E. Integration of Local Ecological Knowledge and Conventional Science: A Study of Seven Community-Based Forestry Organizations in the USA. *Ecol. Soc.* **2008**, *13*, 37. [CrossRef]
159. Dowsley, M. Community Clusters in Wildlife and Environmental Management: Using TEK and Community Involvement to Improve Co-Management in an Era of Rapid Environmental Change. *Polar Res.* **2009**, *28*, 43–59. [CrossRef]
160. McNeely, J.A.; Mooney, H.A.; Neville, L.E.; Schei, P.J.; Waage, J.K. *Global Strategy on Invasive Alien Species*; IUCN: Gland, Switzerland; Cambridge, UK, 2001; ISBN 2-8317-0609-2.

agronomy

MDPI

Article

Is *Ambrosia trifida* L. Preparing for a Wider Invasion? Changes in the Plant Morpho-Functional Traits over a Decade

Aleksandra Savić [1], Maor Matzrafi [2], Sanja Đurović [3], Rodolfo Gentili [4,*] and Sandra Citterio [4]

1 Department of Weed Research, Institute for Plant Protection and Environment, 11040 Belgrade, Serbia; aleksandra.m.savic@gmail.com
2 Department of Plant Pathology and Weed Research, Agricultural Research Organization, Newe Yaar Research Center, Ramat-Yishay IL-30095, Israel; maorm@volcani.agri.gov.il
3 Department of Phytopharmacy and Environmental Protection, Institute for Plant Protection and Environment, 11040 Belgrade, Serbia; stojakovicsm@yahoo.com
4 Department of Earth and Environmental Sciences, University of Milano-Bicocca, 20126 Milano, Italy; sandra.citterio@unimib.it
* Correspondence: rodolfo.gentili@unimib.it; Tel.: +39-0264482700

Abstract: Plastic traits are argued to favor the establishment of invasive alien plants and to evolve during the invasion process, so that invasions may be favored by both plasticity and evolutionary changes in plasticity. Despite this aspect being pivotal to understanding invasion processes, no information is available about the effect of residence time on the evolution of phenotypic plasticity of aliens to produce locally adapted phenotypes. Therefore, we aimed to evaluate changes in the morpho-functional traits of the invasive alien *Ambrosia trifida* L. over time. Specifically, we grew *A. trifida* plants under laboratory (at different temperatures: 10 °C, 20 °C and 25 °C) and field conditions by using seeds collected from 2011 to 2020 from an Eastern European population. Data about seedling emergence, vegetative (laboratory and field) and reproductive (field) traits were collected, e.g., maximum plant height (cm), total dry biomass (g), number of flowers (n) and number of new seeds (n). Analysis of variance (ANOVA), linear regressions and a plasticity index were applied to assess differences in plant performances when grown from seeds of different ages. An auto-regressive integrated moving average (ARIMA) model was applied to predict future trends of some key plant traits (maximum height, total dry biomass and number of new seeds). Overall, the time of seedling emergence was not statistically different among seeds of different age, and regression analyses exhibited a positive relationship between residence time (seed age) and vegetative and reproductive parameters of *A. trifida*. Only male flower production exhibited a negative trend over time. At all temperatures (especially at 25 °C), the vegetative parameters increased with the decrease in residence time. The plasticity index of the measured traits tended to decrease over time for most traits both in the field and in the laboratory trials. The ARIMA model predicted that plant traits will continue to increase in size over time by 2030, suggesting further plastic adaptation of *A. trifida*. The results highlighted that residence time influences the evolution of phenotypic plasticity of *A. trifida*, which has strengthened its adaptability to the new conditions over a decade, also demonstrating a great adaptive and invasion potential for the future.

Keywords: giant ragweed; plasticity; adaptations; invasion; plant traits

1. Introduction

Biological invasions, representing one of the main components of global change, can have negative consequences on the structure, function and dynamics of ecosystems [1]. The success of biological invasions depends on a species' potential to colonize new habitats (e.g., high dispersal efficiency, the persistence of the seed bank, and population density) and on its competitive ability against native plant species (i.e., fast plant growth in terms of higher biomass production and high regeneration capacity; see [2]). In general, the impact of biological invasions on the resident community depends both on the biological characteristics of the invader and on those of the invaded community [2].

In particular, as highlighted by genetic studies, many invasive species show a quite rapid adaptation to novel environments, occurring within less than 20 generations [3,4]. However, the different speed of adaptive evolution across invasive species make it difficult to predict the potential range expansions of species, especially considering that this aspect has been scarcely considered until recently [5].

Recent studies have attempted to clarify how evolutionary changes are influenced by phenotypic plasticity, as the success of an invasive species has often been attributed to an all-purpose genotype that produces a suitable phenotype in different environments, implying a high level of phenotypic plasticity [6,7]. Plastic traits have been reported to evolve during the invasion process, so that invasions may be favored by both plasticity and evolutionary changes in plasticity [7]. Nevertheless, only a few studies have documented the divergence of traits in natural populations, after observing a gradient of plasticity in the same populations [6,8–10]. Specifically, the "plasticity first" hypothesis predicts that phenotypic plasticity may promote adaptive evolutionary changes by producing populations with increased fitness under different environmental conditions across the species' distribution ranges (stressful or novel) [11,12]. Over time, environmentally modulated changes in traits can be optimized through a genetic accommodation (i.e., selection) [13]. In this framework, during the period of climate change, many invasive alien species are predicted to be in the process of adaptation over a short time, which could lead to increased populations with better fitness [14–16]. In turn, trait plasticity could confer a competitive advantage to invasive aliens compared to native species, augmenting their impact on ecosystems [17]. Therefore, some plant traits could be modulated by new environmental conditions (for species colonizing new territories) via phenotypic plasticity and could be responsible for the invasion success of alien species [17]. Considering such important issues, more studies predicting how invasive species will behave and adapt to new environmental conditions are needed. Specifically, during biological invasions, species commonly spread over large and climatically diverse areas [18]. Therefore, predicting how plants will respond to new areas needs to be understood in terms of adaptation of local populations to the new conditions. Temperature in particular can exert effects on plants directly, and also indirectly through environmental factors that covary with temperature [19]. *Ambrosia trifida* L. is one of the most noxious alien plant species in Europe. It is native to North America, and today it occurs in almost all United States and Canadian provinces, with a lesser presence in the western part of the continent [20]. It has been widely introduced into Europe [21] and reported in many regions of Europe and Asia [22,23]. *A. trifida* is mostly in the naturalization phase, with a tendency to spread to new areas. In Central Europe, the species' populations have been recorded in wide areas in a phase of spreading [22,24]. The species is known to be highly variable morphologically and genetically and highly competitive for resources and space [25,26]. Indeed, it is among the first plant species emerging in early spring [27]. By late May, plants can reach a height of up to 4 m [28]. *A. trifida* is highly efficient in utilizing water and nutrients from soil and thus affects the availability of natural resources needed for the growth and development of other plant species [29,30]. As an agricultural weed, *A. trifida* may lead to significant yield losses of certain crops [31]. For example, Baysinger and Sims [32] and Harrison et al. [33] reported losses in maize and soybean yields of 13% and 50%, respectively, with only one *A. trifida* plant/m^2. In addition, Harrison et al. [31] found a yield loss in corn of 60% with 14 plants of *A. trifida*/10 m^2. In addition, in non-agricultural habitats, it can negatively affect the biodiversity of local flora and vegetation by suppressing autochthonous, non-native, rare and endangered species [34]. Finally, it is also an allergenic species. Its pollen is an important source of aeroallergens, and it has become the most serious and persistent cause of allergy-associated respiratory diseases in invaded areas [20,23]. Many researches have also demonstrated that plant diffusion and the prevalence of allergy are increasing worldwide due to all the drivers of global change and in particular to temperature rise [20].

Despite its potential for invasiveness, studies about the adaptation potential and plasticity of *A. trifida* are scarce. Previous research has mainly investigated its negative effects on agricultural production [32,35], herbicide resistance [36,37] and its distribution in Europe [22]. No information is available about the effect of residence time on the evolution of phenotypic plasticity to produce locally adapted phenotypes.

In this paper, considering *A. trifida* ecological characteristics, we hypothesized that the residence time of its populations in the invaded area may have promoted the adaptation of the species to the new environment by changing the morpho-functional traits associated with invasiveness to maximize the plant fitness. Specifically, starting from the scientific evidence of the high level of phenotypic plasticity of *A. trifida*, we assumed that the initial invasive environment had induced a phenotype characterized by traits that allowed the species to satisfy its needs, which, over time and with the persistence of the same environmental conditions, became heritable.

To test the hypothesis, we examined the morpho-functional traits of *A. trifida* plants from 10 seed populations collected in the same agriculture area during 10 consecutive years (from 2011 to 2020). Trait changes detected under different temperatures in controlled conditions and in field were statistically elaborated to determine the effect of the residence time on the evolution of phenotypic plasticity of *A. trifida* and to estimate its local adaptive potential.

2. Materials and Methods

2.1. Sampling

Ambrosia trifida L. seeds were collected from an agriculture area in Central Bačka (Vojvodina Province), Republic of Serbia (45°30′ N, 19°31′ E), each year from 2011 to 2020. The seeds were collected at maturity (in autumn) from at least 100 different mother plants (every year), then brought to the lab and stored at room temperature (20 to 23 °C) until use. To determine the changes in morpho-functional traits of plants grown from the 10 seed populations, a laboratory experiment was performed at the Institute for Plant Protection and Environment (Belgrade), and a field experiment was carried out at a farm near Dobrić, Republic of Serbia (44°41′ N, 19°34′ E) in 2021. Over the period of seed collection (2011–2020), the climatic conditions during *A. trifida* vegetative season (mean temperature and total rainfalls) at the site of seed collection and at the experimental site were similar (about 18 °C and 500 mm).

2.2. Laboratory Experiment

All the 10 stocks of *A. trifida* seeds were tested for emergence and growth under controlled conditions. To break the dormancy of seeds, they were exposed to stratification treatments in three steps: low temperature (4 °C/7 days), high temperature (40 °C/2 days) and low temperature again with the addition of KNO3 (4 °C/7 days).

Emergence tests were conducted in the soil substrate (Cultivo, Zeleni Hit, Belgrade, Serbia) (10 seeds/pot) at three different temperatures for 28 days and repeated three times: 10 °C, 20 °C and 25 °C. Such temperatures were chosen based on literature indicating that *A. trifida* germination occurs within a temperature range of 5–41 °C, with a maximum germination rate at 20 °C and an optimum between 10 and 25 °C [38]. Thus, we chose the three temperatures with the temperature of 20 °C corresponding to the peak of maximum germination and the 10 and 25 °C as the lower and upper optimum. With regards to 10 °C, we specifically chose this temperature also because *A. trifida* has been observed to be one of the first summer annual weeds to emerge under this soil temperature early in spring ([22] and according to the Republic Hydrometeorological Institute of Serbia). In addition, testing different temperatures can highlight the ability of the species to adapt to different climatic conditions, as already observed for *A. artemisiifolia* [17]. The sprouted seeds were then used to test plant growth and development at the three different temperatures (10 °C, 20 °C, 25 °C) in three growth chambers with identical photoperiods and light intensity (16/8 h light/dark). We used 150 pots (140 mm × 110 mm)

(3 temperatures × 5 replications per year × 10 years). Pots were arranged in a completely randomized design and were rotated weekly to minimize spatial variation. The seeds were planted at 0.5 cm depth in each pot and regularly watered to maintain soil moisture. After sprouting, one seedling was grown per pot for about 2 months, for each treatment. The experiment was set up on 12 April, and the following plant traits were recorded on three subsequent dates (3 May, 24 May and 14 June): maximum plant height (cm), plant number of leaves (n), maximum plant width (cm), fresh and dry biomass of roots (g) and total plant dry biomass (g).

2.3. Field Experiment

The experiment was performed between April and October 2021 (Dobrić, Serbia, 44°41′ N, 19°34′ E) in a fallow field where no crop had been sown for more than seven years. The results of the chemical analysis of powdered clay soil regarding the basic agrochemical properties, the experiment timeline and the mean daily air temperature and precipitation during the trial period are reported in the Supplementary Materials: Tables S1–S3; Figure S1.

The experiment was set up on the principle of a completely randomized block design with three replications (Supplementary Figure S2). In particular, 10 treatments, corresponding to seeds collected for 10 successive years, from 2011 to 2020, were considered. Each year, seeds were sown in three parcels of 1 m^2 × 1 m^2 and grown at a density of five plants per square meter. The given plant density was maintained by thinning spontaneous vegetation growing in the parcels every 7 to 10 days over the season. The distribution of *A. trifida* within the parcels was kept uniform to minimize differences due to competition among individuals [34]. During plant growth, the following data about vegetative and reproductive traits were collected: maximum plant height (cm), maximum plant width (cm), number of leaves (n) (20 May, 20 June, 24 July, 24 August, 15 September), total dry biomass (g) at the end of the experiment (15 October), number of inflorescences (n) (24 August), number of male flowers (n) (24 August), number of seeds (n) (15 October) and seed mass (g) (15 October). Plant maximum height and width were measured with a wooden pullout meter, while dry biomass was measured with a precision scale in the laboratory. The seed mass was measured on an analytical balance, and the seed counter determined the exact number of seeds per plant (The Contador: Pfeuffer GmbH, Kitzingen, Germany, an optical seed counter with an integrated vibration channel).

2.4. Statistical Analysis

ANOVA and Bonferroni's test were applied to determine statistical differences in seed mass and time of emergences among populations (2011–2020) during the laboratory test. The relationships between the residence time and the recorded *A. trifida* traits in control and field conditions were analyzed by linear regression. Standard error of parameters and adjusted R^2 were used to show model fitness. Based on the data obtained, the behavior of the future generations of *A. trifida* were predicted by applying the ARIMA model [39]. Phenotypic plasticity of individual traits was calculated for the species for each observation year with the relative distance plasticity index (RDPI), as described by [40]. The RDPI is calculated as the Euclidian distances between the expressed trait values of individuals exposed to different temperatures in both control and field conditions. In particular, the RDPI considered interactions with temperature in lab and field conditions for all traits year by year. All analyses were performed in R version 4.2.2. [41].

3. Results

Overall, although the 10 seed populations of *A. trifida* of different ages showed the same average mass ($F = 0.789$, $p > 0.05$, Figure 1) and the same mean time of seedling emergence at 10 °C ($F = 1.202$, $p > 0.05$, Figure 2), 20 °C ($F = 2.189$, $p > 0.05$, Figure 2) and 25 °C ($F = 1.024$, $p > 0.05$, Figure 2), their derived plants varied in vegetative and generative traits under both laboratory and field conditions.

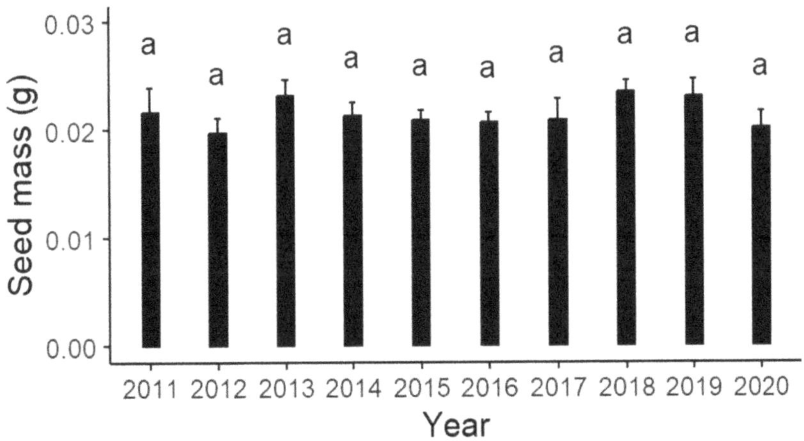

Figure 1. Mean mass of *A. trifida* seeds collected in successive years from 2011 to 2020. Identical letters indicate non-significant differences between years ($p > 0.05$).

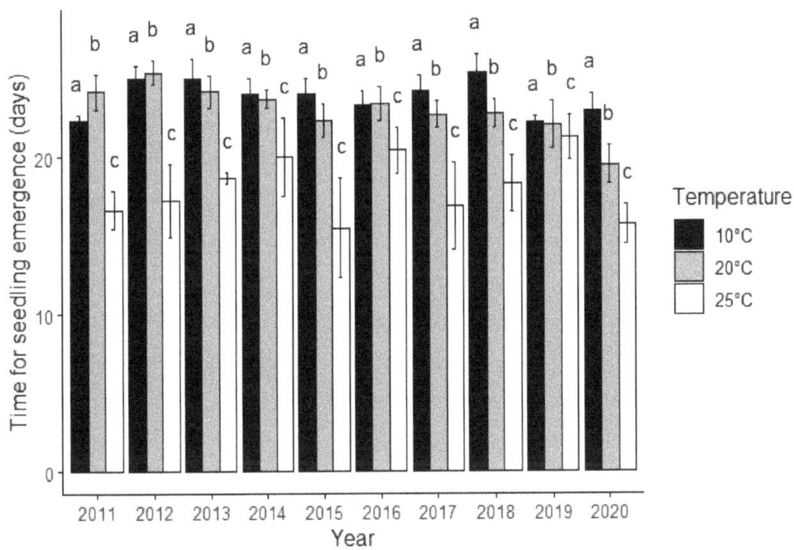

Figure 2. Time required for seedling emergence from the 10 seed populations of *A. trifida* at different temperatures (10 °C, 20 °C and 25 °C) under laboratory conditions. Identical letters indicate non-significant differences ($p > 0.05$) between the years.

3.1. Plant Traits Measurements in Controlled Conditions

Regression analyses exhibited a positive relationship between residence time (seed age) and vegetative parameters of *A. trifida* (Figure 3). At all temperatures, the vegetative parameters increased with the decrease of residence time; the increase was most pronounced at 25 °C (Figure 3). Specifically, a relatively marked positive relationship was recorded between the residence time of *A. trifida* and the trait of maximum height (F = 6.835, $p < 0.01$), which was most pronounced at temperatures of 10 °C ($R^2 = 0.231$, $p < 0.01$) and 25 °C ($R^2 = 0.355$, $p < 0.01$) (Figure 3A). A positive trend was also recorded for the parameters of maximum plant width (F = 17.285, $p < 0.01$) (Figure 3B), number of leaves (F = 21.632, $p < 0.01$) (Figure 3C), root dry biomass (F = 14.34, $p < 0.01$) (Figure 3E) and the total plant dry biomass (F = 72.065, $p < 0.01$) (Figure 3F), which was most pronounced at 25 °C ($R^2 = 0.662$, $p < 0.01$) (Figure 3F). Significant interactions between temperature and year were recorded for number of leaves (F = 3.91, $p < 0.05$), fresh root mass (F = 5.477, $p < 0.05$), dry root mass (F = 8.833, $p < 0.01$) and dry mass (F = 14.04, $p < 0.01$).

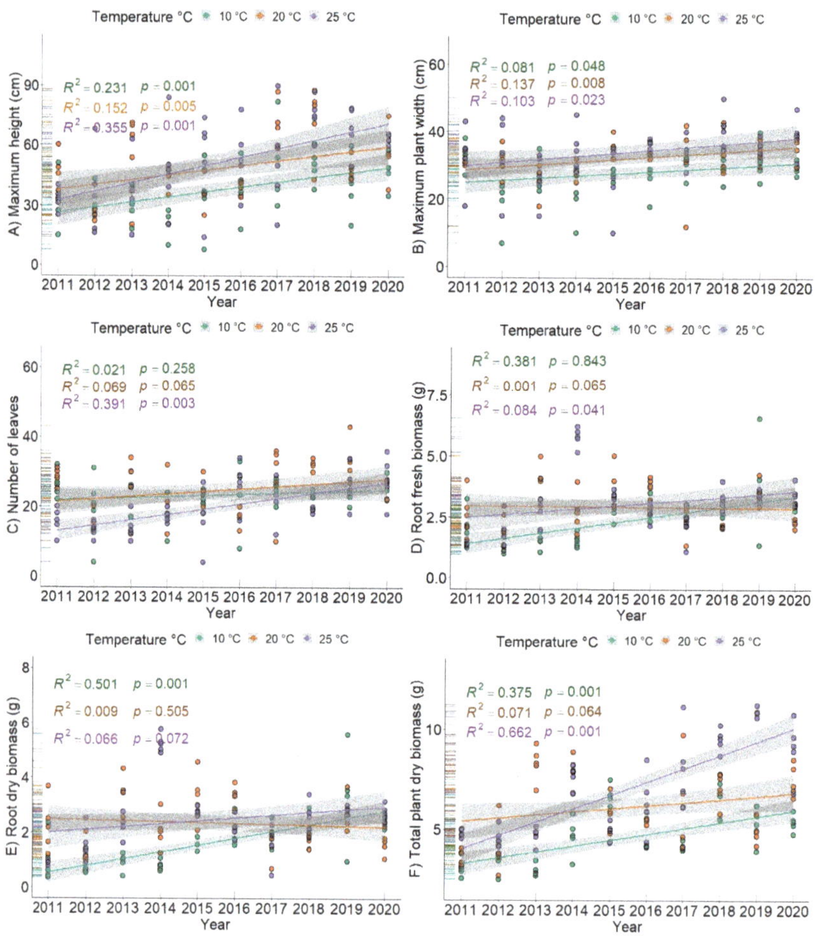

Figure 3. Influence of residence time on vegetative traits (**A**—maximum height; **B**—maximum plant widht; **C**—number of leaves; **D**—root fresh biomass; **E**—root dry biomass; **F**—total plant dry biomass) of *A. trifida* plants grown in controlled conditions at different temperatures.

3.2. Plant Traits Measurements in Field Conditions

In the field experiment, the effect of residence time on vegetative and reproductive traits was similar to that observed in control conditions (Figure 4). With decreasing seed age, vegetative and almost all generative parameters increased. Regression analysis confirmed the significant influence of residence time on the examined parameters. Throughout the growing season, height, width and number of leaves of the plants exhibited a positive trend (Figure 4A–C). Specifically, in the last assessment, a statistically significant positive relationship ($p < 0.01$) was found between residence time and plant maximum height (F = 58.521, $R^2 = 0.283$, $p < 0.01$) (Figure 4A), total dry biomass (F = 141.03, $R^2 = 0.488$, $p < 0.01$) (Figure 4D), number of inflorescences (F = 83.618, $R^2 = 0.487$, $p < 0.01$) (Figure 4E) and number of newly produced seeds (F = 66.467, $R^2 = 0.309$, $p < 0.01$) (Figure 4G).

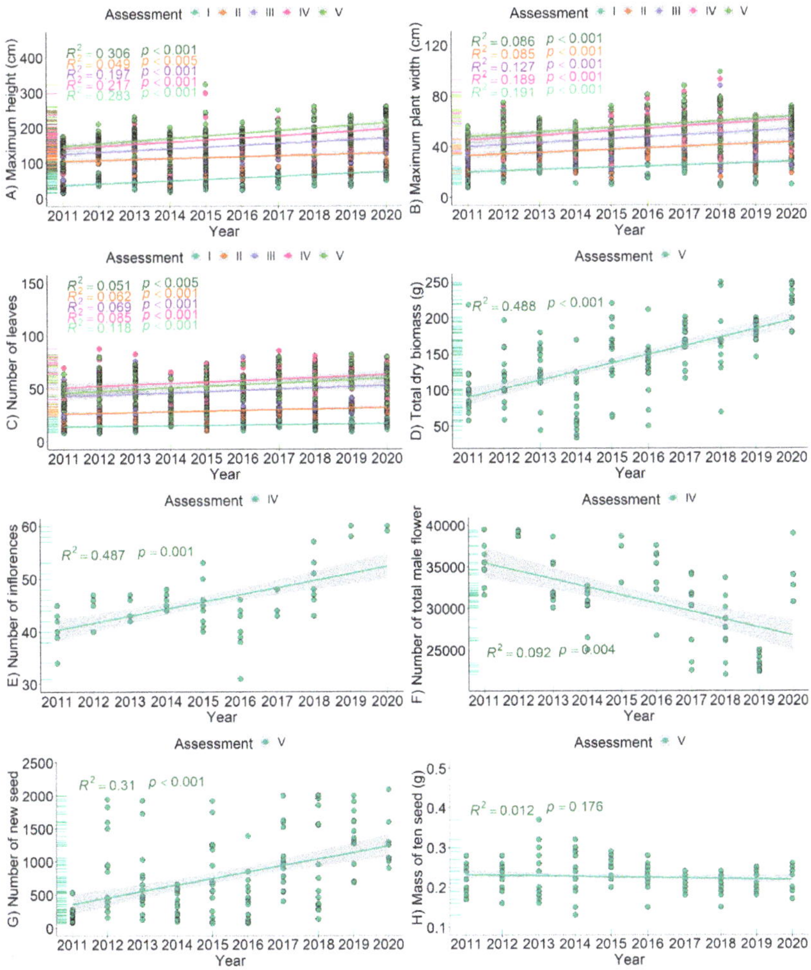

Figure 4. Influence of residence time on morpho-functional vegetative (**A**—maximum height; **B**—maximum plant width; **C**—number of leaves; **D**—total dry biomass) and reproductive traits (**E**—number of inflorences; **F**—number of total male flower; **G**—number of new seed; **H**—mass of ten seed) of *A. trifida* plants grown in field.

A less marked relationship was found for the parameters of plant width (F = 34.812, R^2 = 0.191, $p < 0.01$) (Figure 4B) and number of leaves (F = 19.746, R^2 = 0.118, $p < 0.01$) (Figure 4C). Conversely, a negative correlation in the total production of *A. trifida* male flowers per plant was recorded (F = 8.953, R^2 = 0.092, $p < 0.05$) (Figure 4F), and the relationship between residence time and mass of 10 newly produced seeds was not significant (F = 1.468, R^2 = 0.012, p = 0.176) (Figure 4H). Moreover, significant positive Pearson's correlations between plant mass and seed production were recorded for all 10 consecutive years of seed populations (F = 39.14; R^2 = 0.756, $p < 0.01$ for 2011, F = 130.53; R^2 = 0.911, $p < 0.01$ for 2012, F = 21.76; R^2 = 0.626, $p < 0.01$ for 2013, F = 128.1; R^2 = 0.91, $p < 0.01$ for 2014, F = 48.97; R^2 = 0.791, $p < 0.01$ for 2015, F = 45.26; R^2 = 0.77, $p < 0.01$ for 2016, F = 65.25; R^2 = 0.83, $p < 0.01$ for 2017, F = 154.6; R^2 = 0.92, $p < 0.01$ for 2018, F = 139.13; R^2 = 0.91, $p < 0.01$ for 2019, F = 58.88; R^2 = 0.82, $p < 0.01$ for 2020) (Figure 5).

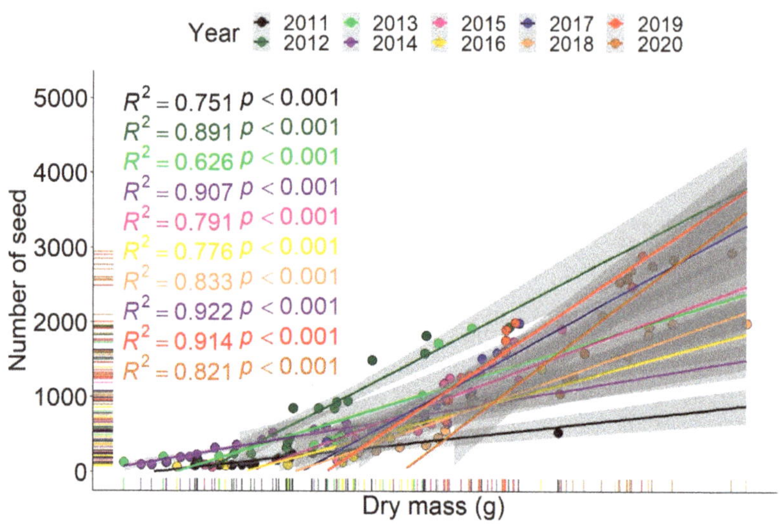

Figure 5. Correlations between vegetative and reproductive traits of *A. trifida*.

3.3. Plasticity of Traits

The phenotypic plasticity according to the RDPI, considering interactions with temperature in lab and field conditions, was calculated for all traits (Figures 6A and 6B, respectively). Overall, all traits exhibited significant differences (all with $p < 0.001$) over the investigated period, most notably for the highest and the lowest RDPI values across years (ANOVA and post hoc tests are shown in the Supplementary Materials). In the lab conditions, under the effects of different temperatures, the traits exhibited RDPI values ranging between 0.074 (plant width) and 0.539 (dry root mass). In the field conditions, the traits exhibited RDPI values ranging between 0.015 (number of inflorescences) and 0.471 (number of new seeds). Overall, both in the lab and the field, the plasticity of most of the traits tended to decrease through two main patterns over time: (a) a linear decreasing trend (e.g., root dry biomass in the lab; plant height in the field) and (b) a downward parabola trend, with a peak in the intermediate years of observation (e.g., plant height in the lab; total dry biomass and number of seeds in the field). A few traits showed a stable trend (total dry biomass in the lab; number of inflorescences in the field).

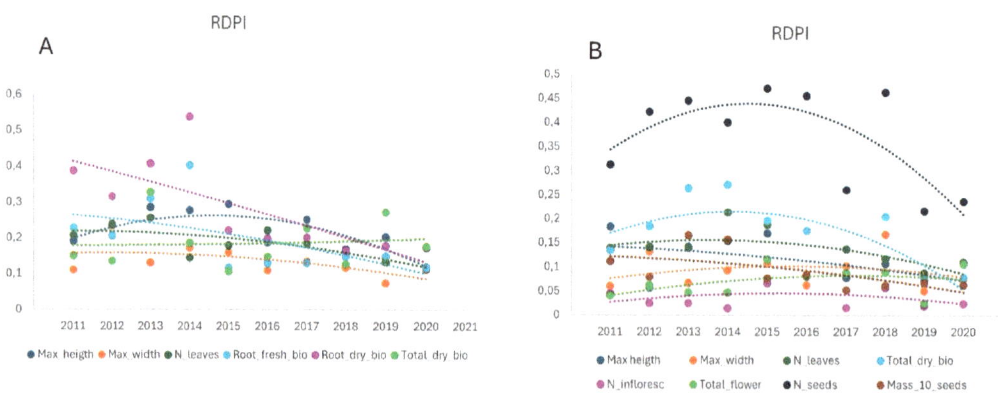

Figure 6. Relative distance plasticity index (RDPI) of *A. trifida* over time and considering the interaction with temperatures in (**A**) laboratory conditions and in (**B**) field conditions.

3.4. Prediction (ARIMA Model) of the Future Performance of A. trifida

Measurements in field of the maximum plant height, plant dry biomass and number of produced seeds were used for predicting the future performance of *A. trifida*. Predictions revealed that all three parameters were statistically significant with positive trends. In the case of the maximum height, the ARIMA model was best fitted, and the expected maximum plant height is estimated to be 287.86 cm at 2030. For the same year, the expected dry mass and number of seeds are estimated to be 359.12 g and 2944.37/plant, respectively (Figure 7, Supplementary Materials: Table S4).

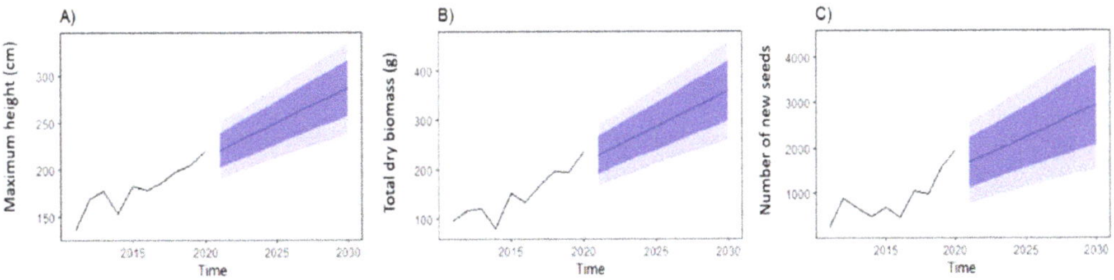

Figure 7. Time forecast (ARIMA model) for parameters (**A**) maximum height, (**B**) total dry biomass and (**C**) number of newly produced seeds. Purple area refers to the prediction of estimated parameters for the future.

4. Discussion

Our results show that in a 10-year experiment in both an agricultural area of Serbia, within its invasive range, and in lab conditions, *A. trifida* has modified most of its vegetative and reproductive traits, increasing its performance over time. This suggests that the species is adapting to local conditions through a genic and/or non-genic control of the main phenotypic traits. In our experiment, the residence time did not affect the germination time but modulated the plant development, generating ever larger individuals capable of producing more and more seeds. This trend was observed in both laboratory (at three different temperatures) and field experiments, suggesting an environmentally induced trait change refined through selection over time. Indeed, when undergoing change to adapt to new environments, populations that express a variety of phenotypes, as shown for *A. trifida* [42], are more likely to be able to adapt quickly, since beneficial phenotypes are more likely to be present and selected for [43]. We point out that the mechanisms enabling plant adaptation are still unclear [43,44] and the analysis of trait variation alone is not sufficient to determine the contribution of genetic and epigenetic factors. However, the reduction of the phenotypic plasticity that we observed in our experiments for most traits within a period of only 10 years suggests the heritability of those traits that allow the species to meet its needs in that specific environment and then its adaptation to local conditions. It is important to underline that we excluded the influence of seed collection and/or storage on our results. The ever larger plants that we obtained from the populations of seeds collected in the subsequent 10 years were not from larger seeds, which are reported to have higher germination and survival rates [45], as seeds were collected from over 100 different plants each year, and no statistical differences in their mean mass, viability and germination (percentage and time) was observed among the seed populations at the beginning of the experiment.

In any case, our results are different from those reported by [46]. They found that young natural populations of the invasive species *Mikania micrantha* were competitively inferior to the old ones, also considering an increasing distance from the source population, suggesting that invasive plants can achieve invasion success by modulating phenotypic plasticity and then changing their traits and genetic characters [46]. On the other hand, in our results in both field and lab conditions, a decrease in seed age corresponded to an increase of vegetative traits at all the tested temperatures (lab conditions), indicating a certain equilibrium between plasticity and trait heritability (genetic or epigenetic mechanisms) over time. Accordingly, Hovick et al. [47] suggested that the increasingly successful spreading of this species in new environments and regions can be attributed, at least in part, to its adaptive potential. Interestingly, so far, the average height of *A. trifida* plants in Europe, its invasive range, is about half of that recorded in America [48], its native range (2 vs. 4 m), and is increasing and expected to reach 3 m in 10 years according to our data and the predictive ARIMA model. It seems that adaptation to the invasive range tends to make the species regain the size and, in general, the better performance it displays in its native range. We can speculate that the better native population performance, at a vegetative but also reproductive level, relates to genotypes that are probably similar to those that characterized the populations of *A. trifida* at the beginning of its introduction into Europe, which did not perform as well in this new environment, and which started a gradual adapting process.

Across the literature, several plants and, in particular, invasive species show evidence of plasticity and genetic adaptation to environmental gradients [17,49,50]. In fact, in variable environments, invasive species are reported to rapidly adjust those traits generally associated with invasive potential, such as generation time, growth rate, and dispersal ability, which result in phenotypic plasticity and new genetic architectures or adaptation abilities to new regions through genetic accommodation that can subsequently occur [4,51]. In our results, plasticity occurs over time with a multi-temporal shift of traits and growth adjustments. Of particular interest in our experiment was the different resource allocations for reproduction. We observed that the residence time of *A. trifida* was negatively correlated with male flower production, whereas it was positively correlated with plant biomass and seed production. In species belonging to the genus *Ambrosia*, the pattern of male and female flowers is peculiar, with a more significant number of male than female inflorescences [52]. This trend can indicate that different conditions may cause different resource allocations in the studied species, likely acting on phenotypic plasticity [42,53].

Although there is not a great amount of data in the literature about morphological differences within the same population at the multi-year age level of the species, some investigations have reported that many biotic and abiotic factors (of which some are implicated in the environmental changes) might result in the rapid selection and evolution of plant phenotypes. Some species can adapt to local climatic conditions after only one generation [54]. Franks et al. [55] found rapid evolution of flowering time in *Brassica rapa* under the influence of climate change, after just three generations. Additionally, Nowak et al. [56] demonstrated the adaptive evolution of *Noccaea caerulescens* via trait variation after one generation of exposure to various levels of Zn contamination in the soil. In the same way, adaptive changes in life-cycle length in response to water stress have been observed over a few generations in *Arabidopsis thaliana* [57]. In the case of *A. trifida*, many different abiotic and biotic environmental factors can have contributed to the species' adaptation to the new environment by driving changes in the morpho-functional traits in order to maximize the plant's fitness. However, at the state of the art, it is difficult to determine what environmental factors have driven this change. Climate matching is a basic requirement for persistence in a new area. According to its global distribution, *A. trifida* comes from temperate areas. Although with several exceptions, this species mainly "moves" from/to warm temperate climate regions, generally avoiding equatorial, arid and snow climates, and apparently shows a limited shift of climatic niche between the native and adventive range [23]. In agreement with this, our laboratory test showed a better performance of the species at 20–25 °C than at 10 °C, regardless of the seeds' age. In addition, a limited shift of climatic niche is also supported by the analysis of mean daily temperatures and precipitations in the *A. trifida* native range and during its residence time in the invasive range, which did not change significantly. However, in addition to climate, many other abiotic and biotic factors can contribute to the difference between native and invasive niches [42]. Thus, further specific studies are needed to understand the factors that have determined the lower performance of *A. trifida* at the early stage of introduction and its need for adaptation by changing the morpho-functional traits generally associated with invasive potential [58]. The concern, of course, is that the acquisition of adaptive traits associated with invasiveness may favor the spread of *A. trifida* in Europe in the next several years with negative impacts at the environmental, social and economic levels. As several studies have highlighted, plasticity across generations would influence the ecological and evolutionary dynamics of weed adaptation [59–61].

Considering that our results were produced in a short observation period, it is reasonable to assume a positive effect of changes in plant size rather than a negative one, as already observed for most invaders, since they are generally highly adaptable and successful in new environments. The cases reporting that invasive species can experience declines in fitness and adaptability are rare across the literature and generally occur over medium to long time periods (see, for instance, [62]).

In conclusion, our results showed that a 10-year residence time of *A. trifida* populations in the invasion area under steady-state conditions has likely promoted a continuous trend of trait adjustment, allowing the adaptation of the species to the new environment through more vigorous vegetative growth and more efficient reproduction. Indeed, while the mechanisms enabling plant adaptation are still unclear, the reduction of the phenotypic plasticity that we observed in our experiments for most traits over time suggests the heritability of the favorable traits that allow adaptation to local conditions. The results seem to confirm the hypothesis that the adaptability potential of *A. trifida* was drastically strengthened over a decade, and that in the future we can expect plants with a larger vegetative performance and fitness that may greatly influence the spread of this species. Nevertheless, further studies about the genetic mechanisms involved in the accommodation of invasive plant traits are needed.

Supplementary Materials: The following supporting information can be downloaded at: https://www.mdpi.com/article/10.3390/agronomy14081601/s1, Figure S1: Weather conditions during the 2021 field experiment (Dobrić); Figure S2: Field experimental design; Table S1: Chemical properties of soil in the experimental field Table S2: Timeline and additional information about the field trials during the season 2021; Table S3: Temperature and Rainfall data for the localities mentioned in the article; Table S4: Time forecast (ARIMA model) for parameters Maximum height, Total dry biomass and Number of new seeds of *A. trifida*.

Author Contributions: A.S.: conceptualization, designed the study and methodology, responsible for the investigation (field and control conditions), resources, funding acquisition, data curation, bioinformatics analysis, visualization, writing—original draft, validation. M.M.: writing—review and editing. S.Đ.: help with the field sampling. R.G.: conceptualization, designing the study and methodology, funding acquisition, resources, supervision, validation, writing—review and editing. S.C.: conceptualization, designing the study and methodology, funding acquisition, resources, supervision, validation, writing—review and editing. All authors revised the manuscript and approved it for submission. All authors have read and agreed to the published version of the manuscript.

Funding: This research was funded by the Italian National Recovery and Resilience Plan (NRRP), Mission 4 Component 2 Investment 1.4—Call for tender No. 3138 of 16 December 2021, rectified by Decree n.3175 of 18 December 2021 of the Italian Ministry of University and Research, funded by the European Union—NextGenerationEU. Award Number: Project code CN_00000033, Concession Decree No. 1034 of 17 June 2022, adopted by the Italian Ministry of University and Research, CUP, H43C22000530001, Project title "National Biodiversity Future Center—NBFC". The Ministry of Science, Technological Development and Innovation Serbia (451-03-66/2024-03/200010) supported a very small piece of this project.

Data Availability Statement: Information about the data presented in this study are available from Aleksandra Savić (aleksandra.m.savic@gmail.com). The data are not publicly available due to privacy.

Acknowledgments: Special thanks go to farmer Vlastimir Savić for his support and assistance during the field experiment.

Conflicts of Interest: The authors declare no conflicts of interest.

References

1. Richardson, D.M.; Pyšek, P. Plant invasions: Merging the concepts of species invasiveness and community invasibility. *Prog. Phys. Geogr.* **2006**, *30*, 409–431. [CrossRef]
2. Pyšek, P.; Richardson, D.M. Invasive species, environmental change and management, and health. *Annu. Rev. Environ. Resour.* **2010**, *35*, 25–55. [CrossRef]
3. Prentis, P.J.; Wilson, J.R.; Dormontt, E.E.; Richardson, D.M.; Lowe, A.J. Adaptive evolution in invasive species. *Trends Plant Sci.* **2008**, *13*, 288–294. [CrossRef] [PubMed]
4. Whitney, K.D.; Gabler, C.A. Rapid evolution in introduced species, invasive traits and recipient communities: Challenges for predicting invasive potential. *Divers. Distrib.* **2008**, *14*, 569–580. [CrossRef]
5. Neve, P.; Villa-Aiub, M.; Roux, R. Evolutionary-thinking in agricultural weed management. *New Phytol.* **2009**, *184*, 783–793. [CrossRef] [PubMed]
6. Ghalambor, C.K.; McKay, J.K.; Carroll, S.P.; Reznick, D.N. Adaptive versus non-adaptive phenotypic plasticity and the potential for contemporary adaptation in new environments. *Funct. Ecol.* **2007**, *21*, 394–407. [CrossRef]
7. Kelly, M. Adaptation to climate change through genetic accommodation and assimilation of plastic phenotypes. *Philos. Trans. R. Soc. B* **2019**, *374*, 20180176. [CrossRef]
8. Ghalambor, C.K.; Hoke, K.L.; Ruell, E.; Fischer, E.K.; Reznick, D.N.; Hughes, K.A. Non-adaptive plasticity potentiates rapid adaptive evolution of gene expression in nature. *Nature* **2015**, *525*, 372–375. [CrossRef] [PubMed]
9. Hendry, A.P. Key questions on the role of phenotypic plasticity in eco-evolutionary dynamics. *J. Hered.* **2016**, *107*, 25–41. [CrossRef]
10. Walter, G.M.; Clark, J.; Cristaudo, A.; Terranova, D.; Nevado, B.; Catara, S.; Paunov, M.; Velikova, V.; Filatov, D.; Cozzolino, S.; et al. Adaptive divergence generates distinct plastic responses in two closely related Senecio species. *Evolution* **2022**, *76*, 1229–1245. [CrossRef]
11. West-Eberhard, M.J. *Developmental Plasticity and Evolution*; Oxford University Press: Oxford, UK, 2003.
12. Levis, N.A.; Pfennig, D.W. Evaluating 'plasticity-first' evolution in nature: Key criteria and empirical approaches. *Trends Ecol. Evol.* **2016**, *31*, 563–574. [CrossRef] [PubMed]
13. Lalejini, A.; Ferguson, A.J.; Grant, N.A.; Ofria, C. Adaptive phenotypic plasticity stabilizes evolution in fluctuating environments. *Front. Ecol. Evol.* **2021**, *9*, 715381. [CrossRef]
14. Clements, D.; Ditommaso, A. Climate change and weed adaptation: Can evolution of invasive plants lead to greater range expansion than forecasted? *Weed Res.* **2011**, *51*, 227–240. [CrossRef]
15. Sun, Y.; Brönnimann, O.; Roderick, G.K.; Poltavsky, A.; Lommen, S.T.; Müller-Schärer, H. Climatic suitability ranking of biological control candidates: A biogeographic approach for ragweed management in Europe. *Ecosphere* **2017**, *8*, e01731. [CrossRef]
16. Sun, Y.; Bossdorf, O.; Grados, R.D.; Liao, Z.; Müller-Schärer, H. Rapid genomic and phenotypic change in response to climate warming in a widespread plant invader. *Glob. Chang. Biol.* **2020**, *26*, 6511–6522. [CrossRef] [PubMed]
17. Gentili, R.; Ambrosini, R.; Augustinus, B.A.; Caronni, S.; Cardarelli, E.; Montagnani, C.; Müller-Schärer, H.; Schaffner, U.; Sandra, C. High phenotypic plasticity in a prominent plant invader along altitudinal and temperature gradients. *Plants* **2021**, *10*, 2144. [CrossRef] [PubMed]
18. van Boheemen, L.A.; Atwater, D.Z.; Hodgins, K.A. Rapid and repeated local adaptation to climate in an invasive plant. *New Phytol.* **2019**, *222*, 614–627. [CrossRef] [PubMed]
19. Kardol, P.; De Long, J.R.; Wardle, D.A. Local plant adaptation across a subarctic elevational gradient. *R. Soc. Open Sci.* **2014**, *1*, 140141. [CrossRef] [PubMed]
20. Montagnani, C.; Gentili, R.; Citterio, S. Ragweed is in the air: *Ambrosia* L. (Asteraceae) and pollen allergens in a changing world. *Curr. Protein Pept. Sci.* **2022**, *24*, 98–111. [CrossRef]
21. Lawalree, A. Les *Ambrosia* adventices en Europe occidentale. *Bull. Jard. Bot. L'état À Brux.* **1947**, *18*, 305–315.
22. Follak, S.; Dullinger, S.; Kleinbauer, I.; Moser, D.; Essl, F. Invasion dynamics of three allergenic invasive Asteraceae (*Ambrosia trifida*, *Artemisia annua*, *Iva xanthiifolia*) in central and eastern Europe. *Preslia* **2013**, *85*, 41–61.
23. Montagnani, C.; Gentili, R.; Smith, M.; Guarino, M.F.; Citterio, S. The Worldwide Spread, Success, and Impact of Ragweed (*Ambrosia* spp.). *Crit. Rev. Plant Sci.* **2017**, *36*, 139–178. [CrossRef]
24. Obradov, D.P.; Radak, B.Đ.; Bokić, B.S.; Anačkov, G.T. Floristic diversity of the central part of the South Bačka loess terrace (Vojvodina, Serbia). *Biol. Serbica* **2021**, *42*, 5–25. [CrossRef]
25. Patzoldt, W.L.; Tranel, P.J. Molecular analysis of cloransulam resistance in a population of giant ragweed. *Weed Sci.* **2002**, *50*, 299–305. [CrossRef]
26. Sako, Y.; Regnier, E.E.; Daoust, T.; Fujimura, K.; Harrison, S.K.; McDonald, M.B. Computer image analysis and classification of giant ragweed seeds. *Weed Sci.* **2001**, *49*, 738–745. [CrossRef]
27. Chauvel, B.; Fried, G.; Follak, S.; Chapman, D.; Kulakova, Y.; Le Bourgeois, T.; Marisavljevic, D.; Monty, A.; Rossi, J.P.; Starfinger, U.; et al. Monographs on invasive plants in Europe N° 5: *Ambrosia trifida* L. *Bot. Lett.* **2021**, *168*, 167–190. [CrossRef]
28. Malidža, G.; Vrbničanin, S. Novo nalazište alohtone korovske vrste *Ambrosia trifida* L. na području Vojvodine. In Proceedings of the VIII Savetovanje o Zaštiti Bilja, Zbornik Rezimea, Vojvodina, Serbia, 2006; pp. 44–45.
29. Abul-Fatih, H.A.; Bazzaz, F.A. The biology of *Ambrosia trifida* L. II. Germination, emergence, growth and survival. *New Phytol.* **1979**, *83*, 817–827. [CrossRef]

30. Mun, S.; Lee, E.J. Litter decomposition rate and nutrient dynamics of giant ragweed (*Ambrosia trifida* L.) in the non-native habitat of South Korea. *Plant Soil* **2020**, *449*, 373–388. [CrossRef]
31. Webster, T.M.; Loux, M.M.; Regnier, E.E.; Harrison, S.K. Giant ragweed (*Ambrosia trifida*) canopy architecture and interference studies in soybean (*Glycine max*). *Weed Technol.* **1994**, *8*, 559–564. [CrossRef]
32. Baysinger, J.A.; Sims, B.D. Giant ragweed (*Ambrosia trifida* L.) interference in soybeans (*Glycine max*). *Weed Sci.* **1991**, *39*, 358–362. [CrossRef]
33. Harrison, S.K.; Regnier, E.E.; Schmoll, J.T.; Webb, J.E. Competition and fecundity of giant ragweed in corn. *Weed Sci.* **2001**, *49*, 224–229. [CrossRef]
34. Savić, A.; Oveisi, M.; Božić, D.; Pavlović, D.; Saulić, M.; Schärer, H.M.; Vrbničanin, S. Competition between Ambrosia artemisiifolia and *Ambrosia trifida*: Is there a threat of a stronger competitor? *Weed Res.* **2021**, *61*, 298–306. [CrossRef]
35. Harrison, S.K.; Regnier, E.E.; Schmoll, J.T. Postdispersal predation of giant ragweed (*Ambrosia trifida*) seed in no-tillage corn. *Weed Sci.* **2003**, *51*, 955–964. [CrossRef]
36. Vink, J.P.; Soltani, N.; Robinson, D.E.; Tardif, F.J.; Lawton, M.B.; Sikkema, P.H. Occurrence and distribution of glyphosate-resistant giant ragweed (*Ambrosia trifida* L.) in southwestern Ontario. *Can. J. Plant Sci.* **2012**, *92*, 533–539. [CrossRef]
37. Kaur, S.; Aulakh, J.; Jhala, A.J. Growth and seed production of glyphosate-resistant giant ragweed (*Ambrosia trifida* L.) in response to water stress. *Can. J. Plant Sci.* **2016**, *96*, 828–836. [CrossRef]
38. Ruziev, F.; Park, K.I.; Umurzokov, M.; Khaitov, B.; Bo, A.B.; Jia, W.Q.; Hien, L.T.; Choi, J.S.; Park, K.W. Seed Germination Ecology of Giant Ragweed (*Ambrosia trifida*) in Korea. *Weed Turfgrass Sci.* **2020**, *9*, 21–28. [CrossRef]
39. Box, G.E.P.; Jenkins, G.M. *Time Series Analysis: Forecasting and Control*; Holden-Day: Oakland, CA, USA, 1976.
40. Valladares, F.; Sanchez-Gomez, D.; Zavala, M.A. Quantitative estimation of phenotypic plasticity: Bridging the gap between the evolutionary concept and its ecological applications. *J. Ecol.* **2006**, *94*, 1103–1116. [CrossRef]
41. R Core Team. *R: A Language and Environment for Statistical Computing*; R Foundation for Statistical Computing: Vienna, Austria, 2022. Available online: https://www.R-project.org/ (accessed on 20 June 2024).
42. Wang, R.; Sun, B.; Li, J.D.; Wang, G.J.; Sun, J.N.; Wang, X.R.; Zhong, R.T. Effects of light intensity on the phenotypic plasticity of invasive species *Ambrosia trifida*. *J. Appl. Ecol.* **2012**, *23*, 1797–1802.
43. Lee, Y.; Chen, A.S.; Tajwar, F.; Kumar, A.; Yao, H.; Liang, P.; Finn, C. Surgical fine-tuning improves adaptation to distribution shifts. *arXiv* **2022**, arXiv:2210.11466.
44. Ng, E.T.H.; Kinjo, A.R. Plasticity-led evolution as an intrinsic property of developmental gene regulatory networks. *Sci. Rep.* **2023**, *13*, 19830. [CrossRef] [PubMed]
45. Meyer, S.E.; Carlson, S.L. Achene mass variation in *Ericameria nauseosus* (Asteraceae) in relation to dispersal ability and seedling fitness. *Funct. Ecol.* **2008**, *15*, 274–281. [CrossRef]
46. Huang, F.; Peng, S.; Chen, B.; Liao, H.; Huang, Q.; Lin, Z.; Liu, G. Rapid evolution of dispersal related traits during range expansion of an invasive vine *Mikania micrantha*. *Oikos* **2015**, *124*, 1023–1030. [CrossRef]
47. Hovick, S.M.; McArdle, A.; Harrison, S.K.; Regnier, E. A mosaic of phenotypic variation in giant ragweed (*Ambrosia trifida*): Local- and continental-scale patterns in a range-expanding agricultural weed. *Evol. Appl.* **2018**, *11*, 995–1009. [CrossRef] [PubMed]
48. Bassett, I.J.; Crompton, C.W. The biology of Canadian weeds. 55. *Ambrosia trifida* L. *Can. J. Plant Sci.* **1982**, *62*, 1003–1010. [CrossRef]
49. Turner, K.G.; Fréville, H.; Rieseberg, L.H. Adaptive plasticity and niche expansion in an invasive thistle. *Ecol. Evol.* **2015**, *5*, 3183–3197. [CrossRef] [PubMed]
50. Gentili, R.; Ferré, C.; Cardarelli, E.; Caronni, S.; Montagnani, C.; Abu El Khair, D.; Citterio, S.; Comolli, R. Performing as a transformer species? The invasive alien *Reynoutria bohemica* changes ecosystem properties in a riparian woodland. *Weed Res.* **2022**, *62*, 446–456. [CrossRef]
51. Fierst, J.L. A history of phenotypic plasticity accelerates adaptation to a new environment. *J. Evol. Biol.* **2011**, *24*, 1992–2001. [CrossRef]
52. Wayne, P.; Foster, S.; Connolly, J.; Bazzaz, F.; Epstein, P. Production of allergenic pollen by ragweed (*Ambrosia artemisiifolia* L.) is increased in CO_2-enriched atmospheres. *Ann. Allergy Asthma Immunol.* **2002**, *88*, 279–282. [CrossRef]
53. Zhao, W.; Liu, T.; Liu, Y.; Wang, H.; Wang, R.; Ma, Q.; Bi, X. The significance of biomass allocation to population growth of the invasive species *Ambrosia artemisiifolia* and *Ambrosia trifida* with different densities. *BMC Ecol. Evol.* **2021**, *21*, 175. [CrossRef]
54. Skrøppa, T.; Kohmann, K. Adaptation to local conditions after one generation in Norway spruce. *Tree Genet. Genomes* **1997**, *4*, 171–177. [CrossRef]
55. Franks, S.J.; Sim, S.; Weis, A.E. Rapid evolution of flowering time by an annual plant in response to a climate fluctuation. *Proc. Natl. Acad. Sci. USA* **2007**, *104*, 1278–1282. [CrossRef] [PubMed]
56. Nowak, J.; Frérot, H.; Faure, N.; Glorieux, C.; Liné, C.; Pourrut, B.; Pauwels, M. Can zinc pollution promote adaptive evolution in plants? Insights from a one-generation selection experiment. *J. Exp. Bot.* **2018**, *69*, 5561–5572. [CrossRef] [PubMed]
57. Brachi, B.; Aimé, C.; Glorieux, C.; Cuguen, J.; Roux, F. Adaptive value of phenological traits in stressful environments: Predictions based on seed production and laboratory natural selection. *PLoS ONE* **2012**, *7*, e32069. [CrossRef] [PubMed]
58. Schutte, B.J.; Regnier, E.E.; Harrison, S.K. The association between seed size and seed longevity among maternal families in *Ambrosia trifida* L. populations. *Seed Sci. Res.* **2008**, *18*, 201–211. [CrossRef]

59. Herman, J.J.; Sultan, S.E. Adaptive transgenerational plasticity in plants: Case studies, mechanisms, and implications for natural populations. *Front. Plant Sci.* **2011**, *2*, 102. [CrossRef] [PubMed]
60. Matzrafi, M.; Osipitan, O.A.; Ohadi, S.; Mesgaran, M.B. Under pressure: Maternal effects promote drought tolerance in progeny seed of Palmer amaranth (*Amaranthus palmeri*). *Weed Sci.* **2020**, *69*, 31–38. [CrossRef]
61. Osipitan, O.A.; Matzrafi, M.; Ohadi, S.; Mesgaran, M.B. Maternal water stress reduces sensitivity of offspring to herbicides in *Amaranthus palmeri* (Palmer amaranth). *Weed Res.* **2021**, *61*, 486–495. [CrossRef]
62. Sádlo, J.; Vítková, M.; Pergl, J.; Pyšek, P. Towards site-specific management of invasive alien trees based on the assessment of their impacts: The case of Robinia pseudoacacia. *NeoBiota* **2017**, *35*, 1–34. [CrossRef]

diversity

MDPI

Article

Are Iron-Rich Calcareous Mine Sites Easily Invaded by Invasive Plant Species?

Jin-Hui Liu [1], Justin S. H. Wan [1,*], Susan Rutherford [1], Ali Al-Namazi [2], Hui Liu [1], Zhi-Cong Dai [1], Jian-Fan Sun [1], Xiao-Qin Sun [3] and Dao-Lin Du [1,*]

1 Institute of Environment and Ecology, School of the Environment and Safety Engineering, Jiangsu University, Zhenjiang 212013, China
2 King Abdulaziz City for Science and Technology (KACST), P.O. Box 6086, Riyadh 11442, Saudi Arabia
3 Institute of Botany, Jiangsu Province and Chinese Academy of Sciences, Nanjing 210014, China
* Correspondence: wanjustinsh@gmail.com (J.S.H.W.); daolindu@163.com (D.-L.D.)

Abstract: Plant diversity in relatively harsh environments, such as metal-polluted areas tends to be relatively low. Invasive plants may invade harsh environments more easily than native plants. However, studies often find fewer invasive species in stressful edaphic habitats (such as serpentine soils). Those examples may represent relatively extreme conditions. Moderately stressful habitats may be more invaded given the advantages of invasive plants. We surveyed the plant diversity in four site pairs across three seasons. Sites consist of abandoned mines and reference sites. The mine sites have calcareous soils with relatively high iron, basic pH, and lower nutrients than reference sites. Results: There were 153 plant species among the four site pairs. Around 80 and 66% of species in calcareous and reference sites were introduced species respectively. Diversity varied across seasons but tended to be lower in the mine sites. One of the mines was significantly more invaded. Across sites, the number of invasive species and their abundances was not different from that of native species. Invasive plants are as capable of invading moderately stressful calcareous sites as native species, with some sites tending to be even more invaded.

Keywords: plant diversity; invasibility; edaphic; stress; abundance

1. Introduction

For over a century, biological invasion is an important part of global environmental change, with increasing research interest worldwide [1]. Many alien species are widely distributed in non-native areas, affecting ecosystems, livelihoods, and having significant economic consequences [2]. Invasive plants are usually widely distributed and have a strong ability to adapt to different environments. These invasive plants can displace native plant species and alter the structure and function of native plant communities [3], with the effect of reducing local or regional biodiversity [4]. Invasive plant species tend to have advantages over native species in disturbed or resource-rich conditions. Even under low resource availability, invasive plants may have higher resource utilization efficiency than native plants [5–8]. Anthropogenic activities such as mining change the structure and function of plant communities, while the species diversity in mining areas could be high, disturbance from mining activities may promote the invasion of alien plants. Studies found that many invasive plants have the ability to settle in sites polluted by heavy metals [9–11]. Some alien invasive plants have a high tolerance to pollutants, such as heavy metals [12–14]. Therefore, it is very necessary to study these questions.

Past studies have found that invasive plants can invade metal-contaminated areas [15]. Toxicity from metal pollution can seriously interfere with the growth and photosynthesis of plants [16,17]. Moreover, adaptive changes to detoxify metals and other stressors present in metal-polluted sites (such as low soil moisture or low nutrients) can reduce the performance of plants inhabiting such sites. Although the toxic effects of heavy metals can hinder the growth and fecundity of most plants, tolerant plants can proliferate in metal-polluted habitats [18,19]. Hyperaccumulator and metallophyte species are commonly found in metal-contaminated environments such as mine sites. Hyperaccumulators are species that can accumulate high concentrations of metals in their tissues without detriment to their growth [20]. Metallophytes are plant species that are able to grow and form sustained populations in polluted soils, with obligatory metallophytes being found only in metal-contaminated sites, and facultative metallophytes found in both non-contaminated and contaminated sites [21]. In their introduced ranges, invasive plants found in mine sites can be assumed to be facultative metallophytes (as they are highly abundant at the regional level and not restricted to mine sites). Studies have confirmed that many invasive plants have a strong tolerance to metals, with some able to hyperaccumulate metals in tissues [22–25]. Moreover, many invasive plants have inherent advantages in growth, fecundity, and other physiological characteristics (such as high phenotypic plasticity and allelopathy). Under competition, invasive plants usually perform better than or even inhibit native plant species [26,27]. Although competition may not be as important as tolerance to stress in stressful sites, few habitats are truly non-competitive.

Mine sites represent relatively new, highly disturbed habitats appearing relatively suddenly in time. Several invasion hypotheses suggest that the richness of invasive plants' metal-polluted sites should be higher compared to normal sites. The ideal weed hypothesis (IWH) posits that successfully established introduced plants have relatively high growth rates, germination rates, and competitive ability [28], and are likely to colonize new areas through long-distance dispersal [29]. Similarly, the empty niche hypothesis (ENH) predicts that ecosystems with low species diversity and more empty niches are more vulnerable to invasion [30,31]. The invasion meltdown hypothesis (IMH) holds that the positive interaction between invasive species will lead to positive group-level feedback, which will intensify the impact of invaders and promote additional invasion by other alien species [32]. Similarly, invasive plants can competitively exclude native plants (e.g., in calcareous grasslands [33]).

Some studies have found that plant communities in abiotically stressful or 'harsh' environments are less often invaded by alien plant species than those in milder environments. A review by Zefferman [34] found that invasive plants do not often invade harsh environments such as those with high soil salinity (e.g., salt marshes) or sodic and/or alkaline soils, which have been observed to be less invaded [28]. The proportion of invasive species may be higher in less stressful sites such as mine sites under restoration (e.g., Bauman et al. [35]) and calcareous sites because stressors are less extreme. For example, some studies have shown that there is a unimodal relationship between the fine-scale species richness and the main vegetation gradient in the dry calcareous grassland in the Baltic Sea [36]. Some studies, such as Bauman et al. [35], show that there is no difference in plant diversity due to the increased coverage of invasive and naturalized species in mine areas under reclamation, and plant diversity has decreased in over five years. However, the alkalinity of the unweathered soil in the reclaimed mining areas is significantly higher (>6.5 pH), which may be conducive to the colonization of non-native herbaceous plants that can adapt to this condition [37]. A survey by Aguiar et al. [38] found that alien species are present in 91% of the survey sites, and are more common in sites at two river types (siliceous and calcareous rivers) than in the reference sites, and calcareous sites have higher invasibility, in the non-reference sites of the two river types, alien species have been more common, and the calcareous background is more sensitive to invasion. Globally, calcareous soil accounts for about 30% of the land area, and its characteristics mainly include high calcium carbonate content, low organic matter content, and low plant nutrient elements absorbed and utilized by plants [39]. Calcareous soil mainly exists in areas with higher altitudes, and the pH value is mainly around 7.8. Compared with most soils, calcareous soil is more vulnerable to heavy metal pollution [40]. Restoration work such as the addition of lime, fertilizer, or mulch used to improve the soil in barren sites may ameliorate stressors and promote colonization by adjacent plant species [41]. It is not yet clear if plant communities in certain harsh habitats such as sites with lower metal contamination and calcareous sites would be less invaded. If such sites have lower native plant diversity then invasive plants should be able to invade more easily [42]. Sites vary widely in harshness. Previous studies finding a lower presence of invaders in harsh sites may represent the most extreme conditions, such as heavily polluted sites. Communities with low species diversity and functional diversity in moderately stressful sites should be more seriously invaded than those with high species diversity and functional diversity.

In this study, we tested the plant diversity in an area with multiple mine sites, and test the following hypothesis: (1) the proportion (by the number of individuals and number of species) of invasive plants in the mine sites will be higher than that of native plants compared to adjacent reference sites, (2) the plant diversity in mine sites will be lower than reference sites; and (3) invasive species are abundant in both mine sites and non-mine areas, but native species will be restricted to either site.

2. Materials and Methods

2.1. Study Area

Four mine sites were located on hills around Zhenjiang in Jiangsu Province, China (Figure 1). These are Xianglushan (119°13′41″ E, 32°07′42.9″ N), Goutoushan (119°20′12″ E, 32°03′48″ N), Majishan (119°38′13″ E, 32°06′39″ N, Dantu District), and Longwangshan (119°36′55″ E, 32°08′12″ N). Henceforth, these ‘Locations’ will be referred to as XLS, GTS, MJS, and LWS, respectively. ‘Sites’ refer to the site type (mine site or reference site) within each ‘Location’. The mine site vegetation consisted of low herbs and grasses, with few trees and shrubs. Adjacent woodland and open sites with sparse or no tree cover were chosen for the reference sites paired with the mine sites (1–1.5 km from the respective mine sites). The soil chemical properties of all sites are presented in Table 1. Mine site soils are calcareous and have relatively high levels of iron and calcium, with a basic pH (7.6–8.2) and lower C and N content. The MJS mine site soil is also relatively high in zinc (>500 mg/kg). Mine soils are very rocky with sparser vegetation cover, particularly during spring (Figure 2). Reference sites had neutral to slightly acidic pH and lower levels of iron.

In the early 1960s to 70s, large-scale quarrying occurred in the areas surrounding Zhenjiang. Mining activities ceased around 2001, after which efforts were made to restore some of the sites. A portion of the mine sites at MJS and XLS was planted with the shrubs *Lespedeza bicolor* and *Swida walteri* in restoration work, so we avoided sampling those areas of the mines. The soil type in the area is highly ferric yellow-brown soil, and occasionally brownish-red soft plastic loam. The rock stratum is mostly covered with 2–3 m of residual soil, and the lithology mainly includes quartz sandstone, basalt, limestone, dolomite, etc. The climate is highly seasonal and is within the subtropical monsoon climate zone, with an annual average temperature of 16 °C. The dominant wind direction is southwest in summer and northwest in winter. The annual average precipitation is 989 mm, and the annual average relative humidity is 76%.

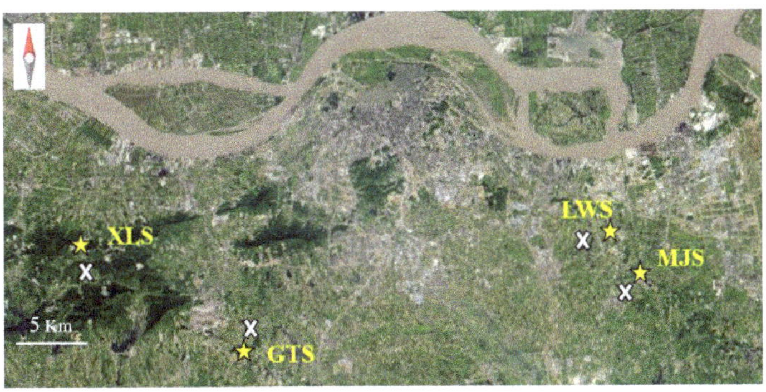

Figure 1. Map of survey site pairs in the areas around Zhenjiang City on the Yangtze River, Jiangsu Province. Yellow stars indicate mine sites and white crosses indicate reference sites.

Figure 2. Photographs detailing the local site conditions: (**a**) MJS mine site, (**b**) GTS reference site, (**c**) MJS mine site vegetation in spring, (**d**) MJS mine site vegetation in autumn, (**e**) GTS mine site, and (**f**) LWS mine site.

Table 1. Soil properties from the sampling locations for plant diversity located around Zhenjiang (all values are total content in mg/kg, except for total carbon and nitrogen in g/kg). The soil was sampled at 10–30 cm depth. Metal concentration values are means ± SD (n = 3). Different lowercase letters represent significant differences at the location × site level ($p < 0.05$).

Location	Zn	Cr	Cu	Cd	Pb	Mn	Ca	Mg	Fe	C (g/kg)	N (g/kg)	pH
Mine sites												
GTS	215 ± 15 a	215 ± 26 a	61 ± 2 a	1.24 ± 0.87 a	52 ± 7 a	1210 ± 253 a	214,219 ± 7837 a	19,024 ± 2191 a	57,294 ± 3398 ab	8.73	0.68	8.13
LWS	263 ± 37 a	129 ± 12 a	46 ± 2 a	0.32 ± 0.05 a	47 ± 10 a	651 ± 267 a	89,221 ± 58,662 ab	80,848 ± 53,213 b	63,587 ± 2924 ab	5.94	0.77	7.65
MJS	544 ± 251 a	217 ± 100 a	82 ± 34 a	0.83 ± 0.31 a	186 ± 67 b	1383 ± 663 a	89,199 ± 80,260 ab	31,249 ± 14,924 ab	85,181 ± 28,443 a	7.84	0.81	8.06
XLS	334 ± 17 a	460 ± 253 a	69 ± 4 a	1.09 ± 0.21 a	79 ± 9 a	1089 ± 140 a	213,298 ± 57,124 a	19,206 ± 5678 a	60,118 ± 9012 ab	6.67	0.66	7.98
Reference sites												
LWS	355 ± 92 a	314 ± 168 a	66 ± 8 a	0.62 ± 0.08 a	93 ± 24 a	916 ± 110 a	37,863 ± 6863 b	22,634 ± 919 a	67,201 ± 10,723 ab	19.22	1.80	7.81
XLS	239 ± 28 a	224 ± 67 a	58 ± 12 a	0.67 ± 0.30 a	68 ± 7 a	759 ± 147 a	133,782 ± 114,978 ab	13,859 ± 452 a	68,089 ± 9087 ab	11.07	1.00	7.56
GTS	211 ± 31 a	113 ± 11 a	34 ± 3 a	1.40 ± 0.47 a	60 ± 5 a	677 ± 27 a	15,179 ± 4579 b	9238 ± 1433 a	45,059 ± 4892 ab	20.82	1.93	6.72
MJS	202 ± 15 a	157 ± 14 a	41 ± 15 a	0.43 ± 0.07 a	61 ± 3 a	501 ± 62 a	6383 ± 1085 b	8309 ± 884 a	50,129 ± 3318 ab	20.83	1.46	5.84

2.2. *Plant Diversity Survey and Soil Sample Collection*

We surveyed the diversity at three site pairs (mine site, and adjacent reference sites across 3 seasons in April and October (i.e., MJS, GTS, and XLS in spring and autumn 2021, and spring 2022). During the last sampling season (spring 2022), one of the site pairs (GTS) was inaccessible due to covid restrictions therefore a substitute site pair was used instead (LWS). Three 5 × 5 m quadrats were randomly placed at each sub-site, and the total number of plant species and number of individuals for each species in each quadrat were recorded. Individuals were counted non-destructively, but for some sprawling or dense clump-forming species (such as *Medicago* sp. or some Poaceae), plants were trimmed for better visibility. Soil samples were taken using a soil auger from a depth of 10–30 cm in three replicates, depending on the penetrability of the substrate. For each replicate, soil from three cores taken from within a 2 × 2 m area was pooled. The soil depth sampled depended on the penetrability of the substrate. A total of 2–3 kg of soil was taken at each site. The soil samples were sieved and air-dried for a week, and then sieved through a 40-mesh sieve to facilitate the measurement of the physical and chemical properties of the soil. After the soil was air-dried, samples were sent to Nanjing Convinced-Test Technology Co., Ltd. (Nanjing, China), which conducted the soil analysis. The soil's total nitrogen content was determined by the Kjeldahl method after digestion with sulfuric acid and an accelerator. Soil pH was measured using the glass electrode method. Soil organic carbon was determined by the potassium dichromate oxidation external heating method. Hydrofluoric acid perchloric acid digestion was used to determine the content of calcium and magnesium by ICP-OES. Cadmium, chromium, lead, zinc, copper, and iron concentrations in soil were determined by four acid digestion inductively coupled plasma spectrometry.

2.3. Statistical Analysis

The proportions of invasive and native species for the number of species and the number of individuals (i.e., abundance) were tested using Chi-square tests. To quantify the species diversity at the sites, we calculated the Shannon-Wiener diversity index (H) and Simpson diversity index (D) of each sampling site. The Simpson evenness index (E) was used to quantify the distribution of invasive and native plants at each sampling site. Species similarity was quantified using the Sørensen similarity index (SI). These are calculated using the following:

$$\text{Shannon} - \text{Wiener diversity index (H)} : -\sum_{i=1}^{s} p_i \ln p_i \quad (1)$$

$$\text{Simpson index (D)} : 1 - \sum_{i=1}^{s} \frac{N_i(N_i - 1)}{N(N-1)} \quad (2)$$

$$\text{Simpson} - \text{Evenness index (E)} : \frac{N\left(N/S - 1\right)}{\sum_{i=1}^{s} [N_i(N_i - 1)]} \quad (3)$$

$$\text{Sorensen similarity index (SI)} : SI = (2 * E_C)/(E_1 + E_2) \quad (4)$$

where N_i is the number of individuals of the ith species, $P_i = N_i/N$, is the proportion of the individual number of the *i*th species, S is the number of plant species in the sample, and N is the total number of individuals. The higher the H and D values, the higher the plant diversity at the site. E represents the species' evenness in the investigation area. The higher the E-value, the more uniform the plant distribution. The Sorensen Index was used to measure the similarity between two samples of data. E_C is the total number of elements in common between the sets, E_1 is the number of elements in site 1 and E_2 is the number of elements in site 2. Hutcheson's *t*-test was used to analyze differences in the Shannon-Weiner index (plant diversity) within mine sites versus reference sites [43]. Linear regression was used to test the relationship between abundance (number of individuals) in the mine sites versus abundance in reference sites for invasive and native species. Shapiro-Wilk and Levene's tests for checking the normality and homogeneity of variance, the results of the W test show that all *p*-values were less than 0.05, so the logarithmic transformation was conducted on the data. All analyses were conducted using SPSS statistical software (ver. 26, IBM, Armonk, NY, USA). A p-value < 0.05 is considered statistically significant.

3. Results

There was a total of 153 species belonging to 127 genera and 52 families of plants were recorded (Supplementary materials Table S1). The distribution in the presence of species across sites is summarized in Figure 3.

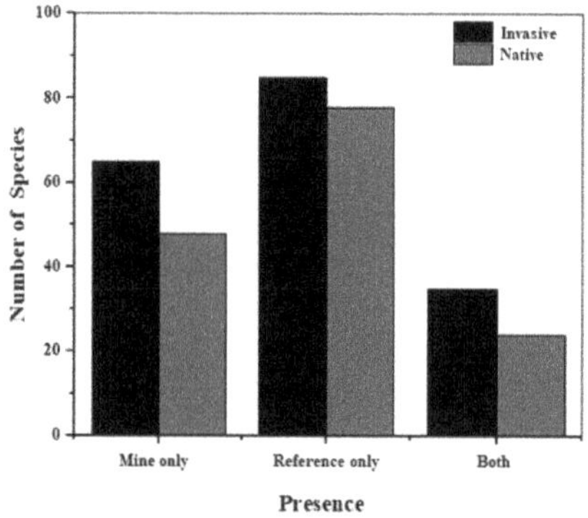

Figure 3. The total number of invasive and native plant species present across site type combinations.

3.1. Abundance and Species Richness of Invasive Plants across Site Pairs

The average abundance of most species across sites at all locations (i.e., XLS, MJS, LWS, and GTS) was not high (Figure 4). Fewer species were present in both mine and reference sites than those present at either of the site types (Figure 3). By focusing on site pairs, the mine sites can be considered to be a filter of the existing species pool in the local area (Figure 5). Interestingly, there was overall little difference between the proportion of invasive plants and the number of native plants at each site, paired by the number of individuals (Table 2) or by the number of species (Table 3). From the average abundance distribution of plant species in mine sites and reference sites in spring 2022 (Figure S1), it can be seen that more scattered points deviate from the coordinate axis. As in the spring of 2021, although most plants seem to prefer one site type over the other (Table 4), there are still more plant species that are present in both site types than in autumn. The difference among mine sites and reference sites in the proportions of invasive plants and native plants among site pairs at each location was not as obvious in spring 2021 and 2022 as in autumn 2021. Basically, there is little difference in the number of plant individuals among site types, but, on the whole, there are more invasive plants (particularly at the MJS site pair). Spring had a significantly higher proportion of invasive species than in autumn in the number of individuals only, but did not differ significantly in the number of species (Table 5).

Table 2. Chi-square test results comparing the number of individuals of invasive versus native species: (a) across all sites, (b) among site pairs (mine sites and adjacent reference sites). Bold p-values indicate statistically significant differences.

(a)				
	Reference	Calcareous	χ^2	p
Total number of individuals across all sites				
Native	1575 (34%)	1997 (35%)		
Invasive	3005 (66%)	3667 (65%)	48.88	0.36
(b)				
	Reference	**Mine site**	χ^2	p
Total number of individuals				
GTS				
Native	307 (40%)	254 (21%)		
Invasive	458 (60%)	946 (79%)	82.37	**<0.001**
MJS				
Native	421 (30%)	964 (39%)		
Invasive	987 (70%)	1504 (61%)	32.76	**<0.001**
XLS				
Native	586 (35%)	664 (44%)		
Invasive	1089 (65%)	842 (56%)	27.56	**<0.001**
LWS				
Native	261 (36%)	115 (23%)		
Invasive	471 (64%)	375 (77%)	20.46	**<0.001**

Table 3. Chi-square test results comparing the number of species of invasive versus native species: (a) across all sites, (b) among site pairs (mine sites and adjacent reference sites). Bold *p*-values indicate statistically significant differences.

(a)				
	Reference	**Mine site**	χ^2	*p*
Total number of species across all sites				
Native	77 (46%)	64 (38%)		
Invasive	88 (54%)	106 (62%)	71.14	**<0.001**
(b)				
	Reference	**Mine site**	χ^2	*p*
Total number of species				
GTS				
Native	24 (52%)	14 (29%)		
Invasive	22 (48%)	34 (71%)	5.17	**< 0.05**
MJS				
Native	21 (38%)	21 (42%)		
Invasive	34 (62%)	29 (58%)	0.16	0.69
XLS				
Native	27 (55%)	25 (42%)		
Invasive	22 (45%)	34 (58%)	1.74	0.19
LWS				
Native	5 (33%)	4 (31%)		
Invasive	10 (67%)	9 (69%)	0.02	0.88

Table 4. Species similarity results among site pairs at each location (mine and adjacent reference sites).

Location	Mine Site total	Reference Total	Species in Common	Sørensen Index
GTS	48	46	16	0.34
MJS	50	55	19	0.36
XLS	59	49	26	0.48
LWS	13	15	10	0.71

Table 5. Chi-square test results for the comparison in the proportion of invasive species to native species during spring vs. autumn across all sites. Bold *p*-values indicate statistically significant differences.

	Spring	**Autumn**	χ^2	*p*
Number of species				
Invasive	54 (54%)	52 (52%)		
Native	46 (46%)	47 (47%)	0.04	0.83
Number of individuals				
Invasive	15,653 (69%)	4367 (55%)		
Native	7184 (31%)	3569 (45%)	473.17	**<0.001**

There was a significant positive relationship between abundance in the mine site and abundance in reference sites for both native and invasive plants, where species that are relatively abundant are also abundant across sites (Figure 6). The total plant number and the plant species richness among the mine sites and reference sites were similar (Figure 3; Table 4). Overall, calcareous mine sites did not differ significantly in the proportion of individuals that are invasive compared to reference sites (Table 2a, $p > 0.05$). However, when the site pairs are considered together, we can find that the difference between the individual numbers of invasive plants and native plants in each sampling area is quite significant, and invasive plants have a higher abundance (Table 2b). The results from the chi-square tests showed that: in general, the abundance of invasive plants is slightly higher than that of native plants, but compared with calcareous areas and reference areas there was no obvious difference between the number of invasive plants and native plant species (Figures 3 and 4; Table 4). Looking at each sampling site separately, we found that, generally, the number of invasive plant species in all mine sites was higher than that of native plants, but in reference sites of GTS and XLS, the number of native plant species is slightly higher than that of invasive plants. Among the four sampling sites, only the number distribution of invasive plants and native plant species in GTS is different, while the number of invasive plants and native plant species in other sampling sites is not significantly different (Table 3).

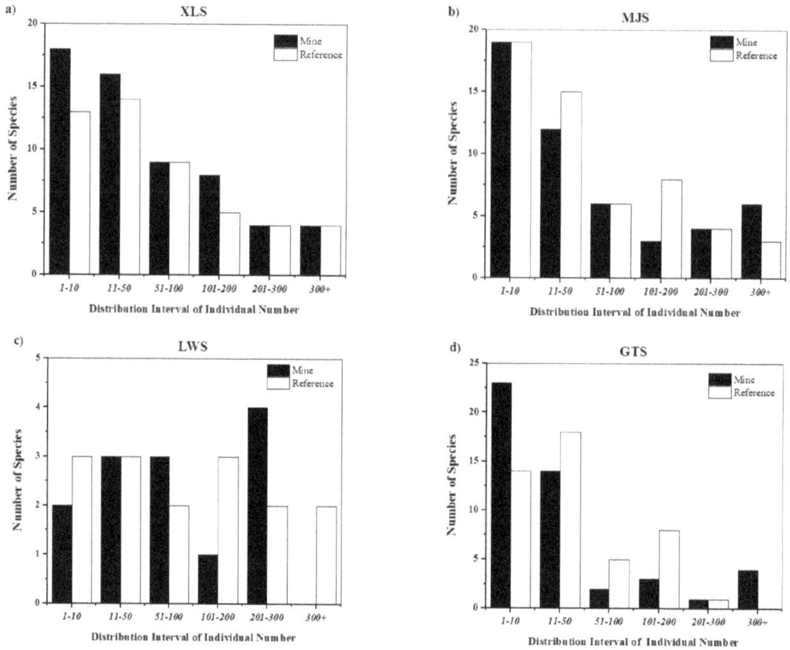

Figure 4. Histogram showing the distribution of plant species across varying levels of abundance at each location: (**a**) XLS, (**b**) MJS, (**c**) LWS, and (**d**) GTS. The sampling at LWS was conducted only during spring 2022 when the GTS site pair was inaccessible.

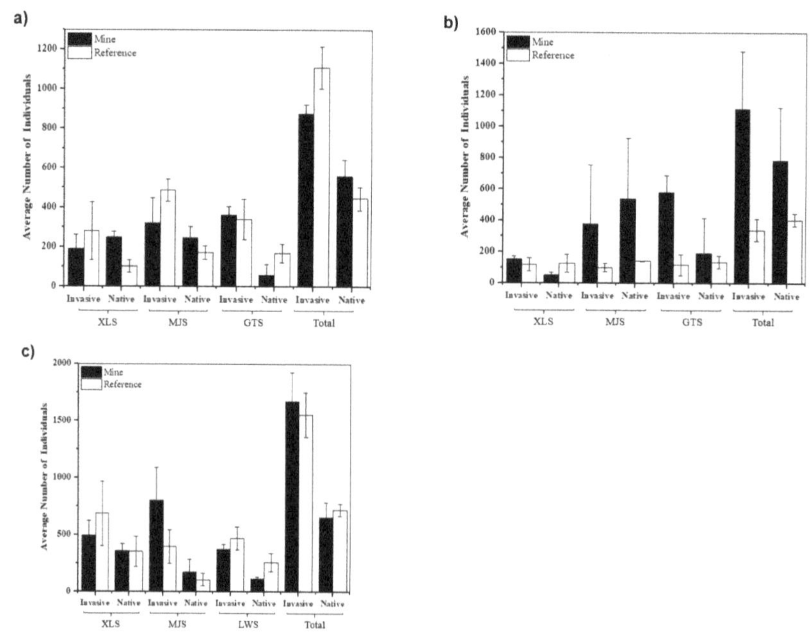

Figure 5. Comparison of average individual numbers of invasive plants and native plants in each sampling area for: (**a**) spring 2021 (April), (**b**) autumn 2021 (October), and (**c**) spring (April) 2022.

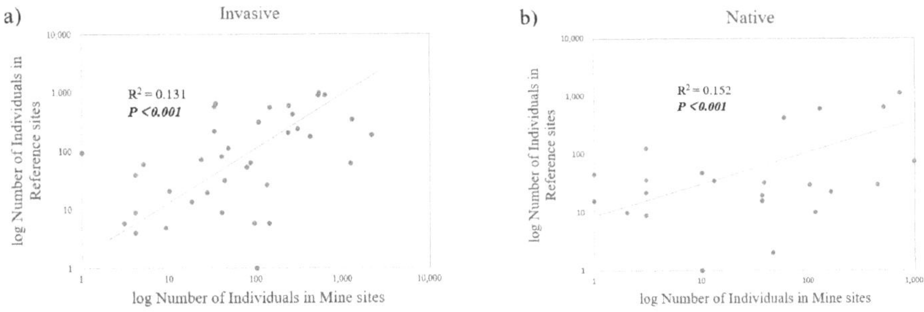

Figure 6. Linear regressions of plant abundance in mine sites versus reference sites for the site pairs at all locations: (**a**) log number of individuals for invasive plants (n = 37) and (**b**) log number of individuals for native plants (n = 23). Each dot point represents an individual species. Values are of the total within each site type, and species absent from either site type were excluded.

3.2. Diversity in Mine Sites Versus Reference Sites

The plant species highly abundant in the reference sites include herbs of Asteraceae, Leguminosae, Rubiaceae, and Geraniaceae, such as *Solidago canadensis, Galium spurium, Vicia sativa, Geranium wilfordii*, etc. Among them, *S. canadensis* accounts for nearly 10% of the total number of individuals. Only a few shrub and tree species were present at both sites (less than 1%), such as *Rubus hirsutus, Paederia cruddasiana, Ginkgo biloba,* and *Celtis sinensis.* Herbs are abundant in reference sites, accounting for 79% of the plant species (mainly Poaceae and Asteraceae). Interestingly, mine sites had relatively low evenness compared to reference sites at GTS and MJS, but only during autumn. There are 62 species of plants present in both mine sites and reference sites, including 38 invasive plants and 24 native species. Plants species were mainly herbs accounting for > 80% (especially Asteraceae and Poaceae which account for 19% and 31%, respectively). In the mine sites, this is similar where Poaceae accounted for 17% and Asteraceae accounted for 27%, including *S. canadensis, Setaria viridis, Erigeron annuus,* and *Elymus dahuricus.* There were only a few shrub species, such as *Rubus hirsutus, Duchesnea indica,* and *Rosa multiflora.* Tree species were common, including 15 species in 12 families, such as *Ginkgo biloba, Populus tomentosa, Pterocarya stenoptera, Celtis sinensis,* and *Broussonetia papyrifera,* all of which are drought-resistant and barren tree species [44]. Patterns in plant diversity varied across seasons. Mine sites tended to have lower diversity than reference sites (i.e., MJS during spring 2021 and 2022, LS during spring 2021, and GTS during autumn 2021), except for the MJS mine site during autumn, which had higher diversity than the reference site (Table 6). The other indices of diversity are presented in Table S2.

3.3. Distribution and Abundance of Native and Invasive Species among Sites

Most species are found only at one of the site types, especially at the MJS site pair (mine and reference sites) in autumn (Figure S1). *S. canadensis* and *Setaria viridis* were the most abundant species across the site pairs at XLS and GTS (Figure S1). In autumn (Figure 5), there were large differences in abundance among site pairs in each sampling location (e.g., GTS mine site *S. canadensis* individuals and GTS reference site *Eleusine indica* individuals; Figure S1). In mine sites, abundance was significantly higher than that of reference sites in spring than in autumn, and most of them are introduced or invasive plants (particularly at the MJS site pair). However, the number of species was similar between autumn and spring ($\chi2 = 0.04$, $p = 0.83$), but invasive species had higher abundance ($\chi2 = 473$, $p < 0.001$). The plant species diversity in spring 2022 was similar to that of spring in the preceding year (Table 3a). Mine sites had similar diversity to reference sites (Table S2).

Table 6. Results from Hutcheson's *t*-tests on the differences in plant diversity among mine versus reference sites (Shannon-Weiner index, H).

Location	Site	H	t	df	*p*
Spring 2021					
GTS	Mine Reference	1.58 1.58	0.05	5.7	0.11
MJS	Mine Reference	1.57 1.58	0.03	5.7	**<0.05**
XLS	Mine Reference	1.55 1.57	0.02	5.7	**<0.05**
Autumn 2021					
GTS	Mine Reference	1.55 1.57	0.02	5.9	**<0.05**
MJS	Mine Reference	1.57 1.54	0.01	5.9	**<0.05**
XLS	Mine site Reference	1.58 1.58	0.18	5.9	0.36
Spring 2022					
LWS	Mine Reference	1.58 1.58	0.07	3.6	0.13
MJS	Mine Reference	1.56 1.58	0.03	3.6	**<0.05**
XLS	Mine Reference	1.58 1.58	0.06	3.6	0.12

4. Discussion

Our results suggest that invasions are commonplace in moderately stressful habitats such as calcareous mine sites, and certain stressful sites may be significantly more invaded than others. In the study region, alien and invasive plants account for the majority (>50%) of all species. Invasive plant abundance in the mine sites was generally higher than that of native plants. Invasive plants still dominate in most cases in reference sites.

Our hypotheses were: (1) the proportion of invasive plants in the mine sites will be higher than that of native plants. On the whole, invasive plants are very prevalent across sites but did not differ significantly between mine sites and reference areas with the exception of the GTS mine site, which had a significantly higher proportion of invasive plant species relative to natives compared to the proportions in the adjacent reference site. The results for the proportion of invasive species by the total number of individuals were not consistent, with some mine sites having a significantly greater proportion of invasive plants compared to that of the adjacent sites (GTS and LWS) and others had an increased proportion of native species in the mine site than in the adjacent sites (XLS and MJS). Overall, the first prediction has only limited support; (2) the plant diversity in mine sites will be lower than in reference sites. Conversely, we found that species diversity, evenness, and richness were similar across sites and seasons. Evenness was relatively low at the GTS and MJS mine sites only during the autumn sampling. The patterns in plant diversity within site pairs varied across seasons. There were no significant differences among sites in some seasons, but mine sites tend to have lower diversity than adjacent sites (except for MJS during autumn 2021). The species similarity tended to be lower in the mine sites. Thus, there was overall slight support for the second prediction; (3) Invasive species are abundant in both mine sites and reference areas, but native species will be restricted to either site. We found that native and invasive species both vary widely in which sites they inhabit and in similar proportions, and that the common species present at both sites tend to be also common in both sites. Therefore, we found no support for the third prediction.

The high presence of invasive species at the calcareous mine sites was not too unexpected. The overall result was similar to findings from other types of calcareous environments. For instance, in a calcareous grassland ecological restoration project [45], the reconstruction of local calcareous grassland plant communities was affected by the high coverage of invasive plants in the calcareous areas. They similarly observed relatively low evenness in the old restoration site compared to reference sites. This may be due to the dominance of invasive grass in these areas. Bart et al. (2011) studied the plant communities in calcareous marshes and found that those environments were mainly composed of woody invasive and weedy native species. Compared with flat land, the area in the marshland that has been plowed or disturbed there were more pioneer native and invasive herbaceous species. Weeds, herbs, and rhizomes are particularly abundant and frequent in these areas. Despite the existence of these invasive weeds and invasive species, the local pioneer species of marsh continue to live in the wing marsh, and the central community is the remains of the local herbaceous calcareous marsh vegetation [46]. There were few invasive species in the center of the calcareous marsh, but more invasive species on the disturbed edges. Thus, the relatively high number of invasive species at some mine sites (e.g., GTS) might be associated with disturbance. The presence of relatively high soil iron or heavy metals (such as high zinc in the MJS mine) did not seem to impact the overall level of plant diversity.

Our investigation into the plant diversity at ferrous calcareous sites found lower similarity among sites within site pairs, particularly between mine sites and adjacent reference sites. This suggests that the stressors present at the mines apply selective pressures on plants. However, these stressors do not limit the number of plant species able to grow in such sites and may promote a different suite of species that are less common outside of the mines. In autumn, the evenness of species in mine sites was much lower than in reference sites, meaning that the plants in reference sites are more evenly distributed in space, but not in the mine sites, where the plants are more clustered. This distribution may be related to the rocky soils found at mine sites. We found that the iron content in the soil of the region is high (i.e., from 45 to 50 g/kg in reference soils, and from 57 to 85 g/kg in mine soils), which may have had an impact on the species diversity of the region. Presently, there is no literature to investigate the impact of iron on plant diversity, Species diversity was much lower in autumn compared to spring (such as at the MJS), but invasive plants are still predominant across the site. This suggests that invasive plants have stronger environmental tolerance than native plants, and have a competitive advantage over plants with poor tolerance in polluted soil [47]. Although invasive plants have a strong tolerance to the environment, there is a sharp contrast between plant species diversity in spring and autumn. It may also be that as resources are reduced during autumn (e.g., lower solar irradiation and reduced rainfall), the growth of many plants becomes slow or stagnant, and invasive plants can make better use of the resources than local plants, which increases the survival probability [6]. However, in general, the richness or number of invasive plants in different seasons and reference sites was not different from that of local plants. In different stages of vegetation restoration, environmental factors have a phased impact on plant diversity. At different stages and locations, they will be more dominant than alien invasive plants, or in most cases, they are equal to invasive plants [48]. On the one hand, local resident plants (both invasive and native plants) may have adapted to this stressful environment for a long time through phenotypic plasticity or ecological differentiation [49–54] has made them completely adapt to environmental pressure like invasive plants. On the other hand, the stressors at the mine sites likely produced a filtering effect, where a portion of the surrounding resident species was excluded from the mines, leaving other species able to tolerate the stressors to become (and not necessarily involving adaptation). Whether plants at the mine sites are adapted to stressors may be addressed in future studies.

The overall diversity is similar across sites, which is not unexpected for moderately stressful habitats such as calcareous sites. Interestingly, the number of species that were only found at mine sites was the greatest (Figure 4). Even in GTS reference sites, there is no species advantage compared with local plants (Table 4). At the same time, due to ore mining, the original vegetation community of the mine has been seriously altered. The survey area belongs to the early stage of natural succession with many Poaceae (especially *Imperata cylindrica, Digitaria sanguinalis, Setaria viridis*) and Asteraceae (especially *Erigeron annuus, E. canadensis* and *S. canadensis*), which are commonly found in wastelands and most of them being invasive species. Poaceae and Asteraceae species are dominant plants in the early stages of natural succession in the restoration of vegetation community [55]. These species have a wide range of distribution, strong adaptability, fast growth, large numbers, and form local small communities. Many Asteraceae and Poaceae plants can adapt to environments with high soil Mn, Cr, Cd, and other heavy metals and other stress factors such as exposure to winds and sun, low nutrients, and drought, so they can be considered pioneer plants for ecological restoration [56,57].

This study aims to explore whether invasive plant species can readily colonize moderately stressful habitats, such as calcareous mine sites, and provide information on species suitable for ecological restoration. The mechanisms of plant invasion are complex and diverse, and an understanding of whether those mechanisms are the same when invasive plants colonize stressful habitats is still developing. The results suggest invasive species are not inhibited from colonizing moderately stressful habitats and may have advantages in some stressful sites such as the GTS mine site which was drastically more invaded than the other mine sites (especially by *Setaria viridis*, Figure S1). The reasons for this are still unknown and warrant further investigation. Environmental protection and ecological restoration of vegetation at abandoned mine sites are important for ecological health. However, the impacts of invasive plant species in polluted sites should also be considered, particularly as they can easily invade many of such sites which can impede ecological restoration. Future work can address the major problems in global ecological protection.

Supplementary Materials: The following supporting information can be downloaded at: https://www.mdpi.com/article/10.3390/d14110986/s1, Figure S1: Average abundance distribution of plant species in the mining area and reference sites: (**a**) April, spring 2021; (**b**) October, autumn 2021; (**c**) April, spring 2022. Each dot point represents an individual plant species; Table S1: The presence and abundance of each individual species at each site. Table S2: Plant diversity, abundance and evenness between mine sites and reference sites across all seasons (H is Shannon's Diversity Index, D is Simpson's Diversity Index, R is Margalef Index, and E is Evenness index). The sampling at LWS was conducted only during spring 2022 when the GTS site pair was inaccessible. Indices were calculated from three quadrats.

Author Contributions: Conceptualization, J.S.H.W. and S.R.; methodology, J.-H.L., S.R. and J.S.H.W.; software, D.-L.D.; validation, Z.-C.D., J.-F.S. and X.-Q.S.; formal analysis, J.-H.L. and J.S.H.W.; investigation, J.-H.L. and H.L.; resources, J.S.H.W.; data curation, J.-H.L.; writing-original draft, J.-H.L.; writing-review and editing, J.-H.L., J.S.H.W., S.R., A.A.-N. and X.-Q.S.; visualization, J.-H.L.; supervision, J.S.H.W. and D.-L.D.; project administration, J.S.H.W.; funding acquisition, J.S.H.W. Additional details: J.-H.L. conducted the surveys, analyzed the data, collected soil, and wrote the first draft of the manuscript. J.S.H.W. designed the study, provided materials and logistics, collected soil, and edited the manuscript. S.R. designed the study and contributed to data collection. H.L., Z.-C.D. and J.-F.S. contributed to data collation. J.-H.L. and X.-Q.S. identified the plant species. All authors have read and agreed to the published version of the manuscript.

Funding: This research was funded by the Jiangsu Natural Science Foundation (BK20200905). JSHW was supported by the Jiangsu University Research Foundation Fund (20JDG056) and the Jiangsu province's 'Double Innovation' talent program. SR was supported by the Jiangsu University Research Foundation Fund (20JDG055), and DZC by the National Natural Science Foundation of China (32271587), carbon peak and carbon neutrality technology innovation foundation of Jiangsu Province (BK20220030).

Institutional Review Board Statement: Not applicable.

Data Availability Statement: The data is available in the Supplementary Materials.

Acknowledgments: We thank Rahil Afzal for his help with soil collection.

Conflicts of Interest: The authors declare no conflict of interest.

References

1. Dawson, W.; Schrama, M.; Austin, A. Identifying the role of soil microbes in plant invasions. *J. Ecol.* **2016**, *104*, 1211–1218. [CrossRef]
2. Pyšek, P.; Hulme, P.E.; Simberloff, D.; Bacher, S.; Blackburn, T.M.; Carlton, J.T.; Dawson, W.; Essl, F.; Foxcroft, L.C.; Genovesi, P.; et al. Scientists' warning on invasive alien species. *Biol. Rev.* **2020**, *95*, 1511–1534. [CrossRef] [PubMed]
3. Xie, W.; Zhong, C.; Li, X.; Guo, Z.; Shi, S. Hybridization with natives augments the threats of introduced species in sonneratia mangroves. *Aquat. Bot.* **2019**, *160*, 103166. [CrossRef]

4. Dai, Z.C.; Zhu, B.; Wan, J.S.H.; Rutherford, S. Editorial: Global Changes and Plant Invasions. *Front. Ecol.* **2022**, *10*, 845816. [CrossRef]
5. Morris, T.L.; Esler, K.J.; Barger, N.N.; Jacobs, S.M.; Cramer, M.D. Ecophysiological traits associated with the competitive ability of invasive Australian acacias. *Divers. Distrib.* **2011**, *17*, 898–910. [CrossRef]
6. Matzek, V. Superior performance and nutrient-use efficiency of invasive plants over non-invasive congeners in a resource-limited environment. *Biol. Invasions* **2011**, *13*, 3005–3014. [CrossRef]
7. Firn, J.; Prober, S.M.; Buckley, Y.M. Plastic traits of an exotic grass con-tribute to its abundance but are not always favorable. *PLoS ONE* **2012**, *7*, e35870. [CrossRef] [PubMed]
8. Funk, J.L. Differences in plasticity between invasive and native plants from a low resource environment. *J. Ecol.* **2008**, *96*, 1162–1174. [CrossRef]
9. Lee, W.G.; Mark, A.F.; Wilson, J.B. Ecotypic differentiation in the ultramafic flora of the South Island, New Zealand. *N. Z. J. Bot.* **1983**, *21*, 141–156. [CrossRef]
10. Gulezian, P.Z.; Jennifer, L.I.; Kelly, J.G. Establishment of an Invasive Plant Species (Conium maculatum) in Contaminated Roadside Soil in Cook County, Illinois. *Am. Midl. Nat.* **2012**, *168*, 375–395. [CrossRef]
11. Prabakaran, K.; Li, J.; Anandkumar, A.; Leng, Z.; Zou, C.B.; Du, D. Managing environmental contamination through phytoremediation by invasive plants: A review. *Ecol. Eng.* **2019**, *138*, 28–37. [CrossRef]
12. Gong, W.; Wang, Y.; Chen, C.; Xiong, Y.; Zhou, Y.; Xiao, F.; Li, B.; Wang, Y. The rapid evolution of an invasive plant due to increased selection pressures throughout its invasive history. *Ecotoxicol Environ. Saf.* **2022**, *233*, 113322. [CrossRef] [PubMed]
13. Yan, W.; Chao, C.; Yxa, B.; Yi, W.; Qla, B. Combination effects of heavy metal and inter-specific competition on the invasiveness of Alternanthera philoxeroides. *Environ. Exp. Bot.* **2021**, *189*, 104532.
14. Cai, H.H. Effects of Heavy Metal cd on Alien Invasive Plants and Their Resistance to Disease. Master's Thesis, Jiangsu University, Zhenjiang, China, 2016.
15. Dai, Z.C.; Cai, H.H.; Qi, S.S.; Li, J.; Du, D.L. Cadmium hyperaccumulation as an inexpensive metal armor against disease in Crofton weed. *Environ. Pollut.* **2020**, *267*, 115649. [CrossRef] [PubMed]
16. Gao, F.F. Effects of Cu, Zn, Cd, Pb on the Growth of Three Leguminous Plants and Their Adsorption Properties. Master's Thesis, Northeastern University, Shenyang, China, 2008.
17. Wang, H.; Jiao, Z.H.; Wu, H.; Wang, Y.G.; Wang, X.X.; Wang, L.S. Review on evaluation methods of heavy metal pollution in soil around lead zinc mine. *J. Shenyang Univ. (Nat. Sci.)* **2021**, *33*, 300–306.
18. Chowdhury, A.; Maiti, S.K. Identification of metal tolerant plant species in mangrove ecosystem by using community study and multivariate analysis: A case study from Indian Sunderban. *Environ. Earth Sci.* **2016**, *75*, 744. [CrossRef]
19. Tovar-Sánchez, E.; Hernández-Plata, I.; Martínez, M.S.; Valencia-Cuevas, L.; Galante, P.M. *'Heavy Metal Pollution as a Biodiversity Threat' in Heavy Metals*; IntechOpen: London, UK, 2018.
20. Jaffré, T.; Brooks, R.R.; Lee, J.; Reeves, R.D. Sebertia acuminata: A Hyperaccumulator of Nickel from New Caledonia. *Science* **1976**, *193*, 579–580. [CrossRef]
21. Li, M.J.; Yu, F.X.; Ding, J.N.; Xiong, Z.T. Water Metabolism Changes of Metallophyte-Elsholtzia haichowensis Treated with Copper and Water Stress. *Jiangxi Sci.* **2014**, *32*, 46–50.
22. Chen, C.; Zhang, H.X.; Wang, A.G.; Lu, M.; Shen, Z.G.; Lian, C.L. Phenotypic plasticity accounts for most of the variation in leaf manganese concentrations in Phytolacca americana growing in manganese-contaminated environments. *Plant Soil* **2015**, *396*, 215–227. [CrossRef]
23. Chen, L.; Gao, J.; Zhu, Q.; Wang, Y.; Yang, Y. Accumulation and output of heavy metals by the invasive plant Spartina alterniflora in a coastal salt marsh. *Pedosphere* **2018**, *28*, 884–894. [CrossRef]
24. Liu, X.W.; Qi, C.M.; Wang, Z.C.; Ouyang, C.B.; Li, Y.; Yan, D.D.; Wang, Q.X.; Guo, M.X.; Yuan, Z.H.; He, F.L. Biochemical and ultrastructural changes induced by lead and cadmium to Crofton weed (*Eupatorium adenophorum* Spreng.). *Int. J. Environ. Res.* **2018**, *12*, 597–607. [CrossRef]
25. Dai, Z.C.; Wan, L.Y.; Qi, S.S.; Rutherford, S.; Du, D.L. Synergy among hypotheses in the invasion process of alien plants: A road map within a timeline. *Perspect. Plant Ecol.* **2020**, *47*, 125575. [CrossRef]
26. Davidson, A.M.; Jennions, M.; Nicotra, A.B. Do invasive species show higher phenotypic plasticity than native species and, if so, is it adaptive? A meta-analysis. *Ecol. Lett.* **2015**, *14*, 419–431. [CrossRef] [PubMed]
27. Zheng, Y.L.; Feng, Y.L.; Zhang, L.K.; Callaway, R.M.; Alfonso, V.B.; Luo, D.Q.; Liao, Z.Y.; Lei, Y.B.; Barclay, G.F.; Carlos, S.P. Integrating novel chemical weapons and evolutionarily increased competitive ability in success of a tropical invader. *New Phytol.* **2015**, *205*, 1350–1359. [CrossRef] [PubMed]
28. Baker, H.G. The Evolution of weeds. *Annu. Rev. Ecol. Evol.* **1974**, *5*, 1–4. [CrossRef]
29. Wang, M.N.; Dai, Z.C.; Qi, S.S.; Wang, X.Y.; Du, D.L. Main hypotheses and research progress of invasion mechanism of exotic plants. *Jiangsu Agric. Sci.* **2014**, *42*, 378–382.
30. Hierro, J.L.; Callaway, M. A biogeographical approach to plant invasions: The importance of studying exotics in their introduced and native range. *J. Ecol.* **2005**, *93*, 5–15. [CrossRef]
31. Harvey, K.J.; Nipperess, D.A.; Britton, D.R.; Hughes, L. Australian family ties: Does a lack of relatives help invasive plants escape natural enemies? *Biol. Invasions* **2012**, *14*, 2423–2434. [CrossRef]

32. Green, P.T.; O'Dowd, D.J.; Abbott, K.L.; Jeffery, M.; Retallick, K.; Mac, N.R. Invasional meltdown: Invader-invader mutualism facilitates a secondary invasion. *Ecology* **2011**, *92*, 1758–1768. [CrossRef]
33. Pitz, C.; Mahy, G.; Harzé, M.; Uyttenbroeck, R.; Monty, A. Comparison of mining spoils to determine the best substrate for rehabilitating limestone quarries by favoring native grassland species over invasive plants. *Ecol. Eng.* **2018**, *127*, 510–518. [CrossRef]
34. Zefferman, E.; Stevens, J.; Charles, G.; Dunbar-Irwin, M.; Emam, T.; Fick, S.; Morales, L.V.; Wolf, K.M.; Young, D.J.; Young, T.P. Plant communities in harsh sites are less invaded: A summary of observations and proposed explanations. *AoB Plants* **2015**, *7*, 1–21. [CrossRef] [PubMed]
35. Bauman, J.M.; Cochran, C.; Chapman, J.; Gilland, K. Plant community development following restoration treatments on a legacy reclaimed mine site. *Ecol. Eng.* **2015**, *83*, 521–528. [CrossRef]
36. Triin, R.; Aveliina, H.; Meelis, P.; Karin, B.; Pille, G.; Ejvind, R.; Krista, T.; Sergey, Z.; Honor, C.P. Determinants of fine-scale plant diversity in dry calcareous grasslands within the Baltic Sea region. *Agric. Ecosyst. Environ.* **2014**, *182*, 9–68.
37. Franklin, J.A.; Zipper, C.E.; Burger, J.A.; Skousen, J.G.; Jacobs, D.F. Influence of herbaceous ground cover on forest restoration of eastern US coal surface mines. *New For.* **2012**, *43*, 905–924. [CrossRef]
38. Aguiar, F.C.; Ferreira, M.T.; Albuquerque, A.; Moreira, I. Alien and endemic flora at reference and non-reference sites in Mediterranean-type streams in Portugal. *Aquat. Conserv.* **2010**, *17*, 335–347. [CrossRef]
39. Hu, Y.; Huang, Y.; Su, J.; Gao, Z.; Nan, Z. Temporal changes of metal bioavailability and extracellular enzyme activities in relation to afforestation of highly contaminated calcareous soil. *Sci. Total Environ.* **2018**, *622*, 1056–1066. [CrossRef]
40. Moore, T.J.; Loeppert, R.H. Significance of potassium chloride pH of calcareous soils. *Soil Sci. Soc. Am. J.* **1987**, *51*, 908–912. [CrossRef]
41. Johnson, C.D.; Skousen, J.G. Minesoil Properties of 15 Abandoned Mine Land Sites in West Virginia. *J. Environ. Qual.* **1995**, *24*, 635–643. [CrossRef]
42. Maron, J.; Marler, M. Native plant diversity resists invasion at both low and high resource levels. *Ecology* **2007**, *88*, 2651–2661. [CrossRef]
43. Hutcheson, K. A test for comparing diversities based on Shannon formula. *J. Theor. Biol.* **1970**, *29*, 151–154. [CrossRef]
44. Hu, S.; Xiao, X.; Jia, H.S.; Zhou, J. Comparison and analysis on dust-retention ability of major deciduous greening species in Xuzhou. *Chin. Agric. Sci. Bull.* **2012**, *28*, 95–98.
45. Piqueray, J.; Bottin, G.; Delescaille, L.M.; Bisteau, E.; Colinet, G.; Mahy, G. Rapid restoration of a species-rich ecosystem assessed from soil and vegetation indicators: The case of calcareous grasslands restored from forest stands. *Ecol. Indic.* **2011**, *11*, 724–733. [CrossRef]
46. Bart, D.; Simon, M.; Carpenter, Q.; Graham, S. Historical Land Use and Plant-Community Variability in a Wisconsin Calcareous Fen. *Rhodora* **2011**, *113*, 160–186. [CrossRef]
47. Zhang, Q.; Yang, R.; Tang, J.; Chen, X. Competitive interaction between the invasive Solidago canadensis and native Kummerowia striata in lead contaminated soil. *Bot. Stud.* **2008**, *49*, 385–391.
48. Liu, X.P.; Zhang, W.J.; Yang, F.; Zhou, X.; Liu, Z.J.; Qu, F.; Lian, S.Q.; Wang, C.L.; Tang, X.G. Changes in vegetation–environment relationships over long-term natural restoration process in Middle Taihang Mountain of North China. *Ecol. Eng.* **2012**, *49*, 193–200. [CrossRef]
49. Schweitzer, J.A.; Larson, K. Greater morphological plasticity of exotic honeysuckle species may make them better invaders than native species. *Torrey Bot. Soc.* **1999**, *126*, 15–23. [CrossRef]
50. Geng, Y.P.; Pan, X.Y.; Xu, C.Y.; Zhang, W.J.; Li, B.; Chen, J.K.; Lu, B.R.; Song, Z.P. Phenotypic plasticity rather than locally adapted ecotypes allows the invasive alligator weed to colonize a wide range of habitats. *Biol. Invasions* **2007**, *9*, 245–256. [CrossRef]
51. Molina-Montenegro, M.A.; Palma-Rojas, C.; Alcayaga-Olivares, Y.; Oses, R.; Corcuera, L.J.; Cavieres, L.A.; Gianoli, E. Ecophysiological traits suggest local adaptation rather than plasticity in the invasive *Taraxacum officinale* (dandelion) from native and introduced habitat range. *Plant Ecol. Divers.* **2011**, *4*, 36–42.
52. Molina-Montenegro, M.A.; Palma-Rojas, C.; Alcayaga-Olivares, Y.; Oses, R.; Corcuera, L.J.; Cavieres, L.A.; Gianoli, E. Ecophysiological plasticity and local differentiation help explain the invasion success of *Taraxacum officinale* (dandelion) in South America. *Ecography* **2012**, *36*, 718–730. [CrossRef]
53. Molina-Montenegro, M.A.; Naya, D.E. Latitudinal patterns in phenotypic plasticity and fitness-related traits: Assessing the climatic variability hypothesis (CVH) with an invasive plant species. *PLoS ONE* **2012**, *7*, e47620. [CrossRef]
54. Molina-Montenegro, M.A.; Peñuelas, J.; Munné-Bosch, S.; Sardans, J. Higher plasticity in ecophysiological traits enhances the performance and invasion success of *Taraxacum officinale* (dandelion) in alpine environments. *Biol. Invasions* **2012**, *14*, 21–33. [CrossRef]
55. Han, F.; Li, C.R.; Sun, M.G.; Fan, Y.X.; Zhao, F.; Lu, S.S.; Fu, Y. Plant community structure at an early ecological restoration stage on an abandoned quarry in Sibao Mount. *J. Cent. South Univ.* **2008**, *28*, 35–39.
56. Yang, S.X.; Li, M.S.; Li, Y.; Huang, H.R. Research status of ecological restoration and pollution of heavy metals by plants in Pingle manganese mine. *Minging Saf. Environ. Prot.* **2006**, *33*, 21–23.
57. Luo, Y.P.; Li, M.S.; Zhang, X.H.; Liu, J.; Huang, H.T.; Cai, X.W. Characteristics of bioaccumulation of heavy metals by dominant plants in Lipu manganese mine, Guangxi. *J. GuangXi Norm. Univ.* **2005**, *23*, 89–93.

plants MDPI

Brief Report

Hydrocharis laevigata in Europe

Pablo Garcia-Murillo

Department of Plant Biology and Ecology, Faculty of Pharmacy, University of Seville, 41012 Seville, Spain; pgarcia@us.es

Abstract: *Hydrocharis laevigata* (Humb. & Bonpl. ex Willd.) Byng & Christenh. [= *Limnobium laevigatum* (Humb. & Bonpl. ex Willd.) Heine], Hydrocharitaceae, is a floating-leaf aquatic plant that is native to inland South America. It is an invasive species in several parts of the world. Reports of its presence in Europe have been recently published: naturalised populations occur in three locations on the Iberian Peninsula. The literature also contains records of the species in Hungary and Poland. In addition, it has been observed in Sweden, Belgium, and the Netherlands. *H. laevigata* is highly adaptable and can profoundly transform habitat conditions in its invasive range, causing major issues for ecosystem conservation and human activities. Until recently, *H. laevigata* was not to be found in natural environments in Europe. Factors explaining its spread include its use as an ornamental plant, the eutrophication of inland waters, and the effects of global warming. With a focus on Europe, this short communication provides information on the species' distribution, taxonomy, biology, habitat, and negative impacts.

Keywords: aquatic plants; spongeplant; AIS; biological invasions

1. Introduction

The major decline in freshwater species biodiversity has become one of the most important alterations in ecosystems around the world [1,2]. Pollution, water flow restrictions, and climate change are transforming conditions in freshwater habitats, favouring the establishment of invasive species [3–5].

There is a common consensus, supported by the Red List of European Habitats [6], that the key threats to European aquatic and wetland habitats are alterations to hydrological systems, climate change, pollution, and invasive species [7]. In Europe, native freshwater biota are facing increasing competitive pressure due to the arrival and spread of floating-leaf aquatic plants (e.g., *Azolla filiculoides* and *Eichhornia crassipes*) in inland water bodies [8–15]. Of particular concern is the recent appearance of an aggressive invasive alien species (AIS), *Hydrocharis laevigata*, whose common names include South American spongeplant, smooth frogbit, or Amazon frogbit. Newly established in a range of European ponds, rivers, and reservoirs, the species serves as yet another warning about the deterioration of Europe's freshwater habitats. Its invasive dynamics are reminiscent of those of *Eichhornia crassipes*, including its striking colonisation abilities. This short communication describes *H. laevigata*'s history in Europe, including its emergence in distinct continental regions. It also provides key information about the species' taxonomy, biology, habitat, and negative impacts. To this end, it will facilitate prevention efforts aimed at invasive species, help control the spread of *H. laevigata*, and bolster the preservation of Europe's native biodiversity.

2. Results

2.1. Taxonomy and Nomenclature

Hydrocharis laevigata (Humb. & Bonpl. ex Willd.) Byng & Christenh. belongs to the family Hydrocharitaceae (superorder Lilianae). This taxon of monocots contains genera whose members are among the world's worst aquatic invasive plants, such as *Eichhornia* Kunth and *Hydrilla* Rich.

Phylogenomic, morphological, and anatomical evidence strongly suggest a new vision of the genus *Hydrocharis* L. [16,17], one that includes the species of *Limnobium* Rich., usually considered to be a separate genus by World Flora Online (WFO) [18]. Assuming the species' reclassification, there are currently five known members of the genus *Hydrocharis* L. Three are spread throughout Afro-Asia, and the other two are found from Mexico to northern Argentina and in the Caribbean. Those previously classified in *Limonium* are native to the Americas and utilise wind pollination. The original members of *Hydrocharis* are autochthonous from the Old World and are insect pollinated [19]. Those differences aside, all five species have a close phylogenetic relationship and display similarities in their vegetative traits [19,20].

Hydrocharis laevigata (Humb. & Bonpl. ex Willd.) Byng & Christenh., Global Fl., 4: 53. 2018.

≡ *Salvinia laevigata* Humb. & Bonpl. ex Willd., Sp. Pl., ed. 4, 5: 537. 1810 (basionym).

≡ *Limnobium laevigatum* (Humb. & Bonpl. ex Willd.) Heine in Adansonia, 8: 315. 1968.

≡ *Hydromystria laevigata* (Humb. & Bonpl. ex Willd.) Hunz., Lorentzia, 4: 5. 1981.

≡ *Limnobium spongia* subsp. *laevigatum* (Humb. & Bonpl. ex Willd.) Lowden, Rhodora, 94: 129. 1992).

Type (lectotype designated by [21] (p. 220): Colombia; Bogotá. Humboldt & Bonpland (holotype: B-Willd. 20251).

= *Hydromystria stolonifera* G. Mey., Prim. Fl. Esseq. 153. 1818.

≡ *Limnobium stoloniferum* (G. Mey.) Griseb., Fl. Brit. W. I. 506. 1864.

This species has traditionally been included in the genus *Limnobioum* Rich., under the name *Limnobium laevigatum* (Humb. & Bonpl. ex Willd.) Heine. Other authors have considered this taxon to be a subspecies of *L. spongia* (Bosc) Steud.

2.2. Species Description

According to [21–24], *H. laevigata* is an aquatic stoloniferous plant whose stems are floating on or suspended in the water. They develop floating or emergent leaf rosettes, which are sometimes rooted. The species' differentiated leaves are petiolate, laminate, cordate to reniform, or orbiculate, stipulate, and dimorphic. The aerial leaves have flat, leathery blades and long erect petioles measuring up to 40 cm; the floating leaves have thick spongy tissue on the lower surface and shorter petioles (minimum of 12 cm). Flowers are about 15 mm wide, unisexual, pedicelled, and enclosed in a spathe. Staminate flowers often occur in clusters of two or three; have three (rarely four) sepals and petals each; possess a variable number of stamens, whose filaments form a single column; and usually exhibit staminodes. Pistillate flowers are solitary; have rudimentary sepals and petals, although the latter may be lacking; and possess a unilocular ovary with six carpels, six styles, and divided linear stigmas that are longer than the perianth segments. Furthermore, female flowers may sometimes exhibit staminodes. The species' fruit consists of globose or ellipsoid fleshy berry-like capsules (~1.5 cm long and 0.4–3.5 mm wide) that contain as many as 100 ellipsoid seeds measuring around 10 mm long.

2.3. Habitat

Found in the tropics and subtropics [22,25], *H. laevigata* grows in a variety of freshwater habitats, such as shallow ponds; lakes; dams; reservoirs; pools; and the margins of slow shady rivers, canals, and ditches; environments with a high degree of eutrophication are particularly hospitable [26,27]. The species occurs at elevations from sea level all the way up to 2800 m. It is absent from fast-flowing watercourses [21,22,26,28], and cold temperatures limit its distribution [25].

2.4. Reproduction

Hydrocharis laevigata reproduces vegetatively via the fragmentation of the stolon segments connecting the leaf rosettes. The floating rosettes send out stolons bearing ramets on their ends [29]. The rosettes are easily dislodged by waterfowl or large mammals in the species' native range [26], as well as by water currents or other types of water movement in general. Consequently, *H. laevigata* is able to rapidly and efficiently colonise suitable new habitats [28].

With regards to sexual reproduction, the species mainly utilises wind pollination [17,22,28]. However, water pollination has also been observed, as has insect pollination, mediated by tiny aphid nymphs (Aphididae) and the aphids' predators, ladybird beetles (Coccinellidae) [24].

After pollination, the pedicel of female flowers tips downwards, such that the fruit develops submerged in water; once it reaches maturity, it breaks open, releasing a mucilaginous mass of about 100 seeds [21].

2.5. Global Distribution

The native range of *H. laevigata* extends across the tropics of Central and South America. Naturalised populations can now be found on all continents, with the exception of Antarctica.

According to [22,24,30], this species is naturally found in water bodies in parts of the Americas with tropical, subtropical, and, sometimes, temperate climates. More specifically, these locations are the CARIBBEAN: Antigua and Barbuda, Cuba, the Dominican Republic, Guadeloupe, St. Lucia, Montserrat, Martinique, Trinidad and Tobago, and Puerto Rico; CENTRAL AMERICA: Mexico, Costa Rica, Guatemala, Nicaragua, Panama, and El Salvador; and SOUTH AMERICA: French Guiana, Guyana, Suriname, Venezuela, Brazil, Colombia, Ecuador, Peru, Argentina, Paraguay, and Uruguay. It should also be noted that there may exist some problematic records for South America. For example, although [31] stated that the species is native to the Southern Cone, this assertion is questionable in the case of Chile, as [26] indicates that the country's oldest herbarium specimen dates back to 1954. The same is likely true for Argentina: while the species' native range is thought to extend across most of the country, it certainly appears to be invasive in the southernmost regions of Argentina, where it has recently appeared for the first time [32].

As noted in [25,29,33–36], *H. laevigata* has become naturalised and spread across ASIA: Japan, Taiwan, and Indonesia; AFRICA: Zambia and Zimbabwe; AUSTRALIA: New South Wales and Queensland; NORTH AMERICA: southern Canada as well as the north-eastern and western US; and EUROPE.

2.6. Populations in Europe

Based on the scientific literature and the GBIF database, *H. laevigata* is known to occur in six European countries: Belgium, Hungary, The Netherlands, Poland, Spain, and Sweden.

The first record of its presence (as *Limnobium laevigatum*) was in hot springs located in Komárom-Esztergom County, northern Hungary [37]. In September 2018, the species was observed in the area of Fényes springs, north of Tata. At this location, the plants were mainly in a vegetative state, although some did bear flowers. One month later, the plants had spread to the Ferencmajori fishponds, traveling a distance of about 1.3 km. In October 2018, the species turned up at another site within the same county: Dunaalmás, a stream fed by warm water [37]. It was observed that the *H. laevigata* populations became established near hot springs (temperatures = 21–23 °C), which do not freeze in the winter. Recently, Dr. Riezing (pers. comm.) has indicated that these populations in northern Hungary have persisted, likely because of mild winter conditions and the warm water temperatures. However, the plants occur in small groups.

On the Iberian Peninsula, the species seems to have rapidly expanded its range. Since 2018, it has been found at three locations that are quite distant from each other. The first published record for this species (as *Limnobioum laevigatum*) was associated with a reservoir near Madrid [38]. A few years later, it was observed in a stream next to a dam near Córdoba [39]. Finally, it was also reported to occur in the Guadaira, a tributary of the

Gualdalquivir, near Seville [40] (Figure 1). There are no indications that *H. laevigata* has disappeared from any of these locations in Spain.

Figure 1. (**A**,**B**) *Hydrocharis laevigata* habitat in the outskirts of Seville, Spain. (**C**) Male flowers, photos taken by P. Garcia Murillo in summer 2021; (**D**,**E**) habitat in the outskirts of Krakow, Poland, photos taken in November and December 2020, courtesy of Dr. Arthur Pliszko.

In 2020, *H. laevigata* was observed in southern Poland, in an artificial pond near Agatowa Street in the district of Bieżanów-Prokocim in eastern Kraków [15]. It was noted [15] that the species (as *L. laevigatum*) is likely a sporadic invader in Poland, given that other similar invasive species display constrained ranges due to the country's cold winter temperatures. Dr. Pliszko (pers. comm.) recently confirmed that this population has disappeared because of the species' vulnerability to low temperatures. It cannot survive winter conditions in Poland, outside of the habitat found in greenhouses or residences (Figure 1).

The GBIF lists nine geographical records for *H. laevigata* in Europe [41], which include Ganshoren, Belgium; the outskirts of Stockholm, Sweden; and Aalsmer and Bleiswijk in the Netherlands. Only in the latter country are the records associated with herbarium specimens. In the other cases, there are only photographs, no herbarium samples (Appendices A and B). Two of these locations represent artificial habitats: a greenhouse in Aalsmeer in the Netherlands and the Bergianska Botanick Garden in Sweden. The remaining records for Belgium and the Netherlands were one-time events. Consequently, they can be viewed as sporadic occurrences that have not resulted in *H. laevigata* becoming naturalised. The same is true for the record in Poland. Therefore, at present, the only naturalised *H. laevigata* populations in Europe are in Hungary and Spain (Figure 2).

Figure 2. Locations where *Hydrocharis laevigata* has been observed in Europe. Blue dots: sporadic presence. Red dots: naturalised populations. When the name of the city has a different spelling of the English alphabet, it appears below in its original spelling. Map created using The Geospatial Conservation Assessment Tool.

2.7. Invasiveness

As noted in the introduction, *H. laevigata* displays invasive dynamics reminiscent of those of *Eichhornia crassipes*, including its marked capacity for colonisation [25,32,42]. Indeed, like other invasive aquatic plant species, it is fast growing and thrives in still waters containing high concentrations of nutrients [43]. Under these conditions, it can outcompete native aquatic plant species and profoundly transforms the habitats it invades. While the impacts of this species have yet to be fully described, past observations [22,25,28,44] indicate that the floating rosettes block light from reaching the water below, indirectly fostering anoxic conditions. *H. laevigata* can thus reduce the abundance and diversity of aquatic plants and animals (e.g., fish, phytoplankton, submersed plants) by rendering water bodies unsuitable for other organisms. The species can also generate mats that cause damage to human infrastructures (e.g., by clogging irrigation pipes or sewers, by damaging pumps or generators in hydroelectric power stations). Consequently, *H. laevigata* is considered to be one of the most harmful weeds in the United States and Australia, given its high degree of invasiveness combined with its negative impacts on water quality and aquatic biodiversity [25,29,42,44].

While *H. spongia* (= *Limnobium spongia*) is rarely cultivated, *H. laevigatum* is commonly found in aquaria and garden ponds, largely because of its ecological breadth and pronounced reproductive capacity [32,35–37,45].

2.8. Control

Typical methods for controlling *H. laevigata* are mechanical harvesting and herbicide treatments [45,46]. In the Sacramento-San Joaquin Delta, diquat dibromide has been used, either on its own or in tandem with low levels of glyphosate [46]; this work yielded unclear results. Another study [47] evaluated the performance of the herbicides imazamox, penoxsulam and topramezone. Imazamox (with 1% methylated seed oil surfactant) was found to control all the growth stages of *H. laevigata*; penoxsulam was particularly effective in targeting seedling and rosette growth. Research is currently lacking on potential biological control methods. However, [36] points out that certain North American herbivores display a dietary specialisation for American frogbit (*H. spongia*). They might therefore be candidate biological control agents.

3. Discussion

In summary, *H. laevigata* is an invasive plant species that poses a significant threat to ecosystems and certain human activities in its non-native range. Its likelihood of establishment is considerably elevated in nutrient-rich, slow-moving freshwater systems subject to mild winter temperatures [25–28], such as those found in southern Europe or the northern Mediterranean. Its invasion dynamics are similar to those of *Eichhornia crassipes* [25,32,42].

As indicated earlier, the ornamental plant trade is the main source of *H. laevigata* (and *Eichhornia crassipes*) introductions. This fact explains how the species has spread so far from its native range to parts of the world as diverse as Australia, Zimbabwe, Japan, and Spain.

The range expansion of *H. laevigata* within Europe underscores the deterioration of the continent's inland aquatic ecosystems. Such is, above all, the result of eutrophication, but other causal factors include changes to hydrological regimes (i.e., reduced levels of water flow), decreasing degrees of water transparency, and global warming, which are simultaneously transforming natural aquatic environments [5,48–53]. These ecological shifts are harmful to native aquatic plant populations while promoting the establishment of invasive species such as *H. laevigata*, leading to profound transformations in European ecosystems [12,14,54–57].

It is curious to note that, while the invasive *H. laevigata* has begun to expand its range in Europe, the native *H. morsus-ranae*, a close relative, is experiencing decline. While the latter has always been patchily distributed across western Europe, populations are shrinking in certain regions [58]. For example, on the Iberian Peninsula, *H. morsus-ranae* now occurs in a single area [59], a range constriction that took place over the last century. Probable causes include habitat degradation due to farming and industrial activities as well as changes in the hydrochemical characteristics and hydrological regimes of water bodies [58]. At the same time, *H. morsus-ranae* has become an AIS in the north-eastern USA and southern Canada [60,61].

Hence, given the plant's recent range expansion in Europe, it is crucial to maximise efforts to rapidly detect its presence. By acting now, during the early stages of the invasion, it is more likely that *H. laevigata* can be controlled [62,63]. At particular risk is the Mediterranean, where conditions in rivers, streams, lakes, ponds, reservoirs, and canals favour the arrival and establishment of *H. laevigata*.

In this regard, it should be added that the introduction and spread of IAS are a major concern for the European Union; for those reasons, Regulation (EU) No 1143/2014 of the European Parliament and of the Council of 22 October 2014 on the prevention and management of the introduction and spread of invasive alien species was published, which is being developed and updated through different implementing regulations and which establishes rules to prevent, minimize, and mitigate the adverse effects of IAS. This regulation is associated with a list of IAS: "the Union list", which is the fundamental instrument for implementing the aforementioned regulation. *Hydrocharis laevigata* is not included in this list. Taking into account the problems that this species causes to native biodiversity and to human activities in other parts of the world, it is hoped that the information contained in this brief note will serve as a starting point to initiate actions to include *Hydrocharis laevigata* in the "the Union list", something that will undoubtedly benefit the conservation of European biodiversity and reduce its harmful effects on human activities.

4. Materials and Methods

This study was based on an extensive literature analysis, data from field surveys, and the examination of herbarium specimens (at MA, SEV, COFC; see Appendix A).

The Geospatial Conservation Assessment Tool (http://geocat.kew.org/, accessed on 1 November 2022) was used to create the map (Figure 2) showing where *H. laevigata* has been observed (location descriptions: Appendices A and B).

Funding: This research was funded by the Andalusian Regional Government Research Group PLACCA (RNM116).

Data Availability Statement: Not applicable.

Acknowledgments: The author thanks Norbert Riezing from Hungary and Artur Pliszko from Poland for providing information about *H. laevigata* occurrence in those two countries. The author also expresses his appreciation to Artur Pliszko for furnishing photos of the Polish population.

Conflicts of Interest: The author declares no conflict of interest.

Appendix A

Table A1. Information on the occurrence of *Hydrocharis laevigata* in Europe compiled by GBIF (accessed: November 2022).

gbifID	Locality	State Province	Decimal Latitude	Decimal Longitude	Event Date	Basis of Record
3400985464	Aalsmeer, commercial greenhouse	Noord-Holland	52.285.699	4.818.572	2019-03-06	PRESERVED_SPECIMEN
3400971442	Bleiswijk	Zuid-Holland	52.012.874	4.549.629	2016-05-13	PRESERVED_SPECIMEN
3400962451	Bleiswijk	Zuid-Holland	52.012.874	4.549.629	2016-05-13	PRESERVED_SPECIMEN
1569965647	Ganshoren	Brussels Capital Region	508.804	430.617	2013-09-01	HUMAN_OBSERVATION
1569965625	Ganshoren	Brussels Capital Region	5.088.037	430.569	2013-08-25	HUMAN_OBSERVATION
1569965614	Ganshoren	Brussels Capital Region	5.088.004	430.411	2013-09-19	HUMAN_OBSERVATION
1569965606	Ganshoren	Brussels Capital Region	5.088.037	430.954	2013-09-11	HUMAN_OBSERVATION
1569965603	Ganshoren	Brussels Capital Region	50.88	430.454	2013-09-01	HUMAN_OBSERVATION
1065300305	Bergianska Botanick Garden	Stockholms stad	59.37	180.471	Not shown	LIVING_SPECIMEN

Appendix B

Table A2. Herbarium specimens of *Hydrocharis laevigata* relating to the European records.

Location	Country/Province	Date	Herbarium Codes
Aalsmeer, commercialgreenhouse	Noord-Holland/The Netherlands	2019-03-06	WAG0452864
Bleiswijk	Zuid-Holland/The Netherlands	2016-05-13	WAG0453625, WAG0453626
Agatowa street	Poland/Krakow	2020-09	KRA 0557636, 0557637, 0557638;
Near Martin Gonzalo Dam	Spain/Cordoba	2018-06-10	COFC 65837
Colmenar Viejo, Grajal Dam	Spain/Madrid	2019-09-28	MA-01-00947148
Colmenar Viejo, Grajal Dam	Spain/Madrid	2021-09-08	MA-01-951386
River Guadaira, near hermitage of the Virgin of Valme	Spain/Seville	2021-06-08	SEV 289200
River Guadaira, near hermitage of the Virgin of Valme	Spain/Seville	2021-08-30	SEV 289203

References

1. Jenkins, M. Prospects for biodiversity. *Science* **2003**, *302*, 1175–1177. [CrossRef]
2. Nel, J.L.; Reyers, B.; Roux, D.J.; Cowling, R.M. Expanding protected areas beyond their terrestrial comfort zone: Identifying spatial options for river conservation. *Biol. Conserv.* **2009**, *142*, 1605–1616. [CrossRef]
3. Markovic, D.; Carrizo, S.; Freyhof, J.; Cid, N.; Lengyel, S.; Scholz, M.; Kasperdius, H.; Darwall, W. Europe's freshwater biodiversity under climate change: Distribution shifts and conservation needs. *Divers. Distrib.* **2014**, *20*, 1097–1107. [CrossRef]
4. Panlasigui, S.; Davis, A.J.; Mangiante, M.J.; Darling, J.A. Assessing threats of non-native species to native freshwater biodiversity: Conservation priorities for the United States. *Biol. Conserv.* **2018**, *224*, 199–208. [CrossRef]
5. Klimaszyk, P.; Gołdyn, R. Water Quality of Freshwater Ecosystems in a Temperate Climate. *Water* **2020**, *12*, 2643. [CrossRef]
6. Janssen, J.A.M.; Rodwell, J.S.; García Criado, M.; Gubbay, S.; Haynes, T.; Nieto, A.; Sanders, N.; Landucci, F.; Loidi, J.; Ssymank, A.; et al. *European Red List of Habitats. Part 2. Terrestrial and Freshwater Habitats*; Publications Office of the European Union: Luxembourg, 2016; pp. 1–34. Available online: https://data.europa.eu/doi/10.2779/091372 (accessed on 1 November 2022).
7. Grzybowski, M.; Glińska-Lewczuk, K. Principal threats to the conservation of freshwater habitats in the continental biogeographical region of Central Europe. *Biodivers. Conserv.* **2019**, *28*, 4065–4097. [CrossRef]
8. Carrapiço, F. *Azolla em Portugal*; Instituto da Água Ministério do Ambiente e do Ordenamento do Território Centro de Biologia Ambiental: Lisbom, Portugal, 2001. Available online: http://azolla.fc.ul.pt/documents/AzollaPortugal.pdf (accessed on 1 November 2022).
9. Monteiro, A.; Moreira, I.; Santos, A.C.; Serrasqueiro, P.M. *Gestão do Jacinto-Aquático (Eichhornia crassipes) na Lezíria Grande de Vila Franca de Xira*; Anais do Instituto Superior de Agronomia: Lisbom, Portugal, 2003. Available online: https://www.repository.utl.pt/bitstream/10400.5/2288/1/REP-ANAIS-297.pdf (accessed on 1 November 2022).
10. Almeida, A.S.R.L. Avaliação da Qualidade da Água na Pateira de Fermentelos: Caso de Estudo na Àrea de Projecto. Master Thesis, Universidade de Aveiro, Aveiro, Portugal, 2006. Available online: http://hdl.handle.net/10773/4662 (accessed on 1 November 2022).
11. García Murillo, P.; Fernández Zamudio, R.; Cirujano Bracamonte, S.; Sousa Martín, A.; Espinar, J.M. The invasion of Donana National Park (SW Spain) by the mosquito fern (*Azolla filiculoides* Lam). *Limnetica* **2007**, *26*, 243–250. [CrossRef]
12. Ruiz Téllez, T.; de Rodrigo López, E.M.; Lorenzo Granado, G.; Albano Pérez, E.; Morán López, R.; Sánchez Guzmán, J.M. The water hyacinth, Eichhornia crassipes: An invasive plant in the Guadiana River Basin (Spain). *Aquat. Invasions* **2008**, *3*, 42–53. [CrossRef]
13. Ministerio para la Transicion Ecológica. *Estrategia de Gestión, Control y Posible Erradicación del Camalote (Eichhornia crassipes). Technical report*; Ministerio para la Transicion Ecológica: Madrid, Spain, 2019. Available online: https://www.miteco.gob.es/va/biodiversidad/publicaciones/estrategiadegestioneichhorniacrassipes3deoctubre2019_tcm39-502314.pdf (accessed on 1 November 2022).
14. Piñero Rodriguez, M.J.; Fernandez Zamudio, R.; Arribas, R.; Gómez Mestre, I.; Diaz Paniagua, C. The invasive aquatic fern *Azolla filiculoides* negatively impacts water quality, aquatic vegetation and amphibian larvae in Mediterranean environments. *Biol. Invasions* **2021**, *23*, 755–769. [CrossRef]
15. Pliszko, A.; Górecki, A. First record of *Limnobium laevigatum* (Humb. & Bonpl. ex Willd.) Heine (Hydrocharitaceae) and Pontederia crassipes Mart. (Pontederiaceae) in Poland. *BioInvasions Rec.* **2021**, *10*, 537–543. [CrossRef]
16. Bernardini, B.; Lucchese, F. New phylogenetic insights into Hydrocharitaceae. *Ann. Bot.* **2018**, *8*, 45–58. [CrossRef]

17. Li, Z.Z.; Lehtonen, S.; Gichira, A.W.; Martins, K.; Efremov, A.; Wang, Q.F.; Chen, J.M. Plastome phylogenomics and historical biogeography of aquatic plant genus *Hydrocharis* (Hydrocharitaceae). *BMC Plant Biol.* **2022**, *22*, 106. [CrossRef] [PubMed]
18. WFO. *Hydrocharis laevigata* (Humb. & Bonpl. ex Willd.) Byng & Christenh. World Flora Online Data. 2021. Available online: http://www.worldfloraonline.org/taxon/wfo-0001427027 (accessed on 1 November 2022).
19. Les, D.H. *Aquatic Plants of North America: Ecology, Life History and Systematics*; CRC Press: Boca Raton, FL, USA, 2020; Volume 2, pp. 56–58.
20. Tanaka, N.; Setoguchi, H.; Murata, J. Phylogeny of the family Hydrocharitaceae inferred from rbcL and matK gene sequence data. *J. Plant Res.* **1997**, *110*, 329–337. [CrossRef]
21. Acevedo-Rodríguez, P.; Strong, M.T. Monocotyledons and gymnosperms of Puerto Rico and the Virgin Islands. *Contrib. U. S. Natl. Herb.* **2005**, *52*, 1–415.
22. Cook, C.D.K.; Urmi-König, K. A revision of the genus *Limnobium* including *Hydromystria* (Hydrocharitaceae). *Aquat. Bot.* **1983**, *17*, 1–27. [CrossRef]
23. Lot, A.; Medina Lemos, R.; Chiang, F. *Plantas Acuáticas Mexicanas, Una Contribución a la Flora de México, Volumen I, Monocotiledóneas*; Universidad Nacional Autónoma de México—Instituto de Biología: Coyoacán, Mexico, 2013; 398p.
24. Lowden, R.M. Floral variation and taxonomy of Limnobium LC Richard (Hydrocharitaceae). *Rhodora* **1992**, *94*, 111–134.
25. APHIS. Weed Risk Assessment for *Limnobium laevigatum* (Humb. & Bonpl. ex Willd.) Heine (Hydrocharitaceae)—South American Spongeplant, United States Department of Agriculture, Animal and Plant Health Inspection Service, Raleigh, USA. 2013. Available online: https://www.aphis.usda.gov/plant_health/plant_pest_info/weeds/downloads/wra/Limnobium_laevigatum_WRA.pdf (accessed on 1 November 2022).
26. Martín, C.S.; Contreras, D.; Vidal, O.; Solís, J.L.; Ramírez, C. Distribución en Chile y colonización del río Cayumapu (Valdivia) por el macrófito acuático invasor *Limnobium laevigatum*. *Gayana Bot.* **2021**, *78*, 65–76. [CrossRef]
27. Hernández, J.; Rangel, J.O. La vegetación del humedal de Jaboque (Bogotá, DC). *Caldasia* **2009**, *31*, 355–379.
28. Ruiz-Merino, M.; Campos-Cuéllar, R.; Germán-Gómez, A.; Aponte, H. Características, historia natural y aplicaciones de *Hydrocharis laevigata*: Una revisión. *Caldasia* **2022**, *44*, 432–441. [CrossRef]
29. SERCUL (2020). *Amazon Frogbit—Limnobium laevigatum Ex-Situ Experiments and Field Investigations*; South East Regional Centre for Urban Landcare Inc. (SERCUL). Perth NRM.: Perth, Australia, 2020. Available online: https://www.sercul.org.au/wp-content/uploads/2021/09/Amazon-Frogbit-Research-Report-2020web.pdf (accessed on 1 November 2022).
30. U.S. National Plant Germplasm System. USDA, Agricultural Research Service, Germplasm Resources Information Network (GRIN Taxonomy). National Germplasm Resources Laboratory, Beltsville, USA. 2022. Available online: https://npgsweb.ars-grin.gov/gringlobal/taxon/taxonomydetail?id=419808 (accessed on 1 November 2022).
31. Zuloaga, F.O.; Morrone, O.; Belgrano, M.J. *Catálogo de las Plantas Vasculares del Cono Sur (Argentina, Sur de Brasil, Chile, Paraguay y Uruguay). Volumen 1: Pteridophyta, Gymnospermae y Monocotyledoneae*; Smithonian: Washington, DC, USA, 2008; 983p.
32. Pérez, C.; Seewald, P. *Limnobium laevigatum* (Humb. & bonpl. ex willd.) Heine (Alismatales, Hydrocharitaceae), una nueva especie invasora en la cuenca del Río Negro, provincia de río negro. *Argent. His. Nat.* **2019**, *9*, 211–218.
33. Kadono, Y. Alien aquatic plants naturalized in Japan: History and present status. *Glob. Environ. Res.* **2004**, *8*, 163–169.
34. Haynes, R.R.; Holm-Nielsen, L.B. The genera of Hydrocharitaceae in the Southeastern United States. *Harv. Pap. Bot.* **2001**, *5*, 201–275.
35. Howard, G.W.; Hyde, M.A.; Bingham, M.G. Alien *Limnobium laevigatum* (Humb. & Bonpl. ex Willd.) Heine (Hydrocharitaceae) becoming prevalent in Zimbabwe and Zambia. *Bioinvasions Rec.* **2016**, *5*, 221–225. [CrossRef]
36. Pratt, P.D.; Moran, P.J.; Pitcairn, M.; Reddy, A.M.; O'Brien, J. Biological control of invasive plants in California's Delta: Past, present, and future. *J. Aquat. Plant Manag.* **2021**, *59*, 55–66.
37. Riezing, N. Újabb adventív vízinövény Magyarországon: Limnobium laevigatum (Hydrocharitaceae). *Kitaibelia* **2019**, *24*, 9–15. [CrossRef]
38. Corro, M.; Izuzquiza, Á.; Cirujano, S. Primera cita de la especie potencialmente invasora *Limnobium laevigatum* (Humb. & Bonpl. ex Willd.) Heine (Hydrocharitaceae) en la *Península Ibérica*. *BV News Publ. Científicas* **2019**, *8*, 75–80.
39. Martínez-Sagarra, G.; García Murillo, P.; Devesa, J.A. *Hydrocharis laevigata* (Humb. & Bonpl. ex Willd.) Byng & Christenh.(Hydrocharitaceae), novedad para la flora de Andalucía (España). *Acta Bot. Malacit.* **2021**, *46*, 93–94. [CrossRef]
40. García Murillo, P.; Boniquito, J.M.; Gutiérrez González, D.; Rodríguez Hiraldo, C. Nueva localidad ibérica de la planta exótica invasora *Hydrocharis laevigata* (Humb. & Bonpl. ex Willd.) Byng & Christenh. *Folia Bot. Extremad.* **2021**, *15*, 141–145.
41. GBIF. *Hydrocharis laevigata* (Humb. & Bonpl. ex Willd.) Byng & Christenh. GBIF Backbone Taxonomy, Checklist Dataset, GBIF Secretariat. 2022. Available online: https://www.gbif.org/dataset/d7dddbf4-2cf0-4f39-9b2a-bb099caae36c (accessed on 1 November 2022). [CrossRef]
42. Larson, L. Spongeplant: A New Aquatic Weed Threat in Delta. Cal-IPC News Spring 2011, Berkeley, USA. 2011. Available online: https://www.cal-ipc.org/wp-content/uploads/2017/03/Cal-IPC_News_2011Spring-6.pdf#page=4 (accessed on 1 November 2022).
43. Aponte, H. Productividad de *Limnobium laevigatum* (Hydrocharitaceae) bajo condiciones de laboratorio. *Polibotánica* **2017**, *44*, 157–166.
44. Van de Witte, Y. *Limnobium laevigatum (South American spongeplant)*; CABI Compendium; CABI International: Wallingford, UK, 2022. [CrossRef]

45. DiTomaso, J.M.; Kyser, G.B.; Oneto, S.R.; Wilson, R.G.; Orloff, S.B.; Anderson, L.W.; Wright, S.D.; Roncoroni, J.A.; Miller, T.L.; Prather, T.S. *Weed Control in Natural Areas in the Western United States*; University of California Weed Research and Information Center: Davis, CA, USA, 2013; 544p.
46. Willis, B.E.; Heilman, M.A.; Bishop, W.M.; Shuler, S.W. Evaluation of Multiple Herbicides for Control of Sponge Plant (Limnobium laevigatum). *J. Geosci. Environ. Prot.* **2018**, *6*, 56–64. [CrossRef]
47. California Department of Parks and Recreation. Water Hyacinth Control Program and Sponge Plant Control Program Aquatic Pesticide Application Plan. 2013. Available online: https://dbw.parks.ca.gov/pages/28702/files/WHCP%20SCP%20Aquatic%20Pesticide%20Application%20Plan%202013.pdf (accessed on 1 November 2022).
48. Bol, R.; Gruau, G.; Mellander, P.E.; Dupas, R.; Bechmann, M.; Skarbøvik, E.; Bieroza, M.; Djodjic, F.; Glendell, M.; Jordan, P.; et al. Challenges of reducing phosphorus based water eutrophication in the agricultural landscapes of Northwest Europe. *Front. Mar. Sci.* **2018**, *5*, 276. [CrossRef]
49. Charlton, M.B.; Bowes, M.J.; Hutchins, M.G.; Orr, H.G.; Soley, R.; Davison, P. Mapping eutrophication risk from climate change: Future phosphorus concentrations in English rivers. *Sci. Total Environ.* **2018**, *613*, 1510–1526. [CrossRef] [PubMed]
50. Paredes, I.; Ramírez, F.; Aragonés, D.; Bravo, M.A.; Forero, M.G.; Green, A.J. Ongoing anthropogenic eutrophication of the catchment area threatens the Doñana World Heritage Site (South-west Spain). *Wetl. Ecol. Manag.* **2021**, *29*, 41–65. [CrossRef]
51. Bonometto, A.; Ponis, E.; Cacciatore, F.; Riccardi, E.; Pigozzi, S.; Parati, P.; Novello, M.; Ungaro, N.; Acquavita, A.; Manconi, P.; et al. A New Multi-Index Method for the Eutrophication Assessment in Transitional Waters: Large-Scale Implementation in Italian Lagoons. *Environments* **2022**, *9*, 41. [CrossRef]
52. Mack, L.; Andersen, H.E.; Beklioğlu, M.; Bucak, T.; Couture, R.M.; Cremona, F.; Ferreira, M.T.; Hutchins, M.G.; Mischke, U.; Molina-Navarro, E.; et al. The future depends on what we do today—Projecting Europe's surface water quality into three different future scenarios. *Sci. Total Environ.* **2019**, *668*, 470–484. [CrossRef] [PubMed]
53. De Vries, W.; Schulte-Uebbing, L.; Kros, H.; Voogd, J.C.; Louwagie, G. Spatially explicit boundaries for agricultural nitrogen inputs in the European Union to meet air and water quality targets. *Sci. Total Environ.* **2021**, *786*, 147283. [CrossRef]
54. Rodríguez-Garlito, E.C.; Paz-Gallardo, A.; Plaza, A. Monitoring the Spatiotemporal Distribution of Invasive Aquatic Plants in the Guadiana River, Spain. *IEEE J. Sel. Top. Appl. Earth Obs. Remote Sens.* **2023**, *16*, 228–241. [CrossRef]
55. Van Rees, C.B.; Waylen, K.A.; Schmidt-Kloiber, A.; Thackeray, S.J.; Kalinkat, G.; Martens, K.; Domisch, S.; Lillebø, A.I.; Hermoso, V.; Grossart, H.P.; et al. Safeguarding freshwater life beyond 2020: Recommendations for the new global biodiversity framework from the European experience. *Conserv. Lett.* **2021**, *14*, e12771. [CrossRef]
56. Troia, A.; Ilardi, V.; Oddo, E. Monitoring of alien aquatic plants in the inland waters of Sicily (Italy). *Webbia* **2020**, *75*, 77–83. [CrossRef]
57. Gross, E.M.; Groffier, H.; Pestelard, C.; Hussner, A. Ecology and Environmental Impact of *Myriophyllum heterophyllum*, an Aggressive Invader in European Waterways. *Diversity* **2020**, *12*, 127. [CrossRef]
58. Efremov, A.N.; Grishina, V.S.; Kislov, D.E.; Mesterházy, A.; Toma, C. The genus Hydrocharis L. (Hydrocharitaceae): Distribution features and conservation status. *Bot. Pac.* **2020**, *9*, 83–94. [CrossRef]
59. García-Murillo, P.; Cirujano, S.; Medina, L.; Sousa, A. ¿Se extinguirá Hydrocharis morsus-ranae L. de la Península Ibérica? *Port. Acta Biol.* **2000**, *19*, 149–158.
60. Catling, P.M.; Dore, W.G. Status and identification of Hydrocharis morsus-ranae and Limnobium spongia (Hydrocharitaceae) in northeastern North America. *Rhodora* **1982**, *84*, 523–545.
61. Zhu, B.; Ottaviani, C.C.; Naddafi, R.; Dai, Z.; Du, D. Invasive European frogbit (*Hydrocharis morsus-ranae* L.) in North America: An updated review 2003–16. *J. Plant Ecol.* **2018**, *11*, 17–25. [CrossRef]
62. Epanchin-Niell, R.; Thompson, A.L.; Treakle, T. Public contributions to early detection of new invasive pests. *Conserv. Sci. Pract.* **2021**, *3*, e422. [CrossRef]
63. Larson, E.R.; Graham, B.M.; Achury, R.; Coon, J.J.; Daniels, M.K.; Gambrell, D.K.; Jonasen, K.L.; King, G.D.; LaRacuente, N.; Perrin-Stowe, T.I.; et al. From eDNA to citizen science: Emerging tools for the early detection of invasive species. *Front. Ecol. Environ.* **2020**, *18*, 194–202. [CrossRef]

plants

MDPI

Article

Lindernia dubia (L.) Pennel as an Alien Weed in Central Spain: A Case Study

María Dolores Curt [1,*], Gema Sánchez [1], Pedro Luis Aguado [1] and Inés Santín-Montanyá [2]

[1] Department of Agrarian Production, Universidad Politécnica de Madrid (UPM), Calle Ramiro de Maeztu 7, 28040 Madrid, Spain; gema.sanchezc@upm.es (G.S.); pl.aguado@upm.es (P.L.A.)

[2] Department of Environment and Agronomy, Instituto Nacional de Investigación y Tecnología Agraria y Alimentaria (INIA-CSIC), 28040 Madrid, Spain; isantin@inia.csic.es

* Correspondence: md.curt@upm.es

Abstract: *Lindernia dubia* (L.) Pennell is a species with invasive behavior outside of its native range of distribution (America), linked mainly to aquatic habitats. This annual species has been acknowledged as a weed in rice paddies in Europe and Asia. Due to the impacts of this invasive plant, some authors have even listed this species as a global invader. The present work focused on spontaneous plant species occurring in seedlings of *Typha domingensis* Pers. grown in central Spain for the establishment of constructed wetlands. Weed inventory revealed the presence of *L. dubia* as a dominant spontaneous species in this crop environment. A suite of mesocosm experiments were designed to study the population density of *L. dubia* versus that of the other dominant plant species, and to determine traits associated with its weedy potential. The results showed that *L. dubia* presents competitive attributes such as morphological variability, early flowering, long seeding time, short growth cycle, small and light seeds and a high seed production and germination rate (25 °C), meaning a high reproductive capacity in a cycle of about three months for plant growth in non-limiting conditions. The data obtained from this work provide a basis for understanding the weedy potential of *L. dubia*, and for management decisions of a potentially invasive species, which has been little investigated in Europe

Keywords: *Lindernia dubia*; invasive species; plant traits; cattail seedlings; constructed wetlands

1. Introduction

Lindernia dubia (L.) Pennell (yellowseed false-pimpernel) is an annual herb of small size (stems 1.5–27 cm), native to America, where it grows in the borders of ponds, river banks and other moist-to-wet habitats [1]. Stems can be erect, ascending or prostrate, and they can root at the lower nodes; leaves are simple, opposite and sessile, and their shape is variable, from lanceolate to ovate or suborbicular [1]. Recently, the distribution of this species has expanded from America to various other regions of the World as distant as Europe, Korea or Taiwan [2,3]. Its flowers are gamopetalous, zygomorphic, chasmogamous or cleistogamous [4]; the fruit is a septicidal capsule with an ellipsoid or globose shape, containing numerous seeds which are ellipsoid or rectangular [1]. According to Lewis [1], this species shows a significant morphological plasticity, especially in terms of its vegetative characteristics.

The Global Biodiversity Information database [3] mentions that there is evidence of the impact of this species in Italy, Romania and Japan. Specifically, in Europe, *L. dubia* (hereafter referred as LIDDU) has been identified as an invasive species in natural aquatic habitats of the Netherlands [5], Romania [6] and Italy [7].

In crop environments, LIDDU has been acknowledged as a weed in rice paddies, where its occurrence seems to be expanding [8–10]. Despite the fact that rice is a staple food crop, there is little information on the traits possessed by LIDDU that promote its spread and on its performance as a weed, probably because other weeds that affect rice, such as

Echinochloa spp. and *Cyperus* spp., are much more harmful [11]. In the case of Europe, the cultivation of rice is concentrated in two countries, Italy (60.3%) and Spain (15.5%) (rice cropped in Russian Federation not included) [12], and in both countries LIDDU has been reported as an alien weed affecting rice [13]. In its native range—as a component of natural vegetation—LIDDU may play a certain role in the conservation of aquatic environments and the restoration of wetlands [14]. Additionally, LIDDU may also occur in other productive moist habitats, such as meadows, plant nurseries or treatment wetlands.

Treatment wetlands are constructed aquatic habitats intended for the restoration of ecosystems or the management of wastewater; they are regarded as nature-based solutions in various contexts [15]. Among the most commonly used plants in constructed wetlands (CW) are *Typha* spp. [16]. *Typha* spp. can be established in constructed wetlands from rhizomes, seeds or seedlings; while the supply of rhizomes requires many resources (rhizome collection, logistics expenses), seeds are easy to collect and use. However, direct sowing is less successful than the establishment of seedlings, due to seed drift from water movement. This fact, along with the need for large number of plants (five to fifty individuals·m^{-2} depending on the type of wetland) [17], has placed the focus on nursery-propagated plants for CW establishment [18]. In this respect, proper sanitation is essential for producing marketable plants and preventing the spread of pests; some authors highlight that weed growth in container-grown nursery stock is a serious problem not only in terms of economic losses but also in terms of the risk of the spread of troublesome plant species, such as alien or invasive plant species [19]. In this regard, we hypothesized that helophyte nurseries may represent an environment prone to weed infestation, due to high availability of growth resources such as water and nutrients. To the best of our knowledge, the occurrence of weeds in helophyte nurseries has not been addressed in the published literature.

In order to shed some light on spontaneous flora occurring in helophyte nurseries and the interspecific competition between dominant species, we set up a mesocosm experiment with the helophyte *Typha domingensis* Pers. (commonly known as cattail) intentionally grown from a seed in a nursery in Spain for constructed wetlands. The primary objectives were to develop an inventory of potential weeds in cattail nurseries and to assess the occurrence of the prevailing taxa and their introduction pathways in order to understand the source and species that should be controlled. In the course of species identification works, the alien species *Lindernia dubia* (L.) Pennel was recorded; this finding has encouraged research on the performance and characterization of this species. Thus, in this paper, (i) a case study of weeds in cattail seedlings is presented, (ii) a new record of the alien species *L. dubia* is reported, (iii) the distribution of dominant weeds is assessed and (iv) traits of *L. dubia*, a dominant alien species, are determined in order to build a dataset for the estimation of its weedy potential.

2. Results

2.1. Weed Inventory and Assessment of Dominant Species

Plant species identified as weeds of cattail seedlings and their constancy values are shown in Table 1. In all, there were five families, Linderniaceae, Brassicaceae, Asteraceae, Cyperaceae, Onagraceae and Poaceae, represented in this particular environment, including, five annual species (in four families) and four perennials (in three families) per life cycle. The Poaceae family presented taxa in both categories; the annuals *Echinochloa crus-galli* (L.) P. Beauv. (ECHCG; common name: barnyardgrass) and *Digitaria sanguinalis* (L.) Scop. (DIGSA; hairy crabgrass) and the perennial *Agrostis stolonifera* L. (AGSST; creeping bentgrass). Per their native range, three alien species were recorded, including *Lindernia dubia* (L.) Pennel (LIDDU; yellowseed false pimpernel), *Cyperus eragrostis* Lam. (CYPER; tall flatsedge) and *Cyperus rotundus* L. (CYPRO; nutgrass), according to the list of alien species in Europe provided by EASIN [20] (CYPRO is considered partly native to Europe). The occurrence of LIDDU represented a new record of this alien species in Madrid (Figure 1) [21].

Table 1. Inventory of plant species present in cattail seedling trays and constancy values.

Taxon	EPPO Code [1]	Family	Life Cycle	Origin	Constancy
Lindernia dubia (L.) Pennel	LIDDU	Linderniaceae	Annual	America	0.9828
Cardamine hirsuta L.	CARHI	Brassicaceae	Winter annual	Europe	0.9425
Cyperus eragrostis Lam.	CYPER	Cyperaceae	Perennial	Tropical America	0.3391
Epilobium roseum Schreb.	EPIRO	Onagraceae	Perennial	Europe to China	0.2471
Agrostis stolonifera L.	AGSST	Poaceae	Perennial	Eurasia, North Africa	0.0862
Digitaria sanguinalis (L.) Scop	DIGSA	Poaceae	Summer annual	Mediterranean region to Central Asia and Malaysia	0.0632
Cyperus rotundus L.	CYPRO	Cyperaceae	Perennial	Indian sub-continent	0.0287
Sonchus oleraceus L.	SONOL	Asteraceae	Summer annual (biennial)	Mediterranean region, Europe	0.0230
Echinochloa crus-galli (L.) P.Beauv.	ECHCG	Poaceae	Summer annual	Europe	0.0172

[1] EPPO = European and Mediterranean Plant Protection Organization https://gd.eppo.int (accessed on 20 September 2023).

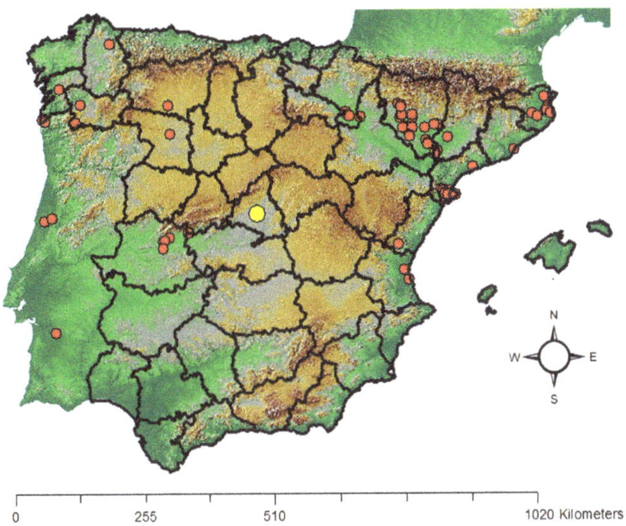

Figure 1. Confirmed records of *Linaria dubia* (L.) Pennel in the Iberian Peninsula (red dots), according to the Information System of the plants of Spain [21]. (Simplified map). Yellow dot = Madrid.

LIDDU was the most frequent species in the flooded seedling trays, showing a rate of constancy of 0.9828 (Table 1), while CARHI was second in the rank (0.9425 constancy). LIDDU and CARHI by far surpassed the occurrence of other species, which presented a <0.34 constancy. The existence of dominance was also confirmed by the estimate of the Simpson index (D = 0.73), while the estimate of the Shannon index (H = 1.53) indicated a low diversity.

Population density by dominant species and tray position are shown in Figure 2. In total, the population density ranged from a minimum of 606 to a maximum of 1543 individuals$\cdot$m^{-2}, showing a mean value of 1024 $\pm$ 263 individuals$\cdot$m^{-2}. Per plant species, LIDDU presented 465 $\pm$ 251 individuals$\cdot$m^{-2} on average, while CARHI represented 560 $\pm$ 223 individuals$\cdot$m^{-2}. The per tray position mean values (LIDDU + CARHI) were 1128 $\pm$ 240 and 929 $\pm$ 256 individuals$\cdot$m^{-2}, for the border and inner trays, respectively. Noteworthily, the variability in density was higher in the inner than in the border trays. No significant differences were found between species nor tray position.

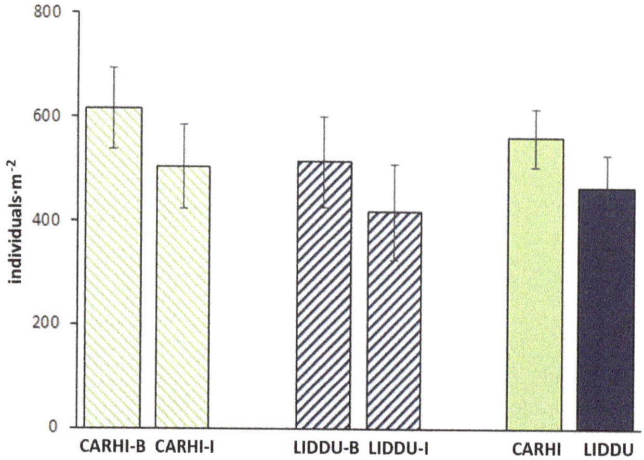

Figure 2. Population density of dominant plant species. CARHI = *Cardamine hirsuta* L.; LIDDU = *Lindernia dubia* (L.) Pennel; B = border position; I = inner position. The bars show ± standard error.

2.2. The Growth Cycle of LIDDU

Throughout the duration of the experiment, no plant species were observed in the control treatment, meaning that the substrate was free of plant propagules. Concerning the sown pots, the chronology of phenological events is shown in Table 2. Seeds germinated within a very short amount of time, and soon after, cotyledons appeared visible to the naked eye. Flowering started a little before a month had passed. As the plants developed leaf whorls, new flower buds arose in the leaf axils; thus, the stages of flowering and fruiting were almost contemporary to the stage of seed dehiscence, which was first noted on DAS 57 (DAS = Days After Sowing). According to these records, the length of the LIDDU reproductive cycle, i.e., from sowing to the start of seed dispersal, was nearly two months. Plants dried up from DAS 83 and were considered dead on DAS 100; that is, the duration of the growth cycle was estimated at 100 days in these experimental conditions. (See Supplementary Material S1).

Table 2. Chronology of phenological observations. *Lindernia dubia* grown from seeds at 25 °C. DAS = Days After Sowing.

DAS	Start of Phenological Stage
6	Visible cotyledons
15	Leaf development
30	Flowering
42	Fruit development
57	Seed dehiscence
83	Senescence
100	Dead plants

Once the cycle was over, the whole plants (shoots + roots) were removed from the pots for growth measurements. Plant counts showed that stem density was equivalent to 6833 individuals·m^{-2} on average (7.1% coefficient of variation, cv), which would be equivalent to 14-times the population recorded for LIDDU in the nursery. The dry matter content of the above ground biomass was 26.5% (cv 1.4%); its dry weight was equivalent to 368.8 g·m^{-2} (12.7% cv). The ratio of the shoots to roots was 6.3 (dry weight basis). Mean growth rate—from sowing to senescence—was equivalent to 19.4 g·week^{-1}·m^{-2}.

2.3. Germination of LIDDU Seeds

The start of the germination (germination time) was observed on Day 2 for the naturally-dehisced seeds in the 25 °C treatment, whereas in the 20 °C treatment the germination started on Day 3. The first observations of open cotyledons were made on Day 4 (19% of the germinated seeds) and Day 6 (1%), respectively. In the 25 °C treatment, the first hypocotyl fibrous root started to grow on Day 10, whereas this observation was recorded on Day 17 for the other treatment (see Supplementary Material S1). The germination period (number of days from the first observed germination to the last one) was 2.3 times higher in the 20 °C treatment than in the 25 °C one.

The time course of germination in the two temperature treatments is shown in Figure 3. Significant differences were observed between the results of both treatments, especially in final cumulative rates: 97.8 ± 1.4% at 25 °C versus 22.8 ± 7.7% at 20 °C. In addition, the variability within replicates was usually higher in the 20 °C treatment. Consistent with this, the fit of data to the Gompertz model gave better coefficient of correlation for the 25 °C treatment (r = 0.9948) than for the 20 °C one (r = 0.8435).

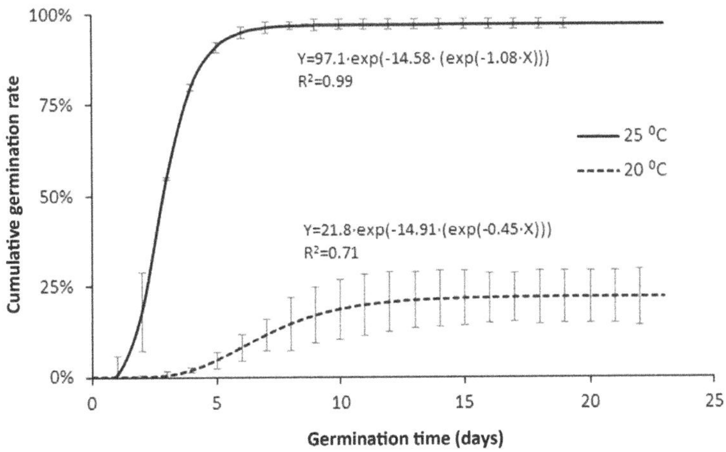

Figure 3. Fitted curves for the germination of naturally-dehisced seeds of *Lindernia dubia*. The bars show ±standard deviation values.

With regard to the whole fruit seeds (WFS), the germination at 25 °C resulted in a 71.8 ± 6.6% final germination rate on average. The asymptote in the fitted germination curve took the value of 69.8%, while the coefficient of correlation was 0.9656.

2.4. LIDDU Traits

The stem length of LIDDU plants that accompanied cattails ranged from 3.6 to 24.7 cm (10.5 cm on average, 41% coefficient of variation) (Figure 4a) and exhibited from five to twenty-one leaf whorls (Figure 4b), which meant about one leaf whorl per cm. The number of capsules per stem also varied greatly, from three to forty-three, in line with the variability found in stem length; on average, there were 17.6 capsules per stem (43% c.v.) (Figure 4c). Nevertheless, there was a close relationship between the number of fruits and stem length (Figure 5), suggesting a regular pattern of flowering and fruiting linked to leaf whorls. Seed production also varied greatly among capsules; it ranged from 34 to 292 seeds, with a mean value of 149.4 seeds·capsule^{-1} (30% c.v.) (Figure 4d).

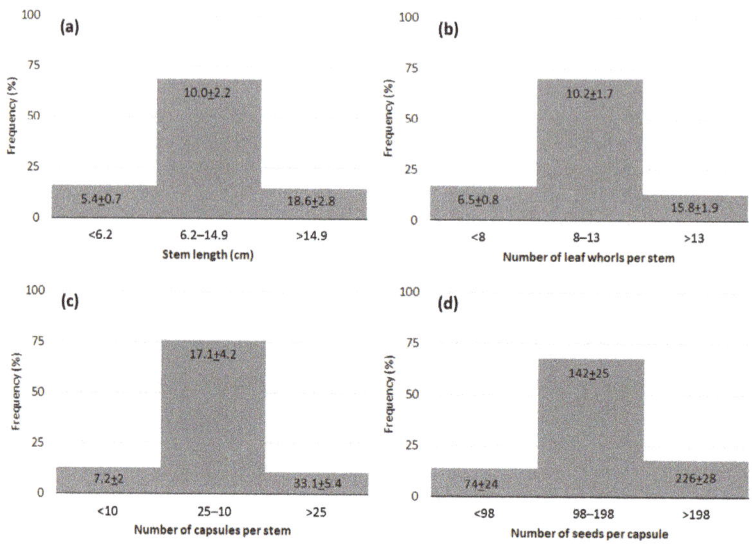

Figure 4. Frequency diagrams for data on traits of *Lindernia dubia*: (**a**) stem length; (**b**) number of leaf whorls per stem; (**c**) number of capsules per stem; (**d**) number of seeds per capsule.

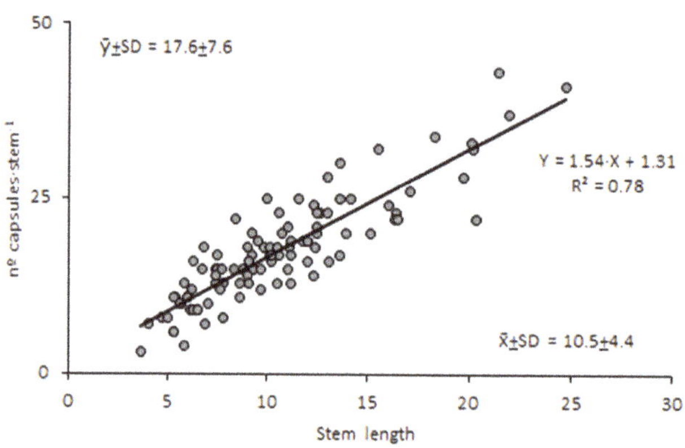

Figure 5. Relationship between the number of capsules and the length of the stem of *Lindernia dubia*.

The variability found in the seed measurements was low, showing that the seeds were rather uniform in size and mass, and tiny. The mean seed length and seed width were 0.35 mm (9% c.v.) (Figure 6a) and 0.14 mm (13% c.v.) (Figure 6b), respectively. The ratio of seed length to width ranged from 1.60 to 4.22, with a mean value of 2.6; this value reflected the rectangular seed shape (Figure 6c). The weight of a thousand seeds varied from 2.86 to 3.27 mg; the NDS lot was slightly heavier (3.09 ± 0.17 mg) than the WHS lot (3.00 ± 0.12 mg), but the difference was not statistically significant. The mean value for 1000-seed weight was 3.04 mg (Figure 6d).

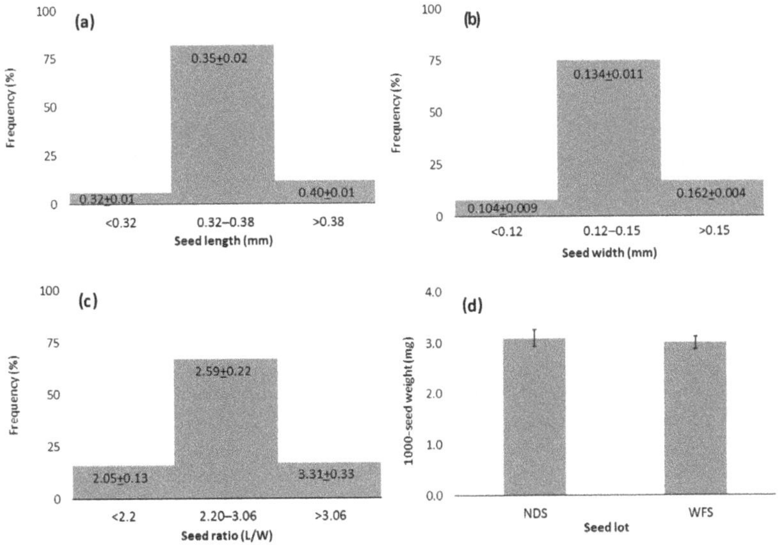

Figure 6. Frequency diagrams for data on traits of *Lindernia dubia*: (**a**) Seed length; (**b**) Seed width; (**c**) Seed ratio length to width; (**d**) 1000-seed weight. NDS = naturally-dehisced seeds; WFS = whole fruit seeds. The bars show ± standard deviation.

The potential seed production of a plant, or plant seed number, was estimated at 2629 seeds·plant^{-1} from the mean values found for the number of capsules per stem and the number of seeds per capsule; this amount would mean that the population seed number in the seedling trays might achieve over 1.2×10^6 seeds·m^{-2} at the end of the growth cycle. In order to assess the reproductive capacity of a plant in the absence of competence, a 97.1% optimum germination rate (value of the asymptote on the germination fitted curve of NDS at 25 °C) and a 72.7% seed dispersal rate (from the asymptote values of NDS and WFS germination fitted curves) were assumed. With these assumptions, the maximum reproductive capacity (sexual propagation) was estimated at 1857 individuals in a cycle of about three months, as long as temperature, substrate and water were not limiting factors.

3. Discussion

Recently, the spread of alien species has become a matter of concern, especially when they present invasive traits. In an agricultural or horticultural environment, the record of an alien plant species may represent a potential weed infestation and/or a risk for nearby ecosystems. One of the main goals of this work was to gain insight into the spontaneous flora occurring in cattail nurseries. Certainly, the use of potted cattail seedlings infested with weeds found in constructed wetlands may result in the spread of undesirable plant species, including potential invaders.

In the present work, the inventory of species accompanying cattail seedlings—grown in a substrate free of plant propagules—revealed a low amount of diversity; it also showed that such species have a preference for moist habitats [22]. Thus, CYPER was reported as the most frequent alien species on the river banks of the Cantabric watershed in Spain [23]. AGSST is widely distributed in Europe and Spain, where it appears in humid meadows [22]. CYPRO and ECHCG have been acknowledged as major weeds of rice fields globally [11], and DIGSA has been reported as a weed in rice fields of India and USA [11]. In Spain, CYPRO, ECHCG and SONOL were reported as the main weeds in sprinkler-irrigated rice [24]. In a study of the success of alien weeds in irrigated Mediterranean orchards, CYPRO and ECHCG were categorized as 'high successful' invaders [25]. Notwithstanding the above, the constancy values of the above-mentioned species were low in our case study; the dominant species were others such as LIDDU (first in the rank) and CARHI.

To the best of our knowledge, the occurrence of the alien species LIDDU in our area of study (Madrid, Spain) has not been reported so far [21,26,27]. According to the Information System of the Plants of Spain [21] the closest record of the occurrence of LIDDU was located in the proximities of the Rosarito reservoir, i.e., 138 km from our location. It is worth noting that this species was cited as a global plant invader by Laginhas and Bradley [28], and that it was reported as an invasive plant species in some regions of Europe [5]. In eastern Spain (provinces of Tarragona and Valencia) it was reported within the list of emergent rice field flora (flooded rice crops), although with little presence [29]. The results for seed weight in this work and the literature data [30,31] showed that LIDDU seeds are lightweight so, the accidental introduction of this alien species into our region might have occurred through a wide range of pathways. It could have even occurred due to waterfowl, a pathway suggested by Lovas-Kiss et al. [32] for LIDDU dispersal in Hungary, and by Soons et al. [33] for small seeds in general. In this respect, it is worth noting that our facilities are about 500 m from the Manzanares river (closest river).

CARHI (hairy bittercress) is a well-known weed species, which is widely studied in the literature. Vaughn et al. [34] stated that CARHI was a major pest in nurseries in USA and highlighted that CARHI occurred "in patches in nursery crops and can solidly fill nursery pots if left uncontrolled" (sic). Warnings about CARHI as a weed of plant nurseries have also been launched in Spain at least since 1997 [35]. Vaughn et al. [34] highlighted large seed production (up to 5000 seeds per plant), quick generation times (30–60 days per generation, in favorable climates) and enhanced mechanisms of seed dispersal (ballistic seed dispersal) as competitive attributes of CARHI.

The occurrence of LIDDU and CARHI across seedling trays seemed to be influenced by the proximity to the corridors, although no statistically significant differences were found due to the high variability in the counts, e.g., LIDDU presented a 63% coefficient of variation for plant counts in the inner trays. Despite the variability, it was noticed that LIDDU performed a little worse in cells all around surrounded by cattails (i.e., in the inner trays). This observation is considered to be in line with the results of Simon and Jansen's [5] work, who inferred that LIDDU would not easily out-compete fast-growing erect plant species; however, these authors suggested that the impact of LIDDU on small or minute plant communities could be adverse. Competition could explain the differences observed between tray positions in the present study. Another explanation could be the unintentional human contribution to seed dispersal (workers walked through the corridors) and wind dispersal; open corridors may result in preferential air flow paths, entailing a higher amount of seed dispersal towards the closest trays.

In vitro germination tests showed that the seeds of LIDDU have the potential to germinate very quickly at 25 °C, and that the germination rate is good, especially for naturally dehisced seeds. Interestingly, the growth of hypocotyl roots was observed at an early plantlet stage (see Supplementary Material S1), an observation that has not been previously reported in the literature. The only datum on LIDDU germination found in published scientific works was the 63% germination rate reported for the control treatment in the study of waterfowl seed dispersal by Lovas-Kiss et al. [32]; their test was conducted at room temperature (20–25 °C) and with a 16 h light/8 h dark photoperiod. Consistent with those temperature conditions, the value reported was between the rates we found at 25 °C and 20 °C in this work. Furthermore, the trends observed in our work seem to be consistent with previous studies of freshwater wetlands and other aquatic habitats that highlighted the capacity of LIDDU to germinate over other species present in seed banks [14,36]. In our tests, the germination rate sharply decreased at 20 °C, a fact that pointed to thermophilic preferences of this species. Although our findings refer to seed lots from a single location, the response of seed germination to temperature obtained in this work suggests that infestation by LIDDU is not likely to happen in the cold season in temperate climates, at least in the short period of time since the introduction of this species.

Nevertheless, Simons and Jansen [5] categorized LIDDU as a naturalized invasive in the Netherlands (temperate Atlantic north-western European climate), but estimated that the transition from being a casual alien plant to that status could have taken about ten years. Additionally, there is evidence of the environmental impact of LIDDU in warmer countries within Europe such as Italy and Romania [3].

The chronology of phenological events and plant traits data presented in this work suggest that LIDDU possesses competitive attributes such as morphological variability, early flowering, long seeding time, short growth cycle, small and light seeds, and high seed production. Peralta and Royuela [37] cited that LIDDU had a summer phenology in North Spain, flowering from July to September, like we observed in central Spain. The botanical description accompanying the data for LIDDU in North America from [38] reads "flowering year-round"; Baker [31] observed that LIDDU flowering ended in autumn (September) in California. With regard to plant size, the values in this work were in line with the botanical description of LIDDU by Lewis [1]. The results for seed length and width were consistent with the taxonomical study of Linderniaceae conducted in Korea by Bazarragchaa et al. [39], and also with data provided in the study of endozoochorous seed dispersal in Hungary [32]. Bazarragchaa et al. [39] described the inflorescence of LIDDU as occurring in axillary flower pairs, solitary axillary or branching in the subtending leaf. Despite the fact that the number of capsules in a single stem was not reported by those authors, it can be inferred that the finding of nearly two fruits (capsules) per leaf whorl in our work is in line with their inflorescence description. Data on capsule seed number or plant seed number were not found in the consulted literature. Some data on 1000-seed weight were found but appeared controversial to a certain extent. Thus, the Seed Information Database [30] reports 0.01 g as the mean value for 1000-seed weight, whereas the Baker Seed Collection database, 0.003 g for the same number of seeds [31]. However, a note on the former value says that minor covering structures could have been included in the weights. Therefore, our results for 1000-seed weight are consistent with Baker's, and support the hypothesis that minor covering structures were included in the weights reported for LIDDU by the Seed Information Database [30].

In the literature, there has been much discussion on what plant traits can be associated with invasiveness, as well as on the dependence of plant traits on the environmental context; however, it is widely accepted that traits related to fecundity and species dispersal are essential to assess the potential success of an invader [40]. Concerning weed fecundity, Norris [41] underlined the risks of extrapolating data from controlled conditions to field conditions, but at the same time acknowledged that many papers report experiments carried out under artificial conditions; in addition, this author encouraged the collection of data for more weed species. In this sense, our estimates of plant seed production and maximum plant reproductive capacity should be taken with caution; however, they provide a basis along with the other results from this work for the estimation of the weedy potential of LIDDU.

4. Materials and Methods

4.1. Site Description

This study was developed at the nursery of aquatic plants of the Technical University of Madrid (UPM, Spain) (40°26′36″ N, 3°44′18″ W, 650 m a.s.l.). The climate at the site is temperate with dry and hot summers; normal values of temperature are 14.6 °C annual mean temperature, −7.4 °C absolute minimum temperature (January) and 40 °C absolute maximum temperature (August) [42].

The aquatic plant facilities included experimental wetlands, a greenhouse for plant propagation and laboratories for plant characterization. These facilities were intended to produce cattail plants for the establishment of externally constructed wetlands, as well as for research on phytodepuration. The greenhouse was endowed with an automated system for ventilation (20 °C set point); in addition, doors remained open during summertime. In the interior of the greenhouse, benches were conventionally arranged: two adjacent rows of benches in the middle (central rows) plus two separate rows at each side of the central rows with two aisle spaces of 1.10 m width in between. Benches were watertight by means of a polyethylene (PE) pond liner. They were periodically filled up with tap water to simulate shadow ponds. Every year in the period from 2017 to 2023, cattails were raised from seed in 96-cell plastic trays (52.5 × 33 × 8 cm) placed in the simulated water ponds; before seeding, trays were filled with a commercial substrate (Pindstrup Mosebrug S.A.E. code 51212, Burgos, Spain). After emergence, seedlings were thinned to one per cell and let to grow until planting in constructed wetlands. The alien LIDDU was noticed for the first time in the summer of 2022.

4.2. Weed Inventory and Assessment of the Dominant Species

The occurrence of spontaneous plant species in cattail seedlings was studied in autumn 2022. At that time, cattails were about 50–65 cm tall and exhibited 5–7 green leaves. Identification of spontaneous plant species growing in the trays was carried out using botanical keys and monographs [22,43]. Species constancy was assessed as the ratio of the number of seedling trays with a determined species to the total number of seedling trays (n = 174). The indexes of Shannon (H) [44] and Simpson (D) [45] were calculated from tray counts in order to estimate species diversity and dominance.

The occurrence of LIDDU was assessed in terms of population density and distribution across the nursery, as compared to CARHI, the other dominant weed species in this study. A randomized block design with four blocks and two replications per block was used. To this end, two categories, border (trays placed next to the aisle) and inner (trays placed on the inner area of the bench table), were made. Population density (individuals·m^{-2}) was calculated from the number of individuals counted in each seedling tray, considering the tray surface area (1536 cm^2·tray^{-1}).

4.3. Growth Chamber Experiments

Two types of experiments were developed in a plant growth chamber (300 L capacity, forced air circulation, 6 × F30W/Gro-Lux T8) to gain insight into LIDDU growth cycle and seed germination.

The first one was a pot experiment, where the pot dimensions were top diameter, 10.5 cm; bottom diameter, 8.5 cm; and height, 9.5 cm. Pots (n = 4) were filled up with the same substrate as the one used in the cattail nursery (Pindstrup Mosebrug S.A.E. code 51212). A control (no sown treatment) was used to check the absence of plant propagules in the substrate. Mature fruits (light-brown capsules, starting to break) were taken from senescent LIDDU plants in the seedling trays; seeds were spread on the pot surface, using 50 capsules·pot^{-1}; capsules were left on the ground. Then, the just-sown pots were placed in a water bath inside the growth chamber at 25 °C constant temperature and in a 14 h day/10 h dark cycle. Air humidity was maintained ≥79.1% throughout the experiment. Water level in the bath was kept at pot surface level (daily water replenishment). Dates of the start of the phenological stages visible cotyledons, flowering, fruiting, seed dehiscence and plant senescence (Supplementary Material S1) were recorded and expressed as DAS (Days After Sowing). Once the seed dehiscence was observed, naturally dehisced seeds from the potted plants were collected on a daily basis; after seed cleaning (debris separation under stereomicroscope) they were gathered in a seed lot named NDS (=naturally dehisced seeds). When plants were fully dry (brown-colored shoots), they were uprooted in order to determine the following parameters: population density (individuals·m^{-2}), shoots dry matter content (%), average growth rate (dry mass accumulated per week and ground area, g·week^{-1}·m^{-2}) and the ratio of shoots/roots (dry weight basis). Dry weights were determined after oven-drying at 105 °C.

The other type of experiments was in vitro seed germination tests. Germination of NDS was performed in covered Petri-dishes filled with filter paper moistened with distilled water; 2 thermal regimes (20 and 25 °C constant temperature) were tested using three replicates of about 100 seeds each (n = 3 × 100). The photoperiodic regime was 14 h light/10 h dark. Observations were made on a daily basis under a Leica EZ4 stereomicroscope (Wetzlar, Germany) 35 × (transmitted illumination). A seed was considered to have germinated when the radicle emerged (see Supplementary Material S1). The germination rate was calculated as the number of germinated seeds divided by the total number of seeds in the Petri dish. Dates of the first observation of visible cotyledons and lateral root growth were annotated as well.

Another seed lot was prepared and tested for in vitro germination. All seeds (naturally dehisced plus not-dehisced seeds) inside 50 mature LIDDU capsules were extracted by shaking and by hand with the help of a thin brush and a spatula; after seed cleaning under stereomicroscope, seeds were gathered in a seed lot named WFS (Whole Fruit Seeds). WFS germination was tested at 25 °C constant thermal regime, following the same procedure as for NDS tests. Final germination rate of WFS was compared to that of NDS in order to assess the seed dispersal rate of LIDDU capsules.

4.4. LIDDU Traits

Plant traits studied were the following: stem length (SL), number of whorls per stem (NW), number of capsules per stem (NC), seed production per capsule (NS), seed size, 1000-seed weight, and maximum reproductive capacity in absence of competence. One hundred shoots in the senescence stage (brown shoots) were randomly sampled from cattail seedling trays in the nursery for SL, NW and NC; from these plants, fifty capsules—dry capsules starting to open—were taken at random for NS. Seed extraction and seed cleaning (separation from debris) were performed by hand with the help of laboratory equipment (thin brushes, tweezers, spatulas, stereomicroscope); afterwards, seeds from each capsule were counted under a stereomicroscope to determine NS. Seed size measurements included seed length (L) and width (W) and the ratio of L/W [39]. The sample size for seed measurements was one hundred seeds. Thousand-seed weight (air-dry seeds) was determined in triplicate using a five decimal precision balance. This trait was studied for the two lots of seeds, NDS and WFS ($n = 3 \times 2 \times 1000$).

The maximum reproductive capacity (individuals per ground area, $pl \cdot m^{-2}$) was calculated as the product of mean NS, mean NC, seed dispersal rate and optimum germination rate. It was assumed that the optimum germination rate was the highest asymptote value found for the germination curves in the germination tests.

4.5. Statistical Analysis

Descriptive statistics (mean, standard deviation, minimum, maximum, coefficients of variation) were calculated. Data for LIDDU traits were graphically presented in terms of frequency (percentage in a category). Population densities of dominant species were analyzed by one-way ANOVA with the location of plants in the nursery as between-subject factor and two weed dominant species (factor with two levels) as a within-subject factor. For the study of seed germination, data were fitted to the S-shaped Gompertz function: $Y = a \times \exp(-b \times \exp(-c \times T)$, using the cumulative germination rate as dependent variable and the germination time (days) as independent variable; parameters a, b and c in the Gompertz function were calculated by successive iterations. Data analyses were conducted using the statistics software Statgraphics 19®.

5. Conclusions

The occurrence of the alien plant species *Lindernia dubia* (L.) Pennel, as a dominant weed in cattail seedlings, has been reported in Madrid (Central Spain) for the first time; its introduction might have occurred due to waterfowl and wind. The population density attained by *L. dubia* has been found to be comparable to that of *Cardamine hirsuta* L., a well-known weed species in plant nurseries. The characterization of *L. dubia* has provided evidence that this species presents the following competitive attributes: variability in stem length, number of capsules per plant and number of seeds per capsule; short growth cycle (assessed at 25 °C); ability to flower at an early stage of growth; long seeding time; small and light seeds; high seed production; and a high germination rate at 25 °C; all of this suggests a high reproductive capacity in a short period of time. On the whole, the dataset obtained in this study has provided a basis for understanding the weedy potential of *L. dubia*, and for management decisions of this potentially invasive species in Europe.

Supplementary Materials: The following supporting information can be downloaded at: https://www.mdpi.com/article/10.3390/plants13131859/s1, Supplementary Material S1: Photographic report.

Author Contributions: Conceptualization, investigation, methodology, formal analysis, writing—original draft preparation, M.D.C.; data curation, investigation, formal analysis, G.S.; resources, visualization, validation, P.L.A.; writing—review and editing, validation, supervision, I.S.-M. All authors have read and agreed to the published version of the manuscript.

Funding: This research received no external funding.

Data Availability Statement: Data are contained within the article.

Acknowledgments: Technical support for the study of LIDDU germination was provided by Eng. J.G. Apodaca. Support from Projects FPA220000PA1402 and Life16 CCM/GR/000044 is also gratefully acknowledged.

Conflicts of Interest: The authors declare no conflicts of interest.

References

1. Lewis, D.Q. A Revision of the New World Species of Lindernia (Scrophulariaceae). *Castanea* **2000**, *65*, 93–122.
2. Seebens, H.; Blackburn, T.M.; Hulme, P.E.; van Kleunen, M.; Liebhold, A.M.; Orlova-Bienkowskaja, M.; Pyšek, P.; Schindler, S.; Essl, F. Around the World in 500 Years: Inter-Regional Spread of Alien Species over Recent Centuries. *Glob. Ecol. Biogeogr.* **2021**, *30*, 1621–1632. [CrossRef]
3. Global Biodiversity Information. GBIF.Org. Available online: https://www.gbif.org/en/ (accessed on 20 April 2024).
4. Ikeda, E.; Miura, R. A Note on the Proportion of Cleistogamous Flowers in *Lindernia dubia* (L.) Pennel (Scrophulariaceae) in a Paddy Field. *Weed Res. Jpn.* **1994**, *39*, 177–179. [CrossRef]
5. Simons, E.L.A.N.; Jansen, M.G.M. Ecology of Naturalized Invasive Species *Lindernia dubia* (L.) Pennell in the Netherlands. *Gorteria Dutch Bot. Arch.* **2018**, *40*, 1–10.
6. Neacșu, A.; Imbrea, I.; Lațo, A.; Arsene, G.G. *Lindernia dubia* (L.) Pennell: A New Invasive in the Romanian Banat. *AgroLife Sci. J.* **2021**, *10*, 129–135. [CrossRef]
7. Galasso, G.; Domina, G.; Ardenghi, N.M.G.; Aristarchi, C.; Bacchetta, G.; Bartolucci, F.; Bonari, G.; Bouvet, D.; Brundu, G.; Buono, S.; et al. Notulae to the Italian Alien Vascular Flora: 7. *Ital. Bot.* **2019**, *7*, 157–182. [CrossRef]
8. Webb, D.A.; Philcox, D. Lindernia All. In *Flora Europaea*; Cambridge University Press: Cambridge, UK, 1972; Volume 3, pp. 203–204.
9. Kumar, V.; Mahajan, G.; Sheng, Q.; Chauhan, B.S. Weed Management in Wet Direct-Seeded Rice (*Oryza sativa* L.): Issues and Opportunities. In *Advances in Agronomy*; Academic Press Inc.: Cambridge, MA, USA, 2023; Volume 179, pp. 91–133. ISBN 9780443192623.
10. Yoshino, N.; Wang, G.X.; Ito, M.; Auld, B.; Kohara, H.; Enomoto, T. Naturalization and Dissemination of Two Subspecies of *Lindernia dubia* (Scrophulariaceae) in Japan. *Weed Biol. Manag.* **2006**, *6*, 174–176. [CrossRef]
11. Kraehmer, H.; Jabran, K.; Mennan, H.; Chauhan, B.S. Global Distribution of Rice Weeds—A Review. *Crop Prot.* **2016**, *80*, 73–86. [CrossRef]
12. Food and Agriculture Organization of the United Nations. FAOSTAT. Available online: https://www.fao.org/faostat/en/#data (accessed on 30 May 2024).
13. Carretero, J.L. La Vegetación Emergente de Los Arrozales Europeos. In *Anales de Biología*; Servicio de Publicaciones de la Universidad de Murcia: Murcia, Spain, 1988; Volume 15, pp. 135–141.
14. Neff, K.P.; Rusello, K.; Baldwin, A.H. Rapid Seed Bank Development in Restored Tidal Freshwater Wetlands. *Restor. Ecol.* **2009**, *17*, 539–548. [CrossRef]
15. Pistocchi, A. *Nature-Based Solutions for Agricultural Water Management—Characteristics and Enabling Factors for a Broader Adoption*; Publications Office of the European Union: Luxembourg, 2022; ISBN 978-92-76-60487-7. JRC131465. [CrossRef]
16. Vymazal, J. Plants in Constructed, Restored and Created Wetlands. *Ecol. Eng.* **2013**, *61*, 501–504. [CrossRef]
17. Kataki, S.; Chatterjee, S.; Vairale, M.G.; Dwivedi, S.K.; Gupta, D.K. Constructed Wetland, an Eco-Technology for Wastewater Treatment: A Review on Types of Wastewater Treated and Components of the Technology (Macrophyte, Biolfilm and Substrate). *J. Environ. Manag.* **2021**, *283*, 111986. [CrossRef] [PubMed]
18. Rogers, S.M.D. Tissue Culture and Wetland Establishment of the Freshwater Monocots Carex, Juncus, Scirpus, and Typha. *Vitr. Cell. Dev. Biol.-Plant* **2003**, *39*, 1–5. [CrossRef]

19. Case, L.T.; Mathers, H.M.; Senesac, A.F. A Review of Weed Control Practices in Container Nurseries. *Horttechnology* **2005**, *15*, 535–545. [CrossRef]
20. European Commission. EASIN—European Alien Species Information Network. Available online: https://easin.jrc.ec.europa.eu/easin (accessed on 20 September 2023).
21. Anthos. Information System of the Plants of Spain. Real Jardín Botánico, CSIC—Fundación Biodiversidad. Available online: www.anthos.es (accessed on 3 June 2024).
22. Castroviejo, S. *Flora Iberica*; Real Jardín Botánico, CSIC: Madrid, Spain, 1986; Volume 22, ISBN 84-00-06221-3.
23. Liendo, D.; Biurrun, I.; Campos, J.A.; Herrera, M.; Loidi, J.; García-Mijangos, I. Invasion Patterns in Riparian Habitats: The Role of Anthropogenic Pressure in Temperate Streams. *Plant Biosyst.* **2015**, *149*, 289–297. [CrossRef]
24. Cavero, J.; Zaragoza, C.; Cirujeda, A.; Anzalone, A.; Faci, J.M.; Blanco, O. Selectivity and Weed Control Efficacy of Some Herbicides Applied to Sprinkler Irrigated Rice (*Oryza sativa* L.). *Span. J. Agric. Res.* **2011**, *9*, 597. [CrossRef]
25. Juárez-Escario, A.; Valls, J.; Solé-Senan, X.O.; Conesa, J.A. A Plant-Traits Approach to Assessing the Success of Alien Weed Species in Irrigated Mediterranean Orchards. *Ann. Appl. Biol.* **2013**, *162*, 200–213. [CrossRef]
26. Madrid Council, Plan de Fomento y Gestión de la Biodiversidad Ciudad de Madrid, Anexo VI Inventarios de Flora y Fauna. Available online: https://madrid.es/portal/site/munimadrid (accessed on 20 September 2023).
27. Casas, I.; Gavilán, R.; Echevarría, J. Catálogo de La Flora Vascular de La Ciudad Universitaria de Madrid (España). *Bot. Complut.* **1993**, *18*, 175–201.
28. Laginhas, B.B.; Bradley, B.A. Global Plant Invaders a Compendium of Invasive Plant Taxa Documented by the Peer-reviewed.Pdf. *Ecology* **2022**, *103*, e03569. [CrossRef] [PubMed]
29. Carretero, J.L. Rice Field Flora and Vegetation in the Provinces of Valencia and Tarragona. *Collect. Bot.* **1987**, *17*, 113–124. [CrossRef]
30. Society for Ecological Restoration; International Network for Seed Based Restoration; Royal Botanic Gardens Kew Seed Information Database (SID). Available online: https://ser-sid.org/ (accessed on 3 June 2024).
31. Baker, H.G. Baker Seed Collection Database. Available online: https://ucjeps.berkeley.edu/baker/ (accessed on 11 June 2024).
32. Lovas-Kiss, Á.; Navarro-Ramos, M.J.; Vincze, O.; Löki, V.; Urgyán, R.; Pallér-Kapusi, F.; van Leeuwen, C.H.A.; Green, A.J.; Lukács, B.A. Traits for Transport: Alien Wetland Plants Gain an Advantage during Endozoochorous Seed Dispersal by Waterfowl. *Freshw. Biol.* **2023**, *68*, 1703–1715. [CrossRef]
33. Soons, M.B.; Van Der Vlugt, C.; Van Lith, B.; Heil, G.W.; Klaassen, M. Small Seed Size Increases the Potential for Dispersal of Wetland Plants by Ducks. *J. Ecol.* **2008**, *96*, 619–627. [CrossRef]
34. Vaughn, K.C.; Bowling, A.J.; Ruel, K.J. The Mechanism for Explosive Seed Dispersal in *Cardamine hirsuta* (Brassicaceae). *Am. J. Bot.* **2011**, *98*, 1276–1285. [CrossRef] [PubMed]
35. Bay, T.; Gracia, I.; Cavero, J.; Zaragoza, C. *Cardamine hirsuta* Infestante de Los Viveros...¿Importada Con La Turba? In Proceedings of the Actas del Congreso 1997 de la Sociedad Española de Malherbología (SEMH), Valencia, Spain, 24–26 November 1997; SEMH: Valencia, Spain, 1997; pp. 177–180.
36. Abernethy, V.J.; Willby, N.J. Changes along a Disturbance Gradient in the Density and Composition of Propagule Banks in Floodplain Aquatic Habitats. *Plant Ecol.* **1999**, *140*, 177–190. [CrossRef]
37. Peralta, J.; Royuela, M. Herbario-Departamento de Ciencias de La Universidad Pública de Navarra. Available online: https://www.unavarra.es/herbario/htm/Lind_dubi.htm (accessed on 28 May 2024).
38. Lewis, D.Q. *Lindernia dubia*. Available online: http://floranorthamerica.org/Lindernia_dubia (accessed on 22 May 2024).
39. Bazarragchaa, B.; Yang, S.; Kim, H.S.; Lee, S.J.; Lee, J. An Updated Taxonomy of the Family Linderniaceae in Korea. *Korean J. Agric. Sci.* **2019**, *46*, 1007–1018. [CrossRef]
40. Lau, J.A.; Funk, J.L. How Ecological and Evolutionary Theory Expanded the 'Ideal Weed' Concept. *Oecologia* **2023**, *203*, 251–266. [CrossRef] [PubMed]
41. Norris, R.F. Weed Fecundity: Current Status and Future Needs. *Crop Prot.* **2007**, *26*, 182–188. [CrossRef]
42. Agencia Estatal de Meteorología; Instituto de Meteorología de Portugal. *Iberian Climate Atlas: Air Temperature and Precipitation (1971–2000)*; Agencia Nacional de Meteorología de España (AEMET), Instituto de Meteorología de Portugal (IM): Lisboa, Portugal, 2011; ISBN 9788478370795.
43. Recasens, J.; Conesa, J.A. *Malas Hierbas en Plántula. Guía de Identificación*; Editions de la Universitat de Lleida, Bayer CropScience: Lleida, Spain, 2009; ISBN 978-84-8409-290-4.
44. Shannon, C.; Wever, W. *The Mathematical Theory of Communication*; University of Illinois Press: Urbana, IL, USA, 1949; ISBN 0-252-72546-8.
45. Simpson, E.H. Measurement of Diversity. *Nature* **1949**, *163*, 688. [CrossRef]

MDPI AG
Grosspeteranlage 5
4052 Basel
Switzerland
Tel.: +41 61 683 77 34

MDPI Books Editorial Office
E-mail: books@mdpi.com
www.mdpi.com/books

www.ingramcontent.com/pod-product-compliance
Lightning Source LLC
Chambersburg PA
CBHW061456220926
51850CB00004B/105
* 9 7 8 3 7 2 5 8 6 1 7 7 4 *